S0-CAS-003

ATOMIC ENERGY-LEVEL
and
GROTRIAN DIAGRAMS

ATOMIC ENERGY-LEVEL and GROTRIAN DIAGRAMS

Volume II. Sulfur I - Titanium XXII

Stanley Bashkin
and
John O. Stoner, Jr.

Department of Physics, University of Arizona, Tucson, Arizona 85721

1978

NORTH-HOLLAND PUBLISHING COMPANY
AMSTERDAM · OXFORD · NEW YORK

ISBN: 0 444 85149 6

Published by:

North-Holland Publishing Company
Amsterdam/New York/Oxford

Distributors for the U.S.A. and Canada:
Elsevier North-Holland, Inc.
52 Vanderbilt Avenue
New York, N.Y. 10017

Library of Congress Cataloging in Publication Data (Revised)

Bashkin, Stanley.
Atomic energy levels and Grotrian diagrams.

Volume 2 has title: Atomic energy level and Grotrian diagrams.
Includes bibliographies.
CONTENTS: v. 1. Hydrogen I-Phosphorus XV.--v. 2. Sulphur I-Calcium XX.
1. Energy levels (Quantum mechanics)--Charts, diagrams, etc. 2. Grotrian diagrams. I. Stoner, John O., joint author. II. Title.
QC795.8.E5B37 539.7'2 74-25897
ISBN 0-444-85149-6

PRINTED IN THE NETHERLANDS

INTRODUCTION

We continue the pictorial representation of energy levels and electronic transitions for atoms and ions. The style followed is similar to that of Vol. I and its addendum volume (see below) which include all systems from H I through P XV. Again, two types of drawings have been prepared, one that shows the energy levels in order of increasing excitation energy, and one that shows the transitions between levels. Again, it has frequently been necessary to subdivide the information so as to present it in a comprehensible manner. Thus, spectroscopic terms of a given multiplicity, parity, or coupling may be sufficiently numerous to warrant individual drawings, although there are also many cases where an entire level system is displayed on a single drawing.

A Grotrian diagram may include all the observed transitions for a particular ion, or it may contain only the observed transitions within a given multiplicity, from one multiplicity to another, or among other subsets of the energy levels. In all cases, we have tried to present the information in a clear and useful manner. Transitions whose identifications are in doubt, or for which the wavelengths have been calculated but not observed, are printed in italics; on a few drawings, question marks are used. The only exceptions to this procedure are the wavelengths of transitions in hydrogenic (one-electron) ions that are all taken from calculations and printed in block numerals.

In Vol. I, we depended largely on previous summaries of spectroscopic data for the basic information we used, and only occasional references to individual research papers were noted. For Vol. II, we have continued to use the general references listed in Vol. I and in the addendum volume. However, so much work has appeared since the publication of the several compendia that we have found it helpful to make many more citations than necessary for Vol. I. In addition, we have gone to some pains to compare the results presented by different authors when there are significant discrepancies in the literature. While authors are generally meticulous in referring to previous papers dealing with their subject, it isn't always clear that the later and earlier data may be discrepant. Where it seemed useful, we have listed the different results; occasionally we have simply called attention to the fact that authors are not in agreement and have given the references.

In the pioneering monographs by C.E. Moore (*Atomic Energy Levels*), the terms of some complicated systems are described by prefixes, "a", "b", "c", for low-lying even-parity terms, "z", "y", "x", ... for odd-parity terms, and "e", "f", ... for high-lying even-parity terms. This notation was intended to represent the fact that a term belonging to a given ionization stage could often be described by the coupling of a running electron to a core which was itself a term in the next-higher ionization stage. For example, the lowest ^{4}F term in Sc I has the designation $3d^2(a\,^3F)4s\;a\,^4F$. This means that a 4s electron has been coupled to a $3d^2\;^3F$ "core" to produce the lowest ^{4}F term. The "core" is actually the second excited term in Sc II, designated "a" since it is the least energetic ^{3}F term. The lowest $^4F^\circ$ term in Sc I, $3d4s(a\,^3D)4p\;z\,^4F^\circ$, decays to the even-parity ground term. Still higher terms that make transitions to z^4F° are denoted by letters "e", "f", and "g".

Such nomenclature is sufficiently helpful that it has been copied in many research papers. Nonetheless, it also has some undesirable features. For one, the early and late letters of the alphabet are mixed in their distribution throughout a given term system. Thus there is an $e\,^4G$ term in Sc I, but no lower ^{4}G terms occur. For another, it is not obvious what connection there is between a prefix and a core. Indeed, the ground term of Sc I, for instance, doesn't have a "core" from Sc II, while $z^2F°$ and $w^2F°$ have different cores. Moreover, the Sc I terms do not serve as "cores" for any other system, so the prefixes in Sc I don't serve the same purpose as the prefixes in Sc II.

We have introduced a different designation which is intended to label the parentage of terms more clearly. For each ion, we have simply numbered the spectroscopic terms in order of increasing excitation, using zero for the ground term. In a level's designation, such a number, followed by *, specifies the parent term (the core) associated with a level in the preceding ion. Thus, the Sc I ^4F term mentioned above is given by 2* $4s\,^4F$. Each energy-level or Grotrian diagram for which this notation is useful includes a definition of the cores. For Sc I, one finds:

Cores (C*) From Sc II
0* = $3d(^2D)4s\,^3D$
1* = $3d(^2D)4s\,^1D$
2* = $3d^2\,^3F$, etc.,

for the parents needed for the Sc I system. This notation is similar to that used by M.W.D. Mansfield, Proc. Roy. Soc. (Lond.) **A346**, 539 and 555 (1975) in his discussion of the structure of K I. We have included in our drawings Moore's letter prefixes in addition to the new symbols.

As discussed in the bibliographical notes for the elements beyond Ca, the level designations in AEL, largely copied by us, are often not accurate. We have referred the reader to particular papers for the available calculations of levels' wavefunctions wherever appropriate.

It should be remembered that our diagrams do not indicate either configuration-mixing or the inadequacies of LS and other coupling schemes.

Errors of omission and commission are virtually unavoidable in publications of any complexity. We have found some in the literature, and we call attention to them at appropriate points in our bibliographical material. Some errors have been found in Vol. I; these are corrected and additional information provided in the addendum volume (available to purchasers of Vol. I). We again solicit critical comments from the scientific community, and express our gratitude to those who have written to us about different features of our compilation. Dr. William Martin of the U.S. National Bureau of Standards has been particularly generous in discussing problems concerning notation and citations.

FUTURE WORK

It is our intention to extend our compilation at least through the iron-group elements, and beyond, if that proves to be practical. In five or ten years, there might be enough new information to warrant a revision of Vol. I, and we are continually adding new data to the original drawings. There have been some requests that we turn to systems that are of immediate importance. We are willing to do so, provided there is enough of a demand from our colleagues to indicate a fundamental need. Communications to this effect are welcome.

ACKNOWLEDGMENTS

We take pleasure in listing the names of those draftsmen, mostly students at The University of Arizona, who joined this project subsequent to the completion of Vol. I: J. Duffield, M. Krapcevic, L. Miller, W. Ram, S. Minghelli, L. Schaefer, W. Smith, and D. Webber. The cover design depicts a precision spectrometer used by Ångstrom. This cover was produced by Mr. John W. Howe, who has general charge of the preparation of the drawings.

Financial support was received from NASA, ONR, and NSF.

Sulfur (S)

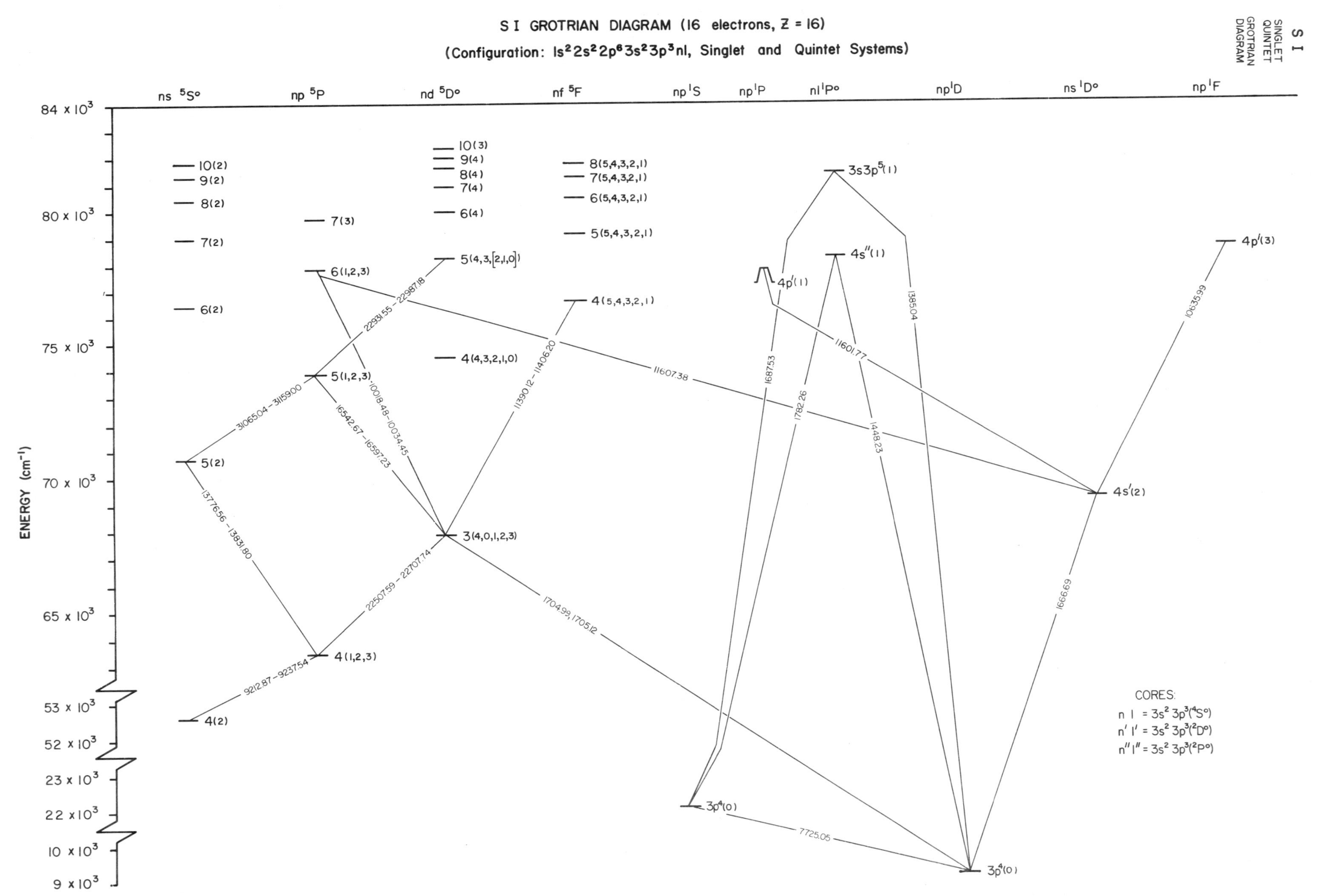
S I
SINGLET
QUINTET
GROTRIAN
DIAGRAM
S I GROTRIAN DIAGRAM (16 electrons, Z = 16)
(Configuration: $1s^2 2s^2 2p^6 3s^2 3p^3 nl$, Singlet and Quintet Systems)
ns $^5S^o$
np ^5P
nd $^5D^o$
nf ^5F
np ^1S
np ^1P
nl $^1P^o$
np ^1D
ns $^1D^o$
np ^1F
ENERGY (cm^{-1})
84 x 10^3
80 x 10^3
75 x 10^3
70 x 10^3
65 x 10^3
53 x 10^3
52 x 10^3
23 x 10^3
22 x 10^3
10 x 10^3
9 x 10^3
10(2)
9(2)
8(2)
7(2)
6(2)
5(2)
4(2)
7(3)
6(1,2,3)
5(1,2,3)
4(1,2,3)
10(3)
9(4)
8(4)
7(4)
6(4)
5(4,3,[2,1,0])
4(4,3,2,1,0)
3(4,0,1,2,3)
8(5,4,3,2,1)
7(5,4,3,2,1)
6(5,4,3,2,1)
5(5,4,3,2,1)
4(5,4,3,2,1)
$3s3p^5(1)$
$4s''(1)$
$4p'(1)$
$4p'(3)$
$4s'(2)$
$3p^4(0)$
$3p^4(0)$
2293.55 – 2298.718
3106.504 – 3115.900
10018.48 – 10034.45
16542.67 – 16597.23
13776.56 – 13831.80
11390.12 – 11406.20
2250.759 – 2270.774
9212.87 – 9237.54
1704.99, 1705.12
11607.38
11601.77
1687.53
1782.26
1448.23
1385.04
1063.599
1666.69
7725.05
CORES:
$n\ l\ = 3s^2\ 3p^3(^4S^o)$
$n'\ l' = 3s^2\ 3p^3(^2D^o)$
$n''\ l'' = 3s^2\ 3p^3(^2P^o)$

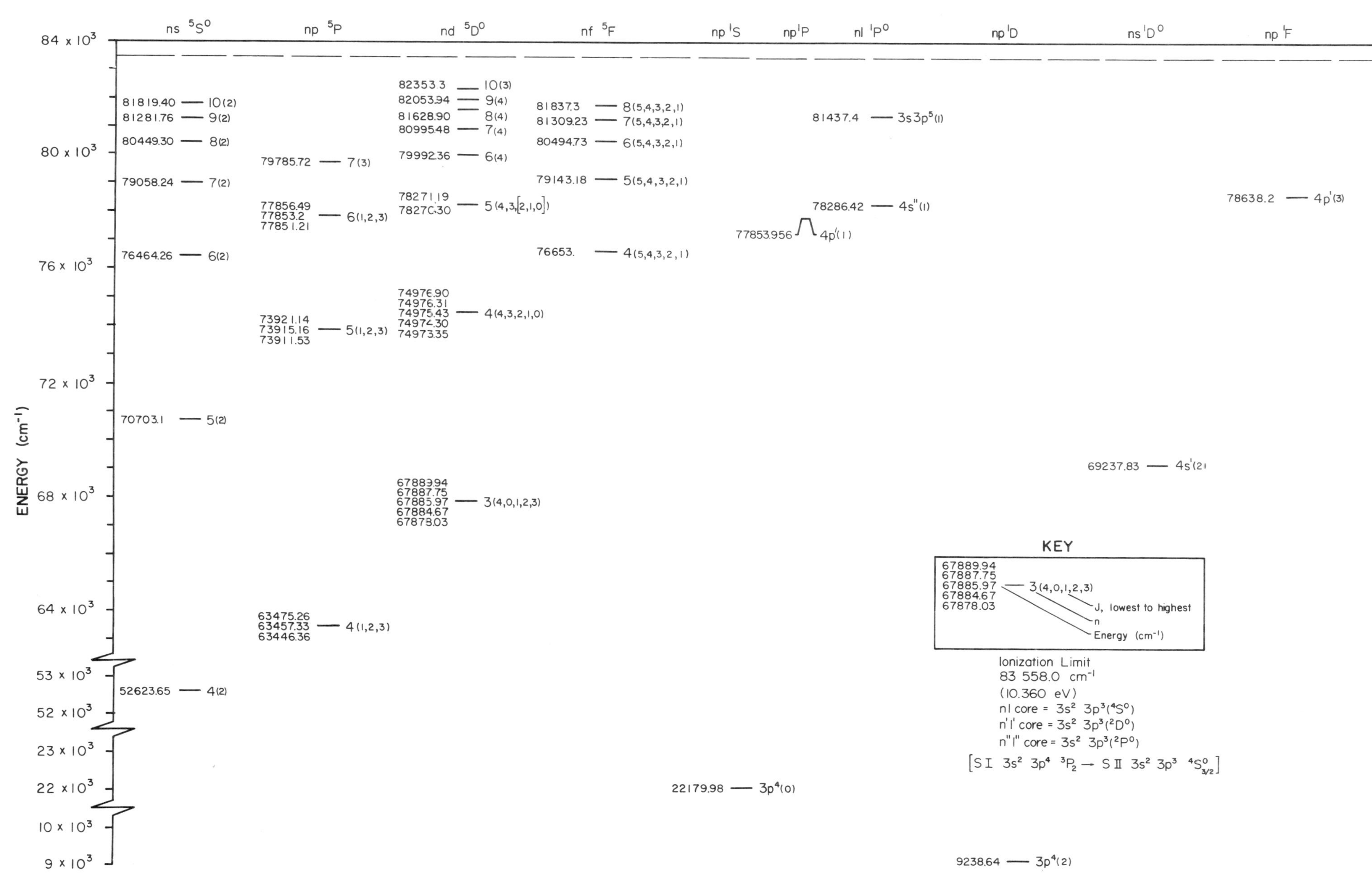
S I ENERGY LEVELS (16 electrons, Z=16)
(Configuration: 1s² 2s² 2p⁶ 3s² 3p³ nl, Singlet and Quintet Systems)
ns ⁵S⁰
np ⁵P
nd ⁵D⁰
nf ⁵F
np ¹S
np ¹P
nl ¹P⁰
np ¹D
ns ¹D⁰
np ¹F
84 x 10³
80 x 10³
76 x 10³
72 x 10³
68 x 10³
64 x 10³
53 x 10³
52 x 10³
23 x 10³
22 x 10³
10 x 10³
9 x 10³
ENERGY (cm⁻¹)
81819.40 — 10(2)
81281.76 — 9(2)
80449.30 — 8(2)
79058.24 — 7(2)
76464.26 — 6(2)
70703.1 — 5(2)
52623.65 — 4(2)
79785.72 — 7(3)
77856.49
77853.2
77851.21 — 6(1,2,3)
73921.14
73915.16
73911.53 — 5(1,2,3)
63475.26
63457.33
63446.36 — 4(1,2,3)
82353.3 — 10(3)
82053.94 — 9(4)
81628.90 — 8(4)
80995.48 — 7(4)
79992.36 — 6(4)
78271.19
78270.30 — 5(4,3,[2,1,0])
74976.90
74976.31
74975.43
74974.30
74973.35 — 4(4,3,2,1,0)
67889.94
67887.75
67885.97
67884.67
67878.03 — 3(4,0,1,2,3)
81837.3 — 8(5,4,3,2,1)
81309.23 — 7(5,4,3,2,1)
80494.73 — 6(5,4,3,2,1)
79143.18 — 5(5,4,3,2,1)
76653. — 4(5,4,3,2,1)
77853.956 4p′(1)
81437.4 — 3s3p⁵(1)
78286.42 — 4s″(1)
22179.98 — 3p⁴(0)
9238.64 — 3p⁴(2)
69237.83 — 4s′(2)
78638.2 — 4p′(3)
KEY
67889.94
67887.75
67885.97
67884.67
67878.03
3(4,0,1,2,3)
J, lowest to highest
n
Energy (cm⁻¹)
Ionization Limit
83 558.0 cm⁻¹
(10.360 eV)
nl core = 3s² 3p³(⁴S⁰)
n′l′ core = 3s² 3p³(²D⁰)
n″l″ core = 3s² 3p³(²P⁰)
[S I 3s² 3p⁴ ³P₂ → S II 3s² 3p³ ⁴S⁰₃/₂]

S I

TRIPLET
GROTRIAN
DIAGRAM

S I GROTRIAN DIAGRAM (16 electrons, Z = 16)
(Configuration: $1s^2 2s^2 2p^6 3s^2 3p^3 nl$, Triplet System)

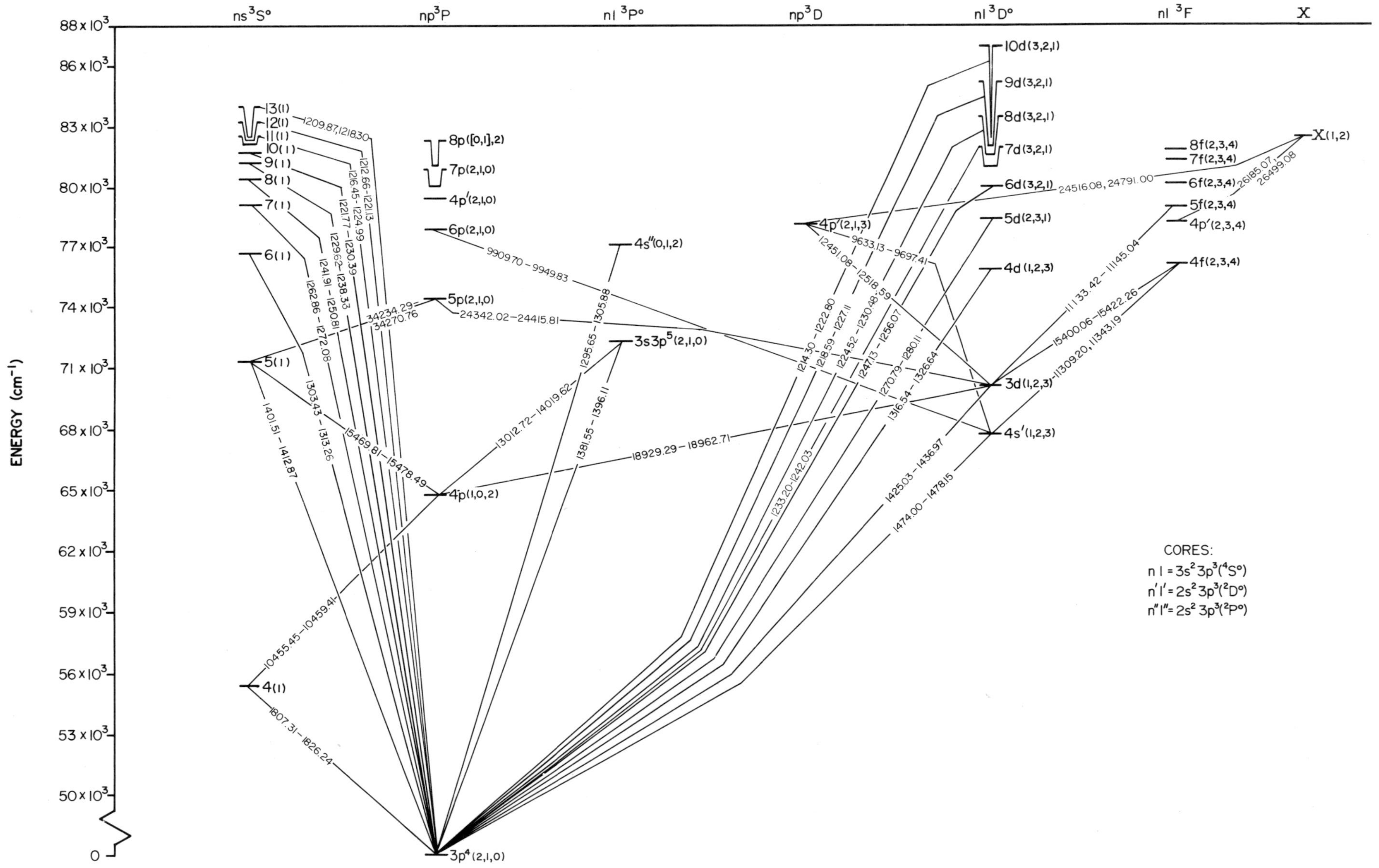

ns $^3S^o$ | np ^3P | nl $^3P^o$ | np ^3D | nl $^3D^o$ | nl ^3F | X

ns $^3S^o$: 4(1), 5(1), 6(1), 7(1), 8(1), 9(1), 10(1), 11(1), 12(1), 13(1)

np ^3P: 4p(1,0,2), 5p(2,1,0), 6p(2,1,0), 4p′(2,1,0), 7p(2,1,0), 8p(0,1,2)

nl $^3P^o$: $3s3p^5$(2,1,0), 4s″(0,1,2)

np ^3D: 4p′(2,1,3)

nl $^3D^o$: 3d(1,2,3), 4s′(1,2,3), 4d(1,2,3), 5d(2,3,1), 6d(3,2,1), 7d(3,2,1), 8d(3,2,1), 9d(3,2,1), 10d(3,2,1)

nl ^3F: 4f(2,3,4), 4p′(2,3,4), 5f(2,3,4), 6f(2,3,4), 7f(2,3,4), 8f(2,3,4)

X: X(1,2)

Ground: $3p^4$(2,1,0)

CORES:
n l = $3s^2 3p^3 (^4S^o)$
n′l′ = $2s^2 3p^3 (^2D^o)$
n″l″ = $2s^2 3p^3 (^2P^o)$

ENERGY (cm^{-1}): 0, 50×10^3, 53×10^3, 56×10^3, 59×10^3, 62×10^3, 65×10^3, 68×10^3, 71×10^3, 74×10^3, 77×10^3, 80×10^3, 83×10^3, 86×10^3, 88×10^3

S I ENERGY LEVELS (16 electrons, Z=16)
(Configuration: $1s^2 2s^2 2p^6 3s^2 3p^3$ nl, Triplet System)

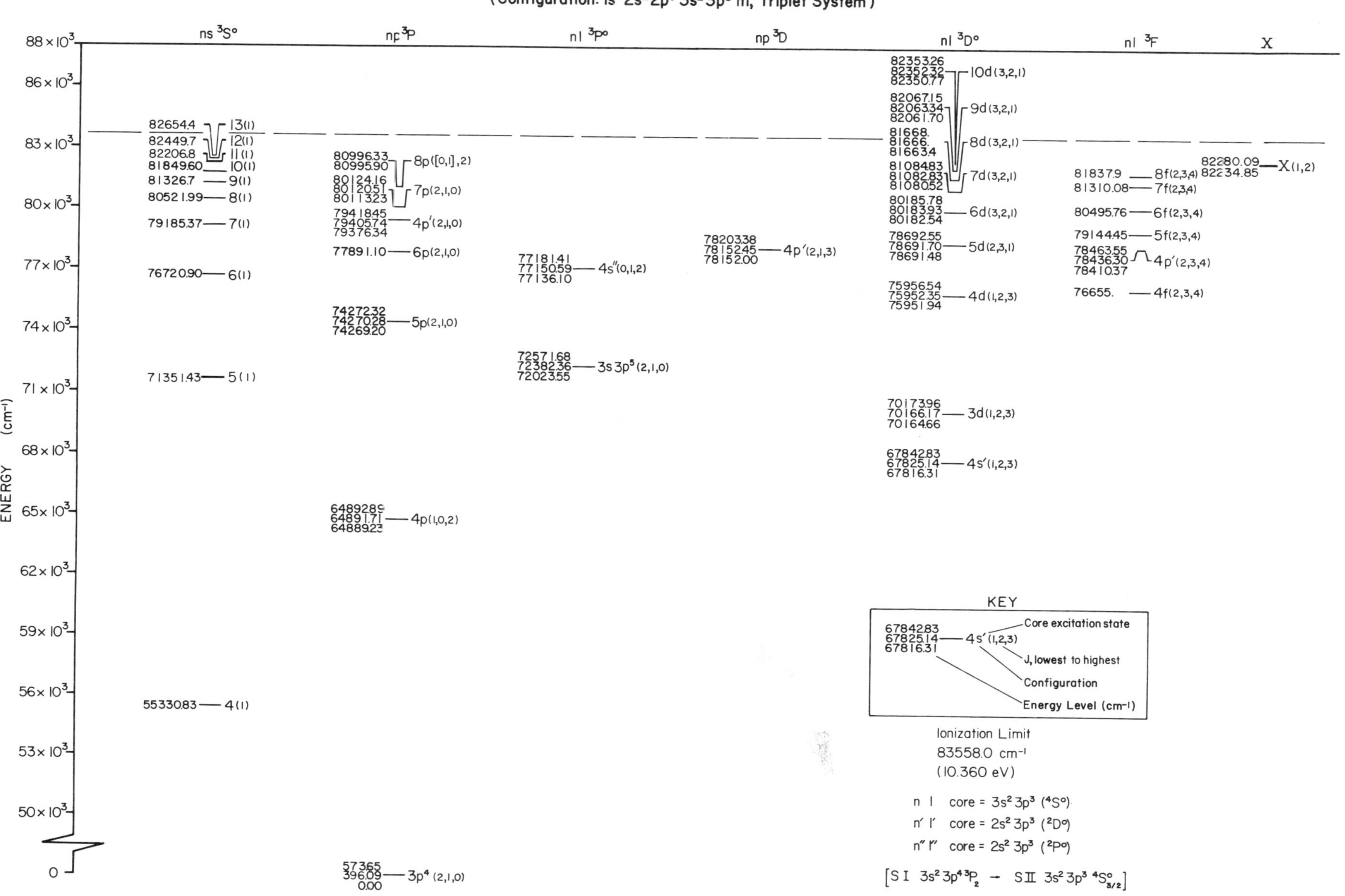

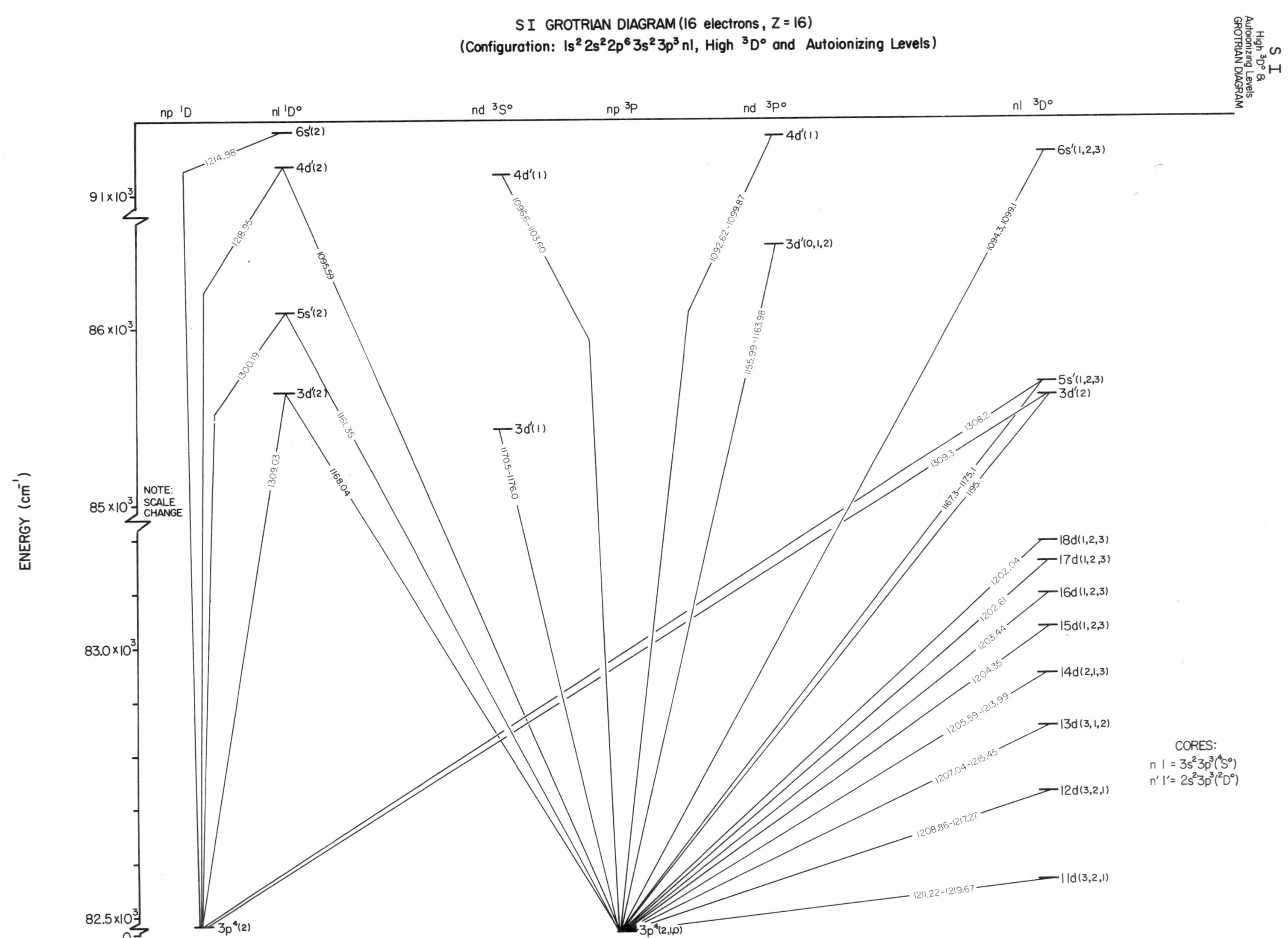

S I
High $^3D^o$ & Autoionizing Levels
GROTRIAN DIAGRAM
S I GROTRIAN DIAGRAM (16 electrons, Z = 16)
(Configuration: $1s^2 2s^2 2p^6 3s^2 3p^3 nl$, High $^3D^o$ and Autoionizing Levels)
np ^1D
nl $^1D^o$
nd $^3S^o$
np ^3P
nd $^3P^o$
nl $^3D^o$
ENERGY (cm^{-1})
91 x 10^3
86 x 10^3
85 x 10^3
83.0 x 10^3
82.5 x 10^3
0
NOTE: SCALE CHANGE
6s'(2)
4d'(2)
5s'(2)
3d'(2)
1214.98
1218.95
1095.59
1300.19
1161.35
1309.03
1168.04
4d'(1)
1096.6-1103.60
3d'(1)
1170.5-1176.0
4d'(1)
1092.62-1099.87
3d'(0,1,2)
1155.99-1163.98
6s'(1,2,3)
1094.3, 1099.1
5s'(1,2,3)
3d'(2)
1308.2
1309.3
1167.3-1175.1
1195.
18d(1,2,3)
17d(1,2,3)
16d(1,2,3)
15d(1,2,3)
14d(2,1,3)
13d(3,1,2)
12d(3,2,1)
11d(3,2,1)
1202.04
1202.61
1203.44
1204.35
1205.59-1213.99
1207.04-1215.45
1208.86-1217.27
1211.22-1219.67
$3p^4(2)$
$3p^4(2,1,0)$
CORES:
$n\ l = 3s^2 3p^3(^4S^o)$
$n'\ l' = 2s^2 3p^3(^2D^o)$

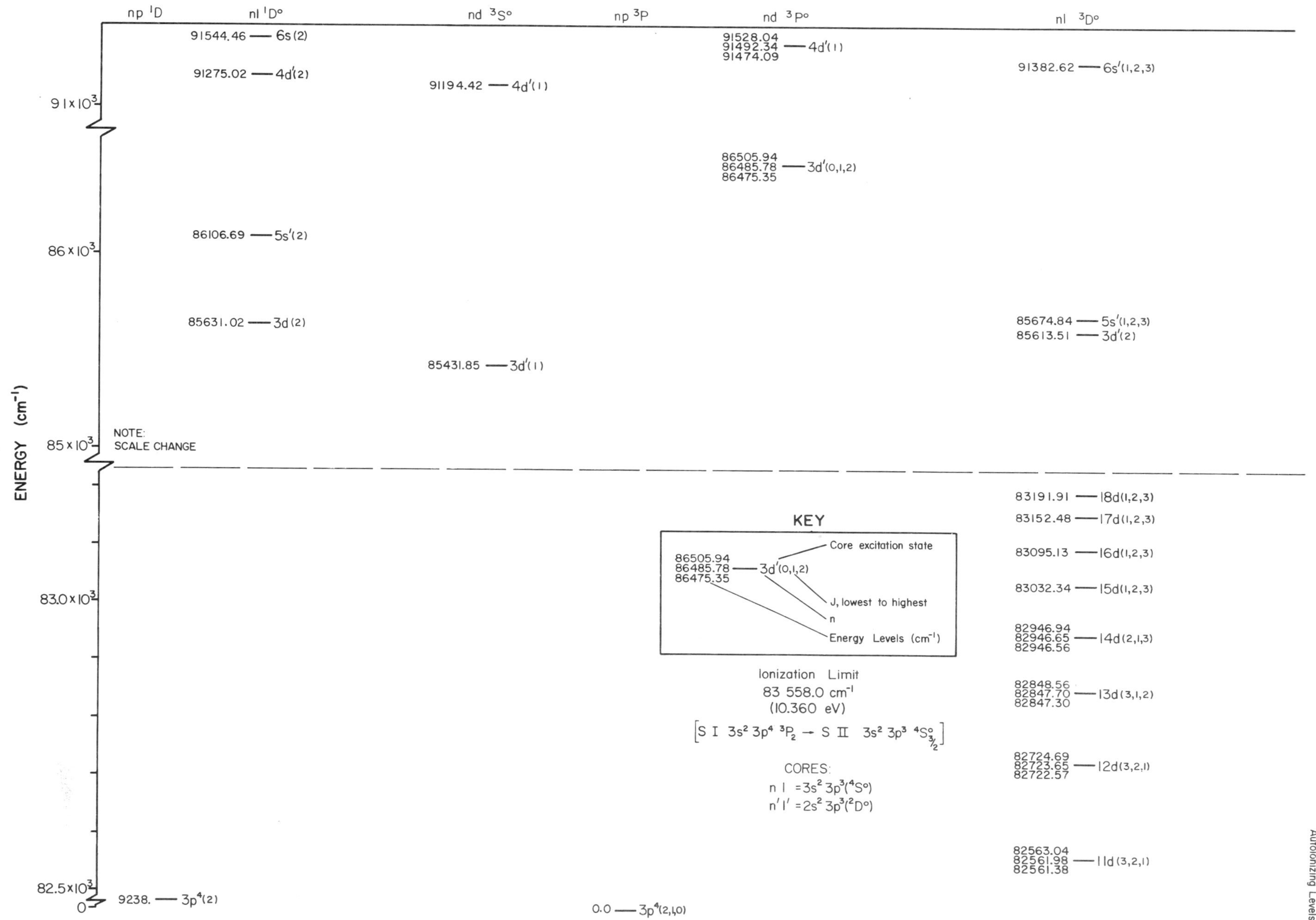
S I ENERGY LEVELS (16 electrons, Z = 16)
(Configuration: $1s^2 2s^2 2p^6 3s^2 3p^3 nl$, High $^3D^o$ and Autoionizing Levels)
np ^1D
nl $^1D^o$
nd $^3S^o$
np ^3P
nd $^3P^o$
nl $^3D^o$
91544.46 — 6s (2)
91275.02 — 4d′(2)
86106.69 — 5s′(2)
85631.02 — 3d (2)
91194.42 — 4d′(1)
85431.85 — 3d′(1)
91528.04 91492.34 91474.09 — 4d′(1)
86505.94 86485.78 86475.35 — 3d′(0,1,2)
91382.62 — 6s′(1,2,3)
85674.84 — 5s′(1,2,3)
85613.51 — 3d′(2)
83191.91 — 18d(1,2,3)
83152.48 — 17d(1,2,3)
83095.13 — 16d(1,2,3)
83032.34 — 15d(1,2,3)
82946.94 82946.65 82946.56 — 14d(2,1,3)
82848.56 82847.70 82847.30 — 13d(3,1,2)
82724.69 82723.65 82722.57 — 12d(3,2,1)
82563.04 82561.98 82561.38 — 11d(3,2,1)
9238. — $3p^4$(2)
0.0 — $3p^4$(2,1,0)
ENERGY (cm^{-1})
91 x 10^3
86 x 10^3
85 x 10^3
83.0 x 10^3
82.5 x 10^3
0
NOTE: SCALE CHANGE
KEY
86505.94 86485.78 86475.35 — 3d′(0,1,2)
Core excitation state
J, lowest to highest
n
Energy Levels (cm^{-1})
Ionization Limit
83 558.0 cm^{-1}
(10.360 eV)
[S I $3s^2 3p^4\ ^3P_2$ → S II $3s^2 3p^3\ ^4S^o_{3/2}$]
CORES:
n l = $3s^2 3p^3(^4S^o)$
n′ l′ = $2s^2 3p^3(^2D^o)$

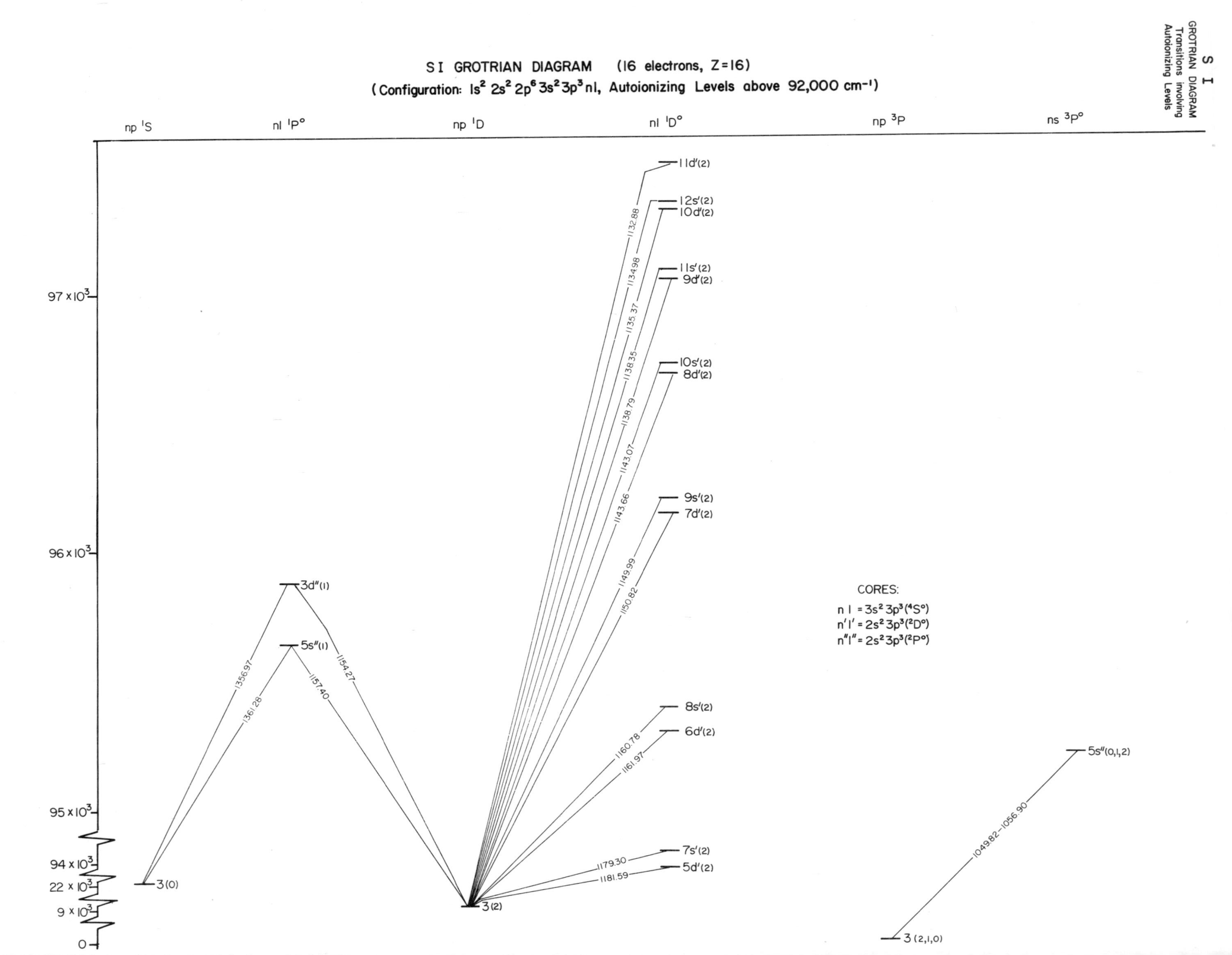
SI GROTRIAN DIAGRAM (16 electrons, Z=16)
(Configuration: $1s^2 2s^2 2p^6 3s^2 3p^3 nl$, Autoionizing Levels above 92,000 cm^{-1})
np ^{1}S
nl ^{1}P°
np ^{1}D
nl ^{1}D°
np ^{3}P
ns ^{3}P°
97 x 10^3
96 x 10^3
95 x 10^3
94 x 10^3
22 x 10^3
9 x 10^3
0
3(0)
3d″(1)
5s″(1)
1356.97
1361.28
1154.27
1157.40
3(2)
11d′(2)
12s′(2)
10d′(2)
11s′(2)
9d′(2)
10s′(2)
8d′(2)
9s′(2)
7d′(2)
8s′(2)
6d′(2)
7s′(2)
5d′(2)
1132.88
1134.98
1135.37
1138.35
1138.79
1143.07
1143.66
1149.99
1150.82
1160.78
1161.97
1179.30
1181.59
CORES:
n l = $3s^2 3p^3 (^4S°)$
n′l′ = $2s^2 3p^3 (^2D°)$
n″l″ = $2s^2 3p^3 (^2P°)$
5s″(0,1,2)
1049.82–1056.90
3 (2,1,0)

S I ENERGY LEVELS (16 electrons, Z=16)

(Configuration: $1s^2 2s^2 2p^6 3s^2 3p^3 nl$, Autoionizing Levels above 92,000 cm^{-1})

ENERGY (cm^{-1})

97×10^3, 96×10^3, 95×10^3, 94×10^3

$nl\ ^1P^o$

Energy	Level
95873.42	$3d''(1)$
95639.13	$5s''(1)$

$nl\ ^1D^o$

Energy	Level
97509.18	$11d'(2)$
97345.86	$12s'(2)$
97315.59	$10d'(2)$
97085.02	$11s'(2)$
97051.08	$9d'(2)$
96722.29	$10s'(2)$
96677.15	$8d'(2)$
96195.86	$9s'(2)$
96133.14	$7d'(2)$
95387.55	$8s'(2)$
95299.32	$6d'(2)$
94034.65	$7s'(2)$
93870.31	$5d'(2)$

$ns\ ^3P^o$

Energy	Level
95254.42 95210.89 95199.89	$5s''(0,1,2)$

KEY

95254.42 / 95210.89 / 95199.89 — $5s''(0,1,2)$

- Core excitation state
- J, lowest to highest
- n
- Energy Levels (cm^{-1})

Ionization Limit
83 558.0 cm^{-1}
(10.360 eV)
[S I $3s^2 3p^4\ ^3P_2 \rightarrow$ S II $3s^2 3p^3\ ^4S^o_{3/2}$]

CORES:
$n\ l = 3s^2 3p^3 (^4S^o)$
$n'l' = 2s^2 3p^3 (^2D^o)$
$n''l'' = 2s^2 3p^3 (^2P^o)$

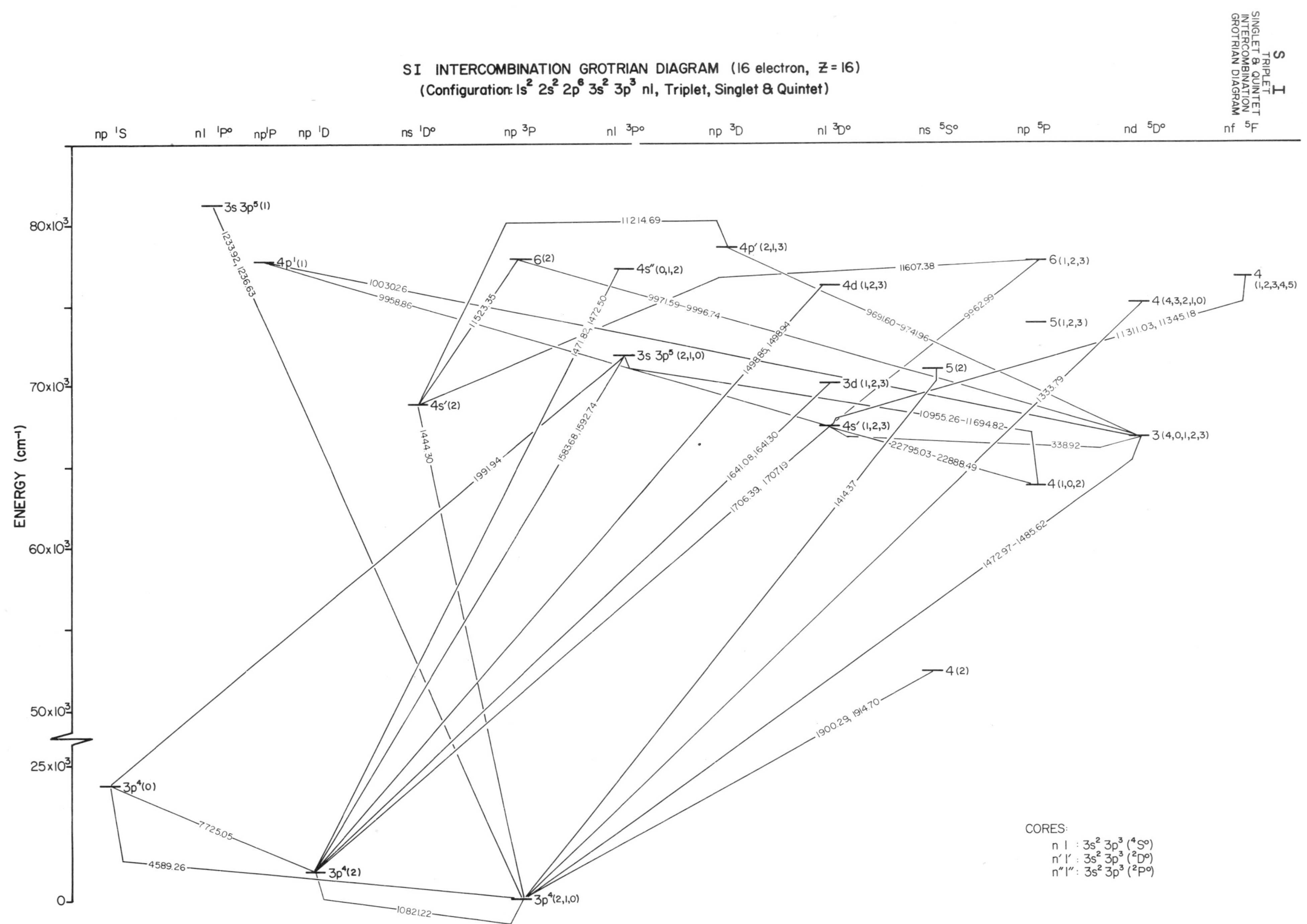
S I
TRIPLET
SINGLET & QUINTET
INTERCOMBINATION
GROTRIAN DIAGRAM
SI INTERCOMBINATION GROTRIAN DIAGRAM (16 electron, Z = 16)
(Configuration: $1s^2\ 2s^2\ 2p^6\ 3s^2\ 3p^3$ nl, Triplet, Singlet & Quintet)
np ^{1}S
nl ^{1}P°
np ^{1}P
np ^{1}D
ns ^{1}D°
np ^{3}P
nl ^{3}P°
np ^{3}D
nl ^{3}D°
ns ^{5}S°
np ^{5}P
nd ^{5}D°
nf ^{5}F
ENERGY (cm^{-1})
80x10^3
70x10^3
60x10^3
50x10^3
25x10^3
0
3s 3p^5(1)
4p'(1)
4p' (2,1,3)
6(2)
4s"(0,1,2)
6(1,2,3)
4d (1,2,3)
4 (1,2,3,4,5)
4 (4,3,2,1,0)
5(1,2,3)
3s 3p^5 (2,1,0)
5(2)
3d (1,2,3)
4s'(2)
4s' (1,2,3)
3 (4,0,1,2,3)
4 (1,0,2)
4 (2)
3p^4(0)
3p^4(2)
3p^4(2,1,0)
11214.69
1233.92, 1236.63
1003026
9958.86
11523.35
1471.82, 1472.50
11607.38
9971.59–9996.74
1498.85, 1498.94
969160–974196
9962.99
11311.03, 11345.18
1333.79
10955.26–11694.82
33892
22795.03–22888.49
1444.30
1583.68, 1592.74
1991.94
1641.08, 1641.30
1706.39, 1707.19
1414.37
1472.97–1485.62
1900.29, 1914.70
7725.05
4589.26
10821.22
CORES:
n l : $3s^2\ 3p^3\ (^4S°)$
n' l' : $3s^2\ 3p^3\ (^2D°)$
n" l" : $3s^2\ 3p^3\ (^2P°)$

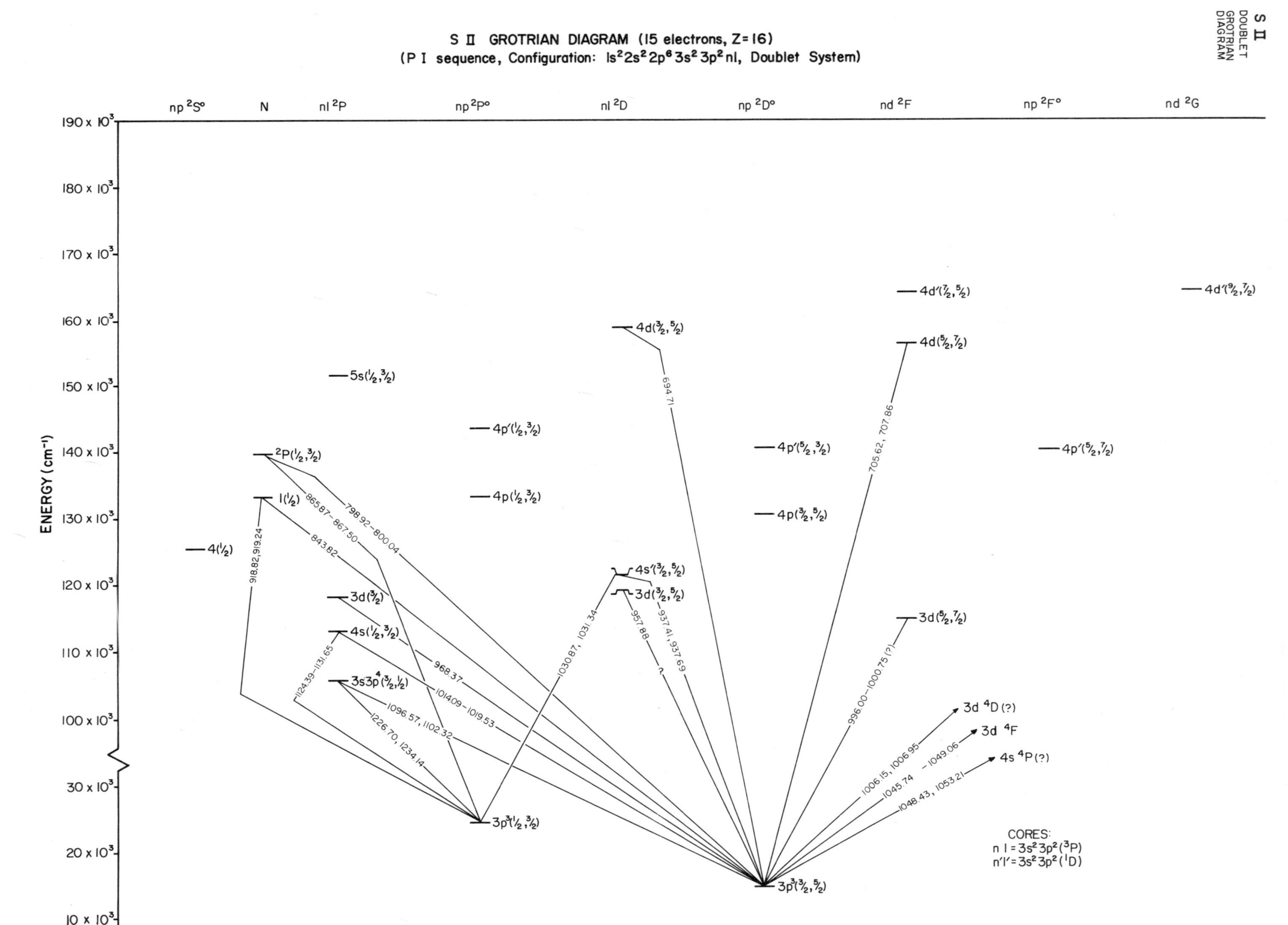
S II
DOUBLET
GROTRIAN
DIAGRAM
S II GROTRIAN DIAGRAM (15 electrons, Z=16)
(P I sequence, Configuration: $1s^2 2s^2 2p^6 3s^2 3p^2 nl$, Doublet System)
np $^2S^o$
N
nl ^2P
np $^2P^o$
nl ^2D
np $^2D^o$
nd ^2F
np $^2F^o$
nd ^2G
ENERGY (cm^{-1})
190 x 10^3
180 x 10^3
170 x 10^3
160 x 10^3
150 x 10^3
140 x 10^3
130 x 10^3
120 x 10^3
110 x 10^3
100 x 10^3
30 x 10^3
20 x 10^3
10 x 10^3
4($\frac{1}{2}$)
^{2}P($\frac{1}{2}$,$\frac{3}{2}$)
1($\frac{1}{2}$)
5s($\frac{1}{2}$,$\frac{3}{2}$)
3d($\frac{3}{2}$)
4s($\frac{1}{2}$,$\frac{3}{2}$)
$3s3p^4$($\frac{3}{2}$,$\frac{1}{2}$)
4p'($\frac{1}{2}$,$\frac{3}{2}$)
4p($\frac{1}{2}$,$\frac{3}{2}$)
$3p^3$($\frac{1}{2}$,$\frac{3}{2}$)
4d($\frac{3}{2}$,$\frac{5}{2}$)
4s'($\frac{3}{2}$,$\frac{5}{2}$)
3d($\frac{3}{2}$,$\frac{5}{2}$)
4p'($\frac{5}{2}$,$\frac{3}{2}$)
4p($\frac{3}{2}$,$\frac{5}{2}$)
$3p^3$($\frac{3}{2}$,$\frac{5}{2}$)
4d'($\frac{7}{2}$,$\frac{5}{2}$)
4d($\frac{5}{2}$,$\frac{7}{2}$)
3d($\frac{5}{2}$,$\frac{7}{2}$)
4p'($\frac{5}{2}$,$\frac{7}{2}$)
4d'($\frac{9}{2}$,$\frac{7}{2}$)
3d ^{4}D (?)
3d ^{4}F
4s ^{4}P (?)
CORES:
n l = $3s^2 3p^2 (^3P)$
n'l' = $3s^2 3p^2 (^1D)$
918.82, 919.24
865.87–867.50
843.82
798.92–800.04
1124.39–1131.65
1226.70, 1234.14
1096.57, 1102.32
1014.09–1019.53
968.37
1030.87, 1031.34
957.88
937.41, 937.69
?
694.71
705.62, 707.86
996.00–1000.75 (?)
1006.15, 1006.95
1045.74 – 1049.06
1048.43, 1053.21

S II ENERGY LEVELS (15 electrons, Z = 16)
(P I sequence, Configuration: $1s^2 2s^2 2p^6 3s^2 3p^2 nl$, Doublet System)

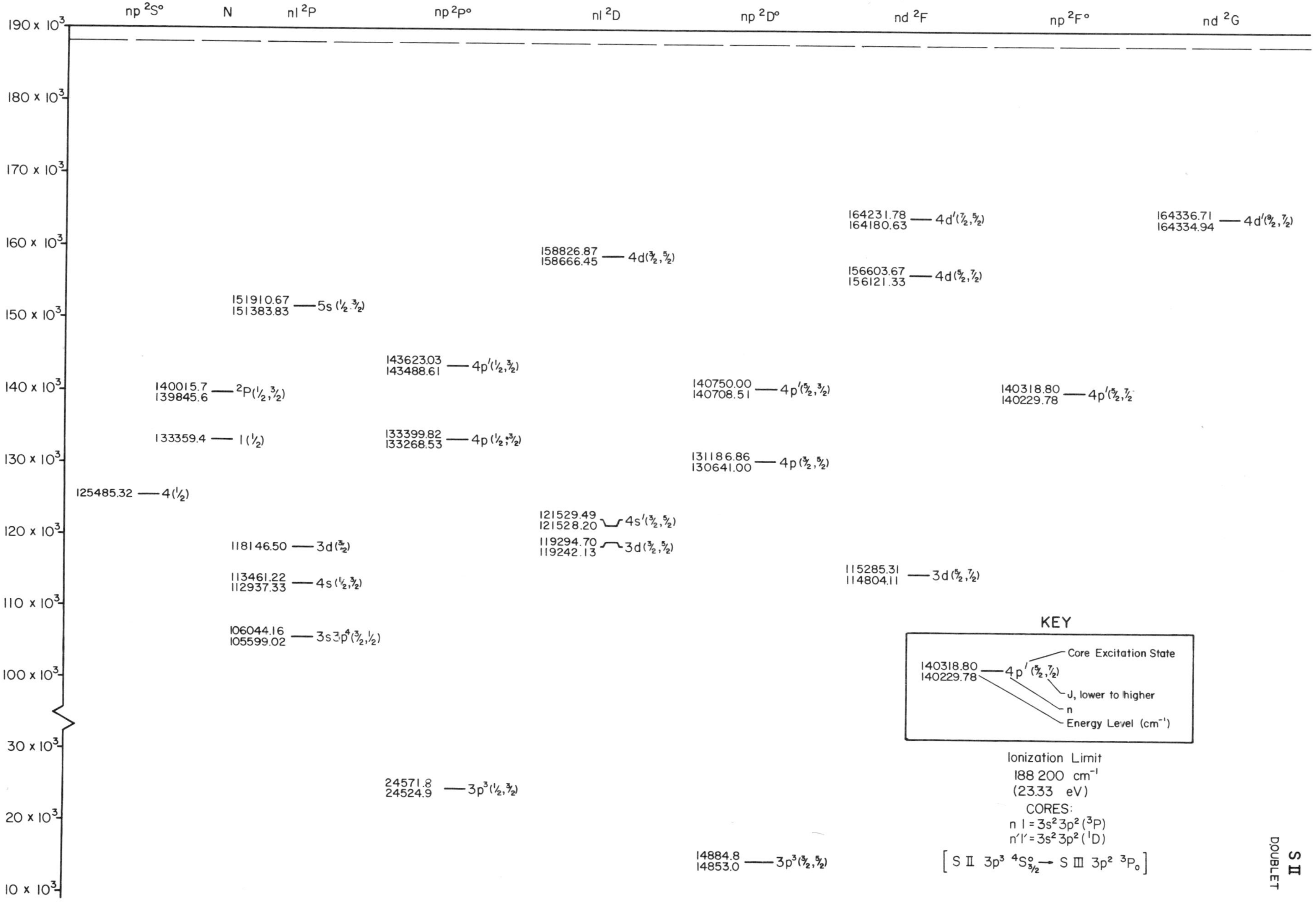

ENERGY (cm^{-1})

190 x 10^3 · 180 x 10^3 · 170 x 10^3 · 160 x 10^3 · 150 x 10^3 · 140 x 10^3 · 130 x 10^3 · 120 x 10^3 · 110 x 10^3 · 100 x 10^3 · 30 x 10^3 · 20 x 10^3 · 10 x 10^3

np $^2S^o$
- 125485.32 — 4(½)

N
- 140015.7, 139845.6 — ^2P(½, 3/2)
- 133359.4 — 1(½)

nl ^2P
- 151910.67, 151383.83 — 5s(½, 3/2)
- 118146.50 — 3d(3/2)
- 113461.22, 112937.33 — 4s(½, 3/2)
- 106044.16, 105599.02 — $3s3p^4$(3/2, ½)

np $^2P^o$
- 143623.03, 143488.61 — 4p′(½, 3/2)
- 133399.82, 133268.53 — 4p(½, 3/2)
- 24571.8, 24524.9 — $3p^3$(½, 3/2)

nl ^2D
- 158826.87, 158666.45 — 4d(3/2, 5/2)
- 121529.49, 121528.20 — 4s′(3/2, 5/2)
- 119294.70, 119242.13 — 3d(3/2, 5/2)

np $^2D^o$
- 140750.00, 140708.51 — 4p′(5/2, 3/2)
- 131186.86, 130641.00 — 4p(3/2, 5/2)
- 14884.8, 14853.0 — $3p^3$(3/2, 5/2)

nd ^2F
- 164231.78, 164180.63 — 4d′(7/2, 5/2)
- 156603.67, 156121.33 — 4d(5/2, 7/2)
- 115285.31, 114804.11 — 3d(5/2, 7/2)

np $^2F^o$
- 140318.80, 140229.78 — 4p′(5/2, 7/2)

nd ^2G
- 164336.71, 164334.94 — 4d′(9/2, 7/2)

KEY

140318.80, 140229.78 — 4p′(5/2, 7/2)
- ′ — Core Excitation State
- (5/2, 7/2) — J, lower to higher
- p — n
- 140318.80, 140229.78 — Energy Level (cm^{-1})

Ionization Limit
188 200 cm^{-1}
(23.33 eV)

CORES:
n l = $3s^2 3p^2 (^3P)$
n′l′ = $3s^2 3p^2 (^1D)$

[S II $3p^3\ ^4S^o_{3/2}$ → S III $3p^2\ ^3P_0$]

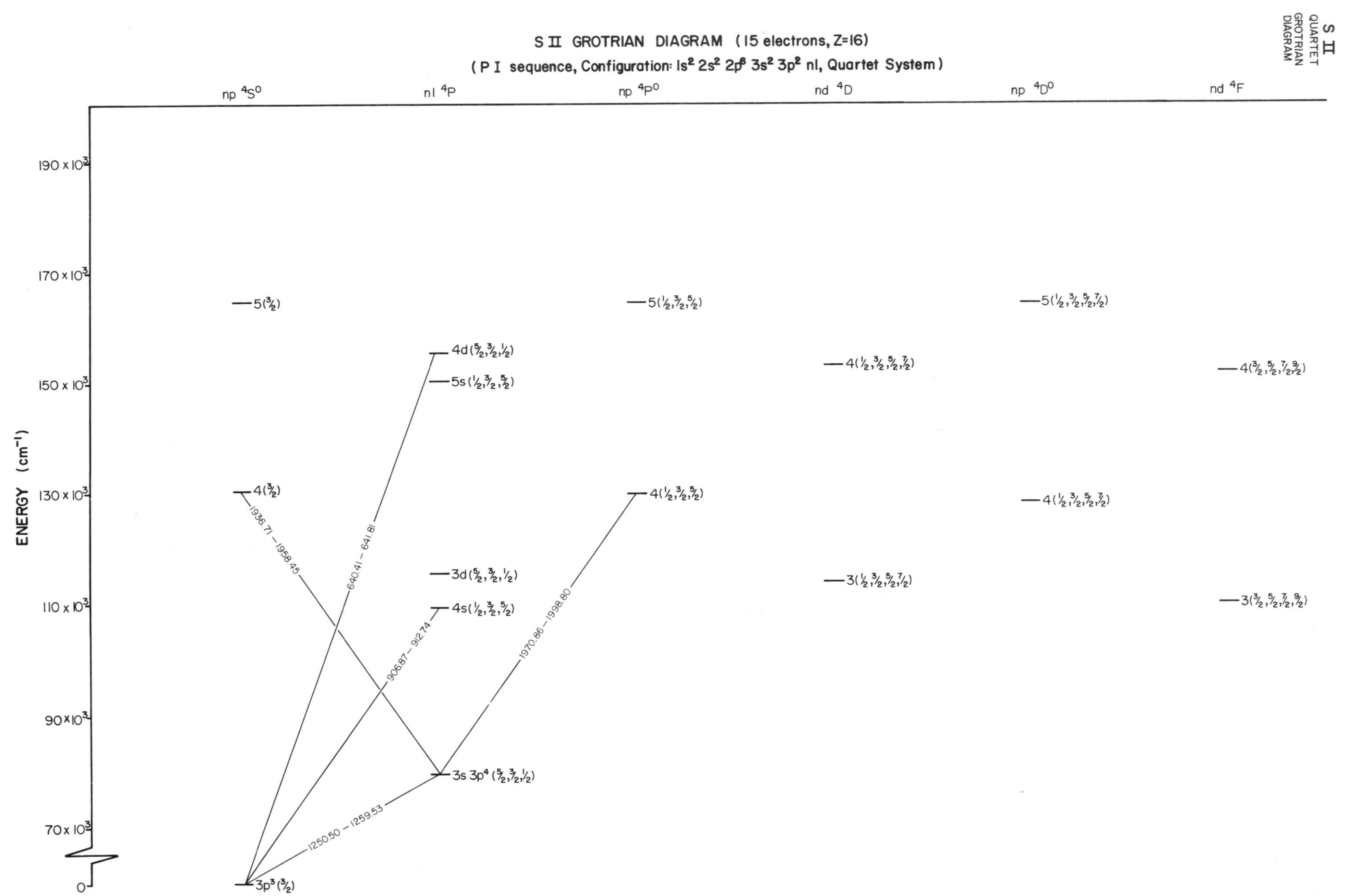
S II
QUARTET
GROTRIAN
DIAGRAM
S II GROTRIAN DIAGRAM (15 electrons, Z=16)
(P I sequence, Configuration: $1s^2\,2s^2\,2p^6\,3s^2\,3p^2\,nl$, Quartet System)
np $^4S^0$
nl ^4P
np $^4P^0$
nd ^4D
np $^4D^0$
nd ^4F
ENERGY (cm^{-1})
190×10^3
170×10^3
150×10^3
130×10^3
110×10^3
90×10^3
70×10^3
0
5(3/2)
4(3/2)
$3p^3$(3/2)
4d(5/2,3/2,1/2)
5s(1/2,3/2,5/2)
3d(5/2,3/2,1/2)
4s(1/2,3/2,5/2)
3s $3p^4$(5/2,3/2,1/2)
5(1/2,3/2,5/2)
4(1/2,3/2,5/2)
4(1/2,3/2,5/2,7/2)
3(1/2,3/2,5/2,7/2)
5(1/2,3/2,5/2,7/2)
4(1/2,3/2,5/2,7/2)
4(3/2,5/2,7/2,9/2)
3(3/2,5/2,7/2,9/2)
1936.71 – 1958.45
640.41 – 641.81
906.87 – 912.74
1970.86 – 1998.80
1250.50 – 1259.53

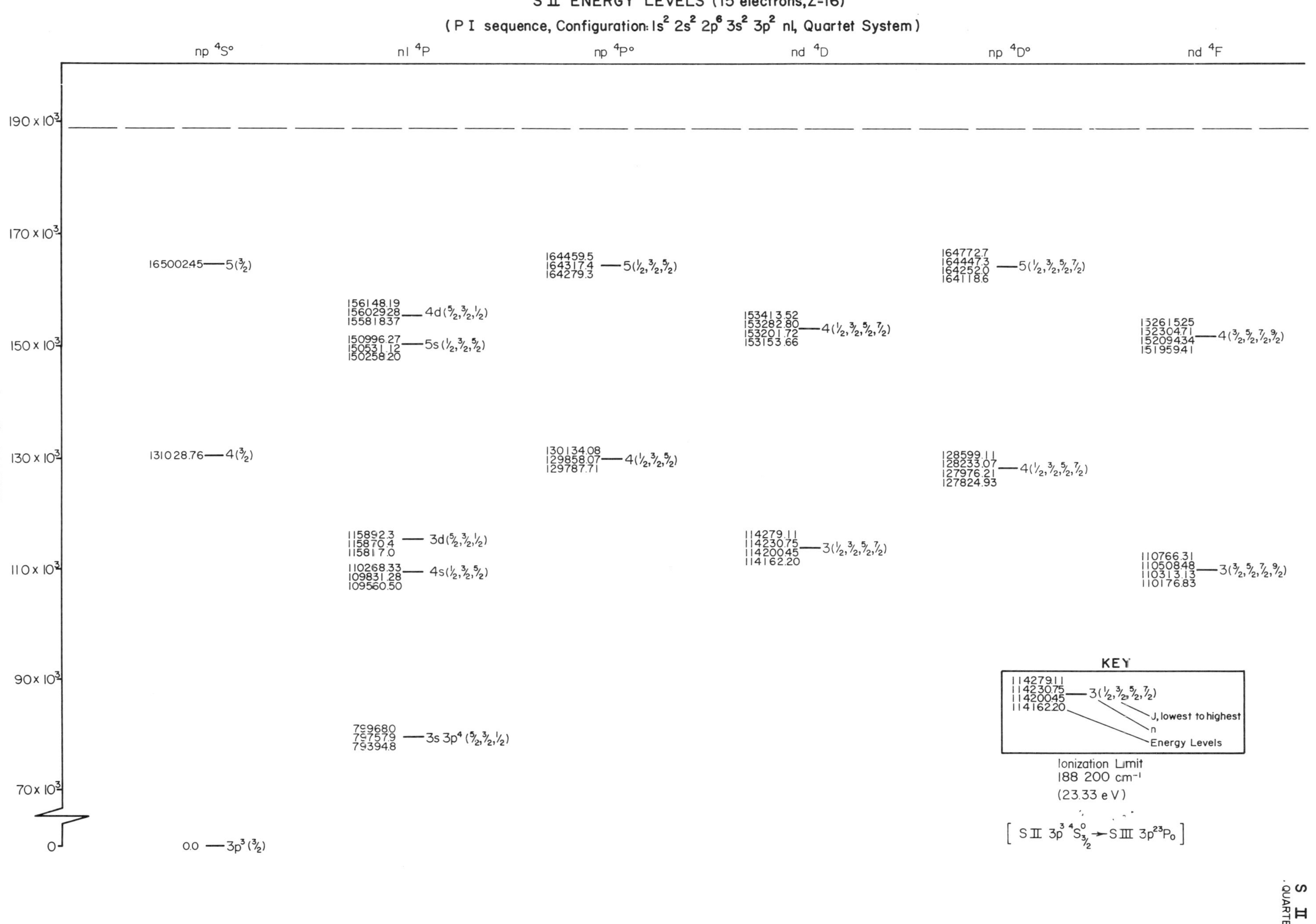
S II ENERGY LEVELS (15 electrons, Z=16)
(P I sequence, Configuration: $1s^2\ 2s^2\ 2p^6\ 3s^2\ 3p^2$ nl, Quartet System)
np $^4S^o$
nl ^4P
np $^4P^o$
nd ^4D
np $^4D^o$
nd ^4F
ENERGY (cm^{-1})
190×10^3
170×10^3
150×10^3
130×10^3
110×10^3
90×10^3
70×10^3
0
16500245 — 5(3/2)
131028.76 — 4(3/2)
0.0 — $3p^3$(3/2)
156148.19
15602928
15581837
— 4d(5/2, 3/2, 1/2)
150996.27
150531.12
150258.20
— 5s(1/2, 3/2, 5/2)
115892.3
115870.4
115817.0
— 3d(5/2, 3/2, 1/2)
110268.33
109831.28
109560.50
— 4s(1/2, 3/2, 5/2)
799680
797579
793948
— 3s $3p^4$(5/2, 3/2, 1/2)
164459.5
164317.4
164279.3
— 5(1/2, 3/2, 5/2)
130134.08
129858.07
129787.71
— 4(1/2, 3/2, 5/2)
153413.52
153282.80
153201.72
153153.66
— 4(1/2, 3/2, 5/2, 7/2)
114279.11
114230.75
11420045
114162.20
— 3(1/2, 3/2, 5/2, 7/2)
164772.7
164447.3
164252.0
164118.6
— 5(1/2, 3/2, 5/2, 7/2)
128599.11
128233.07
127976.21
127824.93
— 4(1/2, 3/2, 5/2, 7/2)
15261525
15230471
15209434
15195941
— 4(3/2, 5/2, 7/2, 9/2)
110766.31
11050848
110313.13
110176.83
— 3(3/2, 5/2, 7/2, 9/2)
KEY
11427911
11423075
11420045
11416220
— 3(1/2, 3/2, 5/2, 7/2)
J, lowest to highest
n
Energy Levels
Ionization Limit
188 200 cm^{-1}
(23.33 eV)
[S II $3p^3\ ^4S^o_{3/2}$ → S III $3p^2\ ^3P_0$]

SIII
SINGLET
TRIPLET
QUINTET
GROTRIAN
DIAGRAM

SIII GROTRIAN DIAGRAM (14 electrons, Z=16)

(Si I sequence, Configuration: $1s^2 2s^2 2p^6 3s^2 3pnl, 3s3p^2np$, Singlet, Triplet, Quintet Systems)

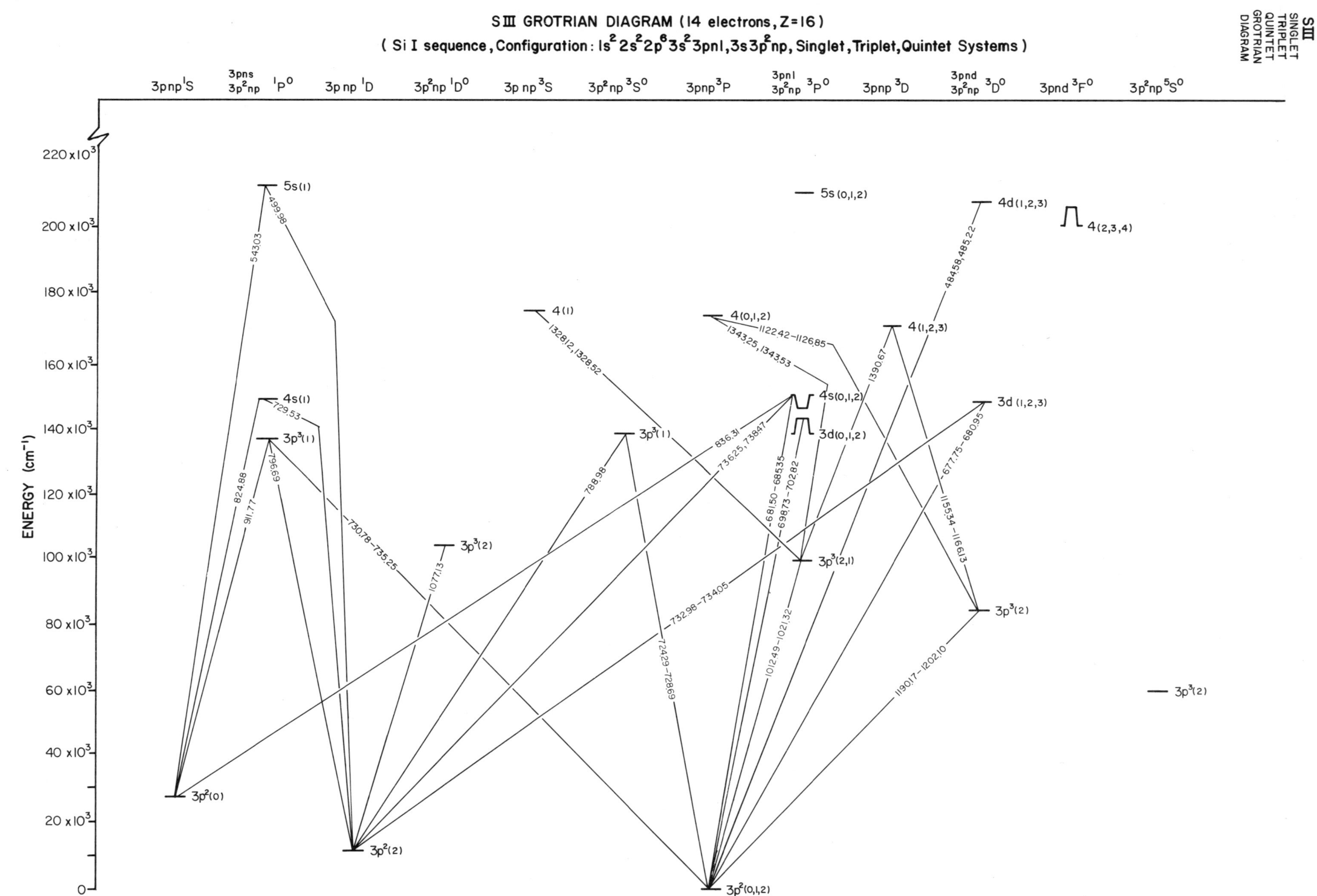

S III ENERGY LEVELS (14 electrons, Z=16)

(Si I sequence, Configuration: $1s^2 2s^2 2p^6 3s^2 3pnl, 3s3p^2np$, Singlet, Triplet, Quintet Systems)

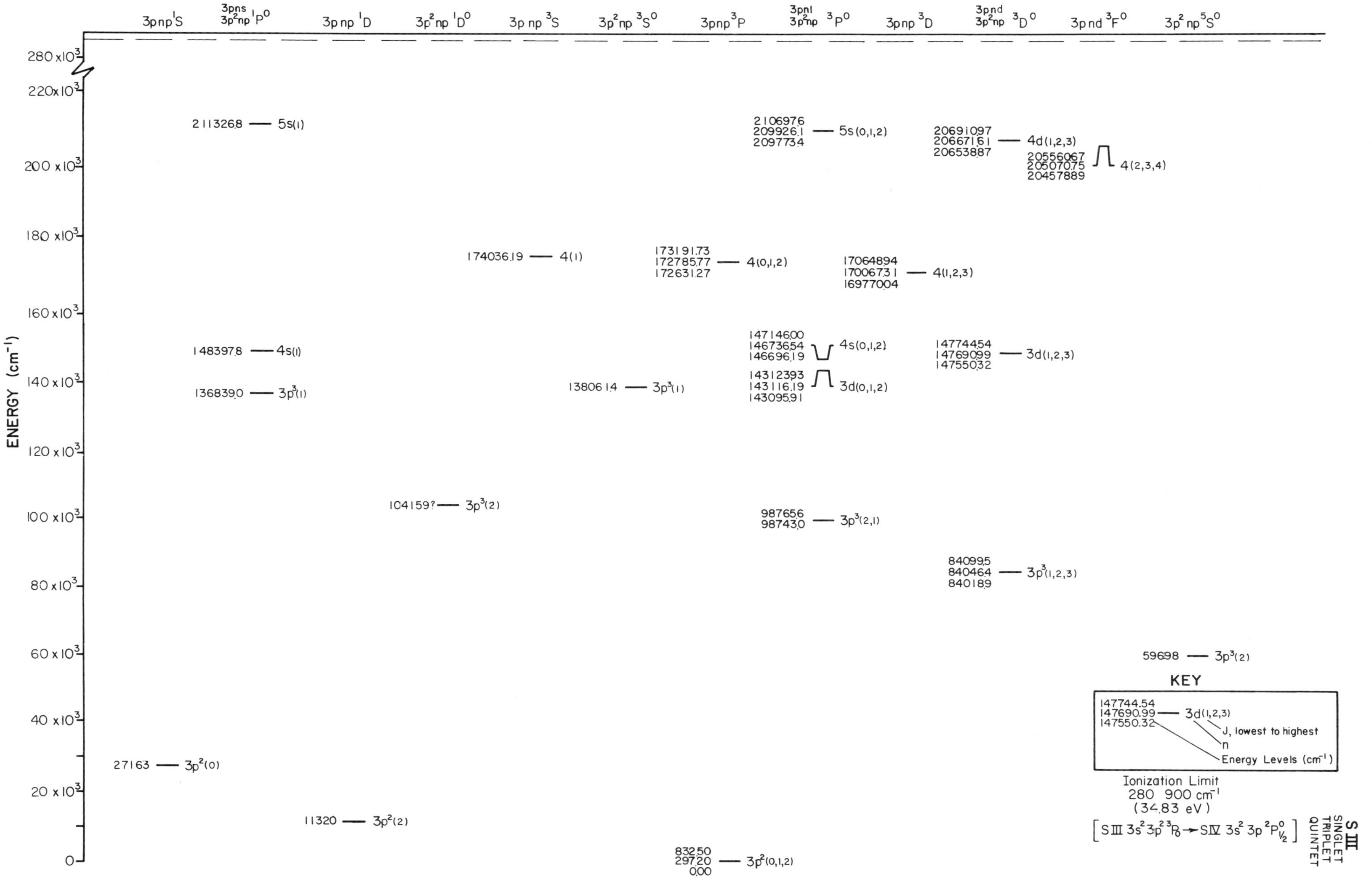

ENERGY (cm^{-1})

Axis: 0, 20 x10³, 40 x10³, 60 x10³, 80 x10³, 100 x10³, 120 x10³, 140 x10³, 160 x10³, 180 x10³, 200 x10³, 220x10³, 280 x10³

$3pnp\ ^1S$
27163 — $3p^2(0)$

$3pns$ / $3p^2np\ ^1P^o$
2113268 — 5s(1)
1483978 — 4s(1)
1368390 — $3p^3(1)$

$3pnp\ ^1D$
11320 — $3p^2(2)$

$3p^2np\ ^1D^o$
104159? — $3p^3(2)$

$3pnp\ ^3S$
174036.19 — 4(1)

$3p^2np\ ^3S^o$
138061.4 — $3p^3(1)$

$3pnp\ ^3P$
173191.73, 172785.77, 172631.27 — 4(0,1,2)
832.50, 297.20, 0.00 — $3p^2(0,1,2)$

$3pnl$ / $3p^2np\ ^3P^o$
2106976, 209926.1, 2097734 — 5s(0,1,2)
147146.00, 146736.54, 146696.19 — 4s(0,1,2)
143123.93, 143116.19, 143095.91 — 3d(0,1,2)
987656, 987430 — $3p^3(2,1)$

$3pnp\ ^3D$
17064894, 17006731, 16977004 — 4(1,2,3)

$3pnd$ / $3p^2np\ ^3D^o$
20691097, 20667161, 20653887 — 4d(1,2,3)
14774454, 14769099, 14755032 — 3d(1,2,3)
840995, 840464, 840189 — $3p^3(1,2,3)$

$3pnd\ ^3F^o$
20556067, 20507075, 20457889 — 4(2,3,4)

$3p^2np\ ^5S^o$
59698 — $3p^3(2)$

KEY

147744.54, 147690.99, 147550.32 — 3d(1,2,3)
J, lowest to highest
n
Energy Levels (cm^{-1})

Ionization Limit
280 900 cm^{-1}
(34.83 eV)
[$S\,III\ 3s^2 3p^2\ ^3P_0 \rightarrow S\,IV\ 3s^2 3p\ ^2P^o_{1/2}$]

S III
SINGLET
TRIPLET
QUINTET

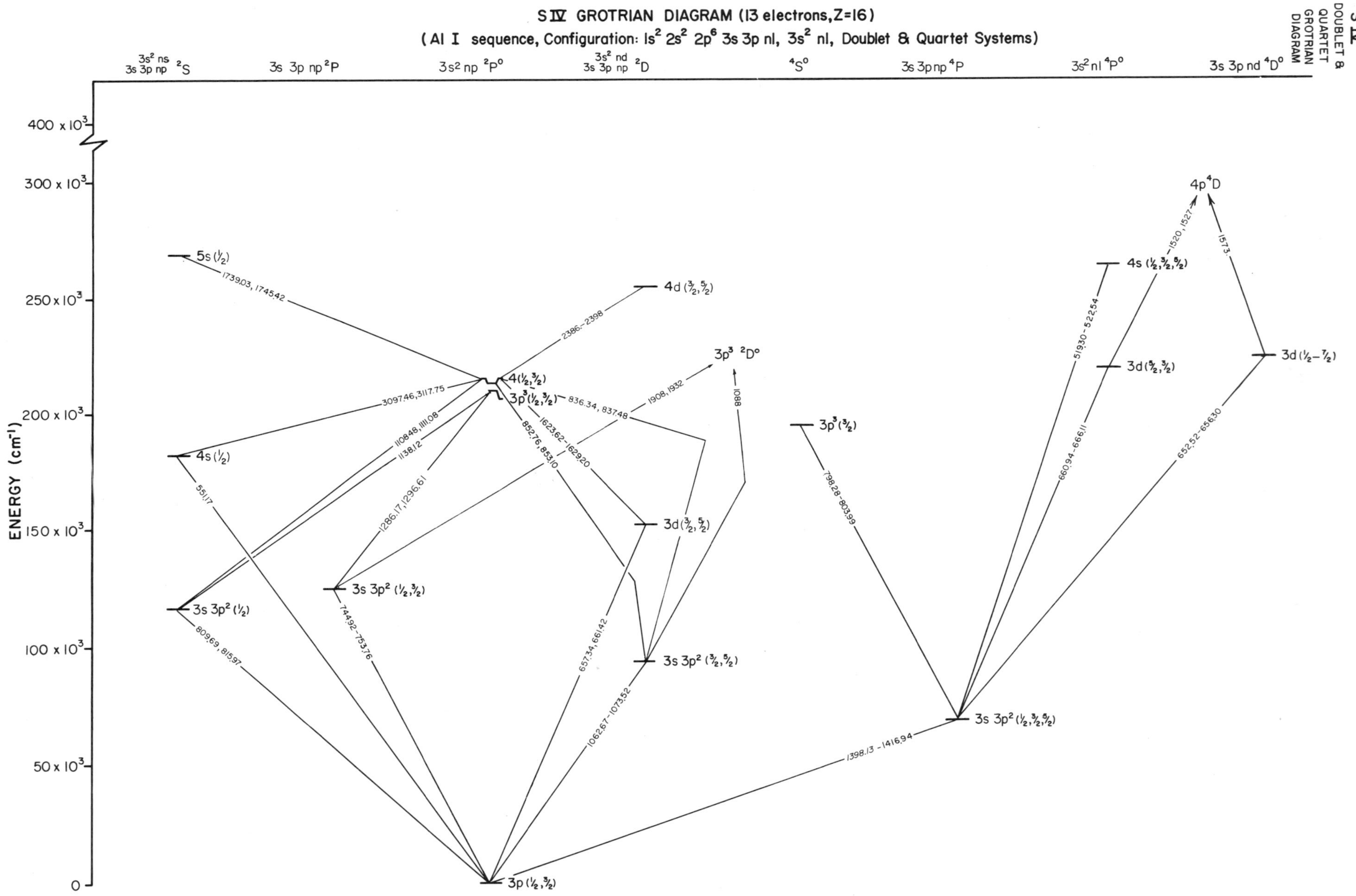

S IV
DOUBLET & QUARTET GROTRIAN DIAGRAM
S IV GROTRIAN DIAGRAM (13 electrons, Z=16)
(Al I sequence, Configuration: 1s² 2s² 2p⁶ 3s 3p nl, 3s² nl, Doublet & Quartet Systems)
3s² ns 3s 3p np ²S
3s 3p np ²P
3s² np ²P°
3s² nd 3s 3p np ²D
⁴S°
3s 3p np ⁴P
3s² nl ⁴P°
3s 3p nd ⁴D°
5s (1/2)
4s (1/2)
3s 3p² (1/2)
3s 3p² (1/2, 3/2)
4(1/2, 3/2)
3p³(1/2, 3/2)
4d (3/2, 5/2)
3p³ ²D°
3d (3/2, 5/2)
3s 3p² (3/2, 5/2)
3p (1/2, 3/2)
3p³ (3/2)
3s 3p² (1/2, 3/2, 5/2)
4s (1/2, 3/2, 5/2)
3d (5/2, 3/2)
4p ⁴D
3d (1/2 - 7/2)
1739.03, 1745.42
551.17
809.69, 815.97
3097.46, 3117.75
1108.48, 1111.08
1138.12
1286.17, 1296.61
744.92 - 753.76
2386 - 2398
1908, 1932
836.34, 837.48
1623.62 - 1629.20
852.76, 853.10
1088
657.34, 661.42
1062.67 - 1073.52
798.28 - 803.99
1398.13 - 1416.94
519.30 - 522.54
660.94 - 666.11
1520, 1527
1573
652.52 - 656.30
ENERGY (cm⁻¹)
400 x 10³
300 x 10³
250 x 10³
200 x 10³
150 x 10³
100 x 10³
50 x 10³
0

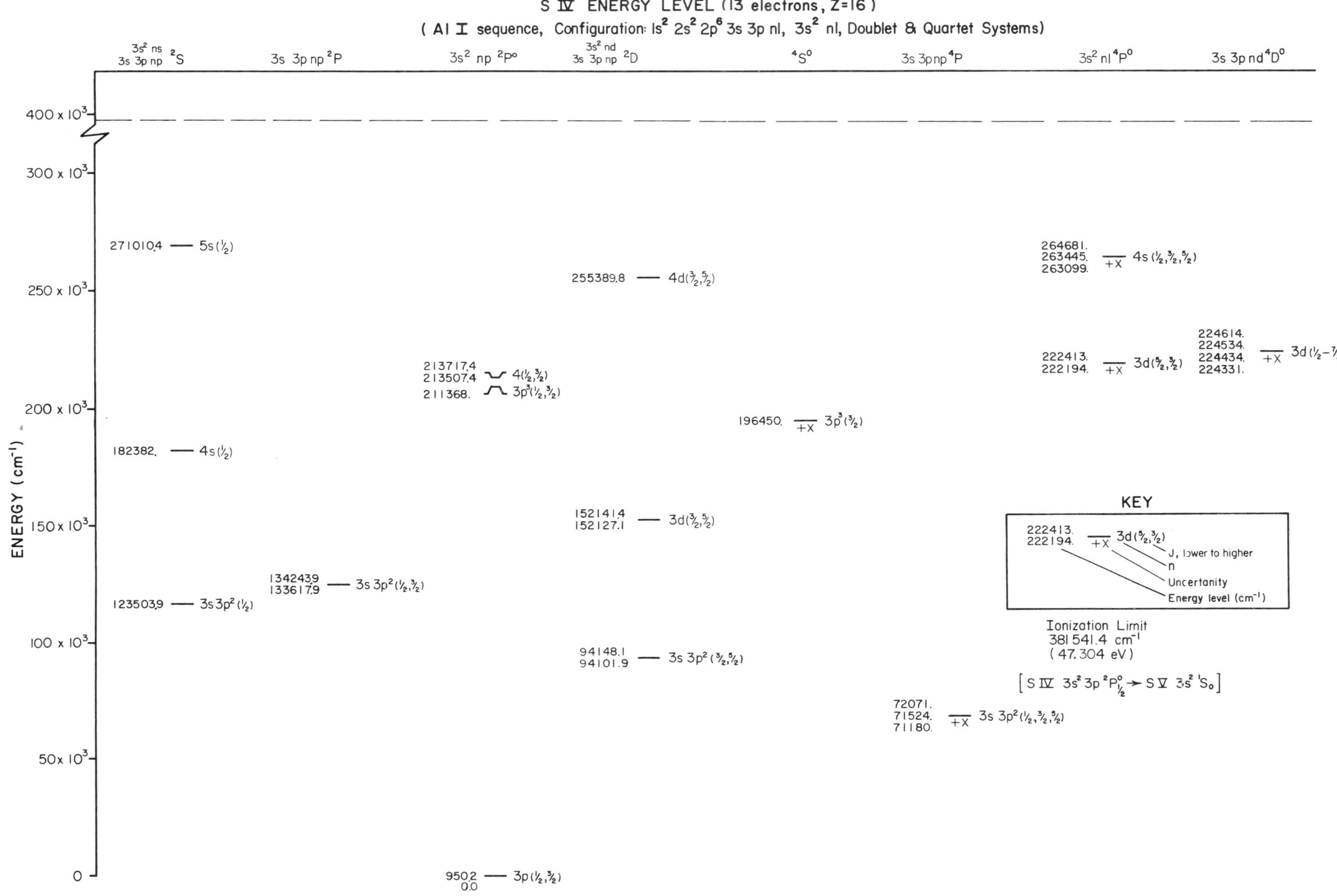
S IV ENERGY LEVEL (13 electrons, Z=16)
(Al I sequence, Configuration: $1s^2 2s^2 2p^6 3s 3p nl$, $3s^2 nl$, Doublet & Quartet Systems)
$3s^2 ns$ $3s 3p np$ ^2S
$3s 3p np$ ^2P
$3s^2 np$ $^2P^o$
$3s^2 nd$ $3s 3p np$ ^2D
$^4S^o$
$3s 3p np$ ^4P
$3s^2 nl$ $^4P^o$
$3s 3p nd$ $^4D^o$
400 x 10^3
300 x 10^3
250 x 10^3
200 x 10^3
150 x 10^3
100 x 10^3
50 x 10^3
0
ENERGY (cm^{-1})
271010.4 — 5s(1/2)
182382. — 4s(1/2)
123503.9 — 3s3p²(1/2)
134243.9
133617.9 — 3s 3p²(1/2,3/2)
213717.4
213507.4 4(1/2,3/2)
211368. 3p³(1/2,3/2)
950.2
0.0 — 3p(1/2,3/2)
255389.8 — 4d(3/2,5/2)
152141.4
152127.1 — 3d(3/2,5/2)
94148.1
94101.9 — 3s 3p²(3/2,5/2)
196450. +X 3p³(3/2)
72071.
71524.
71180. +X 3s 3p²(1/2,3/2,5/2)
264681.
263445.
263099. +X 4s(1/2,3/2,5/2)
222413.
222194. +X 3d(5/2,3/2)
224614.
224534.
224434.
224331. +X 3d(1/2–7/2)
KEY
222413.
222194. +X 3d(5/2,3/2)
J, lower to higher
n
Uncertanity
Energy level (cm^{-1})
Ionization Limit
381 541.4 cm^{-1}
(47.304 eV)
[S IV $3s^2 3p$ $^2P^o_{1/2}$ → S V $3s^2$ $^1S_0$]

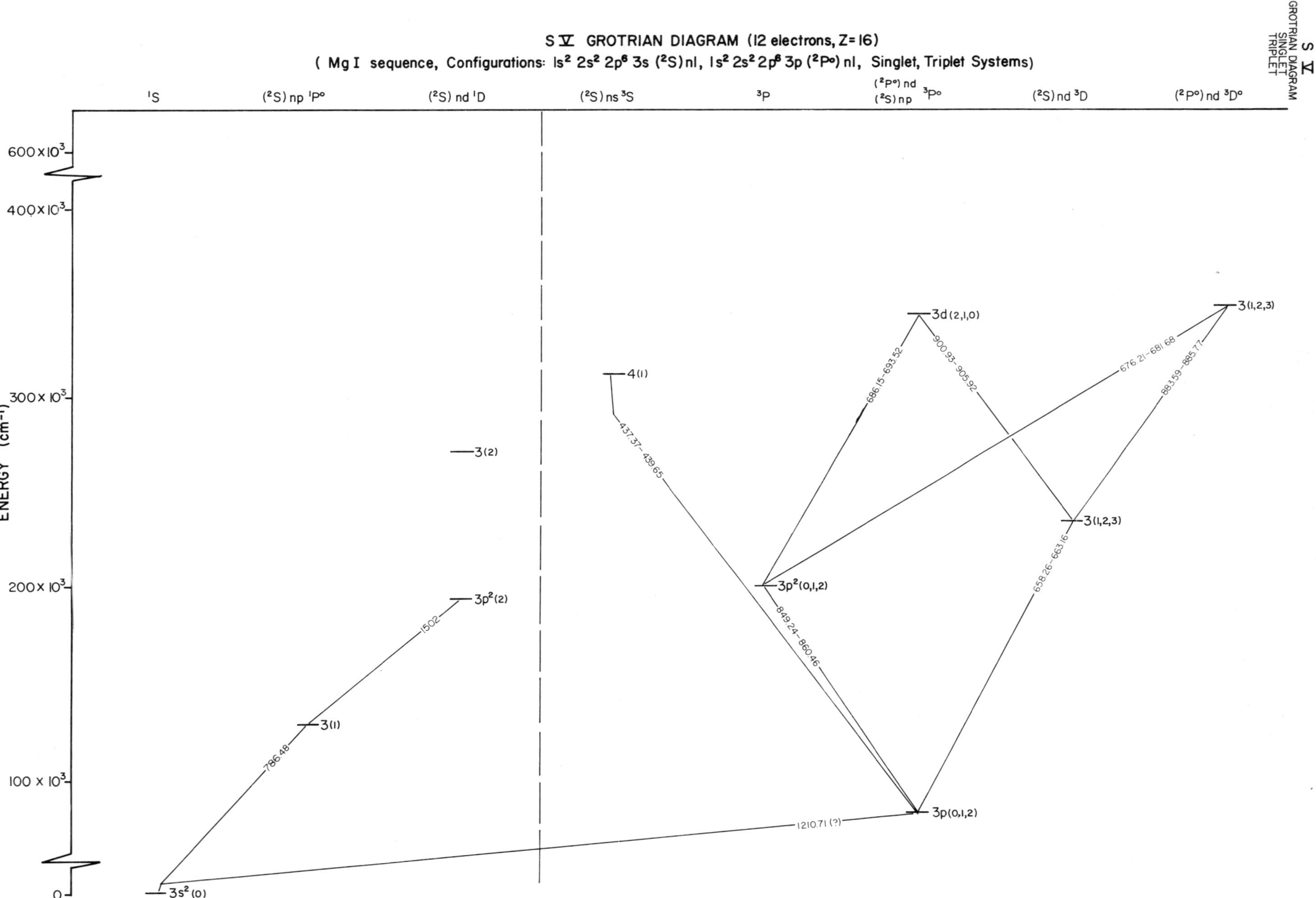
S V GROTRIAN DIAGRAM (12 electrons, Z=16)
(Mg I sequence, Configurations: 1s^2 2s^2 2p^6 3s (^{2}S) nl, 1s^2 2s^2 2p^6 3p (^{2}P°) nl, Singlet, Triplet Systems)
S V
GROTRIAN DIAGRAM
SINGLET
TRIPLET
^{1}S
(^{2}S) np ^{1}P°
(^{2}S) nd ^{1}D
(^{2}S) ns ^{3}S
^{3}P
(^{2}P°) nd
(^{2}S) np ^{3}P°
(^{2}S) nd ^{3}D
(^{2}P°) nd ^{3}D°
600 x 10^3
400 x 10^3
300 x 10^3
200 x 10^3
100 x 10^3
0
ENERGY (cm^{-1})
3(2)
3p^2(2)
3(1)
3s^2 (0)
4(1)
3d (2,1,0)
3(1,2,3)
3(1,2,3)
3p^2(0,1,2)
3p(0,1,2)
1502
786.48
437.37–439.65
686.15–693.52
900.93–905.92
676.21–681.68
883.59–885.77
658.26–663.16
849.24–860.46
1210.71 (?)

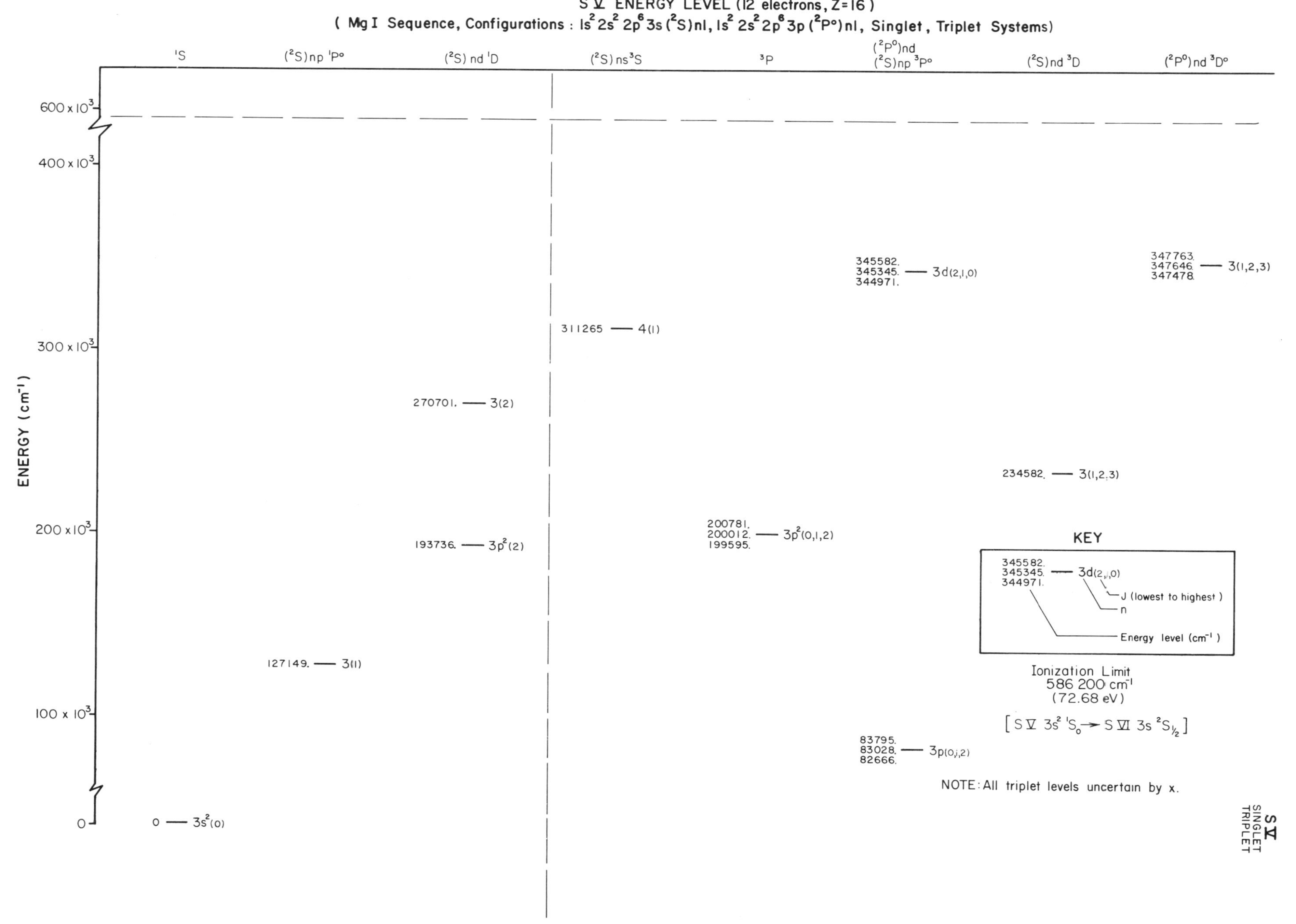

S V ENERGY LEVEL (12 electrons, Z=16)
(Mg I Sequence, Configurations : $1s^2 2s^2 2p^6 3s (^2S) nl$, $1s^2 2s^2 2p^6 3p (^2P^o) nl$, Singlet, Triplet Systems)
^1S
$(^2S) np\ ^1P^o$
$(^2S) nd\ ^1D$
$(^2S) ns\ ^3S$
^3P
$(^2P^0) nd$
$(^2S) np\ ^3P^o$
$(^2S) nd\ ^3D$
$(^2P^0) nd\ ^3D^o$
ENERGY (cm^{-1})
600 x 10^3
400 x 10^3
300 x 10^3
200 x 10^3
100 x 10^3
0
0 — $3s^2(0)$
127149. — 3(1)
270701. — 3(2)
193736. — $3p^2(2)$
311265 — 4(1)
200781. 200012. 199595. — $3p^2(0,1,2)$
345582. 345345. 344971. — 3d(2,1,0)
83795. 83028. 82666. — 3p(0,1,2)
234582. — 3(1,2,3)
347763. 347646. 347478. — 3(1,2,3)
KEY
345582. 345345. 344971. — 3d(2,1,0)
J (lowest to highest)
n
Energy level (cm^{-1})
Ionization Limit
586 200 cm^{-1}
(72.68 eV)
[S V $3s^2\ ^1S_0$ → S VI $3s\ ^2S_{1/2}$]
NOTE: All triplet levels uncertain by x.
S V
SINGLET
TRIPLET

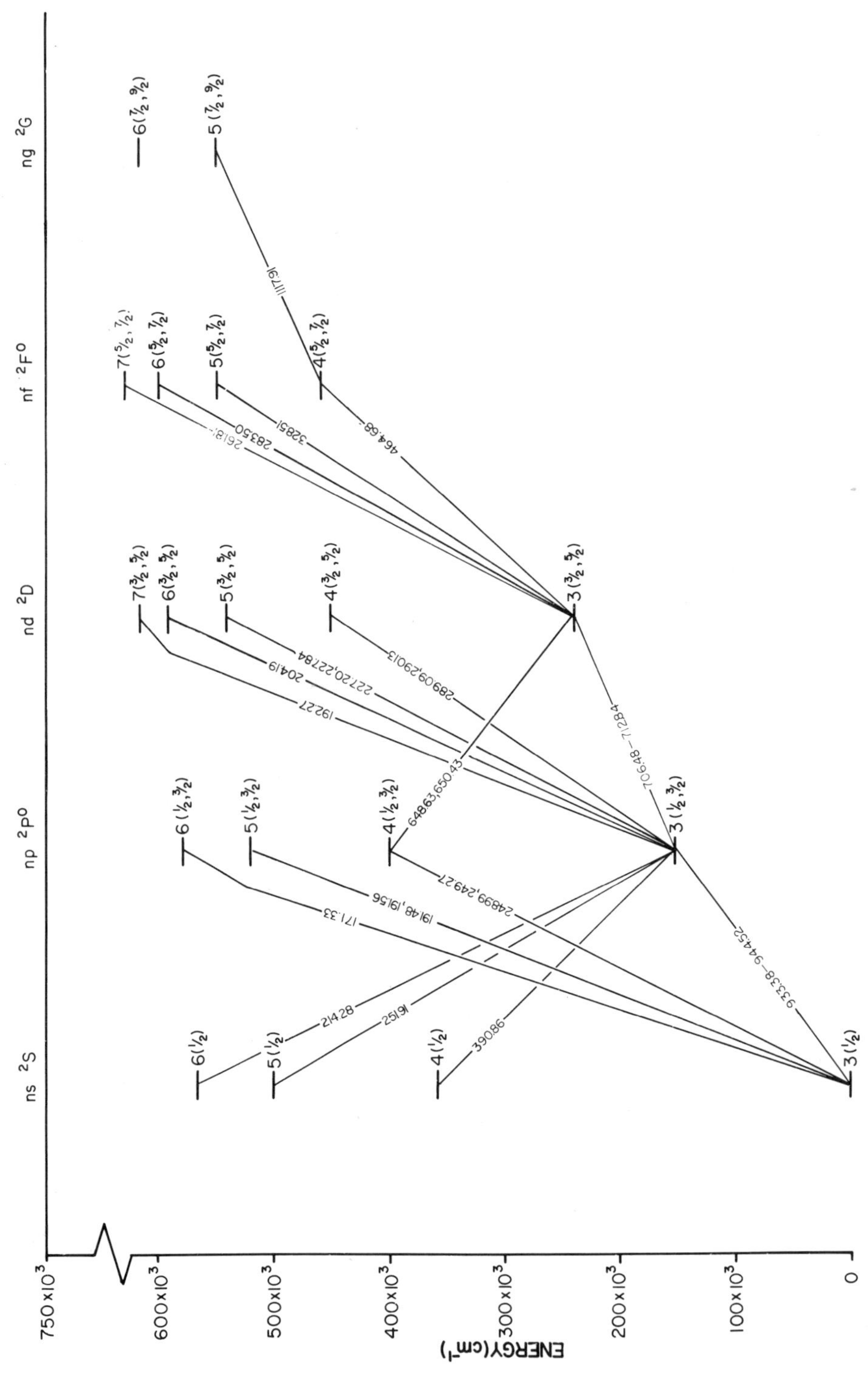
S VI GROTRIAN DIAGRAM (11 electrons, Z=16)
(Na I sequence Configuration: 1s² 2s² 2p⁶ nl, Doublet System)
ns ²S
np ²P°
nd ²D
nf ²F°
ng ²G
ENERGY (cm⁻¹)

S VI ENERGY LEVELS (11 electrons, Z = 16)

(Na I sequence, Configuration: $1s^2\ 2s^2\ 2p^6\ nl$, Doublet System)

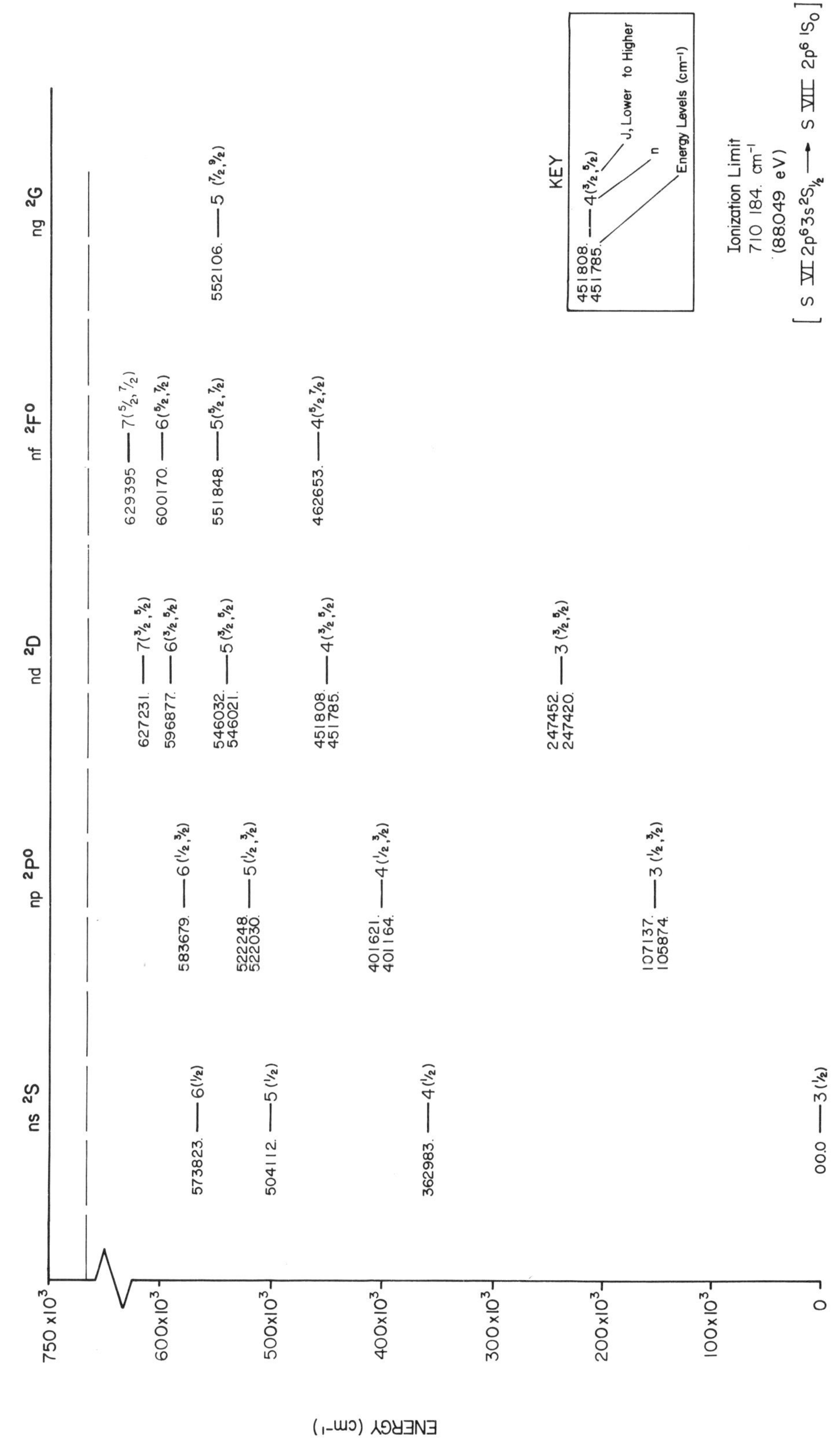

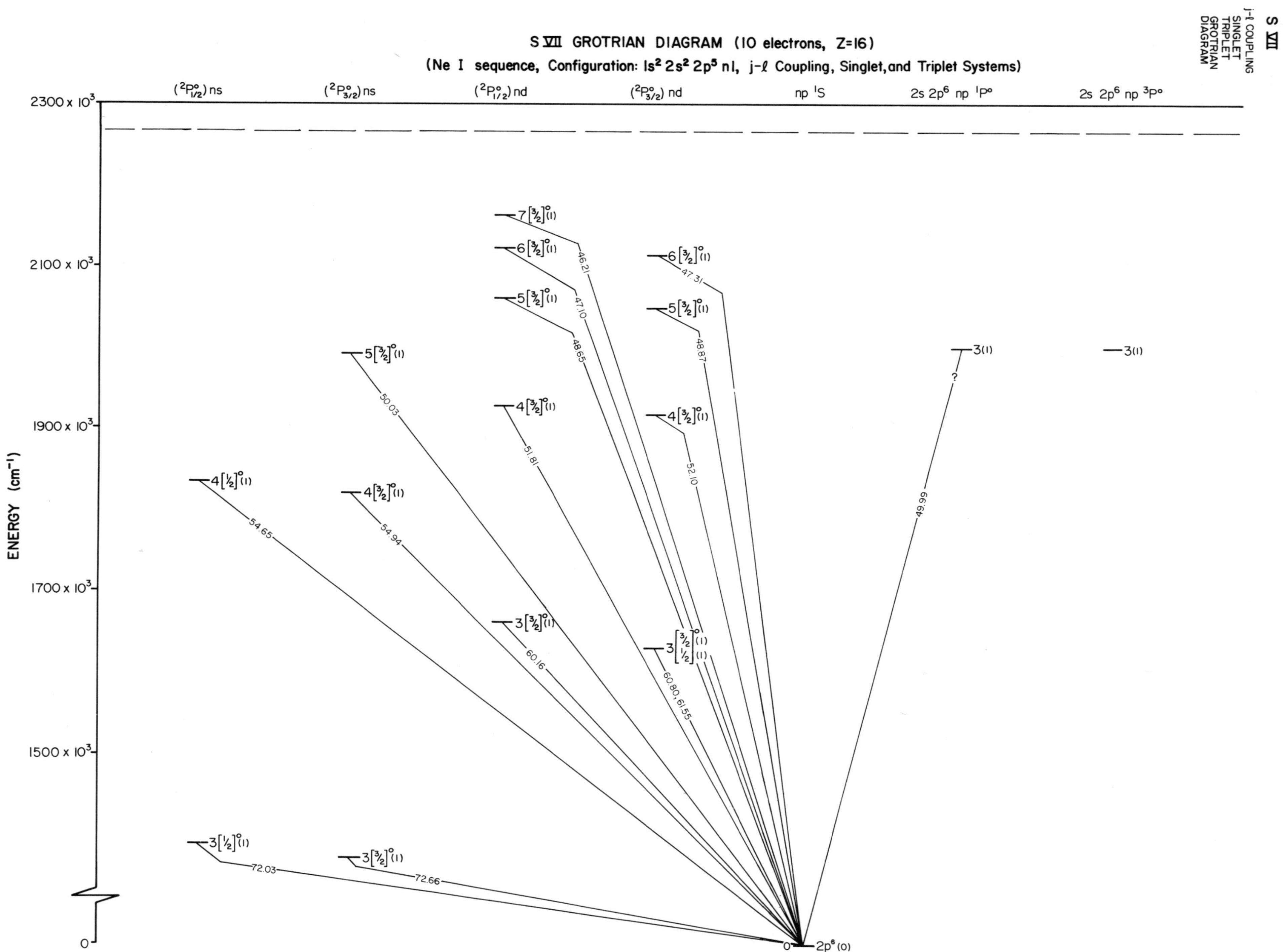

S VII
j-ℓ COUPLING
SINGLET
TRIPLET
GROTRIAN
DIAGRAM
S VII GROTRIAN DIAGRAM (10 electrons, Z=16)
(Ne I sequence, Configuration: $1s^2 2s^2 2p^5 nl$, j-ℓ Coupling, Singlet, and Triplet Systems)
$(^2P^o_{1/2})ns$
$(^2P^o_{3/2})ns$
$(^2P^o_{1/2})nd$
$(^2P^o_{3/2})nd$
$np\ ^1S$
$2s\,2p^6\,np\ ^1P^o$
$2s\,2p^6\,np\ ^3P^o$
$7[3/2]^o(1)$
$6[3/2]^o(1)$
$5[3/2]^o(1)$
$4[3/2]^o(1)$
$3[3/2]^o(1)$
46.21
47.10
48.65
51.81
60.16
$6[3/2]^o(1)$
$5[3/2]^o(1)$
$4[3/2]^o(1)$
$3[3/2, 1/2]^o(1)$
47.31
48.87
52.10
60.80, 61.55
$5[3/2]^o(1)$
$4[3/2]^o(1)$
$3[3/2]^o(1)$
50.03
54.94
72.66
$4[1/2]^o(1)$
$3[1/2]^o(1)$
54.65
72.03
3(1)
?
49.99
3(1)
$2p^6(0)$
0
2300 x 10^3
2100 x 10^3
1900 x 10^3
1700 x 10^3
1500 x 10^3
0
ENERGY (cm^{-1})

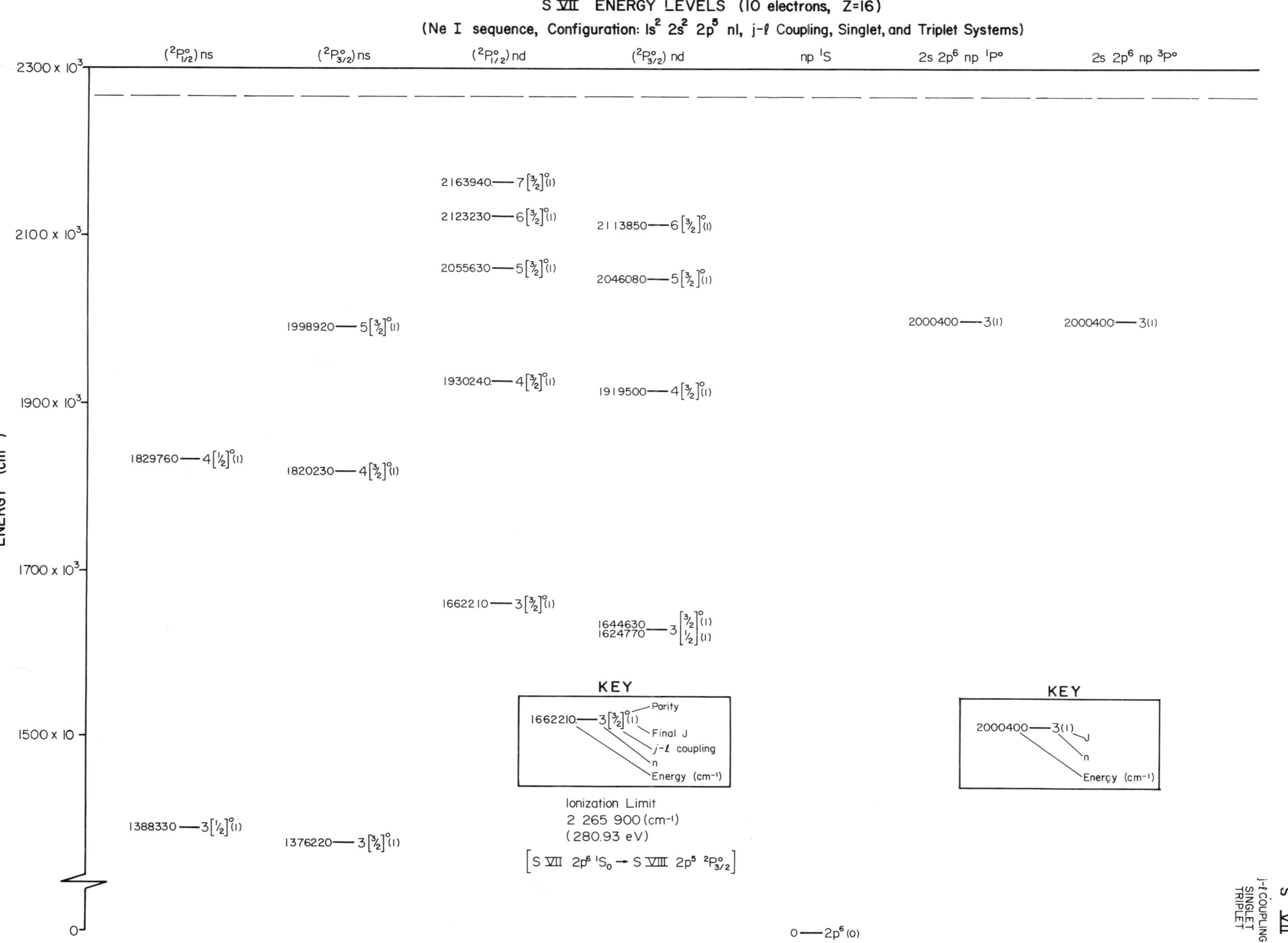

S VII ENERGY LEVELS (10 electrons, Z=16)
(Ne I sequence, Configuration: $1s^2 2s^2 2p^5 nl$, j-ℓ Coupling, Singlet, and Triplet Systems)
$(^2P^o_{1/2})$ ns
$(^2P^o_{3/2})$ ns
$(^2P^o_{1/2})$ nd
$(^2P^o_{3/2})$ nd
np ^1S
2s $2p^6$ np $^1P^o$
2s $2p^6$ np $^3P^o$
2300 x 10^3
2100 x 10^3
1900 x 10^3
1700 x 10^3
1500 x 10
0
ENERGY (cm^{-1})
2163940 — 7[3/2]°(1)
2123230 — 6[3/2]°(1)
2055630 — 5[3/2]°(1)
1930240 — 4[3/2]°(1)
1662210 — 3[3/2]°(1)
2113850 — 6[3/2]°(1)
2046080 — 5[3/2]°(1)
1919500 — 4[3/2]°(1)
1644630 — 3[3/2]°(1)
1624770 — 3[1/2]°(1)
1998920 — 5[3/2]°(1)
1820230 — 4[3/2]°(1)
1376220 — 3[3/2]°(1)
1829760 — 4[1/2]°(1)
1388330 — 3[1/2]°(1)
2000400 — 3(1)
2000400 — 3(1)
0 — $2p^6$(0)
KEY
1662210 — 3[3/2]°(1)
Parity
Final J
j-ℓ coupling
n
Energy (cm^{-1})
KEY
2000400 — 3(1)
J
n
Energy (cm^{-1})
Ionization Limit
2 265 900 (cm^{-1})
(280.93 eV)
[S VII $2p^6$ $^1S_0$ → S VIII $2p^5$ $^2P^o_{3/2}$]
S VII
j-ℓ COUPLING
SINGLET
TRIPLET

S VIII
DOUBLET
QUARTET
GROTRIAN DIAGRAM

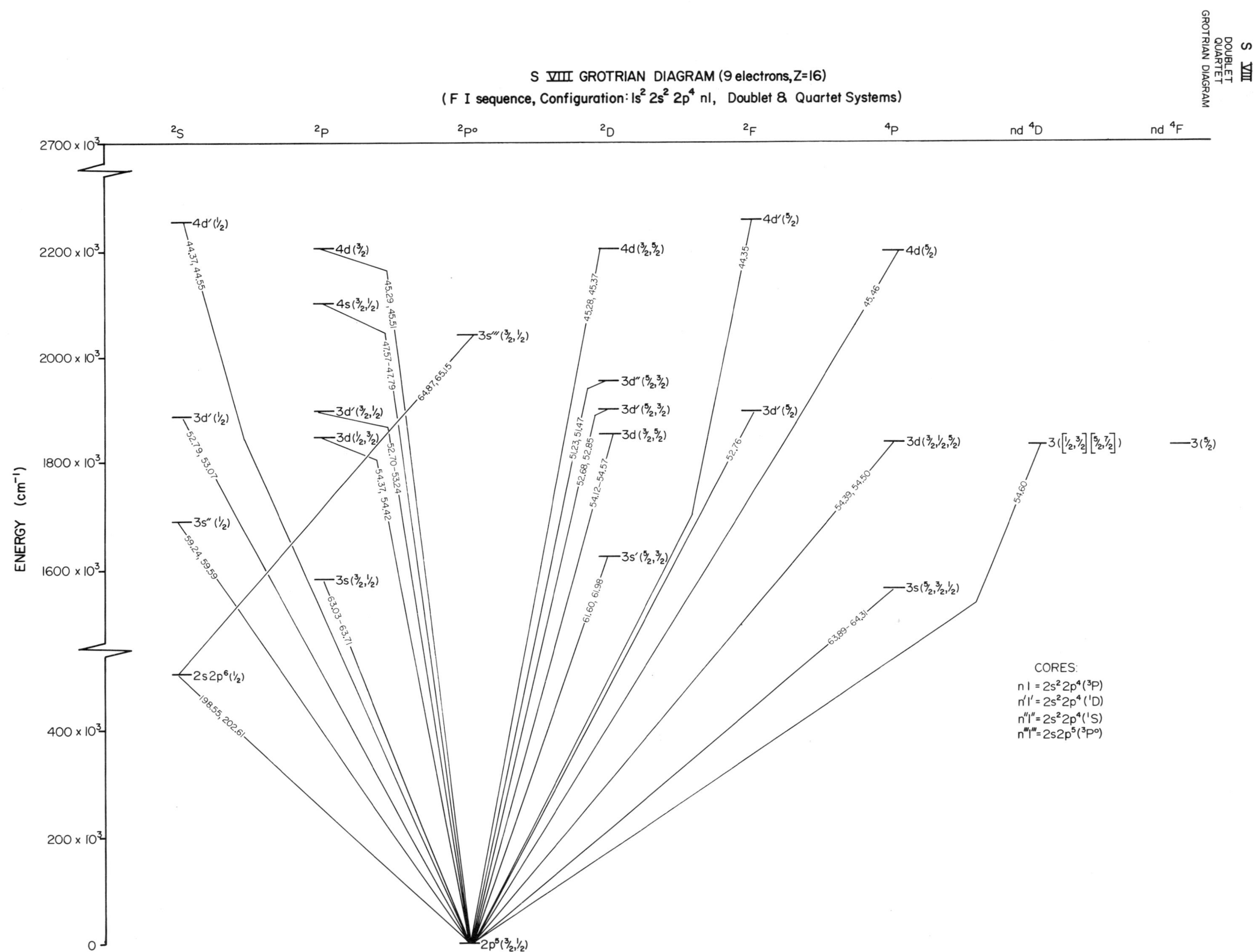

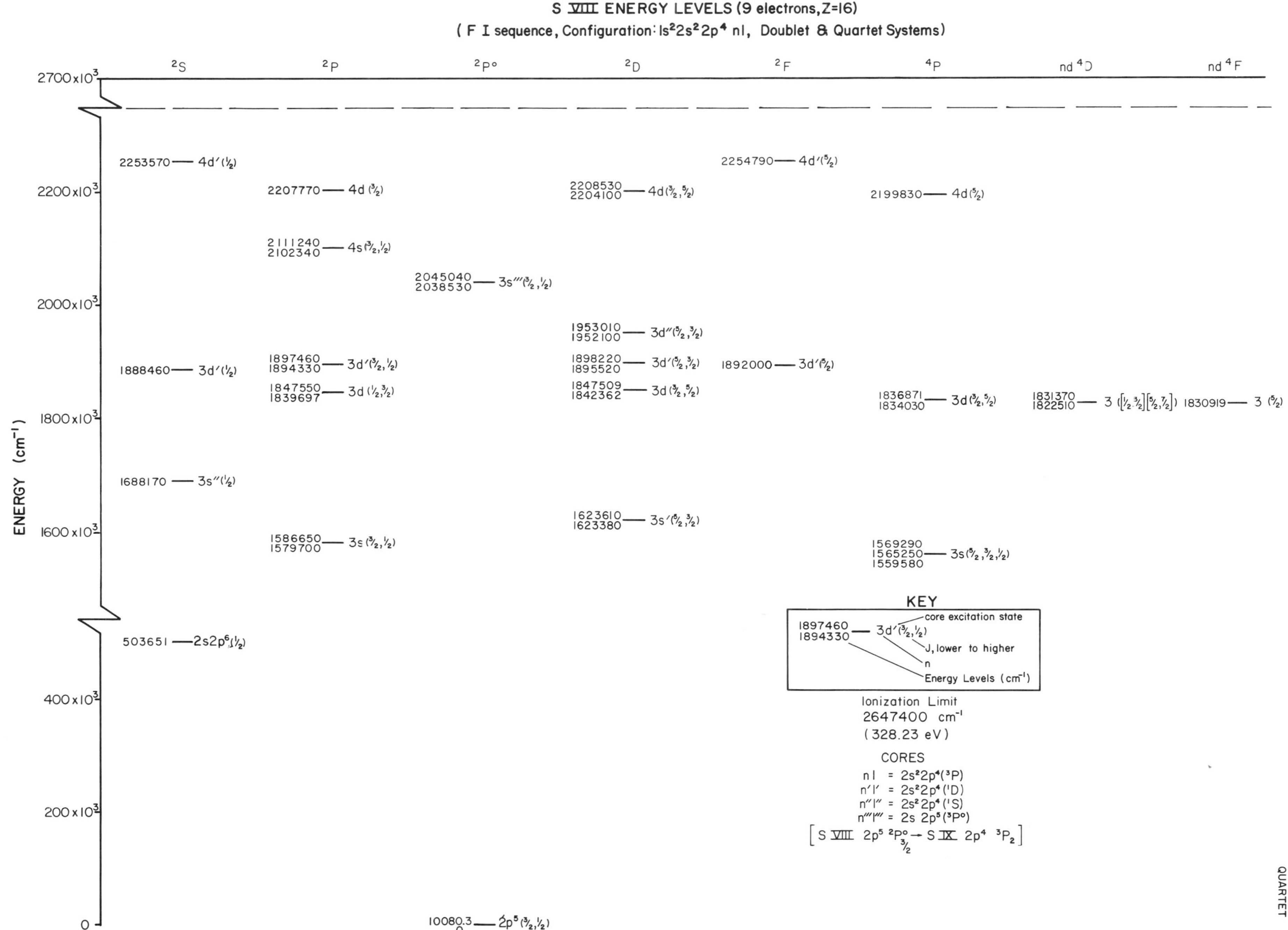

S VIII
DOUBLET
QUARTET

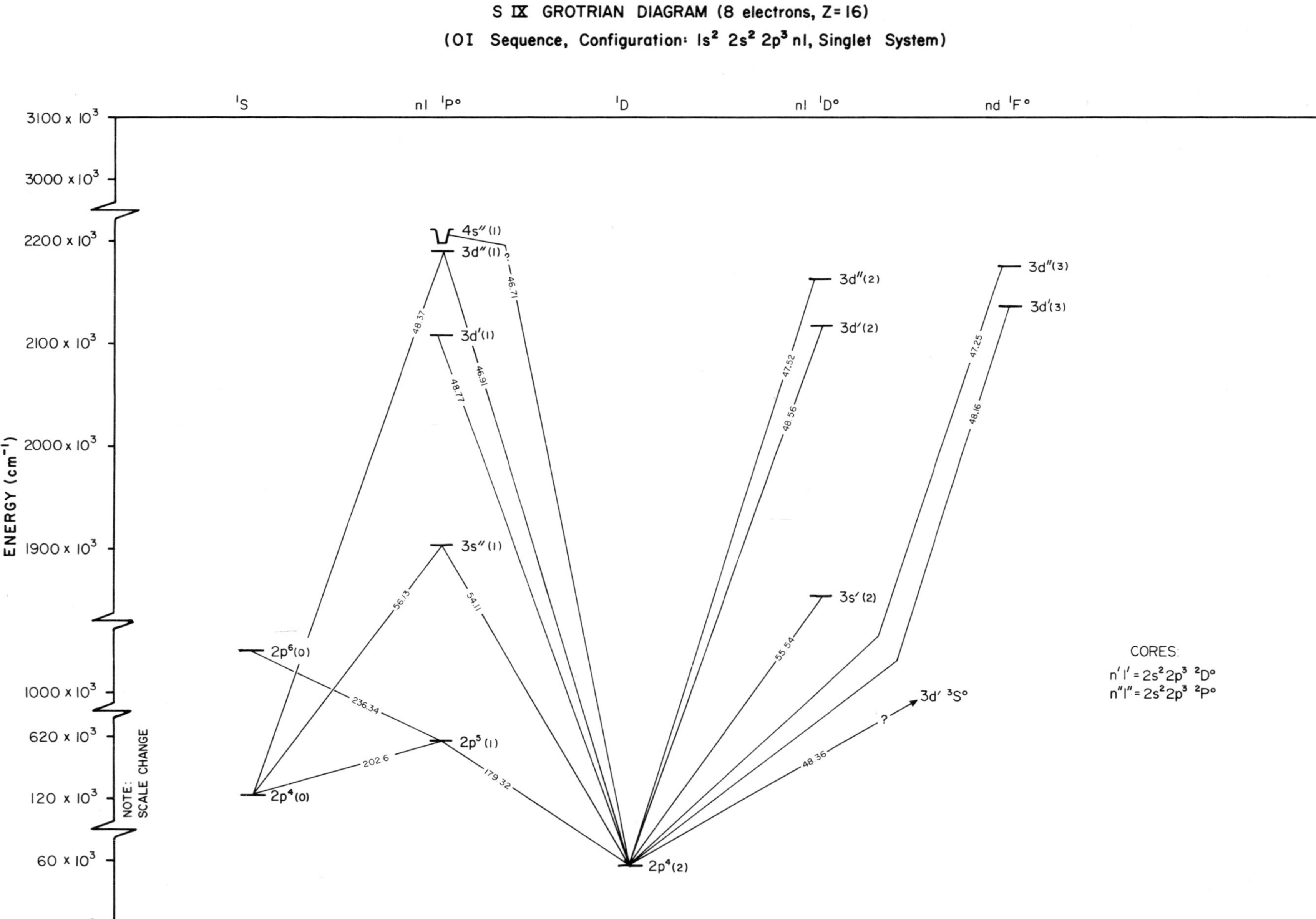
S IX GROTRIAN DIAGRAM (8 electrons, Z= 16)
(OI Sequence, Configuration: $1s^2\ 2s^2\ 2p^3\ nl$, Singlet System)
^1S
nl $^1P°$
^1D
nl $^1D°$
nd $^1F°$
3100 x 10^3
3000 x 10^3
2200 x 10^3
2100 x 10^3
2000 x 10^3
1900 x 10^3
1000 x 10^3
620 x 10^3
120 x 10^3
60 x 10^3
0
ENERGY (cm^{-1})
NOTE:
SCALE CHANGE
4s″(1)
3d″(1)
3d′(1)
3s″(1)
$2p^6$(0)
$2p^5$(1)
$2p^4$(0)
$2p^4$(2)
3d″(2)
3d′(2)
3s′(2)
3d″(3)
3d′(3)
3d′ $^3S°$
?
46.71
48.37
46.91
48.77
56.13
54.11
236.34
202.6
179.32
47.52
48.56
55.54
47.25
48.16
48.36
CORES:
n′l′ = $2s^2 2p^3\ ^2D°$
n″l″ = $2s^2 2p^3\ ^2P°$

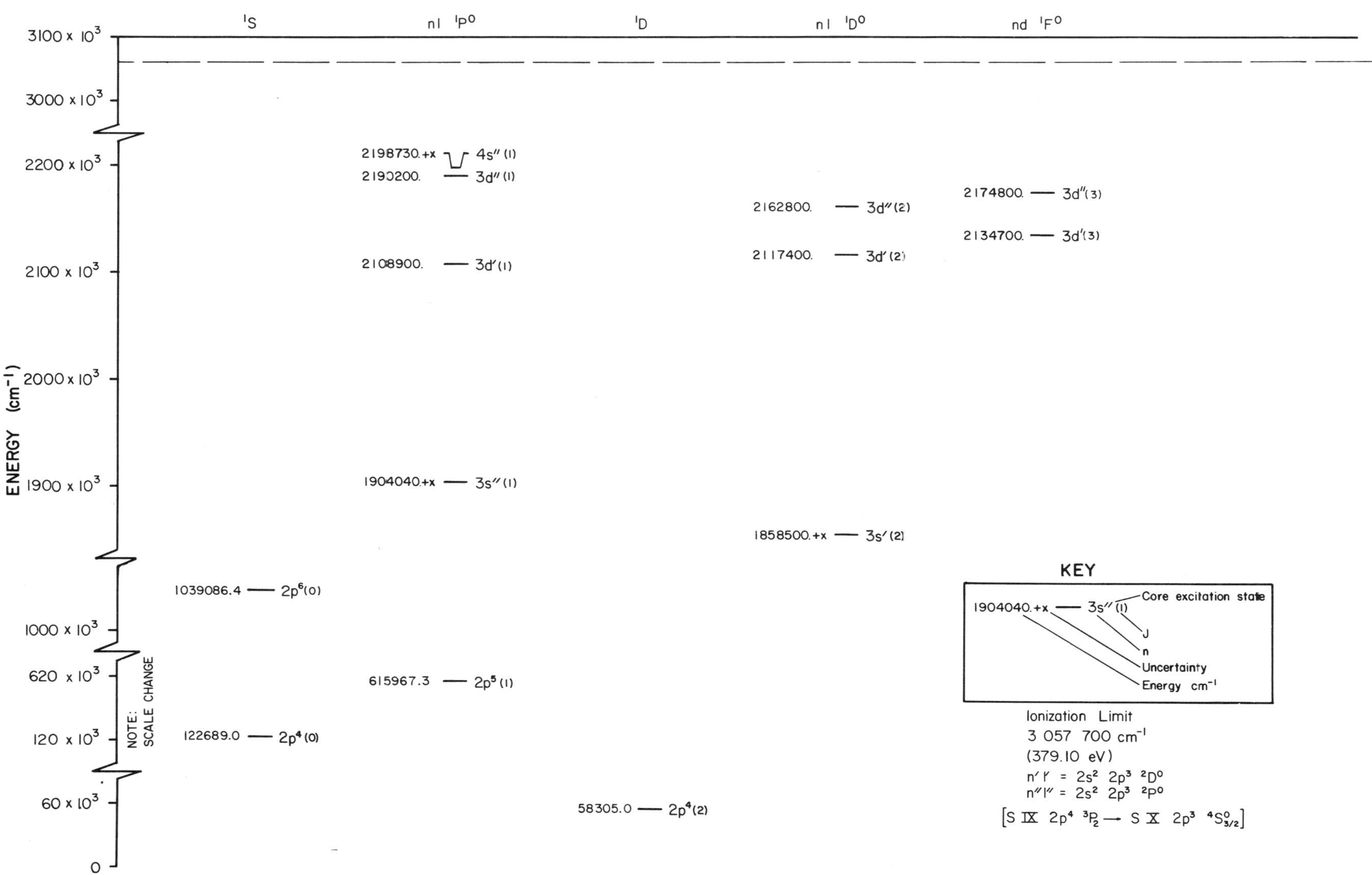
S IX ENERGY LEVELS (8 electrons, Z=16)
(O I Sequence, Configuration: 1s² 2s² 2p³ nl, Singlet System)
¹S
nl ¹P⁰
¹D
nl ¹D⁰
nd ¹F⁰
3100 x 10³
3000 x 10³
2200 x 10³
2100 x 10³
2000 x 10³
1900 x 10³
1000 x 10³
620 x 10³
120 x 10³
60 x 10³
0
ENERGY (cm⁻¹)
NOTE: SCALE CHANGE
2198730.+x 4s''(1)
2190200. 3d''(1)
2108900. — 3d'(1)
1904040.+x — 3s''(1)
615967.3 — 2p⁵(1)
2162800. — 3d''(2)
2117400. — 3d'(2)
1858500.+x — 3s'(2)
2174800. — 3d''(3)
2134700. — 3d'(3)
1039086.4 — 2p⁶(0)
122689.0 — 2p⁴(0)
58305.0 — 2p⁴(2)
KEY
1904040.+x — 3s''(1)
Core excitation state
J
n
Uncertainty
Energy cm⁻¹
Ionization Limit
3 057 700 cm⁻¹
(379.10 eV)
n'l' = 2s² 2p³ ²D⁰
n''l'' = 2s² 2p³ ²P⁰
[S IX 2p⁴ ³P₂ → S X 2p³ ⁴S⁰₃/₂]

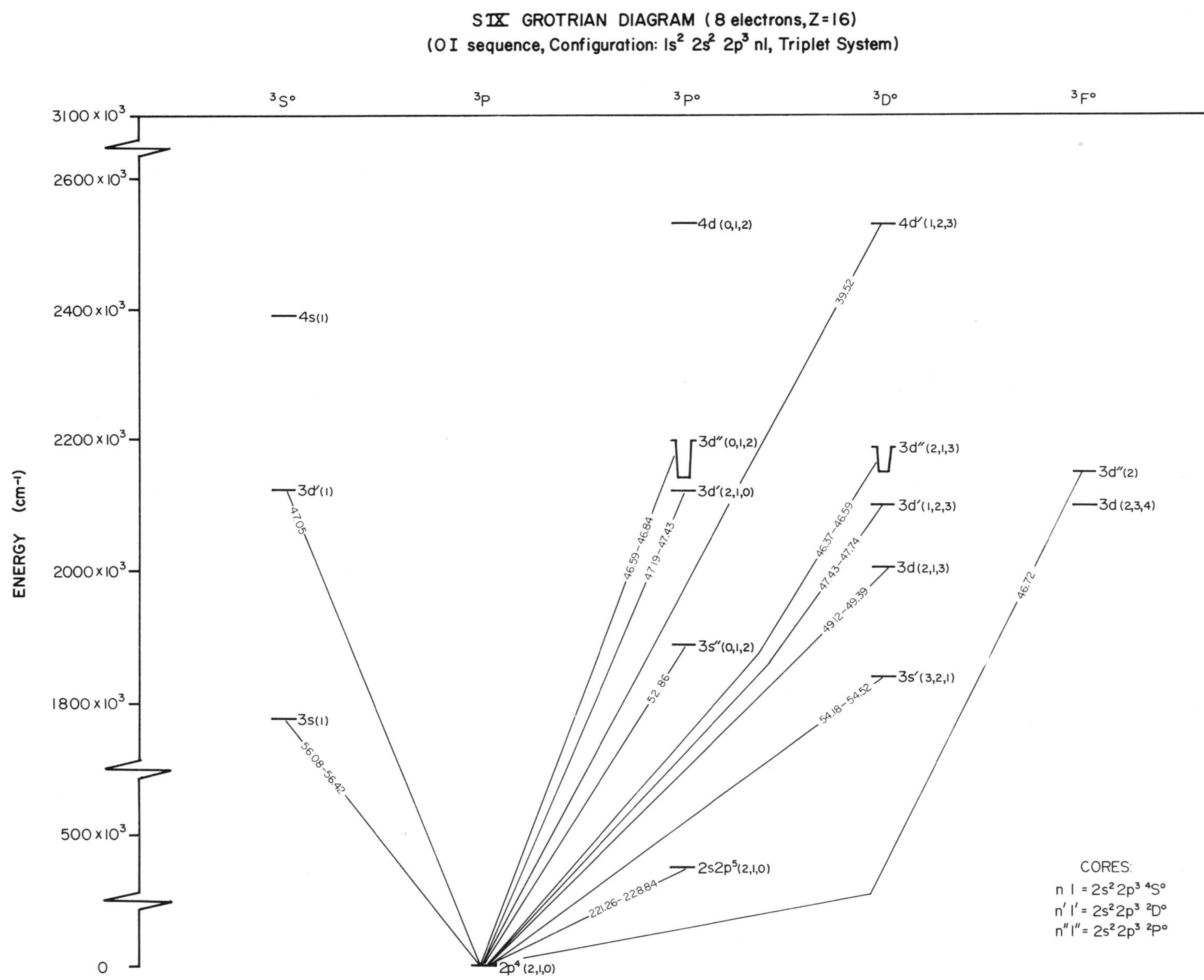
S IX GROTRIAN DIAGRAM (8 electrons, Z = 16)
(O I sequence, Configuration: 1s² 2s² 2p³ nl, Triplet System)
³S°
³P
³P°
³D°
³F°
ENERGY (cm⁻¹)
3100 x 10³
2600 x 10³
2400 x 10³
2200 x 10³
2000 x 10³
1800 x 10³
500 x 10³
0
4s(1)
3d′(1)
3s(1)
47.05
56.08–56.42
4d(0,1,2)
3d″(0,1,2)
3d(2,1,0)
3s″(0,1,2)
2s2p⁵(2,1,0)
46.59–46.84
47.19–47.43
52.86
221.26–228.84
4d′(1,2,3)
3d″(2,1,3)
3d′(1,2,3)
3d(2,1,3)
3s′(3,2,1)
39.52
46.37–46.59
47.43–47.74
49.12–49.39
54.18–54.52
3d″(2)
3d(2,3,4)
46.72
2p⁴(2,1,0)
CORES:
n l = 2s²2p³ ⁴S°
n′ l′ = 2s²2p³ ²D°
n″ l″ = 2s²2p³ ²P°

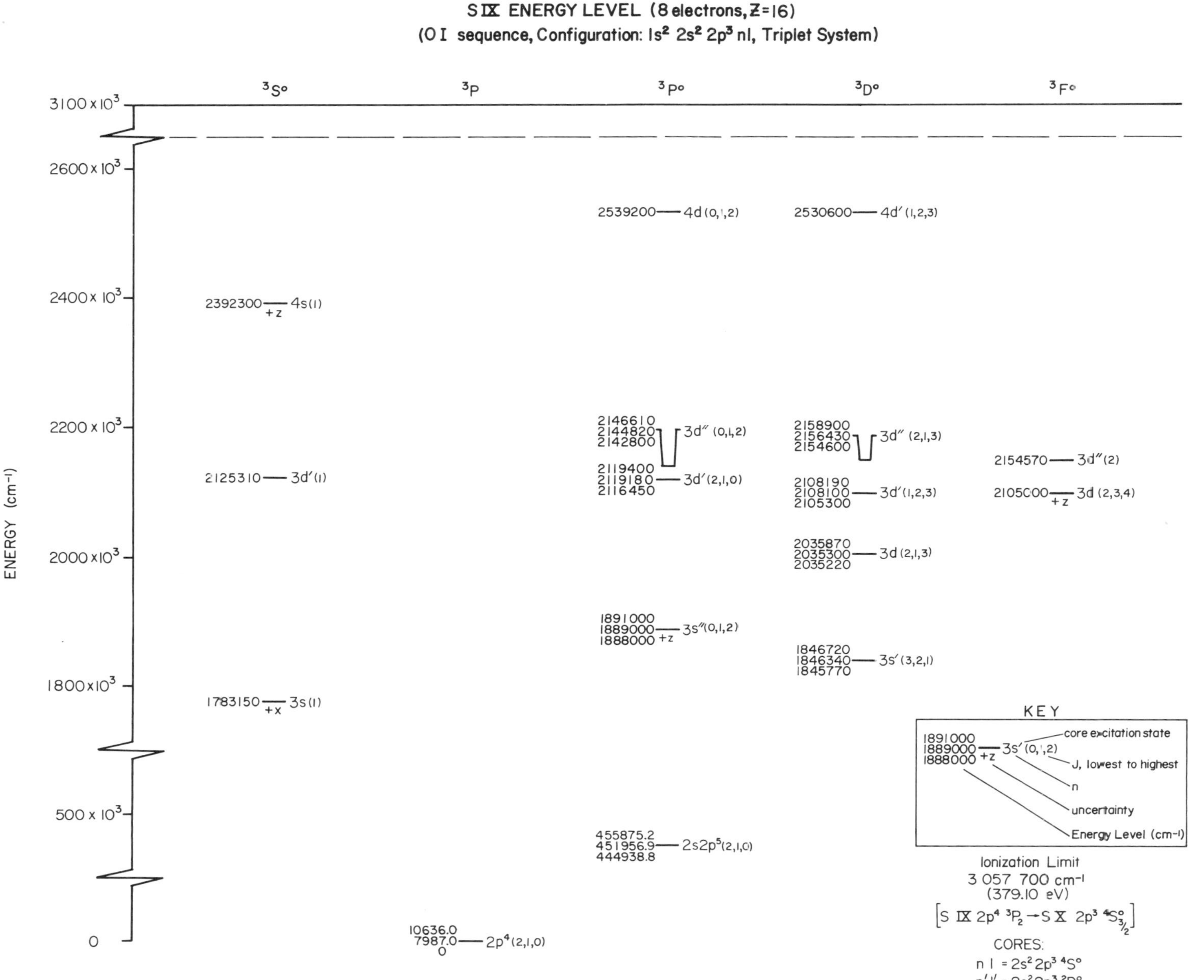
S IX ENERGY LEVEL (8 electrons, Z=16)
(O I sequence, Configuration: 1s² 2s² 2p³ nl, Triplet System)
³S°
³P
³P°
³D°
³F°
ENERGY (cm⁻¹)
3100 x 10³
2600 x 10³
2400 x 10³
2200 x 10³
2000 x 10³
1800 x 10³
500 x 10³
0
2392300 +z 4s(1)
2125310 3d′(1)
1783150 +x 3s(1)
10636.0
7987.0
0 2p⁴(2,1,0)
2539200 4d(0,1,2)
2146610
2144820
2142800 3d″(0,1,2)
2119400
2119180
2116450 3d′(2,1,0)
1891000
1889000
1888000 +z 3s″(0,1,2)
455875.2
451956.9
444938.8 2s2p⁵(2,1,0)
2530600 4d′(1,2,3)
2158900
2156430
2154600 3d″(2,1,3)
2108190
2108100
2105300 3d′(1,2,3)
2035870
2035300
2035220 3d(2,1,3)
1846720
1846340
1845770 3s′(3,2,1)
2154570 3d″(2)
2105000 +z 3d(2,3,4)
KEY
1891000
1889000
1888000 +z 3s′(0,1,2)
core excitation state
J, lowest to highest
n
uncertainty
Energy Level (cm-1)
Ionization Limit
3 057 700 cm⁻¹
(379.10 eV)
[S IX 2p⁴ ³P₂ → S X 2p³ ⁴S°₃/₂]
CORES:
n l = 2s²2p³ ⁴S°
n′l′ = 2s²2p³ ²D°
n″l″ = 2s²2p³ ²P°

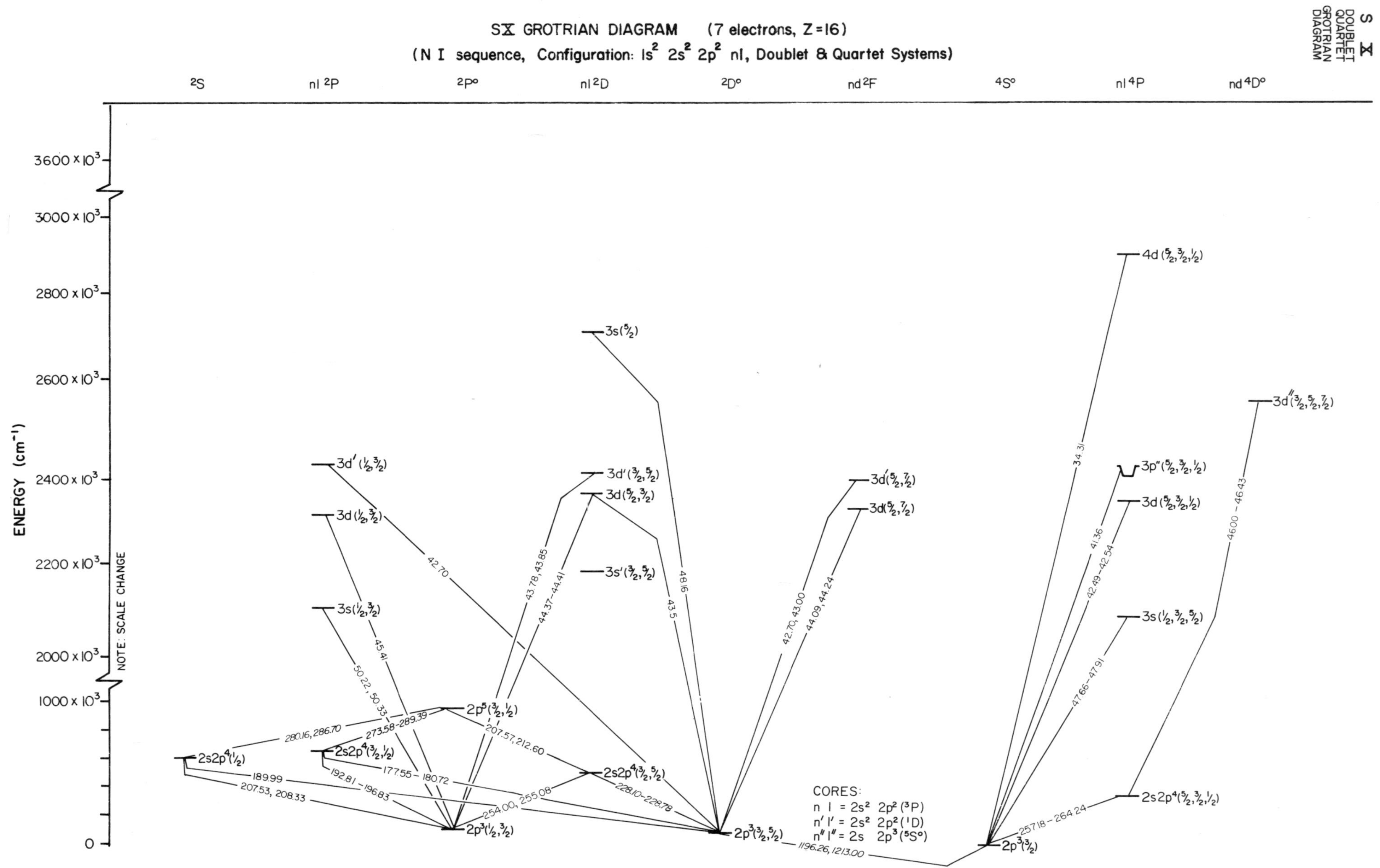
S X
DOUBLET
QUARTET
GROTRIAN
DIAGRAM
S X GROTRIAN DIAGRAM (7 electrons, Z=16)
(N I sequence, Configuration: 1s² 2s² 2p² nl, Doublet & Quartet Systems)
²S
nl ²P
²P°
nl ²D
²D°
nd ²F
⁴S°
nl ⁴P
nd ⁴D°
ENERGY (cm⁻¹)
3600 x 10³
3000 x 10³
2800 x 10³
2600 x 10³
2400 x 10³
2200 x 10³
2000 x 10³
1000 x 10³
0
NOTE: SCALE CHANGE
CORES:
n l = 2s² 2p² (³P)
n′ l′ = 2s² 2p² (¹D)
n″ l″ = 2s 2p³ (⁵S°)
2s2p⁴(½)
3d′(½,3/2)
3d(½,3/2)
3s(½,3/2)
2s2p⁴(3/2,½)
2p⁵(3/2,½)
2p³(½,3/2)
3s(5/2)
3d′(3/2,5/2)
3d(5/2,3/2)
3s′(3/2,5/2)
2s2p⁴(3/2,5/2)
2p³(3/2,5/2)
3d′(5/2,7/2)
3d(5/2,7/2)
2p³(3/2)
4d(5/2,3/2,½)
3p″(5/2,3/2,½)
3d(5/2,3/2,½)
3s(½,3/2,5/2)
2s2p⁴(5/2,3/2,½)
3d″(3/2,5/2,7/2)
280.16, 286.70
273.58-289.39
189.99
207.53, 208.33
192.81-196.83
177.55-180.72
50.22, 50.33
45.41
42.70
207.57, 212.60
254.00, 255.08
228.10-228.78
43.78, 43.85
44.37-44.41
48.16
43.5
42.70, 43.00
44.09, 44.24
1196.26, 1213.00
34.31
41.36
42.49-42.54
47.66-47.91
257.18-264.24
4600-46.43

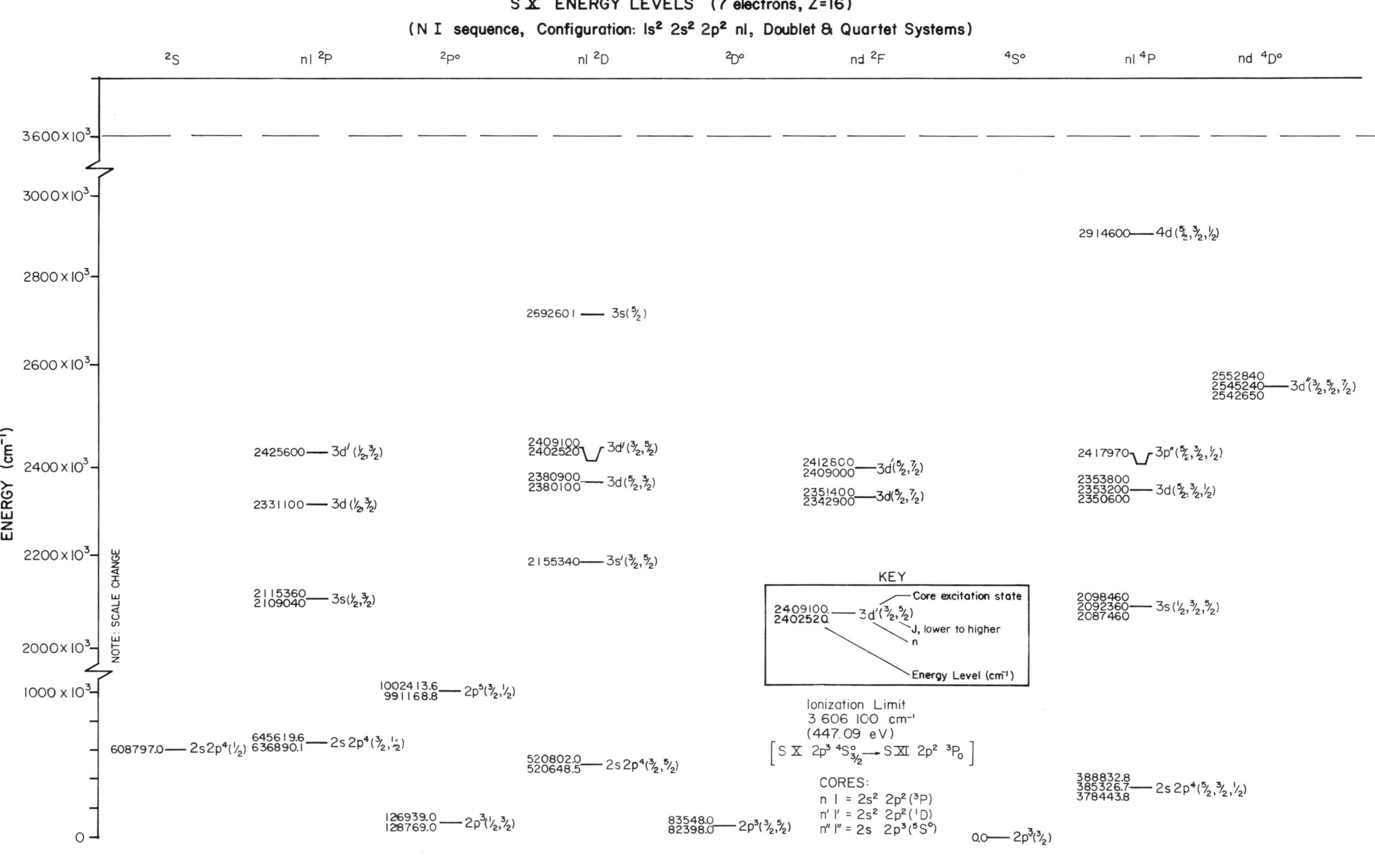
S X ENERGY LEVELS (7 electrons, Z=16)
(N I sequence, Configuration: 1s² 2s² 2p² nl, Doublet & Quartet Systems)
²S
nl ²P
²P°
nl ²D
²D°
nd ²F
⁴S°
nl ⁴P
nd ⁴D°
ENERGY (cm⁻¹)
3600×10³
3000×10³
2800×10³
2600×10³
2400×10³
2200×10³
2000×10³
1000×10³
0
NOTE: SCALE CHANGE
2914600 — 4d(5/2,3/2,1/2)
2692601 — 3s(5/2)
2552840
2545240 — 3d″(3/2,5/2,7/2)
2542650
2425600 — 3d′(1/2,3/2)
2409100
2402520 — 3d′(3/2,5/2)
2417970 — 3p″(5/2,3/2,1/2)
2412600
2409000 — 3d′(5/2,7/2)
2380900
2380100 — 3d(5/2,3/2)
2353800
2353200 — 3d(5/2,3/2,1/2)
2350600
2351400
2342900 — 3d(5/2,7/2)
2331100 — 3d(1/2,3/2)
2155340 — 3s′(3/2,5/2)
2115360
2109040 — 3s(1/2,3/2)
2098460
2092360 — 3s(1/2,3/2,5/2)
2087460
KEY
2409100
2402520 — 3d′(3/2,5/2)
Core excitation state
J, lower to higher
n
Energy Level (cm⁻¹)
1002413.6
991168.8 — 2p⁵(3/2,1/2)
Ionization Limit
3 606 100 cm⁻¹
(447.09 eV)
[S X 2p³ ⁴S°3/2 → S XI 2p² ³P0]
608797.0 — 2s2p⁴(1/2)
645619.6
636890.1 — 2s2p⁴(3/2,1/2)
520802.0
520648.5 — 2s2p⁴(3/2,5/2)
388832.8
385326.7 — 2s2p⁴(5/2,3/2,1/2)
378443.8
CORES:
n l = 2s² 2p²(³P)
n′ l′ = 2s² 2p²(¹D)
n″ l″ = 2s 2p³(⁵S°)
126939.0
128769.0 — 2p³(1/2,3/2)
83548.0
82398.0 — 2p³(3/2,5/2)
0.0 — 2p³(3/2)
NOTE: ALL DOUBLET LEVELS UNCERTAIN BY ±x.

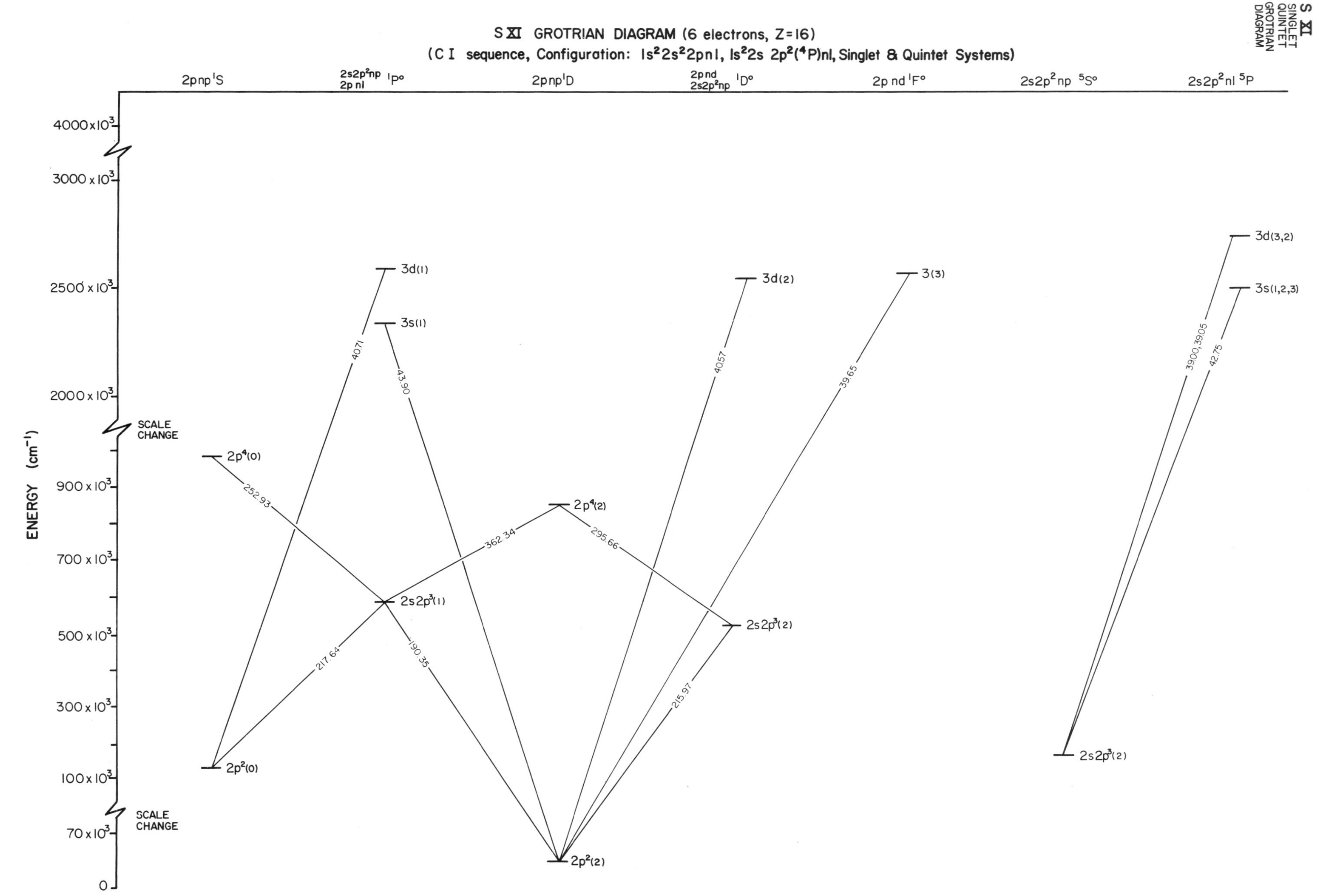
S XI
SINGLET
QUINTET
GROTRIAN
DIAGRAM
S XI GROTRIAN DIAGRAM (6 electrons, Z=16)
(C I sequence, Configuration: 1s²2s²2pnl, 1s²2s 2p²(⁴P)nl, Singlet & Quintet Systems)
2pnp¹S
2s2p²np
2p nl ¹P°
2pnp¹D
2pnd
2s2p²np ¹D°
2p nd ¹F°
2s2p²np ⁵S°
2s2p²nl ⁵P
ENERGY (cm⁻¹)
4000x10³
3000x10³
2500x10³
2000x10³
SCALE CHANGE
900x10³
700x10³
500x10³
300x10³
100x10³
SCALE CHANGE
70x10³
0
3d(1)
3s(1)
3d(2)
3(3)
3d(3,2)
3s(1,2,3)
2p⁴(0)
2p⁴(2)
2s2p³(1)
2s2p³(2)
2p²(0)
2s2p³(2)
2p²(2)
4071
43.90
4057
39.65
3900, 3905
42.75
252.93
362.34
295.66
217.64
190.35
215.97

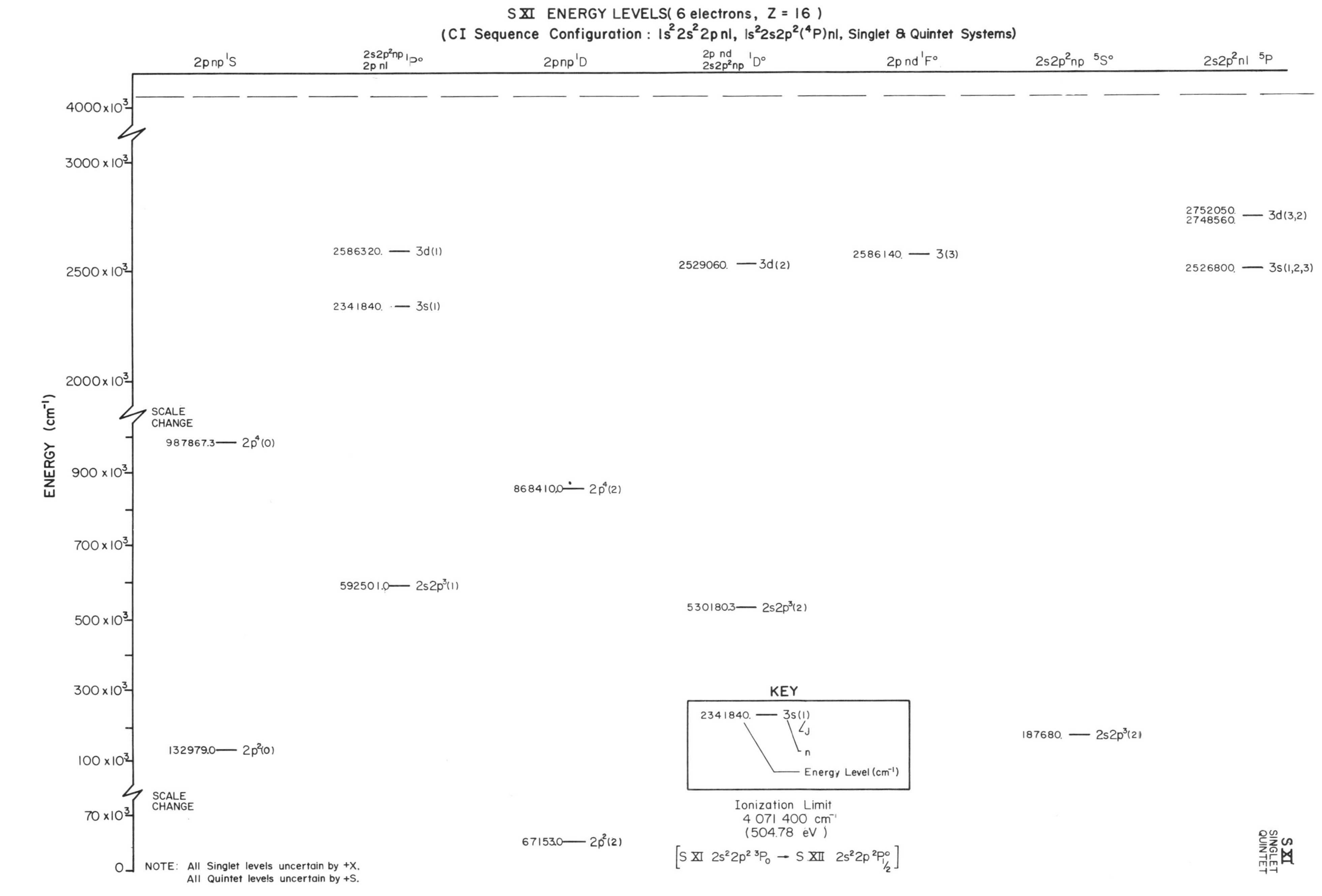
S XI ENERGY LEVELS(6 electrons, Z = 16)
(CI Sequence Configuration : $1s^2 2s^2 2p nl$, $1s^2 2s 2p^2 (^4P) nl$, Singlet & Quintet Systems)
$2pnp\ ^1S$
$2s2p^2np$ $2pnl$ $^1P^o$
$2pnp\ ^1D$
$2pnd$ $2s2p^2np$ $^1D^o$
$2pnd\ ^1F^o$
$2s2p^2np\ ^5S^o$
$2s2p^2nl\ ^5P$
ENERGY (cm^{-1})
4000×10^3
3000×10^3
2500×10^3
2000×10^3
SCALE CHANGE
900×10^3
700×10^3
500×10^3
300×10^3
100×10^3
SCALE CHANGE
70×10^3
0
987867.3 — $2p^4(0)$
132979.0 — $2p^2(0)$
2586320. — 3d(1)
2341840. — 3s(1)
592501.0 — $2s2p^3(1)$
868410.0 — $2p^4(2)$
67153.0 — $2p^2(2)$
2529060. — 3d(2)
530180.3 — $2s2p^3(2)$
2586140. — 3(3)
187680. — $2s2p^3(2)$
2752050.
2748560. — 3d(3,2)
2526800. — 3s(1,2,3)
KEY
2341840. — 3s(1)
J
n
Energy Level (cm^{-1})
Ionization Limit
4 071 400 cm^{-1}
(504.78 eV)
[S XI $2s^2 2p^2\ ^3P_0$ → S XII $2s^2 2p\ ^2P^o_{1/2}$]
NOTE: All Singlet levels uncertain by +X.
All Quintet levels uncertain by +S.
S XI
SINGLET
QUINTET

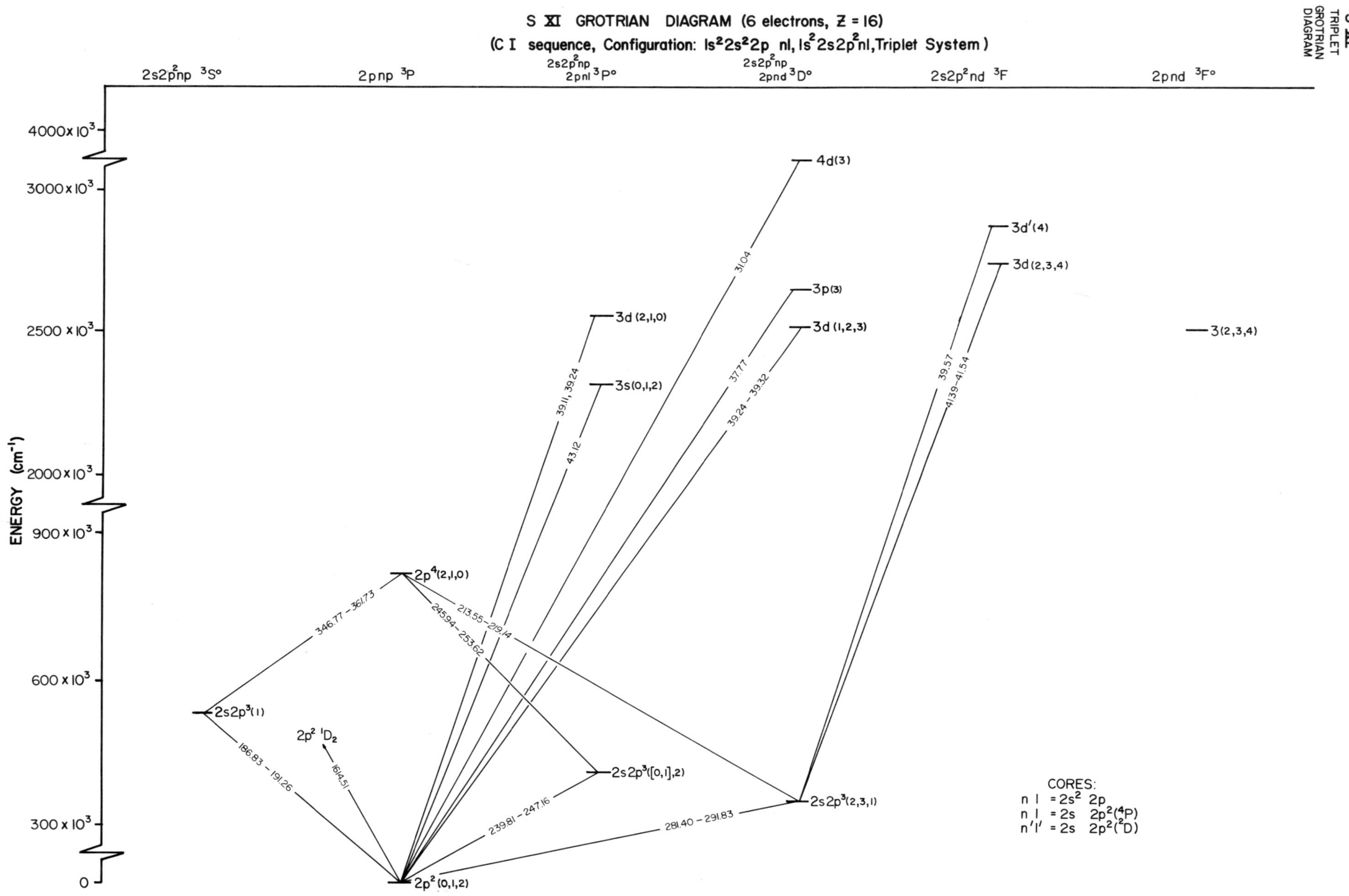
S XI GROTRIAN DIAGRAM (6 electrons, Z = 16)
(C I sequence, Configuration: 1s²2s²2p nl, 1s²2s2p²nl, Triplet System)
2s2p²np ³S°
2pnp ³P
2s2p²np 2pnl ³P°
2s2p²np 2pnd ³D°
2s2p²nd ³F
2pnd ³F°
ENERGY (cm⁻¹)
4000×10³
3000×10³
2500×10³
2000×10³
900×10³
600×10³
300×10³
0
2s2p³(1)
2p⁴(2,1,0)
2p² ¹D₂
2p²(0,1,2)
2s2p³([0,1],2)
2s2p³(2,3,1)
3d(2,1,0)
3s(0,1,2)
4d(3)
3p(3)
3d(1,2,3)
3d′(4)
3d(2,3,4)
3(2,3,4)
186.83 – 191.26
1614.51
346.77 – 361.73
245.94 – 253.62
213.55 – 219.14
239.81 – 247.16
281.40 – 291.83
39.11, 39.24
43.12
31.04
37.77
39.24 – 39.32
39.57
41.39 – 41.54
CORES:
n l = 2s² 2p
n l = 2s 2p²(⁴P)
n′l′ = 2s 2p²(²D)

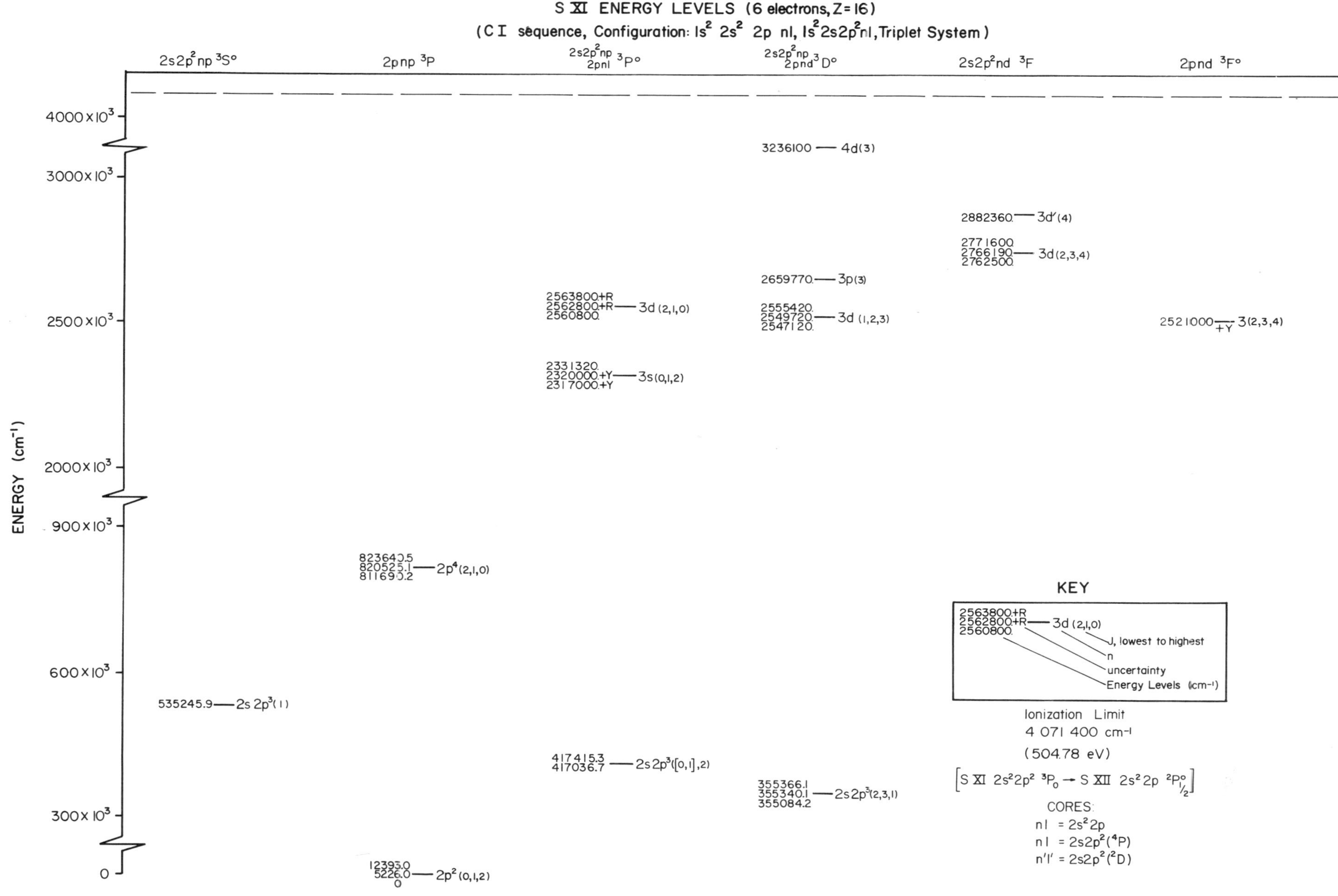
S XI ENERGY LEVELS (6 electrons, Z=16)
(C I sequence, Configuration: 1s² 2s² 2p nl, 1s²2s2p²nl, Triplet System)
2s2p²np ³S°
2pnp ³P
2s2p²np ³P°
2pnl
2s2p²np ³D°
2pnd
2s2p²nd ³F
2pnd ³F°
4000×10³
3000×10³
2500×10³
2000×10³
900×10³
600×10³
300×10³
0
ENERGY (cm⁻¹)
3236100 — 4d(3)
2882360 — 3d'(4)
2771600
2766190
2762500 — 3d(2,3,4)
2659770 — 3p(3)
2563800+R
2562800+R
2560800. — 3d (2,1,0)
2555420.
2549720.
2547120. — 3d (1,2,3)
2521000 +Y — 3(2,3,4)
2331320.
2320000.+Y
2317000.+Y — 3s(0,1,2)
823640.5
820525.1
811690.2 — 2p⁴(2,1,0)
535245.9 — 2s 2p³(1)
417415.3
417036.7 — 2s2p³([0,1],2)
355366.1
355340.1
355084.2 — 2s2p³(2,3,1)
12393.0
5226.0
0 — 2p²(0,1,2)
KEY
2563800+R
2562800+R
2560800. — 3d (2,1,0)
J, lowest to highest
n
uncertainty
Energy Levels (cm⁻¹)
Ionization Limit
4 071 400 cm⁻¹
(504.78 eV)
[S XI 2s²2p² ³P₀ → S XII 2s²2p ²P°1/2]
CORES:
nl = 2s²2p
nl = 2s2p²(⁴P)
n'l' = 2s2p²(²D)

S XII GROTRIAN DIAGRAM (5 electrons, Z = 16)

(B I sequence, Configuration: $1s^2 2s^2 nl$, $1s^2 2s2pnl$, Doublet & Quartet Systems)

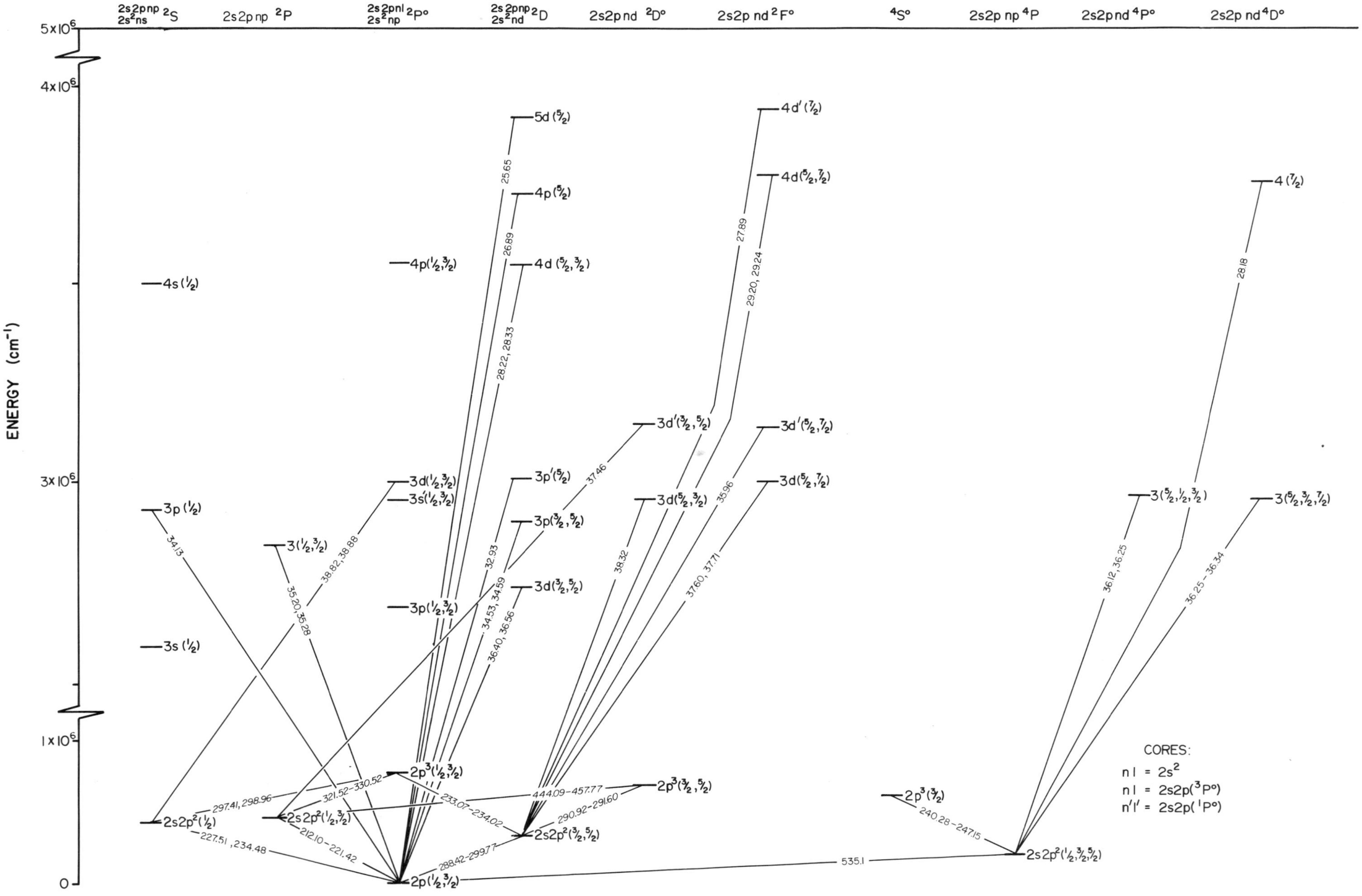

2s2pnp / $2s^2$ns ^{2}S | 2s2pnp ^{2}P | 2s2pnl / $2s^2$np ^{2}P° | 2s2pnp / $2s^2$nd ^{2}D | 2s2pnd ^{2}D° | 2s2pnd ^{2}F° | ^{4}S° | 2s2pnp ^{4}P | 2s2pnd ^{4}P° | 2s2pnd ^{4}D°

ENERGY (cm^{-1})

5x10^6
4x10^6
3x10^6
1x10^6
0

4s(1/2)
3p(1/2)
3s(1/2)
$2s2p^2$(1/2)
3(1/2,3/2)
$2s2p^2$(1/2,3/2)
4p(1/2,3/2)
3d(1/2,3/2)
3s'(1/2,3/2)
3p(1/2,3/2)
$2p^3$(1/2,3/2)
2p(1/2,3/2)
5d(5/2)
4p(5/2)
4d(5/2,3/2)
3p'(5/2)
3p(3/2,5/2)
3d(3/2,5/2)
3d'(3/2,5/2)
3d(5/2,3/2)
$2p^3$(3/2,5/2)
$2s2p^2$(3/2,5/2)
4d'(7/2)
4d(5/2,7/2)
3d'(5/2,7/2)
3d(5/2,7/2)
$2p^3$(3/2)
$2s2p^2$(1/2,3/2,5/2)
3(5/2,1/2,3/2)
4(7/2)
3(5/2,3/2,7/2)

34.13
38.82, 38.88
35.20, 35.28
297.41, 298.96
321.52–330.52
227.51, 234.48
212.10–221.42
25.65
26.89
28.22, 28.33
32.93
34.53, 34.59
36.40, 36.56
37.46
38.32
233.07–234.02
444.09–457.77
290.92–291.60
288.42–299.77
27.89
29.20, 29.24
35.96
37.60, 37.71
240.28–247.15
535.1
36.12, 36.25
28.18
36.25–36.34

CORES:
nl = $2s^2$
nl = 2s2p(^{3}P°)
n'l' = 2s2p(^{1}P°)

S XII ENERGY LEVELS (5 electrons, Z=16)

(B I sequence, Configuration: $1s^2\ 2s^2\ nl$, $1s^2\ 2s\ 2p\ nl$, Doublet & Quartet Systems)

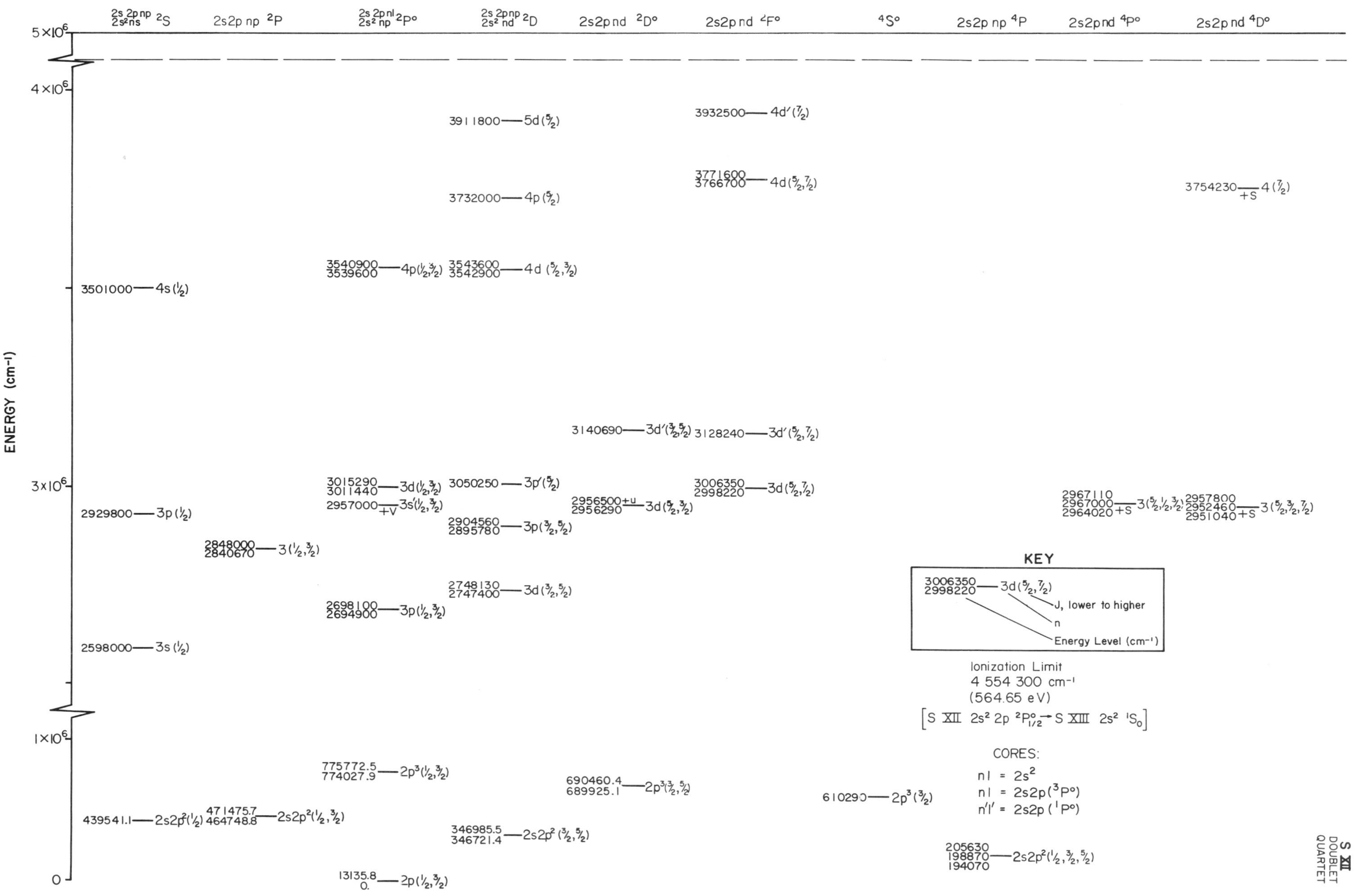

S XII
DOUBLET
QUARTET

ENERGY (cm⁻¹): 0, 1×10⁶, 3×10⁶, 4×10⁶, 5×10⁶

2s2pnp / 2s²ns ²S
3501000 — 4s(1/2)
2929800 — 3p(1/2)
2598000 — 3s(1/2)
439541.1 — 2s2p²(1/2)

2s2p np ²P
2848000, 2840670 — 3(1/2, 3/2)
471475.7, 464748.8 — 2s2p²(1/2, 3/2)

2s2pnl / 2s²np ²P°
3540900, 3539600 — 4p(1/2, 3/2)
3015290, 3011440 — 3d(1/2, 3/2)
2957000+V — 3s(1/2, 3/2)
2698100, 2694900 — 3p(1/2, 3/2)
775772.5, 774027.9 — 2p³(1/2, 3/2)
13135.8, 0. — 2p(1/2, 3/2)

2s2pnp / 2s²nd ²D
3911800 — 5d(5/2)
3732000 — 4p(5/2)
3543600, 3542900 — 4d(5/2, 3/2)
3050250 — 3p′(5/2)
2904560, 2895780 — 3p(3/2, 5/2)
2748130, 2747400 — 3d(3/2, 5/2)
346985.5, 346721.4 — 2s2p²(3/2, 5/2)

2s2pnd ²D°
3140690 — 3d′(3/2, 5/2)
2956500+u, 2956290 — 3d(5/2, 3/2)
690460.4, 689925.1 — 2p³(3/2, 5/2)

2s2pnd ⁴F°
3932500 — 4d′(7/2)
3771600, 3766700 — 4d(5/2, 7/2)
3128240 — 3d′(5/2, 7/2)
3006350, 2998220 — 3d(5/2, 7/2)

⁴S°
610290 — 2p³(3/2)

2s2p np ⁴P
205630, 198870, 194070 — 2s2p²(1/2, 3/2, 5/2)

2s2pnd ⁴P°
2967110, 2967000, 2964020+S — 3(5/2, 1/2, 3/2)

2s2p nd ⁴D°
3754230+S — 4(7/2)
2957800, 2952460, 2951040+S — 3(5/2, 3/2, 7/2)

KEY
3006350, 2998220 — 3d(5/2, 7/2)
J, lower to higher
n
Energy Level (cm⁻¹)

Ionization Limit
4 554 300 cm⁻¹
(564.65 eV)
[S XII $2s^2\ 2p\ ^2P^o_{1/2}$ → S XIII $2s^2\ ^1S_0$]

CORES:
nl = $2s^2$
nl = $2s2p(^3P^o)$
n′l′ = $2s2p(^1P^o)$

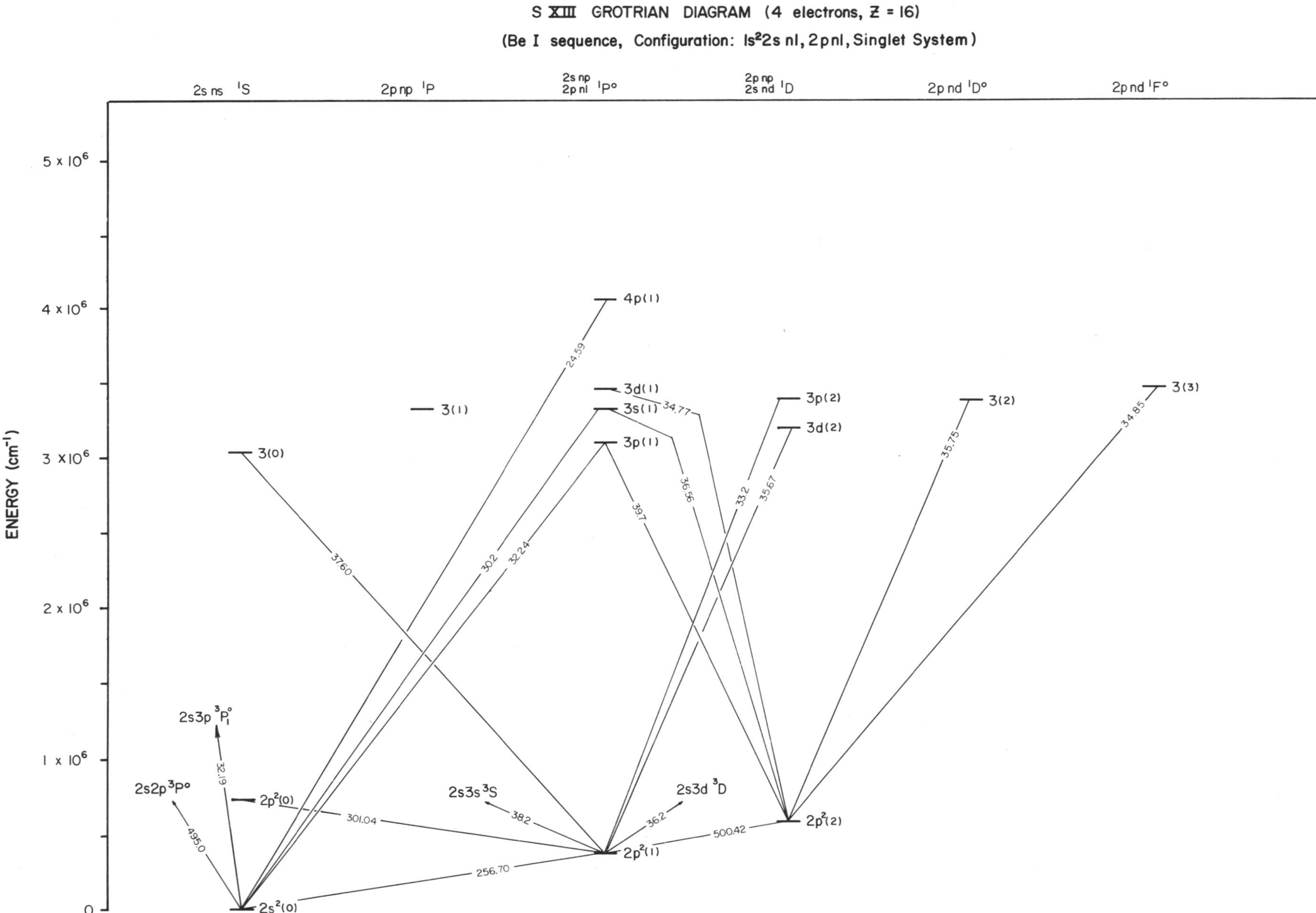
S XIII GROTRIAN DIAGRAM (4 electrons, Z = 16)
(Be I sequence, Configuration: 1s²2s nl, 2pnl, Singlet System)
2s ns ¹S
2p np ¹P
2s np 2p nl ¹P°
2p np 2s nd ¹D
2p nd ¹D°
2p nd ¹F°
ENERGY (cm⁻¹)
0
1 x 10⁶
2 x 10⁶
3 x 10⁶
4 x 10⁶
5 x 10⁶
3(0)
3(1)
4p(1)
3d(1)
3s(1)
3p(1)
3p(2)
3d(2)
3(2)
3(3)
2s3p ³P°₁
2s2p ³P°
2p²(0)
2s3s ³S
2s3d ³D
2p²(1)
2p²(2)
2s²(0)
24.59
34.77
36.56
33.2
35.67
35.75
34.85
37.60
30.2
32.24
39.7
32.19
495.0
301.04
38.2
36.2
500.42
256.70

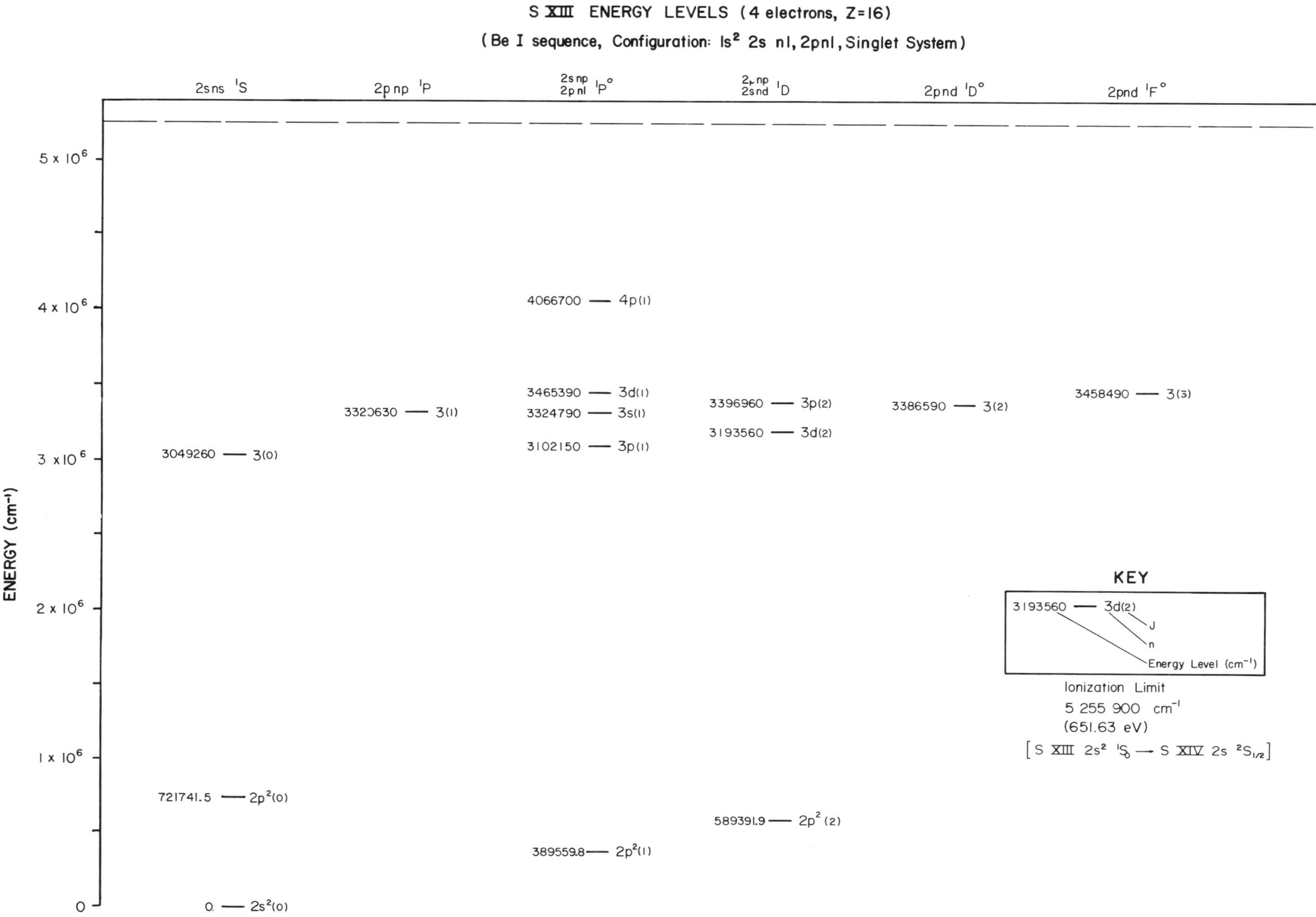
S XIII ENERGY LEVELS (4 electrons, Z=16)
(Be I sequence, Configuration: 1s² 2s nl, 2pnl, Singlet System)
2sns ¹S
2pnp ¹P
2snp 2pnl ¹P°
2pnp 2snd ¹D
2pnd ¹D°
2pnd ¹F°
3049260 — 3(0)
721741.5 — 2p²(0)
0. — 2s²(0)
3320630 — 3(1)
4066700 — 4p(1)
3465390 — 3d(1)
3324790 — 3s(1)
3102150 — 3p(1)
389559.8 — 2p²(1)
3396960 — 3p(2)
3193560 — 3d(2)
589391.9 — 2p² (2)
3386590 — 3(2)
3458490 — 3(3)
KEY
3193560 — 3d(2)
J
n
Energy Level (cm⁻¹)
Ionization Limit
5 255 900 cm⁻¹
(651.63 eV)
[S XIII 2s² ¹S₀ → S XIV 2s ²S₁/₂]
ENERGY (cm⁻¹)
5 x 10⁶
4 x 10⁶
3 x 10⁶
2 x 10⁶
1 x 10⁶
0

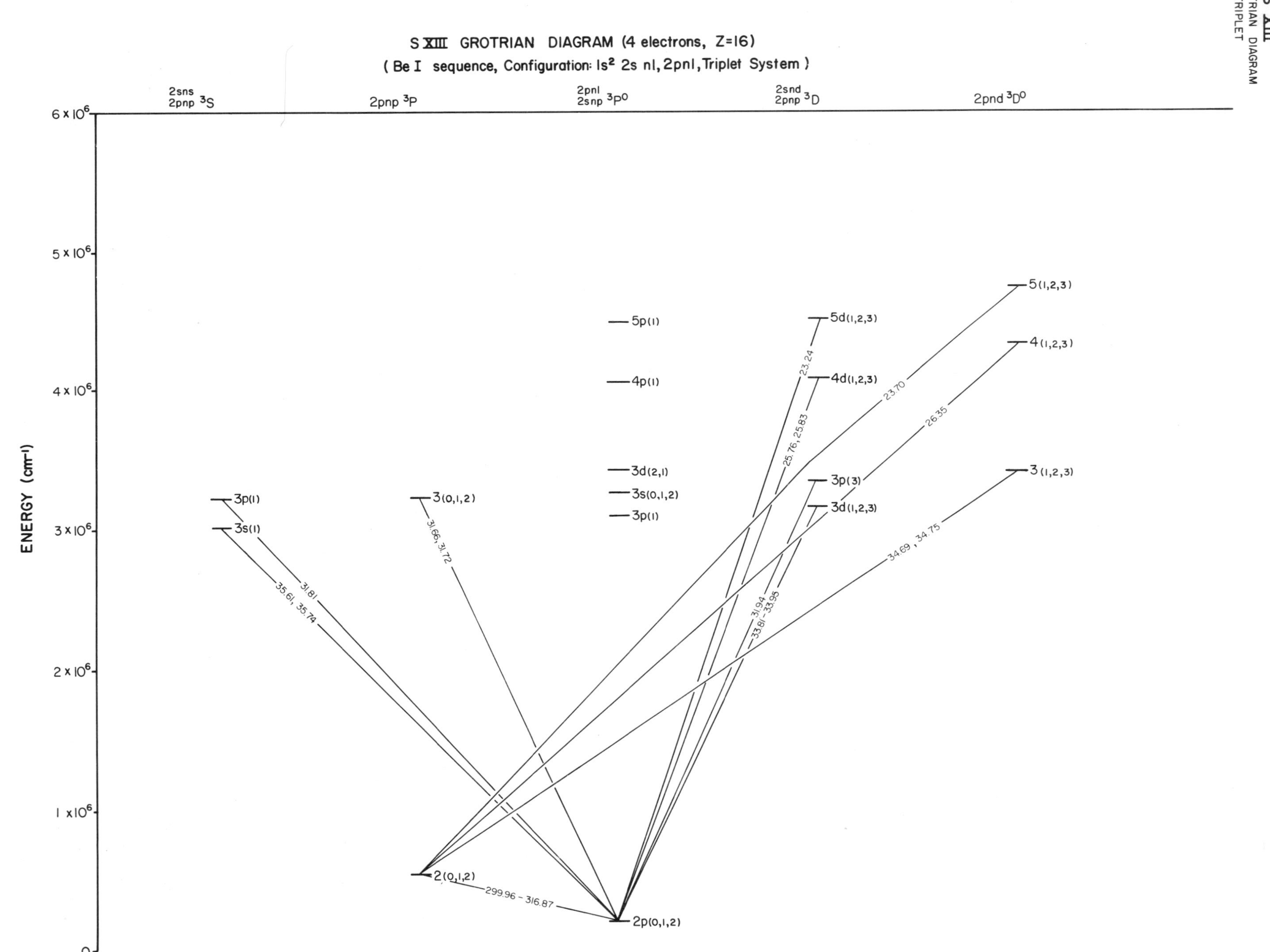
S XIII GROTRIAN DIAGRAM (4 electrons, Z=16)
(Be I sequence, Configuration: 1s^2 2s nl, 2pnl, Triplet System)
2sns 2pnp ^{3}S
2pnp ^{3}P
2pnl 2snp ^{3}P^o
2snd 2pnp ^{3}D
2pnd ^{3}D^o
ENERGY (cm^{-1})
6 x 10^6
5 x 10^6
4 x 10^6
3 x 10^6
2 x 10^6
1 x 10^6
0
3p(1)
3s(1)
31.81
35.61, 35.74
3(0,1,2)
31.66, 31.72
2(0,1,2)
299.96 – 316.87
2p(0,1,2)
5p(1)
4p(1)
3d(2,1)
3s(0,1,2)
3p(1)
5d(1,2,3)
4d(1,2,3)
3p(3)
3d(1,2,3)
23.24
25.76, 25.83
31.94
33.81 – 33.95
5(1,2,3)
4(1,2,3)
3(1,2,3)
23.70
26.35
34.69, 34.75

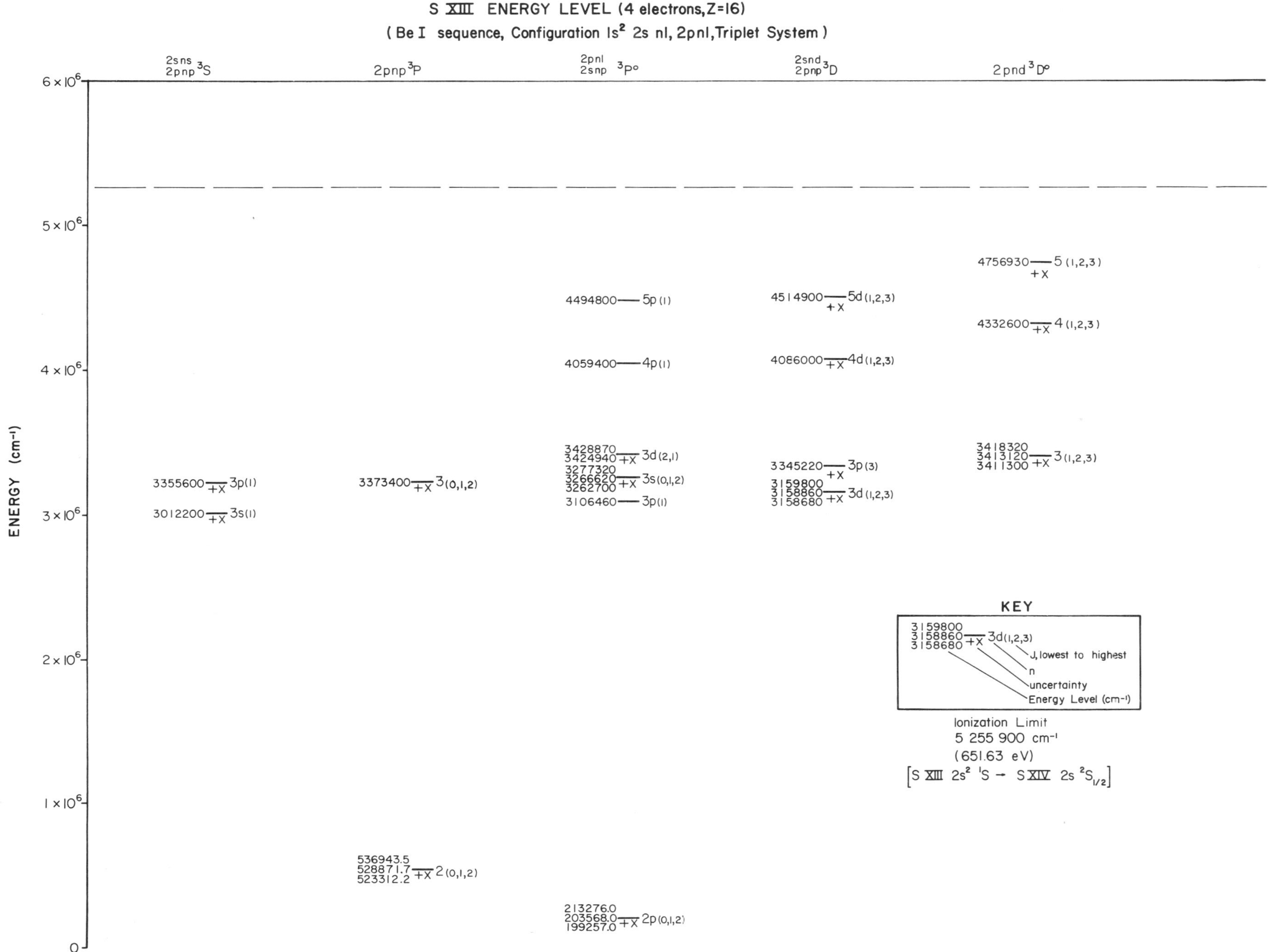
S XIII ENERGY LEVEL (4 electrons, Z=16)
(Be I sequence, Configuration 1s² 2s nl, 2pnl, Triplet System)
2sns ³S
2pnp ³S
2pnp ³P
2pnl 2snp ³P°
2snd ³D
2pnp ³D
2pnd ³D°
ENERGY (cm⁻¹)
6 × 10⁶
5 × 10⁶
4 × 10⁶
3 × 10⁶
2 × 10⁶
1 × 10⁶
0
3355600 +x 3p(1)
3012200 +x 3s(1)
3373400 +x 3(0,1,2)
536943.5
528871.7 +x 2(0,1,2)
523312.2
4494800 5p(1)
4059400 4p(1)
3428870
3424940 +x 3d(2,1)
3277320
3266620 +x 3s(0,1,2)
3262700
3106460 3p(1)
213276.0
203568.0 +x 2p(0,1,2)
199257.0
4514900 +x 5d(1,2,3)
4086000 +x 4d(1,2,3)
3345220 +x 3p(3)
3159800
3158860 +x 3d(1,2,3)
3158680
4756930 +x 5(1,2,3)
4332600 +x 4(1,2,3)
3418320
3413120 +x 3(1,2,3)
3411300
KEY
3159800
3158860 +x 3d(1,2,3)
3158680
J, lowest to highest
n
uncertainty
Energy Level (cm⁻¹)
Ionization Limit
5 255 900 cm⁻¹
(651.63 eV)
[S XIII 2s² ¹S → S XIV 2s ²S1/2]

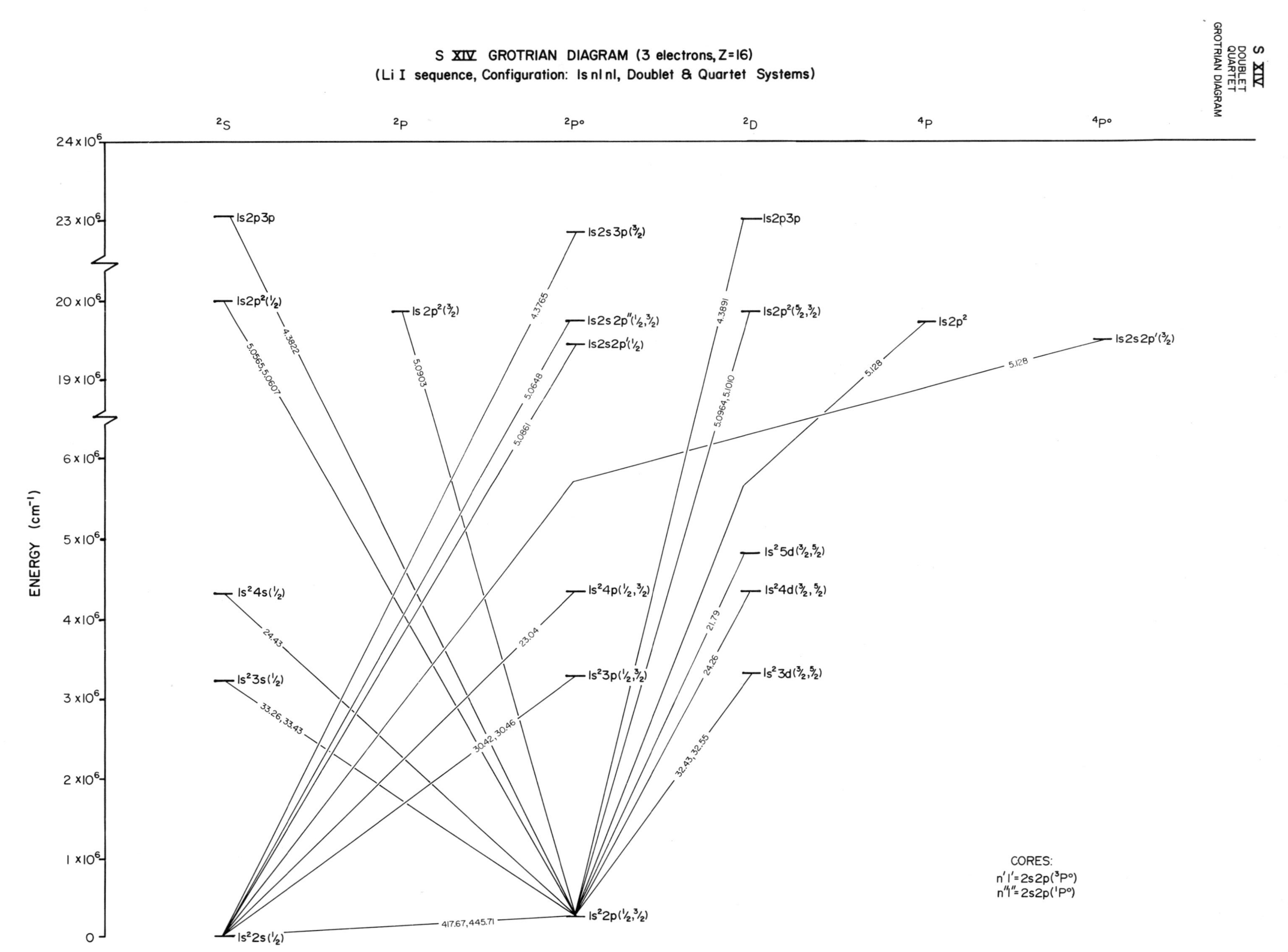

S XIV
DOUBLET
QUARTET
GROTRIAN DIAGRAM
S XIV GROTRIAN DIAGRAM (3 electrons, Z=16)
(Li I sequence, Configuration: 1s nl nl, Doublet & Quartet Systems)
2S
2P
2P°
2D
4P
4P°
ENERGY (cm-1)
24x10^6
23x10^6
20x10^6
19x10^6
6x10^6
5x10^6
4x10^6
3x10^6
2x10^6
1x10^6
0
1s2p3p
1s2s3p(3/2)
1s2p3p
1s2p2(1/2)
1s2p2(3/2)
1s2s2p''(1/2,3/2)
1s2s2p'(1/2)
1s2p2(5/2,3/2)
1s2p2
1s2s2p'(3/2)
1s25d(3/2,5/2)
1s24s(1/2)
1s24p(1/2,3/2)
1s24d(3/2,5/2)
1s23s(1/2)
1s23p(1/2,3/2)
1s23d(3/2,5/2)
1s22s(1/2)
1s22p(1/2,3/2)
4.3822
5.0565,5.0607
5.0903
4.3765
5.0648
5.0861
4.3891
5.0964,5.1010
5.128
5.128
24.43
33.26,33.43
23.04
30.42,30.46
21.79
24.26
32.43,32.55
417.67,445.71
CORES:
n'l'=2s2p(3P°)
n''l''=2s2p(1P°)

S XIV
DOUBLET
QUARTET

S XIV ENERGY LEVELS (3 electrons, Z = 16)

(Li I sequence, Configuration: 1s nl nl, Doublet & Quartet Systems)

ENERGY (cm^{-1})

Axis: 0, 1×10^6, 2×10^6, 3×10^6, 4×10^6, 5×10^6, 6×10^6, 19×10^6, 20×10^6, 23×10^6, 24×10^6

^2S

- 23 051 483. — $1s2p3p$
- 20 000 200 — $1s2p^2(1/2)$
- 4 329 000 +X — $1s^24s(1/2)$
- 3 222 500 — $1s^23s(1/2)$
- 0.0 — $1s^22s(1/2)$

^2P

- 19 884 700 — $1s2p^2(3/2)$

$^2P°$

- 22 849 310 — $1s2s3p(3/2)$
- 19 744 100 — $1s2s2p''(1/2, 3/2)$
- 19 661 400 — $1s2s2p'(1/2)$
- 4 343 000, 4 341 500 +V — $1s^24p(1/2, 3/2)$
- 3 287 000, 3 282 640 — $1s^23p(1/2, 3/2)$
- 239 463.5, 224 325.9 — $1s^22p(1/2, 3/2)$

^2D

- 23 015 608 — $1s2p3p$
- 19 846 000, 19 843 000 — $1s2p^2(5/2, 3/2)$
- 4 830 000 +X — $1s^25d(3/2, 5/2)$
- 4 352 000, 4 349 000 +X — $1s^24d(3/2, 5/2)$
- 3 311 070, 3 307 840 — $1s^23d(3/2, 5/2)$

^4P

- 19 732 700 — $1s2p^2$

$^4P°$

- 19 500 780 — $1s2s2p'(3/2)$

KEY

4 352 000, 4 349 000 +X — $1s^24d(3/2, 5/2)$

- J, lower to higher
- Configuration
- Uncertainty
- Energy Levels (cm^{-1})

Ionization Limit
5 703 600 cm^{-1}
(707.14 eV)

[S XIV $1s^2\,2s\ ^2S_{1/2}$ → S XV $1s^2\ ^1S_0$]

CORES:
$n'l' = 2s2p(^3P°)$
$n''l'' = 2s2p(^1P°)$

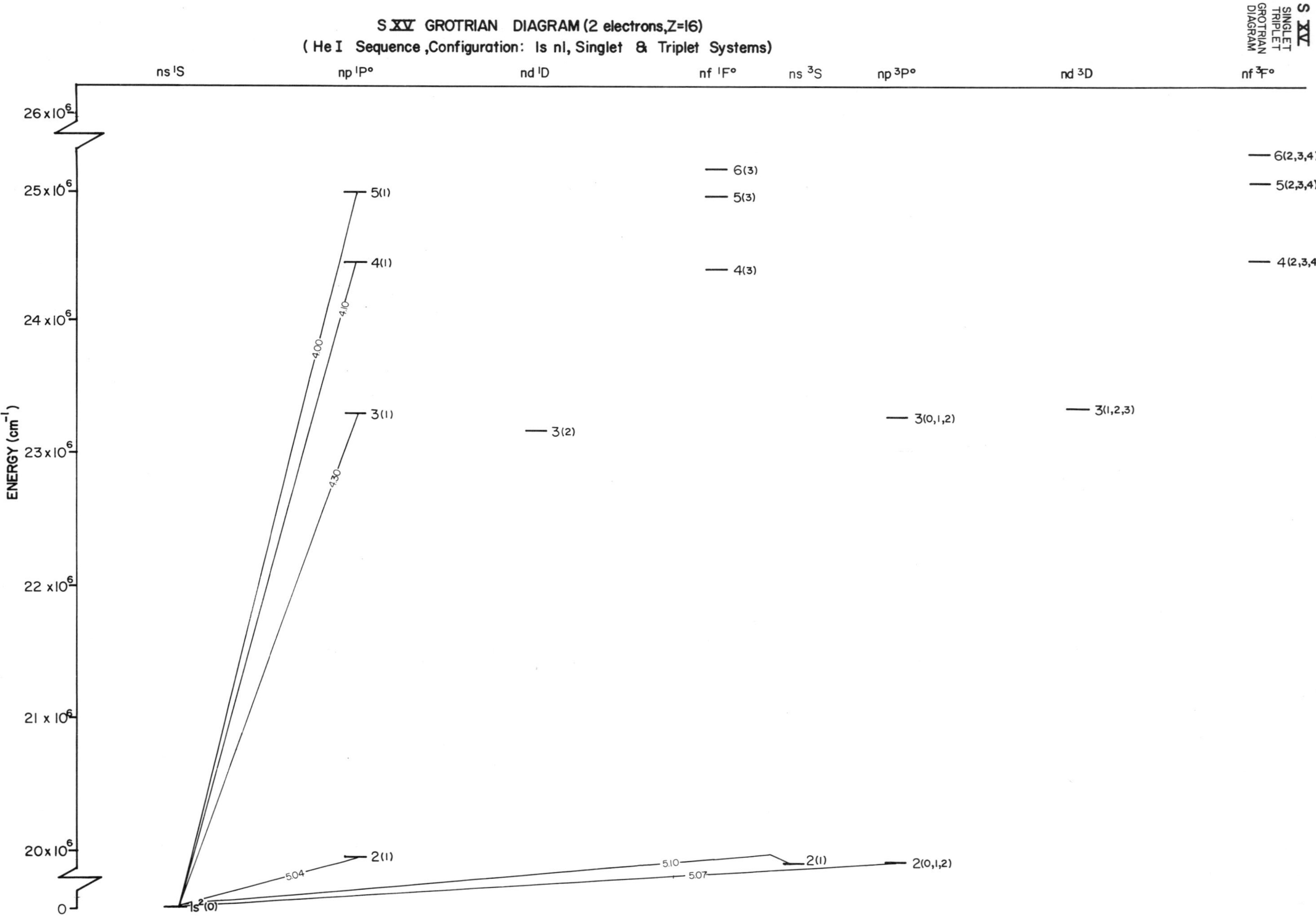
S XV GROTRIAN DIAGRAM (2 electrons, Z=16)
(He I Sequence, Configuration: 1s nl, Singlet & Triplet Systems)
S XV
SINGLET
TRIPLET
GROTRIAN
DIAGRAM
ns ¹S
np ¹P°
nd ¹D
nf ¹F°
ns ³S
np ³P°
nd ³D
nf ³F°
ENERGY (cm⁻¹)
26 x 10⁶
25 x 10⁶
24 x 10⁶
23 x 10⁶
22 x 10⁶
21 x 10⁶
20 x 10⁶
0
1s² (0)
5(1)
4(1)
3(1)
2(1)
3(2)
6(3)
5(3)
4(3)
2(1)
3(0,1,2)
2(0,1,2)
3(1,2,3)
6(2,3,4)
5(2,3,4)
4(2,3,4)
4.00
4.10
4.30
5.04
5.10
5.07

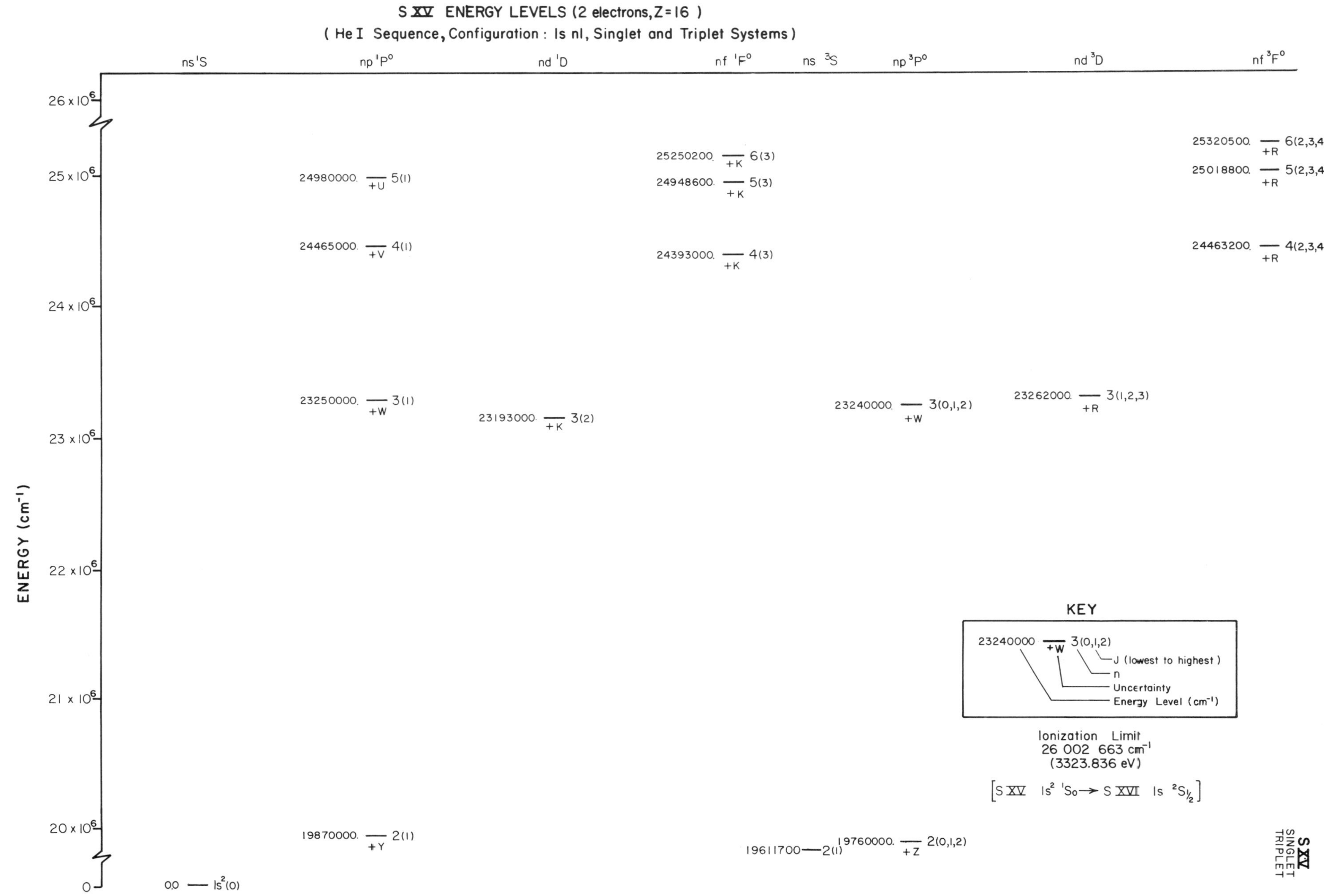

S XV ENERGY LEVELS (2 electrons, Z=16)
(He I Sequence, Configuration : 1s nl, Singlet and Triplet Systems)
ns ¹S
np ¹P°
nd ¹D
nf ¹F°
ns ³S
np ³P°
nd ³D
nf ³F°
ENERGY (cm⁻¹)
26 x 10⁶
25 x 10⁶
24 x 10⁶
23 x 10⁶
22 x 10⁶
21 x 10⁶
20 x 10⁶
0
0.0 — 1s²(0)
24980000. +U 5(1)
24465000. +V 4(1)
23250000. +W 3(1)
19870000. +Y 2(1)
23193000. +K 3(2)
25250200. +K 6(3)
24948600. +K 5(3)
24393000. +K 4(3)
19611700—2(1)
19760000. +Z 2(0,1,2)
23240000. +W 3(0,1,2)
23262000. +R 3(1,2,3)
25320500. +R 6(2,3,4)
25018800. +R 5(2,3,4)
24463200. +R 4(2,3,4)
KEY
23240000 +W 3(0,1,2)
J (lowest to highest)
n
Uncertainty
Energy Level (cm⁻¹)
Ionization Limit
26 002 663 cm⁻¹
(3323.836 eV)
[S XV 1s² ¹S₀ → S XVI 1s ²S½]
S XV
SINGLET
TRIPLET

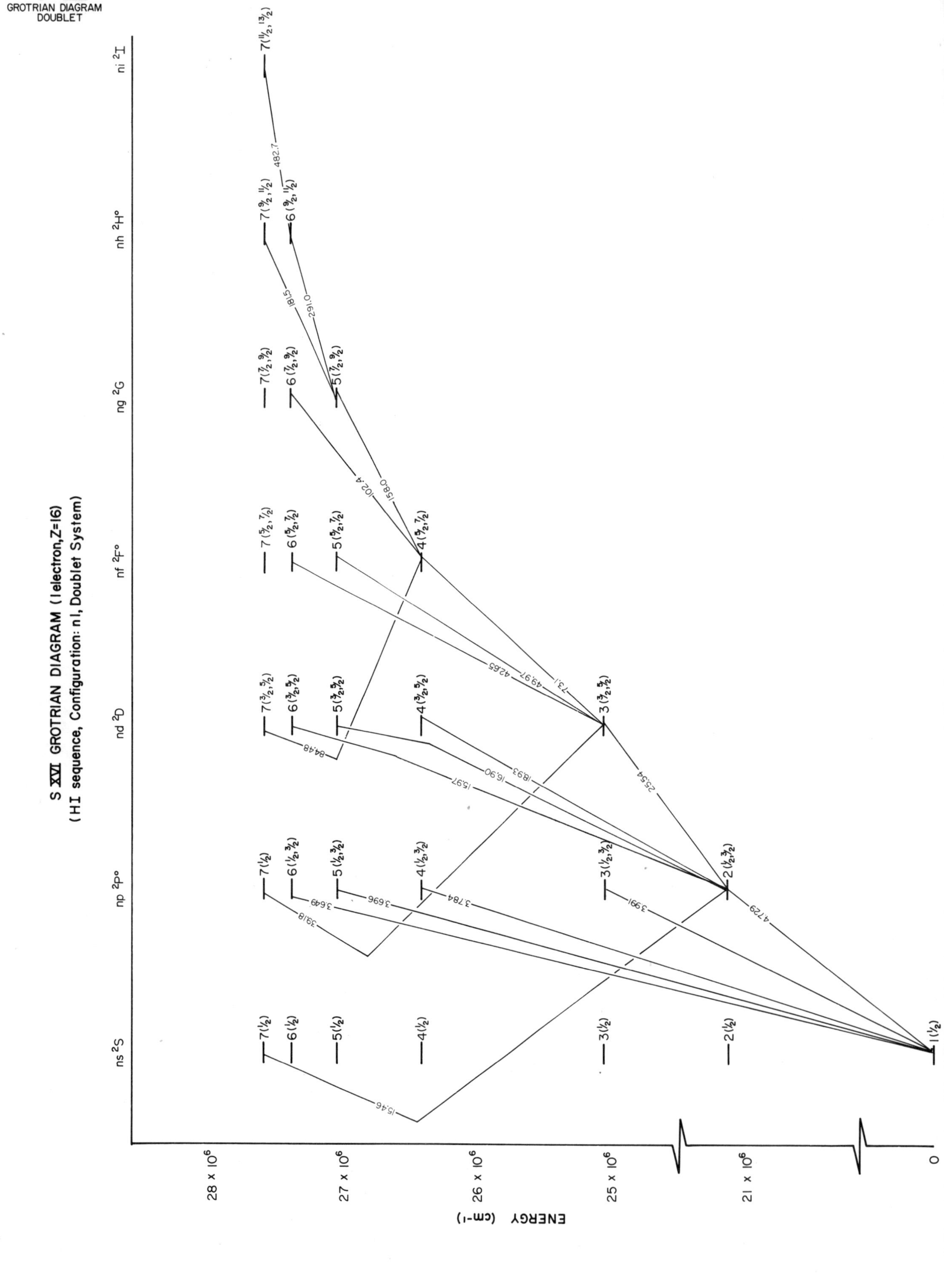

S XVI GROTRIAN DIAGRAM (1 electron, Z=16)
(HI sequence, Configuration: nl, Doublet System)

ns ^2S | np $^2P^o$ | nd ^2D | nf $^2F^o$ | ng ^2G | nh $^2H^o$ | ni ^2I

ENERGY (cm^{-1})

28 x 10^6
27 x 10^6
26 x 10^6
25 x 10^6
21 x 10^6
0

7(½) 6(½) 5(½) 4(½) 3(½) 2(½) 1(½)

7(½) 6(½,³⁄₂) 5(½,³⁄₂) 4(½,³⁄₂) 3(½,³⁄₂) 2(½,³⁄₂)

7(³⁄₂,⁵⁄₂) 6(³⁄₂,⁵⁄₂) 5(³⁄₂,⁵⁄₂) 4(³⁄₂,⁵⁄₂) 3(³⁄₂,⁵⁄₂)

7(⁵⁄₂,⁷⁄₂) 6(⁵⁄₂,⁷⁄₂) 5(⁵⁄₂,⁷⁄₂) 4(⁵⁄₂,⁷⁄₂)

7(⁷⁄₂,⁹⁄₂) 6(⁷⁄₂,⁹⁄₂) 5(⁷⁄₂,⁹⁄₂)

7(⁹⁄₂,¹¹⁄₂) 6(⁹⁄₂,¹¹⁄₂)

7(¹¹⁄₂,¹³⁄₂)

15.46 39.18 3.649 3.696 3.784 3.991 4.729 84.48 15.97 16.90 18.93 25.54 42.65 49.97 73.1 102.4 158.0 181.5 291.0 482.7

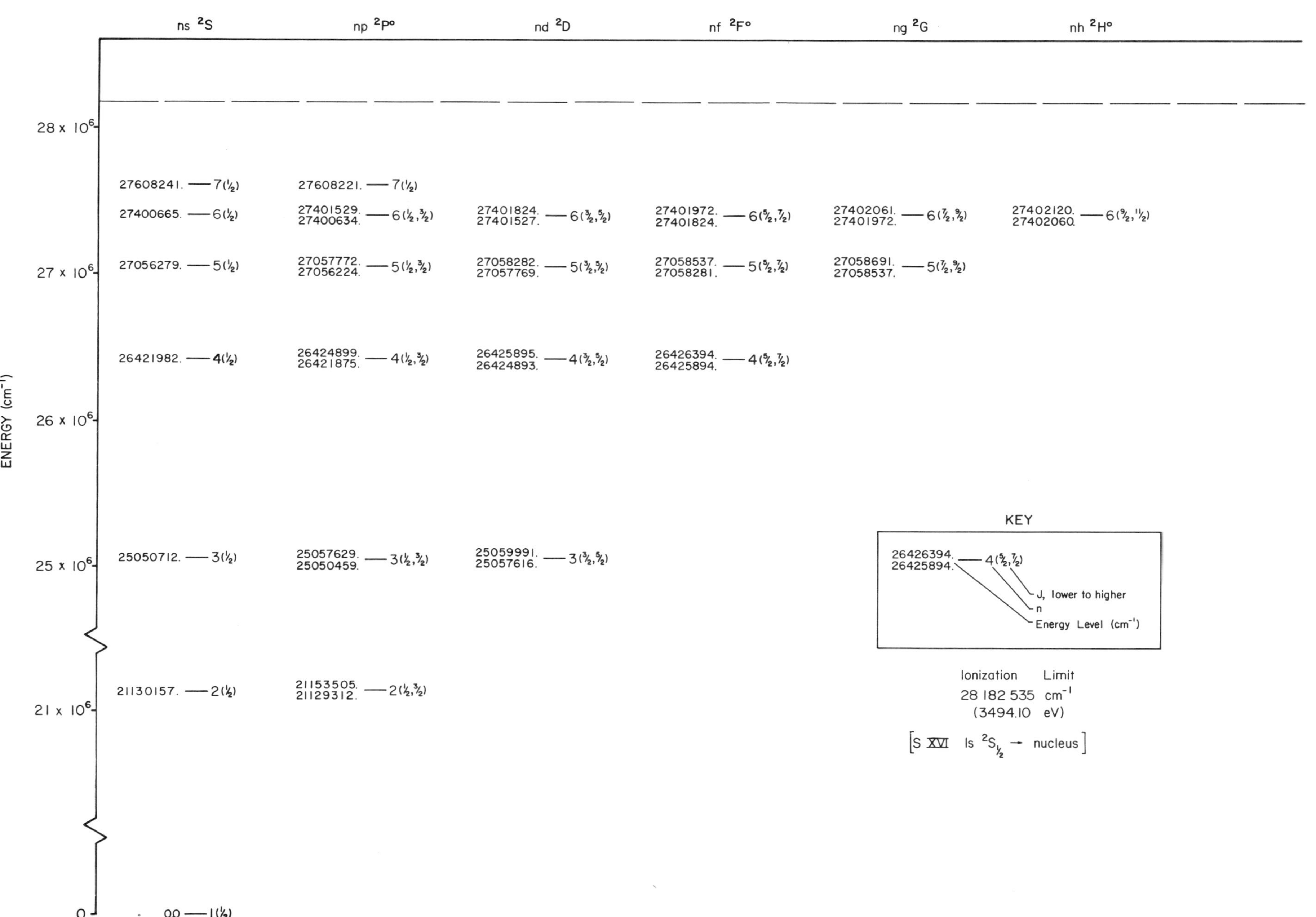
S XVI ENERGY LEVELS (1 electron, Z = 16)
(H I sequence, Configuration: nl, Doublet System)
ns ²S
np ²P°
nd ²D
nf ²F°
ng ²G
nh ²H°
27608241. — 7(½)
27400665. — 6(½)
27056279. — 5(½)
26421982. — 4(½)
25050712. — 3(½)
21130157. — 2(½)
0.0 — 1(½)
27608221. — 7(½)
27401529. 27400634. — 6(½,³⁄₂)
27057772. 27056224. — 5(½,³⁄₂)
26424899. 26421875. — 4(½,³⁄₂)
25057629. 25050459. — 3(½,³⁄₂)
21153505. 21129312. — 2(½,³⁄₂)
27401824. 27401527. — 6(³⁄₂,⁵⁄₂)
27058282. 27057769. — 5(³⁄₂,⁵⁄₂)
26425895. 26424893. — 4(³⁄₂,⁵⁄₂)
25059991. 25057616. — 3(³⁄₂,⁵⁄₂)
27401972. 27401824. — 6(⁵⁄₂,⁷⁄₂)
27058537. 27058281. — 5(⁵⁄₂,⁷⁄₂)
26426394. 26425894. — 4(⁵⁄₂,⁷⁄₂)
27402061. 27401972. — 6(⁷⁄₂,⁹⁄₂)
27058691. 27058537. — 5(⁷⁄₂,⁹⁄₂)
27402120. 27402060. — 6(⁹⁄₂,¹¹⁄₂)
KEY
26426394. 26425894. — 4(⁵⁄₂,⁷⁄₂)
J, lower to higher
n
Energy Level (cm⁻¹)
Ionization Limit
28 182 535 cm⁻¹
(3494.10 eV)
[S XVI 1s ²S½ → nucleus]
28 x 10⁶
27 x 10⁶
26 x 10⁶
25 x 10⁶
21 x 10⁶
0
ENERGY (cm⁻¹)

S I | $Z = 16$ | 16 electrons

Most of the levels are taken from Kelly's compilation. In some cases the levels within a term have not been resolved experimentally, and the order of their *J*-values (normal, inverted, etc.) is unknown. The values of *J* expected to be present are nevertheless shown on the drawings.

I.S. Bowen, Ap. J. **121**, 306 (1955).

Author identifies a wavelength of a forbidden transition observed in a nebula.

I.S. Bowen, Ap. J. **132**, 1 (1960).

Author gives a line table of observed forbidden transitions.

K.B.S. Eriksson, J. Opt. Soc. Am. **63**, 632 (1973).

Authors have remeasured wavelengths of two forbidden spectral lines.

G. Hubner and C. Wittig, J. Opt. Soc. Am. **61**, 415 (1971).

Two laser lines are identified with transitions in S I.

L.R. Jakobsson, Ark. Fys. **34**, 19 (1966).

Author gives line tables, energy level tables, a transition array chart, and a partial energy level diagram based on observations in the extra-photographic infrared region, in the region around 9700 and 10 000 Å and in the region 9212–34 270 Å.

V. Kaufman and L.J. Radziemski, Jr., J. Opt. Soc. Am. **59**, 227 (1969).

Authors give an energy level table based on observations.

J.W. McConkey, D.J. Burns, K.A. Moran, and J.A. Kernahan, Nature **217**, 538 (1968).

Authors give a term table based on observations. The values for the $3p^4$ levels do not fully agree with Kelly's tables.

Y.G. Toresson, Ark. Fys. **18**, 417 (1960).

Author gives line and term tables from lines observed in the vacuum ultraviolet region.

S II | $Z = 16$ | 15 electrons

The levels designated "N" are taken from Moore's AEL tables.

H.G. Berry, R.M. Schectman, I. Martinson, W.S. Bickel, and S. Bashkin, J. Opt. Soc. Am. **60**, 335 (1970).

Many strong lines in the spectrum of S II–S VI have been observed.

L.M. Beyer, W.E. Maddox, and L.B. Bridwell, J. Opt. Soc. Am. **63**, 365 (1973).

Several lines observed in the beam-gas spectra of sulfur, 500–1200 Å, are ascribed to forbidden transitions in S II–S V.

L. Block and E. Block, J. Phys. Rad. **6**, 30 (1935).

Authors give line tables from observations in the region 320–1260 Å.

I.S. Bowen, Ap. J. **121**, 306 (1955).

Author gives wavelengths of forbidden transitions observed in nebulae.

I.S. Bowen, Ap. J. **132**, 1 (1960).

Author gives line tables from nebular observations.

S III | $Z = 16$ | 14 electrons

H.G. Berry, R.M. Schectman, I. Martinson, W.S. Bickel, and S. Bashkin, J. Opt. Soc. Am. **60**, 335 (1970).

Many strong lines in the spectrum of S II–S VI have been observed.

L.M. Beyer, W.E. Maddox, and L.B. Bridwell, K. Opt. Soc. Am. **63**, 365 (1973).

Several lines observed in the spectrum of sulfur, 500–1200 Å, are ascribed to forbidden transitions in S II–S V.

L. Bloch and E. Bloch, J. Phys. Rad. **6**, 30 (1935).

Authors give line tables from observations in the region 320–1260 Å.

I.S. Bowen, Ap. J. **121**, 306 (1955).

Author gives a table of lines observed in nebulae.

I.S. Bowen, Ap. J. **132**, 1 (1960).

Author gives a line table from laboratory and nebular observations.

H.A. Robinson, Phys. Rev. **52**, 724 (1937).

Author gives line and term tables for lines observed in the region 495–4700 Å.

S IV $Z = 16$ 13 electrons

The location of the quartet system relative to the doublet system is known to within a few cm^{-1} (see Ekberg and Svensson). Most wavelengths are taken from Kelly's compilation with, however, a few from Berry's 1971 article.

H.G. Berry, R.M. Schectman, I. Martinson, W.S. Bickel, and S. Bashkin, J. Opt. Soc. Am. **60**, 335 (1970).

Many strong lines in the spectrum of S II–S VI have been observed.

L.M. Beyer, W.E. Maddox, and L.B. Bridwell, J. Opt. Soc. Am. **63**, 365 (1973).

Several lines observed in the spectrum of sulfur, 500–1200 Å, are ascribed to forbidden transitions in S I–S V. These have not been included in this set of drawings.

H.G. Berry, J. Opt. Soc. Am. **61**, 983 (1971).

Author gives a table of lines observed in the spectra of sulfur in the region 1085–3435 Å.

L. Bloch and E. Bloch, J. Phys. Rad. **6**, 30 (1935).

Authors give a table of lines observed in the region 320–1260 Å.

I.S. Bowen, Phys. Rev. **39**, 8 (1932).

Author give line and term tables for lines observed in the region 510–810 Å.

J.O. Ekberg and L.Å. Svensson, Physica Scripta **2**, 283 (1970).

These authors analyze the XUV spectra of K, Ca, Sc, and Ti isoelectronic with P I, Si I, and Al I.

C. Froese Fischer, J. Quant. Spectrosc. Radiat. Transfer **8**, 755 (1968).

This author reports calculations of energies and oscillator strengths in the Al I sequence.

S V $Z = 16$ 12 electrons

H.G. Berry, R.M. Schectman, I. Martinson, W.S. Bickel, and S. Bashkin, J. Opt. Soc. Am. **60**, 335 (1970).

Many strong lines in the spectrum of S II–S VI have been observed.

H.G. Berry, J. Opt. Soc. Am. **61**, 983 (1971).

Author gives a table of lines observed in the spectrum of sulfur in the region 1085–3435 Å.

L.M. Beyer, W.E. Maddox, and L.B. Bridwell, J. Opt. Soc. Am. **63**, 365 (1973).

Several lines observed in the spectrum of sulfur, 500–1200 Å, are ascribed to forbidden transitions in S II–S V.

L. Bloch and E. Bloch, J. Phys. Rad. **6**, 30 (1935).

Authors give line tables from observations in the region 320–1260 Å.

I.S. Bowen, Phys. Rev. **38**, 8 (1937).

Author gives line and term tables for lines observed in the region 435–910 Å.

S VI $Z = 16$ 11 electrons

H.G. Berry, R.M. Schectman, I. Martinson, W.S. Bickel, and S. Bashkin, J. Opt. Soc. Am. **60**, 335 (1970).

Many strong lines in the spectrum of S II–S VI have been observed.

H.B. Berry, J. Opt. Soc. Am. **61**, 983 (1971).

Author gives a line table from observations of spectra in the region 1085–3435 Å.

H.A. Robinson, Phys. Rev. **52**, 724 (1937).

Author gives line and term tables from observations in the region 170–945 Å.

S VII $Z = 16$ 10 electrons

E. Ferner, Ark. Mat. Astr. Fys. **36A**, 1 (1949).

Author gives line and term tables from observations in the region 45–75 Å.

S VIII $Z = 16$ 9 electrons

The energy levels given in Kelly's compilation have been taken in part from *Atomic Energy Levels*, by Moore, and in part from Feldman *et al.* The classifications of some levels differ in these two sources. We have taken the more recent classifications.

U. Feldman, G.A. Doschek, R.D. Cowan, and L. Cohen, J. Opt. Soc. Am. **63**, 1445 (1973).

Authors give experimental and theoretical energy levels and wavelengths for the F I sequence from K XI to Co XIX.

E. Ferner, Ark. Mat. Astr. Fys. **36A**, 1 (1949).

Author gives line and term tables from observations in the region 44–65 Å.

S.O. Kastner, Solar Physics **2**, 196 (1967).

This author calculates oscillator strengths for resonance lines in some Si and S ions.

S IX $Z = 16$ 8 electrons

W.A. Deutschman, Doctorate thesis for the Dept. of Astro-Geophysics, University of Colorado, obtained from University Microfilms, Ann Arbor, Michigan, U.S.A., order number 68–2645 (1970).

Author gives a line table, an energy level table, and term value diagrams from observations in the region 175–225 Å.

W.A. Deutschman and L.L. House, Ap. J. **144**, 435 (1966).

Authors give a line table from observations in the region 175–225 Å.

E. Ferner, Ark. Mat. Astr. Fys. **36A**, 1 (1949).

Authors gives line and term tables based on observations in the region 46–56 Å.

E.Ya. Kononov, Opt. and Spectros. **20**, 303 (1966).

Author gives line and energy levels tables based on observations in the region 175–230 Å.

B.C. Fawcett, R.D. Cowan, and R.W. Hayes; Ap. J. **187**, 377 (1974).

Authors give new S IX identifications and classifications (39–56 Å).

B.C. Fawcett, Atomic Data and Nuc. Data Tables **16**, 135 (1975).

Author gives tables of wavelengths and classifications of observed emission lines due to $2s^2 2p^n - 2s2p^{n+1}$ and $2s2p^n - 2p^{n+1}$ transitions.

S X $Z = 16$ 7 electrons

For wavelengths below 100 Å we have used those presented by Fawcett and Hayes; for others we have used the compilation of Kelly and Palumbo.

W.A. Deutschman, Doctorate thesis for the Dept. of Astro-Geophysics, University of Colorado, obtained from University Microfilms, Ann Arbor, Michigan, U.S.A., order number 68–2645 (1970).

Author gives a line table, an energy level table, and term value diagrams from observations in the region 175–270 Å.

W.A. Deutschman and L.L. House, Ap. J. **144**, 435 (1966).

Authors give a table for lines observed in the region 175–260 Å.

W.A. Deutschman and L.L. House, Ap. J. **149**, 451 (1967).

Authors give a table of lines observed in the region 175–265 Å.

B.C. Fawcett, D.D. Burgess, and N.J. Peacock, Proc. Phys. Soc. **91**, 970 (1967).

Authors give transition wavelengths from observations.

B.C. Fawcett and R.W. Hayes, Physica Scripta **8**, 244 (1973).

Authors give wavelengths below 100 Å. They classify transitions in S X–S XIV and isoelectronic spectra in P, Cl, and Ar.

E. Ferner, Ark. Mat. Astr. Fys. **36A**, 1 (1949).

Author gives observed transition wavelengths in the region 44–48 Å.

S.O. Kastner, J. Opt. Soc. Am. **63**, 738 (1973).

Calculated energies of low-lying terms are presented.

E.Ya. Kononov, Opt. Spectrosc. **20**, 303 (1966).

Author gives line and energy level tables for lines observed in the region 175–230 Å.

S XI $Z = 16$ 6 electrons

In several cases (indicated by different uncertainties within a term) different levels within a term have been located by different authors.

Under columns headed by two configurations, the correct configuration can be identified unambiguously by its parity.

We have taken wavelengths below 100 Å from Fawcett and Hayes, and wavelengths above 100 Å from Kelly and Palumbo.

W.A. Deutschman, Doctorate thesis for the Dept. of Astro-Geophysics, University of Colorado, obtained from University Microfilms, Ann Arbor, Michigan, U.S.A., order number 68–2645 (1970).

Author gives a line table, an energy level table, and a term value diagram from observations in the region 185–250 Å.

W.A. Deutschman and L.L. House, Ap. J. **149**, 451 (1967).

Authors give a table for lines observed in the region 185–250 Å.

B.C. Fawcett and R.W. Hayes, Physica Scripta **8**, 244 (1973).

The authors classify several spectra in the wavelength range below 100 Å. They classify transitions in S X–S XIV and isoelectronic spectra in P, Cl, and Ar.

S XII $Z = 16$ 5 electrons

Uncertainties in the locations of several doublet and quartet levels are indicated by letters on these drawings. Transitions wavelengths below 100 Å are taken from Fawcett and Hayes; others are taken from the compilation by Kelly and Palumbo.

W.A. Deutschman, Doctorate thesis for the Dept. of Astro-Geophysics, University of Colorado, obtained from University Microfilms, Ann Arbor, Michigan, U.S.A., order number 68–2645 (1970).

Author gives a transition wavelength observed for the $2s^22p\ ^2P^\circ - 2s2p^2\ ^2D$ transition.

B.C. Fawcett, D.D. Burgess, and N.J. Peacock, Proc. Phys. Soc. **91**, 970 (1967).

Authors give wavelengths for the observed transitions of a multiplet.

B.C. Fawcett and R.W. Hayes, Physica Scripta **8**, 244 (1973).

Author classify transitions in S X to S XIV, and isoelectronic spectra in P, Cl, and Ar for wavelengths below 100 Å.

L.J. Shamey, J. Opt. Soc. Am. **61**, 942 (1971).

Author gives line and energy level tables from calculations.

S XIII $Z = 16$ 4 electrons

Several authors disagree slightly on the wavelengths of these classified transitions. We have taken those of Fawcett and Hayes first, then included those only listed by Kelly and Palumbo.

B.C. Fawcett, D.D. Burgess, and N.J. Peacock, Proc. Phys. Soc. **91**, 970 (1967).

Authors give wavelengths for the observed transitions of a multiplet.

B.C. Fawcett and R.W. Hayes, Physica Scripta **8**, 244 (1973).

Authors give classifications in S X to XIV and isoelectronic spectra of P, Cl, and Ar.

A. Roldsmith, L. Oren, and L. Cohen, Ap. J. **188**. 197 (1974).

New transitions in the region 25–40 Å are identified for S XIII and S XIV.

There is an error in Table #5 of this reference. The last configurations should be 2p3d, not 2p3p as listed. The 2p3d $^3D_3^\circ$ level should read 3418320.

S XIV $Z = 16$ 3 electrons

Wavelengths in the range 30–34 Å are taken from Fawcett and Hayes; others are taken from Kelly and Palumbo. The energies of the multiply-excited levels are estimated from the wavelengths given in the latter.

B.C. Fawcett and R.W. Hayes, Physica Scripta **8**, 244 (1973).

Authors give classifications of S X to XIV and of isoelectronic spectra of P, Cl, and Ar for wavelengths below 100 Å.

S. Goldsmith, L. Oren, and L. Cohen, Ap. J. **188**, 197 (1974).

New transitions in the region 25–40 Å are identified for S XIII and S XIV.

S XV $Z = 16$ 2 electrons

Please see the general references.

S XVI $Z = 16$ 1 electron

We have taken calculated values rather than experimental values of energies and wavelengths. Since fine-structure and hyperfine-structure splittings are small in hydrogenic systems, and because all levels having the same n are nearly degenerate, we have indicated only the average wavelength for each set of transitions $n - n'$.

G.W. Erickson private communication (1976).

J.D. Garcia and J.E. Mack, J. Opt. Soc. Am. **55**, 654 (1965).

Authors give calculated energy level and line tables for one-electron atomic spectra.

Chlorine (Cl)

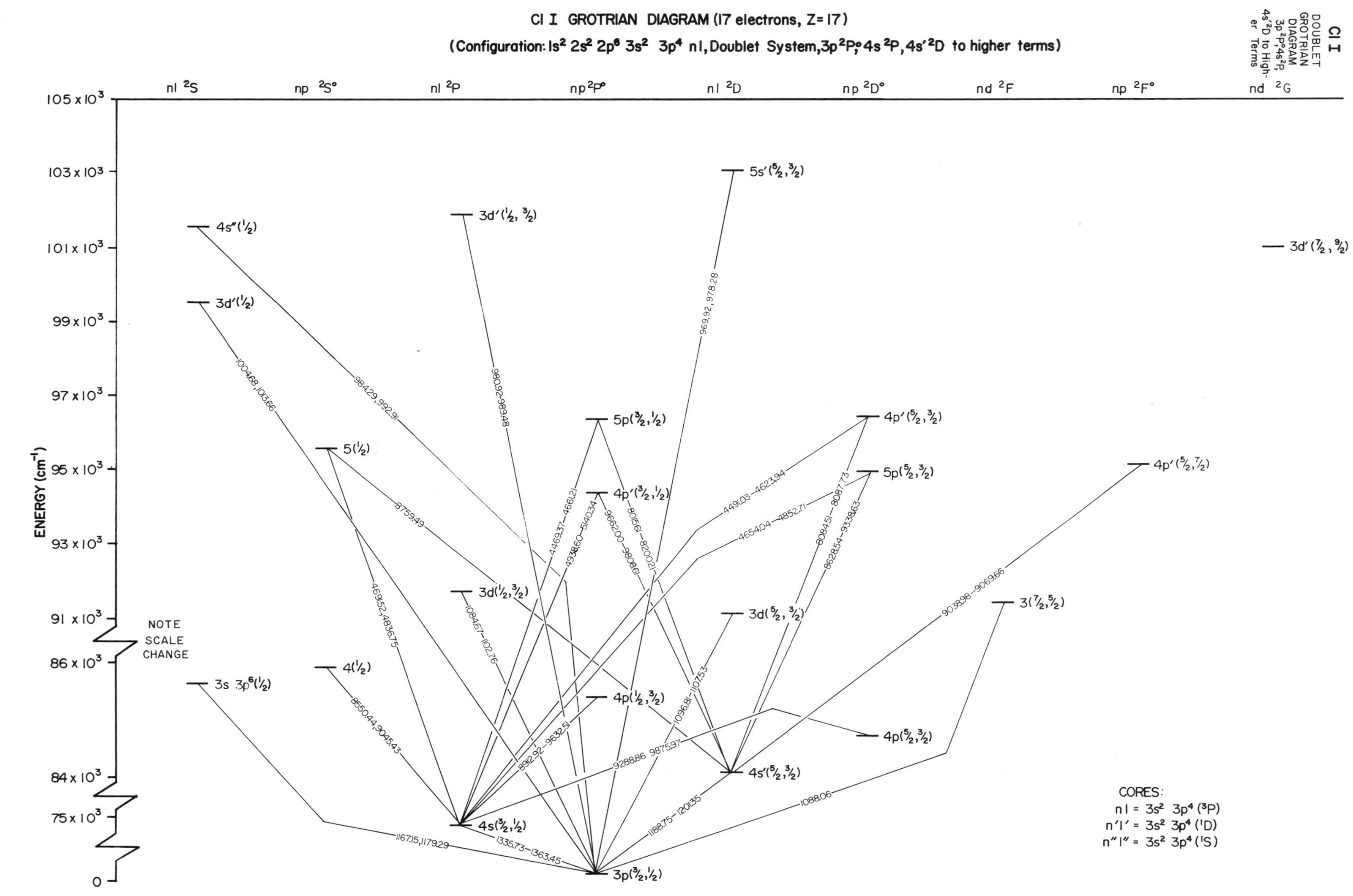

Cl I GROTRIAN DIAGRAM (17 electrons, Z= 17)
(Configuration: 1s² 2s² 2p⁶ 3s² 3p⁴ nl, Doublet System, 3p ²P°, 4s ²P, 4s′ ²D to higher terms)
Cl I
DOUBLET GROTRIAN DIAGRAM 3p ²P°, 4s ²P, 4s′ ²D to Higher Terms
nl ²S
np ²S°
nl ²P
np ²P°
nl ²D
np ²D°
nd ²F
np ²F°
nd ²G
ENERGY (cm⁻¹)
105 x 10³
103 x 10³
101 x 10³
99 x 10³
97 x 10³
95 x 10³
93 x 10³
91 x 10³
86 x 10³
84 x 10³
75 x 10³
0
NOTE SCALE CHANGE
4s″(½)
3d′(½)
3s 3p⁶(½)
5(½)
4(½)
3d′(½, 3/2)
3d(½, 3/2)
4s(3/2, ½)
5p(3/2, ½)
4p′(3/2, ½)
4p(½, 3/2)
3p(3/2, ½)
5s′(5/2, 3/2)
3d(5/2, 3/2)
4s′(5/2, 3/2)
4p′(5/2, 3/2)
5p(5/2, 3/2)
4p(5/2, 3/2)
3(7/2, 5/2)
4p′(5/2, 7/2)
3d′(7/2, 9/2)
984.29, 992.91
1004.68, 1013.66
980.92-989.48
969.92, 978.28
875.949
469.152, 483.675
855.044, 904.543
1084.67-1102.76
446.937-466.121
493.860-514.034
966.200-980.861
801.561-820.021
449.103-462.394
465.404-485.271
808.451-808.773
862.854-933.863
903.898-906.966
1096.81-1107.53
891.292-963.251
928.886, 987.597
1188.75-1201.35
1088.06
1167.15, 1179.29
1335.73-1363.45
CORES:
nl = 3s² 3p⁴ (³P)
n′l′ = 3s² 3p⁴ (¹D)
n″l″ = 3s² 3p⁴ (¹S)

Cl I ENERGY LEVELS (17 electrons, Z=17)

(Configuration: $1s^2\ 2s^2\ 2p^6\ 3s^2\ 3p^4\ nl$, Doublet System)

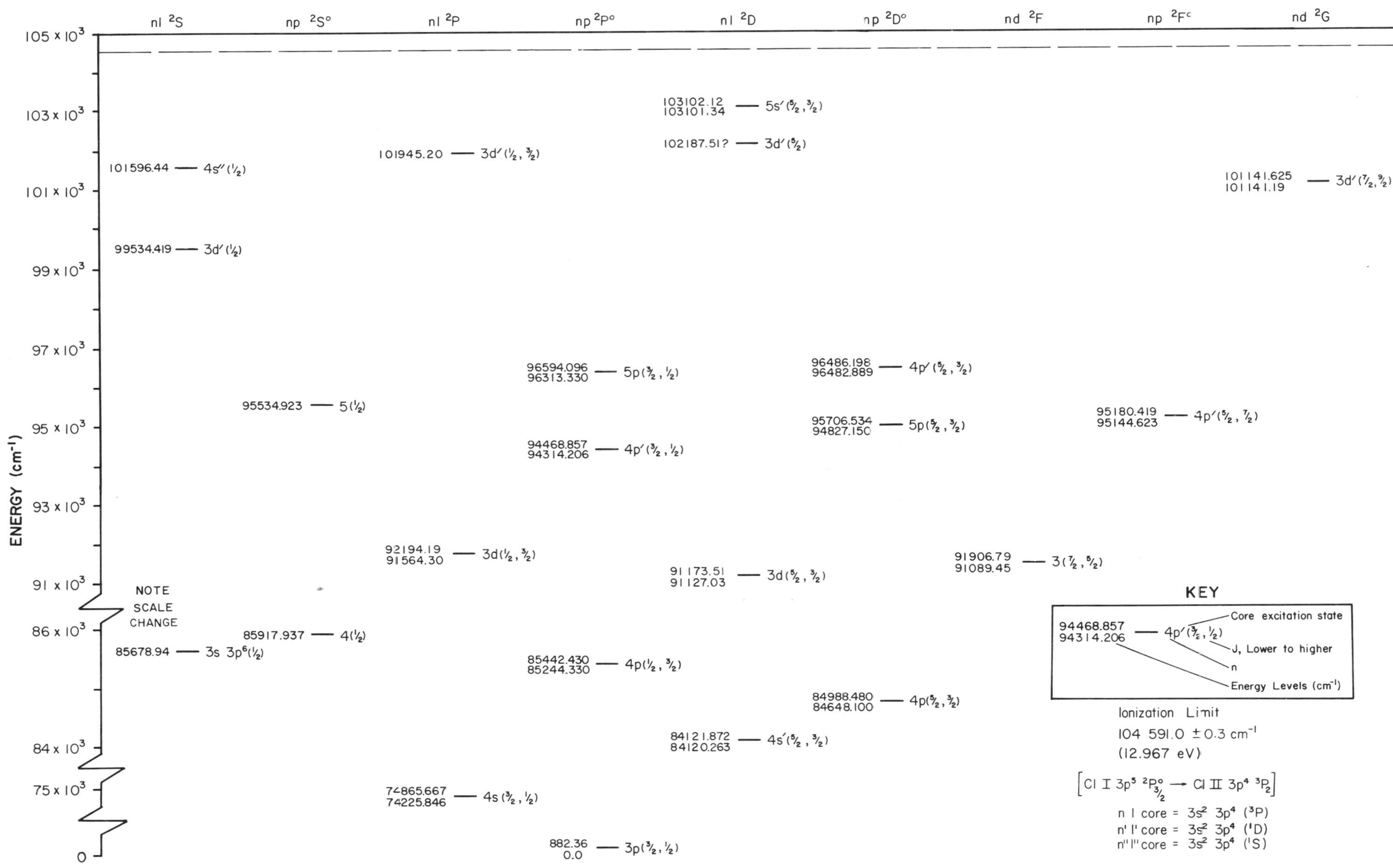

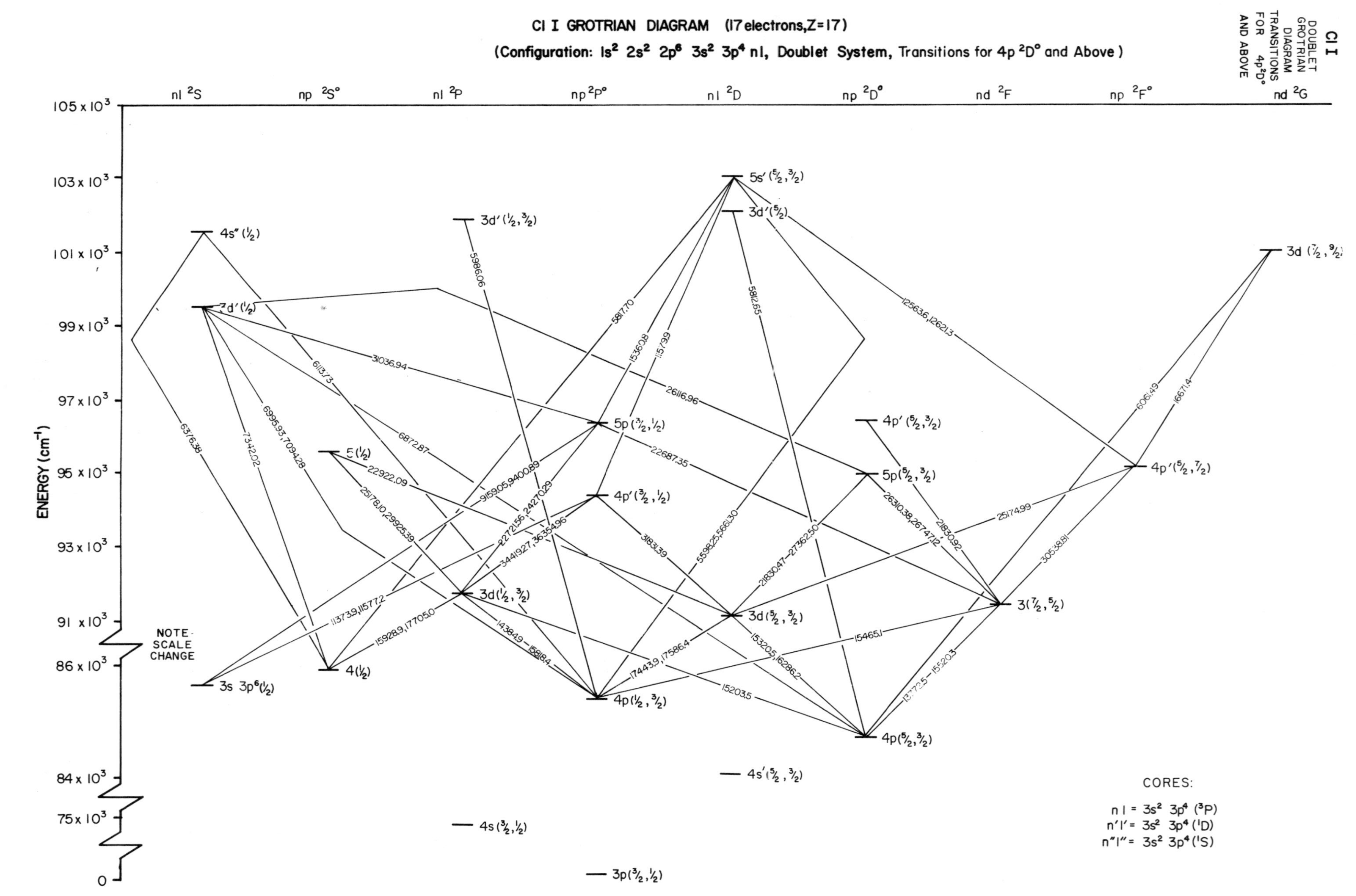

Cl I GROTRIAN DIAGRAM (17 electrons, Z=17)
(Configuration: 1s² 2s² 2p⁶ 3s² 3p⁴ nl, Doublet System, Transitions for 4p ²D° and Above)
Cl I
DOUBLET
GROTRIAN
DIAGRAM
TRANSITIONS
FOR 4p²D°
AND ABOVE
nl ²S
np ²S°
nl ²P
np ²P°
nl ²D
np ²D°
nd ²F
np ²F°
nd ²G
ENERGY (cm⁻¹)
105 x 10³
103 x 10³
101 x 10³
99 x 10³
97 x 10³
95 x 10³
93 x 10³
91 x 10³
86 x 10³
84 x 10³
75 x 10³
0
NOTE
SCALE
CHANGE
4s″(½)
3d′(½)
3s 3p⁶(½)
5(½)
4(½)
3d′(½, ³⁄₂)
3d(½, ³⁄₂)
5p(³⁄₂, ½)
4p′(³⁄₂, ½)
4p(½, ³⁄₂)
4s(³⁄₂, ½)
3p(³⁄₂, ½)
5s′(⁵⁄₂, ³⁄₂)
3d′(⁵⁄₂)
3d(⁵⁄₂, ³⁄₂)
4s′(⁵⁄₂, ³⁄₂)
4p′(⁵⁄₂, ³⁄₂)
5p(⁵⁄₂, ³⁄₂)
4p(⁵⁄₂, ³⁄₂)
3(⁷⁄₂, ⁵⁄₂)
4p′(⁵⁄₂, ⁷⁄₂)
3d (⁷⁄₂, ⁹⁄₂)
6376.38
7342.02
6995.93, 7094.28
6113.73
31036.94
6872.87
22922.09
25178.10, 29925.39
5986.06
9159.05, 9400.89
22721.56, 24270.29
34419.27, 36354.96
11373.9, 11577.2
15928.9, 17705.0
14384.9 – 15818.4
5817.70
15360.8
11517.99
26116.96
22687.35
31833.9
5812.65
55982.5, 55613.0
21830.47 – 27362.50
17443.9, 17586.4
15320.5, 16286.2
15203.5
15465.1
12563.6, 12621.3
26310.38, 26747.12
21830.92
25174.99
30538.81
13772.5 – 15520.3
6061.49
16671.4
CORES:
n l = 3s² 3p⁴ (³P)
n′l′ = 3s² 3p⁴ (¹D)
n″l″ = 3s² 3p⁴ (¹S)

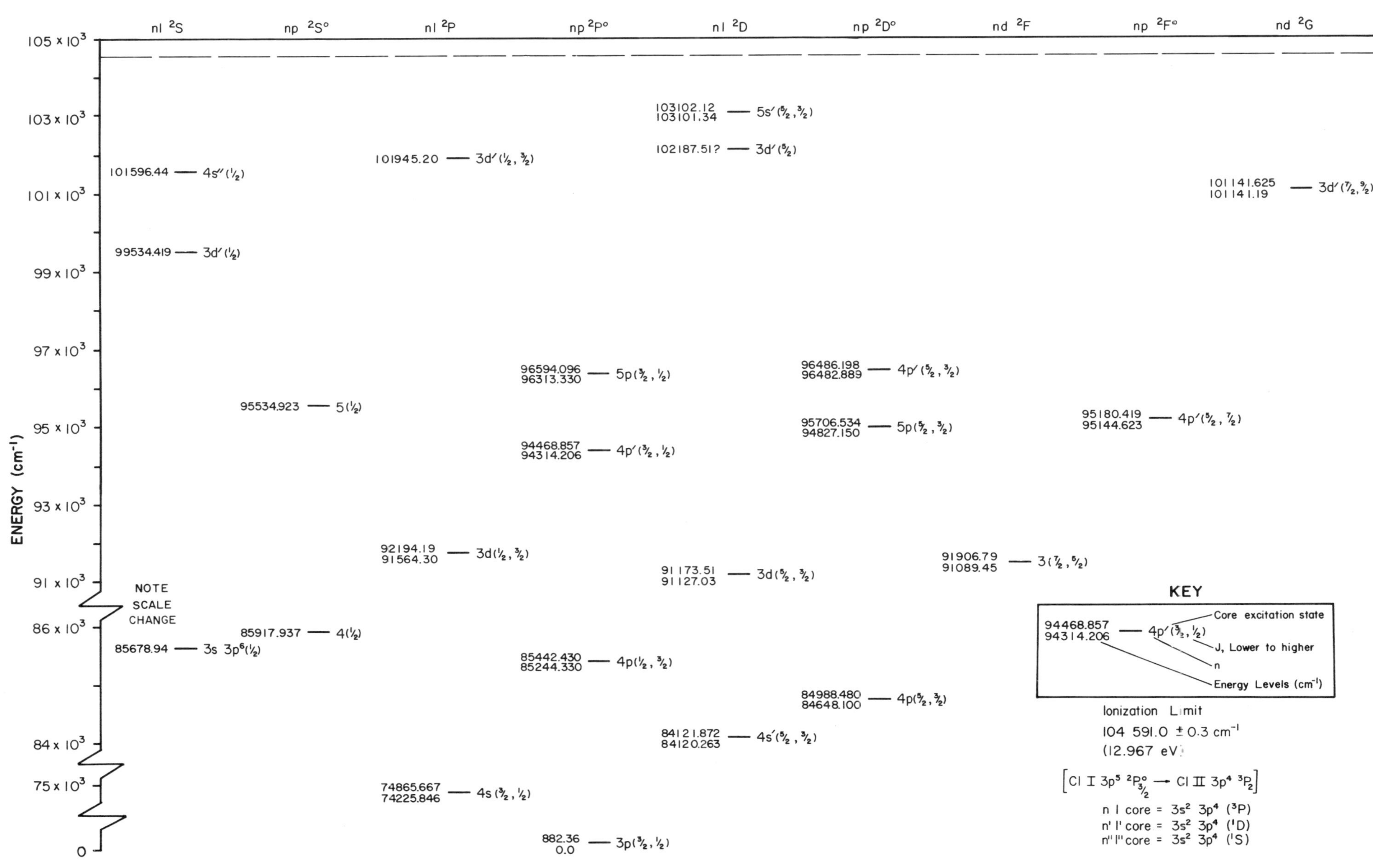
Cl I ENERGY LEVELS (17 electrons, Z=17)
(Configuration: 1s² 2s² 2p⁶ 3s² 3p⁴ nl, Doublet System)
nl ²S
np ²S°
nl ²P
np ²P°
nl ²D
np ²D°
nd ²F
np ²F°
nd ²G
ENERGY (cm⁻¹)
105 x 10³
103 x 10³
101 x 10³
99 x 10³
97 x 10³
95 x 10³
93 x 10³
91 x 10³
86 x 10³
84 x 10³
75 x 10³
0
NOTE
SCALE
CHANGE
101596.44 — 4s″(½)
99534.419 — 3d′(½)
85678.94 — 3s 3p⁶(½)
95534.923 — 5(½)
85917.937 — 4(½)
101945.20 — 3d′(½, 3/2)
92194.19
91564.30 — 3d(½, 3/2)
74865.667
74225.846 — 4s(3/2, ½)
96594.096
96313.330 — 5p(3/2, ½)
94468.857
94314.206 — 4p′(3/2, ½)
85442.430
85244.330 — 4p(½, 3/2)
882.36
0.0 — 3p(3/2, ½)
103102.12
103101.34 — 5s′(5/2, 3/2)
102187.51? — 3d′(5/2)
91173.51
91127.03 — 3d(5/2, 3/2)
84121.872
84120.263 — 4s′(5/2, 3/2)
96486.198
96482.889 — 4p′(5/2, 3/2)
95706.534
94827.150 — 5p(5/2, 3/2)
84988.480
84648.100 — 4p(5/2, 3/2)
91906.79
91089.45 — 3(7/2, 5/2)
95180.419
95144.623 — 4p′(5/2, 7/2)
101141.625
101141.19 — 3d′(7/2, 9/2)
KEY
94468.857
94314.206 — 4p′(3/2, ½)
Core excitation state
J, Lower to higher
n
Energy Levels (cm⁻¹)
Ionization Limit
104 591.0 ± 0.3 cm⁻¹
(12.967 eV)
[Cl I 3p⁵ ²P°3/2 → Cl II 3p⁴ ³P₂]
n l core = 3s² 3p⁴ (³P)
n′ l′ core = 3s² 3p⁴ (¹D)
n″ l″ core = 3s² 3p⁴ (¹S)

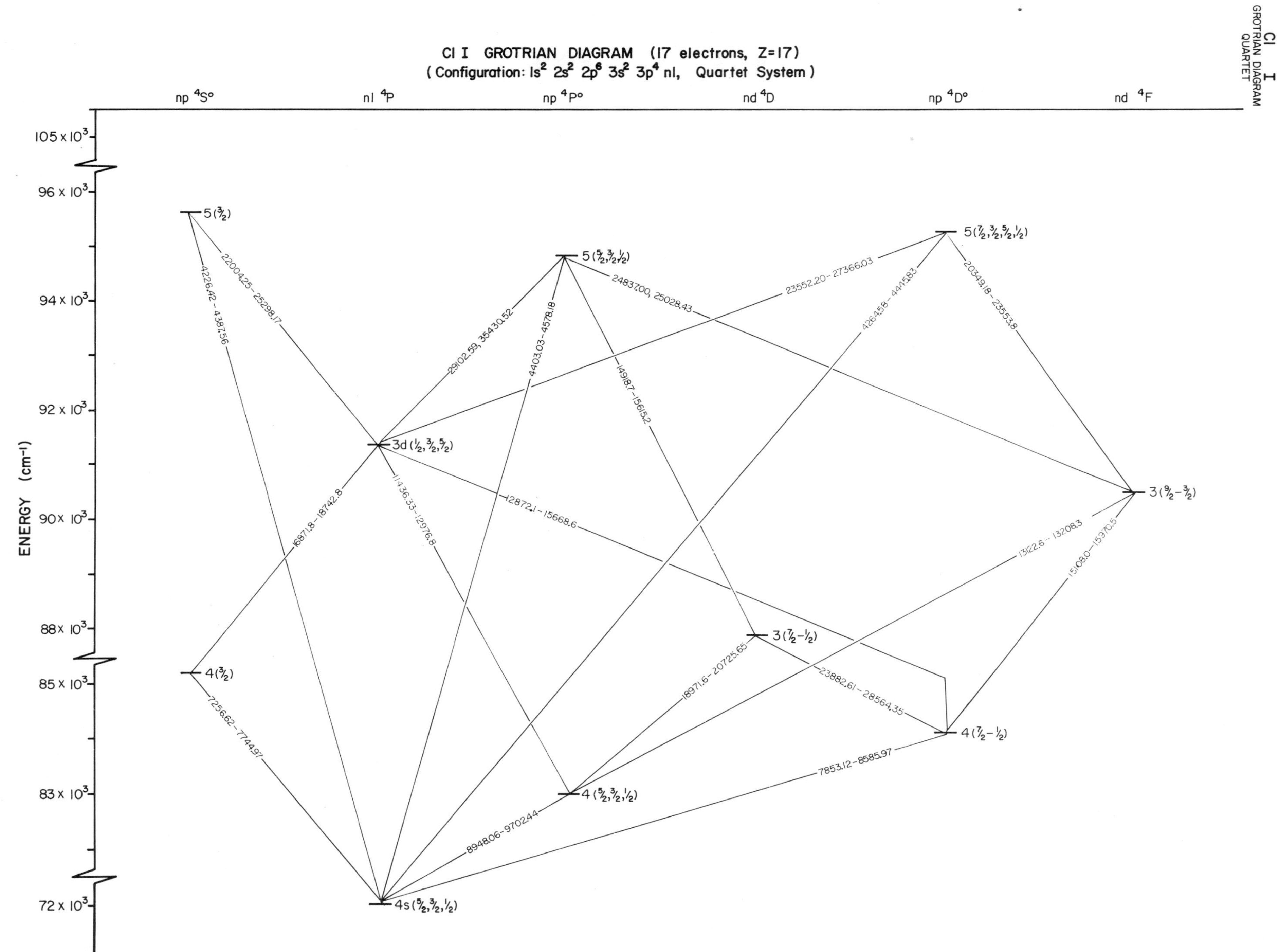
Cl I GROTRIAN DIAGRAM (17 electrons, Z=17)
(Configuration: $1s^2\ 2s^2\ 2p^6\ 3s^2\ 3p^4$ nl, Quartet System)
np $^4S^o$
nl ^4P
np $^4P^o$
nd ^4D
np $^4D^o$
nd ^4F
ENERGY (cm^{-1})
105×10^3
96×10^3
94×10^3
92×10^3
90×10^3
88×10^3
85×10^3
83×10^3
72×10^3
5(3/2)
4(3/2)
3d (1/2, 3/2, 5/2)
4s (5/2, 3/2, 1/2)
5(5/2, 3/2, 1/2)
4 (5/2, 3/2, 1/2)
3 (7/2 – 1/2)
5(7/2, 3/2, 5/2, 1/2)
4 (7/2 – 1/2)
3 (9/2 – 3/2)
2200.4.25 – 25298.17
4226.42 – 4387.56
7256.62 – 7744.97
16871.8 – 18742.8
29102.59, 35430.52
4403.03 – 4578.18
11436.33 – 12976.8
12872.1 – 15668.6
2483.7.00, 25028.43
14918.7 – 15615.2
23552.20 – 27366.03
4264.58 – 4445.83
20349.18 – 23553.8
8948.06 – 9702.44
18971.6 – 20725.65
23882.61 – 28564.35
13122.6 – 13208.3
15108.0 – 15370.5
7853.12 – 8585.97

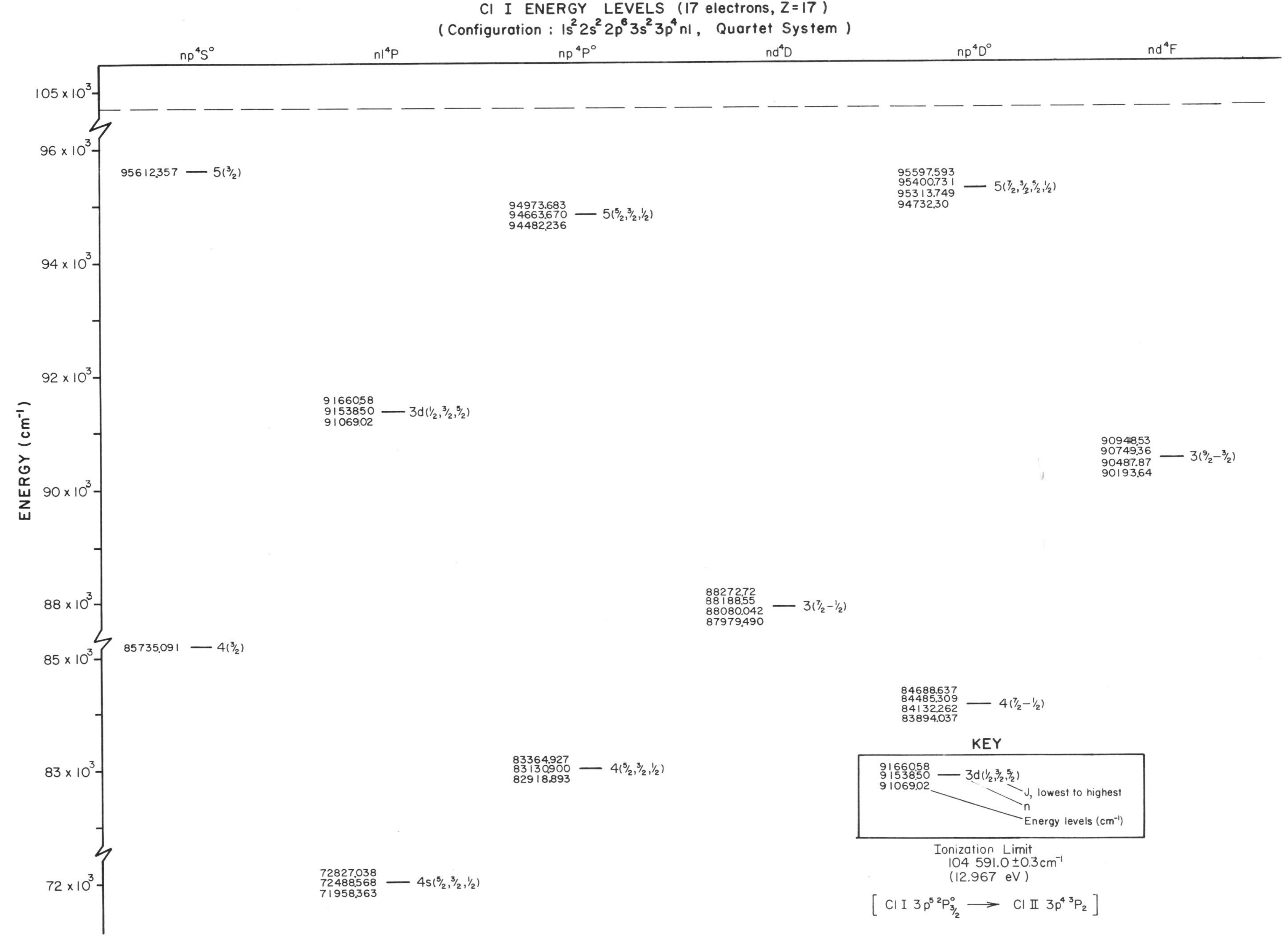
Cl I ENERGY LEVELS (17 electrons, Z=17)
(Configuration : 1s²2s²2p⁶3s²3p⁴nl , Quartet System)
np⁴S°
nl⁴P
np⁴P°
nd⁴D
np⁴D°
nd⁴F
ENERGY (cm⁻¹)
105 x 10³
96 x 10³
94 x 10³
92 x 10³
90 x 10³
88 x 10³
85 x 10³
83 x 10³
72 x 10³
95612.357 — 5(3/2)
85735.091 — 4(3/2)
91660.58
91538.50
91069.02 — 3d(1/2,3/2,5/2)
72827.038
72488.568
71958.363 — 4s(5/2,3/2,1/2)
94973.683
94663.670
94482.236 — 5(5/2,3/2,1/2)
83364.927
83130.900
82918.893 — 4(5/2,3/2,1/2)
88272.72
88188.55
88080.042
87979.490 — 3(7/2–1/2)
95597.593
95400.731
95313.749
94732.30 — 5(7/2,3/2,5/2,1/2)
84688.637
84485.309
84132.262
83894.037 — 4(7/2–1/2)
90948.53
90749.36
90487.87
90193.64 — 3(9/2–3/2)
KEY
91660.58
91538.50
91069.02 — 3d(1/2,3/2,5/2)
J, lowest to highest
n
Energy levels (cm⁻¹)
Ionization Limit
104 591.0 ± 0.3 cm⁻¹
(12.967 eV)
[Cl I 3p⁵ ²P°3/2 ⟶ Cl II 3p⁴ ³P₂]

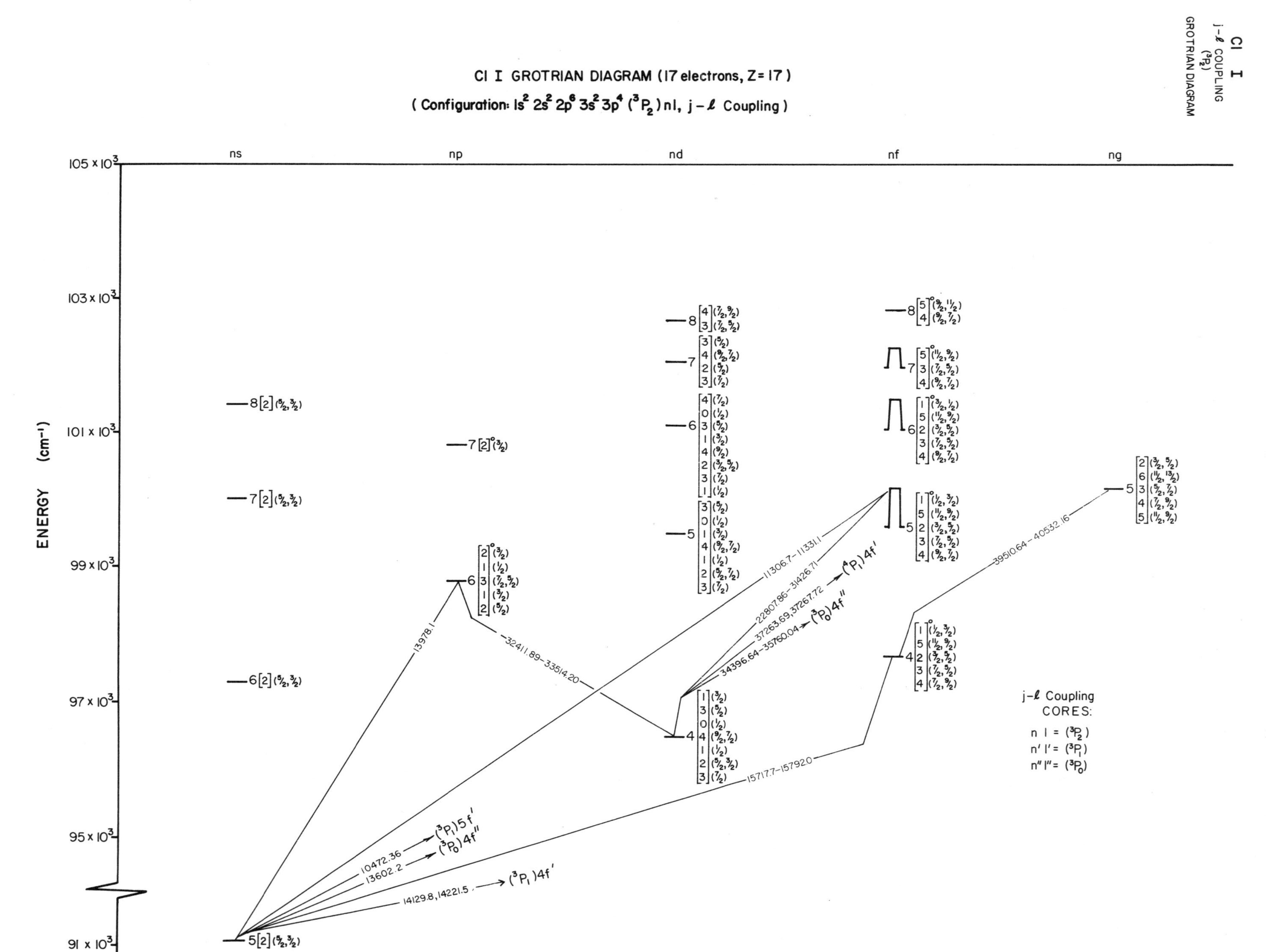
Cl I GROTRIAN DIAGRAM (17 electrons, Z= 17)
(Configuration: 1s² 2s² 2p⁶ 3s² 3p⁴ (³P₂) nl, j – ℓ Coupling)
ns
np
nd
nf
ng
ENERGY (cm⁻¹)
105 x 10³
103 x 10³
101 x 10³
99 x 10³
97 x 10³
95 x 10³
91 x 10³
j–ℓ Coupling
CORES:
n l = (³P₂)
n′ l′ = (³P₁)
n″ l″ = (³P₀)
(³P₁)5f′
(³P₀)4f″
(³P₁)4f′
13978.1
10472.36
13602.2
14129.8, 14221.5
32411.89–33514.20
11306.7–11331.1
22807.86–31426.71
37263.69, 37267.72
34396.64–35760.04
15717.7–157920
39510.64–40532.16

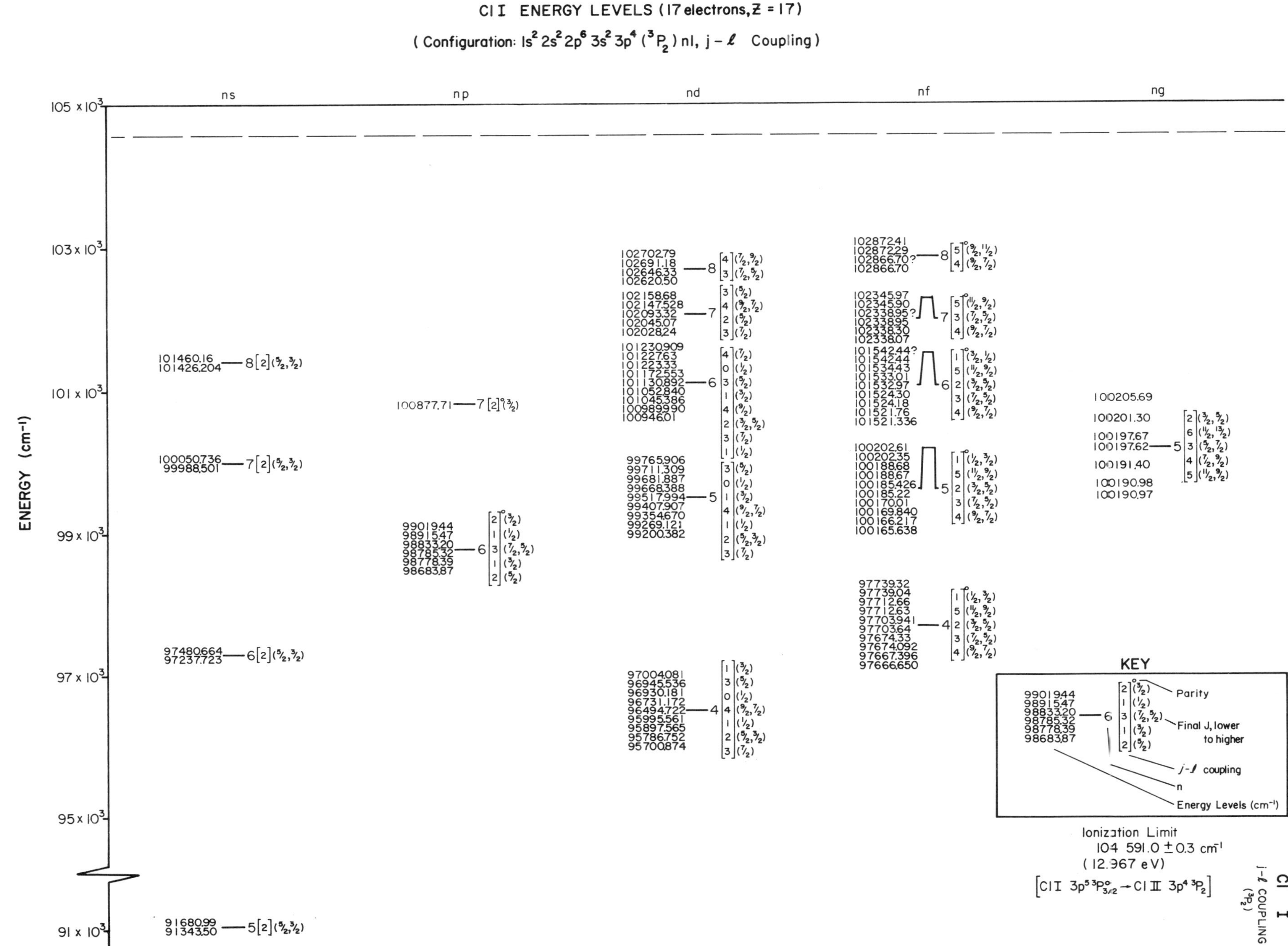
Cl I ENERGY LEVELS (17 electrons, Z = 17)
(Configuration: $1s^2 2s^2 2p^6 3s^2 3p^4 (^3P_2) nl$, j – ℓ Coupling)
ns
np
nd
nf
ng
ENERGY (cm⁻¹)
105 x 10³
103 x 10³
101 x 10³
99 x 10³
97 x 10³
95 x 10³
91 x 10³
KEY
Parity
Final J, lower to higher
j-ℓ coupling
n
Energy Levels (cm⁻¹)
Ionization Limit
104 591.0 ± 0.3 cm⁻¹
(12.967 eV)
[Cl I $3p^5\,^3P^o_{3/2}$ → Cl II $3p^4\,^3P_2$]
Cl I
j-ℓ COUPLING
($^3P_2$)

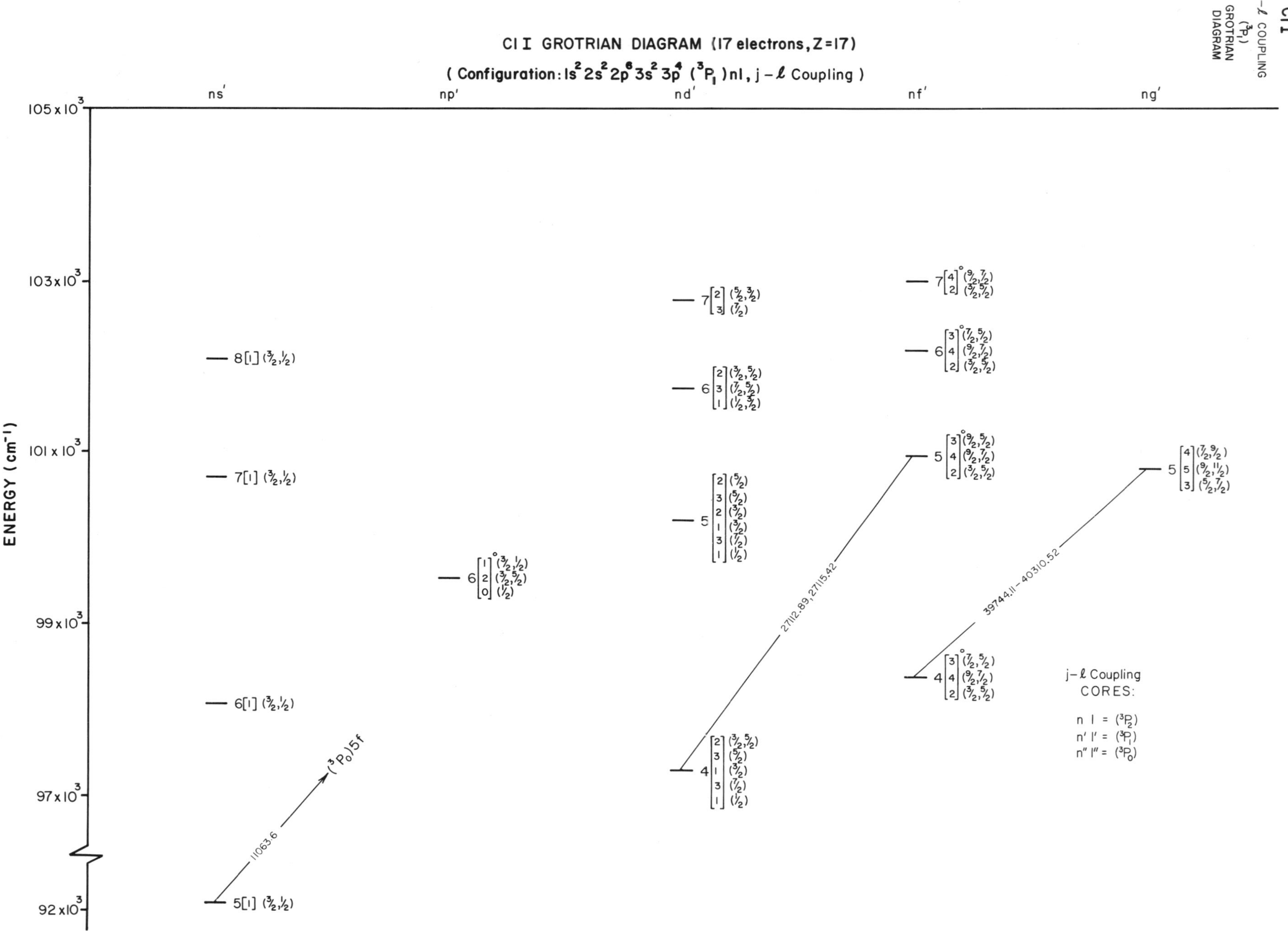
Cl I GROTRIAN DIAGRAM (17 electrons, Z=17)
(Configuration: $1s^2 2s^2 2p^6 3s^2 3p^4$ ($^3P_1$) nl, j – ℓ Coupling)
ns′
np′
nd′
nf′
ng′
ENERGY (cm^{-1})
105 x 10^3
103 x 10^3
101 x 10^3
99 x 10^3
97 x 10^3
92 x 10^3
8[1] (3/2,1/2)
7[1] (3/2,1/2)
6[1] (3/2,1/2)
5[1] (3/2,1/2)
11063.6
($^3P_0$)5f
27112.89, 27115.42
39744.11 – 40310.52
j–ℓ Coupling
CORES:
n l = ($^3P_2$)
n′ l′ = ($^3P_1$)
n″ l″ = ($^3P_0$)

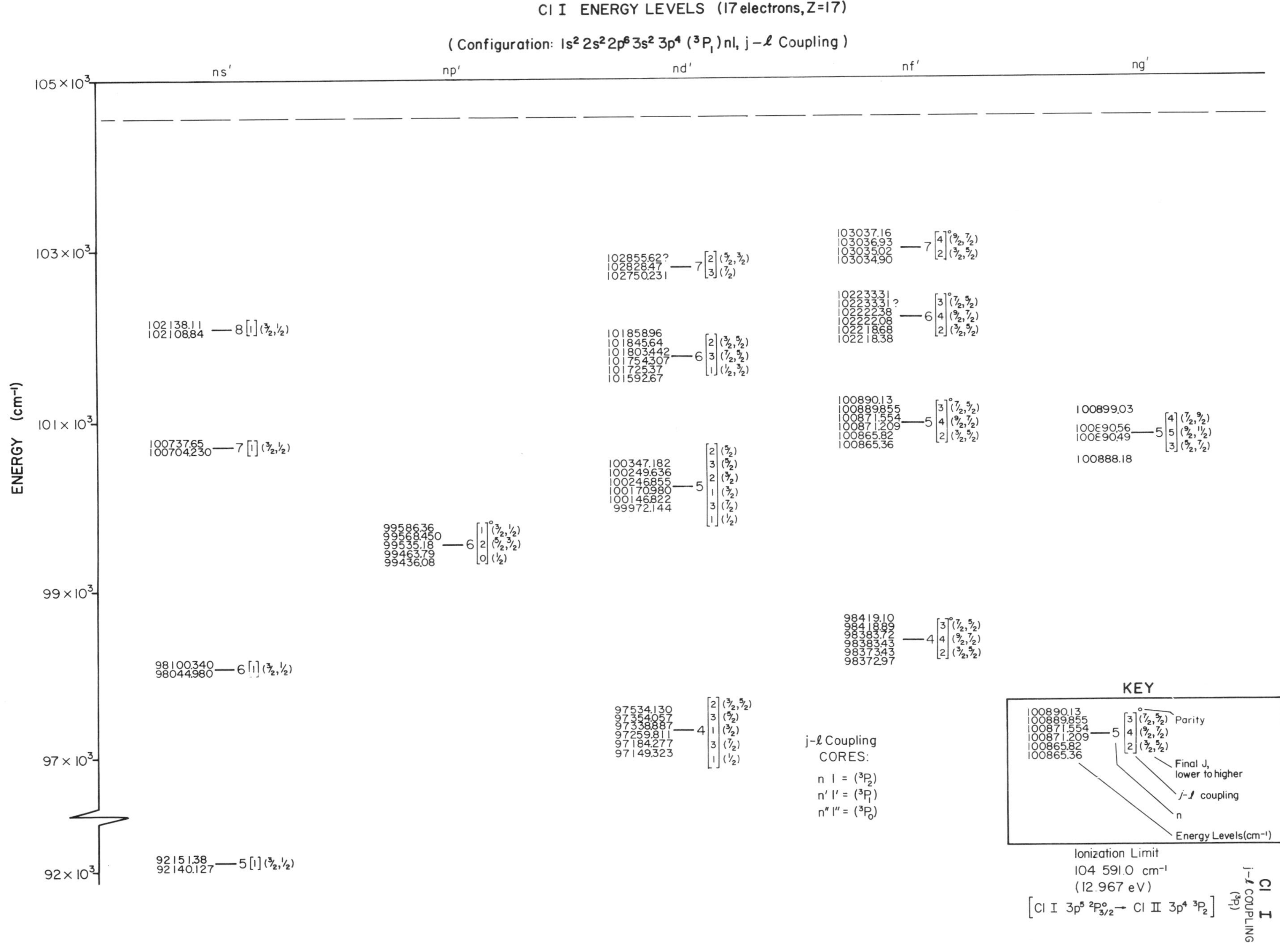
Cl I ENERGY LEVELS (17 electrons, Z=17)
(Configuration: 1s² 2s² 2p⁶ 3s² 3p⁴ (³P₁) nl, j – ℓ Coupling)
ns′
np′
nd′
nf′
ng′
ENERGY (cm⁻¹)
105×10³
103×10³
101×10³
99×10³
97×10³
92×10³
102138.11
102108.84
100737.65
100704.230
98100.340
98044.980
92151.38
92140.127
99586.36
99568.450
99535.18
99463.79
99436.08
102855.62?
102828.47
102750.231
101858.96
101845.64
101803.442
101754.307
101725.37
101592.67
100347.182
100249.636
100246.855
100170.980
100146.822
99972.144
97534.130
97354.057
97338.887
97259.811
97184.277
97149.323
103037.16
103036.93
103035.02
103034.90
102233.31
102233.31?
102222.38
102222.08
102218.68
102218.38
100890.13
100889.855
100871.554
100871.209
100865.82
100865.36
98419.10
98418.89
98383.72
98383.43
98373.43
98372.97
100899.03
100890.56
100890.49
100888.18
j-ℓ Coupling
CORES:
n l = (³P₂)
n′ l′ = (³P₁)
n″ l″ = (³P₀)
KEY
Parity
Final J, lower to higher
j-ℓ coupling
n
Energy Levels (cm⁻¹)
Ionization Limit
104 591.0 cm⁻¹
(12.967 eV)
[Cl I 3p⁵ ²P°₃/₂ → Cl II 3p⁴ ³P₂]
Cl I
j-ℓ COUPLING
(³P₁)

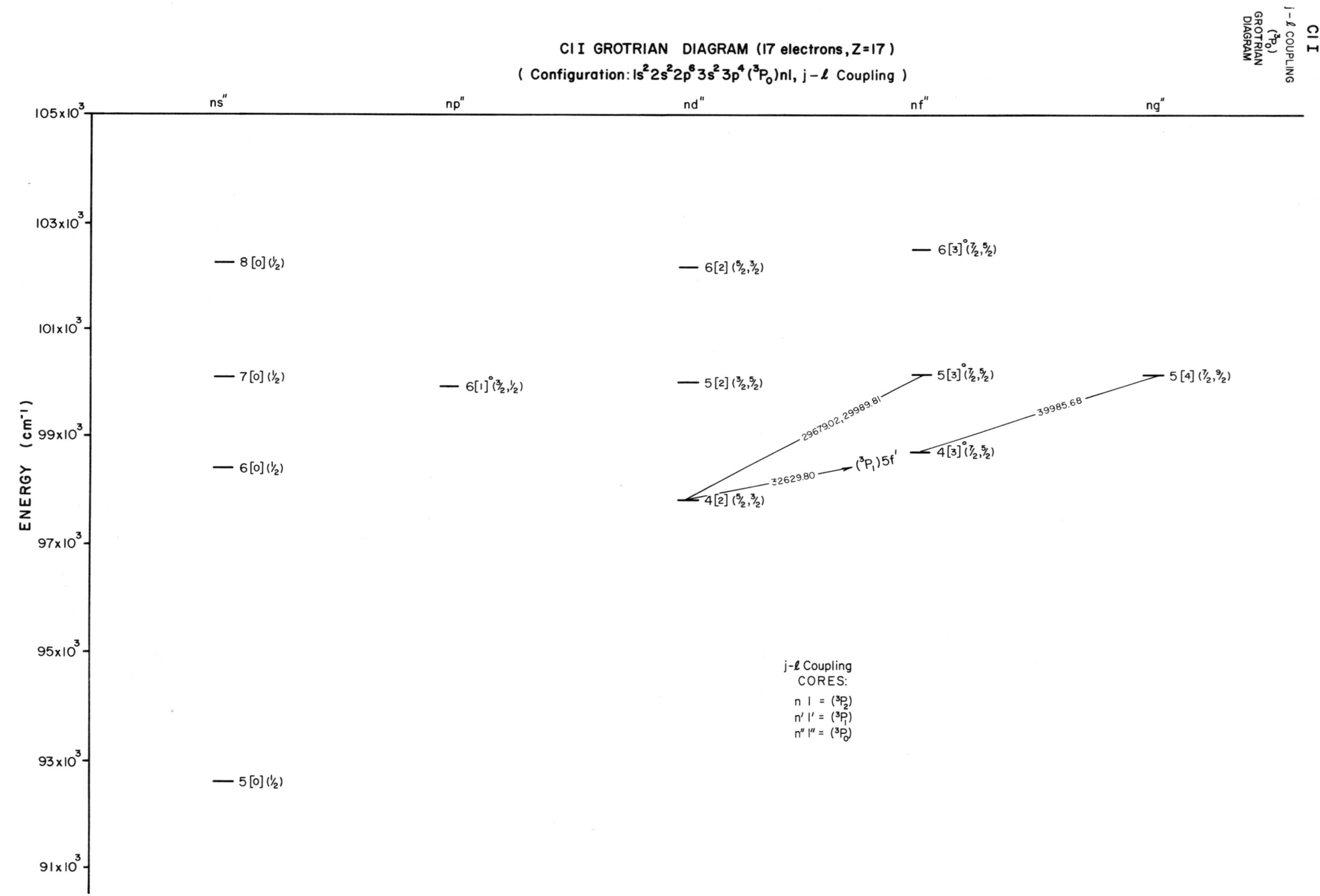

Cl I
j-ℓ COUPLING
($^3P_0$)
GROTRIAN
DIAGRAM
Cl I GROTRIAN DIAGRAM (17 electrons, Z=17)
(Configuration: $1s^2 2s^2 2p^6 3s^2 3p^4 (^3P_0)nl$, j-ℓ Coupling)
ns″
np″
nd″
nf″
ng″
105x10^3
103x10^3
101x10^3
99x10^3
97x10^3
95x10^3
93x10^3
91x10^3
ENERGY (cm^{-1})
8 [0] (½)
7 [0] (½)
6 [0] (½)
5 [0] (½)
6[1]° (3/2, 1/2)
6[2] (5/2, 3/2)
5 [2] (3/2, 5/2)
4[2] (5/2, 3/2)
6[3]° (7/2, 5/2)
5[3]° (7/2, 5/2)
4[3]° (7/2, 5/2)
5 [4] (7/2, 9/2)
29679.02, 29989.81
32629.80
($^3P_1$)5f′
39985.68
j-ℓ Coupling
CORES:
n l = ($^3P_2$)
n′ l′ = ($^3P_1$)
n″ l″ = ($^3P_0$)

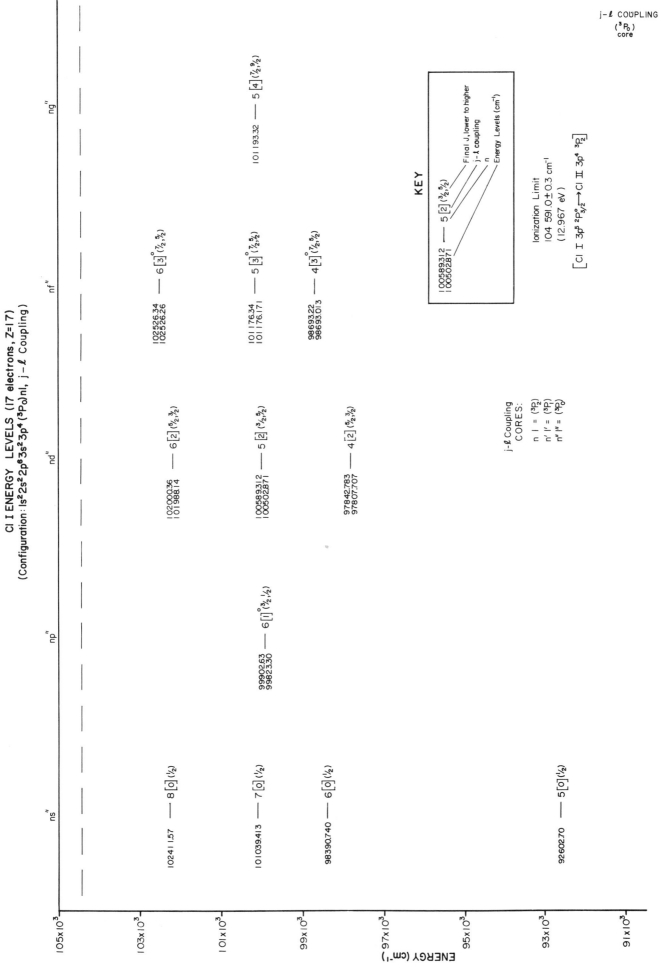

Cl I ENERGY LEVELS (17 electrons, Z=17)

(Configuration: $1s^2 2s^2 2p^6 3s^2 3p^4 (^3P_0)nl$, j-ℓ Coupling)

ENERGY (cm^{-1}): 105×10^3, 103×10^3, 101×10^3, 99×10^3, 97×10^3, 95×10^3, 93×10^3, 91×10^3

ns″

102411.57 — $8[0](1/2)$

101039.413 — $7[0](1/2)$

98390.740 — $6[0](1/2)$

92602.70 — $5[0](1/2)$

np″

99902.63
99823.30 — $6[1]^{\circ}(3/2, 1/2)$

nd″

102000.36
101988.14 — $6[2](5/2, 3/2)$

100589.312
100502.871 — $5[2](3/2, 5/2)$

97842.783
97807.707 — $4[2](5/2, 3/2)$

nf″

102526.34
102526.26 — $6[3]^{\circ}(7/2, 5/2)$

101176.34
101176.171 — $5[3]^{\circ}(7/2, 5/2)$

98693.22
98693.013 — $4[3]^{\circ}(7/2, 5/2)$

ng″

101193.32 — $5[4](7/2, 9/2)$

j-ℓ Coupling
CORES:

n l = ($^3P_2$)
n′ l′ = ($^3P_1$)
n″ l″ = ($^3P_0$)

KEY

100589.312
100502.871 — $5[2](3/2, 5/2)$

Final J, lower to higher
j-ℓ coupling
n
Energy Levels (cm^{-1})

Ionization Limit
104 591.0 ± 0.3 cm^{-1}
(12.967 eV)
[Cl I $3p^5\ ^2P^{\circ}_{3/2}$ → Cl II $3p^4\ ^3P_2$]

Cl I
DOUBLET to QUARTET
INTERCOMBINATION
GROTRIAN DIAGRAM

Cl I INTERCOMBINATION GROTRIAN DIAGRAM (17 electrons, Z = 17)

(Configuration : $1s^2 2s^2 2p^6 3s^2 3p^4 nl$, Doublets to Quartets)

^2S | $nl\,^2P$ | $np\,^2P^o$ | $nl\,^2D$ | $np\,^2D^o$ | $nd\,^2F$ | $np\,^4S^o$ | $nl\,^4P$ | $np\,^4P^o$ | $nd\,^4D$ | $np\,^4D^o$ | $nd\,^4F$

ENERGY (cm^{-1})

96×10^3 · 92×10^3 · 88×10^3 · 84×10^3 · 76×10^3 · 72×10^3 · 0

Levels: $5(\frac{3}{2})$; $5(\frac{5}{2},\frac{3}{2},\frac{1}{2})$; $5(\frac{7}{2}-\frac{1}{2})$; $3d(\frac{1}{2},\frac{3}{2})$; $3d(\frac{5}{2},\frac{3}{2})$; $3(\frac{7}{2},\frac{5}{2})$; $3d(\frac{1}{2},\frac{3}{2},\frac{5}{2})$; $3(\frac{9}{2},\frac{7}{2},\frac{5}{2},\frac{3}{2})$; $3(\frac{7}{2},\frac{5}{2},\frac{3}{2},\frac{1}{2})$; $3s3p^6(\frac{1}{2})$; $4(\frac{1}{2},\frac{3}{2})$; $4(\frac{3}{2})$; $4(\frac{5}{2},\frac{3}{2})$; $4s'(\frac{5}{2},\frac{3}{2})$; $4(\frac{7}{2}-\frac{1}{2})$; $4(\frac{5}{2},\frac{3}{2},\frac{1}{2})$; $4s(\frac{3}{2},\frac{1}{2})$; $4s(\frac{5}{2},\frac{3}{2},\frac{1}{2})$; $3(\frac{3}{2},\frac{1}{2})$

Transitions: 24696.47; 22288.84, 22522.23; 4674.54, 4818.70; 26979.06; 24786.87; 29466.26-38817.64; 9212.52-9483.43; 4818.43-5049.61; 3117.75; 4677.77-4822.13; 28613.17-35382.19; 8862.45-8932.60; 22597.39-27729.61; 10376.16; 28267.71-29796.30; 15883.3-17767.7; 8700.46; 14255.8-15263.0; 10930.98-11088.1; 16385.7, 17119.1; 28237.29-32337.29; 11132.85-11145.39; 10999.52-11110.29; 8686.26, 9197.60; 11226.2; 9555.06-10392.55; 1373.12-1396.53

CORES :
$nl = 3s^2 3p^4(^3P)$
$nl' = 3s^2 3p^4(^1D)$
$nl'' = 3s^2 3p^4(^1S)$

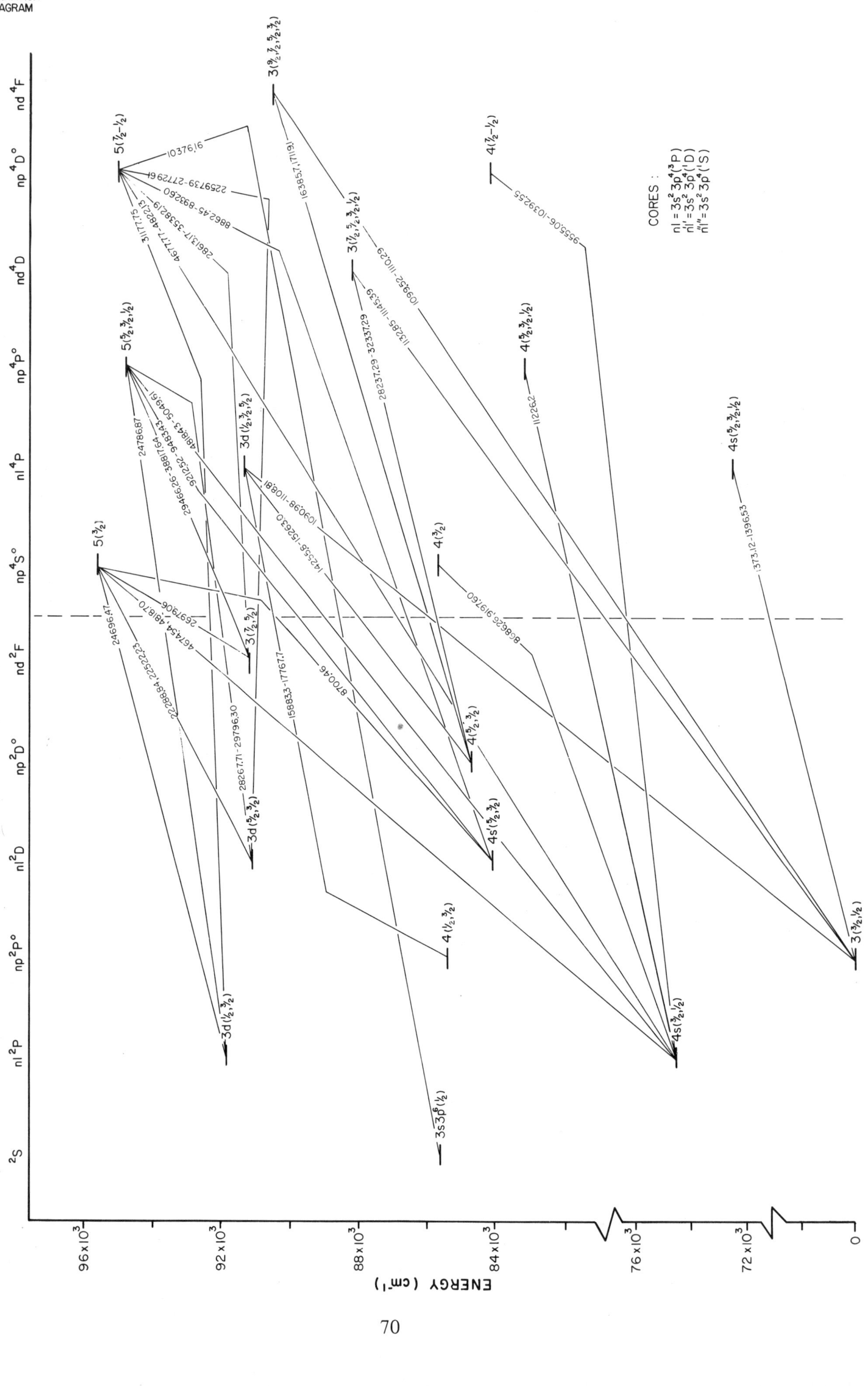

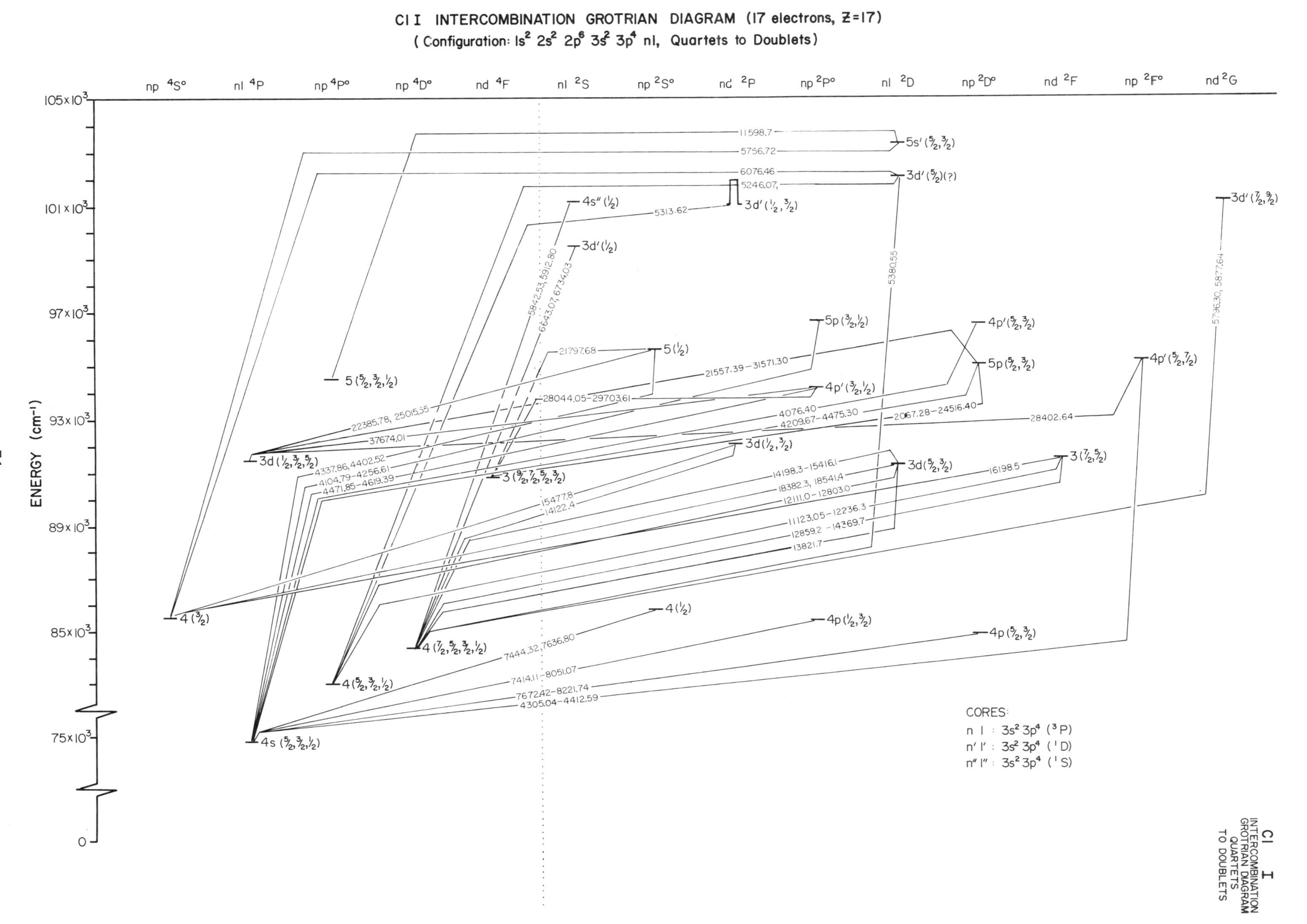
Cl I INTERCOMBINATION GROTRIAN DIAGRAM (17 electrons, Z=17)
(Configuration: $1s^2 2s^2 2p^6 3s^2 3p^4$ nl, Quartets to Doublets)
np $^4S°$
nl ^4P
np $^4P°$
np $^4D°$
nd ^4F
nl ^2S
np $^2S°$
nd ^2P
np $^2P°$
nl ^2D
np $^2D°$
nd ^2F
np $^2F°$
nd ^2G
ENERGY (cm^{-1})
105×10^3
101×10^3
97×10^3
93×10^3
89×10^3
85×10^3
75×10^3
0
5s′ (5/2, 3/2)
3d′ (5/2)(?)
3d′ (7/2, 9/2)
4s″ (1/2)
3d′ (1/2, 3/2)
3d′ (1/2)
5p (3/2, 1/2)
4p′ (5/2, 3/2)
5 (1/2)
5p (5/2, 3/2)
4p′ (5/2, 7/2)
5 (5/2, 3/2, 1/2)
4p′ (3/2, 1/2)
3d (1/2, 3/2, 5/2)
3d (1/2, 3/2)
3d (5/2, 3/2)
3 (7/2, 5/2)
3 (9/2, 7/2, 5/2, 3/2)
4 (3/2)
4 (1/2)
4p (1/2, 3/2)
4p (5/2, 3/2)
4 (7/2, 5/2, 3/2, 1/2)
4 (5/2, 3/2, 1/2)
4s (5/2, 3/2, 1/2)
11598.7
5756.72
6076.46
5246.07,
5313.62
5842.53, 5912.80
6643.07, 6734.03
5380.55
5796.30, 5877.64
21797.68
21557.39 – 31571.30
28044.05 – 29703.61
22385.78, 25015.35
37674.01
4337.86, 4402.52
4104.79 – 4256.61
4471.85 – 4619.39
4076.40
4209.67 – 4475.30
2067.28 – 24516.40
28402.64
14198.3 – 15416.1
18382.3, 18541.4
12111.0 – 12803.0
11123.05 – 12236.3
12859.2 – 14369.7
13821.7
16198.5
15477.8
14122.4
7444.32, 7636.80
7414.11 – 8051.07
7672.42 – 8221.74
4305.04 – 4412.59
CORES:
n l : $3s^2 3p^4$ (^3P)
n′ l′ : $3s^2 3p^4$ (^1D)
n″ l″ : $3s^2 3p^4$ (^1S)
Cl I
INTERCOMBINATION
GROTRIAN DIAGRAM
QUARTETS
TO DOUBLETS

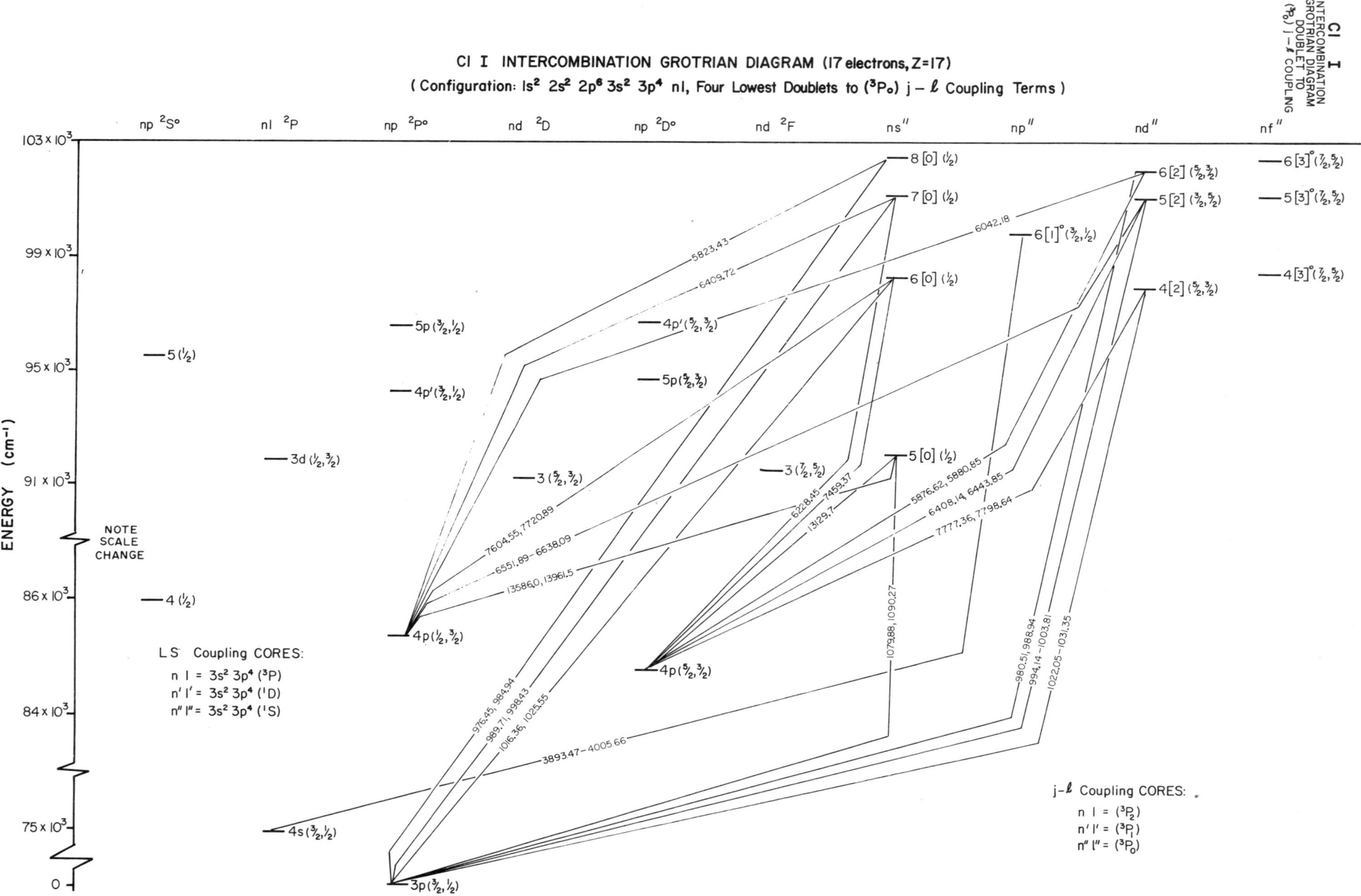

Cl I
INTERCOMBINATION GROTRIAN DIAGRAM DOUBLET TO
($^3P_0$) j – ℓ COUPLING
Cl I INTERCOMBINATION GROTRIAN DIAGRAM (17 electrons, Z=17)
(Configuration: $1s^2\ 2s^2\ 2p^6\ 3s^2\ 3p^4$ nl, Four Lowest Doublets to ($^3P_0$) j – ℓ Coupling Terms)
np $^2S^o$
nl ^2P
np $^2P^o$
nd ^2D
np $^2D^o$
nd ^2F
ns''
np''
nd''
nf''
ENERGY (cm^{-1})
103 x 10^3
99 x 10^3
95 x 10^3
91 x 10^3
NOTE SCALE CHANGE
86 x 10^3
84 x 10^3
75 x 10^3
0
5 (1/2)
4 (1/2)
3d (1/2, 3/2)
4s (3/2, 1/2)
5p (3/2, 1/2)
4p' (3/2, 1/2)
4p (1/2, 3/2)
3p (3/2, 1/2)
3 (5/2, 3/2)
4p' (5/2, 3/2)
5p (5/2, 3/2)
4p (5/2, 3/2)
3 (7/2, 5/2)
8 [0] (1/2)
7 [0] (1/2)
6 [0] (1/2)
5 [0] (1/2)
6 [1]° (3/2, 1/2)
6 [2] (5/2, 3/2)
5 [2] (3/2, 5/2)
4 [2] (5/2, 3/2)
6 [3]° (7/2, 5/2)
5 [3]° (7/2, 5/2)
4 [3]° (7/2, 5/2)
5823.43
6409.72
6042.18
7604.55, 7720.89
6551.89 – 6638.09
13586.0, 13961.5
6228.45
7459.37
13129.7
5876.62, 5880.85
6408.14, 6443.85
7777.36, 7798.64
1079.88, 1090.27
976.45, 984.94
989.71, 998.43
1016.36, 1025.55
3893.47 – 4005.66
980.51, 988.94
994.14 – 1003.81
1022.05 – 1031.35
LS Coupling CORES:
n l = $3s^2\ 3p^4$ (^3P)
n' l' = $3s^2\ 3p^4$ (^1D)
n'' l'' = $3s^2\ 3p^4$ (^1S)
j – ℓ Coupling CORES:
n l = ($^3P_2$)
n' l' = ($^3P_1$)
n'' l'' = ($^3P_0$)

Cl I

DOUBLETS & ($^3P_0$)
j – ℓ COUPLING
INTERCOMBINATION
GROTRIAN DIAGRAM

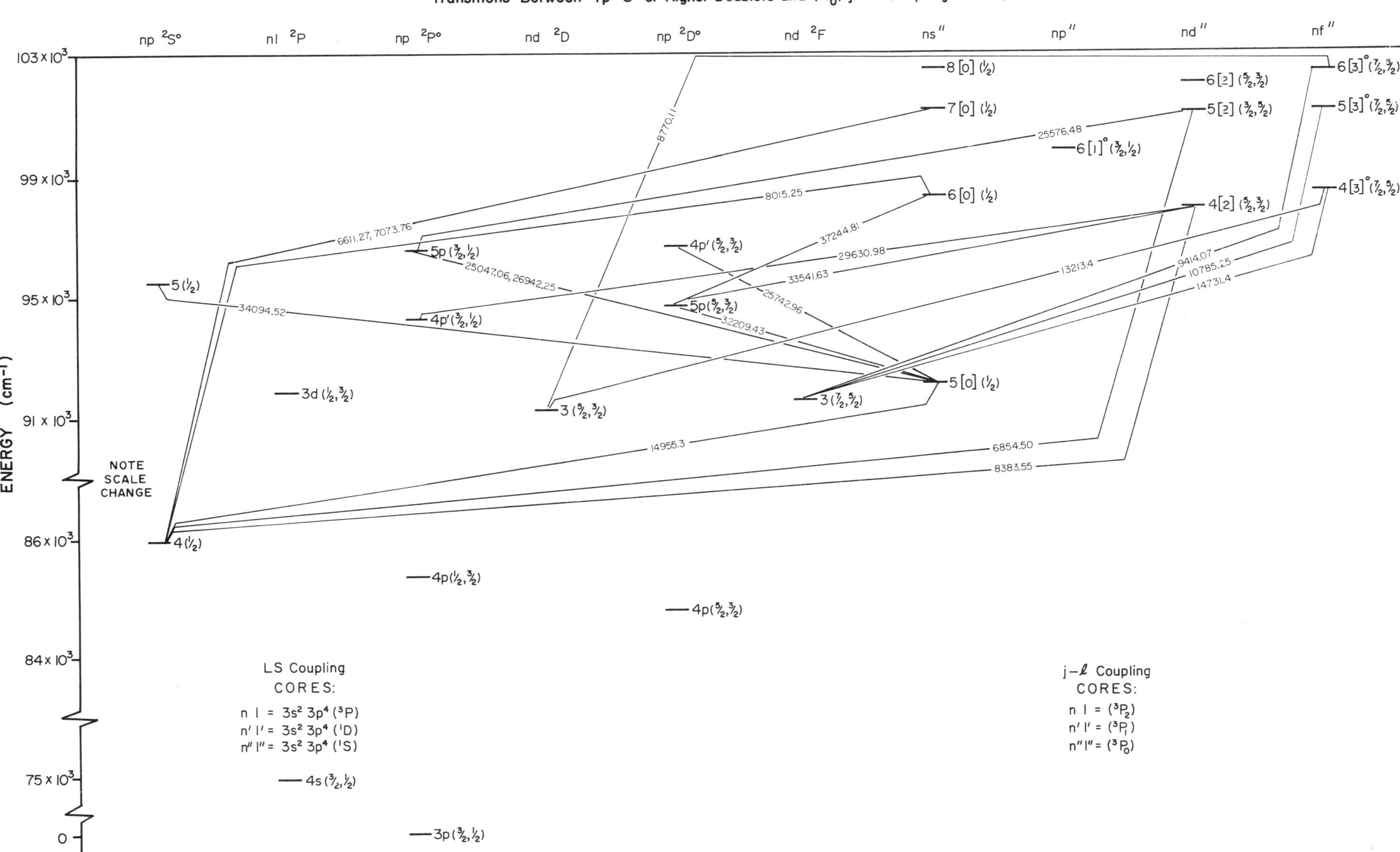

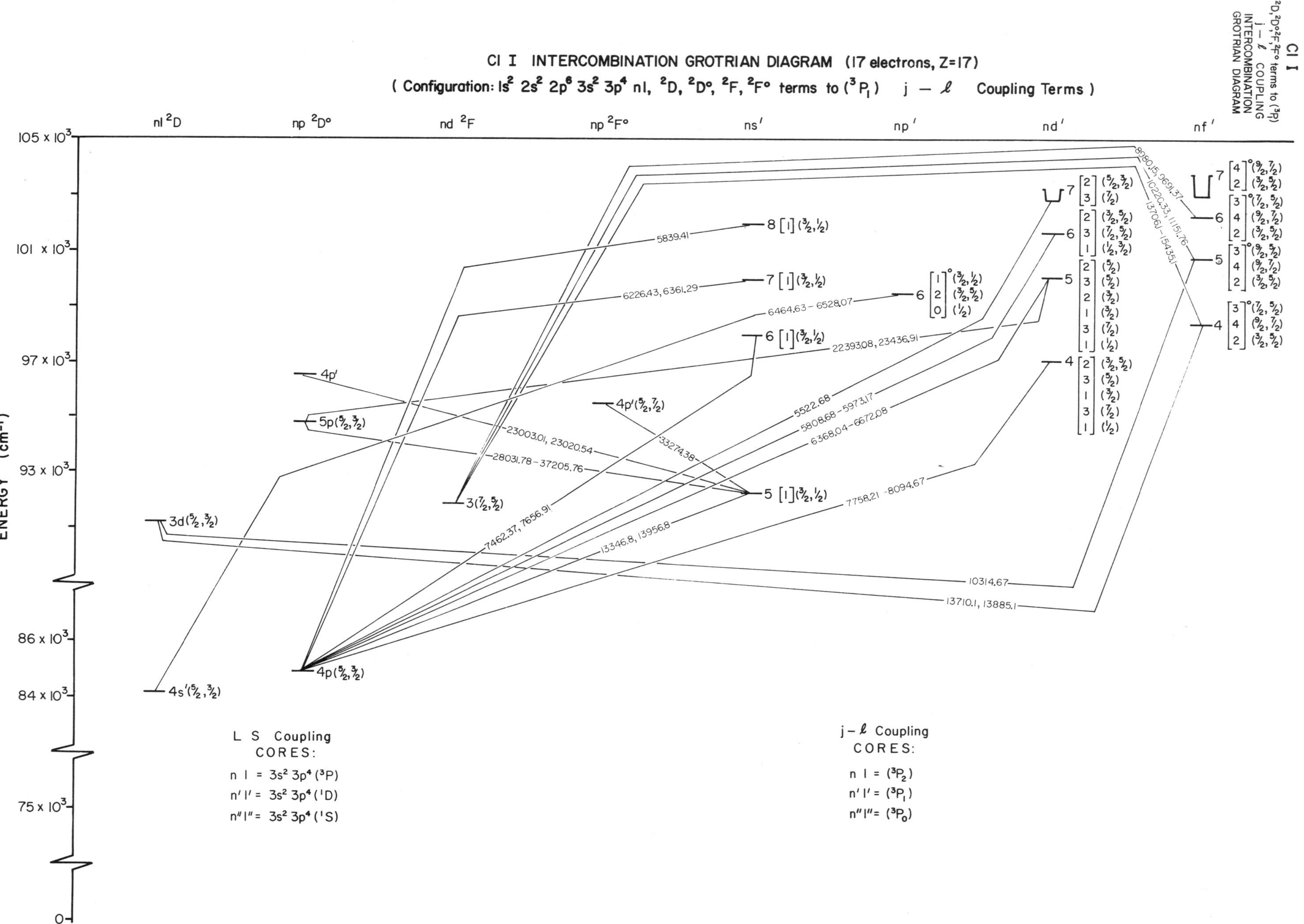

Cl I
$^2D, ^2D^o, ^2F, ^2F^o$ terms to ($^3P_1$)
j – ℓ COUPLING
INTERCOMBINATION
GROTRIAN DIAGRAM
Cl I INTERCOMBINATION GROTRIAN DIAGRAM (17 electrons, Z=17)
(Configuration: $1s^2\ 2s^2\ 2p^6\ 3s^2\ 3p^4$ nl, ^2D, $^2D^o$, ^2F, $^2F^o$ terms to ($^3P_1$) j – ℓ Coupling Terms)
nl ^2D
np $^2D^o$
nd ^2F
np $^2F^o$
ns′
np′
nd′
nf′
ENERGY (cm⁻¹)
105 x 10³
101 x 10³
97 x 10³
93 x 10³
86 x 10³
84 x 10³
75 x 10³
0
3d(5/2, 3/2)
4s′(5/2, 3/2)
4p′
5p(5/2, 3/2)
4p(5/2, 3/2)
3(7/2, 5/2)
4p′(5/2, 7/2)
8 [1](3/2, 1/2)
7 [1](3/2, 1/2)
6 [1](3/2, 1/2)
5 [1](3/2, 1/2)
5839.41
6226.43, 6361.29
6464.63 – 6528.07
22393.08, 23436.91
23003.01, 23020.54
28031.78 – 37205.76
33274.38
7462.37, 7656.91
13346.8, 13956.8
5522.68
5808.68 – 5973.17
6368.04 – 6672.08
7758.21 – 8094.67
10314.67
13710.1, 13885.1
8980.15, 9691.37
10220.33, 11151.76
13706.1 – 15435.1
L S Coupling
CORES:
n l = $3s^2\ 3p^4\ (^3P)$
n′ l′ = $3s^2\ 3p^4\ (^1D)$
n″ l″ = $3s^2\ 3p^4\ (^1S)$
j – ℓ Coupling
CORES:
n l = ($^3P_2$)
n′ l′ = ($^3P_1$)
n″ l″ = ($^3P_0$)

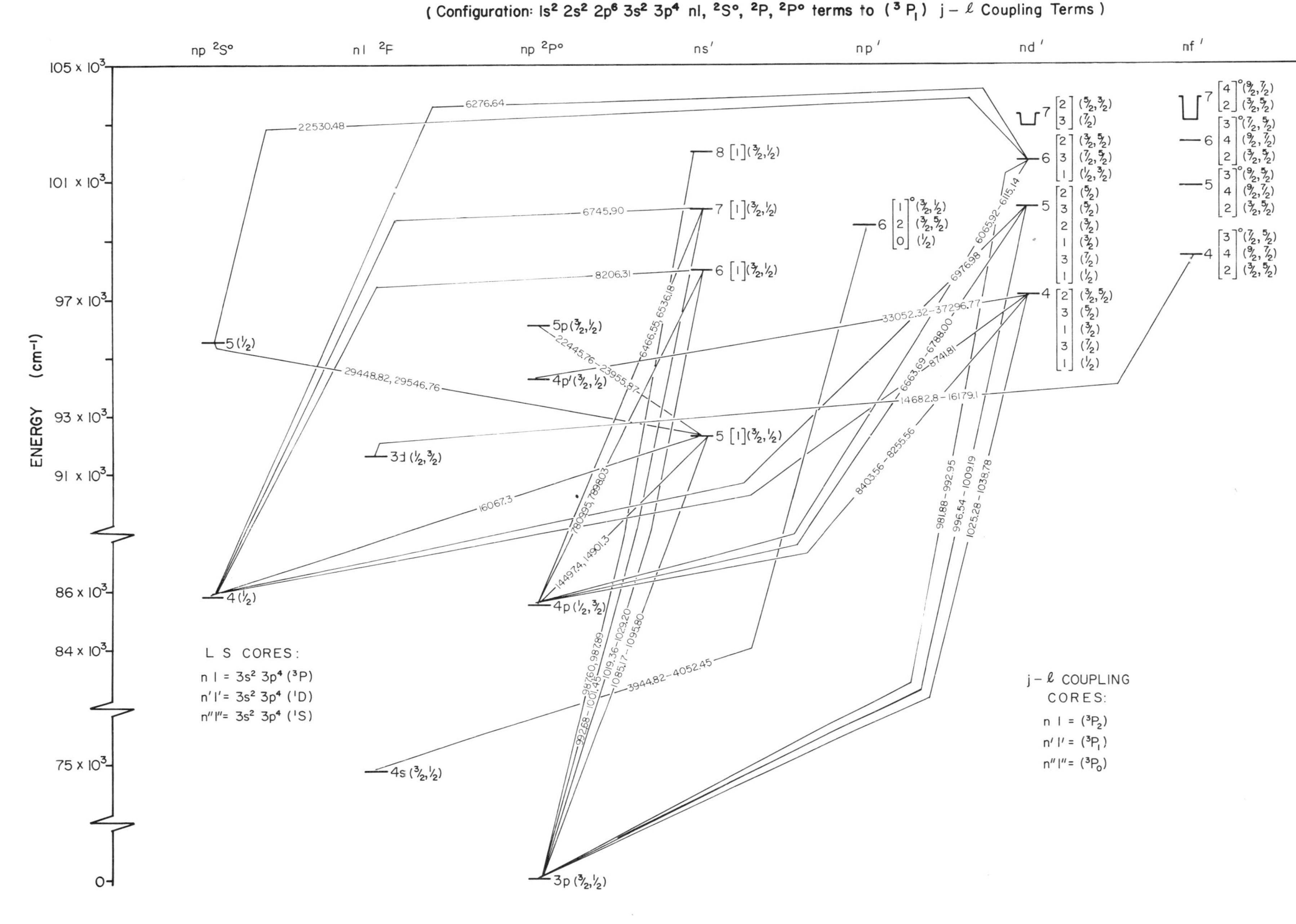
Cl I INTERCOMBINATION GROTRIAN DIAGRAM (17 electrons, Z=17)
(Configuration: 1s² 2s² 2p⁶ 3s² 3p⁴ nl, ²S°, ²P, ²P° terms to (³P₁) j – ℓ Coupling Terms)
np ²S°
nl ²F
np ²P°
ns'
np'
nd'
nf'
ENERGY (cm⁻¹)
105 x 10³
101 x 10³
97 x 10³
93 x 10³
91 x 10³
86 x 10³
84 x 10³
75 x 10³
0
5 (½)
4 (½)
5p (3/2, ½)
4p' (3/2, ½)
4p (½, 3/2)
3p (3/2, ½)
3d (½, 3/2)
4s (3/2, ½)
8 [1] (3/2, ½)
7 [1] (3/2, ½)
6 [1] (3/2, ½)
5 [1] (3/2, ½)
6276.64
22530.48
6745.90
8206.31
29448.82, 29546.76
22445.76 – 23955.87
16067.3
6466.55, 6536.18
780995, 789803
144974, 14901.3
98760, 98789
99268 – 1001.45
1019.36 – 1029.20
1085.17 – 1095.80
3944.82 – 4052.45
33052.32 – 37296.77
14682.8 – 16179.1
6065.92 – 6115.14
6976.98
6663.69 – 6788.00
8741.81
8403.56 – 8255.56
981.88 – 992.95
996.54 – 1009.19
1025.28 – 1038.78
L S CORES:
n l = 3s² 3p⁴ (³P)
n' l' = 3s² 3p⁴ (¹D)
n'' l'' = 3s² 3p⁴ (¹S)
j – ℓ COUPLING
CORES:
n l = (³P₂)
n' l' = (³P₁)
n'' l'' = (³P₀)
Cl I
²S°, ²P, ²P° terms to (³P₁)
j – ℓ COUPLING
INTERCOMBINATION
GROTRIAN DIAGRAM

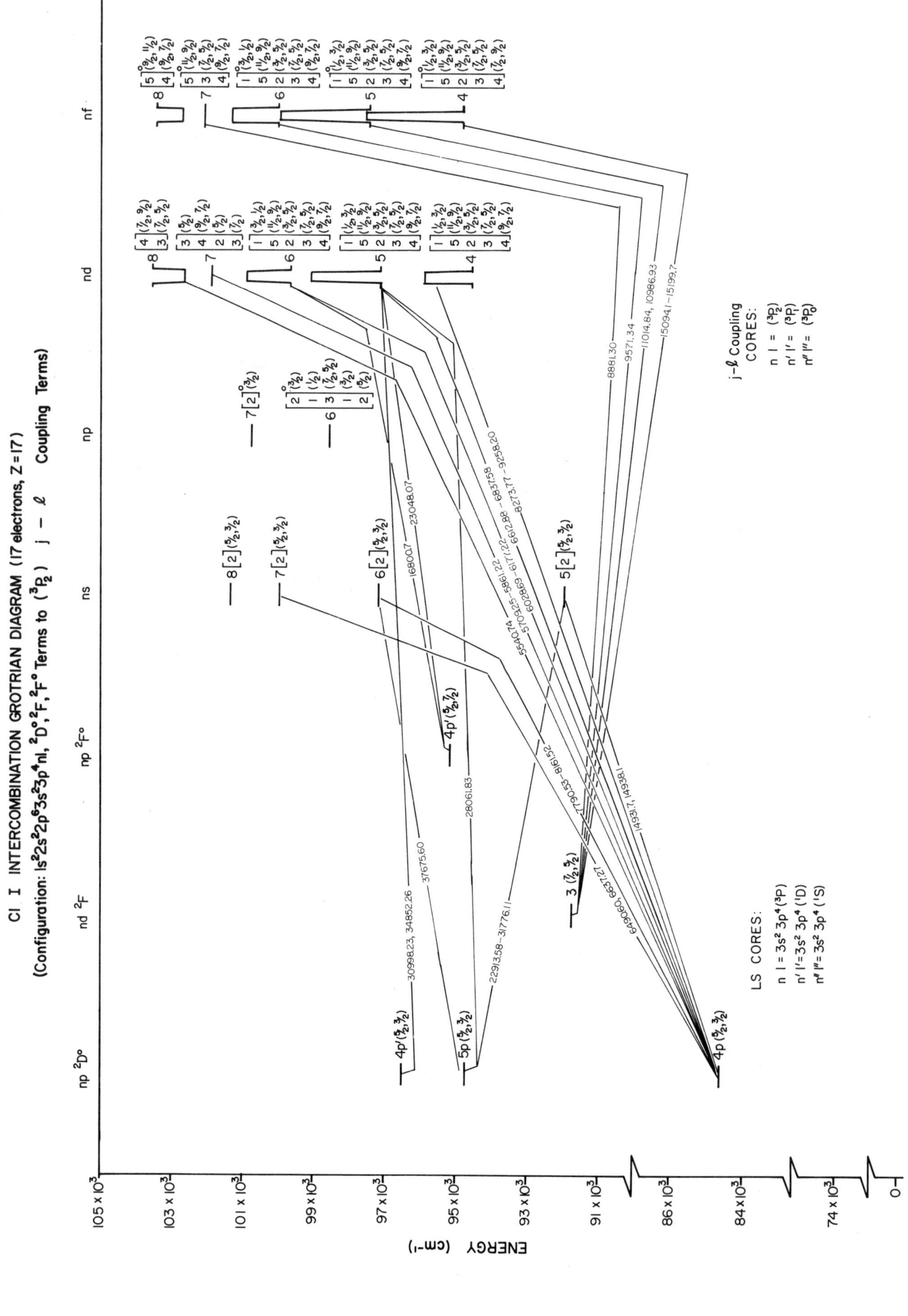

Cl I
$^2D^o, ^2F, ^2F^o$ terms to $(^3P_2)$
j – ℓ COUPLING
INTERCOMBINATION
GROTRIAN DIAGRAM

Cl I INTERCOMBINATION GROTRIAN DIAGRAM (17 electrons, Z = 17)

(Configuration: $1s^2 2s^2 2p^6 3s^2 3p^4 nl$, $^2D^o$, ^2F, $^2F^o$ Terms to $(^3P_2)$ j – ℓ Coupling Terms)

np $^2D^o$ | nd ^2F | np $^2F^o$ | ns | np | nd | nf

ENERGY (cm⁻¹)

105 x 10³
103 x 10³
101 x 10³
99 x 10³
97 x 10³
95 x 10³
93 x 10³
91 x 10³
86 x 10³
84 x 10³
74 x 10³
0

4p′ (5/2, 3/2)
5p (5/2, 3/2)
4p (5/2, 3/2)
3 (7/2, 5/2)
4p′ (5/2, 7/2)
8 [2] (5/2, 3/2)
7 [2] (5/2, 3/2)
6 [2] (5/2, 3/2)
5 [2] (5/2, 3/2)
7 [2]° (3/2)
6 [2]° (3/2), [1] (1/2), [3] (7/2, 5/2), [1] (3/2), [2] (5/2)

nd:
8 [4] (7/2, 9/2), [3] (7/2, 5/2)
7 [3] (5/2), [4] (9/2, 7/2), [2] (5/2), [3] (7/2)
6 [1] (3/2, 1/2), [5] (11/2, 9/2), [2] (3/2, 5/2), [3] (7/2, 5/2), [4] (9/2, 7/2)
5 [1] (1/2, 3/2), [5] (11/2, 9/2), [2] (3/2, 5/2), [3] (7/2, 5/2), [4] (9/2, 7/2)
4 [1] (1/2, 3/2), [5] (11/2, 9/2), [2] (3/2, 5/2), [3] (7/2, 5/2), [4] (9/2, 7/2)

nf:
8 [5]° (9/2, 11/2), [4] (9/2, 7/2)
7 [5]° (11/2, 9/2), [3] (7/2, 5/2), [4] (9/2, 7/2)
6 [1]° (3/2, 1/2), [5] (11/2, 9/2), [2] (3/2, 5/2), [3] (7/2, 5/2), [4] (9/2, 7/2)
5 [1]° (1/2, 3/2), [5] (11/2, 9/2), [2] (3/2, 5/2), [3] (7/2, 5/2), [4] (9/2, 7/2)
4 [1]° (1/2, 3/2), [5] (11/2, 9/2), [2] (3/2, 5/2), [3] (7/2, 5/2), [4] (7/2, 9/2)

30998.23, 34852.26
37675.60
16800.7
23048.07
28061.83
22913.58–31776.11
5540.74
5709.25–5861.22
6028.69–6177.22
6612.88–6837.58
8273.77–9258.20
7790.53–8161.52
6490.60, 6637.27
14931.7, 14938.1
8881.30
9571.34
11014.84, 10986.93
15094.1–15199.7

LS CORES:
n l = $3s^2 3p^4 (^3P)$
n′ l′ = $3s^2 3p^4 (^1D)$
n″ l″ = $3s^2 3p^4 (^1S)$

j–ℓ Coupling
CORES:
n l = $(^3P_2)$
n′ l′ = $(^3P_1)$
n″ l″ = $(^3P_0)$

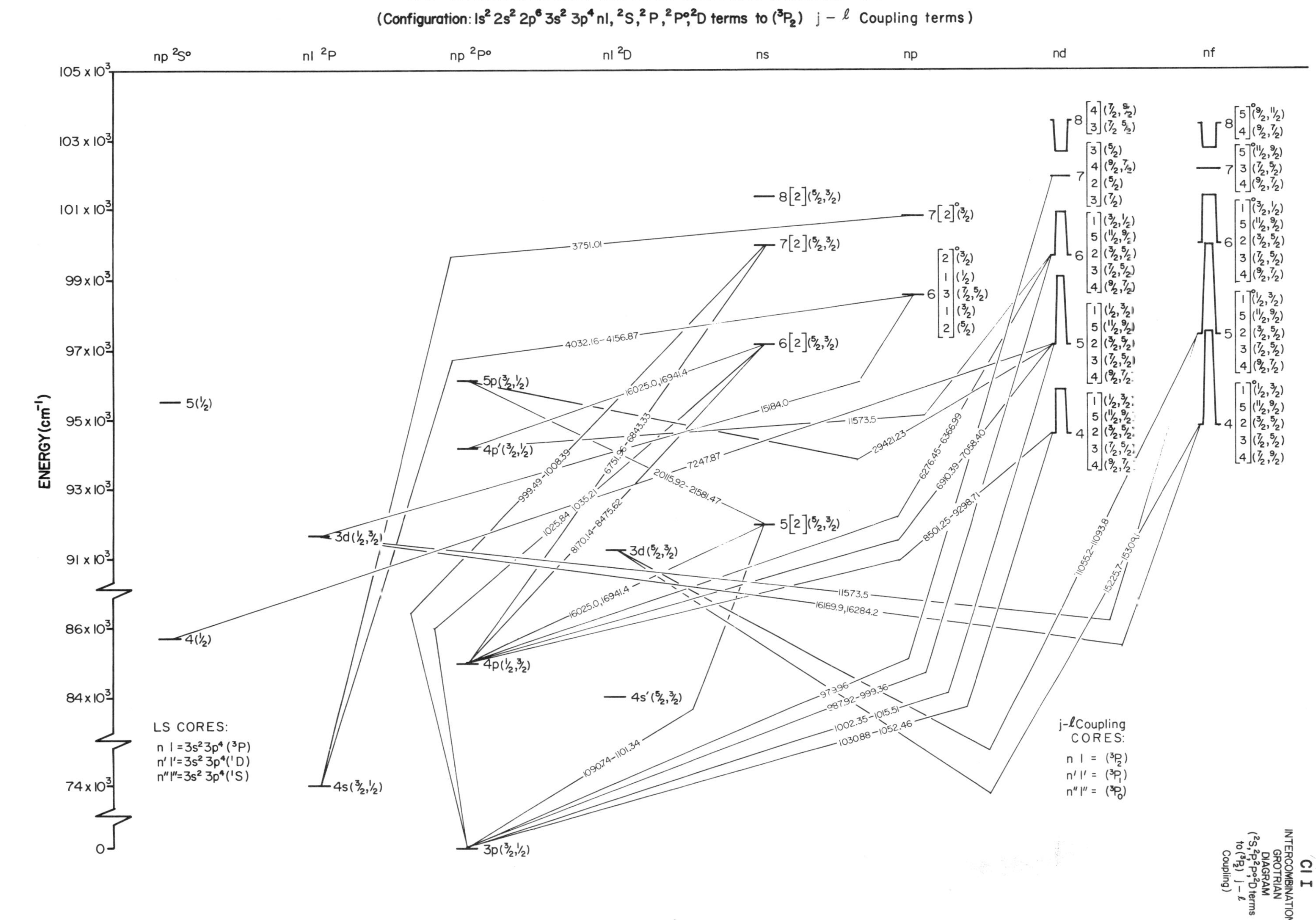

Cl I INTERCOMBINATION GROTRIAN DIAGRAM (17 electrons, Z = 17)
(Configuration: 1s² 2s² 2p⁶ 3s² 3p⁴ nl, ²S, ²P, ²P°, ²D terms to (³P₂) j – ℓ Coupling terms)
np ²S°
nl ²P
np ²P°
nl ²D
ns
np
nd
nf
ENERGY (cm⁻¹)
105 x 10³
103 x 10³
101 x 10³
99 x 10³
97 x 10³
95 x 10³
93 x 10³
91 x 10³
86 x 10³
84 x 10³
74 x 10³
0
5(½)
4(½)
3d(½, 3/2)
4s(3/2, ½)
5p(3/2, ½)
4p′(3/2, ½)
4p(½, 3/2)
3p(3/2, ½)
3d(5/2, 3/2)
4s′(5/2, 3/2)
8[2](5/2, 3/2)
7[2](5/2, 3/2)
6[2](5/2, 3/2)
5[2](5/2, 3/2)
7[2]°(3/2)
6 [2]°(3/2) [1](½) [3](7/2, 5/2) [1](3/2) [2](5/2)
8 [4](7/2, 9/2) [3](7/2, 5/2)
7 [3](5/2) [4](9/2, 7/2) [2](5/2) [3](7/2)
6 [1](3/2, ½) [5](11/2, 9/2) [2](3/2, 5/2) [3](7/2, 5/2) [4](9/2, 7/2)
5 [1](½, 3/2) [5](11/2, 9/2) [2](3/2, 5/2) [3](7/2, 5/2) [4](9/2, 7/2)
4 [1](½, 3/2) [5](11/2, 9/2) [2](3/2, 5/2) [3](7/2, 5/2) [4](9/2, 7/2)
8 [5]°(9/2, 11/2) [4](9/2, 7/2)
7 [5]°(11/2, 9/2) [3](7/2, 5/2) [4](9/2, 7/2)
6 [1]°(3/2, ½) [5](11/2, 9/2) [2](3/2, 5/2) [3](7/2, 5/2) [4](9/2, 7/2)
5 [1]°(½, 3/2) [5](11/2, 9/2) [2](3/2, 5/2) [3](7/2, 5/2) [4](9/2, 7/2)
4 [1]°(½, 3/2) [5](11/2, 9/2) [2](3/2, 5/2) [3](7/2, 5/2) [4](7/2, 9/2)
3751.01
4032.16–4156.87
16025.0, 16941.4
15184.0
11573.5
2942.23
7247.87
999.49–1008.39
6751.96–6843.33
1035.21
1025.84
8170.14–8475.62
20115.92–21581.47
6276.45–6366.99
6910.39–7058.40
8501.25–9298.71
11055.2–11093.8
15225.7–15309.1
11573.5
16169.9, 16284.2
16025.0, 16941.4
973.96
987.92–999.36
1002.35–1015.51
1030.88–1052.46
1090.74–1101.34
LS CORES:
n l = 3s² 3p⁴ (³P)
n′ l′ = 3s² 3p⁴ (¹D)
n″ l″ = 3s² 3p⁴ (¹S)
j-ℓ Coupling CORES:
n l = (³P₂)
n′ l′ = (³P₁)
n″ l″ = (³P₀)
Cl I
INTERCOMBINATION
GROTRIAN
DIAGRAM
(²S, ²P, ²P°, ²D terms
to (³P₂) j – ℓ
Coupling)

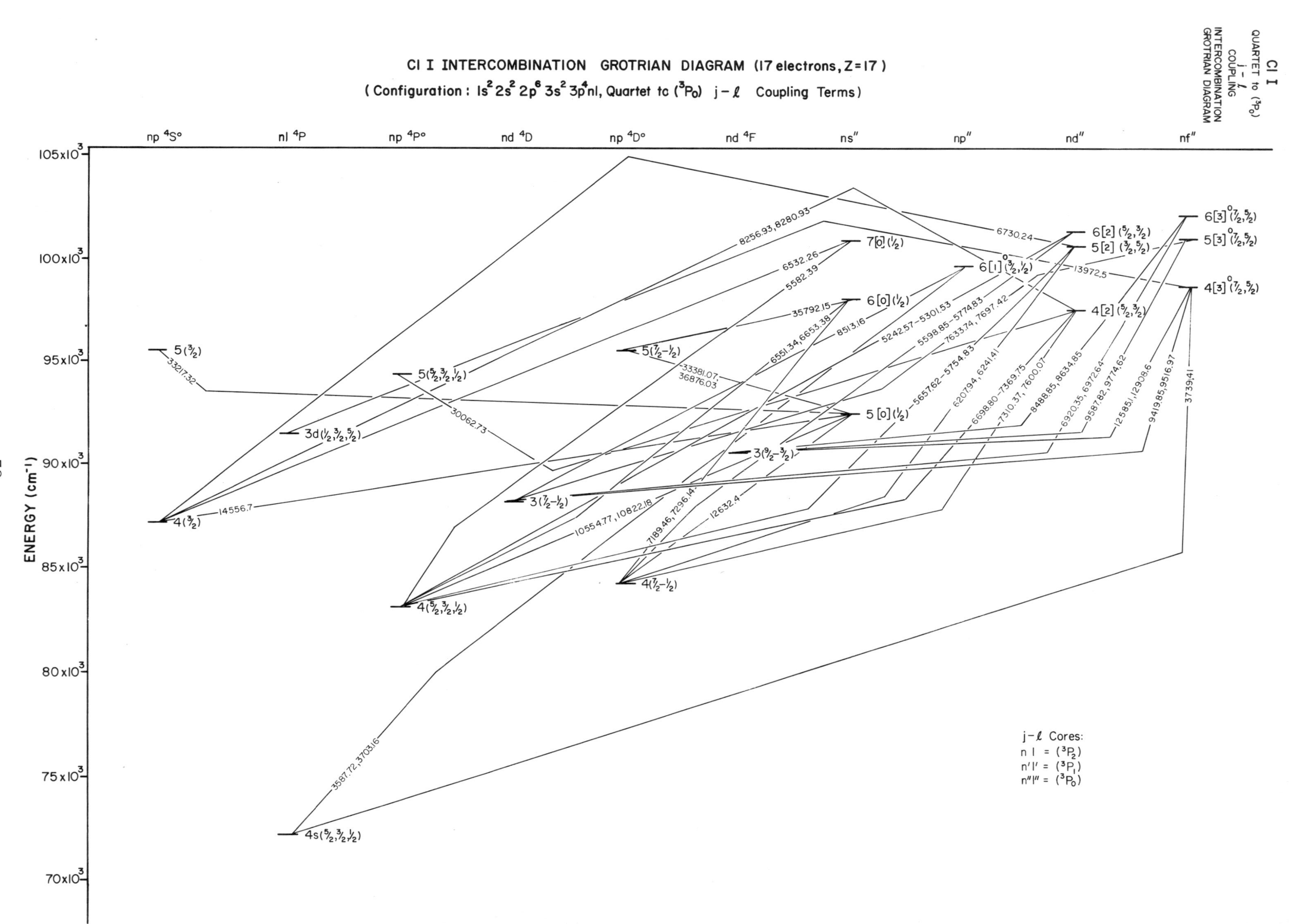

Cl I
QUARTET to ($^3P_0$)
j - ℓ
COUPLING
INTERCOMBINATION
GROTRIAN DIAGRAM
Cl I INTERCOMBINATION GROTRIAN DIAGRAM (17 electrons, Z=17)
(Configuration: $1s^2 2s^2 2p^6 3s^2 3p^4 nl$, Quartet to ($^3P_0$) j - ℓ Coupling Terms)
np $^4S^o$
nl ^4P
np $^4P^o$
nd ^4D
np $^4D^o$
nd ^4F
ns''
np''
nd''
nf''
ENERGY (cm^{-1})
105×10^3
100×10^3
95×10^3
90×10^3
85×10^3
80×10^3
75×10^3
70×10^3
5(3/2)
4(3/2)
3d(1/2,3/2,5/2)
4s(5/2,3/2,1/2)
5(5/2,3/2,1/2)
4(5/2,3/2,1/2)
5(7/2–1/2)
3(7/2–1/2)
4(7/2–1/2)
3(9/2–3/2)
7[0](1/2)
6[0](1/2)
5[0](1/2)
6[1]^o(3/2,1/2)
6[2](5/2,3/2)
5[2](3/2,5/2)
4[2](5/2,3/2)
6[3]^o(7/2,5/2)
5[3]^o(7/2,5/2)
4[3]^o(7/2,5/2)
33217.32
14556.7
30062.73
3587.72, 3703.16
8256.93, 8280.93
6532.26
5582.39
35792.15
6551.34, 6653.38
8513.16
33381.07, 36876.03
10554.77, 10822.18
7189.46, 7296.14
12632.4
5242.57–5301.53
5598.85–5774.83
7633.74, 7697.42
5657.62–5754.83
6207.94, 6241.41
6698.80–7369.75
7310.37, 7600.07
8488.85, 8634.85
6920.35, 6972.64
9587.82, 9774.62
12585.1, 12908.6
9419.85, 9516.97
3739.41
6730.24
13972.5
j-ℓ Cores:
n l = ($^3P_2$)
n'l' = ($^3P_1$)
n''l'' = ($^3P_0$)

Cl I INTERCOMBINATION GROTRIAN DIAGRAM (17 electrons, Z=17)

(Configuration: $1s^2 2s^2 2p^6 3s^2 3p^4 nl$, $^4S^o$ ^4P, $^4P^o$ terms to $(^3P_1)$ j - ℓ Coupling Terms)

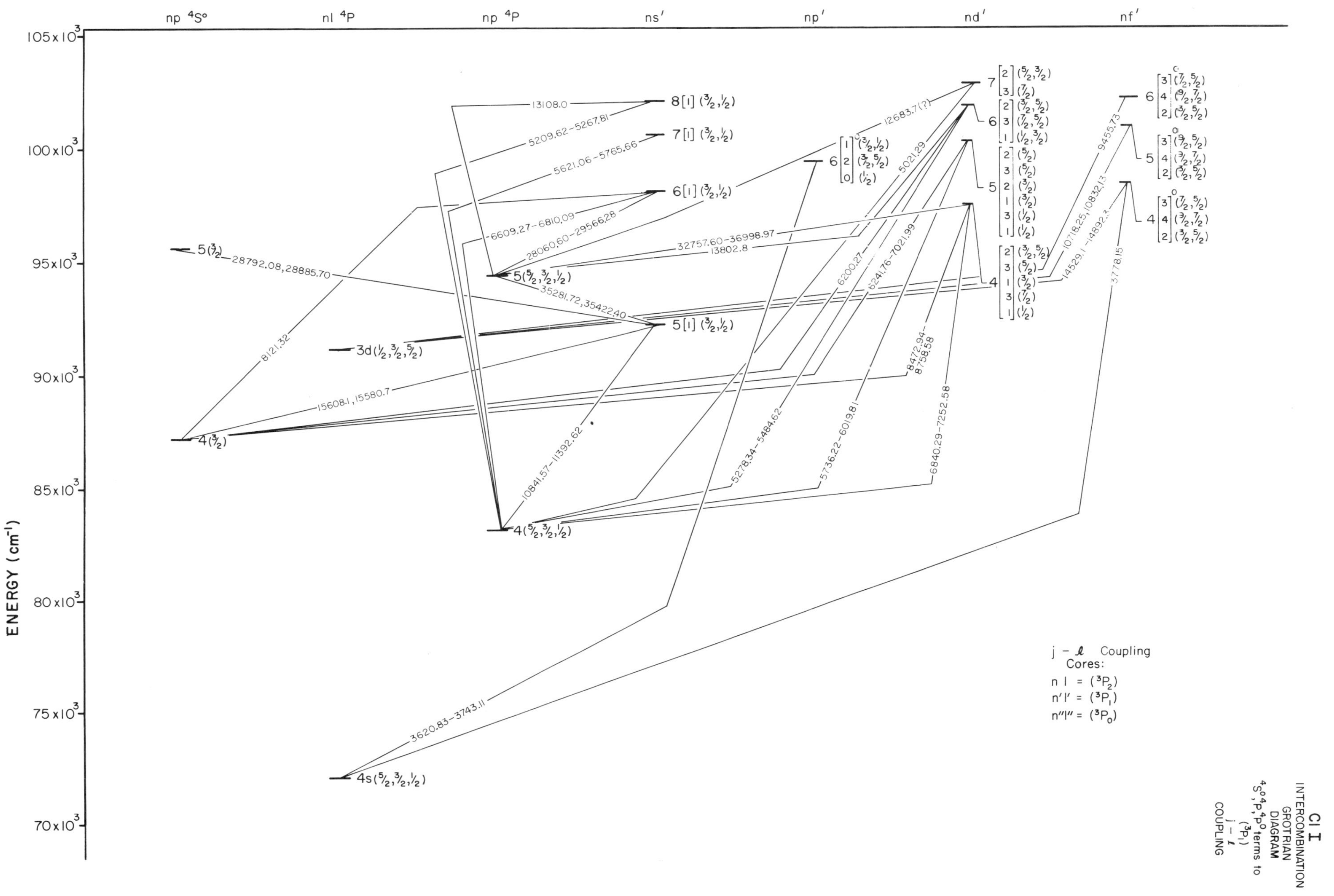

np $^4S^o$ | nl ^4P | np ^4P | ns′ | np′ | nd′ | nf′

ENERGY (cm^{-1})

105×10^3 100×10^3 95×10^3 90×10^3 85×10^3 80×10^3 75×10^3 70×10^3

5(3/2)
4(3/2)
3d(1/2, 3/2, 5/2)
4s(5/2, 3/2, 1/2)
5(5/2, 3/2, 1/2)
4(5/2, 3/2, 1/2)
8[1] (3/2, 1/2)
7[1] (3/2, 1/2)
6[1] (3/2, 1/2)
5[1] (3/2, 1/2)
6 [1] (3/2, 1/2) [2] (3/2, 5/2) [0] (1/2)
7 [2] (5/2, 3/2) [3] (7/2)
6 [2] (3/2, 5/2) [3] (7/2, 5/2) [1] (1/2, 3/2)
5 [2] (5/2) [3] (5/2) [2] (3/2) [1] (3/2) [3] (1/2) [1] (1/2)
4 [2] (3/2, 5/2) [3] (5/2) [1] (3/2) [3] (7/2) [1] (1/2)
6 [3] (7/2, 5/2) [4] (9/2, 7/2) [2] (3/2, 5/2)
5 [3] (9/2, 5/2) [4] (3/2, 7/2) [2] (3/2, 5/2)
4 [3] (7/2, 5/2) [4] (3/2, 7/2) [2] (3/2, 5/2)

13108.0
5209.62–5267.81
5621.06–5765.66
6609.27–6810.09
28060.60–29566.28
32757.60–36998.97
13802.8
35281.72, 35422.40
28792.08, 28885.70
8121.32
15608.1, 15580.7
10841.57–11392.62
12683.7(?)
5021.29
6200.27
6241.76–7021.99
8472.94–
8758.58
5278.34–5484.62
5736.22–6019.81
6840.29–7252.58
9455.73
10718.25, 10832.13
14529.1–14892.3
3778.15
3620.83–3743.11

j – ℓ Coupling
Cores:
n l = $(^3P_2)$
n′l′ = $(^3P_1)$
n″l″ = $(^3P_0)$

Cl I
INTERCOMBINATION
GROTRIAN
DIAGRAM
$^4S^o, ^4P, ^4P^o$ terms to
$(^3P_1)$
j – ℓ
COUPLING

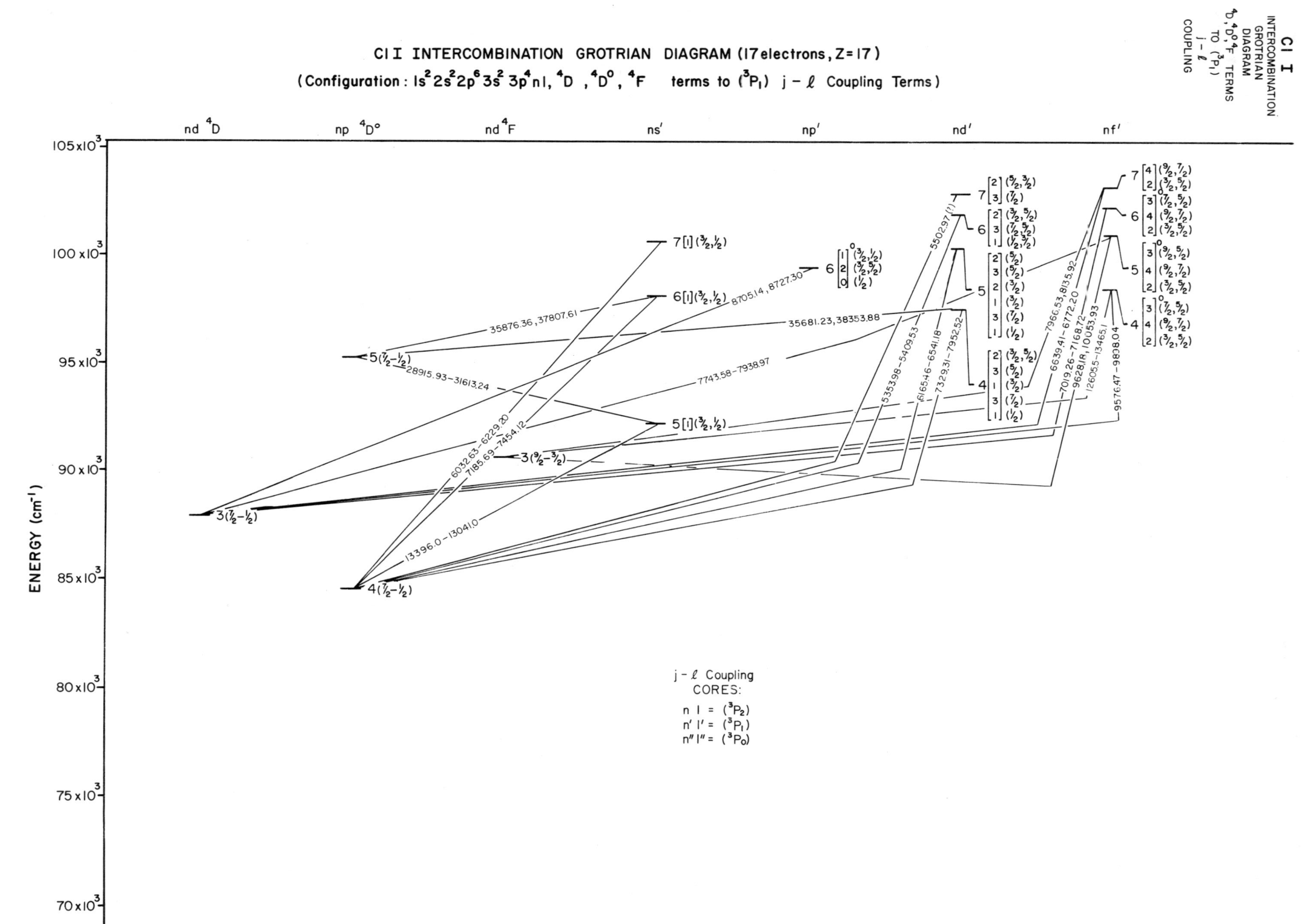

Cl I
INTERCOMBINATION
GROTRIAN
DIAGRAM
$^4D, ^4D^o, ^4F$ TERMS
TO $(^3P_1)$
j - ℓ
COUPLING
Cl I INTERCOMBINATION GROTRIAN DIAGRAM (17 electrons, Z=17)
(Configuration: $1s^2 2s^2 2p^6 3s^2 3p^4 nl$, ^4D, $^4D^o$, ^4F terms to $(^3P_1)$ j - ℓ Coupling Terms)
nd ^4D
np $^4D^o$
nd ^4F
ns'
np'
nd'
nf'
ENERGY (cm^{-1})
105×10^3
100×10^3
95×10^3
90×10^3
85×10^3
80×10^3
75×10^3
70×10^3
3(7/2-1/2)
5(7/2-1/2)
4(7/2-1/2)
3(9/2-3/2)
7[1](3/2,1/2)
6[1](3/2,1/2)
5[1](3/2,1/2)
35876.36, 37807.61
28915.93-31613.24
6032.63-6229.20
7185.69-7454.12
13396.0-13041.0
8705.14, 8727.30
35681.23, 38353.88
7743.58-7938.97
5502.97(?)
5353.98-5409.53
6165.46-6541.18
7329.31-7952.52
7966.53, 8135.92
6639.41-6772.20
7019.26-7168.72
9628.1R, 10053.93
12605.5-13465.1
9576.47-9838.04
j - ℓ Coupling
CORES:
n l = $(^3P_2)$
n' l' = $(^3P_1)$
n'' l'' = $(^3P_0)$

Cl I INTERCOMBINATION GROTRIAN DIAGRAM (17 electrons, Z=17)

(Configuration: $1s^2\ 2s^2\ 2p^6\ 3s^2\ 3p^4$ nl, $^4S^o$, ^4P, $^4P^o$ terms to ($^3P_2$) j – ℓ Coupling Terms)

np $^4S^o$ | nl ^4P | np $^4P^o$ | ns | np | nd | nf

ENERGY (cm^{-1})

105 x 10^3, 100 x 10^3, 95 x 10^3, 90 x 10^3, 85 x 10^3, 80 x 10^3, 75 x 10^3, 70 x 10^3

j-ℓ Coupling
CORES:
n l = ($^3P_2$)
n′ l′ = ($^3P_1$)
n″ l″ = ($^3P_0$)

Cl I

$^4S^o$, ^4P, $^4P^o$ terms to ($^3P_2$)

j – ℓ COUPLING

INTERCOMBINATION GROTRIAN DIAGRAM

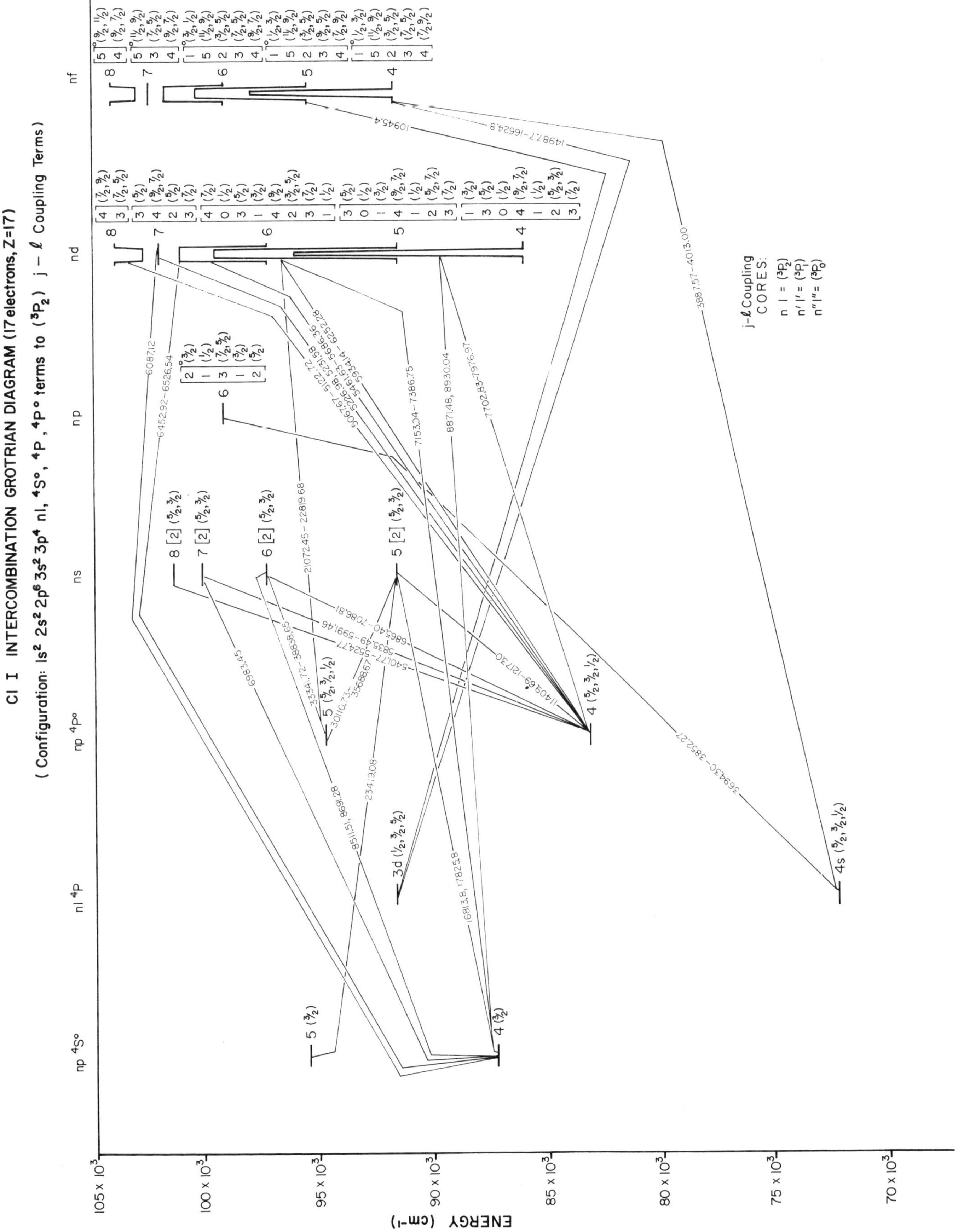

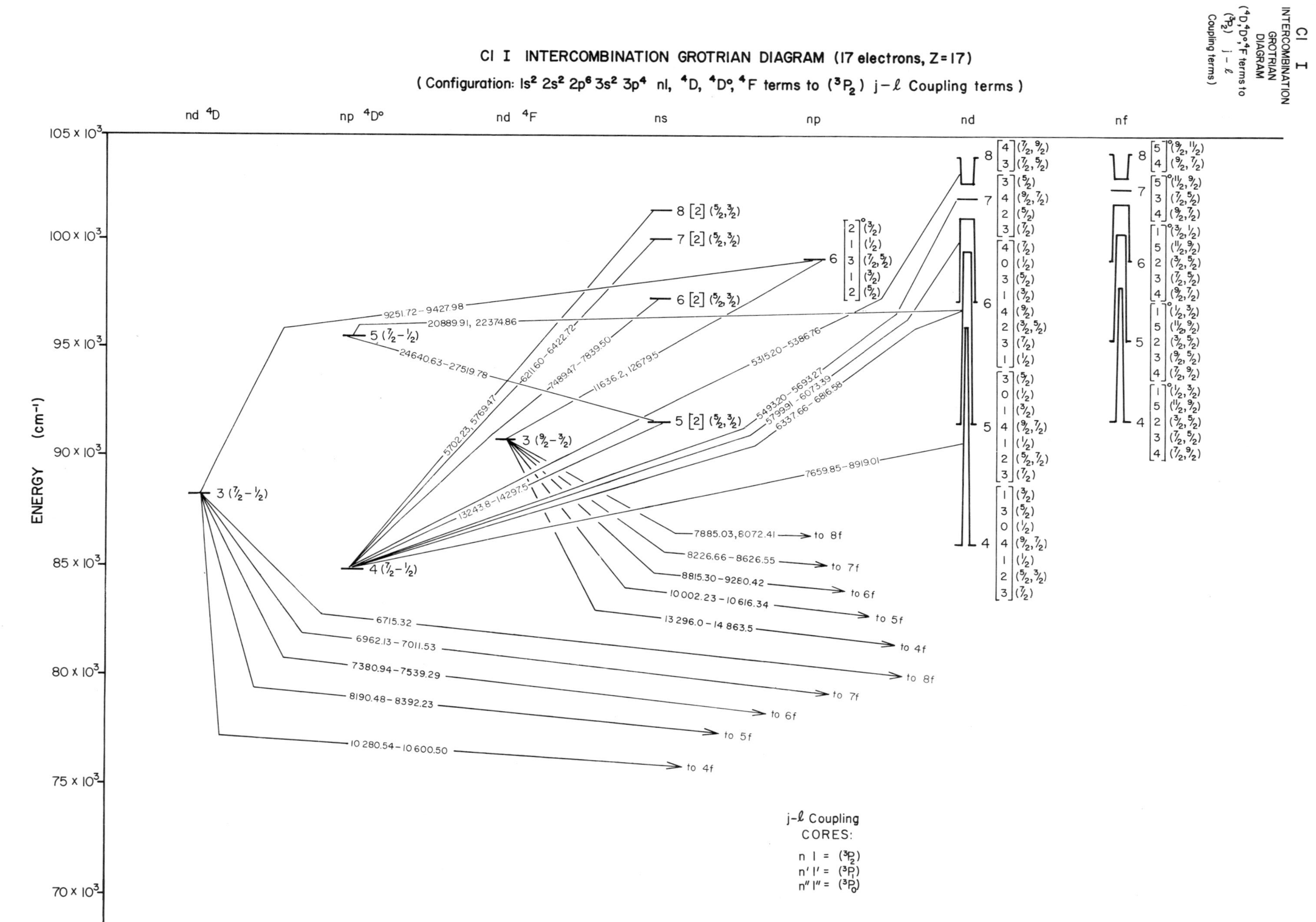
Cl I INTERCOMBINATION GROTRIAN DIAGRAM (17 electrons, Z=17)
(Configuration: 1s² 2s² 2p⁶ 3s² 3p⁴ nl, ⁴D, ⁴D°, ⁴F terms to (³P₂) j−ℓ Coupling terms)
nd ⁴D
np ⁴D°
nd ⁴F
ns
np
nd
nf
ENERGY (cm⁻¹)
105 x 10³
100 x 10³
95 x 10³
90 x 10³
85 x 10³
80 x 10³
75 x 10³
70 x 10³
3 (7/2−1/2)
4 (7/2−1/2)
5 (7/2−1/2)
3 (9/2−3/2)
8 [2] (5/2, 3/2)
7 [2] (5/2, 3/2)
6 [2] (5/2, 3/2)
5 [2] (5/2, 3/2)
9251.72−9427.98
20889.91, 22374.86
24640.63−27519.78
5702.23, 5769.47
6211.60−6422.72
7489.47−7839.50
11636.2, 12679.5
5315.20−5386.76
5493.20−5693.27
5799.91−6073.39
6337.66−6816.58
7659.85−8919.01
13243.8−14297.5
7885.03, 8072.41 to 8f
8226.66−8626.55 to 7f
8815.30−9280.42 to 6f
10 002.23−10 616.34 to 5f
13 296.0−14 863.5 to 4f
6715.32 to 8f
6962.13−7011.53 to 7f
7380.94−7539.29 to 6f
8190.48−8392.23 to 5f
10 280.54−10 600.50 to 4f
j-ℓ Coupling
CORES:
n l = (³P₂)
n' l' = (³P₁)
n'' l'' = (³P₀)
Cl I
INTERCOMBINATION
GROTRIAN
DIAGRAM
(⁴D,⁴D°,⁴F terms to
(³P₂) j − ℓ
Coupling terms)

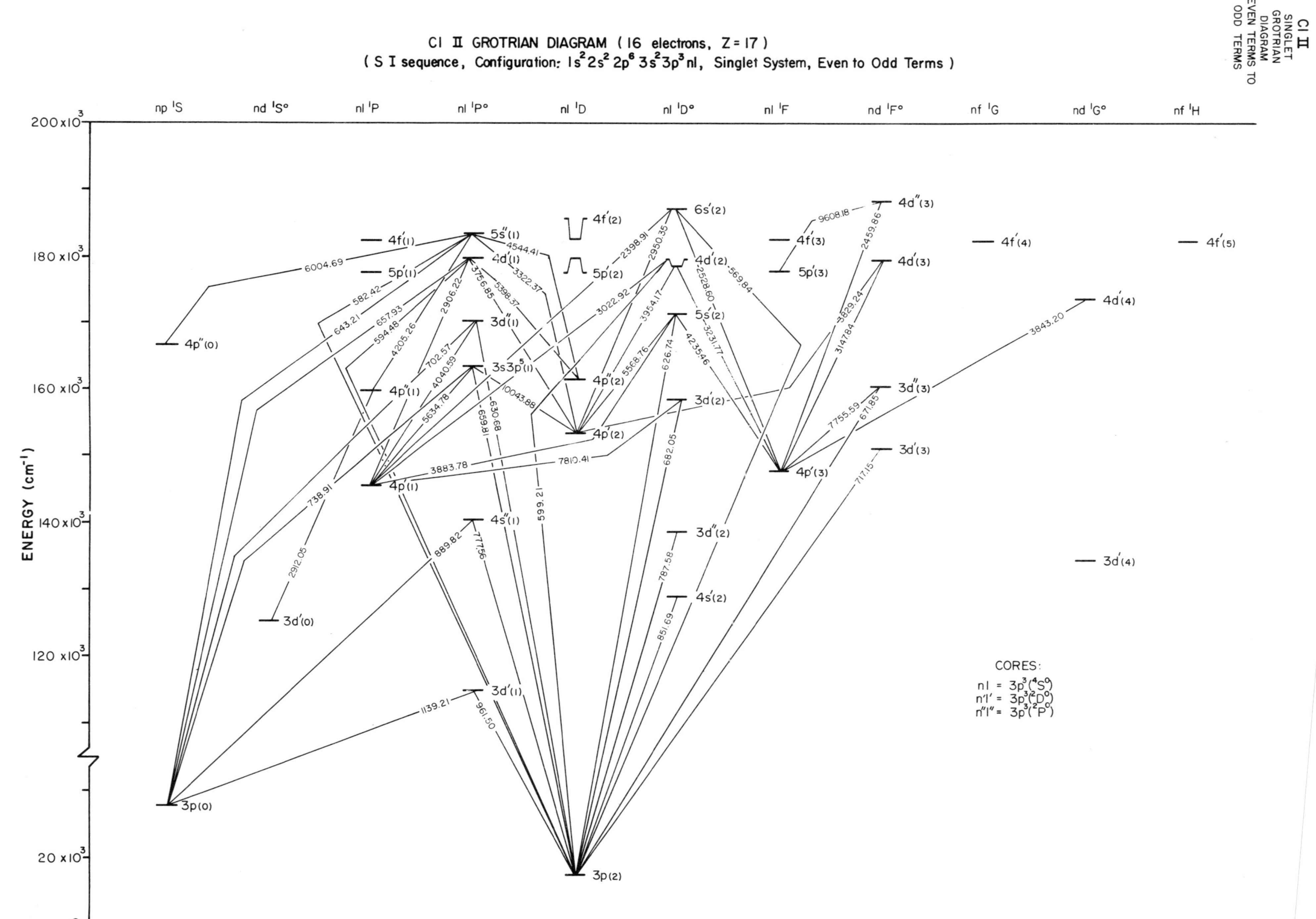
Cl II GROTRIAN DIAGRAM (16 electrons, Z = 17)
(S I sequence, Configuration: $1s^2 2s^2 2p^6 3s^2 3p^3 nl$, Singlet System, Even to Odd Terms)
np ^{1}S
nd ^{1}S°
nl ^{1}P
nl ^{1}P°
nl ^{1}D
nl ^{1}D°
nl ^{1}F
nd ^{1}F°
nf ^{1}G
nd ^{1}G°
nf ^{1}H
ENERGY (cm^{-1})
200x10^3
180x10^3
160x10^3
140x10^3
120x10^3
20x10^3
0
4p"(0)
3p(0)
3d'(0)
4f'(1)
5p'(1)
4p"(1)
4p'(1)
5s"(1)
4d'(1)
3d"(1)
3s3p^5(1)
4s"(1)
3d'(1)
4f'(2)
5p'(2)
4p"(2)
4p'(2)
3p(2)
6s'(2)
4d'(2)
5s'(2)
3d'(2)
3d"(2)
4s'(2)
4f'(3)
5p'(3)
4p'(3)
4d"(3)
4d'(3)
3d"(3)
3d'(3)
4f'(4)
4d'(4)
3d'(4)
4f'(5)
6004.69
582.42
643.21
657.93
594.48
4205.26
2906.22
3756.85
5398.37
3322.37
4544.41
702.57
4040.59
5634.78
10043.88
659.81
630.68
3883.78
7810.41
738.91
2912.05
889.82
777.56
599.21
1139.21
961.50
2398.91
2950.35
3022.92
3954.17
5568.76
626.74
682.05
787.58
851.69
2528.60
569.84
3231.77
4235.46
9608.18
2459.86
3829.24
3147.84
7755.59
671.85
717.15
3843.20
CORES:
nl = $3p^3(^4S^o)$
n'l' = $3p^3(^2D^o)$
n"l" = $3p^3(^2P^o)$

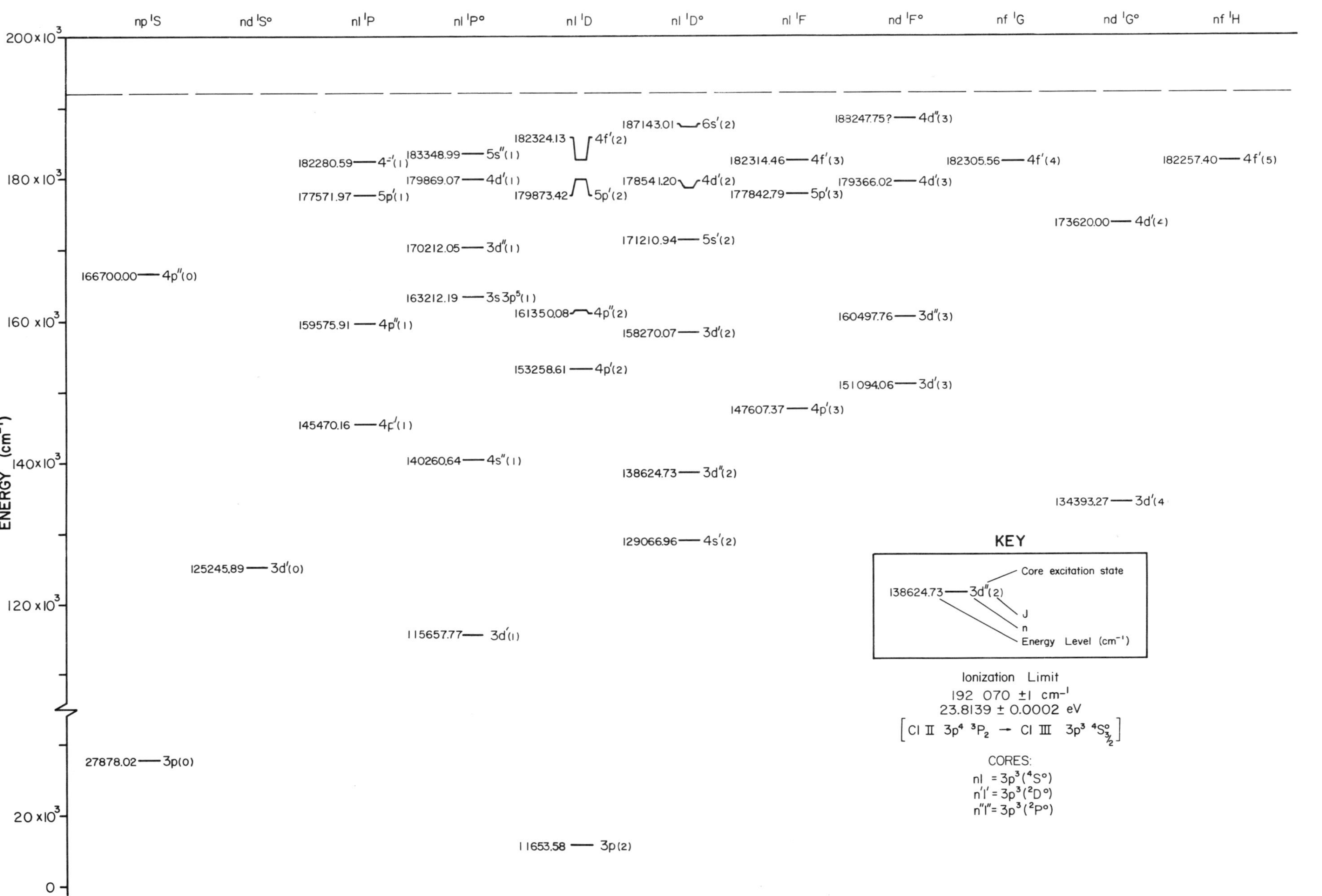
Cl II ENERGY LEVELS (16 electrons, Z = 17)
(S I sequence , Configuration: $1s^2 2s^2 2p^6 3s^2 3p^3 nl$, Singlet System)
np ^1S
nd $^1S°$
nl ^1P
nl $^1P°$
nl ^1D
nl $^1D°$
nl ^1F
nd $^1F°$
nf ^1G
nd $^1G°$
nf ^1H
187143.01 — 6s'(2)
183247.75? — 4d''(3)
182324.13 — 4f'(2)
182280.59 — 4-'(1)
183348.99 — 5s''(1)
182314.46 — 4f'(3)
182305.56 — 4f'(4)
182257.40 — 4f'(5)
179869.07 — 4d'(1)
178541.20 — 4d'(2)
179366.02 — 4d'(3)
177571.97 — 5p'(1)
179873.42 — 5p'(2)
177842.79 — 5p'(3)
173620.00 — 4d'(4)
171210.94 — 5s'(2)
170212.05 — 3d''(1)
166700.00 — 4p''(0)
163212.19 — $3s3p^5$(1)
161350.08 — 4p''(2)
160497.76 — 3d''(3)
159575.91 — 4p''(1)
158270.07 — 3d'(2)
153258.61 — 4p'(2)
151094.06 — 3d'(3)
147607.37 — 4p'(3)
145470.16 — 4p'(1)
140260.64 — 4s''(1)
138624.73 — 3d''(2)
134393.27 — 3d'(4
129066.96 — 4s'(2)
125245.89 — 3d'(0)
115657.77 — 3d'(1)
27878.02 — 3p(0)
11653.58 — 3p(2)
KEY
138624.73 — 3d''(2)
Core excitation state
J
n
Energy Level (cm^{-1})
Ionization Limit
192 070 ±1 cm^{-1}
23.8139 ± 0.0002 eV
[Cl II $3p^4\ ^3P_2$ → Cl III $3p^3\ ^4S°_{3/2}$]
CORES:
nl = $3p^3(^4S°)$
n'l' = $3p^3(^2D°)$
n''l'' = $3p^3(^2P°)$
200×10³
180×10³
160×10³
140×10³
120×10³
20×10³
0
ENERGY (cm^{-1})

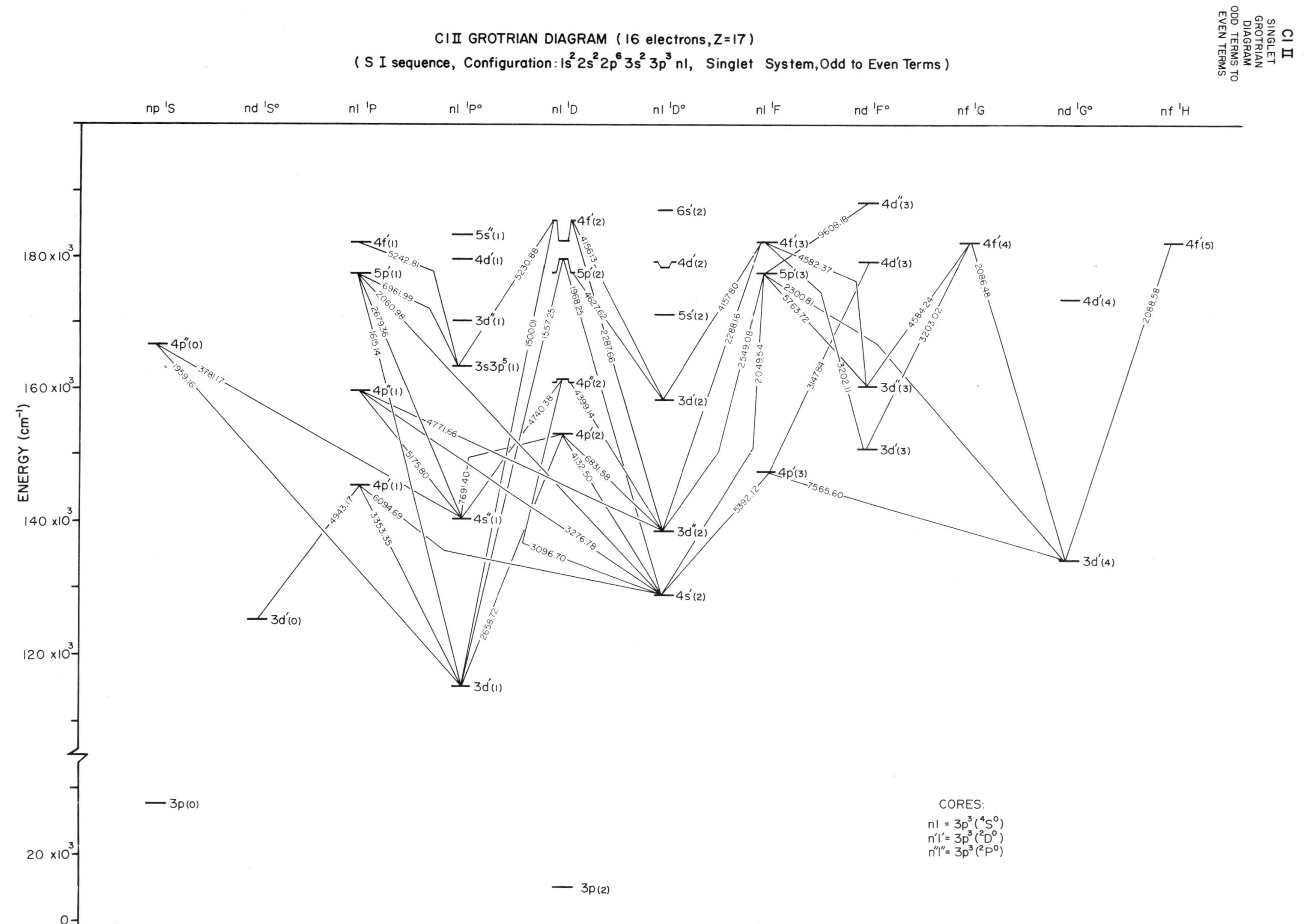
CI II GROTRIAN DIAGRAM (16 electrons, Z=17)
(S I sequence, Configuration: $1s^2 2s^2 2p^6 3s^2 3p^3$ nl, Singlet System, Odd to Even Terms)
np ¹S
nd ¹S°
nl ¹P
nl ¹P°
nl ¹D
nl ¹D°
nl ¹F
nd ¹F°
nf ¹G
nd ¹G°
nf ¹H
ENERGY (cm⁻¹)
180 x10³
160 x10³
140 x10³
120 x10³
20 x10³
0
CORES:
nl = $3p^3(^4S^o)$
n'l' = $3p^3(^2D^o)$
n''l'' = $3p^3(^2P^o)$

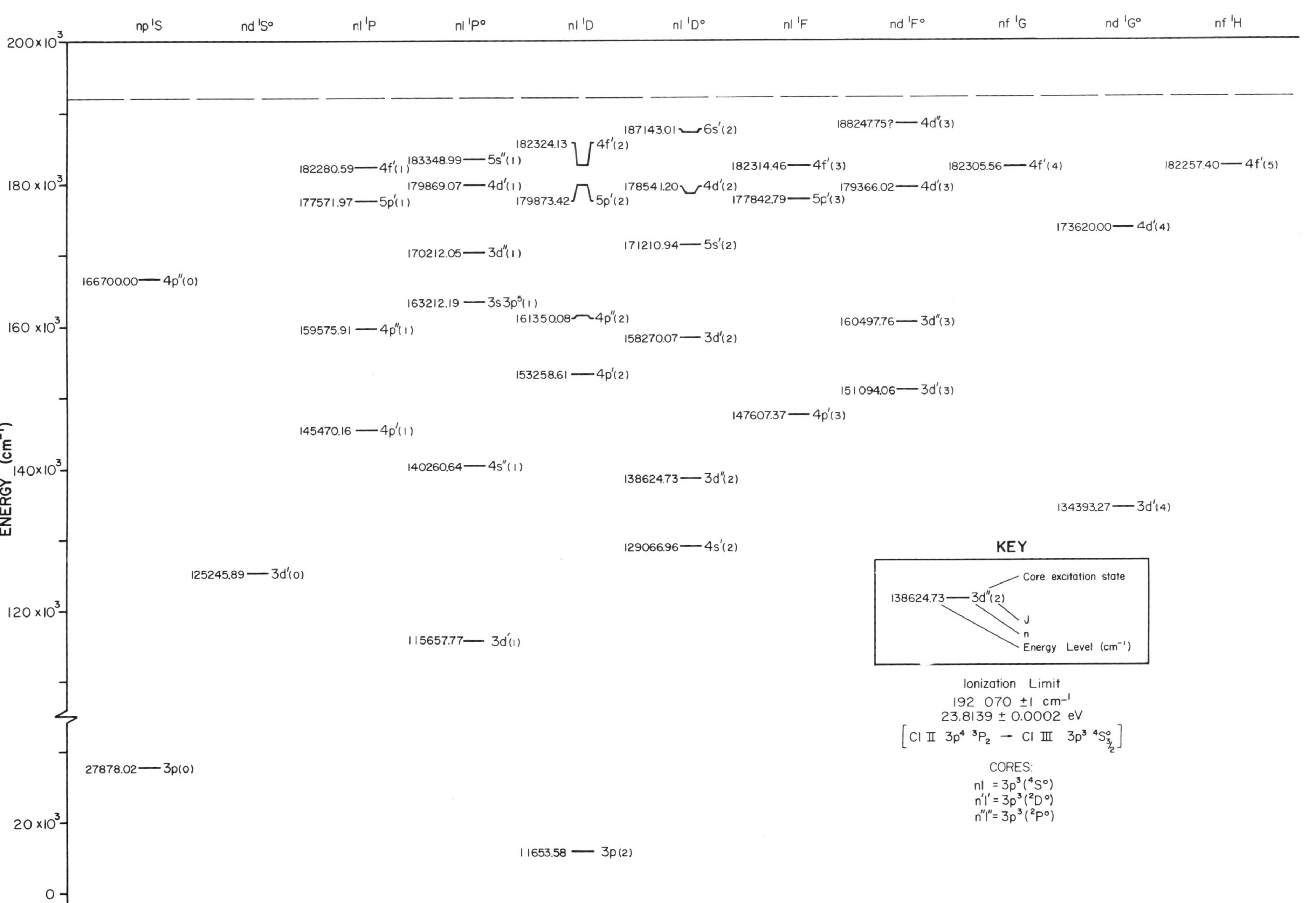
Cl II ENERGY LEVELS (16 electrons, Z = 17)
(S I sequence , Configuration: 1s²2s²2p⁶3s²3p³nl, Singlet System)
np ¹S
nd ¹S°
nl ¹P
nl ¹P°
nl ¹D
nl ¹D°
nl ¹F
nd ¹F°
nf ¹G
nd ¹G°
nf ¹H
187143.01—6s'(2)
188247.75?—4d''(3)
182324.13 4f'(2)
182280.59—4f'(1)
183348.99—5s''(1)
182314.46—4f'(3)
182305.56—4f'(4)
182257.40—4f'(5)
179869.07—4d'(1)
178541.20—4d'(2)
179366.02—4d'(3)
177571.97—5p'(1)
179873.42 5p'(2)
177842.79—5p'(3)
173620.00—4d'(4)
171210.94—5s'(2)
170212.05—3d''(1)
166700.00—4p''(0)
163212.19—3s3p⁵(1)
161350.08—4p''(2)
159575.91—4p''(1)
160497.76—3d''(3)
158270.07—3d'(2)
153258.61—4p'(2)
151094.06—3d'(3)
147607.37—4p'(3)
145470.16—4p'(1)
140260.64—4s''(1)
138624.73—3d''(2)
134393.27—3d'(4)
129066.96—4s'(2)
125245.89—3d'(0)
115657.77—3d'(1)
27878.02—3p(0)
11653.58—3p(2)
KEY
138624.73—3d''(2)
Core excitation state
J
n
Energy Level (cm⁻¹)
Ionization Limit
192 070 ±1 cm⁻¹
23.8139 ± 0.0002 eV
[Cl II 3p⁴ ³P₂ → Cl III 3p³ ⁴S°3/2]
CORES:
nl = 3p³(⁴S°)
n'l' = 3p³(²D°)
n''l'' = 3p³(²P°)
200×10³
180×10³
160×10³
140×10³
120×10³
20×10³
0
ENERGY (cm⁻¹)

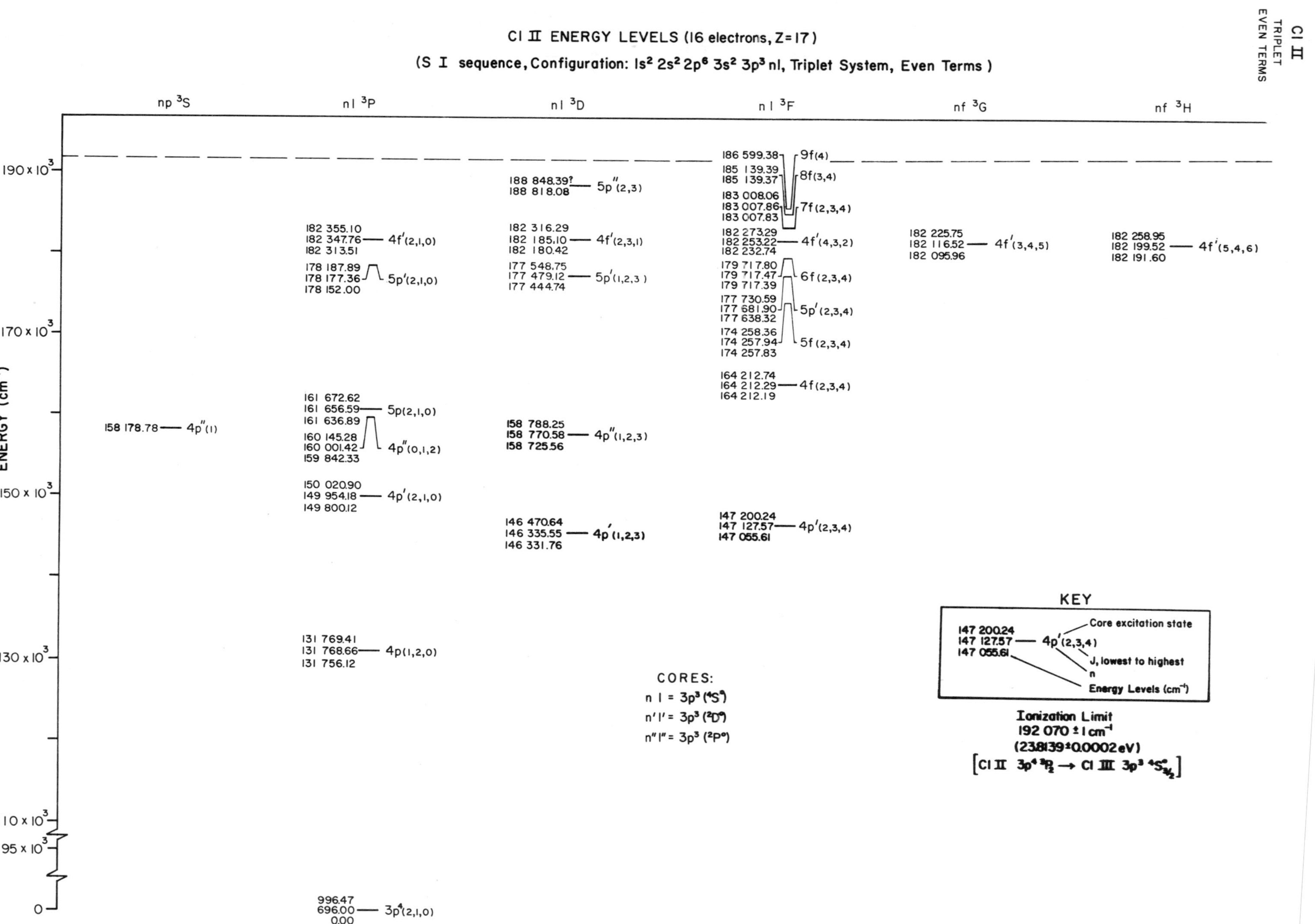
Cl II ENERGY LEVELS (16 electrons, Z=17)
(S I sequence, Configuration: $1s^2 2s^2 2p^6 3s^2 3p^3 nl$, Triplet System, Even Terms)
np ^{3}S
nl ^{3}P
nl ^{3}D
nl ^{3}F
nf ^{3}G
nf ^{3}H
ENERGY (cm^{-1})
190 x 10^3
170 x 10^3
150 x 10^3
130 x 10^3
110 x 10^3
95 x 10^3
0
158 178.78 — 4p″(1)
182 355.10
182 347.76 — 4f′(2,1,0)
182 313.51
178 187.89
178 177.36 — 5p′(2,1,0)
178 152.00
161 672.62
161 656.59 — 5p(2,1,0)
161 636.89
160 145.28
160 001.42 — 4p″(0,1,2)
159 842.33
150 020.90
149 954.18 — 4p′(2,1,0)
149 800.12
131 769.41
131 768.66 — 4p(1,2,0)
131 756.12
996.47
696.00 — 3p^4(2,1,0)
0.00
188 848.39? — 5p″(2,3)
188 818.08
182 316.29
182 185.10 — 4f′(2,3,1)
182 180.42
177 548.75
177 479.12 — 5p′(1,2,3)
177 444.74
158 788.25
158 770.58 — 4p″(1,2,3)
158 725.56
146 470.64
146 335.55 — 4p′(1,2,3)
146 331.76
186 599.38 — 9f(4)
185 139.39
185 139.37 — 8f(3,4)
183 008.06
183 007.86 — 7f(2,3,4)
183 007.83
182 273.29
182 253.22 — 4f′(4,3,2)
182 232.74
179 717.80
179 717.47 — 6f(2,3,4)
179 717.39
177 730.59
177 681.90 — 5p′(2,3,4)
177 638.32
174 258.36
174 257.94 — 5f(2,3,4)
174 257.83
164 212.74
164 212.29 — 4f(2,3,4)
164 212.19
147 200.24
147 127.57 — 4p′(2,3,4)
147 055.61
182 225.75
182 116.52 — 4f′(3,4,5)
182 095.96
182 258.95
182 199.52 — 4f′(5,4,6)
182 191.60
CORES:
n l = $3p^3$ ($^4S^o$)
n′l′ = $3p^3$ ($^2D^o$)
n″l″ = $3p^3$ ($^2P^o$)
KEY
147 200.24
147 127.57 — 4p′(2,3,4)
147 055.61
Core excitation state
J, lowest to highest
n
Energy Levels (cm^{-1})
Ionization Limit
192 070 ± 1 cm^{-1}
(23.8139 ± 0.0002 eV)
[Cl II $3p^4\ ^3P_2$ → Cl III $3p^3\ ^4S^o_{3/2}$]

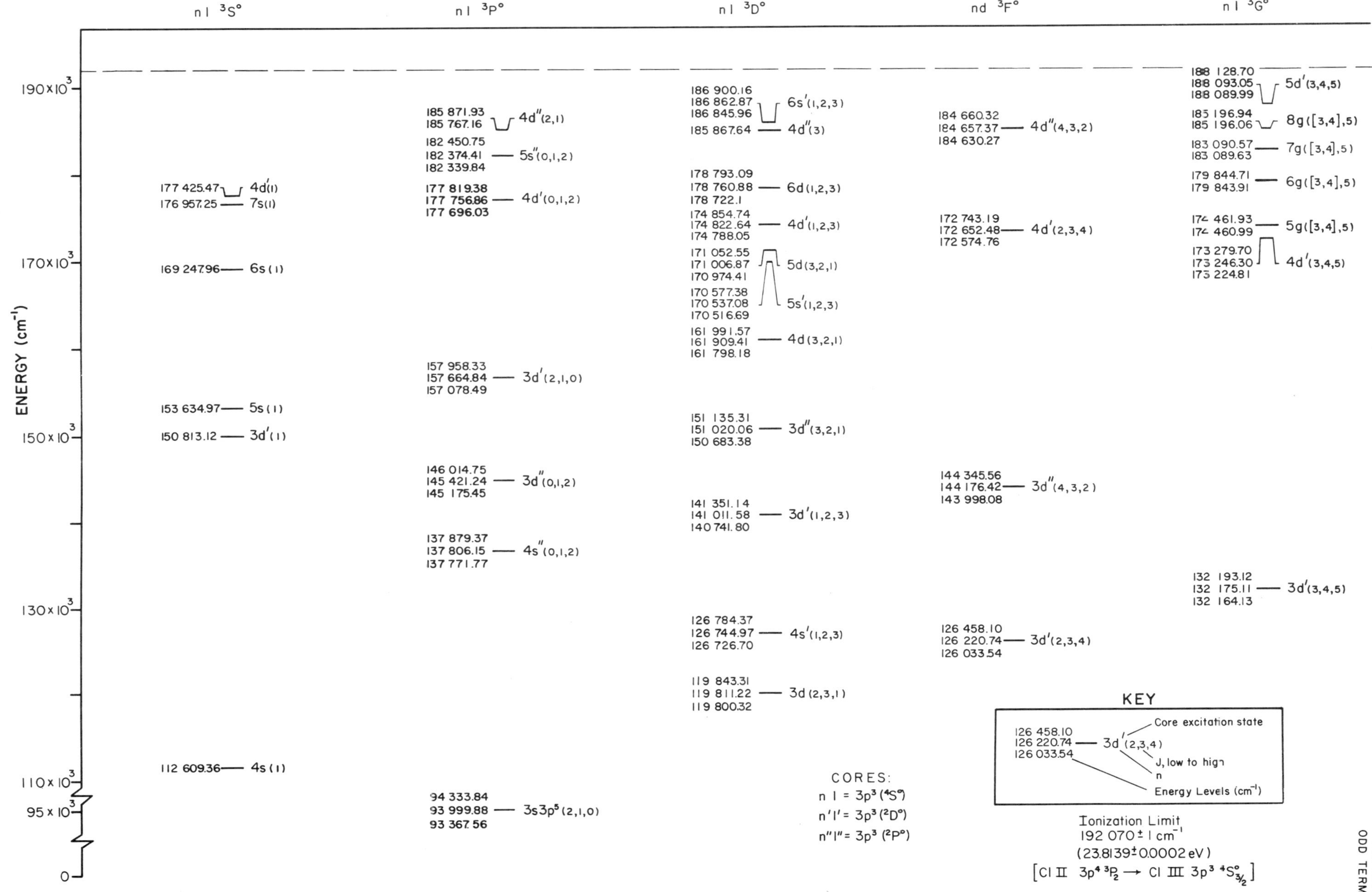
Cl II ENERGY LEVELS (16 electrons, Z=17)
(S I sequence, Configuration: 1s² 2s² 2p⁶ 3s² 3p³ nl, Triplet System, Odd Terms)
nl ³S°
nl ³P°
nl ³D°
nd ³F°
nl ³G°
ENERGY (cm⁻¹)
190×10³
170×10³
150×10³
130×10³
110×10³
95×10³
0
177 425.47 4d′(1)
176 957.25 7s(1)
169 247.96 6s(1)
153 634.97 5s(1)
150 813.12 3d′(1)
112 609.36 4s(1)
185 871.93
185 767.16 4d″(2,1)
182 450.75
182 374.41 5s″(0,1,2)
182 339.84
177 819.38
177 756.86 4d′(0,1,2)
177 696.03
157 958.33
157 664.84 3d′(2,1,0)
157 078.49
146 014.75
145 421.24 3d″(0,1,2)
145 175.45
137 879.37
137 806.15 4s″(0,1,2)
137 771.77
94 333.84
93 999.88 3s3p⁵(2,1,0)
93 367.56
186 900.16
186 862.87 6s′(1,2,3)
186 845.96
185 867.64 4d″(3)
178 793.09
178 760.88 6d(1,2,3)
178 722.1
174 854.74
174 822.64 4d′(1,2,3)
174 788.05
171 052.55
171 006.87 5d(3,2,1)
170 974.41
170 577.38
170 537.08 5s′(1,2,3)
170 516.69
161 991.57
161 909.41 4d(3,2,1)
161 798.18
151 135.31
151 020.06 3d″(3,2,1)
150 683.38
141 351.14
141 011.58 3d′(1,2,3)
140 741.80
126 784.37
126 744.97 4s′(1,2,3)
126 726.70
119 843.31
119 811.22 3d(2,3,1)
119 800.32
184 660.32
184 657.37 4d″(4,3,2)
184 630.27
172 743.19
172 652.48 4d′(2,3,4)
172 574.76
144 345.56
144 176.42 3d″(4,3,2)
143 998.08
126 458.10
126 220.74 3d′(2,3,4)
126 033.54
188 128.70
188 093.05 5d′(3,4,5)
188 089.99
185 196.94
185 196.06 8g([3,4],5)
183 090.57
183 089.63 7g([3,4],5)
179 844.71
179 843.91 6g([3,4],5)
174 461.93
174 460.99 5g([3,4],5)
173 279.70
173 246.30 4d′(3,4,5)
173 224.81
132 193.12
132 175.11 3d′(3,4,5)
132 164.13
CORES:
n l = 3p³ (⁴S°)
n′l′ = 3p³ (²D°)
n″l″ = 3p³ (²P°)
KEY
126 458.10
126 220.74 3d′(2,3,4)
126 033.54
Core excitation state
J, low to high
n
Energy Levels (cm⁻¹)
Ionization Limit
192 070 ± 1 cm⁻¹
(23.8139 ± 0.0002 eV)
[Cl II 3p⁴ ³P₂ → Cl III 3p³ ⁴S°3/2]
Cl II
TRIPLET
ODD TERMS

Cl II GROTRIAN DIAGRAM (16 electrons, Z=17)

(S I sequence, Configuration: $1s^2\ 2s^2\ 2p^6\ 3s^2\ 3p^3\,nl$, $3p^4\ ^3P$ to Odd Terms)

nl $^3S^o$ | nl $^3P^o$ | np ^3P | nl $^3D^o$ | nd $^3F^o$ | nl $^3G^o$

ENERGY (cm^{-1})

90×10^3
170×10^3
150×10^3
130×10^3
110×10^3
95×10^3
0

4d′(1)
7s(1)
6s(1)
5s(1)
3d′(1)
4s(1)

4d″(2,1)
5s″(0,1,2)
4d′(0,1,2)
3d′(2,1,0)
3d″(0,1,2)
4s″(0,1,2)
$3s\,3p^5$(2,1,0)

6s′(1,2,3)
4d″(3)
6d(1,2,3)
4d′(1,2,3)
5d(3,2,1)
5s′(1,2,3)
4d(3,2,1)
3d″(3,2,1)
3d′(1,2,3)
4s′(1,2,3)
3d(2,3,1)

4d″(4,3,2)
4d′(2,3,4)
3d″(4,3,2)
3d′(2,3,4)

5d′(3,4,5)
8g([3,4],5)
7g([3,4],5)
6g([3,4],5)
5g([3,4],5)
4d′(3,4,5)
3d′(3,4,5)

$3p^4$(2,1,0)

538.01–540.90
548.09–551.33
562.37–565.74
563.62–566.80
565.11, 567.34
590.85–594.35
650.89–655.14
663.07–667.48
634.26–639.46
684.86–692.40
725.27–730.94
888.03–895.95
1063.83–1079.08
535.05
559.30–562.66
571.90–575.40
584.61–588.04
586.24–589.90
617.32–621.14
661.66–666.05
707.46–715.59
788.74–795.35
834.42–841.41
579.20, 581.80
692.78–696.14
792.26, 797.85
756.63

CORES:
n l = $3p^3\ (^4S^o)$
n′l′ = $3p^3\ (^2D^o)$
n″l″ = $3p^3\ (^2P^o)$

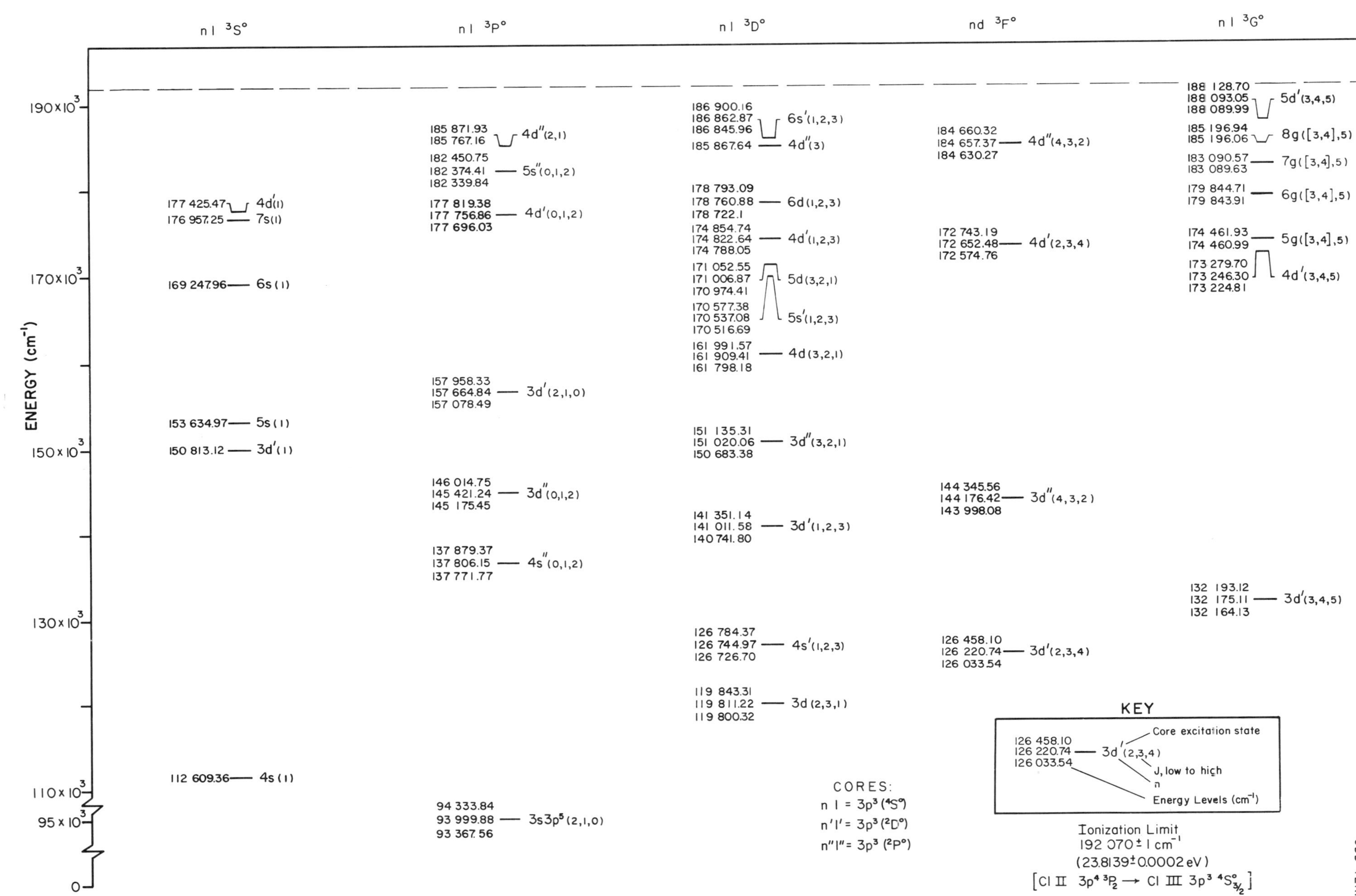
Cl II
TRIPLET
ODD TERMS
Cl II ENERGY LEVELS (16 electrons, Z=17)
(S I sequence, Configuration: 1s² 2s² 2p⁶ 3s² 3p³ nl, Triplet System, Odd Terms)
nl ³S°
nl ³P°
nl ³D°
nd ³F°
nl ³G°
ENERGY (cm⁻¹)
190×10³
170×10³
150×10³
130×10³
110×10³
95×10³
0
177 425.47 4d′(1)
176 957.25 7s(1)
169 247.96 6s(1)
153 634.97 5s(1)
150 813.12 3d′(1)
112 609.36 4s(1)
185 871.93
185 767.16 4d″(2,1)
182 450.75
182 374.41 5s″(0,1,2)
182 339.84
177 819.38
177 756.86 4d′(0,1,2)
177 696.03
157 958.33
157 664.84 3d′(2,1,0)
157 078.49
146 014.75
145 421.24 3d″(0,1,2)
145 175.45
137 879.37
137 806.15 4s″(0,1,2)
137 771.77
94 333.84
93 999.88 3s3p⁵(2,1,0)
93 367.56
186 900.16
186 862.87 6s′(1,2,3)
186 845.96
185 867.64 4d″(3)
178 793.09
178 760.88 6d(1,2,3)
178 722.1
174 854.74
174 822.64 4d′(1,2,3)
174 788.05
171 052.55
171 006.87 5d(3,2,1)
170 974.41
170 577.38
170 537.08 5s′(1,2,3)
170 516.69
161 991.57
161 909.41 4d(3,2,1)
161 798.18
151 135.31
151 020.06 3d″(3,2,1)
150 683.38
141 351.14
141 011.58 3d′(1,2,3)
140 741.80
126 784.37
126 744.97 4s′(1,2,3)
126 726.70
119 843.31
119 811.22 3d(2,3,1)
119 800.32
184 660.32
184 657.37 4d″(4,3,2)
184 630.27
172 743.19
172 652.48 4d′(2,3,4)
172 574.76
144 345.56
144 176.42 3d″(4,3,2)
143 998.08
126 458.10
126 220.74 3d′(2,3,4)
126 033.54
188 128.70
188 093.05 5d′(3,4,5)
188 089.99
185 196.94
185 196.06 8g([3,4],5)
183 090.57
183 089.63 7g([3,4],5)
179 844.71
179 843.91 6g([3,4],5)
174 461.93
174 460.99 5g([3,4],5)
173 279.70
173 246.30 4d′(3,4,5)
173 224.81
132 193.12
132 175.11 3d′(3,4,5)
132 164.13
CORES:
n l = 3p³ (⁴S°)
n′l′ = 3p³ (²D°)
n″l″ = 3p³ (²P°)
KEY
126 458.10
126 220.74 3d′(2,3,4)
126 033.54
Core excitation state
J, low to high
n
Energy Levels (cm⁻¹)
Ionization Limit
192 070 ± 1 cm⁻¹
(23.8139 ± 0.0002 eV)
[Cl II 3p⁴ ³P₂ → Cl III 3p³ ⁴S°₃/₂]

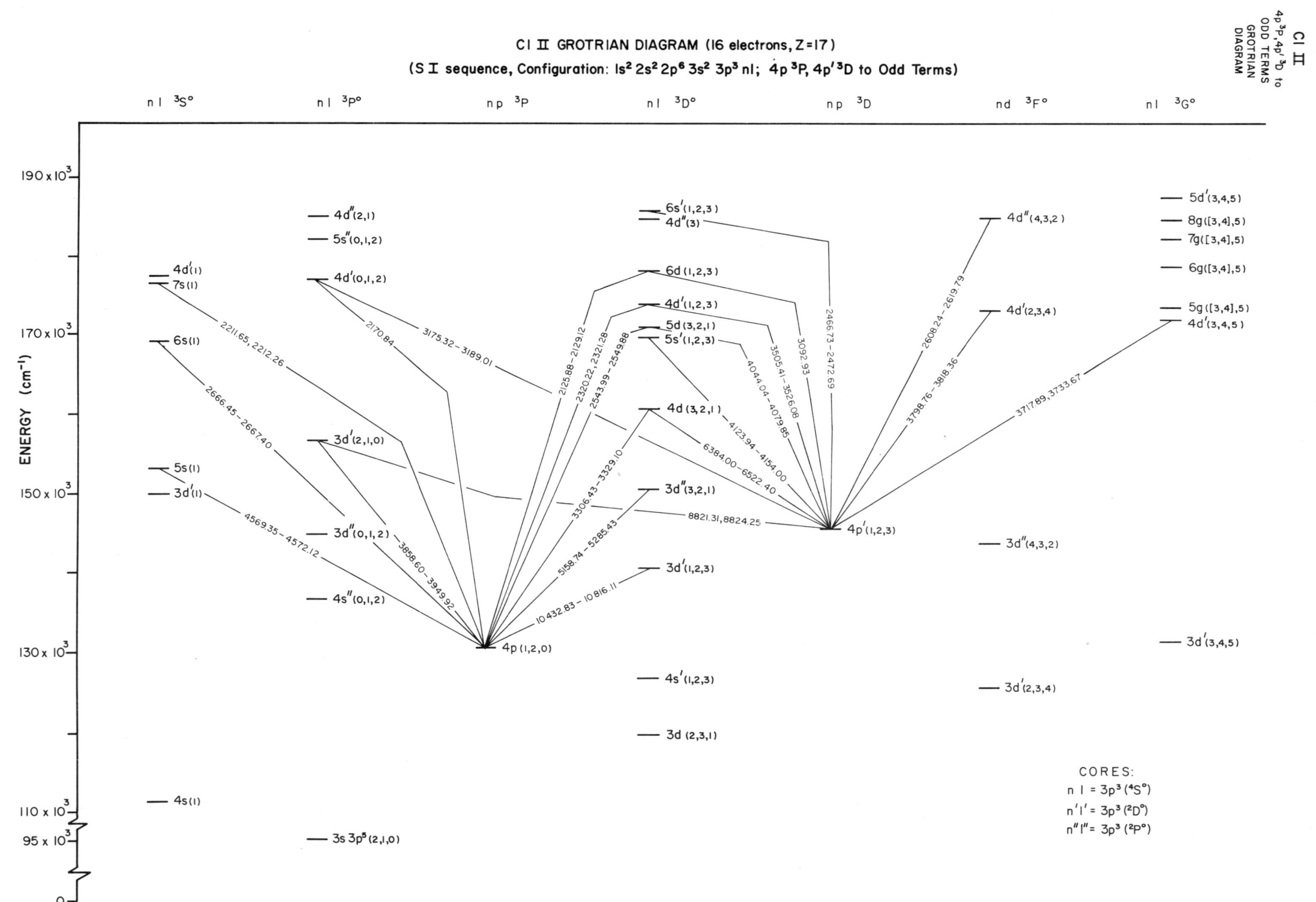

Cl II GROTRIAN DIAGRAM (16 electrons, Z=17)
(S I sequence, Configuration: 1s² 2s² 2p⁶ 3s² 3p³ nl; 4p ³P, 4p′ ³D to Odd Terms)
Cl II
4p ³P, 4p′ ³D to ODD TERMS GROTRIAN DIAGRAM
nl ³S°
nl ³P°
np ³P
nl ³D°
np ³D
nd ³F°
nl ³G°
ENERGY (cm⁻¹)
190 x 10³
170 x 10³
150 x 10³
130 x 10³
110 x 10³
95 x 10³
0
4d′(1)
7s(1)
6s(1)
5s(1)
3d′(1)
4s(1)
4d″(2,1)
5s″(0,1,2)
4d′(0,1,2)
3d′(2,1,0)
3d″(0,1,2)
4s″(0,1,2)
3s 3p⁵(2,1,0)
4p(1,2,0)
6s′(1,2,3)
4d″(3)
6d(1,2,3)
4d′(1,2,3)
5d(3,2,1)
5s′(1,2,3)
4d(3,2,1)
3d″(3,2,1)
3d′(1,2,3)
4s′(1,2,3)
3d(2,3,1)
4p′(1,2,3)
4d″(4,3,2)
4d′(2,3,4)
3d″(4,3,2)
3d′(2,3,4)
5d′(3,4,5)
8g([3,4],5)
7g([3,4],5)
6g([3,4],5)
5g([3,4],5)
4d′(3,4,5)
3d′(3,4,5)
2211.65, 2212.26
2666.45 - 2667.40
4569.35 - 4572.12
2170.84
3175.32 - 3189.01
3858.60 - 3949.92
2125.88 - 2129.12
2320.22, 2321.28
2543.99 - 2549.88
3306.43 - 3329.10
5158.74 - 5285.43
10432.83 - 10816.11
2466.73 - 2472.69
3092.93
3505.41 - 3526.08
4044.04 - 4079.85
4123.94 - 4154.00
6384.00 - 6522.40
8821.31, 8824.25
2608.24 - 2619.79
3798.76 - 3818.36
3717.89, 3733.67
CORES:
n l = 3p³ (⁴S°)
n′l′ = 3p³ (²D°)
n″l″ = 3p³ (²P°)

Cl II ENERGY LEVELS (16 electrons, Z=17)

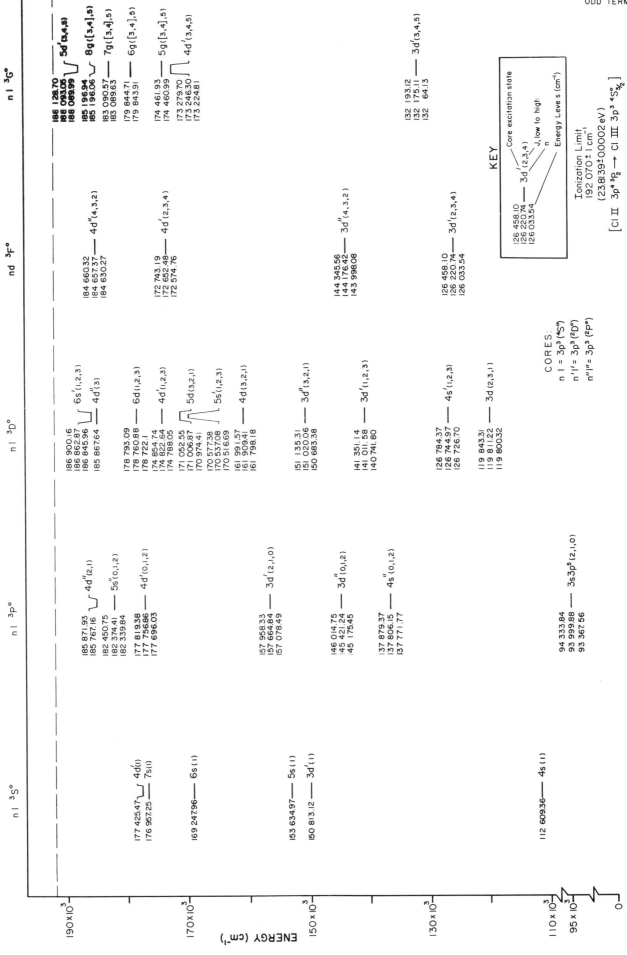

(S I sequence, Configuration: $1s^2\,2s^2\,2p^6\,3s^2\,3p^3\,nl$, Triplet System, Odd Terms)

Cl II
TRIPLET
ODD TERMS

ENERGY (cm^{-1}): 190×10³, 170×10³, 150×10³, 130×10³, 110×10³, 95×10³, 0

nl $^3S^o$

177 425.47	4d′(1)
176 957.25	7s(1)
169 247.96	6s(1)
153 634.97	5s(1)
150 813.12	3d′(1)
112 609.36	4s(1)

nl $^3P^o$

185 871.93 185 767.16	4d″(2,1)
182 450.75 182 374.41 182 339.84	5s″(0,1,2)
177 819.38 177 756.86 177 696.03	4d′(0,1,2)
157 958.33 157 664.84 157 078.49	3d′(2,1,0)
146 014.75 145 421.24 145 175.45	3d″(0,1,2)
137 879.37 137 806.15 137 771.77	4s″(0,1,2)
94 333.84 93 999.88 93 367.56	$3s3p^5$(2,1,0)

nl $^3D^o$

186 900.16 186 862.87 186 845.96	6s′(1,2,3)
185 867.64	4d″(3)
178 793.09 178 760.88 178 722.1	6d(1,2,3)
174 854.74 174 822.64 174 788.05	4d′(1,2,3)
171 052.55 171 006.87 170 974.41	5d(3,2,1)
170 577.38 170 537.08 170 516.69	5s′(1,2,3)
161 991.57 161 909.41 161 798.18	4d(3,2,1)
151 135.31 151 020.06 150 683.38	3d″(3,2,1)
141 351.14 141 011.58 140 741.80	3d′(1,2,3)
126 784.37 126 744.97 126 726.70	4s′(1,2,3)
119 843.31 119 811.22 119 800.32	3d(2,3,1)

nd $^3F^o$

184 660.32 184 657.37 184 630.27	4d″(4,3,2)
172 743.19 172 652.48 172 574.76	4d′(2,3,4)
144 345.56 144 176.42 143 998.08	3d″(4,3,2)
126 458.10 126 220.74 126 033.54	3d′(2,3,4)

nl $^3G^o$

188 128.70 188 093.05 188 089.99	5d′(3,4,5)
185 196.94 185 196.06	8g([3,4],5)
183 090.57 183 089.63	7g([3,4],5)
179 844.71 179 843.91	6g([3,4],5)
174 461.93 174 460.99	5g([3,4],5)
173 279.70 173 246.30 173 224.81	4d′(3,4,5)
132 193.12 132 175.11 132 64.13	3d′(3,4,5)

CORES:
n l = $3p^3$ ($^4S^o$)
n′l′ = $3p^3$ ($^2D^o$)
n″l″ = $3p^3$ ($^2P^o$)

KEY

126 458.10
126 220.74 — 3d′(2,3,4)
126 033.54

Core excitation state; J, low to high; n; Energy Levels (cm^{-1})

Ionization Limit
192 070 ± 1 cm^{-1}
(23.8139 ± 0.0002 eV)
[Cl II $3p^4\,^3P_2$ → Cl III $3p^3\,^4S^o_{3/2}$]

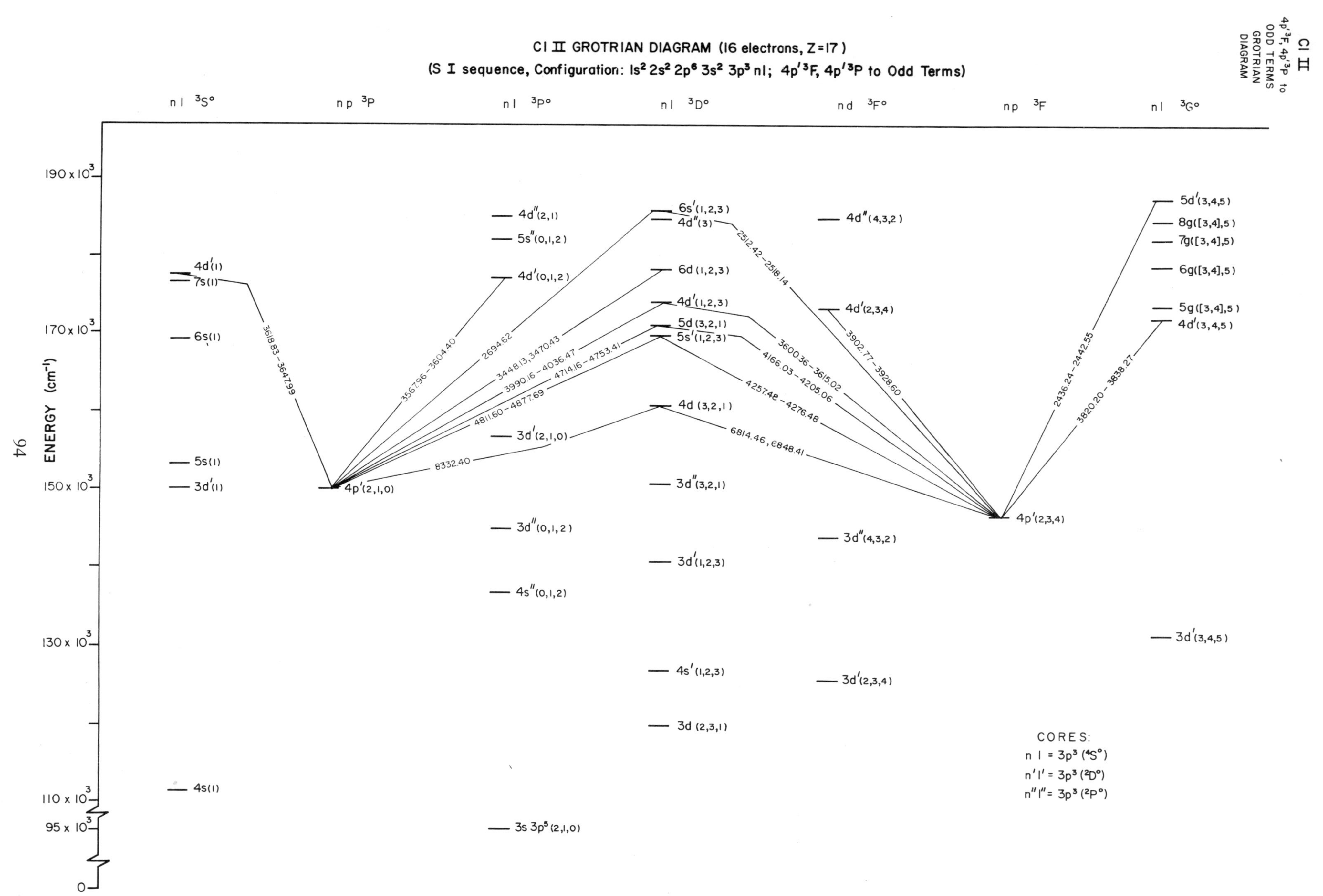

Cl II GROTRIAN DIAGRAM (16 electrons, Z=17)
(S I sequence, Configuration: 1s² 2s² 2p⁶ 3s² 3p³ nl; 4p′ ³F, 4p′ ³P to Odd Terms)
Cl II
4p′ ³F, 4p′ ³P to ODD TERMS GROTRIAN DIAGRAM
nl ³S°
np ³P
nl ³P°
nl ³D°
nd ³F°
np ³F
nl ³G°
ENERGY (cm⁻¹)
190 x 10³
170 x 10³
150 x 10³
130 x 10³
110 x 10³
95 x 10³
0
4d′(1)
7s(1)
6s(1)
5s(1)
3d′(1)
4s(1)
4p′(2,1,0)
4d″(2,1)
5s″(0,1,2)
4d′(0,1,2)
3d′(2,1,0)
3d″(0,1,2)
4s″(0,1,2)
3s 3p⁵(2,1,0)
6s′(1,2,3)
4d″(3)
6d(1,2,3)
4d′(1,2,3)
5d(3,2,1)
5s′(1,2,3)
4d(3,2,1)
3d″(3,2,1)
3d′(1,2,3)
4s′(1,2,3)
3d(2,3,1)
4d″(4,3,2)
4d′(2,3,4)
3d″(4,3,2)
3d′(2,3,4)
4p′(2,3,4)
5d′(3,4,5)
8g([3,4],5)
7g([3,4],5)
6g([3,4],5)
5g([3,4],5)
4d′(3,4,5)
3d′(3,4,5)
3618.83 – 3647.99
3567.96 – 3604.40
2694.62
3448.13, 3470.43
3990.16 – 4036.47
4714.16 – 4753.41
4811.60 – 4877.69
8332.40
2512.42 – 2518.14
3600.36 – 3615.02
4166.03 – 4205.06
4257.48 – 4276.48
6814.46, 6848.41
3902.77 – 3928.60
2436.24 – 2442.55
3820.20 – 3838.27
CORES:
n l = 3p³ (⁴S°)
n′l′ = 3p³ (²D°)
n″l″ = 3p³ (²P°)

Cl II ENERGY LEVELS (16 electrons, Z=17)

(S I sequence, Configuration: $1s^2\ 2s^2\ 2p^6\ 3s^2\ 3p^3\ nl$, Triplet System, Odd Terms)

ENERGY (cm^{-1}): 190×10³, 170×10³, 150×10³, 130×10³, 110×10³, 95×10³, 0

$nl\ ^3S^o$

177 425.47	4d′(1)
176 957.25	7s(1)
169 247.96	6s(1)
153 634.97	5s(1)
150 813.12	3d′(1)
112 609.36	4s(1)

$nl\ ^3P^o$

185 871.93 185 767.16	4d″(2,1)
182 450.75 182 374.41 132 339.84	5s″(0,1,2)
177 819.38 177 756.86 177 696.03	4d′(0,1,2)
157 958.33 157 664.84 157 078.49	3d′(2,1,0)
146 014.75 145 421.24 145 175.45	3d″(0,1,2)
137 879.37 137 806.15 137 771.77	4s″(0,1,2)
94 333.84 93 999.88 93 367.56	$3s3p^5$(2,1,0)

$nl\ ^3D^o$

186 900.16 186 862.87 186 845.96	6s′(1,2,3)
185 867.64	4d″(3)
178 793.09 178 760.88 178 722.1	6d(1,2,3)
174 854.74 174 822.64 174 788.05	4d′(1,2,3)
171 052.55 171 006.87 170 974.41	5d(3,2,1)
170 577.38 170 537.08 170 516.69	5s′(1,2,3)
161 991.57 161 909.41 161 798.18	4d(3,2,1)
151 135.31 151 020.06 150 683.38	3d″(3,2,1)
141 351.14 141 011.58 140 741.80	3d′(1,2,3)
126 784.37 126 744.97 126 726.70	4s′(1,2,3)
119 843.31 119 811.22 119 800.32	3d(2,3,1)

$nd\ ^3F^o$

184 660.32 184 657.37 184 630.27	4d″(4,3,2)
172 743.19 172 652.48 172 574.76	4d′(2,3,4)
144 345.56 144 176.42 143 998.08	3d″(4,3,2)
126 458.10 126 220.74 126 033.54	3d′(2,3,4)

$nl\ ^3G^o$

188 28.70 188 093.05 188 089.99	5d′(3,4,5)
185 196.94 185 196.06	8g([3,4],5)
183 090.57 183 089.63	7g([3,4],5)
179 844.71 179 843.91	6g([3,4],5)
174 461.93 174 460.99	5g([3,4],5)
173 279.70 173 246.30 173 224.81	4d′(3,4,5)
132 193.12 132 175.11 132 164.13	3d′(3,4,5)

CORES:
$n\,l = 3p^3\,(^4S^o)$
$n'l' = 3p^3\,(^2D^o)$
$n''l'' = 3p^3\,(^2P^o)$

KEY

126 458.10
126 220.74 — 3d′(2,3,4)
126 033.54

- ′ — Core excitation state
- (2,3,4) — J, low to high
- 3 — n
- 126 458.10 … — Energy Levels (cm^{-1})

Ionization Limit
192 070 ± 1 cm^{-1}
(23.8139 ± 0.0002 eV)
[Cl II $3p^4\ ^3P_2$ → Cl III $3p^3\ ^4S^o_{3/2}$]

Cl II
TRIPLET
ODD TERMS

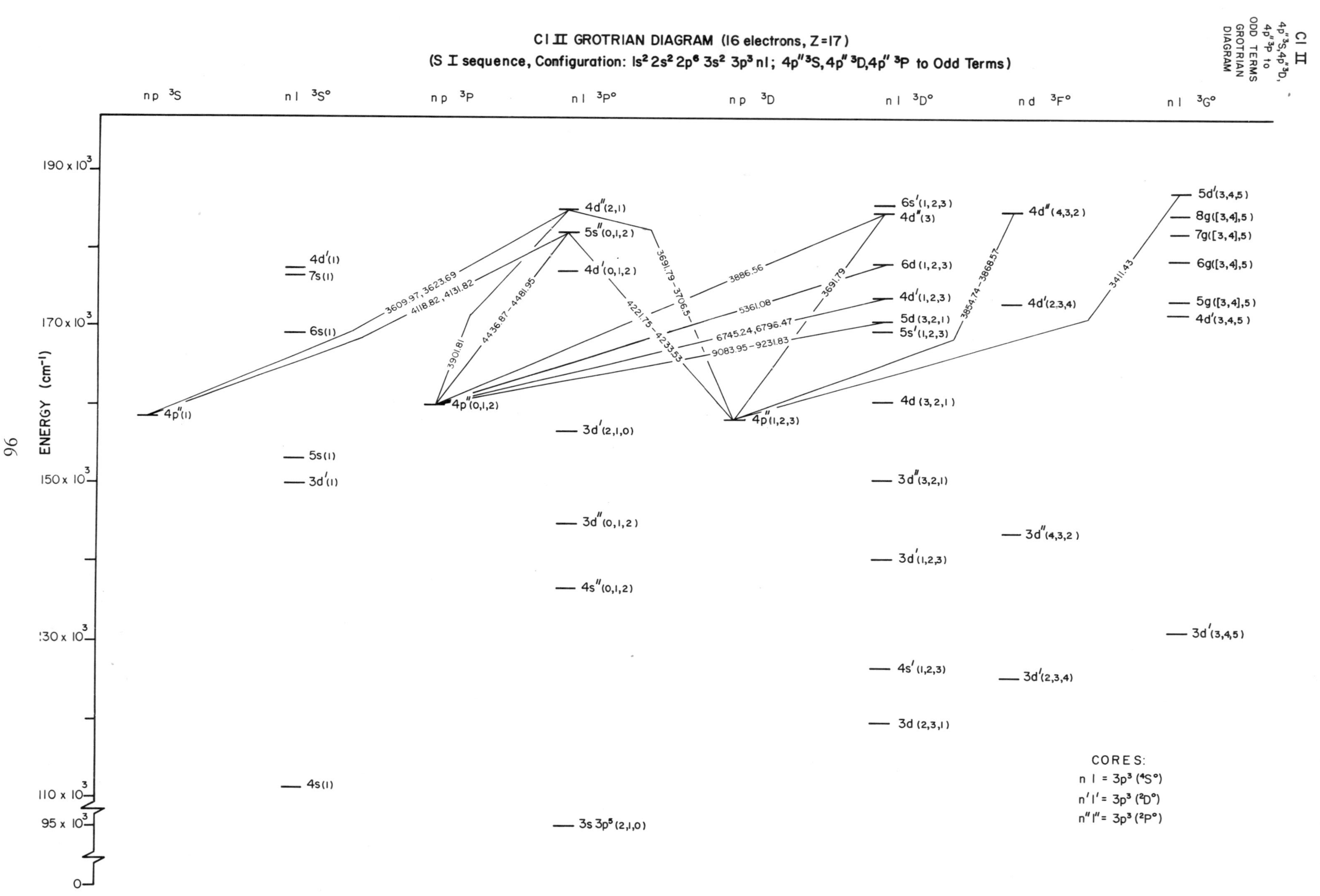

Cl II
4p″³S, 4p″³D, 4p″³P to ODD TERMS GROTRIAN DIAGRAM
Cl II GROTRIAN DIAGRAM (16 electrons, Z=17)
(S I sequence, Configuration: 1s² 2s² 2p⁶ 3s² 3p³ nl; 4p″³S, 4p″³D, 4p″³P to Odd Terms)
np ³S
nl ³S°
np ³P
nl ³P°
np ³D
nl ³D°
nd ³F°
nl ³G°
ENERGY (cm⁻¹)
190 x 10³
170 x 10³
150 x 10³
130 x 10³
110 x 10³
95 x 10³
0
4p″(1)
4d′(1)
7s(1)
6s(1)
5s(1)
3d′(1)
4s(1)
4p″(0,1,2)
4d″(2,1)
5s″(0,1,2)
4d′(0,1,2)
3d′(2,1,0)
3d″(0,1,2)
4s″(0,1,2)
3s 3p⁵ (2,1,0)
4p″(1,2,3)
6s′(1,2,3)
4d″(3)
6d (1,2,3)
4d′(1,2,3)
5d (3,2,1)
5s′(1,2,3)
4d (3,2,1)
3d″(3,2,1)
3d′(1,2,3)
4s′(1,2,3)
3d (2,3,1)
4d″(4,3,2)
4d′(2,3,4)
3d″(4,3,2)
3d′(2,3,4)
5d′(3,4,5)
8g([3,4],5)
7g([3,4],5)
6g([3,4],5)
5g([3,4],5)
4d′(3,4,5)
3d′(3,4,5)
3609.97, 3623.69
4118.82, 4131.82
3901.81
4436.87–4481.95
3691.79–3706.5
4221.75–4233.53
3886.56
5361.08
6745.24, 6796.47
9083.95–9231.83
3691.79
3854.74–3868.57
3411.43
CORES:
n l = 3p³ (⁴S°)
n′l′ = 3p³ (²D°)
n″l″ = 3p³ (²P°)

Cl II ENERGY LEVELS (16 electrons, Z=17)

(S I sequence, Configuration: $1s^2\ 2s^2\ 2p^6\ 3s^2\ 3p^3\ nl$, Triplet System, Odd Terms)

ENERGY (cm^{-1})

Axis: 0; 95 x 10^3; 110 x 10^3; 130 x 10^3; 150 x 10^3; 170 x 10^3; 190 x 10^3

n l $^3S^o$

177 425.47	4d′(1)
176 957.25	7s(1)
169 247.96	6s(1)
153 634.97	5s(1)
150 813.12	3d′(1)
112 609.36	4s(1)

n l $^3P^o$

185 871.93 185 767.16	4d″(2,1)
182 450.75 182 374.41 182 339.84	5s″(0,1,2)
177 819.38 177 756.86 177 696.03	4d′(0,1,2)
157 958.33 157 664.84 157 078.49	3d′(2,1,0)
146 014.75 145 421.24 145 175.45	3d″(0,1,2)
137 879.37 137 806.15 137 771.77	4s″(0,1,2)
94 333.84 93 999.88 93 367.56	$3s3p^5$(2,1,0)

n l $^3D^o$

186 900.16 186 862.87 186 845.96	6s′(1,2,3)
185 867.64	4d″(3)
178 793.09 178 760.88 178 722.1	6d(1,2,3)
174 854.74 174 822.64 174 788.05	4d′(1,2,3)
171 052.55 171 006.87 170 974.41	5d(3,2,1)
170 577.38 170 537.08 170 516.69	5s′(1,2,3)
161 991.57 161 909.41 161 798.18	4d(3,2,1)
151 135.31 151 020.06 150 683.38	3d″(3,2,1)
141 351.14 141 011.58 140 741.80	3d′(1,2,3)
126 784.37 126 744.97 126 726.70	4s′(1,2,3)
119 843.31 119 811.22 119 800.32	3d(2,3,1)

nd $^3F^o$

184 660.32 184 657.37 184 630.27	4d″(4,3,2)
172 743.19 172 652.48 172 574.76	4d′(2,3,4)
144 345.56 144 176.42 143 998.08	3d″(4,3,2)
126 458.10 126 220.74 126 033.54	3d′(2,3,4)

n l $^3G^o$

188 128.70 188 093.05 188 089.99	5d′(3,4,5)
185 196.94 185 196.06	8g([3,4],5)
183 090.57 183 089.63	7g([3,4],5)
179 844.71 179 843.91	6g([3,4],5)
174 461.93 174 460.99	5g([3,4],5)
173 279.70 173 246.30 173 224.81	4d′(3,4,5)
132 193.12 132 175.11 132 164.13	3d′(3,4,5)

CORES:
n l = $3p^3\ (^4S^o)$
n′l′ = $3p^3\ (^2D^o)$
n″l″ = $3p^3\ (^2P^o)$

KEY

126 458.10
126 220.74 — 3d′(2,3,4)
126 033.54

Core excitaticn state
J, low to high
n
Energy Levels (cm^{-1})

Ionization Limit
192 070 ± 1 cm^{-1}
(23.8139 ± 0.0002 eV)
[Cl II $3p^4\ ^3P_2$ → Cl III $3p^3\ ^4S^o_{3/2}$]

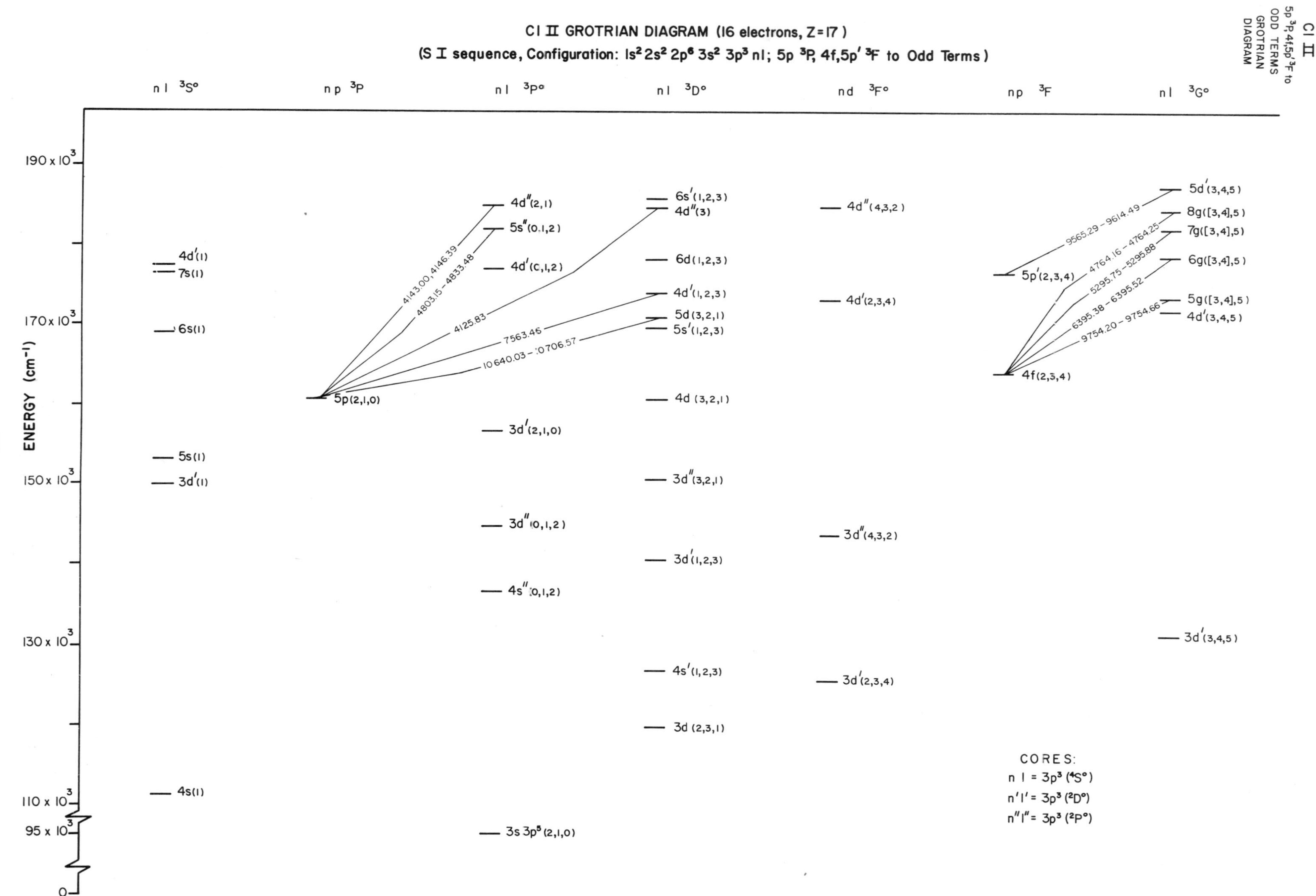
Cl II
5p ³P, 4f,5p′ ³F to
ODD TERMS
GROTRIAN
DIAGRAM
Cl II GROTRIAN DIAGRAM (16 electrons, Z = 17)
(S I sequence, Configuration: 1s² 2s² 2p⁶ 3s² 3p³ nl; 5p ³P, 4f,5p′ ³F to Odd Terms)
nl ³S°
np ³P
nl ³P°
nl ³D°
nd ³F°
np ³F
nl ³G°
ENERGY (cm⁻¹)
190 x 10³
170 x 10³
150 x 10³
130 x 10³
110 x 10³
95 x 10³
0
4d′(1)
7s(1)
6s(1)
5s(1)
3d′(1)
4s(1)
5p(2,1,0)
4d″(2,1)
5s″(0,1,2)
4d′(c,1,2)
3d′(2,1,0)
3d″(0,1,2)
4s″(0,1,2)
3s 3p⁵(2,1,0)
4143.00, 4146.39
4803.15 – 4833.48
4125.83
7563.46
10640.03 – 10706.57
6s′(1,2,3)
4d″(3)
6d(1,2,3)
4d′(1,2,3)
5d(3,2,1)
5s′(1,2,3)
4d (3,2,1)
3d″(3,2,1)
3d′(1,2,3)
4s′(1,2,3)
3d (2,3,1)
4d″(4,3,2)
4d′(2,3,4)
3d″(4,3,2)
3d′(2,3,4)
5p′(2,3,4)
4f(2,3,4)
9565.29 – 9614.49
4764.16 – 4764.25
5295.75 – 5295.88
6395.38 – 6395.52
9754.20 – 9754.66
5d′(3,4,5)
8g([3,4],5)
7g([3,4],5)
6g([3,4],5)
5g([3,4],5)
4d′(3,4,5)
3d′(3,4,5)
CORES:
n l = 3p³ (⁴S°)
n′l′ = 3p³ (²D°)
n″l″ = 3p³ (²P°)

Cl II ENERGY LEVELS (16 electrons, Z=17)

(S I sequence, Configuration: $1s^2\ 2s^2\ 2p^6\ 3s^2\ 3p^3\ nl$, Triplet System, Odd Terms)

ENERGY (cm^{-1}): 190×10^3, 170×10^3, 150×10^3, 130×10^3, 110×10^3, 95×10^3, 0

nl $^3S^o$

- 177 425.47 — 4d′(1)
- 176 957.25 — 7s(1)
- 169 247.96 — 6s(1)
- 153 634.97 — 5s(1)
- 150 813.12 — 3d′(1)
- 112 609.36 — 4s(1)

nl $^3P^o$

- 185 871.93, 185 767.16 — 4d″(2,1)
- 182 450.75, 182 374.41, 182 339.84 — 5s″(0,1,2)
- 177 819.38, 177 756.86, 177 696.03 — 4d′(0,1,2)
- 157 958.33, 157 664.84, 157 078.49 — 3d′(2,1,0)
- 146 014.75, 145 421.24, 145 175.45 — 3d″(0,1,2)
- 137 879.37, 137 806.15, 137 771.77 — 4s″(0,1,2)
- 94 333.84, 93 999.88, 93 367.56 — $3s3p^5$(2,1,0)

nl $^3D^o$

- 186 900.16, 186 862.87, 186 845.96 — 6s′(1,2,3)
- 185 867.64 — 4d″(3)
- 178 793.09, 178 760.88, 178 722.1 — 6d(1,2,3)
- 174 854.74, 174 822.64, 174 788.05 — 4d′(1,2,3)
- 171 052.55, 171 006.87, 170 974.41 — 5d(3,2,1)
- 170 577.38, 170 537.08, 170 516.69 — 5s′(1,2,3)
- 161 991.57, 161 909.41, 161 798.18 — 4d(3,2,1)
- 151 135.31, 151 020.06, 150 683.38 — 3d″(3,2,1)
- 141 351.14, 141 011.58, 140 741.80 — 3d′(1,2,3)
- 126 784.37, 126 744.97, 126 726.70 — 4s′(1,2,3)
- 119 843.31, 119 811.22, 119 800.32 — 3d(2,3,1)

nd $^3F^o$

- 184 660.32, 184 657.37, 184 630.27 — 4d″(4,3,2)
- 172 743.19, 172 652.48, 172 574.76 — 4d′(2,3,4)
- 144 345.56, 144 176.42, 143 998.08 — 3d″(4,3,2)
- 126 458.10, 126 220.74, 126 033.54 — 3d′(2,3,4)

nl $^3G^o$

- 188 128.70, 188 093.05, 188 089.99 — 5d′(3,4,5)
- 185 196.94, 185 196.06 — 8g([3,4],5)
- 183 090.57, 183 089.63 — 7g([3,4],5)
- 179 844.71, 179 843.91 — 6g([3,4],5)
- 174 461.93, 174 460.99 — 5g([3,4],5)
- 173 279.70, 173 246.30, 173 224.81 — 4d′(3,4,5)
- 132 193.12, 132 175.11, 132 154.13 — 3d′(3,4,5)

CORES:
n l = $3p^3\ (^4S^o)$
n′l′ = $3p^3\ (^2D^o)$
n″l″ = $3p^3\ (^2P^o)$

KEY

126 458.10, 126 220.74, 126 033.54 — 3d′(2,3,4)
- Core excitation state
- J, low to high
- n
- Energy Levels (cm^{-1})

Ionization Limit
192 070 ± 1 cm^{-1}
(23.8139 ± 0.0002 eV)
[Cl II $3p^4\ ^3P_2$ → Cl III $3p^3\ ^4S^o_{3/2}$]

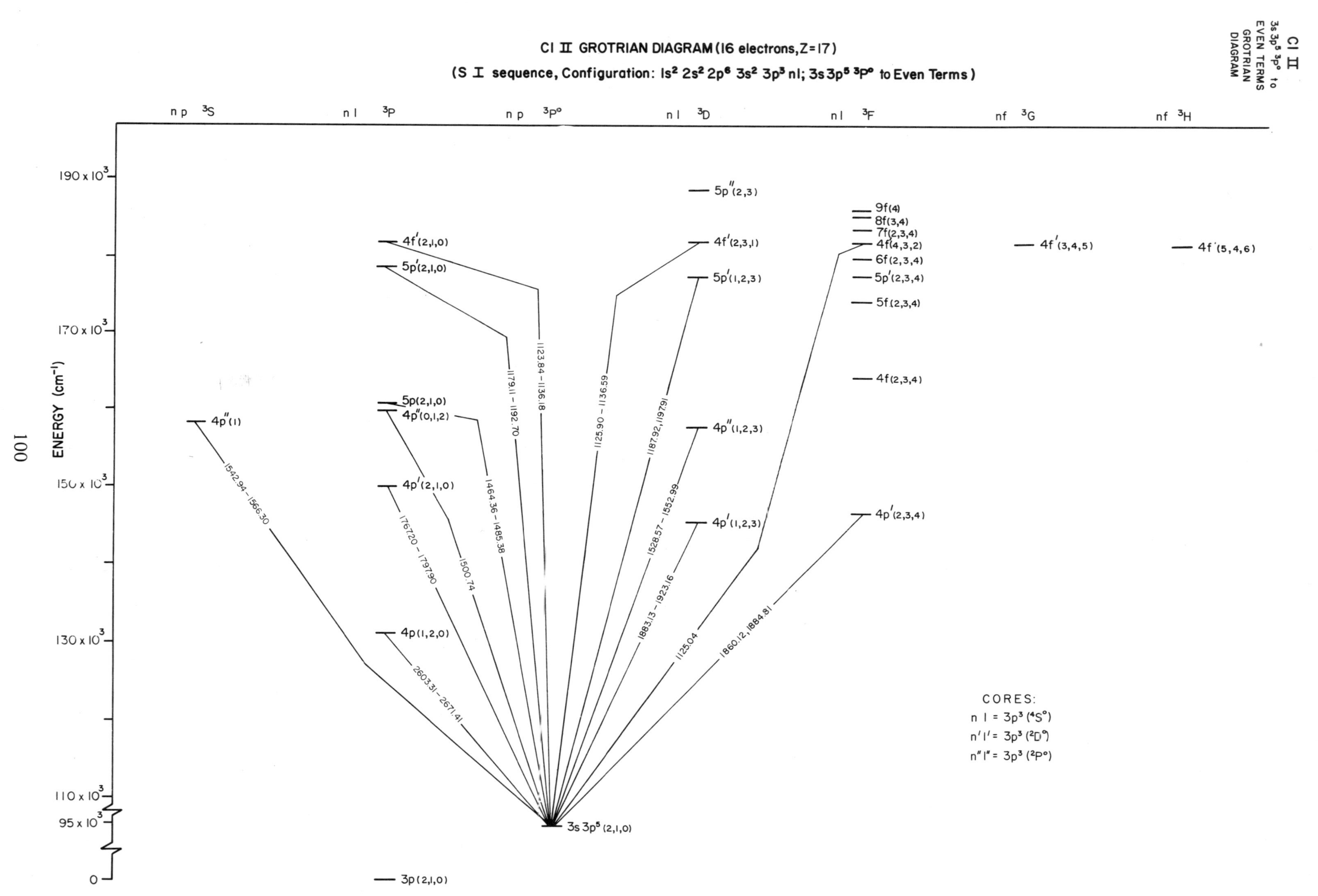
Cl II
3s 3p⁵ ³P° to
EVEN TERMS
GROTRIAN
DIAGRAM
Cl II GROTRIAN DIAGRAM (16 electrons, Z=17)
(S I sequence, Configuration: 1s² 2s² 2p⁶ 3s² 3p³ nl; 3s 3p⁵ ³P° to Even Terms)
np ³S
nl ³P
np ³P°
nl ³D
nl ³F
nf ³G
nf ³H
ENERGY (cm⁻¹)
190 x 10³
170 x 10³
150 x 10³
130 x 10³
110 x 10³
95 x 10³
0
4p″(1)
4f′(2,1,0)
5p′(2,1,0)
5p(2,1,0)
4p″(0,1,2)
4p′(2,1,0)
4p(1,2,0)
3p(2,1,0)
5p″(2,3)
4f′(2,3,1)
5p′(1,2,3)
4p″(1,2,3)
4p′(1,2,3)
9f(4)
8f(3,4)
7f(2,3,4)
4f(4,3,2)
6f(2,3,4)
5p′(2,3,4)
5f(2,3,4)
4f(2,3,4)
4p′(2,3,4)
4f′(3,4,5)
4f′(5,4,6)
3s 3p⁵ (2,1,0)
1542.94 – 1566.30
2603.31 – 2671.41
1767.20 – 1797.90
1500.74
1464.36 – 1485.38
1179.11 – 1192.70
1123.84 – 1136.18
1125.90 – 1136.59
1187.92, 1197.91
1528.57 – 1552.99
1883.13 – 1923.16
1125.04
1860.12, 1884.81
CORES:
n l = 3p³ (⁴S°)
n′l′ = 3p³ (²D°)
n″l″ = 3p³ (²P°)

Cl II
TRIPLET
EVEN TERMS

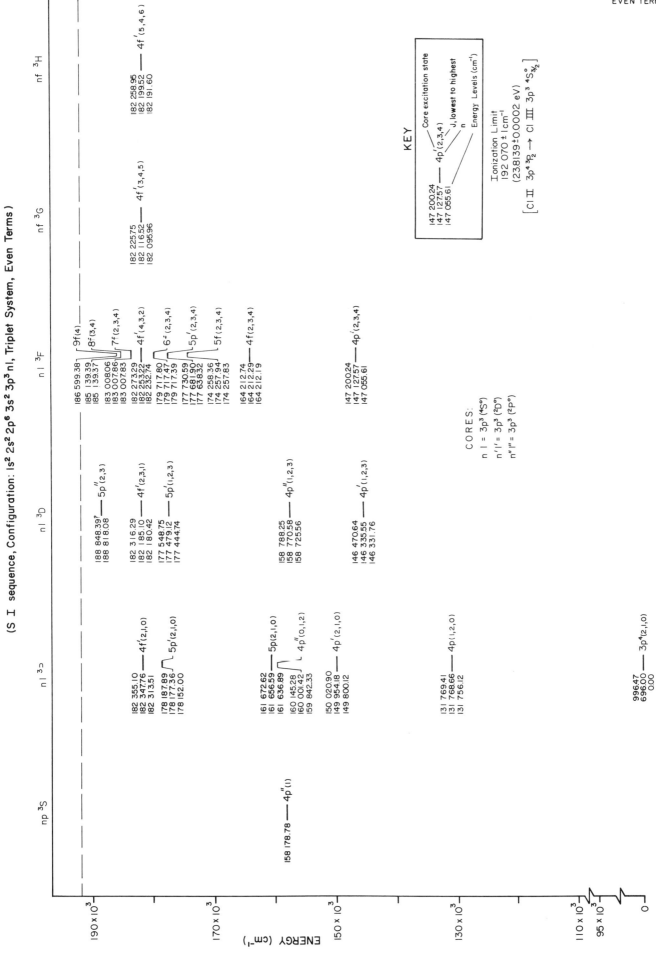

Cl II ENERGY LEVELS (16 electrons, Z = 17)

(S I sequence, Configuration: $1s^2 2s^2 2p^6 3s^2 3p^3 nl$, Triplet System, Even Terms)

np ^3S

158 178.78 — $4p''(1)$

nl ^3P

182 355.10
182 347.76 — $4f'(2,1,0)$
182 313.51

178 187.89
178 177.36 — $5p'(2,1,0)$
178 152.00

161 672.62
161 656.59 — $5p(2,1,0)$
161 636.89

160 145.28
160 001.42 — $4p''(0,1,2)$
159 842.33

150 020.90
149 954.18 — $4p'(2,1,0)$
149 800.12

131 769.41
131 768.66 — $4p(1,2,0)$
131 756.12

996.47
696.00 — $3p^4(2,1,0)$
0.00

nl ^3D

188 848.39?
188 818.08 — $5p''(2,3)$

182 316.29
182 185.10 — $4f'(2,3,1)$
182 180.42

177 548.75
177 479.12 — $5p'(1,2,3)$
177 444.74

158 788.25
158 770.58 — $4p''(1,2,3)$
158 725.56

146 470.64
146 335.55 — $4p'(1,2,3)$
146 331.76

nl ^3F

186 599.38 — $9f(4)$

185 139.39
185 139.37 — $8f(3,4)$

183 008.06
183 007.86 — $7f(2,3,4)$
183 007.83

182 273.29
182 253.22 — $4f'(4,3,2)$
182 232.74

179 717.80
179 717.47 — $6f(2,3,4)$
179 717.39

177 730.59
177 681.90 — $5p'(2,3,4)$
177 638.32

174 258.36
174 257.94 — $5f(2,3,4)$
174 257.83

164 212.74
164 212.29 — $4f(2,3,4)$
164 212.19

147 200.24
147 127.57 — $4p'(2,3,4)$
147 055.61

nf ^3G

182 225.75
182 116.52 — $4f'(3,4,5)$
182 095.96

nf ^3H

182 258.95
182 199.52 — $4f'(5,4,6)$
182 191.60

ENERGY (cm^{-1})

190×10^3
170×10^3
150×10^3
130×10^3
110×10^3
95×10^3
0

CORES:
$n\,l = 3p^3\,(^4S^\circ)$
$n'l' = 3p^3\,(^2D^\circ)$
$n''l'' = 3p^3\,(^2P^\circ)$

KEY

147 200.24
147 127.57 — $4p'(2,3,4)$
147 055.61

Core excitation state
J, lowest to highest
n
Energy Levels (cm^{-1})

Ionization Limit
192 070 ± 1 cm^{-1}
(23.8139 ± 0.0002 eV)
[$Cl\ II\ 3p^4\,^3P_2 \rightarrow Cl\ III\ 3p^3\,^4S^\circ_{3/2}$]

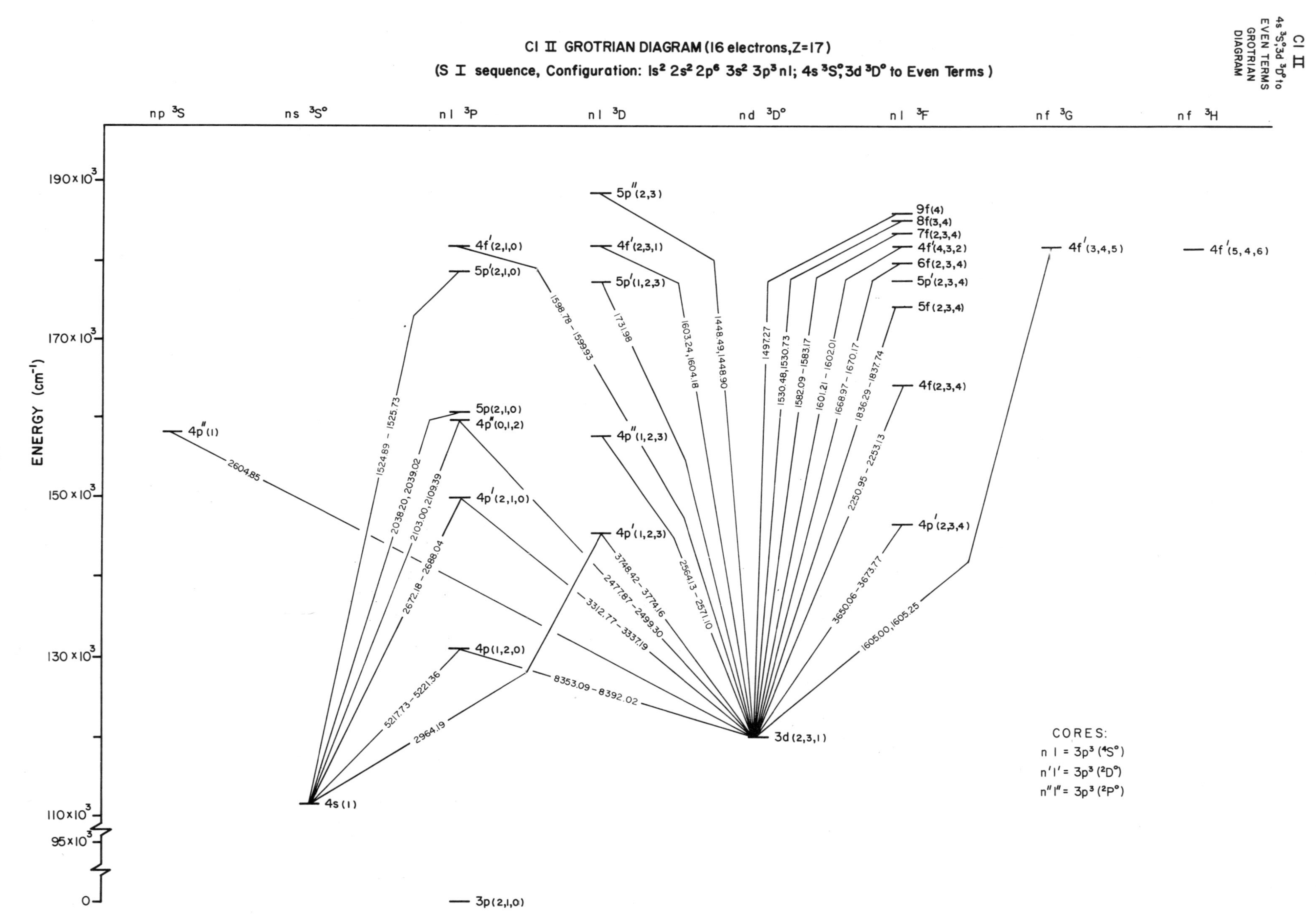
Cl II
4s 3S°,3d 3D° to
EVEN TERMS
GROTRIAN
DIAGRAM
Cl II GROTRIAN DIAGRAM (16 electrons, Z=17)
(S I sequence, Configuration: 1s2 2s2 2p6 3s2 3p3 nl; 4s 3S°, 3d 3D° to Even Terms)
np 3S
ns 3S°
nl 3P
nl 3D
nd 3D°
nl 3F
nf 3G
nf 3H
ENERGY (cm-1)
190 x 10^3
170 x 10^3
150 x 10^3
130 x 10^3
110 x 10^3
95 x 10^3
0
4p''(1)
4s(1)
3p(2,1,0)
4f'(2,1,0)
5p'(2,1,0)
5p(2,1,0)
4p''(0,1,2)
4p'(2,1,0)
4p(1,2,0)
5p''(2,3)
4f'(2,3,1)
5p'(1,2,3)
4p''(1,2,3)
4p'(1,2,3)
3d(2,3,1)
9f(4)
8f(3,4)
7f(2,3,4)
4f'(4,3,2)
6f(2,3,4)
5p'(2,3,4)
5f(2,3,4)
4f(2,3,4)
4p'(2,3,4)
4f'(3,4,5)
4f'(5,4,6)
2604.85
1524.89 - 1525.73
2038.20, 2039.02
2103.00, 2109.39
2672.18 - 2688.04
5217.73 - 5221.36
2964.19
1598.78 - 1599.93
1731.98
1603.24, 1604.18
1448.49, 1448.90
1497.27
1530.48, 1530.73
1582.09 - 1583.17
1601.21 - 1602.01
1668.97 - 1670.17
1836.29 - 1837.74
2250.95 - 2253.13
2564.13 - 2571.10
3748.42 - 3774.16
2477.87 - 2499.30
3312.77 - 3337.19
8353.09 - 8392.02
3650.06 - 3673.77
1605.00, 1605.25
CORES:
n l = 3p3 (4S°)
n'l' = 3p3 (2D°)
n''l'' = 3p3 (2P°)

Cl II ENERGY LEVELS (16 electrons, Z = 17)

(S I sequence, Configuration: $1s^2 2s^2 2p^6 3s^2 3p^3 nl$, Triplet System, Even Terms)

ENERGY (cm^{-1}): 190 x 10^3, 170 x 10^3, 150 x 10^3, 130 x 10^3, 110 x 10^3, 95 x 10^3, 0

np ^3S

158 178.78 — 4p″(1)

nl ^3P

182 355.10, 182 347.76, 182 313.51 — 4f′(2,1,0)

178 187.89, 178 177.36, 178 152.00 — 5p′(2,1,0)

161 672.62, 161 656.59, 161 636.89 — 5p(2,1,0)

160 145.28, 160 001.42, 159 842.33 — 4p″(0,1,2)

150 020.90, 149 954.18, 149 800.12 — 4p′(2,1,0)

131 769.41, 131 768.66, 131 756.12 — 4p(1,2,0)

996.47, 696.00, 0.00 — $3p^4$(2,1,0)

nl ^3D

188 848.39?, 188 818.08 — 5p″(2,3)

182 316.29, 182 185.10, 182 180.42 — 4f′(2,3,1)

177 548.75, 177 479.12, 177 444.74 — 5p′(1,2,3)

158 788.25, 158 770.58, 158 725.56 — 4p″(1,2,3)

146 470.64, 146 335.55, 146 331.76 — 4p′(1,2,3)

nl ^3F

186 599.38 — 9f(4)

185 139.39, 185 139.37 — 8f(3,4)

183 008.06, 183 007.86, 183 007.83 — 7f(2,3,4)

182 273.29, 182 253.22, 182 232.74 — 4f′(4,3,2)

179 717.80, 179 717.47, 179 717.39 — 6f(2,3,4)

177 730.59, 177 681.90, 177 638.32 — 5p′(2,3,4)

174 258.36, 174 257.94, 174 257.83 — 5f(2,3,4)

164 212.74, 164 212.29, 164 212.19 — 4f(2,3,4)

147 200.24, 147 127.57, 147 055.61 — 4p′(2,3,4)

nf ^3G

182 225.75, 182 116.52, 182 095.96 — 4f′(3,4,5)

nf ^3H

182 258.95, 182 199.52, 182 191.60 — 4f′(5,4,6)

CORES:
n l = $3p^3$ ($^4S°$)
n′l′ = $3p^3$ ($^2D°$)
n″l″ = $3p^3$ ($^2P°$)

KEY

147 200.24, 147 127.57, 147 055.61 — 4p′(2,3,4)
- Core excitation state
- J, lowest to highest
- n
- Energy Levels (cm^{-1})

Ionization Limit
192 070 ± 1 cm^{-1}
(23.8139 ± 0.0002 eV)
[Cl II $3p^4\ ^3P_2$ → Cl III $3p^3\ ^4S°_{3/2}$]

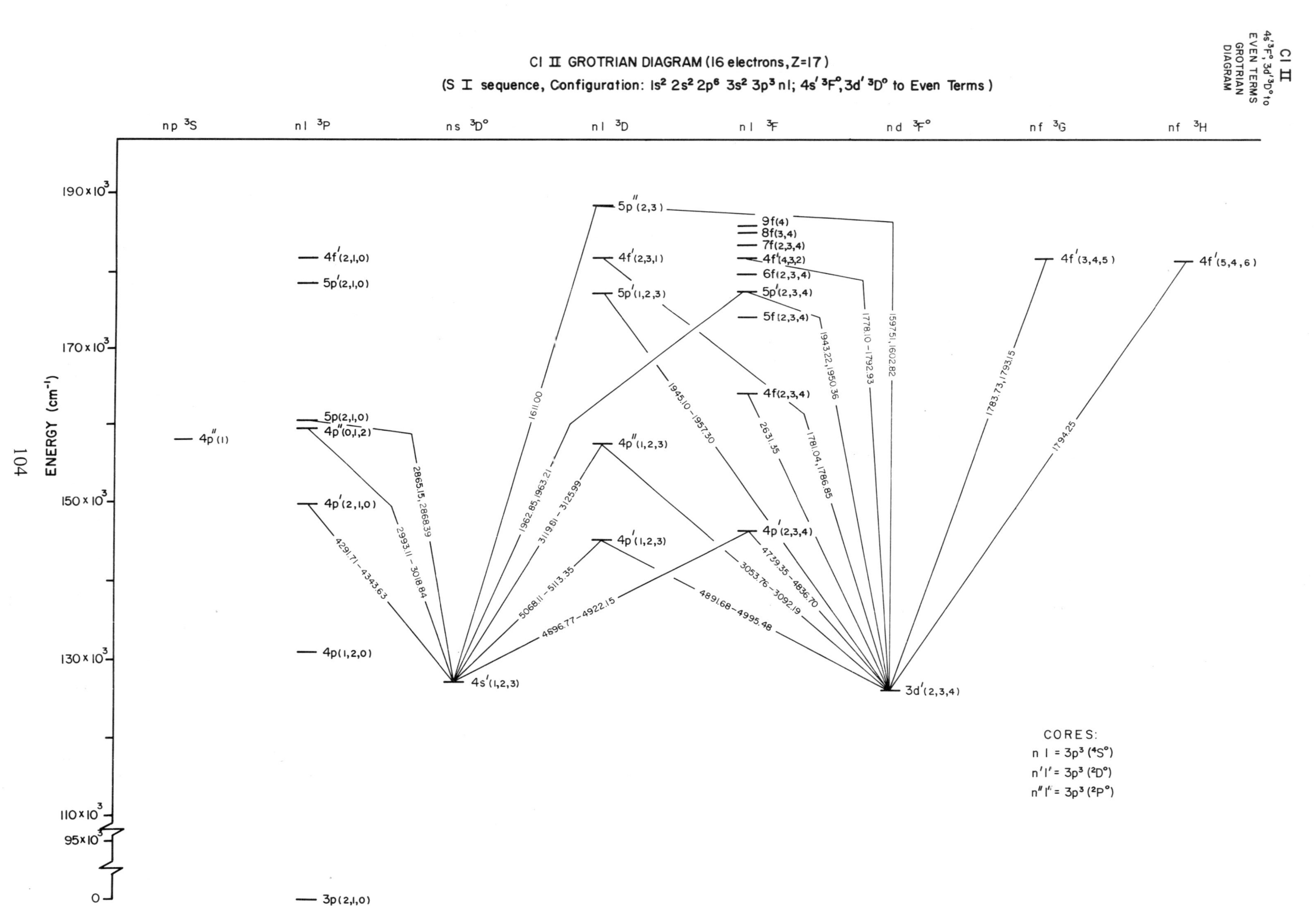

Cl II
4s' ³F°, 3d' ³D° to
EVEN TERMS
GROTRIAN
DIAGRAM
Cl II GROTRIAN DIAGRAM (16 electrons, Z=17)
(S I sequence, Configuration: 1s² 2s² 2p⁶ 3s² 3p³ nl; 4s' ³F°, 3d' ³D° to Even Terms)
np ³S
nl ³P
ns ³D°
nl ³D
nl ³F
nd ³F°
nf ³G
nf ³H
190 x 10³
170 x 10³
150 x 10³
130 x 10³
110 x 10³
95 x 10³
0
ENERGY (cm⁻¹)
5p''(2,3)
9f(4)
8f(3,4)
7f(2,3,4)
4f'(2,1,0)
4f'(2,3,1)
4f'(4,3,2)
4f'(3,4,5)
4f'(5,4,6)
6f(2,3,4)
5p'(2,1,0)
5p'(1,2,3)
5p'(2,3,4)
5f(2,3,4)
4f(2,3,4)
5p(2,1,0)
4p''(0,1,2)
4p''(1)
4p''(1,2,3)
4p'(2,1,0)
4p'(2,3,4)
4p'(1,2,3)
4p(1,2,0)
4s'(1,2,3)
3d'(2,3,4)
3p(2,1,0)
1611.00
1962.85, 1963.21
3119.81 - 3125.99
1945.10 - 1957.30
2631.35
1781.04, 1786.85
1943.22, 1950.36
1778.10 - 1792.93
1597.51, 1602.82
1783.73, 1793.15
1794.25
2865.15, 2868.39
2993.11 - 3018.84
4291.71 - 4343.63
5068.11 - 5113.35
4896.77 - 4922.15
4891.68 - 4995.48
3053.76 - 3092.19
4739.35 - 4836.70
CORES:
n l = 3p³ (⁴S°)
n'l' = 3p³ (²D°)
n''l'' = 3p³ (²P°)

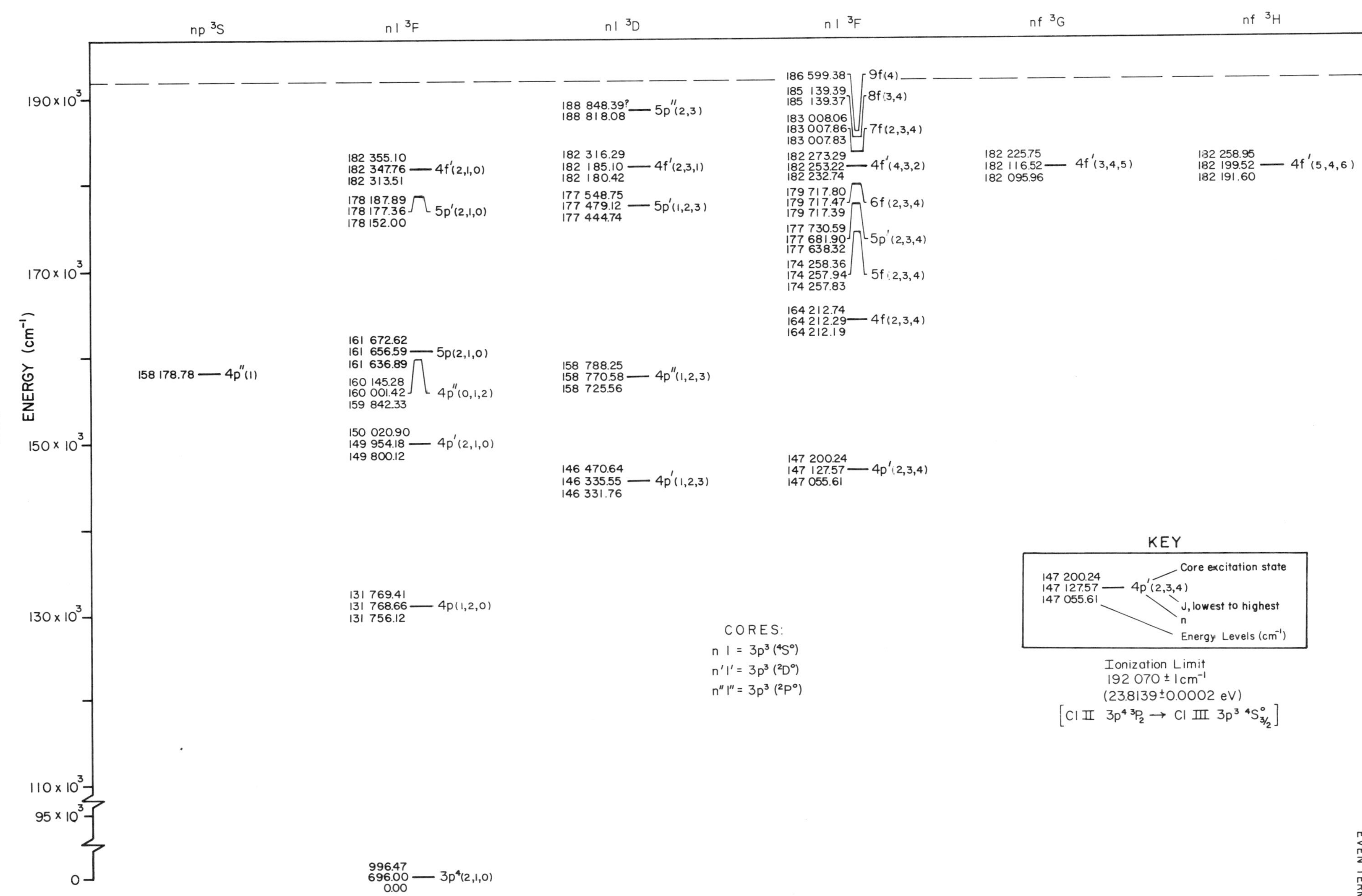
Cl II ENERGY LEVELS (16 electrons, Z= 17)
(S I sequence, Configuration: 1s² 2s² 2p⁶ 3s² 3p³ nl, Triplet System, Even Terms)
np ³S
nl ³F
nl ³D
nl ³F
nf ³G
nf ³H
190 x 10³
170 x 10³
150 x 10³
130 x 10³
110 x 10³
95 x 10³
0
ENERGY (cm⁻¹)
158 178.78 — 4p″(1)
182 355.10
182 347.76 — 4f′(2,1,0)
182 313.51
178 187.89
178 177.36 — 5p′(2,1,0)
178 152.00
161 672.62
161 656.59 — 5p(2,1,0)
161 636.89
160 145.28
160 001.42 — 4p″(0,1,2)
159 842.33
150 020.90
149 954.18 — 4p′(2,1,0)
149 800.12
131 769.41
131 768.66 — 4p(1,2,0)
131 756.12
996.47
696.00 — 3p⁴(2,1,0)
0.00
188 848.39?
188 818.08 — 5p″(2,3)
182 316.29
182 185.10 — 4f′(2,3,1)
182 180.42
177 548.75
177 479.12 — 5p′(1,2,3)
177 444.74
158 788.25
158 770.58 — 4p″(1,2,3)
158 725.56
146 470.64
146 335.55 — 4p′(1,2,3)
146 331.76
186 599.38 — 9f(4)
185 139.39
185 139.37 — 8f(3,4)
183 008.06
183 007.86 — 7f(2,3,4)
183 007.83
182 273.29
182 253.22 — 4f′(4,3,2)
182 232.74
179 717.80
179 717.47 — 6f(2,3,4)
179 717.39
177 730.59
177 681.90 — 5p′(2,3,4)
177 638.32
174 258.36
174 257.94 — 5f(2,3,4)
174 257.83
164 212.74
164 212.29 — 4f(2,3,4)
164 212.19
147 200.24
147 127.57 — 4p′(2,3,4)
147 055.61
182 225.75
182 116.52 — 4f′(3,4,5)
182 095.96
182 258.95
182 199.52 — 4f′(5,4,6)
182 191.60
CORES:
n l = 3p³ (⁴S°)
n′l′ = 3p³ (²D°)
n″l″ = 3p³ (²P°)
KEY
147 200.24
147 127.57 — 4p′(2,3,4)
147 055.61
Core excitation state
J, lowest to highest
n
Energy Levels (cm⁻¹)
Ionization Limit
192 070 ± 1cm⁻¹
(23.8139±0.0002 eV)
[Cl II 3p⁴ ³P₂ → Cl III 3p³ ⁴S°₃/₂]

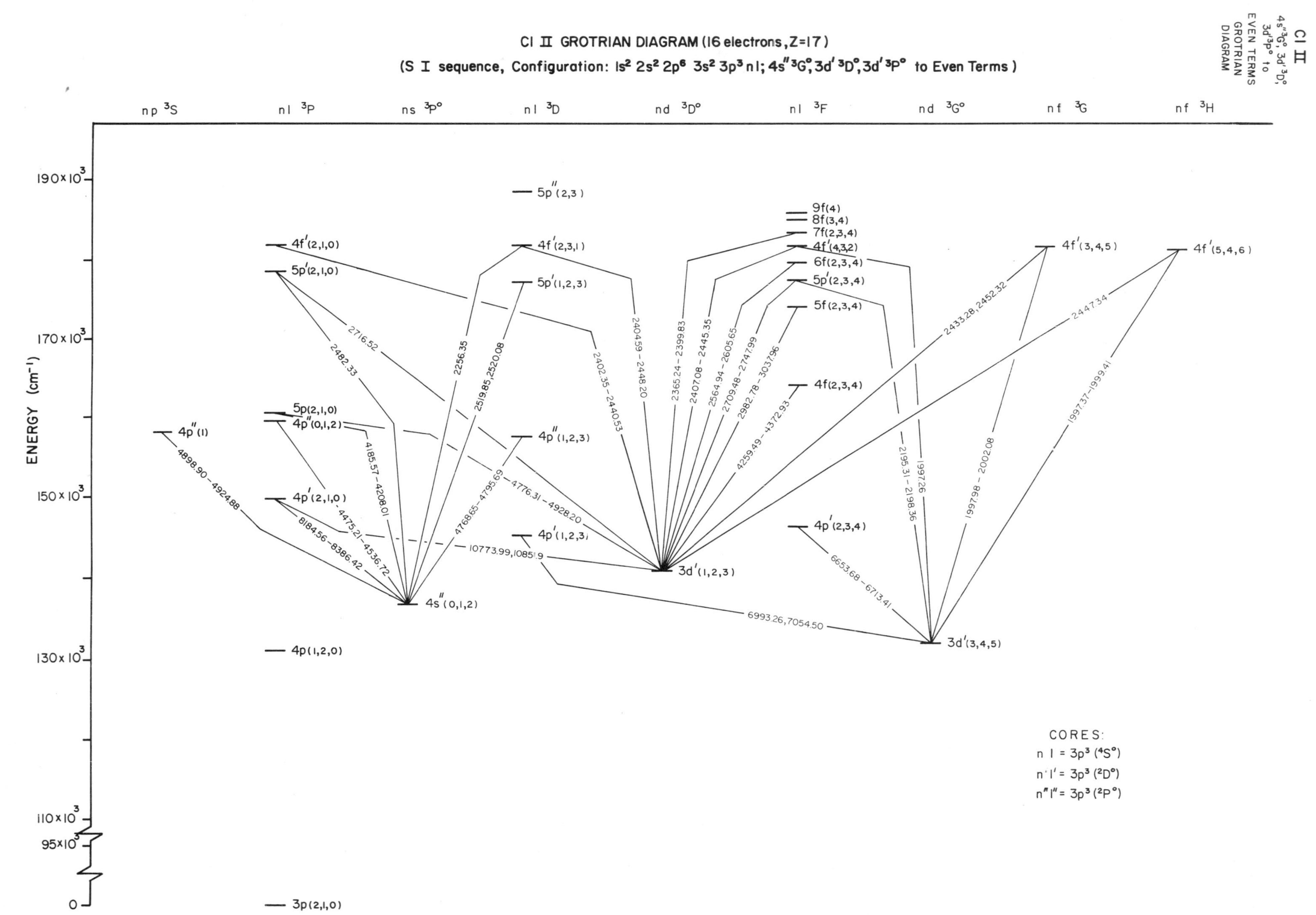
Cl II
4s″³G°, 3d′³D°, 3d′³P° to EVEN TERMS GROTRIAN DIAGRAM
Cl II GROTRIAN DIAGRAM (16 electrons, Z=17)
(S I sequence, Configuration: 1s² 2s² 2p⁶ 3s² 3p³ nl; 4s″³G°, 3d′³D°, 3d′³P° to Even Terms)
np ³S
nl ³P
ns ³P°
nl ³D
nd ³D°
nl ³F
nd ³G°
nf ³G
nf ³H
ENERGY (cm⁻¹)
190×10³
170×10³
150×10³
130×10³
110×10³
95×10³
0
5p″(2,3)
4f′(2,1,0)
5p′(2,1,0)
4f′(2,3,1)
5p′(1,2,3)
9f(4)
8f(3,4)
7f(2,3,4)
4f′(4,3,2)
6f(2,3,4)
5p′(2,3,4)
5f(2,3,4)
4f′(3,4,5)
4f′(5,4,6)
4f(2,3,4)
5p(2,1,0)
4p″(0,1,2)
4p″(1)
4p″(1,2,3)
4p′(2,1,0)
4p′(1,2,3)
4p′(2,3,4)
3d′(1,2,3)
4s″(0,1,2)
3d′(3,4,5)
4p(1,2,0)
3p(2,1,0)
2716.52
2482.33
2256.35
2519.85, 2520.08
2402.35 – 2440.53
2404.59 – 2448.20
2365.24 – 2399.83
2407.08 – 2445.35
2564.94 – 2605.65
2709.48 – 2747.99
2982.78 – 3037.96
4259.49 – 4372.93
2433.28, 2452.32
2447.34
1997.37 – 1999.41
1997.98 – 2002.08
1997.26
2195.31 – 2198.36
4185.57 – 4208.01
4898.90 – 4924.88
4768.65 – 4795.69
4776.31 – 4928.20
4475.21 – 4536.72
8184.56 – 8386.42
10773.99, 10851.9
6653.68 – 6713.41
6993.26, 7054.50
CORES:
n l = 3p³ (⁴S°)
n′l′ = 3p³ (²D°)
n″l″ = 3p³ (²P°)

Cl II ENERGY LEVELS (16 electrons, Z = 17)

(S I sequence, Configuration: $1s^2\,2s^2\,2p^6\,3s^2\,3p^3\,nl$, Triplet System, Even Terms)

ENERGY (cm^{-1}) — axis: 0, 95×10^3, 110×10^3, 130×10^3, 150×10^3, 170×10^3, 190×10^3

np ^3S

158 178.78 — $4p''(1)$

nl ^3P

182 355.10, 182 347.76, 182 313.51 — $4f'(2,1,0)$

178 187.89, 178 177.36, 178 152.00 — $5p'(2,1,0)$

161 672.62, 161 656.59, 161 636.89 — $5p(2,1,0)$

160 145.28, 160 001.42, 159 842.33 — $4p''(0,1,2)$

150 020.90, 149 954.18, 149 800.12 — $4p'(2,1,0)$

131 769.41, 131 768.66, 131 756.12 — $4p(1,2,0)$

996.47, 696.00, 0.00 — $3p^4(2,1,0)$

nl ^3D

188 848.39?, 188 818.08 — $5p''(2,3)$

182 316.29, 182 185.10, 182 180.42 — $4f'(2,3,1)$

177 548.75, 177 479.12, 177 444.74 — $5p'(1,2,3)$

158 788.25, 158 770.58, 158 725.56 — $4p''(1,2,3)$

146 470.64, 146 335.55, 146 331.76 — $4p'(1,2,3)$

nl ^3F

186 599.38 — $9f(4)$

185 139.39, 185 139.37 — $8f(3,4)$

183 008.06, 183 007.86, 183 007.83 — $7f(2,3,4)$

182 273.29, 182 253.22, 182 232.74 — $4f'(4,3,2)$

179 717.80, 179 717.47, 179 717.39 — $6f(2,3,4)$

177 730.59, 177 681.90, 177 638.32 — $5p'(2,3,4)$

174 258.36, 174 257.94, 174 257.83 — $5f(2,3,4)$

164 212.74, 164 212.29, 164 212.19 — $4f(2,3,4)$

147 200.24, 147 127.57, 147 055.61 — $4p'(2,3,4)$

nf ^3G

182 225.75, 182 116.52, 182 095.96 — $4f'(3,4,5)$

nf ^3H

182 258.95, 182 199.52, 182 191.60 — $4f'(5,4,6)$

CORES:
n l = $3p^3\,(^4S^\circ)$
n′l′ = $3p^3\,(^2D^\circ)$
n″l″ = $3p^3\,(^2P^\circ)$

KEY

147 200.24, 147 127.57, 147 055.61 — $4p'(2,3,4)$

Core excitation state
J, lowest to highest
n
Energy Levels (cm^{-1})

Ionization Limit
192 070 ± 1 cm^{-1}
(23.8139 ± 0.0002 eV)
[Cl II $3p^4\,{}^3P_2$ → Cl III $3p^3\,{}^4S^\circ_{3/2}$]

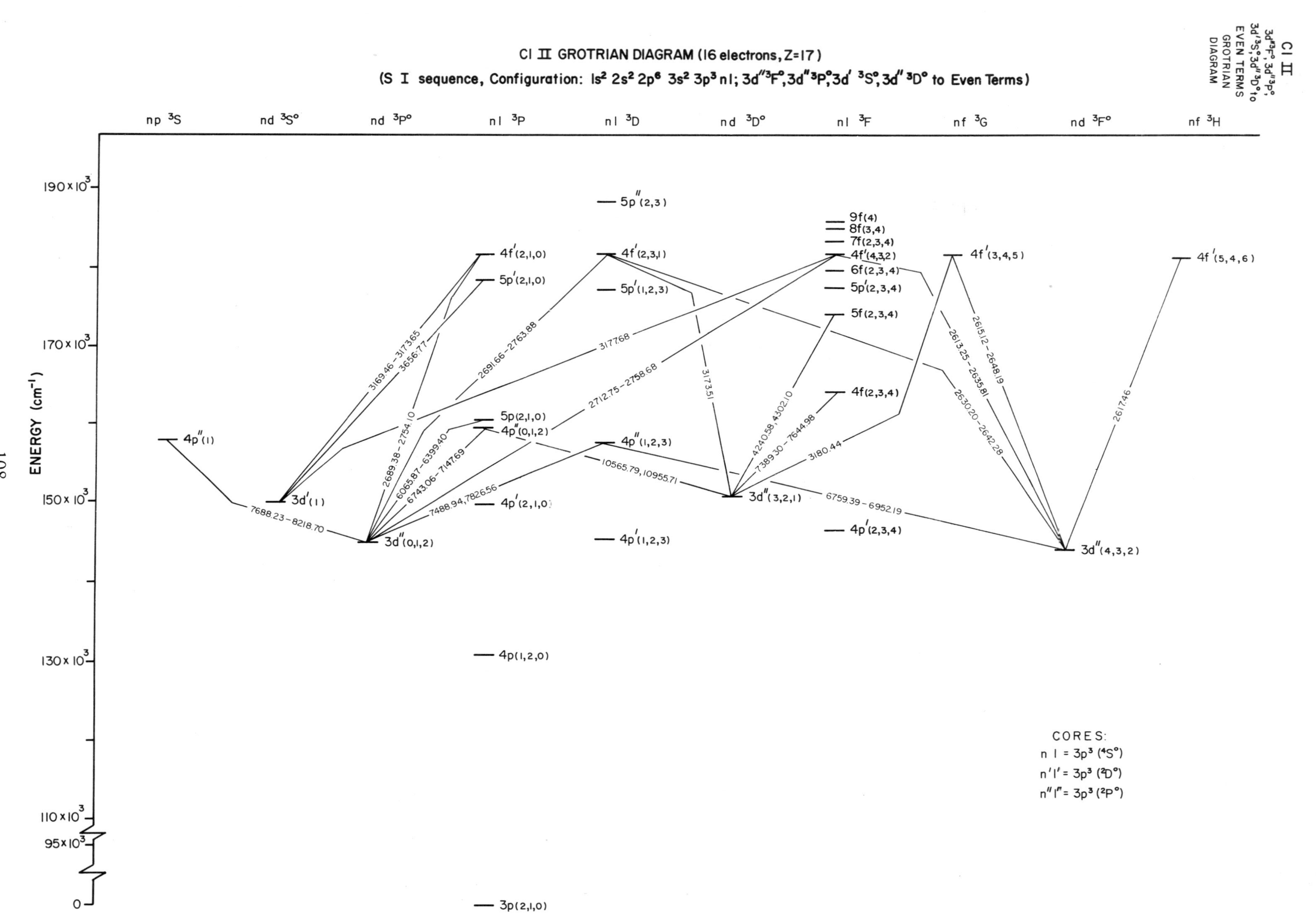

Cl II
3d''3F°, 3d''3P°, 3d'3S°, 3d''3D° to EVEN TERMS GROTRIAN DIAGRAM
Cl II GROTRIAN DIAGRAM (16 electrons, Z=17)
(S I sequence, Configuration: 1s2 2s2 2p6 3s2 3p3 nl; 3d''3F°, 3d''3P°, 3d'3S°, 3d''3D° to Even Terms)
np 3S
nd 3S°
nd 3P°
nl 3P
nl 3D
nd 3D°
nl 3F
nf 3G
nd 3F°
nf 3H
190 x 10^3
170 x 10^3
150 x 10^3
130 x 10^3
110 x 10^3
95 x 10^3
0
ENERGY (cm-1)
5p''(2,3)
9f(4)
8f(3,4)
7f(2,3,4)
4f'(2,1,0)
4f'(2,3,1)
4f'(4,3,2)
4f'(3,4,5)
4f'(5,4,6)
5p'(2,1,0)
6f(2,3,4)
5p'(1,2,3)
5p'(2,3,4)
5f(2,3,4)
4f(2,3,4)
5p(2,1,0)
4p''(0,1,2)
4p''(1)
4p''(1,2,3)
3d'(1)
3d''(3,2,1)
4p'(2,1,0)
3d''(0,1,2)
4p'(1,2,3)
4p'(2,3,4)
3d''(4,3,2)
4p(1,2,0)
3p(2,1,0)
3169.46 - 3173.65
3656.77
2689.38 - 2754.10
6065.87 - 6399.40
6743.06 - 7147.69
7488.94, 7826.56
2691.66 - 2763.88
3177.68
2712.75 - 2758.68
3173.51
10565.79, 10955.71
4240.58, 4302.10
7389.30 - 7644.98
3180.44
6759.39 - 6952.19
2615.12 - 2648.19
2613.25 - 2635.81
2630.20 - 2642.28
2617.46
7688.23 - 8218.70
CORES:
n l = 3p3 (4S°)
n'l' = 3p3 (2D°)
n''l'' = 3p3 (2P°)

Cl II ENERGY LEVELS (16 electrons, Z= 17)

(S I sequence, Configuration: $1s^2\,2s^2\,2p^6\,3s^2\,3p^3\,nl$, Triplet System, Even Terms)

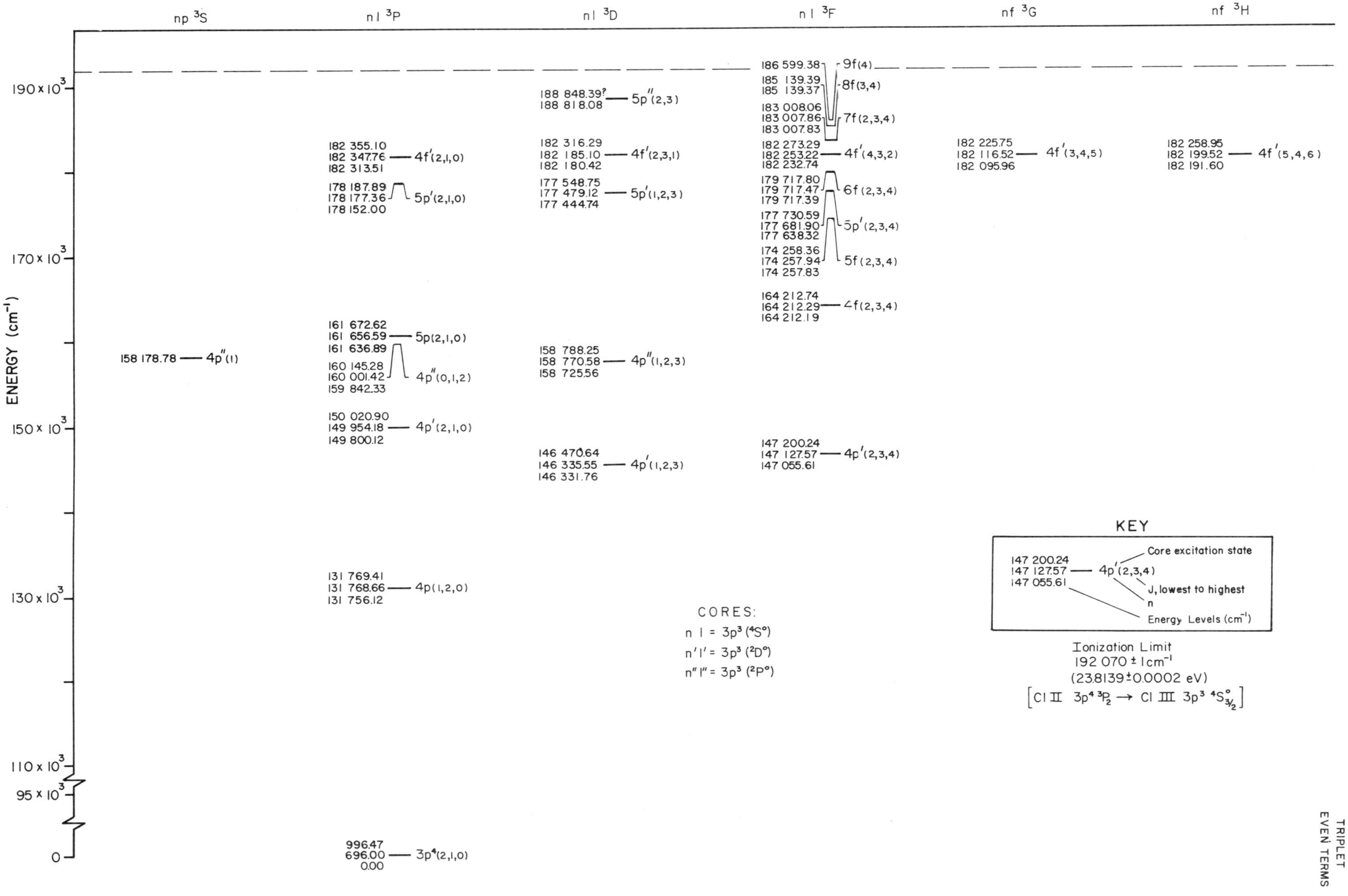

ENERGY (cm^{-1})

Axis: 0, 95 x 10^3, 110 x 10^3, 130 x 10^3, 150 x 10^3, 170 x 10^3, 190 x 10^3

np ^3S

158 178.78 — 4p″(1)

nl ^3P

182 355.10, 182 347.76, 182 313.51 — 4f′(2,1,0)

178 187.89, 178 177.36, 178 152.00 — 5p′(2,1,0)

161 672.62, 161 656.59, 161 636.89 — 5p(2,1,0)

160 145.28, 160 001.42, 159 842.33 — 4p″(0,1,2)

150 020.90, 149 954.18, 149 800.12 — 4p′(2,1,0)

131 769.41, 131 768.66, 131 756.12 — 4p(1,2,0)

996.47, 696.00, 0.00 — $3p^4$(2,1,0)

nl ^3D

188 848.39?, 188 818.08 — 5p″(2,3)

182 316.29, 182 185.10, 182 180.42 — 4f′(2,3,1)

177 548.75, 177 479.12, 177 444.74 — 5p′(1,2,3)

158 788.25, 158 770.58, 158 725.56 — 4p″(1,2,3)

146 470.64, 146 335.55, 146 331.76 — 4p′(1,2,3)

nl ^3F

186 599.38 — 9f(4)

185 139.39, 185 139.37 — 8f(3,4)

183 008.06, 183 007.86, 183 007.83 — 7f(2,3,4)

182 273.29, 182 253.22, 182 232.74 — 4f′(4,3,2)

179 717.80, 179 717.47, 179 717.39 — 6f(2,3,4)

177 730.59, 177 681.90, 177 638.32 — 5p′(2,3,4)

174 258.36, 174 257.94, 174 257.83 — 5f(2,3,4)

164 212.74, 164 212.29, 164 212.19 — 4f(2,3,4)

147 200.24, 147 127.57, 147 055.61 — 4p′(2,3,4)

nf ^3G

182 225.75, 182 116.52, 182 095.96 — 4f′(3,4,5)

nf ^3H

182 258.95, 182 199.52, 182 191.60 — 4f′(5,4,6)

CORES:

n l = $3p^3\,(^4S°)$

n′l′ = $3p^3\,(^2D°)$

n″l″ = $3p^3\,(^2P°)$

KEY

147 200.24, 147 127.57, 147 055.61 — 4p′(2,3,4)

Core excitation state; J, lowest to highest; n; Energy Levels (cm^{-1})

Ionization Limit

192 070 ± 1 cm^{-1}

(23.8139±0.0002 eV)

[Cl II $3p^4\,{}^3P_2 \rightarrow$ Cl III $3p^3\,{}^4S°_{3/2}$]

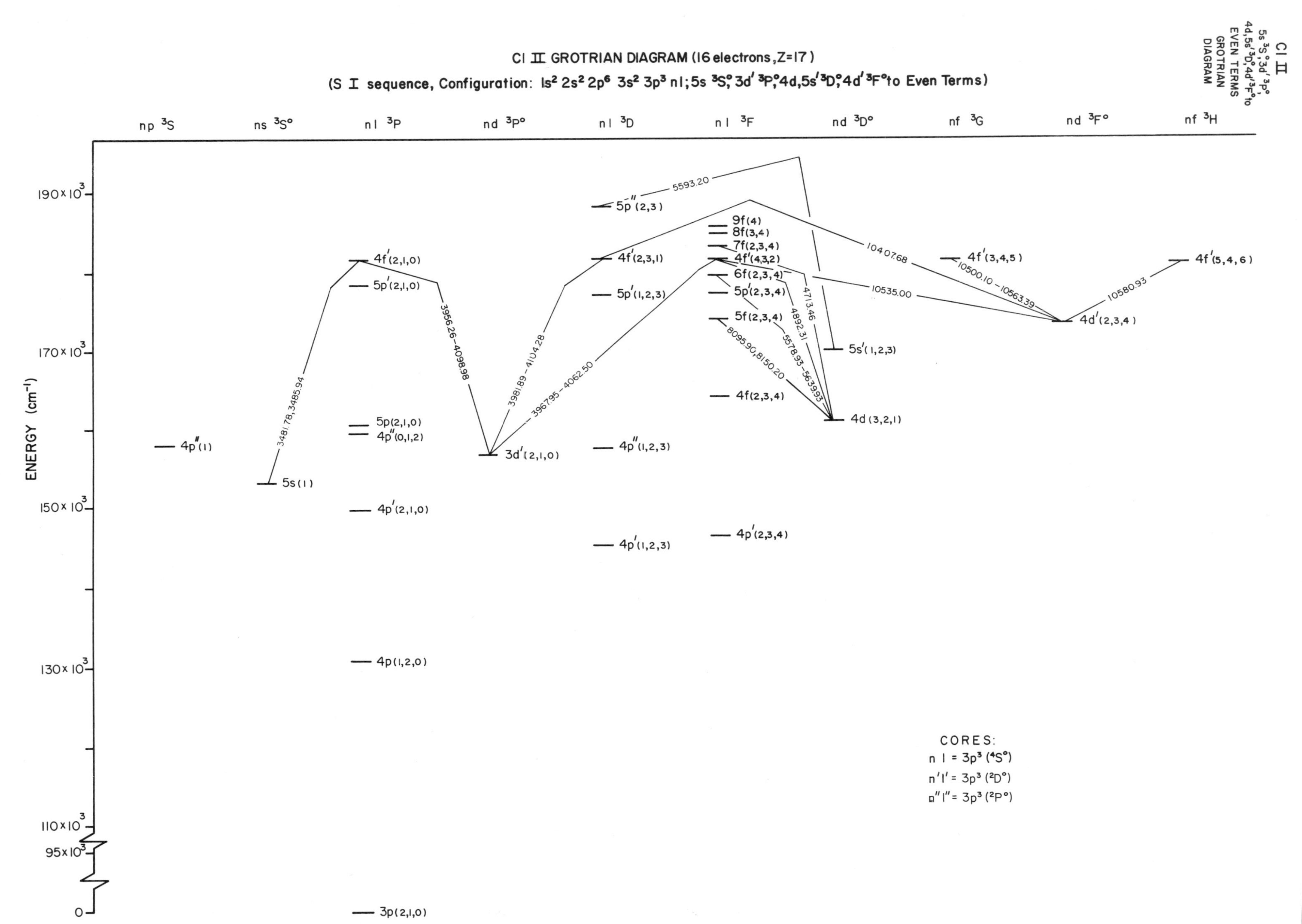

Cl II
5s $^3S^o$, 3d′ $^3P^o$, 4d, 5s′ $^3D^o$, 4d′ $^3F^o$ to
EVEN TERMS
GROTRIAN
DIAGRAM
Cl II GROTRIAN DIAGRAM (16 electrons, Z=17)
(S I sequence, Configuration: $1s^2 2s^2 2p^6 3s^2 3p^3$ nl; 5s $^3S^o$, 3d′ $^3P^o$, 4d, 5s′ $^3D^o$, 4d′ $^3F^o$ to Even Terms)
np ^3S
ns $^3S^o$
nl ^3P
nd $^3P^o$
nl ^3D
nl ^3F
nd $^3D^o$
nf ^3G
nd $^3F^o$
nf ^3H
CORES:
n l = $3p^3$ ($^4S^o$)
n′l′ = $3p^3$ ($^2D^o$)
n″l″ = $3p^3$ ($^2P^o$)
ENERGY (cm^{-1})
190 × 10^3
170 × 10^3
150 × 10^3
130 × 10^3
110 × 10^3
95 × 10^3
0

Cl II ENERGY LEVELS (16 electrons, Z= 17)

(S I sequence, Configuration: $1s^2\,2s^2\,2p^6\,3s^2\,3p^3\,nl$, Triplet System, Even Terms)

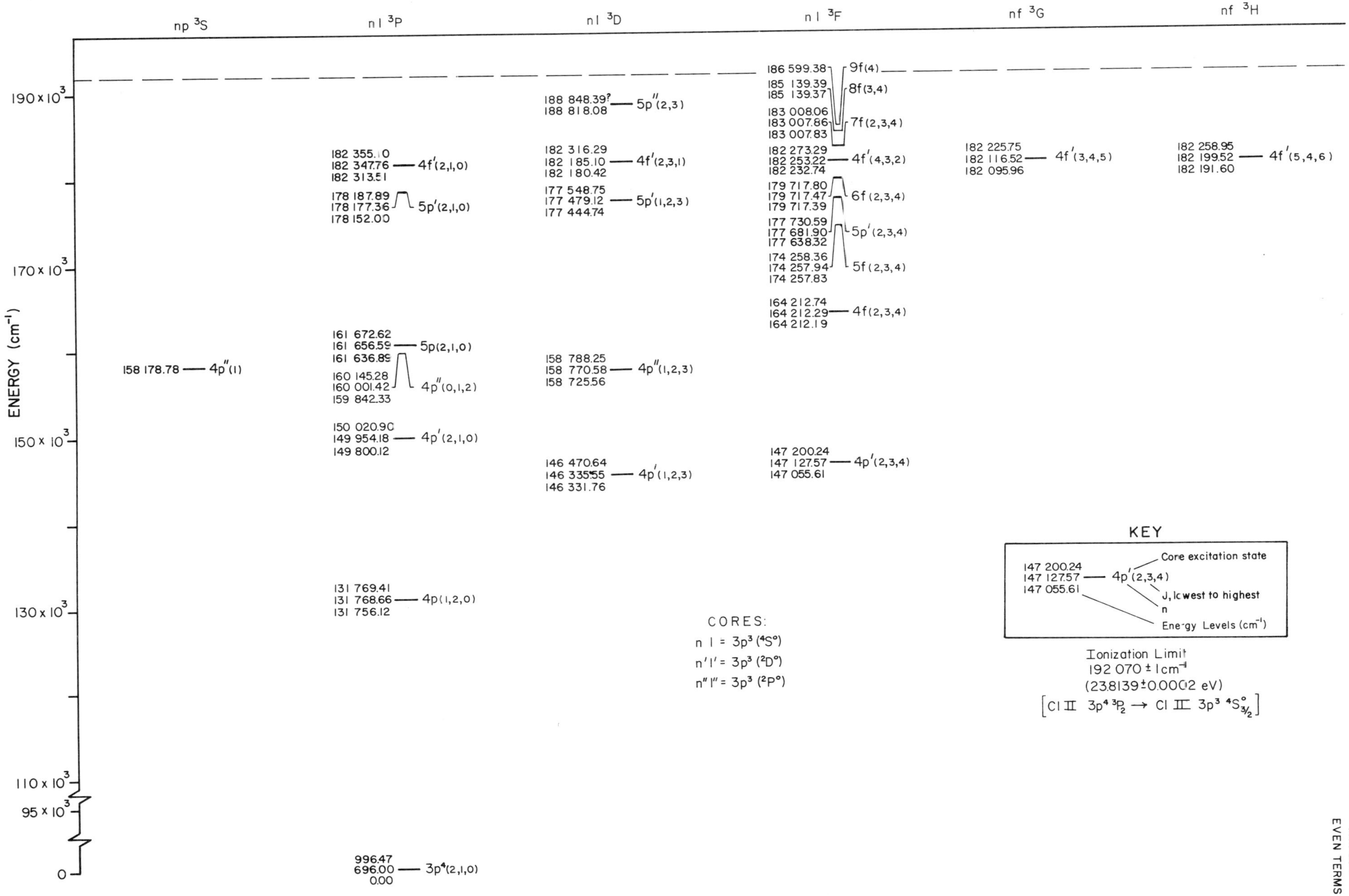

ENERGY (cm^{-1})

190×10^3, 170×10^3, 150×10^3, 130×10^3, 110×10^3, 95×10^3, 0

np ^{3}S

158 178.78 — 4p″(1)

nl ^{3}P

182 355.10
182 347.76 — 4f′(2,1,0)
182 313.51

178 187.89
178 177.36 — 5p′(2,1,0)
178 152.00

161 672.62
161 656.59 — 5p(2,1,0)
161 636.89

160 145.28
160 001.42 — 4p″(0,1,2)
159 842.33

150 020.90
149 954.18 — 4p′(2,1,0)
149 800.12

131 769.41
131 768.66 — 4p(1,2,0)
131 756.12

996.47
696.00 — $3p^4$(2,1,0)
0.00

nl ^{3}D

188 848.39?
188 818.08 — 5p″(2,3)

182 316.29
182 185.10 — 4f′(2,3,1)
182 180.42

177 548.75
177 479.12 — 5p′(1,2,3)
177 444.74

158 788.25
158 770.58 — 4p″(1,2,3)
158 725.56

146 470.64
146 335.55 — 4p′(1,2,3)
146 331.76

nl ^{3}F

186 599.38 — 9f(4)

185 139.39
185 139.37 — 8f(3,4)

183 008.06
183 007.86 — 7f(2,3,4)
183 007.83

182 273.29
182 253.22 — 4f′(4,3,2)
182 232.74

179 717.80
179 717.47 — 6f(2,3,4)
179 717.39

177 730.59
177 681.90 — 5p′(2,3,4)
177 638.32

174 258.36
174 257.94 — 5f(2,3,4)
174 257.83

164 212.74
164 212.29 — 4f(2,3,4)
164 212.19

147 200.24
147 127.57 — 4p′(2,3,4)
147 055.61

nf ^{3}G

182 225.75
182 116.52 — 4f′(3,4,5)
182 095.96

nf ^{3}H

182 258.95
182 199.52 — 4f′(5,4,6)
182 191.60

CORES:
n l = $3p^3$ ($^4S^o$)
n′l′ = $3p^3$ ($^2D^o$)
n″l″ = $3p^3$ ($^2P^o$)

KEY

147 200.24
147 127.57 — 4p′(2,3,4)
147 055.61

4p′: Core excitation state
(2,3,4): J, lowest to highest
4: n
147 200.24, 147 127.57, 147 055.61: Energy Levels (cm^{-1})

Ionization Limit
192 070 ± 1 cm^{-1}
(23.8139 ± 0.0002 eV)
[Cl II $3p^4\,^3P_2$ → Cl III $3p^3\,^4S^o_{3/2}$]

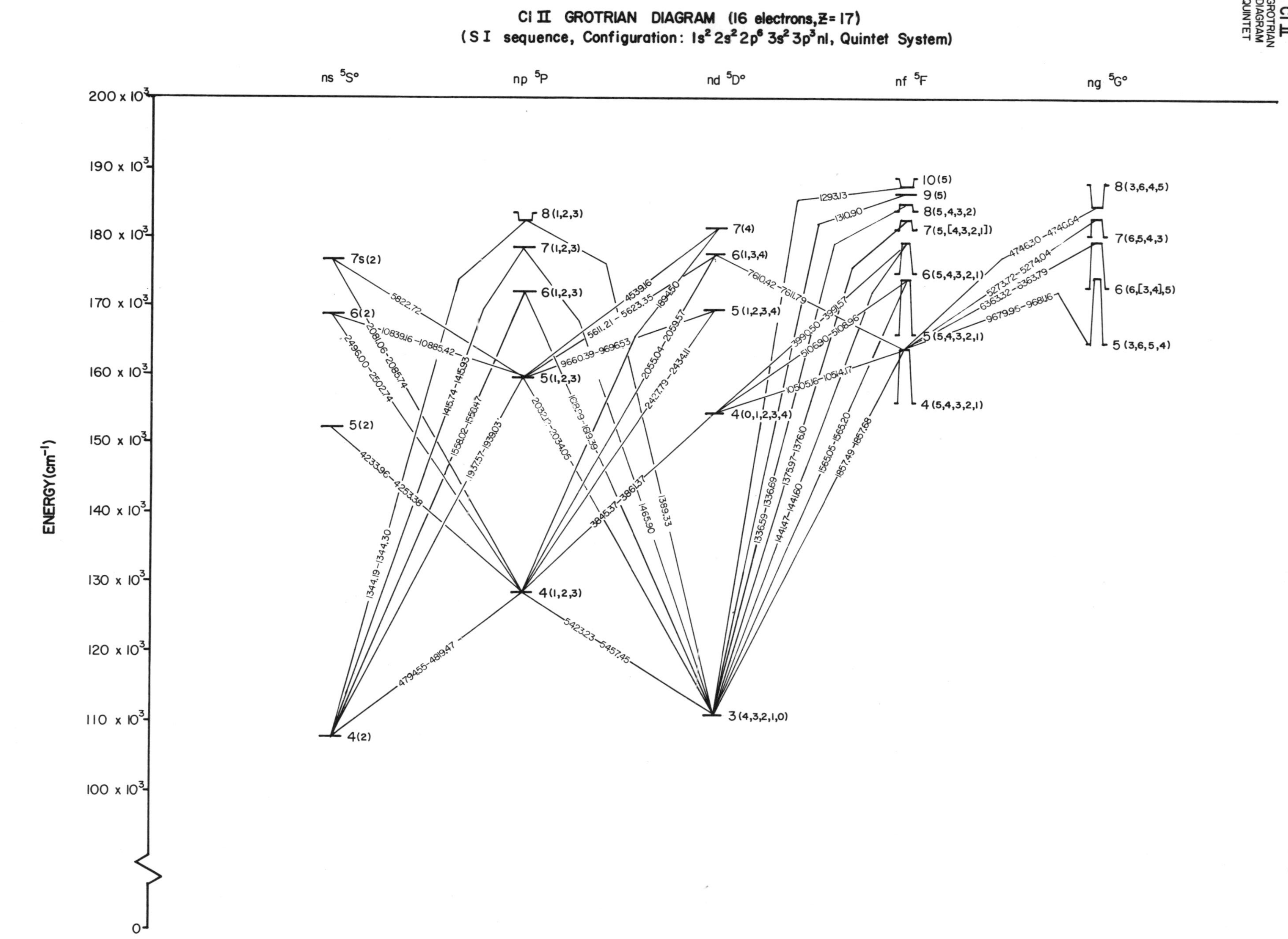
Cl II GROTRIAN DIAGRAM (16 electrons, Z= 17)
(S I sequence, Configuration: 1s^2 2s^2 2p^6 3s^2 3p^3 nl, Quintet System)
ns ^{5}S°
np ^{5}P
nd ^{5}D°
nf ^{5}F
ng ^{5}G°
ENERGY (cm^{-1})
200 x 10^3
190 x 10^3
180 x 10^3
170 x 10^3
160 x 10^3
150 x 10^3
140 x 10^3
130 x 10^3
120 x 10^3
110 x 10^3
100 x 10^3
0
7s(2)
6(2)
5(2)
4(2)
8(1,2,3)
7(1,2,3)
6(1,2,3)
5(1,2,3)
4(1,2,3)
7(4)
6(1,3,4)
5(1,2,3,4)
4(0,1,2,3,4)
3(4,3,2,1,0)
10(5)
9(5)
8(5,4,3,2)
7(5,[4,3,2,1])
6(5,4,3,2,1)
5(5,4,3,2,1)
4(5,4,3,2,1)
8(3,6,4,5)
7(6,5,4,3)
6(6,[3,4],5)
5(3,6,5,4)

Cl II ENERGY LEVELS (16 electrons, Z = 17)

(S I sequence, Configuration: $1s^2 2s^2 2p^6 3s^2 3p^3 nl$, Quintet System)

ENERGY (cm^{-1})

Axis: 200 x 10^3, 190 x 10^3, 180 x 10^3, 170 x 10^3, 160 x 10^3, 150 x 10^3, 140 x 10^3, 130 x 10^3, 120 x 10^3, 110 x 10^3, 100 x 10^3, 0

ns $^5S^o$

Energy Levels	n(J)
176660.14	7(2)
168674.98	6(2)
152234.91	5(2)
107879.66	4(2)

np $^5P^o$

Energy Levels	n(J)
182274.06 182270.12 182267.94	8(1,2,3)
178514.22 178508.59 178504.50	7(1,2,3)
172063.64 172052.10 172045.05	6(1,2,3)
159490.82 159466.66 159451.70	5(1,2,3)
128730.82 128663.57 128622.99	4(1,2,3)

nd $^5D^o$

Energy Levels	n(J)
181515.18	7(4)
177269.25 177268.82 177268.23	6(1,3,4)
169801.79 169800.98 169800.47 169800.20	5(1,2,3,4)
154624.66 154623.57 154620.97 154619.10 154618.09	4(0,1,2,3,4)
110304.52 110303.12 110300.56 110297.72 110296.84	3(4,3,2,1,0)

CORES:

n l = $3p^3(^4S^o)$

n′l′ = $3p^3(^2D^o)$

n″l″ = $3p^3(^2P^o)$

nf ^5F

Energy Levels	n(J)
187628.37	10(5)
186580.60	9(5)
185113.90	8(5,4,3,2)
182973.70 182972.89	7(5,[4,3,2,1])
179671.79 179671.54 179671.18 179670.72 179670.39	6(5,4,3,2,1)
174194.02 174193.79 174193.46 174193.07 174192.65	5(5,4,3,2,1)
164134.61 164134.29 164133.91 164133.48 164133.02	4(5,4,3,2,1)

ng $^5G^o$

Energy Levels	n(J)
185196.25 185196.23 185196.15 185195.95	8(3,6,4,5)
183089.80 183089.74 183089.73 183089.68	7(6,5,4,3)
179843.83 179843.86 179843.75	6(6[3,4]5)
174460.85 174460.84 174460.82 174460.80	5(3,6,5,4)

KEY

110304.52
110303.12
110300.56 — 3(4,3,2,1,0)
110297.72
110296.84

J, lowest to highest
n
Energy Levels (cm^{-1})

Ionization Limit
192 070 ± 1 cm^{-1}
(23.8139 ± 0.0002 eV)
[Cl II $3p^4\,^3P_2$ → Cl III $3p^3\,^4S^o_{3/2}$]

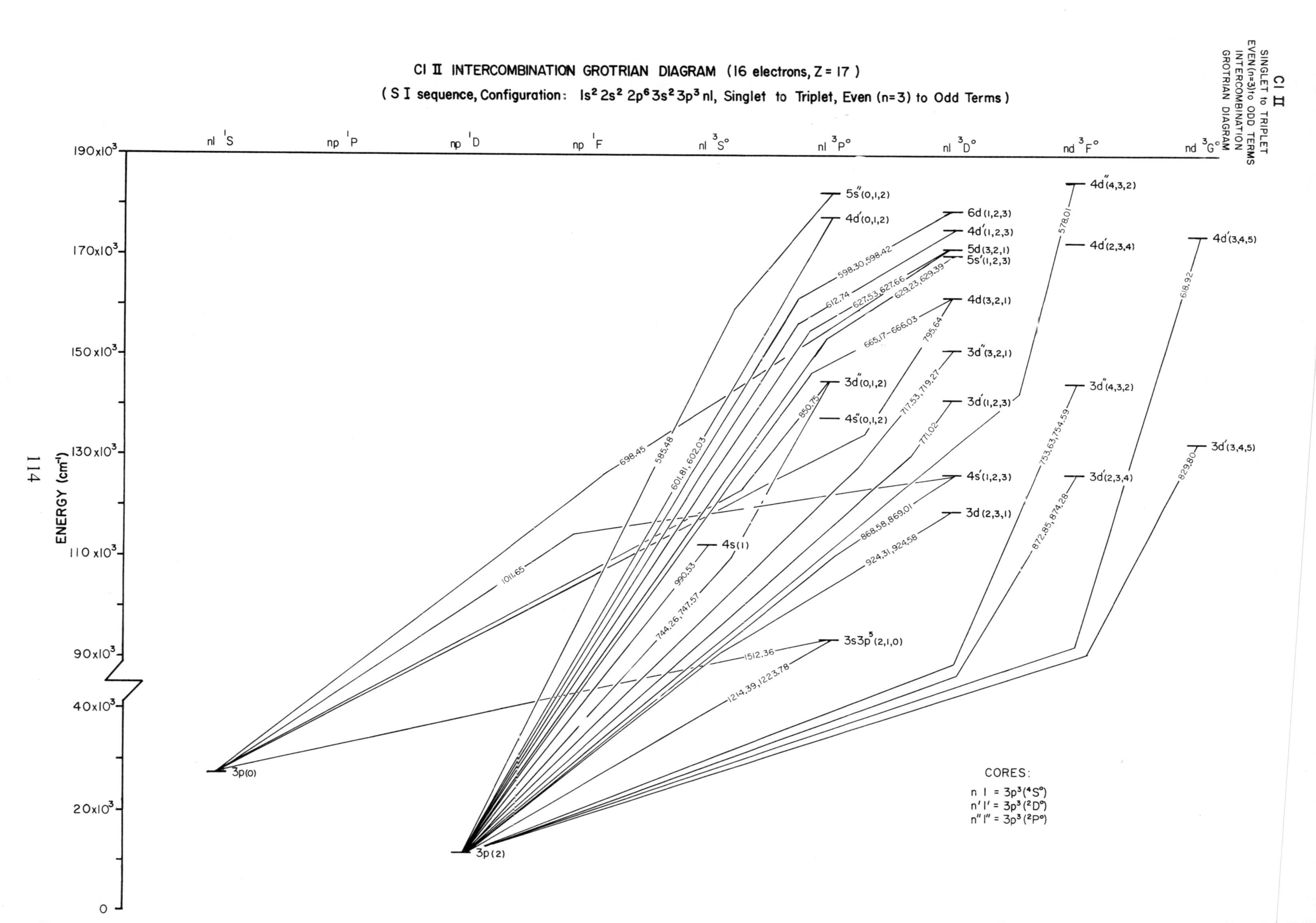
Cl II INTERCOMBINATION GROTRIAN DIAGRAM (16 electrons, Z = 17)
(S I sequence, Configuration: 1s² 2s² 2p⁶ 3s² 3p³ nl, Singlet to Triplet, Even (n=3) to Odd Terms)
Cl II
SINGLET to TRIPLET
EVEN (n=3) to ODD TERMS
INTERCOMBINATION
GROTRIAN DIAGRAM
nl ¹S
np ¹P
np ¹D
np ¹F
nl ³S°
nl ³P°
nl ³D°
nd ³F°
nd ³G°
5s″(0,1,2)
4d′(0,1,2)
3d″(0,1,2)
4s″(0,1,2)
6d(1,2,3)
4d′(1,2,3)
5d(3,2,1)
5s′(1,2,3)
4d(3,2,1)
3d″(3,2,1)
3d′(1,2,3)
4s′(1,2,3)
3d(2,3,1)
4d″(4,3,2)
4d′(2,3,4)
3d″(4,3,2)
3d′(2,3,4)
4d′(3,4,5)
3d′(3,4,5)
4s(1)
3s3p⁵(2,1,0)
3p(0)
3p(2)
598.30,598.42
612.74
627.53,627.66
629.23,629.39
665.17–666.03
795.64
850.75
717.53,719.27
771.02
868.58,869.01
924.31,924.58
578.01
753.63,754.59
872.85,874.28
618.92
829.80
698.45
585.48
601.81,602.03
1011.65
990.53
744.26,747.57
1512.36
1214.39,1223.78
CORES:
n l = 3p³(⁴S°)
n′ l′ = 3p³(²D°)
n″ l″ = 3p³(²P°)
ENERGY (cm⁻¹)
190×10³
170×10³
150×10³
130×10³
110×10³
90×10³
40×10³
20×10³
0

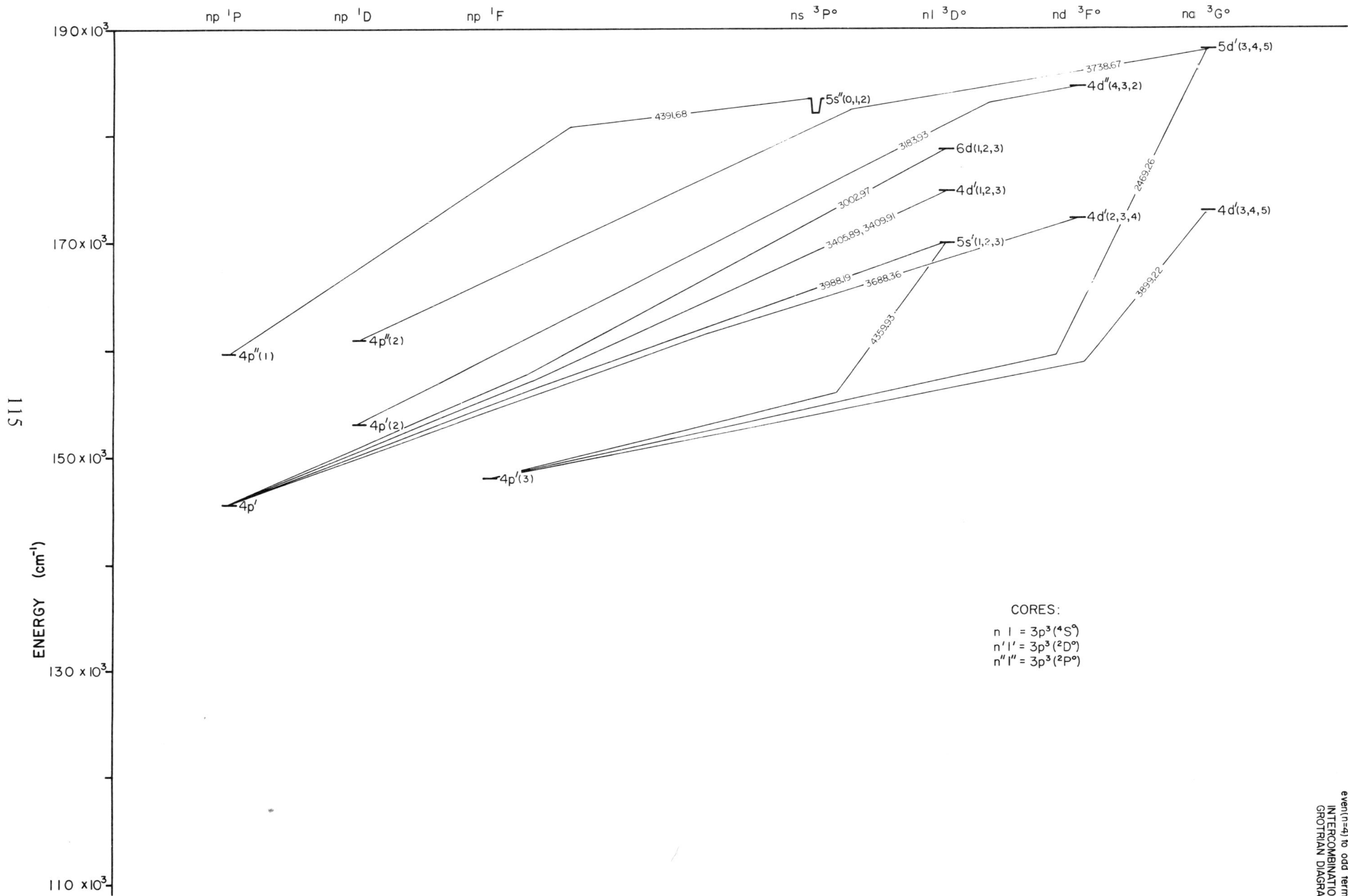
Cl II INTERCOMBINATION GROTRIAN DIAGRAM (16 electrons, Z = 17)
(S I sequence, Configuration: $1s^2 2s^2 2p^6 3s^2 3p^3 nl$, Singlet to Triplet, Even (n = 4) to Odd Terms)
np ^1P
np ^1D
np ^1F
ns $^3P°$
nl $^3D°$
nd $^3F°$
na $^3G°$
190×10^3
170×10^3
150×10^3
130×10^3
110×10^3
ENERGY (cm^{-1})
4p″(1)
4p″(2)
4p′(2)
4p′
4p′(3)
5s″(0,1,2)
5d′(3,4,5)
4d″(4,3,2)
6d(1,2,3)
4d′(1,2,3)
4d′(2,3,4)
5s′(1,2,3)
4d′(3,4,5)
4391.68
3738.67
3183.93
3002.97
3405.89, 3409.91
3988.19
3688.36
4359.93
2469.26
3899.22
CORES:
n l = $3p^3(^4S°)$
n′l′ = $3p^3(^2D°)$
n″l″ = $3p^3(^2P°)$
Cl II
SINGLET to TRIPLET
even(n=4) to odd terms
INTERCOMBINATION
GROTRIAN DIAGRAM

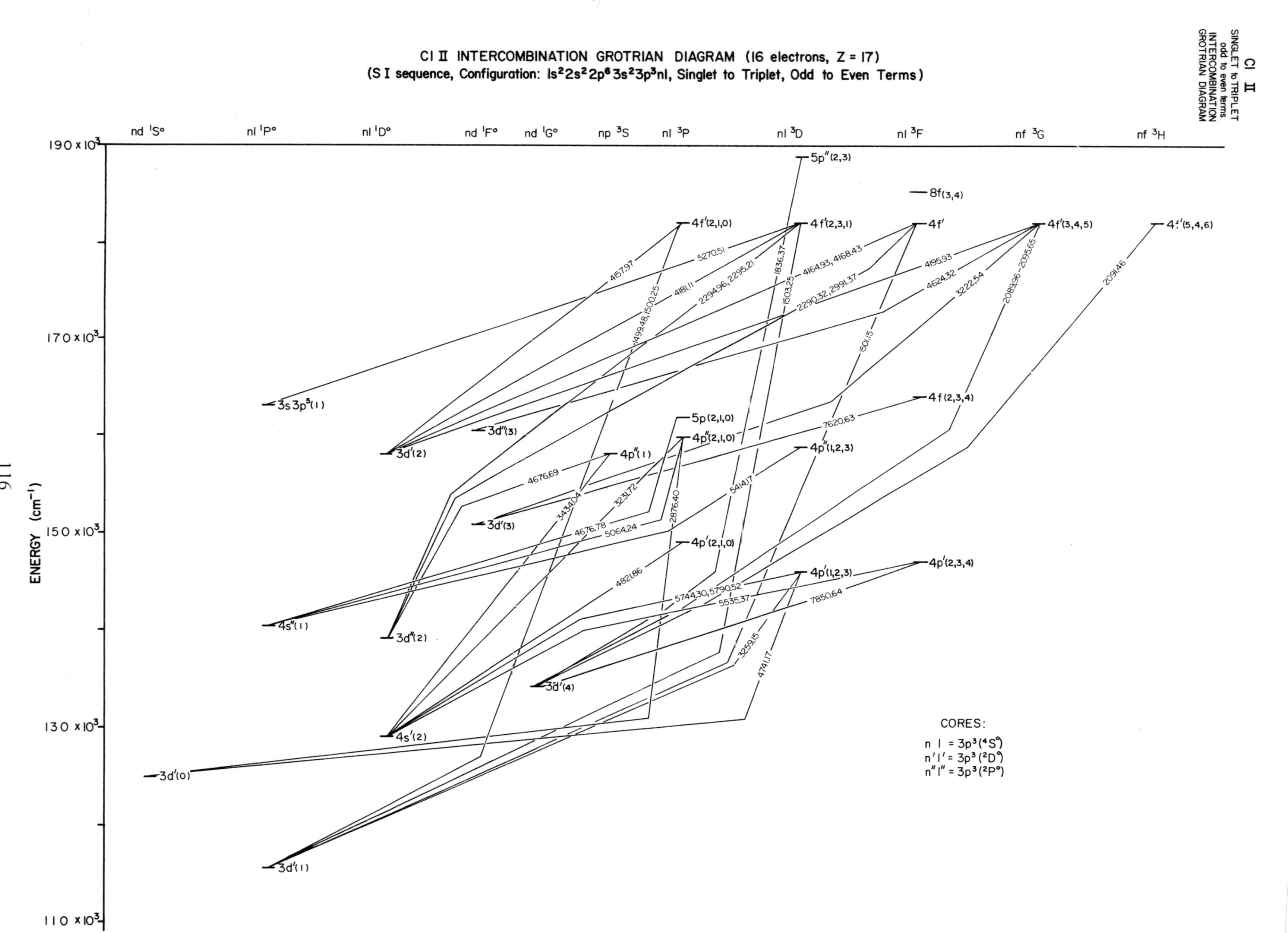

Cl II
SINGLET to TRIPLET
odd to even terms
INTERCOMBINATION
GROTRIAN DIAGRAM
Cl II INTERCOMBINATION GROTRIAN DIAGRAM (16 electrons, Z = 17)
(S I sequence, Configuration: 1s²2s²2p⁶3s²3p³nl, Singlet to Triplet, Odd to Even Terms)
nd ¹S°
nl ¹P°
nl ¹D°
nd ¹F°
nd ¹G°
np ³S
nl ³P
nl ³D
nl ³F
nf ³G
nf ³H
ENERGY (cm⁻¹)
190 x 10³
170 x 10³
150 x 10³
130 x 10³
110 x 10³
5p″(2,3)
8f(3,4)
4f′(2,1,0)
4f′(2,3,1)
4f′
4f′(3,4,5)
4f′(5,4,6)
3s3p⁵(1)
4f(2,3,4)
5p(2,1,0)
4p″(2,1,0)
4p″(1,2,3)
3d″(3)
3d′(2)
4p″(1)
3d′(3)
4p′(2,1,0)
4p′(1,2,3)
4p′(2,3,4)
4s″(1)
3d″(2)
3d′(4)
4s′(2)
3d′(0)
3d′(1)
4157.97
5270.51
1499.48, 1500.25
4181.11
2294.96, 2295.21
1836.37
1503.25
4164.93, 4168.43
2290.32, 2991.37
4195.93
4624.32
3222.54
2089.96–2095.65
2091.46
1501.15
7620.63
4676.69
3434.04
3231.72
2876.40
5414.17
4676.78
5064.24
4821.86
5744.30, 5790.52
5535.37
7850.64
3259.15
4741.17
CORES:
n l = 3p³(⁴S°)
n′l′ = 3p³(²D°)
n″l″ = 3p³(²P°)

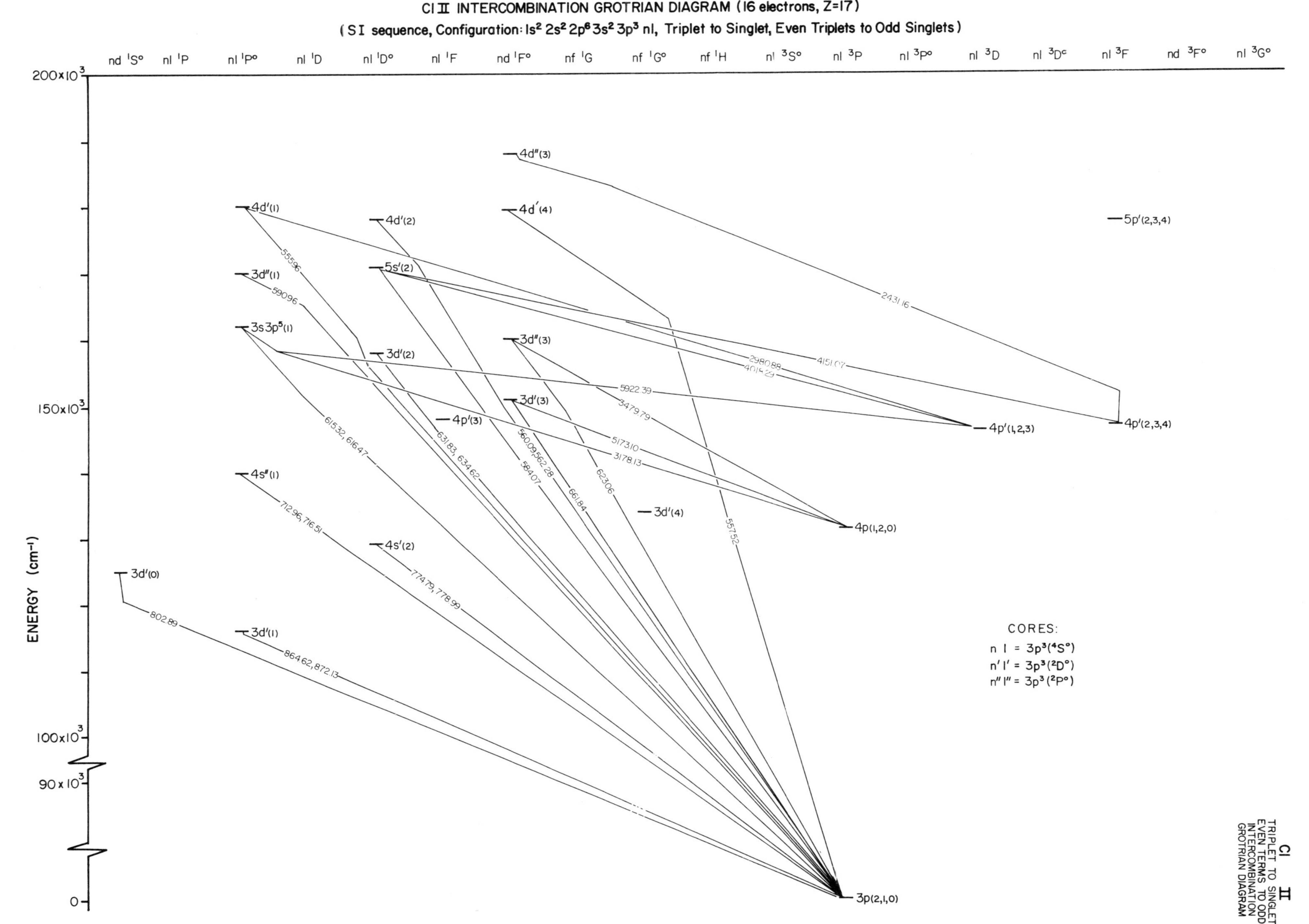
Cl II INTERCOMBINATION GROTRIAN DIAGRAM (16 electrons, Z=17)
(SI sequence, Configuration: 1s² 2s² 2p⁶ 3s² 3p³ nl, Triplet to Singlet, Even Triplets to Odd Singlets)
nd ¹S°
nl ¹P
nl ¹P°
nl ¹D
nl ¹D°
nl ¹F
nd ¹F°
nf ¹G
nf ¹G°
nf ¹H
nl ³S°
nl ³P
nl ³P°
nl ³D
nl ³D°
nl ³F
nd ³F°
nl ³G°
ENERGY (cm⁻¹)
200×10³
150×10³
100×10³
90×10³
0
4d″(3)
4d′(1)
4d′(2)
4d′(4)
5p′(2,3,4)
3d″(1)
5s′(2)
3s3p⁵(1)
3d″(3)
3d′(2)
3d′(3)
4p′(3)
4p′(1,2,3)
4p′(2,3,4)
4s″(1)
3d′(4)
4p(1,2,0)
4s′(2)
3d′(0)
3d′(1)
3p(2,1,0)
555.96
590.96
243.1.16
298.0.88
401.5.29
415.1.07
592.2.39
347.9.79
517.3.10
317.8.13
615.32, 616.47
631.83, 634.62
560.09, 562.28
584.07
661.84
623.06
557.52
712.96, 716.51
774.79, 778.99
802.89
864.62, 872.13
CORES:
n l = 3p³(⁴S°)
n′ l′ = 3p³(²D°)
n″ l″ = 3p³(²P°)
Cl II
TRIPLET TO SINGLET
EVEN TERMS TO ODD
INTERCOMBINATION
GROTRIAN DIAGRAM

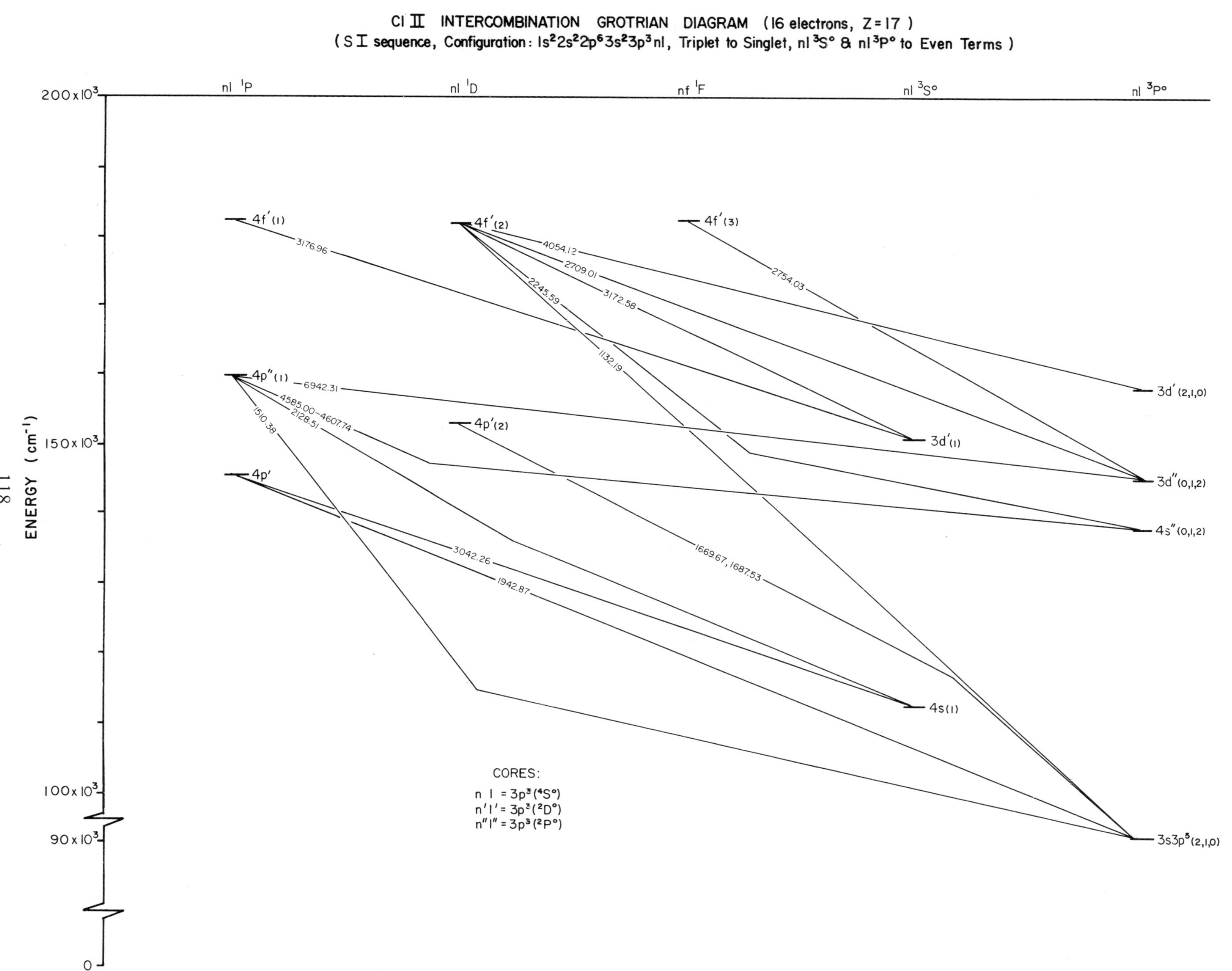
Cl II INTERCOMBINATION GROTRIAN DIAGRAM (16 electrons, Z = 17)
(S I sequence, Configuration: 1s²2s²2p⁶3s²3p³nl, Triplet to Singlet, nl ³S° & nl ³P° to Even Terms)
nl ¹P
nl ¹D
nf ¹F
nl ³S°
nl ³P°
4f′(1)
4f′(2)
4f′(3)
4p″(1)
4p′(2)
4p′
3d′(1)
4s(1)
3d′(2,1,0)
3d″(0,1,2)
4s″(0,1,2)
3s3p⁵(2,1,0)
3176.96
4054.12
2709.01
3172.58
2245.59
1132.19
2754.03
6942.31
4585.00–4607.74
2128.51
1510.38
1669.67, 1687.53
3042.26
1942.87
CORES:
n l = 3p³(⁴S°)
n′l′ = 3p³(²D°)
n″l″ = 3p³(²P°)
200×10³
150×10³
100×10³
90×10³
0
ENERGY (cm⁻¹)

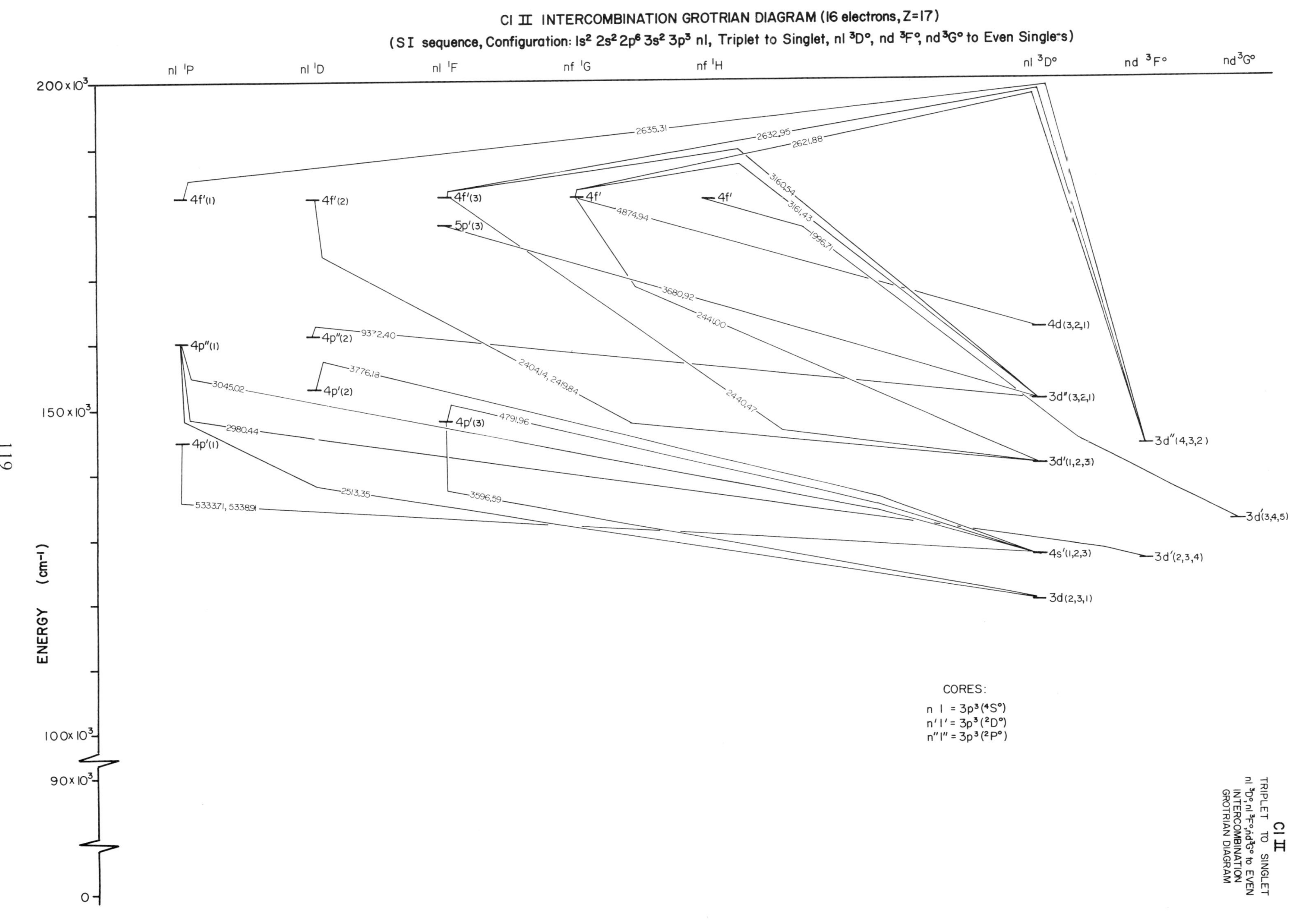
Cl II
TRIPLET TO SINGLET
nl ³D°, nl ³F°, nd³G° to EVEN
INTERCOMBINATION
GROTRIAN DIAGRAM
Cl II INTERCOMBINATION GROTRIAN DIAGRAM (16 electrons, Z=17)
(SI sequence, Configuration: 1s² 2s² 2p⁶ 3s² 3p³ nl, Triplet to Singlet, nl ³D°, nd ³F°, nd³G° to Even Singlets)
nl ¹P
nl ¹D
nl ¹F
nf ¹G
nf ¹H
nl ³D°
nd ³F°
nd³G°
200×10³
150×10³
100×10³
90×10³
0
ENERGY (cm⁻¹)
4f′(1)
4f′(2)
4f′(3)
4f′
4f′
5p′(3)
4p″(1)
4p″(2)
4p′(2)
4p′(3)
4p′(1)
4d(3,2,1)
3d″(3,2,1)
3d″(4,3,2)
3d′(1,2,3)
3d′(3,4,5)
4s′(1,2,3)
3d′(2,3,4)
3d(2,3,1)
2635.31
2632.95
2621.88
3160.54
3161.43
1996.71
4874.94
3680.92
2441.00
9372.40
3776.18
2404.14, 2419.84
2440.47
3045.02
4791.96
2980.44
2513.35
3596.59
533371, 533891
CORES:
n l = 3p³(⁴S°)
n′l′ = 3p³(²D°)
n″l″ = 3p³(²P°)

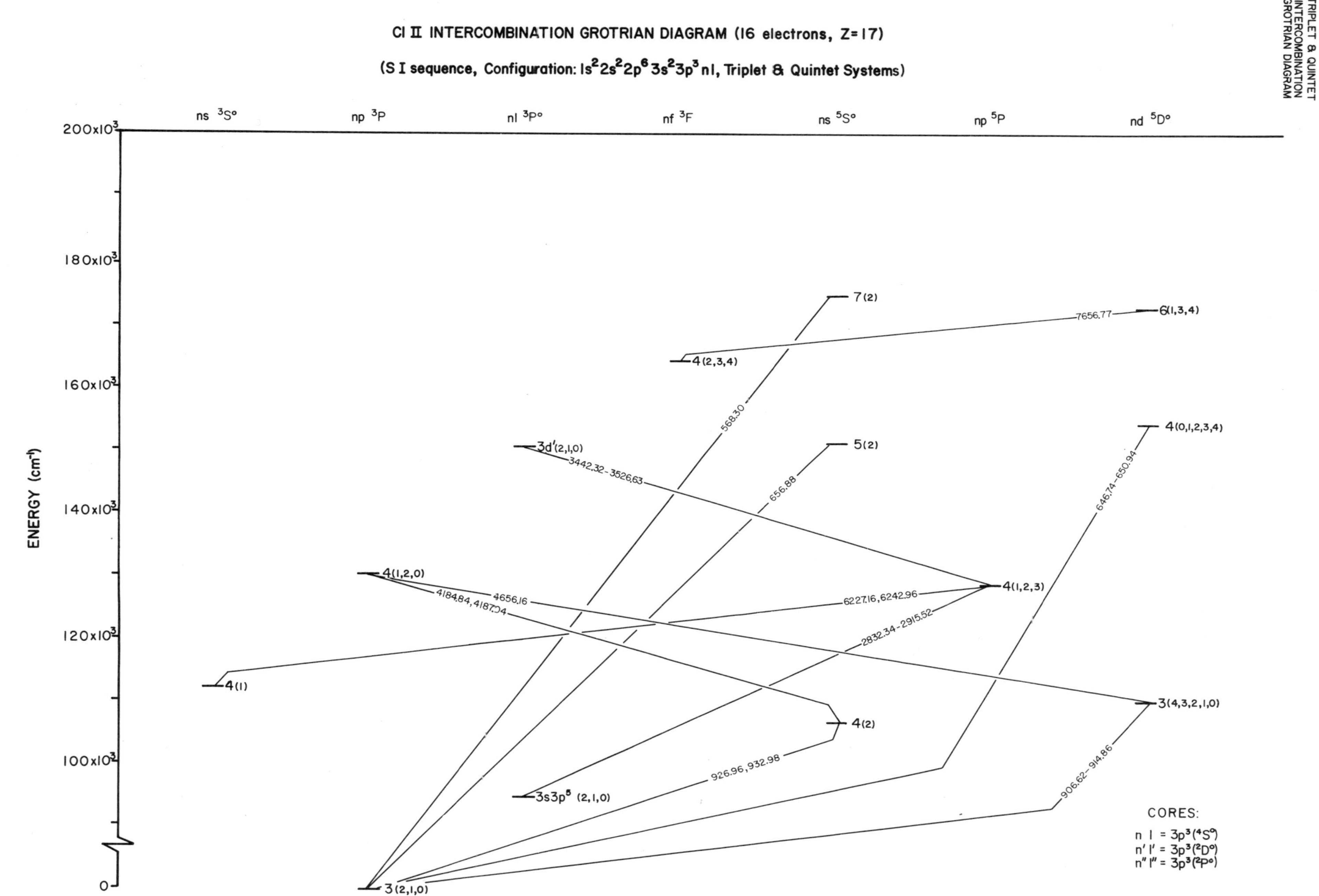
Cl II INTERCOMBINATION GROTRIAN DIAGRAM (16 electrons, Z=17)
(S I sequence, Configuration: 1s²2s²2p⁶3s²3p³nl, Triplet & Quintet Systems)
ns ³S°
np ³P
nl ³P°
nf ³F
ns ⁵S°
np ⁵P
nd ⁵D°
ENERGY (cm⁻¹)
200x10³
180x10³
160x10³
140x10³
120x10³
100x10³
0
CORES:
n l = 3p³(⁴S°)
n′ l′ = 3p³(²D°)
n″ l″ = 3p³(²P°)
4(1)
4(1,2,0)
3(2,1,0)
3d′(2,1,0)
3s3p⁵ (2,1,0)
4(2,3,4)
7(2)
5(2)
4(2)
4(1,2,3)
6(1,3,4)
4(0,1,2,3,4)
3(4,3,2,1,0)
7656.77
568.30
656.88
3442.32-3526.63
4656.16
4184.84, 4187.04
6227.16, 6242.96
2832.34-2915.52
926.96, 932.98
646.74-650.94
906.62-914.86

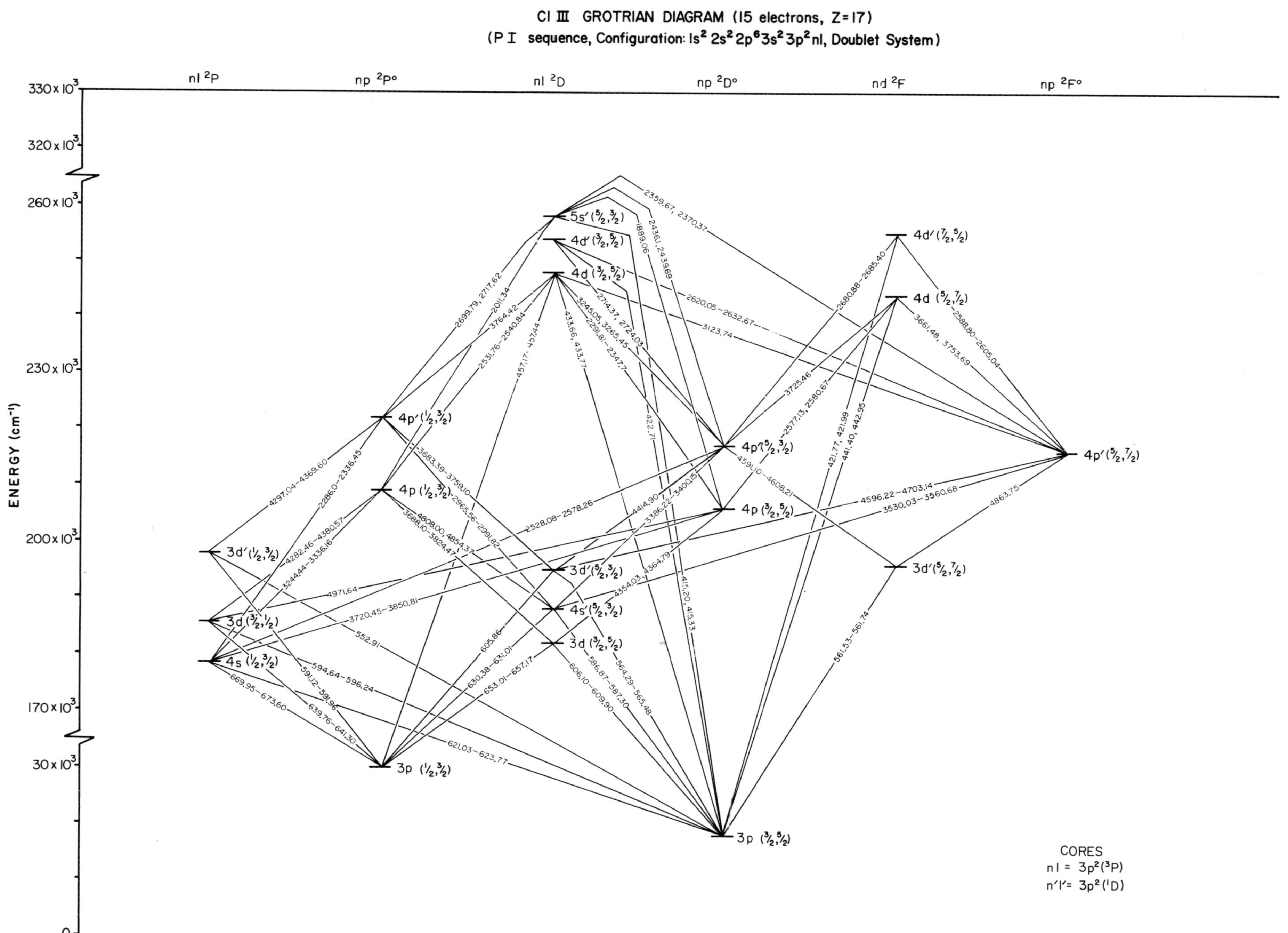
Cl III GROTRIAN DIAGRAM (15 electrons, Z=17)
(P I sequence, Configuration: $1s^2 2s^2 2p^6 3s^2 3p^2 nl$, Doublet System)
nl ^{2}P
np ^{2}P°
nl ^{2}D
np ^{2}D°
nd ^{2}F
np ^{2}F°
ENERGY (cm^{-1})
330 x 10^3
320 x 10^3
260 x 10^3
230 x 10^3
200 x 10^3
170 x 10^3
30 x 10^3
0
5s′(5/2,3/2)
4d′(3/2,5/2)
4d (3/2,5/2)
4d′(7/2,5/2)
4d (5/2,7/2)
4p′(1/2,3/2)
4p (1/2,3/2)
4p′(5/2,3/2)
4p (3/2,5/2)
4p′(5/2,7/2)
3d′(1/2,3/2)
3d (3/2,1/2)
4s (1/2,3/2)
3d′(5/2,3/2)
4s′(5/2,3/2)
3d (3/2,5/2)
3d′(5/2,7/2)
3p (1/2,3/2)
3p (3/2,5/2)
CORES
nl = $3p^2(^3P)$
n′l′= $3p^2(^1D)$

Cl III ENERGY LEVELS (15 electrons, Z=17)

(P I sequence, Configuration: $1s^2 2s^2 2p^6 3s^2 3p^2 nl$, Doublet System)

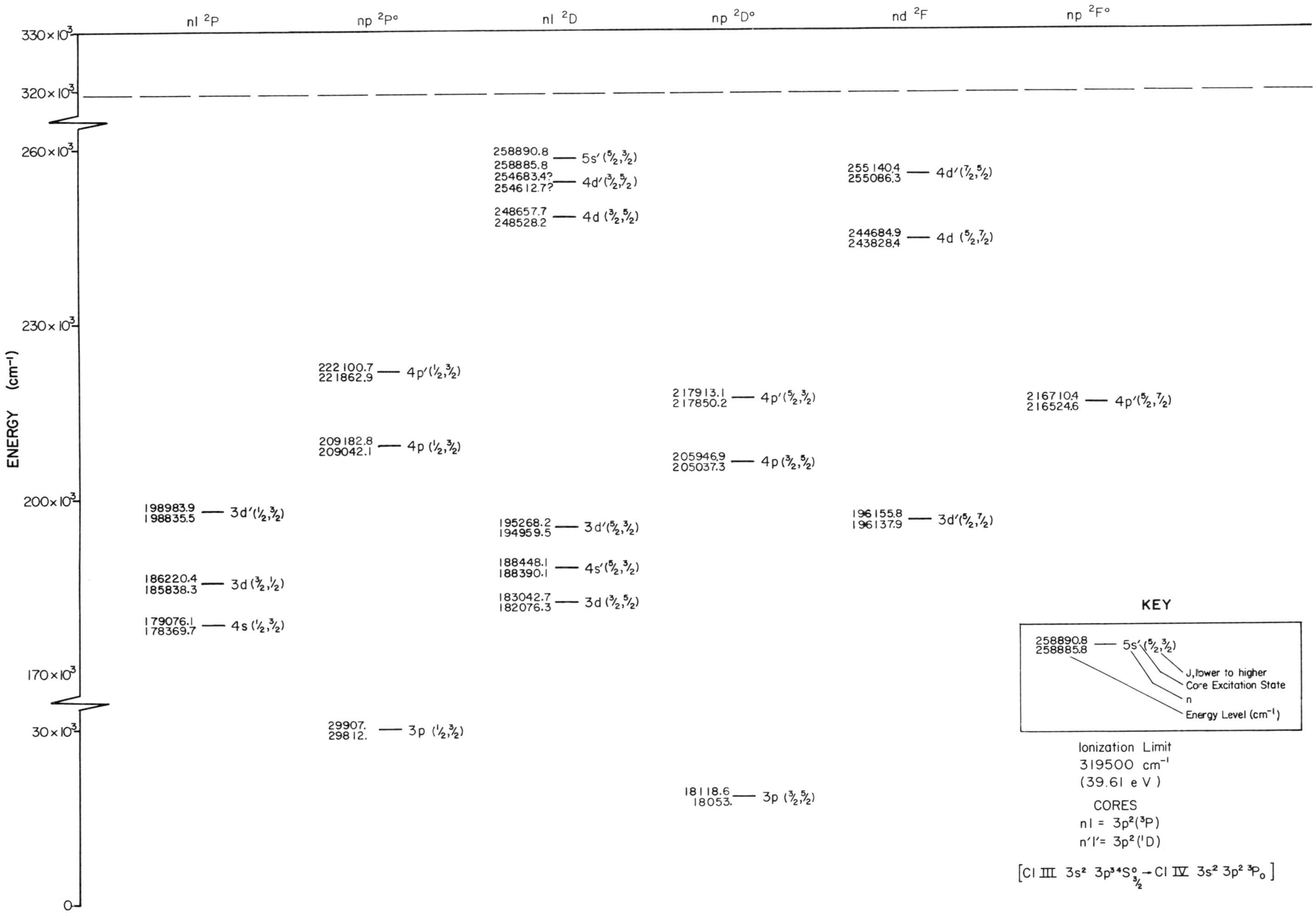

ENERGY (cm^{-1})

330×10^3
320×10^3
260×10^3
230×10^3
200×10^3
170×10^3
30×10^3
0

nl ^{2}P

198983.9, 198835.5 — 3d′(½,3/2)
186220.4, 185838.3 — 3d (3/2,½)
179076.1, 178369.7 — 4s (½,3/2)

np ^{2}P°

222100.7, 221862.9 — 4p′(½,3/2)
209182.8, 209042.1 — 4p (½,3/2)
29907., 29812. — 3p (½,3/2)

nl ^{2}D

258890.8, 258885.8 — 5s′(5/2,3/2)
254683.4?, 254612.7? — 4d′(3/2,5/2)
248657.7, 248528.2 — 4d (3/2,5/2)
195268.2, 194959.5 — 3d′(5/2,3/2)
188448.1, 188390.1 — 4s′(5/2,3/2)
183042.7, 182076.3 — 3d (3/2,5/2)

np ^{2}D°

217913.1, 217850.2 — 4p′(5/2,3/2)
205946.9, 205037.3 — 4p (3/2,5/2)
18118.6, 18053. — 3p (3/2,5/2)

nd ^{2}F

255140.4, 255086.3 — 4d′(7/2,5/2)
244684.9, 243828.4 — 4d (5/2,7/2)
196155.8, 196137.9 — 3d′(5/2,7/2)

np ^{2}F°

216710.4, 216524.6 — 4p′(5/2,7/2)

KEY

258890.8, 258885.8 — 5s′(5/2,3/2)
J, lower to higher
Core Excitation State
n
Energy Level (cm^{-1})

Ionization Limit
319500 cm^{-1}
(39.61 eV)

CORES
nl = $3p^2(^3P)$
n′l′ = $3p^2(^1D)$

[Cl III $3s^2\ 3p^3\ ^4S^o_{3/2}$ → Cl IV $3s^2\ 3p^2\ ^3P_0$]

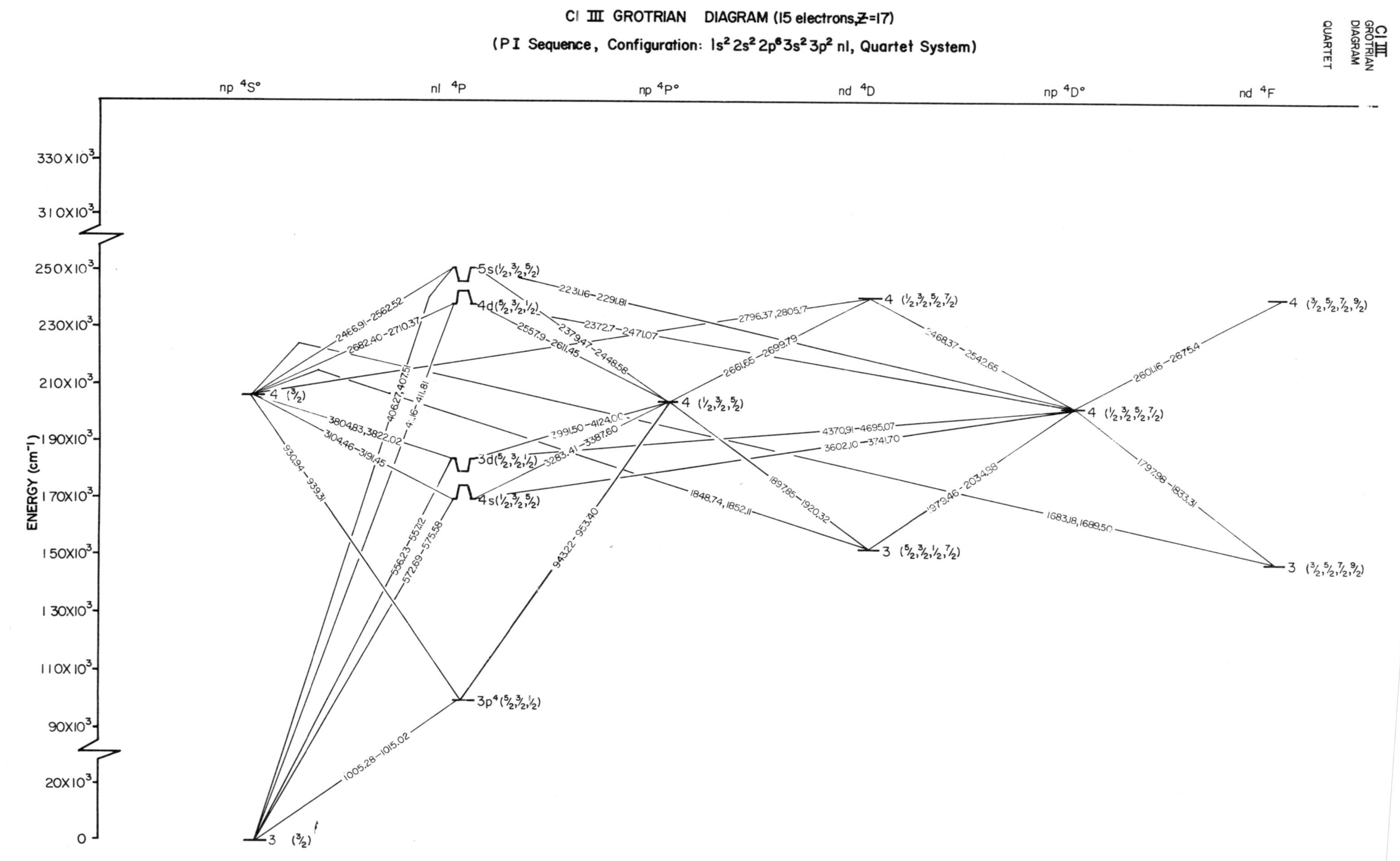

Cl III
GROTRIAN
DIAGRAM
QUARTET
Cl III GROTRIAN DIAGRAM (15 electrons, Z=17)
(P I Sequence, Configuration: 1s² 2s² 2p⁶ 3s² 3p² nl, Quartet System)
np ⁴S°
nl ⁴P
np ⁴P°
nd ⁴D
np ⁴D°
nd ⁴F
ENERGY (cm⁻¹)
330X10³
310X10³
250X10³
230X10³
210X10³
190X10³
170X10³
150X10³
130X10³
110X10³
90X10³
20X10³
0
4 (3/2)
3 (3/2)
5s(1/2,3/2,5/2)
4d(5/2,3/2,1/2)
3d(5/2,3/2,1/2)
4s(1/2,3/2,5/2)
3p⁴(5/2,3/2,1/2)
4 (1/2,3/2,5/2)
4 (1/2,3/2,5/2,7/2)
3 (5/2,3/2,1/2,7/2)
4 (1/2,3/2,5/2,7/2)
4 (3/2,5/2,7/2,9/2)
3 (3/2,5/2,7/2,9/2)
2466.91–2562.52
2682.40–2710.37
3804.83,3822.02
3104.46–3191.45
930.94–939.31
4062.7,4075.1
4.16–411.81
556.23–557.12
572.69–575.58
1005.28–1015.02
2231.16–2291.81
2372.7–2471.07
2557.9–2611.45
2379.47–2448.58
3991.50–4124.00
3283.41–3387.60
943.22–953.40
2796.37,2805.17
2661.65–2699.79
1848.74,1852.11
1897.85–1920.32
2468.37–2542.65
4370.91–4695.07
3602.10–3741.70
1979.46–2034.98
1683.18,1689.50
2601.16–2675.4
1797.98–1833.31

Cl III ENERGY LEVEL (15 electrons, Z=17)

(P I Sequence, Configuration: $1s^2 2s^2 2p^6 3s^2 3p^2 nl$, Quartet System)

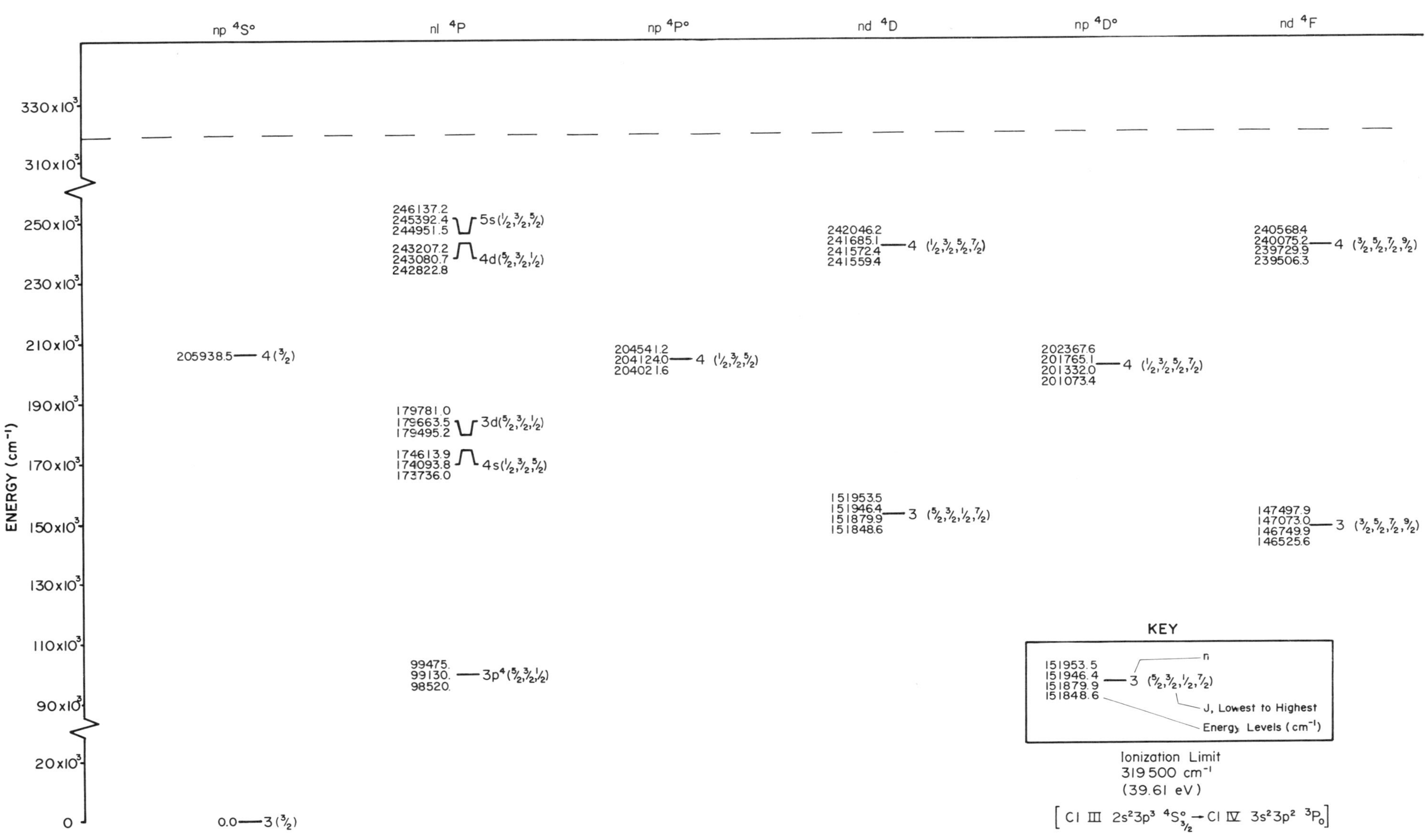

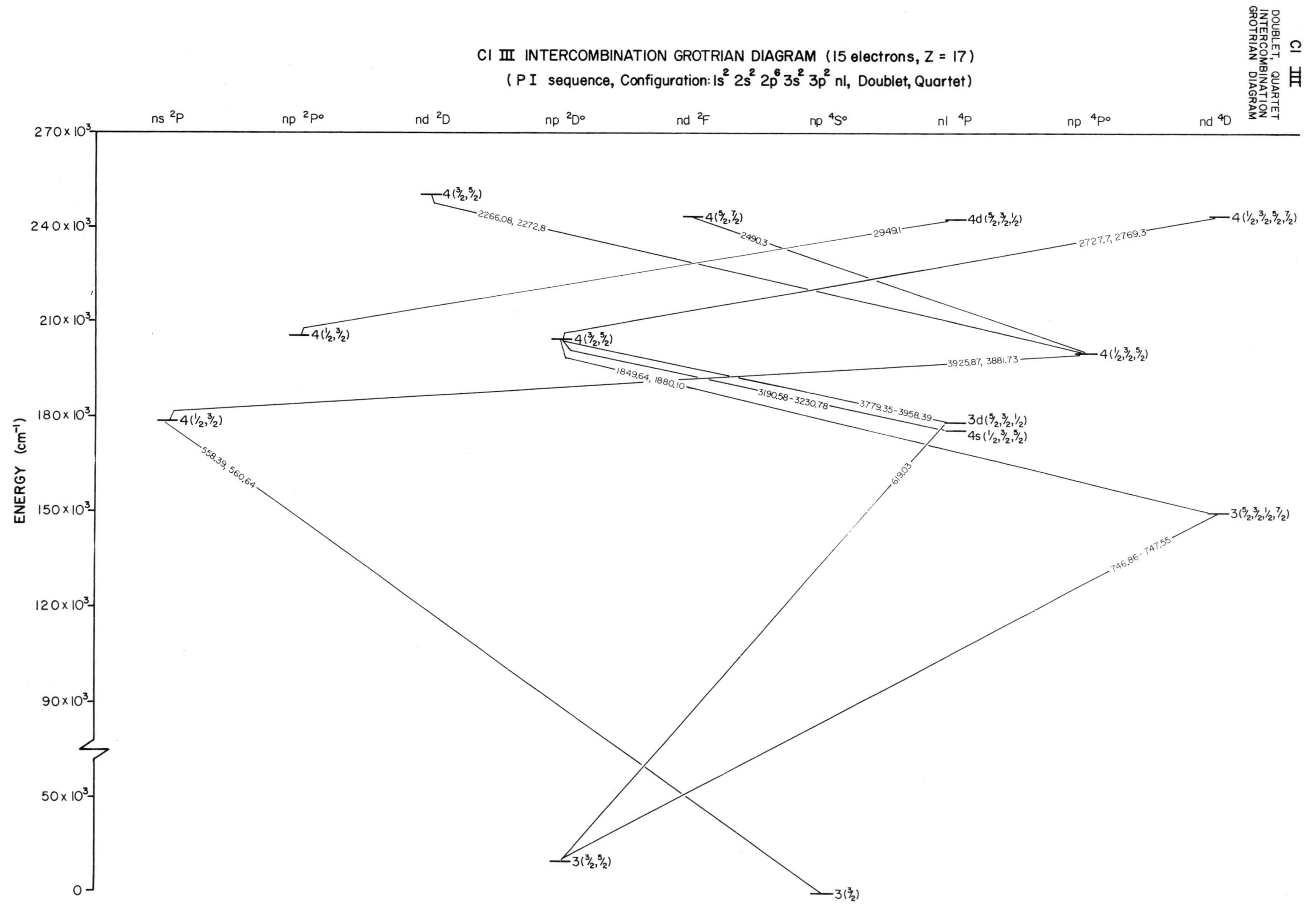

Cl III
DOUBLET, QUARTET
INTERCOMBINATION
GROTRIAN DIAGRAM
Cl III INTERCOMBINATION GROTRIAN DIAGRAM (15 electrons, Z = 17)
(P I sequence, Configuration: 1s² 2s² 2p⁶ 3s² 3p² nl, Doublet, Quartet)
ns ²P
np ²P°
nd ²D
np ²D°
nd ²F
np ⁴S°
nl ⁴P
np ⁴P°
nd ⁴D
270 x 10³
240 x 10³
210 x 10³
180 x 10³
150 x 10³
120 x 10³
90 x 10³
50 x 10³
0
ENERGY (cm⁻¹)
4(3/2,5/2)
4(5/2,7/2)
4d(5/2,3/2,1/2)
4(1/2,3/2,5/2,7/2)
4(1/2,3/2)
4(3/2,5/2)
4(1/2,3/2,5/2)
4(1/2,3/2)
3d(5/2,3/2,1/2)
4s(1/2,3/2,5/2)
3(5/2,3/2,1/2,7/2)
3(3/2,5/2)
3(3/2)
2266.08, 2272.8
2490.3
2949.1
2727.7, 2769.3
3925.87, 3881.73
1849.64, 1880.10
3190.58 - 3230.78
3779.35 - 3958.39
558.39, 560.64
619.03
746.86 - 747.55

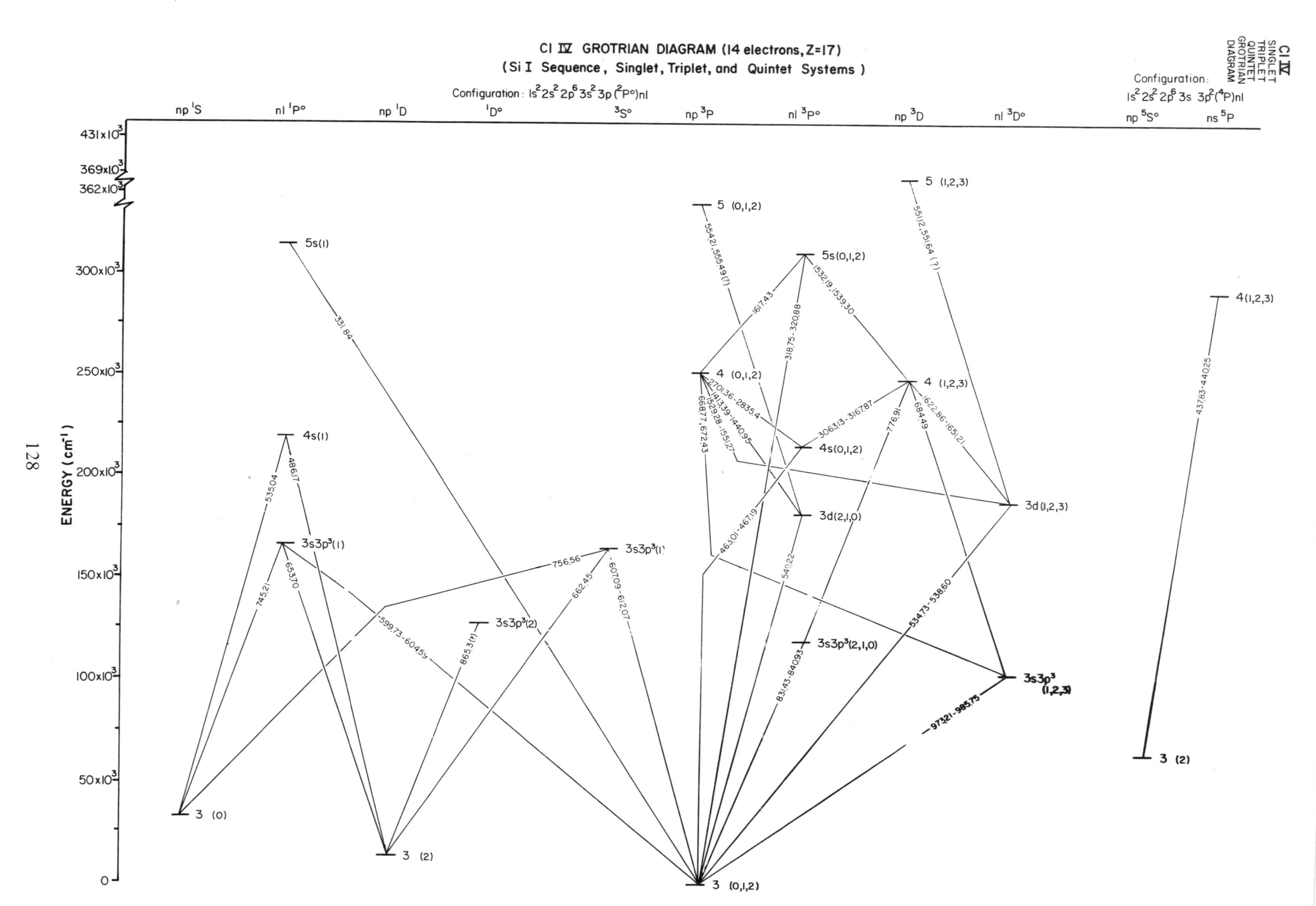

Cl IV GROTRIAN DIAGRAM (14 electrons, Z=17)
(Si I Sequence, Singlet, Triplet, and Quintet Systems)
Configuration: $1s^2 2s^2 2p^6 3s^2 3p(^2P^o)nl$
Cl IV
SINGLET
TRIPLET
QUINTET
GROTRIAN
DIAGRAM
Configuration:
$1s^2 2s^2 2p^6 3s\ 3p^2(^4P)nl$
np ^1S
nl $^1P^o$
np ^1D
$^1D^o$
$^3S^o$
np ^3P
nl $^3P^o$
np ^3D
nl $^3D^o$
np $^5S^o$
ns ^5P
ENERGY (cm^{-1})
431x10^3
369x10^3
362x10^3
300x10^3
250x10^3
200x10^3
150x10^3
100x10^3
50x10^3
0
3 (0)
5s(1)
4s(1)
3s3p^3(1)
3 (2)
3s3p^3(2)
3s3p^3(1)
5 (0,1,2)
4 (0,1,2)
3 (0,1,2)
5s(0,1,2)
4s(0,1,2)
3d(2,1,0)
3s3p^3(2,1,0)
5 (1,2,3)
4 (1,2,3)
3d(1,2,3)
3s3p^3 (1,2,3)
4(1,2,3)
3 (2)
331.84
535.04
486.17
745.21
653.70
599.73-604.59
756.56
865.3(?)
662.45
607.09-612.07
554.21, 555.49(?)
1617.43
318.75-320.88
270.36-283.54
1413.39-1440.95
1529.28-1551.27
668.77, 672.43
463.01-467.19
549.22
831.43-840.93
1532.19, 1539.30
3063.13-3167.87
776.91
684.49
1622.86-1651.21
551.12, 551.64 (?)
534.73-538.60
973.21-985.75
437.83-440.25

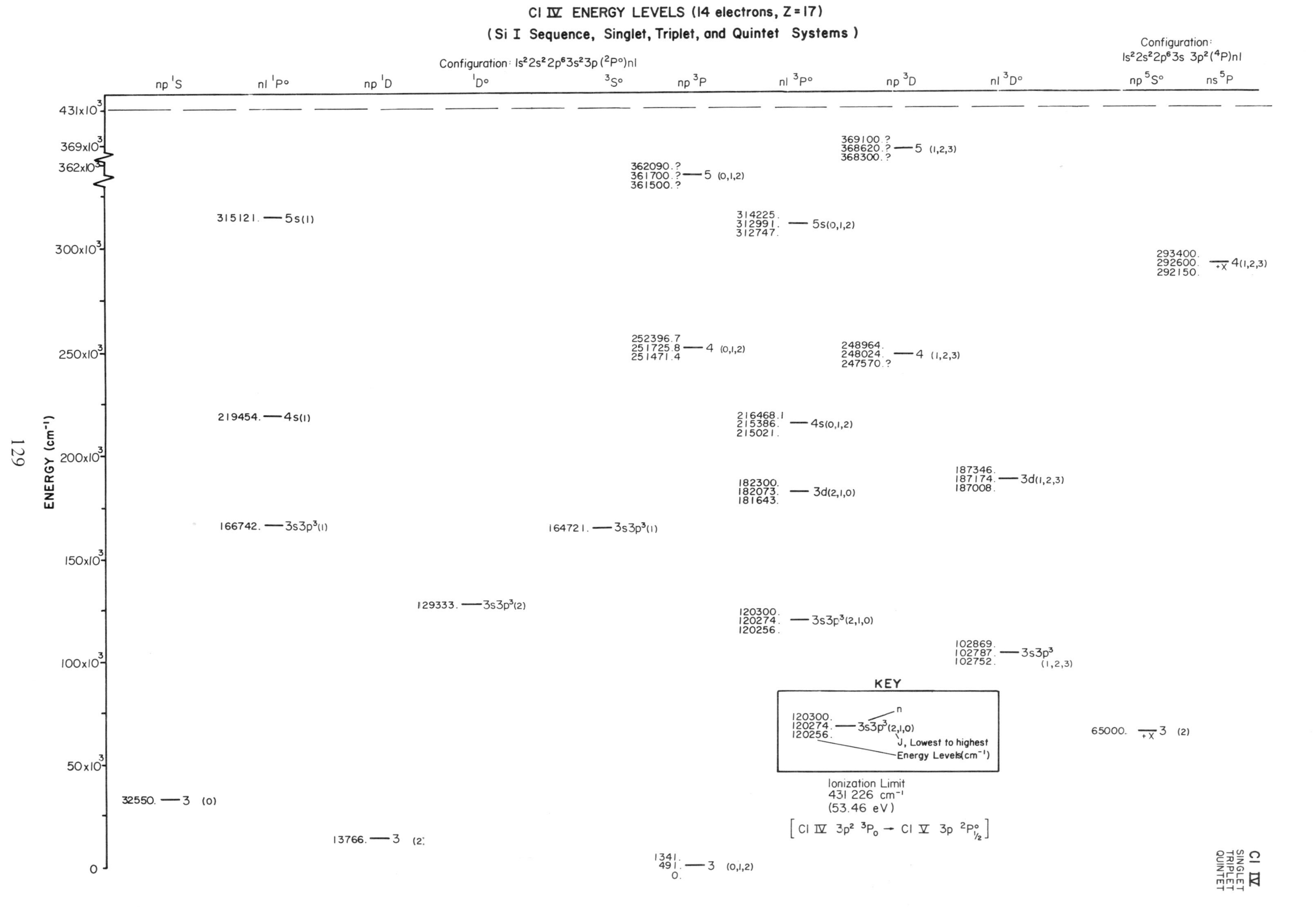

Cl IV ENERGY LEVELS (14 electrons, Z=17)
(Si I Sequence, Singlet, Triplet, and Quintet Systems)
Configuration: 1s²2s²2p⁶3s²3p(²P°)nl
Configuration: 1s²2s²2p⁶3s 3p²(⁴P)nl
np ¹S
nl ¹P°
np ¹D
¹D°
³S°
np ³P
nl ³P°
np ³D
nl ³D°
np ⁵S°
ns ⁵P
369100. ? 368620. ? 368300. ? — 5 (1,2,3)
362090. ? 361700. ? 361500. ? — 5 (0,1,2)
315121. — 5s(1)
314225. 312991. 312747. — 5s(0,1,2)
293400. 292600. 292150. +x 4(1,2,3)
252396.7 251725.8 251471.4 — 4 (0,1,2)
248964. 248024. 247570. ? — 4 (1,2,3)
219454. — 4s(1)
216468.1 215386. 215021. — 4s(0,1,2)
187346. 187174. 187008. — 3d(1,2,3)
182300. 182073. 181643. — 3d(2,1,0)
166742. — 3s3p³(1)
164721. — 3s3p³(1)
129333. — 3s3p³(2)
120300. 120274. 120256. — 3s3p³(2,1,0)
102869. 102787. 102752. — 3s3p³ (1,2,3)
KEY
120300. 120274. 120256. — 3s3p³(2,1,0)
n
J, Lowest to highest
Energy Levels(cm⁻¹)
65000. +x 3 (2)
Ionization Limit
431 226 cm⁻¹
(53.46 eV)
[Cl IV 3p² ³P₀ → Cl V 3p ²P°½]
32550. — 3 (0)
13766. — 3 (2)
1341. 491. 0. — 3 (0,1,2)
Cl IV
SINGLET
TRIPLET
QUINTET
ENERGY (cm⁻¹)
431x10³
369x10³
362x10³
300x10³
250x10³
200x10³
150x10³
100x10³
50x10³
0

Cl V GROTRIAN DIAGRAM (13 electrons, Z=17)

(Al I sequence, Configuration: $1s^2 2s^2 2p^6 3s 3p (^3P^o)nl$, $1s^2 2s^2 2p^6 3s^2 (^1S) nl$, Doublet & Quartet Systems)

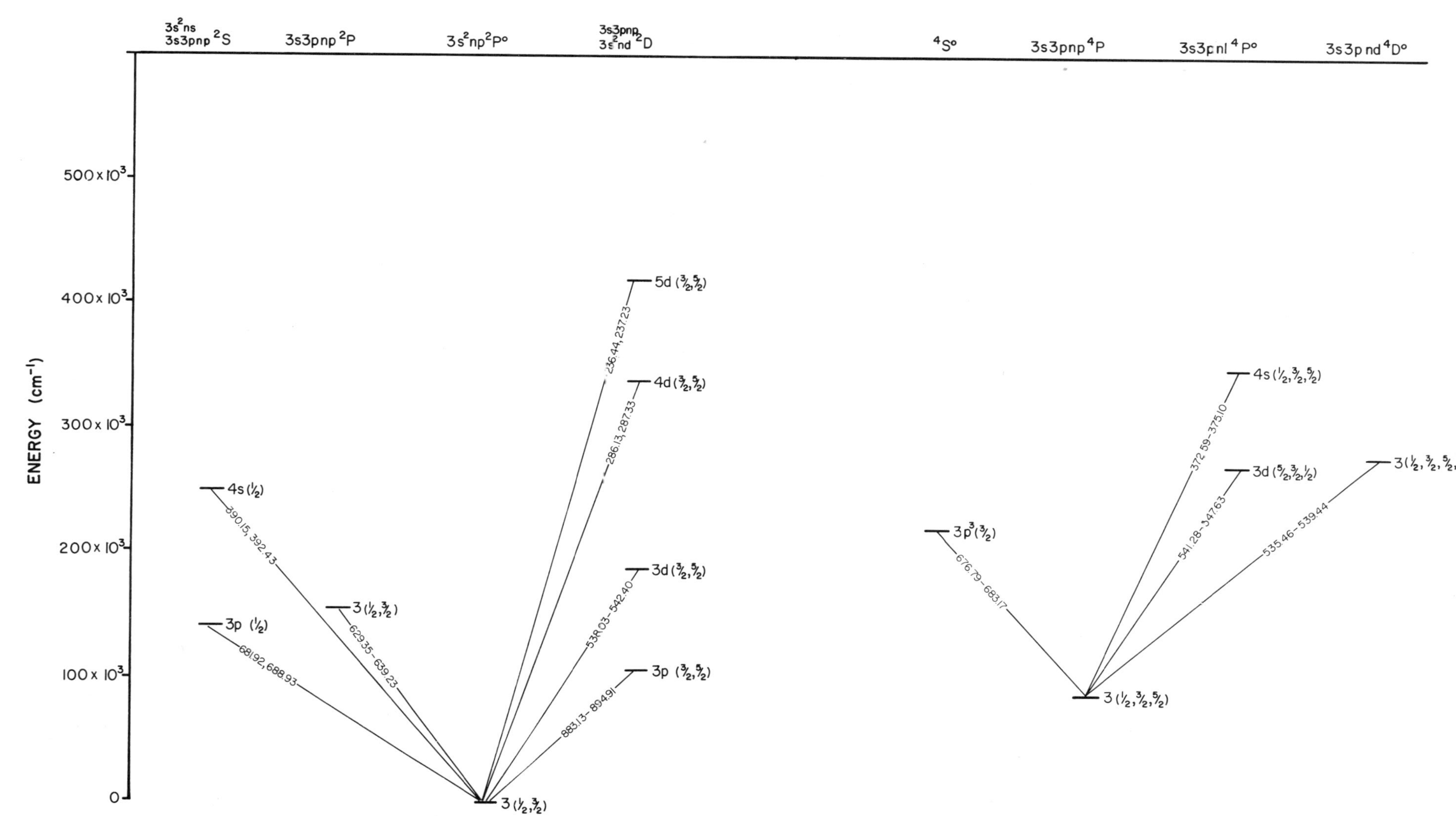

Cl V ENERGY LEVELS (13 electrons, Z=17)

(Al I sequence, Configuration: $1s^2 2s^2 2p^6 3s3p(^3P^0)nl$, $1s^2 2s^2 2p^6 3s^2(^1S)nl$, Doublet & Quartet Systems)

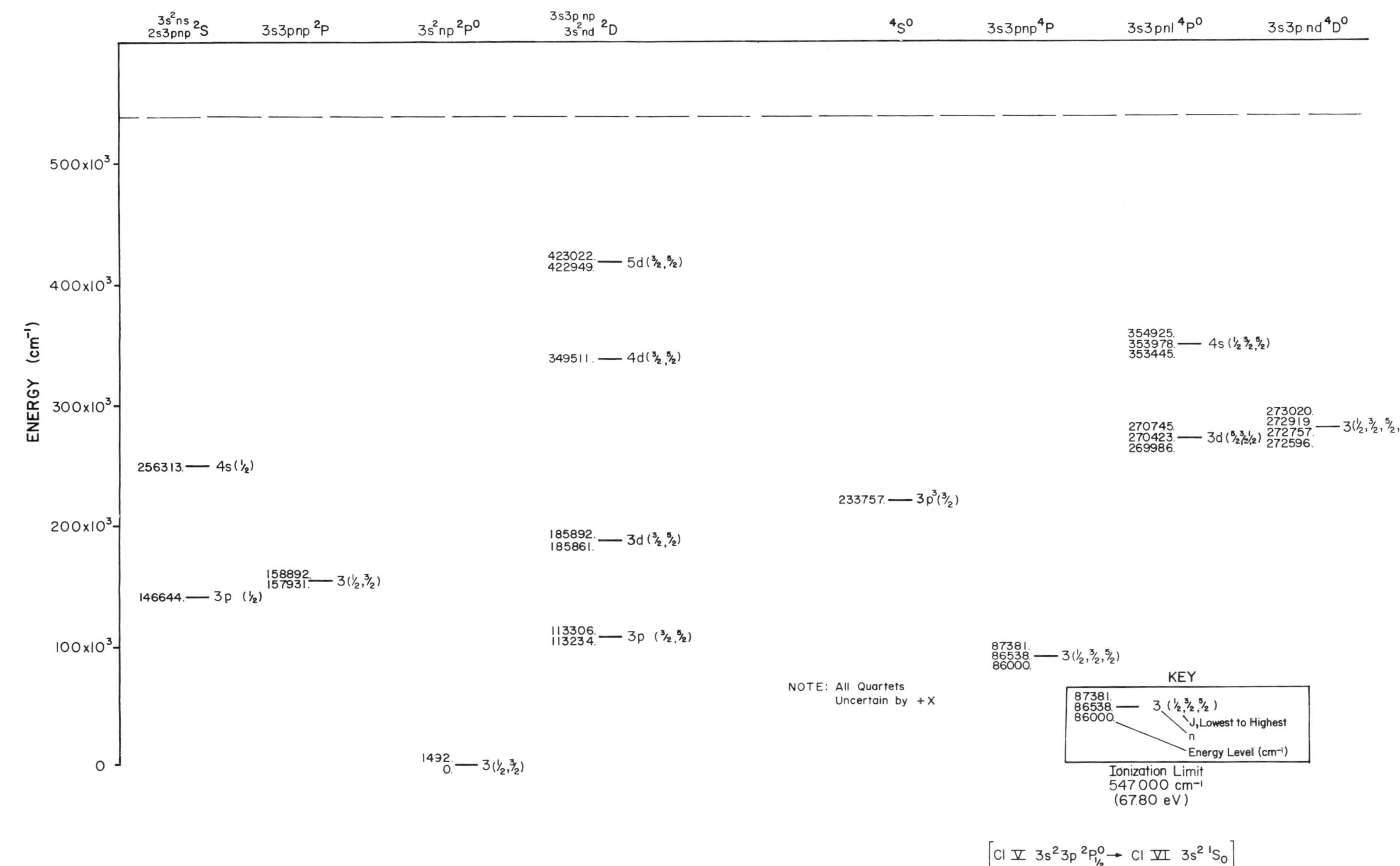

Cl VI GROTRIAN DIAGRAM (12 electrons, Z = 17)

(Mg I sequence, Configurations: $1s^2 2s^2 2p^6 3s (^2S) nl$, $1s^2 2s^2 2p^6 3p (^2P^o) nl$, Singlet & Triplet Systems)

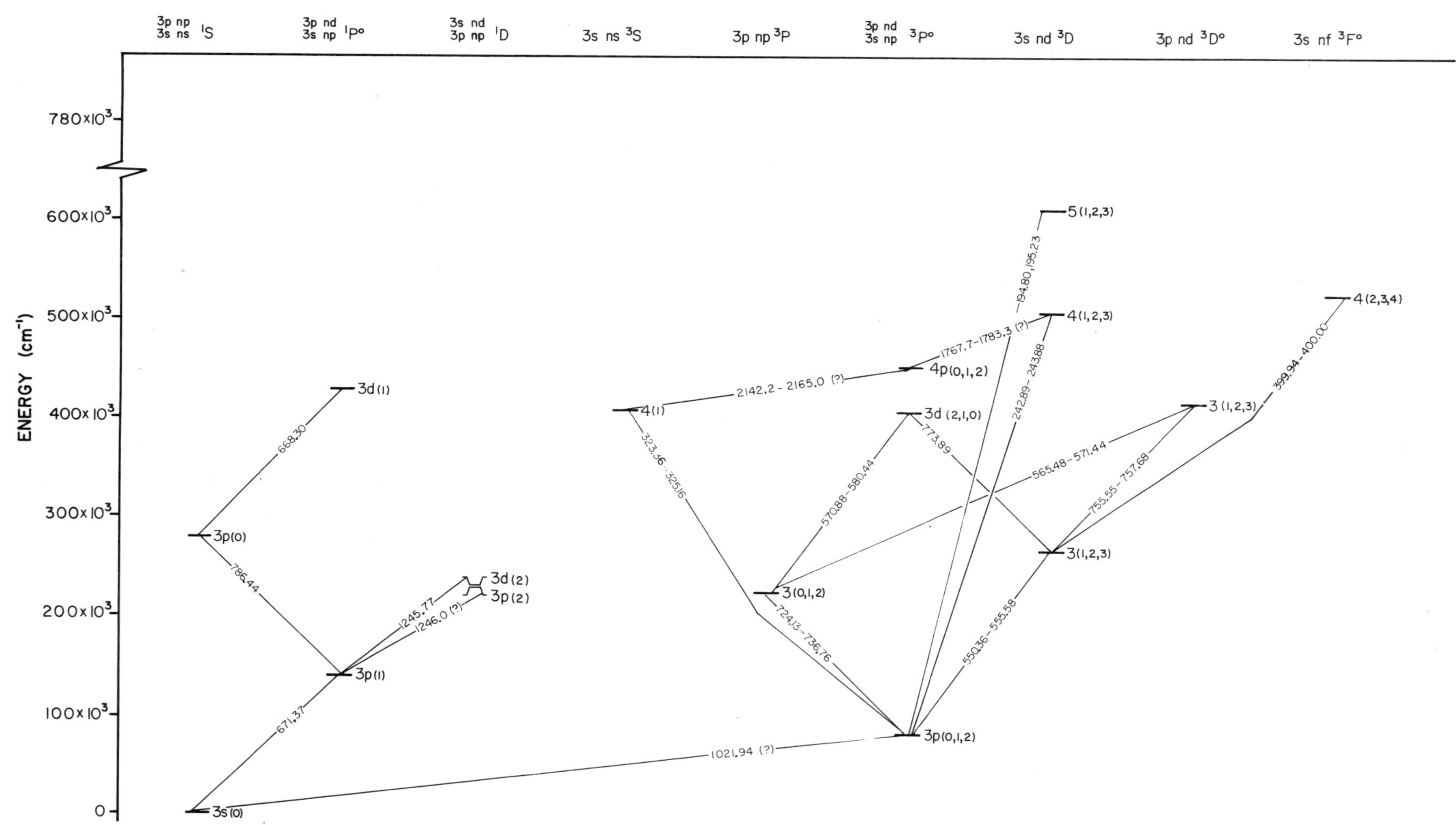

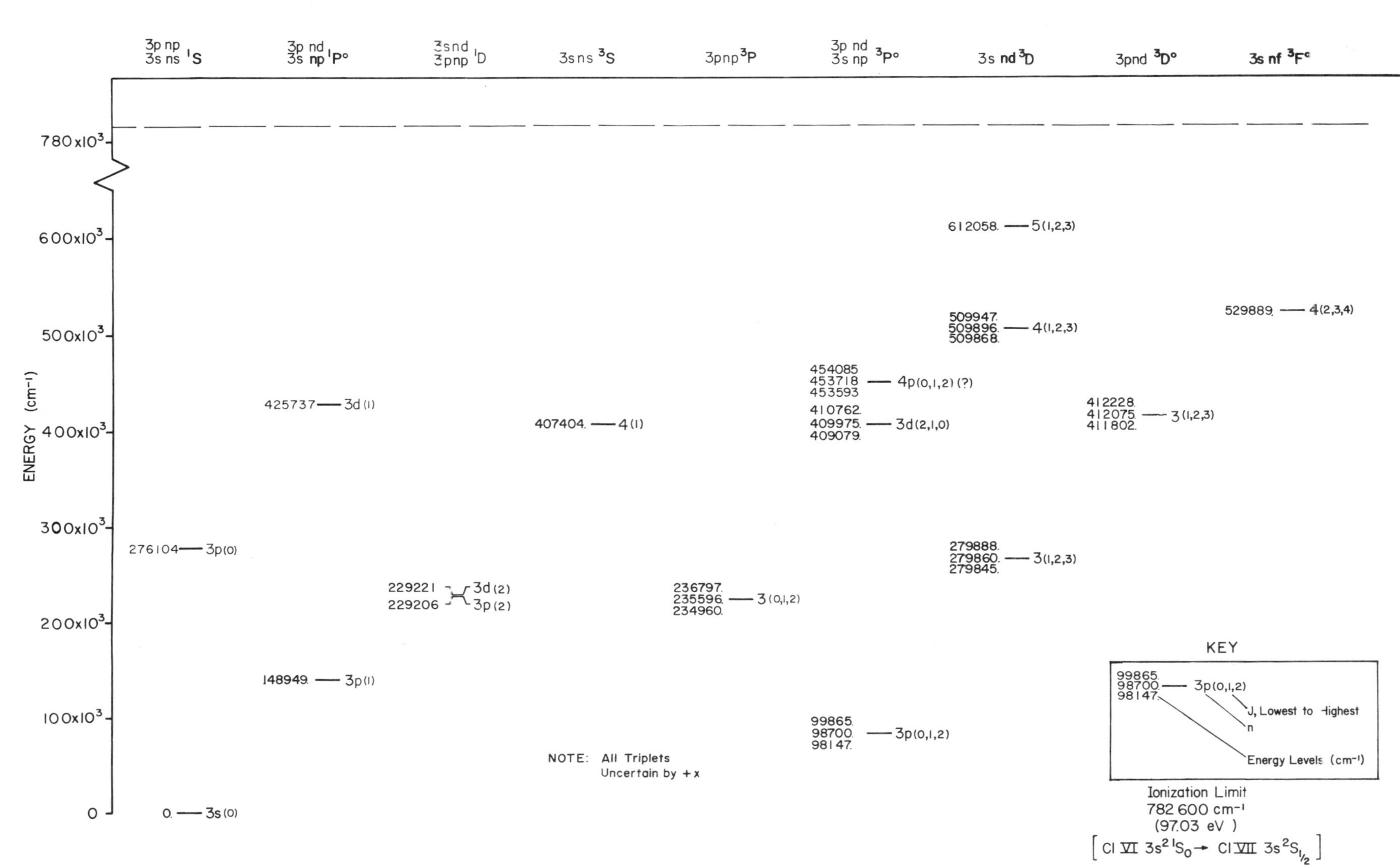
Cl VI ENERGY LEVELS (12 electrons, Z=17)
(Mg I sequence, Configurations: 1s²2s²2p⁶3s(²S)nl, 1s²2s²2p⁶3p(²P⁰)nl, Singlet & Triplet Systems)
3p np 3s ns ¹S
3p nd 3s np ¹P°
3snd 3pnp ¹D
3sns ³S
3pnp ³P
3p nd 3s np ³P°
3s nd ³D
3pnd ³D°
3s nf ³F°
780x10³
600x10³
500x10³
400x10³
300x10³
200x10³
100x10³
0
ENERGY (cm⁻¹)
276104 — 3p(0)
0. — 3s(0)
425737 — 3d(1)
148949. — 3p(1)
229221 3d(2)
229206 3p(2)
407404. — 4(1)
236797.
235596.
234960. — 3(0,1,2)
454085
453718
453593 — 4p(0,1,2) (?)
410762.
409975.
409079. — 3d(2,1,0)
99865.
98700.
98147. — 3p(0,1,2)
612058. — 5(1,2,3)
509947.
509896.
509868. — 4(1,2,3)
279888.
279860.
279845. — 3(1,2,3)
412228.
412075.
411802. — 3(1,2,3)
529889. — 4(2,3,4)
NOTE: All Triplets
Uncertain by +x
KEY
99865.
98700
98147. — 3p(0,1,2)
J, Lowest to Highest
n
Energy Levels (cm⁻¹)
Ionization Limit
782 600 cm⁻¹
(97.03 eV)
[Cl VI 3s² ¹S₀ → Cl VII 3s ²S½]

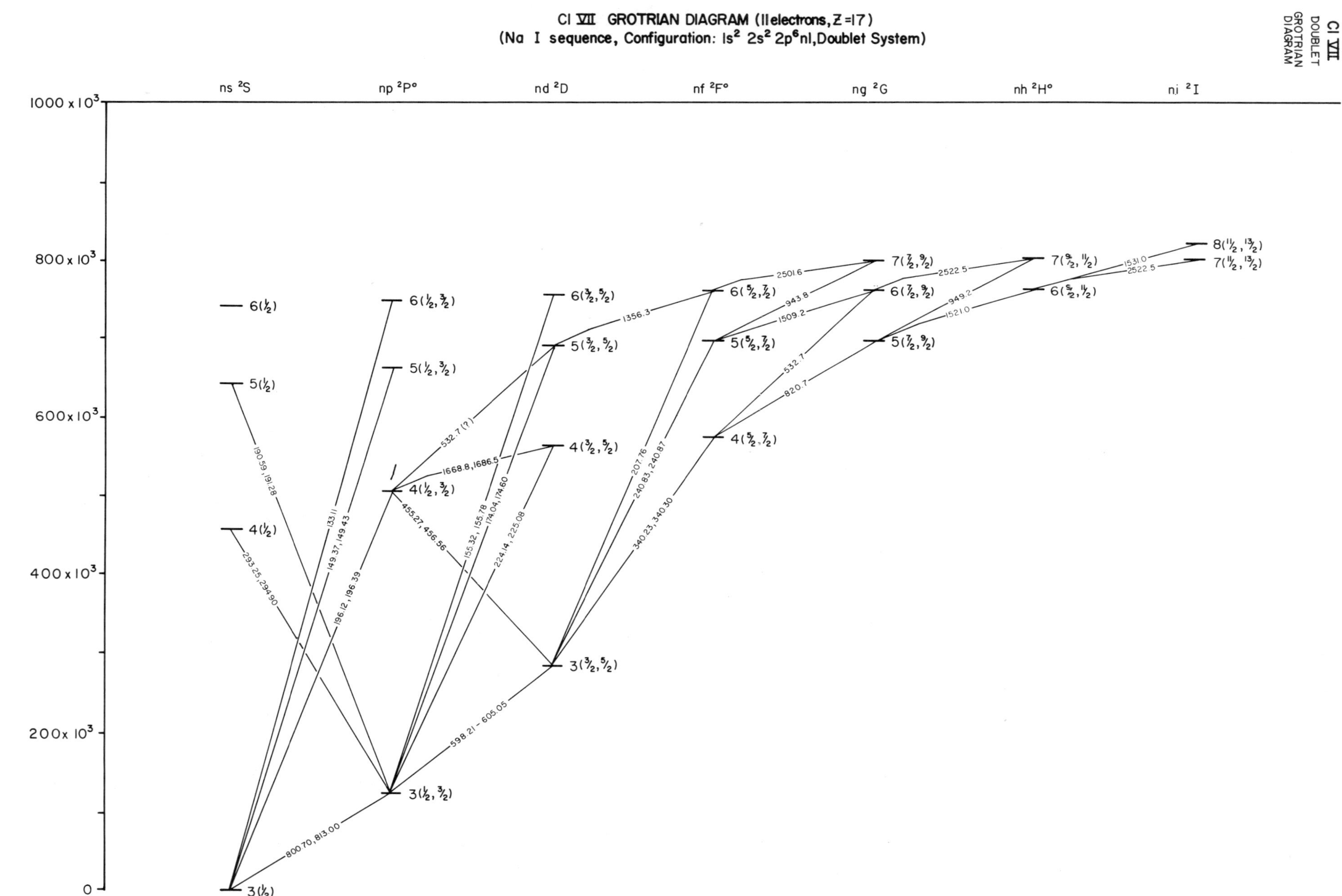
Cl VII GROTRIAN DIAGRAM (11 electrons, Z = 17)
(Na I sequence, Configuration: $1s^2\ 2s^2\ 2p^6 nl$, Doublet System)
ns ^{2}S
np ^{2}P°
nd ^{2}D
nf ^{2}F°
ng ^{2}G
nh ^{2}H°
ni ^{2}I
1000 x 10^3
800 x 10^3
600 x 10^3
400 x 10^3
200 x 10^3
0
6(½)
5(½)
4(½)
3(½)
6(½, 3/2)
5(½, 3/2)
4(½, 3/2)
3(½, 3/2)
6(3/2, 5/2)
5(3/2, 5/2)
4(3/2, 5/2)
3(3/2, 5/2)
6(5/2, 7/2)
5(5/2, 7/2)
4(5/2, 7/2)
7(7/2, 9/2)
6(7/2, 9/2)
5(7/2, 9/2)
7(9/2, 11/2)
6(9/2, 11/2)
8(11/2, 13/2)
7(11/2, 13/2)
190.59, 191.28
293.25, 294.90
133.11
149.37, 149.43
196.12, 196.39
800.70, 813.00
532.7 (?)
1668.8, 1686.5
455.27, 456.56
155.32, 155.78
174.04, 174.60
224.14, 225.08
598.21 - 605.05
1356.3
207.76
240.83, 240.87
340.23, 340.30
2501.6
943.8
1509.2
532.7
820.7
2522.5
949.2
1521.0
1531.0
2522.5

Cl VII ENERGY LEVEL (11 electrons, Z = 17)
(Na I sequence, Configuration: $1s^2\ 2s^2\ 2p^6 nl$, Doublet System)

ENERGY (cm^{-1}): 1000 x 10^3, 800 x 10^3, 600 x 10^3, 400 x 10^3, 200 x 10^3, 0

$ns\ ^2S$	$np\ ^2P^o$	$nd\ ^2D$	$nf\ ^2F^o$	$ng\ ^2G$	$nh\ ^2H^o$	$ni\ ^2I$
						837003 — 8(11/2, 13/2)
				811337 — 7(7/2, 9/2)	811292 — 7(9/2, 11/2)	811329 — 7(11/2, 13/2)
739700 — 6(1/2)	751260 — 6(1/2, 3/2)	766830 — 6(3/2, 5/2)	771549 — 6(5/2, 7/2)	771664 — 6(7/2, 9/2)	771686 — 6(9/2, 11/2)	
647677 — 5(1/2)	669480, 669210 — 5(1/2, 3/2)	697619, 697598 — 5(3/2, 5/2)	705409, 705398 — 5(5/2, 7/2)	705940 — 5(7/2, 9/2)		
464003 — 4(1/2)	509885, 509197 — 4(1/2, 3/2)	569182, 569142 — 4(3/2, 5/2)	584099, 584086 — 4(5/2, 7/2)			
0 — 3(1/2)	124891, 123001 — 3(1/2, 3/2)	290239, 290166 — 3(3/2, 5/2)				

KEY

584099, 584086 — 4(5/2, 7/2)
- J, lower to higher
- n
- Energy Level (cm^{-1})

Ionization Limit
921 051 cm^{-1}
(114.193 eV)
[Cl VII $2p^6 3s\ ^2S_{1/2}$ → Cl VIII $2p^6\ ^1S_0$]

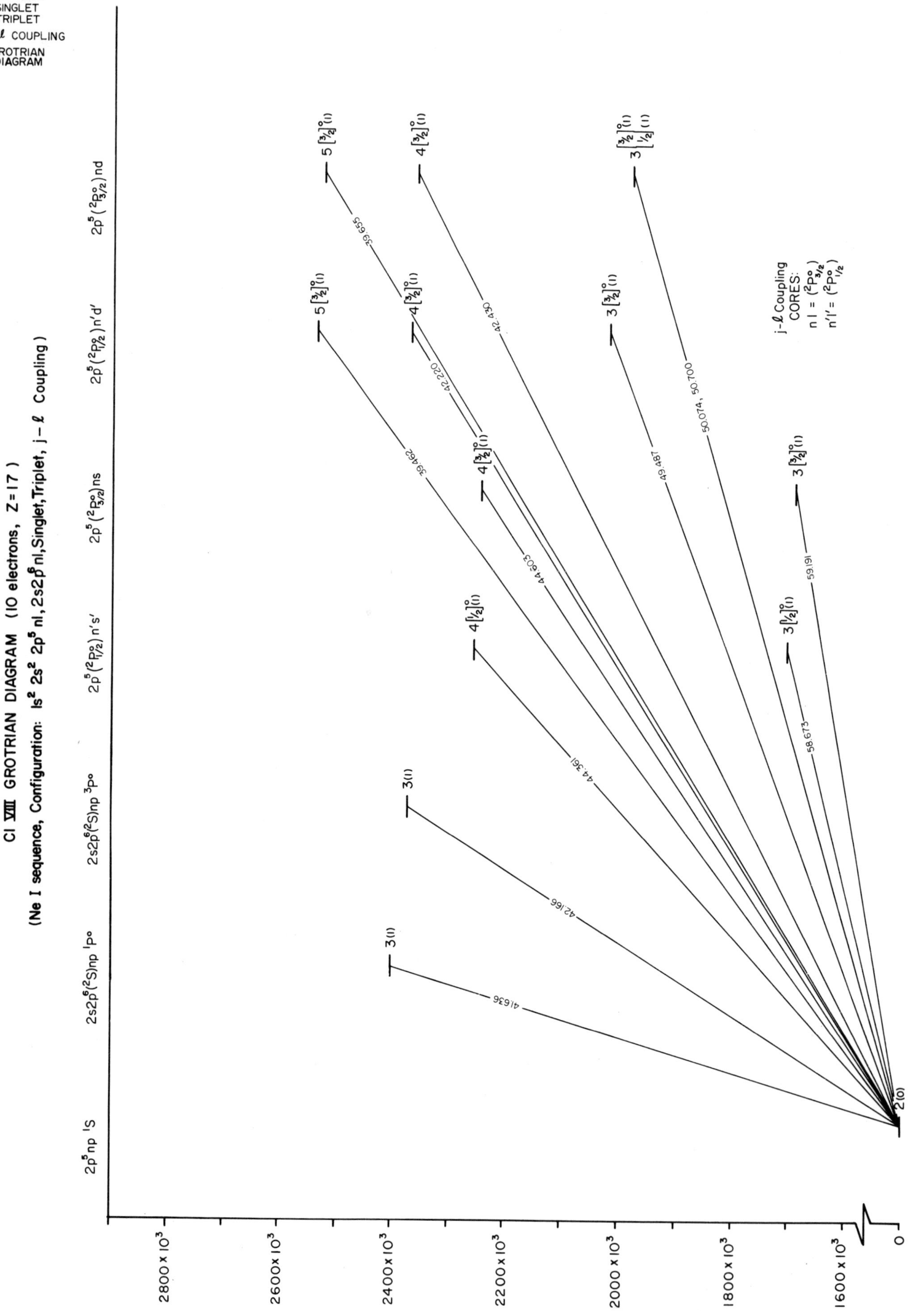

Cl VIII GROTRIAN DIAGRAM (10 electrons, Z=17)

(Ne I sequence, Configuration: $1s^2\ 2s^2\ 2p^5\ nl, 2s2p^6\ nl$, Singlet, Triplet, j – ℓ Coupling)

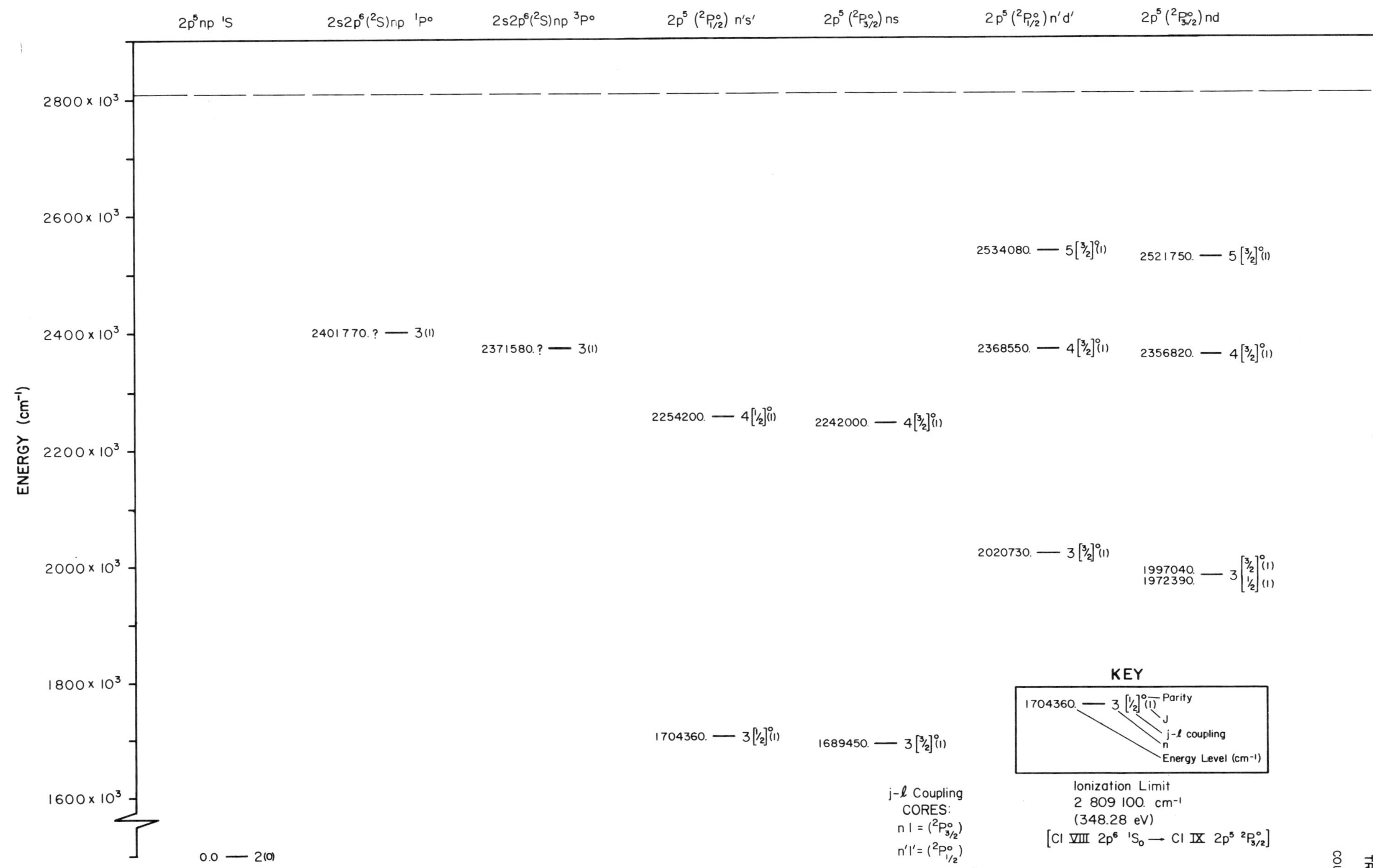
Cl VIII ENERGY LEVELS (10 electrons, Z=17)
(Ne I sequence, Configuration: $1s^2 2s^2 2p^5 nl$, $2s2p^6nl$, Singlet, Triplet, j-ℓ Coupling)
$2p^5np\ ^1S$
$2s2p^6(^2S)np\ ^1P^o$
$2s2p^6(^2S)np\ ^3P^o$
$2p^5(^2P^o_{1/2})n's'$
$2p^5(^2P^o_{3/2})ns$
$2p^5(^2P^o_{1/2})n'd'$
$2p^5(^2P^o_{3/2})nd$
2800×10^3
2600×10^3
2400×10^3
2200×10^3
2000×10^3
1800×10^3
1600×10^3
ENERGY (cm^{-1})
0.0 — 2(0)
2401770.? — 3(1)
2371580.? — 3(1)
2254200. — $4[1/2]^o(1)$
1704360. — $3[1/2]^o(1)$
2242000. — $4[3/2]^o(1)$
1689450. — $3[3/2]^o(1)$
2534080. — $5[3/2]^o(1)$
2368550. — $4[3/2]^o(1)$
2020730. — $3[3/2]^o(1)$
2521750. — $5[3/2]^o(1)$
2356820. — $4[3/2]^o(1)$
1997040. — $3[3/2]^o(1)$
1972390. — $3[1/2](1)$
KEY
1704360. — $3[1/2]^o(1)$
Parity
J
j-ℓ coupling
n
Energy Level (cm^{-1})
j-ℓ Coupling
CORES:
$nl = (^2P^o_{3/2})$
$n'l' = (^2P^o_{1/2})$
Ionization Limit
2 809 100. cm^{-1}
(348.28 eV)
[Cl VIII $2p^6\ ^1S_0$ → Cl IX $2p^5\ ^2P^o_{3/2}$]

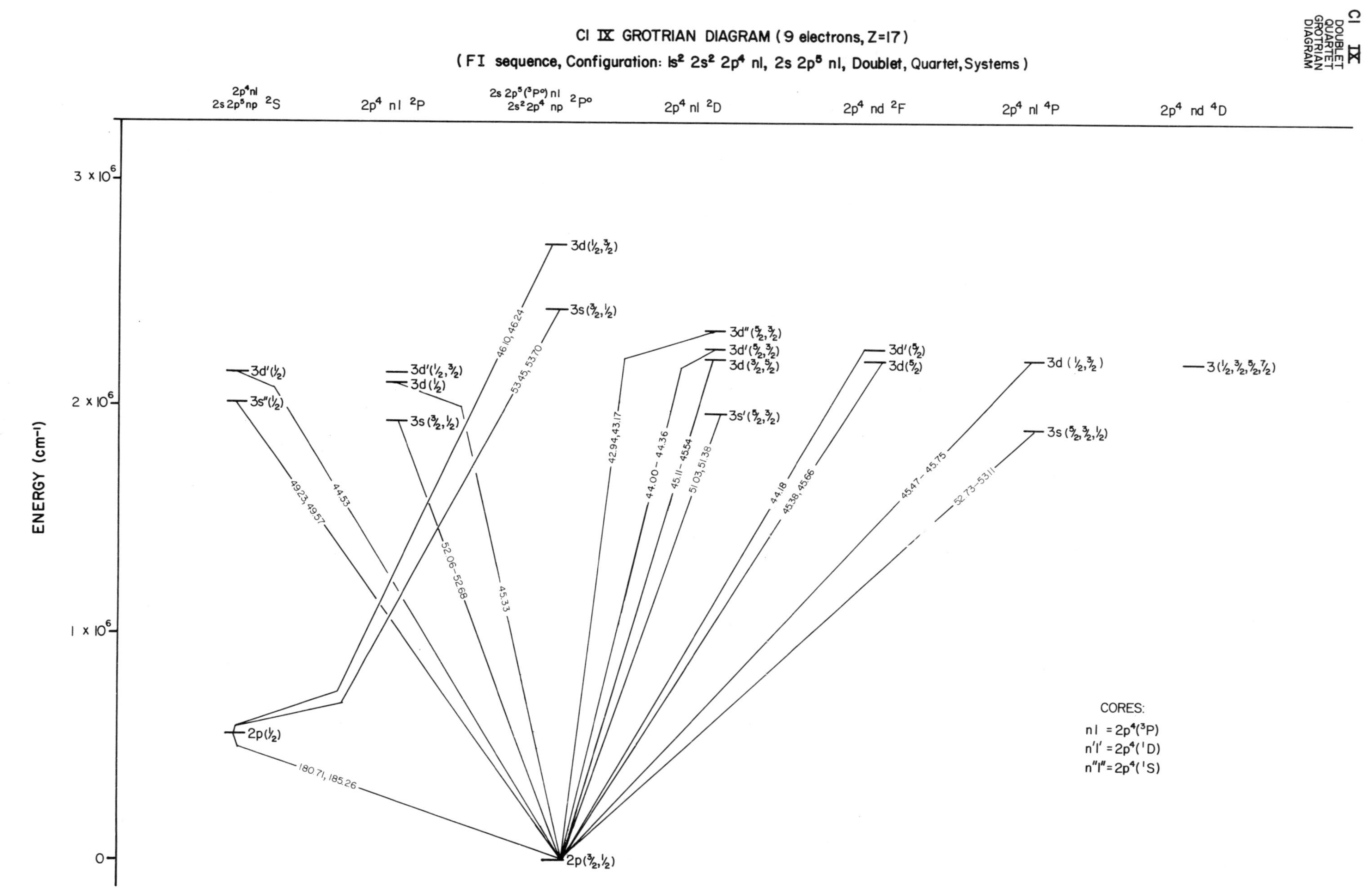

Cl IX
DOUBLET
QUARTET
GROTRIAN
DIAGRAM
Cl IX GROTRIAN DIAGRAM (9 electrons, Z=17)
(FI sequence, Configuration: ls² 2s² 2p⁴ nl, 2s 2p⁵ nl, Doublet, Quartet, Systems)
2p⁴nl 2s2p⁵np ²S
2p⁴ nl ²P
2s 2p⁵(³P°) nl 2s²2p⁴ np ²P°
2p⁴ nl ²D
2p⁴ nd ²F
2p⁴ nl ⁴P
2p⁴ nd ⁴D
ENERGY (cm⁻¹)
3 x 10⁶
2 x 10⁶
1 x 10⁶
0
3d'(½)
3s"(½)
2p(½)
3d'(½, 3/2)
3d(½)
3s(3/2, ½)
3d(½, 3/2)
3s(3/2, ½)
2p(3/2, ½)
3d"(5/2, 3/2)
3d'(5/2, 3/2)
3d(3/2, 5/2)
3s'(5/2, 3/2)
3d'(5/2)
3d(5/2)
3d (½, 3/2)
3s (5/2, 3/2, ½)
3(½, 3/2, 5/2, 7/2)
49.23, 49.57
44.53
180.71, 185.26
52.06–52.68
45.33
46.10, 46.24
53.45, 53.70
42.94, 43.17
44.00 – 44.36
45.11 – 45.54
51.03, 51.38
44.18
45.38, 45.66
45.47 – 45.75
52.73–53.11
CORES:
n l = 2p⁴(³P)
n'l' = 2p⁴(¹D)
n"l" = 2p⁴(¹S)

Cl IX ENERGY LEVELS (9 electrons, Z=17)

(F I sequence, Configuration: $1s^2\ 2s^2\ 2p^4\ nl$, $2s\ 2p^5\ nl$, Doublet, Quartet, Systems)

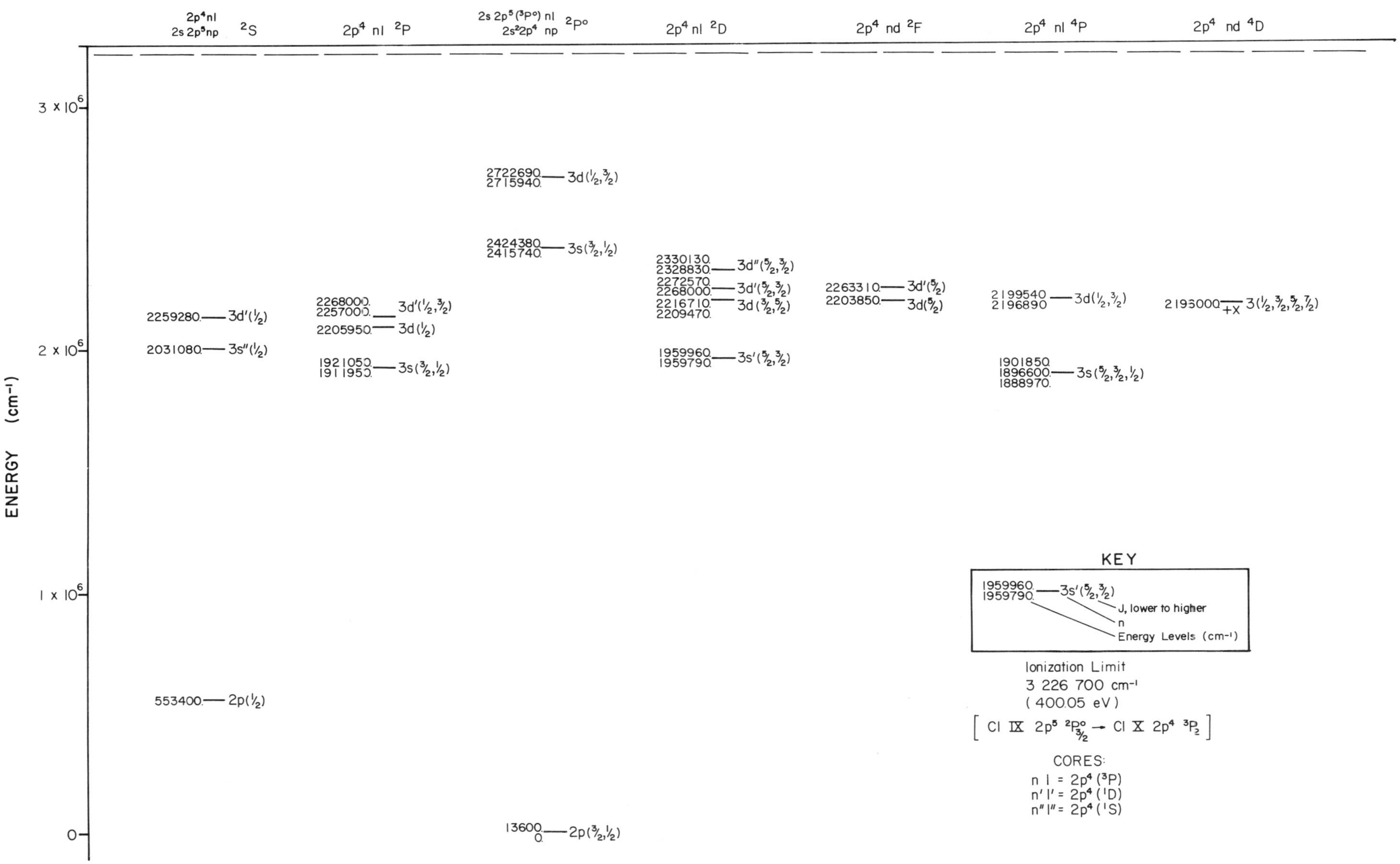

ENERGY (cm^{-1})

Axis: 0 — 1×10^6 — 2×10^6 — 3×10^6

$2p^4nl$ / $2s\,2p^5np$ ^2S

- 2259280 —— $3d'(\frac{1}{2})$
- 2031080 —— $3s''(\frac{1}{2})$
- 553400 —— $2p(\frac{1}{2})$

$2p^4$ nl ^2P

- 2268000, 2257000 —— $3d'(\frac{1}{2},\frac{3}{2})$
- 2205950 —— $3d(\frac{1}{2})$
- 1921050, 1911950 —— $3s(\frac{3}{2},\frac{1}{2})$

$2s\,2p^5(^3P^o)$ nl / $2s^22p^4$ np $^2P^o$

- 2722690, 2715940 —— $3d(\frac{1}{2},\frac{3}{2})$
- 2424380, 2415740 —— $3s(\frac{3}{2},\frac{1}{2})$
- 13600, 0 —— $2p(\frac{3}{2},\frac{1}{2})$

$2p^4$ nl ^2D

- 2330130, 2328830 —— $3d''(\frac{5}{2},\frac{3}{2})$
- 2272570, 2268000 —— $3d'(\frac{5}{2},\frac{3}{2})$
- 2216710, 2209470 —— $3d(\frac{3}{2},\frac{5}{2})$
- 1959960, 1959790 —— $3s'(\frac{5}{2},\frac{3}{2})$

$2p^4$ nd ^2F

- 2263310 —— $3d'(\frac{5}{2})$
- 2203850 —— $3d(\frac{5}{2})$

$2p^4$ nl ^4P

- 2199540, 2196890 —— $3d(\frac{1}{2},\frac{3}{2})$
- 1901850, 1896600, 1888970 —— $3s(\frac{5}{2},\frac{3}{2},\frac{1}{2})$

$2p^4$ nd ^4D

- 2195000 +X —— $3(\frac{1}{2},\frac{3}{2},\frac{5}{2},\frac{7}{2})$

KEY

1959960, 1959790 —— $3s'(\frac{5}{2},\frac{3}{2})$
- J, lower to higher
- n
- Energy Levels (cm^{-1})

Ionization Limit
3 226 700 cm^{-1}
(400.05 eV)
[Cl IX $2p^5\ ^2P^o_{3/2}$ → Cl X $2p^4\ ^3P_2$]

CORES:
n l = $2p^4(^3P)$
n′ l′ = $2p^4(^1D)$
n″ l″ = $2p^4(^1S)$

Cl X
SINGLET
GROTRIAN
DIAGRAM

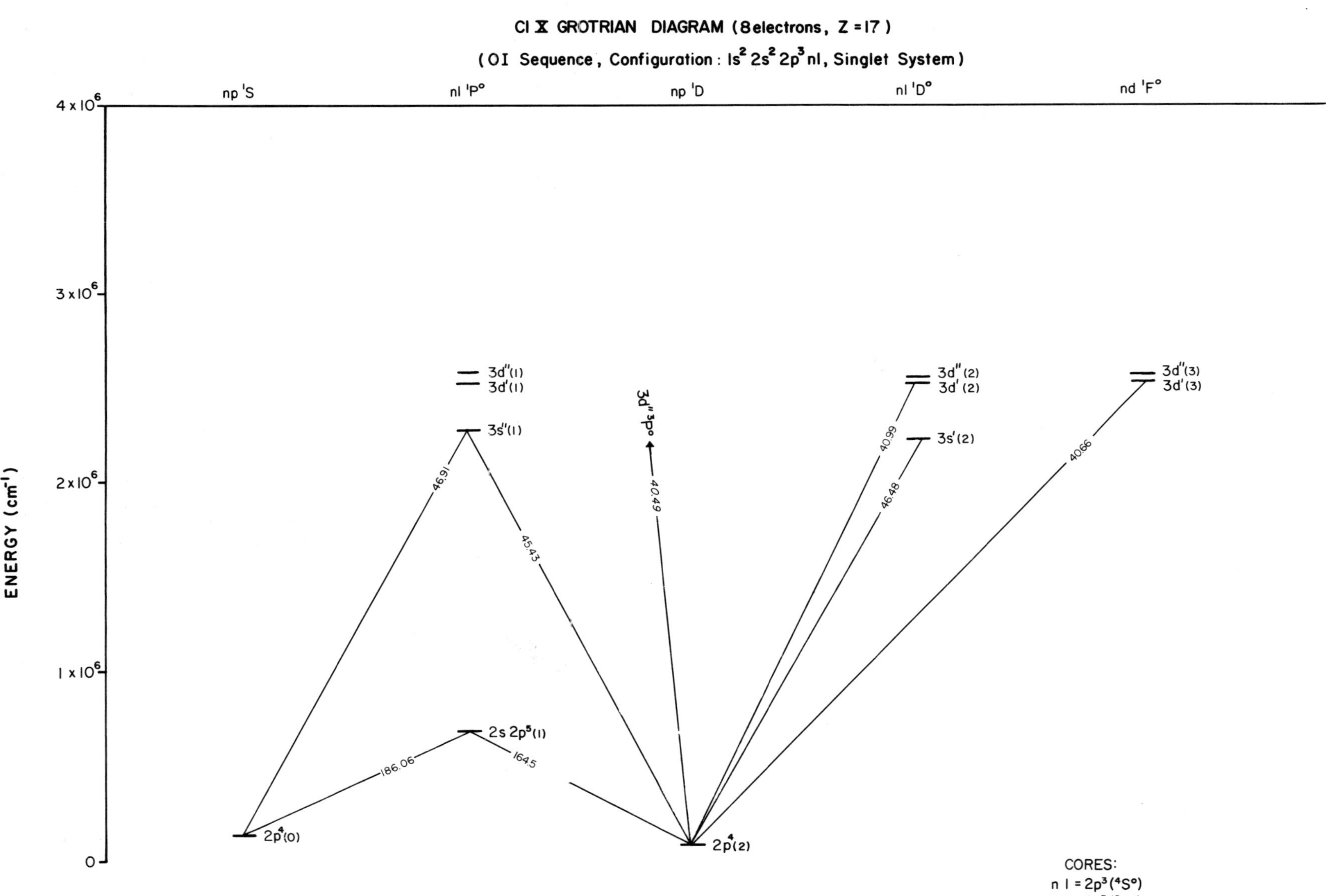

Cl X ENERGY LEVELS (8 electrons, Z = 17)

(OI sequence, Configuration: $1s^2 2s^2 2p^3 nl$, Singlet System)

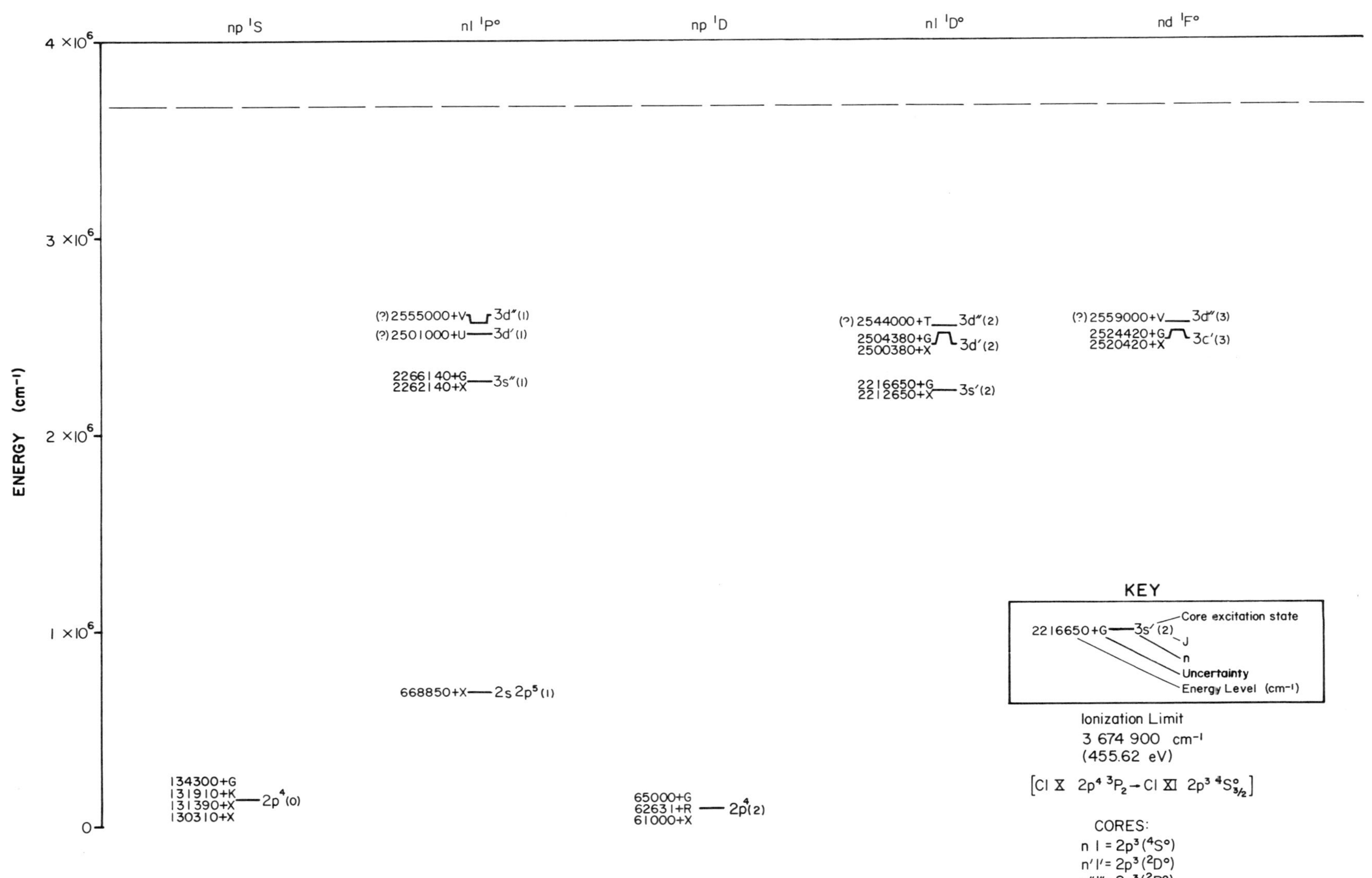

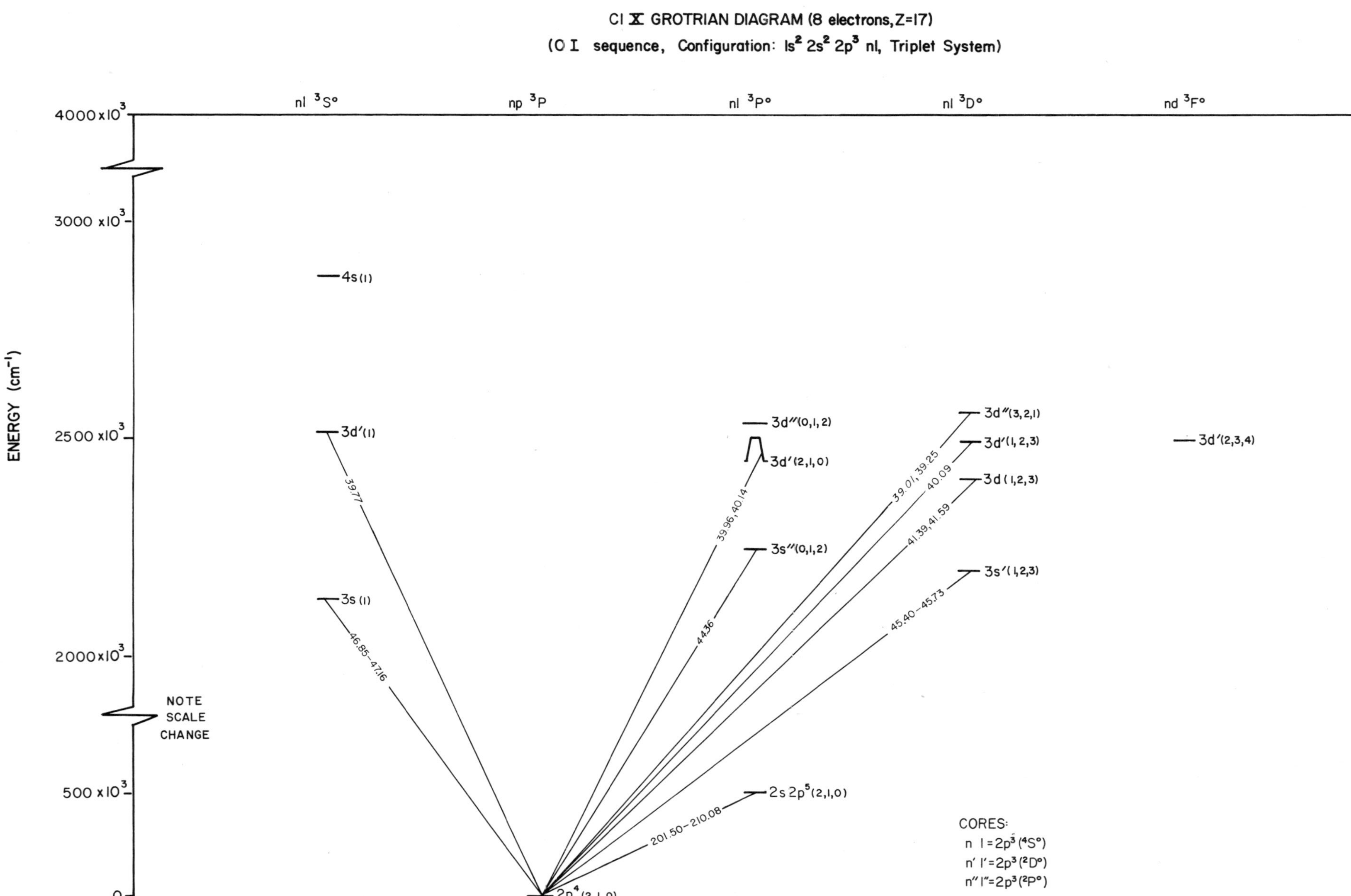
Cl X GROTRIAN DIAGRAM (8 electrons, Z=17)
(O I sequence, Configuration: $1s^2 2s^2 2p^3$ nl, Triplet System)
nl $^3S°$
np ^3P
nl $^3P°$
nl $^3D°$
nd $^3F°$
4000 x10^3
3000 x10^3
2500 x10^3
2000 x10^3
500 x10^3
0
ENERGY (cm^{-1})
NOTE
SCALE
CHANGE
4s (1)
3d′(1)
3s (1)
39.77
46.85–47.16
3d″(0,1,2)
3d′(2,1,0)
3s″(0,1,2)
39.96, 40.14
44.36
3d″(3,2,1)
3d′(1,2,3)
3d (1,2,3)
3s′(1,2,3)
39.01, 39.25
40.09
41.39, 41.59
45.40–45.73
3d′(2,3,4)
2s 2p^5(2,1,0)
201.50–210.08
2p^4(2,1,0)
CORES:
n l = $2p^3(^4S°)$
n′ l′ = $2p^3(^2D°)$
n″ l″ = $2p^3(^2P°)$

Cl X ENERGY LEVELS (8 electrons, Z=17)

(O I sequence, Configuration: $1s^2 2s^2 2p^3$ nl, Triplet System)

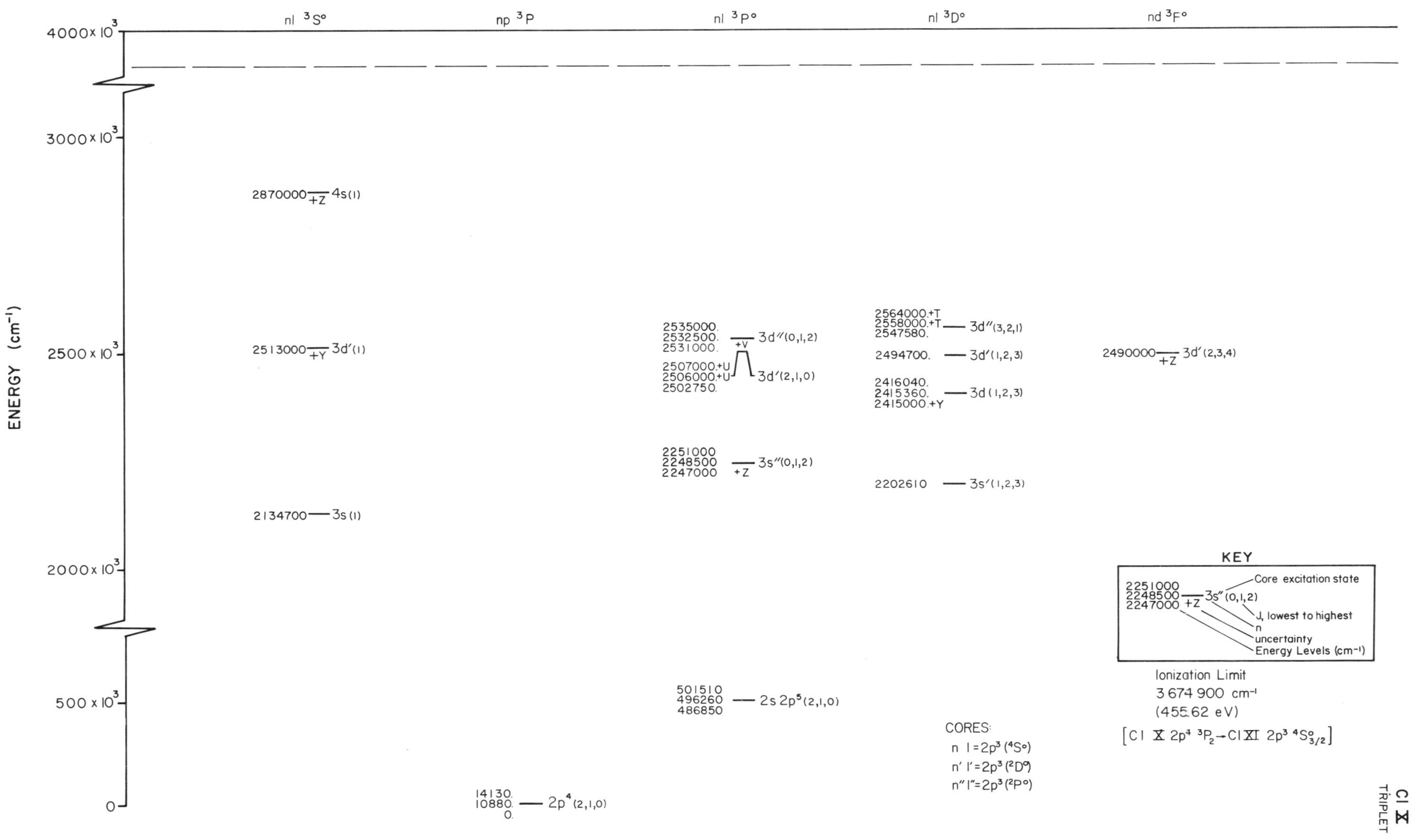

nl $^3S^o$ — np ^3P — nl $^3P^o$ — nl $^3D^o$ — nd $^3F^o$

ENERGY (cm^{-1})

4000×10^3

3000×10^3

2500×10^3

2000×10^3

500×10^3

0

nl $^3S^o$:

2870000 +Z — 4s(1)

2513000 +Y — 3d′(1)

2134700 — 3s(1)

np ^3P:

14130. 10880. 0. — $2p^4$(2,1,0)

nl $^3P^o$:

2535000. 2532500. 2531000. +V — 3d″(0,1,2)

2507000.+U 2506000.+U 2502750. — 3d′(2,1,0)

2251000 2248500 2247000 +Z — 3s″(0,1,2)

501510 496260 486850 — $2s\,2p^5$(2,1,0)

nl $^3D^o$:

2564000.+T 2558000.+T 2547580. — 3d″(3,2,1)

2494700. — 3d′(1,2,3)

2416040. 2415360. 2415000.+Y — 3d(1,2,3)

2202610 — 3s′(1,2,3)

nd $^3F^o$:

2490000 +Z — 3d′(2,3,4)

CORES:

n l = $2p^3$ ($^4S^o$)

n′ l′ = $2p^3$ ($^2D^o$)

n″ l″ = $2p^3$ ($^2P^o$)

KEY

2251000 2248500 2247000 +Z — 3s″(0,1,2)

- Core excitation state
- J, lowest to highest
- n
- uncertainty
- Energy Levels (cm^{-1})

Ionization Limit

3 674 900 cm^{-1}

(455.62 eV)

[Cl X $2p^4\ ^3P_2$ → Cl XI $2p^3\ ^4S^o_{3/2}$]

Cl X TRIPLET

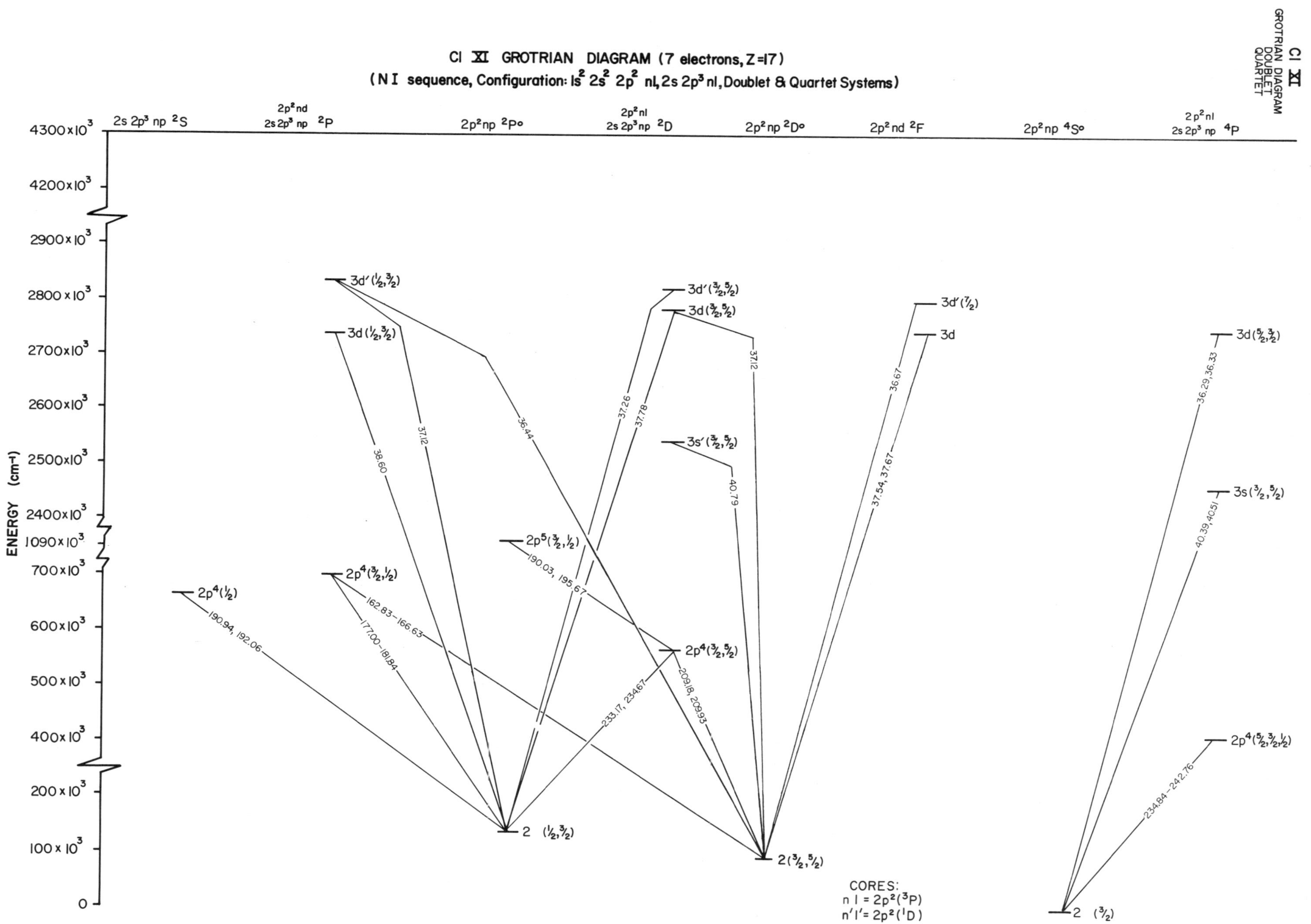
Cl XI
GROTRIAN DIAGRAM
DOUBLET
QUARTET
Cl XI GROTRIAN DIAGRAM (7 electrons, Z=17)
(N I sequence, Configuration: $1s^2 2s^2 2p^2 nl$, $2s 2p^3 nl$, Doublet & Quartet Systems)
$2s 2p^3 np\ ^2S$
$2p^2nd$ $2s 2p^3 np\ ^2P$
$2p^2np\ ^2P^o$
$2p^2nl$ $2s 2p^3 np\ ^2D$
$2p^2np\ ^2D^o$
$2p^2nd\ ^2F$
$2p^2np\ ^4S^o$
$2p^2nl$ $2s 2p^3 np\ ^4P$
ENERGY (cm^{-1})
4300×10^3
4200×10^3
2900×10^3
2800×10^3
2700×10^3
2600×10^3
2500×10^3
2400×10^3
1090×10^3
700×10^3
600×10^3
500×10^3
400×10^3
200×10^3
100×10^3
0
$3d'(1/2,3/2)$
$3d(1/2,3/2)$
$2p^4(3/2,1/2)$
$2p^4(1/2)$
$2p^5(3/2,1/2)$
$3d'(3/2,5/2)$
$3d(3/2,5/2)$
$3s'(3/2,5/2)$
$2p^4(3/2,5/2)$
$3d'(7/2)$
3d
$3d(5/2,3/2)$
$3s(3/2,5/2)$
$2p^4(5/2,3/2,1/2)$
$2\ (1/2,3/2)$
$2(3/2,5/2)$
$2\ (3/2)$
38.60
37.12
36.44
37.26
37.78
37.12
40.79
36.67
37.54, 37.67
36.29, 36.33
40.39, 40.51
234.84–242.76
190.94, 192.06
177.00–181.84
162.83–166.63
190.03, 195.67
233.17, 234.67
209.18, 209.93
CORES:
n l = $2p^2(^3P)$
n′l′ = $2p^2(^1D)$

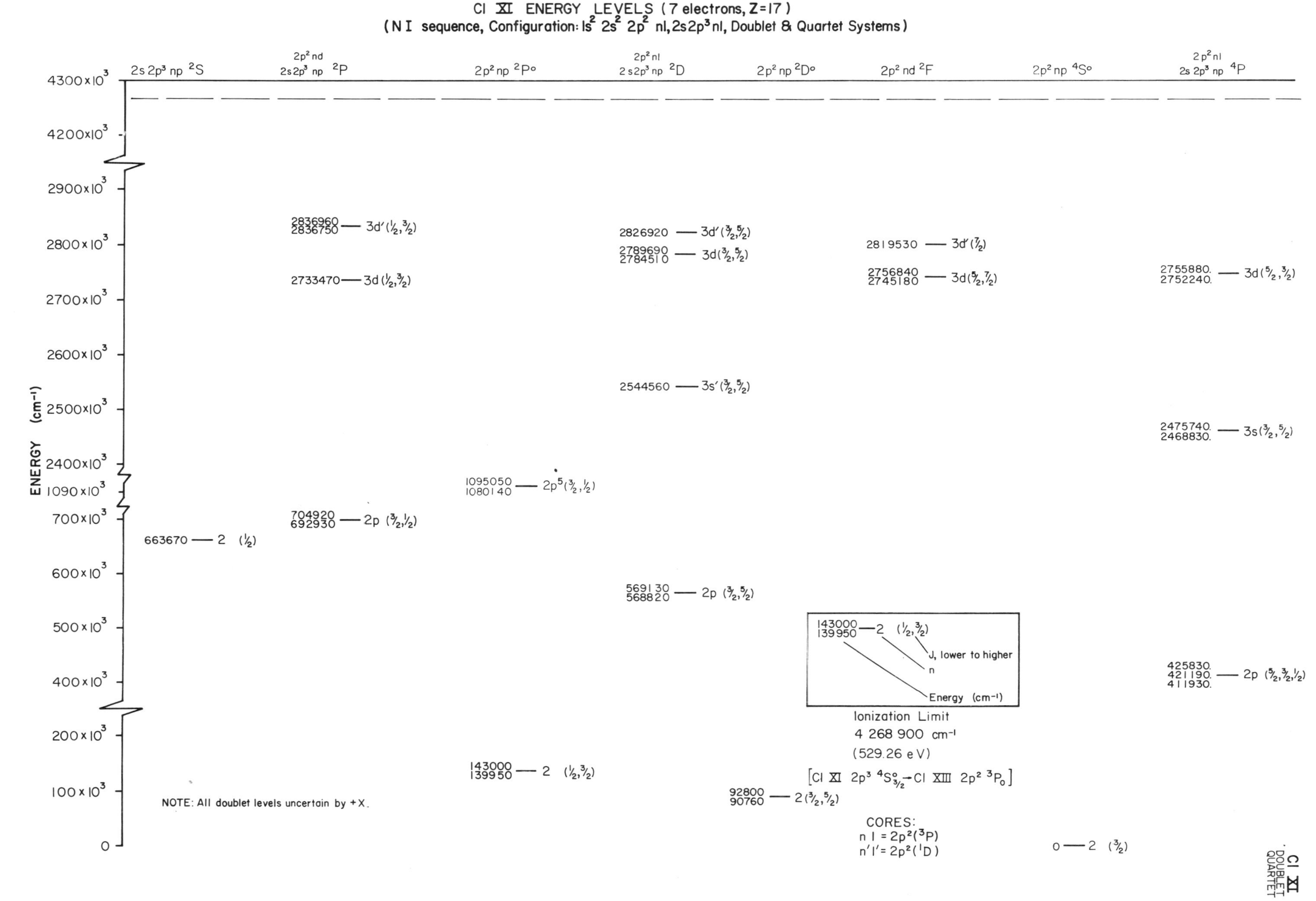

Cl XI ENERGY LEVELS (7 electrons, Z=17)
(N I sequence, Configuration: 1s2 2s2 2p2 nl, 2s2p3 nl, Doublet & Quartet Systems)
2s2p3 np 2S
2p2 nd
2s2p3 np 2P
2p2 np 2Po
2p2 nl
2s2p3 np 2D
2p2 np 2Do
2p2 nd 2F
2p2 np 4So
2p2 nl
2s2p3 np 4P
ENERGY (cm-1)
4300x10^3
4200x10^3
2900x10^3
2800x10^3
2700x10^3
2600x10^3
2500x10^3
2400x10^3
1090x10^3
700x10^3
600x10^3
500x10^3
400x10^3
200x10^3
100x10^3
0
2836960
2836750 — 3d'(½,3/2)
2733470 — 3d(½,3/2)
704920
692930 — 2p (3/2,½)
663670 — 2 (½)
1095050
1080140 — 2p5(3/2,½)
143000
139950 — 2 (½,3/2)
2826920 — 3d'(3/2,5/2)
2789690
2784510 — 3d(3/2,5/2)
2544560 — 3s'(3/2,5/2)
569130
568820 — 2p (3/2,5/2)
92800
90760 — 2(3/2,5/2)
2819530 — 3d'(7/2)
2756840
2745180 — 3d(5/2,7/2)
143000
139950 — 2 (½,3/2)
J, lower to higher
n
Energy (cm-1)
Ionization Limit
4 268 900 cm-1
(529.26 eV)
[Cl XI 2p3 4So3/2 → Cl XIII 2p2 3P0]
CORES:
n l = 2p2(3P)
n'l' = 2p2(1D)
0 — 2 (3/2)
2755880.
2752240. — 3d(5/2,3/2)
2475740.
2468830. — 3s(3/2,5/2)
425830.
421190.
411930. — 2p (5/2,3/2,½)
NOTE: All doublet levels uncertain by +X.
Cl XI
DOUBLET
QUARTET

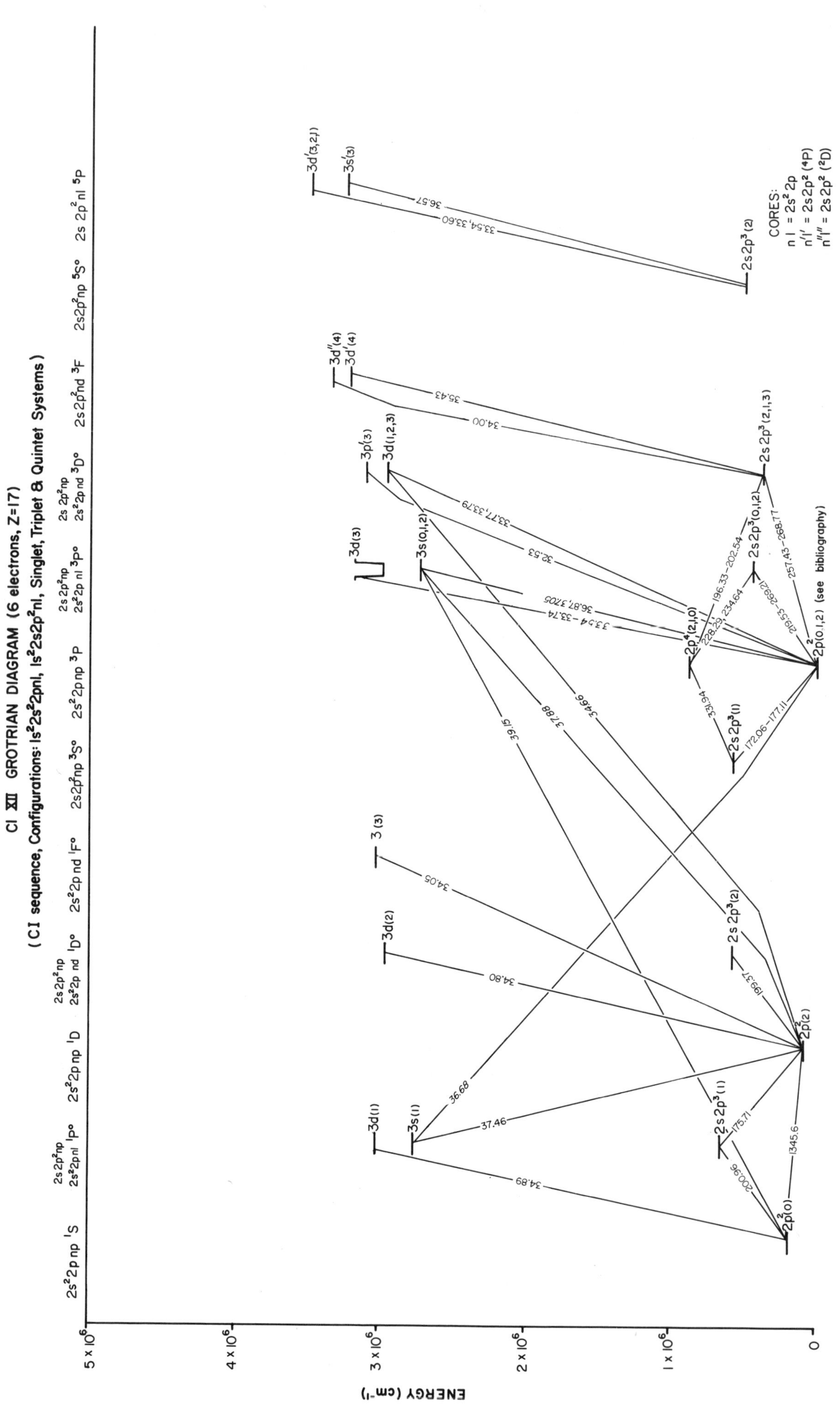

Cl XII GROTRIAN DIAGRAM (6 electrons, Z=17)

(C I sequence, Configurations: $1s^22s^22pnl$, $1s^22s2p^2nl$, Singlet, Triplet & Quintet Systems)

$2s^22p\,np\ {}^1S$
$2s2p^2np$ / $2s^22pnl\ {}^1P^o$
$2s^22p\,np\ {}^1D$
$2s2p^2np$ / $2s^22p\,nd\ {}^1D^o$
$2s^22p\,nd\ {}^1F^o$
$2s2p^2np\ {}^3S^o$
$2s^22p\,np\ {}^3P$
$2s2p^2np$ / $2s^22p\,nl\ {}^3P^o$
$2s2p^2np$ / $2s^22p\,nd\ {}^3D^o$
$2s2p^2nd\ {}^3F$
$2s2p^2np\ {}^5S^o$
$2s2p^2nl\ {}^5P$

ENERGY (cm^{-1})

5×10^6
4×10^6
3×10^6
2×10^6
1×10^6
0

$2p^2(0)$
$2s2p^3(1)$
3s(1)
3d(1)
$2p^2(2)$
$2s2p^3(2)$
3d(2)
3 (3)
$2s2p^3(1)$
$2p^4(2,1,0)$
$2p^2(0,1,2)$ (see bibliography)
$2s2p^3(0,1,2)$
3s(0,1,2)
3d(3)
$2s2p^3(2,1,3)$
3d(1,2,3)
3p′(3)
3d′(4)
3d″(4)
$2s2p^3(2)$
3s′(3)
3d′(3,2,1)

34.89
200.96
175.71
1345.6
37.46
36.68
34.80
199.37
34.05
39.15
37.88
34.66
331.94
172.06–177.11
228.29, 234.64
196.33–202.54
219.53–269.21
257.43–268.77
33.54–33.74
36.87, 37.05
32.53
33.77, 33.79
34.00
35.43
33.54, 33.60
36.57

CORES:
n l = $2s^2\,2p$
n′l′ = $2s2p^2$ (^4P)
n″l″ = $2s2p^2$ (^2D)

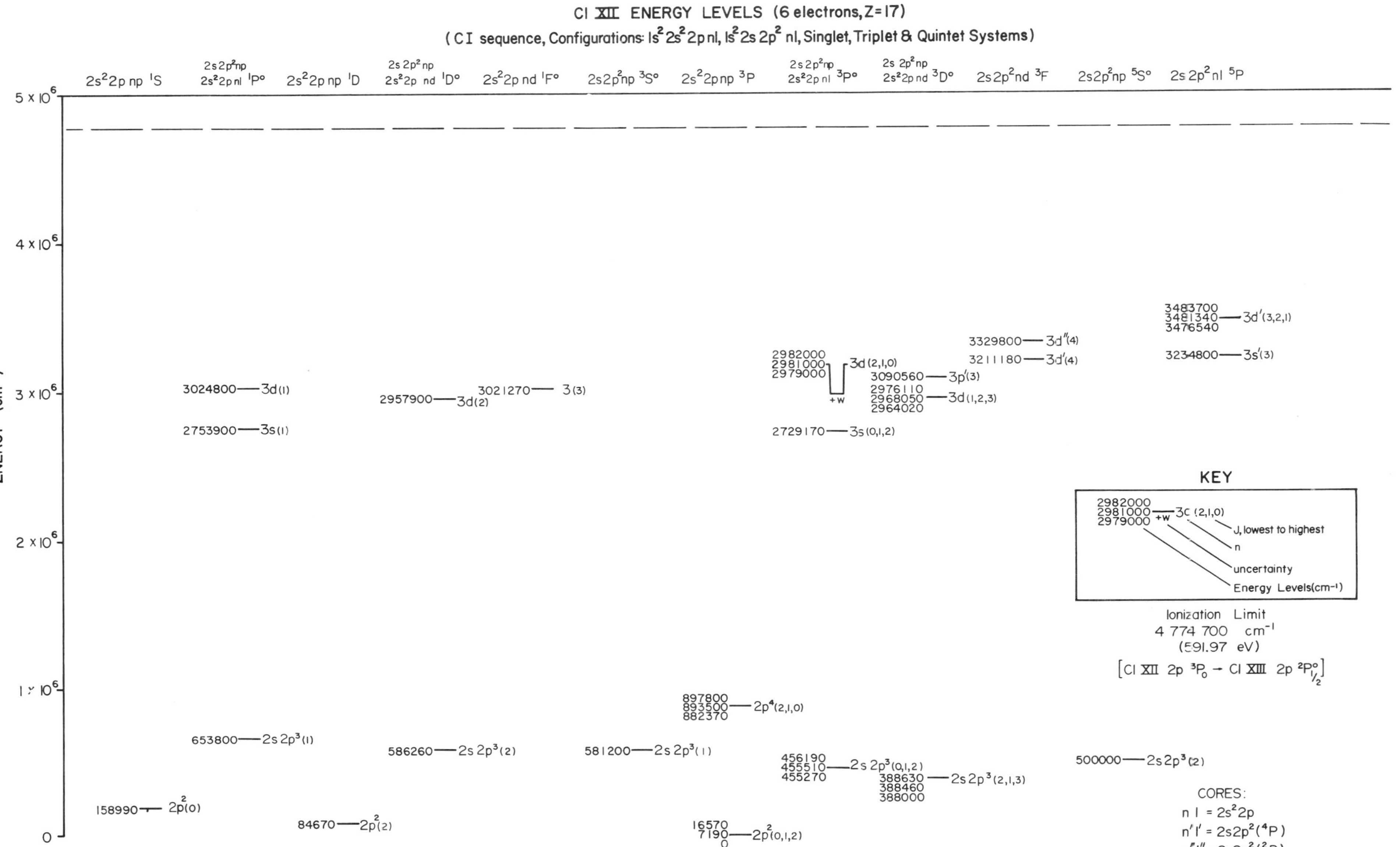
Cl XII ENERGY LEVELS (6 electrons, Z=17)
(CI sequence, Configurations: 1s²2s²2pnl, 1s²2s2p² nl, Singlet, Triplet & Quintet Systems)
2s²2p np ¹S
2s2p²np 2s²2pnl ¹P°
2s²2p np ¹D
2s2p²np 2s²2p nd ¹D°
2s²2p nd ¹F°
2s2p²np ³S°
2s²2pnp ³P
2s2p²np 2s²2pnl ³P°
2s2p²np 2s²2pnd ³D°
2s2p²nd ³F
2s2p²np ⁵S°
2s2p²nl ⁵P
ENERGY (cm⁻¹)
5 x 10⁶
4 x 10⁶
3 x 10⁶
2 x 10⁶
1 x 10⁶
0
158990 — 2p²(0)
3024800 — 3d(1)
2753900 — 3s(1)
653800 — 2s2p³(1)
84670 — 2p²(2)
2957900 — 3d(2)
586260 — 2s2p³(2)
3021270 — 3(3)
581200 — 2s2p³(1)
897800 893500 882370 — 2p⁴(2,1,0)
16570 7190 0 — 2p²(0,1,2)
2982000 2981000 2979000 +w — 3d(2,1,0)
2729170 — 3s(0,1,2)
456190 455510 455270 — 2s2p³(0,1,2)
3090560 — 3p′(3)
2976110 2968050 2964020 — 3d(1,2,3)
388630 388460 388000 — 2s2p³(2,1,3)
3329800 — 3d″(4)
3211180 — 3d′(4)
500000 — 2s2p³(2)
3483700 3481340 3476540 — 3d′(3,2,1)
3234800 — 3s′(3)
KEY
2982000 2981000 2979000 +w — 3c (2,1,0)
J, lowest to highest
n
uncertainty
Energy Levels(cm⁻¹)
Ionization Limit
4 774 700 cm⁻¹
(591.97 eV)
[Cl XII 2p ³P₀ → Cl XIII 2p ²P°½]
CORES:
n l = 2s²2p
n′l′ = 2s2p²(⁴P)
n″l″ = 2s2p²(²D)
NOTE: All singlet levels uncertain by +x, quintet levels by +L.

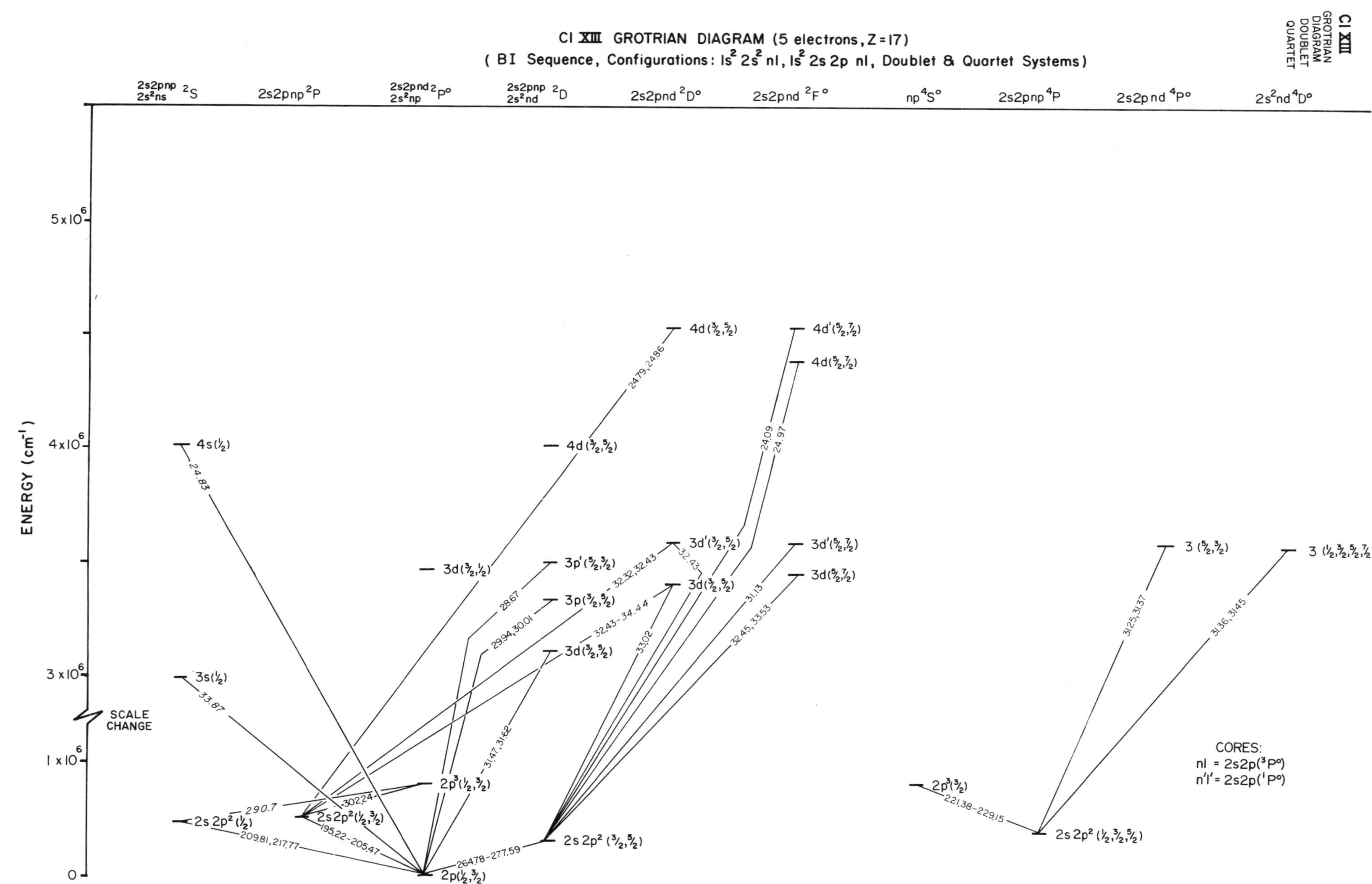
Cl XIII
GROTRIAN
DIAGRAM
DOUBLET
QUARTET
Cl XIII GROTRIAN DIAGRAM (5 electrons, Z=17)
(BI Sequence, Configurations: $1s^2 2s^2 nl$, $1s^2 2s 2p\ nl$, Doublet & Quartet Systems)
2s2pnp 2s²ns ²S
2s2pnp ²P
2s2pnd 2s²np ²P°
2s2pnp 2s²nd ²D
2s2pnd ²D°
2s2pnd ²F°
np ⁴S°
2s2pnp ⁴P
2s2pnd ⁴P°
2s²nd ⁴D°
ENERGY (cm⁻¹)
5x10⁶
4x10⁶
3x10⁶
1x10⁶
0
SCALE CHANGE
4s(1/2)
24.83
3s(1/2)
33.87
2s 2p²(1/2)
290.7
209.81, 217.77
2s 2p²(1/2, 3/2)
302.24
195.22–205.47
2p(1/2, 3/2)
264.78–277.59
2p³(1/2, 3/2)
3d(3/2, 1/2)
28.67
3p'(5/2, 3/2)
29.94, 30.01
3p(3/2, 5/2)
32.32, 32.43
32.43–34.44
3d(3/2, 5/2)
31.47, 31.62
4d(3/2, 5/2)
2479, 2486
4d(3/2, 5/2)
3d'(3/2, 5/2)
32.43
3d(3/2, 5/2)
33.02
2s 2p² (3/2, 5/2)
24.09
24.97
4d'(5/2, 7/2)
4d(5/2, 7/2)
3d'(5/2, 7/2)
3d(5/2, 7/2)
31.13
32.45, 33.53
2p³(3/2)
221.38–229.15
2s 2p² (1/2, 3/2, 5/2)
3 (5/2, 3/2)
31.25, 31.37
3 (1/2, 3/2, 5/2, 7/2)
31.36, 31.45
CORES:
nl = 2s2p(³P°)
n'l' = 2s2p(¹P°)

Cl XIII ENERGY LEVELS (5 electrons, Z=17)

(BI sequence, Configurations: $1s^2\ 2s^2\ nl$, $1s^2\ 2s\ 2p\ nl$, Doublet & Quartet Systems)

Columns: $2s2pnp\ ^2S$ / $2s^2ns$; $2s2pnp\ ^2P$; $2s2pnd\ ^2P^o$ / $2s^2np$; $2s2pnp\ ^2D$ / $2s^2nd$; $2s2pnd\ ^2D^o$; $2s2pnd\ ^2F^o$; $np\ ^4S^o$; $2s2pnp\ ^4P$; $2s2pnd\ ^4P^o$; $2s^2nd\ ^4D^o$

ENERGY (cm^{-1}): 0; 1×10^6; SCALE CHANGE; 3×10^6; 4×10^6; 5×10^6

^2S:
- 4044900 — 4s(1/2)
- 2969700 — 3s(1/2)
- 476620 — $2s2p^2$(1/2)

^2P:
- 512230, 504066 — $2s2p^2$(1/2, 3/2)

$^2P^o$:
- 3476000, 3459000+S — 3d(3/2, 1/2)
- 823000, 822000+Y — $2p^3$(1/2, 3/2)
- 17411, 0 — 2p(1/2, 3/2)

^2D:
- 4116920 — 4d(3/2, 5/2)
- 3506000 — 3p′(5/2, 3/2)
- 3349600, 3340000 — 3p(3/2, 5/2)
- 3180000, 3178130 — 3d(3/2, 5/2)
- 378200, 377670 — $2s2p^2$(3/2, 5/2)

$^2D^o$:
- 4536400 — 4d(3/2, 5/2)
- 3606330, 3587670 — 3d′(3/2, 5/2)
- 3406670 — 3d(3/2, 5/2)

$^2F^o$:
- 4529300 — 4d′(5/2, 7/2)
- 4383000 — 4d(5/2, 7/2)
- 3590950 — 3d′(5/2, 7/2)
- 3460340, 3451660 — 3d(5/2, 7/2)

$^4S^o$:
- 836400 — $2p^3$(3/2)

^4P:
- 400000, 391160, 384680 — $2s2p^2$(1/2, 3/2, 5/2)

$^4P^o$:
- 3590860, 3588270 — 3(5/2, 3/2)

$^4D^o$:
- 3579850, 3575800, 3573660, 3573600 — 3(1/2, 3/2, 5/2, 7/2)

KEY

3606330, 3587670 — 3d′(5/2, 5/2): Core excitation state; J, lower to higher; n; Energy Levels (cm^{-1})

Ionization Limit
5 296 700 cm^{-1}
(656.69 eV)

[Cl XIII $2s^2 2p\ ^2P^o_{1/2}$ Cl XIV $2s^2\ ^1S_0$]

CORES:
nl = 2s2p ($^3P^o$)
n′l′ = 2s2p ($^1P^o$)

NOTE: All quartet levels uncertain by +R.

Cl XIII
DOUBLET
QUARTET

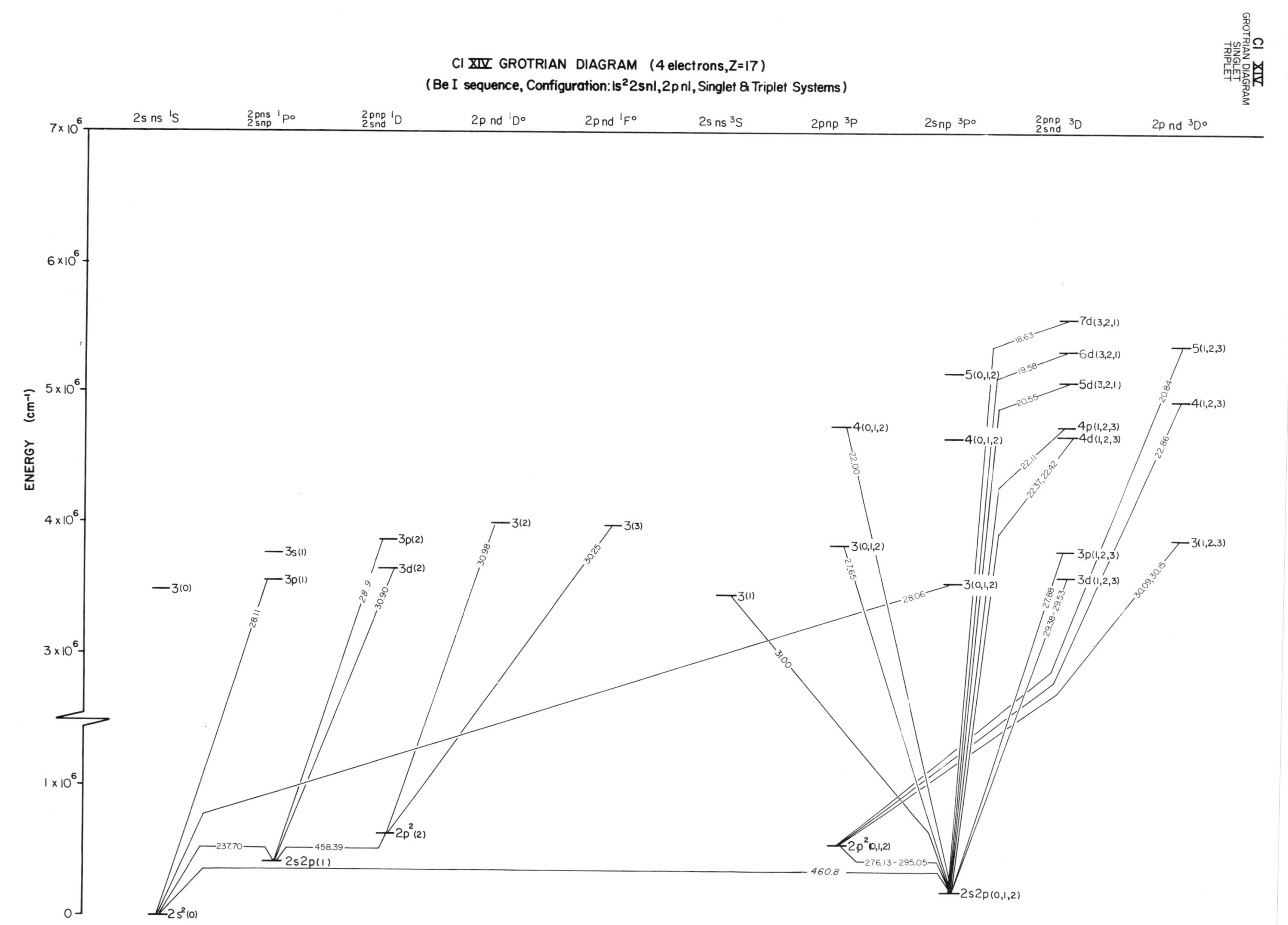
Cl XIV GROTRIAN DIAGRAM (4 electrons, Z=17)
(Be I sequence, Configuration: 1s²2snl, 2pnl, Singlet & Triplet Systems)
ENERGY (cm⁻¹)
7 x 10⁶
6 x 10⁶
5 x 10⁶
4 x 10⁶
3 x 10⁶
1 x 10⁶
0
2s ns ¹S
2pns 2snp ¹P°
2pnp 2snd ¹D
2p nd ¹D°
2p nd ¹F°
2s ns ³S
2pnp ³P
2snp ³P°
2pnp 2snd ³D
2p nd ³D°
3(0)
2s²(0)
3s(1)
3p(1)
2s2p(1)
3p(2)
3d(2)
2p²(2)
3(2)
3(3)
3(1)
4(0,1,2)
3(0,1,2)
2p²(0,1,2)
5(0,1,2)
4(0,1,2)
3(0,1,2)
2s2p(0,1,2)
7d(3,2,1)
6d(3,2,1)
5d(3,2,1)
4p(1,2,3)
4d(1,2,3)
3p(1,2,3)
3d(1,2,3)
5(1,2,3)
4(1,2,3)
3(1,2,3)
28.11
237.70
458.39
28.9
30.90
30.98
30.25
460.8
28.06
31.00
22.00
27.65
276.13 - 295.05
18.63
19.58
20.55
22.11
22.37, 22.42
27.88
29.38 - 29.53
20.84
22.86
30.09, 30.15

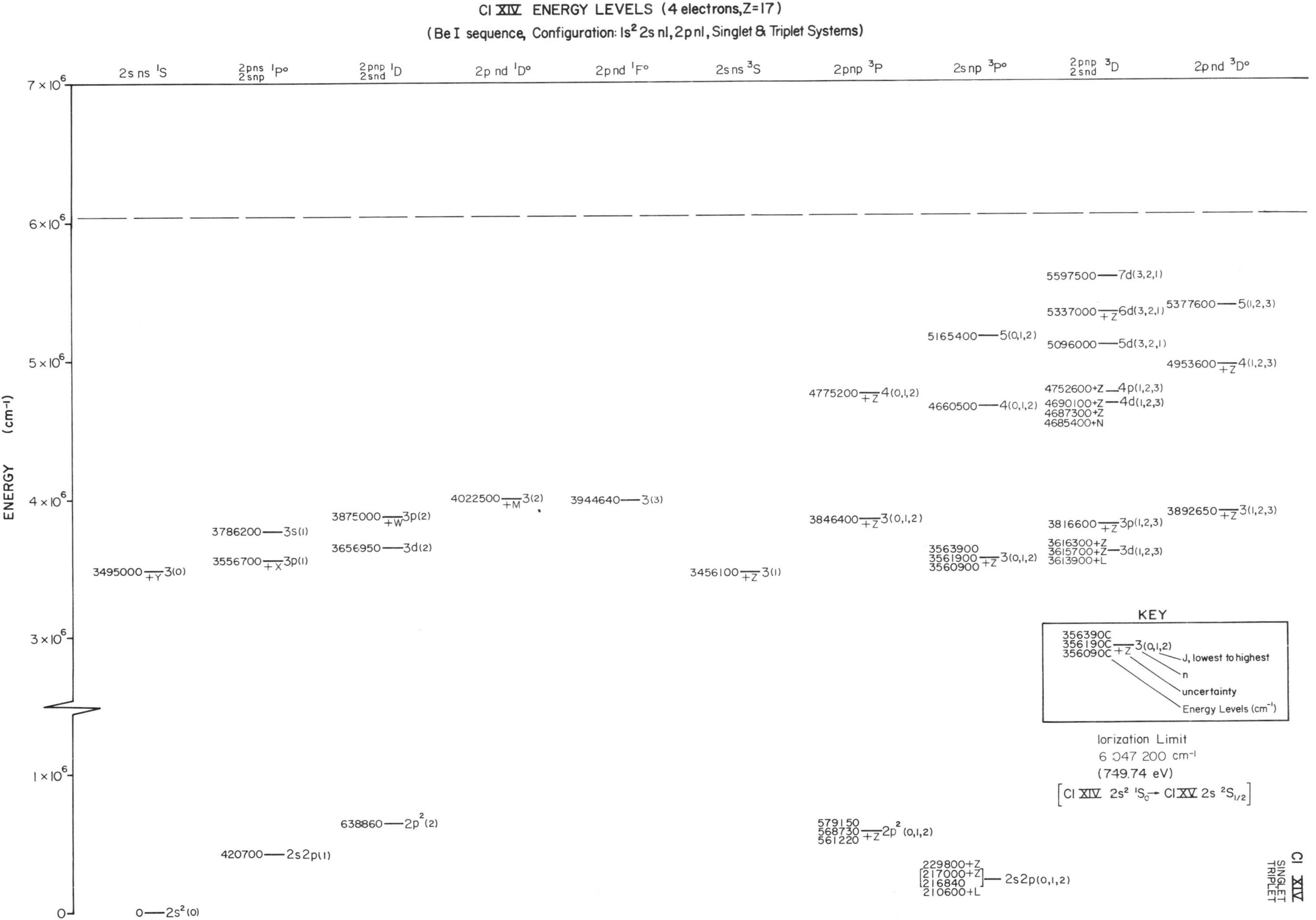

Cl XIV ENERGY LEVELS (4 electrons, Z=17)
(Be I sequence, Configuration: 1s² 2s nl, 2p nl, Singlet & Triplet Systems)
2s ns ¹S
2pns ¹P°
2snp
2pnp ¹D
2snd
2p nd ¹D°
2p nd ¹F°
2s ns ³S
2pnp ³P
2s np ³P°
2pnp ³D
2snd
2p nd ³D°
ENERGY (cm⁻¹)
7×10⁶
6×10⁶
5×10⁶
4×10⁶
3×10⁶
1×10⁶
0
3495000 +Y 3(0)
0 — 2s²(0)
3786200 — 3s(1)
3556700 +X 3p(1)
420700 — 2s2p(1)
3875000 +W 3p(2)
3656950 — 3d(2)
638860 — 2p²(2)
4022500 +M 3(2)
3944640 — 3(3)
3456100 +Z 3(1)
4775200 +Z 4(0,1,2)
3846400 +Z 3(0,1,2)
579150
568730 +Z 2p² (0,1,2)
561220
5165400 — 5(0,1,2)
4660500 — 4(0,1,2)
3563900
3561900
3560900 +Z 3(0,1,2)
229800+Z
217000+Z
216840
210600+L — 2s2p(0,1,2)
5597500 — 7d(3,2,1)
5337000 +Z 6d(3,2,1)
5096000 — 5d(3,2,1)
4752600+Z — 4p(1,2,3)
4690100+Z — 4d(1,2,3)
4687300+Z
4685400+N
3816600 +Z 3p(1,2,3)
3616300+Z
3615700+Z — 3d(1,2,3)
3613900+L
5377600 — 5(1,2,3)
4953600 +Z 4(1,2,3)
3892650 +Z 3(1,2,3)
KEY
356390C
356190C 3(0,1,2)
356090C +Z
J, lowest to highest
n
uncertainty
Energy Levels (cm⁻¹)
Ionization Limit
6 047 200 cm⁻¹
(749.74 eV)
[Cl XIV 2s² ¹S₀ → Cl XV 2s ²S₁/₂]
Cl XIV
SINGLET
TRIPLET

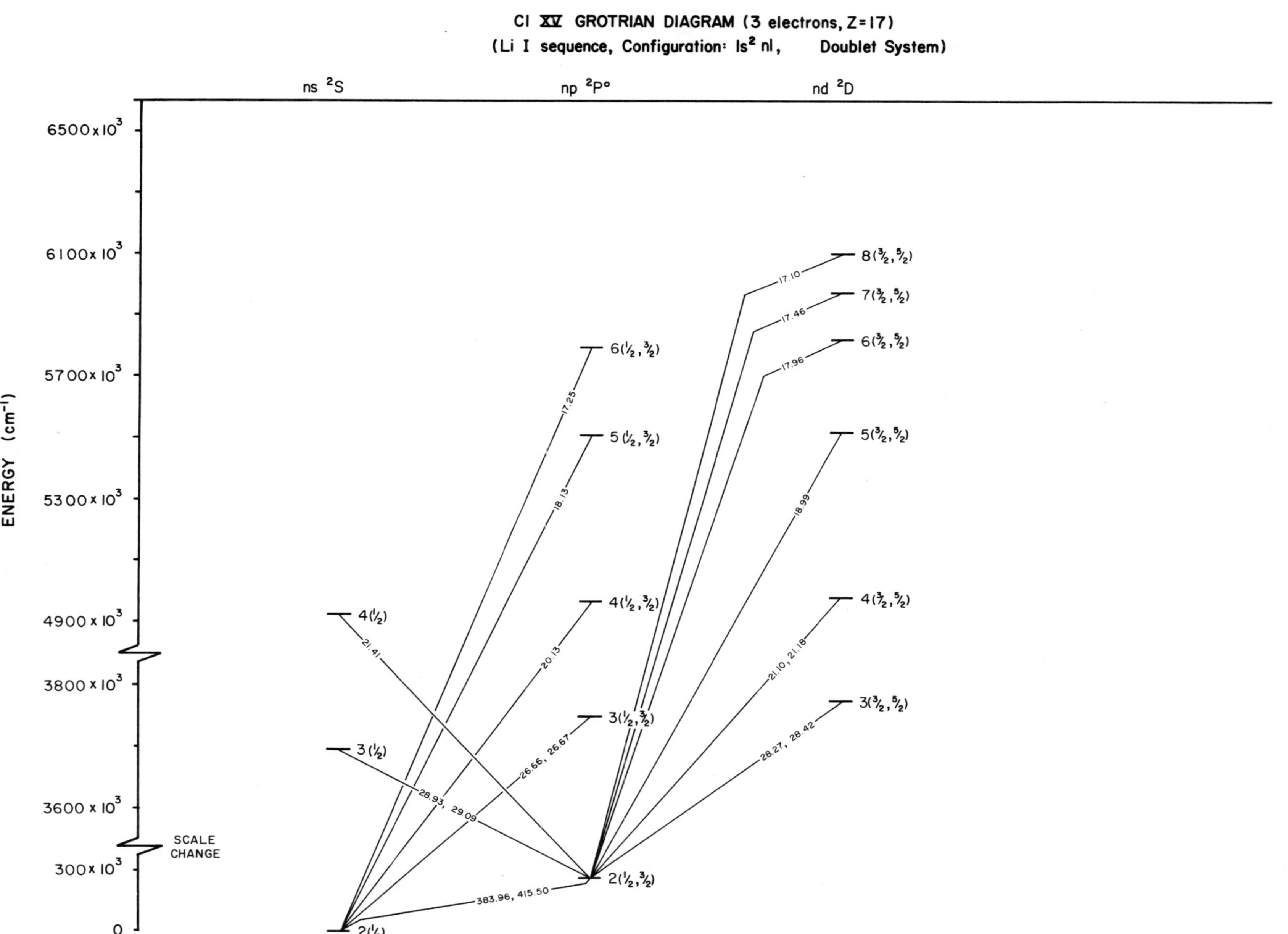
Cl XV GROTRIAN DIAGRAM (3 electrons, Z=17)
(Li I sequence, Configuration: 1s² nl, Doublet System)
ns ²S
np ²P°
nd ²D
6500 x 10³
6100 x 10³
5700 x 10³
5300 x 10³
4900 x 10³
3800 x 10³
3600 x 10³
300 x 10³
0
ENERGY (cm⁻¹)
SCALE CHANGE
4(½)
3(½)
2(½)
6(½, 3/2)
5(½, 3/2)
4(½, 3/2)
3(½, 3/2)
2(½, 3/2)
8(3/2, 5/2)
7(3/2, 5/2)
6(3/2, 5/2)
5(3/2, 5/2)
4(3/2, 5/2)
3(3/2, 5/2)
17.10
17.46
17.96
18.99
21.10, 21.18
28.27, 28.42
17.25
18.13
20.13
26.66, 26.67
21.41
28.93, 29.09
383.96, 415.50

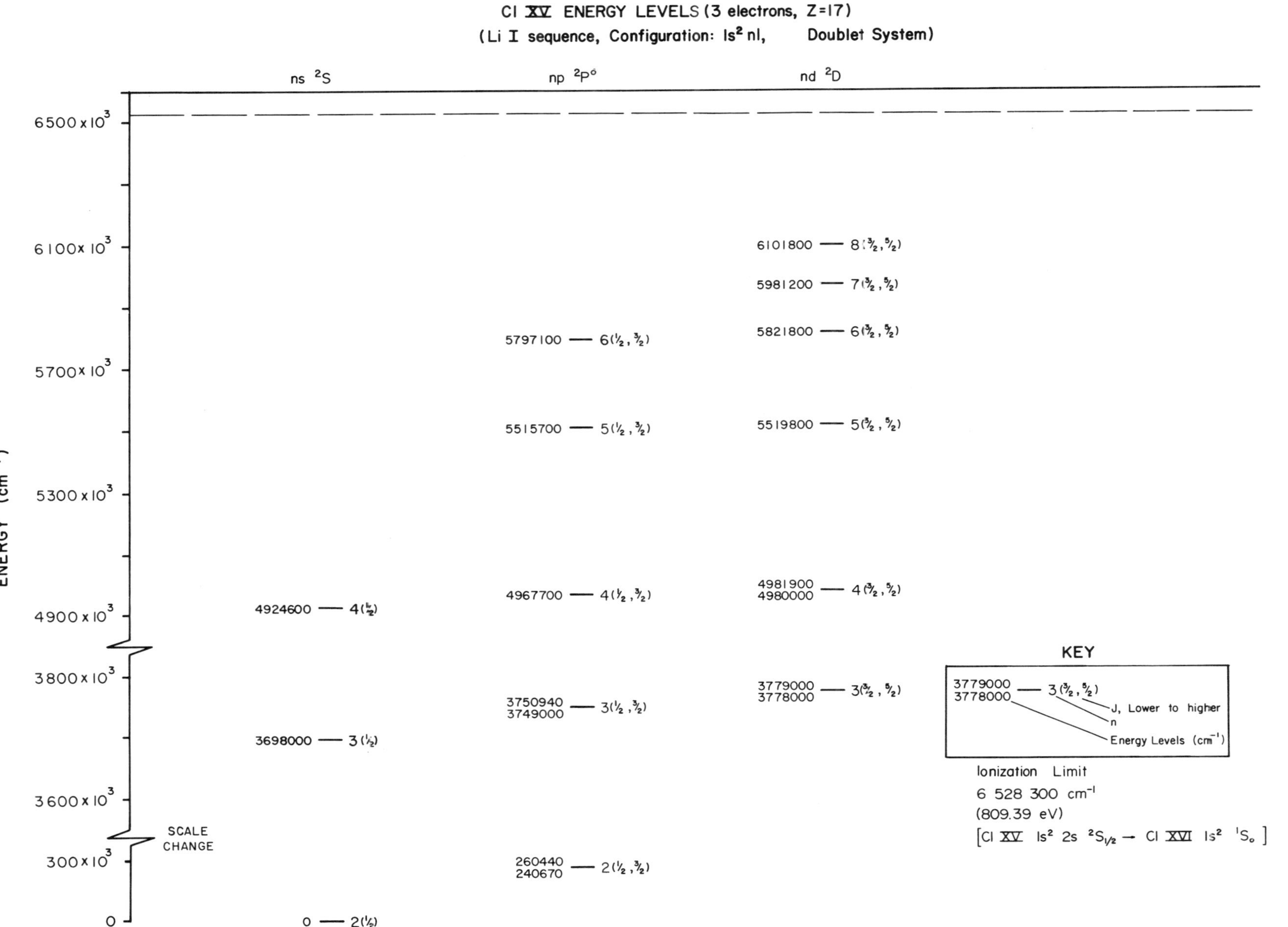
Cl XV ENERGY LEVELS (3 electrons, Z=17)
(Li I sequence, Configuration: 1s² nl, Doublet System)
ns ²S
np ²P°
nd ²D
6500 x 10³
6100 x 10³
5700 x 10³
5300 x 10³
4900 x 10³
3800 x 10³
3600 x 10³
300 x 10³
0
ENERGY (cm⁻¹)
SCALE CHANGE
6101800 — 8(3/2, 5/2)
5981200 — 7(3/2, 5/2)
5821800 — 6(3/2, 5/2)
5797100 — 6(½, 3/2)
5515700 — 5(½, 3/2)
5519800 — 5(3/2, 5/2)
4967700 — 4(½, 3/2)
4981900 4980000 — 4(3/2, 5/2)
4924600 — 4(½)
3779000 3778000 — 3(3/2, 5/2)
3750940 3749000 — 3(½, 3/2)
3698000 — 3(½)
260440 240670 — 2(½, 3/2)
0 — 2(½)
KEY
3779000 3778000 — 3(3/2, 5/2)
J, Lower to higher
n
Energy Levels (cm⁻¹)
Ionization Limit
6 528 300 cm⁻¹
(809.39 eV)
[Cl XV 1s² 2s ²S1/2 → Cl XVI 1s² ¹S0]

Cl XVI
SINGLET
TRIPLET
GROTRIAN
DIAGRAM

Cl XVI GROTRIAN DIAGRAM (2 electrons, Z=17)

(He I sequence, Configuration: 1s nl, Singlet and Triplet Systems)

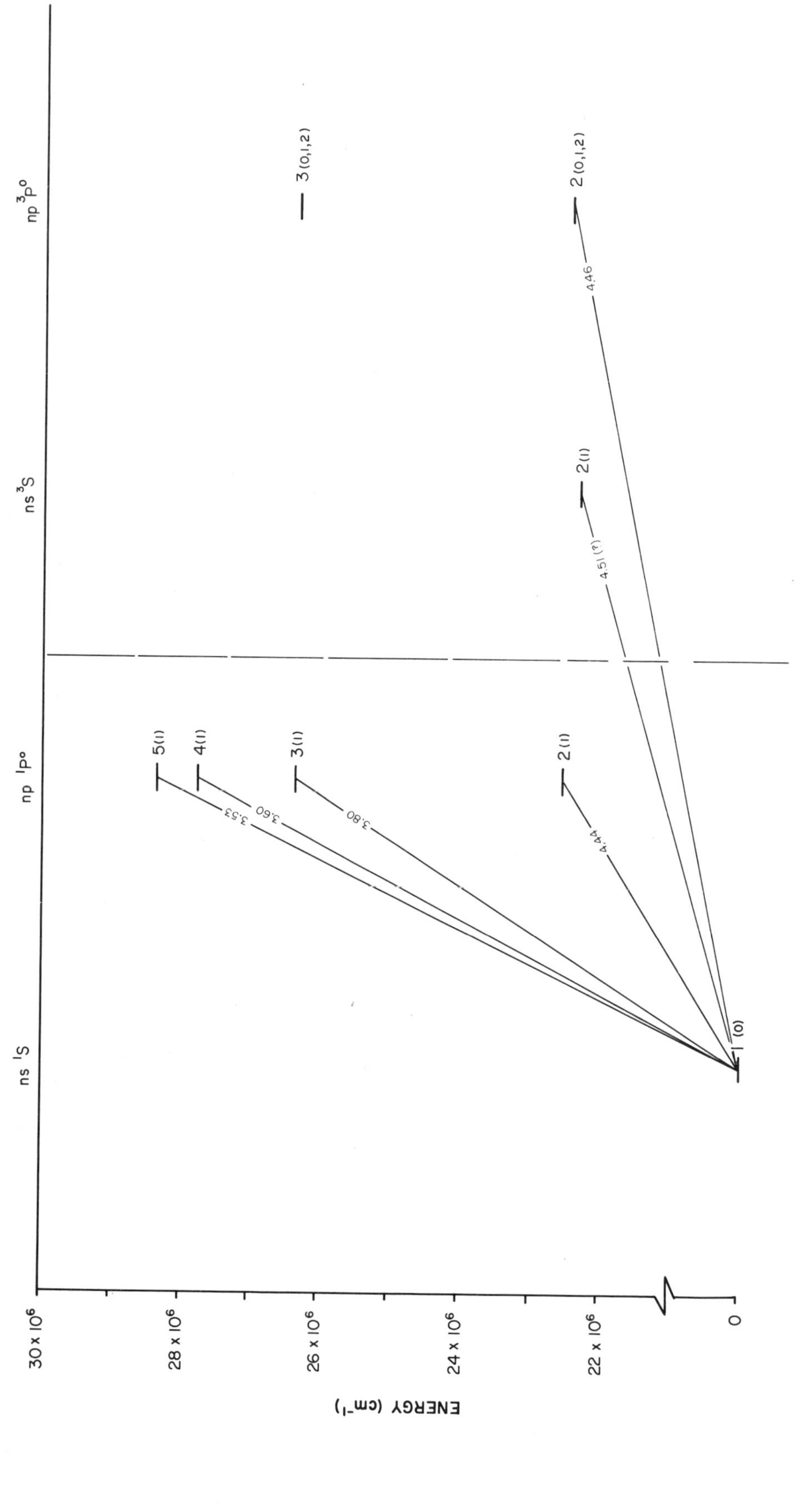

ns ^1S np $^1P^o$ ns ^3S np $^3P^o$

ENERGY (cm^{-1})

30×10^6 28×10^6 26×10^6 24×10^6 22×10^6 0

1(0)

2(1) 3(1) 4(1) 5(1)

4.44 3.80 3.60 3.53

2(1) 4.51(?)

2(0,1,2) 3(0,1,2) 4.46

Cl XVI ENERGY LEVELS (2 electrons, Z = 17)

(He I sequence, Configuration: 1s nl, Singlet and Triplet Systems)

Ionization Limit
29 507 950 cm^{-1}
(3658.425 eV)

[Cl XVI $1s^2\ ^1S_0$ ⟶ Cl XVII $1s\ ^2S_{1/2}$]

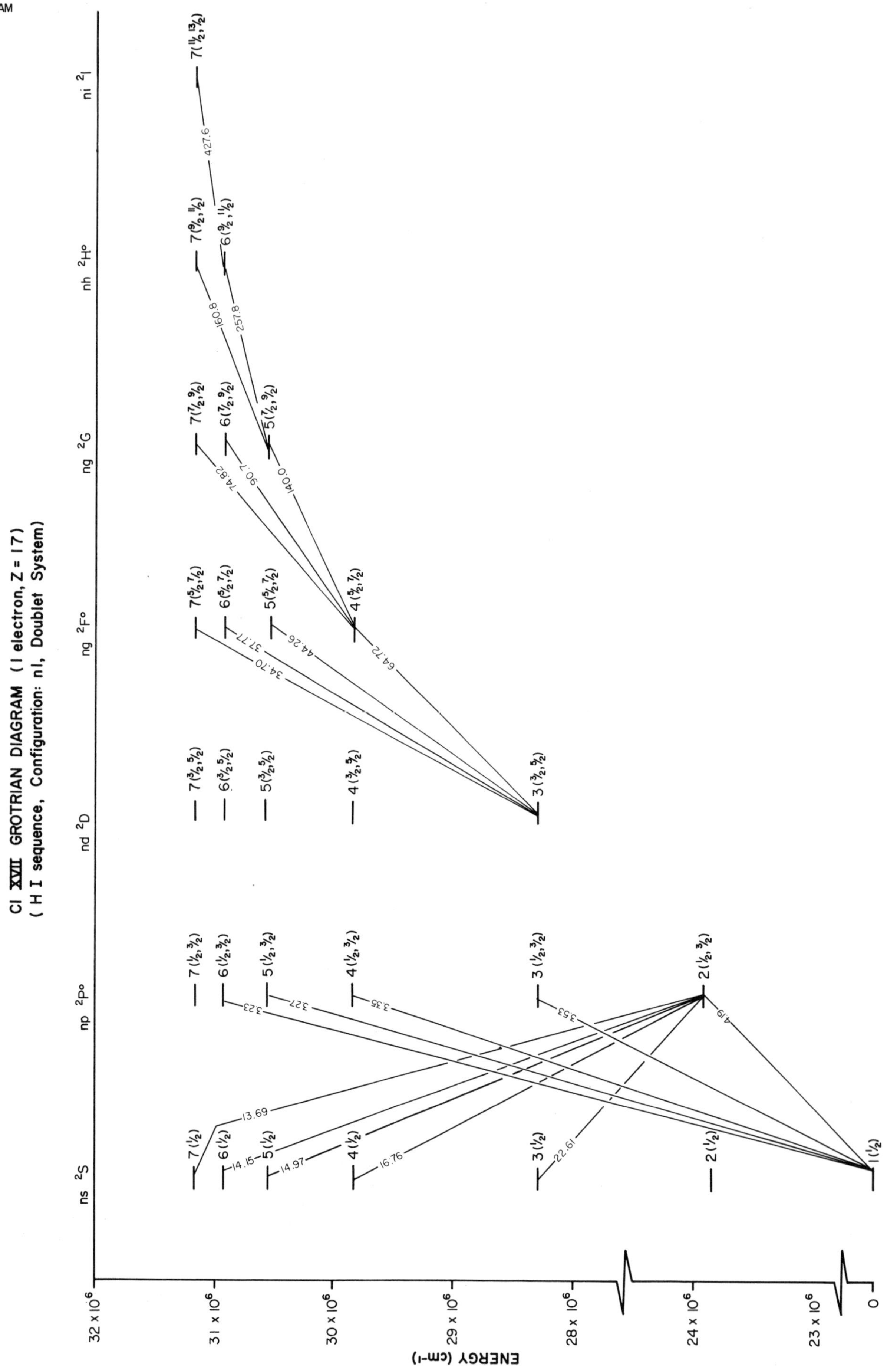

Cl XVII GROTRIAN DIAGRAM (1 electron, Z = 17)
(H I sequence, Configuration: nl, Doublet System)

ns ^2S np $^2P^o$ nd ^2D ng $^2F^o$ ng ^2G nh $^2H^o$ ni ^2I

ENERGY (cm⁻¹)

32×10^6
31×10^6
30×10^6
29×10^6
28×10^6
24×10^6
23×10^6
0

ns ^2S: 7(1/2), 6(1/2), 5(1/2), 4(1/2), 3(1/2), 2(1/2), 1(1/2)

np $^2P^o$: 7(1/2, 3/2), 6(1/2, 3/2), 5(1/2, 3/2), 4(1/2, 3/2), 3(1/2, 3/2), 2(1/2, 3/2)

nd ^2D: 7(3/2, 5/2), 6(3/2, 5/2), 5(3/2, 5/2), 4(3/2, 5/2), 3(3/2, 5/2)

ng $^2F^o$: 7(5/2, 7/2), 6(5/2, 7/2), 5(5/2, 7/2), 4(5/2, 7/2)

ng ^2G: 7(7/2, 9/2), 6(7/2, 9/2), 5(7/2, 9/2)

nh $^2H^o$: 7(9/2, 11/2), 6(9/2, 11/2)

ni ^2I: 7(11/2, 13/2)

13.69 14.15 14.97 16.76 22.61 3.23 3.27 3.35 3.53 4.19 34.70 37.77 44.26 64.72 74.82 90.7 140.0 160.8 257.8 427.6

Cl XVII ENERGY LEVELS (1 electron, Z = 17)
(H I sequence, Configuration: nl, Doublet System)

ENERGY (cm^{-1})

n	ns ^2S	np $^2P^o$	nd ^2D	ng $^2F^o$	ng ^2G	nh $^2H^o$	ni ^2I
7	31180536 — 7(1/2)	31180511 — 7(1/2, 3/2)	— 7(3/2, 5/2)	— 7(5/2, 7/2)	— 7(7/2, 9/2)	— 7(9/2, 11/2)	— 7(11/2, 13/2)
6	30946122 — 6(1/2)	30946083 — 6(1/2, 3/2)	— 6(3/2, 5/2)	— 6(5/2, 7/2)	— 6(7/2, 9/2)	30947978 — 6(9/2, 11/2)	
5	30557189 — 5(1/2)	30557121 — 5(1/2, 3/2)	— 5(3/2, 5/2)	— 5(5/2, 7/2)	30560267 — 5(7/2, 9/2)		
4	29840798 — 4(1/2)	29844524 29840666 — 4(1/2, 3/2)	29845794 29844517 — 4(3/2, 5/2)	29846430 29845792 — 4(5/2, 7/2)			
3	28291923 — 3(1/2)	28300759 28291611 — 3(1/2, 3/2)	28303770 28300742 — 3(3/2, 5/2)				
2	23863187 — 2(1/2)	23893009 23862144 — 2(1/2, 3/2)					
1	0 — 1(1/2)						

Energy axis: 32×10^6, 31×10^6, 30×10^6, 29×10^6, 28×10^6, 24×10^6, 23×10^6, 0

KEY

28303770
28300742 — 3(3/2, 5/2)
J, Lower to higher
n
Energy Levels (cm^{-1})

Ionization Limit
31829006 cm^{-1}
(3946.19 eV)

[Cl XVII 1s $^2S_{1/2}$ → nucleus]

Cl I $Z = 17$ 17 electrons

As pointed out by Radziemski and Kaufman, many levels may be only approximately described as pure *LS* or *JK* coupling; *JK* designations are used except where *LS* coupling is clearly best. We have taken all energies from their tables. Wavelengths for transitions are taken primarily from the references by Radziemski and Kaufman, Humphreys and Paul, and Humphreys, Paul, and Minnhagen.

B. Edlén, Z. Physik, **104**, 407 (1937).

Author gives a line table from observations in the region 6610–7660 Å.

J.E. Hansen, J. Opt. Soc. Am. **67**, 754 (1977).

This author calculates the position and wavefunction for the $sp^6\ ^2S$ term.

C.J. Humphreys and E. Paul, Jr., J. Opt. Soc. Am. **49**, 1180 (1959).

Authors give line and energy tables based on observations in the regions 10 221–25 323 Å and 6920–10 002 Å.

C.J. Humphreys and E. Paul, Jr., J. Opt. Soc. Am. **62**, 432 (1972).

Authors give a line table for observations in the region 19 800–28 570 Å.

C.J. Humphreys, E. Paul, Jr., and L. Minnhagen, J. Opt. Soc. Am. **61**, 110 (1971).

Authors give line and energy level tables based on observations in the region 39 600–40 530 Å.

L. Minnhagen, J. Opt. Soc. Am. **51**, 298 (1961).

Author gives line and energy level tables based on calculations and observations; wavelengths listed are in the region 10 280–16 290 Å.

L.J. Radziemski, Jr., and V. Kaufman, J. Opt. Soc. Am. **59**, 424 (1969).

Authors give line and energy level tables from lines observed in the region 960–40 535 Å.

Cl II $Z = 17$ 16 electrons

The levels and wavelengths are taken from the reference by Radziemski and Kaufman. For most levels, *LS* coupling is a useful approximation.

I.S. Bowen, Ap. J. **132**, 1 (1960).

Author gives a line table of observed and predicted values.

W.B. Bridges and A.N. Chester, IEEE J. Qu. Electronics **1**, 66 (1965).

Authors give a line table of lines observed and calculated in the region 4780–6100 Å in ion lasers.

B. Edlén, Phys. Rev. **61**, 434 (1942).

Author gives a wavelength of an observed transition and term values for the $3s3p^5\ ^1P^o_1$ and $3s^23p^4\ ^1S$ configurations.

V. Kaufman and L.J. Radziemski, Jr., J. Opt. Soc. Am. **59**, 227 (1969).

Authors give a term table from observations.

C.C. Kiess and T.L. de Bruín, J. Research NBS **23**, 443 (1939).

Authors give a term diagram, a term array, and tables of lines observed in the regions 2100–9485 Å and 555–1925 Å.

K. Murakawa, Z. Physik **109**, 162 (1939).

Author gives line and term tables for lines observed in the region 3095–6720 Å.

L.J. Radziemski, Jr., and V. Kaufman, J. Opt. Soc. Am. **64**, 366 (1974).

These authors give observed lines of Cl II from 500 Å to 11 000 Å and analyze them to obtain extensive lists of levels and classified lines.

Cl III $\quad Z = 17 \quad$ 15 electrons

I.S. Bowen, Ap. J. **121**, 306 (1955).

Author gives wavelengths of observed forbidden transitions.

I.S. Bowen, Ap. J. **132**, 1 (1960).

Author gives line tables from nebular observations.

Cl IV $\quad Z = 17 \quad$ 14 electrons

S. Bashkin and I. Martinson, J. Opt. Soc. Am. **61**, 1686 (1971).

Authors identify a transition wavelength and assign a value to an energy level.

I.S. Bowen, Ap. J. **121**, 306, (1955).

Author gives transition wavelengths from nebular observations.

I.S. Bowen, Ap. J. **132**, 1 (1960).

Author gives line tables based on nebular observations.

Cl V $\quad Z = 17 \quad$ 13 electrons

S. Bashkin and I. Martinson, J. Opt. Soc. Am. **61**, 1686 (1971).

Authors identify a transition wavelength and assign a value to an energy level.

L.W. Phillips and W.L. Parker, Phys. Rev. **60**, 301 (1941).

Authors give a line and term table for lines observed in the region 235–550 Å.

Cl VI $\quad Z = 17 \quad$ 12 electrons

According to Ekberg, the uncertainty (x) in the location of the triplet system relative to the singlet system is $-742\ \mathrm{cm}^{-1}$. The transition listed by Kelly and Palumbo at 624.11 Å ($3p^2\ {}^1D$–$3p3d\ {}^1D^\circ$) has been omitted from the present drawings; the transitions involving the $3s4p\ {}^3P^\circ$ levels, together with the suggested positions of those levels (see Bashkin and Martinson) have been included even though these assignments are open to question.

S. Bashkin and I. Martinson, J. Opt. Soc. Am. **61**, 1686 (1971).

Authors give observed transition wavelengths and energy-level values.

J.O. Ekberg, Physica Scripta **4**, 101 (1971).

W.L. Parker and L.W. Phillips, Phys. Rev. **57**, 140 (1940).

Authors give line and term tables for lines observed in the region 190–740 Å.

L.W. Phillips and W.L. Parker, Phys. Rev. **60**, 301 (1941).

Authors give a line and term table for lines observed in the region 565–585 Å.

Cl VII $Z = 17$ 11 electrons

Numerous transitions classified as belonging to hydrogen-like levels have been observed in beam-foil work (Bashkin and Martinson, Hallin *et al.*, Bashkin *et al.*, Bhardwaj *et al.*). The authors of these papers list only a few energies for these transitions; energies have been included on the present drawings when given in the references.

S. Bashkin and I. Martinson, J. Opt. Soc. Am. **61**, 1686 (1971).

Authors give observed transition wavelengths in Cl II–Cl IX.

S. Bashkin, J. Bromander, J.A. Leavitt, and I. Martinson, Physica Scripta **8**, 285 (1973).

Observations were made by these authors of spectra in the range 400–6000 Å belonging to Cl IV–Cl IX.

S.N. Bhardwaj, H.G. Berry, and T. Mossberg, Physica Scripta **9**, 331 (1974).

These authors report lifetime measurements in Cl IV–Cl IX.

W.A. Deutschman and L.L. House, Ap. J. **144**, 435 (1966).

Authors give a line table for observation in the region 133–160 Å.

R. Hallin, J. Lindskog, A. Marelius, J. Pihl, and R. Sjodin, Physica Scripta **8**, 209 (1973).

Authors have made studies of chlorine and oxygen spectra and observe many transitions among hydrogen-like levels in Cl VI–Cl XIV.

L.W. Phillips, Phys. Rev. **53**, 248 (1938).

Author gives line and energy level tables from observations in the region 170–815 Å.

Cl VIII $Z = 17$ 10 electrons

Several authors report observations of hydrogenic transitions.

S. Bashkin, J. Bromander, J.A. Leavitt, and I. Martinson, Physica Scripta **8**, 285 (1973).

S.N. Bhardwaj, H.G. Berry, and T. Mossberg, Physica Scripta **9**, 331 (1974).

R. Hallin, J. Lindskog, A. Marelius, J. Pihl, and R. Sjödin, Physica Scripta **8**, 209 (1973).

B. Edlén, Z. Physik **100**, 726 (1936).

Author gives line and term tables for lines observed in the region 35–60 Å.

Cl IX $Z = 17$ 9 electrons

The designations for the unclassified levels are taken from Moore. Moore primes the entry for $2s2p^6\ ^2S$ and places bars over the other primed terms. We have simplified that notation by omitting the prime and the bars, using primes only to indicate cores with $J = 1/2$.

S. Bashkin, J. Bromander, J.A. Leavitt, and I. Martinson, Physica Scripta **8**, 285 (1973).

S.N. Bhardwaj, H.G. Berry, and T. Mossberg, Physica Scripta **9**, 331 (1974).

R. Hallin, J. Lindskog, A. Marelius, J. Pihl, and R. Sjödin, Physica Scripta **8**, 209 (1973).

The above authors observe some hydrogenic transitions.

W.A. Deutschman, Doctorate thesis for the Dept. of Astro-Geophysics, University of Colorado, obtained from University Microfilms, Ann Arbor, Michigan, U.S.A., order number 68–2645 (1970).

Author gives line and energy level tables and term diagrams from observations.

W.A. Deutschman and L.L. House, Ap. J. **144**, 435 (1966).

Authors give the transition wavelengths of two observed resonance lines.

B. Edlén, Z. Physik **100**, 726 (1936).

Author gives line and term tables from observations in the region 40–60 Å.

U. Feldman, G.A. Doschek, R.D. Cowan, and L. Cohen, J. Opt. Soc. Am. **63**, 1445 (1973).

These authors study the fluorine isoelectronic sequence, obtaining energy levels from grazing-incidence spectra and interpreting them using Hartree–Fock calculations.

Cl X $Z = 17$ 8 electrons

The energy levels listed by Kelly have been taken from several different authors, hence the variety of uncertainties. The most recent seem to be those of Goldsmith *et al.* for which the uncertainty is G = roughly $\pm 300\ \text{cm}^{-1}$.

S. Bashkin, J. Bromander, J.A. Leavitt, and I. Martinson, Physica Scripta **8**, 285 (1973).

R. Hallin, J. Lindskog, A. Marelius, J. Pihl, and R. Sjödin, Physica Scripta **8**, 209 (1973).

These authors observe hydrogenic transitions.

E.U. Condon and H. Odabasi, JILA Report #95, University of Colorado (1968).

This is a report of self-consistent field calculations for 6-, 7-, and 8-electron ions.

W.A. Deutschman, Doctorate thesis for the Dept. of Astro-Geophysics, University of Colorado, obtained from University Microfilms, Ann Arbor, Michigan, U.S.A., order number 68–2645 (1970).

Author gives a line table, an energy level table, and term diagrams based on observations in the region 160–210 Å.

W.A. Deutschman and L.L. House, Ap. J. **144**, 435 (1966).

Authors give a line table of observed resonance lines in the region 160–210 Å.

G.A. Doschek, U. Feldman, and L. Cohen, J. Opt. Soc. Am. **63**, 1463 (1973).

These authors study the $2p^4$–$2p^3 3s$ transitions in the O I isoelectronic sequence.

B. Edlén, Z. Physik **100**, 726 (1936).

Author gives line and term tables from observations in the region 35–60 Å.

S. Goldsmith, U. Feldman, and L. Cohen, J. Opt. Soc. Am. **61**, 615 (1971).

Authors give a table of the ground-configuration energy levels $^1D_2$ and $^1S_0$ from calculations.

Cl XI $Z = 17$ 7 electrons

The levels have been taken from Kelly's unpublished compilation. Some transitions have been taken from Fawcett and Hayes; the remainder are from Kelly and Palumbo.

E.U. Condon and H. Odabasi, JILA Report #95, University of Colorado (1968).

These authors carry out self-consistent field calculations for 6-, 7-, and 8-electron ions.

W.A. Deutschman, Doctorate thesis for the Dept. of Astro-Geophysics, University of Colorada, obtained from University Microfilms, Ann Arbor, Michigan, U.S.A., order number 68–2645 (1970).

Author gives a line table, an energy level table, and term diagrams from observations in the region 160–245 Å.

W.A. Deutschman and L.L. House, Ap. J. **144**, 435 (1966).

Authors give a line table of observed resonance lines in the region 160–245 Å.

W.A. Deutschman and L.L. House, Ap. J. **149**, 451 (1967).

Authors give a line table of observed resonance lines in the region 160–245 Å.

B. Edlén, Z. Physik **100**, 726 (1936).

Author gives two transition wavelengths observed in the vacuum ultraviolet.

B.C. Fawcett and R.W. Hayes, Physica Scripta **8**, 244 (1973).

Theta-pinch spectra are analyzed to provide classification of spectra in the range 36–41 Å in Cl XI.

R. Hallin, J. Lindskog, A. Marelius, J. Pihl, and R. Sjödin, Physica Scripta **8**, 209 (1973).

These authors observe hydrogenic transitions.

S.O. Kastner, J. Opt. Soc. Am. **63**, 738 (1973).

Extrapolations are made from observational data to obtain energy levels for the C I and N I isoelectronic sequences.

Cl XII $Z = 17$ 6 electrons

Only the higher-lying levels have cores with well-defined L and S, and thus can be labeled with primes on the drawings. The transitions between levels having $n = 2$ and $n = 3$ have been taken primarily from Fawcett and Hayes; the rest of the transitions are from Kelly and Palumbo.

E.U. Condon and H. Odabasi, JILA Report #95 (1968).

These authors report self-consistent-field calculations for 6-, 7-, and 8-electron atoms.

W.A. Deutschman, Doctorate thesis for the Dept. of Astro-Geophysics, University of Colorado, obtained from University Microfilms, Ann Arbor, Michigan, U.S.A., order number 68–2645 (1970).

Author gives a line table, an energy level table, and term diagrams from observations in the region 170–230 Å.

W.A. Deutschman and L.L. House, Ap. J. **149**, 451 (1967).

Authors give a table of observed resonance lines in the region 170–230 Å.

B.C. Fawcett, D.D. Burgess, and N.J. Peacock, Proc. Phys. Soc. **91**, 970 (1967).

These authors study inner-shell transitions experimentally.

B.C. Fawcett and R.W. Hayes, Physica Scripta **8**, 244 (1973).

These authors classify observed transitions between levels having $n = 2$ and $n = 3$.

S. Goldsmith, U. Feldman, A. Crooker, and L. Cohen, J. Opt. Soc. Am. **62**, 260 (1972).

These authors classify some lines in this spectrum.

R. Hallin, J. Lindskog, A. Marelius, J. Pihl, and R. Sjödin, Physica Scripta **8**, 209 (1973).

These authors observe hydrogenic transitions.

S.O. Kastner, J. Opt. Soc. Am. **63**, 738 (1973).

This author calculates energies and intervals for the $2s^r2p^k$ configuration from sulfur to scandium.

H. Nussbaumer, Ap. J. **166**, 411 (1971).

This author gives wavelengths for forbidden transitions between levels in $2s^22p^2\ {}^3P$, and between $2s^22p^2\ {}^1S$ and ${}^1D$.

C.F. Moore and H. Wolter, J. Phys. **B6**, L124 (1973).

These authors report X-ray spectra.

Cl XIII $Z = 17$ 5 electrons

Most of the transitions between levels with $n = 2$ and $n = 3$ are taken from Fawcett and Hayes; the others are taken from Kelly and Palumbo. Transitions listed by them at 419.46 Å ($2s2p^2\ {}^2P$–$2p^3\ {}^2D^\circ$) and 24.29 Å ($2s2p^2\ {}^2P$–$2s2p4d'\ {}^2D^\circ$) have been omitted from these drawings.

W.A. Deutschman, Doctorate thesis for the Dept. of Astro-Geophysics, University of Calorado, obtained from University Microfilms, Ann Arbor, Michigan, U.S.A., order number 68–2645 (1970).

Author gives the transition wavelengths of two observed lines.

B.C. Fawcett, D.D. Burgess, and N.J. Peacock, Proc. Phys. Soc. **91**, 970 (1967).

Authors give wavelengths for the observed transitions of a multiplet.

B.C. Fawcett and R.W. Hayes, Physica 8, 244 (1973).

These authors study experimentally transitions between levels of $n = 2$ and $n = 3$.

R. Hallin, J. Lindskog, A. Marelius, J. Pihl, and R. Sjödin, Physica Scripta **8**, 209 (1973).

These authors observe hydrogenic transitions.

C.F. Moore and H. Wolter, J. Phys. **B6**, L124 (1973).

These authors report X-ray spectra.

Cl XIV $Z = 17$ 4 electrons

Transitions between levels with $n = 2$ to $n = 3$ are taken from Fawcett and Hayes; others are taken from Kelly and Palumbo. Note that in the terms' labels we have shown all values of J expected to be present, even though in many cases not all have been observed. The identification of the level 3s3p $^1P_1^\circ$ comes from Kelly (unpublished).

B.C. Fawcett, Atomic Data and Nuclear Data Tables **16**, 135 (1975).

This author presents wavelengths and classifications for transitions of the type $2s^22p^n$–$2s2p^{n+1}$ and $2s2p^n$–$2p^{n+1}$.

B.C. Fawcett, D.D. Burgess, and N.J. Peacock, Proc. Phys. Soc. **91**, 970 (1967)

Authors give the wavelengths for the observed transitions of a multiplet.

B.C. Fawcett and R.W. Hayes, Physics Scripta **8**, 244 (1973).

These authors study experimentally the transitions between levels with $n = 2$ and $n = 3$.

R.H. Garstang and L.J. Shamey, Ap. J. **148**, 665 (1967).

These authors calculate transition rates for 2s $^1S_0$–2p $^3P_1^\circ$ transitions.

R. Hallin, J. Lindskog, A. Marelius, J. Pihl, and R. Sjödin, Physica Scripta **8**, 209 (1973).

These authors observe hydrogenic transitions.

C.F. Moore and H. Wolter, J. Phys. **B6**, L124 (1973).

These authors observe X-ray emissions.

Cl XV $Z = 17$ 3 electrons

The levels 2p $^2P^\circ$ are taken from Fawcett, J. Phys. **B3**, 1152 (1970); Kelly (unpublished) gives $241050 + x$, $261490 + y$ for $^2P^\circ_{1/2,3/2}$ respectively.

B.C. Fawcett, Atomic Data and Nuclear Data Tables **16**, 135 (1975).

This author presents tables of wavelengths and classifications of observed emission lines due to $2s^22p^n$–$2s2p^{n+1}$ and $2s2p^n$–$2p^{n+1}$ transitions.

U. Feldman, G.A. Doschek, D.J. Nagel, R.D. Cowan, and R.R. Whitlock, Ap. J. **192**, 213 (1974).

These authors consider satellite X-ray spectra from laser-produced plasmas.

R. Hallin, J. Lindskog. A. Marelius, J. Pihl, and R. Sjödin, Physica Scripta **8**, 209 (1973).

These authors observe hydrogen-like spectra.

G.A. Martin and W.L. Wiese, J. Phys. Chem. Ref. Data **5**, 537 (1976).

These authors present a critical analysis of oscillator strengths in the lithium isoelectronic sequence.

C.F. Moore and H. Wolter, J. Phys. B6, L124 (1973).

These authors observe X-ray emissions.

Cl XVI | $Z = 17$ | 2 electrons

Please see the general references.

Cl XVII | $Z = 17$ | 1 electron

We have taken calculated values rather than experimental values of energies and wavelengths. Note that the average wavelength is given for each set of transitions $n - n'$. Details can be found in the reference by Garcia and Mack (1965). In those cases for which only one energy value is given for a term, it applies either to the $np\ ^2P^\circ_{1/2}$ level or for the terms having maximum l, to the upper level (having $J = l + 1/2$).

G.W. Erickson, private communication (1976).

J.D. Garcia and J.E. Mack, J. Opt. Soc. Am. **55**, 654 (1965).

Authors give calculated energy level and line tables for one-electron atomic spectra.

C.F. Moore and H. Wolter, J. Phys. B6, L124 (1973).

These authors observe X-ray emissions.

Argon (Ar)

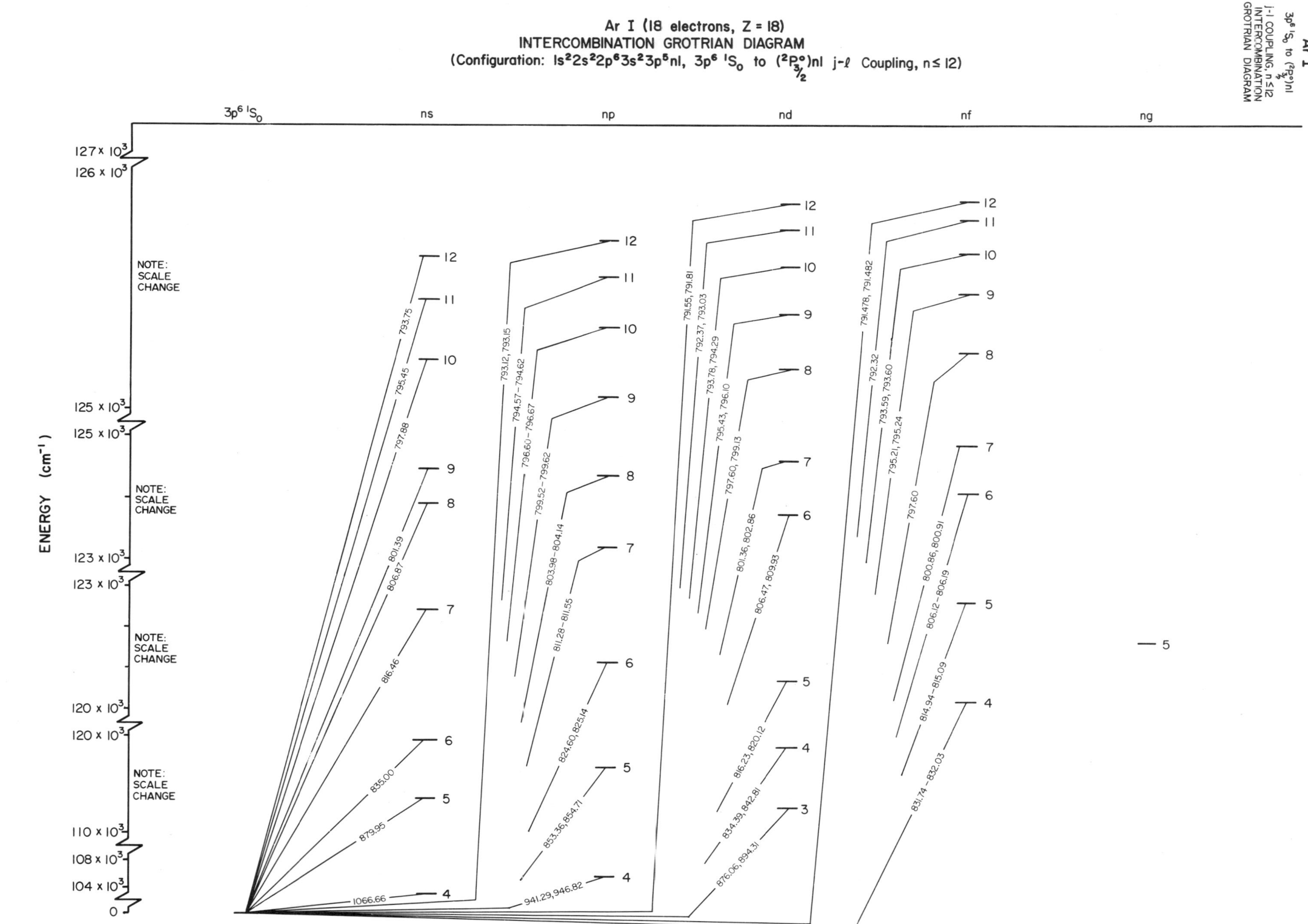

Ar I
$3p^6\ ^1S_0$ to $(^2P^o_{3/2})nl$
j-l COUPLING, n ≤ 12
INTERCOMBINATION
GROTRIAN DIAGRAM
Ar I (18 electrons, Z = 18)
INTERCOMBINATION GROTRIAN DIAGRAM
(Configuration: $1s^2 2s^2 2p^6 3s^2 3p^5 nl$, $3p^6\ ^1S_0$ to $(^2P^o_{3/2})nl$ j-ℓ Coupling, n ≤ 12)
$3p^6\ ^1S_0$
ns
np
nd
nf
ng
ENERGY (cm^{-1})
127 x 10^3
126 x 10^3
125 x 10^3
125 x 10^3
123 x 10^3
123 x 10^3
120 x 10^3
120 x 10^3
110 x 10^3
108 x 10^3
104 x 10^3
0
NOTE: SCALE CHANGE
NOTE: SCALE CHANGE
NOTE: SCALE CHANGE
NOTE: SCALE CHANGE
12
11
10
9
8
7
6
5
4
793.75
795.45
797.88
801.39
806.87
816.46
835.00
879.95
1066.66
793.12, 793.15
794.57 – 794.62
796.60 – 796.67
799.52 – 799.62
803.98 – 804.14
811.28 – 811.55
824.60, 825.14
853.36, 854.71
941.29, 946.82
791.55, 791.81
792.37, 793.03
793.78, 794.29
795.43, 796.10
797.60, 799.13
801.36, 802.86
806.47, 809.93
816.23, 820.12
834.39, 842.81
876.06, 894.31
3
791.478, 791.482
792.32
793.59, 793.60
795.21, 795.24
797.60
800.86, 800.91
806.12 – 806.19
814.94 – 815.09
831.74 – 832.03
5

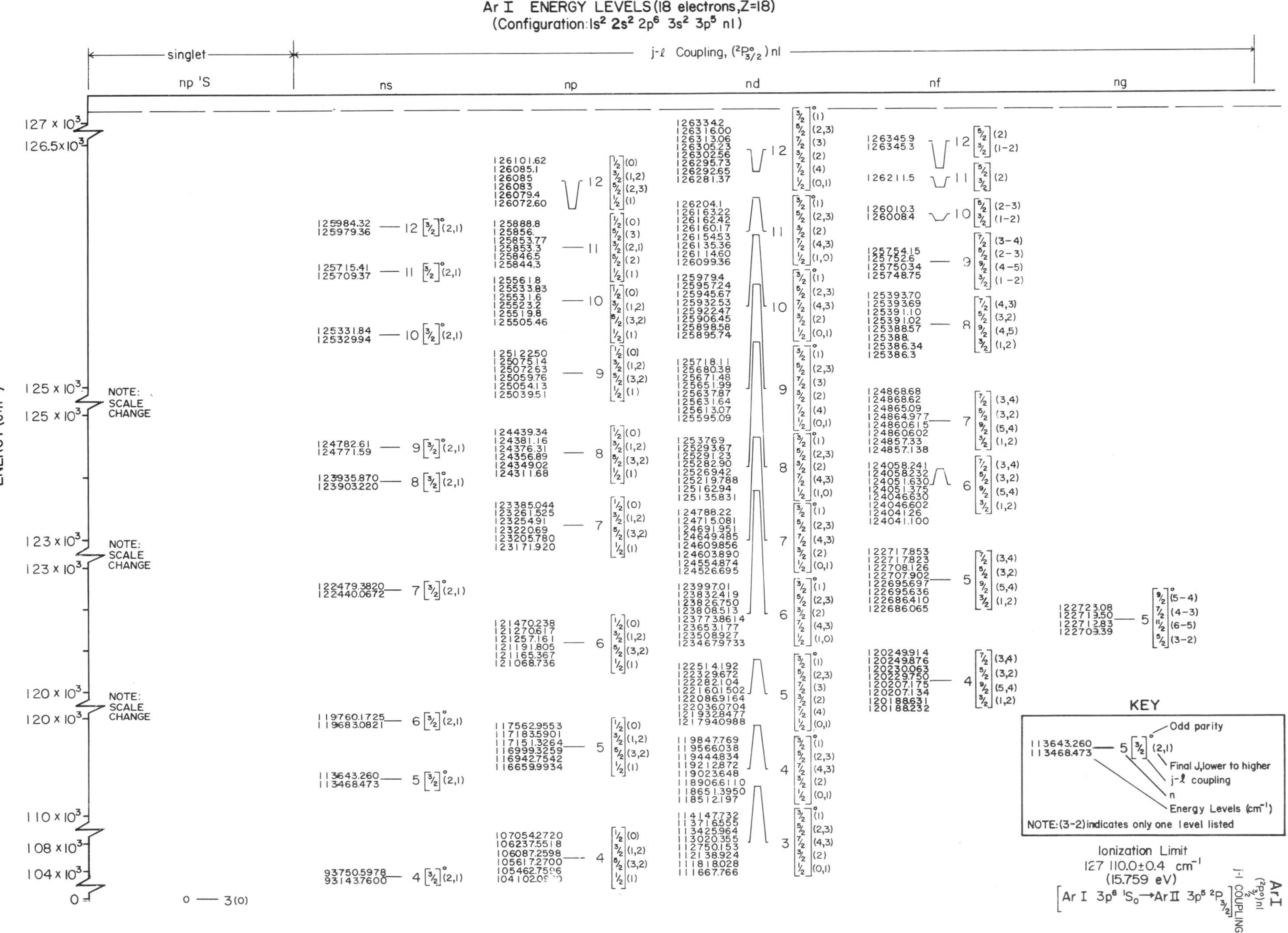
Ar I ENERGY LEVELS (18 electrons, Z=18)
(Configuration: 1s² 2s² 2p⁶ 3s² 3p⁵ nl)
singlet
j-ℓ Coupling, (²P°3/2) nl
np ¹S
ns
np
nd
nf
ng
ENERGY (cm⁻¹)
127 x 10³
126.5x10³
125 x 10³
125 x 10³
123 x 10³
123 x 10³
120 x 10³
120 x 10³
110 x 10³
108 x 10³
104 x 10³
0
NOTE: SCALE CHANGE
NOTE: SCALE CHANGE
NOTE: SCALE CHANGE
0 — 3(0)
KEY
113643.260
113468.473
5 [3/2]° (2,1)
Odd parity
Final J, lower to higher
j-ℓ coupling
n
Energy Levels (cm⁻¹)
NOTE: (3-2) indicates only one level listed
Ionization Limit
127 110.0±0.4 cm⁻¹
(15.759 eV)
[Ar I 3p⁶ ¹S₀→Ar II 3p⁵ ²P3/2]
Ar I
(²P°3/2)nl
j-l COUPLING

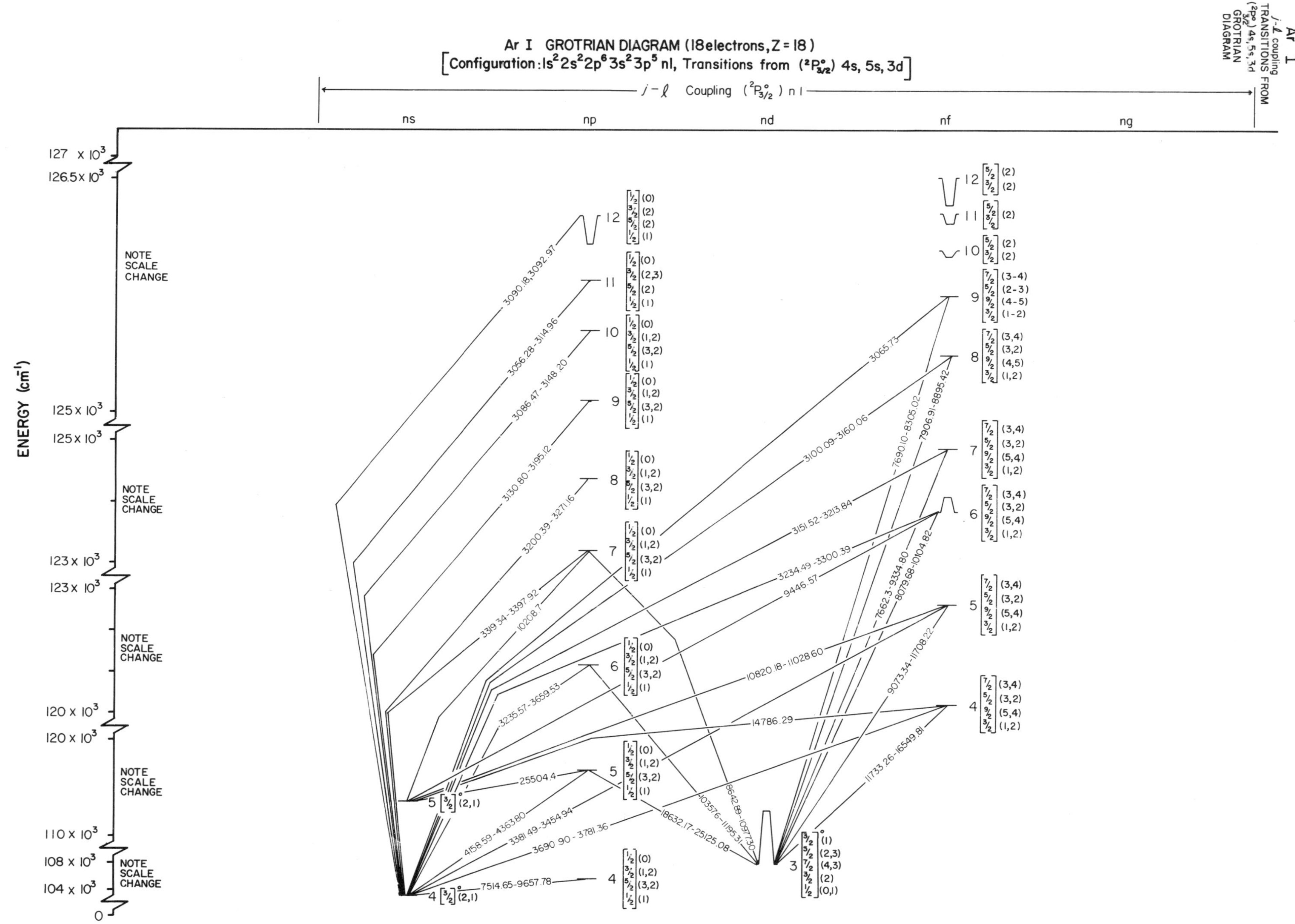

Ar I GROTRIAN DIAGRAM (18 electrons, Z = 18)
[Configuration: $1s^2 2s^2 2p^6 3s^2 3p^5$ nl, Transitions from ($^2P^o_{3/2}$) 4s, 5s, 3d]
j − ℓ Coupling ($^2P^o_{3/2}$) n l
ns
np
nd
nf
ng
Ar I
j-ℓ coupling
TRANSITIONS FROM
($^2P^o_{3/2}$) 4s, 5s, 3d
GROTRIAN
DIAGRAM
ENERGY (cm^{-1})
127 x 10^3
126.5 x 10^3
125 x 10^3
125 x 10^3
123 x 10^3
123 x 10^3
120 x 10^3
120 x 10^3
110 x 10^3
108 x 10^3
104 x 10^3
0
NOTE SCALE CHANGE
NOTE SCALE CHANGE
NOTE SCALE CHANGE
NOTE SCALE CHANGE
NOTE SCALE CHANGE
3090.18, 3092.97
3056.28 - 3114.96
3086.47 - 3148.20
3130.80 - 3195.12
3200.39 - 3271.16
3319.34 - 3397.92
10208.7
3235.57 - 3659.53
25504.4
4158.59 - 4363.80
3381.49 - 3454.94
3690.90 - 3781.36
7514.65 - 9657.78
3065.73
3100.09 - 3160.06
3151.52 - 3213.84
3234.49 - 3300.39
9446.57
10820.18 - 11028.60
14786.29
18632.17 - 25125.08
10357.6 - 11195.31
8642.89 - 10977.30
7690.10 - 8305.02
7906.91 - 8895.42
7662.3 - 9334.80
8079.68 - 10104.82
9073.34 - 11708.22
11733.26 - 16549.81

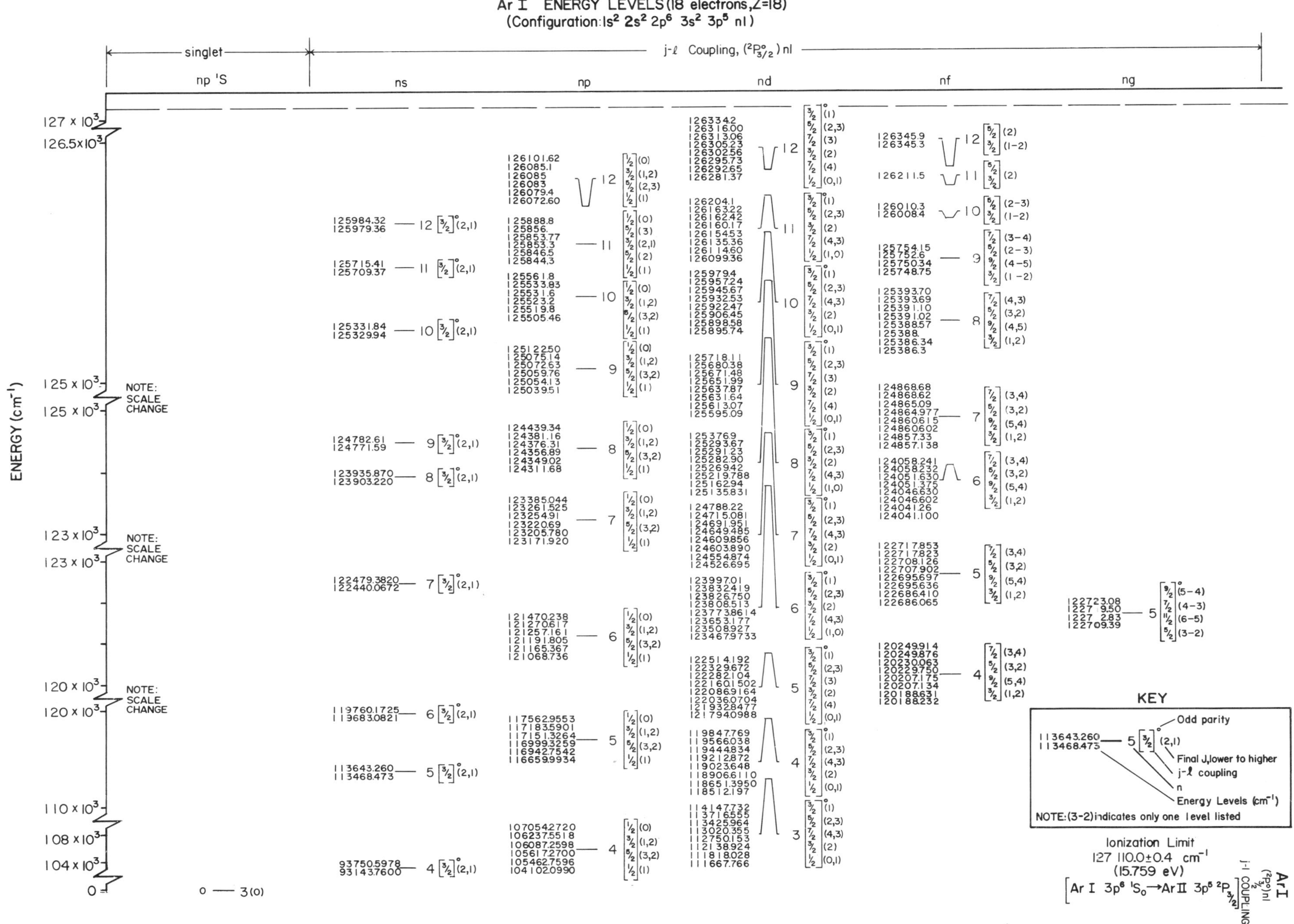
Ar I ENERGY LEVELS (18 electrons, Z=18)
(Configuration: $1s^2\ 2s^2\ 2p^6\ 3s^2\ 3p^5\ nl$)
singlet
j-ℓ Coupling, $(^2P^o_{3/2})nl$
np ¹S
ns
np
nd
nf
ng
ENERGY (cm⁻¹)
NOTE: SCALE CHANGE
KEY
Odd parity
Final J, lower to higher
j-ℓ coupling
n
Energy Levels (cm⁻¹)
NOTE: (3-2) indicates only one level listed
Ionization Limit
127 110.0±0.4 cm⁻¹
(15.759 eV)
$[Ar\ I\ 3p^6\ ^1S_0 \rightarrow Ar\ II\ 3p^5\ ^2P_{3/2}]$
Ar I
$(^2P^o_{3/2})nl$
j-l COUPLING

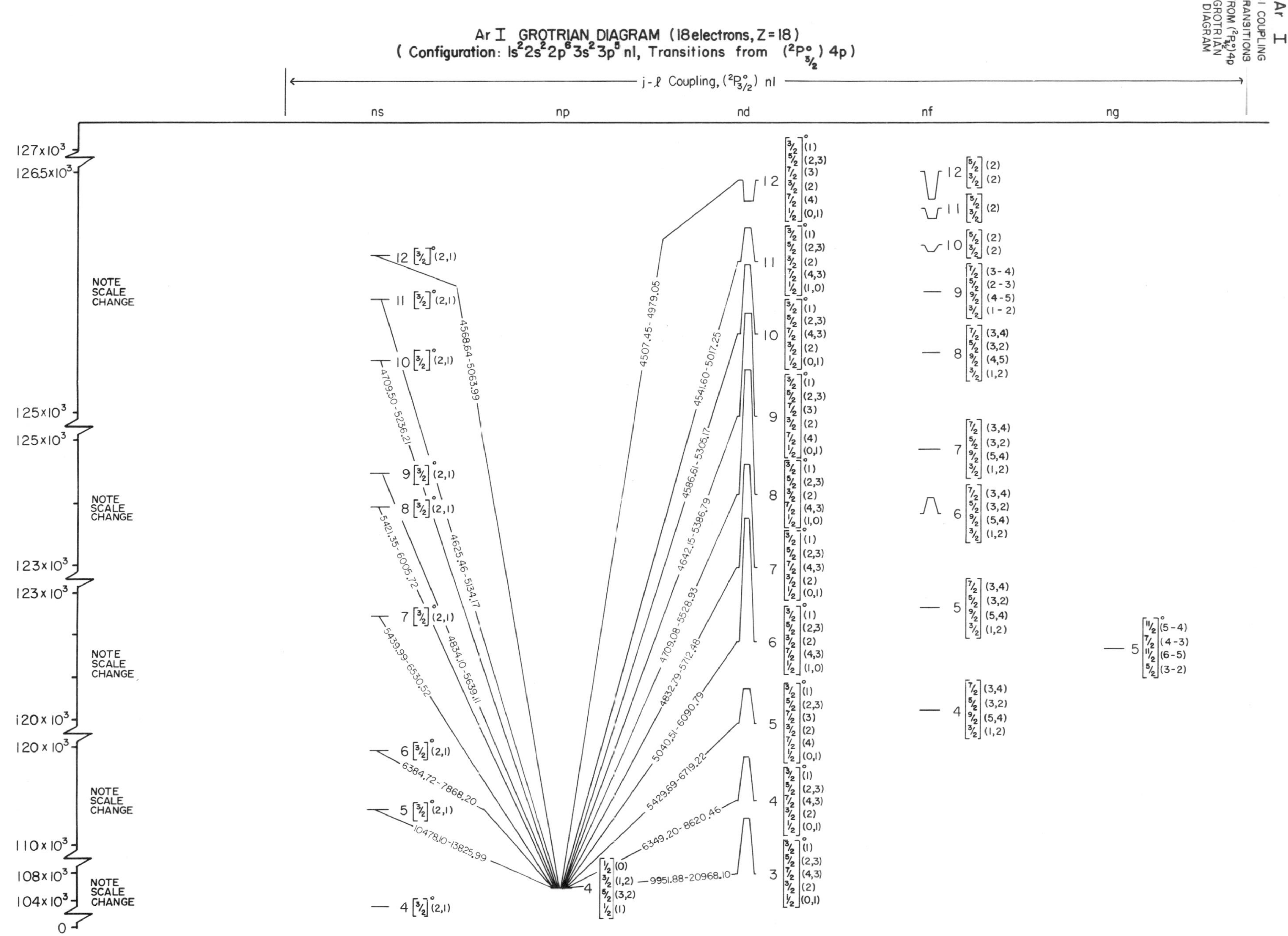
Ar I GROTRIAN DIAGRAM (18 electrons, Z = 18)
(Configuration: 1s² 2s² 2p⁶ 3s² 3p⁵ nl, Transitions from (²P°3/2) 4p)
j-ℓ Coupling, (²P°3/2) nl
ns
np
nd
nf
ng
Ar I
j-l COUPLING
TRANSITIONS
FROM (²P°3/2)4p
GROTRIAN
DIAGRAM
ENERGY (cm⁻¹)
127x10³
126.5x10³
125x10³
125x10³
123x10³
123x10³
120x10³
120x10³
110x10³
108x10³
104x10³
0
NOTE SCALE CHANGE
NOTE SCALE CHANGE
NOTE SCALE CHANGE
NOTE SCALE CHANGE
NOTE SCALE CHANGE
4568.64 - 5063.99
4709.50 - 5236.21
4625.46 - 5134.17
5421.35 - 6005.72
4834.10 - 5639.11
5439.99 - 6530.52
6384.72 - 7868.20
10478.10 - 13825.99
4507.45 - 4979.05
4541.60 - 5017.25
4586.61 - 5305.17
4642.15 - 5386.79
4709.08 - 5528.93
4832.79 - 5712.48
5040.51 - 6090.79
5429.69 - 6719.22
6349.20 - 8620.46
9951.88 - 20968.10

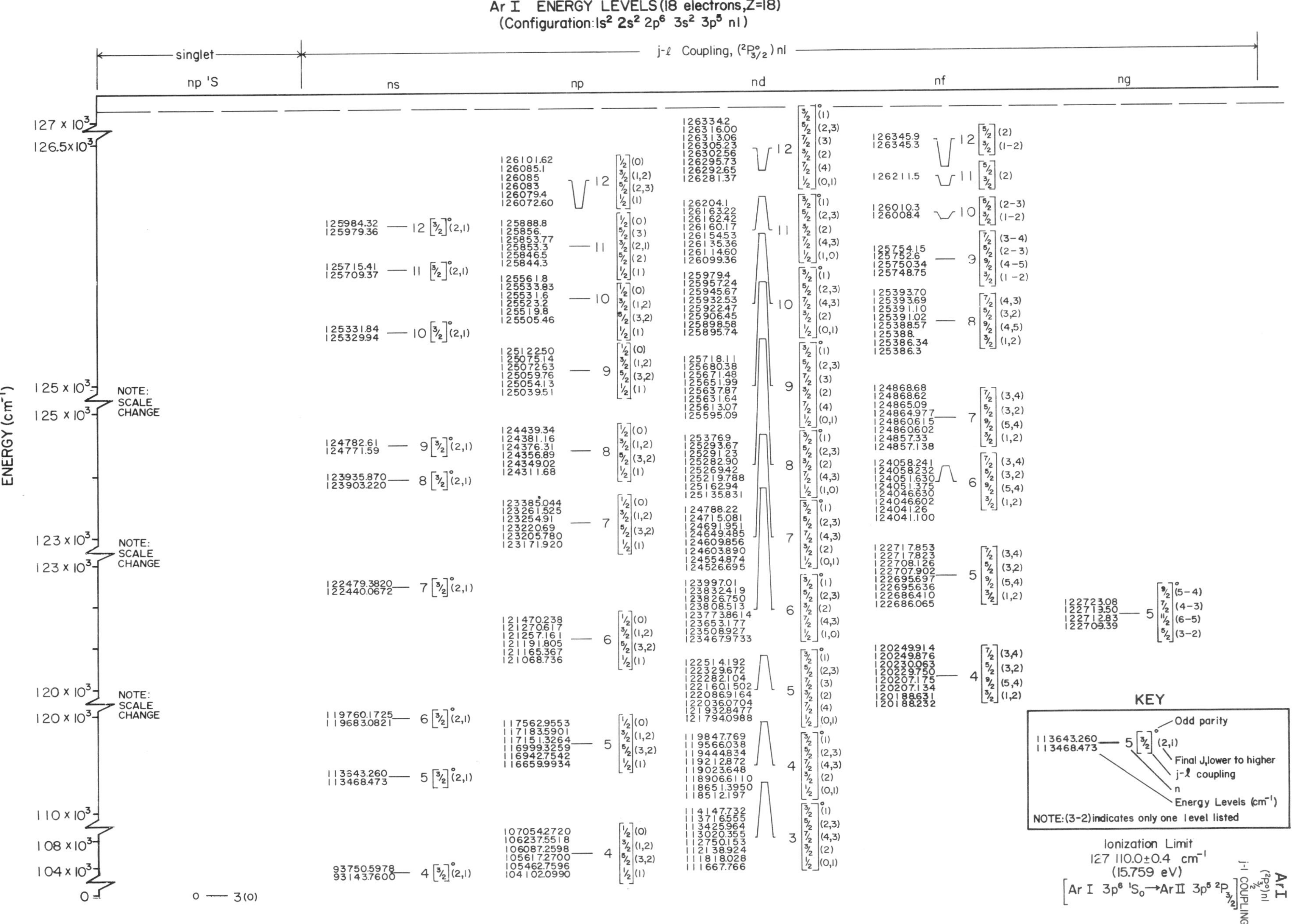

Ar I ENERGY LEVELS (18 electrons, Z=18)
(Configuration: $1s^2\ 2s^2\ 2p^6\ 3s^2\ 3p^5\ nl$)
singlet
j-ℓ Coupling, $(^2P^o_{3/2})\ nl$
np ^1S
ns
np
nd
nf
ng
ENERGY (cm⁻¹)
NOTE: SCALE CHANGE
NOTE: SCALE CHANGE
NOTE: SCALE CHANGE
127 x 10³
126.5x10³
125 x 10³
125 x 10³
123 x 10³
123 x 10³
120 x 10³
120 x 10³
110 x 10³
108 x 10³
104 x 10³
0
KEY
Odd parity
Final J, lower to higher
j-ℓ coupling
n
Energy Levels (cm⁻¹)
NOTE: (3-2) indicates only one level listed
Ionization Limit
127 110.0±0.4 cm⁻¹
(15.759 eV)
$[\text{Ar I } 3p^6\ ^1S_0 \rightarrow \text{Ar II } 3p^5\ ^2P_{3/2}]$
Ar I
$(^2P^o_{3/2})nl$
j-l COUPLING

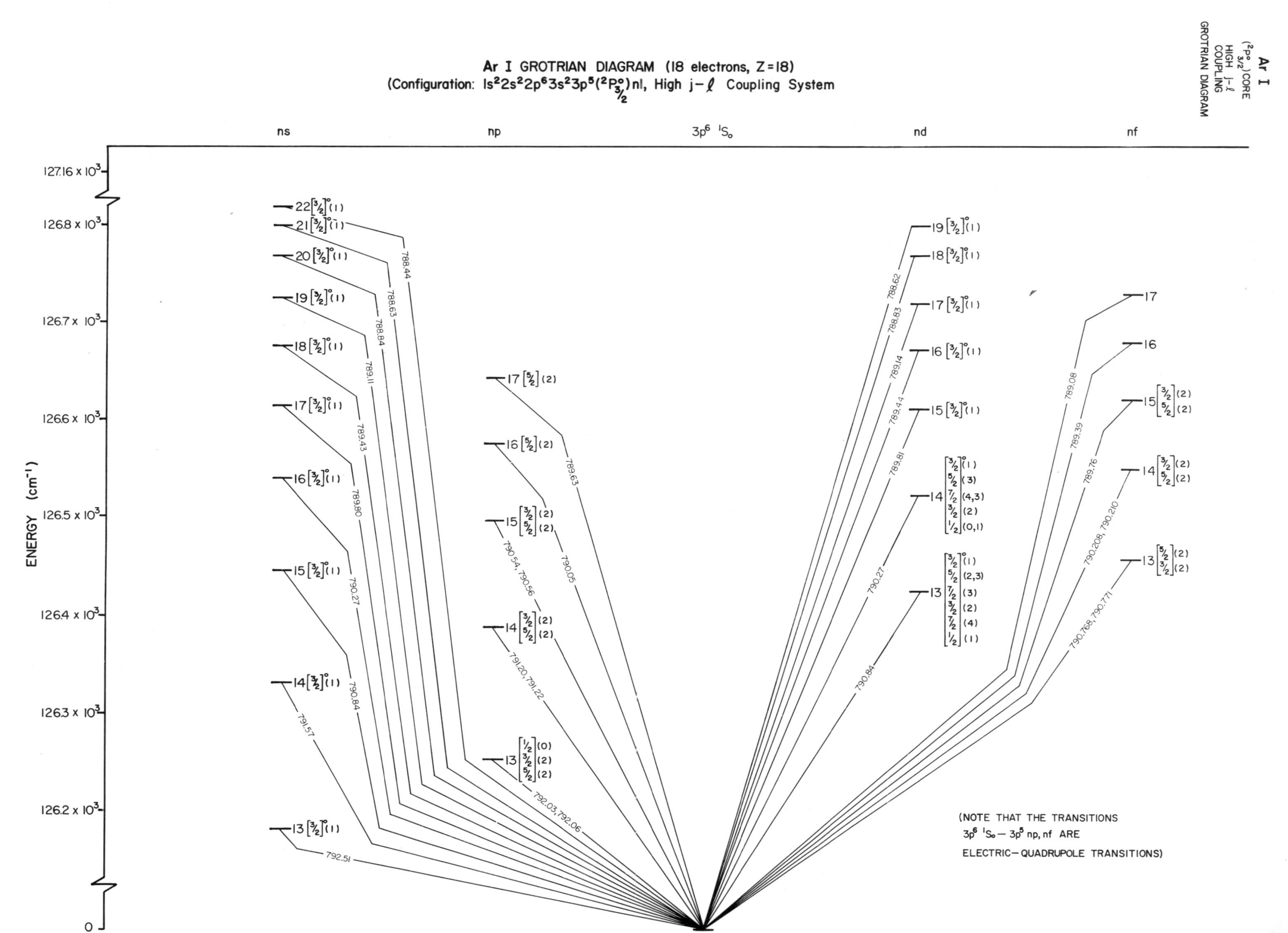

Ar I GROTRIAN DIAGRAM (18 electrons, Z=18)
(Configuration: $1s^2 2s^2 2p^6 3s^2 3p^5 (^2P^\circ_{3/2}) nl$, High j–ℓ Coupling System
Ar I
($^2P^\circ_{3/2}$) CORE
HIGH j-ℓ
COUPLING
GROTRIAN DIAGRAM
ns
np
$3p^6\ ^1S_0$
nd
nf
ENERGY (cm^{-1})
127.16 x 10^3
126.8 x 10^3
126.7 x 10^3
126.6 x 10^3
126.5 x 10^3
126.4 x 10^3
126.3 x 10^3
126.2 x 10^3
0
788.44
788.63
788.84
789.11
789.43
789.80
790.27
790.84
791.57
792.51
789.63
790.05
790.54, 790.56
791.20, 791.22
792.03, 792.06
788.62
788.83
789.14
789.44
789.81
790.27
790.84
789.08
789.39
789.76
790.208, 790.210
790.768, 790.771
(NOTE THAT THE TRANSITIONS
$3p^6\ ^1S_0 - 3p^5$ np, nf ARE
ELECTRIC–QUADRUPOLE TRANSITIONS)

Ar I ENERGY LEVELS (18 electrons, Z=18)

(Configuration: $1s^2 2s^2 2p^6 3s^2 3p^5 (^2P^o_{3/2})nl$, High j-ℓ Coupling System)

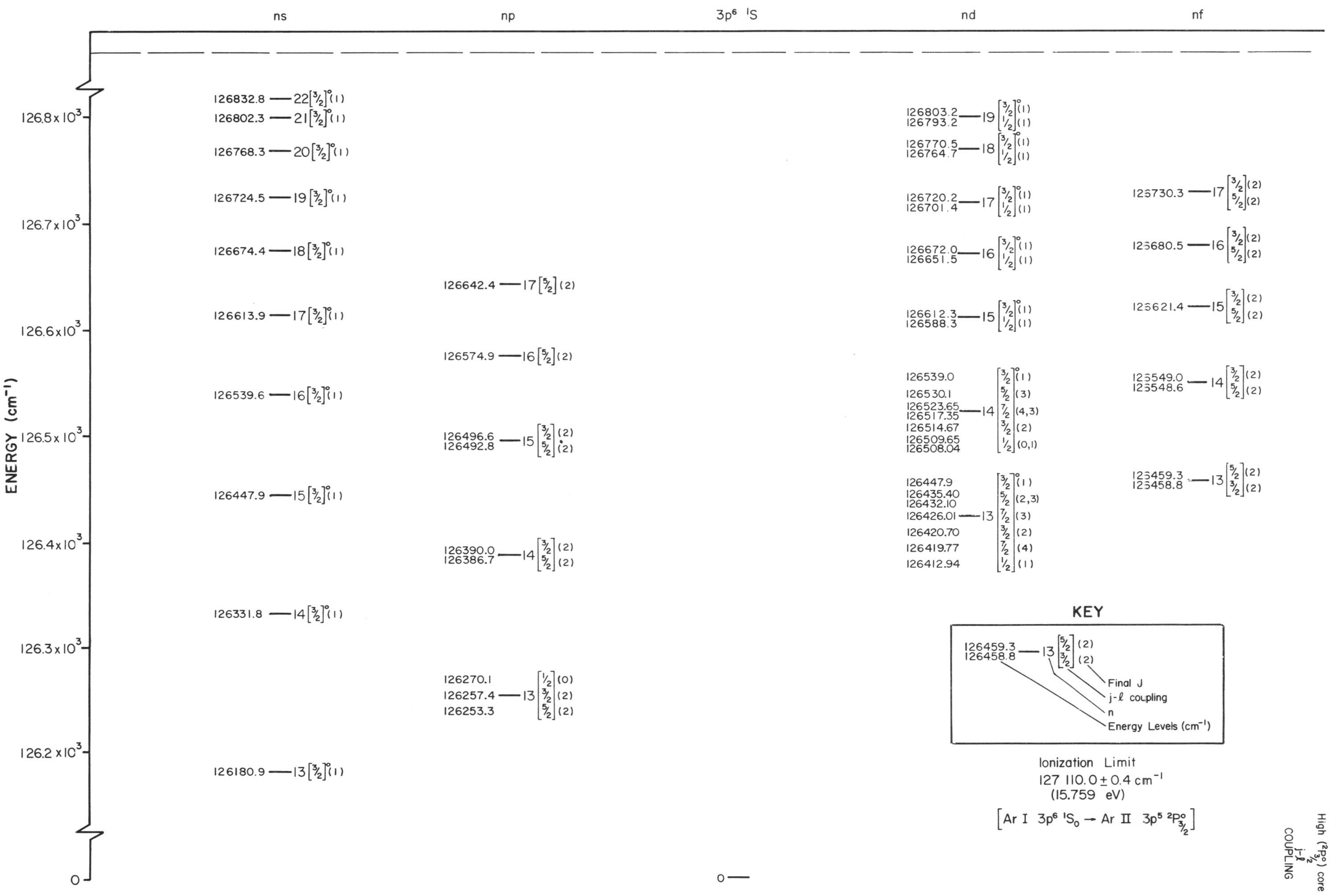

Ar I
High ($^2P^o_{3/2}$) core
j-ℓ
COUPLING

ENERGY (cm^{-1}): 126.8x10³, 126.7x10³, 126.6x10³, 126.5x10³, 126.4x10³, 126.3x10³, 126.2x10³, 0

ns

- 126832.8 — 22 $[3/2]^o$ (1)
- 126802.3 — 21 $[3/2]^o$ (1)
- 126768.3 — 20 $[3/2]^o$ (1)
- 126724.5 — 19 $[3/2]^o$ (1)
- 126674.4 — 18 $[3/2]^o$ (1)
- 126613.9 — 17 $[3/2]^o$ (1)
- 126539.6 — 16 $[3/2]^o$ (1)
- 126447.9 — 15 $[3/2]^o$ (1)
- 126331.8 — 14 $[3/2]^o$ (1)
- 126180.9 — 13 $[3/2]^o$ (1)

np

- 126642.4 — 17 $[5/2]$ (2)
- 126574.9 — 16 $[5/2]$ (2)
- 126496.6, 126492.8 — 15 $[3/2]$ (2), $[5/2]$ (2)
- 126390.0, 126386.7 — 14 $[3/2]$ (2), $[5/2]$ (2)
- 126270.1, 126257.4, 126253.3 — 13 $[1/2]$ (0), $[3/2]$ (2), $[5/2]$ (2)

$3p^6\ ^1S$

- 0 —

nd

- 126803.2, 126793.2 — 19 $[3/2]^o$ (1), $[1/2]$ (1)
- 126770.5, 126764.7 — 18 $[3/2]^o$ (1), $[1/2]$ (1)
- 126720.2, 126701.4 — 17 $[3/2]^o$ (1), $[1/2]$ (1)
- 126672.0, 126651.5 — 16 $[3/2]^o$ (1), $[1/2]$ (1)
- 126612.3, 126588.3 — 15 $[3/2]^o$ (1), $[1/2]$ (1)
- 126539.0, 126530.1, 126523.65, 126517.35, 126514.67, 126509.65, 126508.04 — 14 $[3/2]^o$ (1), $[5/2]$ (3), $[7/2]$ (4,3), $[3/2]$ (2), $[1/2]$ (0,1)
- 126447.9, 126435.40, 126432.10, 126426.01, 126420.70, 126419.77, 126412.94 — 13 $[3/2]^o$ (1), $[5/2]$ (2,3), $[7/2]$ (3), $[3/2]$ (2), $[7/2]$ (4), $[1/2]$ (1)

nf

- 126730.3 — 17 $[3/2]$ (2), $[5/2]$ (2)
- 126680.5 — 16 $[3/2]$ (2), $[5/2]$ (2)
- 126621.4 — 15 $[3/2]$ (2), $[5/2]$ (2)
- 126549.0, 126548.6 — 14 $[3/2]$ (2), $[5/2]$ (2)
- 126459.3, 126458.8 — 13 $[5/2]$ (2), $[3/2]$ (2)

KEY

126459.3, 126458.8 — 13 $[5/2]$ (2), $[3/2]$ (2)
- Final J
- j-ℓ coupling
- n
- Energy Levels (cm^{-1})

Ionization Limit
127 110.0 ± 0.4 cm^{-1}
(15.759 eV)
[Ar I $3p^6\ ^1S_0$ → Ar II $3p^5\ ^2P^o_{3/2}$]

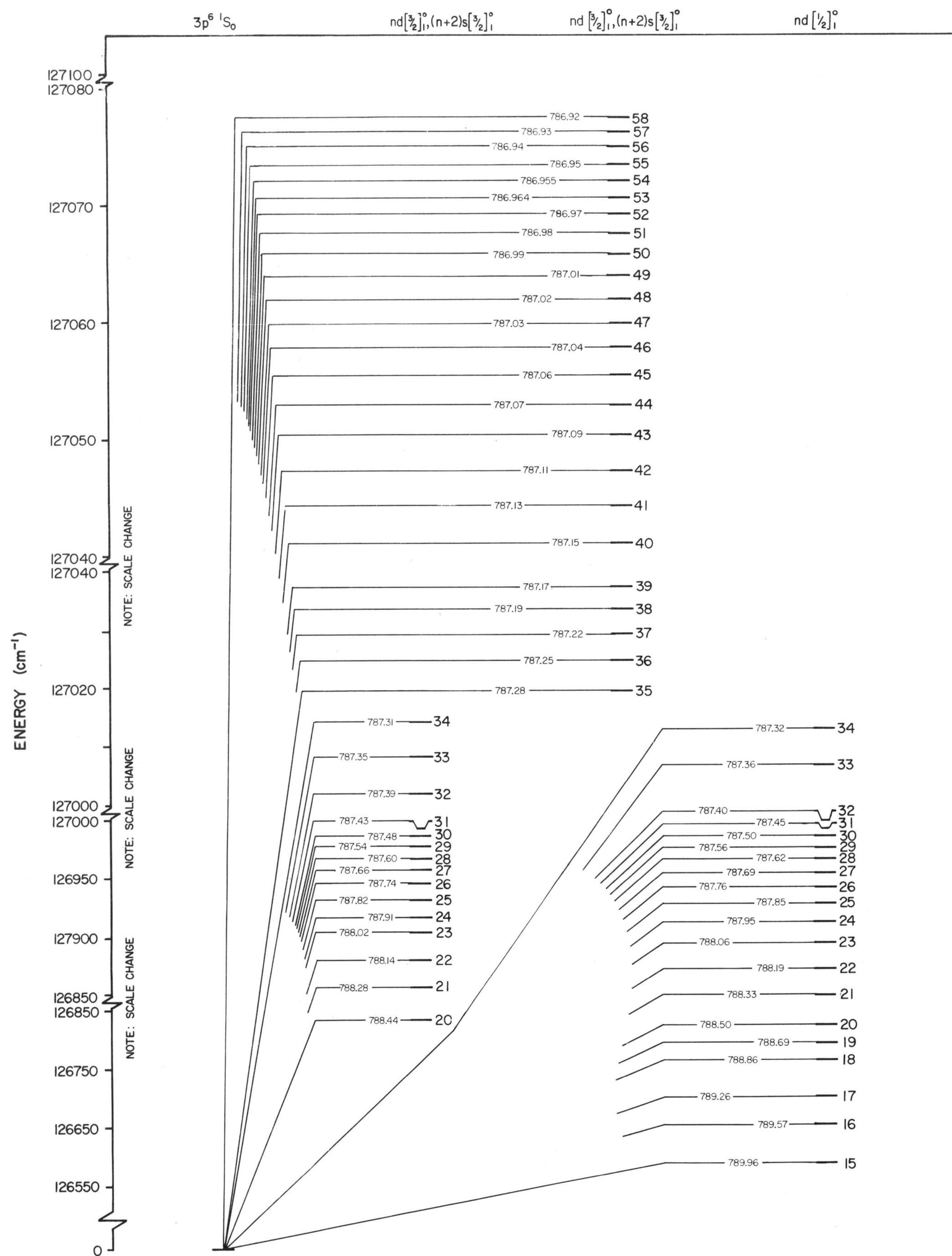
Ar I GROTRIAN DIAGRAM (18 electrons, Z=18)
(Configuration: $1s^22s^22p^63s^23p^5(^2P^o_{3/2})$nd, (n+2)s, Combinations of Higher nd, (n+2)s with Ground Level)
Degenerate Levels
$3p^6\ ^1S_0$
$nd[3/2]^o_1,(n+2)s[3/2]^o_1$
$nd[3/2]^o_1,(n+2)s[3/2]^o_1$
$nd[1/2]^o_1$
ENERGY (cm^{-1})
NOTE: SCALE CHANGE
NOTE: SCALE CHANGE
NOTE: SCALE CHANGE
127100
127080
127070
127060
127050
127040
127040
127020
127000
127000
126950
126900
126850
126850
126750
126650
126550
0
786.92 58
786.93 57
786.94 56
786.95 55
786.955 54
786.964 53
786.97 52
786.98 51
786.99 50
787.01 49
787.02 48
787.03 47
787.04 46
787.06 45
787.07 44
787.09 43
787.11 42
787.13 41
787.15 40
787.17 39
787.19 38
787.22 37
787.25 36
787.28 35
787.31 34
787.35 33
787.39 32
787.43 31
787.48 30
787.54 29
787.60 28
787.66 27
787.74 26
787.82 25
787.91 24
788.02 23
788.14 22
788.28 21
788.44 20
787.32 34
787.36 33
787.40 32
787.45 31
787.50 30
787.56 29
787.62 28
787.69 27
787.76 26
787.85 25
787.95 24
788.06 23
788.19 22
788.33 21
788.50 20
788.69 19
788.86 18
789.26 17
789.57 16
789.96 15

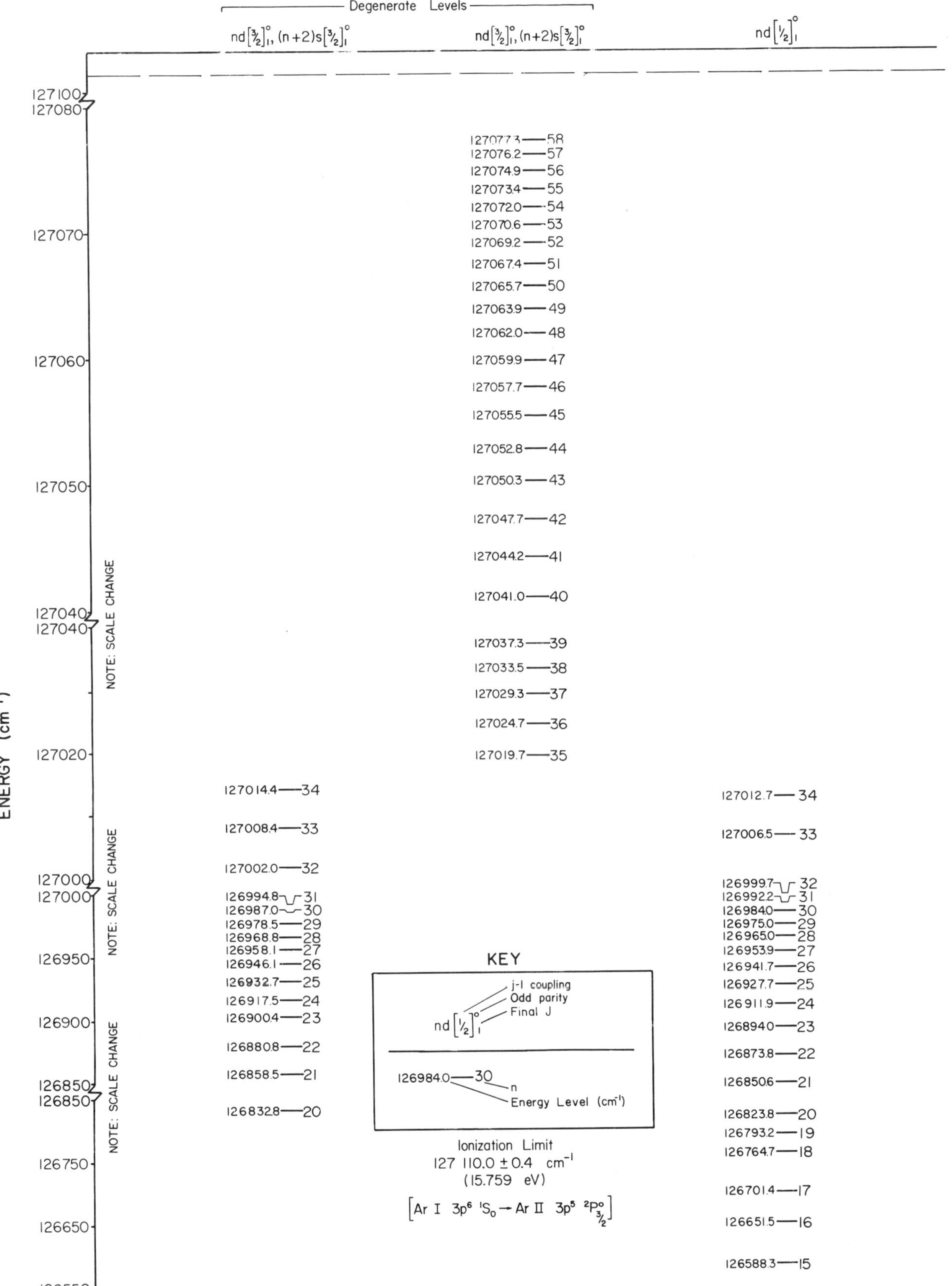
Ar I ENERGY LEVELS (18 electrons, Z=18)
(Configuration: $1s^2 2s^2 2p^6 3s^2 3p^5 (^2P^o_{3/2})$ nd, (n+2)s, Higher j-ℓ Coupling Systems)
Degenerate Levels
$nd[3/2]^o_1$, $(n+2)s[3/2]^o_1$
$nd[3/2]^o_1$, $(n+2)s[3/2]^o_1$
$nd[1/2]^o_1$
ENERGY (cm⁻¹)
127100
127080
127070
127060
127050
127040
127040
127020
127000
127000
126950
126900
126850
126850
126750
126650
126550
NOTE: SCALE CHANGE
NOTE: SCALE CHANGE
NOTE: SCALE CHANGE
127077.3—58
127076.2—57
127074.9—56
127073.4—55
127072.0—54
127070.6—53
127069.2—52
127067.4—51
127065.7—50
127063.9—49
127062.0—48
127059.9—47
127057.7—46
127055.5—45
127052.8—44
127050.3—43
127047.7—42
127044.2—41
127041.0—40
127037.3—39
127033.5—38
127029.3—37
127024.7—36
127019.7—35
127014.4—34
127008.4—33
127002.0—32
126994.8—31
126987.0—30
126978.5—29
126968.8—28
126958.1—27
126946.1—26
126932.7—25
126917.5—24
126900.4—23
126880.8—22
126858.5—21
126832.8—20
127012.7—34
127006.5—33
126999.7—32
126992.2—31
126984.0—30
126975.0—29
126965.0—28
126953.9—27
126941.7—26
126927.7—25
126911.9—24
126894.0—23
126873.8—22
126850.6—21
126823.8—20
126793.2—19
126764.7—18
126701.4—17
126651.5—16
126588.3—15
KEY
j-l coupling
Odd parity
Final J
$nd[1/2]^o_1$
126984.0—30
n
Energy Level (cm⁻¹)
Ionization Limit
127 110.0 ± 0.4 cm⁻¹
(15.759 eV)
[Ar I $3p^6$ $^1S_0$ → Ar II $3p^5$ $^2P^o_{3/2}$]

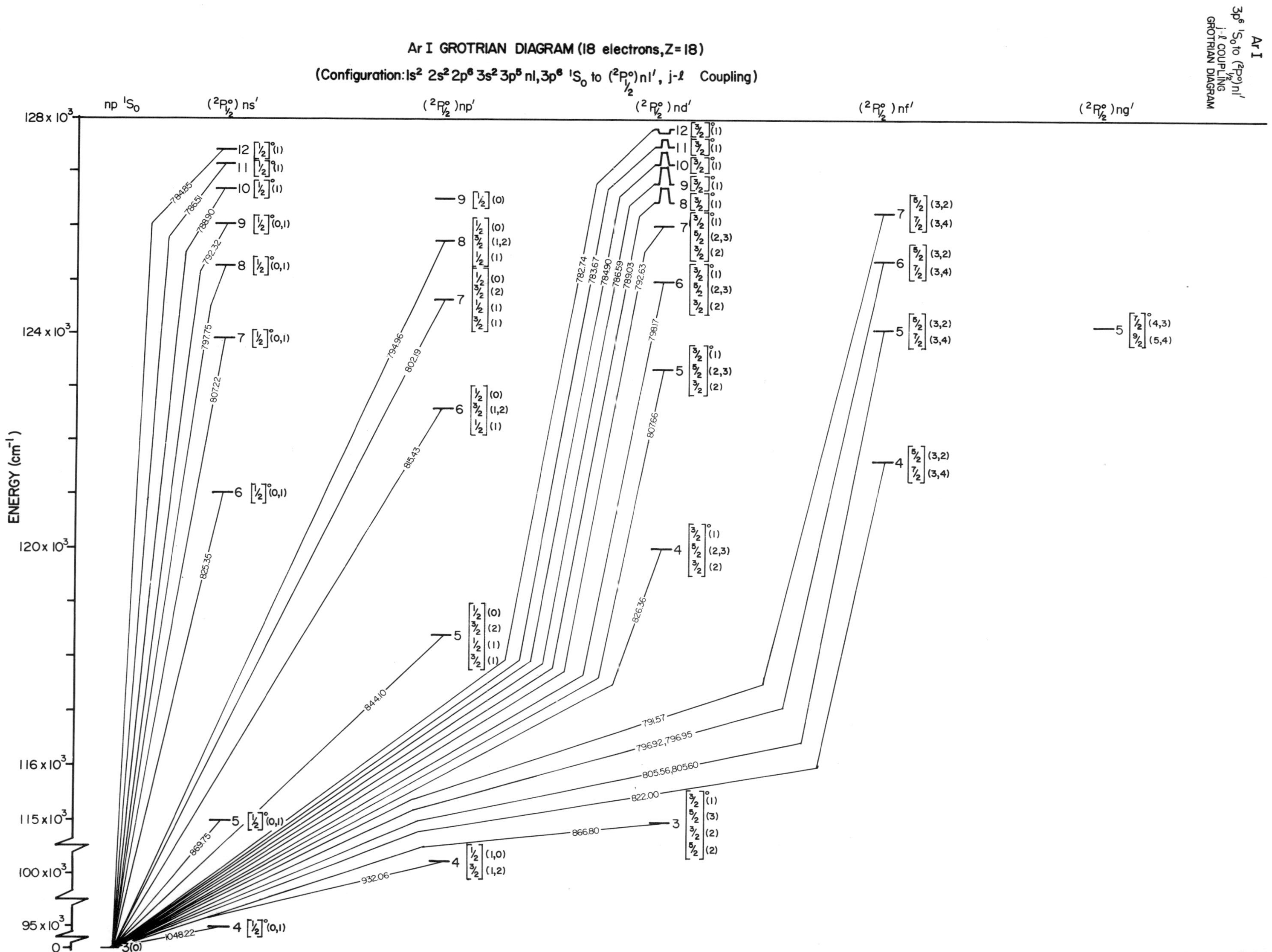

Ar I GROTRIAN DIAGRAM (18 electrons, Z=18)
(Configuration: 1s² 2s² 2p⁶ 3s² 3p⁵ nl, 3p⁶ ¹S₀ to (²P°₁/₂)nl', j-ℓ Coupling)
Ar I
3p⁶ ¹S₀ to (²P°₁/₂)nl'
j-ℓ COUPLING
GROTRIAN DIAGRAM
np ¹S₀
(²P°₁/₂) ns'
(²P°₁/₂) np'
(²P°₁/₂) nd'
(²P°₁/₂) nf'
(²P°₁/₂) ng'
ENERGY (cm⁻¹)
128 x 10³
124 x 10³
120 x 10³
116 x 10³
115 x 10³
100 x 10³
95 x 10³
0
784.85
786.51
788.90
792.32
797.75
807.22
825.35
869.75
1048.22
794.96
802.19
815.43
844.10
932.06
782.74
783.67
784.90
786.59
789.03
792.63
798.17
807.66
826.36
866.80
791.57
796.92,796.95
805.56,805.60
822.00

Ar I ENERGY LEVELS (18 electrons, Z=18)

(Configuration: $1s^2\ 2s^2\ 2p^6\ 3s^2\ 3p^5\ nl$, $(^2P^o_{1/2})\ nl'$, j-ℓ Coupling)

ENERGY (cm^{-1})

128 x 10³
124 x 10³
120 x 10³
116 x 10³
115 x10
100 x10
95 x 10³

$(^2P^o_{1/2})ns'$

127413.7, 127410. — 12 $[\frac{1}{2}]^o$ (1)
127144.6, 127140. — 11 $[\frac{1}{2}]^o$ (0,1)
126758.5, 126750. — 10 $[\frac{1}{2}]^o$ (0,1)
126211.52, 126202.78 — 9 $[\frac{1}{2}]^o$ (0,1)
125353.23, 125334.69 — 8 $[\frac{1}{2}]^o$ (0,1)
123882.203, 123873.000 — 7 $[\frac{1}{2}]^o$ (0,1)
121161.3135, 121096.605 — 6 $[\frac{1}{2}]^o$ (0,1)
114975.019, 114861.635 — 5 $[\frac{1}{2}]^o$ (0,1)
95399.8276, 94553.6652 — 4 $[\frac{1}{2}]^o$ (0,1)

$(^2P^o_{1/2})np'$

126524.19 — 9 $[\frac{1}{2}]$ (0)
125831.53, 125791.95, 125783.71, 125777.32 — 8 $[\frac{1}{2}]$ (0), $[\frac{3}{2}]$ (1,2), $[\frac{1}{2}]$ (1)
124749.805, 124658.44, 124650.95, 124643.45 — 7 $[\frac{1}{2}]$ (0), $[\frac{3}{2}]$ (2), $[\frac{1}{2}]$ (1), $[\frac{3}{2}]$ (1)
122790.547, 122635.054, 122609.637, 122601.225 — 6 $[\frac{1}{2}]$ (0), $[\frac{3}{2}]$ (1,2), $[\frac{1}{2}]$ (1)
118870.9170, 118469.0508, 118459.5975, 118407.4303 — 5 $[\frac{1}{2}]$ (0), $[\frac{3}{2}]$ (2), $[\frac{1}{2}]$ (1), $[\frac{3}{2}]$ (1)
108722.6194, 107496.4166, 107289.7001, 107131.7086 — 4 $[\frac{1}{2}]$ (1,0), $[\frac{3}{2}]$ (1,2)

$(^2P^o_{1/2})nd'$

127756. — 12 $[\frac{3}{2}]^o$ (1)
127605. — 11 $[\frac{3}{2}]^o$ (1)
127404. — 10 $[\frac{3}{2}]^o$ (1)
127131. — 9 $[\frac{3}{2}]^o$ (1)
126737.7 — 8 $[\frac{3}{2}]^o$ (1)
126161.9, 126089.54, 126064.44, 126053.10 — 7 $[\frac{3}{2}]^o$ (1), $[\frac{5}{2}]$ (2,3), $[\frac{3}{2}]$ (2)
125286.26, 125149.937, 125113.376, 125066.432 — 6 $[\frac{3}{2}]^o$ (1), $[\frac{5}{2}]$ (2,3), $[\frac{3}{2}]$ (2)
123815.560, 123557.4227, 123505.4896, 123372.9253 — 5 $[\frac{3}{2}]^o$ (1), $[\frac{5}{2}]$ (2,3), $[\frac{3}{2}]$ (2)
121011.9198, 120753.4714, 120619.0162, 120600.8905 — 4 $[\frac{3}{2}]^o$ (1), $[\frac{5}{2}]$ (2,3), $[\frac{3}{2}]$ (2)
115366.866, 114821.939, 114805.135, 114640.992 — 3 $[\frac{3}{2}]^o$ (1), $[\frac{5}{2}]$ (3), $[\frac{3}{2}]$ (2), $[\frac{5}{2}]$ (2)

$(^2P^o_{1/2})nf'$

126295., 126294.97, 126294.84, 126294.83 — 7 $[\frac{5}{2}]$ (3,2), $[\frac{7}{2}]$ (3,4)
125483.362, 125483.145, 125482.68, 125482.641 — 6 $[\frac{5}{2}]$ (3,2), $[\frac{7}{2}]$ (3,4)
124137.436, 124137.214, 124135.698, 124135.669 — 5 $[\frac{5}{2}]$ (3,2), $[\frac{7}{2}]$ (3,4)
121654.580, 121654.234, 121653.280, 121653.238 — 4 $[\frac{5}{2}]$ (3,2), $[\frac{7}{2}]$ (3,4)

$(^2P^o_{1/2})ng'$

124148.48, 124148.26 — 5 $[\frac{7}{2}]^o$ (4,3), $[\frac{9}{2}]$ (5,4)

KEY

Core
$(^2P^o_{1/2})nd'$
L
parity
outer electron
J
multiplicity

121011.9198, 120753.4714, 120619.0162, 120600.8905 — 4 $[\frac{3}{2}]$ (1), $[\frac{5}{2}]$ (2,3), $[\frac{3}{2}]$ (2)
Final J, lower to higher
j-ℓ coupling
n
Energy Levels (cm^{-1})

Ionization Limit
127 110.0±0.4 cm^{-1}
(15.759 eV)
[Ar I $3p^6\ ^1S_0$ → Ar II $3p^5\ ^2P^o_{3/2}$]

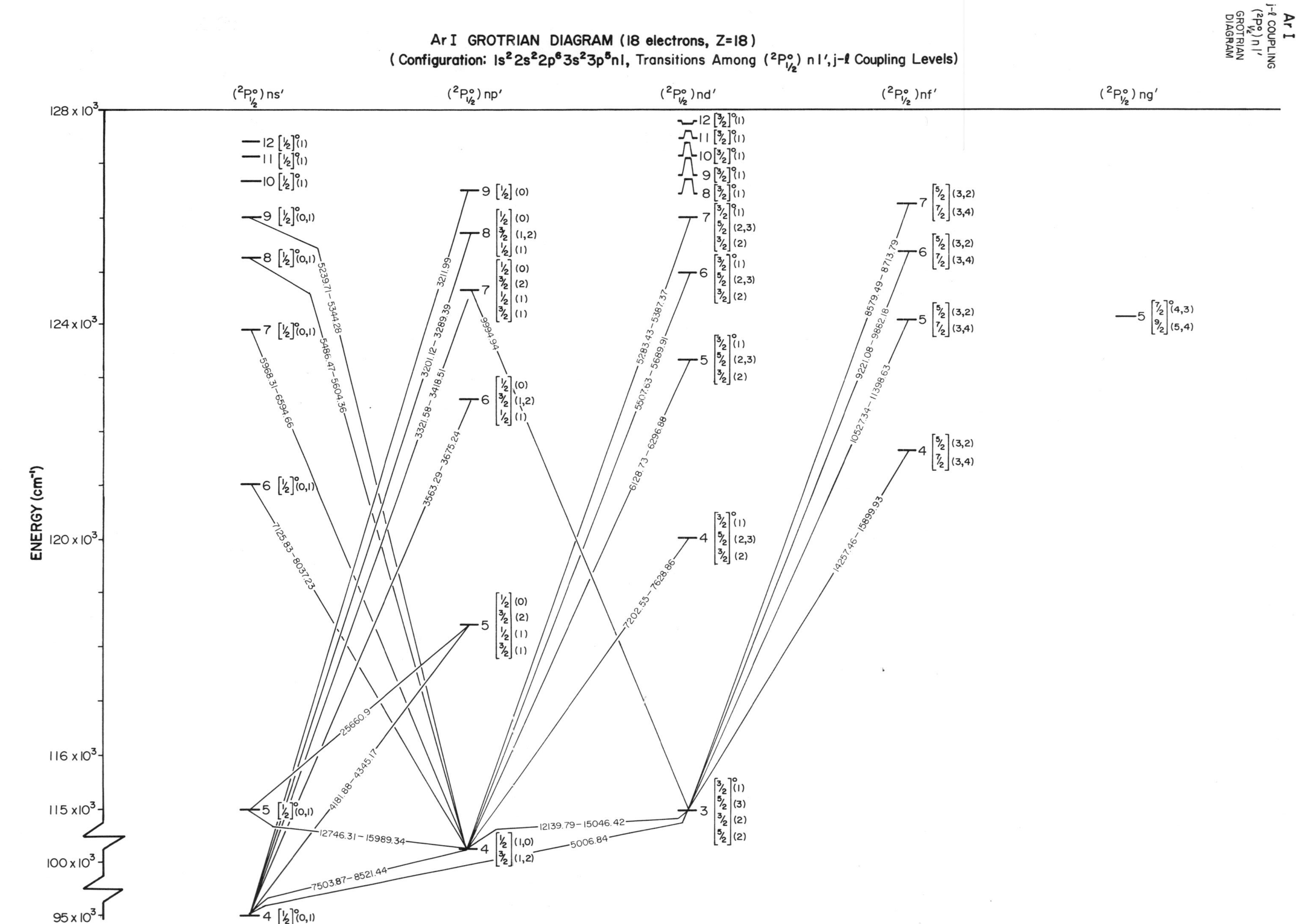

Ar I
j-ℓ COUPLING
(²P°1/2) n l′
GROTRIAN
DIAGRAM
Ar I GROTRIAN DIAGRAM (18 electrons, Z=18)
(Configuration: 1s²2s²2p⁶3s²3p⁵nl, Transitions Among (²P°1/2) n l′, j-ℓ Coupling Levels)
(²P°1/2) ns′
(²P°1/2) np′
(²P°1/2) nd′
(²P°1/2) nf′
(²P°1/2) ng′
ENERGY (cm⁻¹)
128 x 10³
124 x 10³
120 x 10³
116 x 10³
115 x 10³
100 x 10³
95 x 10³
5239.71 – 5344.28
5486.47 – 5604.36
5968.31 – 6594.66
7125.83 – 8037.23
3211.99
3201.12 – 3289.39
3321.58 – 3418.51
3563.29 – 3675.24
9994.94
25660.9
4181.88 – 4345.17
12746.31 – 15989.34
7503.87 – 8521.44
5283.43 – 5387.37
5507.63 – 5689.91
6128.73 – 6296.88
7202.55 – 7628.86
12139.79 – 15046.42
5006.84
8579.49 – 8713.79
9221.08 – 9882.18
10527.34 – 11398.63
14257.46 – 15899.93

Ar I
j-ℓ Coupling
$(^2P^o_{1/2})nl'$

Ar I ENERGY LEVELS (18 electrons, Z = 18)

(Configuration: $1s^2\ 2s^2\ 2p^6\ 3s^2\ 3p^5\ nl$, $(^2P^o_{1/2})\ nl'$, j-ℓ Coupling)

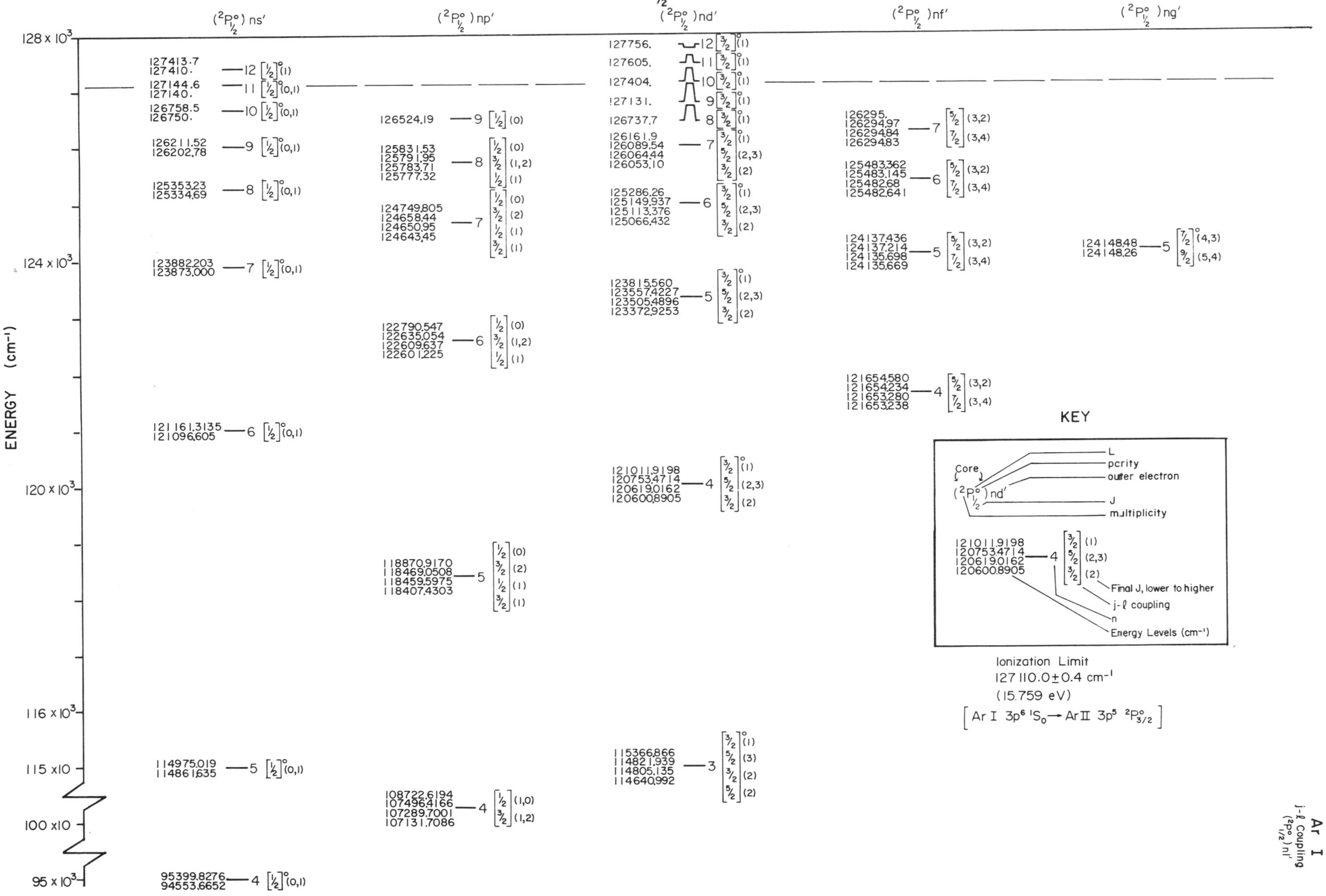

ENERGY (cm⁻¹) axis: 128 x 10³, 124 x 10³, 120 x 10³, 116 x 10³, 115 x10, 100 x10, 95 x 10³

$(^2P^o_{1/2})$ ns'

Energy Levels	n	j-ℓ coupling	Final J
127413.7 127410.	12	$[\frac{1}{2}]^o$	(1)
127144.6 127140.	11	$[\frac{1}{2}]^o$	(0,1)
126758.5 126750.	10	$[\frac{1}{2}]^o$	(0,1)
126211.52 126202.78	9	$[\frac{1}{2}]^o$	(0,1)
125353.23 125334.69	8	$[\frac{1}{2}]^o$	(0,1)
123882.203 123873.000	7	$[\frac{1}{2}]^o$	(0,1)
121161.3135 121096.605	6	$[\frac{1}{2}]^o$	(0,1)
114975.019 114861.635	5	$[\frac{1}{2}]^o$	(0,1)
95399.8276 94553.6652	4	$[\frac{1}{2}]^o$	(0,1)

$(^2P^o_{1/2})$ np'

Energy Levels	n	j-ℓ coupling	Final J
126524.19	9	$[\frac{1}{2}]$	(0)
125831.53 125791.95 125783.71 125777.32	8	$[\frac{1}{2}]$ $[\frac{3}{2}]$ $[\frac{1}{2}]$	(0) (1,2) (1)
124749.805 124658.44 124650.95 124643.45	7	$[\frac{1}{2}]$ $[\frac{3}{2}]$ $[\frac{1}{2}]$ $[\frac{3}{2}]$	(0) (2) (1) (1)
122790.547 122635.054 122609.637 122601.225	6	$[\frac{1}{2}]$ $[\frac{3}{2}]$ $[\frac{1}{2}]$	(0) (1,2) (1)
118870.9170 118469.0508 118459.5975 118407.4303	5	$[\frac{1}{2}]$ $[\frac{3}{2}]$ $[\frac{1}{2}]$ $[\frac{3}{2}]$	(0) (2) (1) (1)
108722.6194 107496.4166 107289.7001 107131.7086	4	$[\frac{1}{2}]$ $[\frac{3}{2}]$	(1,0) (1,2)

$(^2P^o_{1/2})$ nd'

Energy Levels	n	j-ℓ coupling	Final J
127756.	12	$[\frac{3}{2}]^o$	(1)
127605.	11	$[\frac{3}{2}]^o$	(1)
127404.	10	$[\frac{3}{2}]^o$	(1)
127131.	9	$[\frac{3}{2}]^o$	(1)
126737.7	8	$[\frac{3}{2}]^o$	(1)
126161.9 126089.54 126064.44 126053.10	7	$[\frac{3}{2}]^o$ $[\frac{5}{2}]$ $[\frac{3}{2}]$	(1) (2,3) (2)
125286.26 125149.937 125113.376 125066.432	6	$[\frac{3}{2}]^o$ $[\frac{5}{2}]$ $[\frac{3}{2}]$	(1) (2,3) (2)
123815.560 123557.4227 123505.4896 123372.9253	5	$[\frac{3}{2}]^o$ $[\frac{5}{2}]$ $[\frac{3}{2}]$	(1) (2,3) (2)
121011.9198 120753.4714 120619.0162 120600.8905	4	$[\frac{3}{2}]^o$ $[\frac{5}{2}]$ $[\frac{3}{2}]$	(1) (2,3) (2)
115366.866 114821.939 114805.135 114640.992	3	$[\frac{3}{2}]^o$ $[\frac{5}{2}]$ $[\frac{3}{2}]$ $[\frac{5}{2}]$	(1) (3) (2) (2)

$(^2P^o_{1/2})$ nf'

Energy Levels	n	j-ℓ coupling	Final J
126295. 126294.97 126294.84 126294.83	7	$[\frac{5}{2}]$ $[\frac{7}{2}]$	(3,2) (3,4)
125483.362 125483.145 125482.68 125482.641	6	$[\frac{5}{2}]$ $[\frac{7}{2}]$	(3,2) (3,4)
124137.436 124137.214 124135.698 124135.669	5	$[\frac{5}{2}]$ $[\frac{7}{2}]$	(3,2) (3,4)
121654.580 121654.234 121653.280 121653.238	4	$[\frac{5}{2}]$ $[\frac{7}{2}]$	(3,2) (3,4)

$(^2P^o_{1/2})$ ng'

Energy Levels	n	j-ℓ coupling	Final J
124148.48 124148.26	5	$[\frac{7}{2}]^o$ $[\frac{9}{2}]$	(4,3) (5,4)

KEY

$(^2P^o_{1/2})nd'$: Core; L; parity; outer electron; J; multiplicity

Energy Levels (cm⁻¹)	n	j-ℓ coupling	Final J, lower to higher
121011.9198 120753.4714 120619.0162 120600.8905	4	$[\frac{3}{2}]$ $[\frac{5}{2}]$ $[\frac{3}{2}]$	(1) (2,3) (2)

Ionization Limit
127 110.0 ± 0.4 cm⁻¹
(15.759 eV)
[Ar I $3p^6\ ^1S_0$ → Ar II $3p^5\ ^2P^o_{3/2}$]

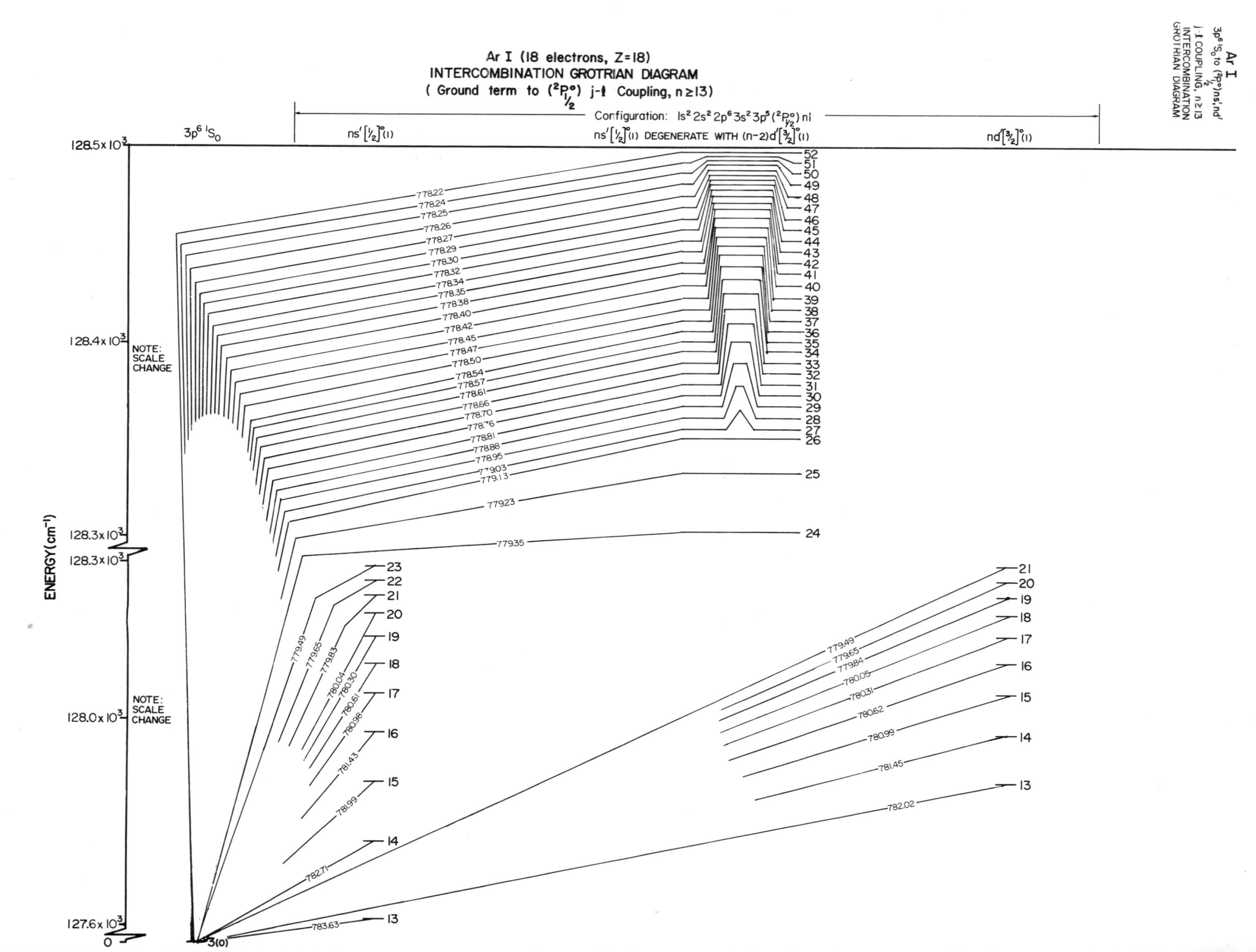
Ar I (18 electrons, Z=18)
INTERCOMBINATION GROTRIAN DIAGRAM
(Ground term to (²P°½) j-l Coupling, n ≥ 13)
Configuration: 1s²2s²2p⁶3s²3p⁵(²P°½) nl
3p⁶ ¹S₀
ns′[½]°(1)
ns′[½]°(1) DEGENERATE WITH (n-2)d′[3/2]°(1)
nd′[3/2]°(1)
Ar I
3p⁶ ¹S₀ to (²P°½)ns′,nd′
j-l COUPLING, n ≥ 13
INTERCOMBINATION
GROTRIAN DIAGRAM
ENERGY(cm⁻¹)
128.5x10³
128.4x10³
128.3x10³
128.3x10³
128.0x10³
127.6x10³
0
NOTE: SCALE CHANGE
NOTE: SCALE CHANGE
3(0)
778.22
778.24
778.25
778.26
778.27
778.29
778.30
778.32
778.34
778.35
778.38
778.40
778.42
778.45
778.47
778.50
778.54
778.57
778.61
778.66
778.70
778.81
778.88
778.95
779.13
779.23
779.35
779.49
779.65
779.83
780.04
780.30
780.61
780.98
781.43
781.99
782.71
783.63
779.49
779.65
779.84
780.05
780.31
780.62
780.99
781.45
782.02

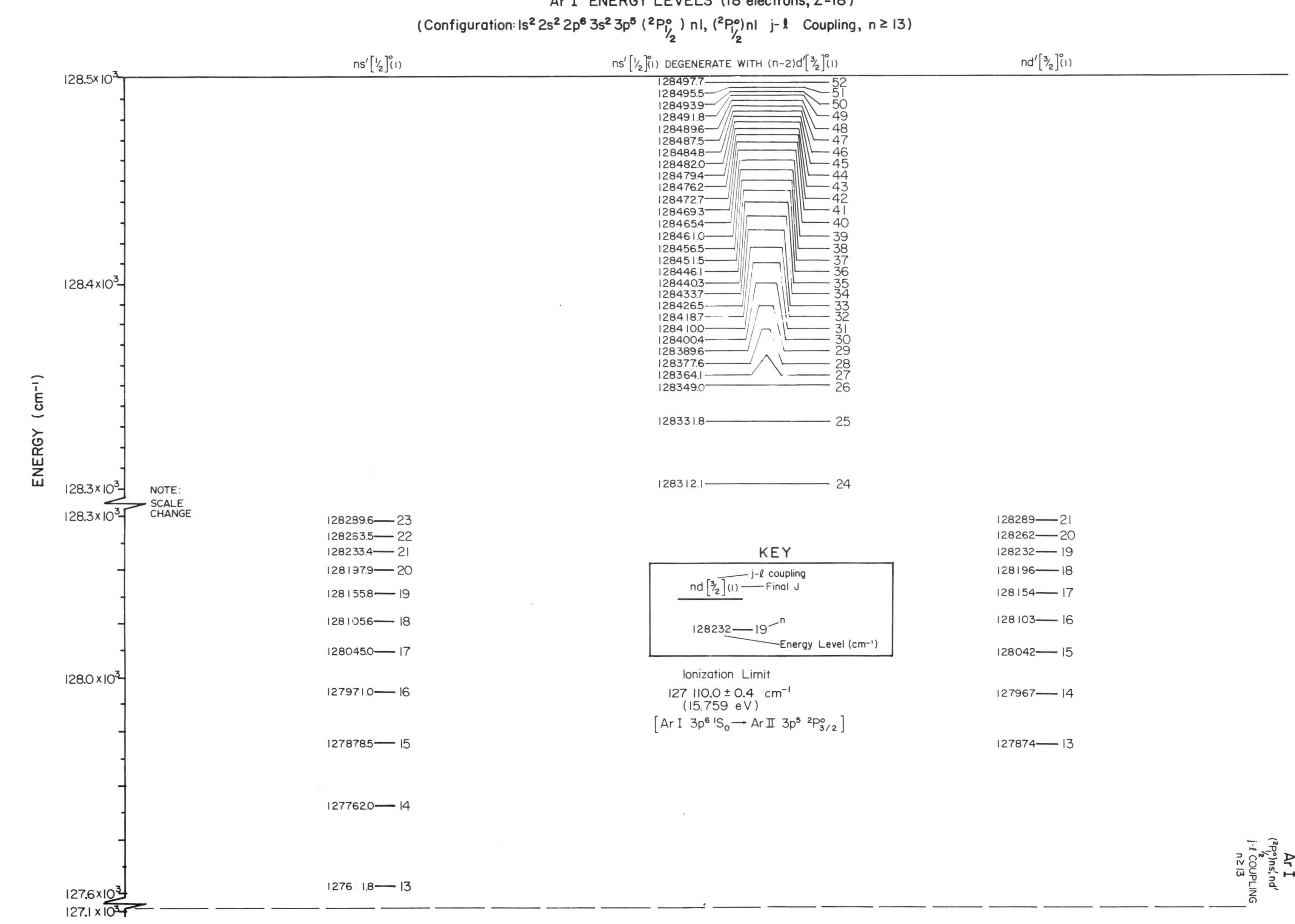

Ar I ENERGY LEVELS (18 electrons, Z=18)
(Configuration: $1s^2 2s^2 2p^6 3s^2 3p^5 (^2P^o_{1/2})nl$, $(^2P^o_{1/2})nl$ j-l Coupling, n ≥ 13)
$ns'[1/2]^o_{(1)}$
$ns'[1/2]^o_{(1)}$ DEGENERATE WITH $(n-2)d'[3/2]^o_{(1)}$
$nd'[3/2]^o_{(1)}$
128497.7 — 52
128495.5 — 51
128493.9 — 50
128491.8 — 49
128489.6 — 48
128487.5 — 47
128484.8 — 46
128482.0 — 45
128479.4 — 44
128476.2 — 43
128472.7 — 42
128469.3 — 41
128465.4 — 40
128461.0 — 39
128456.5 — 38
128451.5 — 37
128446.1 — 36
128440.3 — 35
128433.7 — 34
128426.5 — 33
128418.7 — 32
128410.0 — 31
128400.4 — 30
128389.6 — 29
128377.6 — 28
128364.1 — 27
128349.0 — 26
128331.8 — 25
128312.1 — 24
128289.6 — 23
128253.5 — 22
128233.4 — 21
128197.9 — 20
128155.8 — 19
128105.6 — 18
128045.0 — 17
127971.0 — 16
127878.5 — 15
127762.0 — 14
1276 1.8 — 13
128289 — 21
128262 — 20
128232 — 19
128196 — 18
128154 — 17
128103 — 16
128042 — 15
127967 — 14
127874 — 13
KEY
$nd[3/2]_{(1)}$
j-l coupling
Final J
128232 — 19
n
Energy Level (cm⁻¹)
Ionization Limit
127 110.0 ± 0.4 cm⁻¹
(15.759 eV)
[Ar I $3p^6\ ^1S_0$ → Ar II $3p^5\ ^2P^o_{3/2}$]
NOTE: SCALE CHANGE
128.5×10³
128.4×10³
128.3×10³
128.3×10³
128.0×10³
127.6×10³
127.1×10³
ENERGY (cm⁻¹)
Ar I
$(^2P^o_{1/2})ns', nd'$
j-l COUPLING
n ≥ 13

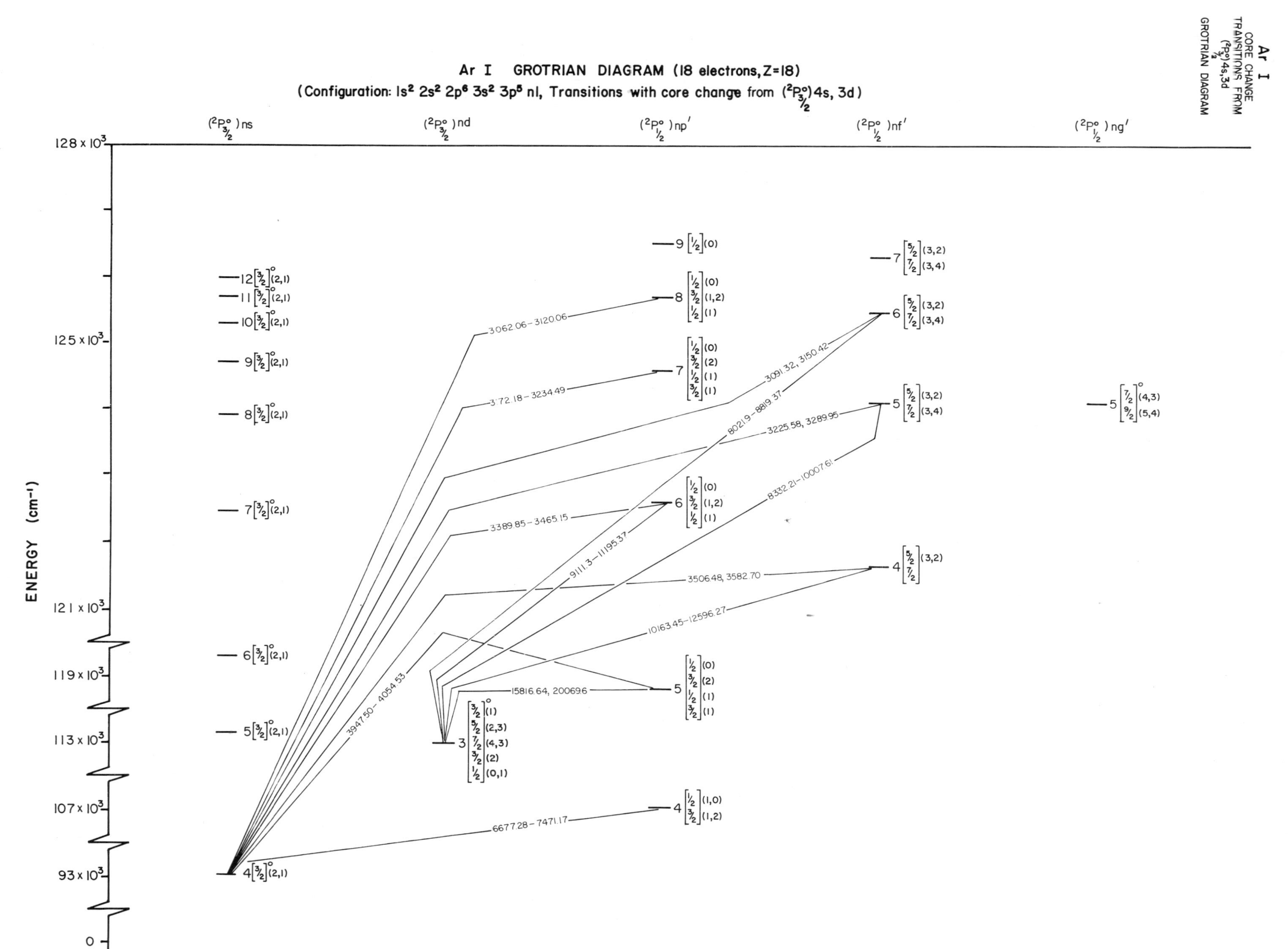

Ar I GROTRIAN DIAGRAM (18 electrons, Z=18)
(Configuration: 1s² 2s² 2p⁶ 3s² 3p⁵ nl, Transitions with core change from (²P°3/2)4s, 3d)
(²P°3/2)ns
(²P°3/2)nd
(²P°1/2)np'
(²P°1/2)nf'
(²P°1/2)ng'
3062.06–3120.06
3172.18–3234.49
3091.32, 3150.42
8021.9–8819.37
3225.58, 3289.95
8332.21–10007.61
3389.85–3465.15
9111.3–11195.37
3506.48, 3582.70
10163.45–12596.27
15816.64, 20069.6
3947.50–4054.53
6677.28–7471.17
ENERGY (cm⁻¹)
128 x 10³
125 x 10³
121 x 10³
119 x 10³
113 x 10³
107 x 10³
93 x 10³
0
Ar I
CORE CHANGE
TRANSITIONS FROM
(²P°3/2)4s,3d
GROTRIAN DIAGRAM

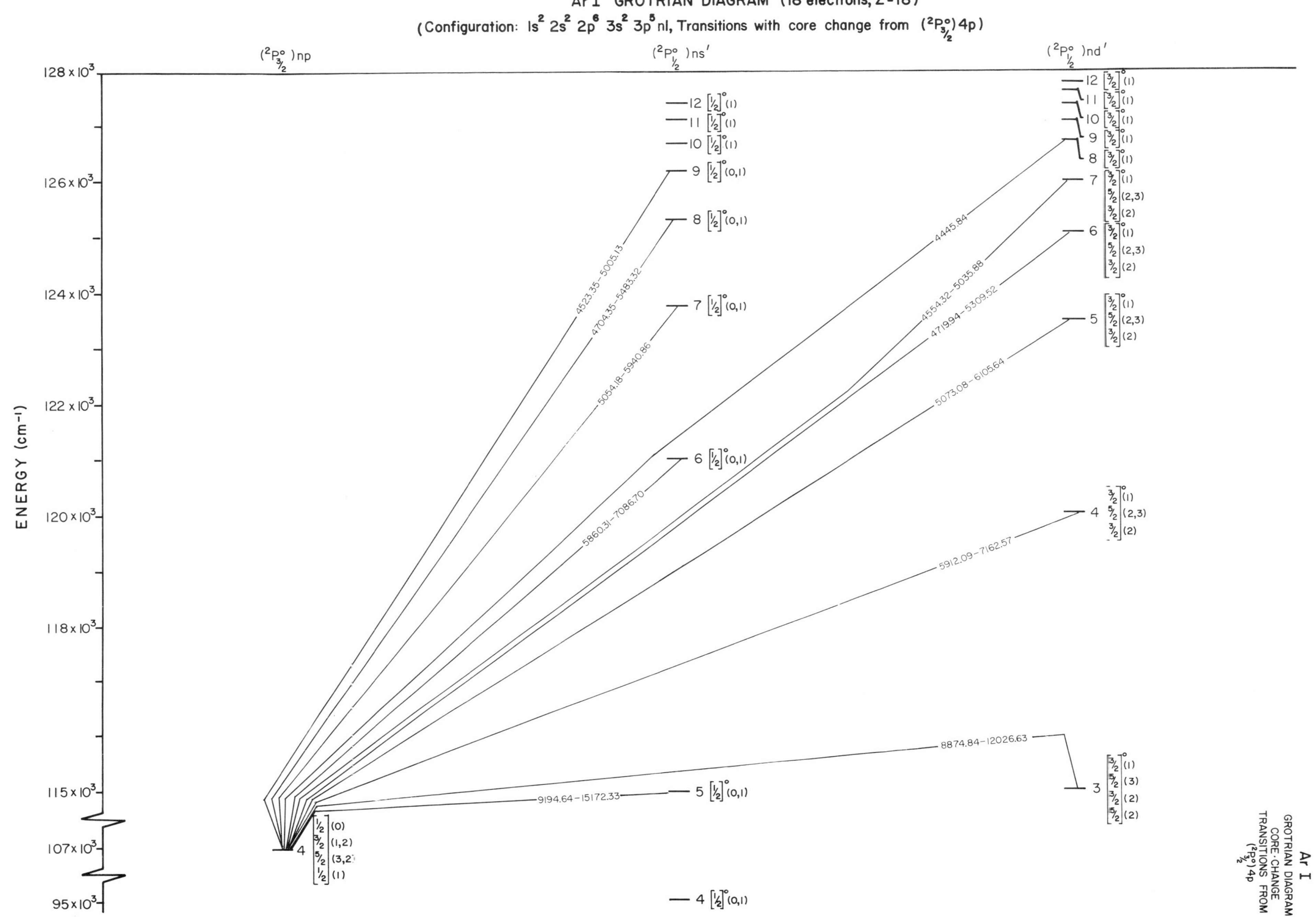

Ar I GROTRIAN DIAGRAM (18 electrons, Z=18)
(Configuration: $1s^2\ 2s^2\ 2p^6\ 3s^2\ 3p^5\ nl$, Transitions with core change from $(^2P^o_{3/2})4p$)
$(^2P^o_{3/2})np$
$(^2P^o_{1/2})ns'$
$(^2P^o_{1/2})nd'$
4523.35–5005.13
4704.35–5483.32
5054.18–5940.86
5860.31–7086.70
9194.64–15172.33
4445.84
4554.32–5035.88
4719.94–5309.52
5073.08–6105.64
5912.09–7162.57
8874.84–12026.63
ENERGY (cm⁻¹)
128 x 10³
126 x 10³
124 x 10³
122 x 10³
120 x 10³
118 x 10³
115 x 10³
107 x 10³
95 x 10³
Ar I
GROTRIAN DIAGRAM
CORE-CHANGE
TRANSITIONS FROM
$(^2P^o_{3/2})4p$

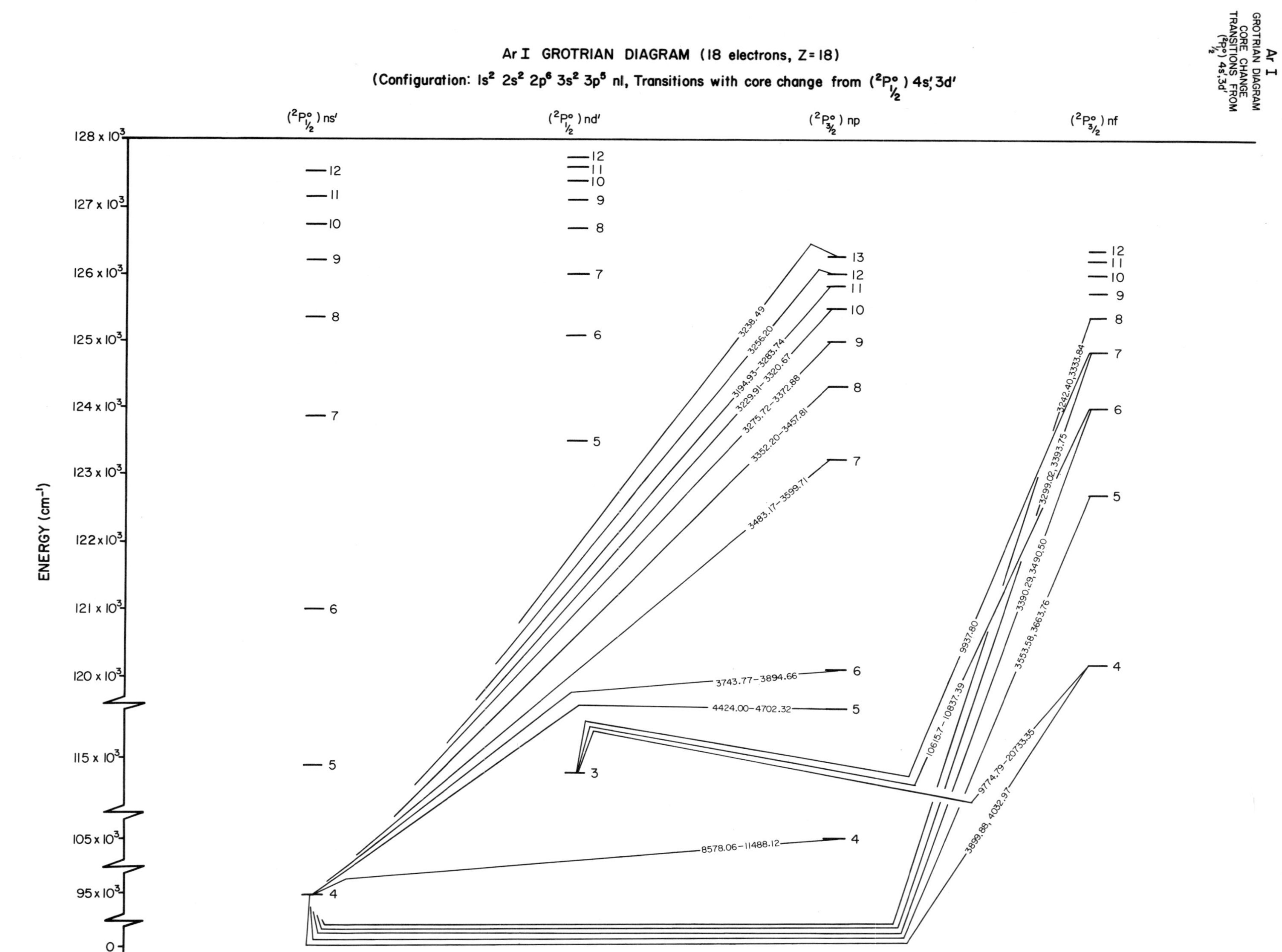

Ar I
GROTRIAN DIAGRAM
CORE CHANGE
TRANSITIONS FROM
($^2P^o_{1/2}$) 4s',3d'
Ar I GROTRIAN DIAGRAM (18 electrons, Z = 18)
(Configuration: $1s^2\ 2s^2\ 2p^6\ 3s^2\ 3p^5$ nl, Transitions with core change from ($^2P^o_{1/2}$) 4s',3d'
($^2P^o_{1/2}$) ns'
($^2P^o_{1/2}$) nd'
($^2P^o_{3/2}$) np
($^2P^o_{3/2}$) nf
3238.49
3256.20
3194.93–3283.74
3229.91–3320.67
3275.72–3372.88
3352.20–3457.81
3483.17–3599.71
3743.77–3894.66
4424.00–4702.32
8578.06–11488.12
3242.40, 3333.84
3299.02, 3393.75
3390.29, 3490.50
3553.58, 3663.76
9937.80
10615.7–10837.39
9774.79–20733.35
3899.88, 4032.97
ENERGY (cm^{-1})
128 x 10^3
127 x 10^3
126 x 10^3
125 x 10^3
124 x 10^3
123 x 10^3
122 x 10^3
121 x 10^3
120 x 10^3
115 x 10^3
105 x 10^3
95 x 10^3
0

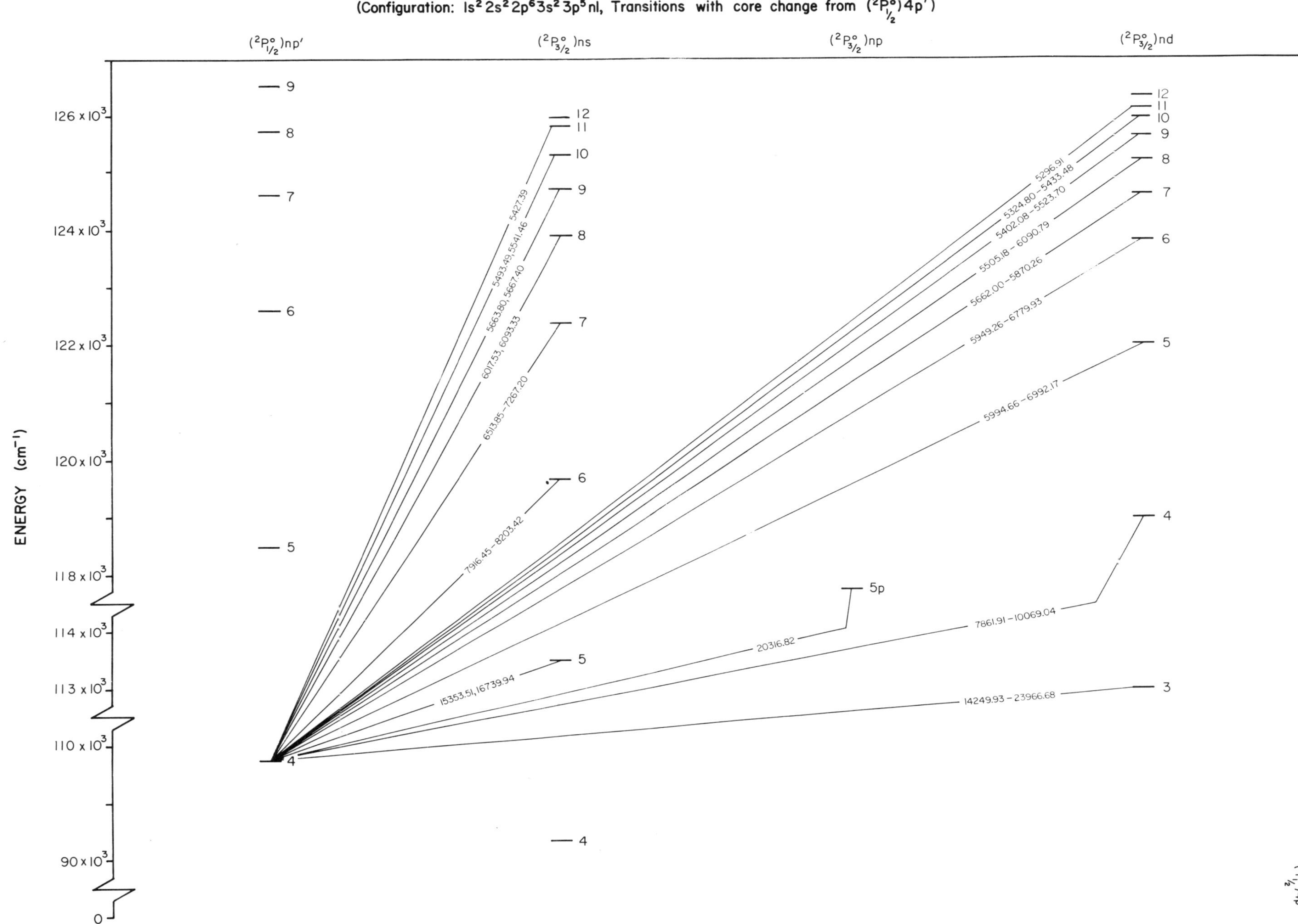
Ar I GROTRIAN DIAGRAM (18 electrons, Z=18)
(Configuration: $1s^2 2s^2 2p^6 3s^2 3p^5 nl$, Transitions with core change from $(^2P^o_{1/2})4p'$)
$(^2P^o_{1/2})np'$
$(^2P^o_{3/2})ns$
$(^2P^o_{3/2})np$
$(^2P^o_{3/2})nd$
5p
5427.39
5493.49, 5541.46
5663.80, 5667.40
6017.53, 6093.33
6513.85 – 7267.20
7916.45 – 8203.42
15353.51, 16739.94
5296.91
5324.80 – 5433.48
5402.08 – 5523.70
5505.18 – 6090.79
5662.00 – 5870.26
5949.26 – 6779.93
5994.66 – 6992.17
20316.82
7861.91 – 10069.04
14249.93 – 23966.68
ENERGY (cm^{-1})
126×10^3
124×10^3
122×10^3
120×10^3
118×10^3
114×10^3
113×10^3
110×10^3
90×10^3
0

Ar I GROTRIAN DIAGRAM (18 electrons, Z=18)

(Configuration: $1s^2 2s^2 2p^6 3s3p^6 nl$, $3s^2 3p^4(^1S)nl(^m L_j)n'l'$, High $^1P^o_1$ Levels and Doubly-excited Levels)

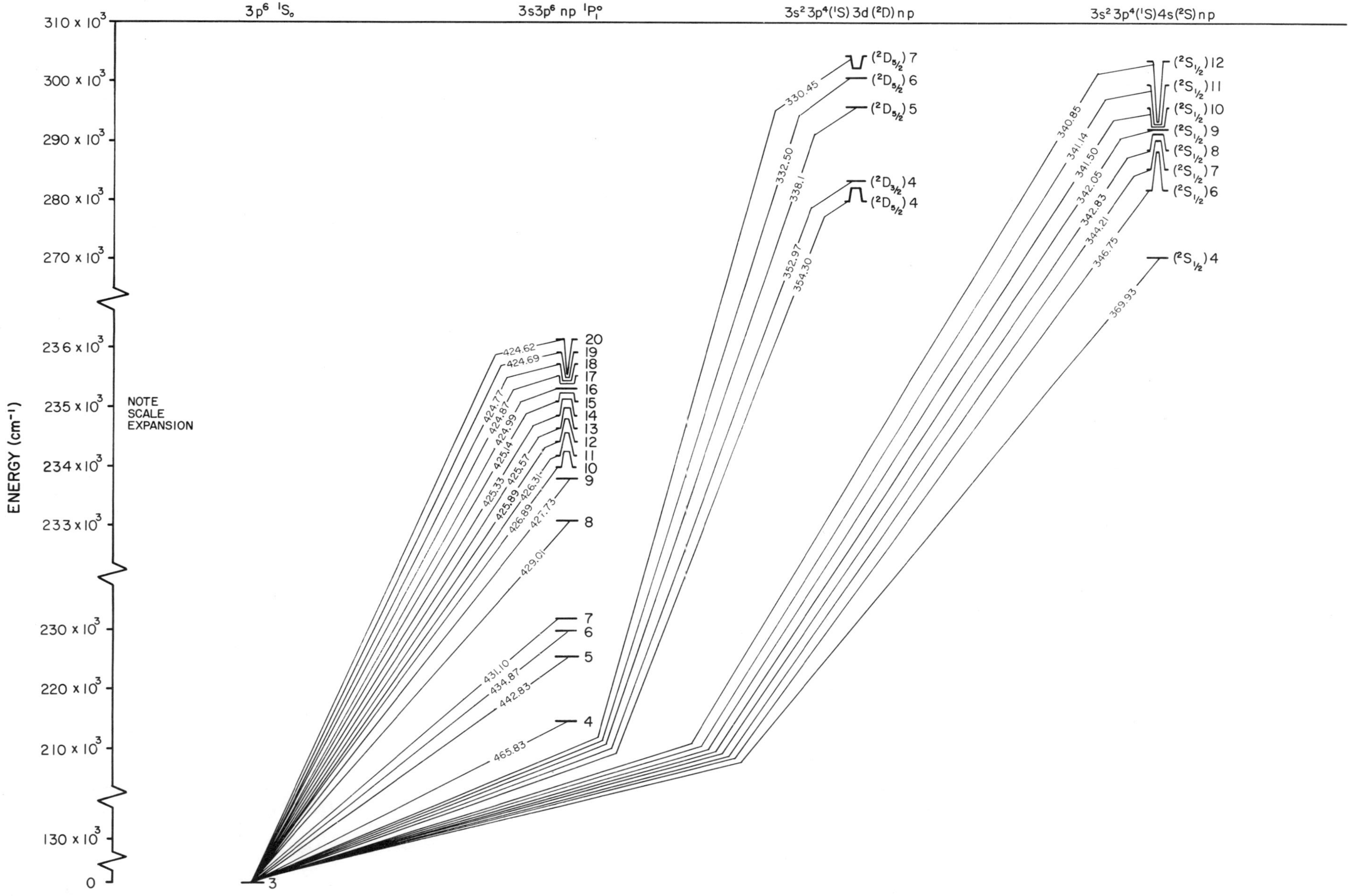

$3p^6\ ^1S_0$

$3s3p^6\ np\ ^1P^o_1$

$3s^2 3p^4(^1S)3d(^2D)np$

$3s^2 3p^4(^1S)4s(^2S)np$

ENERGY (cm^{-1})

310×10^3
300×10^3
290×10^3
280×10^3
270×10^3
236×10^3
235×10^3
234×10^3
233×10^3
230×10^3
220×10^3
210×10^3
130×10^3
0

NOTE
SCALE
EXPANSION

3

20
19
18
17
16
15
14
13
12
11
10
9
8
7
6
5
4

424.62
424.69
424.77
424.87
424.99
425.14
425.33
425.57
425.89
426.31
426.89
427.73
429.01
431.10
434.87
442.83
465.83

$(^2D_{5/2})$ 7
$(^2D_{5/2})$ 6
$(^2D_{5/2})$ 5
$(^2D_{3/2})$ 4
$(^2D_{5/2})$ 4

330.45
332.50
338.1
352.97
354.30

$(^2S_{1/2})$ 12
$(^2S_{1/2})$ 11
$(^2S_{1/2})$ 10
$(^2S_{1/2})$ 9
$(^2S_{1/2})$ 8
$(^2S_{1/2})$ 7
$(^2S_{1/2})$ 6
$(^2S_{1/2})$ 4

340.85
341.14
341.50
342.05
342.83
344.21
346.75
369.93

Ar I ENERGY LEVELS (18 electrons, Z=18)

(Configuration: $1s^2 2s^2 2p^6 3s3p^6 nl, 3s^2 3p^4 (^1S) nl (^m L_j) n'l'$, High $^1P^o_1$ Levels and Doubly Excited Levels)

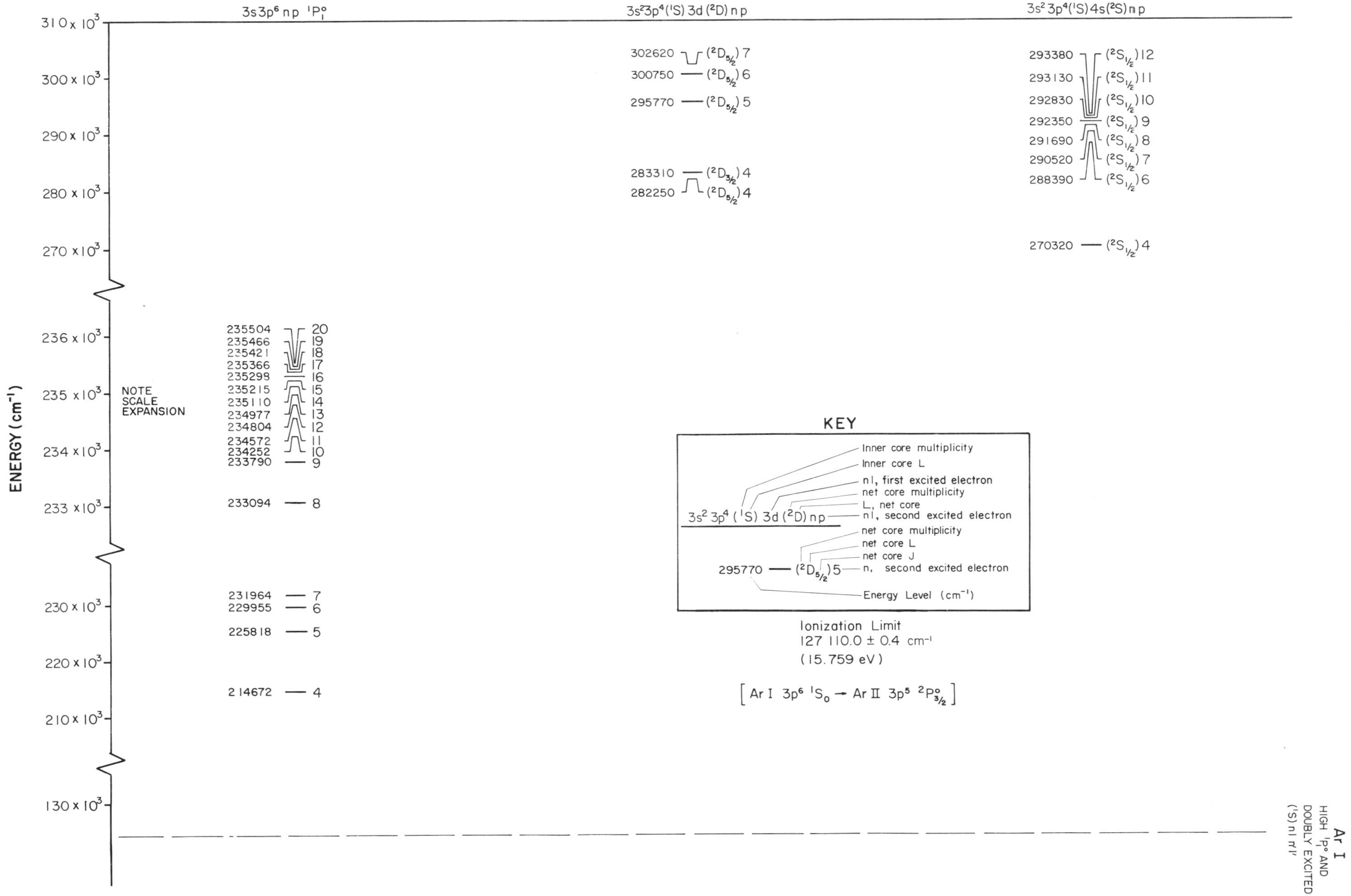

ENERGY (cm^{-1})

Axis: 310 x 10^3, 300 x 10^3, 290 x 10^3, 280 x 10^3, 270 x 10^3, 236 x 10^3, 235 x 10^3, 234 x 10^3, 233 x 10^3, 230 x 10^3, 220 x 10^3, 210 x 10^3, 130 x 10^3

NOTE SCALE EXPANSION

$3s3p^6 np\ ^1P^o_1$

Energy	n
235504	20
235466	19
235421	18
235366	17
235298	16
235215	15
235110	14
234977	13
234804	12
234572	11
234252	10
233790	9
233094	8
231964	7
229955	6
225818	5
214672	4

$3s^2 3p^4 (^1S) 3d (^2D) np$

Energy	Level
302620	$(^2D_{5/2})$ 7
300750	$(^2D_{5/2})$ 6
295770	$(^2D_{5/2})$ 5
283310	$(^2D_{3/2})$ 4
282250	$(^2D_{5/2})$ 4

$3s^2 3p^4 (^1S) 4s (^2S) np$

Energy	Level
293380	$(^2S_{1/2})$ 12
293130	$(^2S_{1/2})$ 11
292830	$(^2S_{1/2})$ 10
292350	$(^2S_{1/2})$ 9
291690	$(^2S_{1/2})$ 8
290520	$(^2S_{1/2})$ 7
288390	$(^2S_{1/2})$ 6
270320	$(^2S_{1/2})$ 4

KEY

$3s^2 3p^4 (^1S) 3d (^2D) np$
- Inner core multiplicity
- Inner core L
- nl, first excited electron
- net core multiplicity
- L, net core
- nl, second excited electron

295770 — $(^2D_{5/2})$ 5
- net core multiplicity
- net core L
- net core J
- n, second excited electron
- Energy Level (cm^{-1})

Ionization Limit
127 110.0 ± 0.4 cm^{-1}
(15.759 eV)

[Ar I $3p^6\ ^1S_0$ → Ar II $3p^5\ ^2P^o_{3/2}$]

Ar I
HIGH $^1P^o_1$ AND
DOUBLY EXCITED
$(^1S) nl\ n'l'$

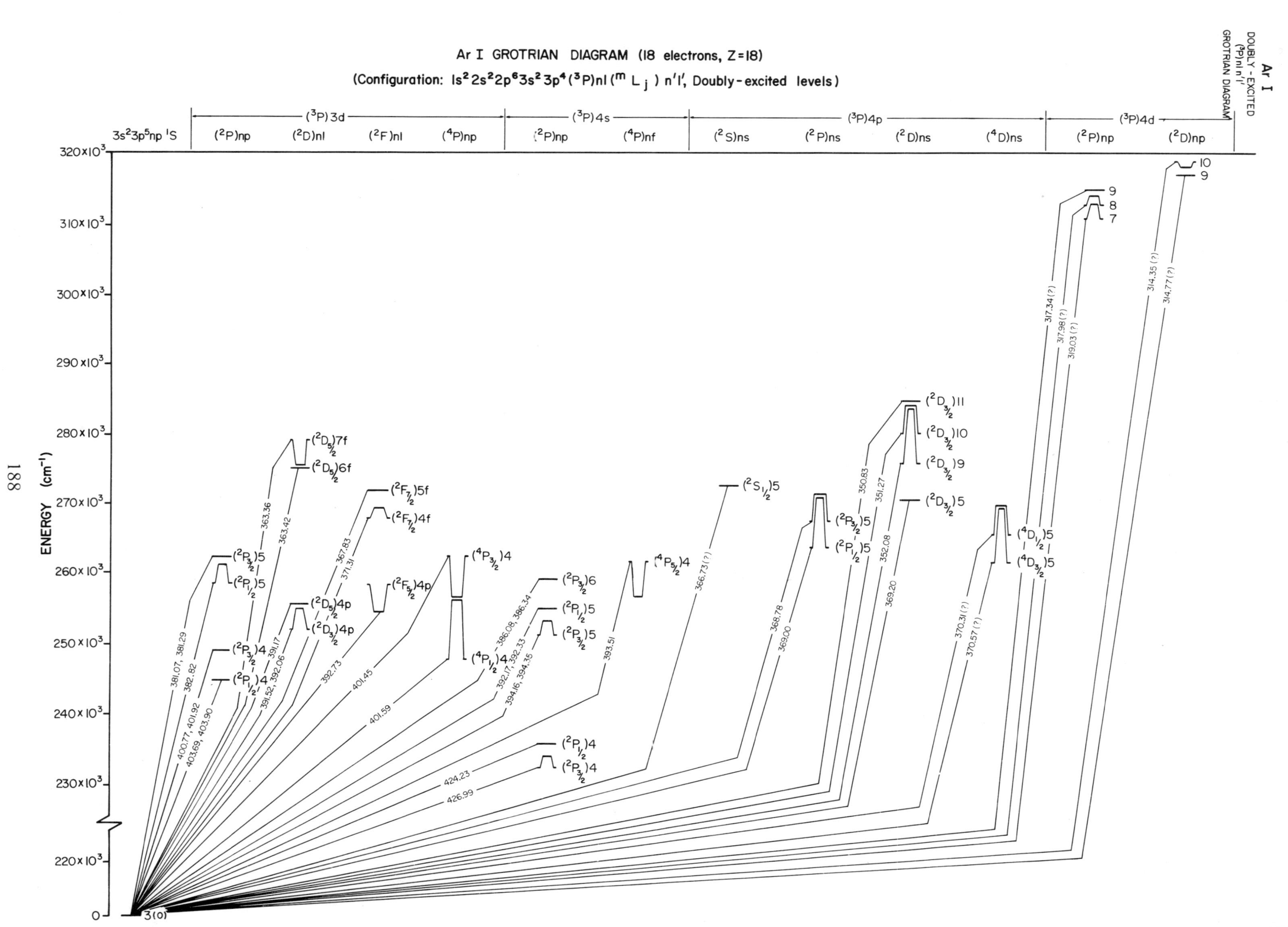
Ar I
DOUBLY - EXCITED
(^{3}P)nl n'l'
GROTRIAN DIAGRAM
Ar I GROTRIAN DIAGRAM (18 electrons, Z=18)
(Configuration: $1s^2 2s^2 2p^6 3s^2 3p^4 (^3P) nl (^m L_j) n'l'$, Doubly-excited levels)
(^{3}P)3d
(^{3}P)4s
(^{3}P)4p
(^{3}P)4d
$3s^2 3p^5 np\ ^1S$
(^{2}P)np
(^{2}D)nl
(^{2}F)nl
(^{4}P)np
(^{2}P)np
(^{4}P)nf
(^{2}S)ns
(^{2}P)ns
(^{2}D)ns
(^{4}D)ns
(^{2}P)np
(^{2}D)np
ENERGY (cm^{-1})
320x10^3
310x10^3
300x10^3
290x10^3
280x10^3
270x10^3
260x10^3
250x10^3
240x10^3
230x10^3
220x10^3
0
3(0)
381.07, 381.29
382.82
400.77, 401.92
403.69, 403.90
363.36
363.42
391.17
391.52, 392.06
367.83
371.31
392.73
401.45
401.59
386.08, 386.34
392.17, 392.33
394.16, 394.35
424.23
426.99
393.51
366.73 (?)
368.78
369.00
350.83
351.27
352.08
369.20
370.31 (?)
370.57 (?)
317.34 (?)
317.98 (?)
319.03 (?)
314.35 (?)
314.77 (?)

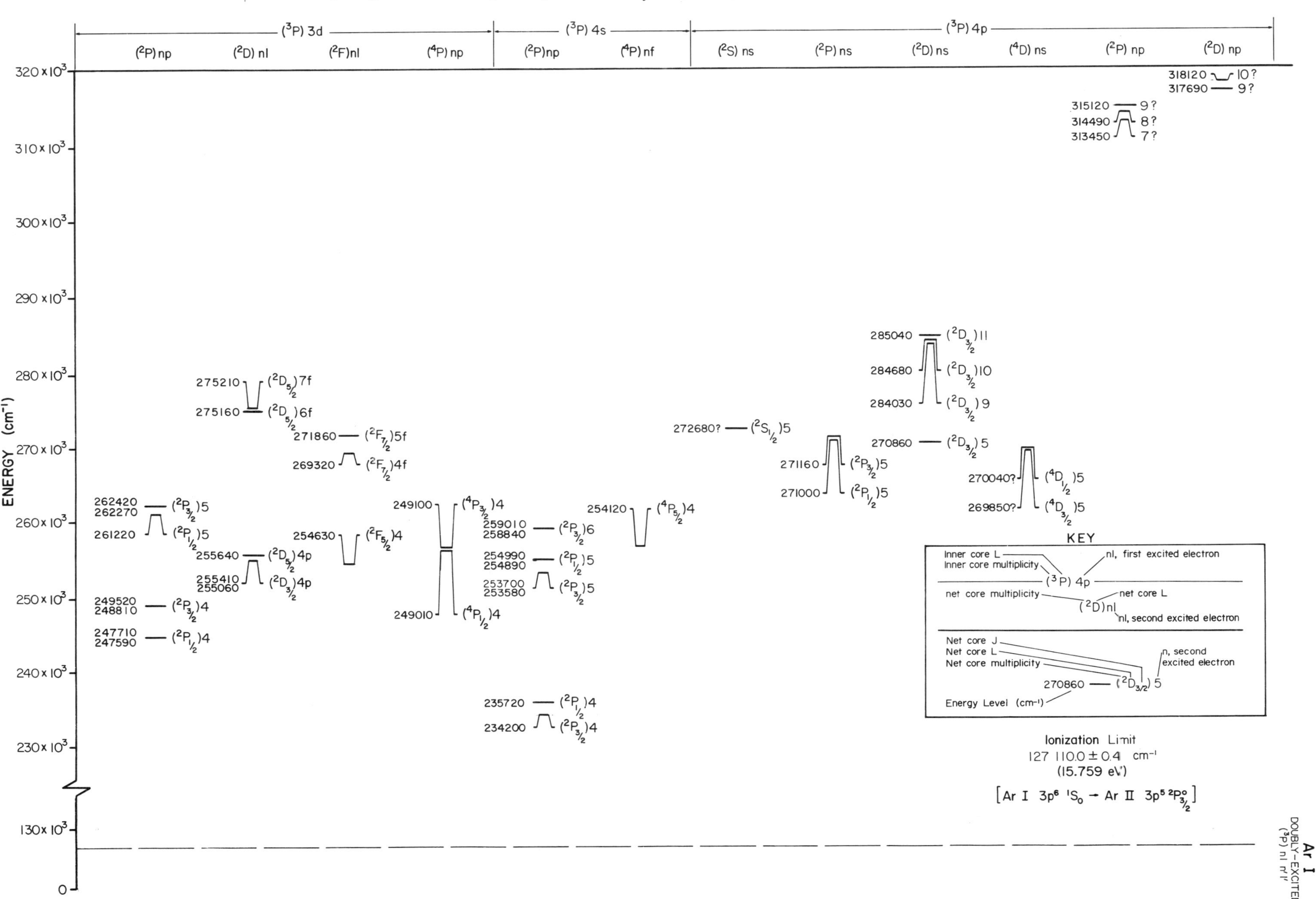
Ar I ENERGY LEVELS (18 electrons, Z = 18)
(Configuration: $1s^2 2s^2 2p^6 3s^2 3p^4 (^3P) nl (^mL_j) n'l'$, Doubly-excited levels)
(^3P) 3d
(^3P) 4s
(^3P) 4p
(^2P)np
(^2D) nl
(^2F)nl
(^4P) np
(^2P)np
(^4P) nf
(^2S) ns
(^2P) ns
(^2D) ns
(^4D) ns
(^2P) np
(^2D) np
ENERGY (cm^{-1})
320 x 10^3
310 x 10^3
300 x 10^3
290 x 10^3
280 x 10^3
270 x 10^3
260 x 10^3
250 x 10^3
240 x 10^3
230 x 10^3
130 x 10^3
0
262420
262270 $(^2P_{3/2})5$
261220 $(^2P_{1/2})5$
249520
248810 $(^2P_{3/2})4$
247710
247590 $(^2P_{1/2})4$
275210 $(^2D_{5/2})7f$
275160 $(^2D_{5/2})6f$
255640 $(^2D_{5/2})4p$
255410
255060 $(^2D_{3/2})4p$
271860 $(^2F_{7/2})5f$
269320 $(^2F_{7/2})4f$
254630 $(^2F_{5/2})4$
249100 $(^4P_{3/2})4$
249010 $(^4P_{1/2})4$
259010
258840 $(^2P_{3/2})6$
254990
254890 $(^2P_{1/2})5$
253700
253580 $(^2P_{3/2})5$
235720 $(^2P_{1/2})4$
234200 $(^2P_{3/2})4$
254120 $(^4P_{5/2})4$
272680? $(^2S_{1/2})5$
271160 $(^2P_{3/2})5$
271000 $(^2P_{1/2})5$
285040 $(^2D_{3/2})11$
284680 $(^2D_{3/2})10$
284030 $(^2D_{3/2})9$
270860 $(^2D_{3/2})5$
270040? $(^4D_{1/2})5$
269850? $(^4D_{3/2})5$
315120 9?
314490 8?
313450 7?
318120 10?
317690 9?
KEY
Inner core L
Inner core multiplicity
nl, first excited electron
(^3P) 4p
net core multiplicity
net core L
(^2D)nl
nl, second excited electron
Net core J
Net core L
Net core multiplicity
n, second excited electron
270860 $(^2D_{3/2})$ 5
Energy Level (cm^{-1})
Ionization Limit
127 110.0 ± 0.4 cm^{-1}
(15.759 eV)
[Ar I $3p^6\ ^1S_0$ → Ar II $3p^5\ ^2P^o_{3/2}$]
Ar I
DOUBLY-EXCITED
(^3P) nl n'l'

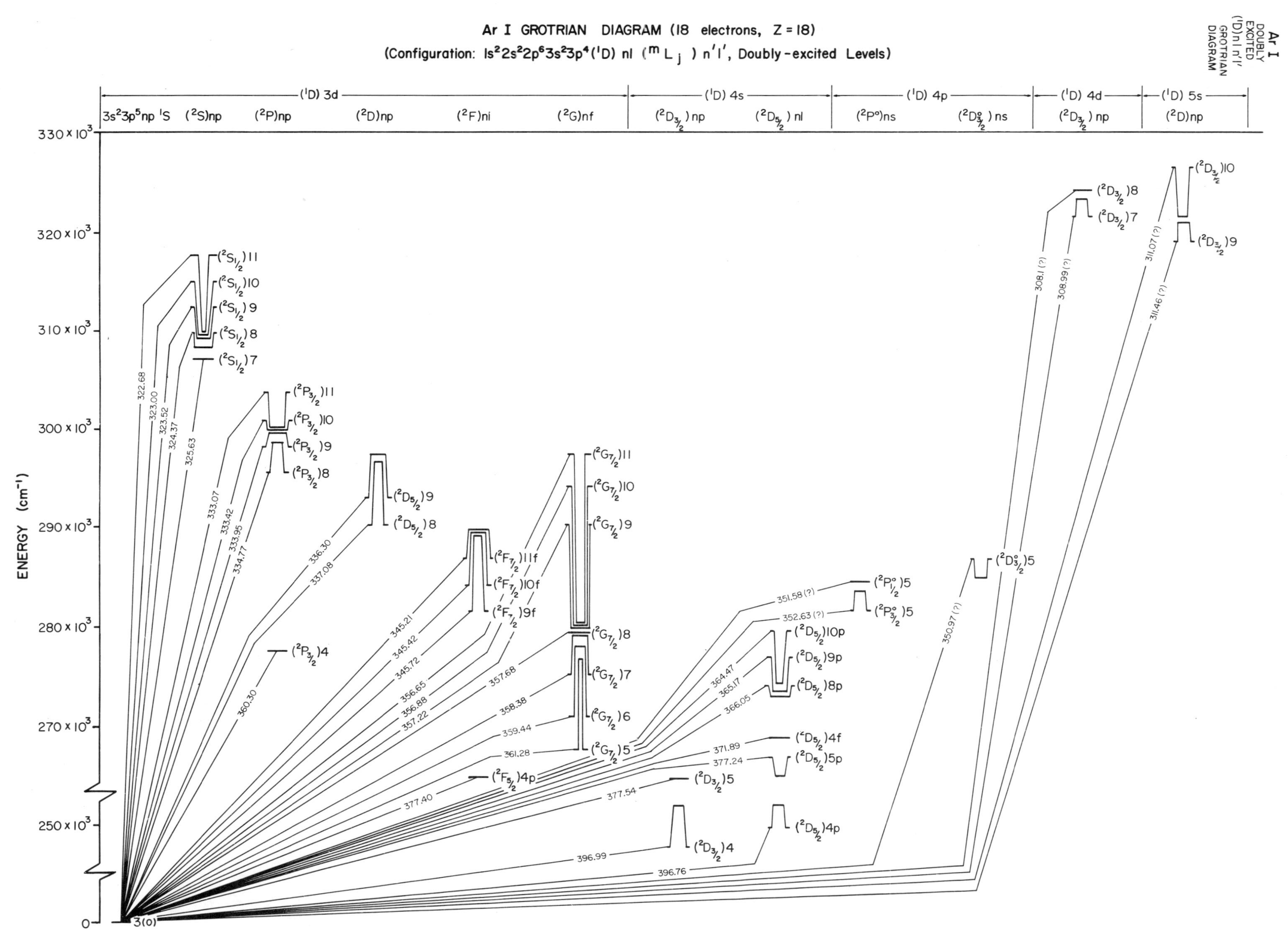
Ar I GROTRIAN DIAGRAM (18 electrons, Z = 18)
(Configuration: 1s²2s²2p⁶3s²3p⁴(¹D) nl (ᵐLⱼ) n′l′, Doubly-excited Levels)
Ar I
DOUBLY
EXCITED
(¹D)nl n′l′
GROTRIAN
DIAGRAM
(¹D) 3d
(¹D) 4s
(¹D) 4p
(¹D) 4d
(¹D) 5s
3s²3p⁵np ¹S
(²S)np
(²P)np
(²D)np
(²F)nl
(²G)nf
(²D3/2)np
(²D5/2)nl
(²P°)ns
(²D°3/2)ns
(²D3/2)np
(²D)np
ENERGY (cm⁻¹)
330×10³
320×10³
310×10³
300×10³
290×10³
280×10³
270×10³
250×10³
0
3(0)
(²S1/2)11
(²S1/2)10
(²S1/2)9
(²S1/2)8
(²S1/2)7
(²P3/2)11
(²P3/2)10
(²P3/2)9
(²P3/2)8
(²P3/2)4
(²D5/2)9
(²D5/2)8
(²F7/2)11f
(²F7/2)10f
(²F7/2)9f
(²F5/2)4p
(²G7/2)11
(²G7/2)10
(²G7/2)9
(²G7/2)8
(²G7/2)7
(²G7/2)6
(²G7/2)5
(²D3/2)5
(²D3/2)4
(²P°1/2)5
(²P°3/2)5
(²D5/2)10p
(²D5/2)9p
(²D5/2)8p
(²D5/2)4f
(²D5/2)5p
(²D5/2)4p
(²D°3/2)5
(²D3/2)8
(²D3/2)7
(²D3/2)10
(²D3/2)9
322.68
323.00
323.52
324.37
325.63
333.07
333.42
333.95
334.77
336.30
337.08
360.30
345.21
345.42
345.72
356.65
356.88
357.22
357.68
358.38
359.44
361.28
377.40
377.54
396.99
396.76
351.58 (?)
352.63 (?)
364.47
365.17
366.05
371.89
377.24
350.97 (?)
308.1 (?)
308.99 (?)
311.07 (?)
311.46 (?)

Ar I ENERGY LEVELS (18 electrons, Z = 18)

(Configuration: $1s^2\ 2s^2\ 2p^6\ 3s^2 3p^4\ (^1D)\ nl\ (^mL_j)\ n'l'$, Doubly-excited Levels)

Ar I

DOUBLY-EXCITED

$(^1D)\ nl\ n'l'$

ENERGY (cm^{-1})

330×10^3, 320×10^3, 310×10^3, 300×10^3, 290×10^3, 280×10^3, 270×10^3, 250×10^3, 130×10^3, 120×10^3

(^1D) 3d

(^2S) np
309900 $(^2S_{1/2})11$
309600 $(^2S_{1/2})10$
309100 $(^2S_{1/2})9$
308290 $(^2S_{1/2})8$
307100 $(^2S_{1/2})7$

(^2P) np
300240 $(^2P_{3/2})11$
299920 $(^2P_{3/2})10$
299450 $(^2P_{3/2})9$
298710 $(^2P_{3/2})8$
277550 $(^2P_{3/2})4$

(^2D) np
297350 $(^2D_{5/2})9$
296670 $(^2D_{5/2})8$

(^2F) nl
289680 $(^2F_{7/2})11f$
289500 $(^2F_{7/2})10f$
289250 $(^2F_{7/2})9f$
264970 $(^2F_{5/2})4p$

(^2G) nf
280390 $(^2G_{7/2})11$
2802 $(^2G_{7/2})10$
279940 $(^2G_{7/2})9$
279580 $(^2G_{7/2})8$
279030 $(^2G_{7/2})7$
278210 $(^2G_{7/2})6$
276790 $(^2G_{7/2})5$

(^1D) 4s

$(^2D_{3/2})$ np
264870 $(^2D_{3/2})5$
251900 $(^2D_{3/2})4$

$(^2D_{5/2})$ nl
274370 $(^2D_{5/2})10p$
273850 $(^2D_{5/2})9p$
273190 $(^2D_{5/2})8p$
268900 $(^2D_{5/2})4f$
265080 $(^2D_{5/2})5p$ *
252040 $(^2D_{5/2})4p$

(^1D) 4p

$(^2P^o)$ns
284430 $(^2P^o_{1/2})5?$
283580 $(^2P^o_{3/2})5?$

$(^2D^o_{3/2})$ ns
284920 $(^2D^o_{3/2})5?$

(^1D) 4d

$(^2D_{3/2})$ np
324570 $(^2D_{3/2})8?$
323640 $(^2D_{3/2})7?$

(^1D) 5s

(^2D) np
321470 $(^2D_{3/2})10?$
321070 $(^2D_{3/2})9?$

KEY

(^1D) 3d $(^2D_{5/2})$ nl — Inner core multiplicity; Inner core L; nl, first excited electron; Net core multiplicity; Net core L; Net core J; nl, second excited electron

268900 — $(^2D_{5/2})$ 4f — Net core multiplicity; nl, second excited electron; Net core J; Net core L; Energy Level (cm^{-1})

Ionization Limit
127110.0 ± 0.4 cm^{-1}
(15.759 eV)
[Ar I $3p^6\ ^1S_0$ ⟶ Ar II $3p^5\ ^2P^o_{3/2}$]

Ar II
DOUBLET
GROTRIAN
DIAGRAM
GROUND TERM
TO HIGHER
DOUBLETS

Ar II GROTRIAN DIAGRAM (17 electrons, Z=18)

(Cl I sequence, Configuration: $1s^2 2s^2 2p^6 3s^2 3p^4 nl$, Ground Term to Higher Doublets)

nl ^2S | np $^2S^o$ | nl ^2P | np $^2P^o$ | nl ^2D | np $^2D^o$ | nd ^2F | np $^2F^o$ | nd ^2G

ENERGY (cm^{-1})

220x10^3
200x10^3
180x10^3
160x10^3
140x10^3
100x10^3
0

nl ^2S: 5s″(½); 5d′(½); 4d′(½); 3d′(½); 4s″(½); 3s3p^6(½)
487.23, 490.65
543.20, 547.46
597.70, 602.86
919.78, 932.05

np $^2S^o$: 6(½); 5(½); 4(½)

nl ^2P: 6d(3/2); 7s(3/2, 1/2); 5d(1/2, 3/2); 6s(3/2, 1/2); 4d′(1/2, 3/2); 4d(1/2, 3/2); 5s(3/2, 1/2); 3d′(3/2, 1/2); 3d(1/2, 3/2); 4s(3/2, 1/2)
477.10
488.96, 492.41
499.92, 502.03
501.19–505.01
524.68–530.50
543.73–550.48
572.01–578.11
686.49–697.94
718.09–730.93

np $^2P^o$: 6p(3/2); 5p(3/2); 4p″(3/2, 1/2); 5p(1/2, 3/2); 4p′(3/2, 1/2); 4p(1/2, 3/2); 3p^5(3/2, 1/2)

nl ^2D: 5d′(5/2); 6s′(3/2, 5/2); 6d(3/2, 5/2); 5d(3/2, 5/2); 4d′(5/2, 3/2); 5s′(5/2, 3/2); 4d(5/2, 3/2); 3d″(5/2, 3/2); 3d′(5/2, 3/2); 3d(3/2, 5/2); 4s′(3/2, 5/2)
465.84
469.63, 472.81
475.91, 480.81
488.79–492.65
500.80, 503.65
510.55–514.31
518.91–522.79
555.77–560.22
578.60–583.44
661.87–670.95
671.85–679.40

np $^2D^o$: 5p′(3/2, 5/2); 6p(3/2, 5/2); 5p(5/2, 3/2); 4p′(3/2, 5/2); 4p(5/2, 3/2)

nd ^2F: 5d′(7/2, 5/2); 7d(9/2); 6d(7/2); 5d(7/2, 5/2); 4d′(7/2, 5/2); 4d(7/2, 5/2); 3d′(5/2, 7/2); 3d(5/2, 7/2)
477.61
494.67
533.08
612.37
666.01

np $^2F^o$: 5p′(5/2, 7/2); 4p′(5/2, 7/2)

nd ^2G: 5d′(7/2, 9/2); 4d′(7/2, 9/2); 3d′(9/2, 7/2)

CORES:
nl = 3p^4(^{3}P)
n′l′ = 3p^4(^{1}D)
n″l″ = 3p^4(^{1}S)

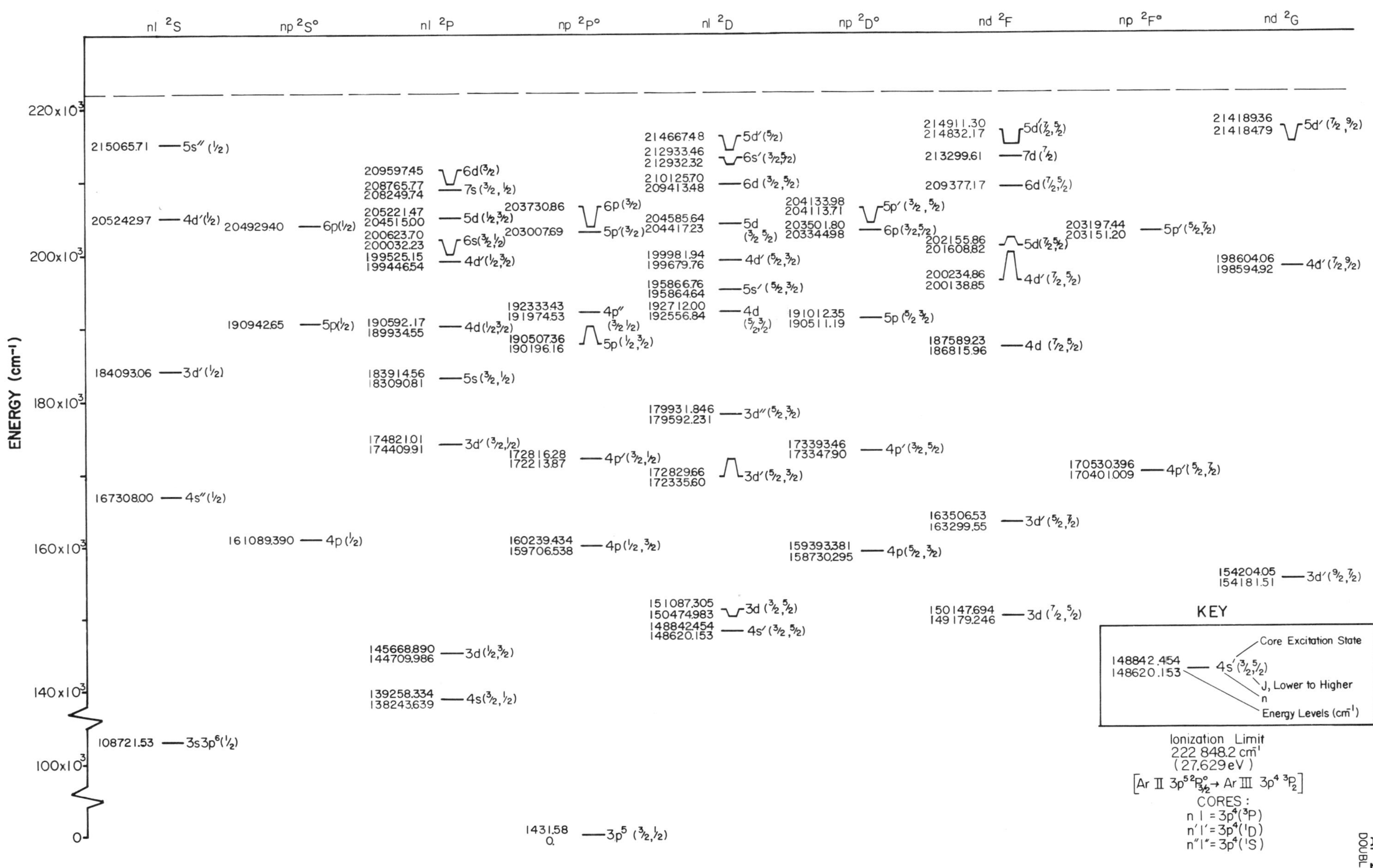

Ar II ENERGY LEVELS (17 electrons, Z=18)
(Cl I sequence, Configuration: $1s^2 2s^2 2p^6 3s^2 3p^4 nl$, Doublet System)
ENERGY (cm⁻¹)
220x10³
200x10³
180x10³
160x10³
140x10³
100x10³
0
nl ²S
215065.71 — 5s''(1/2)
205242.97 — 4d'(1/2)
184093.06 — 3d'(1/2)
167308.00 — 4s''(1/2)
108721.53 — 3s3p⁶(1/2)
np ²S°
204929.40 — 6p(1/2)
190942.65 — 5p(1/2)
161089.390 — 4p(1/2)
nl ²P
209597.45 — 6d(3/2)
208765.77 208249.74 — 7s(3/2,1/2)
205221.47 204515.00 — 5d(1/2,3/2)
200623.70 200032.23 — 6s(3/2,1/2)
199525.15 199446.54 — 4d'(1/2,3/2)
190592.17 189934.55 — 4d(1/2,3/2)
183914.56 183090.81 — 5s(3/2,1/2)
174821.01 174409.91 — 3d'(3/2,1/2)
145668.890 144709.986 — 3d(1/2,3/2)
139258.334 138243.639 — 4s(3/2,1/2)
np ²P°
203730.86 — 6p(3/2)
203007.69 — 5p'(3/2)
192333.43 191974.53 — 4p''(3/2,1/2)
190507.36 190196.16 — 5p(1/2,3/2)
172816.28 172213.87 — 4p'(3/2,1/2)
160239.434 159706.538 — 4p(1/2,3/2)
1431.58 0. — 3p⁵(3/2,1/2)
nl ²D
214667.48 — 5d'(5/2)
212933.46 212932.32 — 6s'(3/2,5/2)
210125.70 209413.48 — 6d(3/2,5/2)
204585.64 204417.23 — 5d(3/2,5/2)
199981.94 199679.76 — 4d'(5/2,3/2)
195866.76 195864.64 — 5s'(5/2,3/2)
192712.00 192556.84 — 4d(5/2,3/2)
179931.846 179592.231 — 3d''(5/2,3/2)
172829.66 172335.60 — 3d'(5/2,3/2)
151087.305 150474.983 — 3d(3/2,5/2)
148842.454 148620.153 — 4s'(3/2,5/2)
np ²D°
204133.98 204113.71 — 5p'(3/2,5/2)
203501.80 203344.98 — 6p(3/2,5/2)
191012.35 190511.19 — 5p(5/2,3/2)
173393.46 173347.90 — 4p'(3/2,5/2)
159393.381 158730.295 — 4p(5/2,3/2)
nd ²F
214911.30 214832.17 — 5d'(7/2,5/2)
213299.61 — 7d(7/2)
209377.17 — 6d(7/2,5/2)
202155.86 201608.82 — 5d(7/2,5/2)
200234.86 200138.85 — 4d'(7/2,5/2)
187589.23 186815.96 — 4d(7/2,5/2)
163506.53 163299.55 — 3d'(5/2,7/2)
150147.694 149179.246 — 3d(7/2,5/2)
np ²F°
203197.44 203151.20 — 5p'(5/2,7/2)
170530.396 170401.009 — 4p'(5/2,7/2)
nd ²G
214189.36 214184.79 — 5d'(7/2,9/2)
198604.06 198594.92 — 4d'(7/2,9/2)
154204.05 154181.51 — 3d'(9/2,7/2)
KEY
148842.454 148620.153 — 4s'(3/2,5/2)
Core Excitation State
J, Lower to Higher
n
Energy Levels (cm⁻¹)
Ionization Limit
222 848.2 cm⁻¹
(27.629 eV)
[Ar II $3p^5\,^2P^o_{3/2}$ → Ar III $3p^4\,^3P_2$]
CORES:
n l = 3p⁴(³P)
n' l' = 3p⁴(¹D)
n'' l'' = 3p⁴(¹S)
Ar II
DOUBLET

Ar II
DOUBLET
GROTRIAN
DIAGRAM
^{2}P TERMS
TO HIGHER
DOUBLETS

Ar II GROTRIAN DIAGRAM (17 electrons, Z=18)

(Cl I sequence, Configuration: $1s^2 2s^2 2p^6 3s^2 3p^4 nl$, ^2P Terms to Higher Doublets)

nl ^{2}S | np ^{2}S° | nl ^{2}P | np ^{2}P° | nl ^{2}D | np ^{2}D° | nd ^{2}F | np ^{2}F° | nd ^{2}G

ENERGY (cm^{-1})

220x10^3
200x10^3
180x10^3
160x10^3
140x10^3
100x10^3
0

nl ^{2}S: 5s″(½), 5d′(½), 4d′(½), 3d′(½), 4s″(½), 3s3p^6(½)

np ^{2}S°: 6(½), 5(½), 4(½)

nl ^{2}P: 6d(3/2), 7s(3/2,1/2), 5d(1/2,3/2), 6s(3/2,1/2), 4d′(1/2,3/2), 4d(1/2,3/2), 5s(3/2,1/2), 3d′(3/2,1/2), 3d(1/2,3/2), 4s(3/2,1/2)

np ^{2}P°: 6p(3/2), 5p′(3/2), 4p″(3/2,1/2), 5p(1/2,3/2), 4p′(3/2,1/2), 4p(1/2,3/2), 3p^5(3/2,1/2)

nl ^{2}D: 5d′(5/2), 6s′(3/2,5/2), 6d(3/2,5/2), 5d(3/2,5/2), 4d′(5/2,3/2), 5s′(5/2,3/2), 4d(5/2,3/2), 3d″(5/2,3/2), 3d′(5/2,3/2), 3d(3/2,5/2), 4s′(3/2,5/2)

np ^{2}D°: 5p′(3/2,5/2), 6p(3/2,5/2), 5p(5/2,3/2), 4p′(3/2,5/2), 4p(5/2,3/2)

nd ^{2}F: 5d(7/2,5/2), 7d(9/2), 6d(7/2), 5d(7/2,5/2), 4d′(7/2,5/2), 4d (7/2,5/2), 3d′(5/2,7/2), 3d (5/2,7/2)

np ^{2}F°: 5p′(5/2,7/2), 4p′(5/2,7/2)

nd ^{2}G: 5d′(7/2,9/2), 4d′(7/2,9/2), 3c′(9/2,7/2)

4757.22
3275.64, 3320.38
6046.89, 6201.10
2162.29
6103.55, 6483.08
4375.95, 4579.35
3495.77
11253.50
5577.69 - 5828.06
6210.42 - 6502.16
3363.30
6174.38, 6208.94
2159.05, 2204.70
2197.79, 2245.12
2115.09 - 2158.88
3556.91 - 3766.12
2891.61 - 3033.51
6437.60 - 7121.74
4545.05 - 4889.03
3490.88, 3611.81
2844.13 - 2932.59
3108.80
4042.19
6808.53 - 7654.03
4726.86 - 4965.07

CORES:
nl = 3p^4(^{3}P)
n′l′ = 3p^4(^{1}D)
n″l″ = 3p^4(^{1}S)

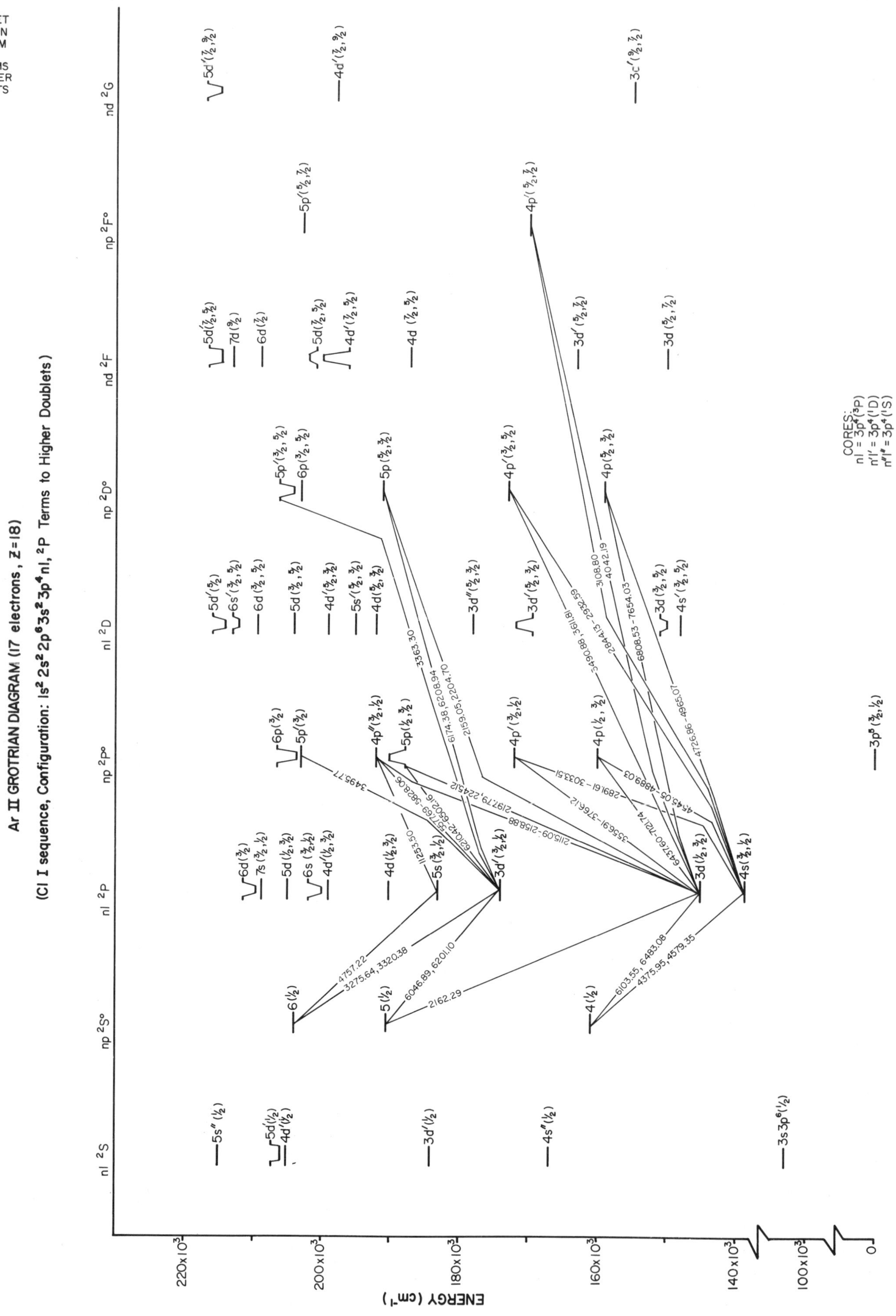

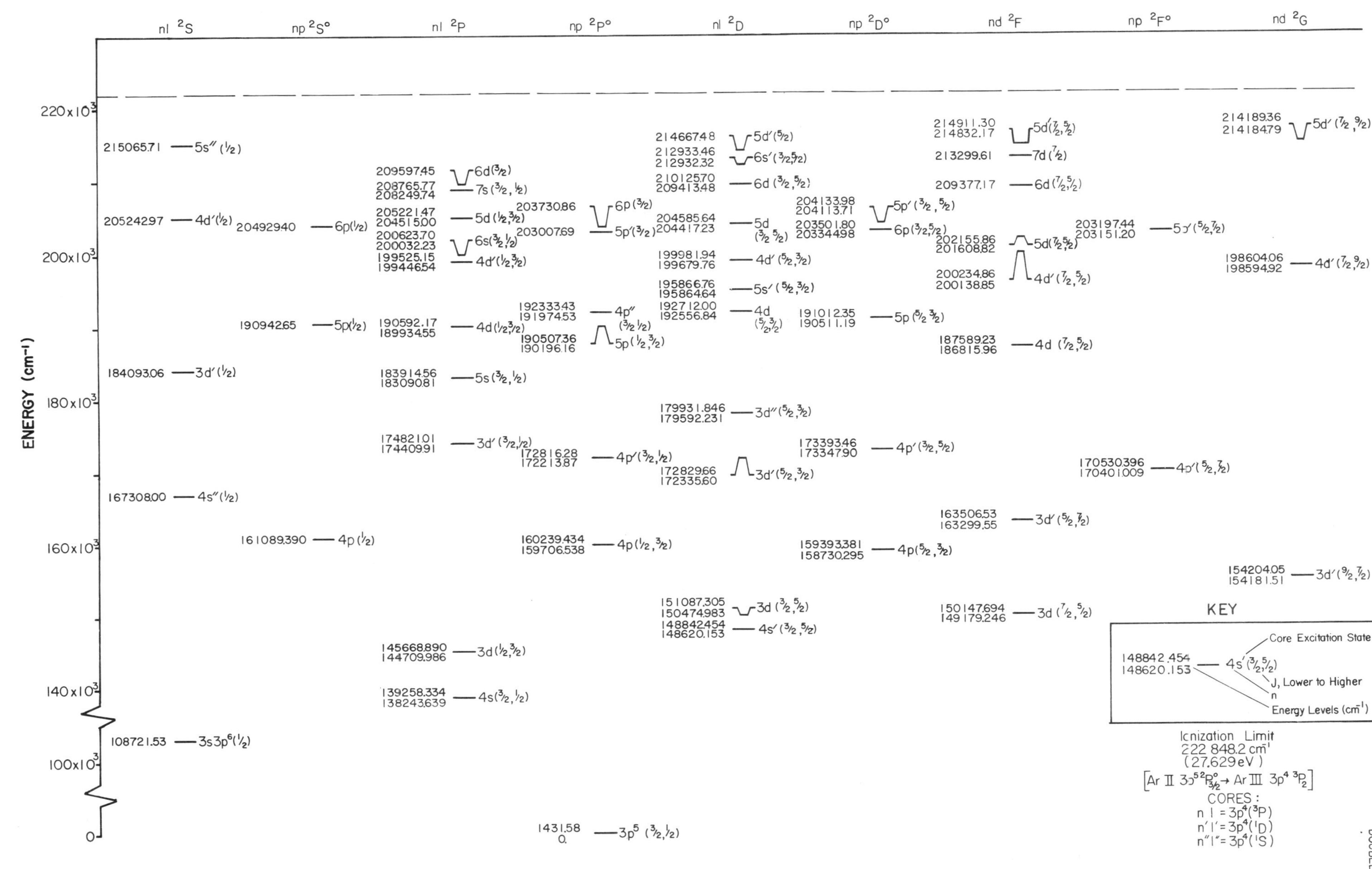

Ar II ENERGY LEVELS (17 electrons, Z=18)
(Cl I sequence, Configuration: 1s2 2s2 2p6 3s2 3p4 nl, Doublet System)
nl 2S
np 2S°
nl 2P
np 2P°
nl 2D
np 2D°
nd 2F
np 2F°
nd 2G
ENERGY (cm-1)
220x10³
200x10³
180x10³
160x10³
140x10³
100x10³
0
215065.71 — 5s″(1/2)
205242.97 — 4d′(1/2)
184093.06 — 3d′(1/2)
167308.00 — 4s″(1/2)
108721.53 — 3s3p6(1/2)
204929.40 — 6p(1/2)
190942.65 — 5p(1/2)
161089.390 — 4p(1/2)
209597.45 6d(3/2)
208765.77 208249.74 7s(3/2, 1/2)
205221.47 204515.00 5d(1/2, 3/2)
200623.70 200032.23 6s(3/2, 1/2)
199525.15 199446.54 4d′(1/2, 3/2)
190592.17 189934.55 4d(1/2, 3/2)
183914.56 183090.81 5s(3/2, 1/2)
174821.01 174409.91 3d′(3/2, 1/2)
145668.890 144709.986 3d(1/2, 3/2)
139258.334 138243.639 4s(3/2, 1/2)
203730.86 6p(3/2)
203007.69 5p′(3/2)
192333.43 191974.53 4p″(3/2, 1/2)
190507.36 190196.16 5p(1/2, 3/2)
172816.28 172213.87 4p′(3/2, 1/2)
160239.434 159706.538 4p(1/2, 3/2)
1431.58 0. 3p5 (3/2, 1/2)
214667.48 5d′(5/2)
212933.46 212932.32 6s′(3/2, 5/2)
210125.70 209413.48 6d (3/2, 5/2)
204585.64 204417.23 5d (3/2, 5/2)
199981.94 199679.76 4d′(5/2, 3/2)
195866.76 195864.64 5s′(5/2, 3/2)
192712.00 192556.84 4d (5/2, 3/2)
179931.846 179592.231 3d″(5/2, 3/2)
172829.66 172335.60 3d′(5/2, 3/2)
151087.305 150474.983 3d (3/2, 5/2)
148842.454 148620.153 4s′(3/2, 5/2)
204133.98 204113.71 5p′(3/2, 5/2)
203501.80 203344.98 6p(3/2, 5/2)
191012.35 190511.19 5p (5/2, 3/2)
173393.46 173347.90 4p′(3/2, 5/2)
159393.381 158730.295 4p(5/2, 3/2)
214911.30 214832.17 5d′(7/2, 5/2)
213299.61 7d (7/2)
209377.17 6d (7/2, 5/2)
202155.86 201608.82 5d(7/2, 5/2)
200234.86 200138.85 4d′(7/2, 5/2)
187589.23 186815.96 4d (7/2, 5/2)
163506.53 163299.55 3d′(5/2, 7/2)
150147.694 149179.246 3d (7/2, 5/2)
203197.44 203151.20 5p′(5/2, 7/2)
170530.396 170401.009 4p′(5/2, 7/2)
214189.36 214184.79 5d′(7/2, 9/2)
198604.06 198594.92 4d′(7/2, 9/2)
154204.05 154181.51 3d′(9/2, 7/2)
KEY
148842.454 148620.153 — 4s′(3/2, 5/2)
Core Excitation State
J, Lower to Higher
n
Energy Levels (cm-1)
Ionization Limit
222 848.2 cm-1
(27.629 eV)
[Ar II 3s2 3p5 2P°3/2 → Ar III 3p4 3P2]
CORES:
n l = 3p4(3P)
n′l′ = 3p4(1D)
n″l″ = 3p4(1S)
Ar II
DOUBLET

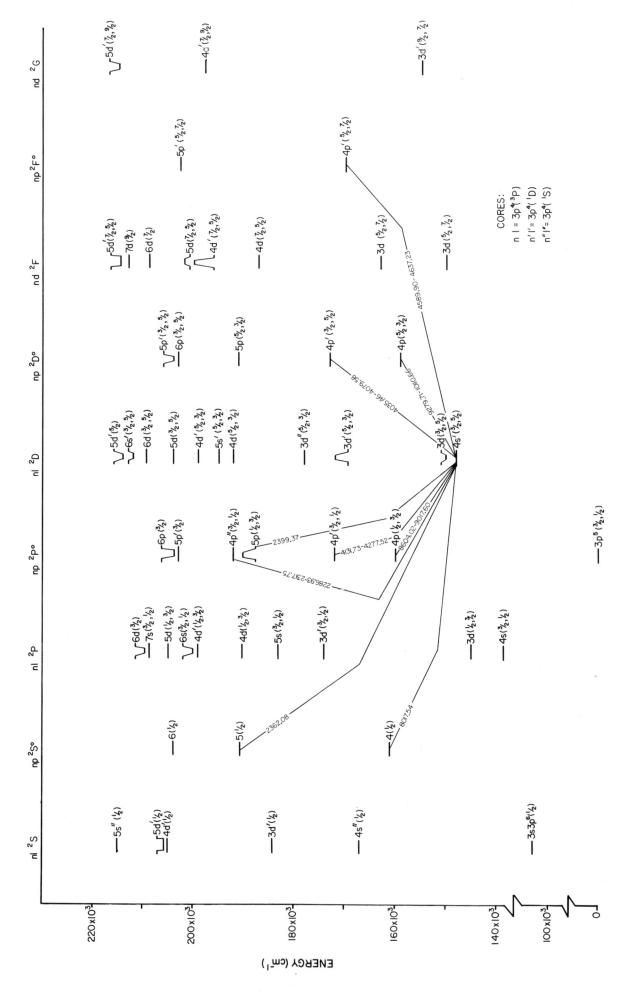

Ar II
DOUBLET
GROTRIAN
DIAGRAM
4s′ ²D TO
HIGHER
DOUBLETS

Ar II GROTRIAN DIAGRAM (17 electrons, Z=18)

(Cl I sequence, Configuration: 1s² 2s² 2p⁶ 3s² 3p⁴ nl, 4s′ ²D Terms to Higher Doublets)

ENERGY (cm⁻¹)

220x10³
200x10³
180x10³
160x10³
140x10³
100x10³
0

nl ²S | np ²S° | nl ²P | np ²P° | nl ²D | np ²D° | nd ²F | np ²F° | nd ²G

CORES:
n l = 3p⁴(³P)
n′l′ = 3p⁴(¹D)
n″l″ = 3p⁴(¹S)

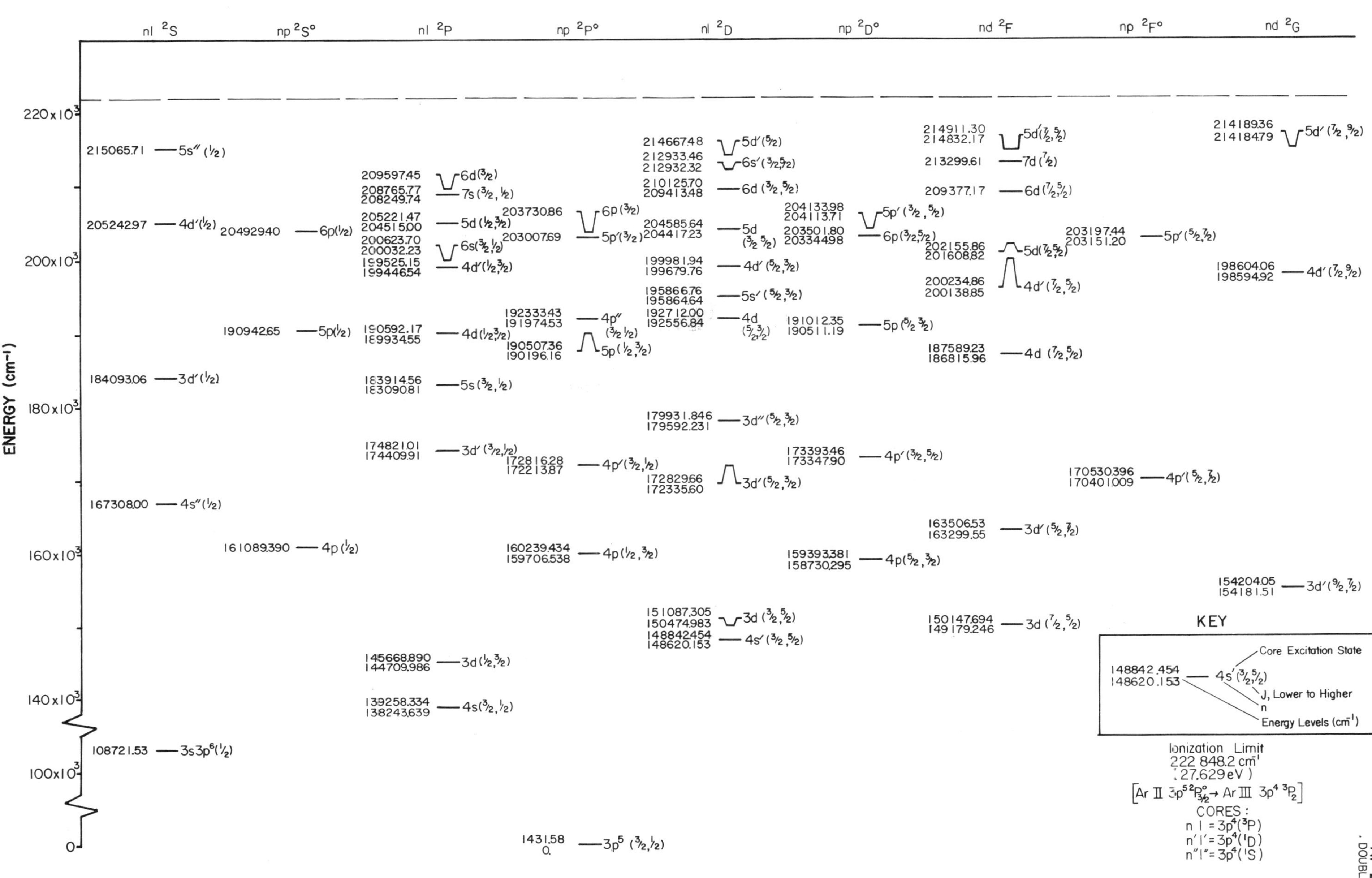
Ar II ENERGY LEVELS (17 electrons, Z=18)
(Cl I sequence, Configuration: 1s² 2s² 2p⁶ 3s² 3p⁴ nl, Doublet System)
ENERGY (cm⁻¹)
220x10³
200x10³
180x10³
160x10³
140x10³
100x10³
0
nl ²S
np ²S°
nl ²P
np ²P°
nl ²D
np ²D°
nd ²F
np ²F°
nd ²G
215065.71 — 5s″ (1/2)
205242.97 — 4d′(1/2)
184093.06 — 3d′(1/2)
167308.00 — 4s″(1/2)
108721.53 — 3s3p⁶(1/2)
204929.40 — 6p(1/2)
190942.65 — 5p(1/2)
161089.390 — 4p (1/2)
209597.45 6d(3/2)
208765.77 208249.74 7s (3/2, 1/2)
205221.47 204515.00 5d (1/2,3/2)
200623.70 200032.23 6s(3/2,1/2)
199525.15 199446.54 4d′(1/2,3/2)
190592.17 189934.55 4d(1/2,3/2)
183914.56 183090.81 5s (3/2, 1/2)
174821.01 174409.91 3d′ (3/2,1/2)
145668.890 144709.986 3d (1/2,3/2)
139258.334 138243.639 4s(3/2, 1/2)
203730.86 6p (3/2)
203007.69 5p′(3/2)
192333.43 191974.53 4p″ (3/2 1/2)
190507.36 190196.16 5p (1/2,3/2)
172816.28 172213.87 4p′(3/2,1/2)
160239.434 159706.538 4p (1/2, 3/2)
1431.58 0. 3p⁵ (3/2,1/2)
214667.48 5d′(5/2)
212933.46 212932.32 6s′(3/2,5/2)
210125.70 209413.48 6d (3/2, 5/2)
204585.64 204417.23 5d (3/2 5/2)
199981.94 199679.76 4d′ (5/2,3/2)
195866.76 195864.64 5s′ (5/2,3/2)
192712.00 192556.84 4d (5/2,3/2)
179931.846 179592.231 3d″(5/2,3/2)
172829.66 172335.60 3d′(5/2,3/2)
151087.305 150474.983 3d (3/2, 5/2)
148842.454 148620.153 4s′ (3/2, 5/2)
204133.98 204113.71 5p′ (3/2, 5/2)
203501.80 203344.98 6p(3/2,5/2)
191012.35 190511.19 5p (5/2 3/2)
173393.46 173347.90 4p′ (3/2,5/2)
159393.381 158730.295 4p(5/2, 3/2)
214911.30 214832.17 5d(7/2,5/2)
213299.61 7d (7/2)
209377.17 6d(7/2,5/2)
202155.86 201608.82 5d(7/2,5/2)
200234.86 200138.85 4d′(7/2, 5/2)
187589.23 186815.96 4d (7/2,5/2)
163506.53 163299.55 3d′ (5/2,7/2)
150147.694 149179.246 3d (7/2, 5/2)
203197.44 203151.20 5p′ (5/2,7/2)
170530.396 170401.009 4p′(5/2, 7/2)
214189.36 214184.79 5d′ (7/2, 9/2)
198604.06 198594.92 4d′ (7/2,9/2)
154204.05 154181.51 3d′(9/2, 7/2)
KEY
148842.454 148620.153 — 4s′ (3/2,5/2)
Core Excitation State
J, Lower to Higher
n
Energy Levels (cm⁻¹)
Ionization Limit
222 848.2 cm⁻¹
(27.629 eV)
[Ar II 3p⁵ ²P°3/2 → Ar III 3p⁴ ³P2]
CORES:
n l = 3p⁴(³P)
n′l′ = 3p⁴(¹D)
n″l″ = 3p⁴(¹S)
Ar II
. DOUBLET

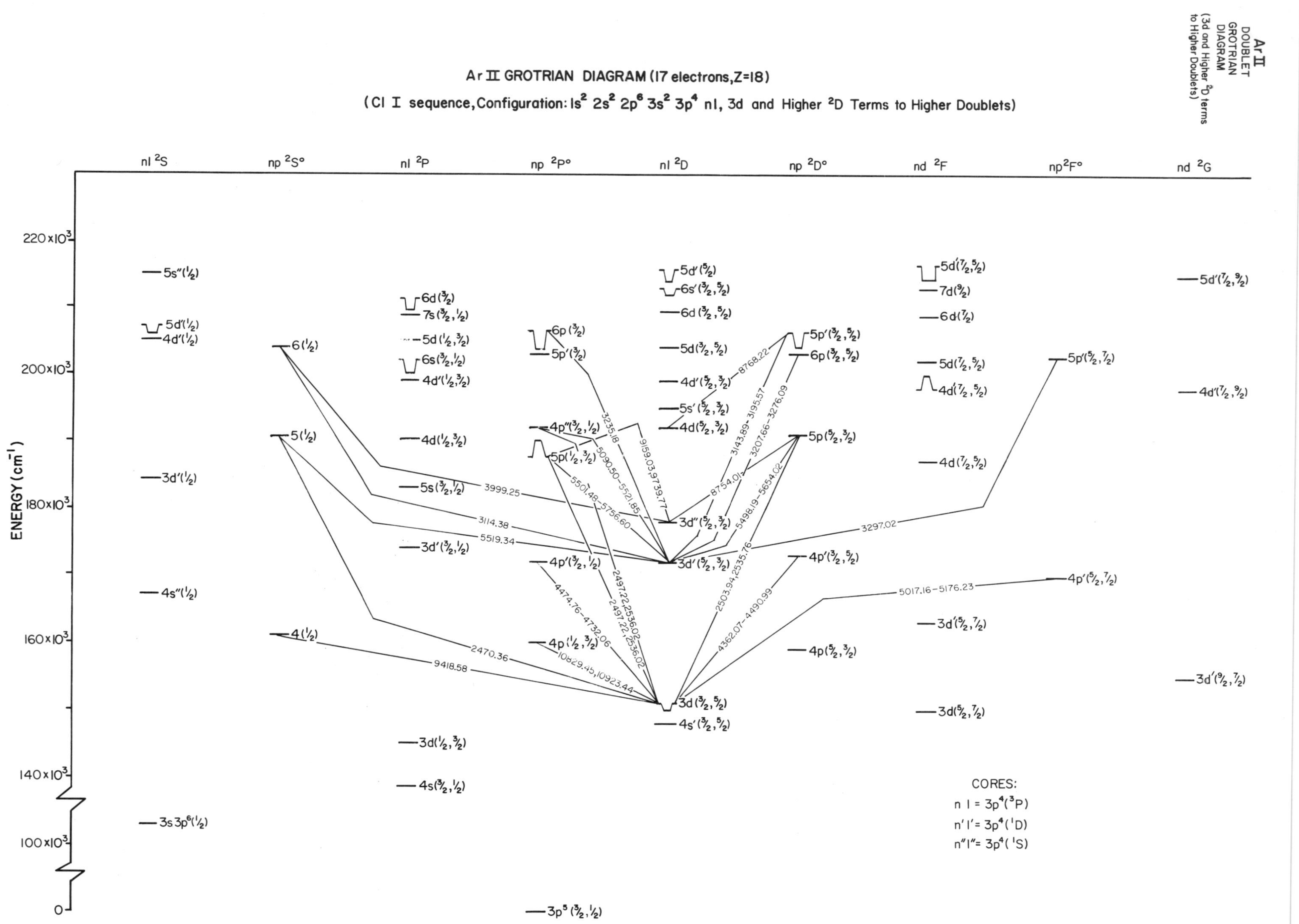
Ar II
DOUBLET
GROTRIAN
DIAGRAM
(3d and Higher ²D terms to Higher Doublets)
Ar II GROTRIAN DIAGRAM (17 electrons, Z=18)
(Cl I sequence, Configuration: 1s² 2s² 2p⁶ 3s² 3p⁴ nl, 3d and Higher ²D Terms to Higher Doublets)
nl ²S
np ²S°
nl ²P
np ²P°
nl ²D
np ²D°
nd ²F
np²F°
nd ²G
ENERGY (cm⁻¹)
220×10³
200×10³
180×10³
160×10³
140×10³
100×10³
0
CORES:
n l = 3p⁴(³P)
n′l′ = 3p⁴(¹D)
n″l″ = 3p⁴(¹S)

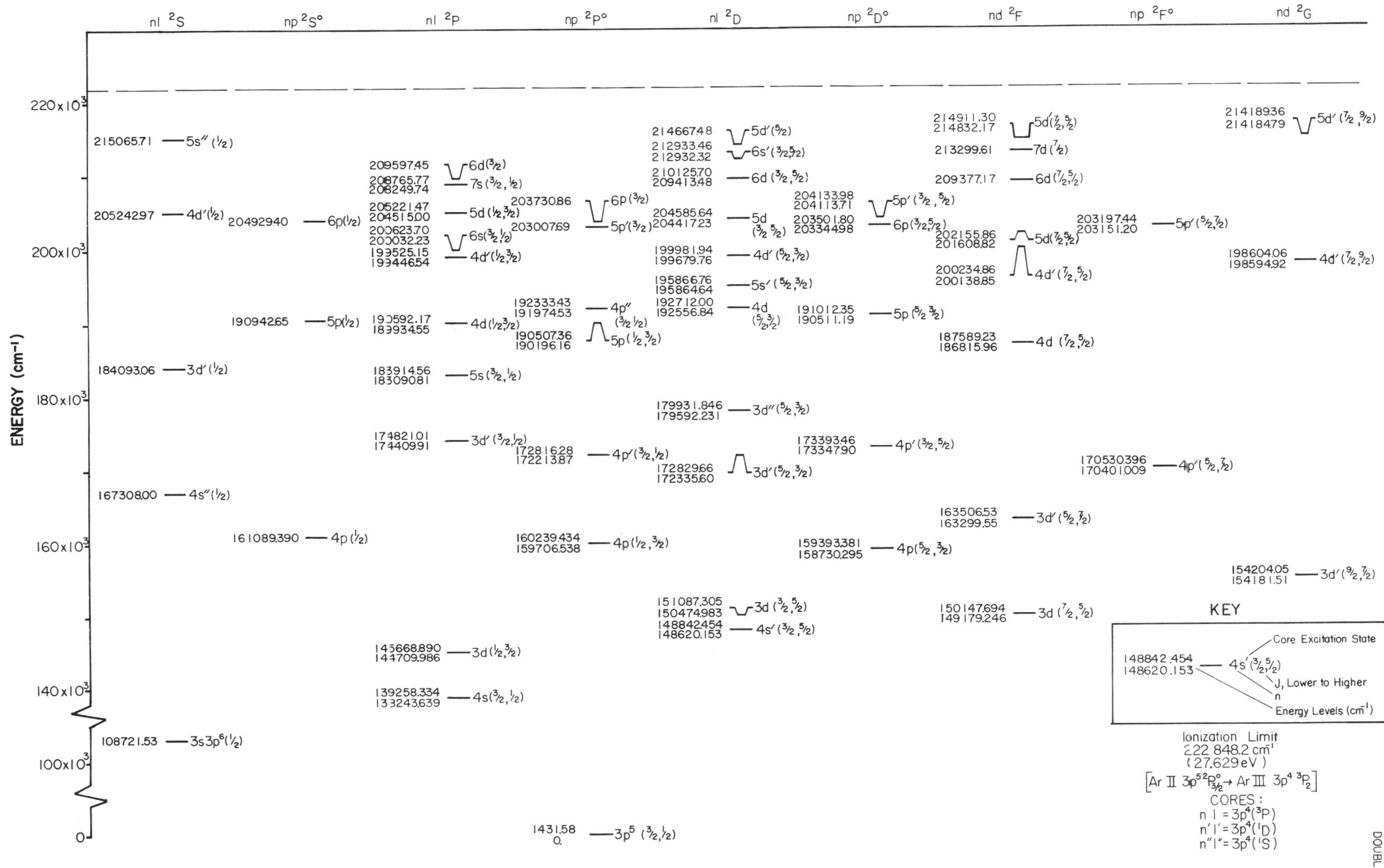

Ar II ENERGY LEVELS (17 electrons, Z=18)
(Cl I sequence, Configuration: $1s^2 2s^2 2p^6 3s^2 3p^4 nl$, Doublet System)
nl ^2S
np $^2S^o$
nl ^2P
np $^2P^o$
nl ^2D
np $^2D^o$
nd ^2F
np $^2F^o$
nd ^2G
ENERGY (cm^{-1})
220x10^3
200x10^3
180x10^3
160x10^3
140x10^3
100x10^3
0
215065.71 — 5s″(½)
205242.97 — 4d′(½)
184093.06 — 3d′(½)
167308.00 — 4s″(½)
108721.53 — 3s3p^6(½)
204929.40 — 6p(½)
190942.65 — 5p(½)
161089.390 — 4p(½)
209597.45 6d(3/2)
208765.77 208249.74 7s(3/2, 1/2)
205221.47 204515.00 5d(1/2, 3/2)
200623.70 200032.23 6s(3/2, 1/2)
199525.15 199446.54 4d′(1/2, 3/2)
190592.17 189934.55 4d(1/2, 3/2)
183914.56 183090.81 5s(3/2, 1/2)
174820.01 174409.91 3d′(3/2, 1/2)
145668.890 144709.986 3d(1/2, 3/2)
139258.334 133243.639 4s(3/2, 1/2)
203730.86 6p(3/2)
203007.69 5p′(3/2)
192333.43 191974.53 4p″(3/2, 1/2)
190507.36 190196.16 5p(1/2, 3/2)
172816.28 172213.87 4p′(3/2, 1/2)
160239.434 159706.538 4p(1/2, 3/2)
1431.58 0. 3p^5(3/2, 1/2)
214667.48 5d′(5/2)
212933.46 212932.32 6s′(3/2, 5/2)
210125.70 209413.48 6d(3/2, 5/2)
204585.64 204417.23 5d(3/2, 5/2)
199981.94 199679.76 4d′(5/2, 3/2)
195866.76 195866.64 5s′(5/2, 3/2)
192712.00 192556.84 4d(5/2, 3/2)
179931.846 179592.231 3d″(5/2, 3/2)
172829.66 172335.60 3d′(5/2, 3/2)
151087.305 150474.983 3d(3/2, 5/2)
148842.454 148620.153 4s′(3/2, 5/2)
204133.98 204113.71 5p′(3/2, 5/2)
203501.80 203344.98 6p(3/2, 5/2)
191012.35 190511.19 5p(5/2, 3/2)
173393.46 173347.90 4p′(3/2, 5/2)
159393.381 158730.295 4p(5/2, 3/2)
214911.30 214832.17 5d′(7/2, 5/2)
213299.61 7d(7/2)
209377.17 6d(7/2, 5/2)
202155.86 201608.82 5d(7/2, 5/2)
200234.86 200138.85 4d′(7/2, 5/2)
187589.23 186815.96 4d(7/2, 5/2)
163506.53 163299.55 3d′(5/2, 7/2)
150147.694 149179.246 3d(7/2, 5/2)
203197.44 203151.20 5p′(5/2, 7/2)
170530.396 170401.009 4p′(5/2, 7/2)
214189.36 214184.79 5d′(7/2, 9/2)
198604.06 198594.92 4d′(7/2, 9/2)
154204.05 154181.51 3d′(9/2, 7/2)
KEY
148842.454 148620.153 — 4s′(3/2, 5/2)
Core Excitation State
J, Lower to Higher
n
Energy Levels (cm^{-1})
Ionization Limit
222 848.2 cm^{-1}
(27.629 eV)
[Ar II $3p^5\,^2P^o_{3/2}$ → Ar III $3p^4\,^3P_2$]
CORES:
n l = $3p^4(^3P)$
n′l′ = $3p^4(^1D)$
n″l″ = $3p^4(^1S)$
Ar II
DOUBLET

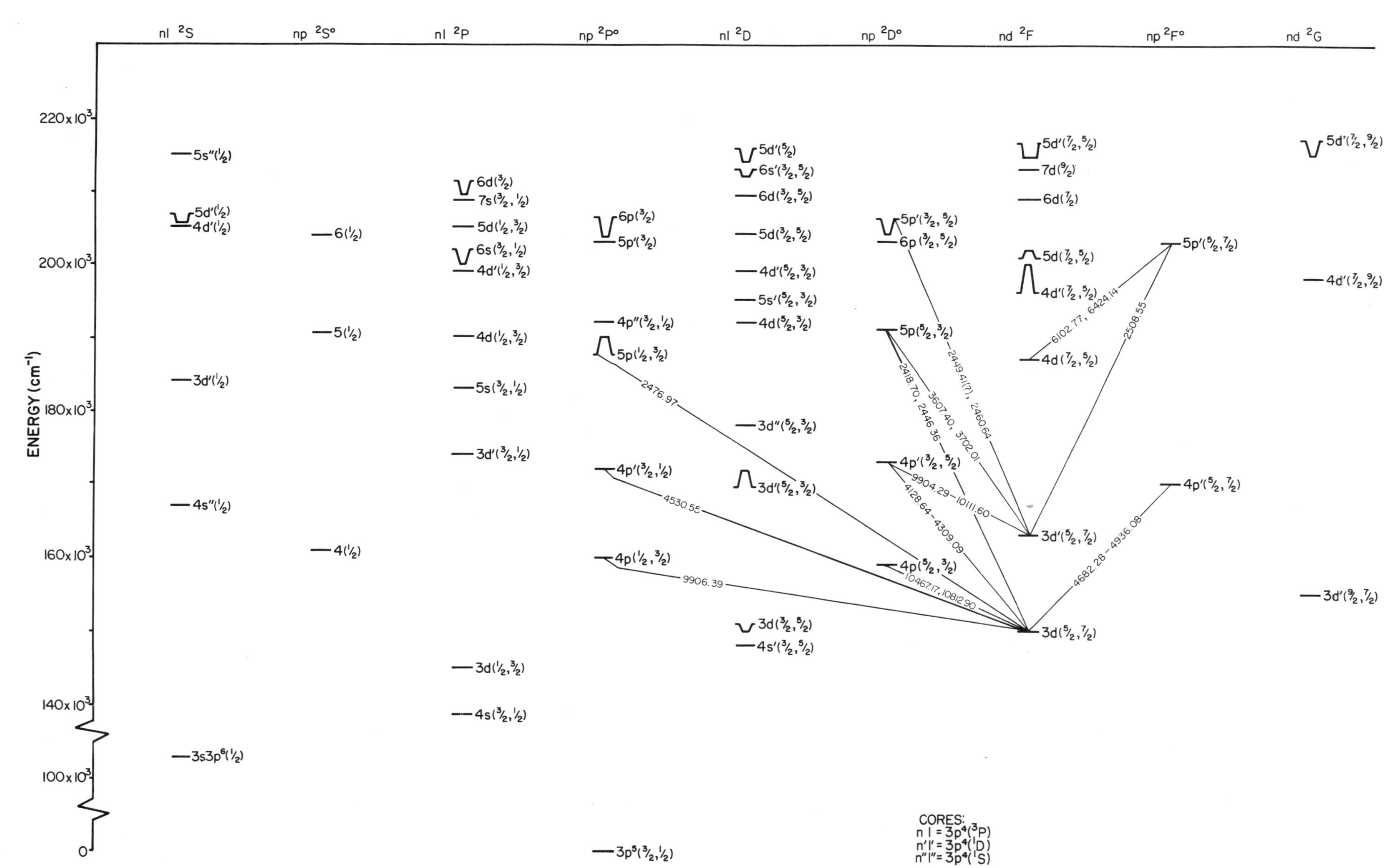
Ar II
DOUBLET
GROTRIAN
DIAGRAM
²F TERMS
TO HIGHER
DOUBLETS
Ar II GROTRIAN DIAGRAM (17 electrons, Z=18)
(Cl I sequence, Configuration: 1s²2s²2p⁶3s²3p⁴nl, ²F Terms to Higher Doublets)
nl ²S
np ²S°
nl ²P
np ²P°
nl ²D
np ²D°
nd ²F
np ²F°
nd ²G
ENERGY (cm⁻¹)
220x10³
200x10³
180x10³
160x10³
140x10³
100x10³
0
5s''(1/2)
5d'(1/2)
4d'(1/2)
3d'(1/2)
4s''(1/2)
3s3p⁶(1/2)
6(1/2)
5(1/2)
4(1/2)
6d(3/2)
7s(3/2, 1/2)
5d(1/2, 3/2)
6s(3/2, 1/2)
4d'(1/2, 3/2)
4d(1/2, 3/2)
5s(3/2, 1/2)
3d'(3/2, 1/2)
3d(1/2, 3/2)
4s(3/2, 1/2)
6p(3/2)
5p(3/2)
4p''(3/2, 1/2)
5p(1/2, 3/2)
4p'(3/2, 1/2)
4p(1/2, 3/2)
3p⁵(3/2, 1/2)
5d'(5/2)
6s'(3/2, 5/2)
6d(3/2, 5/2)
5d(3/2, 5/2)
4d'(5/2, 3/2)
5s'(5/2, 3/2)
4d(5/2, 3/2)
3d''(5/2, 3/2)
3d'(5/2, 3/2)
3d(3/2, 5/2)
4s'(3/2, 5/2)
5p'(3/2, 5/2)
6p(3/2, 5/2)
5p(5/2, 3/2)
4p'(3/2, 5/2)
4p(5/2, 3/2)
5d'(7/2, 5/2)
7d(9/2)
6d(7/2)
5d(7/2, 5/2)
4d'(7/2, 5/2)
4d(7/2, 5/2)
3d'(5/2, 7/2)
3d(5/2, 7/2)
5p'(5/2, 7/2)
4p'(5/2, 7/2)
5d'(7/2, 9/2)
4d'(7/2, 9/2)
3d'(9/2, 7/2)
2476.97
4530.55
9906.39
2449.41(?), 2460.64
2418.70, 2446.36
3607.40, 3702.01
9904.29–10111.60
4128.64–4309.09
10467.17, 10812.90
6102.77, 6424.14
2508.55
4682.28–4936.08
CORES:
n l = 3p⁴(³P)
n'l' = 3p⁴(¹D)
n''l'' = 3p⁴(¹S)

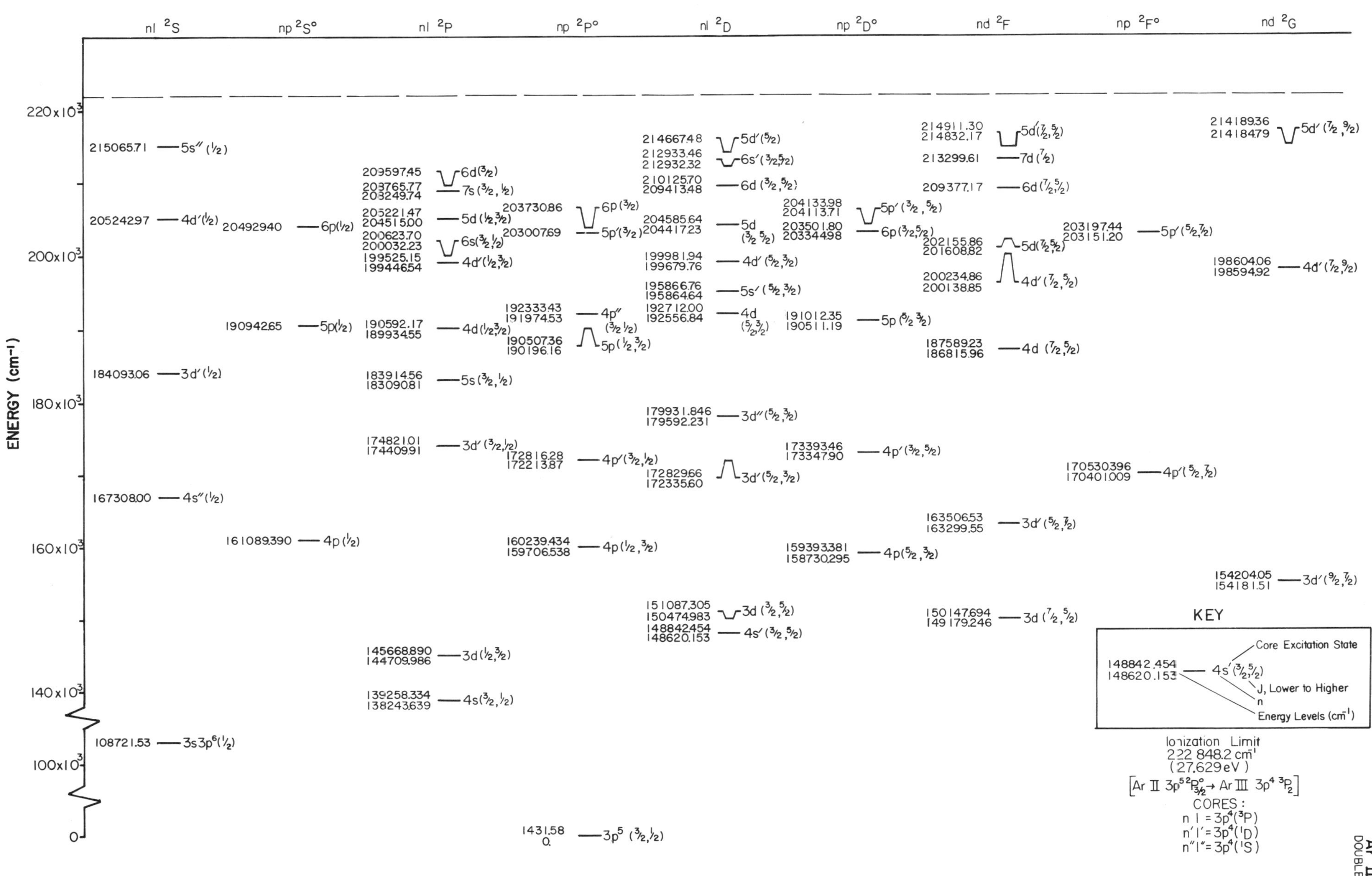

Ar II ENERGY LEVELS (17 electrons, Z=18)
(Cl I sequence, Configuration: 1s2 2s2 2p6 3s2 3p4 nl, Doublet System)
nl 2S
np 2S°
nl 2P
np 2P°
nl 2D
np 2D°
nd 2F
np 2F°
nd 2G
ENERGY (cm-1)
220x10^3
200x10^3
180x10^3
160x10^3
140x10^3
100x10^3
0
215065.71 — 5s″(1/2)
205242.97 — 4d′(1/2)
184093.06 — 3d′(1/2)
167308.00 — 4s″(1/2)
108721.53 — 3s3p6(1/2)
204929.40 — 6p(1/2)
190942.65 — 5p(1/2)
161089.390 — 4p(1/2)
209597.45 — 6d(3/2)
203765.77 203249.74 — 7s(3/2, 1/2)
205221.47 204515.00 — 5d(1/2, 3/2)
200623.70 200032.23 — 6s(3/2, 1/2)
199525.15 199446.54 — 4d′(1/2, 3/2)
190592.17 189934.55 — 4d(1/2, 3/2)
183914.56 183090.81 — 5s(3/2, 1/2)
174821.01 174409.91 — 3d′(3/2, 1/2)
145668.890 144709.986 — 3d(1/2, 3/2)
139258.334 138243.639 — 4s(3/2, 1/2)
203730.86 — 6p(3/2)
203007.69 — 5p′(3/2)
192333.43 191974.53 — 4p″(3/2, 1/2)
190507.36 190196.16 — 5p(1/2, 3/2)
172816.28 172213.87 — 4p′(3/2, 1/2)
160239.434 159706.538 — 4p(1/2, 3/2)
1431.58 0. — 3p5(3/2, 1/2)
214667.48 — 5d′(5/2)
212933.46 212932.32 — 6s′(3/2, 5/2)
210125.70 209413.48 — 6d(3/2, 5/2)
204585.64 204417.23 — 5d(3/2, 5/2)
199981.94 199679.76 — 4d′(5/2, 3/2)
195866.76 195864.64 — 5s′(5/2, 3/2)
192712.00 192556.84 — 4d(5/2, 3/2)
179931.846 179592.231 — 3d″(5/2, 3/2)
172829.66 172335.60 — 3d′(5/2, 3/2)
151087.305 150474.983 — 3d(3/2, 5/2)
148842.454 148620.153 — 4s′(3/2, 5/2)
204133.98 204113.71 — 5p′(3/2, 5/2)
203501.80 203344.98 — 6p(3/2, 5/2)
191012.35 190511.19 — 5p(5/2, 3/2)
173393.46 173347.90 — 4p′(3/2, 5/2)
159393.381 158730.295 — 4p(5/2, 3/2)
214911.30 214832.17 — 5d′(7/2, 5/2)
213299.61 — 7d(7/2)
209377.17 — 6d(7/2, 5/2)
202155.86 201608.82 — 5d(7/2, 5/2)
200234.86 200138.85 — 4d′(7/2, 5/2)
187589.23 186815.96 — 4d(7/2, 5/2)
163506.53 163299.55 — 3d′(5/2, 7/2)
150147.694 149179.246 — 3d(7/2, 5/2)
203197.44 203151.20 — 5p′(5/2, 7/2)
170530.396 170401.009 — 4p′(5/2, 7/2)
214189.36 214184.79 — 5d′(7/2, 9/2)
198604.06 198594.92 — 4d′(7/2, 9/2)
154204.05 154181.51 — 3d′(9/2, 7/2)
KEY
148842.454 148620.153 — 4s′(3/2, 5/2)
Core Excitation State
J, Lower to Higher
n
Energy Levels (cm-1)
Ionization Limit
222 848.2 cm-1
(27.629 eV)
[Ar II 3p5 2P°3/2 → Ar III 3p4 3P2]
CORES:
n l = 3p4(3P)
n′l′ = 3p4(1D)
n″l″ = 3p4(1S)
Ar II
DOUBLET

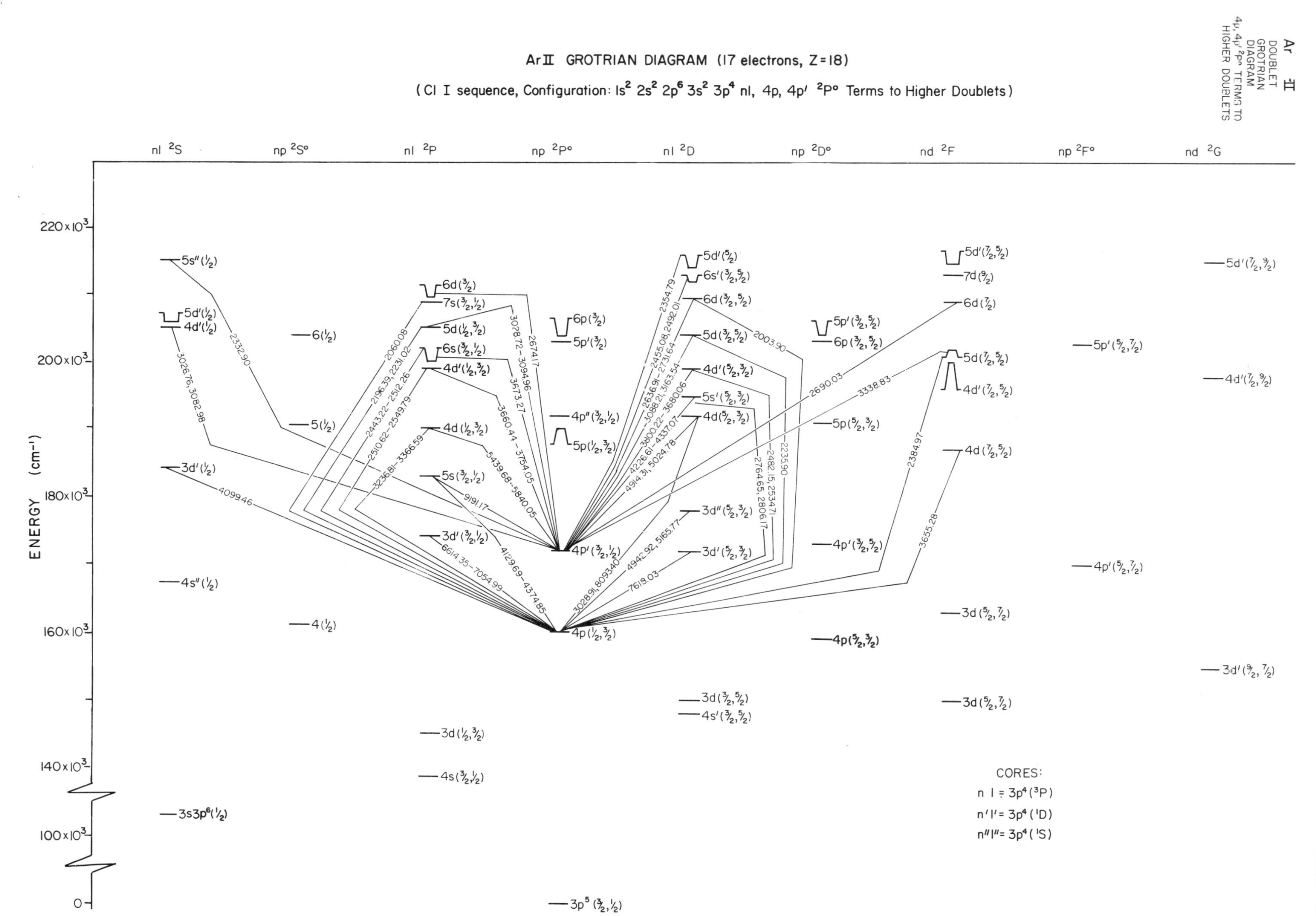
Ar II
DOUBLET
GROTRIAN
DIAGRAM
4p, 4p′ ²P° TERMS TO
HIGHER DOUBLETS
ArII GROTRIAN DIAGRAM (17 electrons, Z=18)
(Cl I sequence, Configuration: 1s² 2s² 2p⁶ 3s² 3p⁴ nl, 4p, 4p′ ²P° Terms to Higher Doublets)
nl ²S
np ²S°
nl ²P
np ²P°
nl ²D
np ²D°
nd ²F
np ²F°
nd ²G
ENERGY (cm⁻¹)
220x10³
200x10³
180x10³
160x10³
140x10³
100x10³
0
CORES:
n l = 3p⁴(³P)
n′l′ = 3p⁴(¹D)
n″l″ = 3p⁴(¹S)

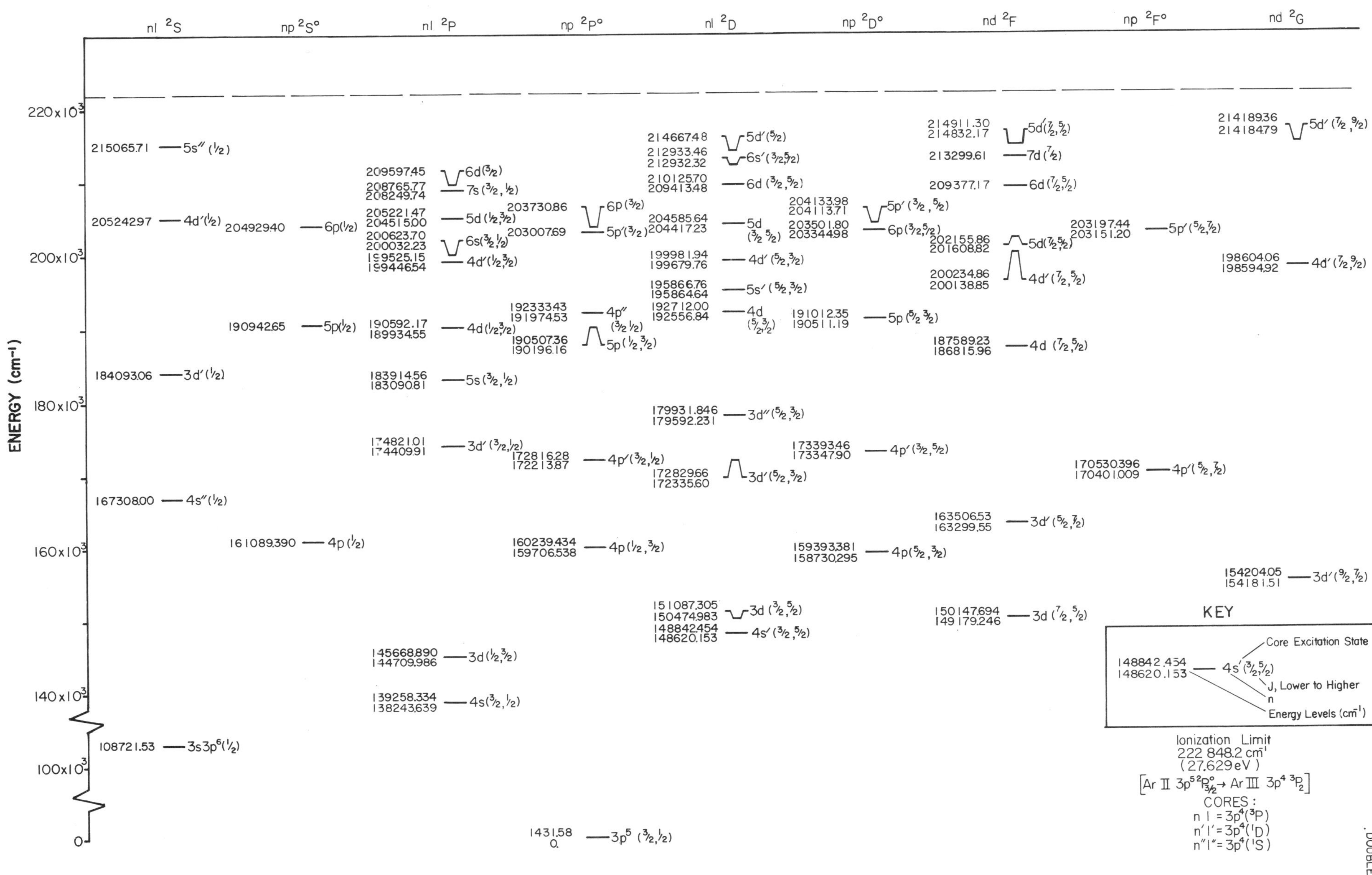

Ar II ENERGY LEVELS (17 electrons, Z=18)
(Cl I sequence, Configuration: 1s² 2s² 2p⁶ 3s² 3p⁴ nl, Doublet System)
nl ²S
np ²S°
nl ²P
np ²P°
nl ²D
np ²D°
nd ²F
np ²F°
nd ²G
ENERGY (cm⁻¹)
220x10³
200x10³
180x10³
160x10³
140x10³
100x10³
0
215065.71 — 5s″ (1/2)
205242.97 — 4d′ (1/2)
184093.06 — 3d′ (1/2)
167308.00 — 4s″ (1/2)
108721.53 — 3s3p⁶ (1/2)
204929.40 — 6p (1/2)
190942.65 — 5p (1/2)
161089.390 — 4p (1/2)
209597.45 6d (3/2)
208765.77 208249.74 7s (3/2, 1/2)
205221.47 204515.00 5d (1/2, 3/2)
200623.70 200032.23 6s (3/2, 1/2)
199525.15 199446.54 4d′ (1/2, 3/2)
190592.17 189934.55 4d (1/2, 3/2)
183914.56 183090.81 5s (3/2, 1/2)
174821.01 174409.91 3d′ (3/2, 1/2)
145668.890 144709.986 3d (1/2, 3/2)
139258.334 138243.639 4s (3/2, 1/2)
203730.86 6p (3/2)
203007.69 5p′ (3/2)
192333.43 191974.53 4p″ (3/2, 1/2)
190507.36 190196.16 5p (1/2, 3/2)
172816.28 172213.87 4p′ (3/2, 1/2)
160239.434 159706.538 4p (1/2, 3/2)
1431.58 0. 3p⁵ (3/2, 1/2)
214667.48 5d′ (5/2)
212933.46 212932.32 6s′ (3/2, 5/2)
210125.70 209413.48 6d (3/2, 5/2)
204585.64 204417.23 5d (3/2, 5/2)
199981.94 199679.76 4d′ (5/2, 3/2)
195866.76 195864.64 5s′ (5/2, 3/2)
192712.00 192556.84 4d (5/2, 3/2)
179931.846 179592.231 3d″ (5/2, 3/2)
172829.66 172335.60 3d′ (5/2, 3/2)
151087.305 150474.983 3d (3/2, 5/2)
148842.454 148620.153 4s′ (3/2, 5/2)
204133.98 204113.71 5p′ (3/2, 5/2)
203501.80 203344.98 6p (3/2, 5/2)
191012.35 190511.19 5p (5/2, 3/2)
173393.46 173347.90 4p′ (3/2, 5/2)
159393.381 158730.295 4p (5/2, 3/2)
214911.30 214832.17 5d (7/2, 5/2)
213299.61 7d (7/2)
209377.17 6d (7/2, 5/2)
202155.86 201608.82 5d (7/2, 5/2)
200234.86 200138.85 4d′ (7/2, 5/2)
187589.23 186815.96 4d (7/2, 5/2)
163506.53 163299.55 3d′ (5/2, 7/2)
150147.694 149179.246 3d (7/2, 5/2)
203197.44 203151.20 5p′ (5/2, 7/2)
170530.396 170401.009 4p′ (5/2, 7/2)
214189.36 214184.79 5d′ (7/2, 9/2)
198604.06 198594.92 4d′ (7/2, 9/2)
154204.05 154181.51 3d′ (9/2, 7/2)
KEY
148842.454 148620.153 — 4s′ (3/2, 5/2)
Core Excitation State
J, Lower to Higher
n
Energy Levels (cm⁻¹)
Ionization Limit
222 848.2 cm⁻¹
(27.629 eV)
[Ar II 3p⁵ ²P°3/2 → Ar III 3p⁴ ³P2]
CORES:
n l = 3p⁴(³P)
n′ l′ = 3p⁴(¹D)
n″ l″ = 3p⁴(¹S)
Ar II
DOUBLET

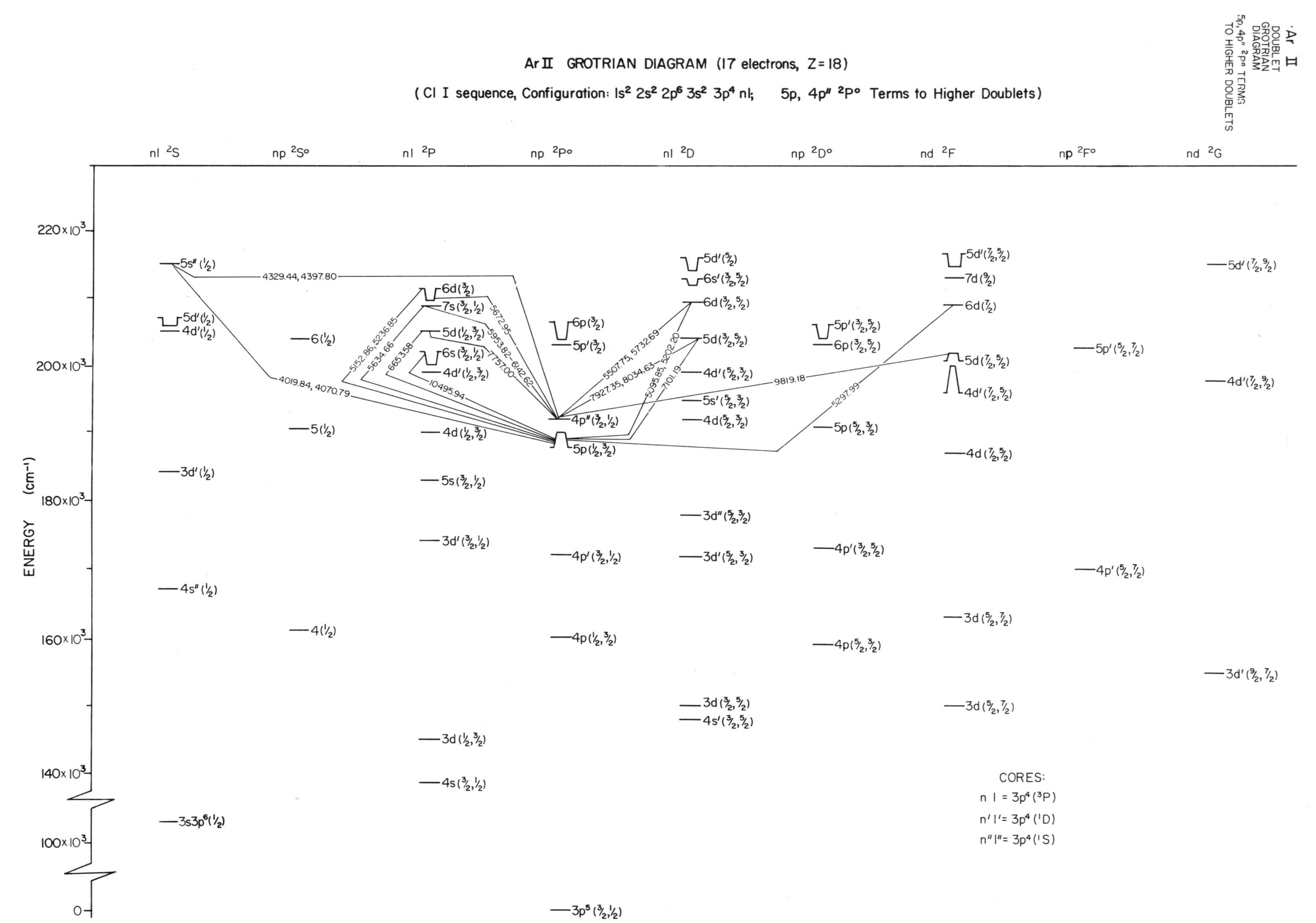
Ar II GROTRIAN DIAGRAM (17 electrons, Z = 18)
(Cl I sequence, Configuration: 1s² 2s² 2p⁶ 3s² 3p⁴ nl; 5p, 4p'' ²P° Terms to Higher Doublets)
nl ²S
np ²S°
nl ²P
np ²P°
nl ²D
np ²D°
nd ²F
np ²F°
nd ²G
ENERGY (cm⁻¹)
220×10³
200×10³
180×10³
160×10³
140×10³
100×10³
0
5s''(1/2)
5d'(1/2)
4d'(1/2)
3d'(1/2)
4s''(1/2)
3s3p⁶(1/2)
6(1/2)
5(1/2)
4(1/2)
6d(3/2)
7s(3/2,1/2)
5d(1/2,3/2)
6s(3/2,1/2)
4d'(1/2,3/2)
4d(1/2,3/2)
5s(3/2,1/2)
3d'(3/2,1/2)
3d(1/2,3/2)
4s(3/2,1/2)
6p(3/2)
5p'(3/2)
4p''(3/2,1/2)
5p(1/2,3/2)
4p'(3/2,1/2)
4p(1/2,3/2)
3p⁵(3/2,1/2)
5d'(5/2)
6s'(3/2,5/2)
6d(3/2,5/2)
5d(3/2,5/2)
4d'(5/2,3/2)
5s'(5/2,3/2)
4d(5/2,3/2)
3d''(5/2,3/2)
3d'(5/2,3/2)
3d(3/2,5/2)
4s'(3/2,5/2)
5p'(3/2,5/2)
6p(3/2,5/2)
5p(5/2,3/2)
4p'(3/2,5/2)
4p(5/2,3/2)
5d'(7/2,5/2)
7d(9/2)
6d(7/2)
5d(7/2,5/2)
4d'(7/2,5/2)
4d(7/2,5/2)
3d(5/2,7/2)
3d(5/2,7/2)
5p'(5/2,7/2)
4p'(5/2,7/2)
5d'(7/2,9/2)
4d'(7/2,9/2)
3d'(9/2,7/2)
4329.44, 4397.80
4019.84, 4070.79
5152.86, 5236.85
5634.66
6653.58
10495.94
5672.95
5953.82–6142.62
7757.00
5507.75, 5732.69
7927.35, 8034.63
5095.85, 5202.20
7101.19
9819.18
5297.99
CORES:
n l = 3p⁴(³P)
n' l' = 3p⁴(¹D)
n'' l'' = 3p⁴(¹S)

Ar II ENERGY LEVELS (17 electrons, Z=18)

(Cl I sequence, Configuration: $1s^2 2s^2 2p^6 3s^2 3p^4 nl$, Doublet System)

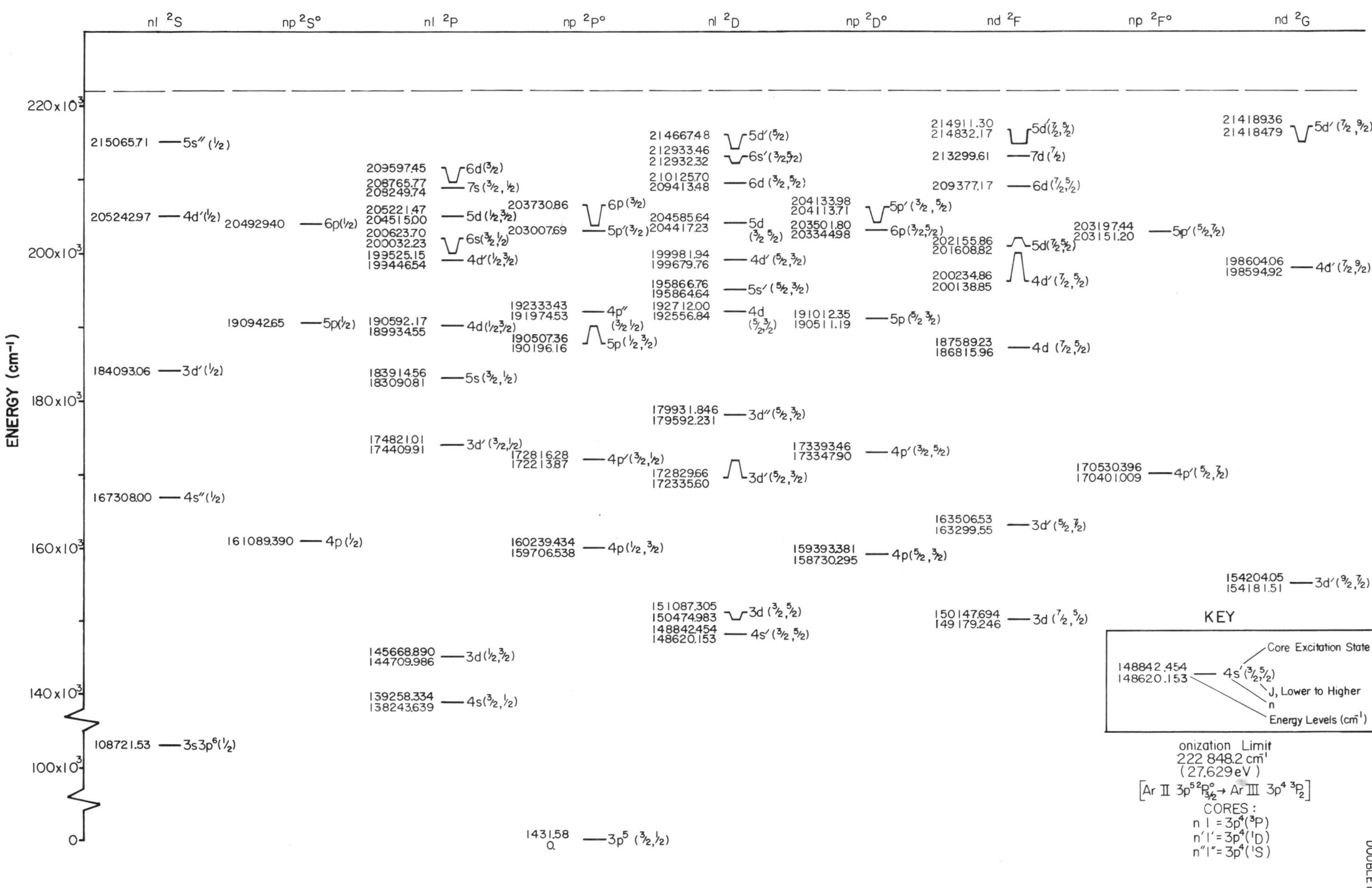

Ar II
DOUBLET
GROTRIAN
DIAGRAM
$^2D^o$ TERMS
TO HIGHER
DOUBLETS

Ar II GROTRIAN DIAGRAM (17 electrons, Z=18)

(Cl I sequence, Configuration: $1s^2 2s^2 2p^6 3s^2 3p^4 nl$, $^2D^o$ Terms to Higher Doublets)

nl ^2S | np ^2S | nl ^2P | np $^2P^o$ | nl ^2D | np $^2D^o$ | nd ^2F | np $^2F^o$ | nd ^2G

ENERGY (cm^{-1})

220x10³
200x10³
180x10³
160x10³
140x10³
100x10³
0

nl ^2S: 5s''(1/2); 5d'(1/2); 4d'(1/2); 3d'(1/2); 4s''(1/2); 3s3p⁶(1/2)

np ^2S: 6(1/2); 5(1/2); 4(1/2)

nl ^2P: 6d(3/2); 7s(3/2, 1/2); 5d(1/2, 3/2); 6s(3/2, 1/2); 4d''(1/2, 3/2); 4d(1/2, 3/2); 5s(3/2, 1/2); 3d''(3/2, 1/2); 3d(1/2, 3/2); 4s(3/2, 1/2)

np $^2P^o$: 6p(3/2); 5p'(3/2); 4p''(3/2, 1/2); 5p(1/2, 3/2); 4p'(3/2, 1/2); 4p(1/2, 3/2); 3p⁵(3/2, 1/2)

nl ^2D: 5d'(5/2); 6s'(3/2, 5/2); 6d(3/2, 5/2); 5d(3/2, 5/2); 4d''(5/2, 3/2); 5s(5/2, 3/2); 4d(5/2, 3/2); 3d''(5/2, 3/2); 3d'(5/2, 3/2); 3d(3/2, 5/2); 4s'(3/2, 5/2)

np $^2D^o$: 5p'(3/2, 5/2); 6p(3/2, 5/2); 5p(5/2, 3/2); 4p'(3/2, 5/2); 4p(5/2, 3/2)

nd ^2F: 5d'(7/2, 5/2); 7d(9/2); 6d(7/2); 5d(7/2, 5/2); 4d'(7/2, 5/2); 4d(7/2, 5/2); 3d'(5/2, 7/2); 3d(5/2, 7/2)

np $^2F^o$: 5p'(5/2, 7/2); 4p'(5/2, 7/2)

nd ^2G: 5d'(7/2, 9/2); 4d'(7/2, 9/2); 3d'(9/2, 7/2)

5379.16
5631.16–5799.73
2018.75, 2024.73
2420.46–2459.95
2491.04, 2495.92
10401.51, 10500.21
3746.45
3136.48–3207.58
3819.02–3830.52
5812.75, 6027.25
3137.63–3273.32
4076.94–4218.67
6375.95–6657.50
4047.48
2422.09
2525.48, 2528.32
5096.85–5432.94
2180.09–2220.35
10555.90
2423.53(?)–2481.48
2740.91, 2741.07
5162.74–5216.82
4439.46–4448.88
3753.52–3803.17
3205.00–3222.39
4792.09–4949.40
2941.89–3014.48
7090.56–7440.49
9258.78, 9344.79
2406.23–2412.46
4386.96(?)
5299.08–5443.68
1988.62, 2000.00
8585.26–9008.46
10383.90
3470.26, 3543.15
3718.21–3737.89
3464.13–3559.51
2414.22, 2447.74
2302.08–2337.78

CORES:
n l = 3p⁴(³P)
n'l' = 3p⁴(¹D)
n''l'' = 3p⁴(¹S)

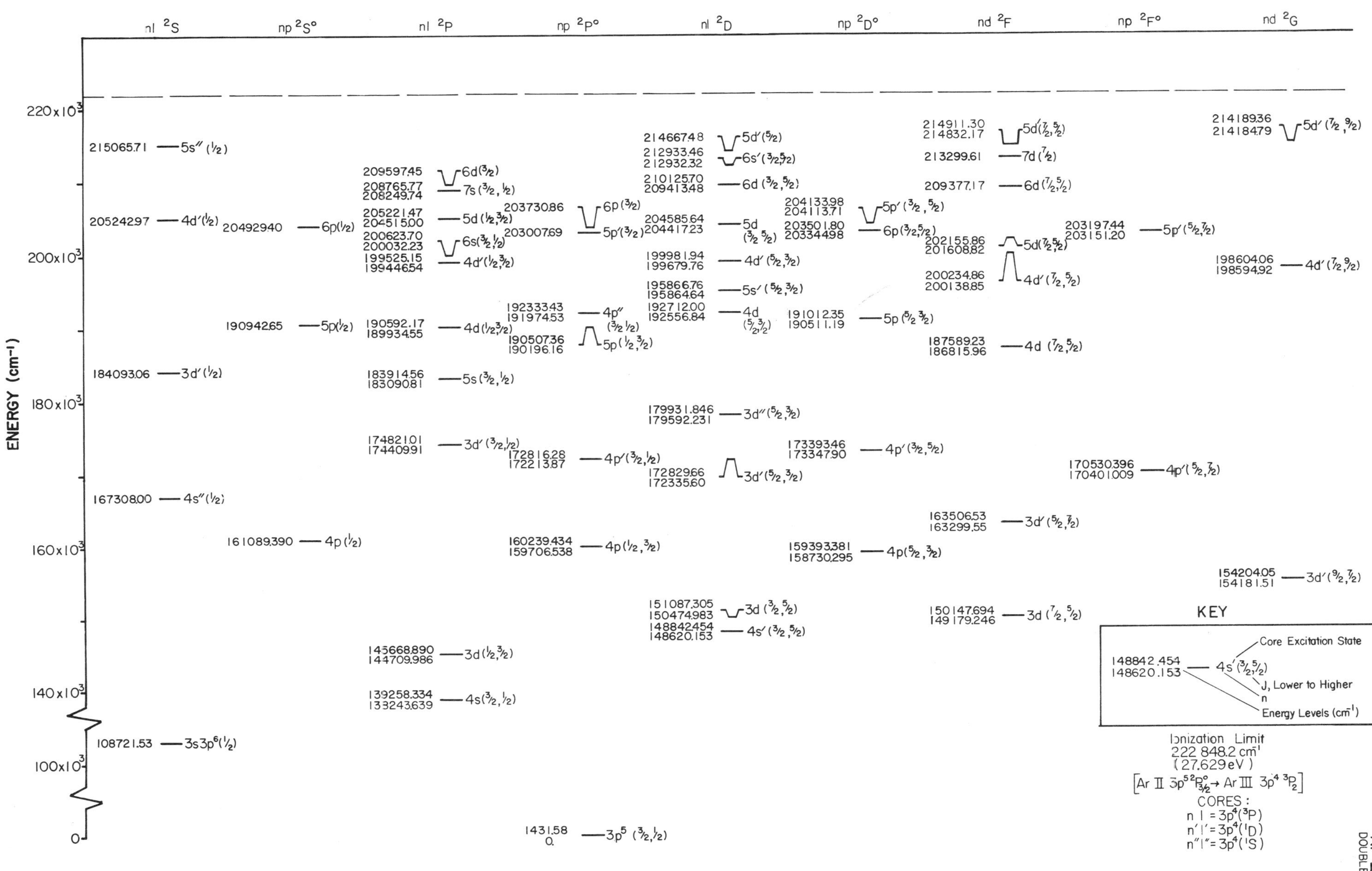
Ar II ENERGY LEVELS (17 electrons, Z=18)
(Cl I sequence, Configuration: 1s² 2s² 2p⁶ 3s² 3p⁴ nl, Doublet System)
nl ²S
np ²S°
nl ²P
np ²P°
nl ²D
np ²D°
nd ²F
np ²F°
nd ²G
ENERGY (cm⁻¹)
220x10³
200x10³
180x10³
160x10³
140x10³
100x10³
0
215065.71 — 5s″(1/2)
205242.97 — 4d′(1/2)
184093.06 — 3d′(1/2)
167308.00 — 4s″(1/2)
108721.53 — 3s3p⁶(1/2)
204929.40 — 6p(1/2)
190942.65 — 5p(1/2)
161089.390 — 4p(1/2)
209597.45 — 6d(3/2)
208765.77 208249.74 — 7s(3/2, 1/2)
205221.47 204515.00 — 5d(1/2, 3/2)
200623.70 200032.23 — 6s(3/2, 1/2)
199525.15 199446.54 — 4d′(1/2, 3/2)
190592.17 189934.55 — 4d(1/2, 3/2)
183914.56 183090.81 — 5s(3/2, 1/2)
174821.01 174409.91 — 3d′(3/2, 1/2)
145668.890 144709.986 — 3d(1/2, 3/2)
139258.334 133243.639 — 4s(3/2, 1/2)
203730.86 — 6p(3/2)
203007.69 — 5p′(3/2)
192333.43 191974.53 — 4p″(3/2, 1/2)
190507.36 190196.16 — 5p(1/2, 3/2)
172816.28 172213.87 — 4p′(3/2, 1/2)
160239.434 159706.538 — 4p(1/2, 3/2)
1431.58 0. — 3p⁵(3/2, 1/2)
214667.48 — 5d′(5/2)
212933.46 212932.32 — 6s′(3/2, 5/2)
210125.70 209413.48 — 6d(3/2, 5/2)
204585.64 204417.23 — 5d(3/2, 5/2)
199981.94 199679.76 — 4d′(5/2, 3/2)
195866.76 195864.64 — 5s′(5/2, 3/2)
192712.00 192556.84 — 4d(5/2, 3/2)
179931.846 179592.231 — 3d″(5/2, 3/2)
172829.66 172335.60 — 3d′(5/2, 3/2)
151087.305 150474.983 — 3d(3/2, 5/2)
148842.454 148620.153 — 4s′(3/2, 5/2)
204133.98 204113.71 — 5p′(3/2, 5/2)
203501.80 203344.98 — 6p(3/2, 5/2)
191012.35 190511.19 — 5p(5/2, 3/2)
173393.46 173347.90 — 4p′(3/2, 5/2)
159393.381 158730.295 — 4p(5/2, 3/2)
214911.30 214832.17 — 5d(7/2, 5/2)
213299.61 — 7d(7/2)
209377.17 — 6d(7/2, 5/2)
202155.86 201608.82 — 5d(7/2, 5/2)
200234.86 200138.85 — 4d′(7/2, 5/2)
187589.23 186815.96 — 4d(7/2, 5/2)
163506.53 163299.55 — 3d′(5/2, 7/2)
150147.694 149179.246 — 3d(7/2, 5/2)
203197.44 203151.20 — 5p′(5/2, 7/2)
170530.396 170401.009 — 4p′(5/2, 7/2)
214189.36 214184.79 — 5d′(7/2, 9/2)
198604.06 198594.92 — 4d′(7/2, 9/2)
154204.05 154181.51 — 3d′(9/2, 7/2)
KEY
148842.454 148620.153 — 4s′(3/2, 5/2)
Core Excitation State
J, Lower to Higher
n
Energy Levels (cm⁻¹)
Ionization Limit
222 848.2 cm⁻¹
(27.629 eV)
[Ar II 3p⁵ ²P°3/2 → Ar III 3p⁴ ³P2]
CORES:
n l = 3p⁴(³P)
n′l′ = 3p⁴(¹D)
n″l″ = 3p⁴(¹S)
Ar II
DOUBLET

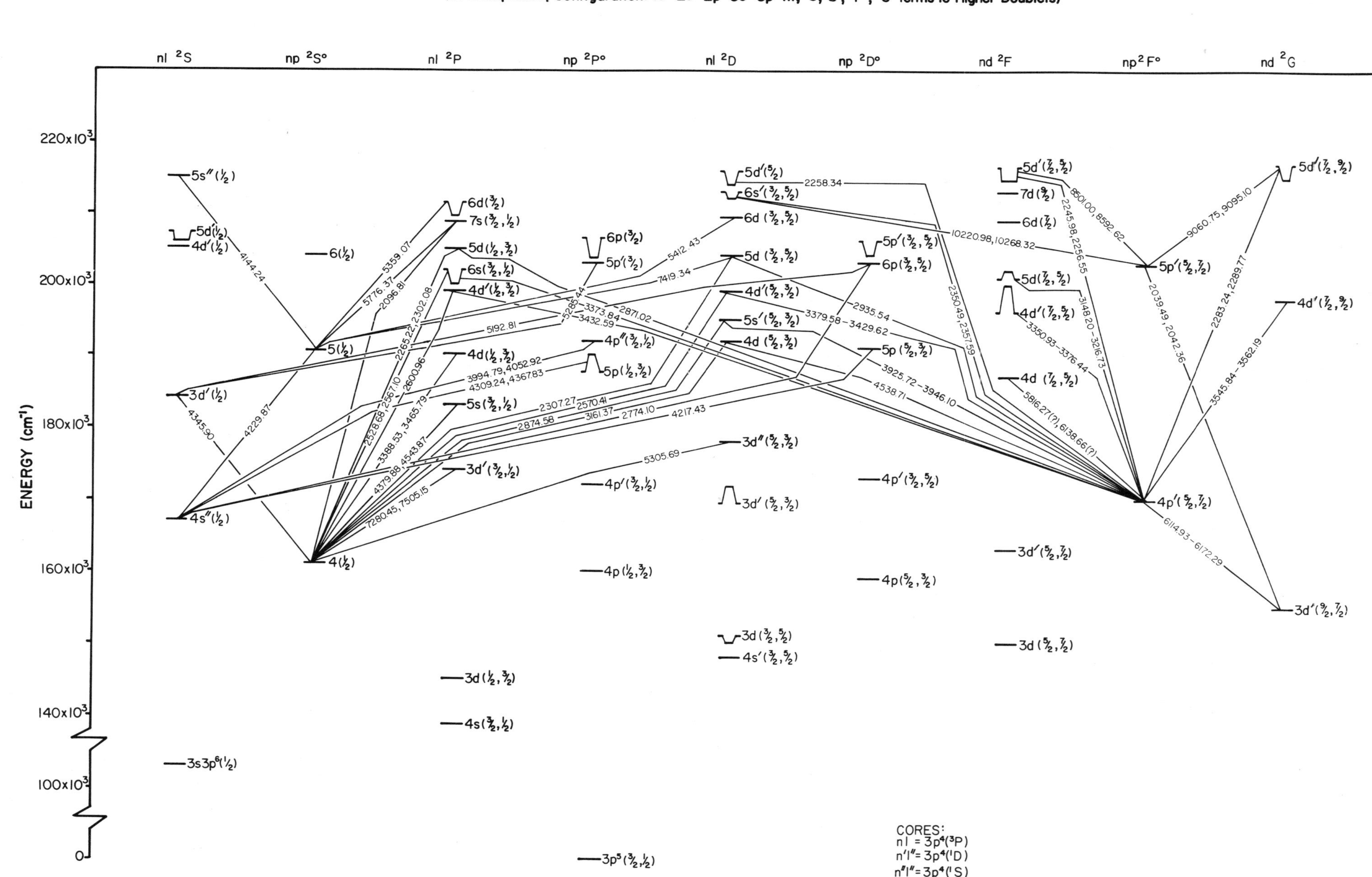

Ar II
DOUBLET
GROTRIAN
DIAGRAM
²S,²S°,²F°,²G
Terms to Higher
Doublets
Ar II GROTRIAN DIAGRAM (17 electrons, Z = 18)
(Cl I sequence, Configuration: 1s² 2s² 2p⁶ 3s² 3p⁴ nl, ²S,²S°,²F°,²G Terms to Higher Doublets)
nl ²S
np ²S°
nl ²P
np ²P°
nl ²D
np ²D°
nd ²F
np ²F°
nd ²G
ENERGY (cm⁻¹)
220x10³
200x10³
180x10³
160x10³
140x10³
100x10³
0
CORES:
nl = 3p⁴(³P)
n′l′ = 3p⁴(¹D)
n″l″ = 3p⁴(¹S)

Ar II ENERGY LEVELS (17 electrons, Z=18)

(Cl I sequence, Configuration: $1s^2 2s^2 2p^6 3s^2 3p^4 nl$, Doublet System)

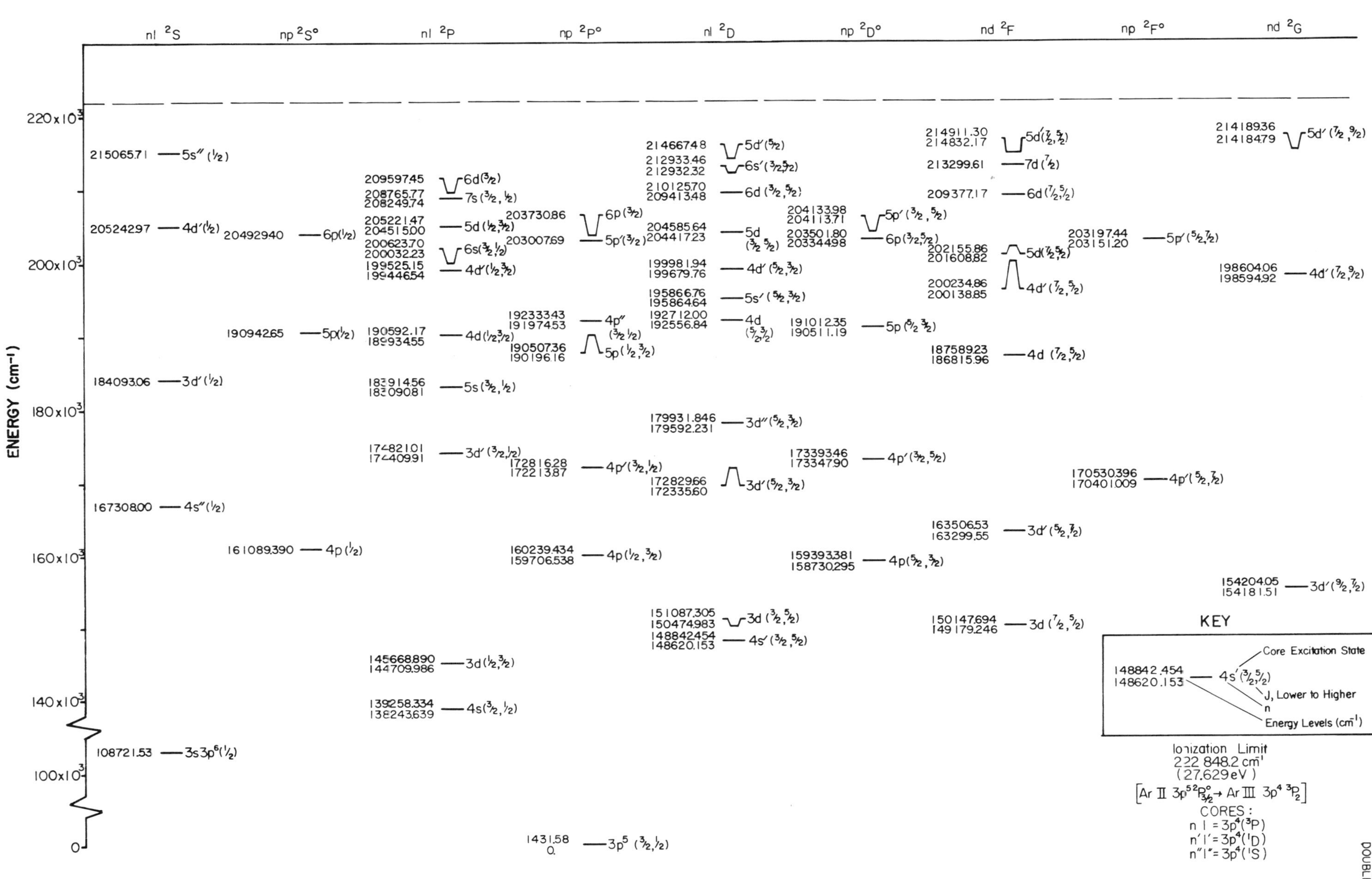

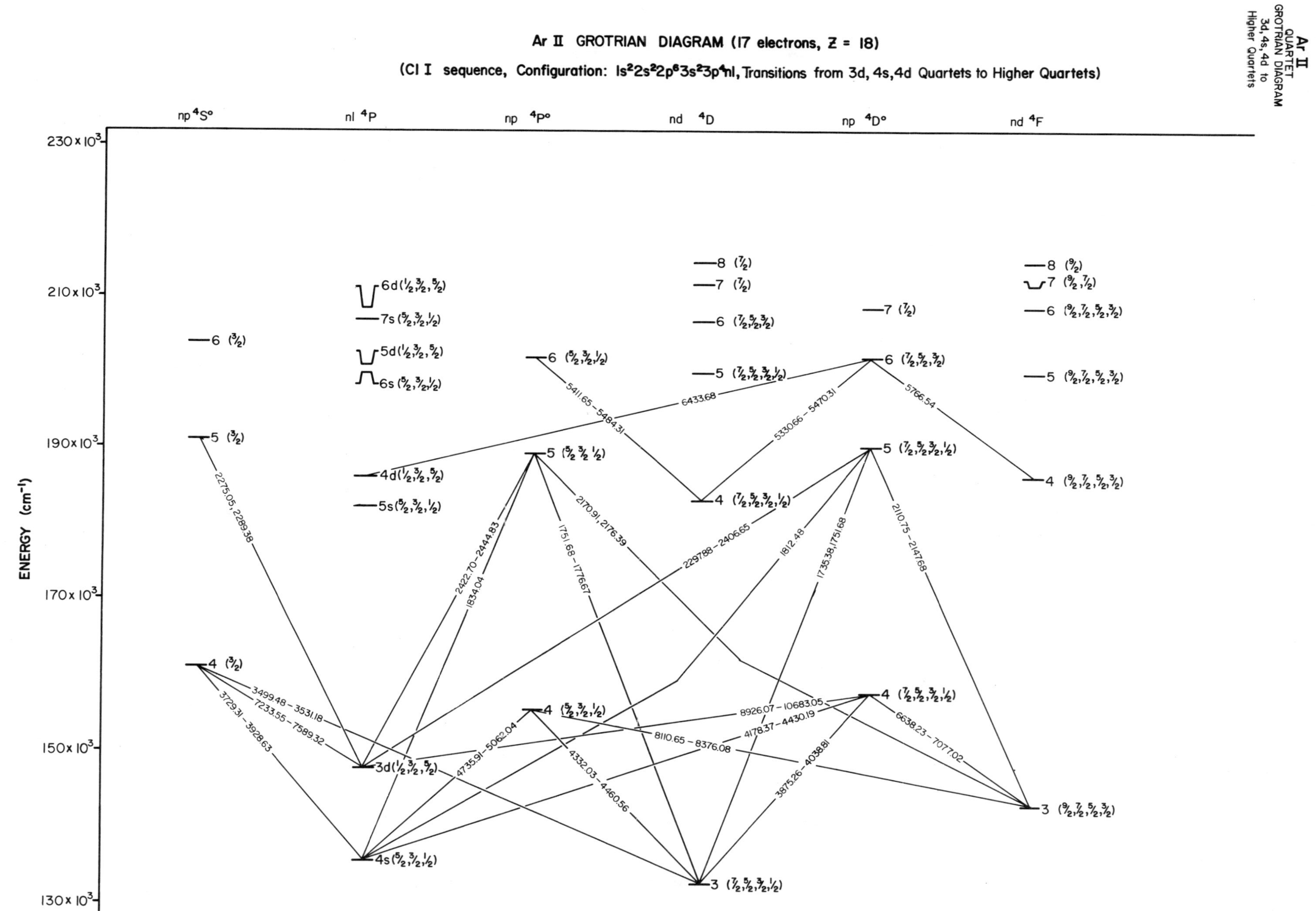
Ar II GROTRIAN DIAGRAM (17 electrons, Z = 18)
(Cl I sequence, Configuration: $1s^22s^22p^63s^23p^4nl$, Transitions from 3d, 4s, 4d Quartets to Higher Quartets)
np ⁴S°
nl ⁴P
np ⁴P°
nd ⁴D
np ⁴D°
nd ⁴F
ENERGY (cm⁻¹)
230×10³
210×10³
190×10³
170×10³
150×10³
130×10³
2275.05, 2289.38
3499.48 – 3531.18
7233.55 – 7589.32
3729.31 – 3928.63
2422.70 – 2444.83
1834.04
1751.68 – 1776.67
2170.91, 2176.39
5411.65 – 5484.31
6433.68
5330.66 – 5470.31
5766.54
2297.88 – 2406.65
1812.48
1735.38, 1751.68
2110.75 – 2147.68
4332.03 – 4460.56
4735.91 – 5062.04
8110.65 – 8376.08
8926.07 – 10683.05
4178.37 – 4430.19
3875.26 – 4038.81
6638.23 – 7077.02

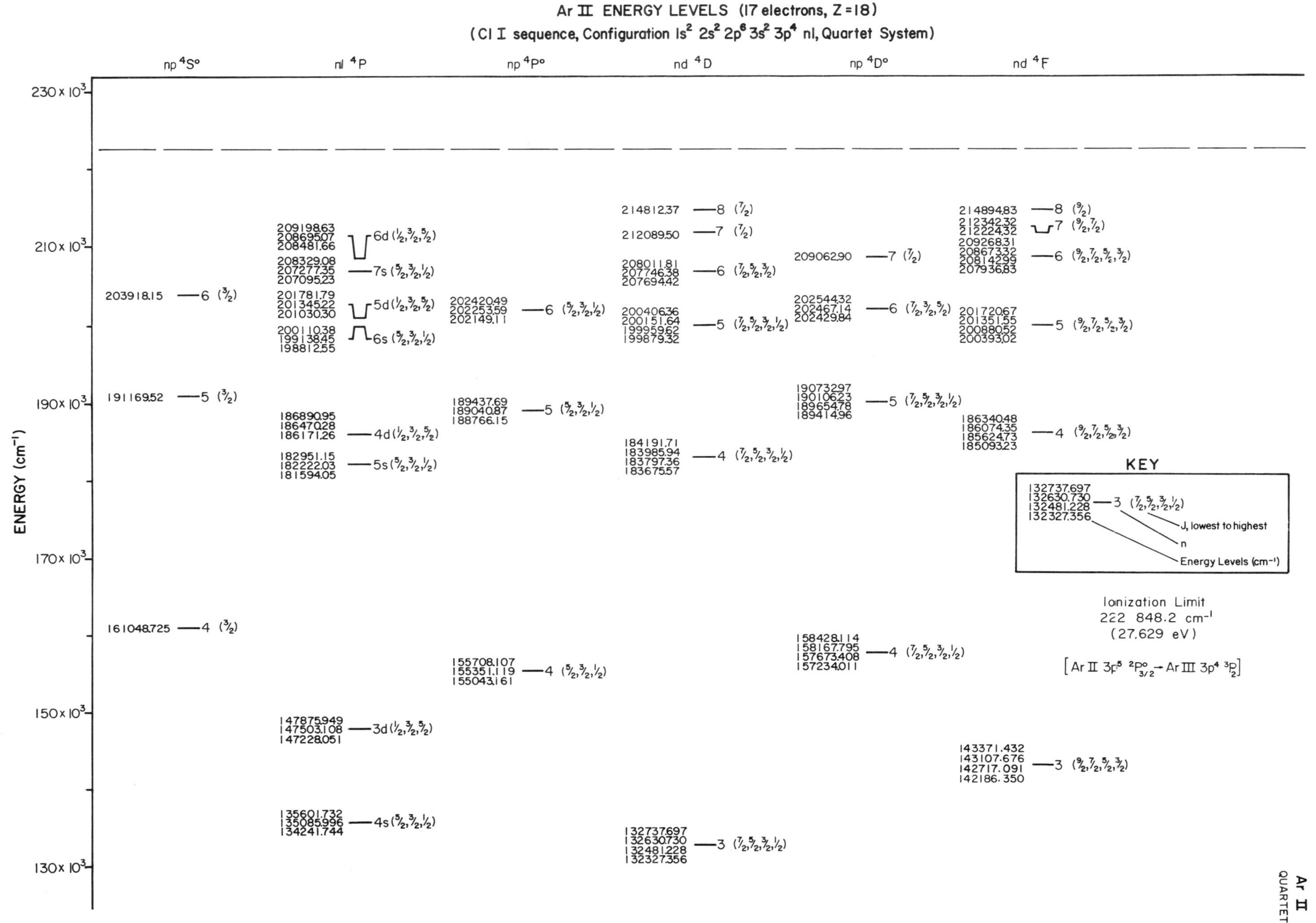
Ar II ENERGY LEVELS (17 electrons, Z=18)
(Cl I sequence, Configuration 1s² 2s² 2p⁶ 3s² 3p⁴ nl, Quartet System)
np ⁴S°
nl ⁴P
np ⁴P°
nd ⁴D
np ⁴D°
nd ⁴F
203918.15 —6 (3/2)
191169.52 —5 (3/2)
161048.725 —4 (3/2)
209198.63 208695.07 208481.66 —6d (1/2, 3/2, 5/2)
208329.08 207277.35 207095.23 —7s (5/2, 3/2, 1/2)
201781.79 201345.22 201030.30 —5d (1/2, 3/2, 5/2)
200110.38 199138.45 198812.55 —6s (5/2, 3/2, 1/2)
186890.95 186470.28 186171.26 —4d (1/2, 3/2, 5/2)
182951.15 182222.03 181594.05 —5s (5/2, 3/2, 1/2)
147875.949 147503.108 147228.051 —3d (1/2, 3/2, 5/2)
135601.732 135085.996 134241.744 —4s (5/2, 3/2, 1/2)
202420.49 202253.59 202149.11 —6 (5/2, 3/2, 1/2)
189437.69 189040.87 188766.15 —5 (5/2, 3/2, 1/2)
155708.107 155351.119 155043.161 —4 (5/2, 3/2, 1/2)
214812.37 —8 (7/2)
212089.50 —7 (7/2)
208011.81 207746.38 207694.42 —6 (7/2, 5/2, 3/2)
200406.36 200151.64 199959.62 199879.32 —5 (7/2, 5/2, 3/2, 1/2)
184191.71 183985.94 183797.36 183675.57 —4 (7/2, 5/2, 3/2, 1/2)
132737.697 132630.730 132481.228 132327.356 —3 (7/2, 5/2, 3/2, 1/2)
209062.90 —7 (7/2)
202544.32 202467.14 202429.84 —6 (7/2, 3/2, 5/2)
190732.97 190106.23 189654.78 189414.96 —5 (7/2, 5/2, 3/2, 1/2)
158428.114 158167.795 157673.408 157234.011 —4 (7/2, 5/2, 3/2, 1/2)
214894.83 —8 (9/2)
212342.32 212224.32 —7 (9/2, 7/2)
209268.31 208673.32 208142.99 207936.83 —6 (9/2, 7/2, 5/2, 3/2)
201720.67 201351.55 200880.52 200393.02 —5 (9/2, 7/2, 5/2, 3/2)
186340.48 186074.35 185624.73 185093.23 —4 (9/2, 7/2, 5/2, 3/2)
143371.432 143107.676 142717.091 142186.350 —3 (9/2, 7/2, 5/2, 3/2)
KEY
132737.697 132630.730 132481.228 132327.356 —3 (7/2, 5/2, 3/2, 1/2)
J, lowest to highest
n
Energy Levels (cm⁻¹)
Ionization Limit
222 848.2 cm⁻¹
(27.629 eV)
[Ar II 3p⁵ ²P°3/2 → Ar III 3p⁴ ³P2]
ENERGY (cm⁻¹)
230 x 10³
210 x 10³
190 x 10³
170 x 10³
150 x 10³
130 x 10³

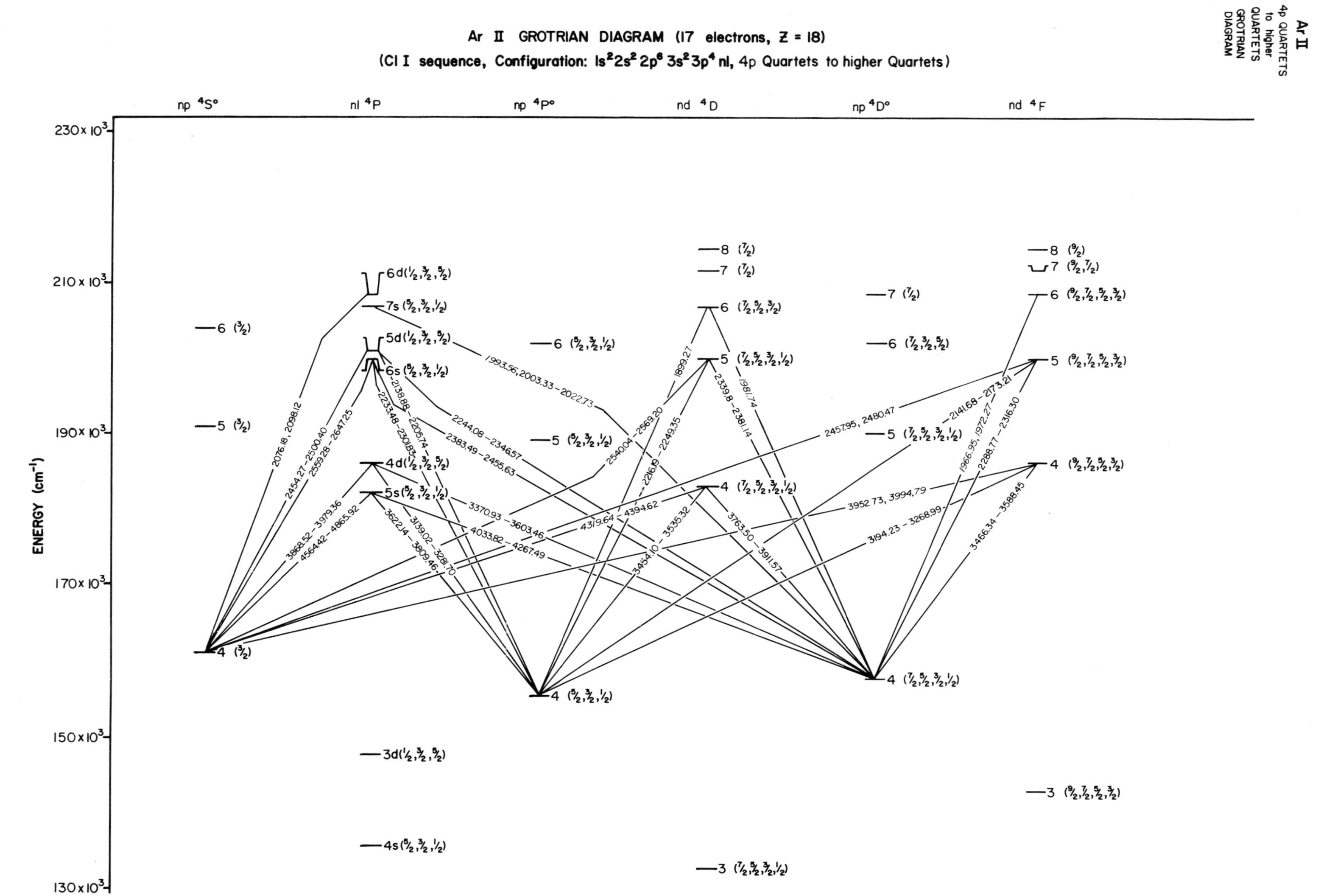
Ar II GROTRIAN DIAGRAM (17 electrons, Z = 18)
(Cl I sequence, Configuration: 1s²2s²2p⁶3s²3p⁴nl, 4p Quartets to higher Quartets)
np ⁴S°
nl ⁴P
np ⁴P°
nd ⁴D
np ⁴D°
nd ⁴F
ENERGY (cm⁻¹)
230x10³
210x10³
190x10³
170x10³
150x10³
130x10³

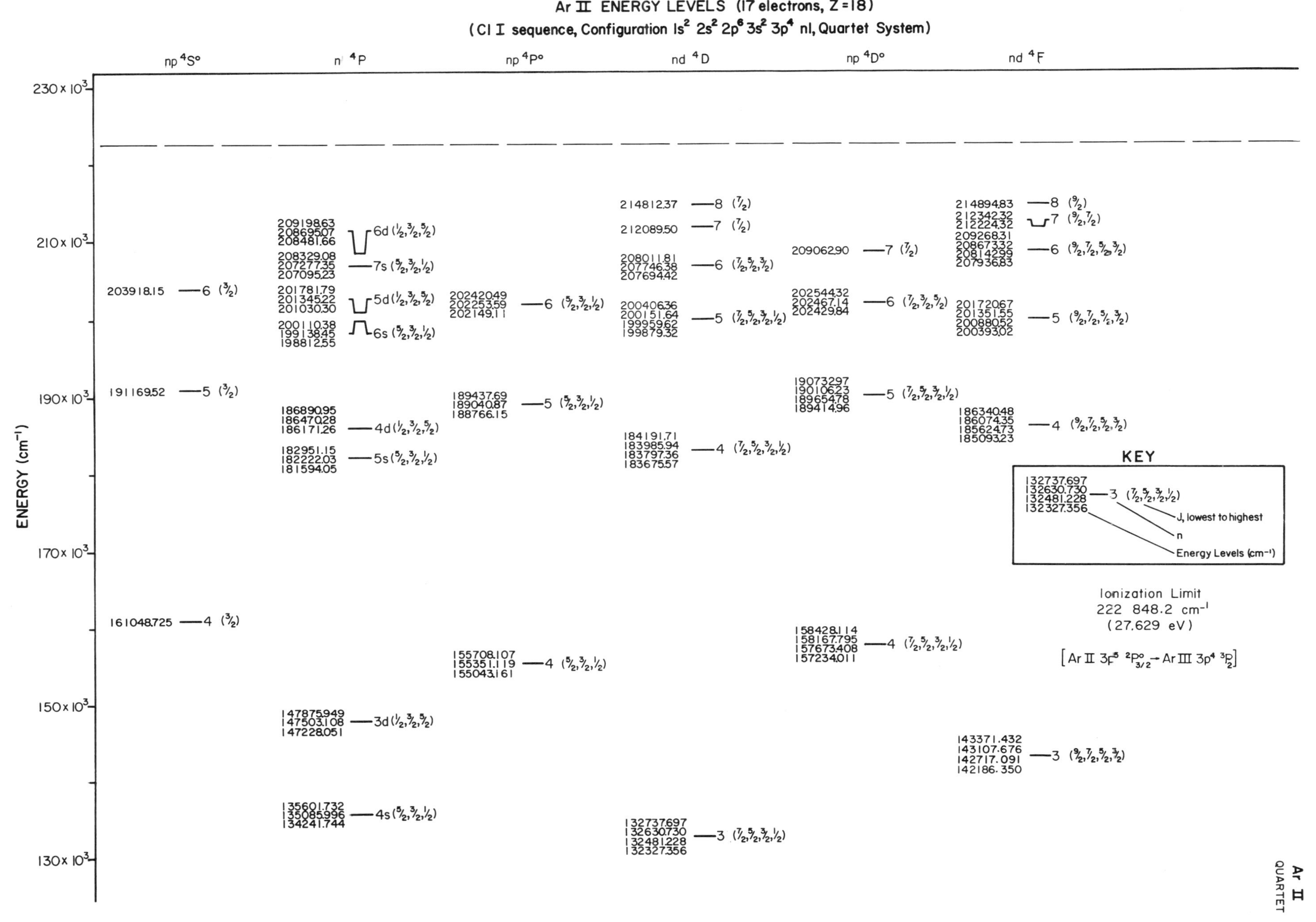
Ar II ENERGY LEVELS (17 electrons, Z=18)
(Cl I sequence, Configuration $1s^2\,2s^2\,2p^6\,3s^2\,3p^4$ nl, Quartet System)
np ^{4}S°
n ^{4}P
np ^{4}P°
nd ^{4}D
np ^{4}D°
nd ^{4}F
ENERGY (cm^{-1})
230 x 10^3
210 x 10^3
190 x 10^3
170 x 10^3
150 x 10^3
130 x 10^3
KEY
132737.697
132630.730
132481.228
132327.356
J, lowest to highest
n
Energy Levels (cm^{-1})
Ionization Limit
222 848.2 cm^{-1}
(27.629 eV)
[Ar II $3p^5\ ^2P^o_{3/2}$ → Ar III $3p^4\ ^3P_2$]
Ar II
QUARTET

Ar II

n=5 QUARTETS
to higher
QUARTETS
GROTRIAN
DIAGRAM

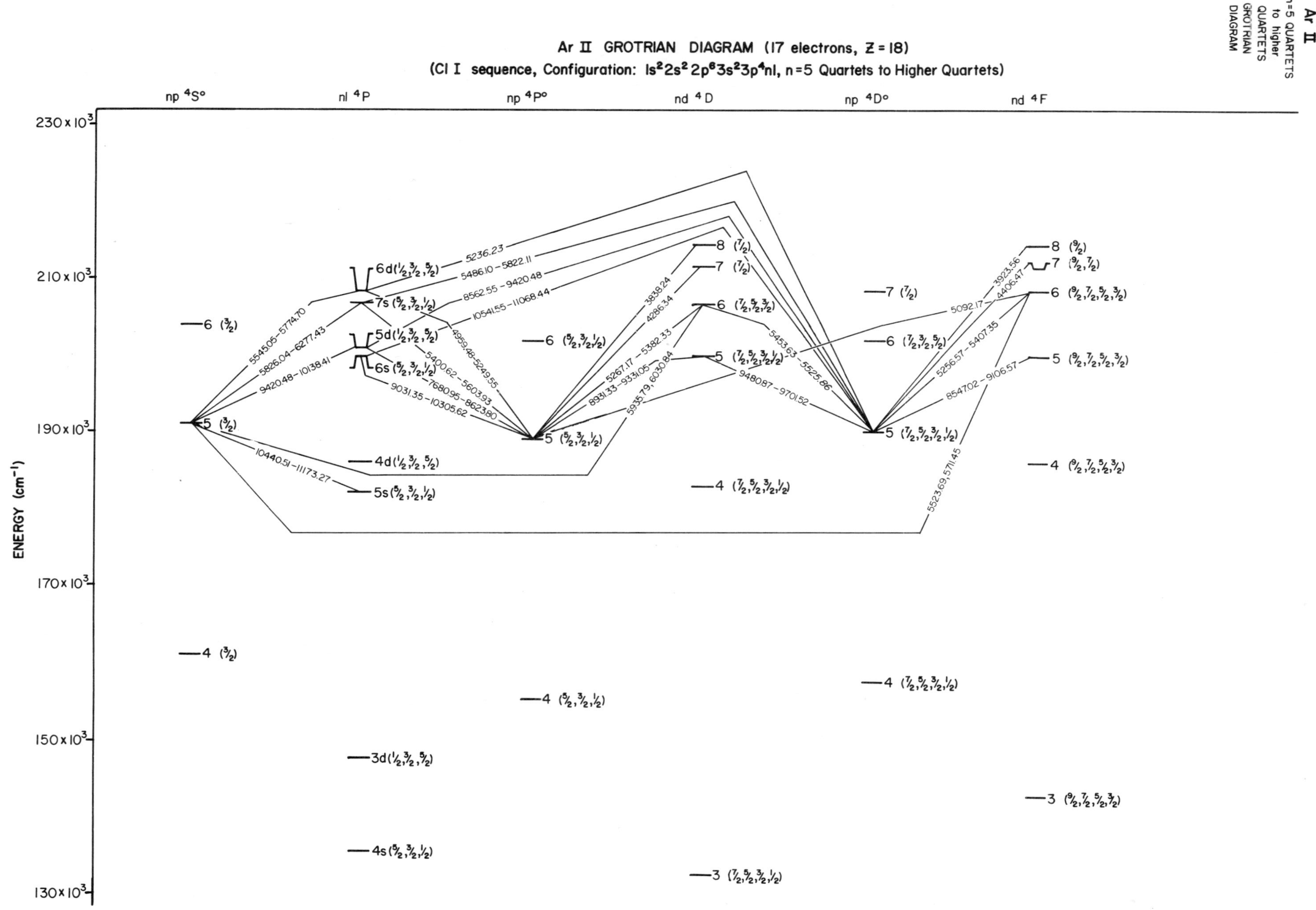

Ar II ENERGY LEVELS (17 electrons, Z=18)
(Cl I sequence, Configuration $1s^2\ 2s^2\ 2p^6\ 3s^2\ 3p^4$ nl, Quartet System)

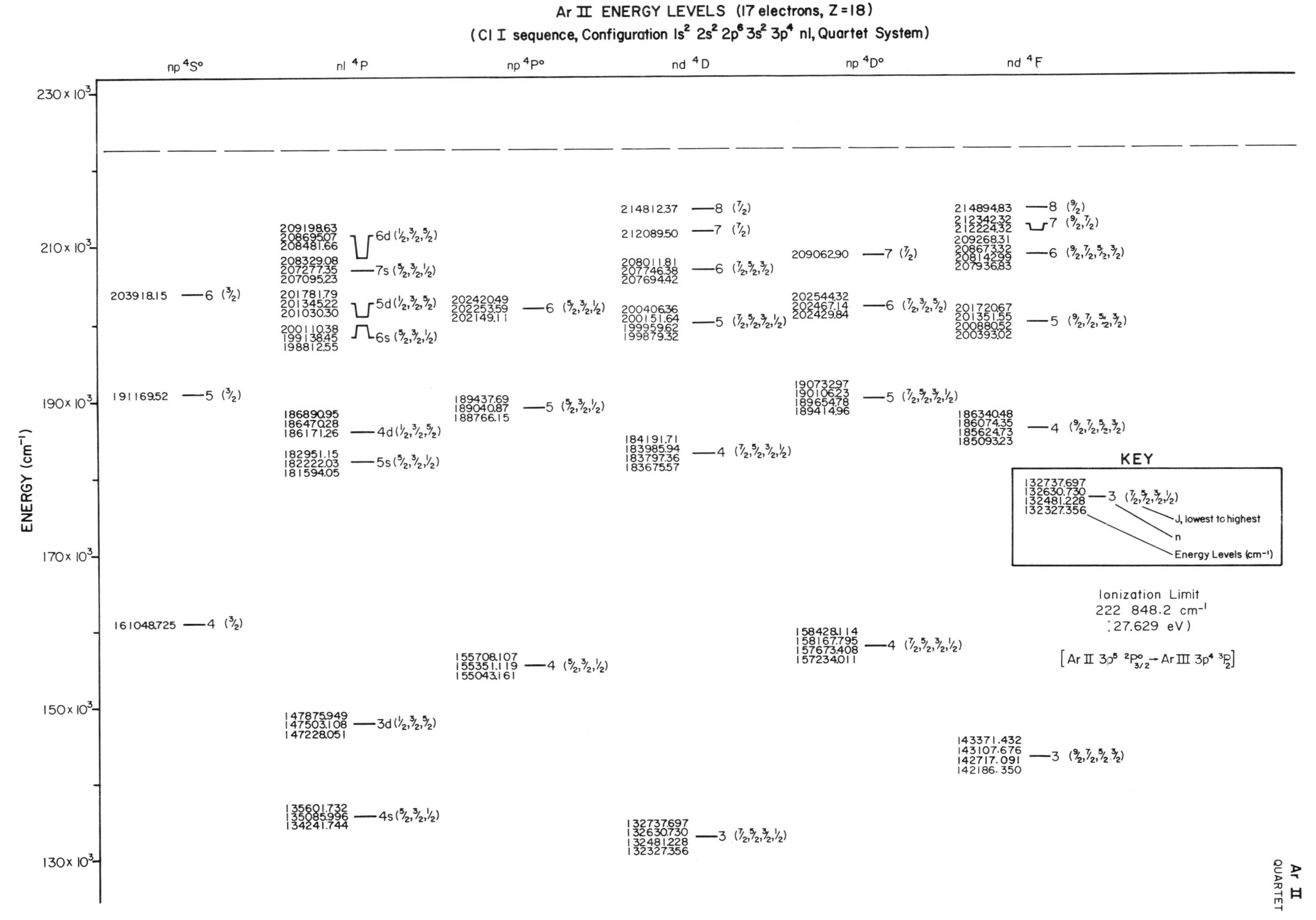

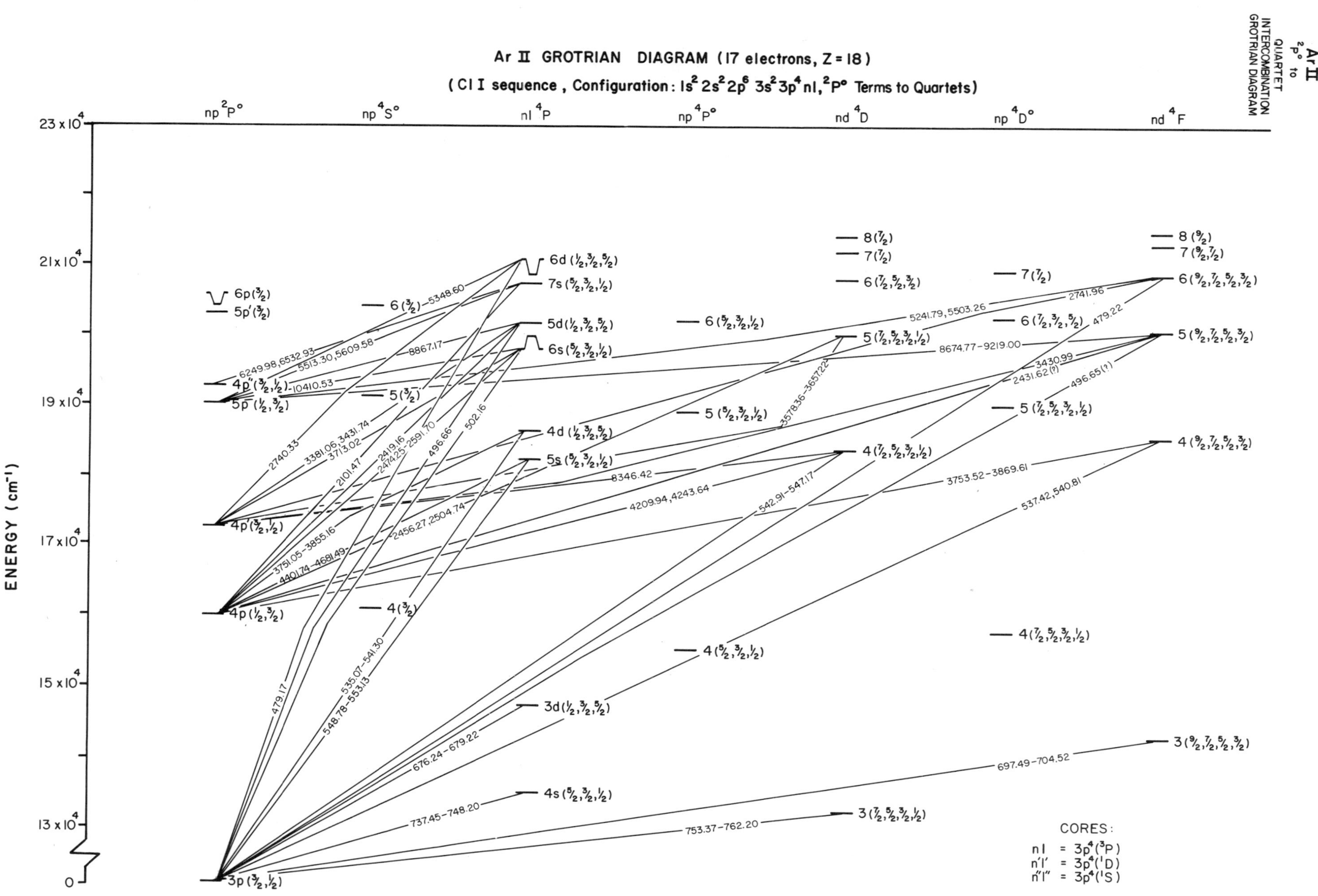
Ar II
²P° to
QUARTET
INTERCOMBINATION
GROTRIAN DIAGRAM
Ar II GROTRIAN DIAGRAM (17 electrons, Z = 18)
(Cl I sequence, Configuration: 1s² 2s² 2p⁶ 3s² 3p⁴ nl, ²P° Terms to Quartets)
np ²P°
np ⁴S°
nl ⁴P
np ⁴P°
nd ⁴D
np ⁴D°
nd ⁴F
ENERGY (cm⁻¹)
CORES:
nl = 3p⁴(³P)
n′l′ = 3p⁴(¹D)
n″l″ = 3p⁴(¹S)

Ar II GROTRIAN DIAGRAM (17 electrons, Z = 18)

(Cl I sequence, Configuration : $1s^2 2s^2 2p^6 3s^2 3p^4 nl$, ^2S, $^2S^o$, ^2P terms to Quartets)

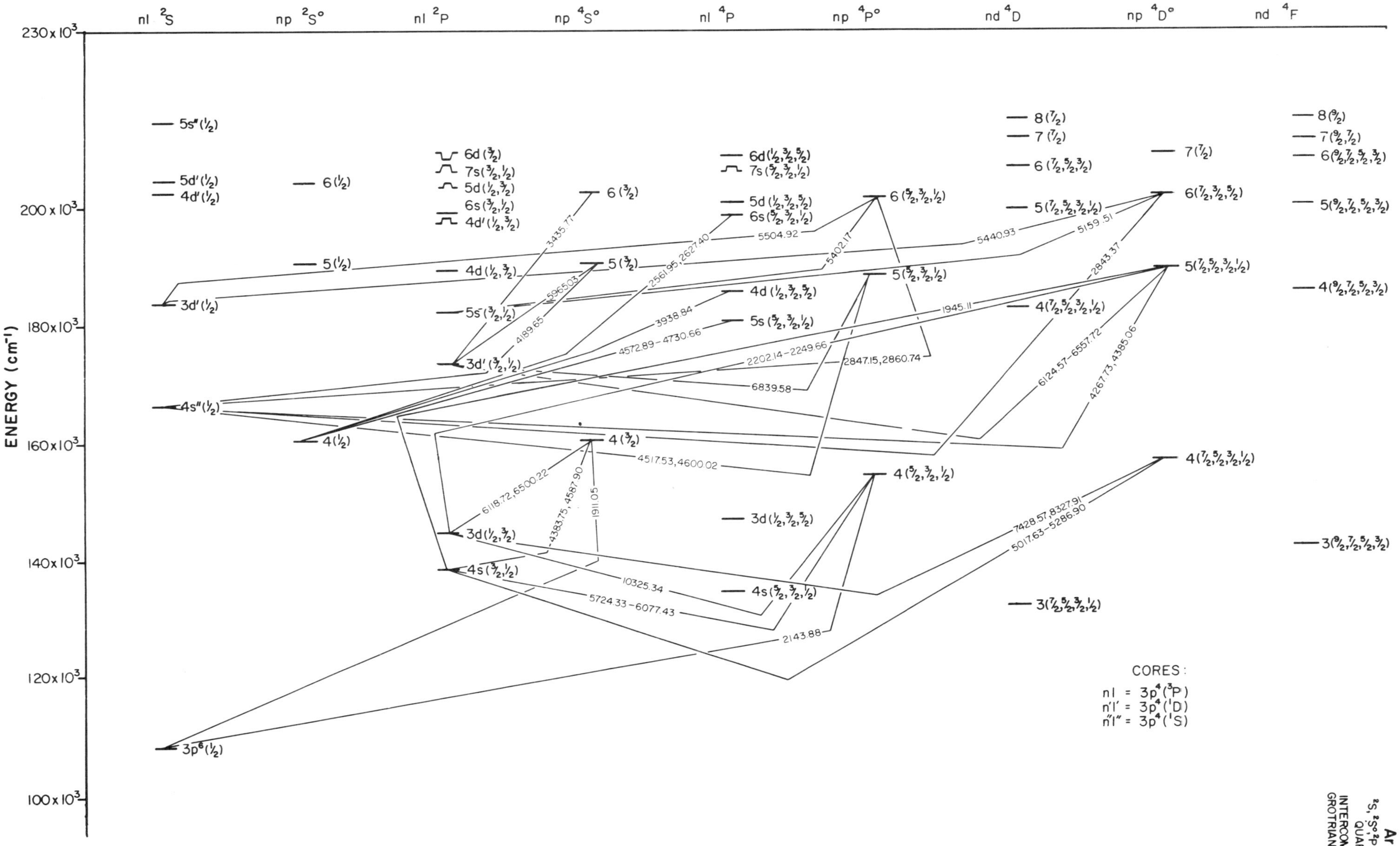

Ar II
^2S, $^2S^o$, ^2P terms to QUARTET INTERCOMBINATION GROTRIAN DIAGRAM

nl ^2S | np $^2S^o$ | nl ^2P | np $^4S^o$ | nl ^4P | np $^4P^o$ | nd ^4D | np $^4D^o$ | nd ^4F

ENERGY (cm^{-1})

230x10^3
200x10^3
180x10^3
160x10^3
140x10^3
120x10^3
100x10^3

nl ^2S: 5s″(1/2); 5d′(1/2); 4d′(1/2); 3d′(1/2); 4s″(1/2); $3p^6$(1/2)

np $^2S^o$: 6(1/2); 5(1/2); 4(1/2)

nl ^2P: 6d(3/2); 7s(3/2,1/2); 5d(1/2,3/2); 6s(3/2,1/2); 4d′(1/2,3/2); 4d(1/2,3/2); 5s(3/2,1/2); 3d′(3/2,1/2); 3d(1/2,3/2); 4s(3/2,1/2)

np $^4S^o$: 6(3/2); 5(3/2); 4(3/2)

nl ^4P: 6d(1/2,3/2,5/2); 7s(5/2,3/2,1/2); 5d(1/2,3/2,5/2); 6s(5/2,3/2,1/2); 4d(1/2,3/2,5/2); 5s(5/2,3/2,1/2); 3d(1/2,3/2,5/2); 4s(5/2,3/2,1/2)

np $^4P^o$: 6(5/2,3/2,1/2); 5(5/2,3/2,1/2); 4(5/2,3/2,1/2)

nd ^4D: 8(7/2); 7(7/2); 6(7/2,5/2,3/2); 5(7/2,5/2,3/2,1/2); 4(7/2,5/2,3/2,1/2); 3(7/2,5/2,3/2,1/2)

np $^4D^o$: 7(7/2); 6(7/2,3/2,5/2); 5(7/2,5/2,3/2,1/2); 4(7/2,5/2,3/2,1/2)

nd ^4F: 8(9/2); 7(9/2,7/2); 6(9/2,7/2,5/2,3/2); 5(9/2,7/2,5/2,3/2); 4(9/2,7/2,5/2,3/2); 3(9/2,7/2,5/2,3/2)

3435.77; 2561.95, 2627.40; 5504.92; 5440.93; 5159.51; 5402.17; 2843.37; 5965.03; 4189.65; 3938.84; 4572.89–4730.66; 1945.11; 2202.14–2249.66; 2847.15, 2860.74; 6839.58; 6124.57–6557.72; 4267.73, 4385.06; 4517.53, 4600.02; 6118.72, 6500.22; 4383.75, 4587.90; 1911.05; 10325.34; 5724.33–6077.43; 7428.57, 8327.91; 5017.63–5286.90; 2143.88

CORES:
nl = $3p^4(^3P)$
n′l′ = $3p^4(^1D)$
n″l″ = $3p^4(^1S)$

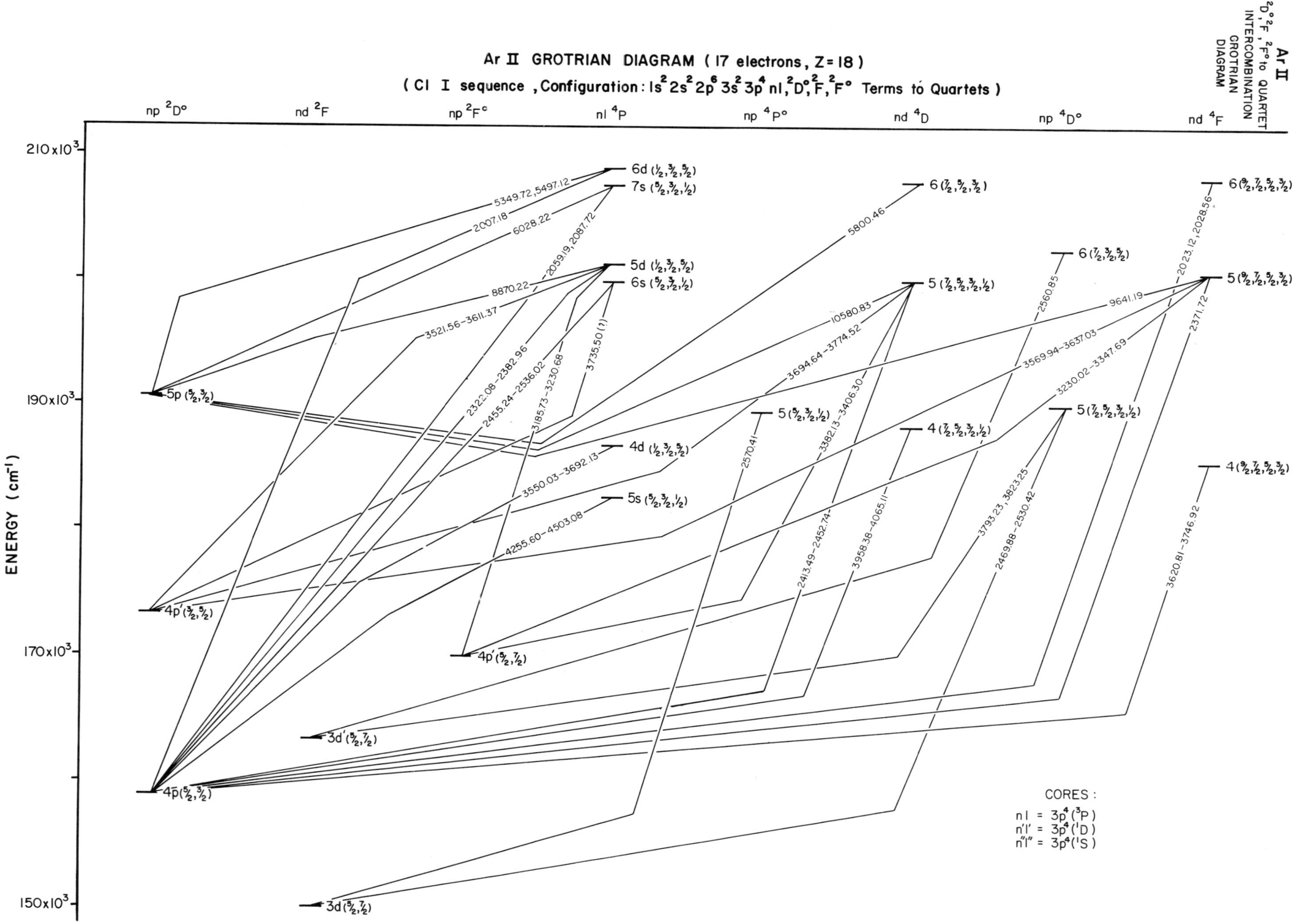
Ar II GROTRIAN DIAGRAM (17 electrons, Z = 18)
(Cl I sequence , Configuration : 1s^2 2s^2 2p^6 3s^2 3p^4 nl, ^{2}D°, ^{2}F, ^{2}F° Terms to Quartets)
Ar II
^{2}D°, ^{2}F , ^{2}F° to QUARTET INTERCOMBINATION GROTRIAN DIAGRAM
np ^{2}D°
nd ^{2}F
np ^{2}F°
nl ^{4}P
np ^{4}P°
nd ^{4}D
np ^{4}D°
nd ^{4}F
ENERGY (cm^{-1})
210x10^3
190x10^3
170x10^3
150x10^3
5349.72, 5497.12
2007.18
6028.22
2059.19, 2087.72
8870.22
3521.56-3611.37
2322.08-2382.96
2455.24-2536.02
3185.73-3230.68
3735.50 (?)
3550.03-3692.13
4255.60-4503.08
5800.46
10580.83
3694.64-3774.52
3382.13-3406.30
2413.49-2452.74
3958.38-4065.11
2570.41
2560.85
9641.19
3569.94-3637.03
3230.02-3347.69
3793.23, 3823.25
2469.88-2530.42
2023.12, 2028.56
2371.72
3620.81-3746.92
CORES :
nl = 3p^4(^{3}P)
n'l' = 3p^4(^{1}D)
n''l'' = 3p^4(^{1}S)

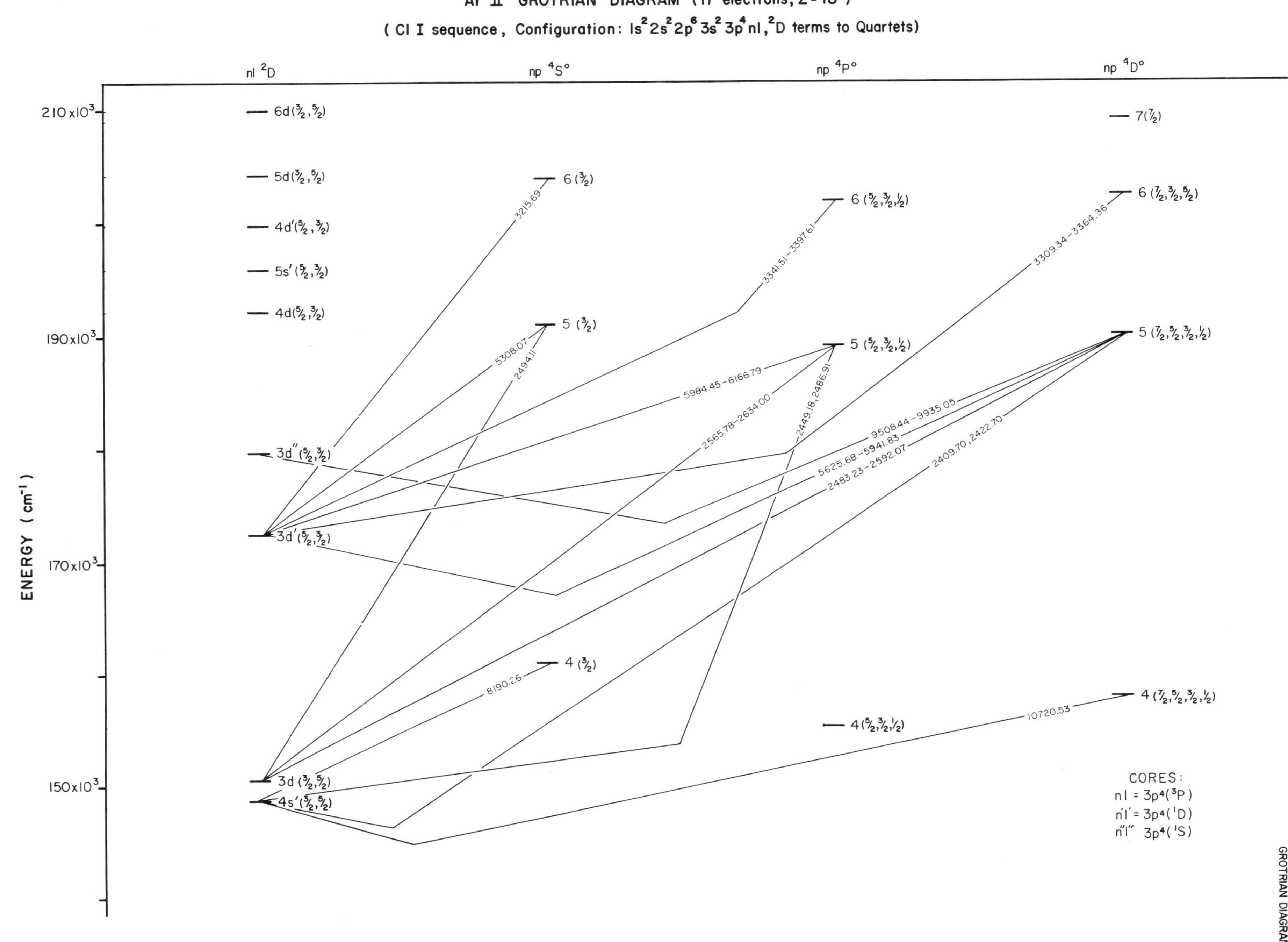
Ar II GROTRIAN DIAGRAM (17 electrons, Z = 18)
(Cl I sequence, Configuration: 1s²2s²2p⁶3s²3p⁴nl, ²D terms to Quartets)
nl ²D
np ⁴S°
np ⁴P°
np ⁴D°
6d(3/2, 5/2)
5d(3/2, 5/2)
4d'(5/2, 3/2)
5s'(5/2, 3/2)
4d(5/2, 3/2)
3d''(5/2, 3/2)
3d'(5/2, 3/2)
3d (3/2, 5/2)
4s'(3/2, 5/2)
6 (3/2)
5 (3/2)
4 (3/2)
6 (5/2, 3/2, 1/2)
5 (5/2, 3/2, 1/2)
4 (5/2, 3/2, 1/2)
7 (7/2)
6 (7/2, 3/2, 5/2)
5 (7/2, 5/2, 3/2, 1/2)
4 (7/2, 5/2, 3/2, 1/2)
CORES:
nl = 3p⁴(³P)
n'l' = 3p⁴(¹D)
n''l'' 3p⁴(¹S)
3215.69
5308.07
2494.11
8190.26
3341.51-3397.61
5984.45-6166.79
2565.78-2634.00
2449.18, 2486.91
3309.34-3364.36
9508.44-9935.05
5625.68-5941.83
2483.23-2592.07
2409.70, 2422.70
10720.53
210×10³
190×10³
170×10³
150×10³
ENERGY (cm⁻¹)

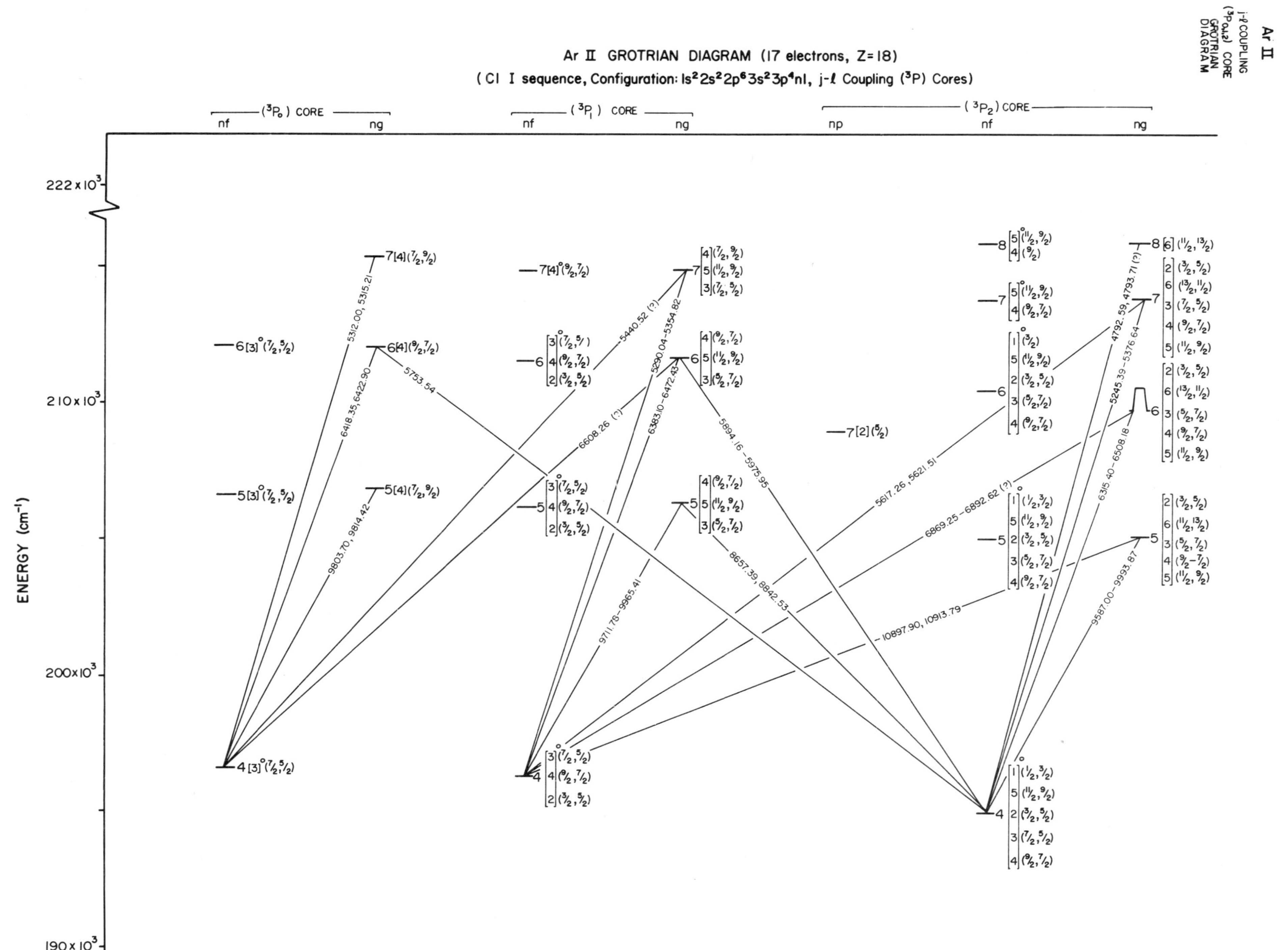

Ar II
j-ℓ COUPLING
(³P₀,₁,₂) CORE
GROTRIAN
DIAGRAM
Ar II GROTRIAN DIAGRAM (17 electrons, Z = 18)
(Cl I sequence, Configuration: 1s²2s²2p⁶3s²3p⁴nl, j-ℓ Coupling (³P) Cores)
(³P₀) CORE
nf
ng
(³P₁) CORE
nf
ng
(³P₂) CORE
np
nf
ng
ENERGY (cm⁻¹)
222 x 10³
210 x 10³
200 x 10³
190 x 10³
5312.00, 5315.21
6418.35, 6422.90
9803.70, 9814.42
5753.54
5440.52 (?)
6608.26 (?)
5290.04 – 5354.82
6383.10 – 6472.43
9711.78 – 9965.41
5894.16 – 5975.95
8657.39, 8842.53
5617.26, 5621.51
6869.25 – 6892.62 (?)
10897.90, 10913.79
4792.59, 4793.71 (?)
5245.39 – 5376.64
6315.40 – 6508.18
9587.00 – 9993.87

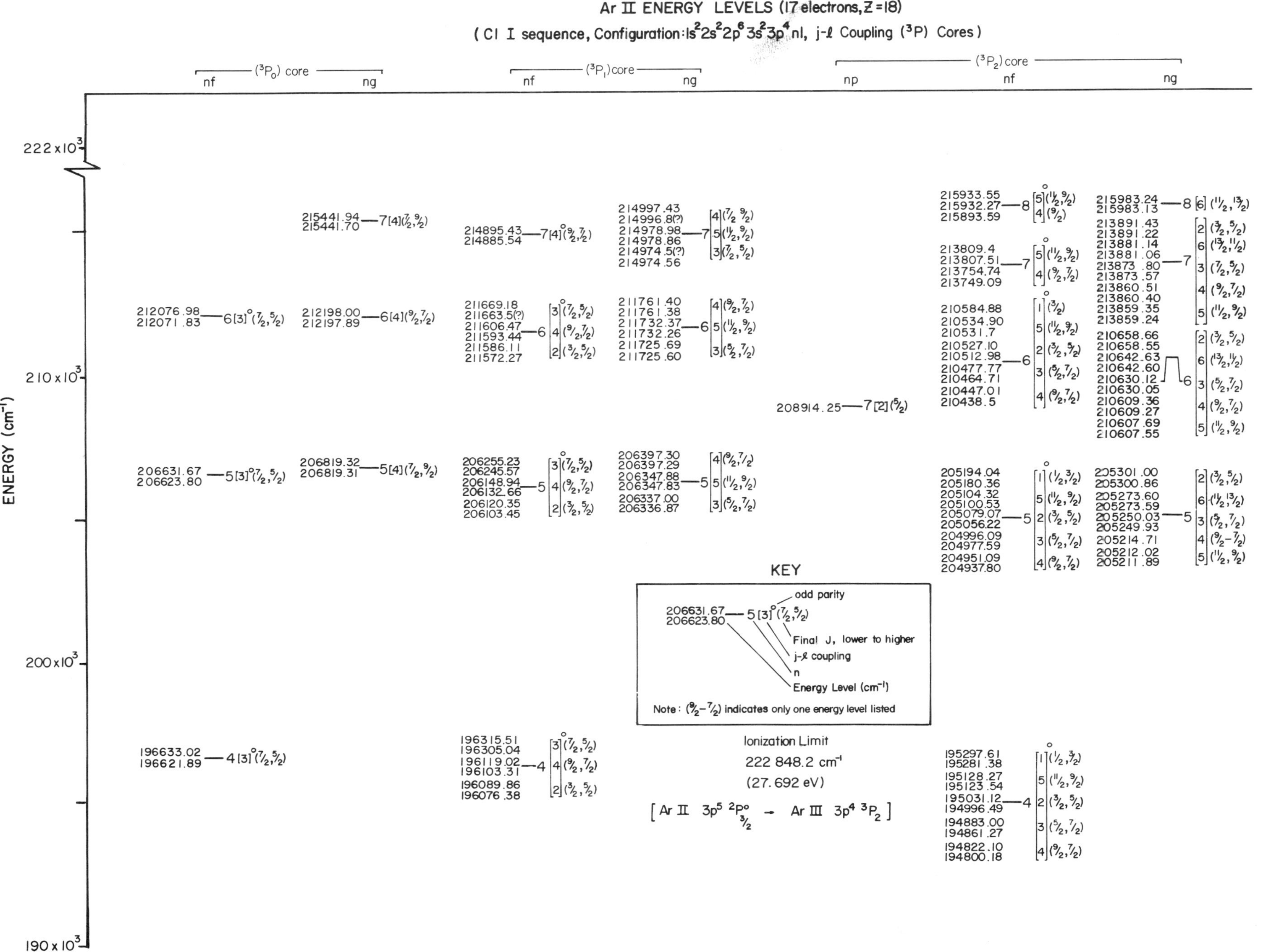
Ar II ENERGY LEVELS (17 electrons, Z = 18)
(Cl I sequence, Configuration: 1s^{2}2s^{2}2p^{6}3s^{2}3p^4 nl, j-ℓ Coupling (^{3}P) Cores)
(^{3}P$_0$) core
(^{3}P$_1$) core
(^{3}P$_2$) core
nf
ng
np
ENERGY (cm^{-1})
222 x 10^3
210 x 10^3
200 x 10^3
190 x 10^3
KEY
206631.67
206623.80
5 [3]° (7/2, 5/2)
odd parity
Final J, lower to higher
j-ℓ coupling
n
Energy Level (cm^{-1})
Note : (9/2 – 7/2) indicates only one energy level listed
Ionization Limit
222 848.2 cm^{-1}
(27.692 eV)
[Ar II 3p^5 ^{2}P°$_{3/2}$ → Ar III 3p^4 ^{3}P$_2$]

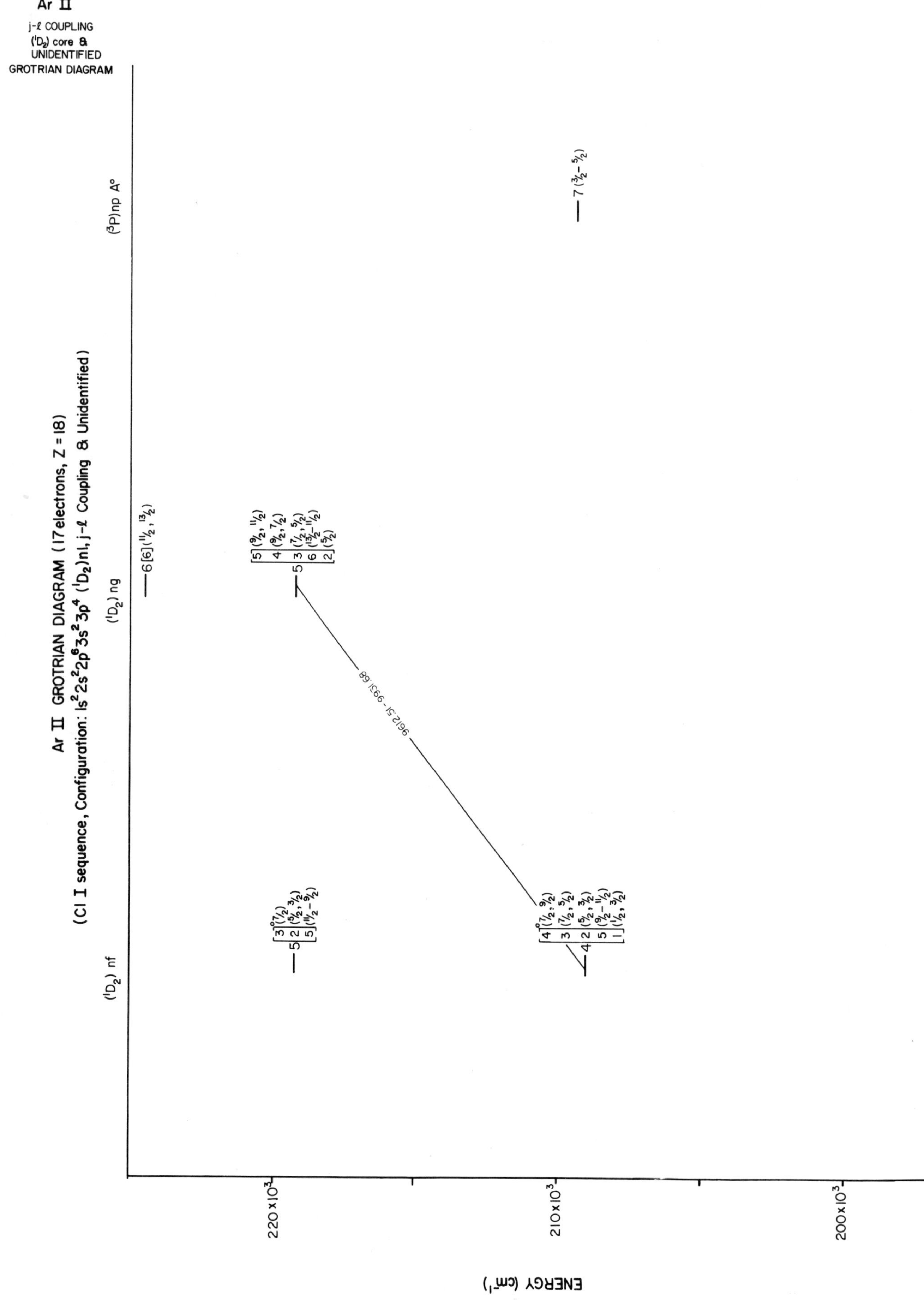
Ar II
j-ℓ COUPLING
($^1D_2$) core &
UNIDENTIFIED
GROTRIAN DIAGRAM
Ar II GROTRIAN DIAGRAM (17 electrons, Z = 18)
(Cl I sequence, Configuration: $1s^2 2s^2 2p^6 3s^2 3p^4$ ($^1D_2$)nl, j-ℓ Coupling & Unidentified)
($^1D_2$) nf
($^1D_2$) ng
(^3P)np A°
ENERGY (cm^{-1})
220×10^3
210×10^3
200×10^3

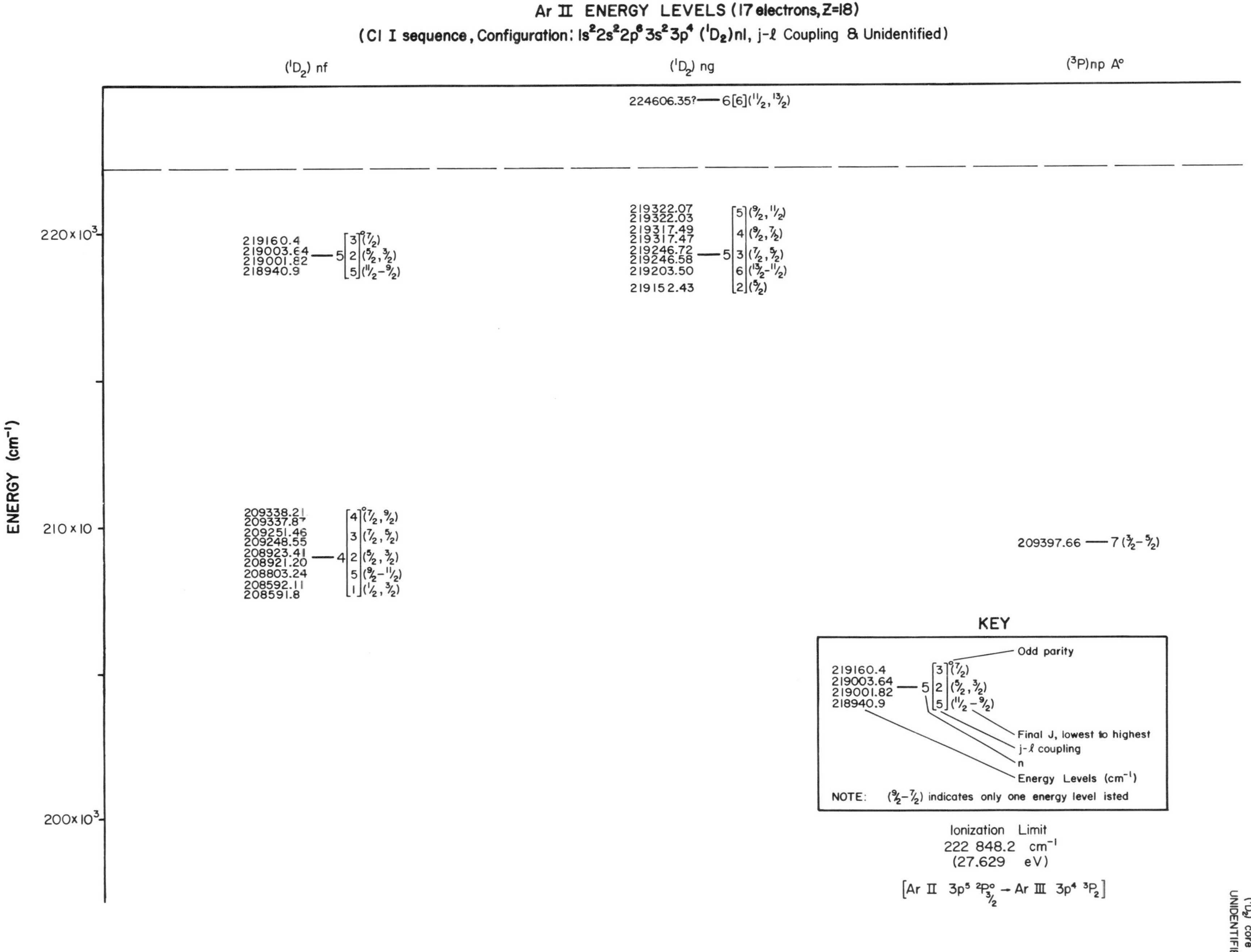
Ar II ENERGY LEVELS (17 electrons, Z=18)
(Cl I sequence, Configuration: 1s²2s²2p⁶3s²3p⁴ (¹D₂)nl, j-ℓ Coupling & Unidentified)
(¹D₂) nf
(¹D₂) ng
(³P)np A°
224606.35? — 6[6](11/2, 13/2)
219160.4
219003.64
219001.82
218940.9
— 5 [3°(7/2) 2 (5/2, 3/2) 5 (11/2 – 9/2)]
219322.07
219322.03
219317.49
219317.47
219246.72
219246.58
219203.50
219152.43
— 5 [5 (9/2, 11/2) 4 (9/2, 7/2) 3 (7/2, 5/2) 6 (13/2 – 11/2) 2 (5/2)]
209338.21
209337.87
209251.46
209248.55
208923.41
208921.20
208803.24
208592.11
208591.8
— 4 [4°(7/2, 9/2) 3 (7/2, 5/2) 2 (5/2, 3/2) 5 (9/2 – 11/2) 1 (1/2, 3/2)]
209397.66 — 7 (3/2 – 5/2)
KEY
219160.4
219003.64
219001.82
218940.9
— 5 [3°(7/2) 2 (5/2, 3/2) 5 (11/2 – 9/2)]
Odd parity
Final J, lowest to highest
j-ℓ coupling
n
Energy Levels (cm⁻¹)
NOTE: (9/2 – 7/2) indicates only one energy level isted
Ionization Limit
222 848.2 cm⁻¹
(27.629 eV)
[Ar II 3p⁵ ²P°3/2 → Ar III 3p⁴ ³P₂]
ENERGY (cm⁻¹)
220×10³
210×10
200×10³

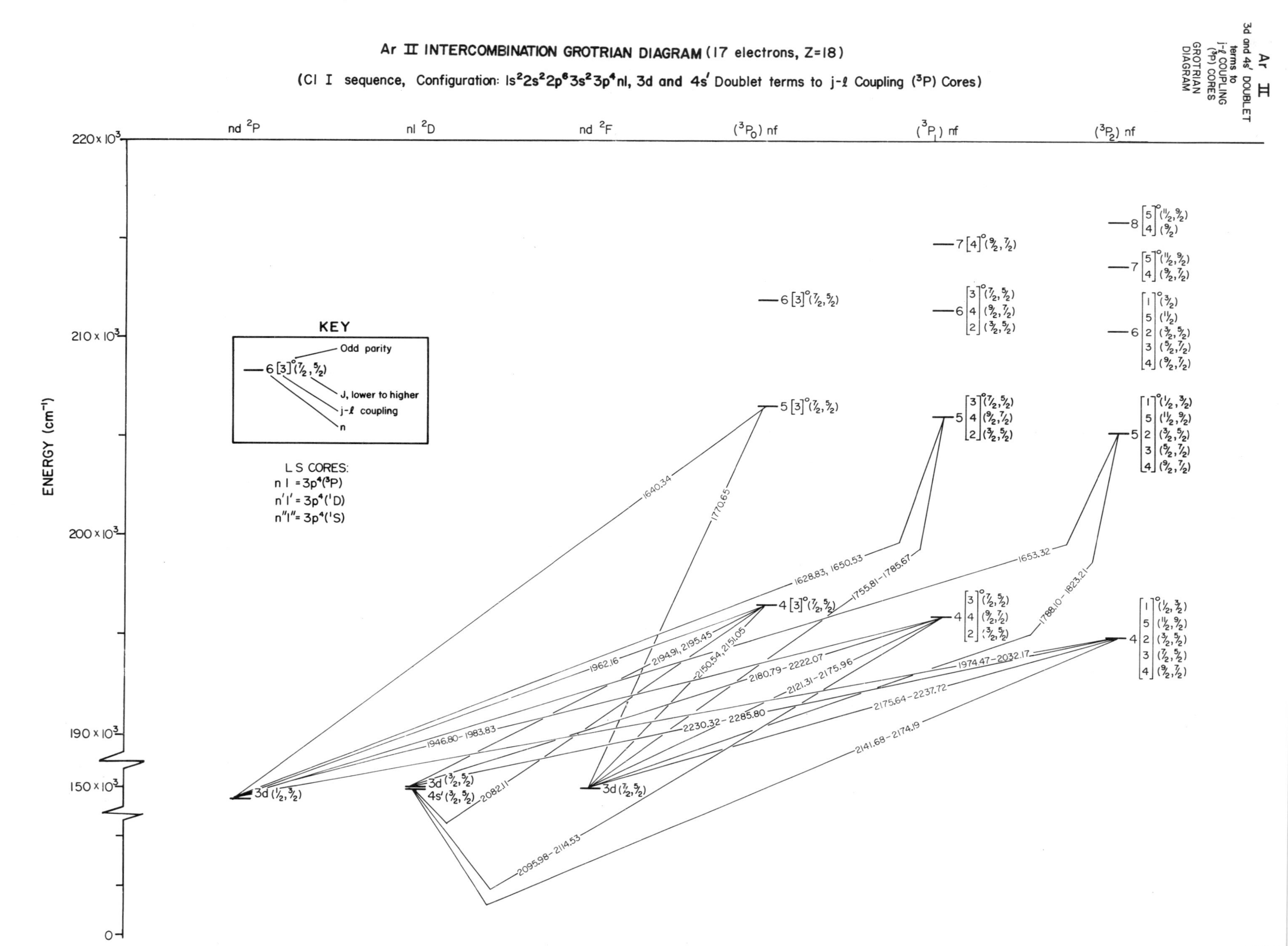

Ar II INTERCOMBINATION GROTRIAN DIAGRAM (17 electrons, Z=18)
(Cl I sequence, Configuration: 1s²2s²2p⁶3s²3p⁴nl, 3d and 4s' Doublet terms to j-ℓ Coupling (³P) Cores)
Ar II
3d and 4s' DOUBLET
terms to
j-ℓ COUPLING
(³P) CORES
GROTRIAN
DIAGRAM
nd ²P
nl ²D
nd ²F
(³P₀) nf
(³P₁) nf
(³P₂) nf
ENERGY (cm⁻¹)
220 x 10³
210 x 10³
200 x 10³
190 x 10³
150 x 10³
0
KEY
Odd parity
J, lower to higher
j-ℓ coupling
n
L S CORES:
n l = 3p⁴(³P)
n'l' = 3p⁴(¹D)
n''l'' = 3p⁴(¹S)
1640.34
1770.65
1628.83, 1650.53
1755.81–1785.67
1653.32
1788.10–1823.21
1962.16
2194.91, 2195.45
2150.54, 2151.05
2180.79–2222.07
2121.31–2175.96
1974.47–2032.17
1946.80–1983.83
2230.32–2285.80
2175.64–2237.72
2141.68–2174.19
2082.11
2095.98–2114.53

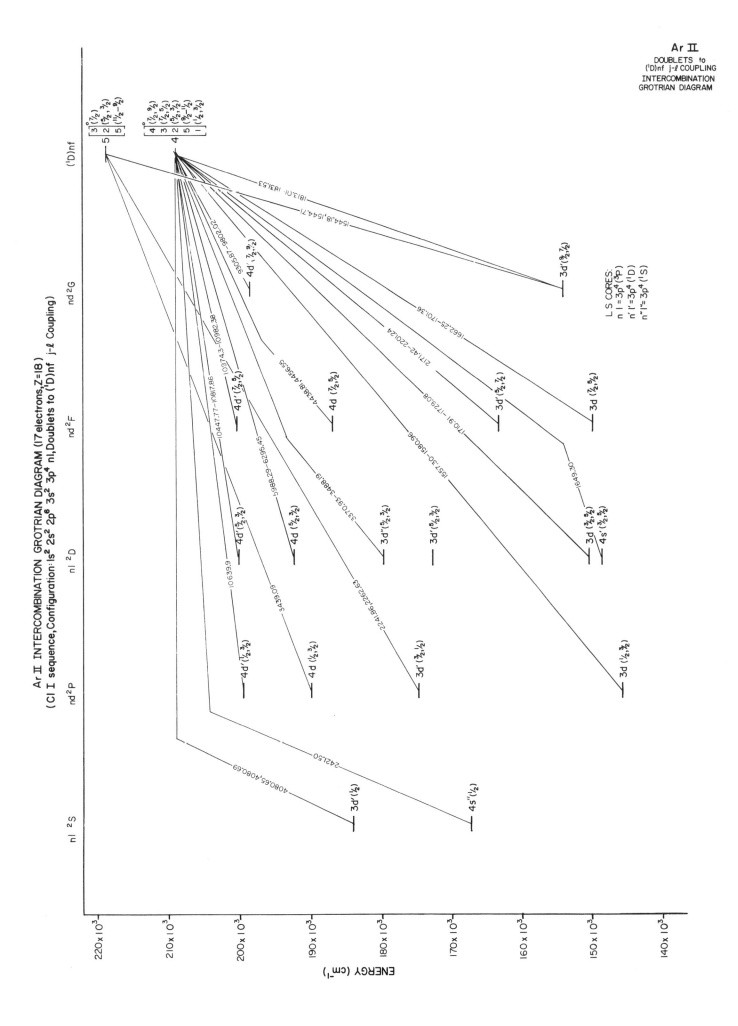

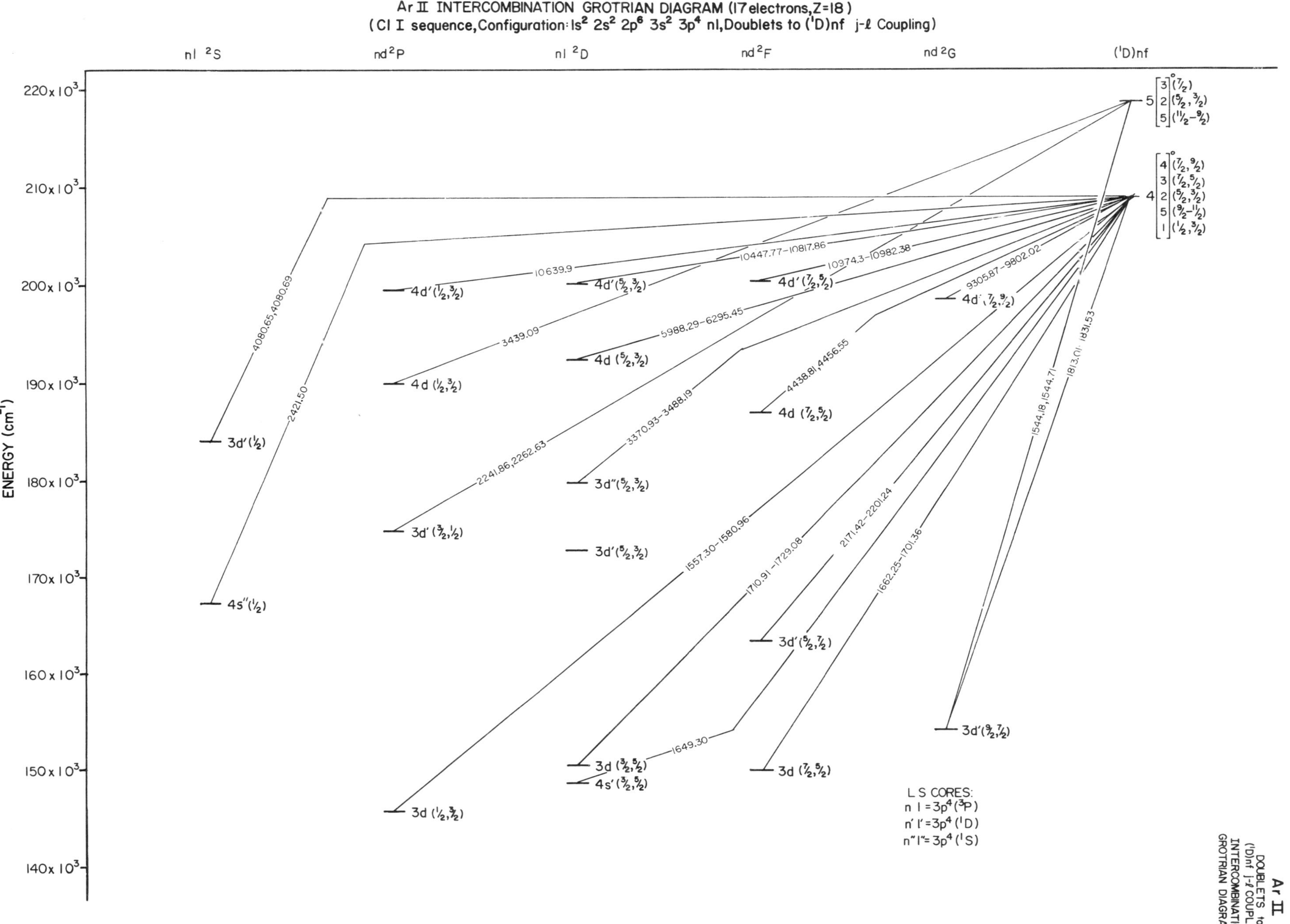
Ar II INTERCOMBINATION GROTRIAN DIAGRAM (17 electrons, Z=18)
(Cl I sequence, Configuration: 1s² 2s² 2p⁶ 3s² 3p⁴ nl, Doublets to (¹D)nf j-ℓ Coupling)
nl ²S
nd²P
nl ²D
nd²F
nd²G
(¹D)nf
ENERGY (cm⁻¹)
220x10³
210x10³
200x10³
190x10³
180x10³
170x10³
160x10³
150x10³
140x10³
5 [3](7/2) 2(5/2, 3/2) 5(11/2–9/2)
4 [4](7/2, 9/2) 3(7/2, 5/2) 2(5/2, 3/2) 5(9/2–11/2) 1(1/2, 3/2)
3d′(1/2)
4s″(1/2)
4d′(1/2, 3/2)
4d (1/2, 3/2)
3d′(3/2, 1/2)
3d (1/2, 3/2)
4d′(5/2, 3/2)
4d (5/2, 3/2)
3d″(5/2, 3/2)
3d′(5/2, 3/2)
3d (3/2, 5/2)
4s′(3/2, 5/2)
4d′(7/2, 5/2)
4d (7/2, 5/2)
3d′(5/2, 7/2)
3d (7/2, 5/2)
4d′(7/2, 9/2)
3d′(9/2, 7/2)
4080.65, 4080.69
2421.50
10639.9
3439.09
2241.86, 2262.63
10447.77–10817.86
5988.29–6295.45
3370.93–3488.19
1557.30–1580.96
1710.91–1729.08
1649.30
10974.3–10982.38
4438.81, 4456.55
2171.42–2201.24
1662.25–1701.36
9305.87–9802.02
1544.18, 1544.71
1813.01–1831.53
L S CORES:
n l = 3p⁴(³P)
n′ l′ = 3p⁴(¹D)
n″ l″ = 3p⁴(¹S)
Ar II
DOUBLETS to
(¹D)nf j-ℓ COUPLING
INTERCOMBINATION
GROTRIAN DIAGRAM

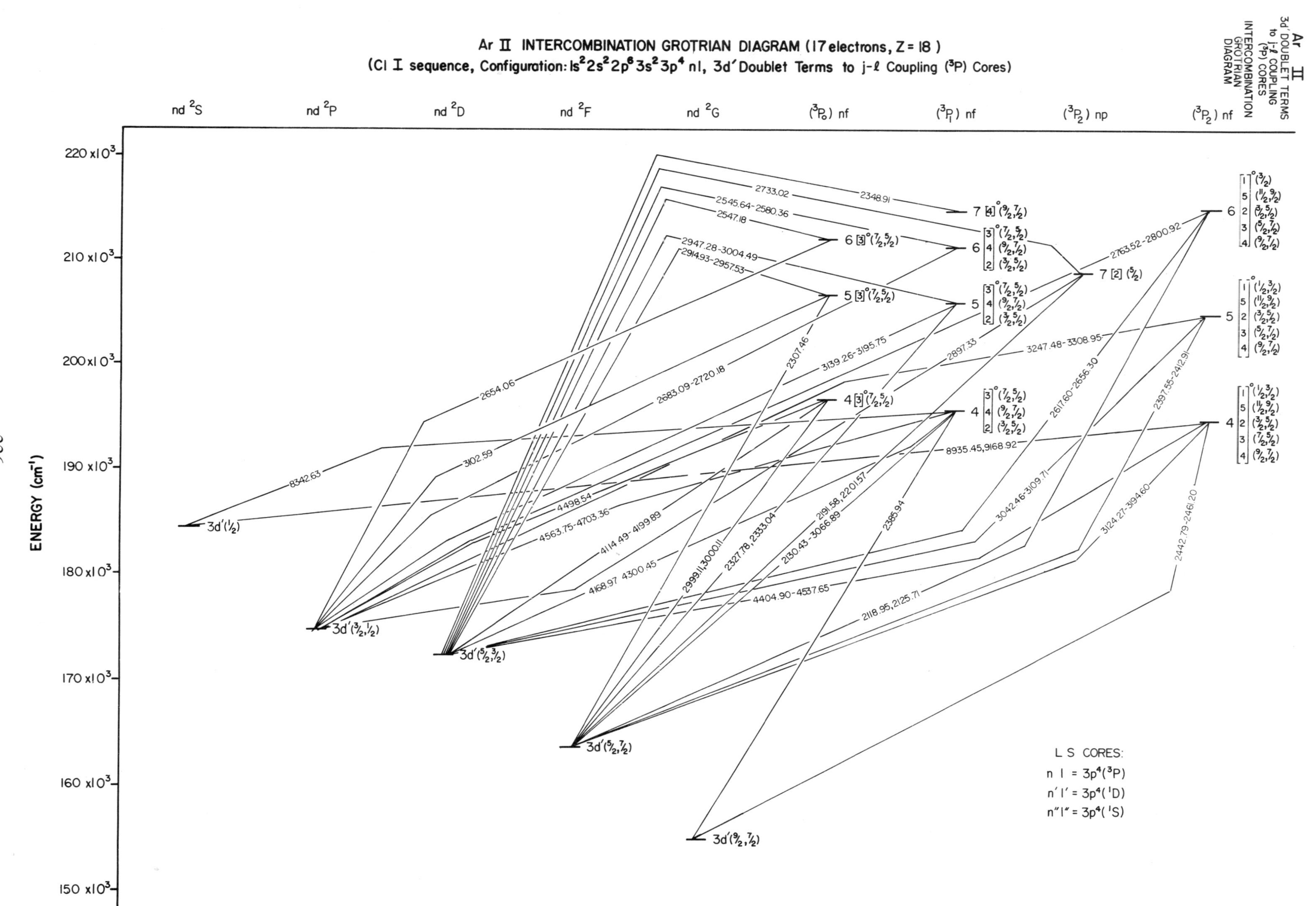

Ar II INTERCOMBINATION GROTRIAN DIAGRAM (17 electrons, Z = 18)
(Cl I sequence, Configuration: 1s²2s²2p⁶3s²3p⁴ nl, 3d′ Doublet Terms to j-ℓ Coupling (³P) Cores)
Ar II
3d′ DOUBLET TERMS
to j-ℓ COUPLING
(³P) CORES
INTERCOMBINATION
GROTRIAN
DIAGRAM
nd ²S
nd ²P
nd ²D
nd ²F
nd ²G
(³P₀) nf
(³P₁) nf
(³P₂) np
(³P₂) nf
ENERGY (cm⁻¹)
220 x10³
210 x10³
200 x10³
190 x10³
180 x10³
170 x10³
160 x10³
150 x10³
3d′(1/2)
3d′(3/2,1/2)
3d′(5/2,3/2)
3d′(5/2,7/2)
3d′(9/2,7/2)
2733.02
2348.91
2545.64-2580.36
2547.18
2947.28-3004.49
2914.93-2957.53
2654.06
2683.09-2720.18
3139.26-3195.75
2307.46
2897.33
3247.48-3308.95
2763.52-2800.92
2617.60-2656.30
2397.55-2412.91
8935.45,9168.92
8342.63
3102.59
4498.54
4563.75-4703.36
4114.49-4199.89
4168.97 4300.45
2999.11,3000.11
2327.78,2333.04
2130.43-3066.89
2191.58,2201.57
2385.94
3042.46-3109.71
3124.27-3194.60
2442.79-2461.20
4404.90-4537.65
2118.95,2125.71
L S CORES:
n l = 3p⁴(³P)
n′ l′ = 3p⁴(¹D)
n″ l″ = 3p⁴(¹S)

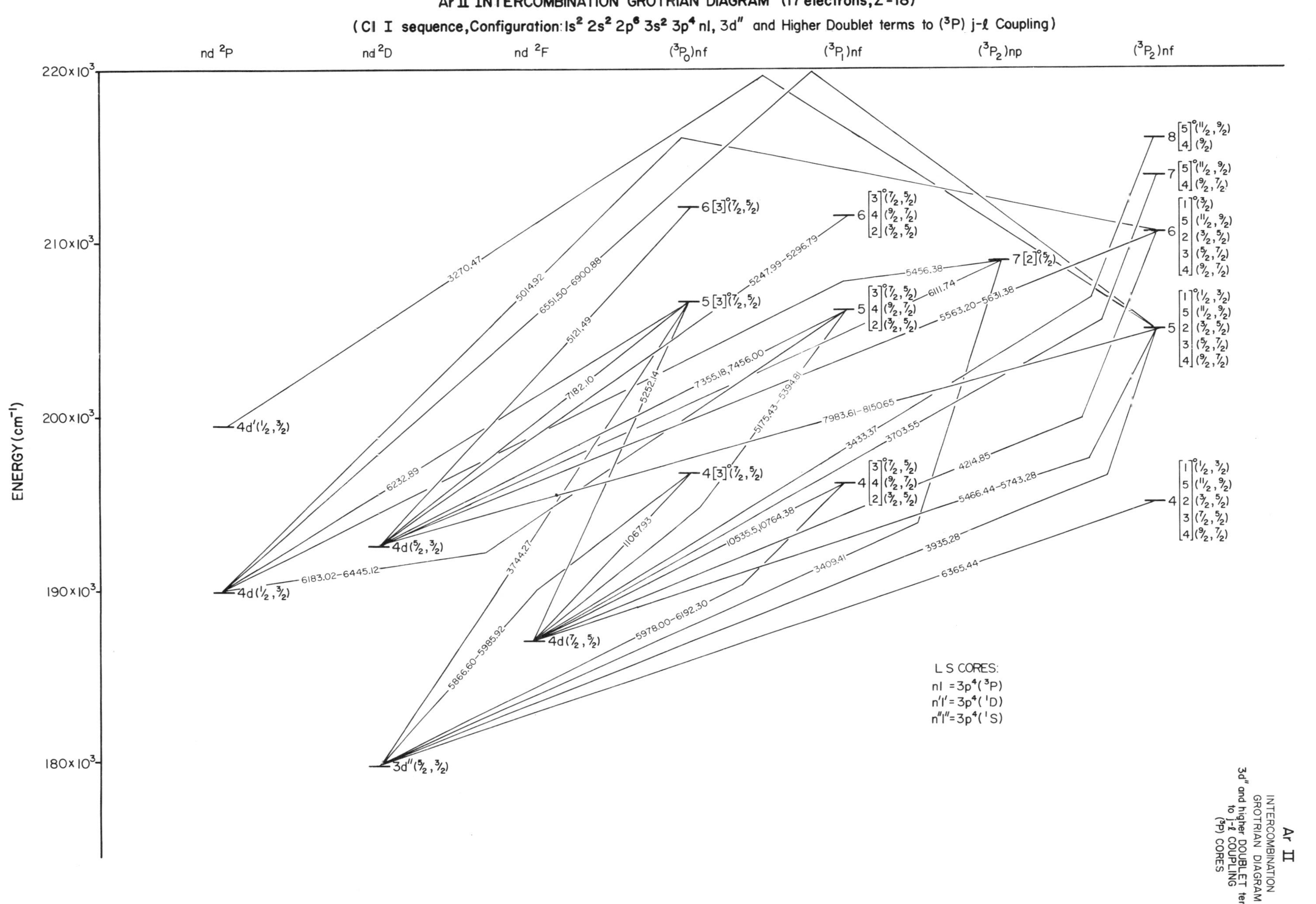

Ar II INTERCOMBINATION GROTRIAN DIAGRAM (17 electrons, Z=18)
(Cl I sequence, Configuration: 1s² 2s² 2p⁶ 3s² 3p⁴ nl, 3d″ and Higher Doublet terms to (³P) j-ℓ Coupling)
nd ²P
nd ²D
nd ²F
(³P₀)nf
(³P₁)nf
(³P₂)np
(³P₂)nf
ENERGY (cm⁻¹)
220×10³
210×10³
200×10³
190×10³
180×10³
4d′(1/2, 3/2)
4d(1/2, 3/2)
4d(5/2, 3/2)
4d(7/2, 5/2)
3d″(5/2, 3/2)
6[3]°(7/2, 5/2)
5[3]°(7/2, 5/2)
4[3]°(7/2, 5/2)
7[2]°(5/2)
3270.47
5014.92
6551.50–6900.88
5121.49
7182.10
6232.89
6183.02–6445.12
3744.27
5866.60–5985.92
5252.14
1106.93
5247.99–5296.79
5456.38
6111.74
5563.20–5631.38
7355.18, 7456.00
5175.43–5394.81
7983.61–8150.65
3433.37
3703.55
4214.85
5466.44–5743.28
10535.5, 10764.38
3409.41
3935.28
5978.00–6192.30
6365.44
L S CORES:
nl = 3p⁴(³P)
n′l′ = 3p⁴(¹D)
n″l″ = 3p⁴(¹S)
Ar II
INTERCOMBINATION
GROTRIAN DIAGRAM
3d″ and higher DOUBLET terms
to j-ℓ COUPLING
(³P) CORES

Ar II INTERCOMBINATION GROTRIAN DIAGRAM (17 electrons, Z = 18)

(Cl I sequence, Configuration: $1s^2 2s^2 2p^6 3s^2 3p^4 nl$, 3d Quartets to (^{3}P) j-ℓ Coupling)

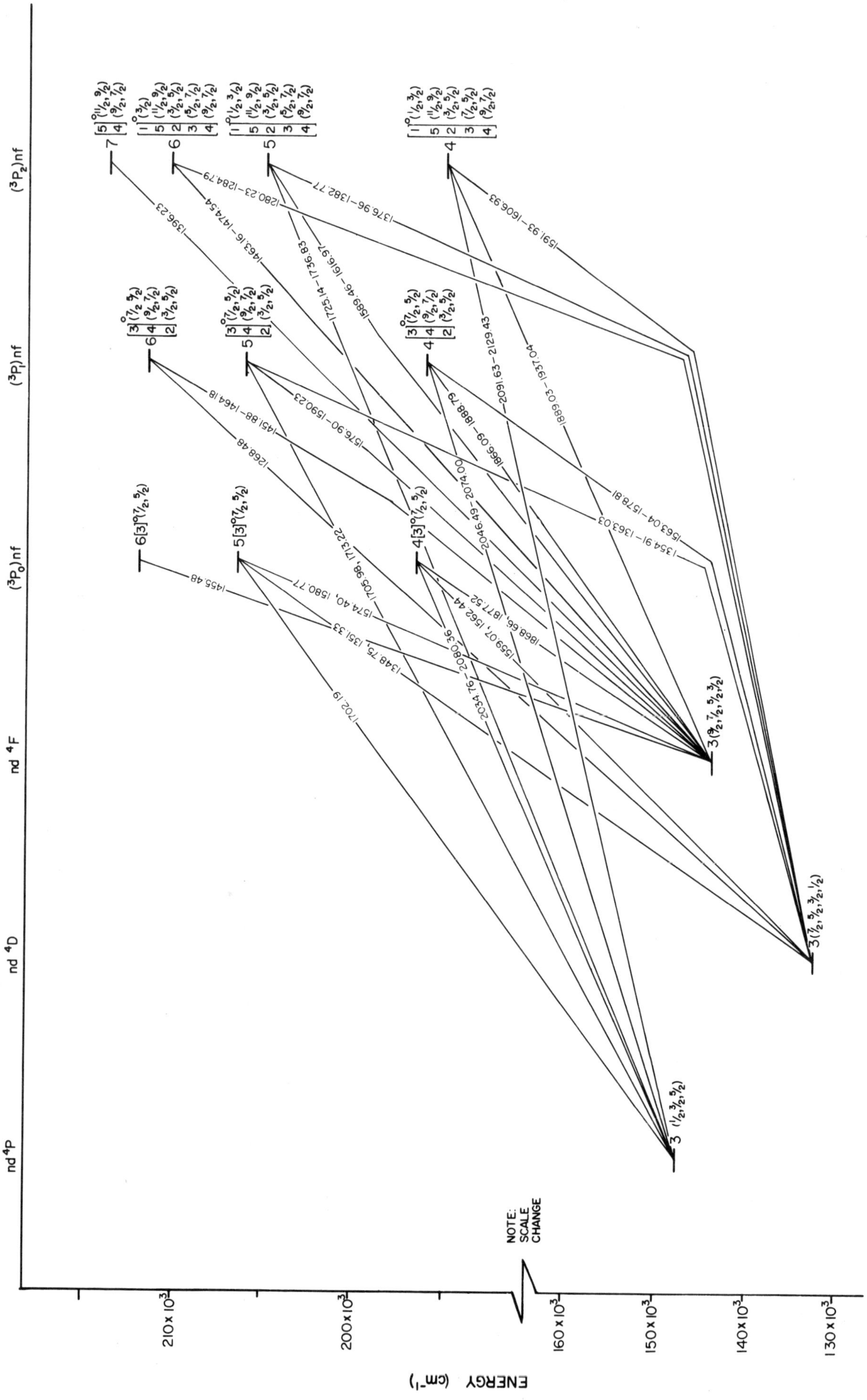

nd ^{4}P
nd ^{4}D
nd ^{4}F
($^3P_0$)nf
($^3P_1$)nf
($^3P_2$)nf

ENERGY (cm^{-1})
210 x 10^3
200 x 10^3
160 x 10^3
150 x 10^3
140 x 10^3
130 x 10^3

NOTE: SCALE CHANGE

3 (1/2, 3/2, 5/2)
3 (7/2, 5/2, 3/2, 1/2)
3 (9/2, 7/2, 5/2, 3/2)
6[3]°(7/2, 5/2)
5[3]°(7/2, 5/2)
4[3]°(7/2, 5/2)
6 [3]° (7/2, 5/2) [4] (9/2, 7/2) [2] (3/2, 5/2)
5 [3]° (7/2, 5/2) [4] (9/2, 7/2) [2] (3/2, 5/2)
4 [3]° (7/2, 5/2) [4] (9/2, 7/2) [2] (3/2, 5/2)
7 [5]° (11/2, 9/2) [4] (9/2, 7/2)
6 [1]° (3/2) [5] (11/2, 9/2) [2] (3/2, 5/2) [3] (5/2, 7/2) [4] (9/2, 7/2)
5 [1]° (1/2, 3/2) [5] (11/2, 9/2) [2] (3/2, 5/2) [3] (5/2, 7/2) [4] (9/2, 7/2)
4 [1]° (1/2, 3/2) [5] (11/2, 9/2) [2] (3/2, 5/2) [3] (7/2, 5/2) [4] (9/2, 7/2)

1455.48
1702.19
1348.75, 1351.33
1574.40, 1580.77
1705.98, 1713.22
1268.48
1451.88–1464.18
1576.90–1590.23
1396.23
1463.16–1474.54
1280.23–1284.79
1725.14–1736.83
1589.46–1616.97
1376.96–1382.77
2034.76–2080.36
1559.07, 1562.44
1868.66, 1877.52
2046.49–2074.00
1866.09–1888.79
2091.63–2129.43
1889.03–1937.04
1591.93–1606.93
1354.91–1363.03
1563.04–1578.81

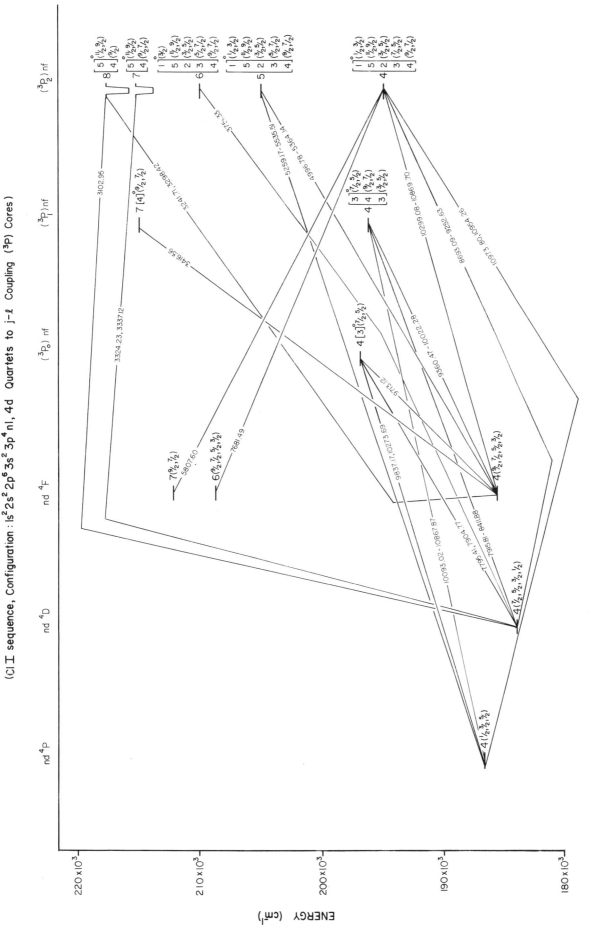

Ar II INTERCOMBINATION GROTRIAN DIAGRAM (17 electrons, Z = 18)

(Cl I sequence, Configuration: $1s^2 2s^2 2p^6 3s^2 3p^4 nl$, 4d Quartets to j-ℓ Coupling (^{3}P) Cores)

nd ^{4}P
nd ^{4}D
nd ^{4}F
($^3P_0$) nf
($^3P_1$) nf
($^3P_2$) nf

ENERGY (cm^{-1})
220×10^3
210×10^3
200×10^3
190×10^3
180×10^3

$4(1/2, 3/2, 5/2)$
$4(7/2, 5/2, 3/2, 1/2)$
$4(9/2, 7/2, 5/2, 3/2)$
$7(9/2, 7/2)$
$6(9/2, 7/2, 5/2, 3/2)$
$4[3]^\circ(7/2, 5/2)$
$7[4]^\circ(9/2, 7/2)$
$4\ [3]^\circ(7/2, 5/2)\ [4](9/2, 7/2)\ [3](3/2, 5/2)$
$8\ [5]^\circ(11/2, 9/2)\ [4](9/2)$
$7\ [5]^\circ(11/2, 9/2)\ [4](9/2, 7/2)$
$6\ [1]^\circ(3/2)\ [5](11/2, 9/2)\ [2](3/2, 5/2)\ [3](5/2, 7/2)\ [4](9/2, 7/2)$
$5\ [1]^\circ(1/2, 3/2)\ [5](11/2, 9/2)\ [2](3/2, 5/2)\ [3](5/2, 7/2)\ [4](9/2, 7/2)$
$4\ [1]^\circ(1/2, 3/2)\ [5](11/2, 9/2)\ [2](3/2, 5/2)\ [3](7/2, 5/2)\ [4](9/2, 7/2)$

3102.95
3324.23, 3337.12
3416.56
3241.71, 3298.42
5807.60
7681.49
3751.33
5259.17 - 5535.51
4996.78 - 5364.14
9837.17, 10273.69
9713.12
10093.02 - 10867.87
7795.41, 7904.77
7915.81 - 8411.88
9360.47 - 10022.28
10299.08 - 10869.70
8693.09 - 9252.63
10973.80, 10954.26

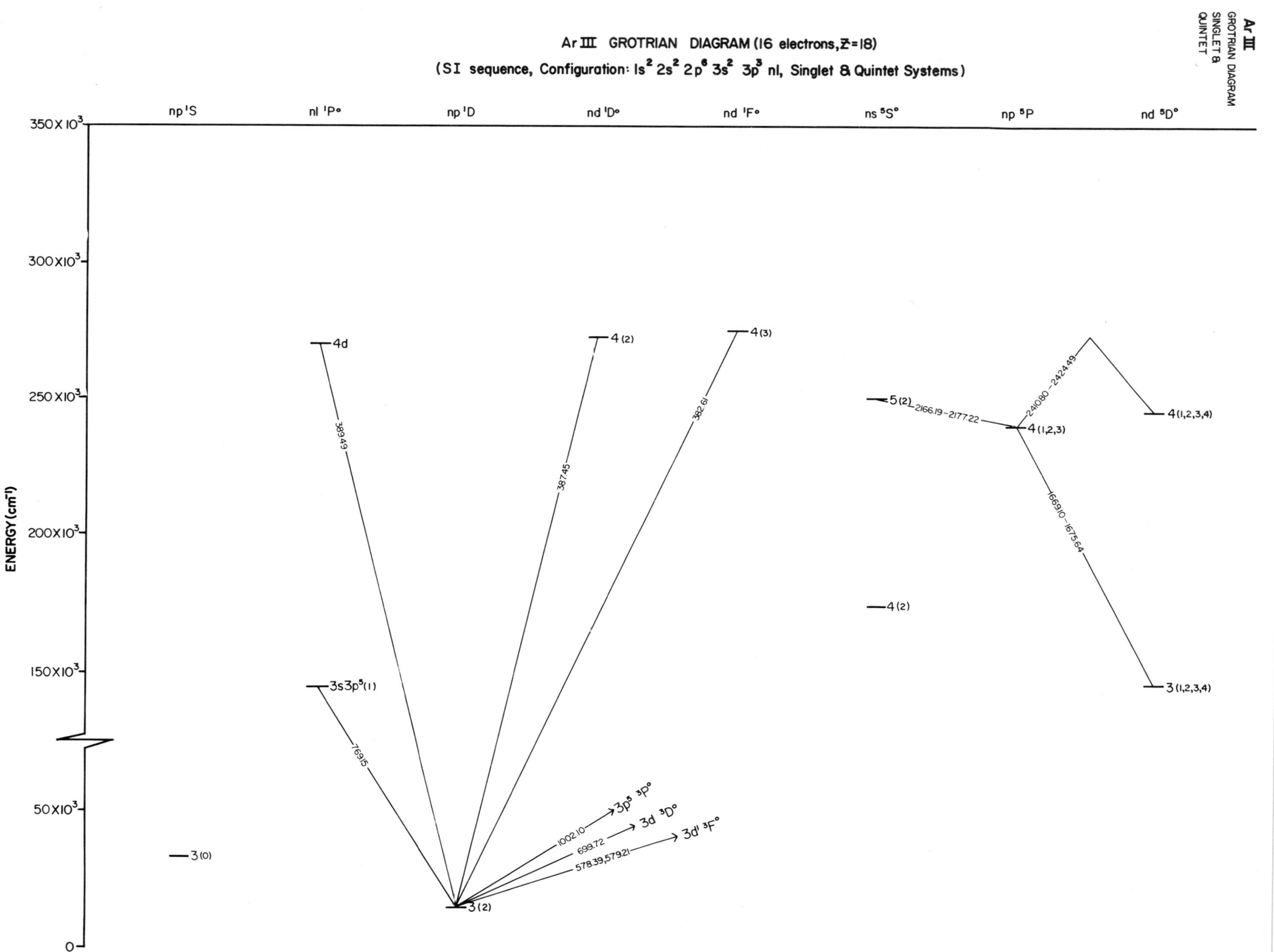

Ar III
GROTRIAN DIAGRAM
SINGLET & QUINTET
Ar III GROTRIAN DIAGRAM (16 electrons, Z=18)
(SI sequence, Configuration: 1s2 2s2 2p6 3s2 3p3 nl, Singlet & Quintet Systems)
np 1S
nl 1P°
np 1D
nd 1D°
nd 1F°
ns 5S°
np 5P
nd 5D°
ENERGY (cm-1)
350 X 10^3
300 X 10^3
250 X 10^3
200 X 10^3
150 X 10^3
50 X 10^3
0
3 (0)
4d
389.49
3s3p5 (1)
769.15
4 (2)
387.45
4 (3)
382.61
3 (2)
1002.10
3p5 3P°
699.72
3d 3D°
578.39,579.21
3d' 3F°
5 (2)
2166.19–2177.22
4 (2)
4 (1,2,3)
2410.80–2424.49
1669.10–1675.64
4 (1,2,3,4)
3 (1,2,3,4)

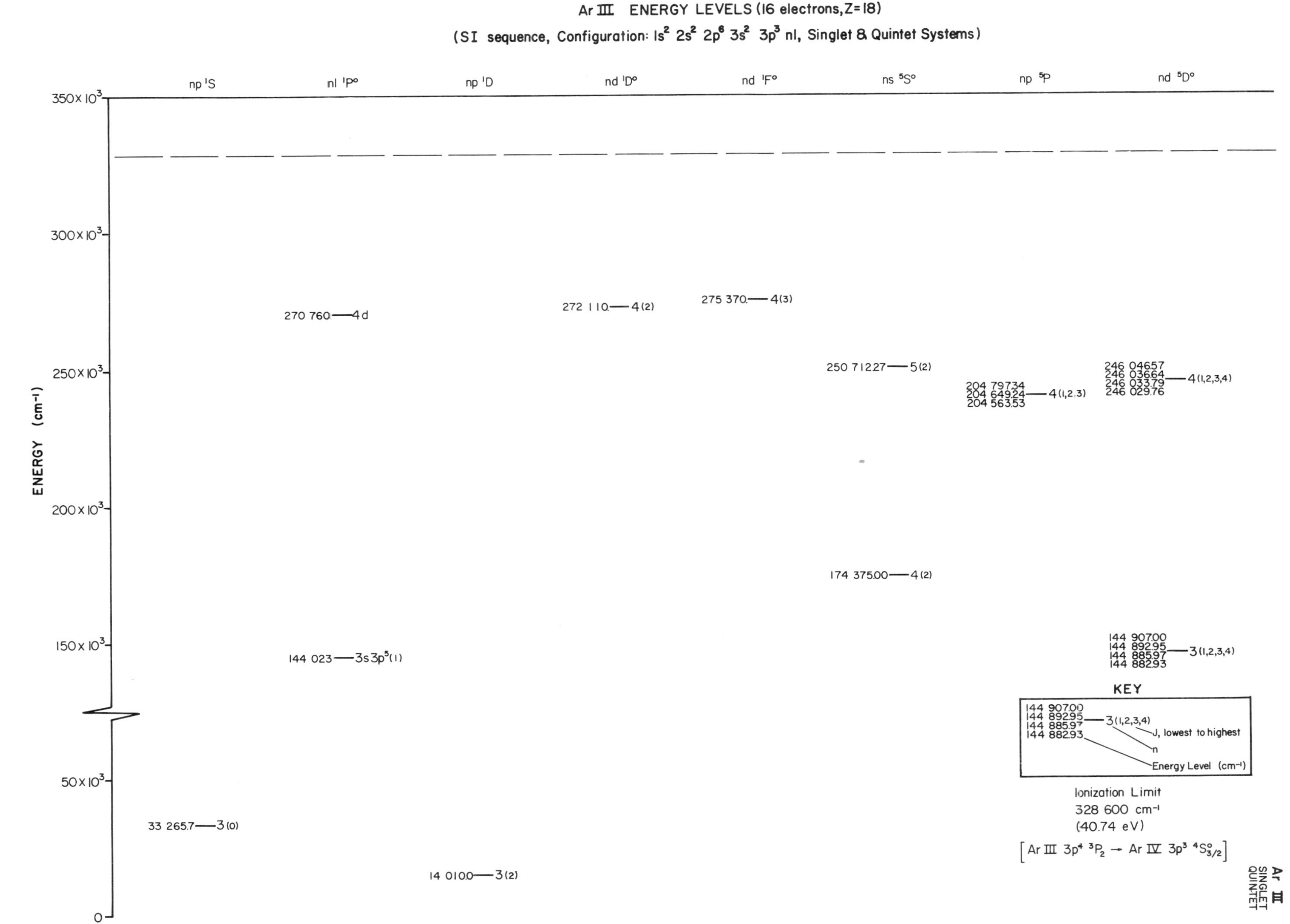
Ar III ENERGY LEVELS (16 electrons, Z=18)
(SI sequence, Configuration: $1s^2\ 2s^2\ 2p^6\ 3s^2\ 3p^3\ nl$, Singlet & Quintet Systems)
np ^{1}S
nl ^{1}P°
np ^{1}D
nd ^{1}D°
nd ^{1}F°
ns ^{5}S°
np ^{5}P
nd ^{5}D°
350 x 10^3
300 x 10^3
250 x 10^3
200 x 10^3
150 x 10^3
50 x 10^3
0
ENERGY (cm^{-1})
33 265.7 — 3(0)
270 760 — 4d
144 023 — 3s3p^5(1)
14 010.0 — 3(2)
272 110. — 4(2)
275 370. — 4(3)
250 712.27 — 5(2)
174 375.00 — 4(2)
204 797.34
204 649.24 — 4(1,2,3)
204 563.53
246 046.57
246 036.64 — 4(1,2,3,4)
246 033.79
246 029.76
144 907.00
144 892.95 — 3(1,2,3,4)
144 885.97
144 882.93
KEY
144 907.00
144 892.95 — 3(1,2,3,4)
144 885.97
144 882.93
J, lowest to highest
n
Energy Level (cm^{-1})
Ionization Limit
328 600 cm^{-1}
(40.74 eV)
[Ar III $3p^4\ ^3P_2$ → Ar IV $3p^3\ ^4S^o_{3/2}$]
Ar III
SINGLET
QUINTET

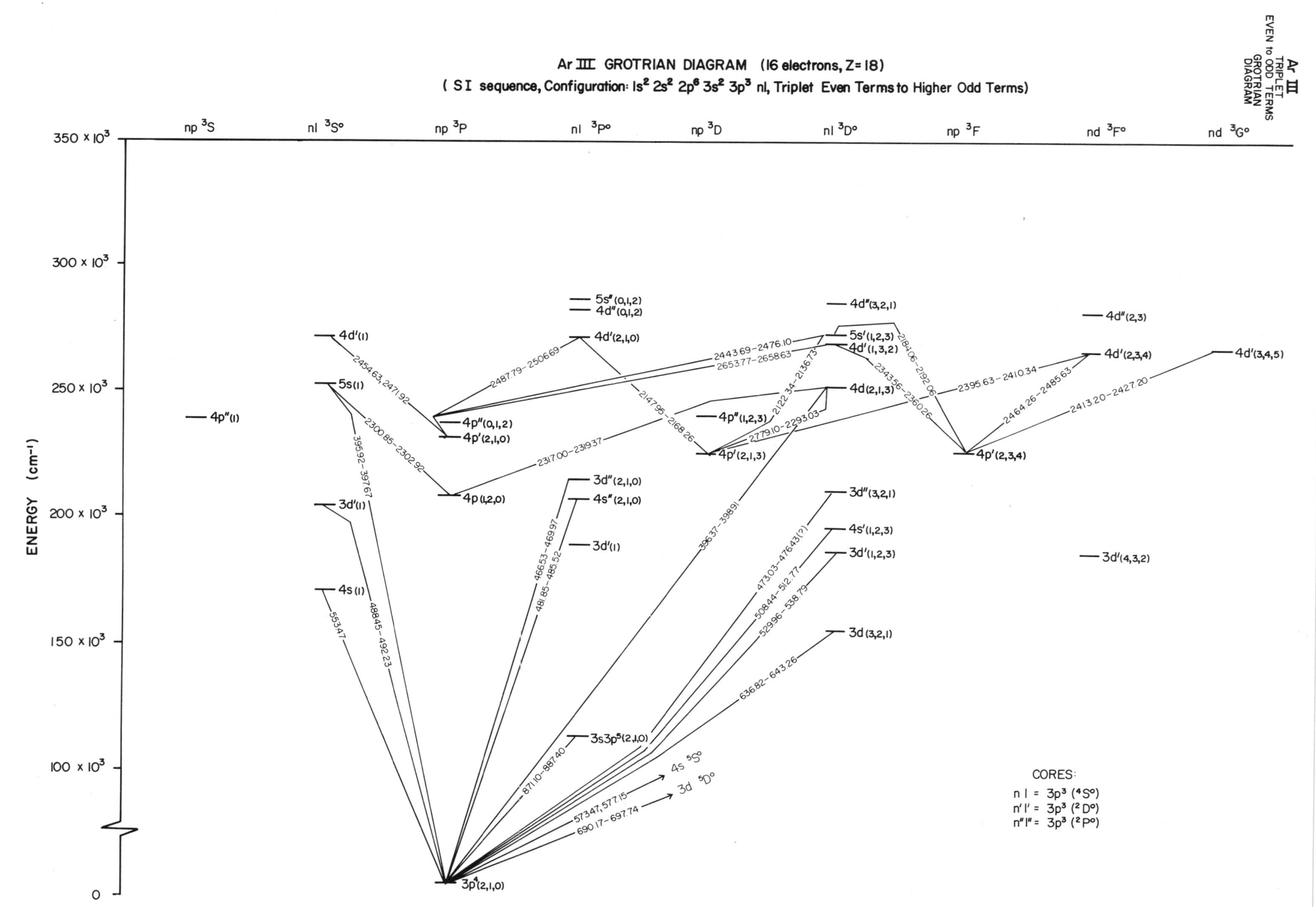

Ar III
TRIPLET
EVEN to ODD TERMS
GROTRIAN
DIAGRAM
Ar III GROTRIAN DIAGRAM (16 electrons, Z= 18)
(SI sequence, Configuration: $1s^2\ 2s^2\ 2p^6\ 3s^2\ 3p^3$ nl, Triplet Even Terms to Higher Odd Terms)
np ^{3}S
nl ^{3}S°
np ^{3}P
nl ^{3}P°
np ^{3}D
nl ^{3}D°
np ^{3}F
nd ^{3}F°
nd ^{3}G°
350 x 10^3
300 x 10^3
250 x 10^3
200 x 10^3
150 x 10^3
100 x 10^3
0
ENERGY (cm^{-1})
4p''(1)
4d'(1)
5s(1)
3d'(1)
4s(1)
4p''(0,1,2)
4p'(2,1,0)
4p(1,2,0)
3p^4(2,1,0)
5s''(0,1,2)
4d''(0,1,2)
4d'(2,1,0)
3d''(2,1,0)
4s''(2,1,0)
3d'(1)
3s3p^5(2,1,0)
4p''(1,2,3)
4p'(2,1,3)
4d''(3,2,1)
5s'(1,2,3)
4d'(1,3,2)
4d(2,1,3)
3d''(3,2,1)
4s'(1,2,3)
3d'(1,2,3)
3d(3,2,1)
4p'(2,3,4)
4d''(2,3)
4d'(2,3,4)
3d'(4,3,2)
4d'(3,4,5)
4s ^{5}S°
3d ^{5}D°
2454.63, 2471.92
2300.85 - 2302.92
395.92 - 397.67
488.45 - 492.23
553.47
2487.79 - 2506.69
2317.00 - 2319.37
466.53 - 469.97
481.85 - 485.52
871.10 - 887.40
2147.95 - 2168.26
2443.69 - 2476.10
2653.77 - 2658.63
2122.34 - 2136.73
2779.10 - 2293.03
396.37 - 398.91
473.03 - 476.43(?)
508.44 - 512.77
529.96 - 538.79
636.82 - 643.26
573.47, 577.15
690.17 - 697.74
2184.06 - 2192.06
2343.56 - 2360.26
2395.63 - 2410.34
2464.26 - 2485.63
2413.20 - 2427.20
CORES:
n l = $3p^3$ (^{4}S°)
n'l' = $3p^3$ (^{2}D°)
n''l'' = $3p^3$ (^{2}P°)

Ar III
TRIPLET

Ar III ENERGY LEVELS (16 electrons, Z=18)
(S I sequence, Configuration: $1s^2 2s^2 2p^6 3s^2 3p^3$ nl, Triplet System)

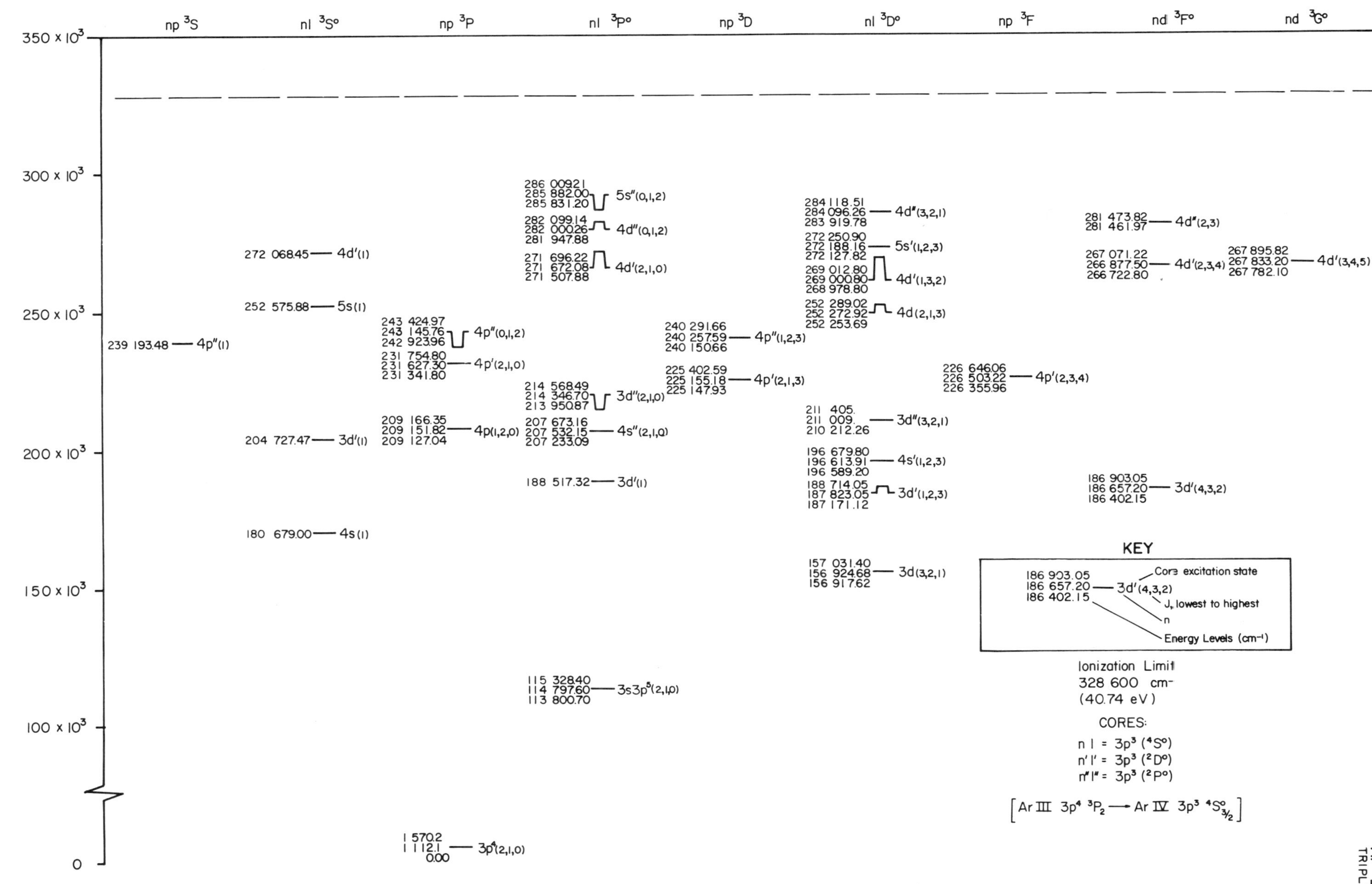

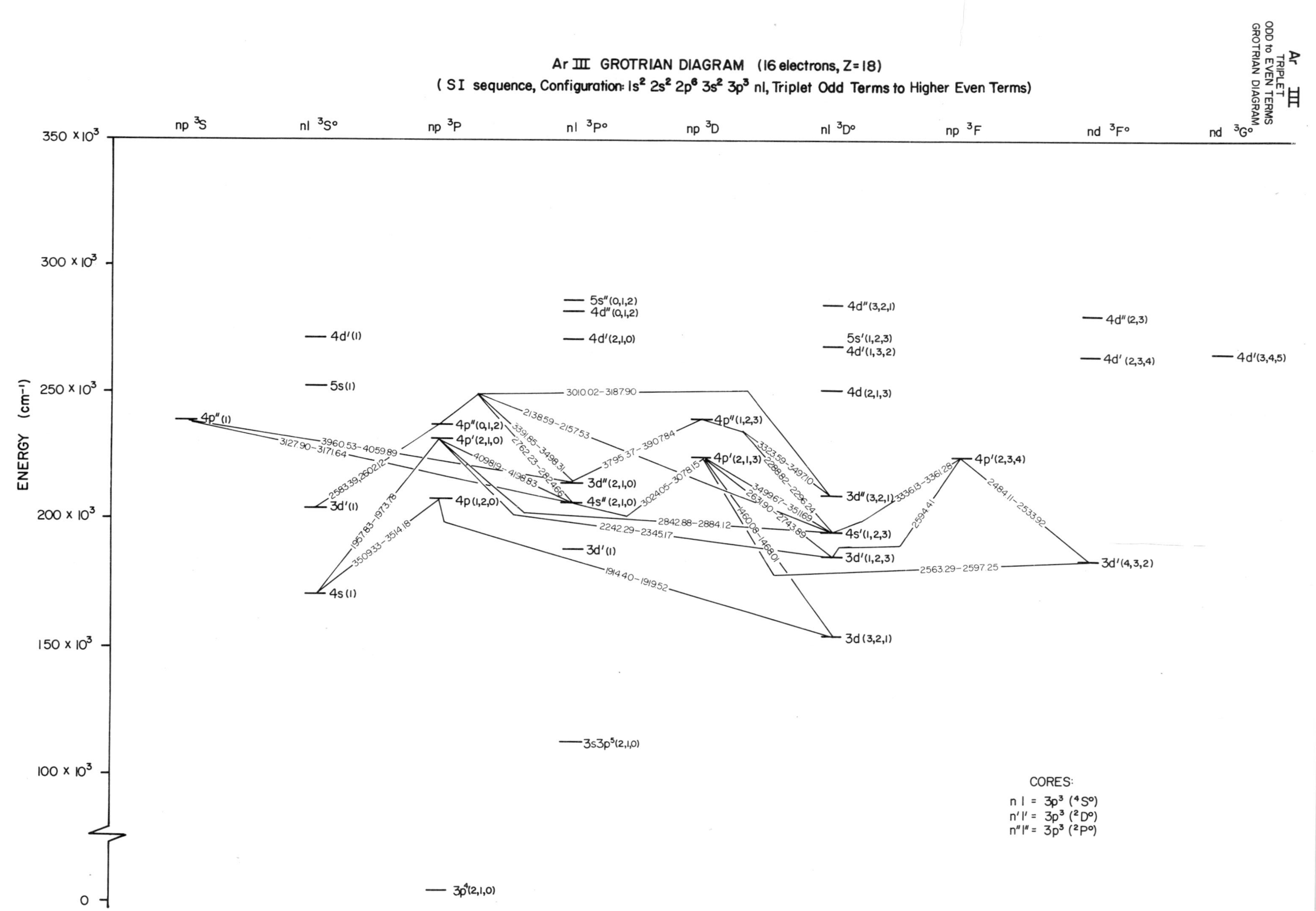

Ar III
TRIPLET
ODD to EVEN TERMS
GROTRIAN DIAGRAM
Ar III GROTRIAN DIAGRAM (16 electrons, Z=18)
(S I sequence, Configuration: 1s² 2s² 2p⁶ 3s² 3p³ nl, Triplet Odd Terms to Higher Even Terms)
np ³S
nl ³S°
np ³P
nl ³P°
np ³D
nl ³D°
np ³F
nd ³F°
nd ³G°
ENERGY (cm⁻¹)
350 x 10³
300 x 10³
250 x 10³
200 x 10³
150 x 10³
100 x 10³
0
4p″(1)
4d′(1)
5s(1)
3d′(1)
4s(1)
4p″(0,1,2)
4p′(2,1,0)
4p(1,2,0)
3p⁴(2,1,0)
5s″(0,1,2)
4d″(0,1,2)
4d′(2,1,0)
3d″(2,1,0)
4s″(2,1,0)
3d′(1)
3s3p⁵(2,1,0)
4p″(1,2,3)
4p′(2,1,3)
4d″(3,2,1)
5s′(1,2,3)
4d′(1,3,2)
4d(2,1,3)
3d″(3,2,1)
4s′(1,2,3)
3d′(1,2,3)
3d(3,2,1)
4p′(2,3,4)
4d″(2,3)
4d′(2,3,4)
3d′(4,3,2)
4d′(3,4,5)
3010.02–3187.90
2138.59–2157.53
3391.85–3498.31
2762.23–2824.66
4098.19–4138.83
3960.53–4059.89
3127.90–3171.64
2583.39,2602.12
1957.83–1973.78
3509.33–3514.18
3795.37–3907.84
3024.05–3078.15
3323.59–3497.10
2288.82–2296.24
3499.67–3511.69
2631.90–2743.89
1460.08–1468.01
2842.88–2884.12
2242.29–2345.17
1914.40–1919.52
3336.13–3361.28
2594.41
2484.11–2533.92
2563.29–2597.25
CORES:
n l = 3p³ (⁴S°)
n′l′ = 3p³ (²D°)
n″l″ = 3p³ (²P°)

Ar III
TRIPLET

Ar III ENERGY LEVELS (16 electrons, Z=18)

(S I sequence, Configuration: $1s^2\,2s^2\,2p^6\,3s^2\,3p^3$ nl, Triplet System)

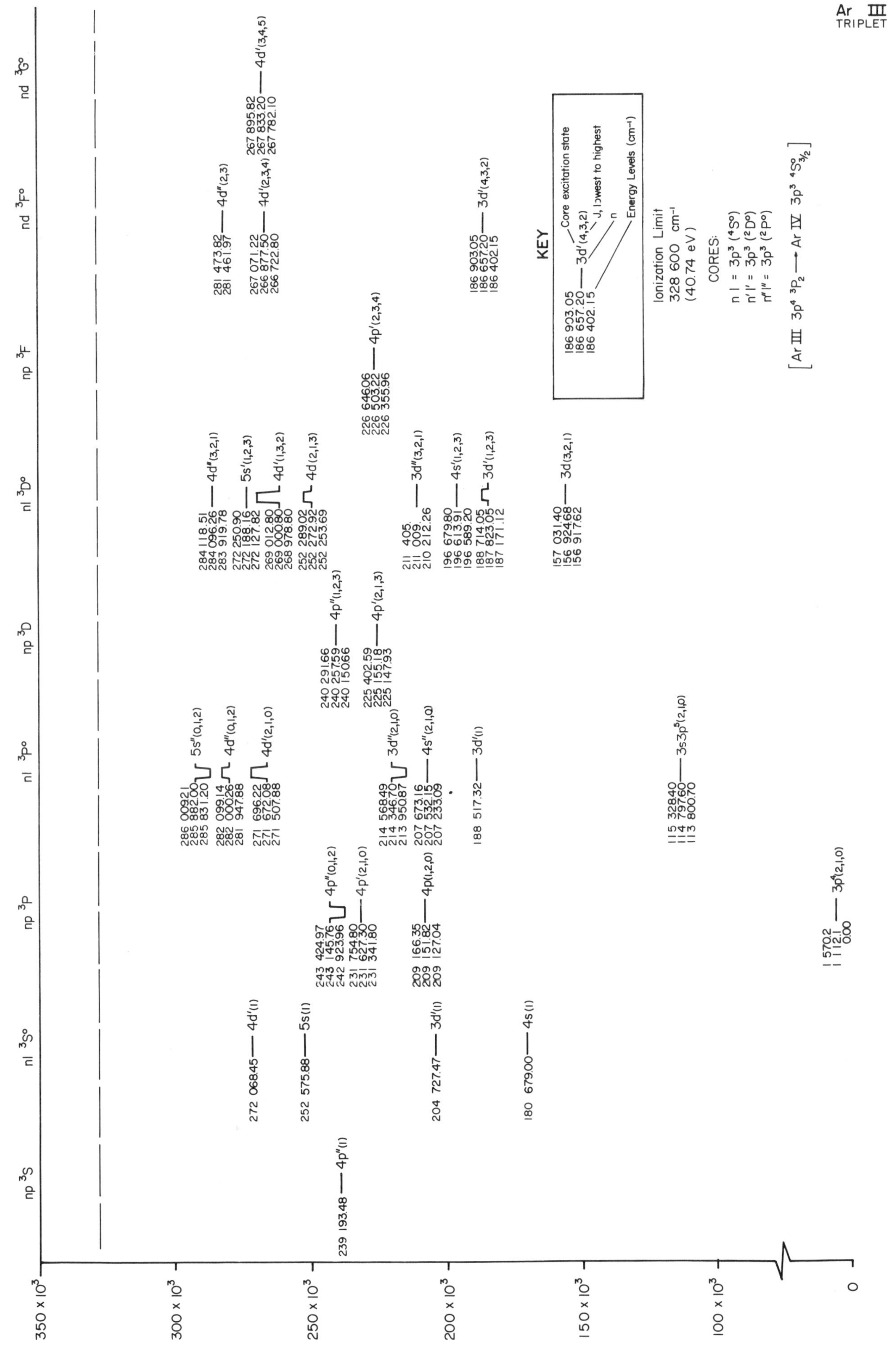

Energy axis: 0, 100 x 10³, 150 x 10³, 200 x 10³, 250 x 10³, 300 x 10³, 350 x 10³

np ³S

239 193.48 — 4p″(1)

nl ³S°

272 068.45 — 4d′(1)
252 575.88 — 5s(1)
204 727.47 — 3d′(1)
180 679.00 — 4s(1)

np ³P

243 424.97
243 145.76
242 923.96 — 4p″(0,1,2)

231 754.80
231 627.30
231 341.80 — 4p′(2,1,0)

209 166.35
209 151.82
209 127.04 — 4p(1,2,0)

1 570.2
1 112.1
0.00 — $3p^4$(2,1,0)

nl ³P°

286 009.21
285 882.00
285 831.20 — 5s″(0,1,2)

282 099.14
282 000.26
281 947.88 — 4d″(0,1,2)

271 696.22
271 672.08
271 507.88 — 4d′(2,1,0)

214 568.49
214 346.70
213 950.87 — 3d″(2,1,0)

207 673.16
207 532.15
207 233.09 — 4s″(2,1,0)

188 517.32 — 3d′(1)

115 328.40
114 797.60
113 800.70 — $3s3p^5$(2,1,0)

np ³D

240 291.66
240 257.59
240 150.66 — 4p″(1,2,3)

225 402.59
225 155.18
225 147.93 — 4p′(2,1,3)

nl ³D°

284 118.51
284 096.26
283 919.78 — 4d″(3,2,1)

272 250.90
272 188.16
272 127.82 — 5s′(1,2,3)

269 012.80
269 000.80
268 978.80 — 4d′(1,3,2)

252 289.02
252 272.92
252 253.69 — 4d(2,1,3)

211 405.
211 009.
210 212.26 — 3d″(3,2,1)

196 679.80
196 613.91
196 589.20 — 4s′(1,2,3)

188 714.05
187 823.05
187 171.12 — 3d′(1,2,3)

157 031.40
156 924.68
156 917.62 — 3d(3,2,1)

np ³F

226 646.06
226 503.22
226 355.96 — 4p′(2,3,4)

nd ³F°

281 473.82
281 461.97 — 4d″(2,3)

267 071.22
266 877.50
266 722.80 — 4d′(2,3,4)

186 903.05
186 657.20
186 402.15 — 3d′(4,3,2)

nd ³G°

267 895.82
267 833.20
267 782.10 — 4d′(3,4,5)

KEY

186 903.05
186 657.20 — 3d′(4,3,2)
186 402.15

Core excitation state
J, lowest to highest
n
Energy Levels (cm⁻¹)

Ionization Limit
328 600 cm⁻¹
(40.74 eV)

CORES:
n l = $3p^3$ ($^4S°$)
n′ l′ = $3p^3$ ($^2D°$)
n″ l″ = $3p^3$ ($^2P°$)

[Ar III $3p^4\ ^3P_2$ → Ar IV $3p^3\ ^4S°_{3/2}$]

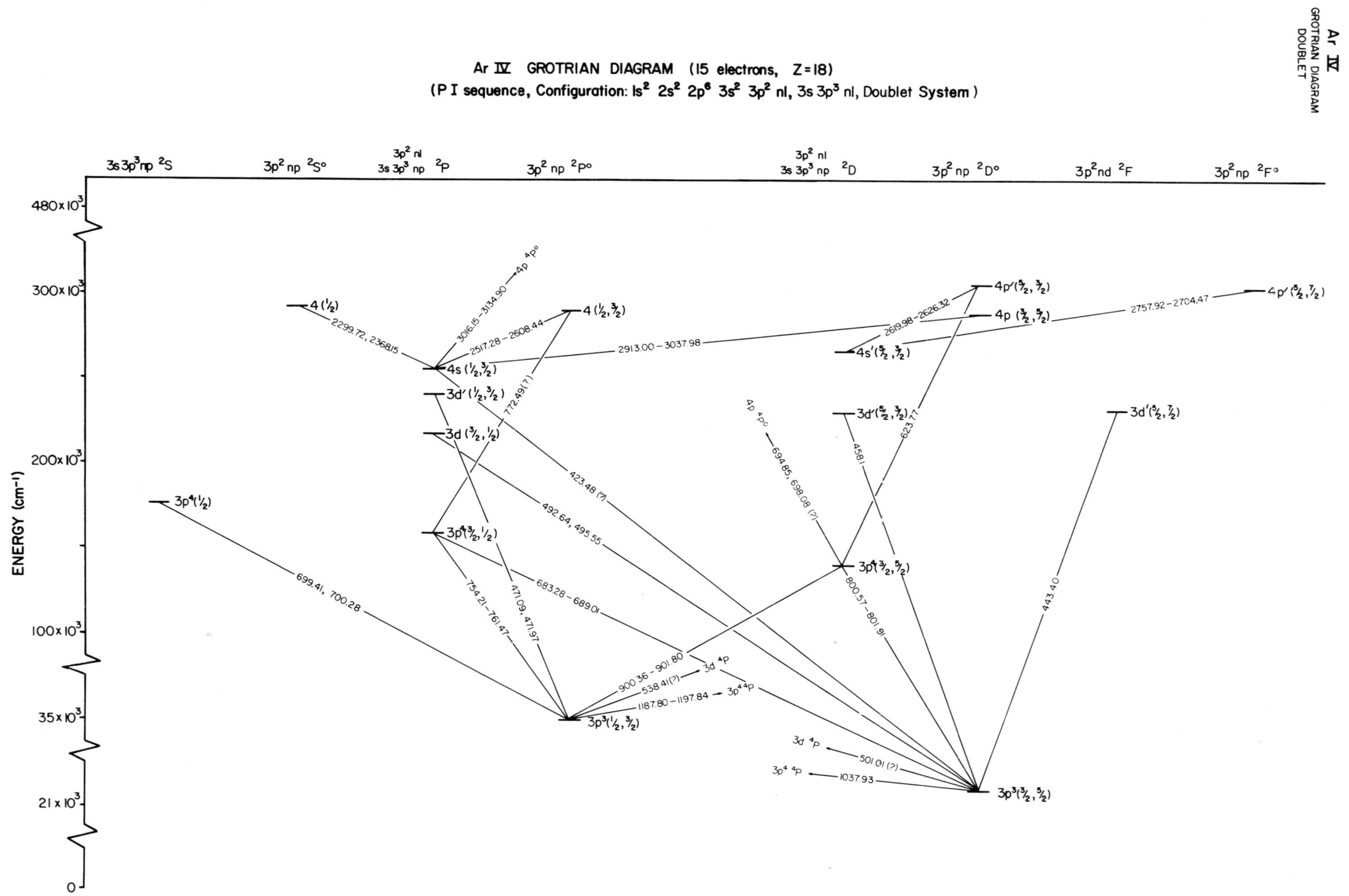
Ar IV GROTRIAN DIAGRAM (15 electrons, Z=18)
(P I sequence, Configuration: $1s^2\ 2s^2\ 2p^6\ 3s^2\ 3p^2\ nl$, $3s\ 3p^3\ nl$, Doublet System)
$3s3p^3np\ ^2S$
$3p^2np\ ^2S^o$
$3p^2nl$ / $3s3p^3np\ ^2P$
$3p^2np\ ^2P^o$
$3p^2nl$ / $3s3p^3np\ ^2D$
$3p^2np\ ^2D^o$
$3p^2nd\ ^2F$
$3p^2np\ ^2F^o$
ENERGY (cm^{-1})
480×10^3
300×10^3
200×10^3
100×10^3
35×10^3
21×10^3
0
$3p^4(1/2)$
$4(1/2)$
$4s(1/2,3/2)$
$3d'(1/2,3/2)$
$3d(3/2,1/2)$
$3p^4(3/2,1/2)$
$4(1/2,3/2)$
$3p^3(1/2,3/2)$
$4s'(5/2,3/2)$
$3d'(5/2,3/2)$
$3p^4(3/2,5/2)$
$4p'(5/2,3/2)$
$4p(3/2,5/2)$
$3d'(5/2,7/2)$
$4p'(5/2,7/2)$
$3p^3(3/2,5/2)$
$4p\ ^4P^o$
$3d\ ^4P$
$3p^4\ ^4P$
699.41, 700.28
2299.72, 2368.15
3016.15 – 3134.90
2517.28 – 2608.44
2913.00 – 3037.98
772.49 (?)
754.21 – 761.47
471.09, 471.97
683.28 – 689.01
492.64, 495.55
423.48 (?)
900.36 – 901.80
538.41 (?)
1187.80 – 1197.84
694.85, 698.08 (?)
4581
623.77
2619.98 – 2626.32
2757.92 – 2704.47
800.57 – 801.91
443.40
501.01 (?)
1037.93

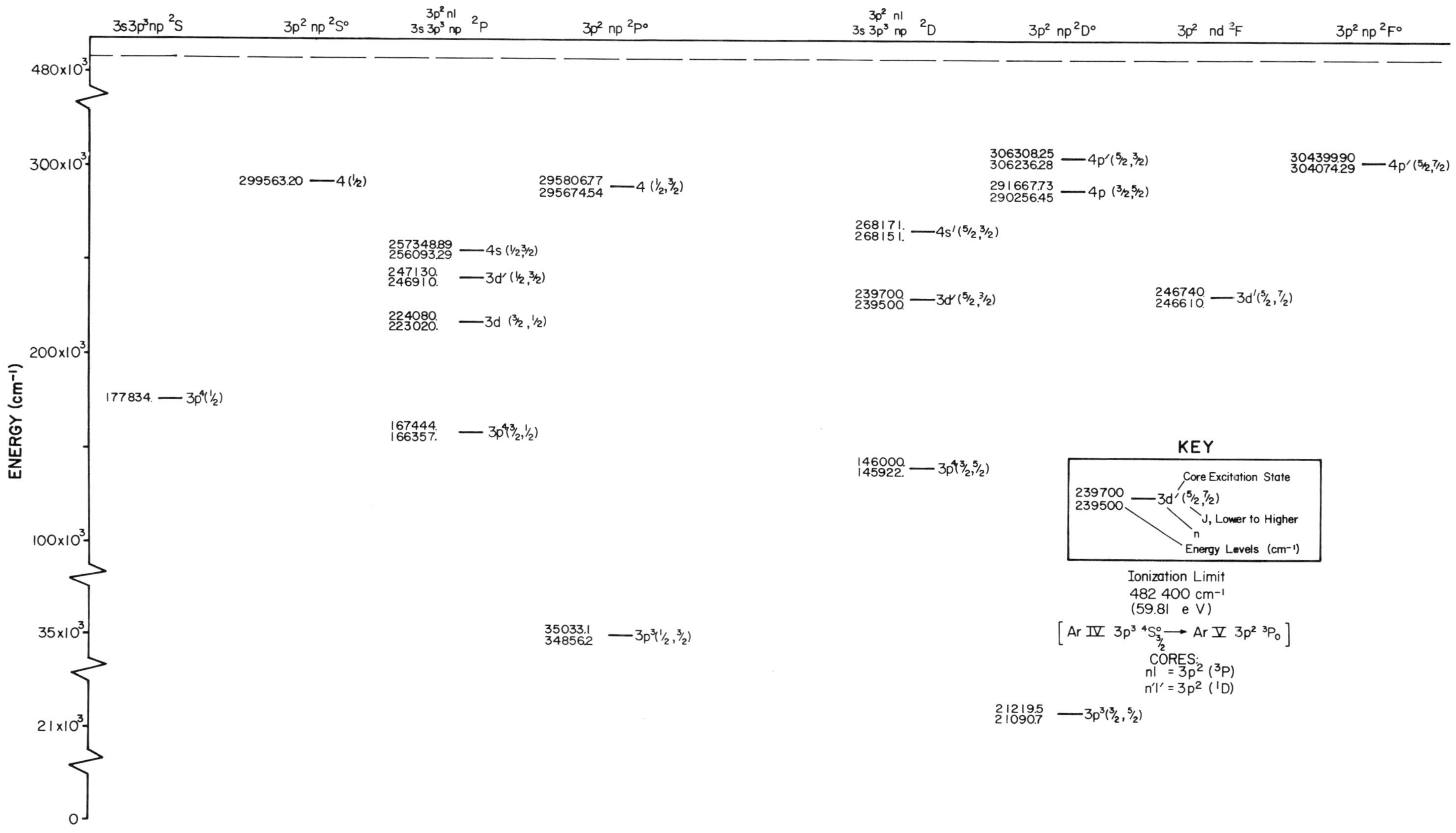
Ar IV ENERGY LEVELS (15 electrons, Z = 18)
(P I sequence, Configuration: 1s²2s²2p⁶3s² 3p² nl, 3s 3p³ nl, Doublet System)
3s3p³np ²S
3p² np ²S°
3p² nl 3s 3p³ np ²P
3p² np ²P°
3p² nl 3s 3p³ np ²D
3p² np ²D°
3p² nd ²F
3p² np ²F°
480x10³
300x10³
200x10³
100x10³
35x10³
21x10³
0
ENERGY (cm⁻¹)
177834. — 3p⁴(½)
299563.20 — 4(½)
257348.89 256093.29 — 4s (½,3/2)
247130. 246910. — 3d′ (½,3/2)
224080. 223020. — 3d (3/2,½)
167444. 166357. — 3p⁴3(3/2,½)
295806.77 295674.54 — 4 (½,3/2)
35033.1 34856.2 — 3p³(½,3/2)
268171. 268151. — 4s′(5/2,3/2)
239700. 239500. — 3d′(5/2,3/2)
146000. 145922. — 3p⁴3(3/2,5/2)
306308.25 306236.28 — 4p′(5/2,3/2)
291667.73 290256.45 — 4p (3/2,5/2)
246740. 246610. — 3d′(5/2,7/2)
304399.90 304074.29 — 4p′(5/2,7/2)
212195 210907 — 3p³(3/2,5/2)
KEY
239700 239500 — 3d′(5/2,7/2)
Core Excitation State
J, Lower to Higher
n
Energy Levels (cm⁻¹)
Ionization Limit
482 400 cm⁻¹
(59.81 eV)
[Ar IV 3p³ ⁴S°3/2 → Ar V 3p² ³P₀]
CORES:
nl = 3p² (³P)
n′l′ = 3p² (¹D)

Ar IV GROTRIAN DIAGRAM (15 electrons, Z=18)

(P I sequence, Configuration: $1s^2 2s^2 2p^6 3s^2 3p^2\, nl$, Quartet System)

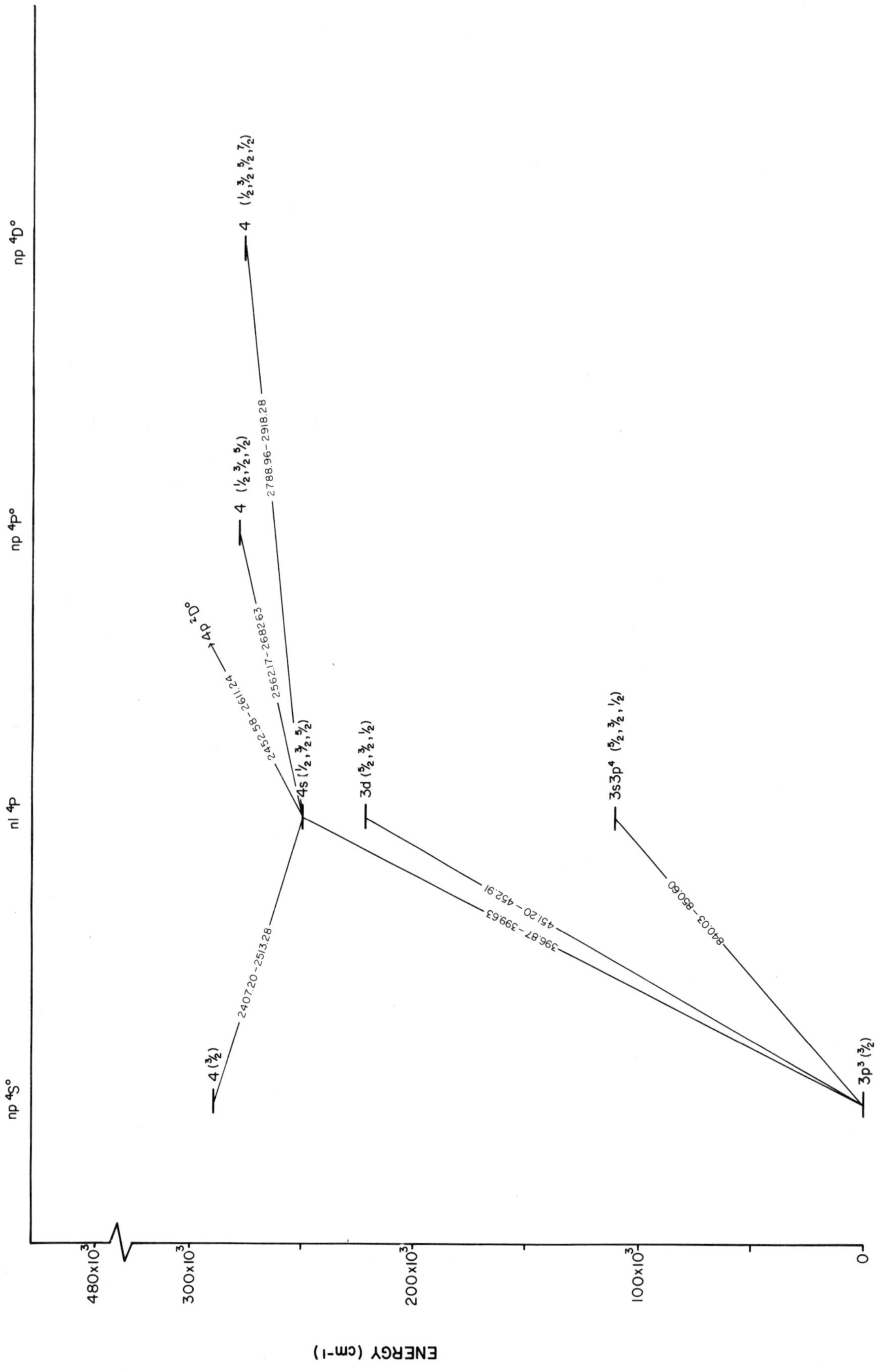

np $^4S^o$

nl ^4P

np $^4P^o$

np $^4D^o$

480x10^3

300x10^3

200x10^3

100x10^3

0

ENERGY (cm^{-1})

4 (3/2)

2407.20 – 2513.28

4p $^2D^o$

2452.58 – 2611.24

4 (1/2, 3/2, 5/2)

2562.17 – 2682.63

4 (1/2, 3/2, 5/2, 7/2)

2788.96 – 2918.28

4s (1/2, 3/2, 5/2)

3d (5/2, 3/2, 1/2)

396.87 – 399.63

451.20 – 452.91

3s3p^4 (5/2, 3/2, 1/2)

840.03 – 850.60

3p^3 (3/2)

Ar IV ENERGY LEVELS (15 electrons, Z=18)

(P I sequence, Configuration: $1s^2 2s^2 2p^6 3s^2 3p^2 nl$, Quartet System)

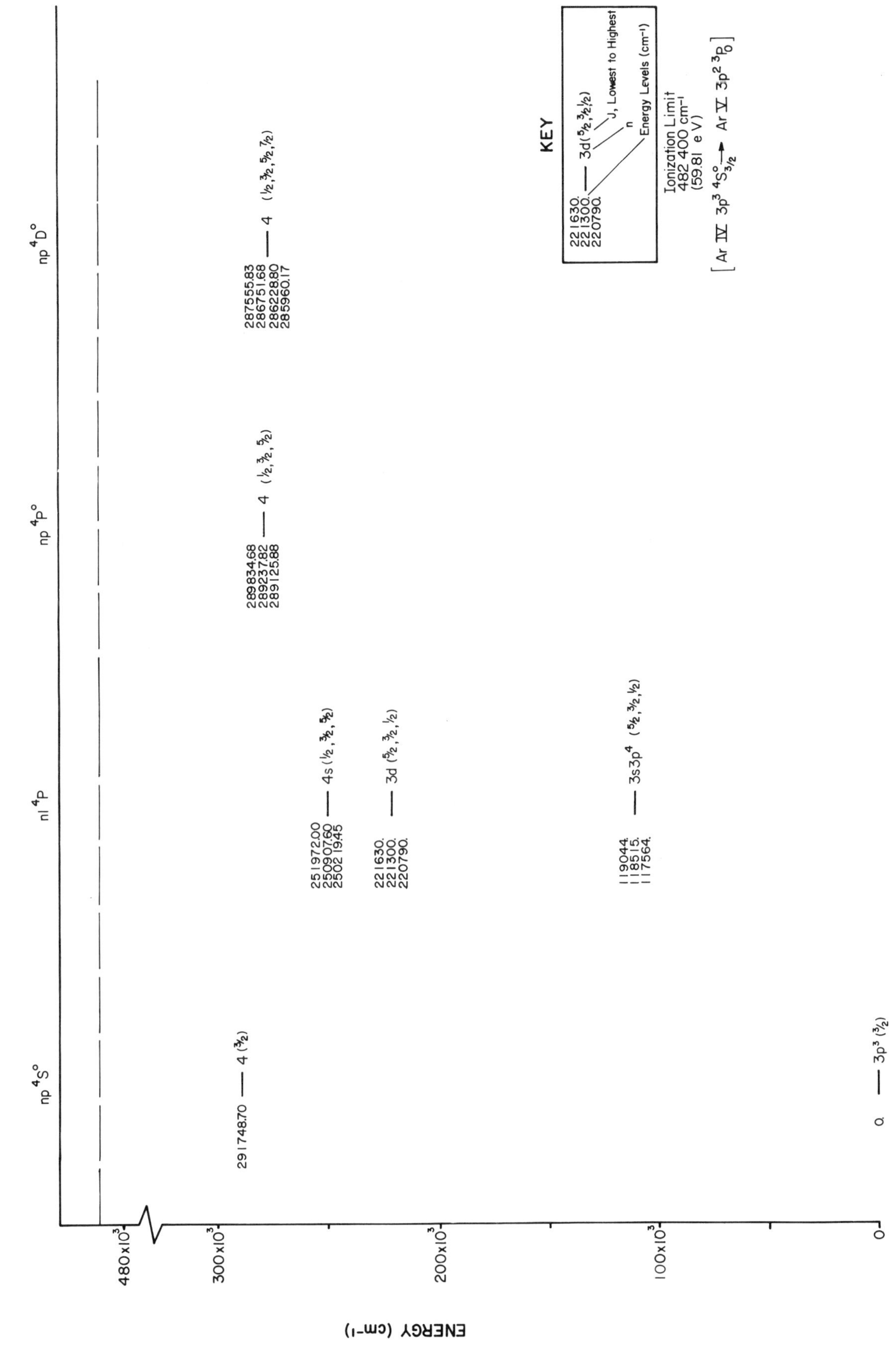

np $^4S^o$
nl ^4P
np $^4P^o$
np $^4D^o$

ENERGY (cm^{-1})

480x10^3
300x10^3
200x10^3
100x10^3
0

291748.70 — 4 (3/2)

0. — $3p^3$ (3/2)

251972.00
250907.60
250219.45 — 4s (1/2, 3/2, 5/2)

221630.
221300.
220790. — 3d (5/2, 3/2, 1/2)

119044.
118515.
117564. — $3s3p^4$ (5/2, 3/2, 1/2)

289834.68
289237.82
289125.88 — 4 (1/2, 3/2, 5/2)

287555.83
286751.68
286228.80
285960.17 — 4 (1/2, 3/2, 5/2, 7/2)

KEY

221630.
221300.
220790. — 3d(5/2, 3/2, 1/2)

J, Lowest to Highest
n
Energy Levels (cm^{-1})

Ionization Limit
482 400 cm^{-1}
(59.81 eV)
[Ar IV $3p^3$ $^4S^o_{3/2}$ → Ar V $3p^2$ $^3P_0$]

Ar V
SINGLET
TRIPLET
GROTRIAN
DIAGRAM

Ar V GROTRIAN DIAGRAM (14 electrons, Z=18)

(Si I Sequence, Configuration: $1s^2\ 2s^2\ 2p^6\ 3s^2\ 3p\ nl, 3s3p^2nl$, Singlet & Triplet Systems)

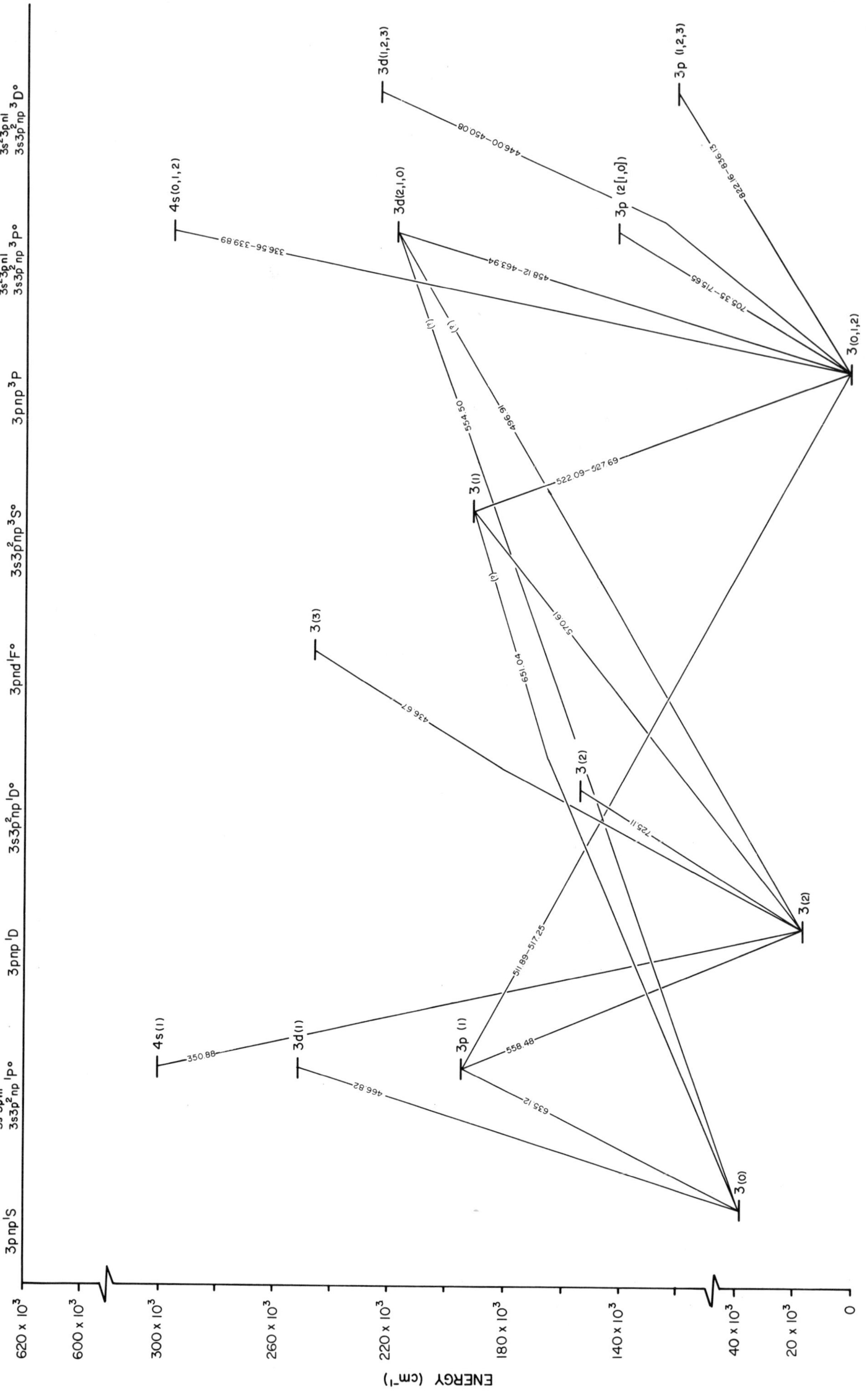

$3pnp\ ^1S$
$3s^23pnl$ $3s3p^2np\ ^1P^o$
$3pnp\ ^1D$
$3s3p^2np\ ^1D^o$
$3pnd\ ^1F^o$
$3s3p^2np\ ^3S^o$
$3pnp\ ^3P$
$3s^23pnl$ $3s3p^2np\ ^3P^o$
$3s^23pnl$ $3s3p^2np\ ^3D^o$

ENERGY (cm^{-1})
620×10^3
600×10^3
300×10^3
260×10^3
220×10^3
180×10^3
140×10^3
40×10^3
20×10^3
0

4s(1)
3d(1)
3p (1)
3(0)
3(2)
3(3)
3(2)
3(1)
4s(0,1,2)
3d(2,1,0)
3d(1,2,3)
3p (2[1,0])
3p (1,2,3)
3(0,1,2)

350.88
466.82
635.12
558.48
511.89–517.25
436.67
725.11
651.04
(?)
554.50
(?)
(?)
496.91
570.61
522.09–527.69
336.56–339.89
458.12–463.94
446.00–450.08
705.35–715.65
822.16–836.13

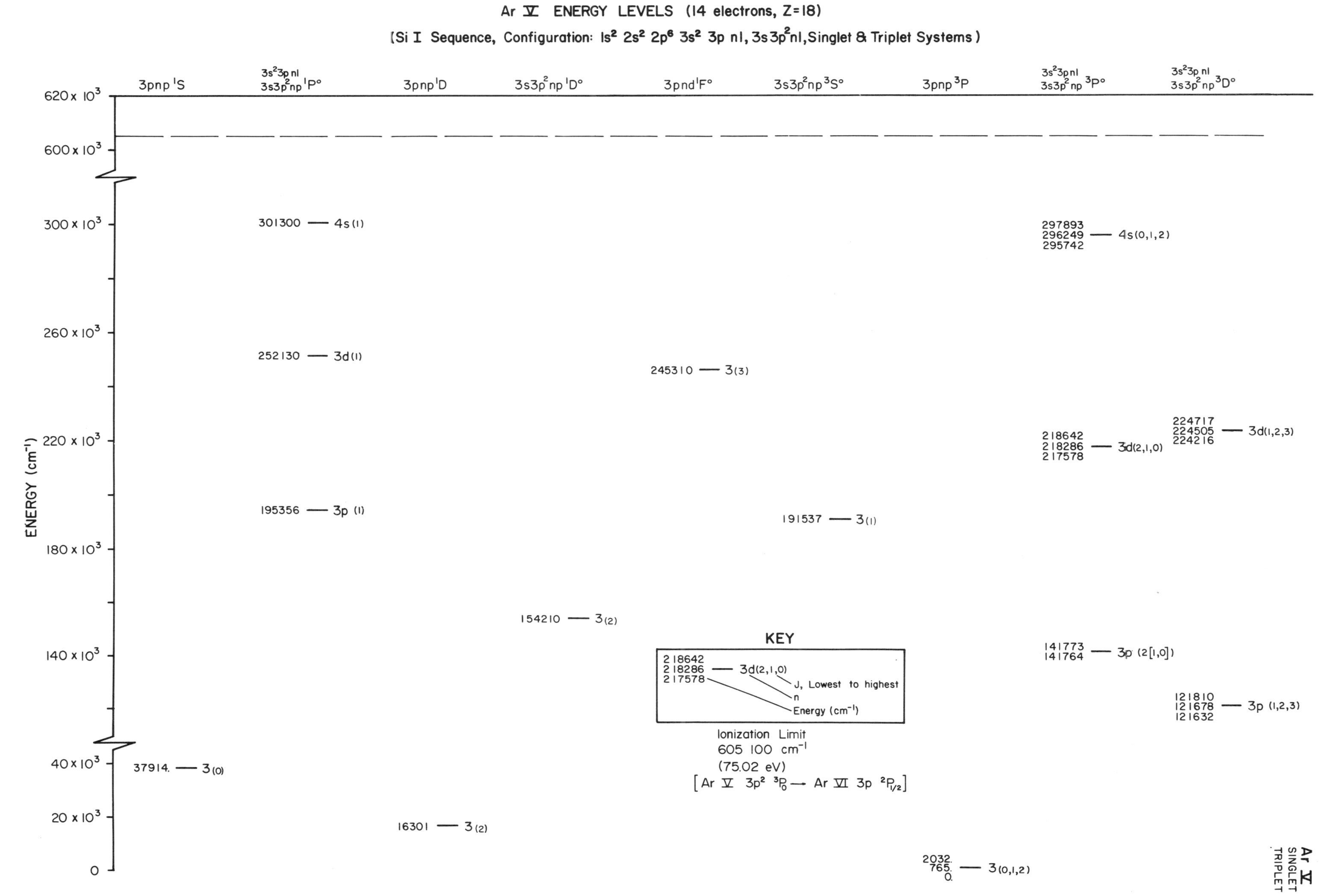

Ar V ENERGY LEVELS (14 electrons, Z=18)
(Si I Sequence, Configuration: 1s² 2s² 2p⁶ 3s² 3p nl, 3s3p²nl, Singlet & Triplet Systems)
3pnp ¹S
3s²3p nl 3s3p²np ¹P°
3pnp ¹D
3s3p²np ¹D°
3pnd ¹F°
3s3p²np ³S°
3pnp ³P
3s²3pnl 3s3p²np ³P°
3s²3p nl 3s3p²np ³D°
620 x 10³
600 x 10³
300 x 10³
260 x 10³
220 x 10³
180 x 10³
140 x 10³
40 x 10³
20 x 10³
0
ENERGY (cm⁻¹)
301300 — 4s(1)
252130 — 3d(1)
195356 — 3p (1)
37914. — 3(0)
16301 — 3(2)
154210 — 3(2)
245310 — 3(3)
191537 — 3(1)
2032. 765. 0. — 3(0,1,2)
297893 296249 295742 — 4s(0,1,2)
218642 218286 217578 — 3d(2,1,0)
141773 141764 — 3p (2[1,0])
224717 224505 224216 — 3d(1,2,3)
121810 121678 121632 — 3p (1,2,3)
KEY
218642 218286 217578 — 3d(2,1,0)
J, Lowest to highest
n
Energy (cm⁻¹)
Ionization Limit
605 100 cm⁻¹
(75.02 eV)
[Ar V 3p² ³P₀ → Ar VI 3p ²P₁/₂]
Ar V
SINGLET
TRIPLET

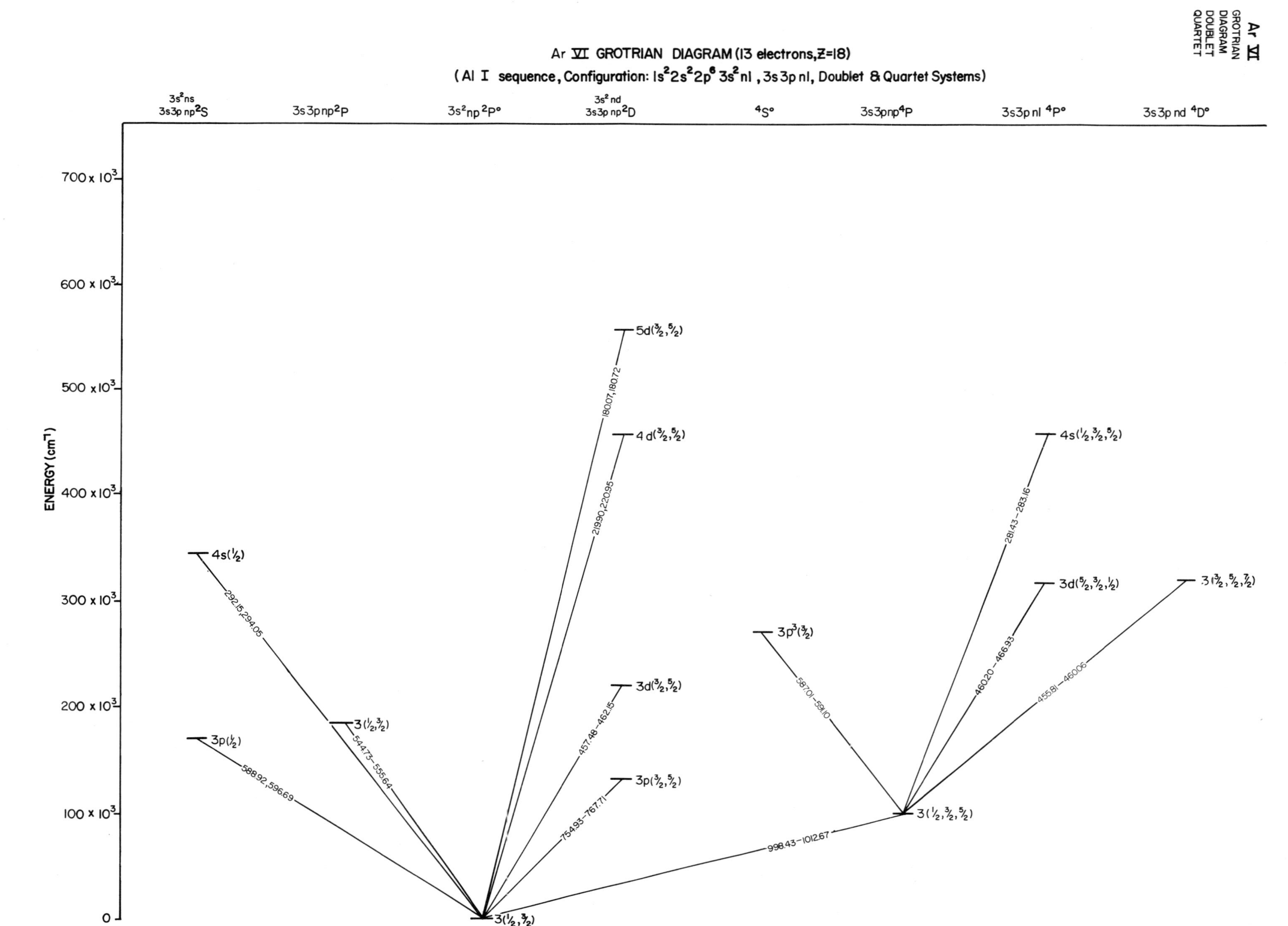
Ar VII GROTRIAN DIAGRAM (13 electrons, Z=18)
(Al I sequence, Configuration: 1s²2s²2p⁶3s²nl, 3s3p nl, Doublet & Quartet Systems)
3s²ns 3s3p np²S
3s3pnp²P
3s²np²P°
3s²nd 3s3p np²D
⁴S°
3s3pnp⁴P
3s3p nl ⁴P°
3s3p nd ⁴D°
ENERGY (cm⁻¹)
700 x 10³
600 x 10³
500 x 10³
400 x 10³
300 x 10³
200 x 10³
100 x 10³
0
4s(½)
292.15,294.05
3p(½)
588.92,596.69
3(½,³⁄₂)
544.73–555.64
5d(³⁄₂,⁵⁄₂)
180.07,180.72
4d(³⁄₂,⁵⁄₂)
219.90,220.95
3d(³⁄₂,⁵⁄₂)
457.48–462.15
3p(³⁄₂,⁵⁄₂)
754.93–767.71
3(½,³⁄₂)
3p³(³⁄₂)
587.01–591.10
998.43–1012.67
3(½,³⁄₂,⁵⁄₂)
4s(½,³⁄₂,⁵⁄₂)
281.43–283.16
3d(⁵⁄₂,³⁄₂,½)
460.20–466.93
3(³⁄₂,⁵⁄₂,⁷⁄₂)
455.81–460.06

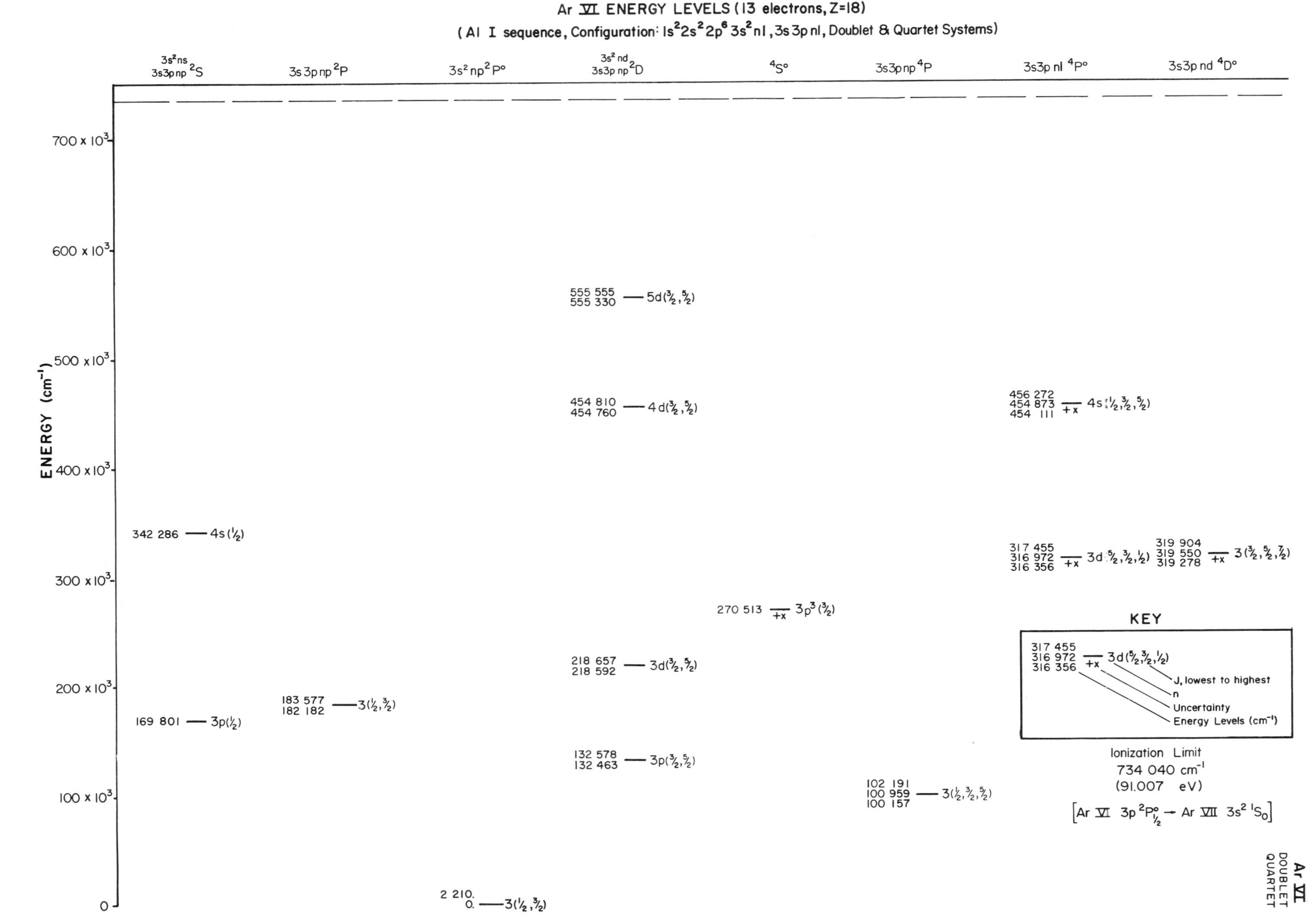

Ar VI ENERGY LEVELS (13 electrons, Z=18)
(Al I sequence, Configuration: 1s²2s²2p⁶3s²nl, 3s3pnl, Doublet & Quartet Systems)
3s²ns 3s3pnp ²S
3s3pnp ²P
3s²np ²P°
3s²nd 3s3pnp ²D
⁴S°
3s3pnp ⁴P
3s3pnl ⁴P°
3s3pnd ⁴D°
700 x 10³
600 x 10³
500 x 10³
400 x 10³
300 x 10³
200 x 10³
100 x 10³
0
ENERGY (cm⁻¹)
555 555 555 330 —— 5d(3/2, 5/2)
454 810 454 760 —— 4d(3/2, 5/2)
456 272 454 873 454 111 +x 4s(½, 3/2, 5/2)
342 286 —— 4s(½)
317 455 316 972 316 356 +x 3d(5/2, 3/2, ½)
319 904 319 550 319 278 +x 3(3/2, 5/2, 7/2)
270 513 +x 3p³(3/2)
218 657 218 592 —— 3d(3/2, 5/2)
183 577 182 182 —— 3(½, 3/2)
169 801 —— 3p(½)
132 578 132 463 —— 3p(3/2, 5/2)
102 191 100 959 100 157 —— 3(½, 3/2, 5/2)
2 210. 0. —— 3(½, 3/2)
KEY
317 455 316 972 316 356 +x 3d(5/2, 3/2, ½)
J, lowest to highest
n
Uncertainty
Energy Levels (cm⁻¹)
Ionization Limit
734 040 cm⁻¹
(91.007 eV)
[Ar VI 3p ²P°½ → Ar VII 3s² ¹S₀]
Ar VI
DOUBLET
QUARTET

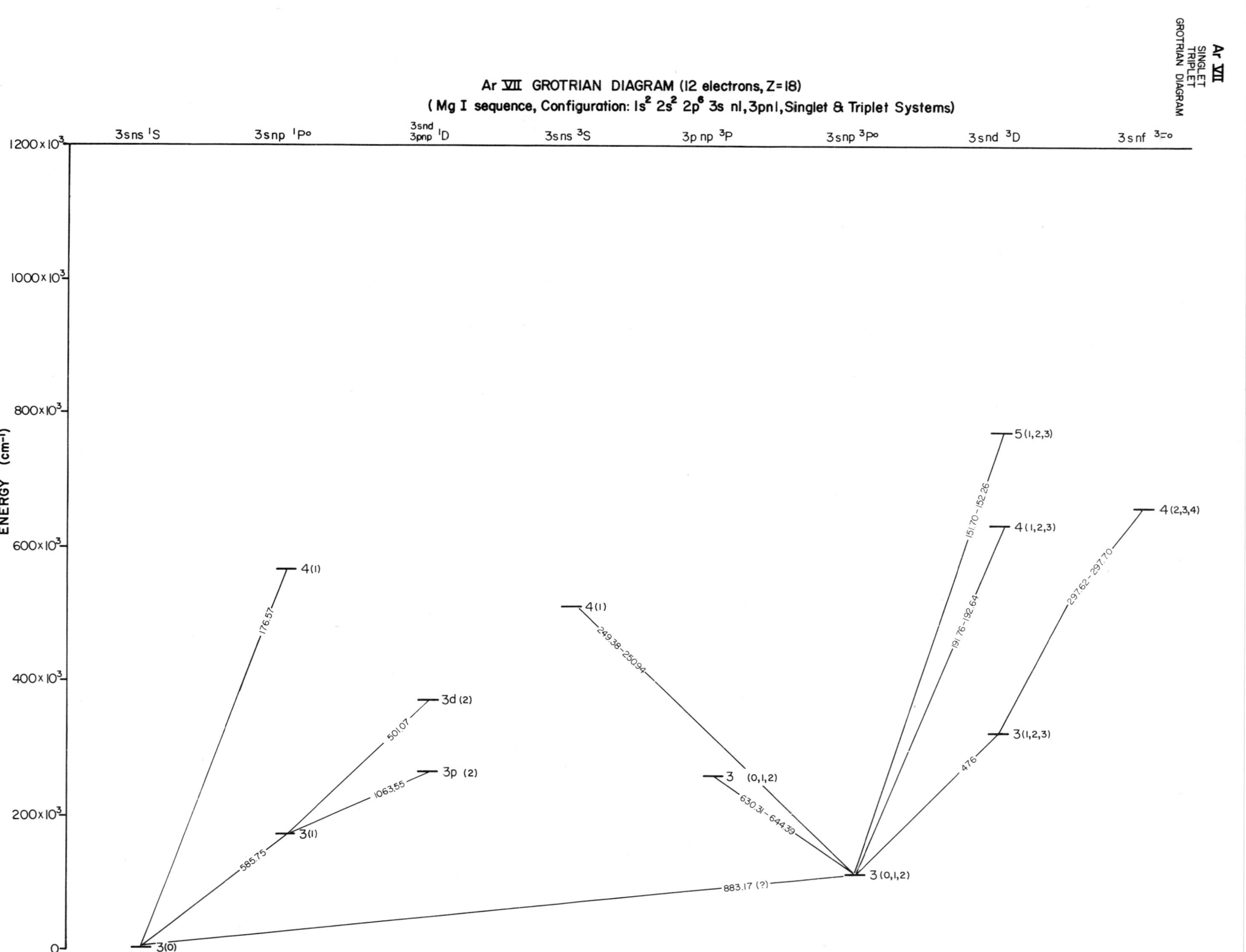
Ar VII
SINGLET
TRIPLET
GROTRIAN DIAGRAM
Ar VII GROTRIAN DIAGRAM (12 electrons, Z=18)
(Mg I sequence, Configuration: 1s² 2s² 2p⁶ 3s nl, 3pnl, Singlet & Triplet Systems)
3sns ¹S
3snp ¹P°
3snd
3pnp ¹D
3sns ³S
3p np ³P
3snp ³P°
3snd ³D
3snf ³F°
ENERGY (cm⁻¹)
1200x10³
1000x10³
800x10³
600x10³
400x10³
200x10³
0
4(1)
176.57
3d (2)
501.07
3p (2)
1063.55
3(1)
585.75
3(0)
4(1)
249.38-250.94
3 (0,1,2)
630.31-644.39
883.17 (?)
3 (0,1,2)
5 (1,2,3)
151.70-152.26
4 (1,2,3)
191.76-192.64
4 (2,3,4)
297.62-297.70
3 (1,2,3)
476

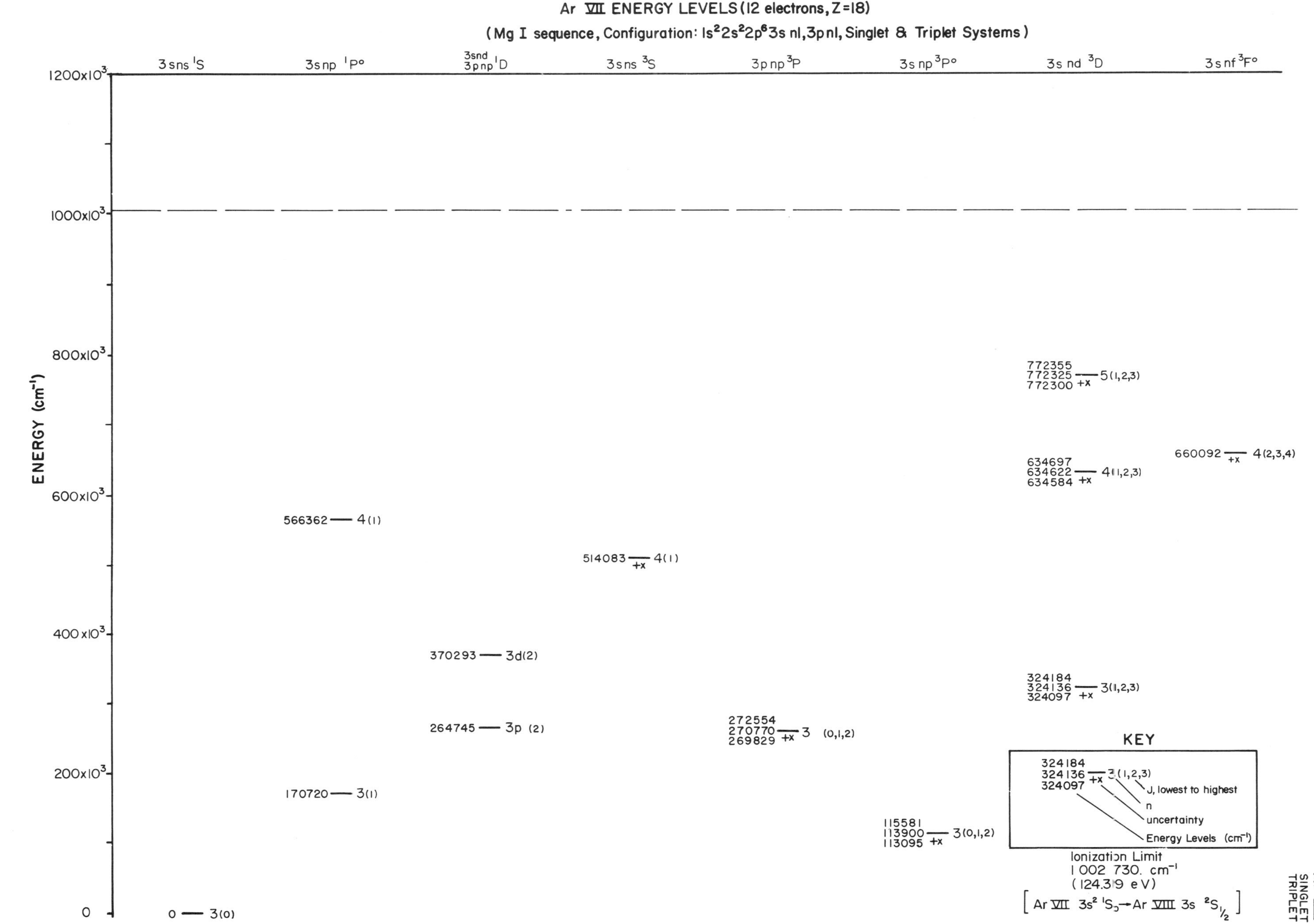

Ar VII ENERGY LEVELS (12 electrons, Z=18)
(Mg I sequence, Configuration: 1s²2s²2p⁶3s nl, 3pnl, Singlet & Triplet Systems)
3sns ¹S
3snp ¹P°
3snd ¹D
3pnp ¹D
3sns ³S
3pnp ³P
3snp ³P°
3snd ³D
3snf ³F°
ENERGY (cm⁻¹)
1200x10³
1000x10³
800x10³
600x10³
400x10³
200x10³
0
0 — 3(0)
170720 — 3(1)
566362 — 4(1)
264745 — 3p (2)
370293 — 3d(2)
514083 — 4(1)
+x
272554
270770 — 3 (0,1,2)
269829
+x
115581
113900 — 3(0,1,2)
113095
+x
324184
324136 — 3(1,2,3)
324097
+x
634697
634622 — 4(1,2,3)
634584
+x
772355
772325 — 5(1,2,3)
772300
+x
660092 — 4(2,3,4)
+x
KEY
324184
324136 — 3(1,2,3)
324097
+x
J, lowest to highest
n
uncertainty
Energy Levels (cm⁻¹)
Ionization Limit
1 002 730. cm⁻¹
(124.319 eV)
[Ar VII 3s² ¹S₀ → Ar VIII 3s ²S½]
Ar VII
SINGLET
TRIPLET

Ar VIII GROTRIAN DIAGRAM (11 electrons, Z=18)

(Na I sequence, Configuration: $1s^2 2s^2 2p^6 nl$, Doublet System)

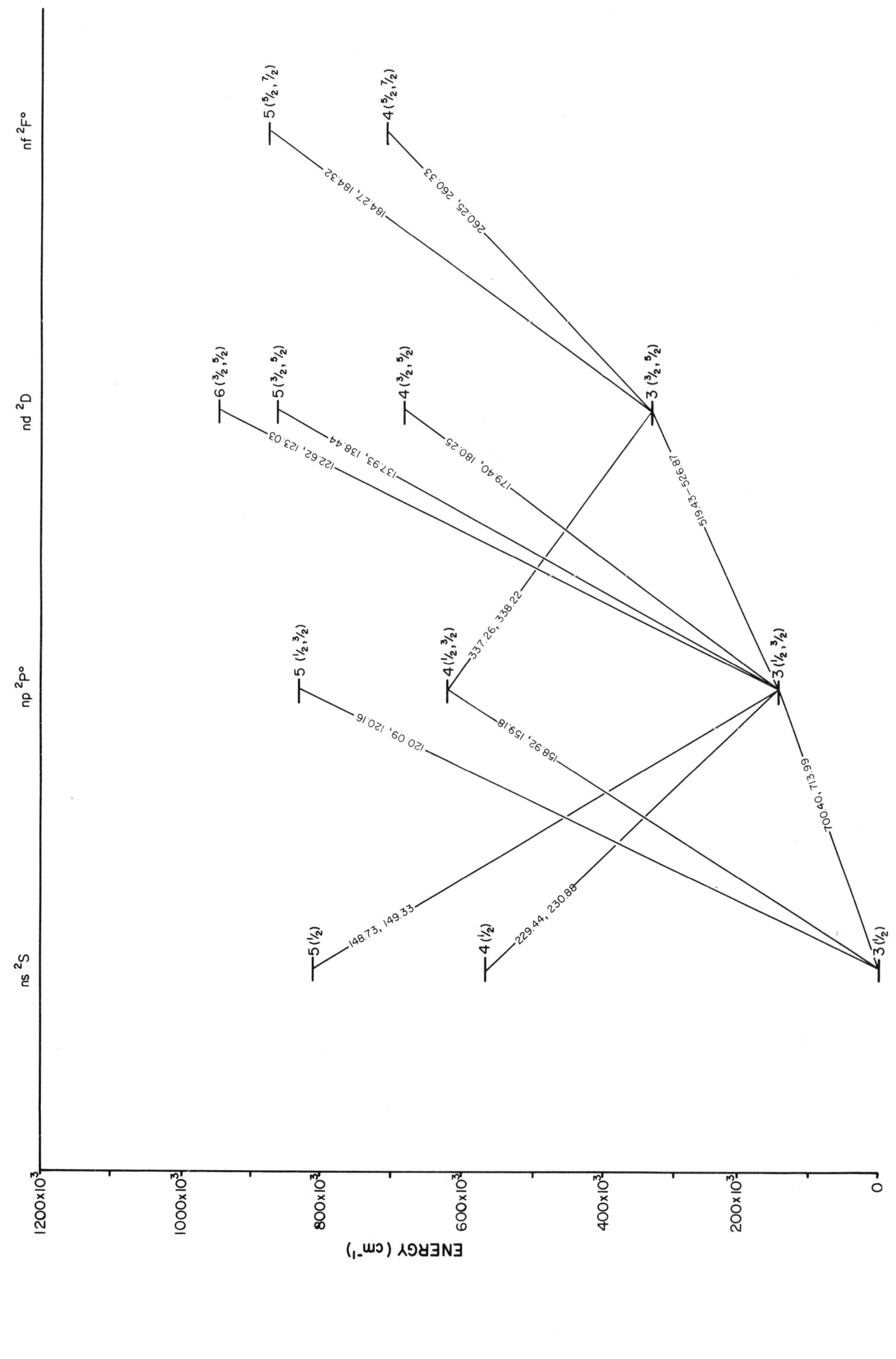

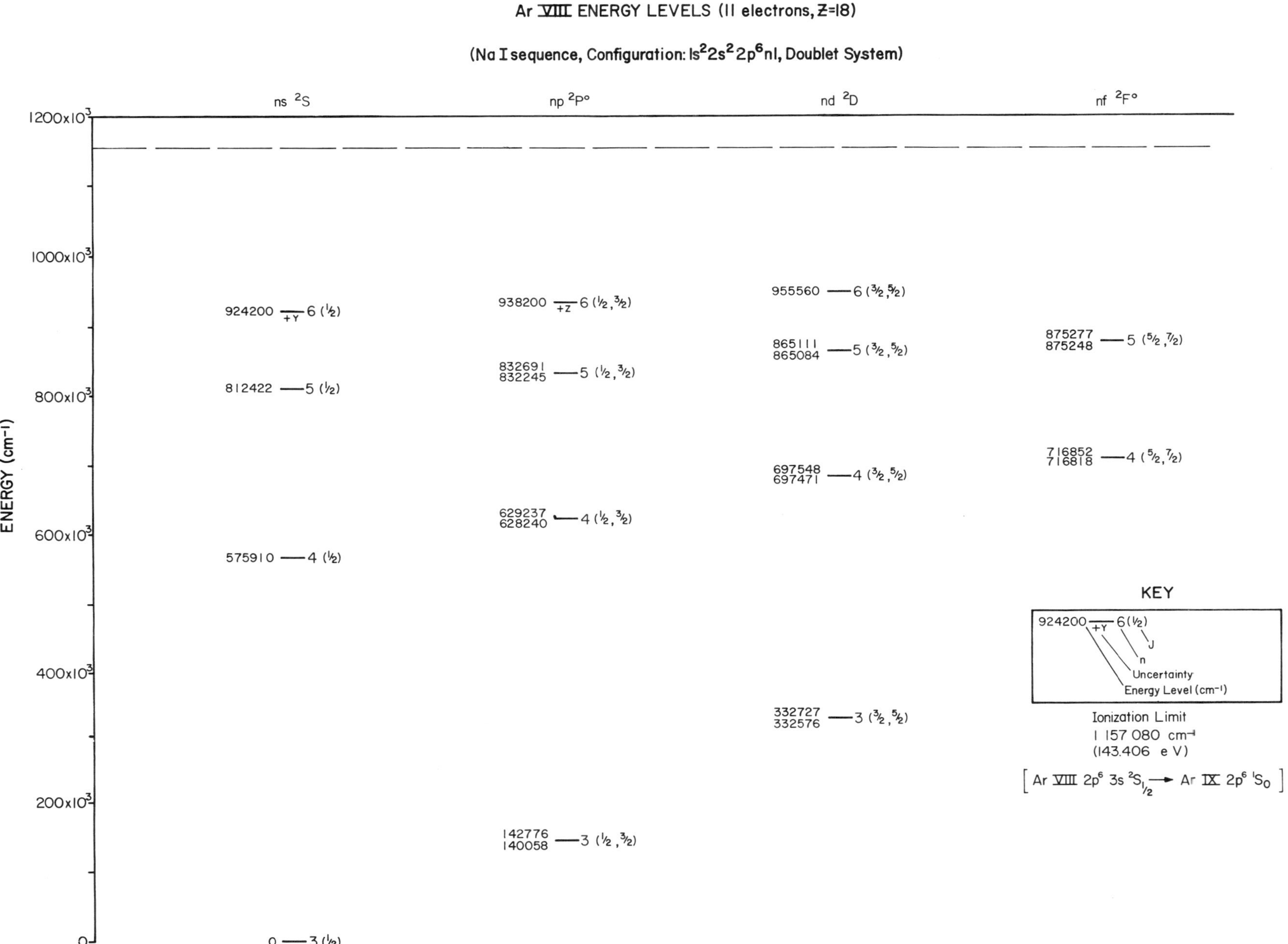
Ar VIII ENERGY LEVELS (11 electrons, Z=18)
(Na I sequence, Configuration: 1s²2s²2p⁶nl, Doublet System)
ns ²S
np ²P°
nd ²D
nf ²F°
924200 +Y 6 (½)
812422 — 5 (½)
575910 — 4 (½)
0 — 3 (½)
938200 +Z 6 (½, 3/2)
832691
832245 — 5 (½, 3/2)
629237
628240 — 4 (½, 3/2)
142776
140058 — 3 (½, 3/2)
955560 — 6 (3/2, 5/2)
865111
865084 — 5 (3/2, 5/2)
697548
697471 — 4 (3/2, 5/2)
332727
332576 — 3 (3/2, 5/2)
875277
875248 — 5 (5/2, 7/2)
716852
716818 — 4 (5/2, 7/2)
KEY
924200 +Y 6 (½)
J
n
Uncertainty
Energy Level (cm⁻¹)
Ionization Limit
1 157 080 cm⁻¹
(143.406 eV)
[Ar VIII 2p⁶ 3s ²S½ → Ar IX 2p⁶ ¹S₀]
1200x10³
1000x10³
800x10³
600x10³
400x10³
200x10³
0
ENERGY (cm⁻¹)

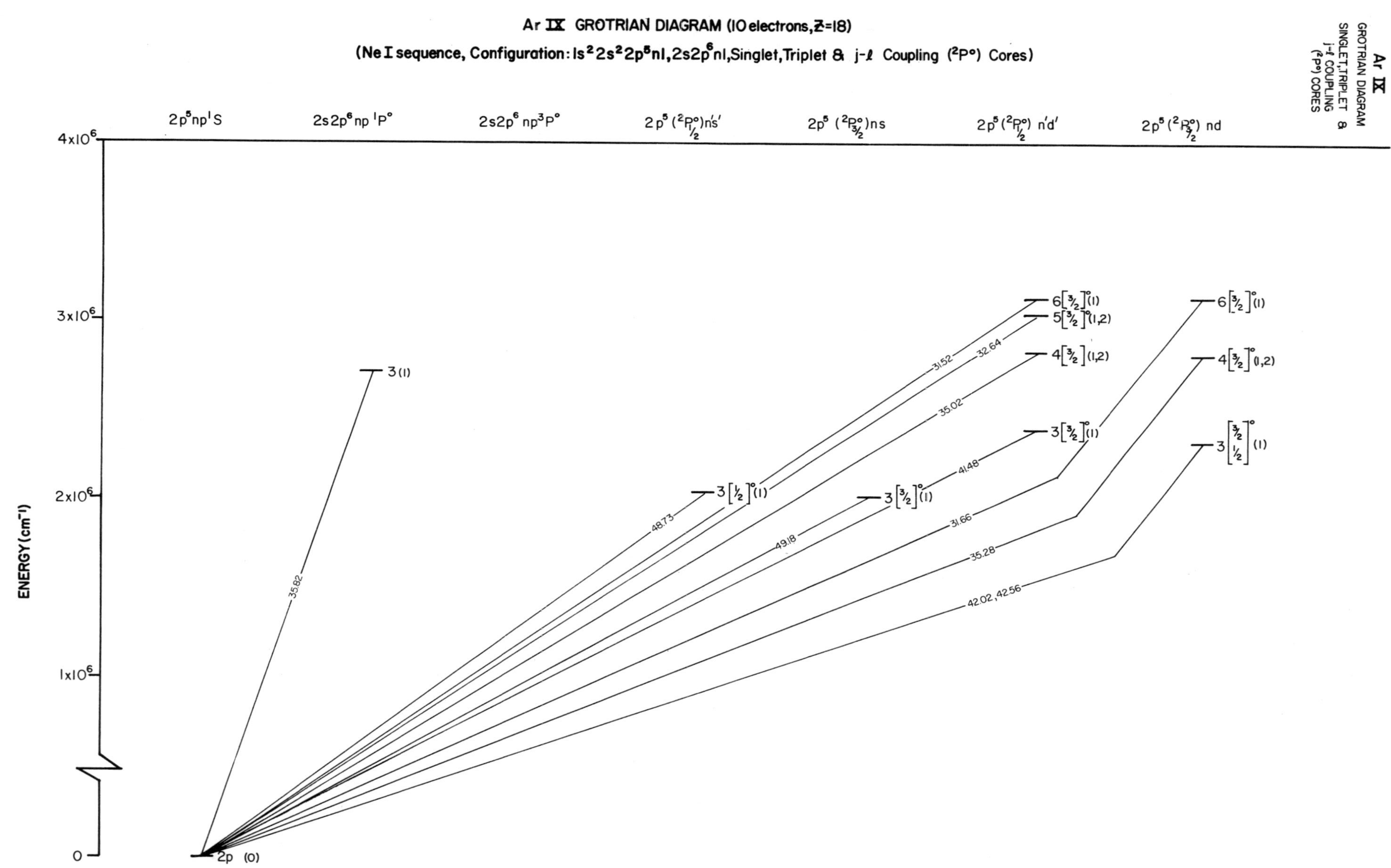

Ar IX GROTRIAN DIAGRAM (10 electrons, Z=18)
(Ne I sequence, Configuration: 1s²2s²2p⁵nl, 2s2p⁶nl, Singlet, Triplet & j-ℓ Coupling (²P°) Cores)
Ar IX
GROTRIAN DIAGRAM
SINGLET, TRIPLET &
j-ℓ COUPLING
(²P°) CORES
2p⁵np ¹S
2s2p⁶np ¹P°
2s2p⁶np ³P°
2p⁵(²P°1/2)n's'
2p⁵(²P°3/2)ns
2p⁵(²P°1/2)n'd'
2p⁵(²P°3/2)nd
ENERGY (cm⁻¹)
4x10⁶
3x10⁶
2x10⁶
1x10⁶
0
2p (0)
3 (1)
35.82
3[1/2]°(1)
48.73
3[3/2]°(1)
49.18
41.48
6[3/2]°(1)
5[3/2]°(1,2)
4[3/2](1,2)
3[3/2]°(1)
31.52
32.64
35.02
6[3/2]°(1)
4[3/2]°(1,2)
3[3/2 1/2]°(1)
31.66
35.28
42.02, 42.56

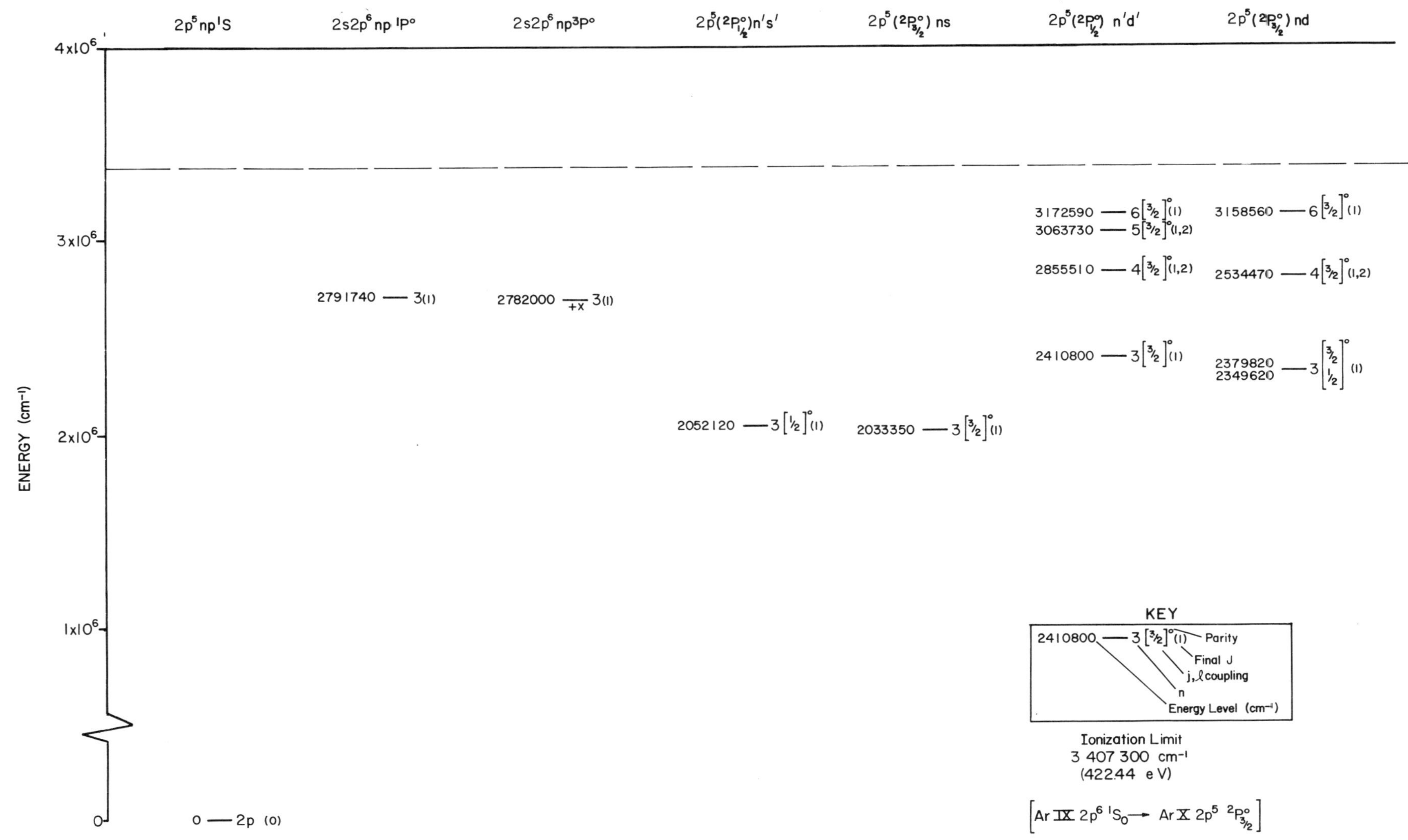
Ar IX ENERGY LEVELS (10 electrons, Z=18)
(Ne I sequence, Configuration: 1s²2s²2p⁵nl, 2s2p⁶nl, Singlet, Triplet & j-ℓ Coupling (²P°) Cores)
2p⁵np ¹S
2s2p⁶ np ¹P°
2s2p⁶ np ³P°
2p⁵(²P°1/2)n′s′
2p⁵(²P°3/2) ns
2p⁵(²P°1/2) n′d′
2p⁵(²P°3/2) nd
4x10⁶
3x10⁶
2x10⁶
1x10⁶
0
ENERGY (cm⁻¹)
3172590 — 6 [3/2]° (1)
3063730 — 5 [3/2]° (1,2)
3158560 — 6 [3/2]° (1)
2855510 — 4 [3/2]° (1,2)
2534470 — 4 [3/2]° (1,2)
2791740 — 3(1)
2782000 +x 3(1)
2410800 — 3 [3/2]° (1)
2379820
2349620 — 3 [3/2 1/2]° (1)
2052120 — 3 [1/2]° (1)
2033350 — 3 [3/2]° (1)
0 — 2p (0)
KEY
2410800 — 3 [3/2]° (1)
Parity
Final J
j, ℓ coupling
n
Energy Level (cm⁻¹)
Ionization Limit
3 407 300 cm⁻¹
(422.44 eV)
[Ar IX 2p⁶ ¹S₀ → Ar X 2p⁵ ²P°3/2]

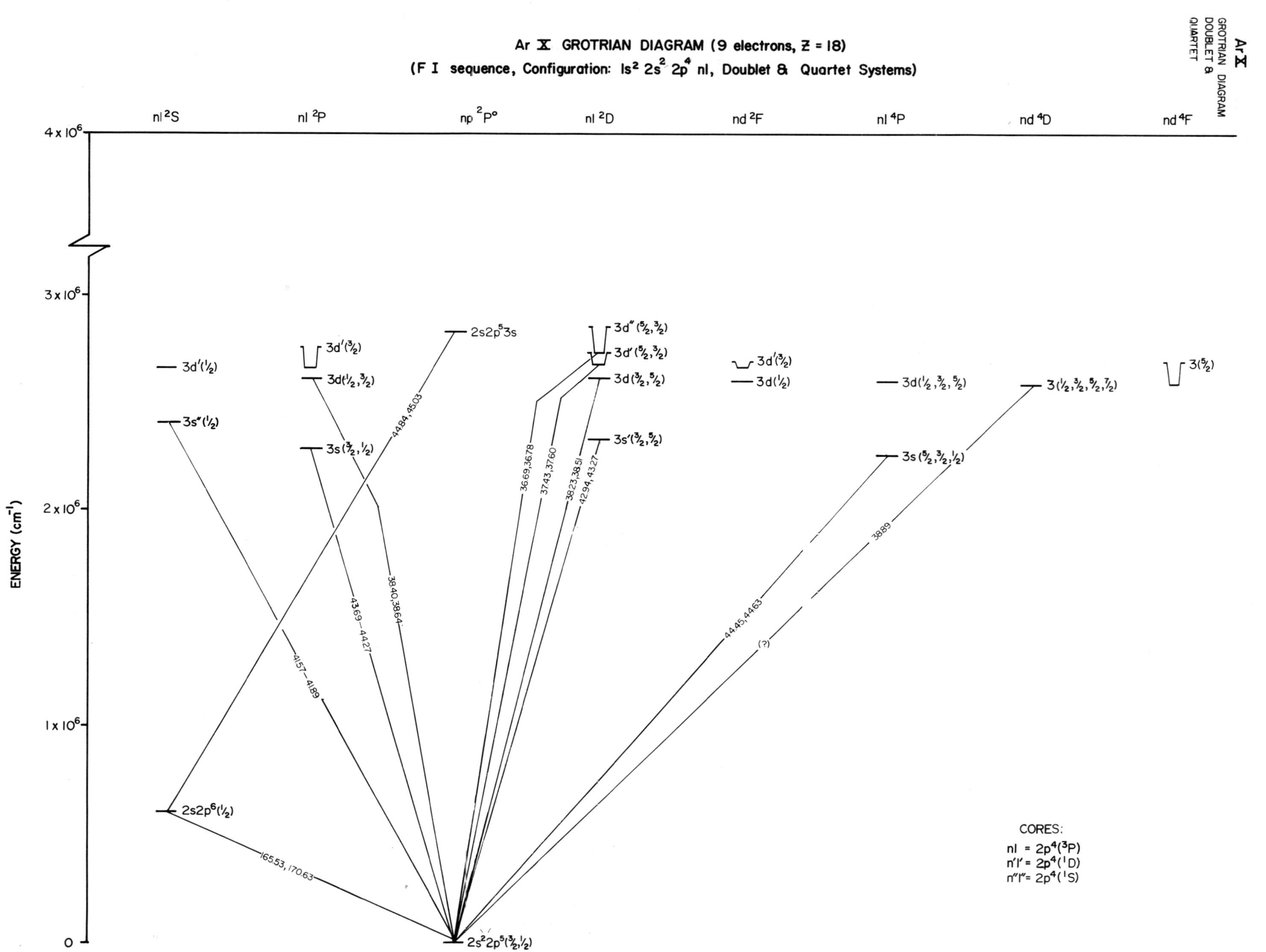
Ar X GROTRIAN DIAGRAM (9 electrons, Z = 18)
(F I sequence, Configuration: $1s^2 2s^2 2p^4$ nl, Doublet & Quartet Systems)
Ar X
GROTRIAN DIAGRAM
DOUBLET &
QUARTET
$nl\,^2S$
$nl\,^2P$
$np\,^2P^o$
$nl\,^2D$
$nd\,^2F$
$nl\,^4P$
$nd\,^4D$
$nd\,^4F$
$3d'(1/2)$
$3s''(1/2)$
$2s2p^6(1/2)$
$3d'(3/2)$
$3d(1/2, 3/2)$
$3s(3/2, 1/2)$
$2s2p^5 3s$
$3d''(5/2, 3/2)$
$3d'(5/2, 3/2)$
$3d(3/2, 5/2)$
$3s'(3/2, 5/2)$
$3d'(3/2)$
$3d(1/2)$
$3d(1/2, 3/2, 5/2)$
$3s(5/2, 3/2, 1/2)$
$3(1/2, 3/2, 5/2, 7/2)$
$3(5/2)$
$2s^2 2p^5(3/2, 1/2)$
165.53, 170.63
4157–4189
4369–4427
3840, 3864
4484, 4503
3669, 3678
3743, 3760
3823, 3851
4294, 4327
4445, 4463
(?)
3889
CORES:
$nl = 2p^4(^3P)$
$n'l' = 2p^4(^1D)$
$n''l'' = 2p^4(^1S)$
ENERGY (cm^{-1})
0
1×10^6
2×10^6
3×10^6
4×10^6

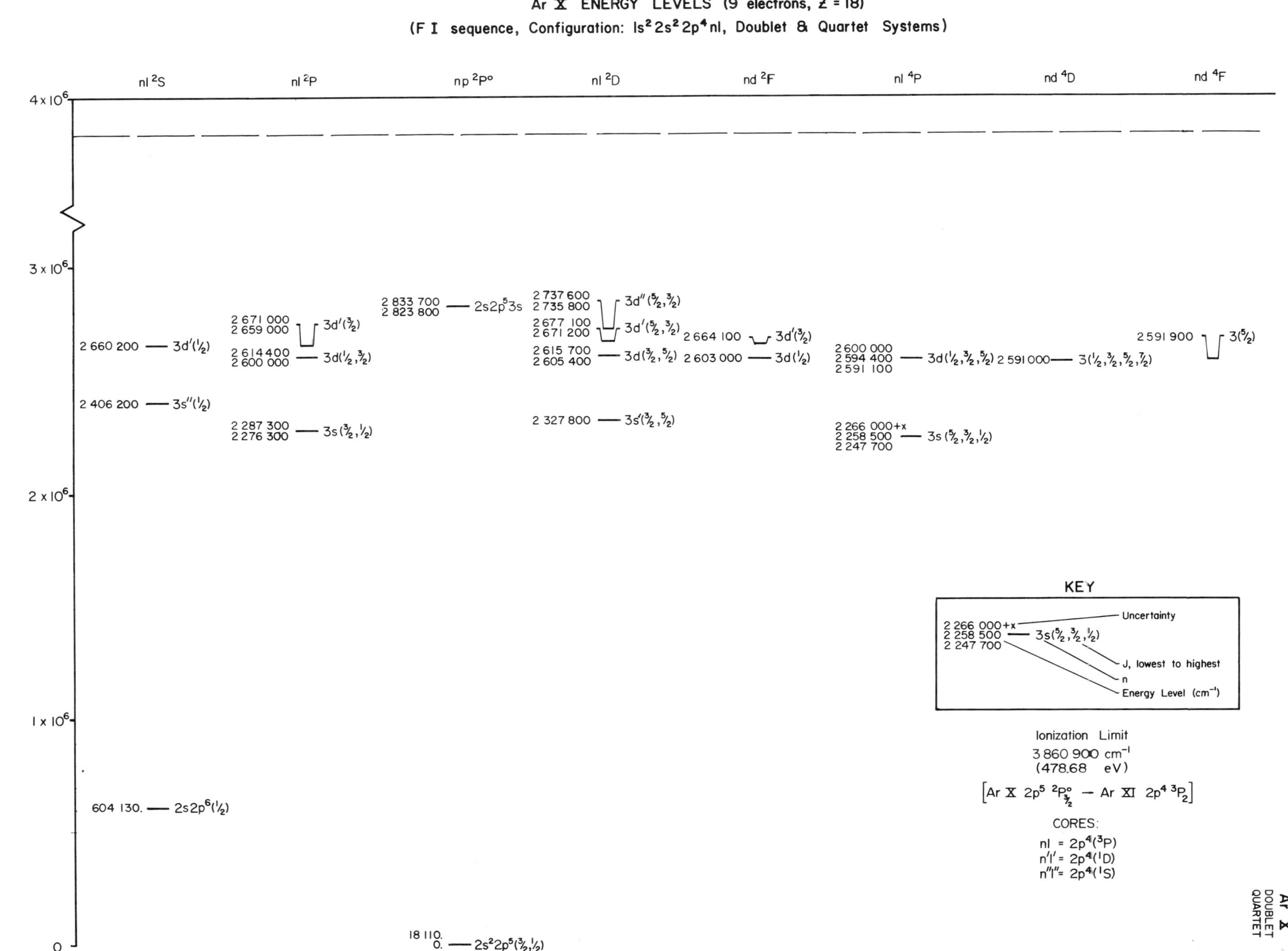
Ar X ENERGY LEVELS (9 electrons, Z = 18)
(F I sequence, Configuration: 1s^{2}2s^{2}2p^{4}nl, Doublet & Quartet Systems)
nl ^{2}S
nl ^{2}P
np ^{2}P°
nl ^{2}D
nd ^{2}F
nl ^{4}P
nd ^{4}D
nd ^{4}F
ENERGY (cm^{-1})
4 x 10^6
3 x 10^6
2 x 10^6
1 x 10^6
0
2 660 200 — 3d'(1/2)
2 406 200 — 3s''(1/2)
604 130. — 2s2p^6(1/2)
2 671 000
2 659 000 3d'(3/2)
2 614 400
2 600 000 — 3d(1/2, 3/2)
2 287 300
2 276 300 — 3s(3/2, 1/2)
2 833 700
2 823 800 — 2s2p^{5}3s
18 110.
0. — 2s^{2}2p^5(3/2, 1/2)
2 737 600
2 735 800 3d''(5/2, 3/2)
2 677 100
2 671 200 3d'(5/2, 3/2)
2 615 700
2 605 400 — 3d(3/2, 5/2)
2 327 800 — 3s'(3/2, 5/2)
2 664 100 3d'(3/2)
2 603 000 — 3d(1/2)
2 600 000
2 594 400
2 591 100 — 3d(1/2, 3/2, 5/2)
2 266 000+x
2 258 500
2 247 700 — 3s(5/2, 3/2, 1/2)
2 591 000 — 3(1/2, 3/2, 5/2, 7/2)
2 591 900 3(5/2)
KEY
2 266 000+x
2 258 500
2 247 700 — 3s(5/2, 3/2, 1/2)
Uncertainty
J, lowest to highest
n
Energy Level (cm^{-1})
Ionization Limit
3 860 900 cm^{-1}
(478.68 eV)
[Ar X 2p^5 ^{2}P°$_{3/2}$ — Ar XI 2p^4 ^{3}P$_2$]
CORES:
nl = 2p^4(^{3}P)
n'l' = 2p^4(^{1}D)
n''l'' = 2p^4(^{1}S)
Ar X
DOUBLET
QUARTET

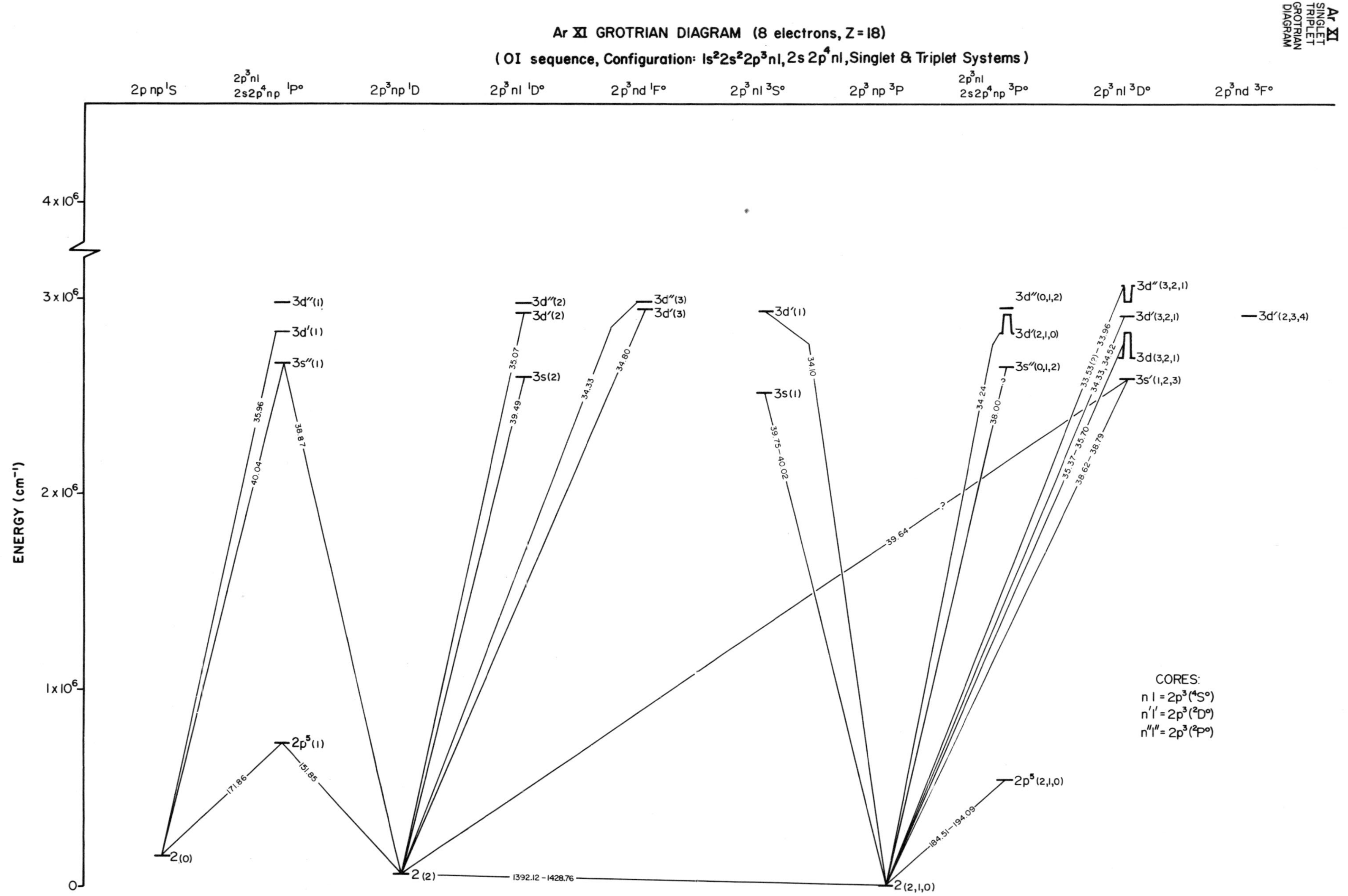

Ar XI
SINGLET
TRIPLET
GROTRIAN
DIAGRAM
Ar XI GROTRIAN DIAGRAM (8 electrons, Z = 18)
(OI sequence, Configuration: $1s^2 2s^2 2p^3 nl, 2s 2p^4 nl$, Singlet & Triplet Systems)
$2p\,np\,^1S$
$2p^3nl$ $2s2p^4np$ $^1P^o$
$2p^3np\,^1D$
$2p^3nl\,^1D^o$
$2p^3nd\,^1F^o$
$2p^3nl\,^3S^o$
$2p^3np\,^3P$
$2p^3nl$ $2s2p^4np$ $^3P^o$
$2p^3nl\,^3D^o$
$2p^3nd\,^3F^o$
4×10^6
3×10^6
2×10^6
1×10^6
0
ENERGY (cm^{-1})
3d″(1)
3d′(1)
3s″(1)
$2p^5$(1)
2(0)
2(2)
2(2,1,0)
3d″(2)
3d′(2)
3s(2)
3d″(3)
3d′(3)
3d′(1)
3s(1)
3d″(0,1,2)
3d′(2,1,0)
3s″(0,1,2)
3d″(3,2,1)
3d′(3,2,1)
3d(3,2,1)
3s′(1,2,3)
3d′(2,3,4)
$2p^5$(2,1,0)
35.96
40.04
38.87
171.86
151.85
35.07
39.49
34.33
34.80
34.10
39.75 – 40.02
34.24
38.00 – ?
33.53(?) – 33.96
34.33, 34.52
35.37 – 35.70
38.62 – 38.79
39.64 – ?
1392.12 – 1428.76
184.51 – 194.09
CORES:
$n\,l = 2p^3(^4S^o)$
$n'l' = 2p^3(^2D^o)$
$n''l'' = 2p^3(^2P^o)$

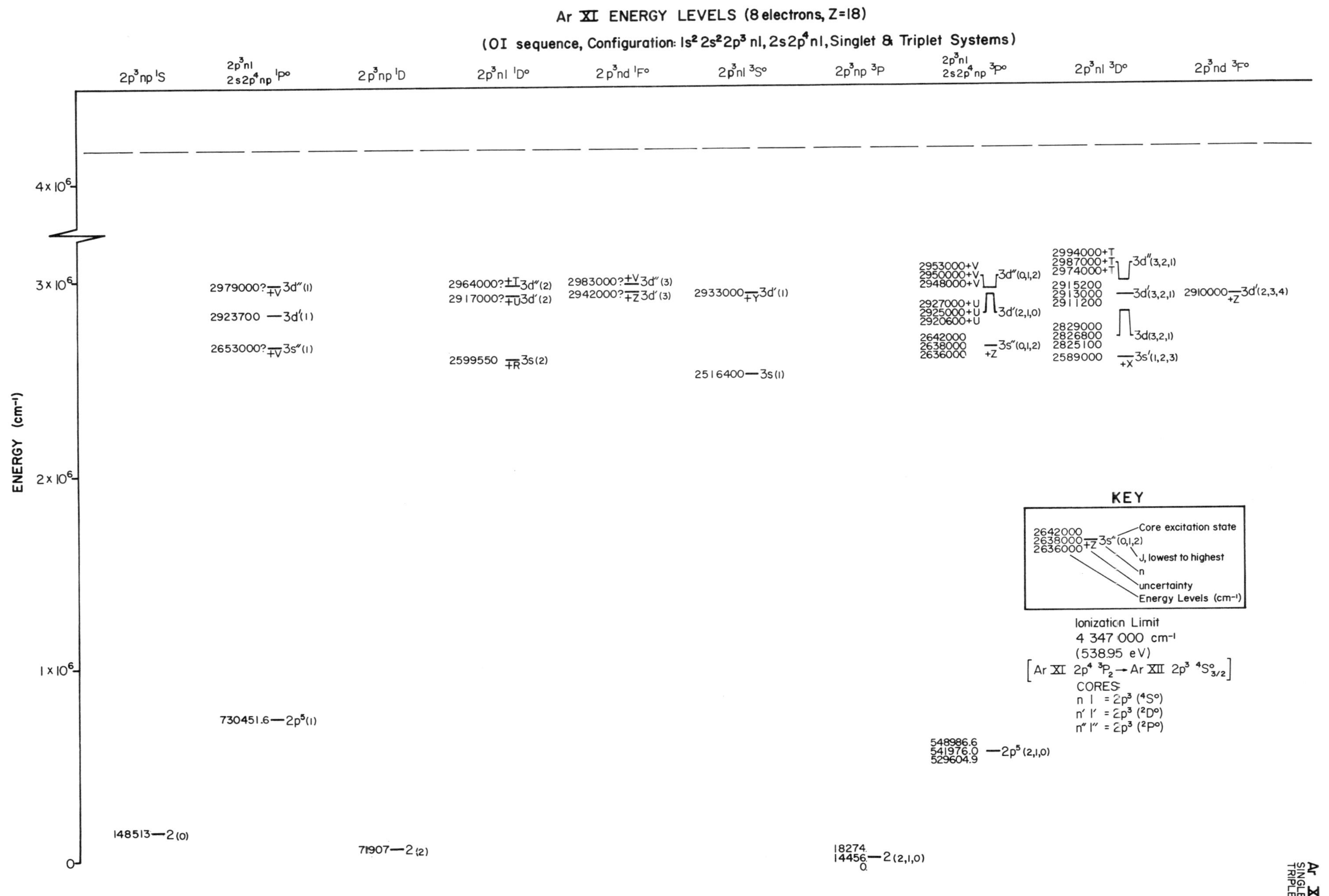
Ar XI ENERGY LEVELS (8 electrons, Z=18)
(OI sequence, Configuration: 1s² 2s² 2p³ nl, 2s2p⁴ nl, Singlet & Triplet Systems)
2p³np ¹S
2p³nl 2s2p⁴np ¹P°
2p³np ¹D
2p³nl ¹D°
2p³nd ¹F°
2p³nl ³S°
2p³np ³P
2p³nl 2s2p⁴np ³P°
2p³nl ³D°
2p³nd ³F°
ENERGY (cm⁻¹)
4 x 10⁶
3 x 10⁶
2 x 10⁶
1 x 10⁶
0
148513—2(0)
2979000? ∓V 3d″(1)
2923700 —3d′(1)
2653000? ∓V 3s″(1)
730451.6—2p⁵(1)
71907—2(2)
2964000? ±T 3d″(2)
2917000? ∓U 3d′(2)
2599550 ∓R 3s(2)
2983000? ±V 3d″(3)
2942000? ∓Z 3d′(3)
2933000 ∓Y 3d′(1)
2516400—3s(1)
18274
14456—2(2,1,0)
0
2953000+V
2950000+V
2948000+V 3d″(0,1,2)
2927000+U
2925000+U
2920600+U 3d′(2,1,0)
2642000
2638000
2636000 ∓Z 3s″(0,1,2)
548986.6
541976.0
529604.9 —2p⁵(2,1,0)
2994000+T
2987000+T
2974000+T 3d″(3,2,1)
2915200
2913000
2911200 —3d′(3,2,1)
2829000
2826800
2825100 3d(3,2,1)
2589000 ∓X 3s′(1,2,3)
2910000 ∓Z 3d′(2,3,4)
KEY
2642000
2638000
2636000 ∓Z 3s″(0,1,2)
Core excitation state
J, lowest to highest
n
uncertainty
Energy Levels (cm⁻¹)
Ionization Limit
4 347 000 cm⁻¹
(538.95 eV)
[Ar XI 2p⁴ ³P₂ → Ar XII 2p³ ⁴S°₃/₂]
CORES
n l = 2p³ (⁴S°)
n′ l′ = 2p³ (²D°)
n″ l″ = 2p³ (²P°)

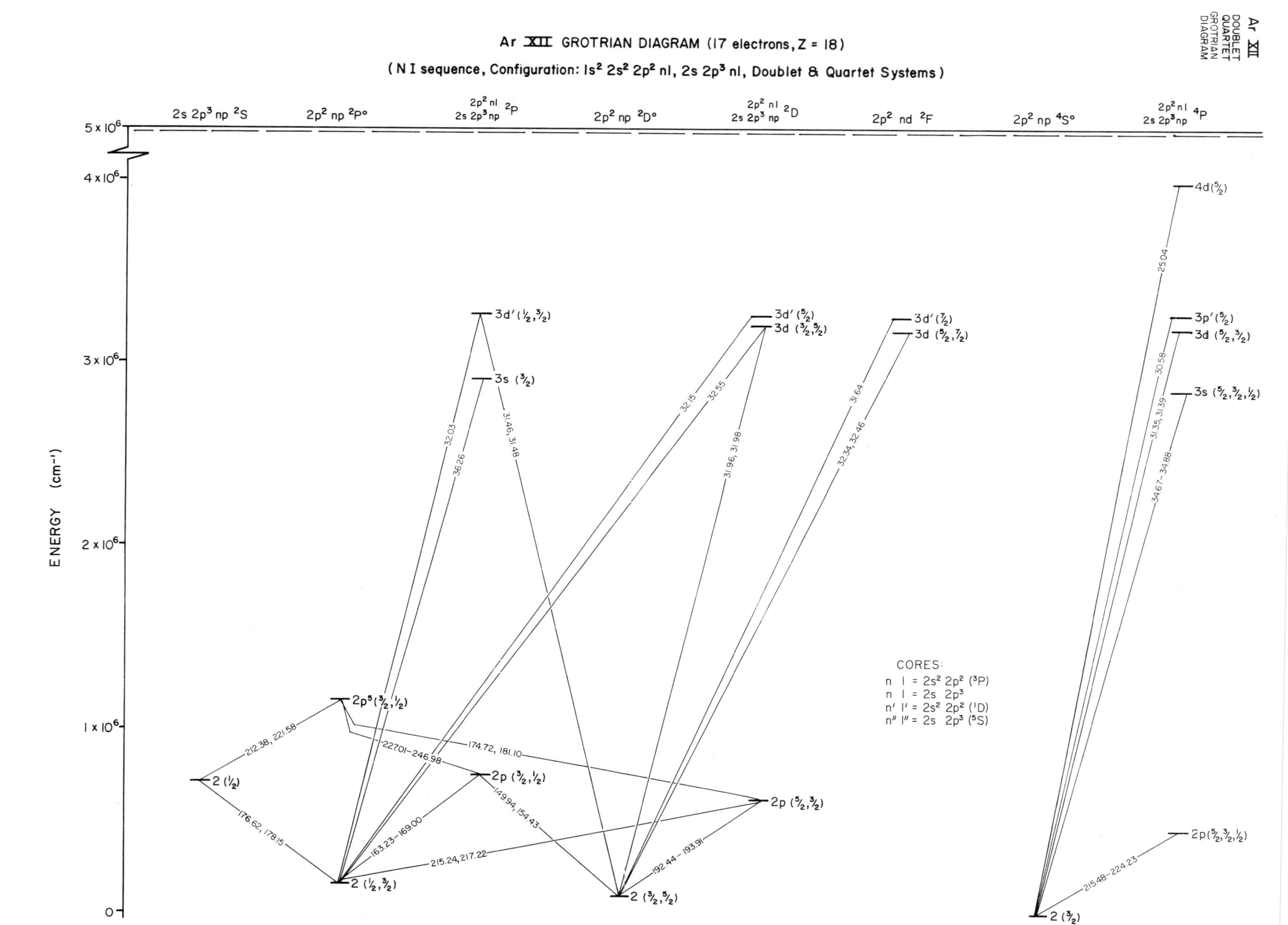
Ar XII GROTRIAN DIAGRAM (17 electrons, Z = 18)
(N I sequence, Configuration: 1s² 2s² 2p² nl, 2s 2p³ nl, Doublet & Quartet Systems)
Ar XII
DOUBLET
QUARTET
GROTRIAN
DIAGRAM
2s 2p³ np ²S
2p² np ²P°
2p² nl ²P
2s 2p³ np
2p² np ²D°
2p² nl ²D
2s 2p³ np
2p² nd ²F
2p² np ⁴S°
2p² nl ⁴P
2s 2p³ np
5 x 10⁶
4 x 10⁶
3 x 10⁶
2 x 10⁶
1 x 10⁶
0
ENERGY (cm⁻¹)
3d′(¹⁄₂, ³⁄₂)
3s (³⁄₂)
32.03
36.26
31.46, 31.48
3d′(⁵⁄₂)
3d (³⁄₂, ⁵⁄₂)
32.15
32.55
31.96, 31.98
3d′(⁷⁄₂)
3d (⁵⁄₂, ⁷⁄₂)
31.64
32.34, 32.46
4d(⁵⁄₂)
25.04
3p′(⁵⁄₂)
3d (⁵⁄₂, ³⁄₂)
30.58
3s (⁵⁄₂, ³⁄₂, ¹⁄₂)
31.35, 31.39
34.67–34.88
CORES:
n l = 2s² 2p² (³P)
n l = 2s 2p³
n′ l′ = 2s² 2p² (¹D)
n″ l″ = 2s 2p³ (⁵S)
2p⁵(³⁄₂, ¹⁄₂)
212.38, 221.58
174.72, 181.10
227.01–246.98
2 (¹⁄₂)
2p (³⁄₂, ¹⁄₂)
2p (⁵⁄₂, ³⁄₂)
176.62, 178.15
163.23–169.00
149.94, 154.43
2p(⁵⁄₂, ³⁄₂, ¹⁄₂)
192.44–193.91
215.24, 217.22
215.48–224.23
2 (¹⁄₂, ³⁄₂)
2 (³⁄₂, ⁵⁄₂)
2 (³⁄₂)

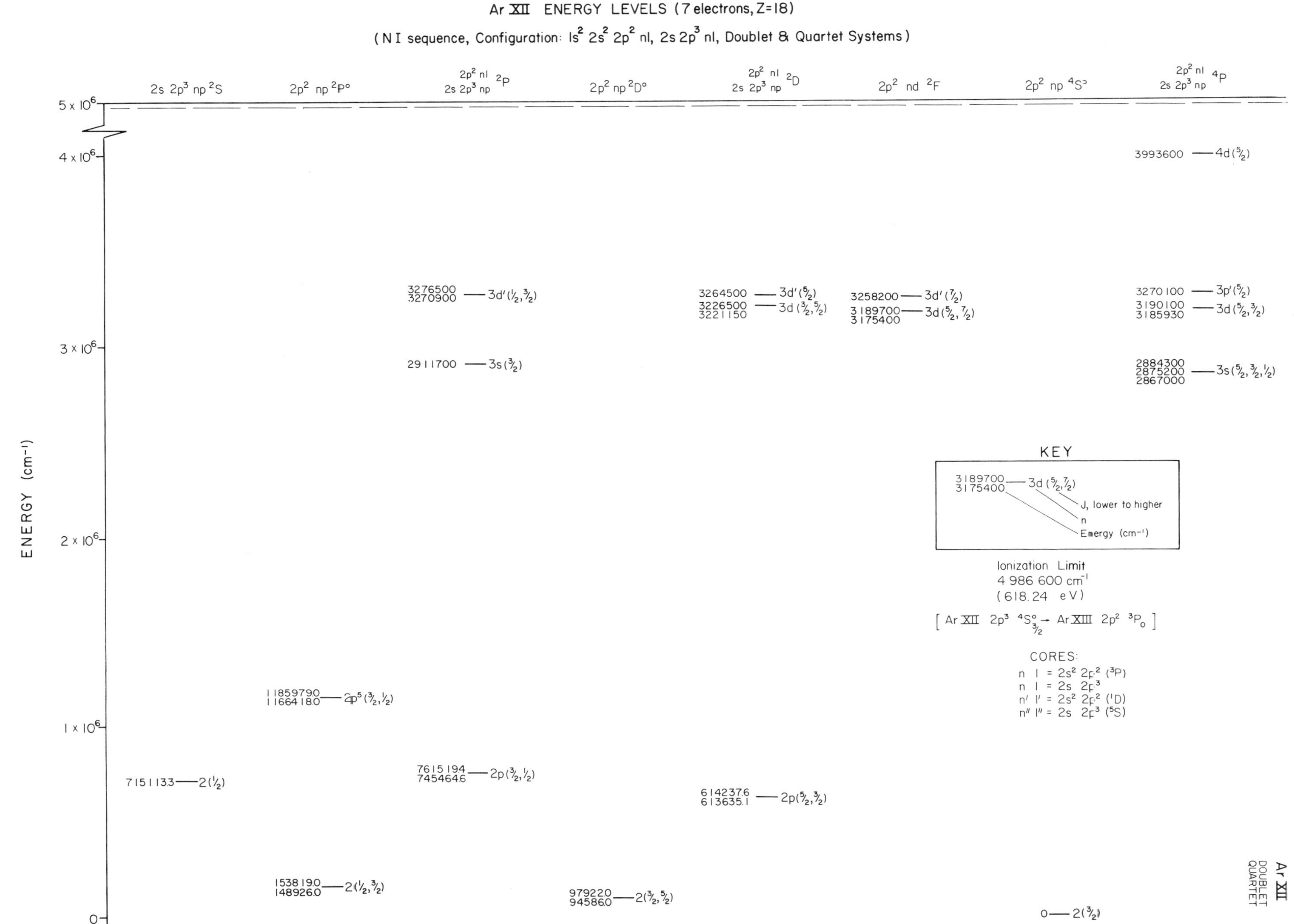

Ar XII ENERGY LEVELS (7 electrons, Z=18)
(N I sequence, Configuration: $1s^2\,2s^2\,2p^2$ nl, $2s\,2p^3$ nl, Doublet & Quartet Systems)
$2s\,2p^3\,np\,^2S$
$2p^2\,np\,^2P^o$
$2p^2$ nl / $2s\,2p^3$ np ^2P
$2p^2\,np\,^2D^o$
$2p^2$ nl / $2s\,2p^3$ np ^2D
$2p^2\,nd\,^2F$
$2p^2\,np\,^4S^o$
$2p^2$ nl / $2s\,2p^3$ np ^4P
5 x 10⁶
4 x 10⁶
3 x 10⁶
2 x 10⁶
1 x 10⁶
0
ENERGY (cm⁻¹)
7151133 — 2(1/2)
11859790 / 11664180 — $2p^5$(3/2, 1/2)
1538190 / 1489260 — 2(1/2, 3/2)
3276500 / 3270900 — 3d′(1/2, 3/2)
2911700 — 3s(3/2)
7615194 / 7454646 — 2p(3/2, 1/2)
979220 / 945860 — 2(3/2, 5/2)
3264500 — 3d′(5/2)
3226500 / 3221150 — 3d(3/2, 5/2)
6142376 / 613635.1 — 2p(5/2, 3/2)
3258200 — 3d′(7/2)
3189700 / 3175400 — 3d(5/2, 7/2)
0 — 2(3/2)
3993600 — 4d(5/2)
3270100 — 3p′(5/2)
3190100 / 3185930 — 3d(5/2, 3/2)
2884300 / 2875200 / 2867000 — 3s(5/2, 3/2, 1/2)
KEY
3189700 / 3175400 — 3d(5/2, 7/2)
J, lower to higher
n
Energy (cm⁻¹)
Ionization Limit
4 986 600 cm⁻¹
(618.24 eV)
[Ar XII $2p^3\,^4S^o_{3/2}$ → Ar XIII $2p^2\,^3P_0$]
CORES:
n l = $2s^2\,2p^2$ (^3P)
n l = $2s\,2p^3$
n′ l′ = $2s^2\,2p^2$ (^1D)
n″ l″ = $2s\,2p^3$ (^5S)
Ar XII
DOUBLET
QUARTET

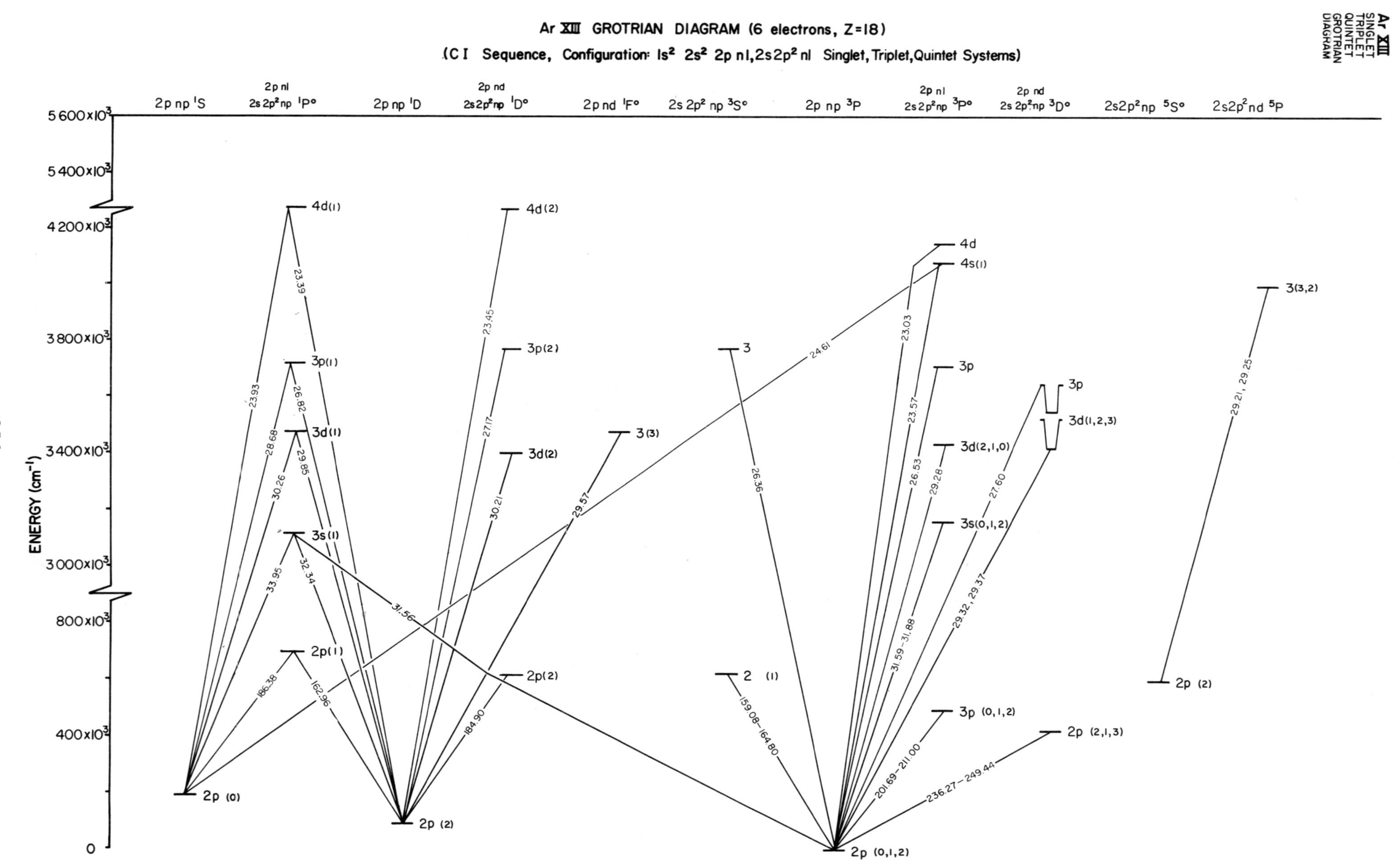

Ar XIII
SINGLET
TRIPLET
QUINTET
GROTRIAN
DIAGRAM
Ar XIII GROTRIAN DIAGRAM (6 electrons, Z=18)
(C I Sequence, Configuration: 1s² 2s² 2p nl, 2s2p² nl Singlet, Triplet, Quintet Systems)
2p np ¹S
2p nl 2s 2p² np ¹P°
2p np ¹D
2p nd 2s 2p² np ¹D°
2p nd ¹F°
2s 2p² np ³S°
2p np ³P
2p nl 2s 2p² np ³P°
2p nd 2s 2p² np ³D°
2s2p² np ⁵S°
2s2p² nd ⁵P
ENERGY (cm⁻¹)
5 600x10³
5 400x10³
4 200x10³
3 800x10³
3 400x10³
3 000x10³
800x10³
400x10³
0
4d(1)
3p(1)
3d(1)
3s(1)
2p(1)
2p (0)
4d(2)
3p(2)
3d(2)
2p(2)
2p (2)
3(3)
3
2 (1)
4d
4s(1)
3p
3d(2,1,0)
3s(0,1,2)
3p (0,1,2)
2p (0,1,2)
3p
3d(1,2,3)
2p (2,1,3)
3(3,2)
2p (2)
23.39
23.93
26.82
28.68
29.85
30.26
32.34
33.95
186.38
162.96
23.45
27.17
30.21
184.90
29.57
31.56
24.61
26.36
159.08–164.80
23.03
23.57
26.53
29.28
31.59–31.88
201.69–211.00
27.60
29.32, 29.37
236.27–249.44
29.21, 29.25

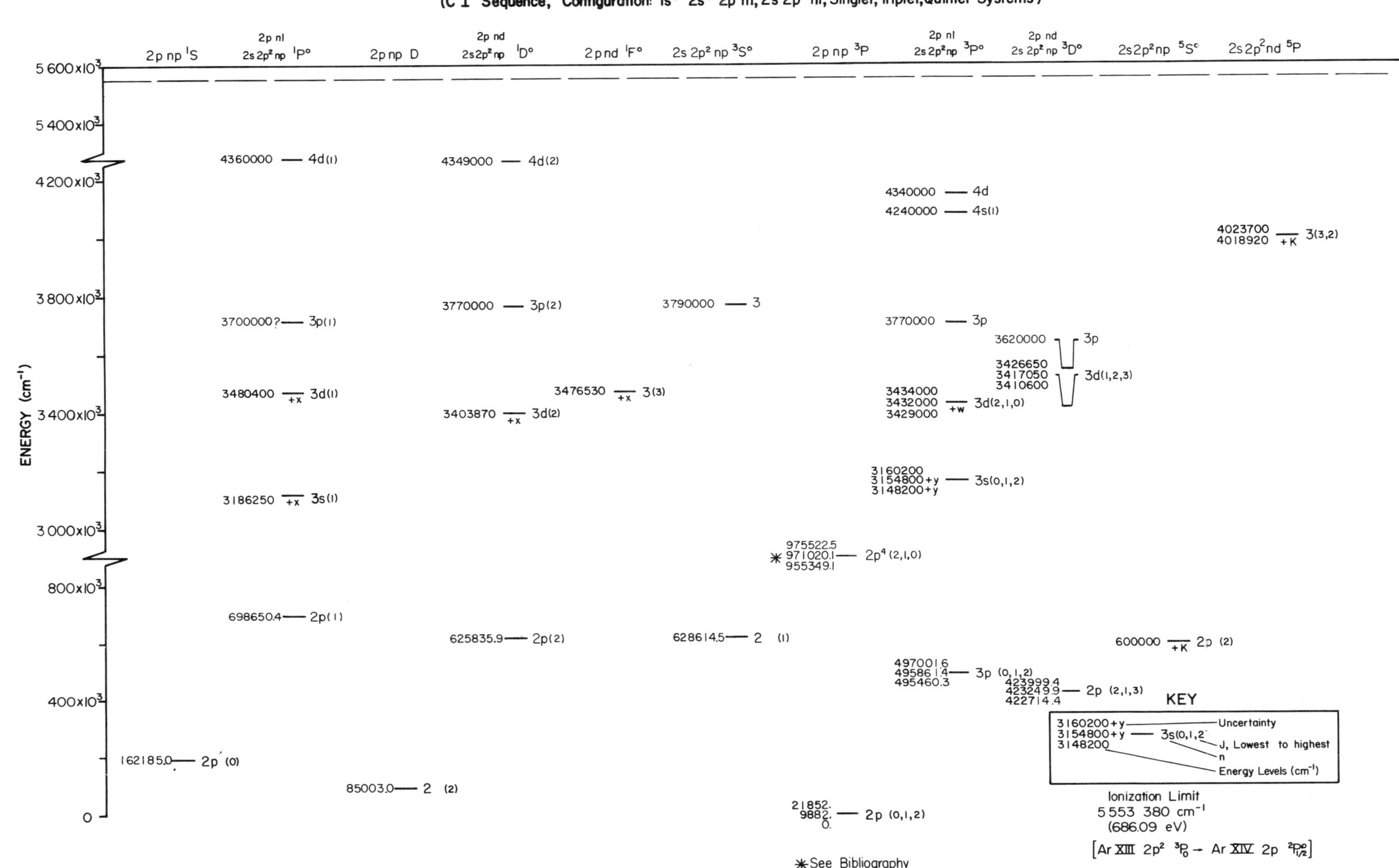
Ar XIII ENERGY LEVELS (6 electrons, Z=18)
(C I Sequence, Configuration: 1s² 2s² 2p nl, 2s 2p² nl, Singlet, Triplet, Quintet Systems)
2p np ¹S
2p nl 2s 2p² np ¹P°
2p np D
2p nd 2s 2p² np ¹D°
2p nd ¹F°
2s 2p² np ³S°
2p np ³P
2p nl 2s 2p² np ³P°
2p nd 2s 2p² np ³D°
2s 2p² np ⁵S°
2s 2p² nd ⁵P
ENERGY (cm⁻¹)
5 600x10³
5 400x10³
4 200x10³
3 800x10³
3 400x10³
3 000x10³
800x10³
400x10³
0
4360000 — 4d(1)
4349000 — 4d(2)
4340000 — 4d
4240000 — 4s(1)
4023700
4018920 +K 3(3,2)
3770000 — 3p(2)
3790000 — 3
3700000? — 3p(1)
3770000 — 3p
3620000 3p
3426650
3417050
3410600 3d(1,2,3)
3480400 +x 3d(1)
3476530 +x 3(3)
3434000
3432000
3429000 +w 3d(2,1,0)
3403870 +x 3d(2)
3160200
3154800+y
3148200+y — 3s(0,1,2)
3186250 +x 3s(1)
975522.5
✳971020.1 — 2p⁴(2,1,0)
955349.1
698650.4 — 2p(1)
625835.9 — 2p(2)
628614.5 — 2 (1)
600000 +K 2p (2)
497001.6
495861.4 — 3p (0,1,2)
495460.3
423999.4
423249.9 — 2p (2,1,3)
422714.4
KEY
3160200+y — Uncertainty
3154800+y — 3s(0,1,2 — J, Lowest to highest
3148200 — n
Energy Levels (cm⁻¹)
162185.0 — 2p′ (0)
85003.0 — 2 (2)
21852.
9882. — 2p (0,1,2)
0.
✳See Bibliography
Ionization Limit
5 553 380 cm⁻¹
(686.09 eV)
[Ar XIII 2p² ³P₀ → Ar XIV 2p ²P°₁/₂]

Ar XIV
DOUBLET
QUARTET
GROTRIAN
DIAGRAM

Ar XIV GROTRIAN DIAGRAM (5 electrons, Z=18)

(B I sequence, Configurations: $1s^22s^2nl$, $1s^22s2pnl$, Doublet & Quartet Systems)

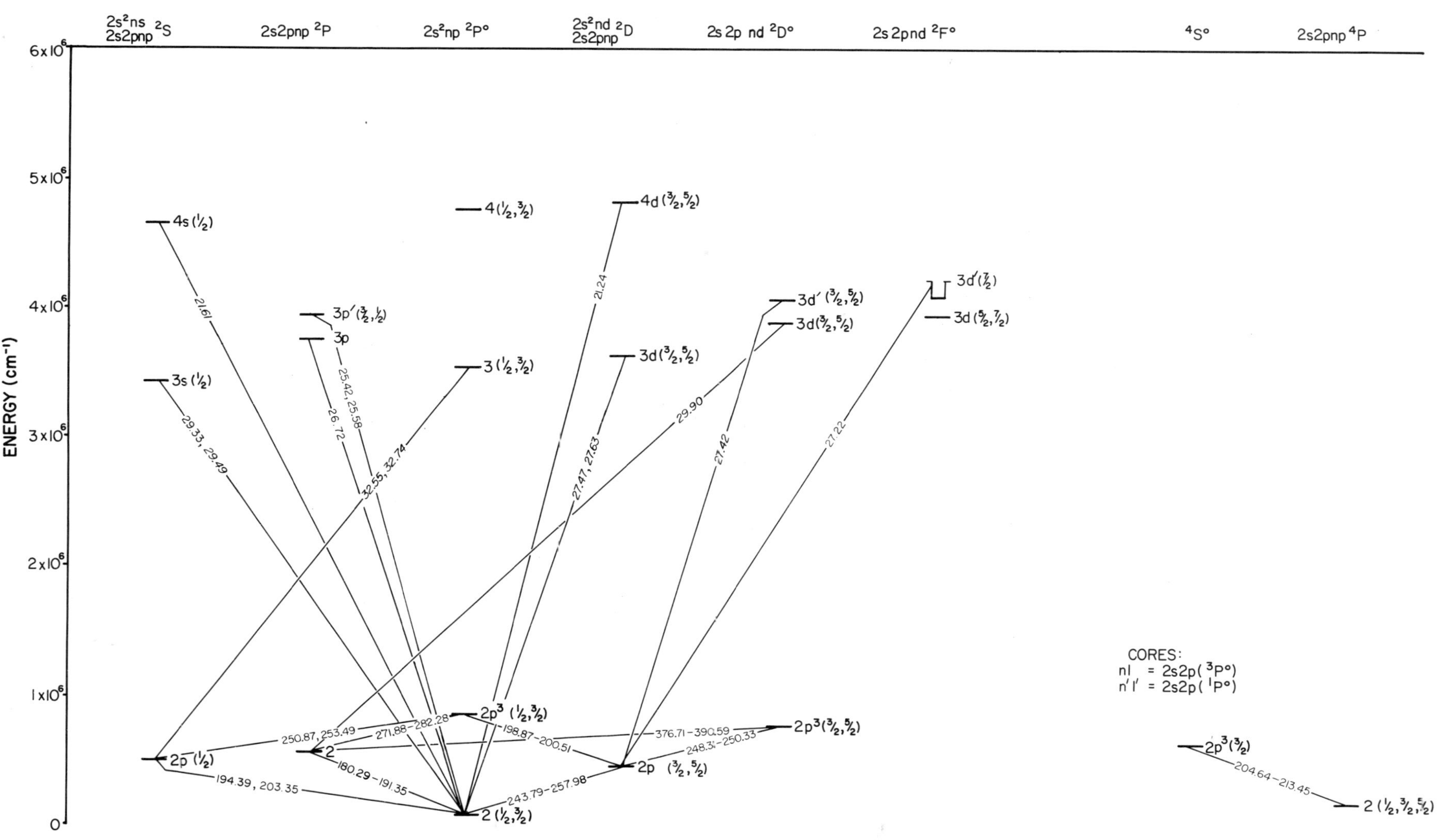

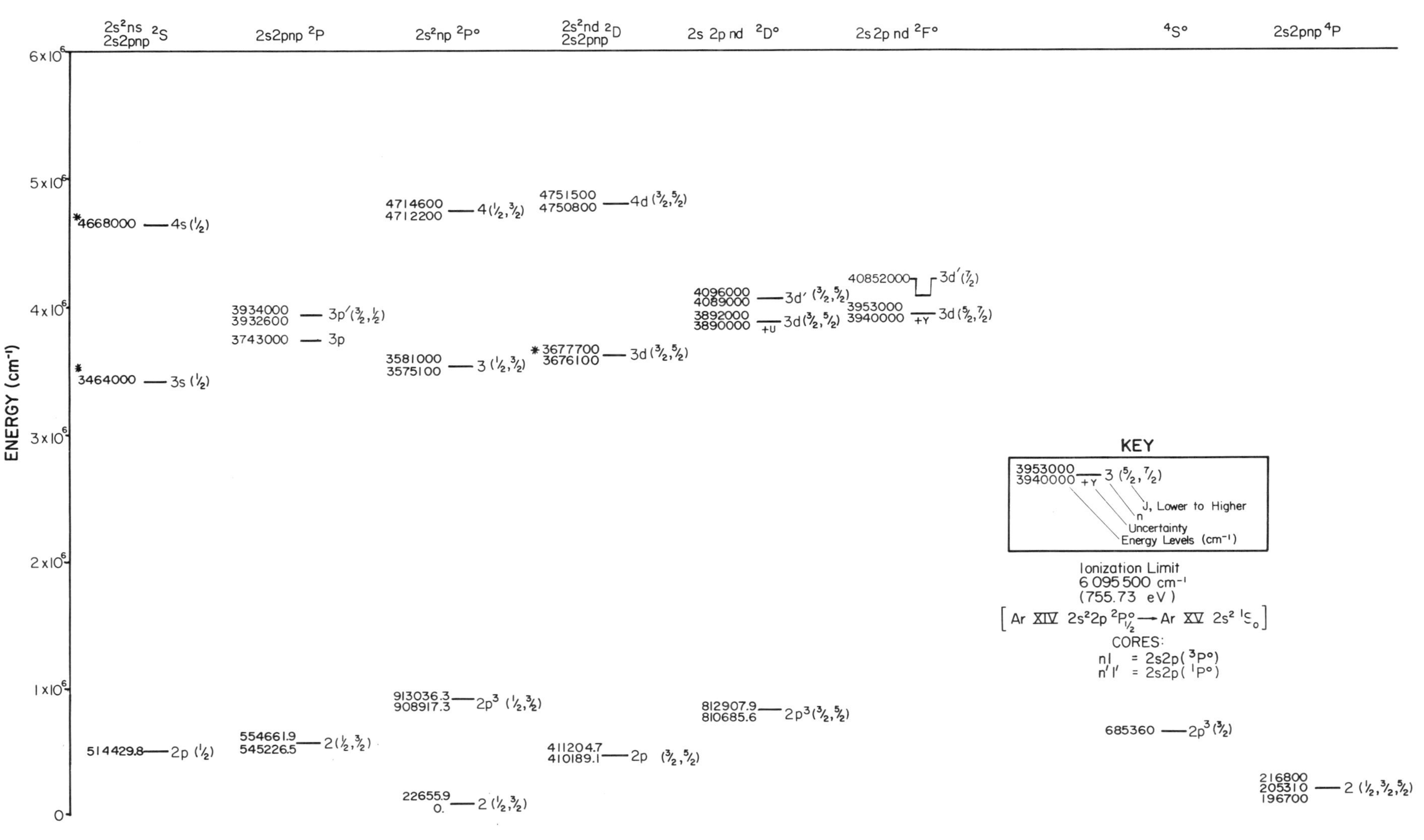
Ar XIV ENERGY LEVELS (5 electrons, Z=18)
(B I sequence, Configurations: 1s²2s²nl, 1s²2s2pnl, Doublet & Quartet Systems)
ENERGY (cm⁻¹)
6x10⁶
5x10⁶
4x10⁶
3x10⁶
2x10⁶
1x10⁶
0
2s²ns ²S
2s2pnp
2s2pnp ²P
2s²np ²P°
2s²nd ²D
2s2pnp
2s 2p nd ²D°
2s2p nd ²F°
⁴S°
2s2pnp ⁴P
*4668000 — 4s(1/2)
*3464000 — 3s(1/2)
514429.8 — 2p(1/2)
3934000
3932600 — 3p′(3/2,1/2)
3743000 — 3p
554661.9
545226.5 — 2(1/2,3/2)
4714600
4712200 — 4(1/2,3/2)
3581000
3575100 — 3(1/2,3/2)
913036.3
908917.3 — 2p³ (1/2,3/2)
22655.9
0. — 2(1/2,3/2)
4751500
4750800 — 4d(3/2,5/2)
*3677700
3676100 — 3d(3/2,5/2)
411204.7
410189.1 — 2p (3/2,5/2)
4096000
4089000 — 3d′(3/2,5/2)
3892000
3890000 +U 3d(3/2,5/2)
812907.9
810685.6 — 2p³(3/2,5/2)
40852000 — 3d′(7/2)
3953000
3940000 +Y 3d(5/2,7/2)
KEY
3953000
3940000 +Y 3(5/2,7/2)
J, Lower to Higher
n
Uncertainty
Energy Levels (cm⁻¹)
Ionization Limit
6 095 500 cm⁻¹
(755.73 eV)
[Ar XIV 2s²2p ²P°1/2 → Ar XV 2s² ¹S0]
CORES:
nl = 2s2p(³P°)
n′l′ = 2s2p(¹P°)
685360 — 2p³(3/2)
216800
205310
196700 — 2(1/2,3/2,5/2)
* see bibliography

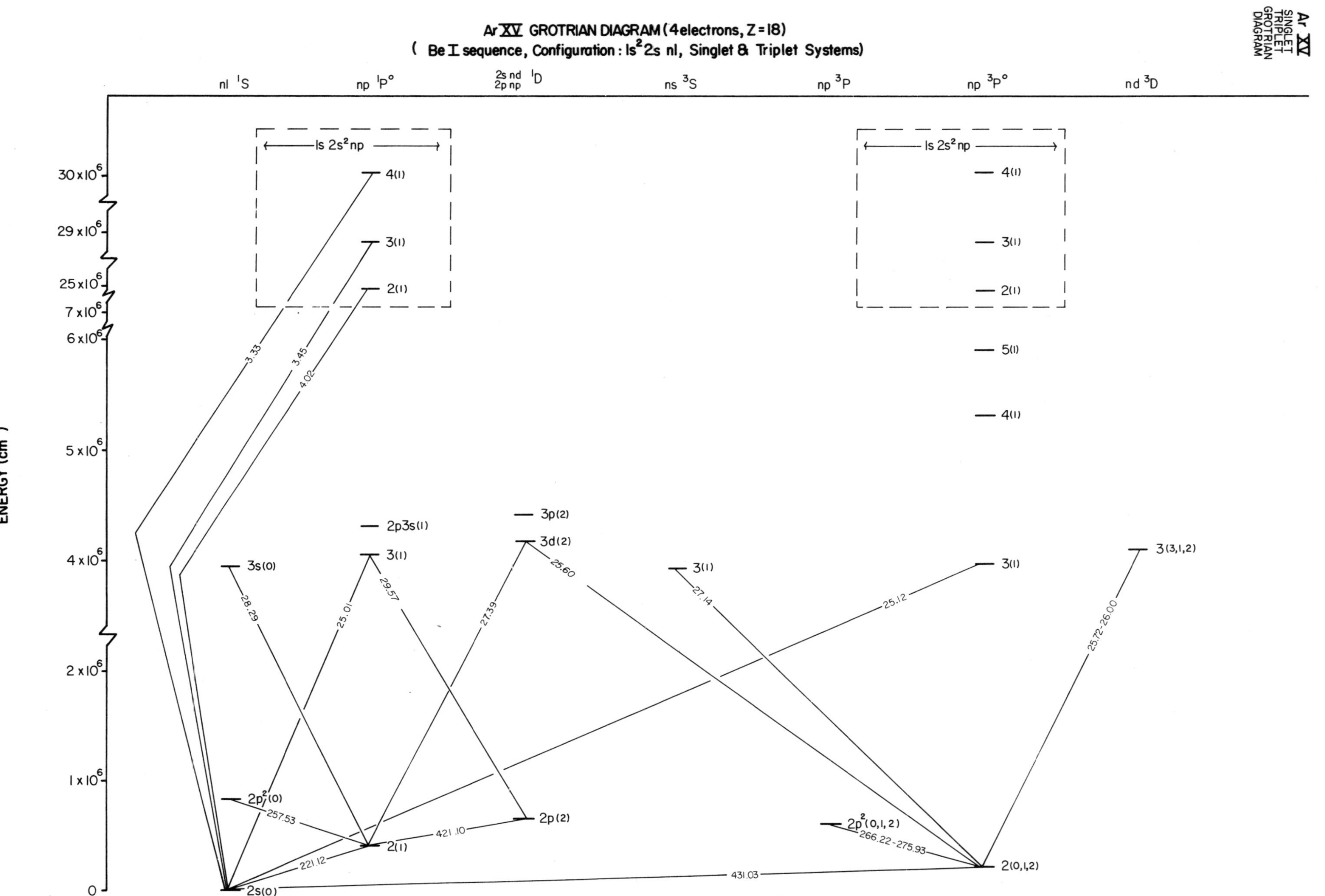

Ar XV
SINGLET
TRIPLET
GROTRIAN
DIAGRAM
Ar XV GROTRIAN DIAGRAM (4electrons, Z=18)
(Be I sequence, Configuration: 1s^{2}2s nl, Singlet & Triplet Systems)
nl ^{1}S
np ^{1}P°
2snd 2pnp ^{1}D
ns ^{3}S
np ^{3}P
np ^{3}P°
nd ^{3}D
1s 2s^{2}np
4(1)
3(1)
2(1)
1s 2s^{2}np
4(1)
3(1)
2(1)
5(1)
4(1)
2p3s(1)
3p(2)
3d(2)
3s(0)
3(1)
3(1)
3(1)
3(3,1,2)
2p^2(0)
2p(2)
2(1)
2p^2(0,1,2)
2(0,1,2)
2s(0)
3.33
3.45
4.02
28.29
25.01
29.57
27.39
25.60
27.14
25.12
25.72-26.00
257.53
421.10
221.12
266.22-275.93
431.03
ENERGY (cm^{-1})
30x10^6
29x10^6
25x10^6
7x10^6
6x10^6
5x10^6
4x10^6
2x10^6
1x10^6
0

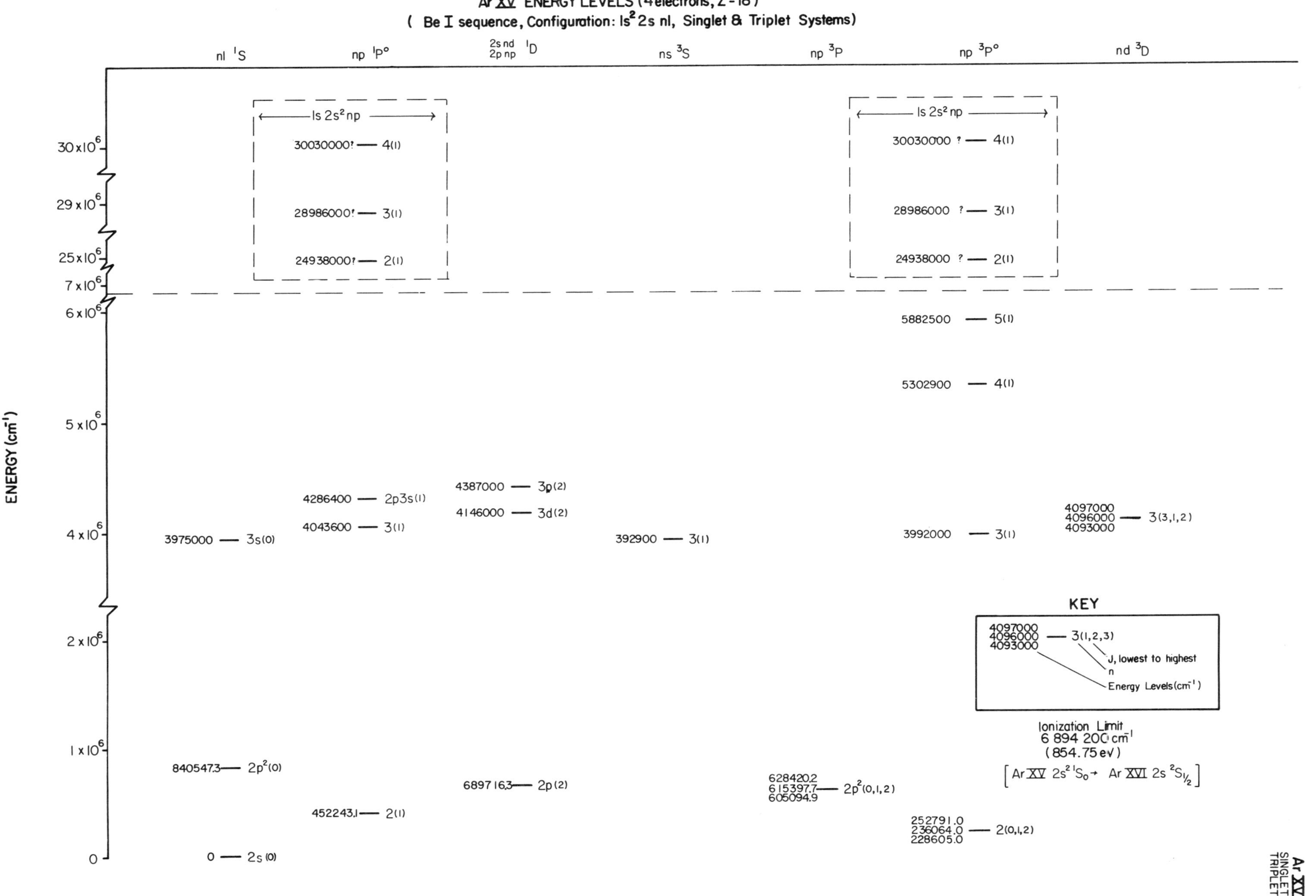
Ar XV ENERGY LEVELS (4 electrons, Z = 18)
(Be I sequence, Configuration: $1s^2 2s\,nl$, Singlet & Triplet Systems)
nl ^1S
np $^1P°$
2s nd
2p np ^1D
ns ^3S
np ^3P
np $^3P°$
nd ^3D
ENERGY (cm^{-1})
30 x 10^6
29 x 10^6
25 x 10^6
7 x 10^6
6 x 10^6
5 x 10^6
4 x 10^6
2 x 10^6
1 x 10^6
0
1s $2s^2$ np
30030000? — 4(1)
28986000? — 3(1)
24938000? — 2(1)
1s $2s^2$ np
30030000 ? — 4(1)
28986000 ? — 3(1)
24938000 ? — 2(1)
5882500 — 5(1)
5302900 — 4(1)
4387000 — 3p(2)
4146000 — 3d(2)
4286400 — 2p3s(1)
4043600 — 3(1)
3975000 — 3s(0)
392900 — 3(1)
3992000 — 3(1)
4097000
4096000
4093000 — 3(3,1,2)
KEY
4097000
4096000
4093000 — 3(1,2,3)
J, lowest to highest
n
Energy Levels (cm^{-1})
Ionization Limit
6 894 200 cm^{-1}
(854.75 eV)
[Ar XV $2s^2\,^1S_0$ → Ar XVI $2s\,^2S_{1/2}$]
840547.3 — $2p^2$(0)
689716.3 — 2p(2)
452243.1 — 2(1)
628420.2
615397.7
605094.9 — $2p^2$(0,1,2)
252791.0
236064.0
228605.0 — 2(0,1,2)
0 — 2s(0)

Ar XVI
DOUBLET
GROTRIAN DIAGRAM

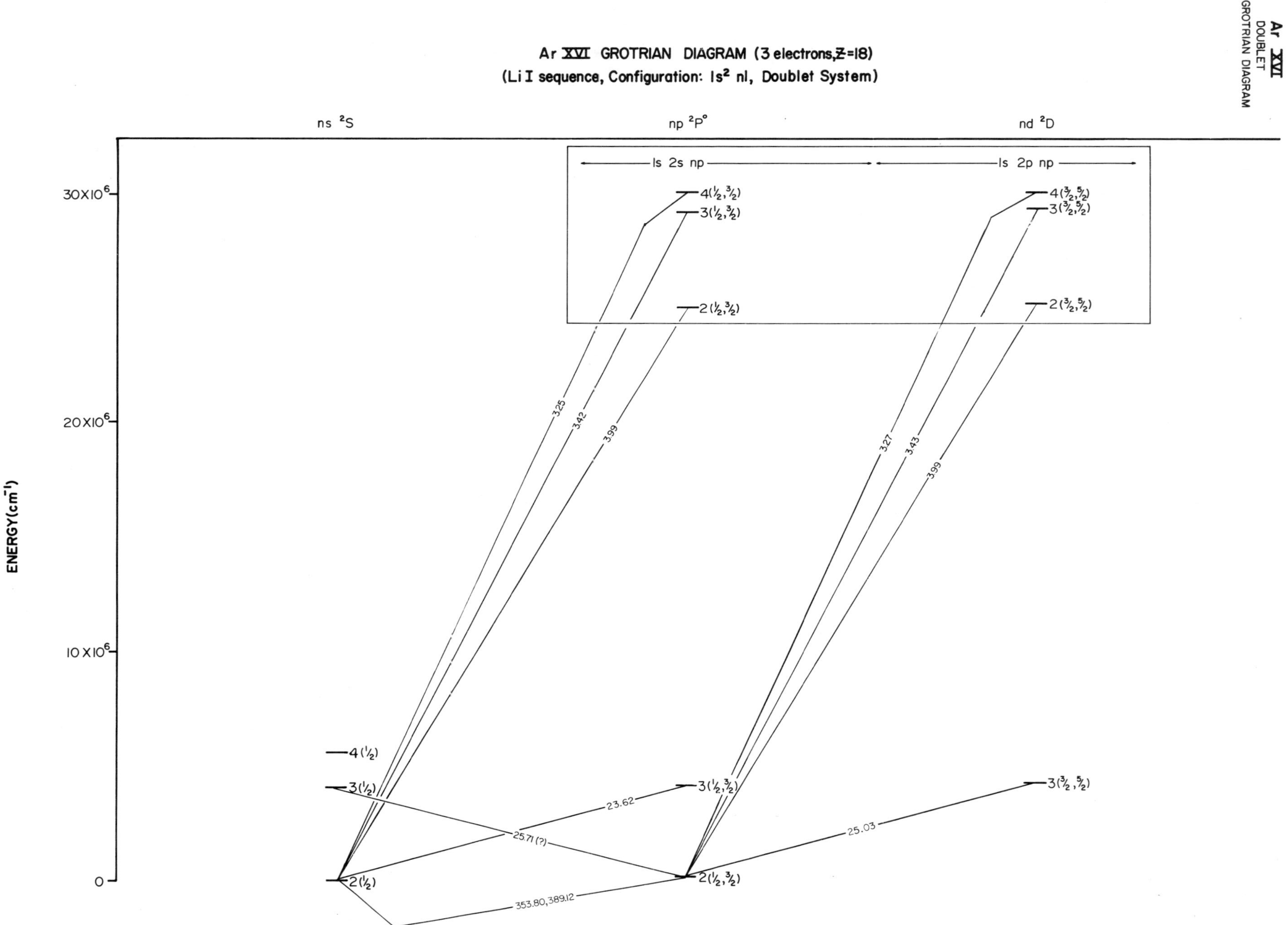

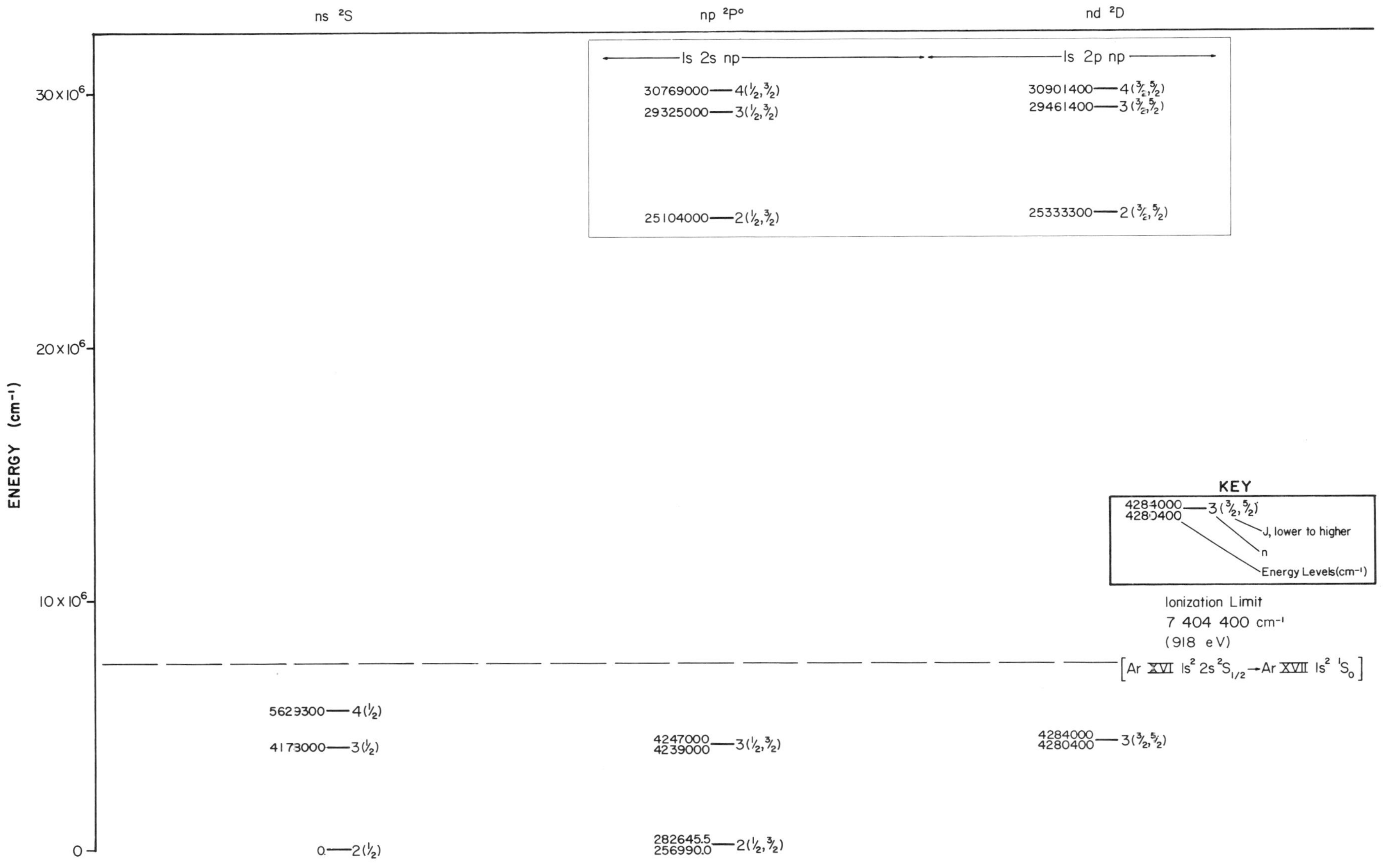
Ar XVI ENERGY LEVELS (3electrons, Z=18)
(Li I sequence, Configuration: 1s² nl, Doublet System)
ns ²S
np ²P°
nd ²D
ENERGY (cm⁻¹)
30×10⁶
20×10⁶
10×10⁶
0
1s 2s np
1s 2p np
30769000—4(1/2,3/2)
29325000—3(1/2,3/2)
25104000—2(1/2,3/2)
30901400—4(3/2,5/2)
29461400—3(3/2,5/2)
25333300—2(3/2,5/2)
KEY
4284000
4280400—3(3/2,5/2)
J, lower to higher
n
Energy Levels(cm⁻¹)
Ionization Limit
7 404 400 cm⁻¹
(918 eV)
[Ar XVI 1s² 2s ²S1/2 → Ar XVII 1s² ¹S0]
5623300—4(1/2)
4173000—3(1/2)
0—2(1/2)
4247000
4239000—3(1/2,3/2)
282645.5
256990.0—2(1/2,3/2)
4284000
4280400—3(3/2,5/2)

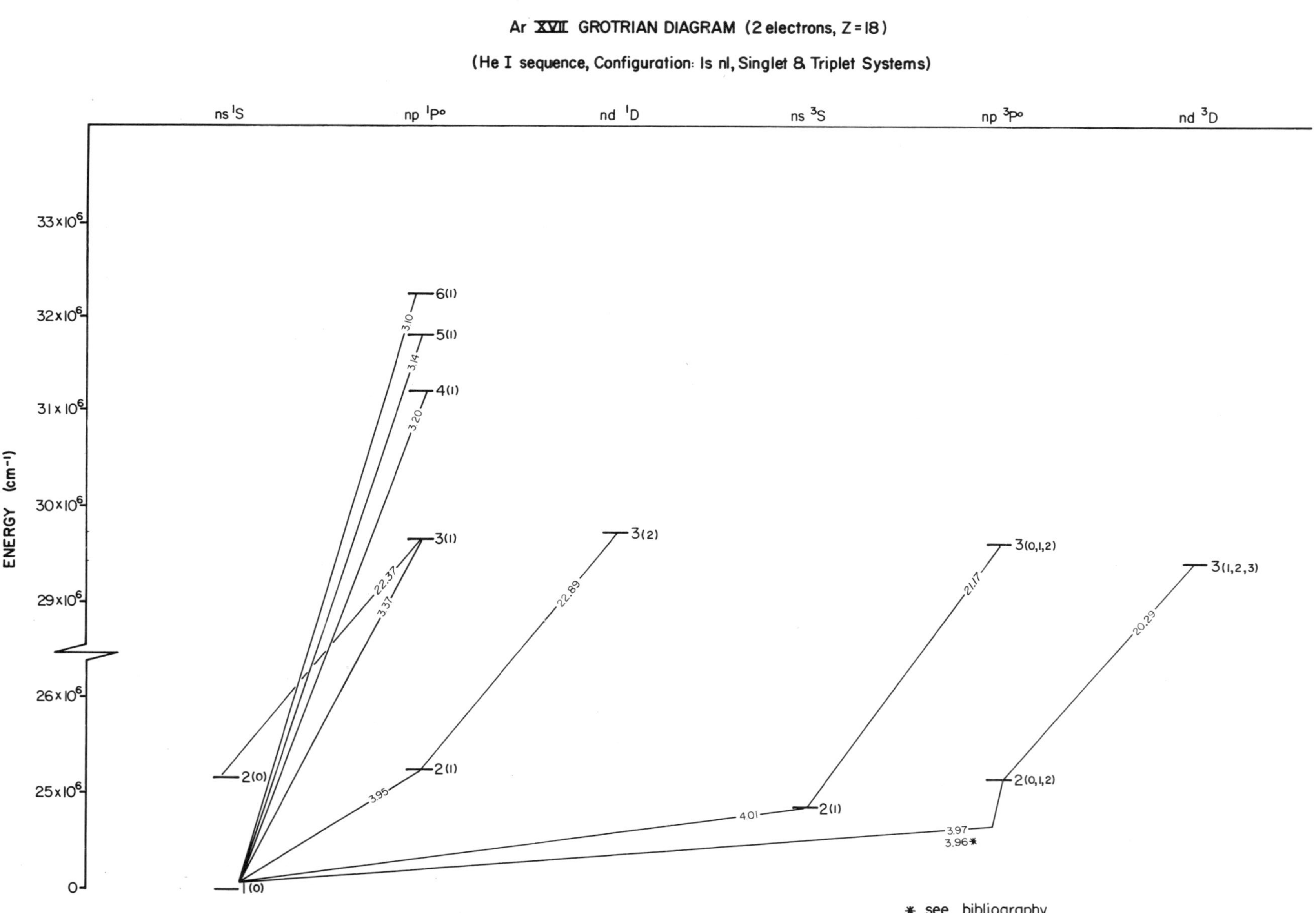
Ar XVII GROTRIAN DIAGRAM (2 electrons, Z = 18)
(He I sequence, Configuration: 1s nl, Singlet & Triplet Systems)
ns ¹S
np ¹P°
nd ¹D
ns ³S
np ³P°
nd ³D
ENERGY (cm⁻¹)
33x10⁶
32x10⁶
31x10⁶
30x10⁶
29x10⁶
26x10⁶
25x10⁶
0
2(0)
1(0)
6(1)
5(1)
4(1)
3(1)
2(1)
3(2)
2(1)
3(0,1,2)
2(0,1,2)
3(1,2,3)
3.10
3.14
3.20
22.37
3.37
3.95
22.89
4.01
21.17
3.97
3.96*
20.29
* see bibliography

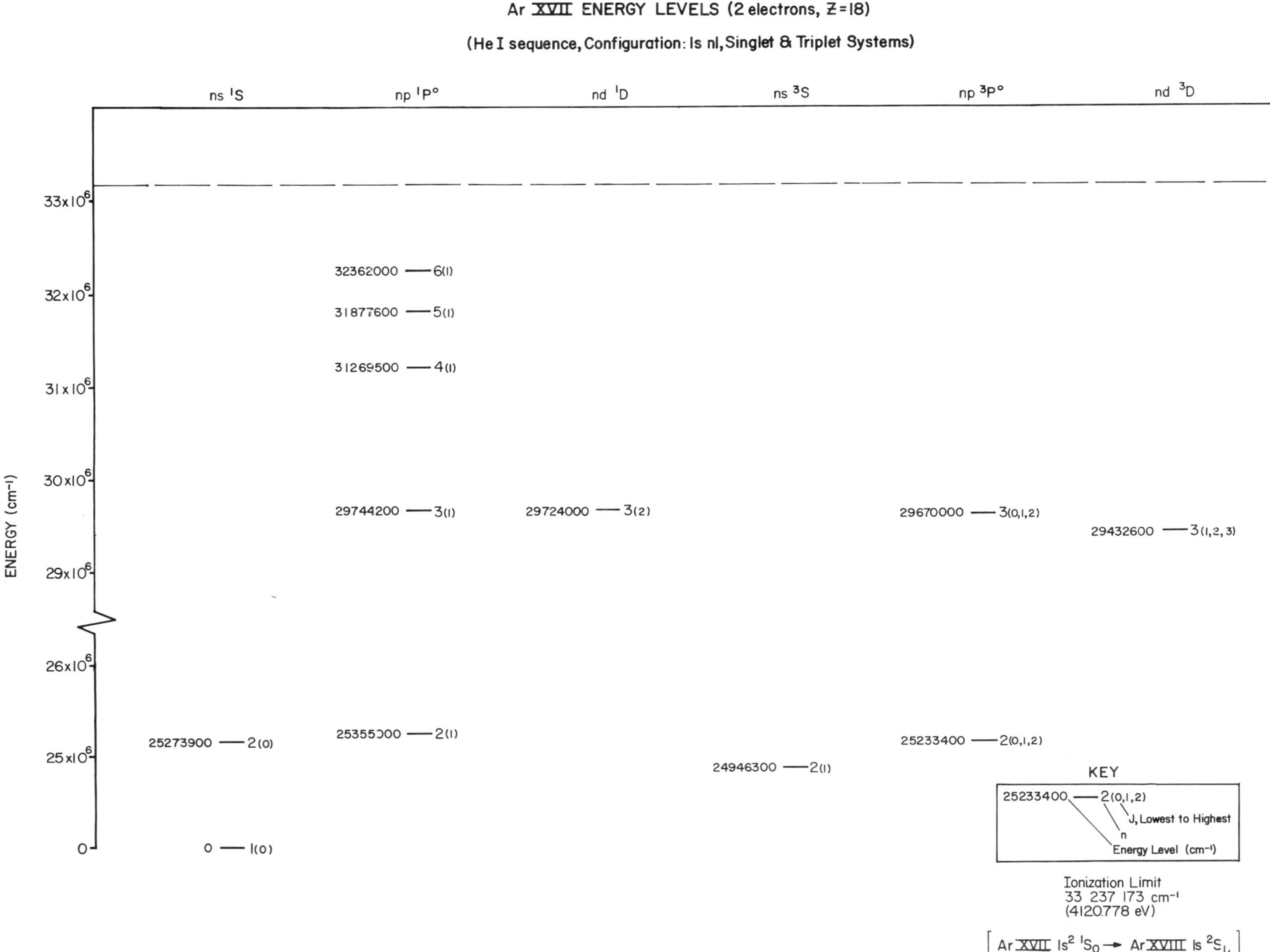
Ar XVII ENERGY LEVELS (2 electrons, Z=18)
(He I sequence, Configuration: 1s nl, Singlet & Triplet Systems)
ns ^1S
np $^1P^o$
nd ^1D
ns ^3S
np $^3P^o$
nd ^3D
33x10⁶
32362000 — 6(1)
32x10⁶
31877600 — 5(1)
31269500 — 4(1)
31x10⁶
30x10⁶
29744200 — 3(1)
29724000 — 3(2)
29670000 — 3(0,1,2)
29432600 — 3(1,2,3)
ENERGY (cm⁻¹)
29x10⁶
26x10⁶
25273900 — 2(0)
25355000 — 2(1)
25233400 — 2(0,1,2)
25x10⁶
24946300 — 2(1)
KEY
25233400 — 2(0,1,2)
J, Lowest to Highest
n
Energy Level (cm⁻¹)
0
0 — 1(0)
Ionization Limit
33 237 173 cm⁻¹
(4120.778 eV)
[Ar XVII 1s² $^1S_0$ → Ar XVIII 1s $^2S_{1/2}$]

Ar XVIII
DOUBLET
GROTRIAN
DIAGRAM

Ar XVIII GROTRIAN DIAGRAM (1 electron, Z = 18)
(H I sequence, Configuration: nl, Doublet System)

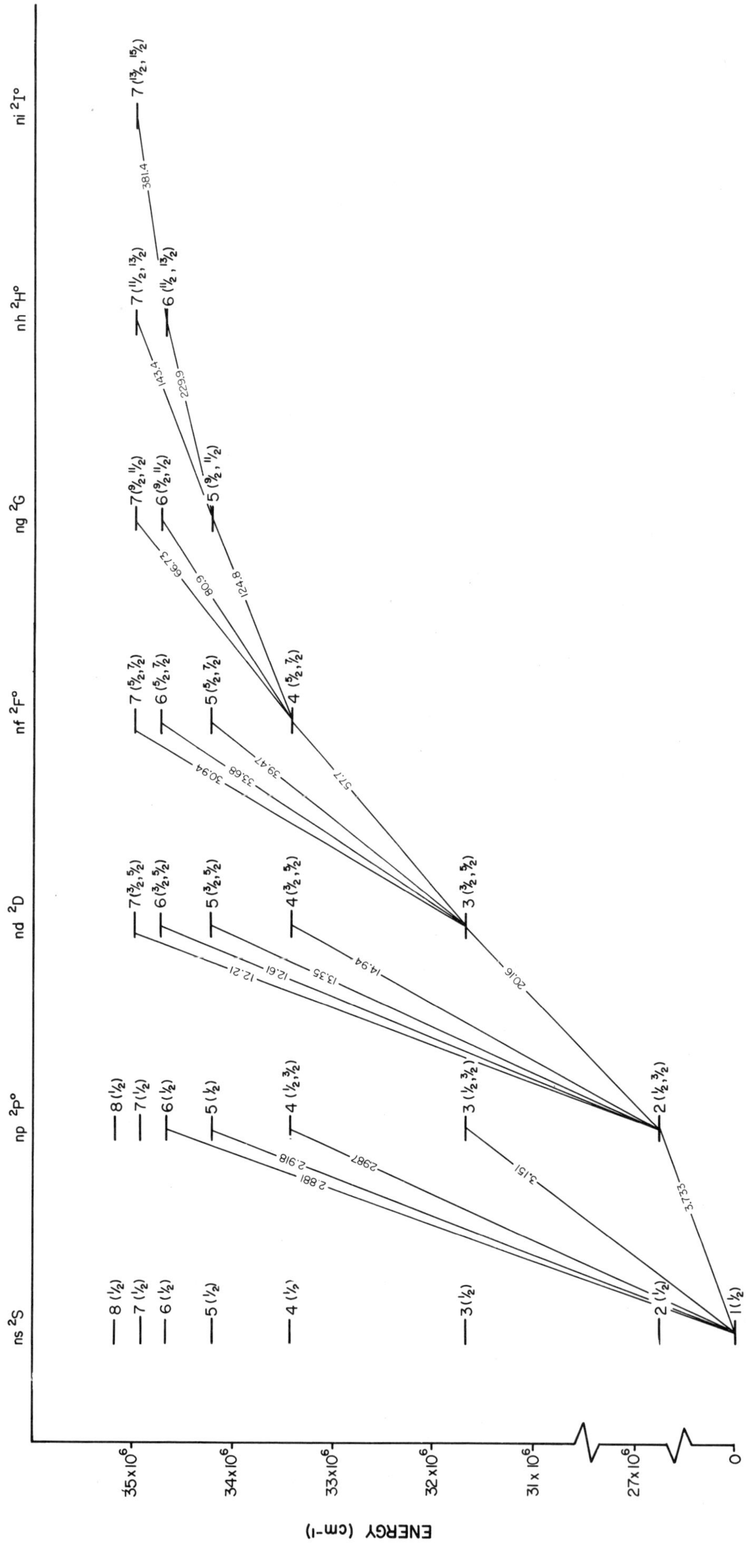

ns ^{2}S — np ^{2}P° — nd ^{2}D — nf ^{2}F° — ng ^{2}G — nh ^{2}H° — ni ^{2}I°

ns ^{2}S: 1(½), 2(½), 3(½), 4(½), 5(½), 6(½), 7(½), 8(½)

np ^{2}P°: 2(½, 3/2), 3(½, 3/2), 4(½, 3/2), 5(½), 6(½), 7(½), 8(½)

nd ^{2}D: 3(3/2, 5/2), 4(3/2, 5/2), 5(3/2, 5/2), 6(3/2, 5/2), 7(3/2, 5/2)

nf ^{2}F°: 4(5/2, 7/2), 5(5/2, 7/2), 6(5/2, 7/2), 7(5/2, 7/2)

ng ^{2}G: 5(9/2, 11/2), 6(9/2, 11/2), 7(9/2, 11/2)

nh ^{2}H°: 6(11/2, 13/2), 7(11/2, 13/2)

ni ^{2}I°: 7(13/2, 15/2)

Transition wavelengths: 3.733, 3.151, 2.987, 2.918, 2.881, 20.16, 14.94, 13.35, 12.61, 12.21, 57.7, 39.47, 33.68, 30.94, 124.8, 80.9, 66.73, 229.9, 143.4, 381.4

ENERGY (cm-1): 0, 27x10^6, 31 x 10^6, 32x10^6, 33x10^6, 34x10^6, 35x10^6

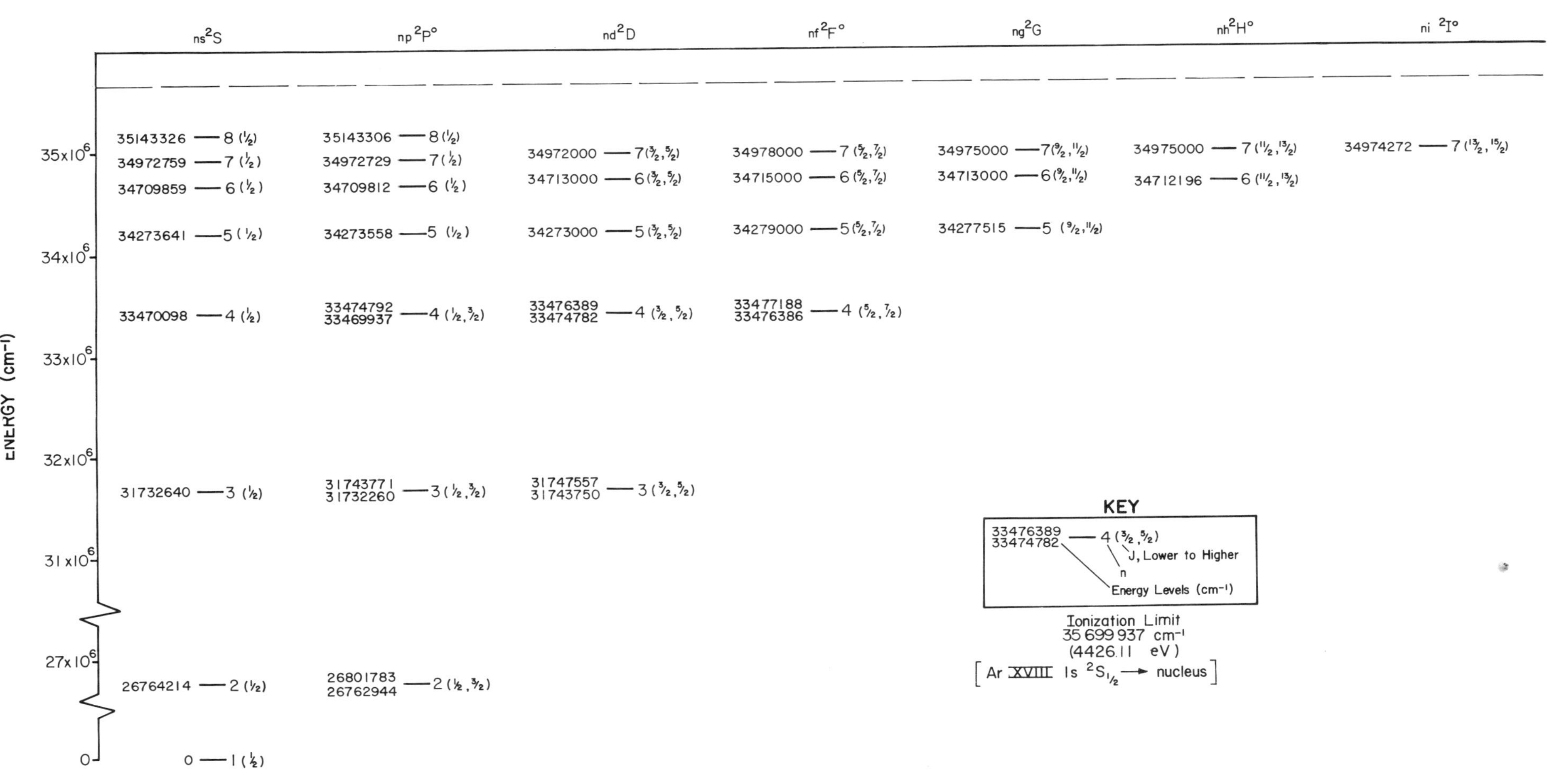
Ar XVIII ENERGY LEVELS (1 electron, Z=18)
(H I sequence, Configuration: nl, Doublet System)
ns²S
np²P°
nd²D
nf²F°
ng²G
nh²H°
ni²I°
35143326 — 8 (½)
34972759 — 7 (½)
34709859 — 6 (½)
34273641 — 5 (½)
33470098 — 4 (½)
31732640 — 3 (½)
26764214 — 2 (½)
0 — 1 (½)
35143306 — 8 (½)
34972729 — 7 (½)
34709812 — 6 (½)
34273558 — 5 (½)
33474792 33469937 — 4 (½, 3/2)
31743771 31732260 — 3 (½, 3/2)
26801783 26762944 — 2 (½, 3/2)
34972000 — 7 (3/2, 5/2)
34713000 — 6 (3/2, 5/2)
34273000 — 5 (3/2, 5/2)
33476389 33474782 — 4 (3/2, 5/2)
31747557 31743750 — 3 (3/2, 5/2)
34978000 — 7 (5/2, 7/2)
34715000 — 6 (5/2, 7/2)
34279000 — 5 (5/2, 7/2)
33477188 33476386 — 4 (5/2, 7/2)
34975000 — 7 (9/2, 11/2)
34713000 — 6 (9/2, 11/2)
34277515 — 5 (9/2, 11/2)
34975000 — 7 (11/2, 13/2)
34712196 — 6 (11/2, 13/2)
34974272 — 7 (13/2, 15/2)
KEY
33476389 33474782 — 4 (3/2, 5/2)
J, Lower to Higher
n
Energy Levels (cm⁻¹)
Ionization Limit
35 699 937 cm⁻¹
(4426.11 eV)
[Ar XVIII 1s ²S½ → nucleus]
35x10⁶
34x10⁶
33x10⁶
32x10⁶
31x10⁶
27x10⁶
0
ENERGY (cm⁻¹)

Ar I $Z = 18$ 18 electrons

The low-lying levels are taken from Kelly's unpublished compilation, in turn largely from Minnhagen (1973). Many levels lying close to the ionization limit were found by Yoshino, and his ionization limit has been used in these diagrams. Doubly-excited levels high in the continuum were found by Madden *et al.* (1969). Transitions are taken largely from Striganov *et al.*, Yoshino, Minnhagen, and Madden *et al.*.

Numerous transitions forbidden in the dipole approximation occur in this spectrum. The most prominent of these are electric-quadrupole transitions seen in absorption in the far ultraviolet (Madden *et al.*).

O. Andrade, M. Gallardo, and K. Bockasten, Appl. Phys. Letters **11**, 99 (1967).

Authors give transition wavelengths of observed superradiant lines in lasers.

K. Bockasten and O. Andrade, Nature **215**, 382 (1967).

Authors identify two lines observed by A.D. Brisbane [Nature **214**, 75 (1967)].

K. Bockasten, T. Lundholm, and O. Andrade, Phys. Letters **22**, 145 (1966).

Authors give three observed transition wavelengths in the near infrared.

J.C. Boyce, Phys. Rev. **48**, 396 (1935).

Author gives a table of observed and calculated lines in the region 800–1070 Å.

K. Burns and K.B. Adams, J. Opt. Soc. Am. **43**, 1020 (1953).

Authors give line and energy level tables for lines observed in the region 5450–9657 Å.

R.D. Cowan, J. Opt. Soc. Am. **58**, 924 (1968).

Author gives a line table and an energy level table of experimental and calculated values.

K.G. Ericsson and L.R. Lidholt, IEEE J. Qu. Electronics **3**, 94 (1967).

Authors observe and identify a line at 7067.2 Å.

W.L. Faust, R.A. McFarlane, C.K.N. Patel, and C.G.B. Garrett, Phys. Rev. **133A**, 1476 (1964).

Authors give a line table of observed stimulated emission spectral lines.

G. Hepner, Compt. Rend. **248**, 1142 (1959).

Author gives a line table of observations in the region 18 420–25 660 Å.

C.J. Humphreys and H.J. Kostkowski, J. Research NBS **49**, 73 (1952).

Authors give a line table of observations in the region 12 110–16 940 Å.

C.J. Humpreys, E. Paul, Jr., R.D. Cowan, and K.L. Andrew. J. Opt. Soc. Am. **57**, 855 (1967).

Authors give a line table and a supermultiplet array based on observations in the region 39 650–40 480 Å.

P.G. Kruger and S.G. Weissberg, Phys. Rev. **48**, 659 (1935).

Authors give a term table from observations.

P.G. Kruger, S.G. Weissberg, and L.W. Phillips, Phys. Rev. **51**, 1090 (1937).

Authors give a term table from observations.

H.H. Li and C.J. Humphreys, J. Opt. Soc. Am. **64**, 1072 (1974).

These authors observe this spectrum interferometrically in the range 3173–11672 Å.

S. Liberman, Compt. Rend. **261**, 2601 (1965).

Author gives a table of lines observed in the region 20 620–72 170 Å.

T.A. Littlefield and W.R.C. Rowley, Proc. Roy. Soc. **276A**, 502 (1963).

Authors give line and energy level tables from lines observed in the region 12 405–16 950 Å and calculated for 2p–3d and 2p–3s transitions (in the region 9195–32 310 Å).

T.A. Littlefield and D.T. Turnball, Proc. Roy. Soc. **218A**, 577 (1954).

Authors give line and energy level tables of lines observed in the region 3550–6540 Å.

R.P. Madden, D.L. Ederer, and K. Codling, Phys. Rev. **177**, 136 (1969).

Authors give tables of lines observed in the regions 290–430 Å and 210–280 Å.

L. Minnhagen, J. Opt. Soc. Am. **63**, 1185 (1973).

Author reobserves and reanalyzes this spectrum over the range 3000–12 356 Å.

G. Norlén, Ark. Fys. **35**, 119 (1967).

Author gives line and energy level tables from lines observed in the region 5150–6960 Å.

G. Norlén, Physica Scripta **8**, 249 (1973).

This author reobserves and reanalyzes this spectrum in the range 3400–9800 Å.

E. Paul, Jr. and C.J. Humphreys, J. Opt. Soc. Am. **49**, 1186 (1959).

Authors give a table of lines observed in the region 13 860–25 125 Å.

E.R. Peck, B.N. Khanna, and N.C. Anderholm, J. Opt. Soc. Am. **52**, 536 (1962).

Authors give wavelengths in the near infrared as secondary standards.

B. Petersson, Ark. Fys. **27**, 317 (1964).

Author gives two transition wavelengths in the vacuum ultraviolet and a table of Ritz standards in the vacuum ultraviolet region.

E.K. Plyler, L.R. Blaine, and E.D. Tidwell, J. Research NBS **55**, 279 (1955).

Authors give wavelengths for standards in the infrared region.

T. Sasaki, N. Kaifu, N. Ito, K. Shimada, and I. Sakai, Sci. of Light **13**, 115 (1964).

Authors give a Grotrian diagram.

W.R. Sittner and E.R. Peck, J. Opt. Soc. Am. **39**, 474 (1949).

Authors give a table of wave numbers from lines observed in the region 12 00–22 000 Å.

Ar II $Z = 18$ 17 electrons

Energy levels are taken largely from Kelly (unpublished). Wavelengths have been taken primarily from Striganov and Sventitskii, Norlén, and from Kelly and Palumbo. There are some discrepancies in the classifications and wavelengths taken from different authors.

Although a substantial part of the spectrum of Ar II may be approximately classified as belonging to transitions within doublet and quartet systems, many intercombinations are known.

J.C. Boyce, Phys. Rev. **48**, 396 (1935).

Author gives line and term tables for lines observed in the region 480–1980 Å.

W.B. Bridges and A.N. Chester, IEEE J. Qu. Electronics **1**, 66 (1965).

Authors give a line table of lines observed and calculated in ion lasers.

G. Convert, M. Armand, and P. Martinot-Lagarde, Compt. Rend. **258**, 4467 (1961).

Authors give a line table and a partial Grotrian diagram for lines observed in the region 4545–4765 Å.

B. Edlén, Z. Physik **104**, 407 (1937).

Author gives a line table for lines observed in the region near 2530 Å.

K.G. Ericsson and L.R. Lidholt, IEEE J. Qu. Electronics **3**, 94 (1967).

Authors give two wavelengths from superradiant transitions observed in lasers.

U. Fink, S. Bashkin, and W.S. Bickel, J. Quant. Spectrosc. Radiat. Transfer **10**, 1241 (1970).

Authors give a line table of calculated and observed values in the region 3375–4965 Å.

J.E. Hansen, J. Opt. Soc. Am. **67**, 754 (1977).

This author calculates the position and wavefunction for the $sp^6\ {}^2S$ term.

G. Herzberg, Proc. Phys. Soc. **248A**, 309 (1958).

Author gives an energy level diagram and line tables for lines observed in the near vacuum ultraviolet regions, and for lines calculated in the far ultraviolet region.

B. Kjöllerström, N.H. Möller, and H. Svensson, Ark. Fys. **29**, 167 (1965).

Authors calculate the energy levels for three terms.

H.H. Li and C.J. Humphreys, J. Opt. Soc. Am. **64**, 1072 (1974).

Authors study the spectrum of Ar II in the photographic infrared.

L. Maissel, J. Opt. Soc. Am. **48**, 853 (1958).

Author gives a table of lines observed in the region 3460–3660 Å in a study of the Stark shift.

L. Minnhagen, Ark. Fys. **14**, 483 (1958).

Author gives transition and energy level arrays and line tables for lines observed in the regions 1260–2000 Å and 480–1000 Å.

L. Minnhagen, Ark. Fys. **14**, 123 (1958).

Author gives a partial energy level diagram and an energy level table.

L. Minnhagen, Ark. Fys. **18**, 97 (1960).

Author gives energy level tables and arrays, and a partial energy level diagram from empirical data and from calculations for the *n*f and *n*g levels.

L. Minnhagen, Ark. Fys. **25**, 203 (1963).

Author gives line tables for lines observed in the region 2000–12 500 Å and 1400–2000 Å, and gives energy level tables, term tables, and a partial energy level diagram from observations and calculations.

L. Minnhagen, J. Opt. Soc. Am. **61**, 1257 (1971), and **63**, 1185 (1973).

Author gives line and energy level tables and a partial energy level diagram from observations and calculations.

G. Norlén, Ark. Fys. **35**, 119 (1967).

Author gives a line table for lines observed in the region 4880–6870 Å.

G. Norlén, Physica Scripta **8**, 249 (1973).

Author gives interferometrically-determined wavelengths and levels.

E.H. Pinnington, B. Curnette, and M. Dufay, J. Opt. Soc. Am. **61**, 978 (1971).

Authors give a table of lines observed in the region 520–1000 Å.

T. Sasaki, N. Kaifu, N. Ito, K. Shimada, and I. Sakai, Sci. of Light **13**, 115 (1964).

Authors give a table of unidentified lines and a Grotrian diagram.

Ar III $\quad\quad$ $Z = 18$ $\quad\quad$ 16 electrons

The levels are taken mostly from Kelly's unpublished compilation.

I.S. Bowen, Ap. J. **121**, 306 (1955).

Author gives transition wavelengths from nebular observations.

I.S. Bowen, Ap. J. **132**, 1 (1960).

Author gives line tables from laboratory and nebular observations.

J.C. Boyce, Phys. Rev. **48**, 396 (1935).

Author gives line and term tables from lines observed in the region 395–1980 Å. These have been mostly supplanted by later authors' work.

W.B. Bridges and A.N. Chester, IEEE J. Qu. Electronics **1**, 66 (1965).

Authors give a table of lines observed in ion lasers in the region 2620–5510 Å. These differ slightly in some cases from wavelengths given in Striganov and Sventitskii.

B. Edlén, Phys. Rev. **61**, 434 (1942).

Author gives a wavelength of an observed transition and term values for the $3s^2 3p^4\ ^1S$ and $3s3p^5\ ^1P^\circ$ terms.

K.G. Ericsson and L.R. Lidholt, IEEE J. Qu. Electronics **3**, 94 (1967).

Authors give an observed superradiant transition.

B.C. Fawcett, N.J. Peacock, and R.D. Cowan, J. Phys. B (Proc. Phys. Soc.) **1**, 295 (1968).

Authors give a line table from observations in the region 380–400 Å.

U. Fink, S. Bashkin, and W.S. Bickel, J. Quant. Spectrosc. Radiat. Transfer. **10**, 1241 (1970).

Authors give a line table of calculated and observed values in the region 3285–3515 Å.

R.A. MacFarlane, Appl. Opt. **3**, 1196 (1964).

Author gives transition wavelengths observed in an optical maser for the $4p\ ^3P_{1,2}$–$4s\ ^3S^\circ_1$ transitions.

E.H. Pinnington, B. Curnutte, and M. Dufay, J. Opt. Soc. Am. **61**, 978 (1971).

Authors give a table of observed lines in the region 520–1000 Å.

G.S. Rostovikov, V.P. Samoilov, and Yu.M. Smirnov, Opt. Spectrosc. **35**, 222 (1973).

These authors calculate transition probabilities for some transitions in Ar III.

T. Sasaki, N. Kaifu, N. Ito, K. Shimada, and I. Sakai, Sci. of Light **13**, 115 (1964).

These authors give a table of unidentified lines and a Grotrian diagram.

Ar IV $\quad Z = 18 \quad$ 15 electrons

The levels are taken from Kelly's unpublished compilation, and from Moore's tables. Wavelengths are taken from Kelly and Palumbo, and from Striganov and Sventitskii.

I.S. Bowen, Ap. J. **121**, 306 (1955).

Author gives a line table from nebular observations.

I.S. Bowen, Ap. J. **132**, 1 (1960).

Author gives a line table from nebular observations.

J.C. Boyce, Phys. Rev. **48**, 396 (1935).

Authors give line and term tables from observations in the region 390–1200 Å.

B.C. Fawcett, A.H. Gabriel, and P.A.H. Saunders, Proc. Phys. Soc. **90**, 863 (1967).

Authors give wavelengths of the observed transitions of a multiplet.

B.C. Fawcett, N.J. Peacock, and R.D. Cowan, J. Phys. B (Proc. Phys. Soc.) **1**, 295 (1968).

Authors give wavelengths of observed transitions in the region 405–430 Å.

A.H. Gabriel, B.C. Fawcett, and C. Jordan, Proc. Phys. Soc. **87**, 825 (1966).

Authors give a wavelength of an observed transition in the vacuum ultraviolet region.

E.H. Pinnington, B. Curnutte, and M. Dufay, J. Opt. Soc. Am. **61**, 978 (1971).

Authors give a line table for observations in the region 530–1000 Å.

Ar V $Z = 18$ 14 electrons

I.S. Bowen, Ap. J. **121**, 306 (1955).

Author gives transition wavelengths from nebular observations.

I.S. Bowen, Ap. J. **132**, 1 (1960).

Author gives a line table from laboratory and nebular observations.

J.C. Boyce, Phys. Rev. **48**, 396 (1935).

Author gives line and term tables from observations in the region 700–840 Å.

J.O. Ekberg and L.Å. Svensson, Physica Scripta **2**, 283 (1970).

These authors analyze the spectra of K, Ca, Sc, and Ti isoelectronic with P I, Si I, and Al I, and in this process identify lines in Ar V.

B.C. Fawcett, A.H. Gabriel, and P.A.H. Saunders, Proc. Phys. Soc. **90**, 863 (1967).

Authors observe a line at 436.60 Å.

L.W. Phillips and W.L. Parker, Phys. Rev. **60**, 301 (1941).

Authors give line and term tables from observations in the region 330–840 Å.

E.H. Pinnington, B. Curnutte, and M. Dufay, J. Opt. Soc. Am. **61**, 978 (1971).

Authors give a table of lines observed in the region 530–1000 Å.

Ar VI $Z = 18$ 13 electrons

B.C. Fawcett, B.B. Jones, amd R. Wilson, Proc. Phys. Soc. **78**, 1223 (1961).

Authors give the wavelengths of observed transitions in the vacuum ultraviolet.

L.W. Phillips and W.L. Parker, Phys. Rev. **60**, 301 (1941).

Authors give line and term tables from observations in the region 180–600 Å.

E. Schönheit, Optik **23**, 409 (1965–1966).

This author finds numerous lines in highly-excited Ar that have not yet been classified.

Ar VII $Z = 18$ 12 electrons

J.O. Ekberg, Physica Scripta **4**, 101 (1971).

This author analyzes several Mg-I-like spectra and makes some new identifications in Ar VII.

L.W. Phillips and W.L. Parker, Phys. Rev. **60**, 301 (1941).

Authors give line and term tables from observations in the region 150–650 Å.

E. Schönheit, Optik **23**, 409 (1965–1966).

This author finds numerous lines in highly-excited Ar that have not yet been classified.

Ar VIII Z = 18 11 electrons

R.U. Datla, H.-J. Kunze, and D. Petrini, Phys. Rev. **A6**, 38 (1972).

These authors study collision rates and transition probabilities in Ar VIII and show a Grotrian diagram.

B.C. Fawcett, B.B. Jones, and R. Wilson, Proc. Phys. Soc. **78**, 1223 (1961).

Authors give observed wavelengths for the 3s $^2S_{1/2}$–3p $^2P^\circ_{3/2,1/2}$ transitions in the vacuum ultraviolet region.

L.W. Phillips and W.L. Parker, Phys. Rev. **60**, 301 (1941).

Authors give line and term tables from observations in the region 120–720 Å.

E. Schönheit, Optik **23**, 409 (1965–1966).

This author finds numerous lines in highly-excited Ar that have not yet been classified.

Ar IX Z = 18 10 electrons

B.C. Fawcett, A.H. Gabriel, W.G. Griffin, B.B. Jones, and R. Wilson, Nature **200**, 1303 (1963).

Authors give wavelengths of the observed transitions in the vacuum ultraviolet region.

B.C. Fawcett, A.H. Gabriel, B.B. Jones, and N.J. Peacock, Proc. Phys. Soc. **84**, 257 (1964).

Authors give a table of lines observed in the range 30–45 Å.

B.C. Fawcett, Proc. Phys. Soc. **86**, 1087 (1965).

Author identifies a line at 35.82 Å.

L.W. Phillips and W.L. Parker, Phys. Rev. **60**, 301 (1941).

Authors give a line and term table from empirical data.

Ar X Z = 18 9 electrons

J.P. Buchet, M.C. Buchet-Poulizac, A. Denis, J. Désesquelles, and G. DoCao, Physica Scripta **9**, 221 (1974).

These authors observe hydrogenic transitions and other transitions in Ar X–XV by beam-foil spectroscopy.

J.P. Connerade, N.J. Peacock, and R.J. Speer, Solar Phys. **18**, 63 (1971), and private communication from J.P. Connerade (1977).

These authors identify observed transitions in Ar X–XVIII using Hartree–Fock calculations. These identifications have in some cases been supplanted by more-recent investigations.

W.A. Deutschman and L.L. House, Ap. J. **144**, 435 (1966).

Authors give wavelengths for the observed transitions of a multiplet.

B.C. Fawcett, Proc. Phys. Soc. **86**, 1087 (1965).

Author gives transition wavelengths for lines observed below 40 Å.

B.C. Fawcett, A.H. Gabriel, W.G. Griffin, B.B. Jones, and R. Wilson, Nature **200**, 1303 (1963).

Authors give transition wavelengths of observed spectral lines in the vacuum ultraviolet region.

B.C. Fawcett, A.H. Gabriel, B.B. Jones, and N.J. Peacock, Proc. Phys. Soc. **84**, 257 (1964).

Authors give a table of lines observed in the region 40–45 Å and include the wavelengths 166 Å and 171 Å.

B.C. Fawcett and R.W. Hayes, Physica Scripta **8**, 244 (1973).

These authors classify transitions in S X–XIV and some isoelectronic spectra in P, Cl, and Ar.

B.C. Fawcett, A.H. Gabriel, and T.M. Paget, J. Phys. B4, 986 (1971).

These authors report classifications of new Ar X–XV lines from a θ pinch.

U. Feldman, G.A. Doschek, R.D. Cowan, and L. Cohen, J. Opt. Soc. Am. **63**, 1445 (1973).

These authors give levels and lines for members of the F I isoelectronic sequence.

Ar XI | $Z = 18$ | 8 electrons

The lowest five levels are taken from B.C. Fawcett, Atomic Data and Nuclear Data Tables **16**, 135 (1975). The same paper gives wavelengths for the transitions $2\,^3P$–$2\,^3P^\circ$, $2p^4\,^1D$–$2p^5\,^1P^\circ$, and $2p^2\,^1S$–$2p^5\,^1P^\circ$. The high levels $[2s\,2p^4(^3P,\,^1D)3p]$ are from J.P. Connerade, N.J. Peacock, and R.J. Speer, Solar Physics **18**, 63 (1971). There are small differences between the two foregoing papers; we have generally followed the listings in Kelly and Palumbo. However, we used Connerade *et al.* for $2p^3(^2P^\circ)3s''\,^1P^\circ$.

J.P. Buchet, M.C. Buchet-Poulizac, A. Denis, J. Désesquelles, and G. DoCao, Physica Scripta **9**, 221 (1974).

The authors observe transitions in Ar X–XV by beam-foil spectroscopy.

J.P. Connerade, N.J. Peacock, and R.J. Speer, Solar Phys. **18**, 63 (1971), and private communication from J.P. Connerade (1977).

These authors identify observed transitions in Ar X–XVIII using Hartree–Fock calculations. These identifications have in some cases been supplanted by more-recent investigations.

W.A. Deutschman, Doctorate thesis from the Dept. of Astro-Geophysics, University of Colorado, obtained from University Microfilms, Ann Arbor, Michigan, U.S.A., order number 68–2645 (1970).

Author gives a line table, an energy level table, and term diagrams from lines observed in the region 150–195 Å.

W.A. Deutschman and L.L. House, Ap. J. **144**, 435 (1966).

Authors give a line table of observed lines in the region 150–195 Å.

B.C. Fawcett and A.H. Gabriel, Proc. Phys. Soc. **84**, 1038 (1964).

Authors give a table of observed lines in the region 150–195 Å.

B.C. Fawcett, A.H. Gabriel, and T.M. Paget, J. Phys. B4, 986 (1971).

These authors report the classification of lines in Ar X–XV from θ-pinch and laser-produced plasmas.

B.C. Fawcett and R.W. Hayes, Physica Scripta **8**, 244 (1973).

These authors classify transitions in S X–XIV and some isoelectronic spectra of P, Cl, and Ar.

B.C. Fawcett, A.H. Gabriel, B.B. Jones, and N.J. Peacock, Proc. Phys. Soc. (London) **84**, 257 (1964).

These authors report new spectra from Ar IX–XII from laboratory plasma sources.

A.E. Goertz, J. Opt. Soc. Am. **55**, 742 (1965).

This author predicts the positions for two levels in Ar XI.

Ar XII | $Z = 18$ | 7 electrons

J.P. Buchet, M.C. Buchet-Poulizac, A. Denis, J. Désesquelles and G. DoCao, Physica Scripta **9**, 221 (1974).

These authors observe transitions in Ar X–XV by beam-foil spectroscopy.

J.P. Connerade, N.J. Peacock, and R.J. Speer, Solar Phys. **18**, 63 (1971), and private communication (1977).

These authors identify observed transitions in Ar X–XVIII using Hartree–Fock calculations. These identifications have been supplanted in som cases by more recent investigations, especially by Fawcett and Hayes (1973) for Ar XII and Ar XIII.

W.A. Deutschman, Doctorate thesis for the Dept. of Astro-Geophysics, University of Colorado, obtained from University Microfilms, Ann Arbor, Michigan, U.S.A., order number 68–2645 (1970).

Author gives a line table, an energy level table, and term diagrams based on observed lines in the region 145–225 Å.

W.A. Deutschman and L.L. House, Ap. J. **144**, 435 (1966).

Authors give a table of lines observed in the region 150–225 Å.

W.A. Deutschman and L.L. House, Ap. J. **149**, 451 (1967).

Authors give a table of lines observed in the region 150–225 Å.

B.C. Fawcett and A.H. Gabriel, Proc. Phys. Soc. **84**, 1038 (1964).

Authors give a table of observed resonance lines in the region 150–225 Å.

B.C. Fawcett, A.H. Gabriel, and T.M. Paget, J. Phys. B4, 986 (1971).

These authors report the classification of lines in Ar X–XV from θ-pinch and laser-produced plasmas.

B.C. Fawcett, A.H. Gabriel, B.B. Jones, and N.J. Peacock, Proc. Phys. Soc. **84**, 257 (1964).

Authors give a table of lines observed in the region 30–50 Å.

B.C. Fawcett and R.W. Hayes, Physica Scripta **8**, 244 (1973).

These authors classify transitions in S X–XIV and some isoelectronic spectra of P, Cl, and Ar.

S.O. Kastner, J. Opt. Soc. Am. **63**, 738 (1973).

This author reports calculations of levels and intervals in $2s^r 2p^k$ configurations for S through Sc.

J.D. Purcell and K.G. Widing, Ap. J. **176**, 239 (1972).

These authors report observations of high-ionized spectra of a solar flare.

Ar XIII $Z = 18$ 6 electrons

B.C. Fawcett, Atomic Data and Nuclear Data Tables **16**, 135 (1975) gives the level $2p^4\ {}^3P$; however, no transitions to that level have been seen. We list the level but not the predicted wavelengths.

J.P. Buchet, M.C. Buchet-Poulizac, A. Denis, J. Désesquelles, and G. DoCao, Physica Scripta **9**, 221 (1974).

These authors observe transitions in Ar X–XV by beam-foil spectroscopy.

J.P. Connerade, N.J. Peacock, and R.J. Speer, Solar Phys. **18**, 63 (1971), and private communication (1977).

These authors identify observed transitions in Ar X–XVIII using Hartree–Fock calculations. These identifications have been supplanted in some cases by more-recent investigations, especially by Fawcett and Hayes (1973) for Ar XII and Ar XIII.

W.A. Deutschman, Doctorate thesis for the Dept. of Astro-Geophysics, University of Colorado, obtained from University of Microfilms, Ann Arbor, Michigan, U.S.A., order number 68–2645 (1970).

Author gives a line table, an energy level table, and a term diagram based on observations in the region 160–200 Å.

W.A. Deutschman and L.L. House, Ap. J. **149**, 451 (1967).

Authors give a table of observed lines in the region 150–210 Å.

B.C. Fawcett, A.H. Gabriel, and T.M. Paget, J. Phys. B4, 986 (1971). B.C. Fawcett and R.W. Hayes, Physica Scripta **8**, 244 (1973).

These authors classify transitions in S X–XIV and some isoelectronic spectra in P, Cl, and Ar.

S. Goldsmith, U. Feldman, A. Crooker, and L. Cohen, J. Opt. Soc. Am. **62**, 260 (1972).

These authors find a few levels' energies in Ar XIII.

S.O. Kastner, J. Opt. Soc. Am. **63**, 738 (1973).

This author reports calculations of levels and intervals for configurations $2s^r\, 2p^k$ in S through Sc.

J.D. Purcell and K.G. Widing, Ap. J. **176**, 239 (1972).

These authors report observations of a few Ar XIII lines in a solar flare.

Ar XIV | $Z = 18$ | 5 electrons

Levels labeled with asterisks (*) in the drawings are taken from Shamey (1971). Kelly's unpublished compilation gives somewhat different values that agree with Fawcett's (1975) calculations.

J.P. Buchet, M.C. Buchet-Poulizac, A. Denis, J. Désesquelles, and G. DoCao, Physica Scripta **9**, 221 (1974).

These authors observe transitions in Ar X–XV by beam-foil spectroscopy.

J.P. Connerade, N.J. Peacock, and R.J. Speer, Solar Phys. **18**, 63 (1971), and private communication (1977).

These authors identify observed transitions in Ar X–XVIII using Hartree–Fock calculations. These identifications have been supplanted in some cases by more-recent investigations.

G.A. Doschek, Space Sci. Rev. **13**, 765 (1972).

This author presents identifications for lines in Ar XIV–XVIII seen in solar flares.

B. Edlén, Solar Phys. **9**, 439 (1969).

This author identifies a solar coronal line at λ 4412 Å in Ar XIV ($2s^2 2p\ ^2P^\circ_{3/2} - ^2P^\circ_{1/2}$).

B.C. Fawcett, Atomic Data and Nuclear Data Tables **16**, 135 (1975).

This author tabulates and in some cases calculates intervals in the transitions $2s^2 2p^n - 2s2p^{n+1}$ and $2s2p^n - 2p^{n+1}$.

B.C. Fawcett, A.H. Gabriel, and T.M. Paget, J. Phys. B4, 986 (1971).

These authors report the classification of lines in Ar X–XV from θ-pinch and laser-produced plasmas.

J.D. Purcell and K.G. Widing Ap. J. **176**, 239 (1972).

These authors consider the problems involved in identifying Ar XIV lines in the spectrum of a solar flare.

L.J. Shamey, J. Opt. Soc. Am. **61**, 942 (1971).

This author gives calculated tables of lines (180–500 Å) and energy levels.

Ar XV | $Z = 18$ | 4 electrons

J.P. Buchet, M.C. Buchet-Poulizac, A. Denis, J. Désesquelles, and G. DoCao, Physica Scripta **9**, 221 (1974).

These authors observe transitions in Ar X–XV by beam-foil spectroscopy.

J.P. Connerade, N.J. Peacock, and R.J. Speer, Solar Phys. **18**, 63 (1971), and private communication (1977).

These authors identify observed transitions in Ar X–XVIII using Hartree–Fock calculations. These identifications have been supplanted in some cases by more-recent investigations.

B.C. Fawcett, Atomic Data and Nuclear Data Tables **16**, 135 (1975).

This author tabulates, and in some cases calculates intervals in the transitions $2s^2 2p^n - 2s2p^{n+1}$ and $2s2p^n - 2p^{n+1}$.

B.C. Fawcett, A.H. Gabriel, and T.M. Paget, J. Phys. **B4**, 986 (1971).

These authors report the classification of lines in Ar X–XV from θ-pinch and laser-produced plasmas.

R.H. Garstang and L.J. Shamey, Ap. J. **148**, 665 (1967).

These authors predict the oscillator strength for the forbidden transition $2s\ ^1S_0 - 2p\ ^3P_1^\circ$ in Ar XV.

J.D. Purcell and K.G. Widing, Ap. J. **176**, 239 (1972).

These authors discuss an Ar XV line observed in the spectrum of a solar flare [see also K.G. Widing, Ap. J. **197**, L33 (1975), and G.A. Doschek, Space Sci. Rev. **13**, 765 (1972)].

Ar XVI $Z = 18$ 3 electrons

J.P. Connerade, N.J. Peacock, and R.J. Speer, Solar Phys. **18**, 63 (1971), and private communication (1977).

These authors calculate wavelengths for the transitions $2s\ ^2S - 3p\ ^2P^\circ$ and $2p\ ^2P^\circ - 3d\ ^2D$; we include these. Some of the identifications of observed transitions presented in this paper have been supplanted by more-recent investigations.

G.A. Doschek, Space Sci. Rev. **13**, 765 (1972).

This author presents identifications for lines in Ar XIV–XVIII seen in solar flares.

B.C. Fawcett, Atomic Data and Nuclear Data Tables **16**, 135 (1975).

This author tabulates, and in some cases calculates intervals in the transitions $2s^2 2p^n - 2s2p^{n+1}$ and $2s2p^n - 2p^{n+1}$.

J.D. Purcell and K.G. Widing, Ap. J. **176**, 239 (1972).

These authors identify lines of Ar XVI in the spectrum of a solar flare.

Ar XVII $Z = 18$ 2 electrons

L. Cohen, U. Feldman, M. Swartz, and J.H. Underwood, J. Opt. Soc. Am. **58**, 843 (1968).

Authors give wavelengths of two observed transitions in the X-ray region.

J.P. Connerade, N.J. Peacock, and R.J. Speer, Solar Phys. **18**, 63 (1971), and private communication (1977).

These authors calculate wavelengths for the transitions $2p\ ^1P^\circ - 3d\ ^1D$ and $2p\ ^3P^\circ - 3d\ ^3D$; we have assigned levels on that basis. Some of the identifications presented in this paper have been supplanted by more-recent investigations.

G.A. Doschek, Space Sci. Rev. **13**, 765 (1972).

This author presents identifications for lines in Ar XIV–XVIII seen in solar flares.

R. Marrus and R.W. Schmieder, Phys. Rev. **A5**, 1160 (1972).

These authors report a non-dispersive measurement of the $1\ ^1S_0 - 2\ ^3S_1$ transition energy with a solid-state X-ray detector.

For hydrogenic systems we show calculated levels and wavelengths: Since the wavelengths for $nl - n'l'$ are nearly independent of l, l', we have shown only the mean wavelength for each such complex.

J.P. Connerade, N.J. Peacock, and R.J. Speer, Solar Physics **18**, 63 (1971) give 20.16 Å for the transition 2–3; we have shown it as 2p ^{2}P° –3d ^{2}D.

G.A. Doschek, Space Sci. Rev. **13**, 765 (1972).

This author presents identifications for lines in Ar XIV–XVIII seen in solar flares.

G.W. Erickson, private communication (1976); in press, J. Phys. Chem. Ref. Data.

This author calculates energies of levels for hydrogenic species up to $Z = 105$.

J.D. Garcia and J.E. Mack, J. Opt. Soc. Am. **55**, 654 (1965).

Authors give calculated energy level and line tables for one-electron atomic spectra.

Note added in proof

An additional reference for Ar I reads:

K. Yoshino, J. Opt. Soc. Am. **60**, 1220 (1970).

This author studies in detail the absorption spectrum in the vacuum ultraviolet.

Potassium (K)

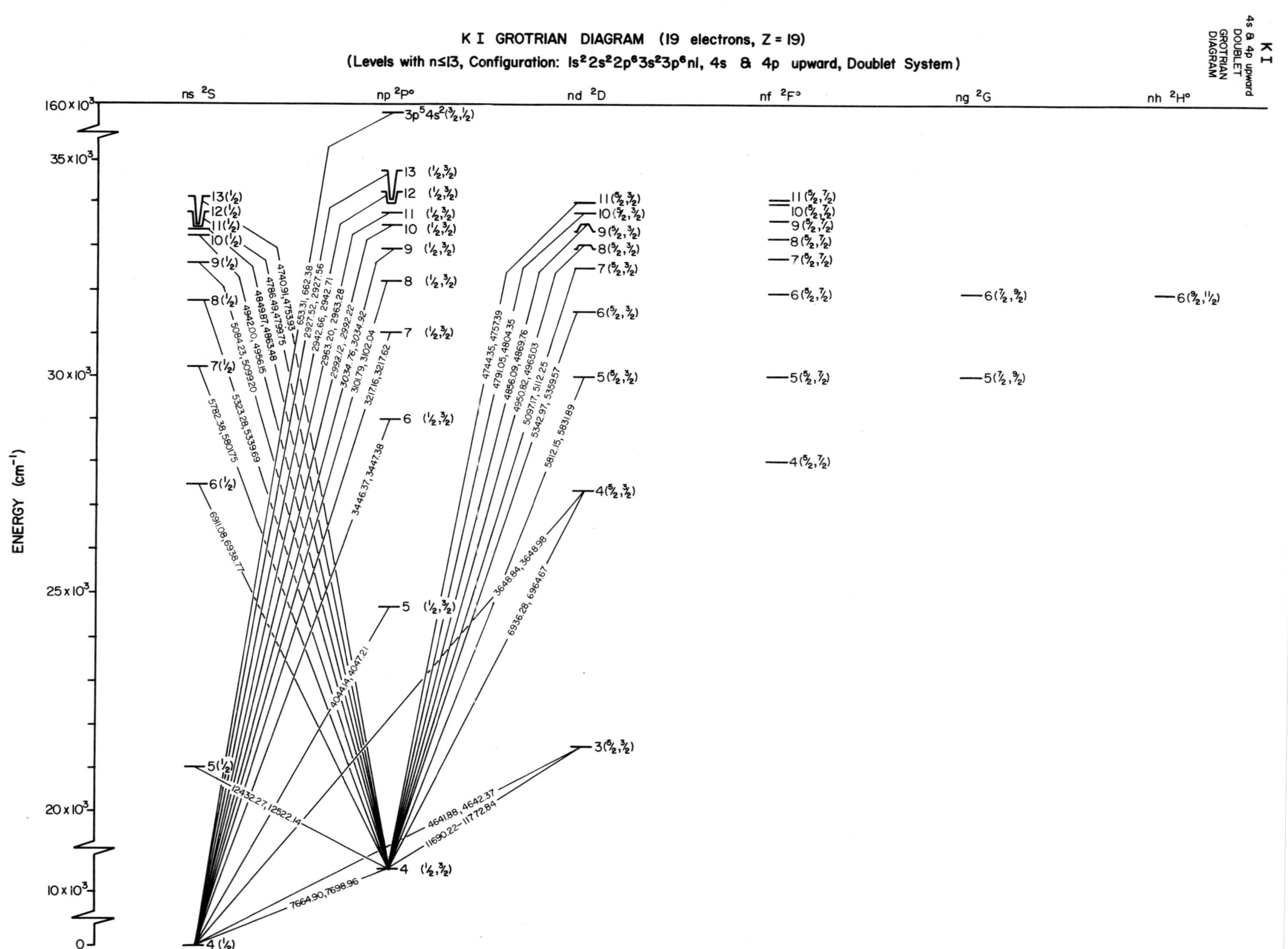
K I GROTRIAN DIAGRAM (19 electrons, Z = 19)
(Levels with n≤13, Configuration: 1s²2s²2p⁶3s²3p⁶nl, 4s & 4p upward, Doublet System)
K I
4s & 4p upward
DOUBLET
GROTRIAN
DIAGRAM
ns ²S
np ²P°
nd ²D
nf ²F°
ng ²G
nh ²H°
3p⁵4s²(3/2,1/2)
ENERGY (cm⁻¹)
160×10³
35×10³
30×10³
25×10³
20×10³
10×10³
0
7664.90, 7698.96
4044.14, 4047.21
3446.37, 3447.38
3217.16, 3217.62
3101.79, 3102.04
3034.76, 3034.92
2992.12, 2992.22
2963.20, 2963.28
2942.66, 2942.71
2927.52, 2927.56
653.31, 662.38
4740.91, 4753.93
4786.49, 4799.75
4849.87, 4863.48
4942.00, 4956.15
5084.23, 5099.20
5323.28, 5339.69
5782.38, 5801.75
6911.08, 6938.77
12432.27, 12522.14
4641.88, 4642.37
11690.22–11772.84
6936.28, 6964.67
3648.84, 3648.98
5812.15, 5831.89
5342.97, 5359.57
5097.17, 5112.25
4950.82, 4965.03
4856.09, 4869.76
4791.05, 4804.35
4744.35, 4757.39

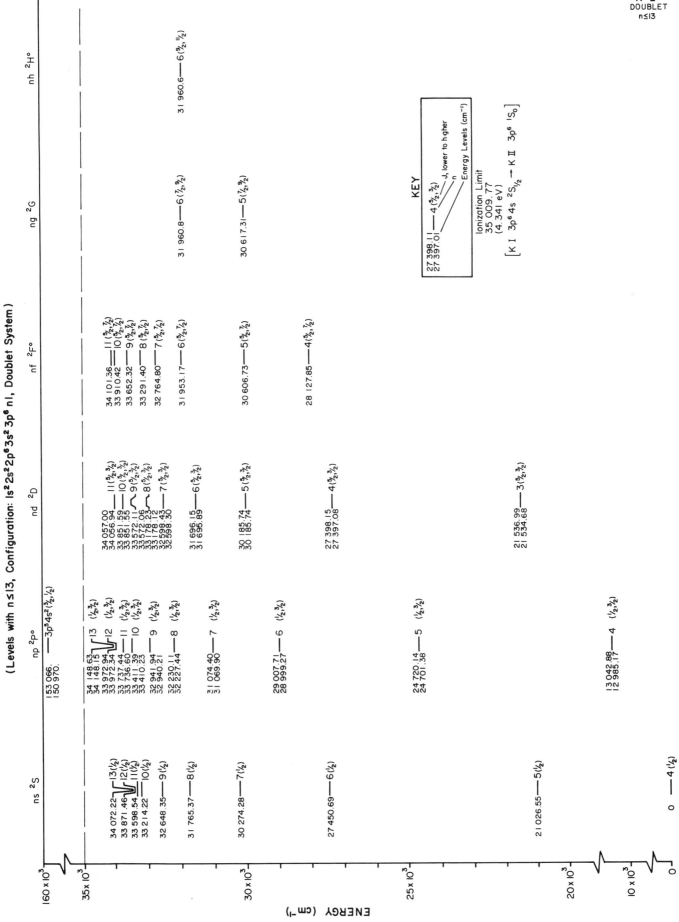

K I ENERGY LEVELS (19 electrons, Z=19)

(Levels with n≤13, Configuration: $1s^2 2s^2 2p^6 3s^2 3p^6 nl$, Doublet System)

ENERGY (cm^{-1}): 160 x 10^3, 35 x 10^3, 30 x 10^3, 25 x 10^3, 20 x 10^3, 10 x 10^3, 0

ns ^2S

34 072.22 — 13(1/2)
33 871.46 — 12(1/2)
33 598.54 — 11(1/2)
33 214.22 — 10(1/2)
32 648.35 — 9(1/2)
31 765.37 — 8(1/2)
30 274.28 — 7(1/2)
27 450.69 — 6(1/2)
21 026.55 — 5(1/2)
0 — 4(1/2)

np $^2P^o$

153 066., 150 970. — $3p^5 4s^2$(3/2, 1/2)
34 148.63, 34 148.15 — 13 (1/2, 3/2)
33 972.94, 33 972.34 — 12 (1/2, 3/2)
33 737.44, 33 736.60 — 11 (1/2, 3/2)
33 411.39, 33 410.23 — 10 (1/2, 3/2)
32 941.94, 32 940.21 — 9 (1/2, 3/2)
32 230.11, 32 227.44 — 8 (1/2, 3/2)
31 074.40, 31 069.90 — 7 (1/2, 3/2)
29 007.71, 28 999.27 — 6 (1/2, 3/2)
24 720.14, 24 701.38 — 5 (1/2, 3/2)
13 042.88, 12 985.17 — 4 (1/2, 3/2)

nd ^2D

34 057.00, 34 056.94 — 11(5/2, 3/2)
33 851.59, 33 851.55 — 10(5/2, 3/2)
33 572.11, 33 572.06 — 9(5/2, 3/2)
33 178.23, 33 178.12 — 8(5/2, 3/2)
32 598.43, 32 598.30 — 7(5/2, 3/2)
31 696.15, 31 695.89 — 6(5/2, 3/2)
30 185.74, 30 185.74 — 5(5/2, 3/2)
27 398.15, 27 397.08 — 4(5/2, 3/2)
21 536.99, 21 534.68 — 3(5/2, 3/2)

nf $^2F^o$

34 101.36 — 11(5/2, 7/2)
33 910.42 — 10(5/2, 7/2)
33 652.32 — 9(5/2, 7/2)
33 291.40 — 8(5/2, 7/2)
32 764.80 — 7(5/2, 7/2)
31 953.17 — 6(5/2, 7/2)
30 606.73 — 5(5/2, 7/2)
28 127.85 — 4(5/2, 7/2)

ng ^2G

31 960.8 — 6(7/2, 9/2)
30 617.31 — 5(7/2, 9/2)

nh $^2H^o$

31 960.6 — 6(9/2, 11/2)

KEY

27 398.11, 27 397.01 — 4(5/2, 3/2)
J, lower to higher
n
Energy Levels (cm^{-1})

Ionization Limit
35 009.77
(4.341 eV)
[K I $3p^6 4s$ $^2S_{1/2}$ → K II $3p^6$ $^1S_0$]

K I
DOUBLET
n≤13

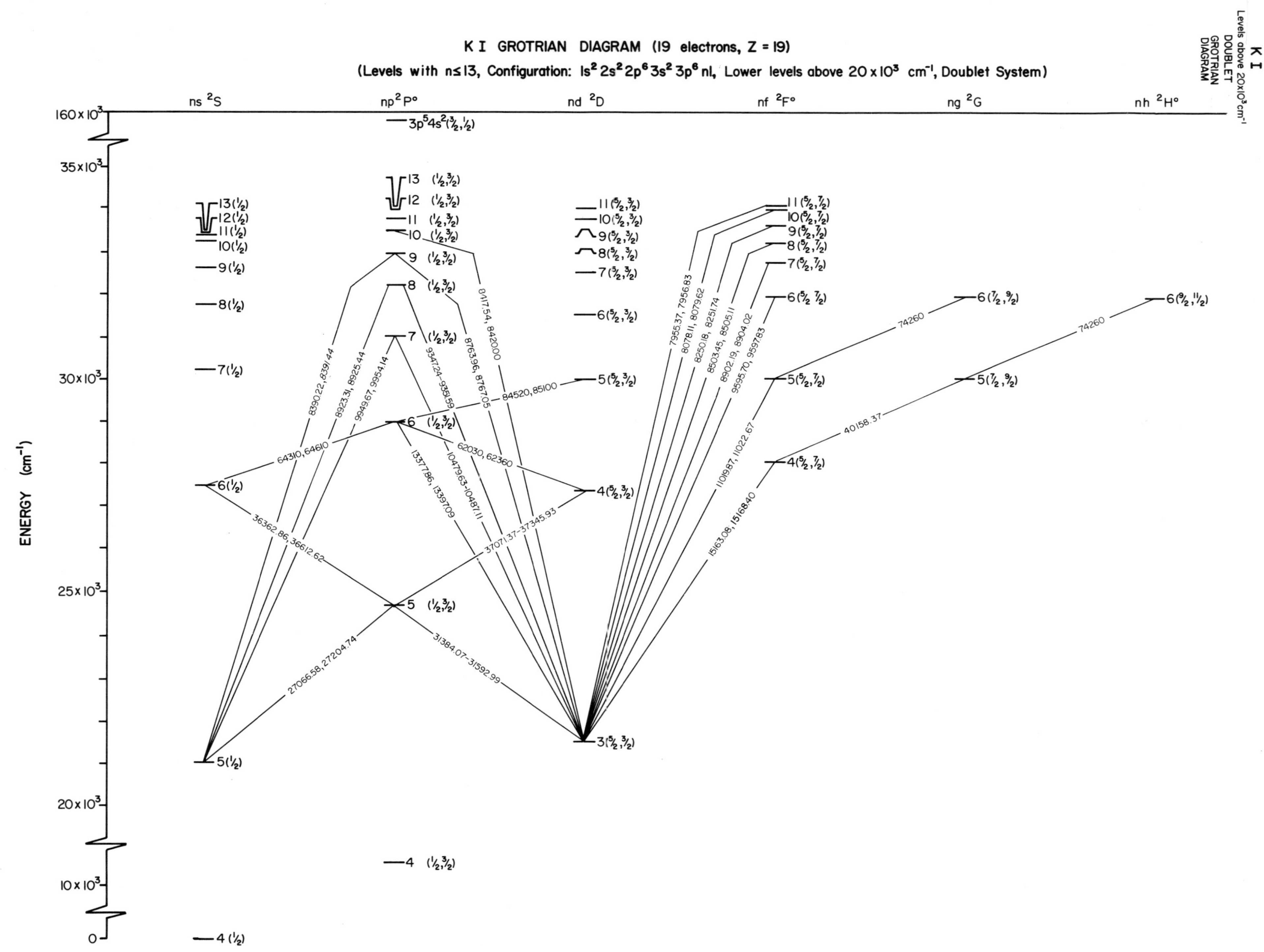
K I GROTRIAN DIAGRAM (19 electrons, Z = 19)
(Levels with n≤13, Configuration: 1s² 2s² 2p⁶ 3s² 3p⁶ nl, Lower levels above 20 x 10³ cm⁻¹, Doublet System)
K I
Levels above 20x10³cm⁻¹
DOUBLET
GROTRIAN
DIAGRAM
ns ²S
np ²P°
nd ²D
nf ²F°
ng ²G
nh ²H°
3p⁵4s²(3/2,1/2)
ENERGY (cm⁻¹)
160x10³
35x10³
30x10³
25x10³
20x10³
10x10³
0
8390.22, 8391.44
8923.31, 8925.44
9949.67, 9954.14
8417.54, 8420.00
8763.96, 8767.05
9347.24-9351.59
84520, 85100
64310, 64610
62030, 62360
13377.86, 13397.09
10479.63-10487.11
36362.86, 36612.62
37071.37-37345.93
27066.58, 27204.74
31384.07-31592.99
7955.37, 7956.83
8078.11, 8079.62
8250.18, 8251.74
8503.45, 8505.11
8902.19, 8904.02
9595.70, 9597.83
11019.87, 11022.67
15163.08, 15168.40
74260
74260
40158.37

K I DOUBLET n≤13

K I ENERGY LEVELS (19 electrons, Z=19)

(Levels with n≤13, Configuration: $1s^2 2s^2 2p^6 3s^2 3p^6 nl$, Doublet System)

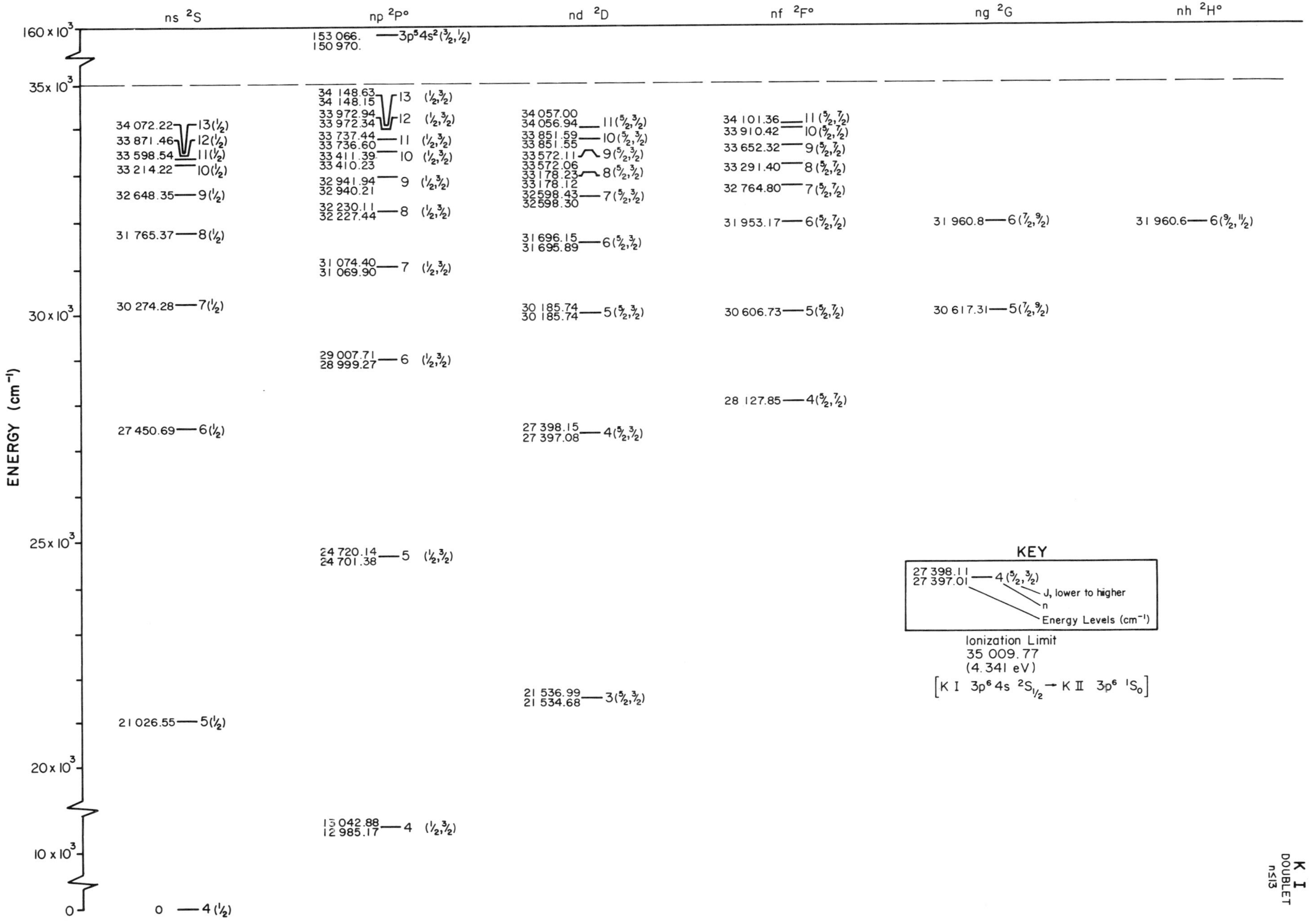

ENERGY (cm^{-1}) — axis: 0, 10×10^3, 20×10^3, 25×10^3, 30×10^3, 35×10^3, 160×10^3

ns ^2S

Energy Levels	n (J)
34 072.22	13(1/2)
33 871.46	12(1/2)
33 598.54	11(1/2)
33 214.22	10(1/2)
32 648.35	9(1/2)
31 765.37	8(1/2)
30 274.28	7(1/2)
27 450.69	6(1/2)
21 026.55	5(1/2)
0	4(1/2)

np $^2P^o$

Energy Levels	n	J
153 066. 150 970.	$3p^5 4s^2$	(3/2, 1/2)
34 148.63 34 148.15	13	(1/2, 3/2)
33 972.94 33 972.34	12	(1/2, 3/2)
33 737.44 33 736.60	11	(1/2, 3/2)
33 411.39 33 410.23	10	(1/2, 3/2)
32 941.94 32 940.21	9	(1/2, 3/2)
32 230.11 32 227.44	8	(1/2, 3/2)
31 074.40 31 069.90	7	(1/2, 3/2)
29 007.71 28 999.27	6	(1/2, 3/2)
24 720.14 24 701.38	5	(1/2, 3/2)
13 042.88 12 985.17	4	(1/2, 3/2)

nd ^2D

Energy Levels	n (J)
34 057.00 34 056.94	11(5/2, 3/2)
33 851.59 33 851.55	10(5/2, 3/2)
33 572.11 33 572.06	9(5/2, 3/2)
33 178.23 33 178.12	8(5/2, 3/2)
32 598.43 32 598.30	7(5/2, 3/2)
31 696.15 31 695.89	6(5/2, 3/2)
30 185.74 30 185.74	5(5/2, 3/2)
27 398.15 27 397.08	4(5/2, 3/2)
21 536.99 21 534.68	3(5/2, 3/2)

nf $^2F^o$

Energy Levels	n (J)
34 101.36	11(5/2, 7/2)
33 910.42	10(5/2, 7/2)
33 652.32	9(5/2, 7/2)
33 291.40	8(5/2, 7/2)
32 764.80	7(5/2, 7/2)
31 953.17	6(5/2, 7/2)
30 606.73	5(5/2, 7/2)
28 127.85	4(5/2, 7/2)

ng ^2G

Energy Levels	n (J)
31 960.8	6(7/2, 9/2)
30 617.31	5(7/2, 9/2)

nh $^2H^o$

Energy Levels	n (J)
31 960.6	6(9/2, 11/2)

KEY

27 398.11
27 397.01 — 4(5/2, 3/2)

- J, lower to higher
- n
- Energy Levels (cm^{-1})

Ionization Limit
35 009.77
(4.341 eV)
[K I $3p^6 4s$ $^2S_{1/2}$ → K II $3p^6$ $^1S_0$]

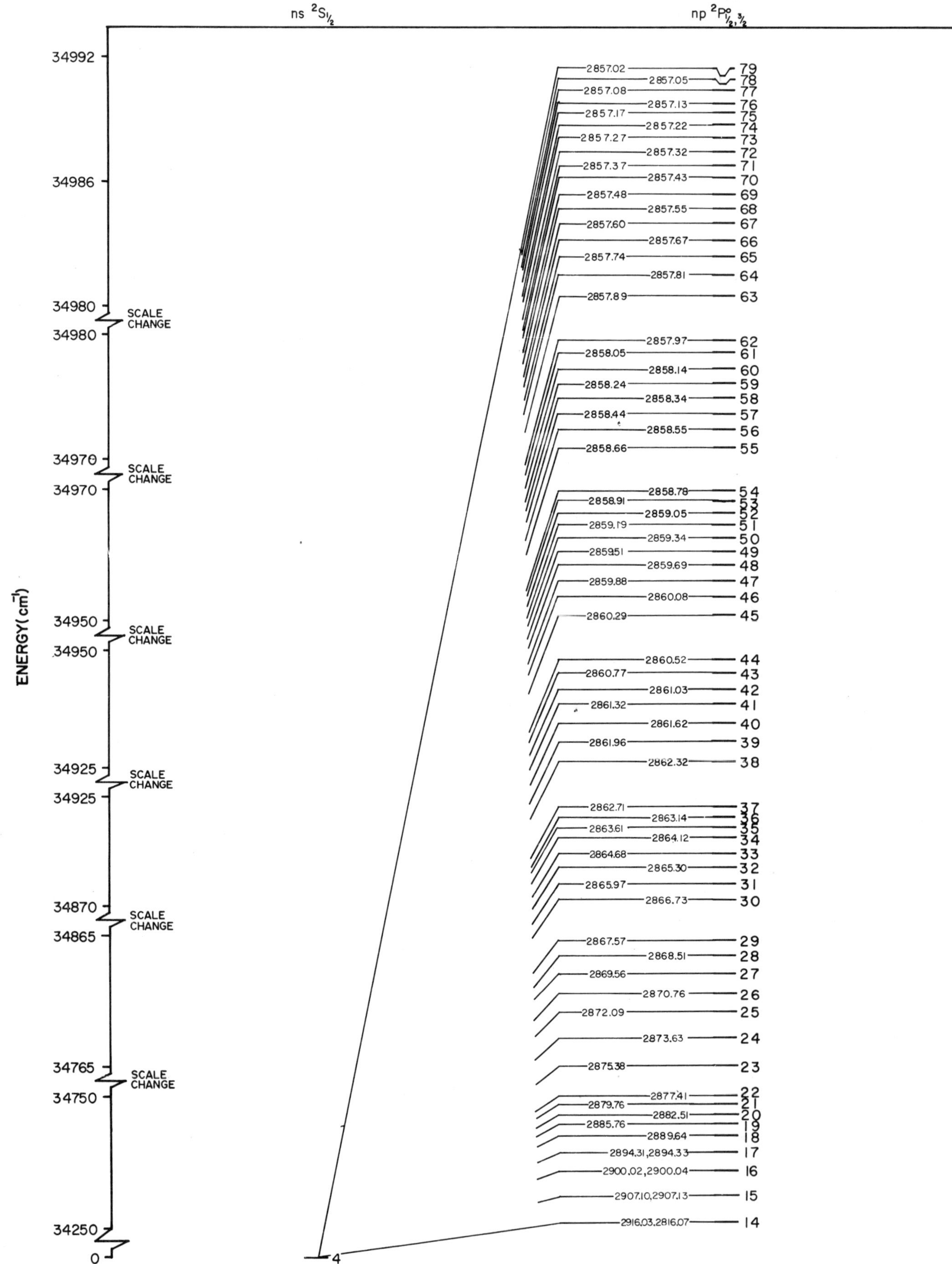
K I GROTRIAN DIAGRAM (19 electrons, Z=19)
(²P° Levels (n ≥ 14), Configuration: 1s² 2s² 2p⁶ 3s² 3p⁶np, Doublet System)
ns ²S1/2
np ²P°1/2,3/2
ENERGY (cm⁻¹)
34992
34986
34980
34980
34970
34970
34950
34950
34925
34925
34870
34865
34765
34750
34250
0
SCALE CHANGE
4
2857.02
2857.05
2857.08
2857.13
2857.17
2857.22
2857.27
2857.32
2857.37
2857.43
2857.48
2857.55
2857.60
2857.67
2857.74
2857.81
2857.89
2857.97
2858.05
2858.14
2858.24
2858.34
2858.44
2858.55
2858.66
2858.78
2858.91
2859.05
2859.19
2859.34
2859.51
2859.69
2859.88
2860.08
2860.29
2860.52
2860.77
2861.03
2861.32
2861.62
2861.96
2862.32
2862.71
2863.14
2863.61
2864.12
2864.68
2865.30
2865.97
2866.73
2867.57
2868.51
2869.56
2870.76
2872.09
2873.63
2875.38
2877.41
2879.76
2882.51
2885.76
2889.64
2894.31,2894.33
2900.02,2900.04
2907.10,2907.13
2916.03,2816.07
79
78
77
76
75
74
73
72
71
70
69
68
67
66
65
64
63
62
61
60
59
58
57
56
55
54
53
52
51
50
49
48
47
46
45
44
43
42
41
40
39
38
37
36
35
34
33
32
31
30
29
28
27
26
25
24
23
22
21
20
19
18
17
16
15
14

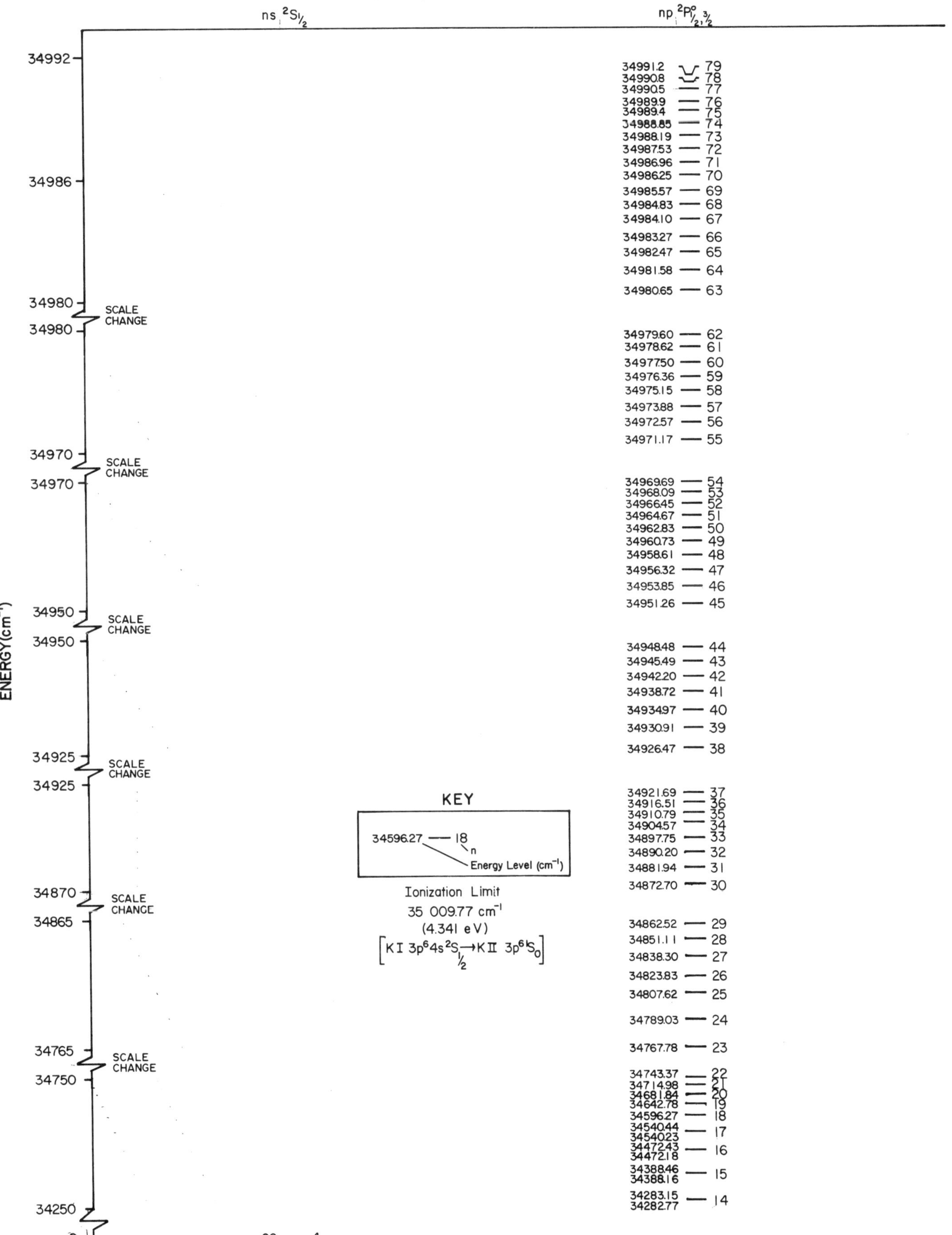
K I ENERGY LEVELS (19 electrons, Z=19)
($^2P^o$ Levels (n≥14), Configuration: $1s^2\ 2s^2\ 2p^6\ 3s^2\ 3p^6 np$, Doublet System)
ns $^2S_{1/2}$
np $^2P^o_{1/2,3/2}$
ENERGY (cm^{-1})
34992
34986
34980
34980
SCALE CHANGE
34970
34970
SCALE CHANGE
34950
34950
SCALE CHANGE
34925
34925
SCALE CHANGE
34870
34865
SCALE CHANGE
34765
34750
SCALE CHANGE
34250
0
0.0 — 4
34991.2 — 79
34990.8 — 78
34990.5 — 77
34989.9 — 76
34989.4 — 75
34988.85 — 74
34988.19 — 73
34987.53 — 72
34986.96 — 71
34986.25 — 70
34985.57 — 69
34984.83 — 68
34984.10 — 67
34983.27 — 66
34982.47 — 65
34981.58 — 64
34980.65 — 63
34979.60 — 62
34978.62 — 61
34977.50 — 60
34976.36 — 59
34975.15 — 58
34973.88 — 57
34972.57 — 56
34971.17 — 55
34969.69 — 54
34968.09 — 53
34966.45 — 52
34964.67 — 51
34962.83 — 50
34960.73 — 49
34958.61 — 48
34956.32 — 47
34953.85 — 46
34951.26 — 45
34948.48 — 44
34945.49 — 43
34942.20 — 42
34938.72 — 41
34934.97 — 40
34930.91 — 39
34926.47 — 38
34921.69 — 37
34916.51 — 36
34910.79 — 35
34904.57 — 34
34897.75 — 33
34890.20 — 32
34881.94 — 31
34872.70 — 30
34862.52 — 29
34851.11 — 28
34838.30 — 27
34823.83 — 26
34807.62 — 25
34789.03 — 24
34767.78 — 23
34743.37 — 22
34714.98 — 21
34681.84 — 20
34642.78 — 19
34596.27 — 18
34540.44
34540.23 — 17
34472.43
34472.18 — 16
34388.46
34388.16 — 15
34283.15
34282.77 — 14
KEY
34596.27 — 18
n
Energy Level (cm^{-1})
Ionization Limit
35 009.77 cm^{-1}
(4.341 eV)
[K I $3p^6 4s\ ^2S_{1/2}$ → K II $3p^6\ ^1S_0$]

K I INTERCOMBINATION GROTRIAN DIAGRAM (19 electrons, Z=19)

(Configuration: $1s^2 2s^2 2p^6 3s^2 3p^5 nl(L^o)nl$, Autoionizing Doublet and Quartet Systems)

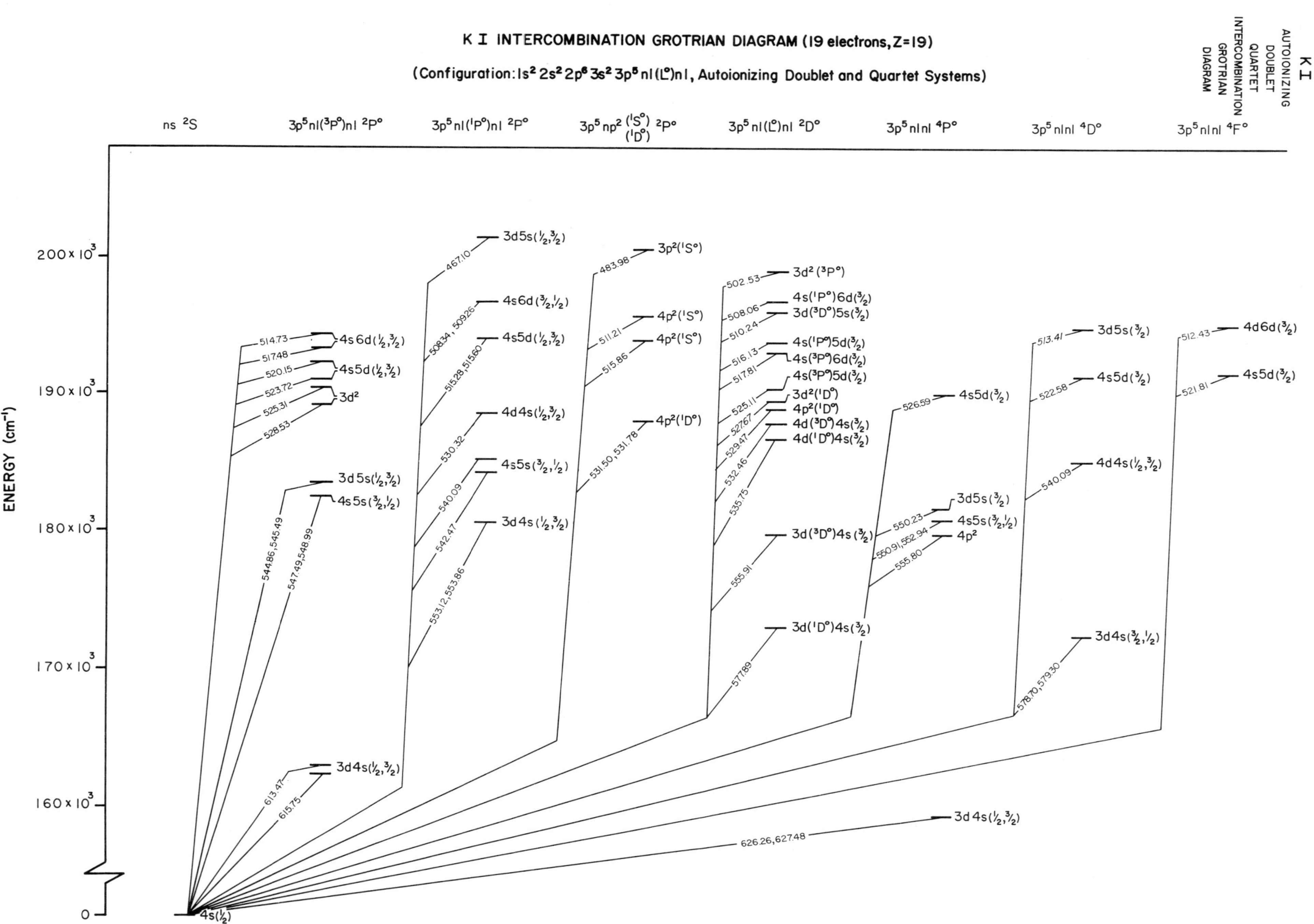

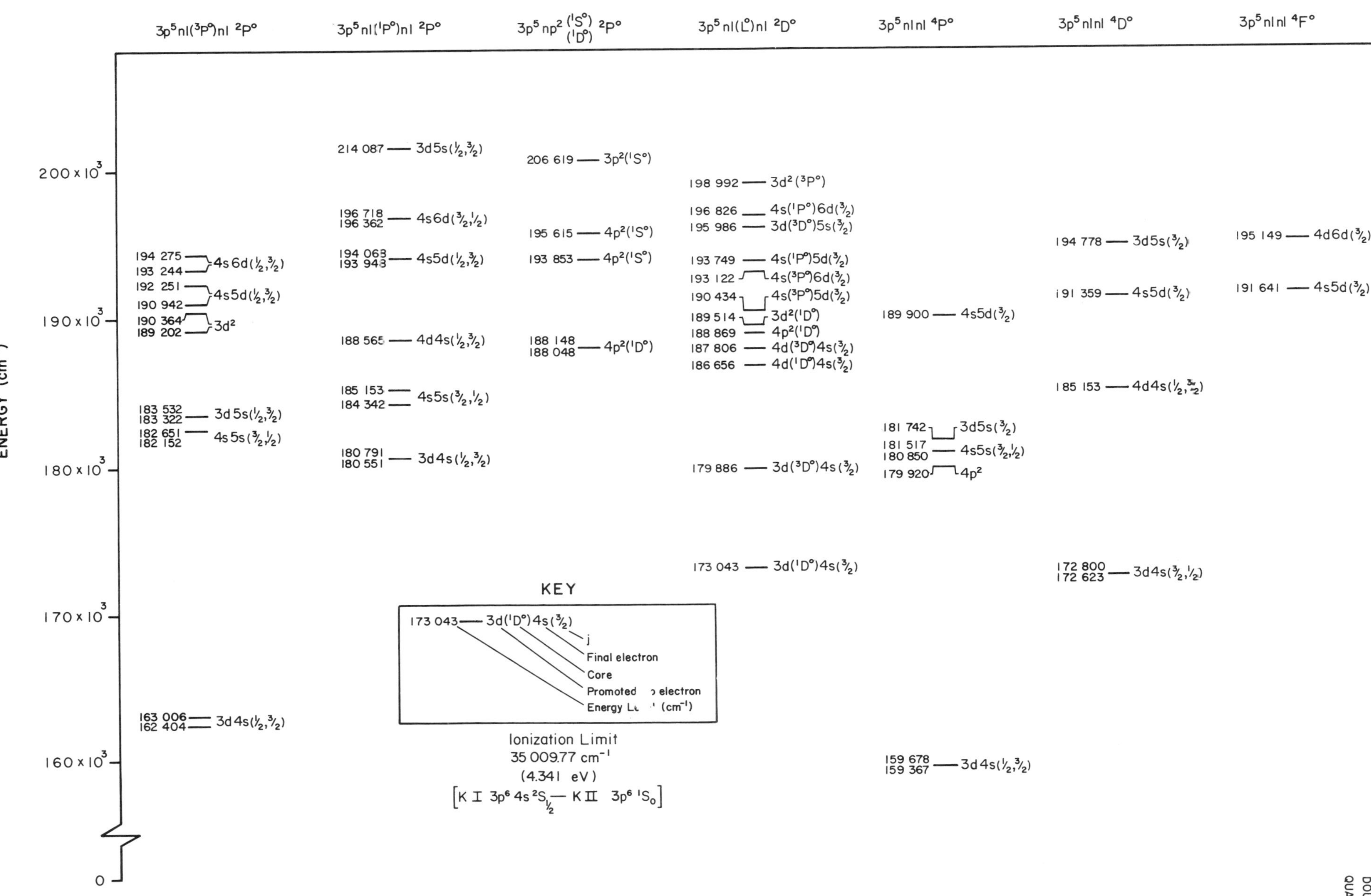
K I ENERGY LEVELS (19 electrons, Z=19)
(Configuration: $1s^2 2s^2 2p^6 3s^2 3p^5 nl(L^o)nl$, Autoionizing Doublet and Quartet Systems)
$3p^5nl(^3P^o)nl\ ^2P^o$
194 275 — 193 244 — 4s6d(1/2,3/2)
192 251 — 190 942 — 4s5d(1/2,3/2)
190 364 — 189 202 — $3d^2$
183 532 183 322 — 3d5s(1/2,3/2)
182 651 182 152 — 4s5s(3/2,1/2)
163 006 162 404 — 3d4s(1/2,3/2)
$3p^5nl(^1P^o)nl\ ^2P^o$
214 087 — 3d5s(1/2,3/2)
196 718 196 362 — 4s6d(3/2,1/2)
194 063 193 943 — 4s5d(1/2,3/2)
188 565 — 4d4s(1/2,3/2)
185 153 — 184 342 — 4s5s(3/2,1/2)
180 791 180 551 — 3d4s(1/2,3/2)
$3p^5np^2(^1S^o)(^1D^o)\ ^2P^o$
206 619 — $3p^2(^1S^o)$
195 615 — $4p^2(^1S^o)$
193 853 — $4p^2(^1S^o)$
188 148 188 048 — $4p^2(^1D^o)$
$3p^5nl(L^o)nl\ ^2D^o$
198 992 — $3d^2(^3P^o)$
196 826 — $4s(^1P^o)6d(3/2)$
195 986 — $3d(^3D^o)5s(3/2)$
193 749 — $4s(^1P^o)5d(3/2)$
193 122 — $4s(^3P^o)6d(3/2)$
190 434 — $4s(^3P^o)5d(3/2)$
189 514 — $3d^2(^1D^o)$
188 869 — $4p^2(^1D^o)$
187 806 — $4d(^3D^o)4s(3/2)$
186 656 — $4d(^1D^o)4s(3/2)$
179 886 — $3d(^3D^o)4s(3/2)$
173 043 — $3d(^1D^o)4s(3/2)$
$3p^5nlnl\ ^4P^o$
189 900 — 4s5d(3/2)
181 742 — 3d5s(3/2)
181 517 180 850 — 4s5s(3/2,1/2)
179 920 — $4p^2$
159 678 159 367 — 3d4s(1/2,3/2)
$3p^5nlnl\ ^4D^o$
194 778 — 3d5s(3/2)
191 359 — 4s5d(3/2)
185 153 — 4d4s(1/2,3/2)
172 800 172 623 — 3d4s(3/2,1/2)
$3p^5nlnl\ ^4F^o$
195 149 — 4d6d(3/2)
191 641 — 4s5d(3/2)
KEY
173 043 — $3d(^1D^o)4s(3/2)$
j
Final electron
Core
Promoted
electron
(cm^{-1})
Ionization Limit
35 009.77 cm^{-1}
(4.341 eV)
[K I $3p^6 4s\ ^2S_{1/2}$ — K II $3p^6\ ^1S_0$]
ENERGY (cm^{-1})
200×10^3
190×10^3
180×10^3
170×10^3
160×10^3
0

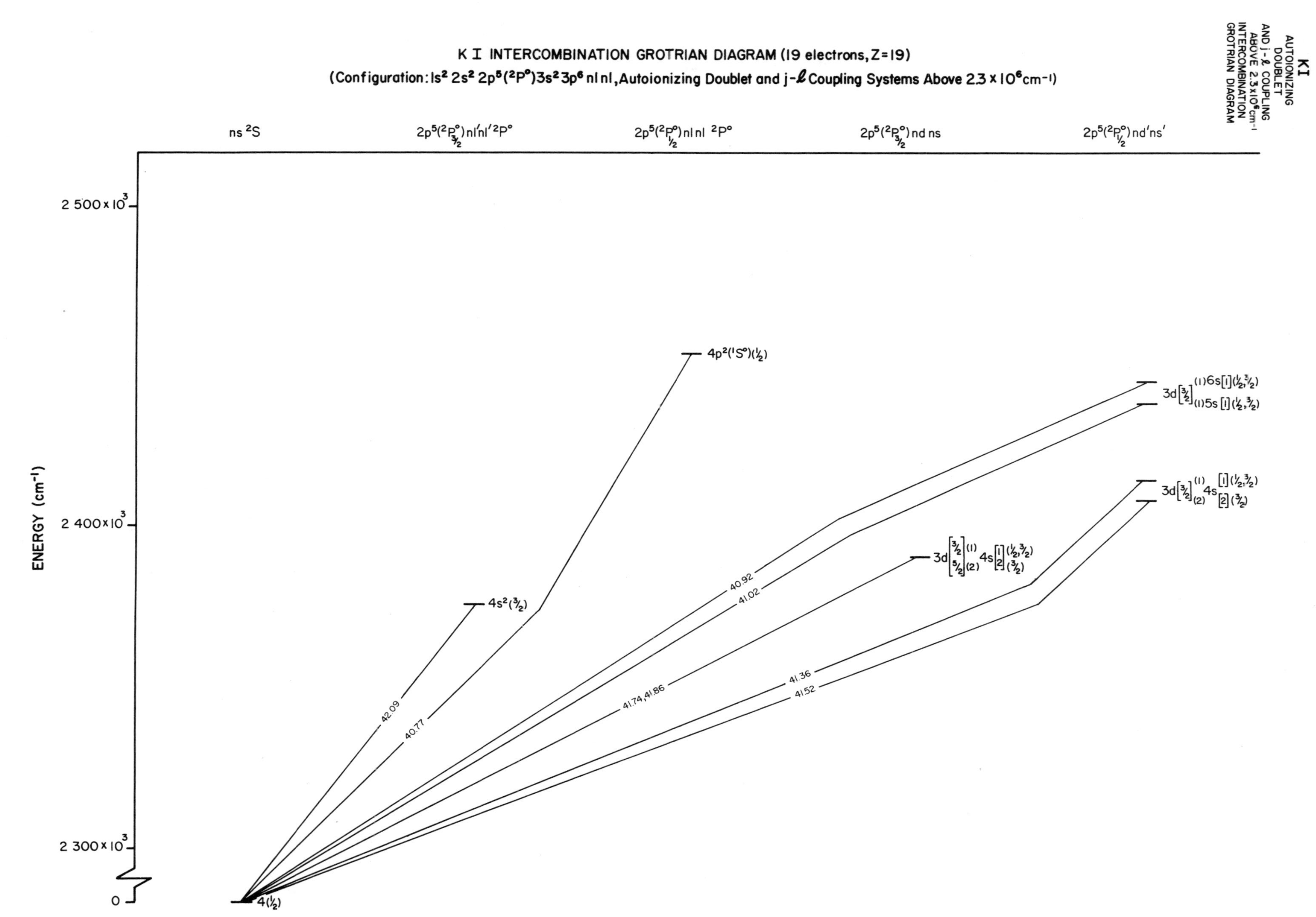
KI
AUTOIONIZING
DOUBLET
AND j-ℓ COUPLING
ABOVE 2.3x10⁶cm⁻¹
INTERCOMBINATION
GROTRIAN DIAGRAM
K I INTERCOMBINATION GROTRIAN DIAGRAM (19 electrons, Z=19)
(Configuration: $1s^2 2s^2 2p^5(^2P^o)3s^2 3p^6$ nl nl, Autoionizing Doublet and j-ℓ Coupling Systems Above $2.3 \times 10^6 cm^{-1}$)
ns ^2S
$2p^5(^2P^o_{3/2})nl'nl'\,^2P^o$
$2p^5(^2P^o_{1/2})nl\,nl\,^2P^o$
$2p^5(^2P^o_{3/2})nd\,ns$
$2p^5(^2P^o_{1/2})nd'ns'$
ENERGY (cm^{-1})
2 500 x 10³
2 400 x 10³
2 300 x 10³
0
4(½)
4s²(3/2)
4p²(¹S°)(½)
3d[3/2](1)6s[1](½,3/2)
3d[3/2](1)5s[1](½,3/2)
3d[3/2](1)4s[1](½,3/2)
3d[3/2](2)4s[2](3/2)
3d[3/2](1)[5/2](2) 4s[1](½,3/2)[2](3/2)
42.09
40.77
40.92
41.02
41.74, 41.86
41.36
41.52

K I ENERGY LEVELS (19 electrons, Z=19)

(Configuration: $1s^2 2s^2 2p^5(^2P^o)3s^2 3p^6 nl nl$, Autoionizing Doublet and j-ℓ Coupling Systems Above 2.3×10^6 cm^{-1})

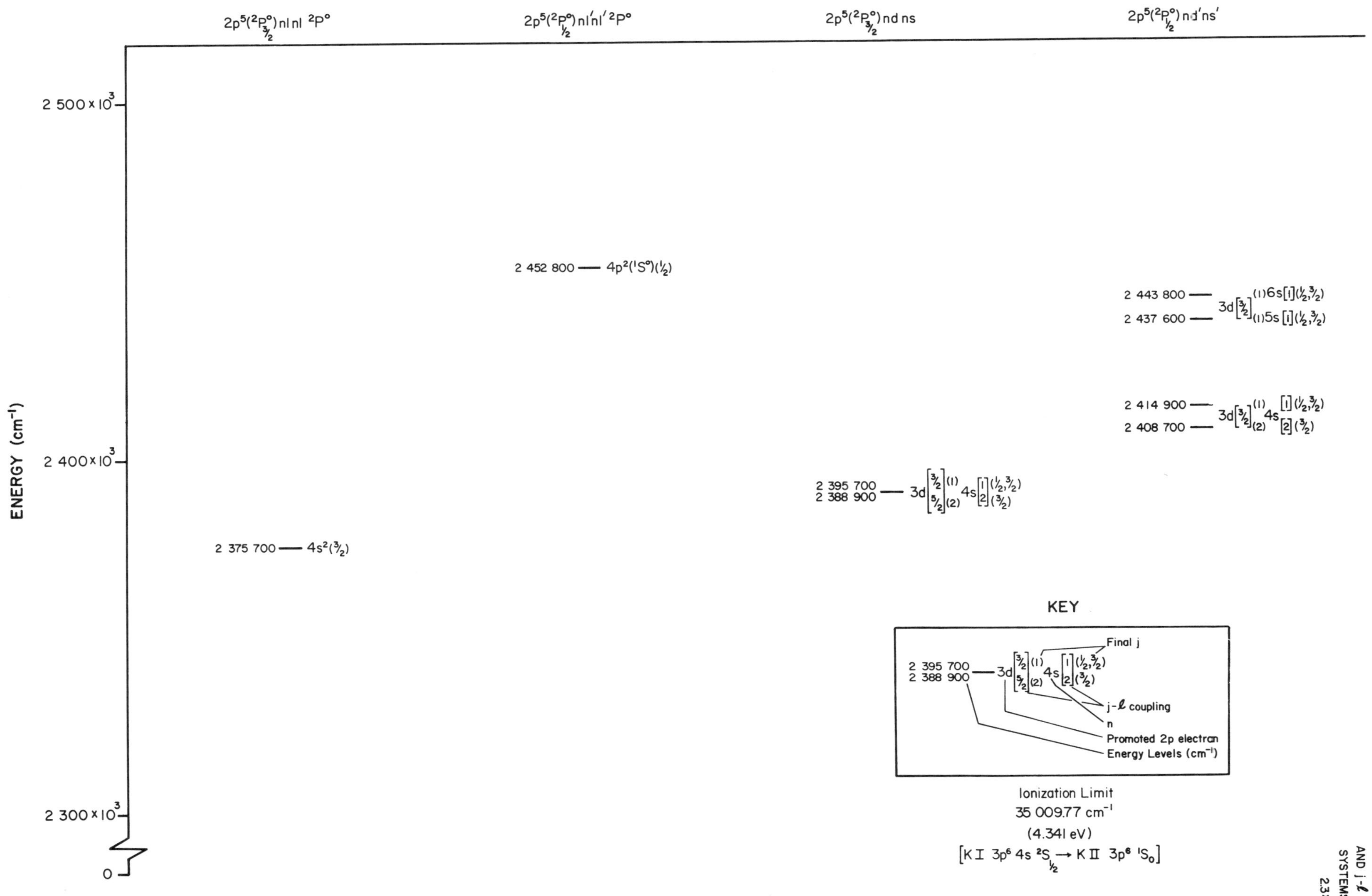

$2p^5(^2P^o_{3/2})nl\,nl\ ^2P^o$

2 375 700 — $4s^2(3/2)$

$2p^5(^2P^o_{1/2})nl\,n'l'\ ^2P^o$

2 452 800 — $4p^2(^1S^o)(1/2)$

$2p^5(^2P^o_{3/2})nd\,ns$

2 395 700, 2 388 900 — $3d\left[\begin{smallmatrix}3/2\\5/2\end{smallmatrix}\right]\begin{smallmatrix}(1)\\(2)\end{smallmatrix}4s\left[\begin{smallmatrix}1\\2\end{smallmatrix}\right]\begin{smallmatrix}(1/2,3/2)\\(3/2)\end{smallmatrix}$

$2p^5(^2P^o_{1/2})nd'ns'$

2 443 800 — $3d[3/2]^{(1)}6s[1](1/2,3/2)$

2 437 600 — $3d[3/2]_{(1)}5s[1](1/2,3/2)$

2 414 900 — $3d[3/2]^{(1)}4s[1](1/2,3/2)$

2 408 700 — $3d[3/2]_{(2)}4s[2](3/2)$

ENERGY (cm^{-1}): 2 500 × 10^3, 2 400 × 10^3, 2 300 × 10^3, 0

KEY

2 395 700, 2 388 900 — $3d\left[\begin{smallmatrix}3/2\\5/2\end{smallmatrix}\right]\begin{smallmatrix}(1)\\(2)\end{smallmatrix}4s\left[\begin{smallmatrix}1\\2\end{smallmatrix}\right]\begin{smallmatrix}(1/2,3/2)\\(3/2)\end{smallmatrix}$

Final j
j-ℓ coupling
n
Promoted 2p electron
Energy Levels (cm^{-1})

Ionization Limit
35 009.77 cm^{-1}
(4.341 eV)
[K I $3p^6 4s\ ^2S_{1/2}$ → K II $3p^6\ ^1S_0$]

K I ENERGY LEVELS (19 electrons, Z=19)

(Configuration: $1s^2 2s^2 2p^6 3s^2 3p^6 nl$, ^2S ($n \geq 13$) and ^2D ($n \geq 11$) Terms, Doublet System)

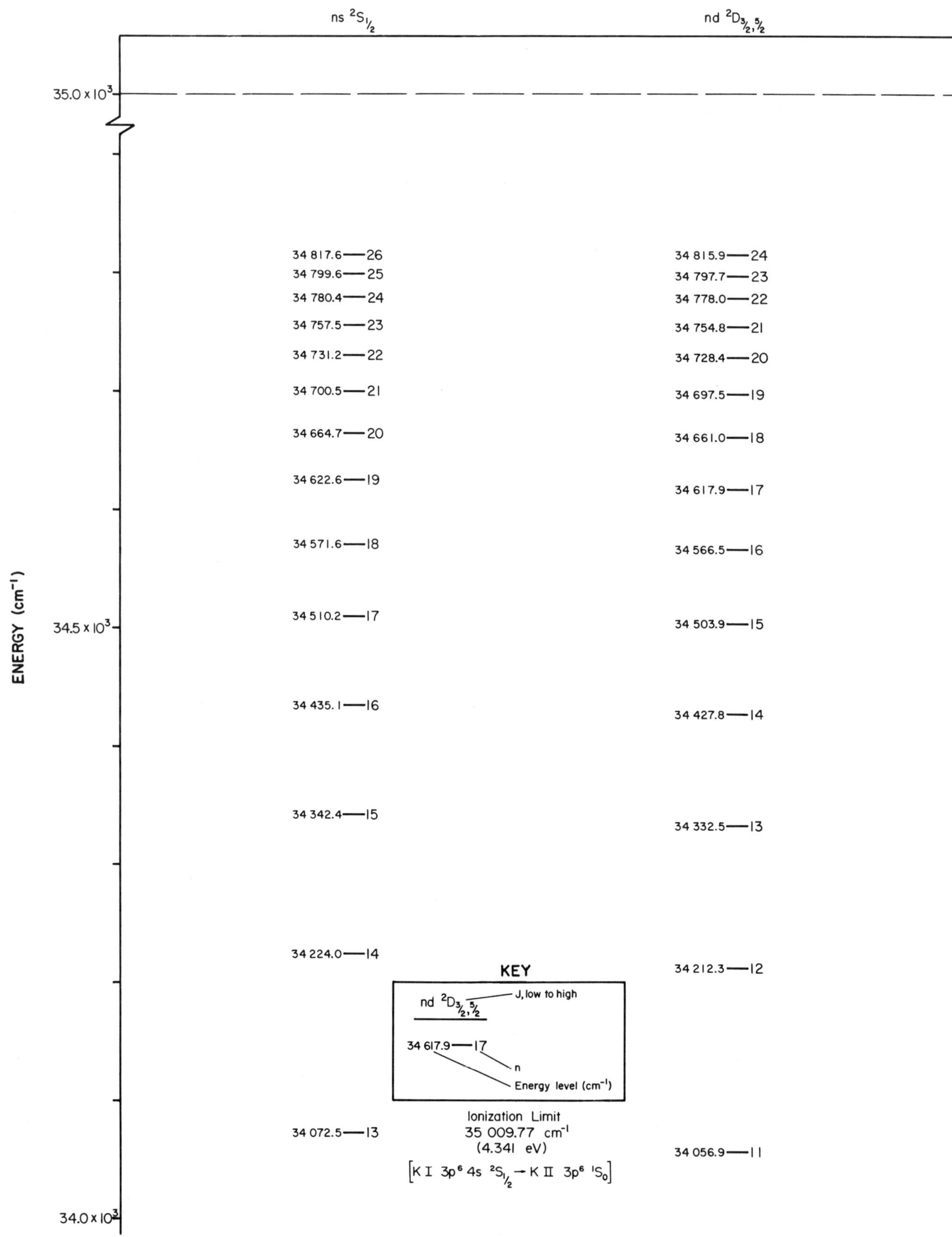

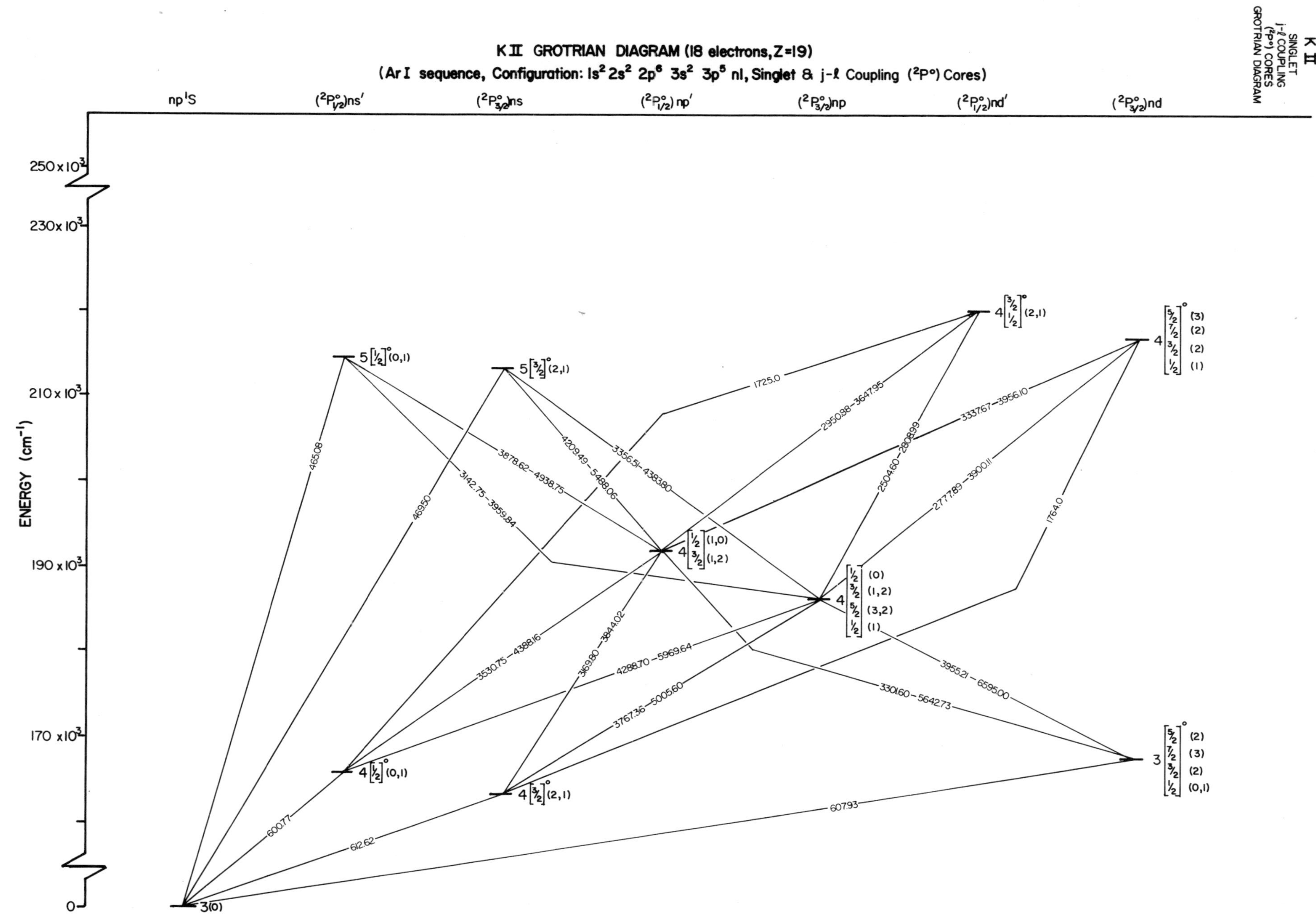
K II
SINGLET
j-ℓ COUPLING
(²P°) CORES
GROTRIAN DIAGRAM
K II GROTRIAN DIAGRAM (18 electrons, Z=19)
(Ar I sequence, Configuration: 1s² 2s² 2p⁶ 3s² 3p⁵ nl, Singlet & j-ℓ Coupling (²P°) Cores)
np ¹S
(²P°1/2)ns′
(²P°3/2)ns
(²P°1/2)np′
(²P°3/2)np
(²P°1/2)nd′
(²P°3/2)nd
ENERGY (cm⁻¹)
250 x 10³
230 x 10³
210 x 10³
190 x 10³
170 x 10³
0
3(0)
5[1/2]°(0,1)
5[3/2]°(2,1)
4[1/2]°(0,1)
4[3/2]°(2,1)
4[1/2](1,0) [3/2](1,2)
4[1/2](0) [3/2](1,2) [5/2](3,2) [1/2](1)
4[3/2 1/2]°(2,1)
4[5/2 (3) 7/2 (2) 3/2 (2) 1/2 (1)]°
3[5/2 (2) 7/2 (3) 3/2 (2) 1/2 (0,1)]°
465.08
469.50
600.77
612.62
607.93
3142.75–3959.84
3878.62–4938.75
4209.49–5488.06
3356.51–4383.80
3530.75–4388.16
3169.80–3844.02
4288.70–5969.64
3767.36–5005.60
1725.0
2950.88–3647.95
2504.60–2808.99
3337.67–3956.10
2777.89–3900.11
1764.0
3955.21–6595.00
3301.60–5642.73

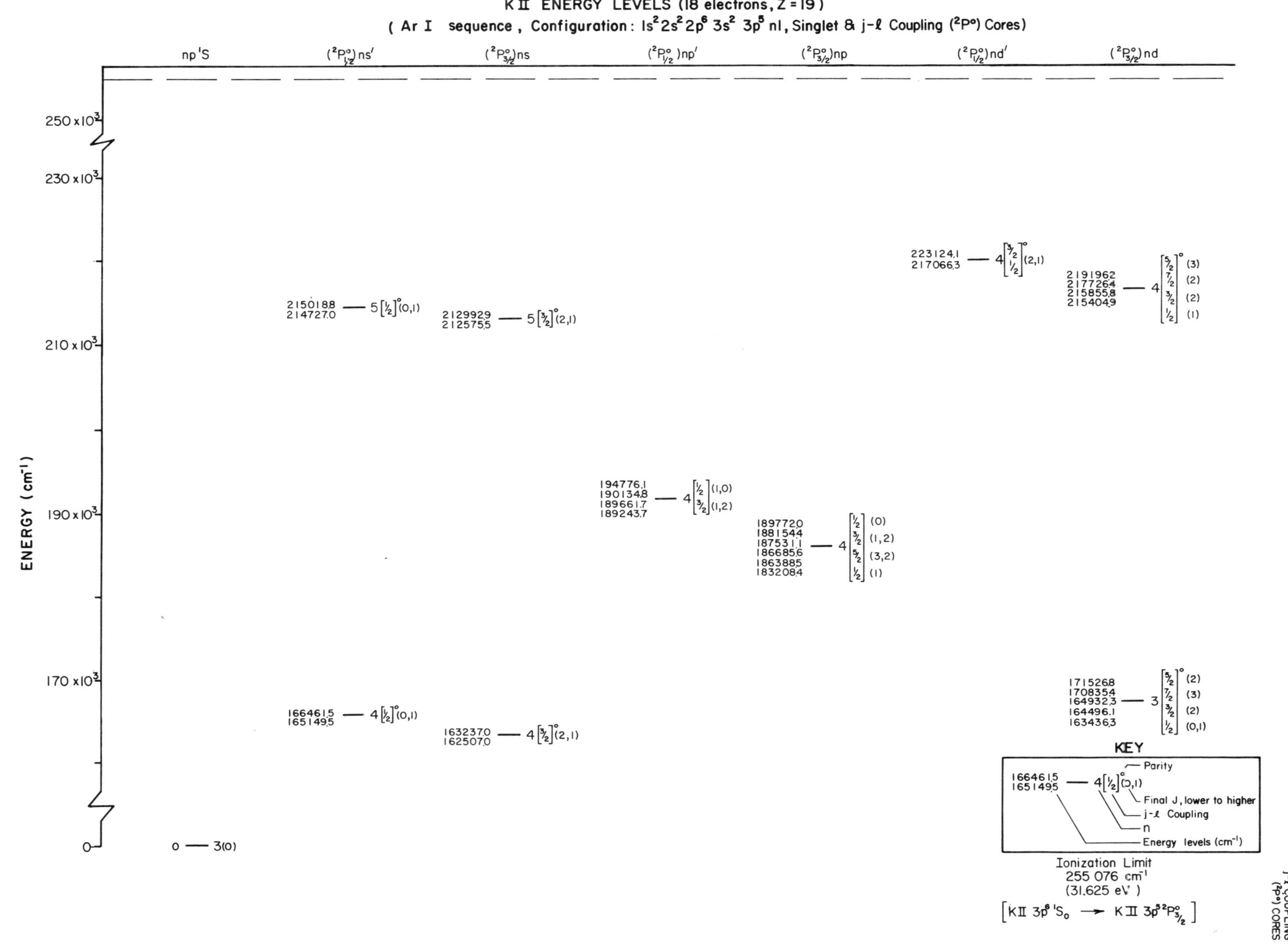
K II ENERGY LEVELS (18 electrons, Z = 19)
(Ar I sequence , Configuration: $1s^2 2s^2 2p^6 3s^2 3p^5 nl$, Singlet & j-ℓ Coupling ($^2P^o$) Cores)
np ^1S
($^2P^o_{1/2}$)ns′
($^2P^o_{3/2}$)ns
($^2P^o_{1/2}$)np′
($^2P^o_{3/2}$)np
($^2P^o_{1/2}$)nd′
($^2P^o_{3/2}$)nd
250 x10³
230 x10³
210 x10³
190 x10³
170 x10³
0
ENERGY (cm⁻¹)
0 — 3(0)
2150188
214727.0
— 5[½]°(0,1)
1664615
165149.5
— 4[½]°(0,1)
2129929
212575.5
— 5[3/2]°(2,1)
1632370
162507.0
— 4[3/2]°(2,1)
194776.1
190134.8
189661.7
189243.7
— 4 [½] (1,0) [3/2] (1,2)
1897720
1881544
187531.1
1866856
1863885
1832084
— 4 [½] (0) 3/2 (1,2) 5/2 (3,2) ½ (1)
223124.1
217066.3
— 4 [3/2 ½]° (2,1)
2191962
2177264
215855.8
215404.9
— 4 [5/2]° (3) 7/2 (2) 3/2 (2) ½ (1)
1715268
1708354
164932.3
164496.1
163436.3
— 3 [5/2]° (2) 7/2 (3) 3/2 (2) ½ (0,1)
KEY
1664615
165149.5
— 4[½]°(0,1)
Parity
Final J, lower to higher
j-ℓ Coupling
n
Energy levels (cm⁻¹)
Ionization Limit
255 076 cm⁻¹
(31.625 eV)
[K II $3p^6\ ^1S_0$ → K II $3p^5\ ^2P^o_{3/2}$]
K II
SINGLET
j-ℓ COUPLING
($^2P^o$) CORES

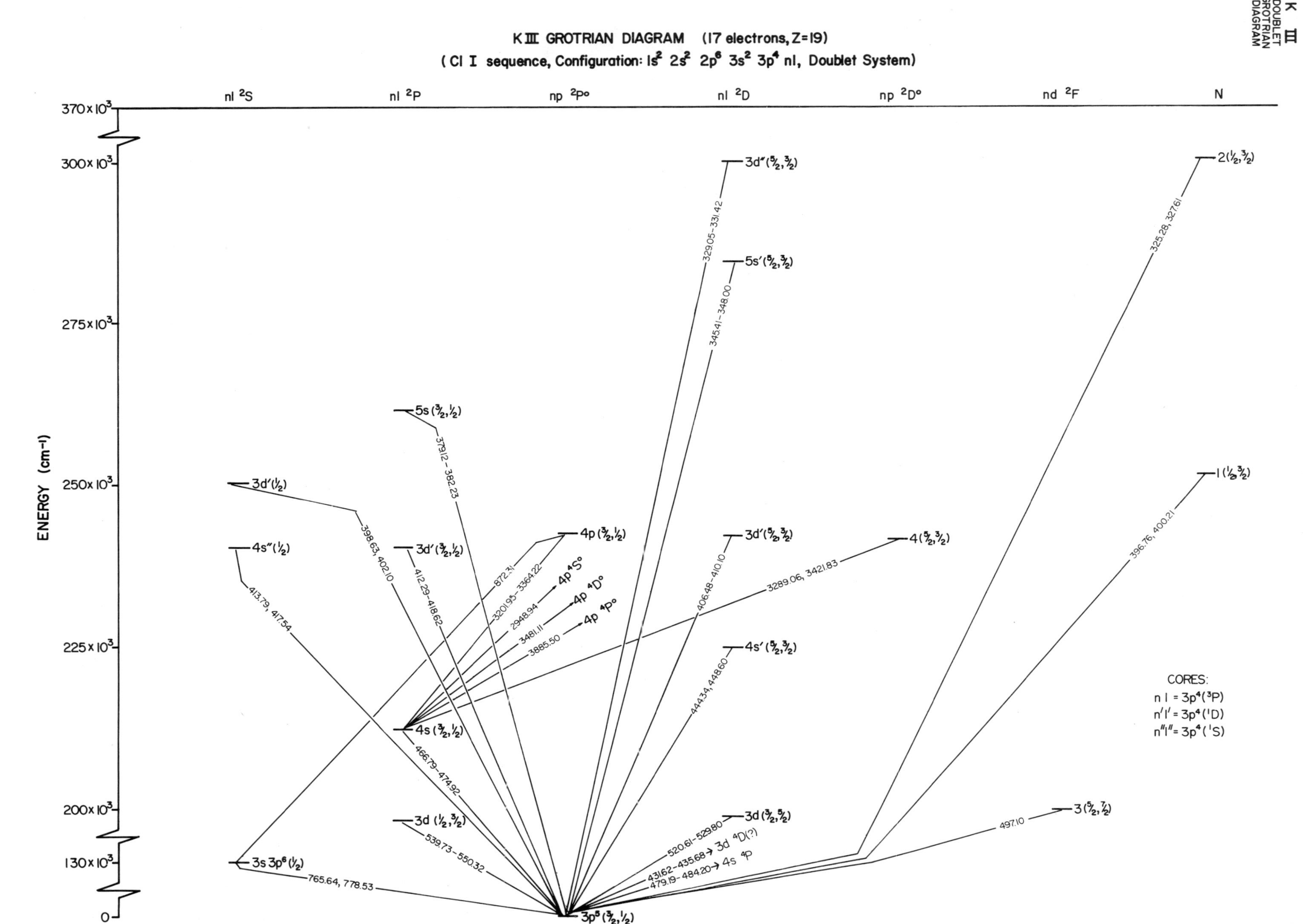
K III
DOUBLET
GROTRIAN
DIAGRAM
K III GROTRIAN DIAGRAM (17 electrons, Z=19)
(Cl I sequence, Configuration: 1s² 2s² 2p⁶ 3s² 3p⁴ nl, Doublet System)
nl ²S
nl ²P
np ²P°
nl ²D
np ²D°
nd ²F
N
ENERGY (cm⁻¹)
370x10³
300x10³
275x10³
250x10³
225x10³
200x10³
130x10³
0
3d″(5/2,3/2)
2(1/2,3/2)
329.05–331.42
325.28, 327.61
5s′(5/2,3/2)
345.41–348.00
5s(3/2,1/2)
379.12–382.23
3d′(1/2)
1(1/2,3/2)
396.76, 400.21
4p(3/2,1/2)
3d′(5/2,3/2)
4s″(1/2)
3d′(3/2,1/2)
4(5/2,3/2)
398.63, 402.10
412.29–418.62
872.31
3201.95–3364.22
2948.94
4p ⁴S°
3481.11
4p ⁴D°
3885.50
4p ⁴P°
406.48–410.10
3289.06, 3421.83
413.79, 417.54
4s′(5/2,3/2)
444.34, 448.60
CORES:
n l = 3p⁴(³P)
n′l′ = 3p⁴(¹D)
n″l″ = 3p⁴(¹S)
4s(3/2,1/2)
466.79–474.92
3d(1/2,3/2)
3d(3/2,5/2)
3(5/2,7/2)
539.73–550.32
520.61–529.80
497.10
3s 3p⁶(1/2)
431.62–435.68 → 3d ⁴D(?)
479.19–484.20 → 4s ⁴P
765.64, 778.53
3p⁵(3/2,1/2)

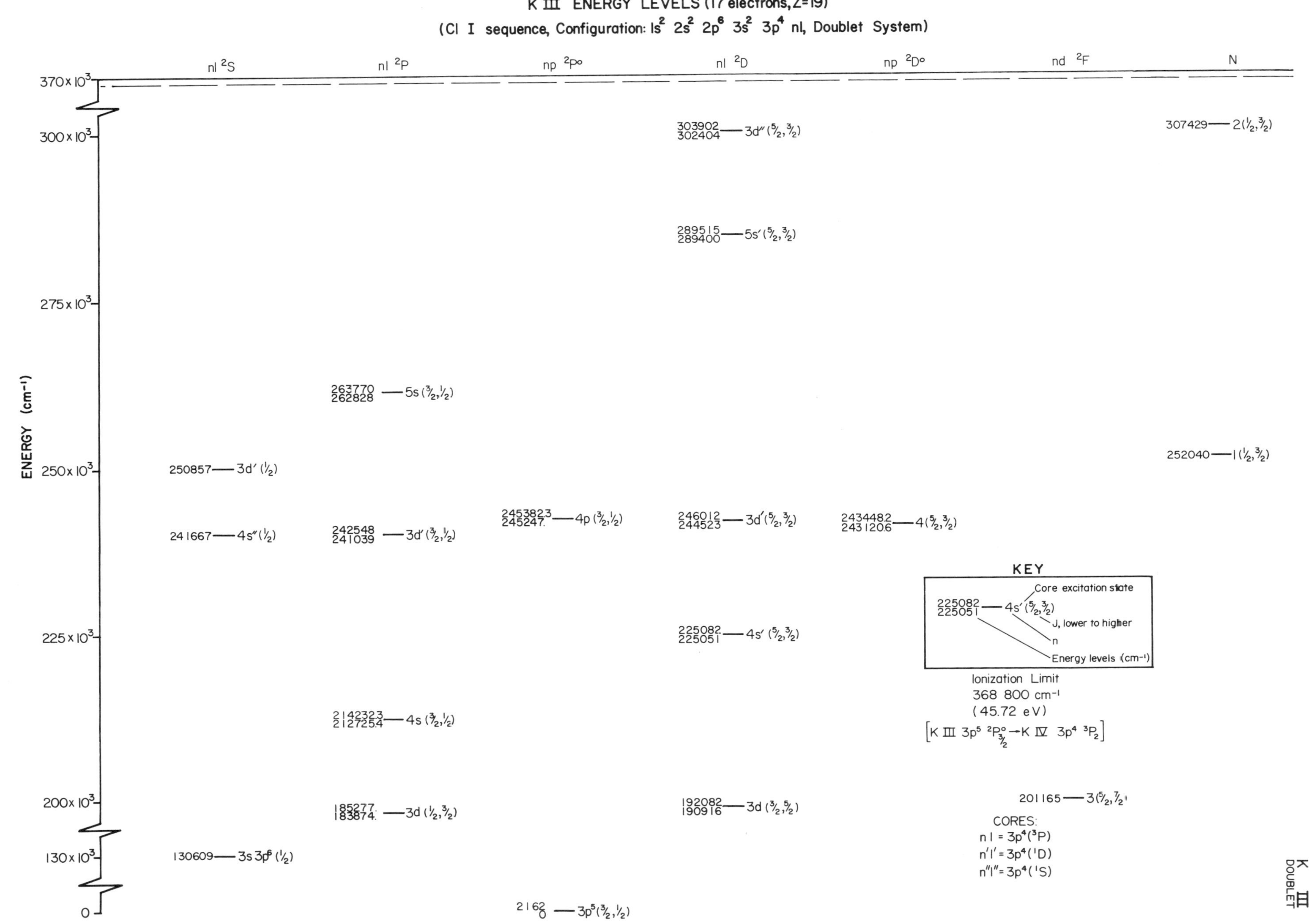
K III ENERGY LEVELS (17 electrons, Z=19)
(Cl I sequence, Configuration: 1s^2 2s^2 2p^6 3s^2 3p^4 nl, Doublet System)
nl ^{2}S
nl ^{2}P
np ^{2}P°
nl ^{2}D
np ^{2}D°
nd ^{2}F
N
370 x 10^3
300 x 10^3
275 x 10^3
250 x 10^3
225 x 10^3
200 x 10^3
130 x 10^3
0
ENERGY (cm^{-1})
303902
302404 — 3d''(5/2,3/2)
307429 — 2(1/2,3/2)
289515
289400 — 5s'(5/2,3/2)
263770
262828 — 5s(3/2,1/2)
250857 — 3d'(1/2)
252040 — 1(1/2,3/2)
2453823
245247 — 4p(3/2,1/2)
246012
244523 — 3d'(5/2,3/2)
2434482
2431206 — 4(5/2,3/2)
241667 — 4s''(1/2)
242548
241039 — 3d'(3/2,1/2)
KEY
225082
225051 — 4s'(5/2,3/2)
Core excitation state
J, lower to higher
n
Energy levels (cm^{-1})
225082
225051 — 4s'(5/2,3/2)
Ionization Limit
368 800 cm^{-1}
(45.72 eV)
[K III 3p^5 ^{2}P°$_{3/2}$ → K IV 3p^4 ^{3}P$_2$]
2142323
2127254 — 4s(3/2,1/2)
201165 — 3(5/2,7/2)
185277
183874 — 3d(1/2,3/2)
192082
190916 — 3d(3/2,5/2)
CORES:
n l = 3p^4(^{3}P)
n'l' = 3p^4(^{1}D)
n''l'' = 3p^4(^{1}S)
130609 — 3s3p^6(1/2)
2162
0 — 3p^5(3/2,1/2)
K III
DOUBLET

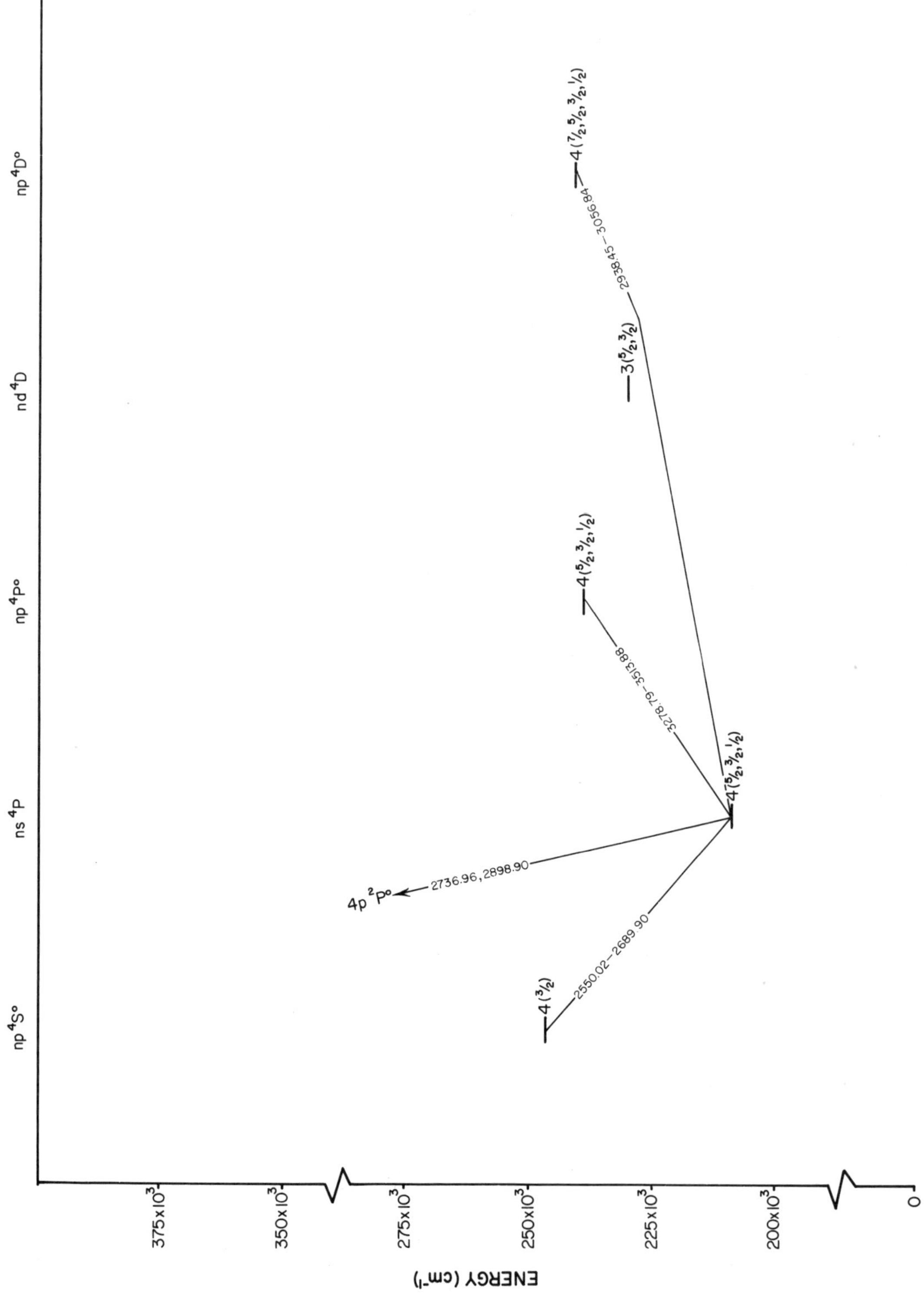
K III GROTRIAN DIAGRAM (17 electrons, Z=19)
(Cl I sequence, Configuration: 1s^{2}2s^{2}2p^{6}3s^{2}3p^{4}nl, Quartet System)
np^{4}S°
ns^{4}P
np^{4}P°
nd^{4}D
np^{4}D°
4(3/2)
2550.02–2689.90
4p ^{2}P°
2736.96, 2898.90
4(5/2, 3/2, 1/2)
3278.79–3513.88
4(5/2, 3/2, 1/2)
3(5/2, 3/2)
2938.45–3056.84
4(7/2, 5/2, 3/2, 1/2)
ENERGY (cm^{-1})
375x10^3
350x10^3
275x10^3
250x10^3
225x10^3
200x10^3
0

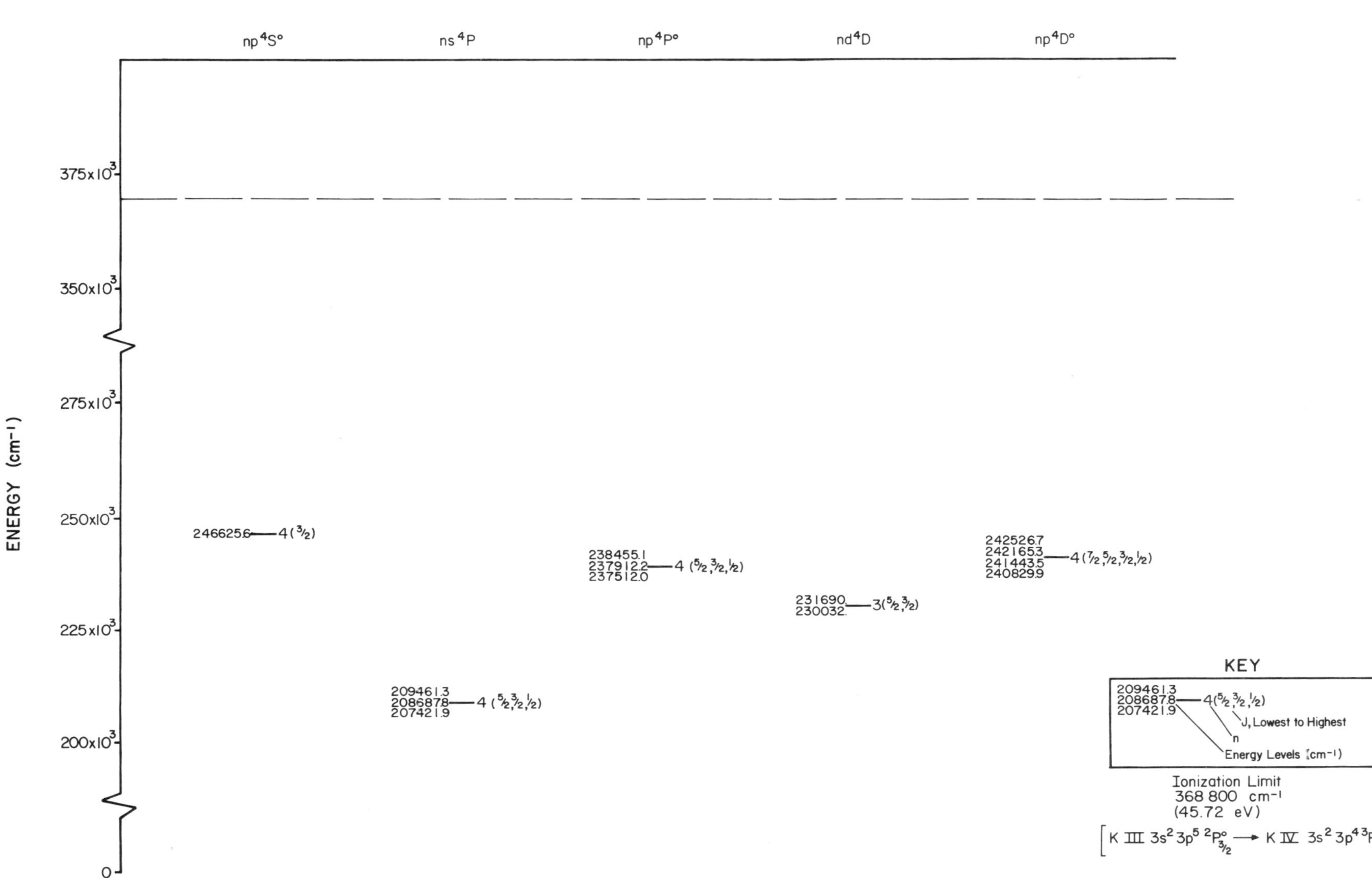
K III ENERGY LEVELS (17 electrons, Z=19)
(Cl I sequence, Configuration: $1s^2 2s^2 2p^6 3s^2 3p^4 nl$, Quartet System)
$np\,^4S^o$
$ns\,^4P$
$np\,^4P^o$
$nd\,^4D$
$np\,^4D^o$
375x10³
350x10³
275x10³
250x10³
225x10³
200x10³
0
ENERGY (cm⁻¹)
246625.6 — 4(3/2)
209461.3 208687.8 207421.9 — 4(5/2,3/2,1/2)
238455.1 237912.2 237512.0 — 4(5/2,3/2,1/2)
231690 230032 — 3(5/2,3/2)
242526.7 242165.3 241443.5 240829.9 — 4(7/2,5/2,3/2,1/2)
KEY
209461.3 208687.8 207421.9 — 4(5/2,3/2,1/2)
J, Lowest to Highest
n
Energy Levels (cm⁻¹)
Ionization Limit
368 800 cm⁻¹
(45.72 eV)
[K III $3s^2 3p^5\,^2P^o_{3/2}$ → K IV $3s^2 3p^4\,^3P_2$]

K IV GROTRIAN DIAGRAM (16 electrons, Z=19)

(S I sequence, Configuration: $1s^2 2s^2 2p^6 3s^2 3p^3 nl$, Singlet and Triplet Systems)

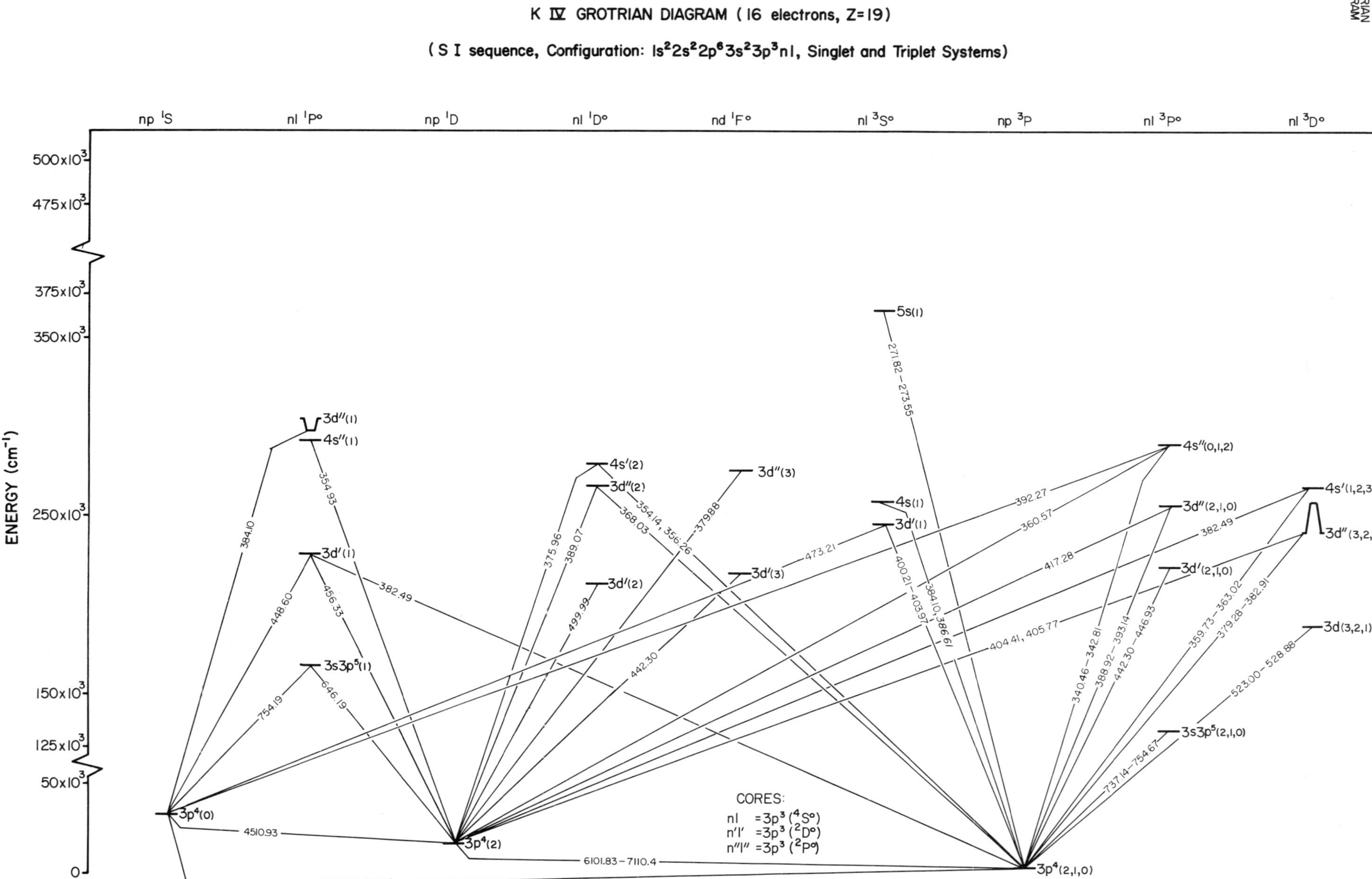

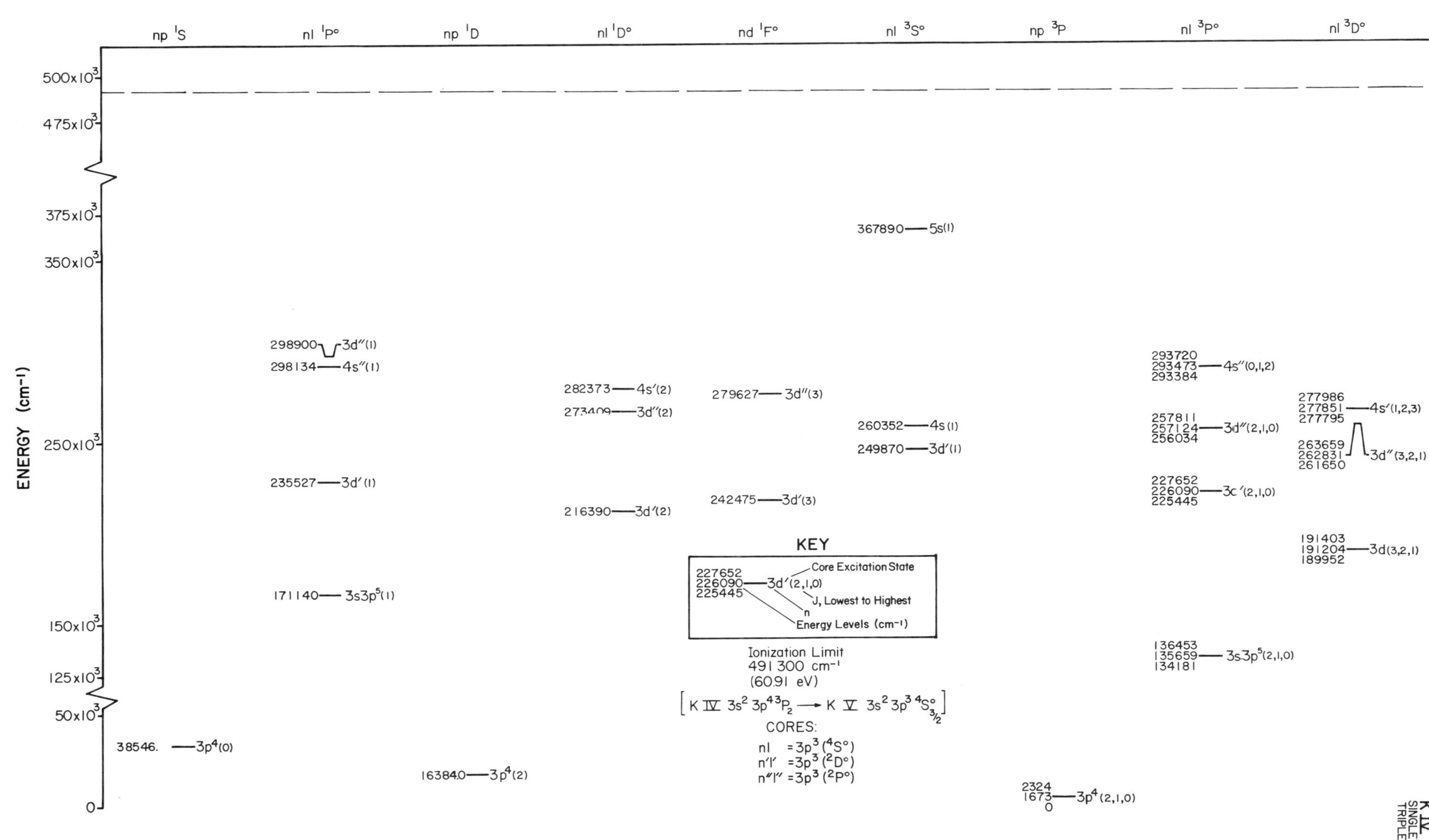
K IV ENERGY LEVELS (16 electrons, Z=19)
(S I sequence, Configuration: 1s²2s²2p⁶3s²3p³nl, Singlet & Triplet Systems)
np ¹S
nl ¹P°
np ¹D
nl ¹D°
nd ¹F°
nl ³S°
np ³P
nl ³P°
nl ³D°
ENERGY (cm⁻¹)
500x10³
475x10³
375x10³
350x10³
250x10³
150x10³
125x10³
50x10³
0
38546. —3p⁴(0)
298900—3d″(1)
298134—4s″(1)
235527—3d′(1)
171140—3s3p⁵(1)
16384.0—3p⁴(2)
282373—4s′(2)
273409—3d″(2)
216390—3d′(2)
279627—3d″(3)
242475—3d′(3)
367890—5s(1)
260352—4s(1)
249870—3d′(1)
2324
1673—3p⁴(2,1,0)
0
293720
293473—4s″(0,1,2)
293384
257811
257124—3d″(2,1,0)
256034
227652
226090—3c′(2,1,0)
225445
136453
135659—3s3p⁵(2,1,0)
134181
277986
277851—4s′(1,2,3)
277795
263659
262831—3d″(3,2,1)
261650
191403
191204—3d(3,2,1)
189952
KEY
227652
226090—3d′(2,1,0)
225445
Core Excitation State
J, Lowest to Highest
n
Energy Levels (cm⁻¹)
Ionization Limit
491 300 cm⁻¹
(60.91 eV)
[K IV 3s² 3p⁴ ³P₂ → K V 3s² 3p³ ⁴S°₃/₂]
CORES:
nl = 3p³(⁴S°)
n′l′ = 3p³(²D°)
n″l″ = 3p³(²P°)
K IV
SINGLET
TRIPLET

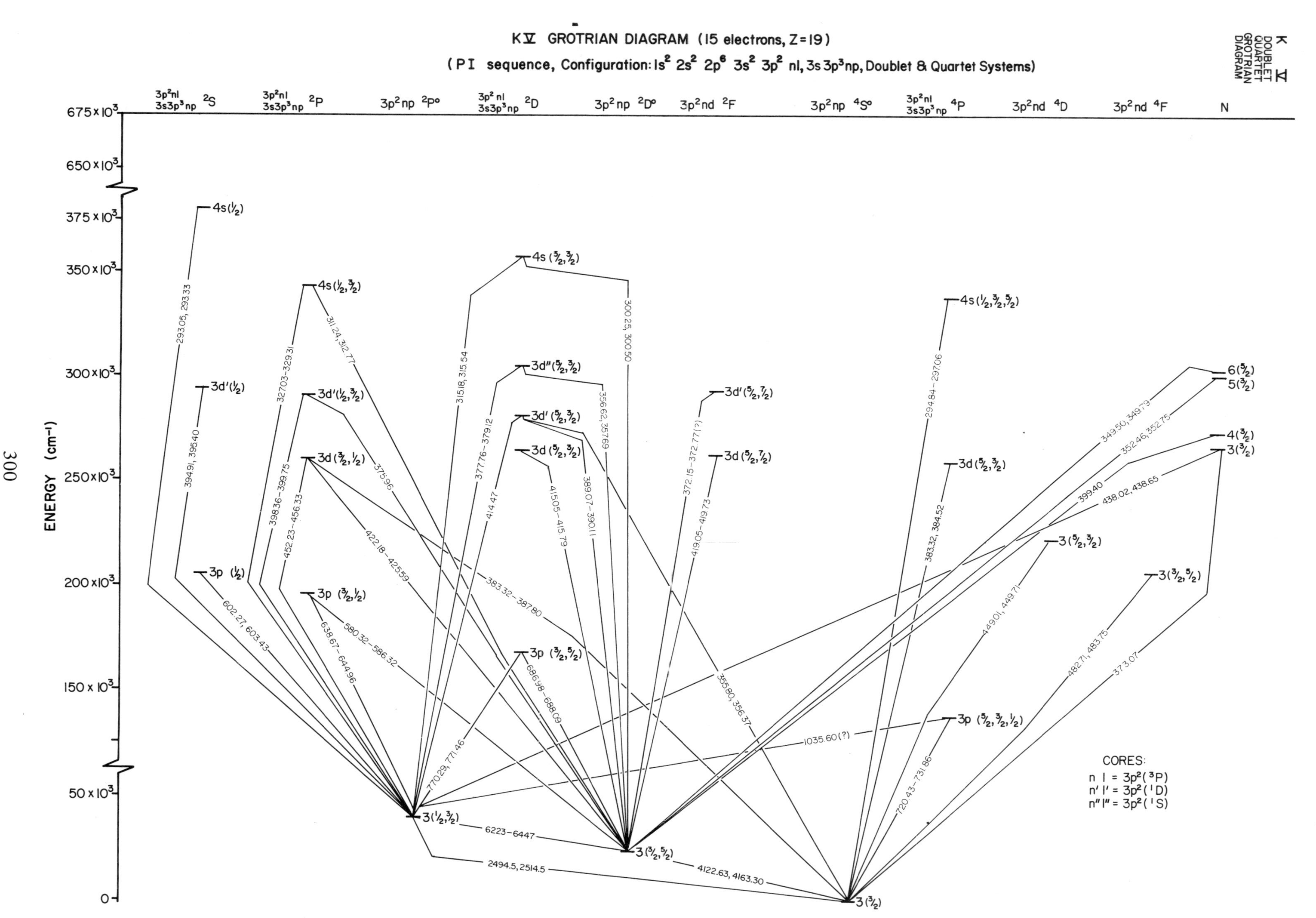

K V
DOUBLET
QUARTET
GROTRIAN
DIAGRAM
K V GROTRIAN DIAGRAM (15 electrons, Z=19)
(P I sequence, Configuration: 1s^2 2s^2 2p^6 3s^2 3p^2 nl, 3s 3p^{3}np, Doublet & Quartet Systems)
ENERGY (cm^{-1})
675×10^3
650×10^3
375×10^3
350×10^3
300×10^3
250×10^3
200×10^3
150×10^3
50×10^3
0
3p^{2}nl 3s3p^{3}np ^{2}S
3p^{2}nl 3s3p^{3}np ^{2}P
3p^{2}np ^{2}P°
3p^{2}nl 3s3p^{3}np ^{2}D
3p^{2}np ^{2}D°
3p^{2}nd ^{2}F
3p^{2}np ^{4}S°
3p^{2}nl 3s3p^{3}np ^{4}P
3p^{2}nd ^{4}D
3p^{2}nd ^{4}F
N
CORES:
n l = 3p^2(^{3}P)
n′ l′ = 3p^2(^{1}D)
n″ l″ = 3p^2(^{1}S)

K V ENERGY LEVELS (15 electrons, Z=19)

(P I sequence, Configuration: $1s^2 2s^2 2p^6 3s^2 3p^2 nl$, $3s3p^3np$, Doublet & Quartet Systems)

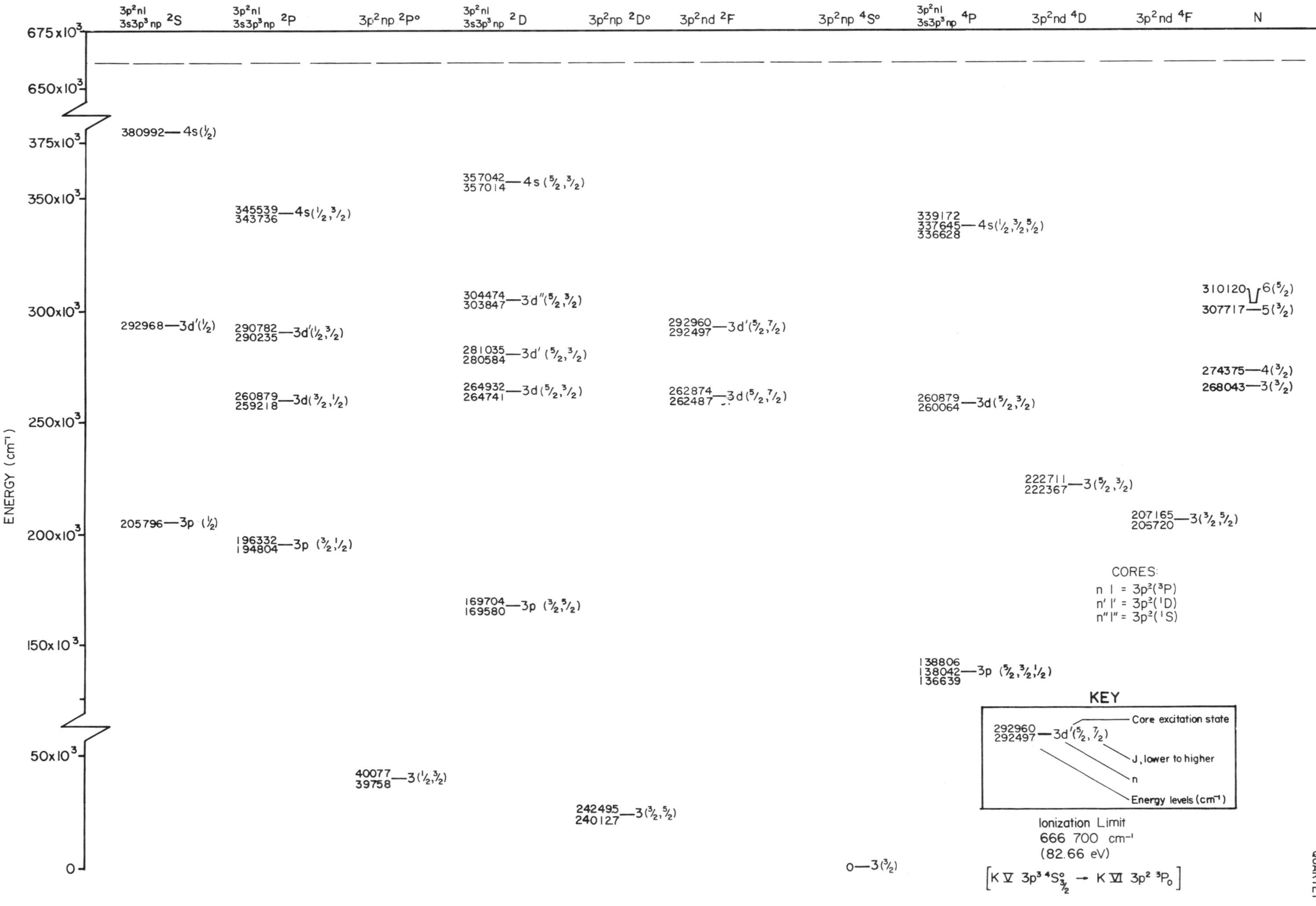

K V
DOUBLET
QUARTET

ENERGY (cm^{-1})

Energy axis: 675x10^3, 650x10^3, 375x10^3, 350x10^3, 300x10^3, 250x10^3, 200x10^3, 150x10^3, 50x10^3, 0

$3p^2nl$ / $3s3p^3np$ ^2S
- 380992—4s($\frac{1}{2}$)
- 292968—3d′($\frac{1}{2}$)
- 205796—3p ($\frac{1}{2}$)

$3p^2nl$ / $3s3p^3np$ ^2P
- 345539, 343736—4s($\frac{1}{2}$,$\frac{3}{2}$)
- 290782, 290235—3d′($\frac{1}{2}$,$\frac{3}{2}$)
- 260879, 259218—3d($\frac{3}{2}$,$\frac{1}{2}$)
- 196332, 194804—3p ($\frac{3}{2}$,$\frac{1}{2}$)

$3p^2np$ $^2P^o$
- 40077, 39758—3($\frac{1}{2}$,$\frac{3}{2}$)

$3p^2nl$ / $3s3p^3np$ ^2D
- 357042, 357014—4s ($\frac{5}{2}$,$\frac{3}{2}$)
- 304474, 303847—3d″($\frac{5}{2}$,$\frac{3}{2}$)
- 281035, 280584—3d′ ($\frac{5}{2}$,$\frac{3}{2}$)
- 264932, 264741—3d ($\frac{5}{2}$,$\frac{3}{2}$)
- 169704, 169580—3p ($\frac{3}{2}$,$\frac{5}{2}$)

$3p^2np$ $^2D^o$
- 242495, 240127—3($\frac{3}{2}$,$\frac{5}{2}$)

$3p^2nd$ ^2F
- 292960, 292497—3d′($\frac{5}{2}$,$\frac{7}{2}$)
- 262874, 262487—3d ($\frac{5}{2}$,$\frac{7}{2}$)

$3p^2np$ $^4S^o$
- 0—3($\frac{3}{2}$)

$3p^2nl$ / $3s3p^3np$ ^4P
- 339172, 337645, 336628—4s($\frac{1}{2}$,$\frac{3}{2}$,$\frac{5}{2}$)
- 260879, 260064—3d ($\frac{5}{2}$,$\frac{3}{2}$)
- 138806, 138042, 136639—3p ($\frac{5}{2}$,$\frac{3}{2}$,$\frac{1}{2}$)

$3p^2nd$ ^4D
- 222711, 222367—3($\frac{5}{2}$,$\frac{3}{2}$)

$3p^2nd$ ^4F
- 207165, 205720—3($\frac{3}{2}$,$\frac{5}{2}$)

N
- 310120—6($\frac{5}{2}$)
- 307717—5($\frac{3}{2}$)
- 274375—4($\frac{3}{2}$)
- 268043—3($\frac{3}{2}$)

CORES:
n l = $3p^2(^3P)$
n′ l′ = $3p^2(^1D)$
n″ l″ = $3p^2(^1S)$

KEY

292960, 292497—3d′($\frac{5}{2}$, $\frac{7}{2}$)
- Core excitation state
- J, lower to higher
- n
- Energy levels (cm^{-1})

Ionization Limit
666 700 cm^{-1}
(82.66 eV)
[K V $3p^3\,^4S^o_{3/2}$ → K VI $3p^2\,^3P_0$]

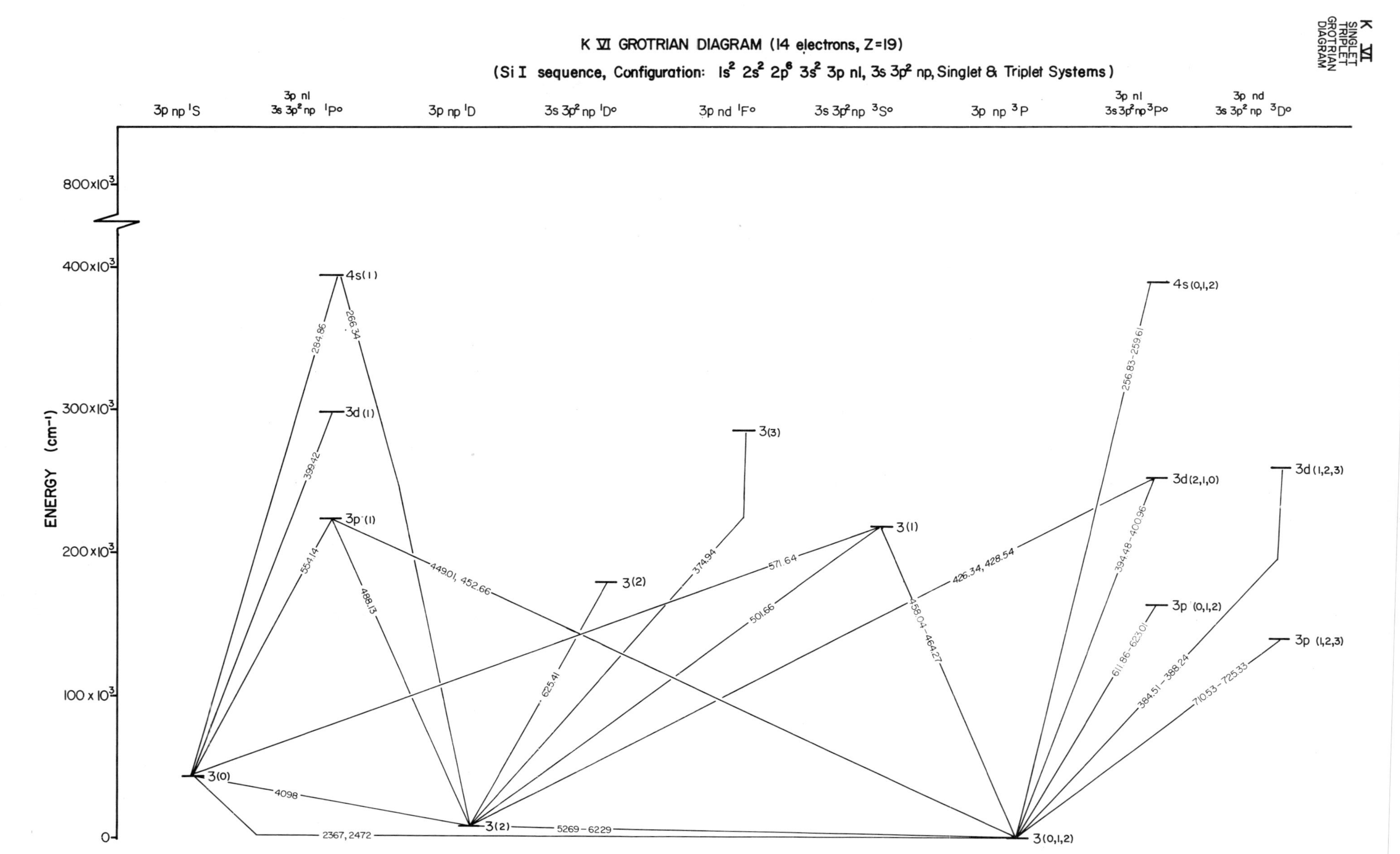
K VI
SINGLET
TRIPLET
GROTRIAN
DIAGRAM
K VI GROTRIAN DIAGRAM (14 electrons, Z=19)
(Si I sequence, Configuration: 1s² 2s² 2p⁶ 3s² 3p nl, 3s 3p² np, Singlet & Triplet Systems)
ENERGY (cm⁻¹)
800x10³
400x10³
300x10³
200x10³
100 x 10³
0
3p np ¹S
3p nl 3s 3p² np ¹P°
3p np ¹D
3s 3p² np ¹D°
3p nd ¹F°
3s 3p² np ³S°
3p np ³P
3p nl 3s 3p² np ³P°
3p nd 3s 3p² np ³D°
3(0)
4s(1)
3d(1)
3p′(1)
3(2)
3(3)
3(1)
3(0,1,2)
4s(0,1,2)
3d(2,1,0)
3p′(0,1,2)
3d(1,2,3)
3p′(1,2,3)
284.86
266.34
399.42
554.14
488.13
449.01, 452.66
4098
2367, 2472
5269–6229
625.41
374.94
571.64
501.66
458.04–464.27
426.34, 428.54
256.83–259.61
394.48–400.96
611.86–623.01
384.51–388.24
710.53–725.33

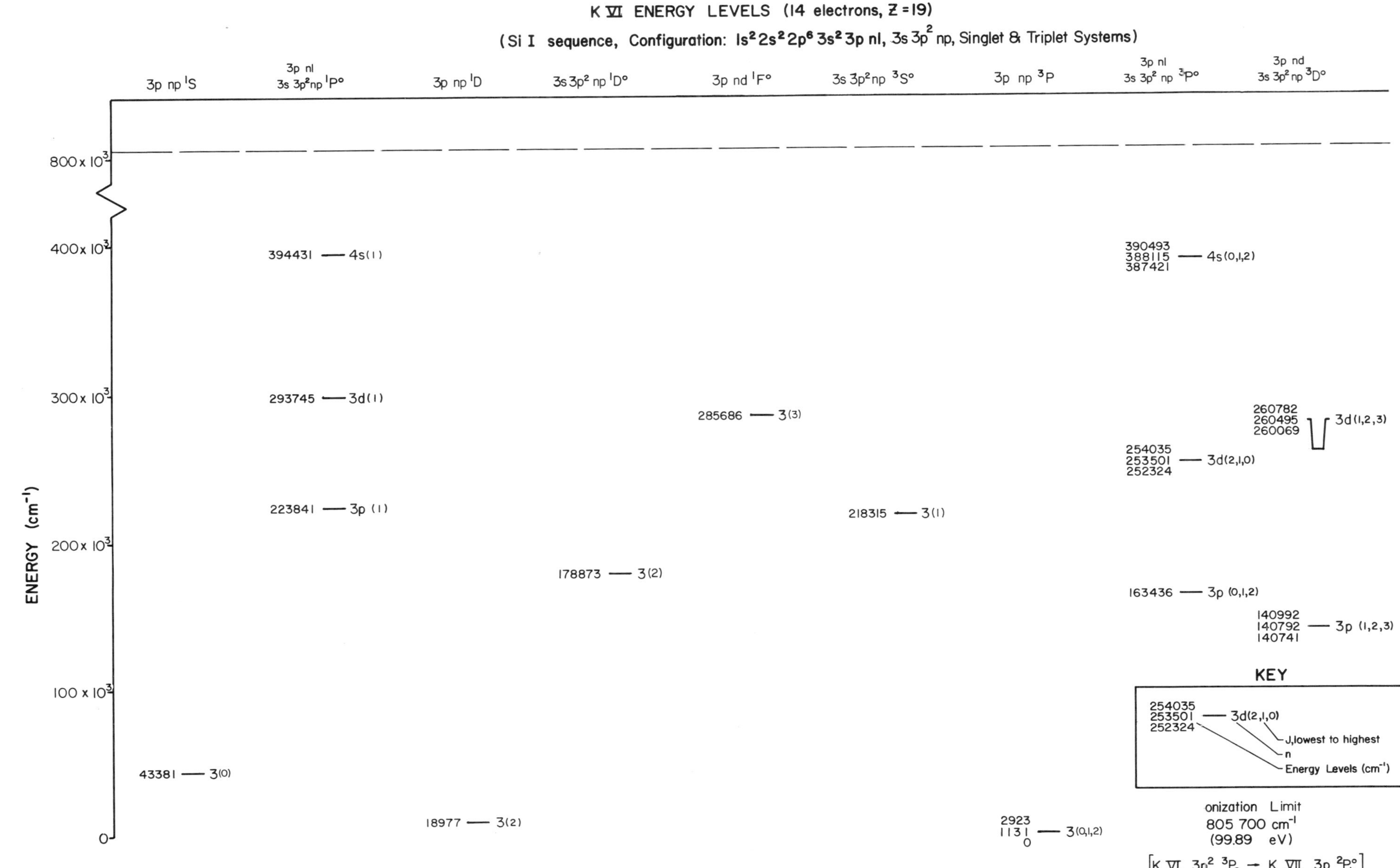

K VI ENERGY LEVELS (14 electrons, Z = 19)
(Si I sequence, Configuration: $1s^2 2s^2 2p^6 3s^2 3p\,nl$, $3s\,3p^2\,np$, Singlet & Triplet Systems)
3p np ^{1}S
3p nl 3s $3p^2$np ^{1}P°
3p np ^{1}D
3s $3p^2$ np ^{1}D°
3p nd ^{1}F°
3s $3p^2$np ^{3}S°
3p np ^{3}P
3p nl 3s $3p^2$ np ^{3}P°
3p nd 3s $3p^2$ np ^{3}D°
800 x 10^3
400 x 10^3
300 x 10^3
200 x 10^3
100 x 10^3
0
ENERGY (cm^{-1})
394431 — 4s(1)
293745 — 3d(1)
223841 — 3p (1)
43381 — 3(0)
18977 — 3(2)
178873 — 3(2)
285686 — 3(3)
218315 — 3(1)
2923 1131 0 — 3(0,1,2)
390493 388115 387421 — 4s(0,1,2)
254035 253501 252324 — 3d(2,1,0)
163436 — 3p (0,1,2)
260782 260495 260069 — 3d(1,2,3)
140992 140792 140741 — 3p (1,2,3)
KEY
254035 253501 252324 — 3d(2,1,0)
J, lowest to highest
n
Energy Levels (cm^{-1})
onization Limit
805 700 cm^{-1}
(99.89 eV)
[K VI $3p^2\ ^3P_0$ → K VII $3p\ ^2P^o_{1/2}$]

K VII
DOUBLET
QUARTET
GROTRIAN
DIAGRAM

K VII GROTRIAN DIAGRAM (13 electrons, Z=19)

(Al I sequence, Configuration: $1s^2\ 2s^2\ 2p^6\ 3s^2\ nl$, $1s^2\ 2s^2\ 2p^6\ 3s\ 3p\ nl$, Doublet & Quartet Systems)

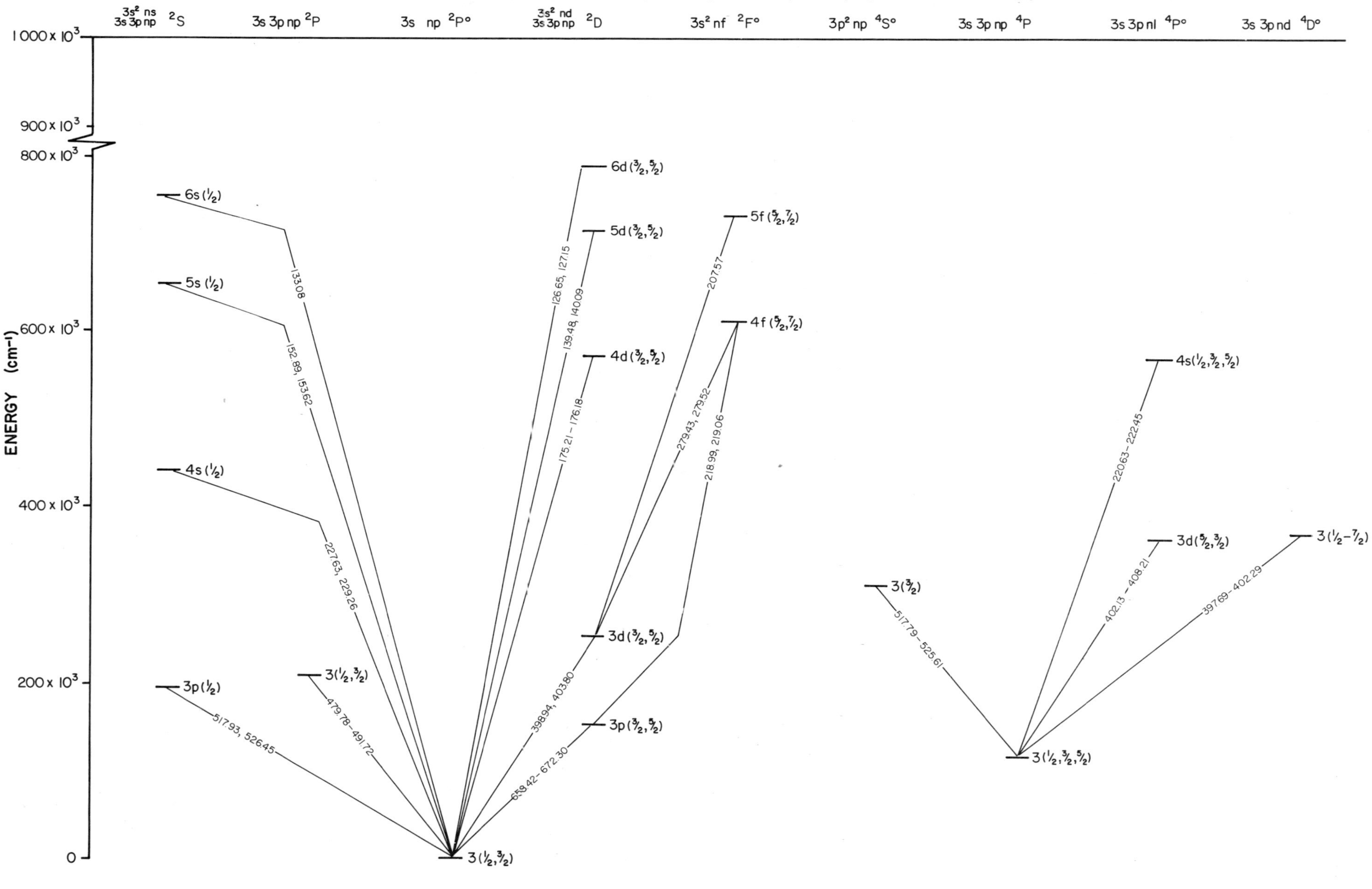

$3s^2\ ns$ / $3s\,3p\,np$ ^2S | $3s\,3p\,np$ ^2P | $3s\ np$ $^2P^o$ | $3s^2\ nd$ / $3s\,3p\,np$ ^2D | $3s^2\ nf$ $^2F^o$ | $3p^2\ np$ $^4S^o$ | $3s\,3p\,np$ ^4P | $3s\,3p\,nl$ $^4P^o$ | $3s\,3p\,nd$ $^4D^o$

ENERGY (cm^{-1})

1000 x 10^3
900 x 10^3
800 x 10^3
600 x 10^3
400 x 10^3
200 x 10^3
0

6s(1/2)
5s(1/2)
4s(1/2)
3p(1/2)
3(1/2,3/2)
133.08
152.89, 153.62
227.63, 229.26
517.93, 526.45
479.78–491.72
6d(3/2,5/2)
5d(3/2,5/2)
4d(3/2,5/2)
3d(3/2,5/2)
3p(3/2,5/2)
126.65, 127.15
139.48, 140.09
175.21 – 176.18
398.94, 403.80
659.42–672.30
5f(5/2,7/2)
4f(5/2,7/2)
207.57
279.43, 279.52
218.99, 219.06
3(1/2,3/2)
3(3/2)
517.79–525.61
3(1/2,3/2,5/2)
4s(1/2,3/2,5/2)
3d(5/2,3/2)
3(1/2–7/2)
220.63–222.45
402.13 – 408.21
397.69–402.29

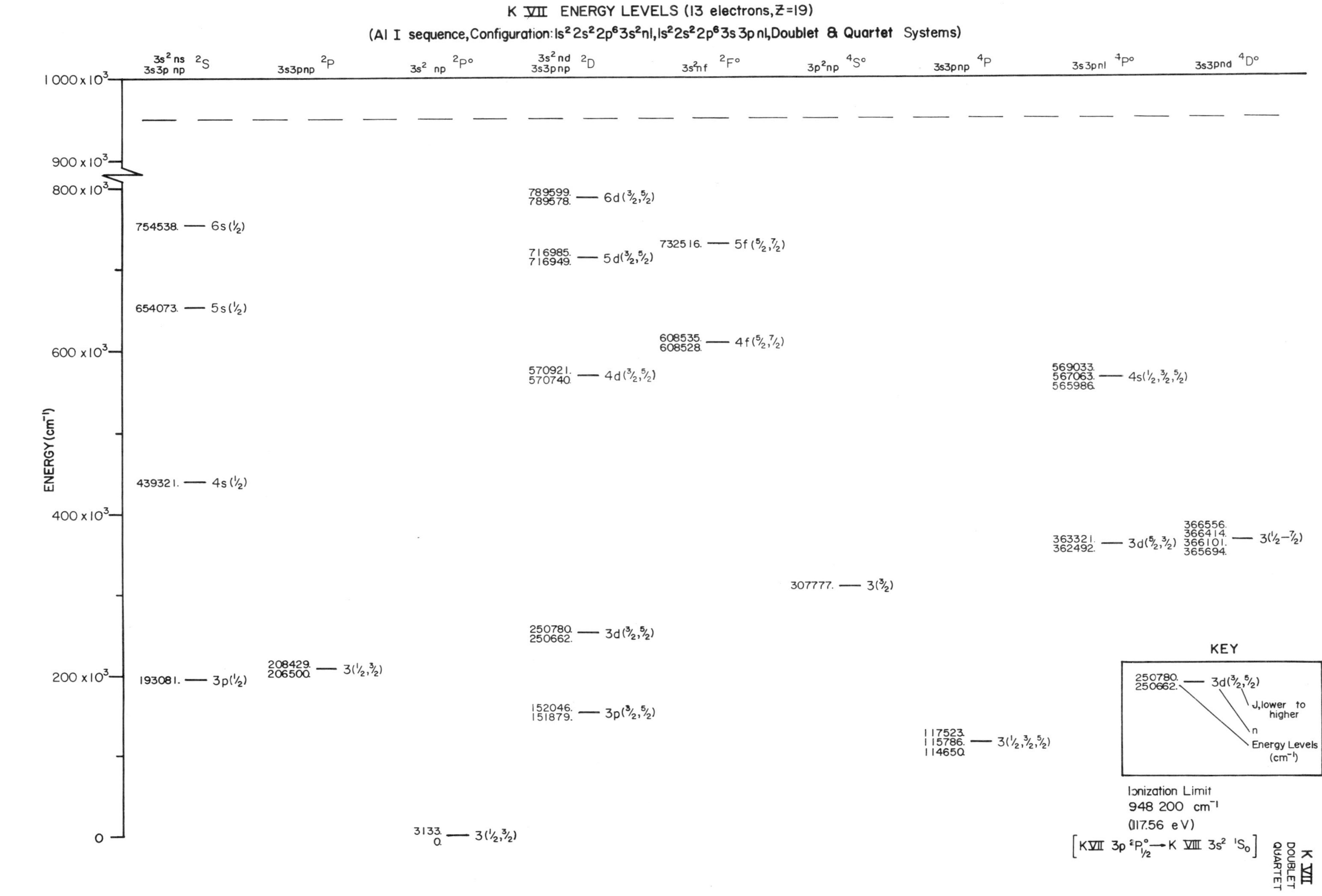
K VII ENERGY LEVELS (13 electrons, Z=19)
(Al I sequence, Configuration: 1s²2s²2p⁶3s²nl, 1s²2s²2p⁶3s3pnl, Doublet & Quartet Systems)
3s² ns 3s3p np ²S
3s3pnp ²P
3s² np ²P°
3s² nd 3s3pnp ²D
3s²nf ²F°
3p²np ⁴S°
3s3pnp ⁴P
3s3pnl ⁴P°
3s3pnd ⁴D°
1000 x 10³
900 x 10³
800 x 10³
600 x 10³
400 x 10³
200 x 10³
0
ENERGY (cm⁻¹)
754538. — 6s(½)
654073. — 5s(½)
439321. — 4s(½)
193081. — 3p(½)
208429. 206500. — 3(½,3/2)
3133. 0. — 3(½,3/2)
789599. 789578. — 6d(3/2,5/2)
716985. 716949. — 5d(3/2,5/2)
570921. 570740. — 4d(3/2,5/2)
250780. 250662. — 3d(3/2,5/2)
152046. 151879. — 3p(3/2,5/2)
732516. — 5f(5/2,7/2)
608535. 608528. — 4f(5/2,7/2)
307777. — 3(3/2)
117523. 115786. 114650. — 3(½,3/2,5/2)
569033. 567063. 565986. — 4s(½,3/2,5/2)
363321. 362492. — 3d(5/2,3/2)
366556. 366414. 366101. 365694. — 3(½–7/2)
KEY
250780. 250662. — 3d(3/2,5/2)
J, lower to higher
n
Energy Levels (cm⁻¹)
Ionization Limit
948 200 cm⁻¹
(117.56 eV)
[K VII 3p ²P°½ → K VIII 3s² ¹S₀]
K VII
DOUBLET
QUARTET

K VIII GROTRIAN DIAGRAM (12 electrons, Z = 19)
(Mg I sequence, Configuration: $1s^2 2s^2 2p^6 3s\,nl$, Singlet System)

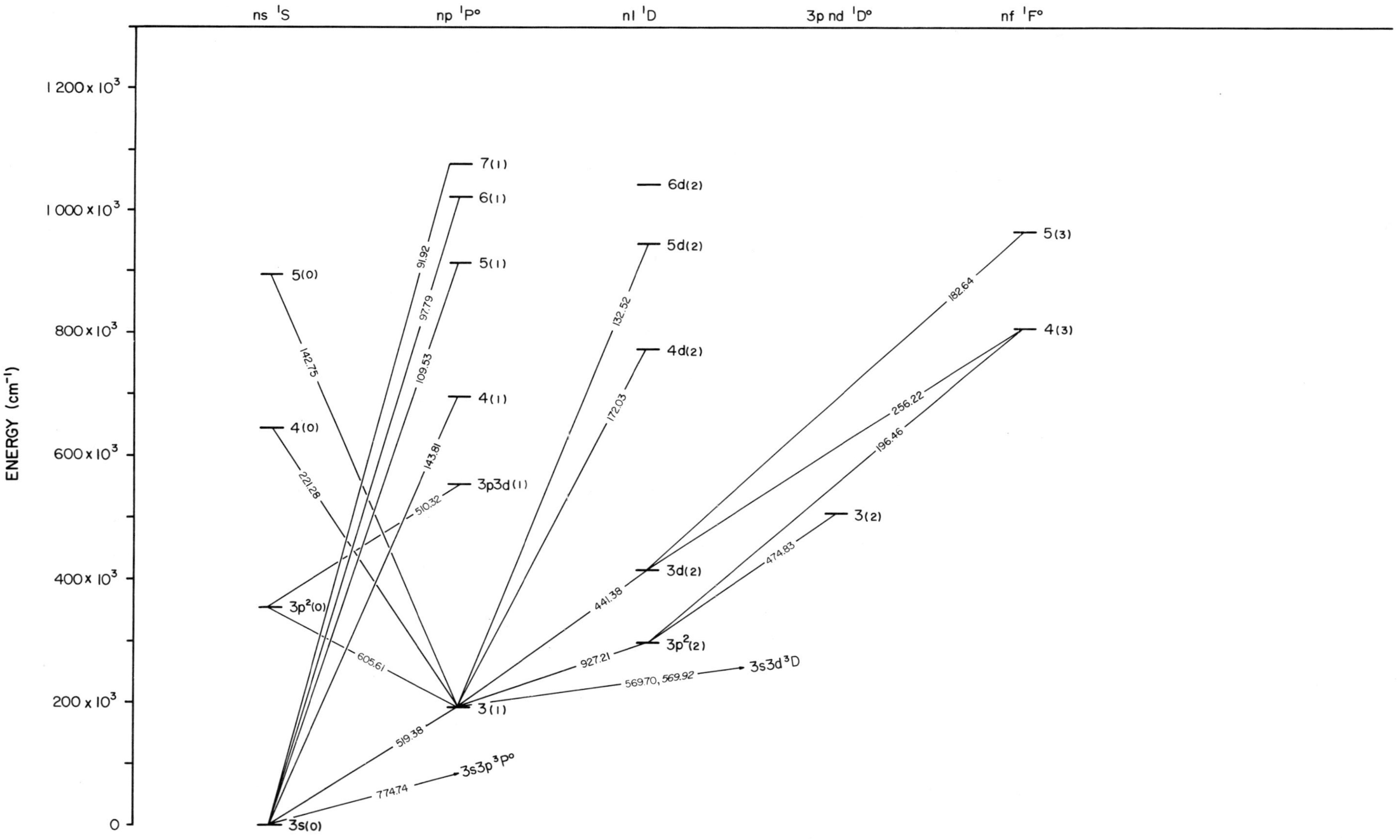

ns ^1S
np $^1P^o$
nl ^1D
3p nd $^1D^o$
nf $^1F^o$

5(0)
4(0)
$3p^2(0)$
3s(0)
7(1)
6(1)
5(1)
4(1)
3p3d(1)
3(1)
$3s3p\,^3P^o$
6d(2)
5d(2)
4d(2)
3d(2)
$3p^2(2)$
$3s3d\,^3D$
3(2)
5(3)
4(3)

142.75
221.28
91.92
97.79
109.53
143.81
510.32
605.61
569.38
774.74
132.52
172.03
441.38
927.21
569.70, 569.92
474.83
182.64
256.22
196.46

ENERGY (cm^{-1})

0
200×10^3
400×10^3
600×10^3
800×10^3
$1\,000 \times 10^3$
$1\,200 \times 10^3$

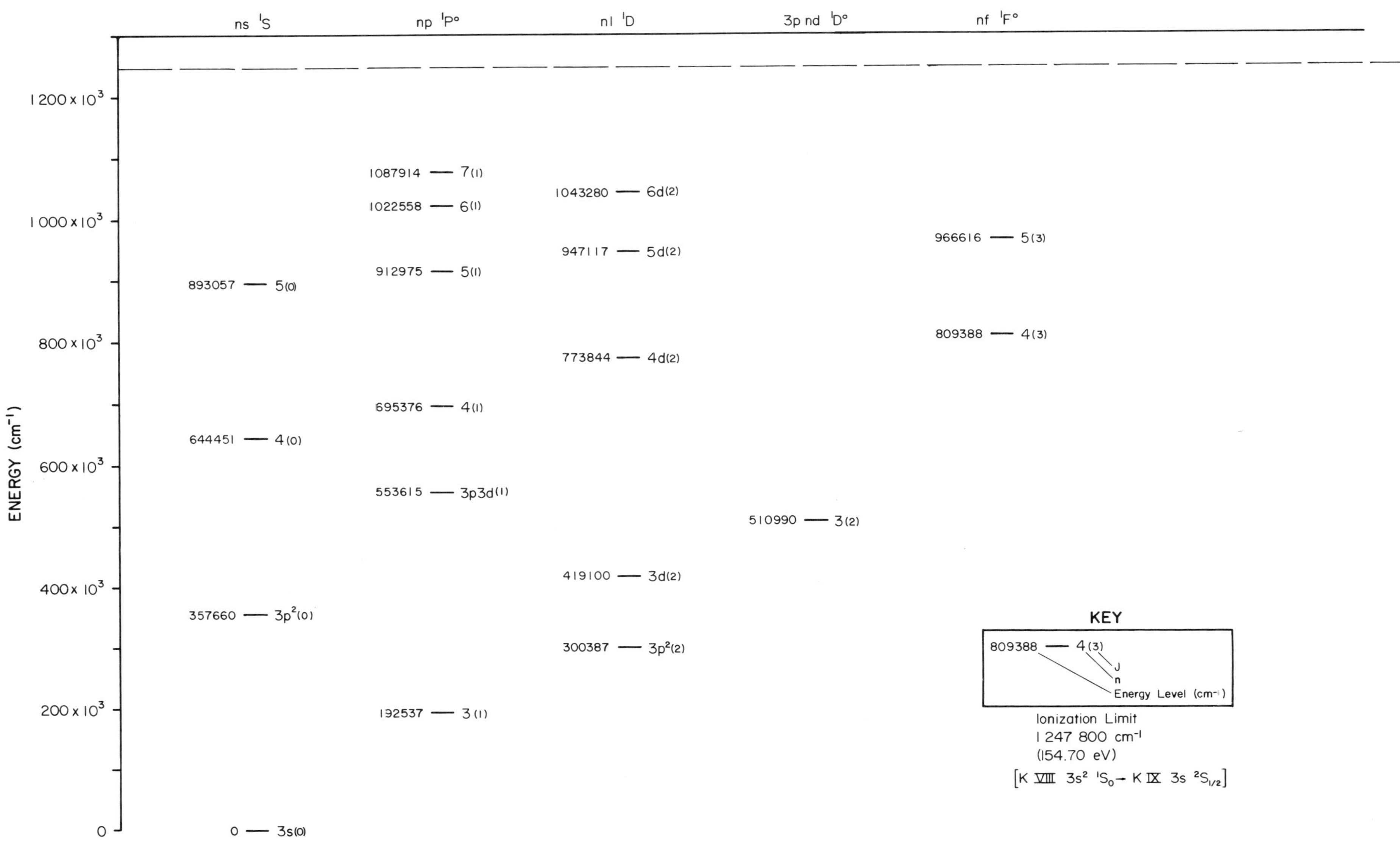
K VIII ENERGY LEVELS (12 electrons, Z = 19)
(Mg I sequence, Configuration: 1s² 2s² 2p⁶ 3s nl, Singlet System)
ns ¹S
np ¹P°
nl ¹D
3p nd ¹D°
nf ¹F°
1200×10³
1000×10³
800×10³
600×10³
400×10³
200×10³
0
ENERGY (cm⁻¹)
893057 — 5(0)
644451 — 4(0)
357660 — 3p²(0)
0 — 3s(0)
1087914 — 7(1)
1022558 — 6(1)
912975 — 5(1)
695376 — 4(1)
553615 — 3p3d(1)
192537 — 3(1)
1043280 — 6d(2)
947117 — 5d(2)
773844 — 4d(2)
419100 — 3d(2)
300387 — 3p²(2)
510990 — 3(2)
966616 — 5(3)
809388 — 4(3)
KEY
809388 — 4(3)
J
n
Energy Level (cm⁻¹)
Ionization Limit
1 247 800 cm⁻¹
(154.70 eV)
[K VIII 3s² ¹S₀ – K IX 3s ²S₁/₂]

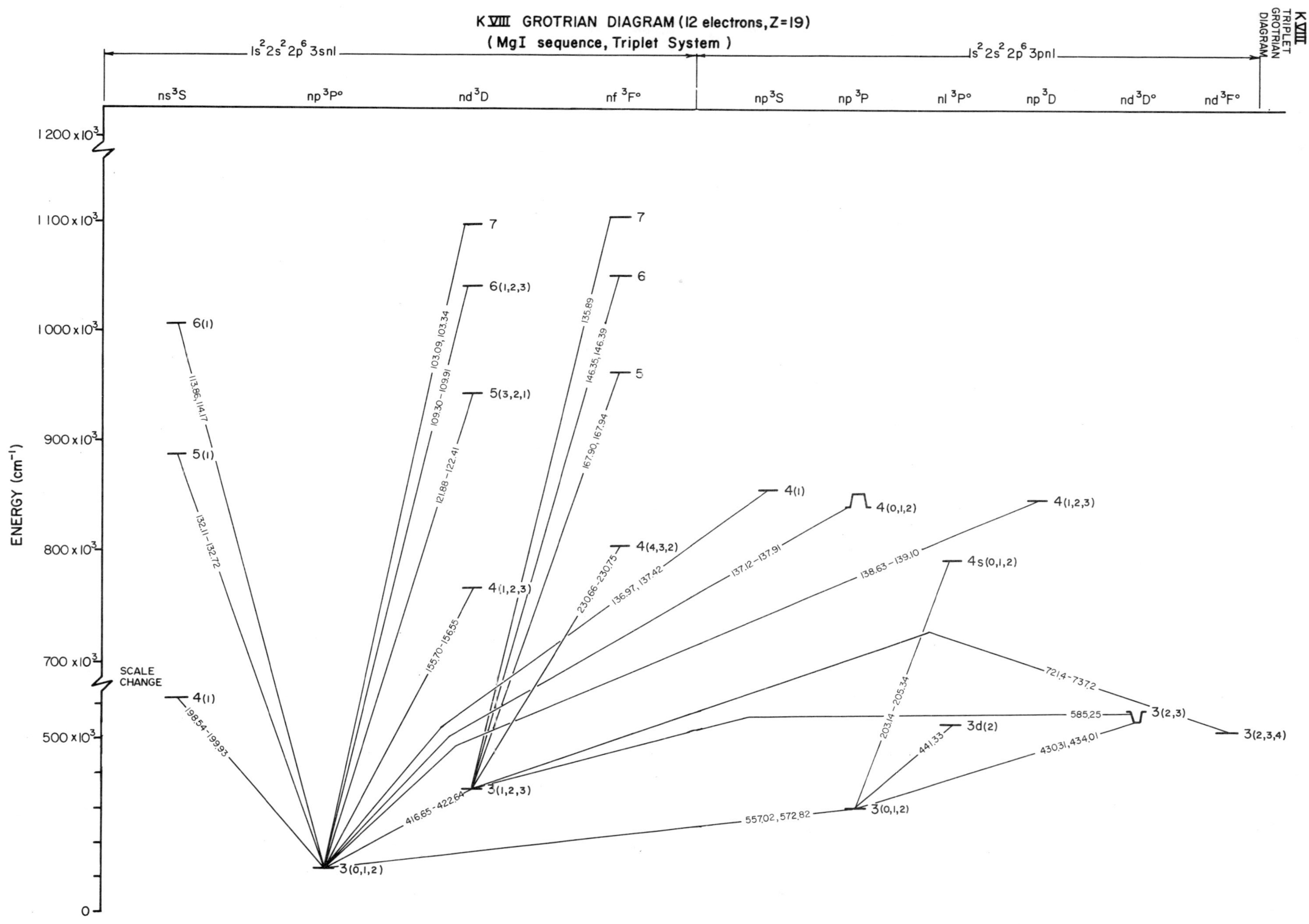
K VIII GROTRIAN DIAGRAM (12 electrons, Z=19)
(MgI sequence, Triplet System)
K VIII
TRIPLET
GROTRIAN
DIAGRAM
$1s^2 2s^2 2p^6 3snl$
$1s^2 2s^2 2p^6 3pnl$
ns^3S
np^3P^o
nd^3D
nf^3F^o
np^3S
np^3P
nl^3P^o
np^3D
nd^3D^o
nd^3F^o
ENERGY (cm^{-1})
1 200 x 10^3
1 100 x 10^3
1 000 x 10^3
900 x 10^3
800 x 10^3
700 x 10^3
500 x 10^3
0
SCALE CHANGE
6(1)
5(1)
4(1)
113.86, 114.17
132.11 – 132.72
198.54 - 199.93
3(0,1,2)
7
6(1,2,3)
5(3,2,1)
4(1,2,3)
103.09, 103.34
109.30 – 109.91
121.88 – 122.41
155.70 - 156.55
416.65 - 422.64
3(1,2,3)
7
6
5
4(4,3,2)
135.89
146.35, 146.39
167.90, 167.94
230.66 - 230.75
4(1)
136.97, 137.42
4(0,1,2)
137.12 - 137.91
4(1,2,3)
138.63 - 139.10
4s(0,1,2)
203.14 - 205.34
3(0,1,2)
557.02, 572.82
3d(2)
441.33
430.31, 434.01
585.25
721.4 - 737.2
3(2,3)
3(2,3,4)

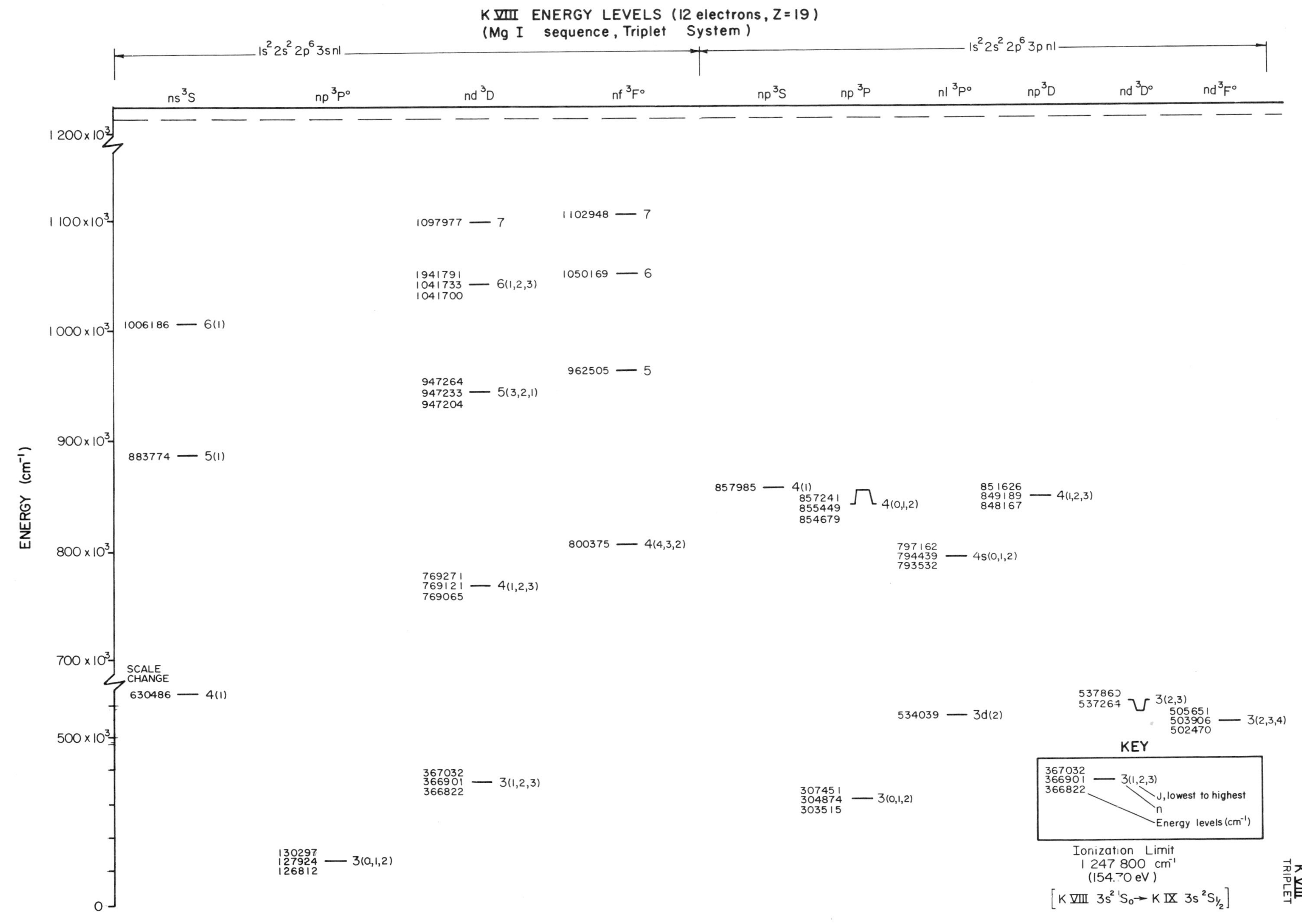

K VIII ENERGY LEVELS (12 electrons, Z = 19)
(Mg I sequence, Triplet System)
1s²2s²2p⁶3snl
1s²2s²2p⁶3pnl
ns³S
np³P°
nd³D
nf³F°
np³S
np³P
nl ³P°
np³D
nd³D°
nd³F°
ENERGY (cm⁻¹)
1 200 x 10³
1 100 x 10³
1 000 x 10³
900 x 10³
800 x 10³
700 x 10³
500 x 10³
0
1006186 — 6(1)
883774 — 5(1)
SCALE CHANGE
630486 — 4(1)
130297
127924
126812 — 3(0,1,2)
1097977 — 7
1941791
1041733
1041700 — 6(1,2,3)
947264
947233
947204 — 5(3,2,1)
769271
769121
769065 — 4(1,2,3)
367032
366901
366822 — 3(1,2,3)
1102948 — 7
1050169 — 6
962505 — 5
800375 — 4(4,3,2)
857985 — 4(1)
857241
855449
854679 — 4(0,1,2)
307451
304874
303515 — 3(0,1,2)
797162
794439
793532 — 4s(0,1,2)
534039 — 3d(2)
851626
849189
848167 — 4(1,2,3)
537863
537264 — 3(2,3)
505651
503906
502470 — 3(2,3,4)
KEY
367032
366901
366822 — 3(1,2,3)
J, lowest to highest
n
Energy levels (cm⁻¹)
Ionization Limit
1 247 800 cm⁻¹
(154.70 eV)
[K VIII 3s² ¹S₀ → K IX 3s ²S½]
K VIII
TRIPLET

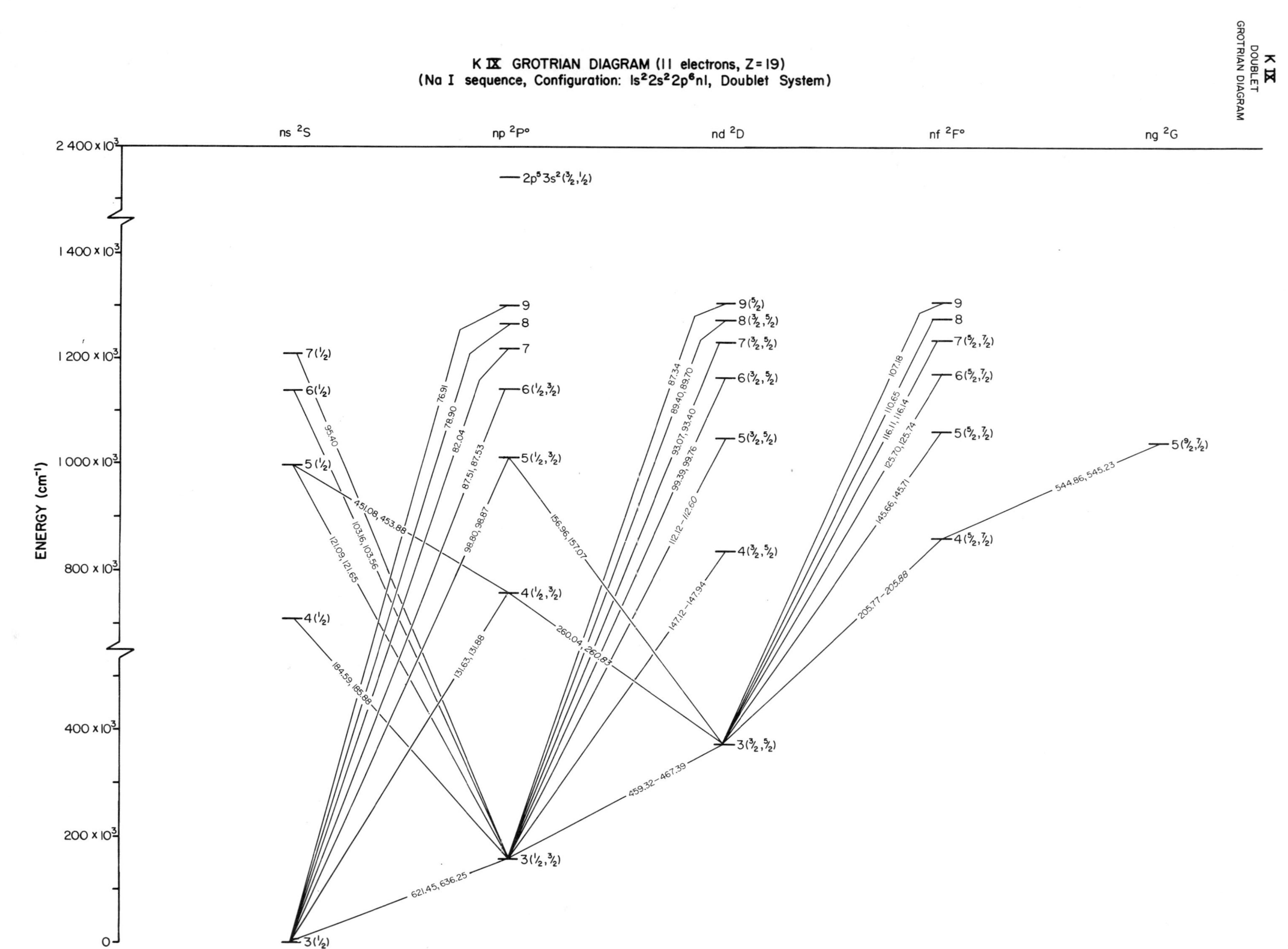
K IX GROTRIAN DIAGRAM (11 electrons, Z=19)
(Na I sequence, Configuration: 1s²2s²2p⁶nl, Doublet System)
ns ²S
np ²P°
nd ²D
nf ²F°
ng ²G
2p⁵3s²(³/₂,¹/₂)
ENERGY (cm⁻¹)
2 400 x 10³
1 400 x 10³
1 200 x 10³
1 000 x 10³
800 x 10³
400 x 10³
200 x 10³
0
7(¹/₂)
6(¹/₂)
5(¹/₂)
4(¹/₂)
3(¹/₂)
9
8
7
6(¹/₂,³/₂)
5(¹/₂,³/₂)
4(¹/₂,³/₂)
3(¹/₂,³/₂)
9(⁵/₂)
8(³/₂,⁵/₂)
7(³/₂,⁵/₂)
6(³/₂,⁵/₂)
5(³/₂,⁵/₂)
4(³/₂,⁵/₂)
3(³/₂,⁵/₂)
9
8
7(⁵/₂,⁷/₂)
6(⁵/₂,⁷/₂)
5(⁵/₂,⁷/₂)
4(⁵/₂,⁷/₂)
5(⁹/₂,⁷/₂)
95.40
451.08, 453.88
103.16, 103.56
121.09, 121.65
184.59, 185.88
76.91
78.90
82.04
87.51, 87.53
98.80, 98.87
131.63, 131.88
621.45, 636.25
156.96, 157.07
260.04, 260.83
87.34
89.40, 89.70
93.07, 93.40
99.39, 99.76
112.12 – 112.60
147.12 – 147.94
459.32 – 467.39
107.18
110.65
116.11, 116.14
125.70, 125.74
145.66, 145.71
205.77 – 205.88
544.86, 545.23

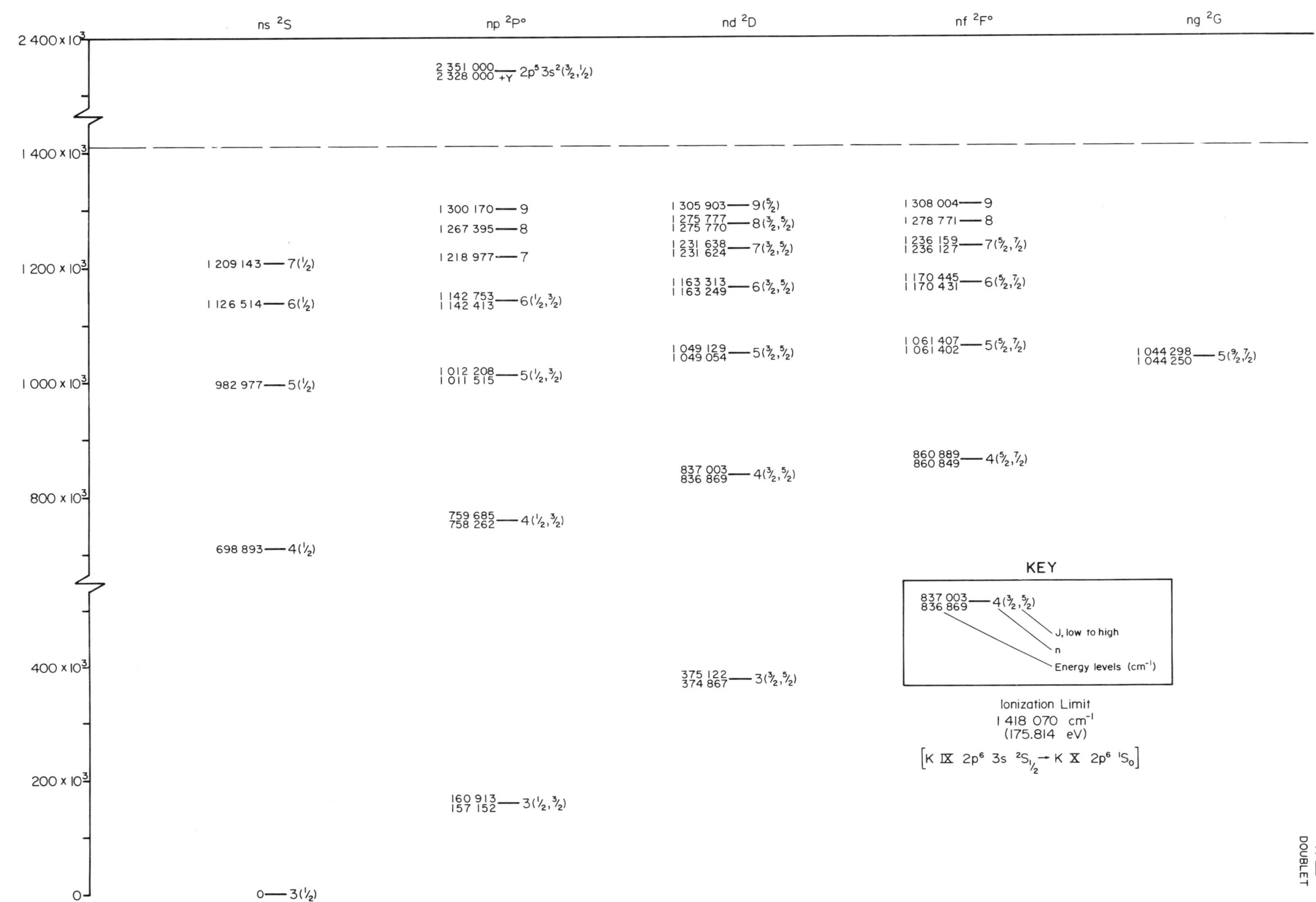
K IX ENERGY LEVELS (11 electrons, Z = 19)
(Na I sequence, Configuration: 1s²2s²2p⁶nl, Doublet System)
ns ²S
np ²P°
nd ²D
nf ²F°
ng ²G
ENERGY (cm⁻¹)
2 400 x 10³
1 400 x 10³
1 200 x 10³
1 000 x 10³
800 x 10³
400 x 10³
200 x 10³
0
2 351 000 / 2 328 000 +Y — 2p⁵3s²(3/2, 1/2)
1 300 170 — 9
1 267 395 — 8
1 218 977 — 7
1 142 753 / 1 142 413 — 6(1/2, 3/2)
1 012 208 / 1 011 515 — 5(1/2, 3/2)
759 685 / 758 262 — 4(1/2, 3/2)
160 913 / 157 152 — 3(1/2, 3/2)
1 209 143 — 7(1/2)
1 126 514 — 6(1/2)
982 977 — 5(1/2)
698 893 — 4(1/2)
0 — 3(1/2)
1 305 903 — 9(5/2)
1 275 777 / 1 275 770 — 8(3/2, 5/2)
1 231 638 / 1 231 624 — 7(3/2, 5/2)
1 163 313 / 1 163 249 — 6(3/2, 5/2)
1 049 129 / 1 049 054 — 5(3/2, 5/2)
837 003 / 836 869 — 4(3/2, 5/2)
375 122 / 374 867 — 3(3/2, 5/2)
1 308 004 — 9
1 278 771 — 8
1 236 159 / 1 236 127 — 7(5/2, 7/2)
1 170 445 / 1 170 431 — 6(5/2, 7/2)
1 061 407 / 1 061 402 — 5(5/2, 7/2)
860 889 / 860 849 — 4(5/2, 7/2)
1 044 298 / 1 044 250 — 5(9/2, 7/2)
KEY
837 003 / 836 869 — 4(3/2, 5/2)
J, low to high
n
Energy levels (cm⁻¹)
Ionization Limit
1 418 070 cm⁻¹
(175.814 eV)
[K IX 2p⁶ 3s ²S1/2 → K X 2p⁶ ¹S₀]
K IX
DOUBLET

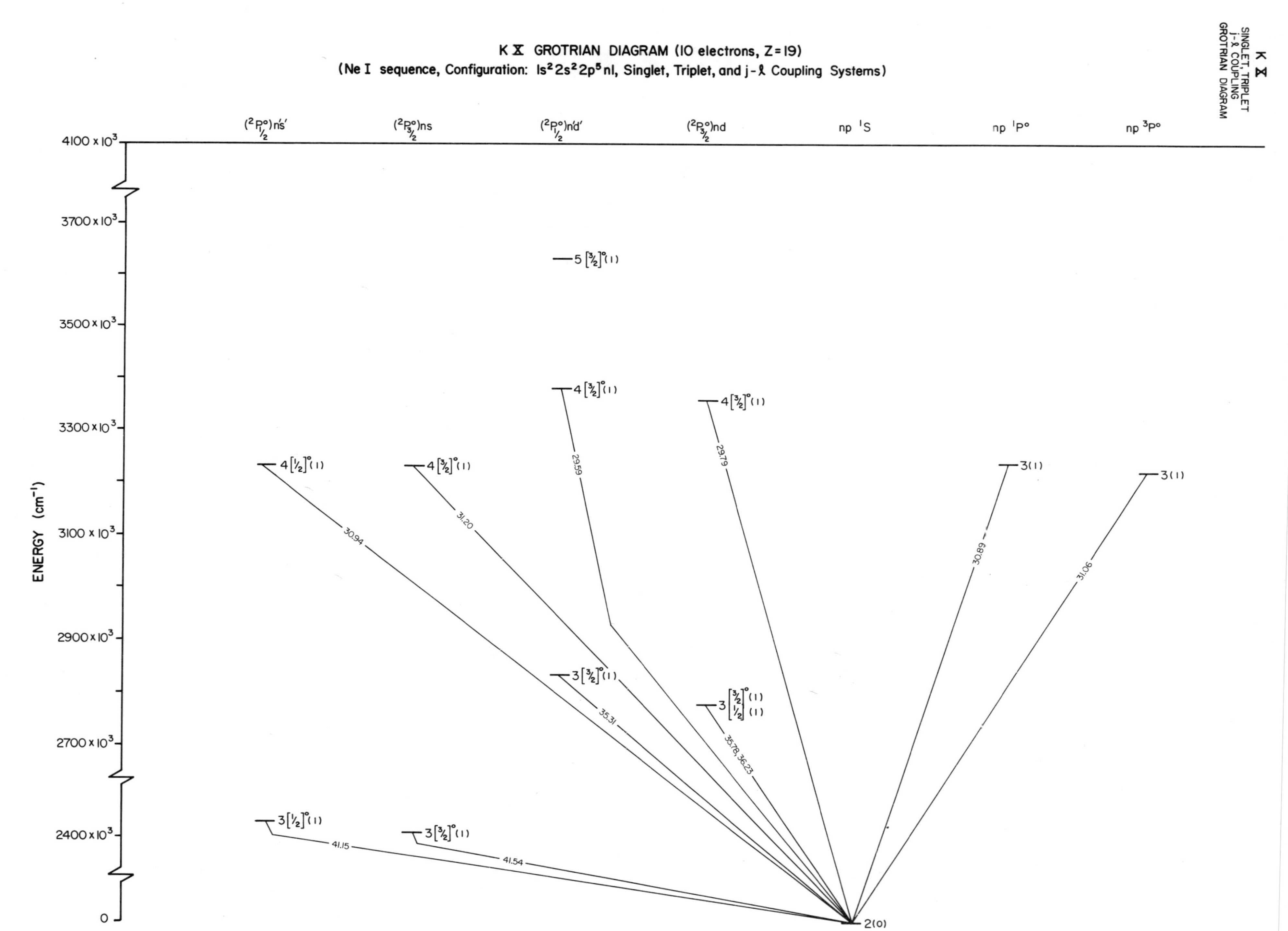

K X
SINGLET, TRIPLET
j-ℓ COUPLING
GROTRIAN DIAGRAM
K X GROTRIAN DIAGRAM (10 electrons, Z = 19)
(Ne I sequence, Configuration: $1s^2 2s^2 2p^5 nl$, Singlet, Triplet, and j-ℓ Coupling Systems)
$(^2P^o_{1/2})n's'$
$(^2P^o_{3/2})ns$
$(^2P^o_{1/2})n'd'$
$(^2P^o_{3/2})nd$
np ^1S
np $^1P^o$
np $^3P^o$
ENERGY (cm^{-1})
4100×10^3
3700×10^3
3500×10^3
3300×10^3
3100×10^3
2900×10^3
2700×10^3
2400×10^3
0
$5[3/2]^o(1)$
$4[3/2]^o(1)$
$4[3/2]^o(1)$
$4[1/2]^o(1)$
$4[3/2]^o(1)$
3(1)
3(1)
$3[3/2]^o(1)$
$3[3/2]^o(1)$ $[1/2](1)$
$3[1/2]^o(1)$
$3[3/2]^o(1)$
2(0)
29.59
29.79
30.94
31.20
30.89
31.06
35.31
35.78, 36.23
41.15
41.54

K X ENERGY LEVELS (10 electrons, Z=19)

(Ne I sequence, Configuration: $1s^2 2s^2 2p^5 nl$, Singlet, Triplet, and j-ℓ Coupling Systems)

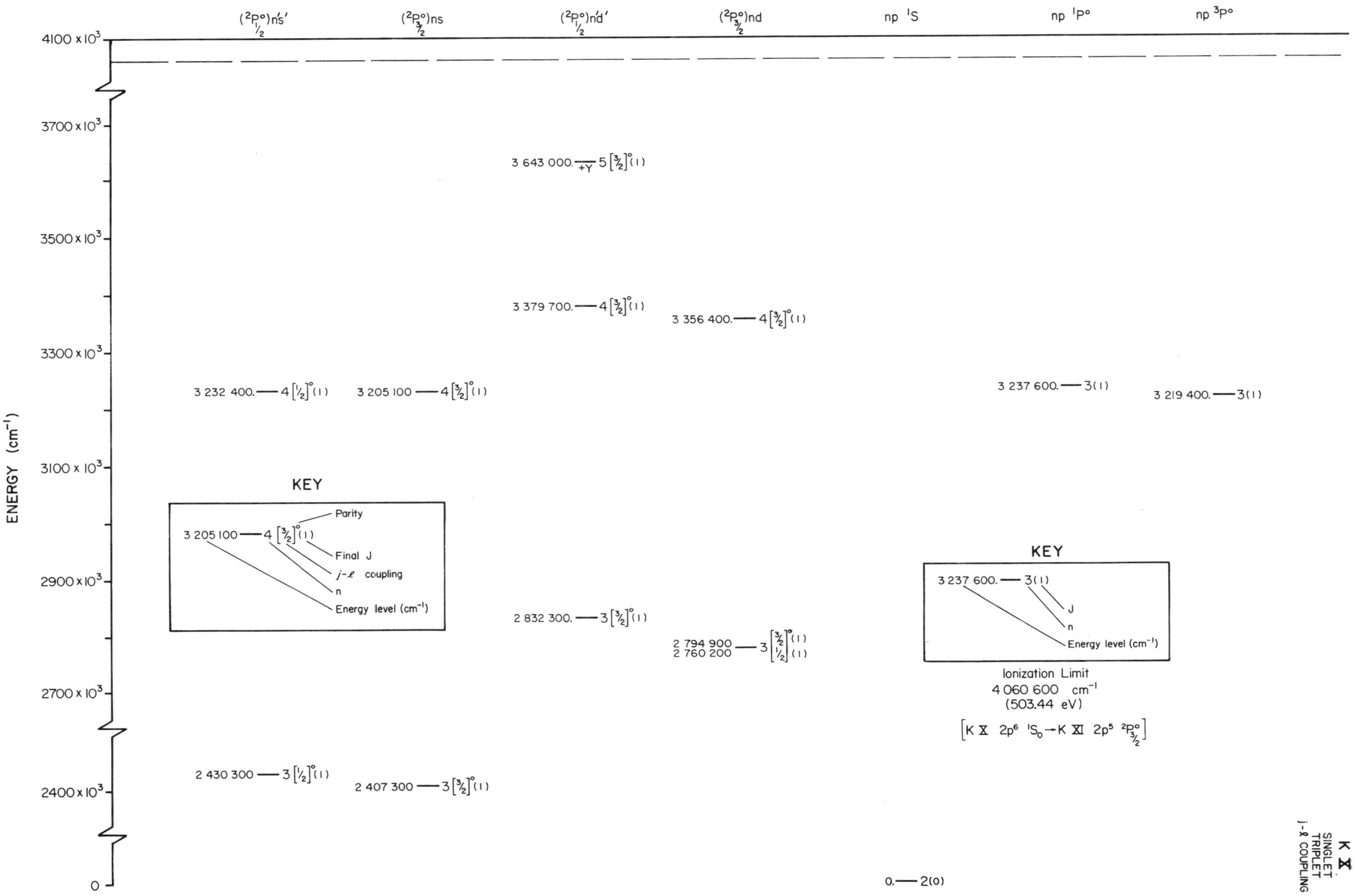

$(^2P^\circ_{1/2})ns'$	$(^2P^\circ_{3/2})ns$	$(^2P^\circ_{1/2})nd'$	$(^2P^\circ_{3/2})nd$	np ^1S	np $^1P^\circ$	np $^3P^\circ$
		3 643 000.+Y — $5[3/2]^\circ(1)$				
		3 379 700. — $4[3/2]^\circ(1)$	3 356 400. — $4[3/2]^\circ(1)$			
3 232 400. — $4[1/2]^\circ(1)$	3 205 100 — $4[3/2]^\circ(1)$				3 237 600. — 3(1)	3 219 400. — 3(1)
		2 832 300. — $3[3/2]^\circ(1)$	2 794 900 — $3[3/2]^\circ(1)$; 2 760 200 — $3[1/2]^\circ(1)$			
2 430 300 — $3[1/2]^\circ(1)$	2 407 300 — $3[3/2]^\circ(1)$					
				0. — 2(0)		

ENERGY (cm^{-1})

4100×10^3, 3700×10^3, 3500×10^3, 3300×10^3, 3100×10^3, 2900×10^3, 2700×10^3, 2400×10^3, 0

KEY

3 205 100 — $4[3/2]^\circ(1)$: Parity; Final J; j-ℓ coupling; n; Energy level (cm^{-1})

KEY

3 237 600. — 3(1): J; n; Energy level (cm^{-1})

Ionization Limit
4 060 600 cm^{-1}
(503.44 eV)
[K X $2p^6\ ^1S_0$ → K XI $2p^5\ ^2P^\circ_{3/2}$]

K X
SINGLET
TRIPLET
j-ℓ COUPLING

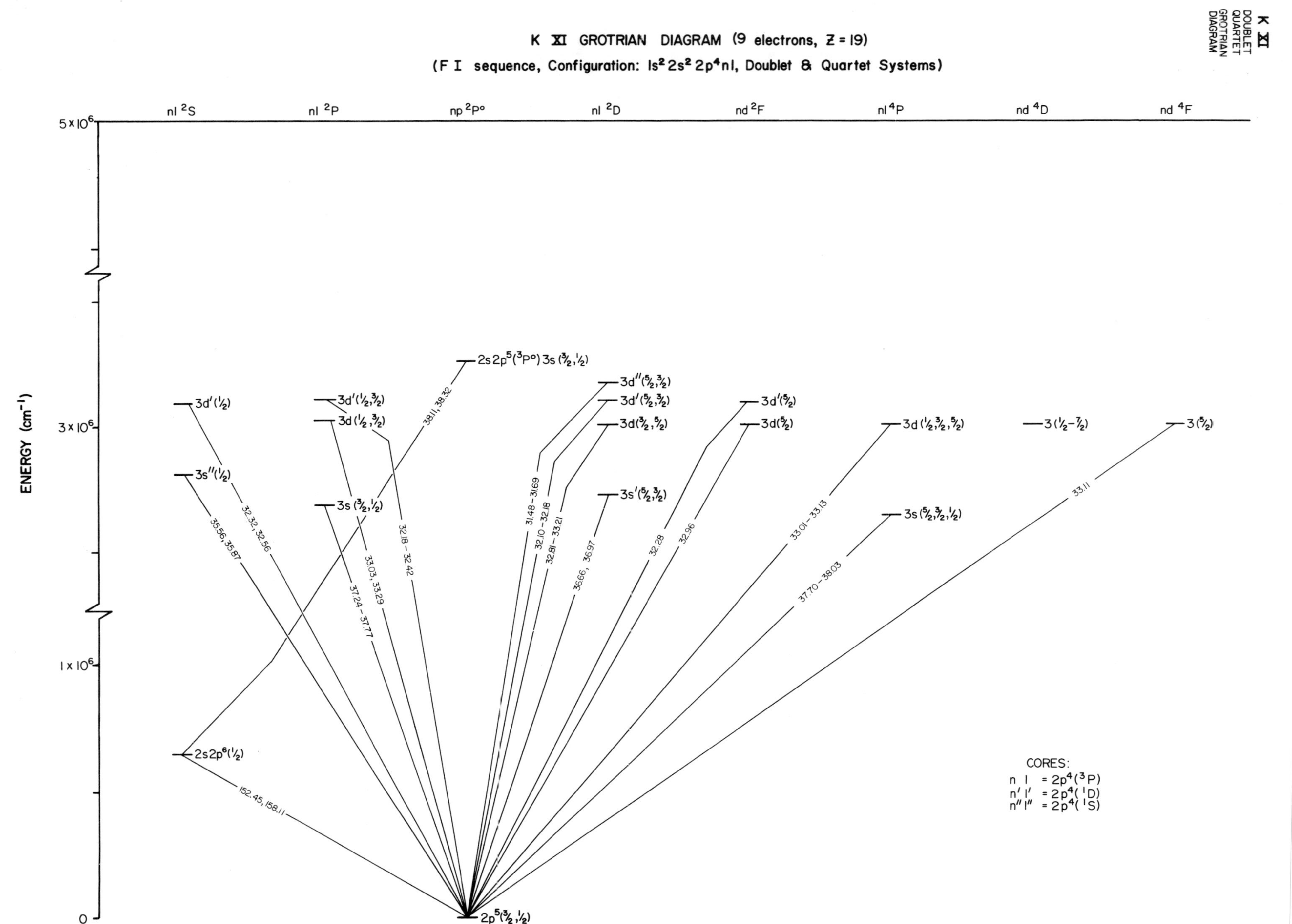
K XI
DOUBLET
QUARTET
GROTRIAN
DIAGRAM
K XI GROTRIAN DIAGRAM (9 electrons, Z = 19)
(F I sequence, Configuration: 1s²2s²2p⁴nl, Doublet & Quartet Systems)
nl ²S
nl ²P
np ²P°
nl ²D
nd ²F
nl ⁴P
nd ⁴D
nd ⁴F
ENERGY (cm⁻¹)
5 x 10⁶
3 x 10⁶
1 x 10⁶
0
2p⁵(³⁄₂,¹⁄₂)
2s2p⁶(¹⁄₂)
3d′(¹⁄₂)
3s″(¹⁄₂)
3d′(¹⁄₂,³⁄₂)
3d(¹⁄₂,³⁄₂)
3s(³⁄₂,¹⁄₂)
2s2p⁵(³P°)3s(³⁄₂,¹⁄₂)
3d″(⁵⁄₂,³⁄₂)
3d′(⁵⁄₂,³⁄₂)
3d(³⁄₂,⁵⁄₂)
3s′(⁵⁄₂,³⁄₂)
3d′(⁵⁄₂)
3d(⁵⁄₂)
3d(¹⁄₂,³⁄₂,⁵⁄₂)
3s(⁵⁄₂,³⁄₂,¹⁄₂)
3(¹⁄₂–⁷⁄₂)
3(⁵⁄₂)
35.56, 35.87
32.32, 32.56
152.45, 158.11
37.24 – 37.77
33.03, 33.29
32.18 – 32.42
38.11, 38.32
31.48 – 31.69
32.10 – 32.18
32.81 – 33.21
36.66, 36.97
32.28
32.96
33.01 – 33.13
37.70 – 38.03
33.11
CORES:
n l = 2p⁴(³P)
n′ l′ = 2p⁴(¹D)
n″ l″ = 2p⁴(¹S)

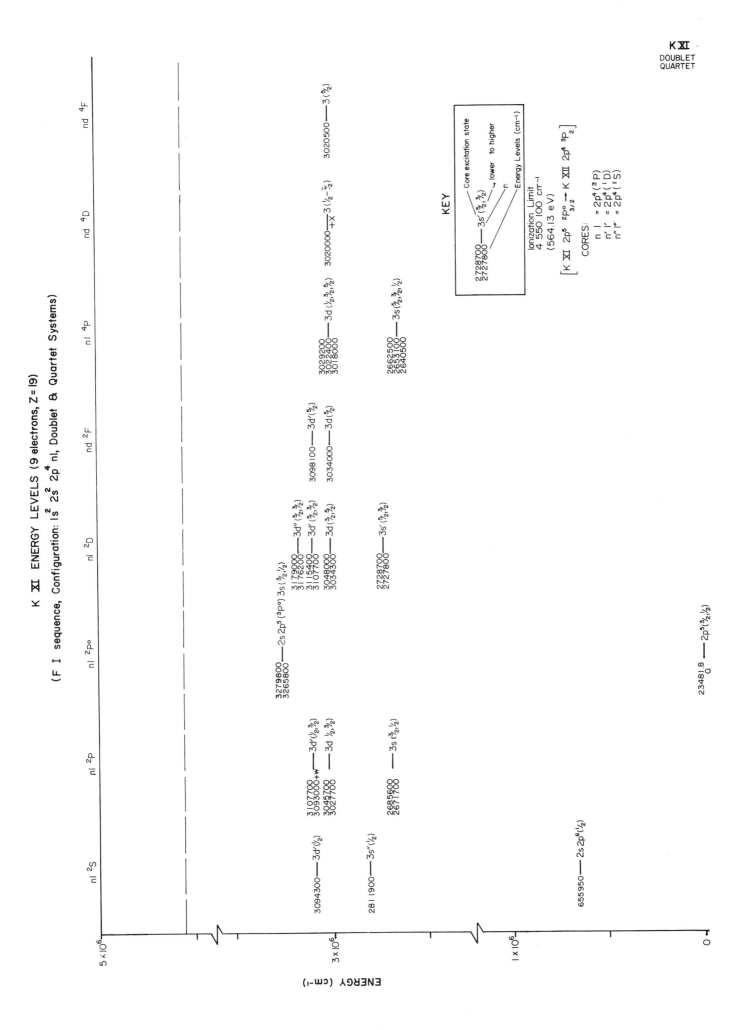

K XI ENERGY LEVELS (9 electrons, Z = 19)

(F I sequence, Configuration: $1s^2 2s^2 2p^4$ nl, Doublet & Quartet Systems)

ENERGY (cm^{-1}): 5×10^6, 3×10^6, 1×10^6, 0

nl ^2S
- 3094300 — 3d′(1/2)
- 2811900 — 3s″(1/2)
- 655950 — $2s2p^6$(1/2)

nl ^2P
- 3107700, 3093000+w — 3d′(1/2, 3/2)
- 3045700, 3027700 — 3d (1/2, 3/2)
- 2685600, 2671700 — 3s (3/2, 1/2)

nl $^2P^o$
- 3279800, 3265800 — $2s2p^5(^3P^o)$ 3s (3/2, 1/2)
- 23481.8, 0. — $2p^5$(3/2, 1/2)

nl ^2D
- 3179000, 3176200 — 3d″(5/2, 3/2)
- 3115400, 3107700 — 3d′(5/2, 3/2)
- 3048000, 3034300 — 3d (3/2, 5/2)
- 2728700, 2727800 — 3s′(5/2, 3/2)

nd ^2F
- 3098100 — 3d′(5/2)
- 3034000 — 3d(5/2)

nl ^4P
- 3029200, 3022400, 3018000 — 3d (1/2, 3/2, 5/2)
- 2662500, 2653100, 2640500 — 3s (5/2, 3/2, 1/2)

nd ^4D
- 3020000 +X — 3 (1/2 – 7/2)

nd ^4F
- 3020500 — 3 (5/2)

KEY

2728700, 2727800 — 3s′(5/2, 3/2): Core excitation state; ↳ lower to higher; n; Energy Levels (cm^{-1})

Ionization Limit
4 550 100 cm^{-1}
(564.13 eV)

[K XI $2p^5$ $^2P^o_{3/2}$ → K XII $2p^4$ $^3P_2$]

CORES:
- n l = $2p^4(^3P)$
- n′ l′ = $2p^4(^1D)$
- n″ l″ = $2p^4(^1S)$

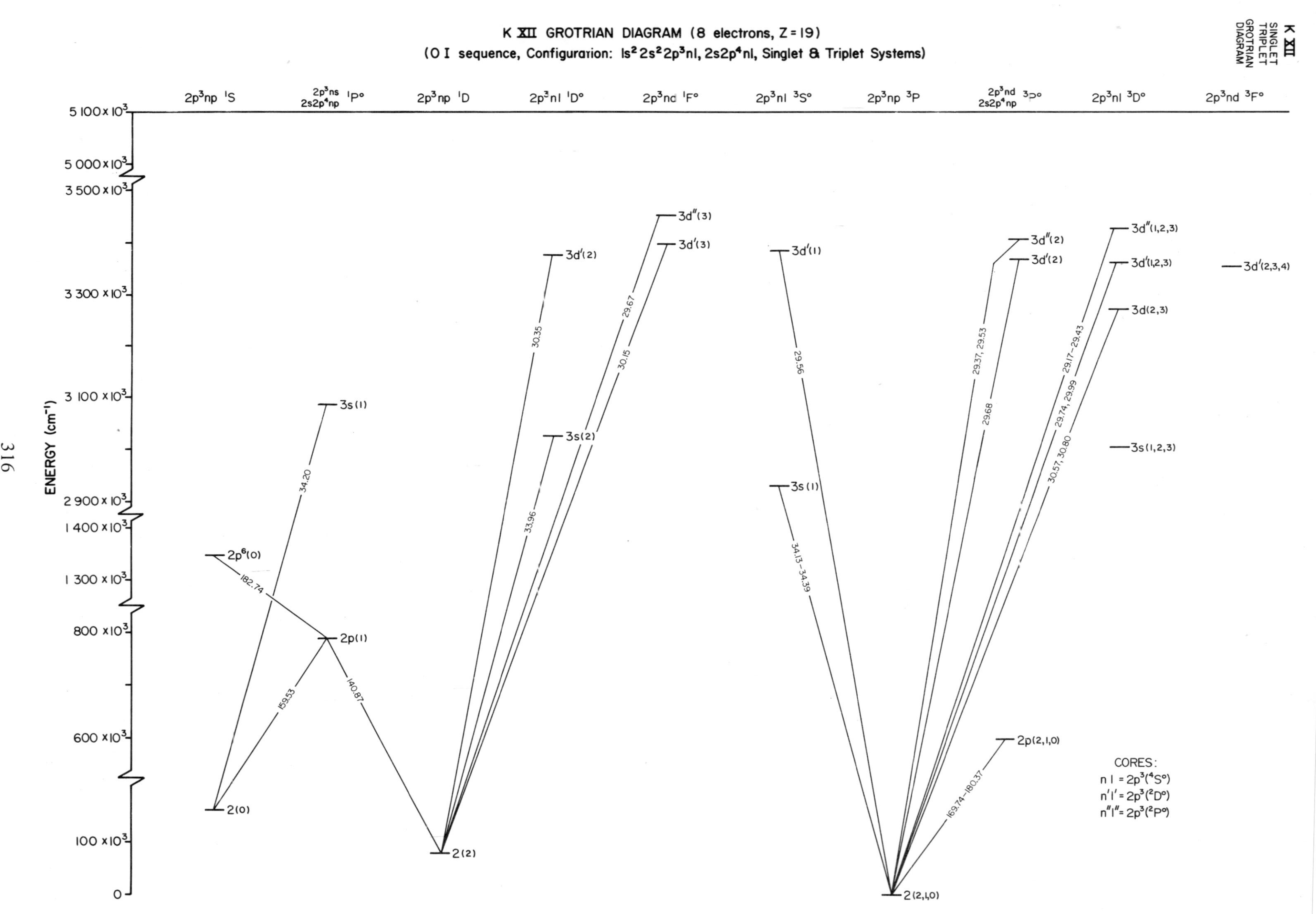

K XII
SINGLET
TRIPLET
GROTRIAN
DIAGRAM
K XII GROTRIAN DIAGRAM (8 electrons, Z = 19)
(O I sequence, Configuration: $1s^2 2s^2 2p^3 nl$, $2s2p^4 nl$, Singlet & Triplet Systems)
$2p^3np\ ^1S$
$2p^3ns$ / $2s2p^4np$ $^1P^o$
$2p^3np\ ^1D$
$2p^3nl\ ^1D^o$
$2p^3nd\ ^1F^o$
$2p^3nl\ ^3S^o$
$2p^3np\ ^3P$
$2p^3nd$ / $2s2p^4np$ $^3P^o$
$2p^3nl\ ^3D^o$
$2p^3nd\ ^3F^o$
ENERGY (cm^{-1})
$5\,100 \times 10^3$
$5\,000 \times 10^3$
$3\,500 \times 10^3$
$3\,300 \times 10^3$
$3\,100 \times 10^3$
$2\,900 \times 10^3$
$1\,400 \times 10^3$
$1\,300 \times 10^3$
800×10^3
600×10^3
100×10^3
0
2(0)
$2p^6(0)$
182.74
3s(1)
34.20
2p(1)
159.53
140.87
2(2)
3d′(2)
30.35
3s(2)
33.96
3d″(3)
3d′(3)
29.67
30.15
3d′(1)
29.56
3s(1)
34.13–34.39
2(2,1,0)
3d″(2)
3d′(2)
29.37, 29.53
29.68
2p(2,1,0)
169.74–180.37
3d″(1,2,3)
3d′(1,2,3)
3d(2,3)
29.17–29.43
29.74, 29.99
30.57, 30.80
3s(1,2,3)
3d′(2,3,4)
CORES:
$nl = 2p^3(^4S^o)$
$n'l' = 2p^3(^2D^o)$
$n''l'' = 2p^3(^2P^o)$

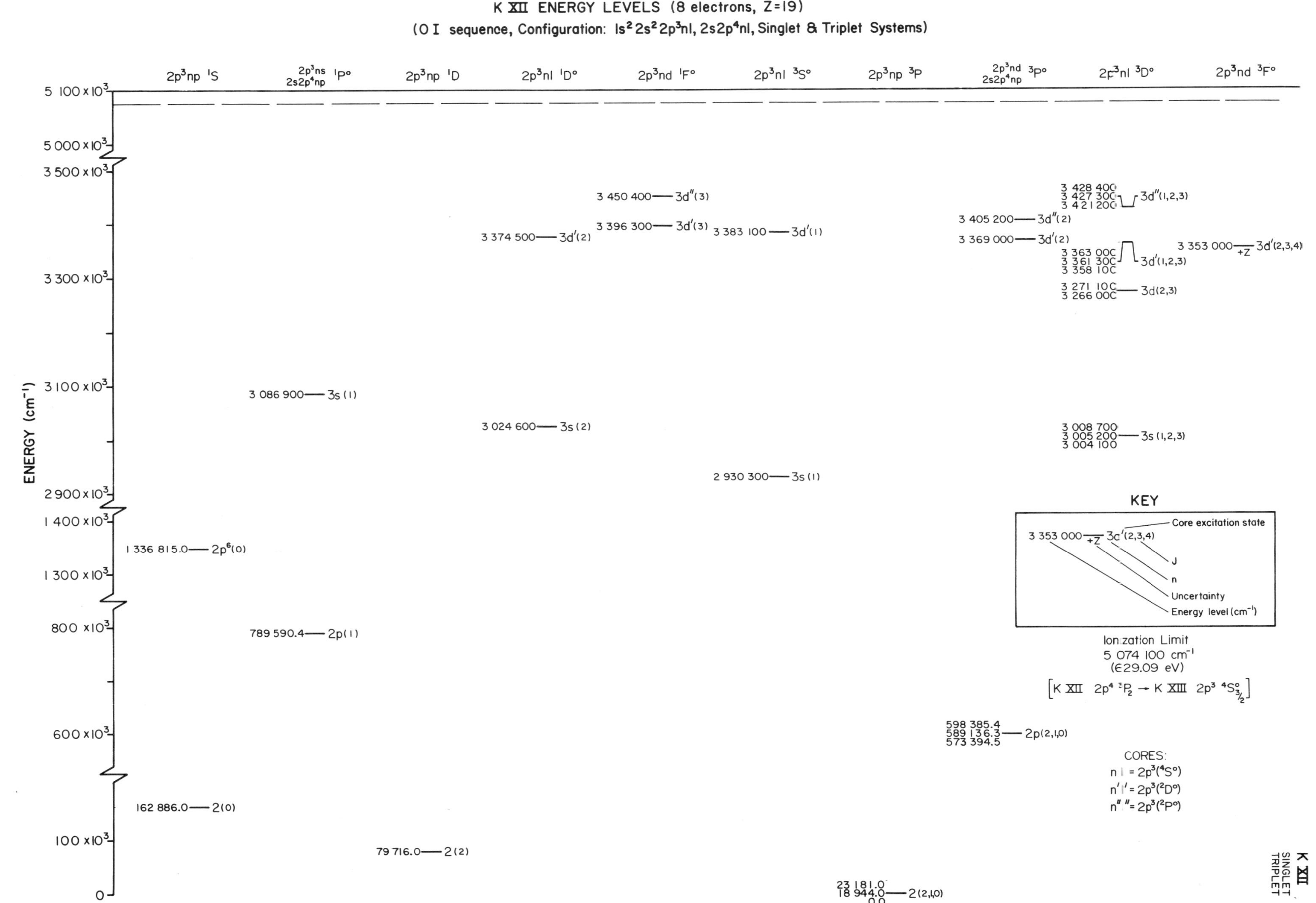
K XII ENERGY LEVELS (8 electrons, Z=19)
(O I sequence, Configuration: 1s² 2s² 2p³nl, 2s2p⁴nl, Singlet & Triplet Systems)
2p³np ¹S
2p³ns 2s2p⁴np ¹P°
2p³np ¹D
2p³nl ¹D°
2p³nd ¹F°
2p³nl ³S°
2p³np ³P
2p³nd 2s2p⁴np ³P°
2p³nl ³D°
2p³nd ³F°
ENERGY (cm⁻¹)
5 100 x10³
5 000 x10³
3 500 x10³
3 300 x10³
3 100 x10³
2 900 x10³
1 400 x10³
1 300 x10³
800 x10³
600 x10³
100 x10³
0
3 450 400 — 3d″(3)
3 396 300 — 3d′(3)
3 374 500 — 3d′(2)
3 383 100 — 3d′(1)
3 405 200 — 3d″(2)
3 369 000 — 3d′(2)
3 428 400
3 427 300
3 421 200 — 3d″(1,2,3)
3 363 000
3 361 300
3 358 100 — 3d′(1,2,3)
3 353 000 +Z — 3d′(2,3,4)
3 271 100
3 266 000 — 3d(2,3)
3 086 900 — 3s(1)
3 024 600 — 3s(2)
3 008 700
3 005 200
3 004 100 — 3s(1,2,3)
2 930 300 — 3s(1)
1 336 815.0 — 2p⁶(0)
789 590.4 — 2p(1)
598 385.4
589 136.3
573 394.5 — 2p(2,1,0)
162 886.0 — 2(0)
79 716.0 — 2(2)
23 181.0
18 944.0
0.0 — 2(2,1,0)
KEY
3 353 000 +Z — 3d′(2,3,4)
Core excitation state
J
n
Uncertainty
Energy level (cm⁻¹)
Ionization Limit
5 074 100 cm⁻¹
(629.09 eV)
[K XII 2p⁴ ³P₂ → K XIII 2p³ ⁴S°₃/₂]
CORES:
n l = 2p³(⁴S°)
n′l′ = 2p³(²D°)
n″l″ = 2p³(²P°)
K XII
SINGLET
TRIPLET

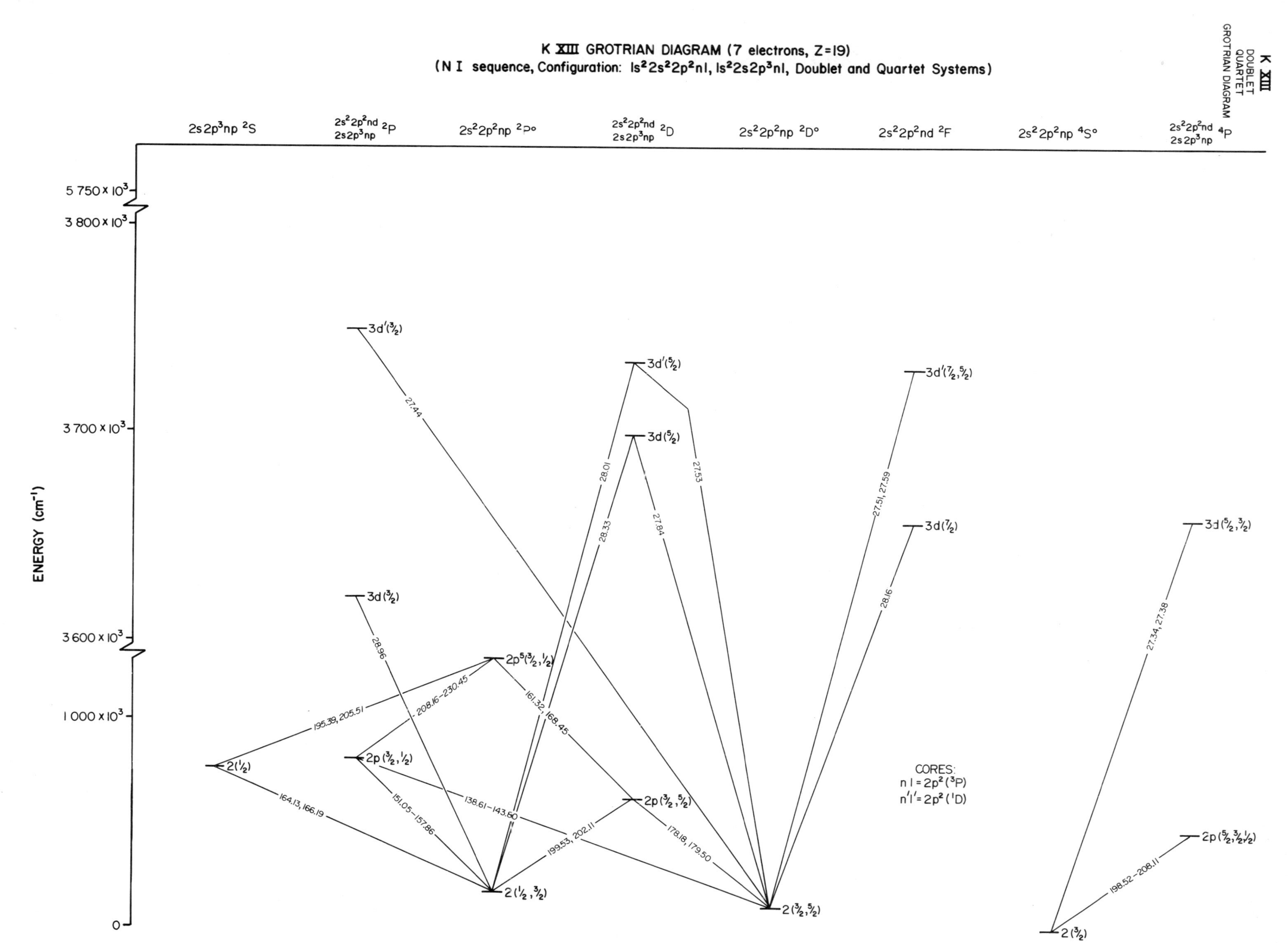
K XIII GROTRIAN DIAGRAM (7 electrons, Z=19)
(N I sequence, Configuration: 1s²2s²2p²nl, 1s²2s2p³nl, Doublet and Quartet Systems)
K XIII
DOUBLET
QUARTET
GROTRIAN DIAGRAM
2s2p³np ²S
2s²2p²nd 2s2p³np ²P
2s²2p²np ²P°
2s²2p²nd 2s2p³np ²D
2s²2p²np ²D°
2s²2p²nd ²F
2s²2p²np ⁴S°
2s²2p²nd 2s2p³np ⁴P
5 750 x 10³
3 800 x 10³
3 700 x 10³
3 600 x 10³
1 000 x 10³
0
ENERGY (cm⁻¹)
3d′(3/2)
3d′(5/2)
3d′(7/2, 5/2)
3d(5/2)
3d(7/2)
3d(5/2, 3/2)
3d(3/2)
2p⁵(3/2, 1/2)
2p(3/2, 1/2)
2(1/2)
2p(3/2, 5/2)
2p(5/2, 3/2, 1/2)
2(1/2, 3/2)
2(3/2, 5/2)
2(3/2)
27.44
28.01
28.33
27.84
27.53
27.51, 27.59
28.16
27.34, 27.38
28.96
195.39, 205.51
208.16–230.45
161.32, 168.45
164.13, 166.19
151.05–157.86
138.61–143.60
199.53, 202.11
178.18, 179.50
198.52–208.11
CORES:
n l = 2p² (³P)
n′l′ = 2p² (¹D)

K XIII ENERGY LEVELS (7 electrons, Z=19)

(N I sequence, Configuration: $1s^2 2s^2 2p^2 nl$, $1s^2 2s 2p^3 nl$, Doublet and Quartet Systems)

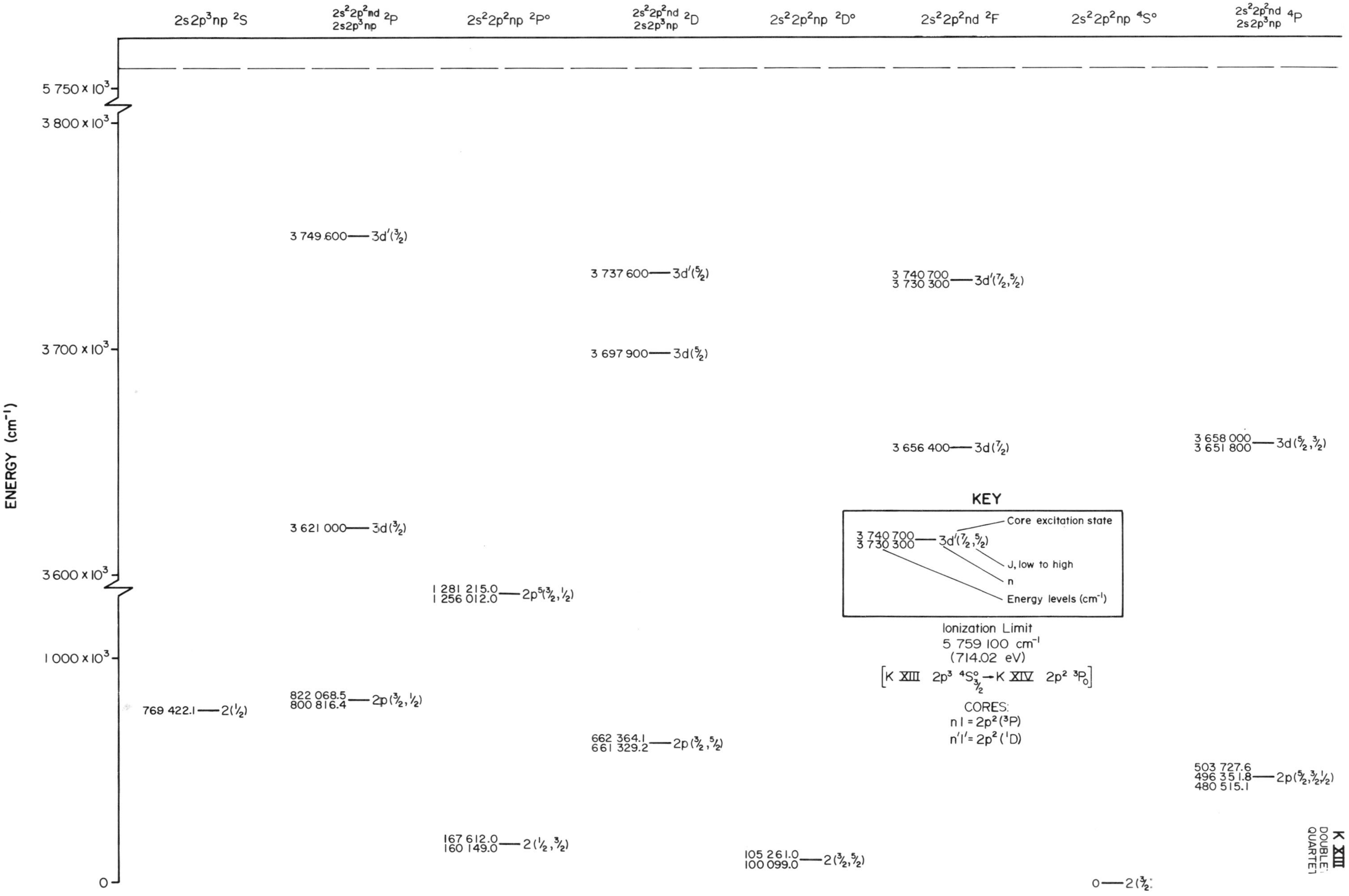

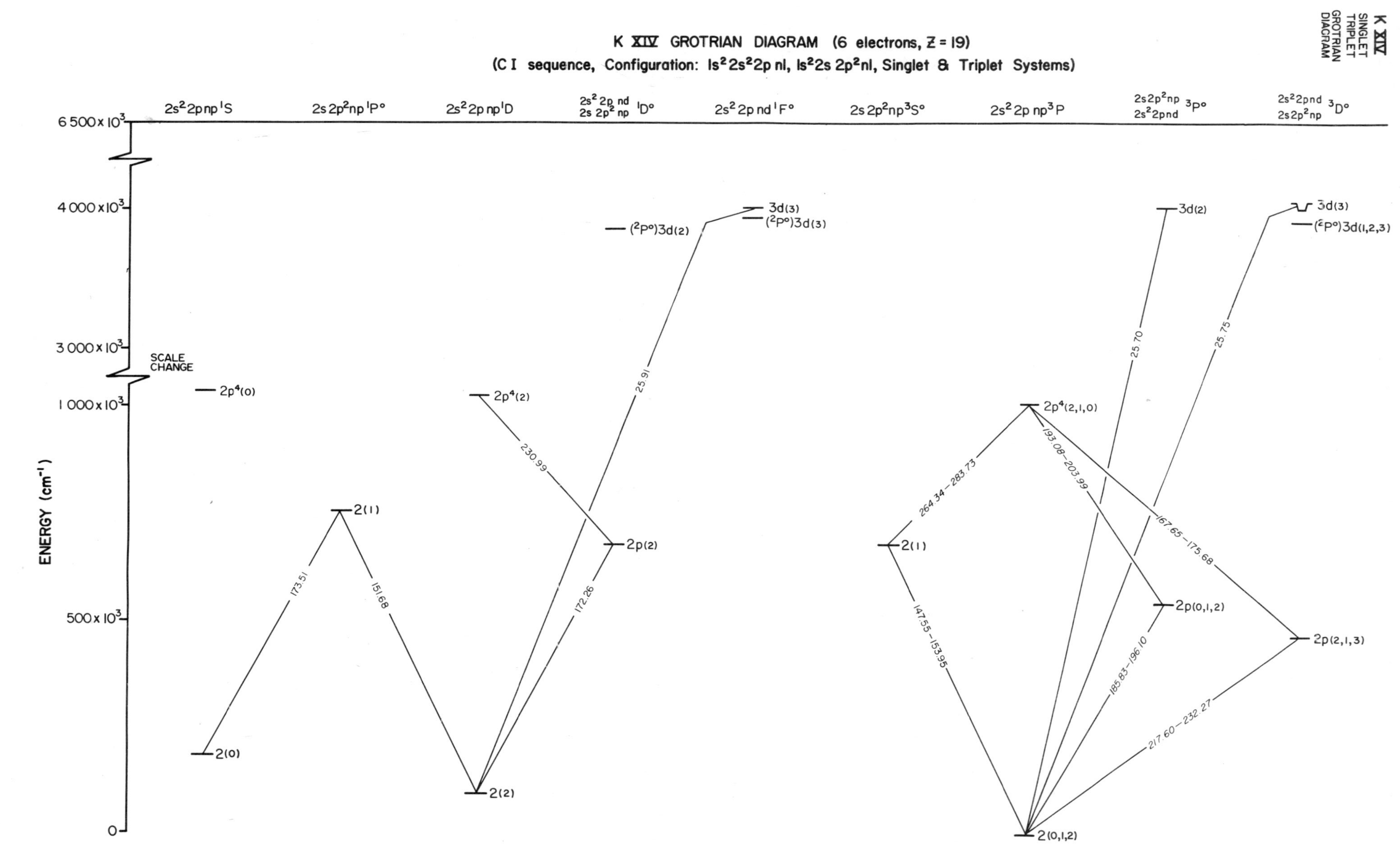
K XIV
SINGLET
TRIPLET
GROTRIAN
DIAGRAM
K XIV GROTRIAN DIAGRAM (6 electrons, Z = 19)
(C I sequence, Configuration: $1s^2 2s^2 2p\,nl$, $1s^2 2s\,2p^2 nl$, Singlet & Triplet Systems)
$2s^2 2p\,np\,^1S$
$2s\,2p^2 np\,^1P^o$
$2s^2 2p\,np\,^1D$
$2s^2 2p\,nd$ / $2s\,2p^2\,np$ $^1D^o$
$2s^2 2p\,nd\,^1F^o$
$2s\,2p^2 np\,^3S^o$
$2s^2 2p\,np\,^3P$
$2s2p^2np$ / $2s^22pnd$ $^3P^o$
$2s^22pnd$ / $2s2p^2np$ $^3D^o$
ENERGY (cm^{-1})
6 500 x 10^3
4 000 x 10^3
3 000 x 10^3
1 000 x 10^3
500 x 10^3
0
SCALE CHANGE
$2p^4(0)$
2(0)
173.51
2(1)
151.68
$2p^4(2)$
230.99
2(2)
2p(2)
172.26
25.91
$(^2P^o)3d(2)$
3d(3)
$(^2P^o)3d(3)$
2(1)
264.34–283.73
147.55–153.95
$2p^4(2,1,0)$
193.08–203.99
167.65–175.68
2(0,1,2)
2p(0,1,2)
185.83–196.10
2p(2,1,3)
217.60–232.27
3d(2)
25.70
3d(3)
$(^2P^o)3d(1,2,3)$
25.75

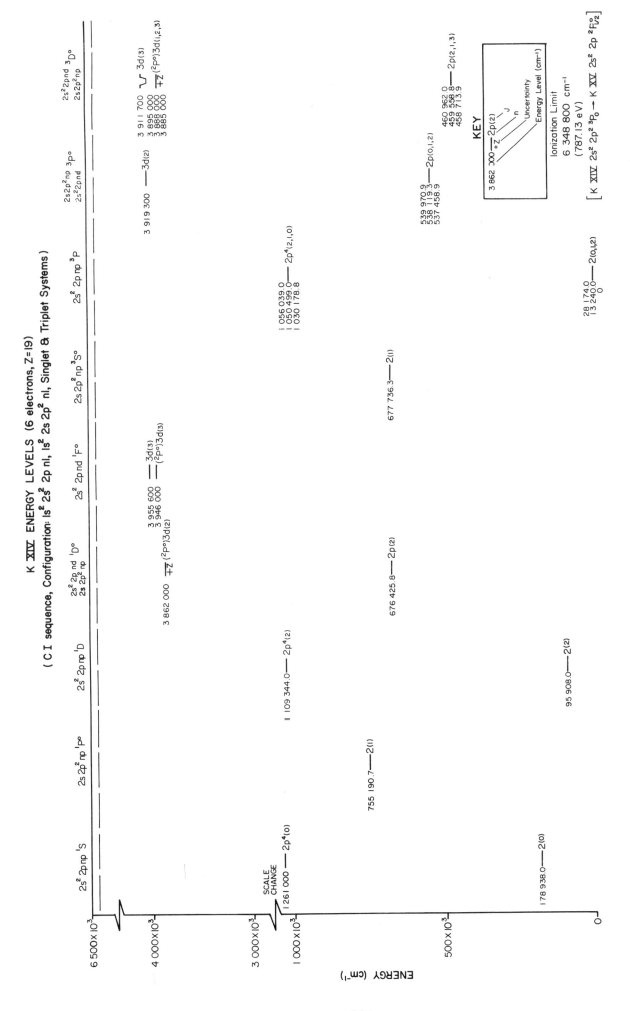

K XIV ENERGY LEVELS (6 electrons, Z=19)

(C I sequence, Configuration: $1s^2 2s^2 2p\,nl$, $1s^2 2s\,2p^2 nl$, Singlet & Triplet Systems)

$2s^2 2p\,np\ ^1S$
$2s\,2p^2 np\ ^1P^o$
$2s^2 2p\,np\ ^1D$
$2s^2 2p\,nd\ ^1D^o$ / $2s\,2p^2 np$
$2s^2 2p\,nd\ ^1F^o$
$2s\,2p^2 np\ ^3S^o$
$2s^2 2p\,np\ ^3P$
$2s\,2p^2 np\ ^3P^o$ / $2s^2 2p\,nd$
$2s^2 2p\,nd\ ^3D^o$ / $2s\,2p^2 np$

6 500 X 10³
4 000 X 10³
3 000 X 10³
SCALE CHANGE
1 000 X 10³
500 X 10³
0
ENERGY (cm⁻¹)

1 261 000 — $2p^4(0)$
178 938.0 — 2(0)
755 190.7 — 2(1)
1 109 344.0 — $2p^4(2)$
95 908.0 — 2(2)
3 862 000 ±Z ($^2P^o$)3d(2)
676 425.8 — 2p(2)
3 955 600 — 3d(3)
3 946 000 — ($^2P^o$)3d(3)
677 736.3 — 2(1)
1 056 039.0
1 050 499.0 — $2p^4(2,1,0)$
1 030 178.8
28 174.0
13 240.0 — 2(0,1,2)
0
3 919 300 — 3d(2)
539 970.9
538 119.3 — 2p(0,1,2)
537 458.9
3 911 700 — 3d(3)
3 895 000
3 888 000 ±Z ($^2P^o$)3d(1,2,3)
3 885 000
460 962.0
459 558.8 — 2p(2,1,3)
458 713.9

KEY
3 862 000 ±Z 2p(2)
J
n
Uncertainty
Energy Level (cm⁻¹)

Ionization Limit
6 348 800 cm⁻¹
(787.13 eV)
[K XIV $2s^2 2p^2\ ^3P_0$ → K XV $2s^2 2p\ ^2F^o_{1/2}$]

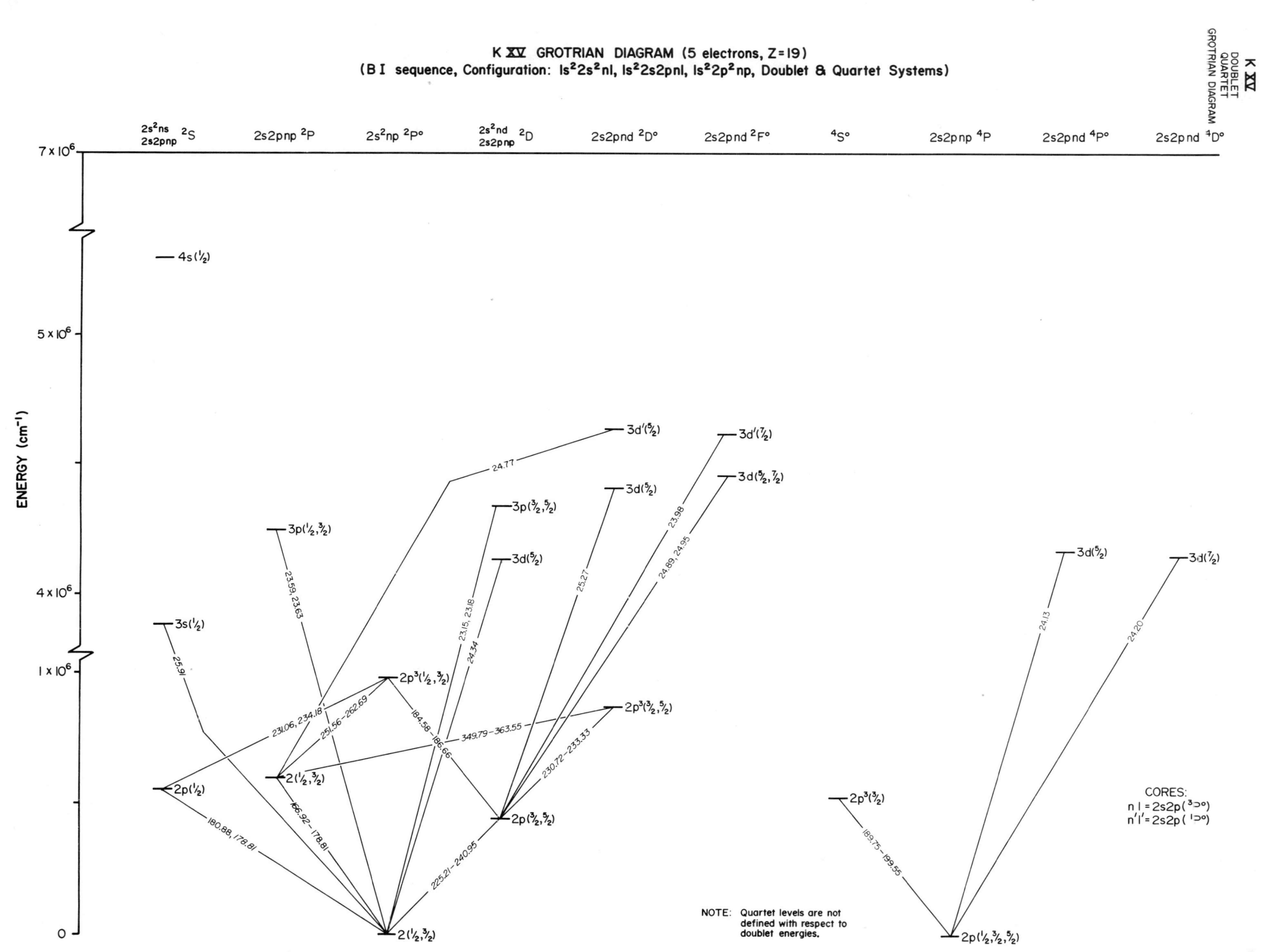
K XV
DOUBLET
QUARTET
GROTRIAN DIAGRAM
K XV GROTRIAN DIAGRAM (5 electrons, Z=19)
(B I sequence, Configuration: ls²2s²nl, ls²2s2pnl, ls²2p²np, Doublet & Quartet Systems)
2s²ns 2s2pnp ²S
2s2pnp ²P
2s²np ²P°
2s²nd 2s2pnp ²D
2s2pnd ²D°
2s2pnd ²F°
⁴S°
2s2pnp ⁴P
2s2pnd ⁴P°
2s2pnd ⁴D°
ENERGY (cm⁻¹)
7 x 10⁶
5 x 10⁶
4 x 10⁶
1 x 10⁶
0
4s(½)
3s(½)
2p(½)
3p(½,³⁄₂)
2(½,³⁄₂)
2p³(½,³⁄₂)
3d′(⁵⁄₂)
3p(³⁄₂,⁵⁄₂)
3d(⁵⁄₂)
3d(⁵⁄₂)
2p(³⁄₂,⁵⁄₂)
2p³(³⁄₂,⁵⁄₂)
3d′(⁷⁄₂)
3d(⁵⁄₂,⁷⁄₂)
2p³(³⁄₂)
2p(½,³⁄₂,⁵⁄₂)
3d(⁵⁄₂)
3d(⁷⁄₂)
25.91
23.59, 23.63
180.88, 178.81
166.92 – 178.81
231.06, 234.18
251.56 – 262.69
24.77
184.58 – 186.66
23.15, 23.18
24.34
349.79 – 363.55
225.21 – 240.95
230.72 – 233.33
25.27
23.98
24.89, 24.95
189.75 – 199.55
24.13
24.20
CORES:
n l = 2s2p(³P°)
n′l′ = 2s2p(¹P°)
NOTE: Quartet levels are not defined with respect to doublet energies.

K XV ENERGY LEVELS (5 electrons, Z = 19)

(B I sequence, Configuration: $1s^2 2s^2 nl$, $1s^2 2s2p\,nl$, $1s^2 2p^2 np$, Doublet & Quartet Systems)

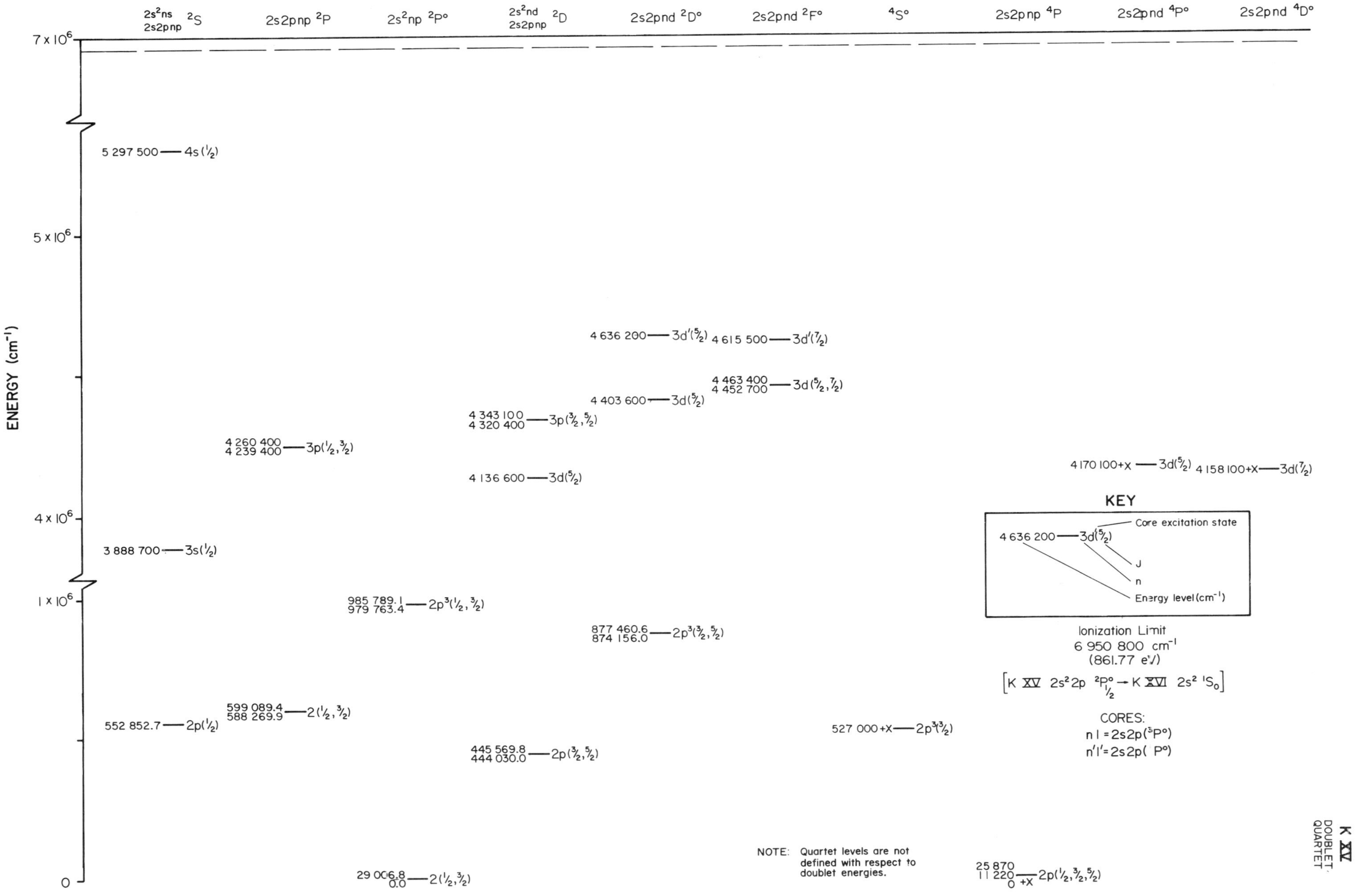

K XV
DOUBLET
QUARTET

ENERGY (cm⁻¹)

7×10^6
5×10^6
4×10^6
1×10^6
0

$2s^2ns$ / $2s2pnp$ ^2S
5 297 500 — 4s(1/2)
3 888 700 — 3s(1/2)
552 852.7 — 2p(1/2)

2s2pnp ^2P
4 260 400, 4 239 400 — 3p(1/2, 3/2)
599 089.4, 588 269.9 — 2(1/2, 3/2)

$2s^2np$ $^2P^o$
985 789.1, 979 763.4 — $2p^3$(1/2, 3/2)
29 006.8, 0.0 — 2(1/2, 3/2)

$2s^2nd$ / $2s2pnp$ ^2D
4 343 100, 4 320 400 — 3p(3/2, 5/2)
4 136 600 — 3d(5/2)
445 569.8, 444 030.0 — 2p(3/2, 5/2)

2s2pnd $^2D^o$
4 636 200 — 3d′(5/2)
4 403 600 — 3d(5/2)
877 460.6, 874 156.0 — $2p^3$(3/2, 5/2)

2s2pnd $^2F^o$
4 615 500 — 3d′(7/2)
4 463 400, 4 452 700 — 3d(5/2, 7/2)

$^4S^o$
527 000+X — $2p^3$(3/2)

2s2pnp ^4P
25 870, 11 220, 0 +X — 2p(1/2, 3/2, 5/2)

2s2pnd $^4P^o$
4 170 100+X — 3d(5/2)

2s2pnd $^4D^o$
4 158 100+X — 3d(7/2)

KEY
4 636 200 — 3d(5/2)
Core excitation state
J
n
Energy level (cm⁻¹)

Ionization Limit
6 950 800 cm⁻¹
(861.77 eV)
[K XV $2s^2 2p\ ^2P^o_{1/2}$ → K XVI $2s^2\ ^1S_0$]

CORES:
nl = 2s2p($^3P^o$)
n′l′ = 2s2p(P^o)

NOTE: Quartet levels are not defined with respect to doublet energies.

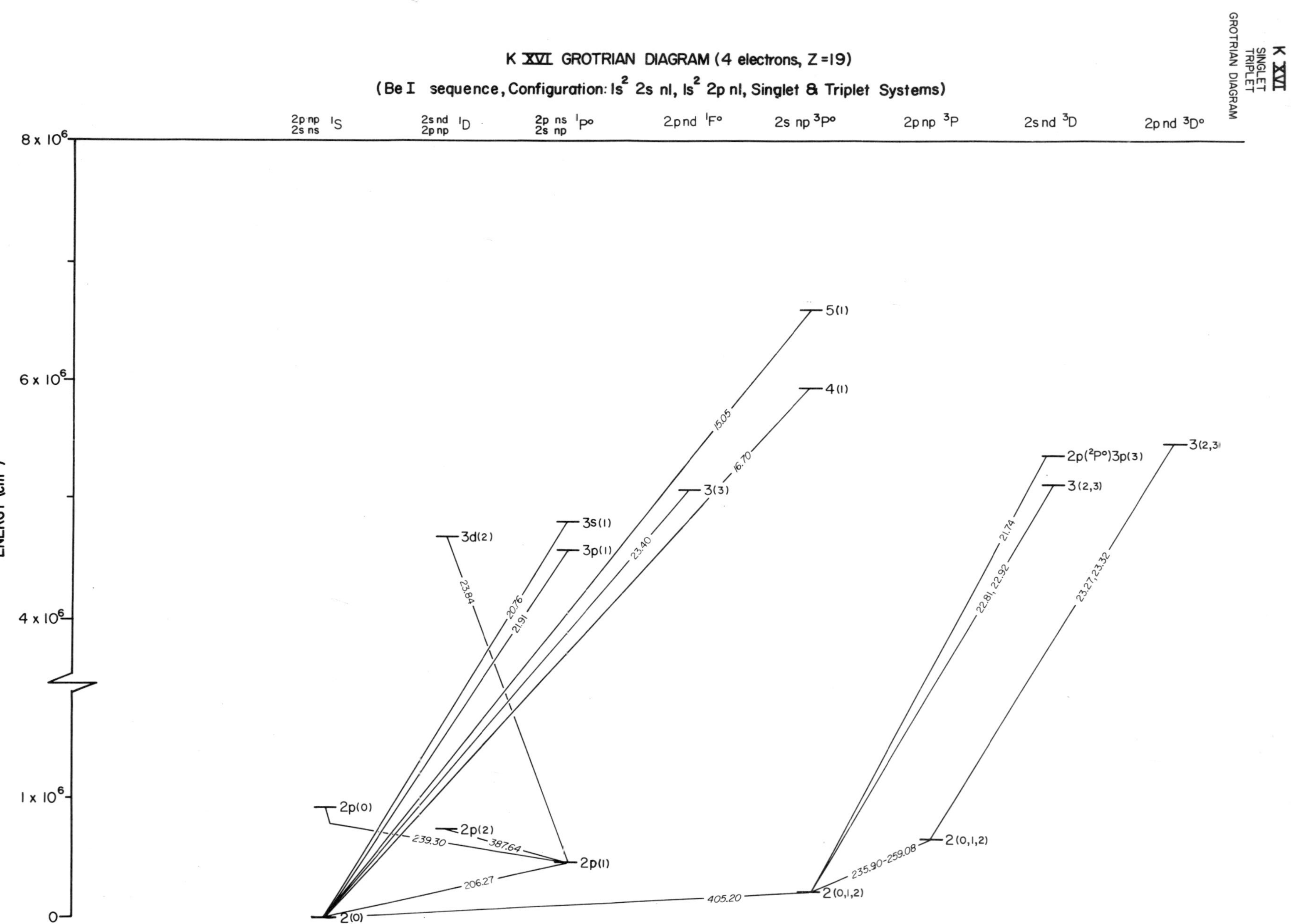
K XVI GROTRIAN DIAGRAM (4 electrons, Z = 19)
(Be I sequence, Configuration: 1s² 2s nl, 1s² 2p nl, Singlet & Triplet Systems)
2p np 2s ns ¹S
2s nd 2p np ¹D
2p ns 2s np ¹P°
2p nd ¹F°
2s np ³P°
2p np ³P
2s nd ³D
2p nd ³D°
ENERGY (cm⁻¹)
8 x 10⁶
6 x 10⁶
4 x 10⁶
1 x 10⁶
0
2(0)
2p(0)
2p(2)
2p(1)
3d(2)
3s(1)
3p(1)
3(3)
4(1)
5(1)
2(0,1,2)
2(0,1,2)
2p(²P°)3p(3)
3(2,3)
3(2,3)
239.30
387.64
206.27
23.84
20.76
21.91
23.40
15.05
16.70
405.20
235.90–259.08
21.74
22.81, 22.92
23.27, 23.32

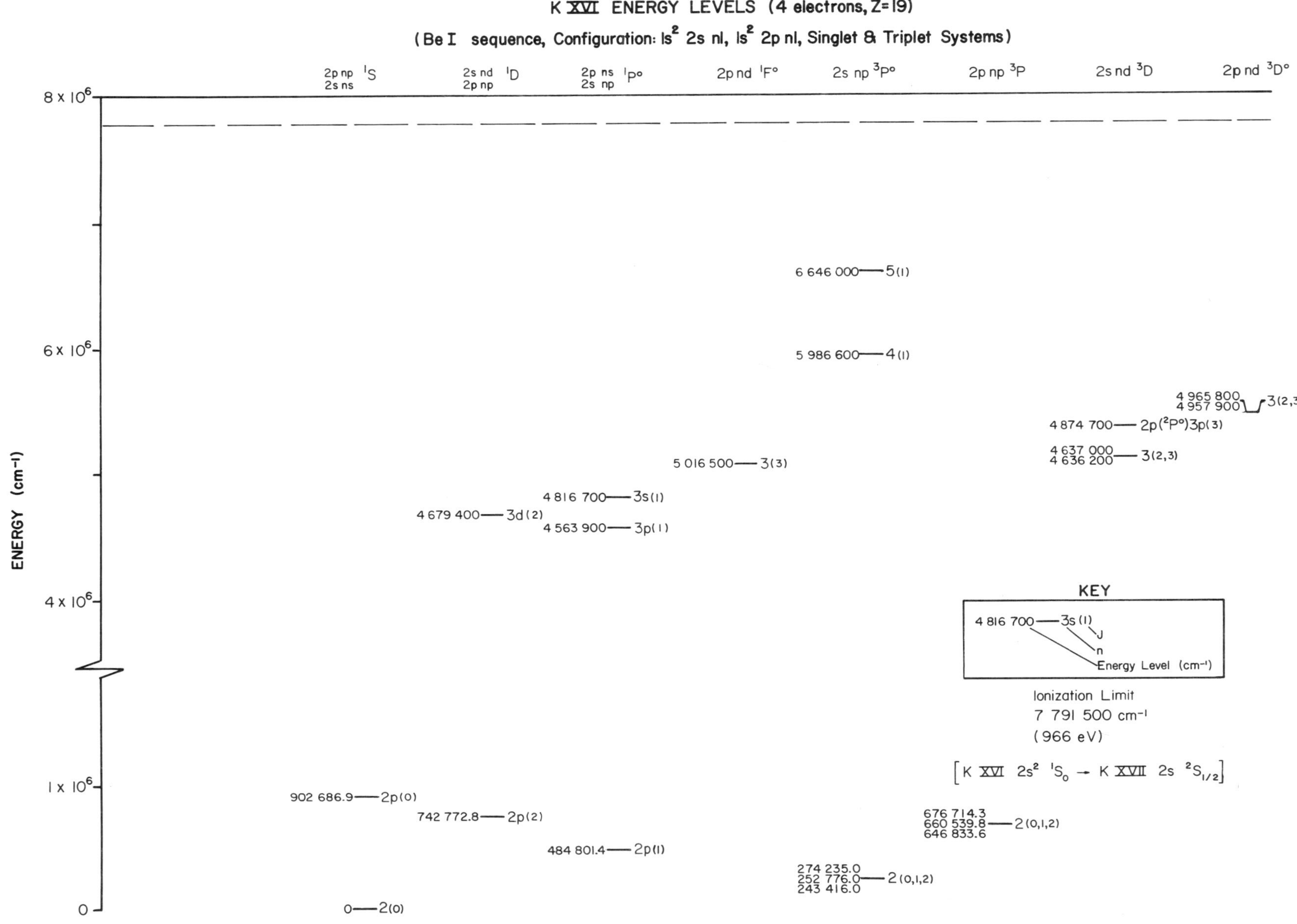
K XVI ENERGY LEVELS (4 electrons, Z=19)
(Be I sequence, Configuration: 1s² 2s nl, 1s² 2p nl, Singlet & Triplet Systems)
2p np ¹S
2s ns
2s nd ¹D
2p np
2p ns ¹P°
2s np
2p nd ¹F°
2s np ³P°
2p np ³P
2s nd ³D
2p nd ³D°
8 x 10⁶
6 x 10⁶
4 x 10⁶
1 x 10⁶
0
ENERGY (cm⁻¹)
6 646 000—5(1)
5 986 600—4(1)
4 965 800
4 957 900—3(2,3)
4 874 700—2p(²P°)3p(3)
4 637 000
4 636 200—3(2,3)
5 016 500—3(3)
4 816 700—3s(1)
4 679 400—3d(2)
4 563 900—3p(1)
KEY
4 816 700—3s(1)
J
n
Energy Level (cm⁻¹)
Ionization Limit
7 791 500 cm⁻¹
(966 eV)
[K XVI 2s² ¹S₀ → K XVII 2s ²S₁/₂]
902 686.9—2p(0)
742 772.8—2p(2)
484 801.4—2p(1)
676 714.3
660 539.8
646 833.6—2(0,1,2)
274 235.0
252 776.0
243 416.0—2(0,1,2)
0—2(0)

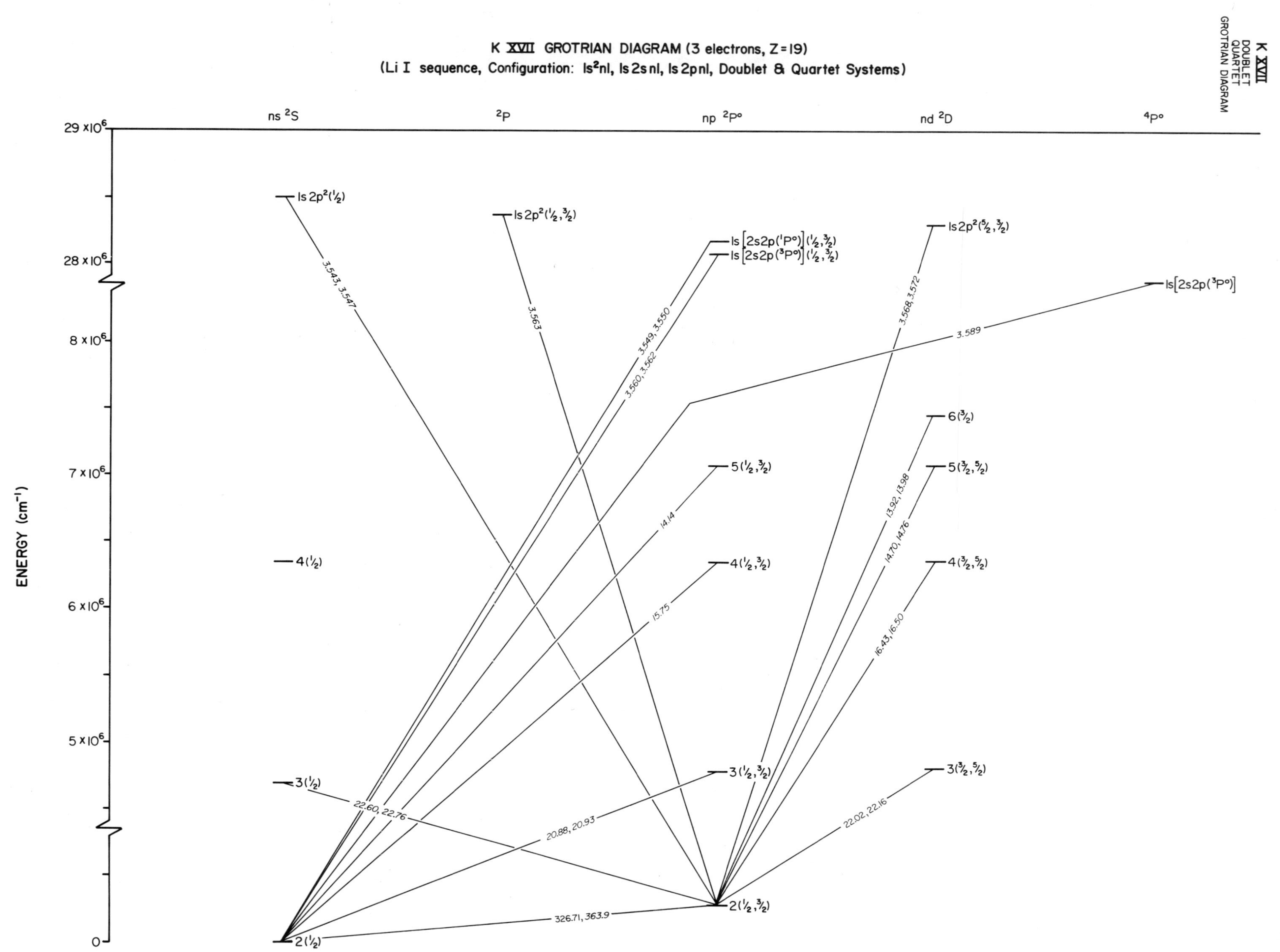
K XVII
DOUBLET
QUARTET
GROTRIAN DIAGRAM
K XVII GROTRIAN DIAGRAM (3 electrons, Z=19)
(Li I sequence, Configuration: $1s^2nl$, $1s2snl$, $1s2pnl$, Doublet & Quartet Systems)
ns ^2S
^2P
np $^2P^o$
nd ^2D
$^4P^o$
ENERGY (cm^{-1})
29 x10^6
28 x10^6
8 x10^6
7 x10^6
6 x10^6
5 x10^6
0
$1s2p^2(1/2)$
$1s2p^2(1/2, 3/2)$
$1s[2s2p(^1P^o)](1/2, 3/2)$
$1s[2s2p(^3P^o)](1/2, 3/2)$
$1s2p^2(5/2, 3/2)$
$1s[2s2p(^3P^o)]$
4(1/2)
3(1/2)
2(1/2)
5(1/2, 3/2)
4(1/2, 3/2)
3(1/2, 3/2)
2(1/2, 3/2)
6(3/2)
5(3/2, 5/2)
4(3/2, 5/2)
3(3/2, 5/2)
3.543, 3.547
3.563
3.549, 3.550
3.560, 3.562
3.568, 3.572
3.589
14.14
15.75
13.92, 13.98
14.70, 14.76
16.43, 16.50
22.60, 22.76
20.88, 20.93
22.02, 22.16
326.71, 363.9

K XVII ENERGY LEVELS (3 electrons, Z=19)

(Li I sequence, Configuration: $1s^2nl$, $1s2snl$, $1s2pnl$, Doublet & Quartet Systems)

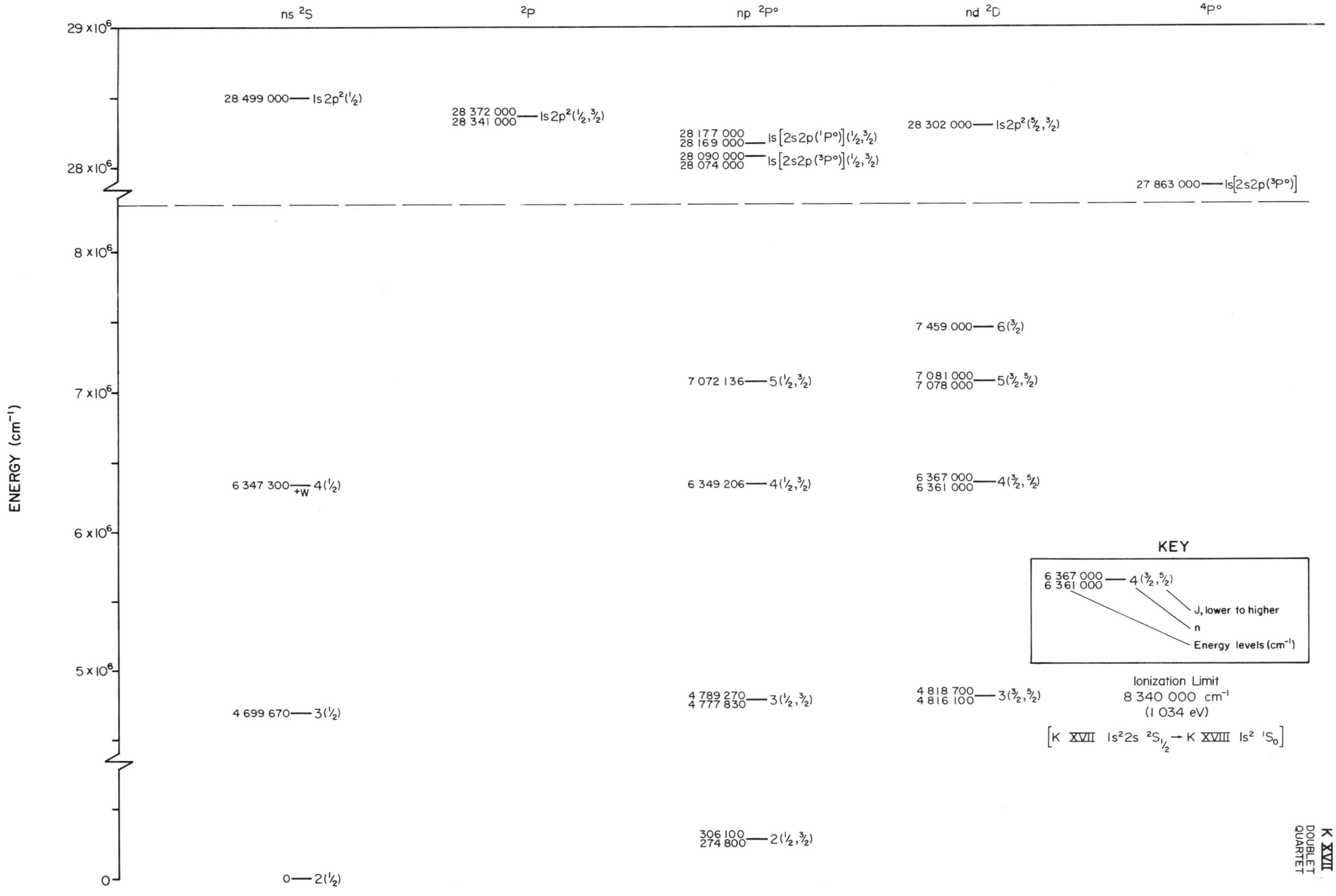

ENERGY (cm^{-1})

Axis: 0, 5×10^6, 6×10^6, 7×10^6, 8×10^6, 28×10^6, 29×10^6

ns ^2S

28 499 000 — $1s2p^2(\frac{1}{2})$

6 347 300 +W — $4(\frac{1}{2})$

4 699 670 — $3(\frac{1}{2})$

0 — $2(\frac{1}{2})$

^2P

28 372 000, 28 341 000 — $1s2p^2(\frac{1}{2},\frac{3}{2})$

np $^2P^o$

28 177 000, 28 169 000 — $1s[2s2p(^1P^o)](\frac{1}{2},\frac{3}{2})$

28 090 000, 28 074 000 — $1s[2s2p(^3P^o)](\frac{1}{2},\frac{3}{2})$

7 072 136 — $5(\frac{1}{2},\frac{3}{2})$

6 349 206 — $4(\frac{1}{2},\frac{3}{2})$

4 789 270, 4 777 830 — $3(\frac{1}{2},\frac{3}{2})$

306 100, 274 800 — $2(\frac{1}{2},\frac{3}{2})$

nd ^2D

28 302 000 — $1s2p^2(\frac{5}{2},\frac{3}{2})$

7 459 000 — $6(\frac{3}{2})$

7 081 000, 7 078 000 — $5(\frac{3}{2},\frac{5}{2})$

6 367 000, 6 361 000 — $4(\frac{3}{2},\frac{5}{2})$

4 818 700, 4 816 100 — $3(\frac{3}{2},\frac{5}{2})$

$^4P^o$

27 863 000 — $1s[2s2p(^3P^o)]$

KEY

6 367 000, 6 361 000 — $4(\frac{3}{2},\frac{5}{2})$

- J, lower to higher
- n
- Energy levels (cm^{-1})

Ionization Limit
8 340 000 cm^{-1}
(1 034 eV)

[K XVII $1s^22s\ ^2S_{1/2}$ → K XVIII $1s^2\ ^1S_0$]

K XVII
DOUBLET
QUARTET

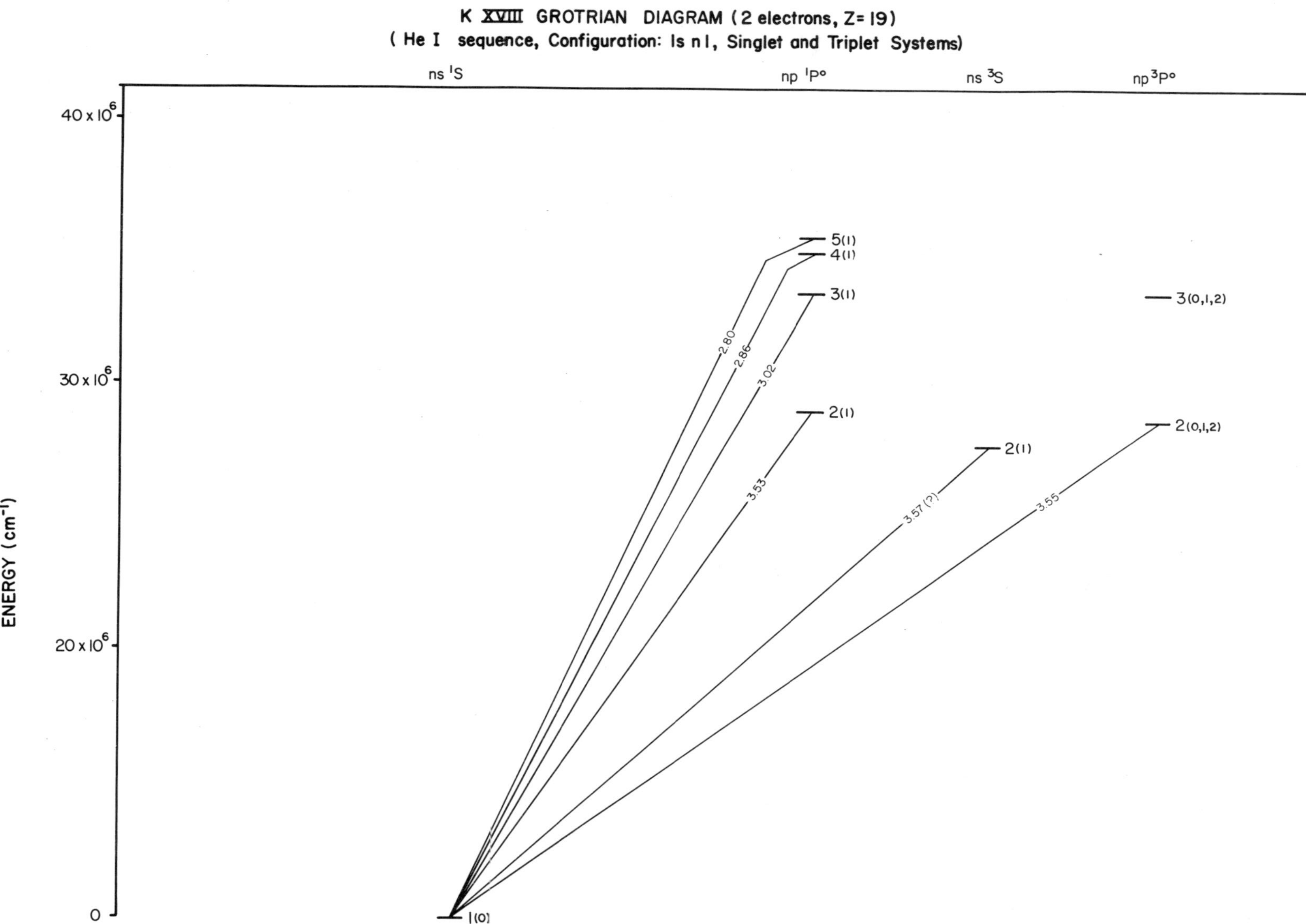
K XVIII GROTRIAN DIAGRAM (2 electrons, Z= 19)
(He I sequence, Configuration: 1s n l, Singlet and Triplet Systems)
ns 1S
np 1P°
ns 3S
np 3P°
40 x 10^6
30 x 10^6
20 x 10^6
0
ENERGY (cm-1)
5(1)
4(1)
3(1)
2(1)
3(0,1,2)
2(0,1,2)
2(1)
1(0)
2.80
2.86
3.02
3.53
3.57 (?)
3.55

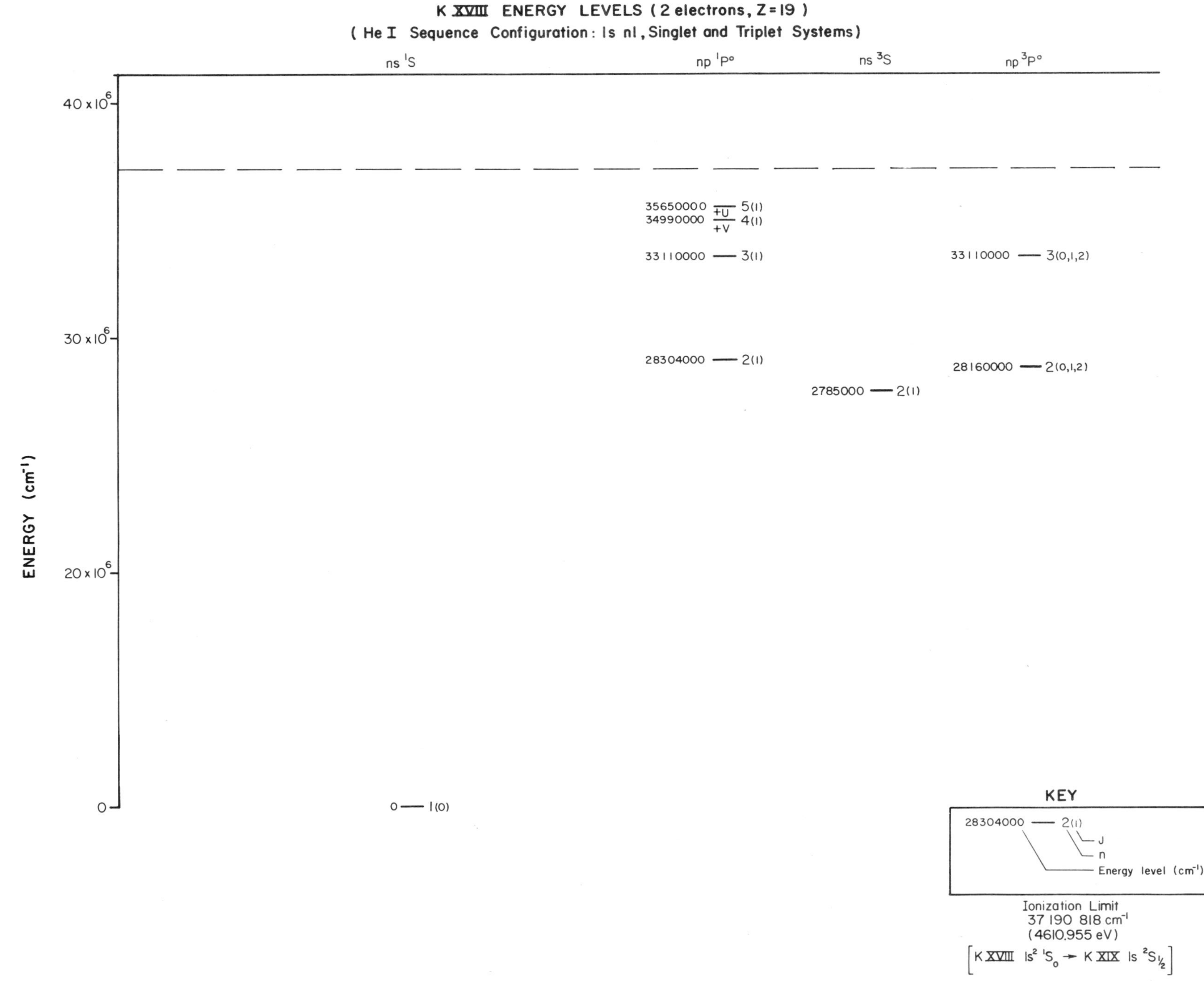
K XVIII ENERGY LEVELS (2 electrons, Z = 19)
(He I Sequence Configuration: 1s nl, Singlet and Triplet Systems)
ns ^1S
np $^1P^o$
ns ^3S
np $^3P^o$
40×10^6
30×10^6
20×10^6
0
ENERGY (cm^{-1})
35650000 — 5(1)
34990000 — 4(1)
+U
+V
33110000 — 3(1)
28304000 — 2(1)
2785000 — 2(1)
33110000 — 3(0,1,2)
28160000 — 2(0,1,2)
0 — 1(0)
KEY
28304000 — 2(1)
J
n
Energy level (cm^{-1})
Ionization Limit
37 190 818 cm^{-1}
(4610.955 eV)
[K XVIII $1s^2\ ^1S_0$ → K XIX $1s\ ^2S_{1/2}$]
K XVIII
SINGLET
TRIPLET

K XIX
GROTRIAN
DIAGRAM
DOUBLET

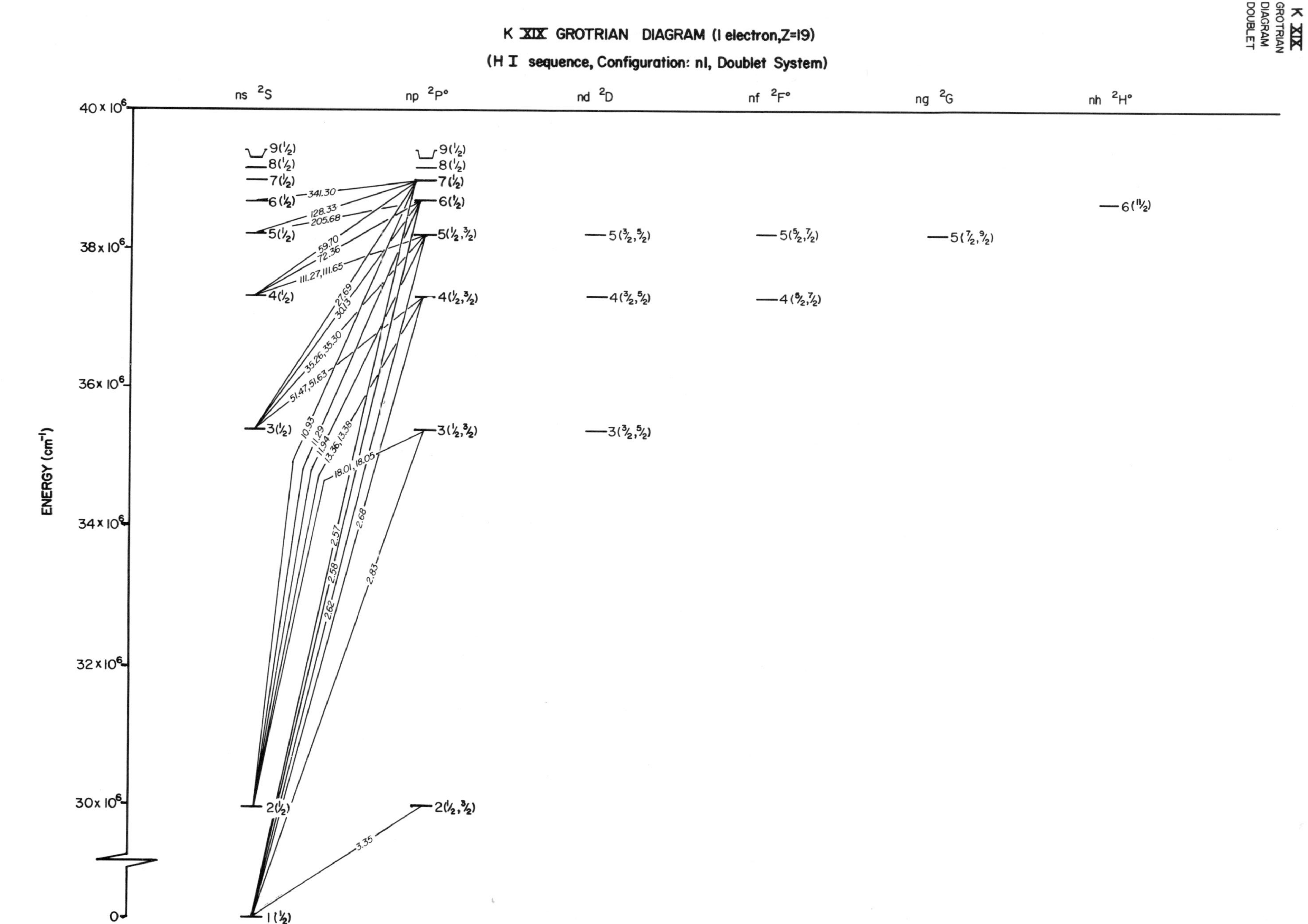

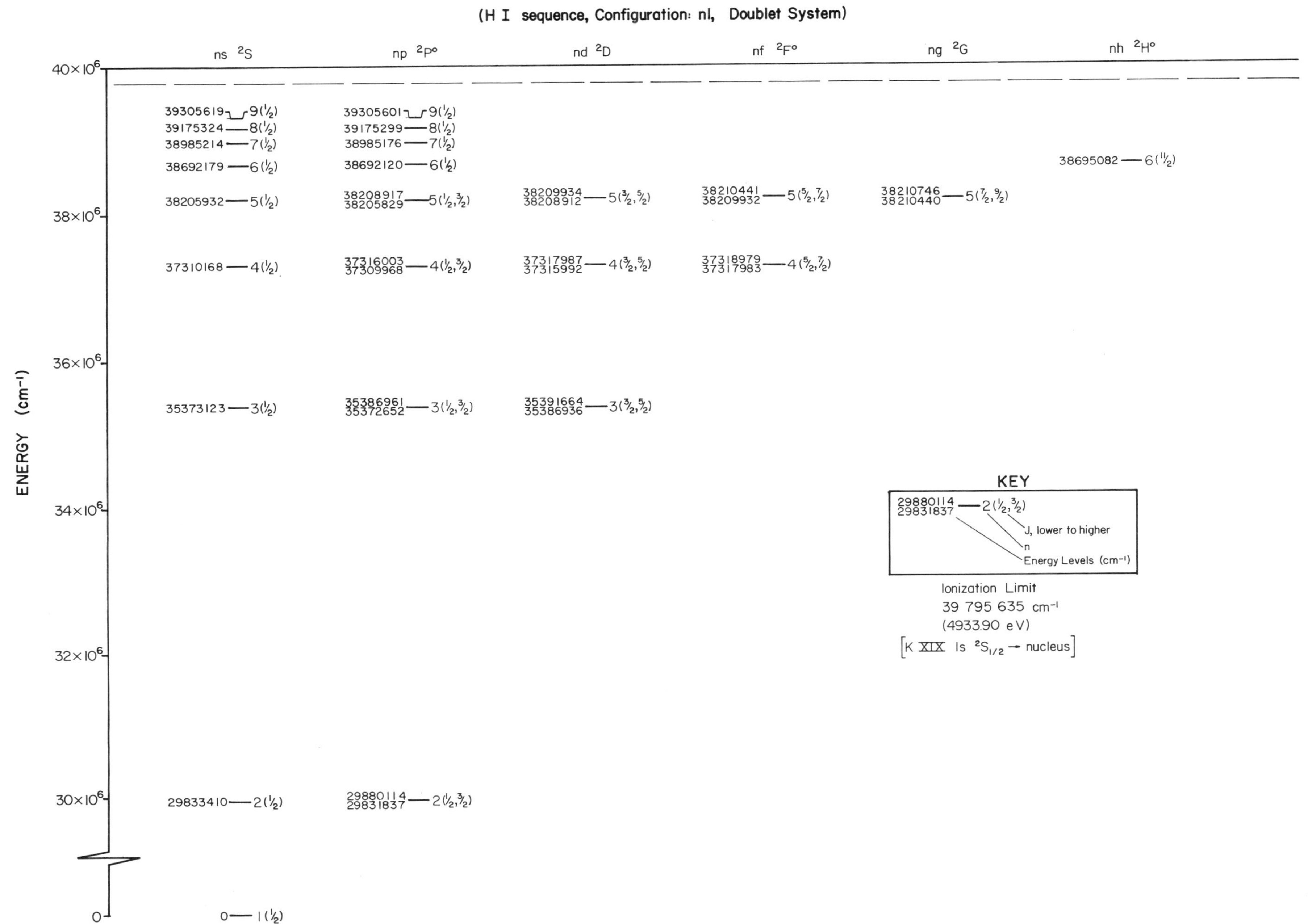
K XIX ENERGY LEVELS (1 electron, Z=19)
(H I sequence, Configuration: nl, Doublet System)
ns 2S
np 2P°
nd 2D
nf 2F°
ng 2G
nh 2H°
39305619 — 9(1/2)
39175324 — 8(1/2)
38985214 — 7(1/2)
38692179 — 6(1/2)
38205932 — 5(1/2)
37310168 — 4(1/2)
35373123 — 3(1/2)
29833410 — 2(1/2)
0 — 1(1/2)
39305601 — 9(1/2)
39175299 — 8(1/2)
38985176 — 7(1/2)
38692120 — 6(1/2)
38208917 38205829 — 5(1/2,3/2)
37316003 37309968 — 4(1/2,3/2)
35386961 35372652 — 3(1/2,3/2)
29880114 29831837 — 2(1/2,3/2)
38209934 38208912 — 5(3/2,5/2)
37317987 37315992 — 4(3/2,5/2)
35391664 35386936 — 3(3/2,5/2)
38210441 38209932 — 5(5/2,7/2)
37318979 37317983 — 4(5/2,7/2)
38210746 38210440 — 5(7/2,9/2)
38695082 — 6(11/2)
KEY
29880114 29831837 — 2(1/2,3/2)
J, lower to higher
n
Energy Levels (cm⁻¹)
Ionization Limit
39 795 635 cm⁻¹
(4933.90 eV)
[K XIX 1s 2S1/2 → nucleus]
ENERGY (cm⁻¹)
40×10⁶
38×10⁶
36×10⁶
34×10⁶
32×10⁶
30×10⁶
0

K I $Z = 19$ 19 electrons

P. Feldman and R. Novick, Phys. Rev. Letters **11**, 278 (1963).

These authors report the existence of autoionizing levels in K I, although no energies are given.

R.A. Fisher, W.C. Knopf, Jr., and F.E. Kinney, Ap. J. **130**, 683 (1959).

Authors give a line table of observed lines in the region 11 690–12 530 Å.

C.D. Harper, S.E. Wheatley, and M.D. Levenson, J. Opt. Soc. Am. **67**, 579 (1977).

These authors locate the nd levels ($11 \leqslant n \leqslant 24$) and the ns levels ($13 \leqslant n \leqslant 26$) by two-photon absorption. (We have corrected two apparent misprints in their Table I.)

I. Johansson, Ark. Fys. **20**, 135 (1961).

Author gives a line table of observed lines in the region 12 430–15 170 Å.

I. Johansson and N. Svendenius, Physica Scripta **5**, 129 (1972).

These authors study the relative intensities of several infrared line of K I and improve some levels' locations.

H.R. Kratz, Phys. Rev. **75**, 1844 (1949).

Author gives a line table for the principal series in absorption, up to $n = 79$.

U. Litzén, Physica Scripta **1**, 253 (1970).

This author studies the 5g levels in K I.

P. Risberg, Ark. Fys. **10**, 583 (1956).

Author gives a line table, a term table, and a Grotrian diagram based on lines observed in the region 15 170–2990 Å.

K II $Z = 19$ 18 electrons

R.D. Cowan, J. Opt. Soc. Am. **58**, 924 (1968).

Author gives a table of calculated lines.

E. Ekefors, Z. Physik **71**, 53 (1931).

This author studies the vacuum-spark spectra of K and presents an extensive line list.

P.G. Kruger and S.G. Weissberg, Phys. Rev. **48**, 659 (1935).

Author give a term table from observations.

P.G. Kruger, S.G. Weissberg, and L.W. Phillips, Phys. Rev. **51**, 1090 (1937).

Authors give a term table from observations.

M.W.D. Mansfield, Proc. Roy. Soc. (London) **A341**, 277 (1974).

This author improves the location and identification of levels in K II, using Hartree–Fock calculations.

K III $Z = 19$ 17 electrons

B. Edlén, Z. Physik **104**, 407 (1937).

Author gives line and term tables based on lines observed in the regions 440–490 Å and 2730–3890 Å.

E. Ekefors, Z. Physik **104**, 407 (1937).

This author studies the vacuum-spark spectrum of K and presents an extensive line list.

A.H. Gabriel, B.C. Fawcett, and Carole Jordan, Nature **206**, 390 (1965).

Authors give the wavelengths of observed transitions in the region 405–415 Å.

A.H. Gabriel, B.C. Fawcett, and Carole Jordan, Proc. Phys. Soc. **87**, 825 (1966).

Authors give wavelengths of the observed transitions in the vacuum ultraviolet region.

J.E. Hansen, J. Opt. Soc. Am. **67**, 754 (1977).

This author calculates the position and wavefunction for the $sp^6\ ^2S$ term.

P.G. Kruger and L.W. Phillips, Phys. Rev. **51**, 1087 (1937).

Authors give an array of terms and observed lines in the vacuum ultraviolet region.

L.Å. Svensson and J.O. Ekberg, Ark. Fys. **37**, 65 (1967).

Authors identify three lines of observed transitions in the region 539–550 Å.

K IV Z = 19 16 electrons

Kelly's tabulation lists 279.877 Å for the transition $3p^4\ ^1D–3d''\ ^1F°$; this assignment is probably incorrent. See B.C. Fawcett, Culham Laboratory Report #ARU-R4 (Sept. 1971, unpublished).

I.S. Bowen, Ap. J. **121**, 306 (1955).

Author gives wavelengths of two forbidden transitions from nebular observations.

I.S. Bowen, Ap. J. **132**, 1 (1960).

Author gives wavelengths of forbidden transitions from laboratory and nebular observations.

B. Edlén, Phys. Rev. **62**, 434 (1942).

Author gives a table of observed and calculated values for the $s^2p^4\ ^1S$ and $sp^5\ ^1P°$ levels.

A.H. Gabriel, B.C. Fawcett, and Carole Jordan, Proc. Phys. Soc. **87**, 825 (1966).

Authors give wavelengths of two observed multiplet transitions.

L.Å. Svensson and J.O. Ekberg, Ark. Fys. **37**, 65 (1967).

Authors given a term table from empirical data.

K V Z = 19 15 electrons

I.S. Bowen, Ap. J. **121**, 306 (1955).

Author gives transition wavelengths observed in nebulae for forbidden transitions.

I.S. Bowen, Ap. J. **132**, 1 (1960).

Author gives a line table for forbidden transitions from laboratory and nebular observations.

B.C. Fawcett, A.H. Gabriel, and P.A.H. Saunders, Proc. Phys. Soc. **90**, 863 (1967).

Authors give wavelengths of observed transitions in one multiplet.

A.H. Gabriel, B.C. Fawcett, and Carole Jordan, Proc. Phys. Soc. **87**, 825 (1966).

Authors give observed transition wavelengths.

J.O. Ekberg and L.Å. Svensson, Physica Scripta **2**, 283 (1970).

These authors classify this spectrum in detail.

K VI Z = 19 14 electrons

I.S. Bowen, Ap. J. **132**, 1 (1960).

Author gives a line table from observations and calculations for forbidden transitions.

B.C. Fawcett, A.H. Gabriel, and P.A.H. Saunders, Proc. Phys. Soc. **90**, 863 (1967).

Authors give wavelengths of two observed transitions.

H.A. Robinson, Phys. Rev. **52**, 724 (1937).

Author gives line and term tables from lines observed in the region 450–500 Å.

J.O. Ekberg and L.Å. Svensson, Physica Scripta **2**, 283 (1970).

These authors classify this spectrum in detail.

K VII $Z = 19$ 13 electrons

J.O. Ekberg and L.Å. Svensson, Physica Scripta **2**, 283 (1970).

These authors classify lines and levels in this spectrum.

B.C. Fawcett, J. Phys. B**3**, 1732 (1970).

This author classifies transitions among low-lying levels in this spectrum.

L.W. Phillips, Phys. Rev. **55**, 708 (1939).

Author gives line and term tables from lines observed in the region 175–405 Å.

K VIII $Z = 19$ 12 electrons

W.L. Parker and L.W. Phillips, Phys. Rev. **57**, 140 (1940).

Authors give line and term tables from lines observed in the region 155–575 Å.

J.O. Ekberg, Physica Scripta **4**, 101 (1971).

This author reanalyzes this spectrum, providing new line and level lists.

B.C. Fawcett, J. Phys. B**3**, 1732 (1970).

This author classifies transitions among low-lying levels in this spectrum.

K IX $Z = 19$ 11 electrons

B. Edlén and E. Bodén have provided (1976) a table of lines for K IX; these usually agree with the results of Cohen and Behring.

D.D. Burgess, B.C. Fawcett, and N.J. Peacock, Proc. Phys. Soc. **92**, 805 (1967).

These authors observe spectra of highly-ionized emitters by excitation of solid targets with lasers.

L. Cohen and W.E. Behring, J. Opt. Soc. Am. **66**, 899 (1976).

These authors report analysis of spectra from sliding sparks.

B. Edlén, Z. Physik **100**, 621 (1936).

Author gives line and term tables from lines observed in the region 180–640 Å.

U. Feldman, G.A. Doschek, D.K. Prinz, and D.J. Nagel, J. Appl. Phys. **47**, 1341 (1976).

These authors observe laser-excited spectra in the region 200–550 Å.

P.G. Kruger and L.W. Phillips, Phys. Rev. **55**, 352 (1939).

Authors give line and term tables from lines observed in the region 110–640 Å.

K X $Z = 19$ 10 electrons

B. Edlén and F. Tyrén, Z. Physik **101**, 206 (1936).

Authors give line and term tables from lines observed in the region 25–45 Å.

K XI Z = 19 9 electrons

W.A. Deutschman, Doctorate thesis for the Dept. of Astro-Geophysics, University of Colorado, obtained from University of Microfilms, Ann Arbor, Michigan, U.S.A., order number 68–2645 (1970).

Author gives a line table, an energy level table, and a term diagram from observations.

W.A. Deutschman and L.L. House, Ap. J. **149**, 451 (1967).

Authors give two wavelengths of observed transitions.

B. Edlén and F. Tyrén, Z. Physik **101**, 206 (1936).

Authors give line table from observations in the region 30–40 Å.

B.C. Fawcett, D.D. Burgess, and N.J. Peacock, Proc. Phys. Soc. **91**, 970 (1967).

Authors give two wavelengths of observed transitions.

U. Feldman, G.A. Doschek, R.D. Cowan, and L. Cohen, J. Opt. Soc. Am. **63**, 1445 (1973).

These authors give energy levels and wavelengths for transitions in the fluorine isoelectronic sequence from K XI to Cr XIX.

B.C. Fawcett, Atomic Data and Nucl. Data Tables, **16**, 135 (1975).

This author presents data on low-lying levels in this spectrum.

K XII Z = 19 8 electrons

W.A. Deutschman, Doctorate thesis for the Dept. of Astro-Feophysics, University of Colorado, obtained from University Microfilms, Ann Arbor , Michigan, U.S.A., order number 68–2645 (1970).

Author gives a line table, and term diagrams from observations in the region 140–180 Å.

W.A. Deutschman and L.L. House, Ap. J. **149**, 451 (1967).

Authors give line table of observations in the region 140–180 Å.

G.A. Doschek, U. Feldman, and L. Cohen, J. Opt. Soc. Am. **63**, 1463 (1973).

These authors observe and analyze the transitions $2p^4-2p^33s$ in K XII.

B.C. Fawcett, Atomic Data and Nuclear Data Tables **16**, 135 (1975).

This author has compiled and calculated levels and wavelengths for transitions of the sort $2s^22p^n-2s2p^{n+1}$ and $2s2p^n-2p^{n+1}$.

B.C. Fawcett, D.D. Burgess, and N.J. Peacock, Proc. Phys. Soc. **91**, 970 (1967).

Authors give a table of lines observed in the region 150–175 Å.

B.C. Fawcett and R.W. Hayes, Mon. Not. Roy. Astron. Soc. **170**, 185 (1975).

These authors classify lines in this spectrum.

A.E. Goertz, J. Opt. Soc. Am. **55**, 742 (1965).

This author predicts two energies for levels in this ion.

K XIII Z = 19 7 electrons

V.A. Boiko, Yu.P. Voinov, V.A. Gribkov, and G.V. Sklizkov, Opt. Spectrosc. **29**, 545 (1970).

These authors classify sixteen lines in this spectrum.

W.A. Deutschman, Doctorate thesis for the Sept. of Astro-Geophysics, University of Colorado, obtained from University Microfilms, Ann Arbor, Michigan, U.S.A., order number 68–2645 (1970).

Author gives wavelengths of observed transitions in the region 135–145 Å.

W.A. Deutschman and L.L. House, Ap. J. **149**, 451 (1967).

Authors give wavelengths of observed transitions in the region 135–145 Å.

B.C. Fawcett, Atomic Data and Nuclear Data Tables **16**, 135 (1975).

This author has compiled and/or calculated levels and wavelengths for the transitions $2s^2 2p^n$ – $2s2p^{n+1}$ and $2s2p^n$ – $2p^{n+1}$.

B.C. Fawcett, J. Phys. B3, 1152 (1970).

This author classifies two lines in this spectrum.

B.C. Fawcett, D.D. Burgess, and N.J. Peacock, Proc. Phys. Soc. **91**, 970 (1967).

Authors give a table of lines observed in the region 140–210 Å, observed from laser-heated plasmas.

B.C. Fawcett and R.W. Hayes, Mon. Not. Roy. Astron. Soc. **170**, 185 (1975).

These authors classify lines in this spectrum.

S.O. Kastner, J. Opt. Soc. Am. **63**, 738 (1973).

This author calculates energies in several isoelectronic sequences.

K XIV $Z = 19$ 6 electrons

V.A. Boiko, Yu.P. Voinov, V.A. Gribkov, and G.V. Sklizkov, Opt. Spectrosc. **29**, 545 (1970).

These authors classify six lines in this spectrum.

B.C. Fawcett, Atomic Data and Nuclear Data Tables **16**, 135 (1975).

This author tabulates classified lines and levels in this spectrum involving configurations $2s^2 2p^n$–$2s2p^{n+1}$ and $2s2p^n$–$2p^{n+1}$.

B.C. Fawcett and R.D. Cowan, Mon. Not. Roy. Astron. Soc. **171**, 1 (1975).

These authors calculate a few energies and intervals in this spectrum.

B.C. Fawcett and R.W. Hayes, Mon. Not. Roy. Astron. Soc. **170**, 185 (1975).

These authors classify lines in this spectrum.

S.O. Kastner, J. Opt. Soc. Am. **63**, 738 (1973).

This author calculates levels and intervals in the configurations $2s^r 2p^k$.

K XV $Z = 19$ 5 electrons

V.A. Boiko, Yu.P. Voinov, V.A. Gribkov, and G.V. Sklizkov, Opt. Spectrosc. **29**, 545 (1970).

These authors classify one line in this spectrum.

B.C. Fawcett, Atomic Data and Nuclear Data Tables **16**, 135 (1975).

This author tabulates classified lines and levels in this spectrum involving configurations $2s^2 2p^n$–$2s2p^{n+1}$ and $2s2p^n$–$2p^{n+1}$.

B.C. Fawcett and R.D. Cowan, Mon. Not. Roy. Astron. Soc. **171**, 1 (1975).

These authors calculate one line in this spectrum.

B.C. Fawcett and R.W. Hayes, Mon. Not. Roy. Astron. Soc. **170**, 185 (1975).

These authors classify several lines in this spectrum.

K XVI $Z = 19$ 4 electrons

B.C. Fawcett, Atomic Data and Nuclear Data Table **16**, 135 (1975).

This author presents wavelengths and classifications for transitions of the type $2s^2 2p^n - 2s2p^{n+1}$ and $2s2p^n - 2p^{n+1}$.

B.C. Fawcett and R.W. Hayes, Mon. Not. Roy. Astron. Soc. **170**, 185 (1975).

These authors classify several lines in this spectrum.

R.H. Garstang and L.J. Shamey, Ap. J. **148**, 665 (1967).

These authors calculate transition probabilities for $2s\ ^1S_0 - 2p\ ^3P_1^{\circ}$ transitions.

K XVII $Z = 19$ 3 electrons

E.V. Aglitskii, V.A. Boiko, S.M. Zakharov, S.A. Pikuz, and A. Ya. Faenov, Sov. J. Quant. Electron. **4**, 500 (1974).

These authors tentatively identify one transition in this system.

B.C. Fawcett, Atomic Data and Nuclear Data Tables **16**, 135 (1975).

This author presents wavelengths and classification for transitions of the type $2s^2 2p^n - 2s2p^{n+1}$ and $2s2p^n - 2p^{n+1}$.

U. Feldman, G.A. Doschek, D.J. Nagel, R.D. Cowan, and R.R. Whitlock, Ap. J. **192**, 213 (1974).

These authors identify satellite lines in laser-produced plasmas.

U. Feldman, G.A. Doschek, D.K. Prinz, and D.J. Nagel, J. Appl. Phys. **47**, 1341 (1976).

These authors identify two lines in this spectrum.

S. Goldsmith, V. Feldman, L. Oren (Katz), and L. Cohen, Ap. J. **174**, 209 (1972).

These authors present new lines and classification in this spectrum.

K XVIII $Z = 19$ 2 electrons

E.V. Aglitskii, V.A. Boiko, S.M. Zakharov, S.A. Pikuz, and A.Ya. Faenov, Sov. J. Quant. Electron. **4**, 500 (1974).

These authors identify numerous transitions in this spectrum.

L. Cohen, U. Feldman, M. Swartz, and J.H. Underwood, J. Opt. Soc. Am. **58**, 843 (1968).

These authors observe one line in this spectrum with a vacuum spark as source.

G.A. Doschek, Space Sci. Rev. **13**, 765 (1972).

This author reports lines in this spectrum observed in solar-flare plasmas.

K XIX $Z = 19$ 1 electron

We have taken calculated values rather than experimental values for energies and wavelengths. Note that each wavelength is an average over l, l'.

G.W. Erickson, private communication (1976), and in press, J. Phys. Chem. Ref. Data.

This author provides improved calculations for one-electron atomic spectra.

J.D. Garcia and J.E. Mack, J. Opt. Soc. Am. **55**, 654 (1965).

Authors give calculated energy level and line tables for one-electron atomic spectra.

Calcium (Ca)

Ca I GROTRIAN DIAGRAM (20 electrons, Z = 20)
(Configuration: $1s^2 2s^2 2p^6 3s^2 3p^6 4s\,nl$, Singlet System)

ns ^{1}S | nl ^{1}P° | nl ^{1}D | np ^{1}D° | nf ^{1}F°

ns ^{1}S levels: 4s(0), 5(0), 6(0), 4p²(0), 7(0), 8(0), 9(0), 10(0), 11(0), 12(0)

nl ^{1}P° levels: 4p(1), 5p(1), 4p′(1), 6p(1), 7p(1), 8p(1), 9p(1), 10p(1), 11p(1), 12p(1), 13p(1)

nl ^{1}D levels: 3d(2), 4d(2), 4p²(2), 5d(2), 6d(2), 7d(2), 3d²(2), 5s′(2)

np ^{1}D° levels: 4p′(2)

nf ^{1}F° levels: 4p′(3), 4(3), 5(3), 6(3), 7(3), 8(3), 9(3), 10(3), 11(3), 12(3)

Transitions: 4226.73, 4575.45, 10343.81, 5867.56, 5512.98, 4847.31, 4496.17, 4312.34, 4203.12, 4132.53, 9338.09, 7595.22, 8256.67, 9416.99, 11955.96, 2398.56, 2721.65, 2118.67, 2150.80, 2200.73, 2275.46, 5041.62, 4526.94, 4240.46, 4058.91, 10888.95, 3946.04, 6717.69, 3328.00, 4092.00, 4201.07, 4412.61, 16157.36, 4685.27, 5188.85, 5857.45, 7326.15, 2680.29, 8162.37, 8606.19, 7148.15, 19815.02, 5349.47, 4878.13, 4355.08, 4108.53, 3972.57, 3889.10, 3833.96, 3795.52, 3767.62, 3761.70

CORES:
nl = 4s(^{2}S)
n′l′ = 3d(^{2}D)

ENERGY (cm^{-1}): 0, 22 x 10^3, 30 x 10^3, 35 x 10^3, 40 x 10^3, 45 x 10^3, 50 x 10^3, 58 x 10^3

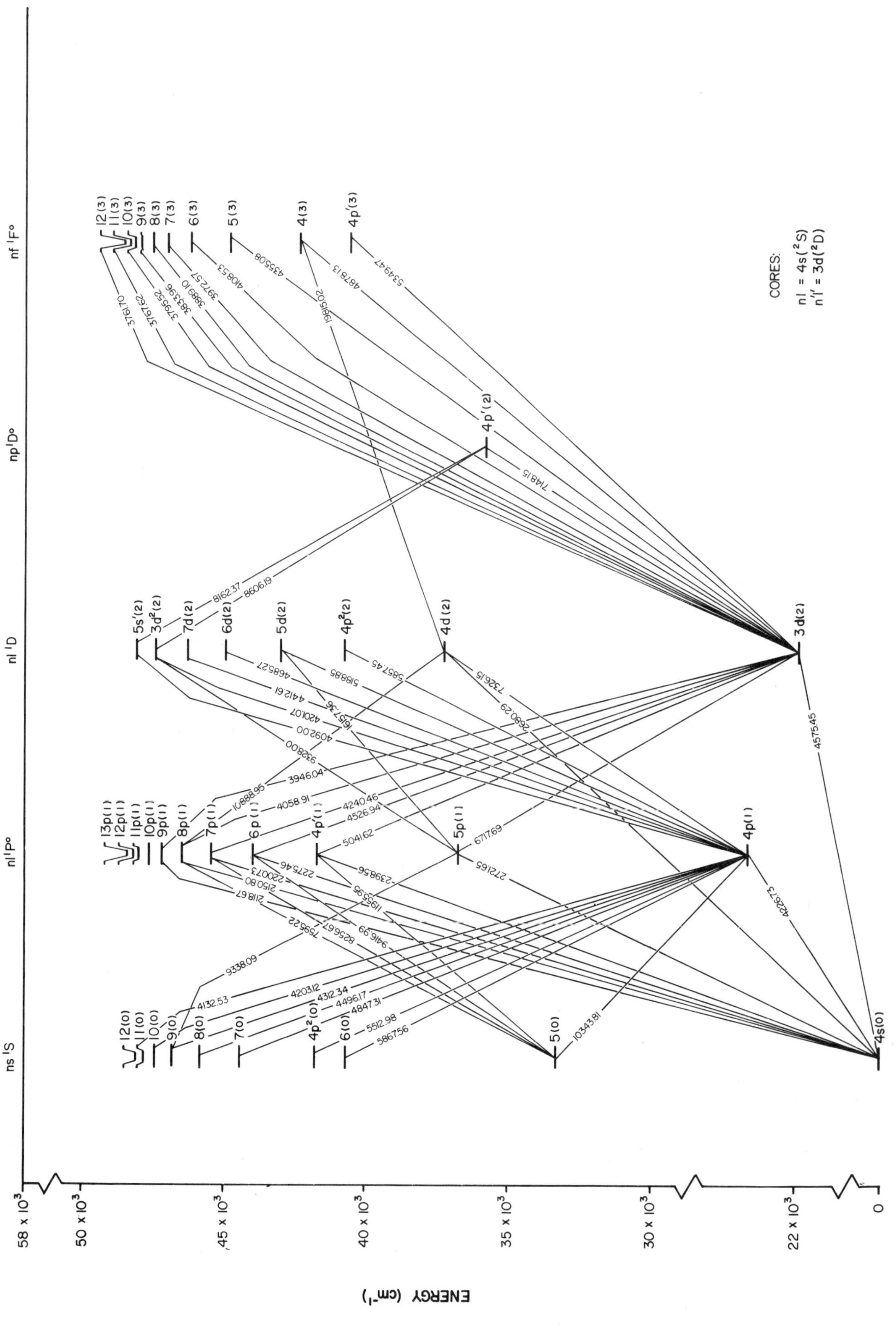

Ca I ENERGY LEVELS (20 electrons, Z = 20)
(Configuration: $1s^2 2s^2 2p^6 3s^2 3p^6 4s\,nl$, Singlet System)

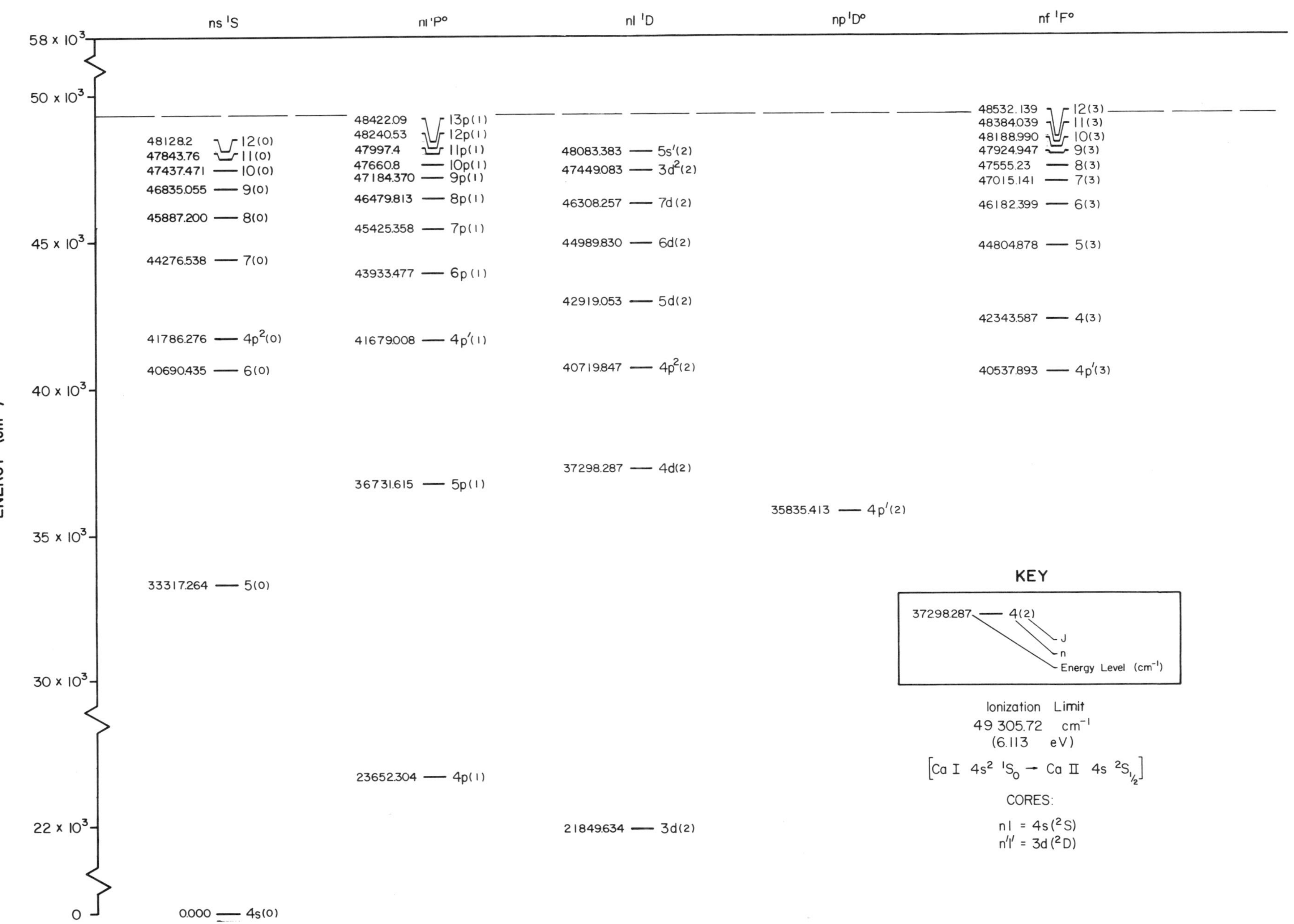

Ca I GROTRIAN DIAGRAM (20 electrons, Z = 20)

(Configuration: $1s^2 2s^2 2p^6 3s^2 3p^6 4s\,nl$, Ground to $^1P^o$ levels, $11 \leq n \leq 79$, Singlet System)

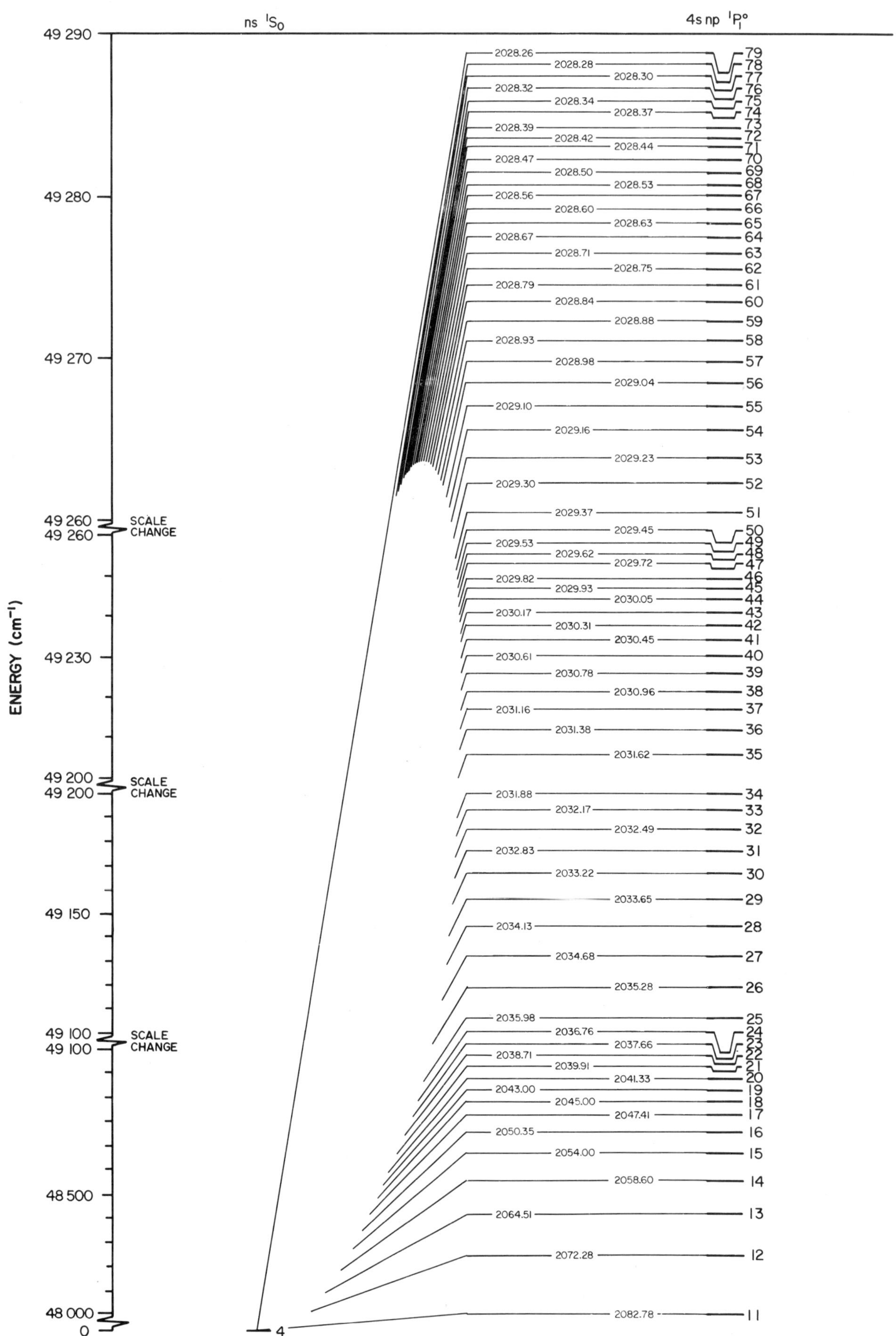

Ca I ENERGY LEVELS (20 electrons, Z=20)

(Configuration: $1s^2\ 2s^2\ 2p^6\ 3s^2\ 3p^6\ 4s\ nl$, Singlet, 11≤n≤79)

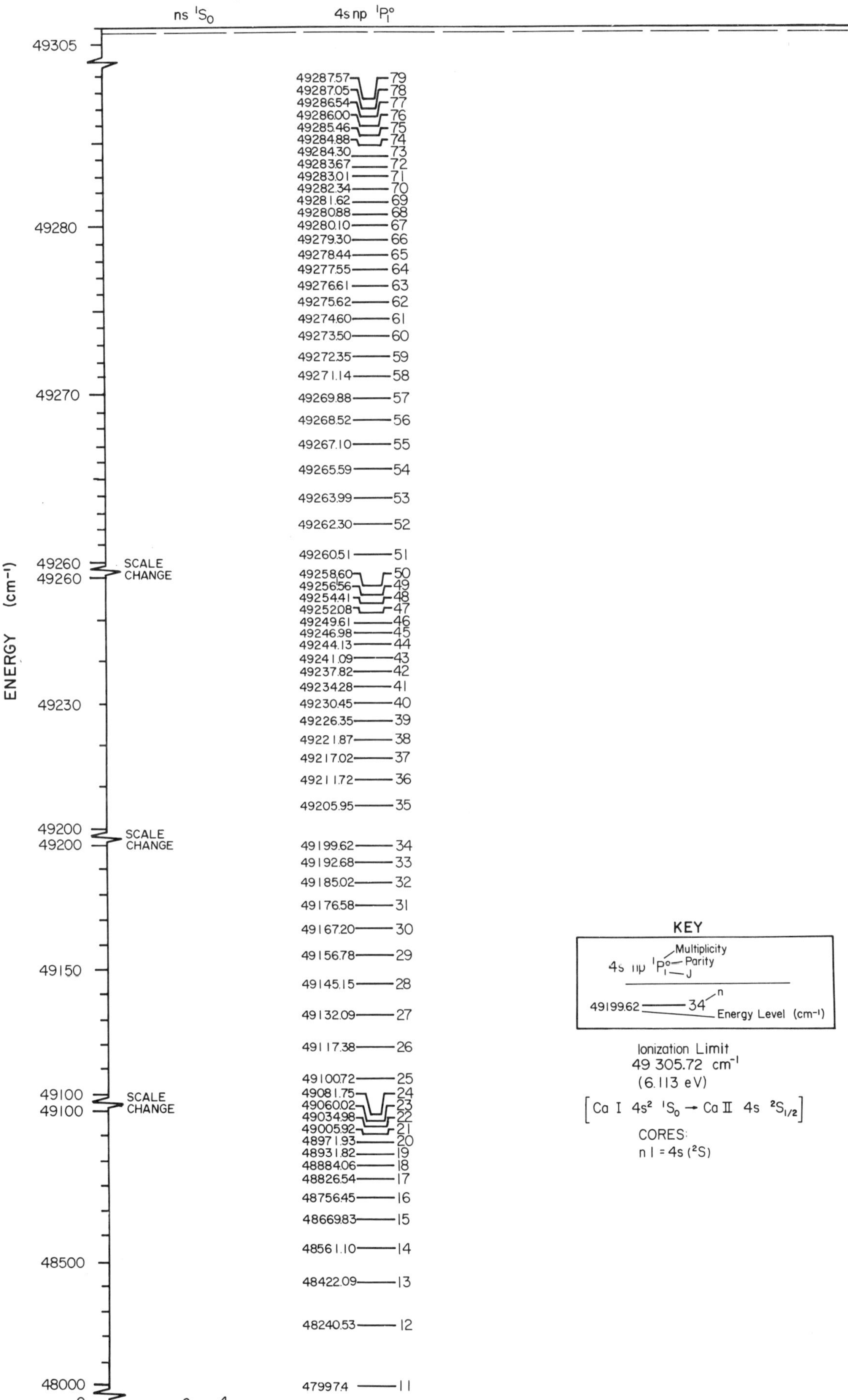

Ca I
SINGLET
Ground to
AUTOIONIZING
GROTRIAN DIAGRAM

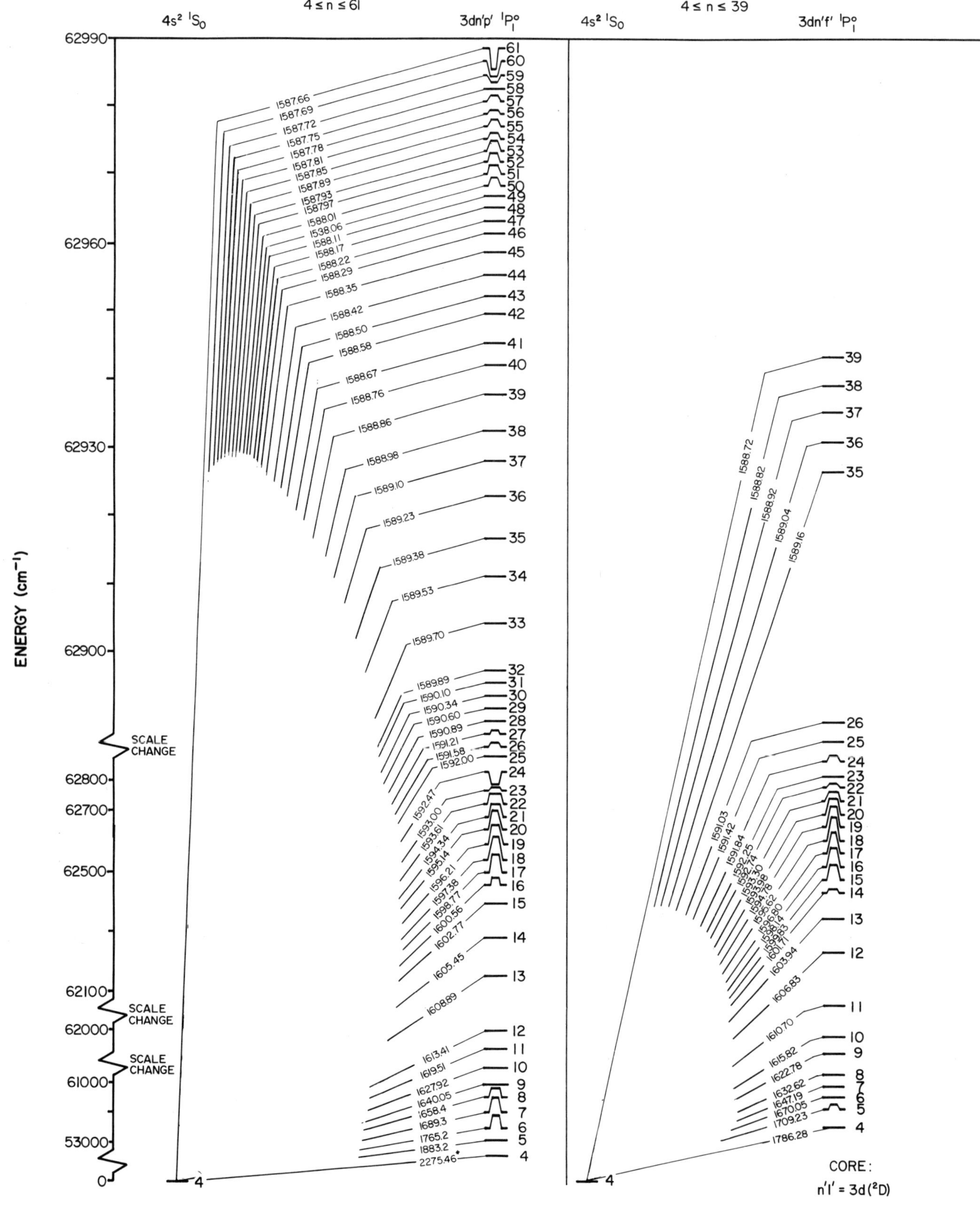

* air wavelength

Ca I ENERGY LEVELS (20 electrons, Z=20)

(Configuration: $1s^2 2s^2 2p^6 3s^2 3p^6 3dn'l'$, Autoionizing Levels, Singlet System)

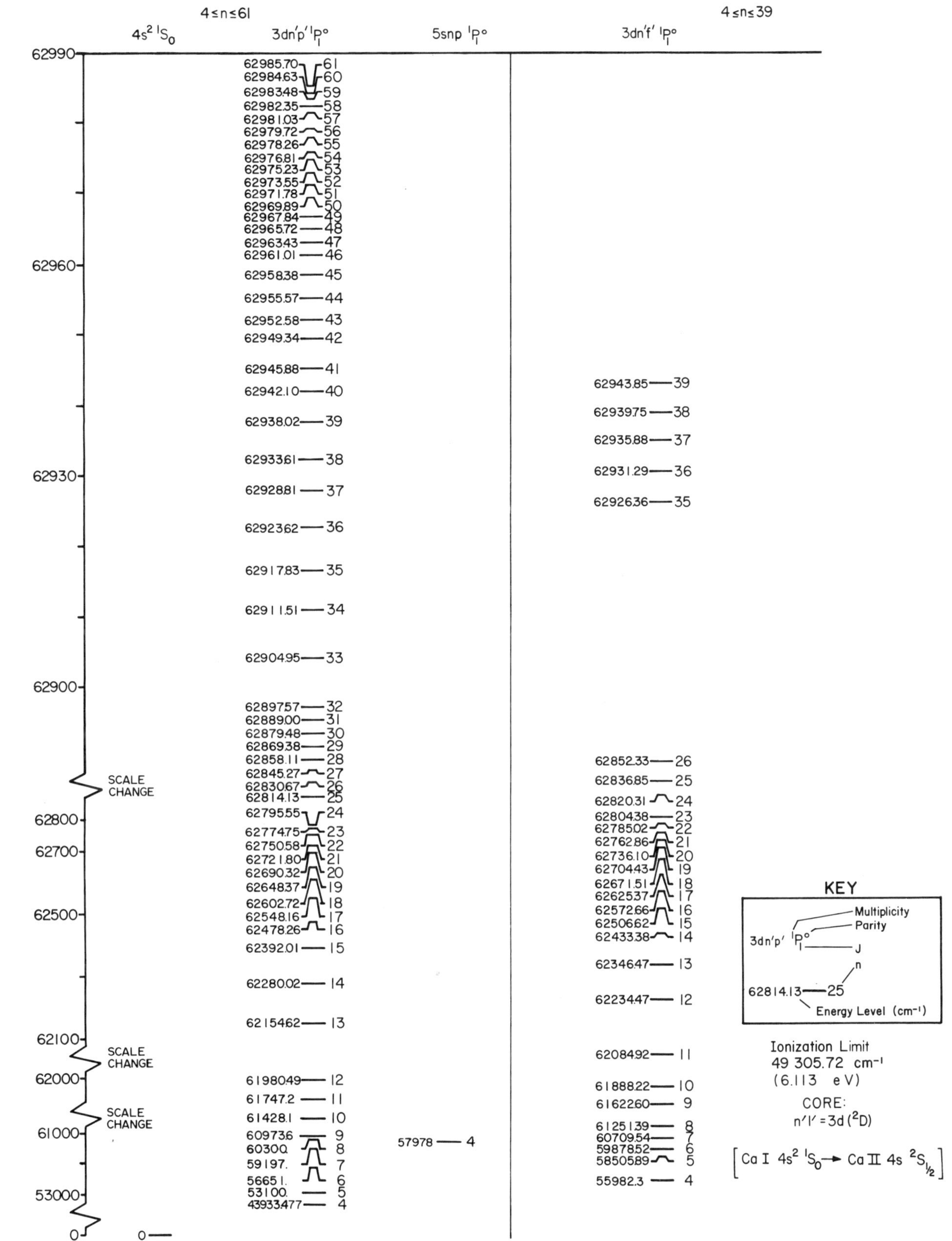

Ca I ENERGY LEVELS (20 electrons, Z = 20)

(Configuration: $1s^2 2s^2 2p^6 3s^2 3p^6 4s\,nl, 3d\,nl$, ^{1}S and ^{1}D levels above 46 800 cm^{-1})

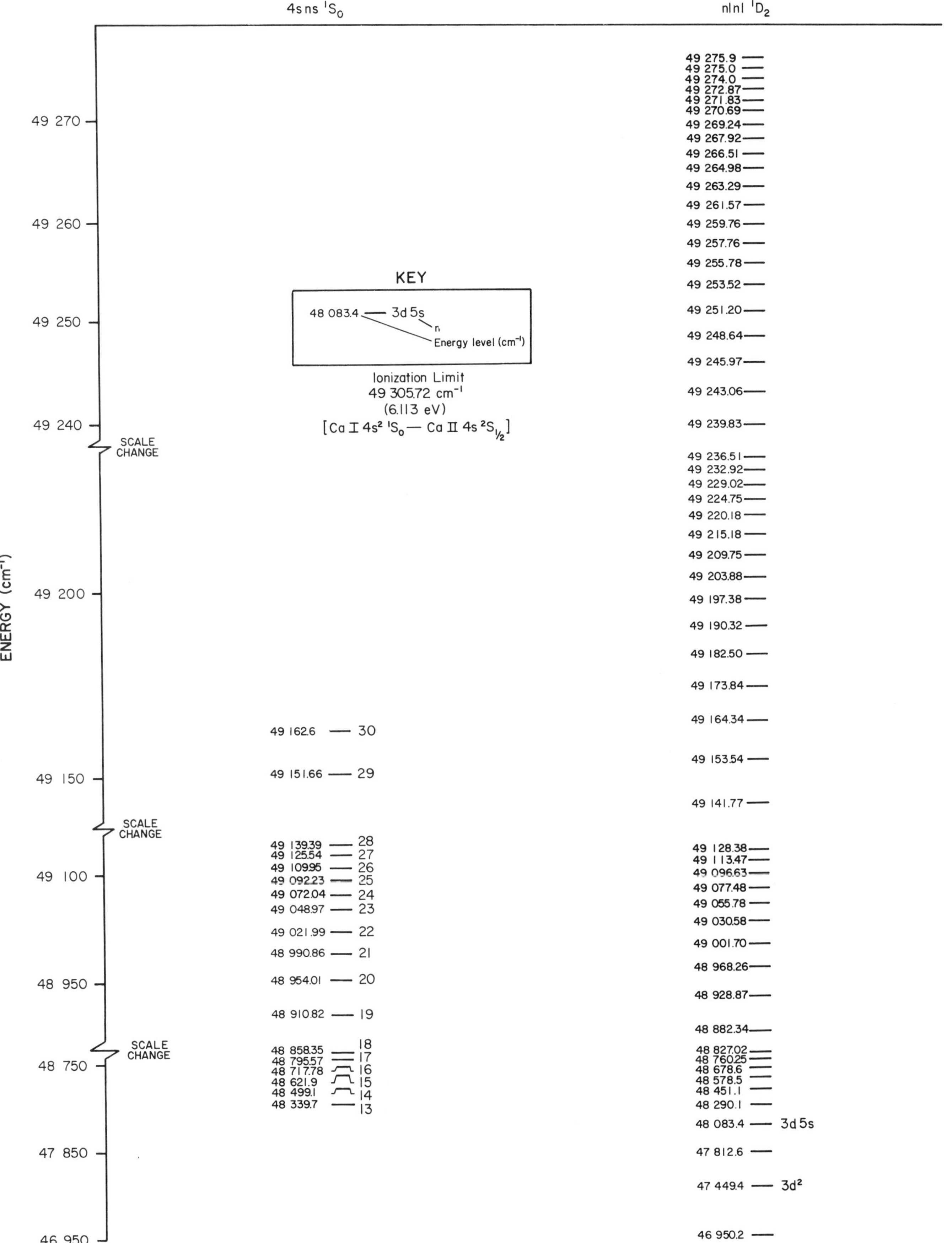

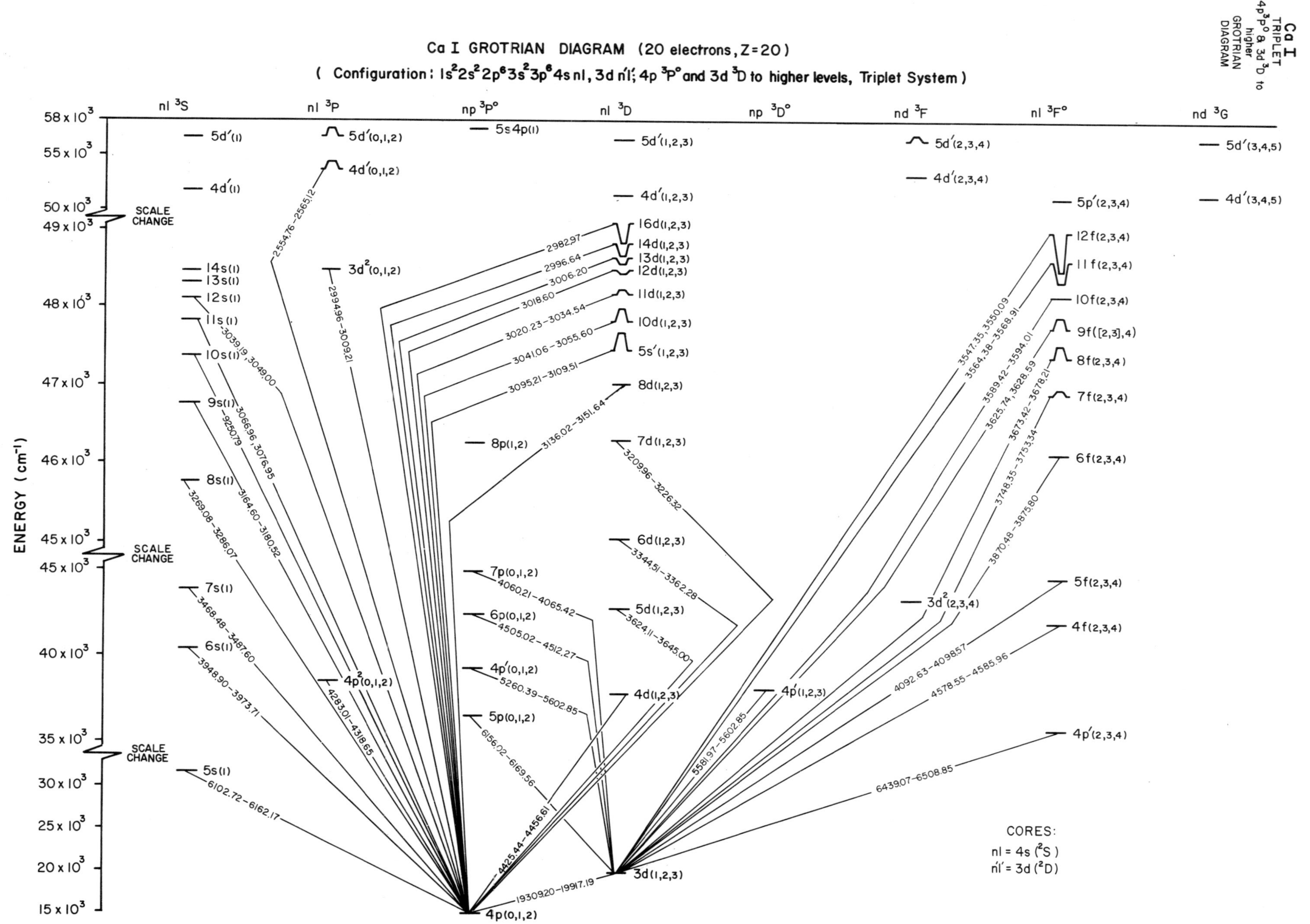

Ca I
TRIPLET
4p ³P° & 3d ³D to higher
GROTRIAN DIAGRAM
Ca I GROTRIAN DIAGRAM (20 electrons, Z=20)
(Configuration: 1s²2s²2p⁶3s²3p⁶4s nl, 3d n'l'; 4p ³P° and 3d ³D to higher levels, Triplet System)
nl ³S
nl ³P
np ³P°
nl ³D
np ³D°
nd ³F
nl ³F°
nd ³G
ENERGY (cm⁻¹)
58 x 10³
55 x 10³
50 x 10³
49 x 10³
48 x 10³
47 x 10³
46 x 10³
45 x 10³
45 x 10³
40 x 10³
35 x 10³
30 x 10³
25 x 10³
20 x 10³
15 x 10³
SCALE CHANGE
SCALE CHANGE
SCALE CHANGE
5d'(1)
4d'(1)
14s(1)
13s(1)
12s(1)
11s(1)
10s(1)
9s(1)
8s(1)
7s(1)
6s(1)
5s(1)
5d'(0,1,2)
4d'(0,1,2)
3d²(0,1,2)
4p²(0,1,2)
5s4p(1)
8p(1,2)
7p(0,1,2)
6p(0,1,2)
4p'(0,1,2)
5p(0,1,2)
4p(0,1,2)
5d'(1,2,3)
4d'(1,2,3)
16d(1,2,3)
14d(1,2,3)
13d(1,2,3)
12d(1,2,3)
11d(1,2,3)
10d(1,2,3)
5s'(1,2,3)
8d(1,2,3)
7d(1,2,3)
6d(1,2,3)
5d(1,2,3)
4d(1,2,3)
3d(1,2,3)
4p'(1,2,3)
5d'(2,3,4)
4d'(2,3,4)
3d²(2,3,4)
5p'(2,3,4)
12f(2,3,4)
11f(2,3,4)
10f(2,3,4)
9f([2,3],4)
8f(2,3,4)
7f(2,3,4)
6f(2,3,4)
5f(2,3,4)
4f(2,3,4)
4p'(2,3,4)
5d'(3,4,5)
4d'(3,4,5)
2554.76–2565.12
2994.96–3009.21
3039.19, 3049.00
3066.96, 3076.95
9250.79
3164.60–3180.52
3269.08–3286.07
3468.48–3487.60
3948.90–3973.71
6102.72–6162.17
4283.01–4318.65
2982.97
2996.64
3006.20
3018.60
3020.23–3034.54
3041.06–3055.60
3095.21–3109.51
3136.02–3151.64
3209.96–3226.32
3344.51–3362.28
3624.11–3645.00
4060.21–4065.42
4505.02–4512.27
5260.39–5602.85
6156.02–6169.56
4425.44–4456.61
19309.20–19917.19
5581.97–5602.85
3547.35, 3550.09
3564.38–3568.91
3589.42–3594.01
3625.74, 3628.59
3673.42–3678.21
3748.35–3753.34
3870.48–3875.80
4092.63–4098.57
4578.55–4585.96
6439.07–6508.85
CORES:
nl = 4s (²S)
n'l' = 3d (²D)

Ca I
TRIPLET

Ca I ENERGY LEVELS (20 electrons, Z=20)

(Configuration: $1s^2\ 2s^2\ 2p^6\ 3s^2\ 3p^6\ 4s\ nl,\ 3d\ n'l'$, Triplet System)

ENERGY (cm^{-1})

Axis: 58×10^3, 55×10^3, 50×10^3, SCALE CHANGE, 49×10^3, 48×10^3, 47×10^3, 46×10^3, 45×10^3, SCALE CHANGE, 45×10^3, 40×10^3, 35×10^3, SCALE CHANGE, 30×10^3, 25×10^3, 20×10^3, 15×10^3

nl ^3S

- 56558.8 — 5d′(1)
- 51571.4 — 4d′(1)
- 48484.7 — 14s(1)
- 48320.4 — 13s(1)
- 48104.02 — 12s(1)
- 47806.20 — 11s(1)
- 47382.048 — 10s(1)
- 46748238 — 9s(1)
- 45738.684 — 8s(1)
- 43980.767 — 7s(1)
- 40474.241 — 6s(1)
- 31539.495 — 5s(1)

nl ^3P

- 57638.2, 57617.8, 57601.0 — 5d′(0,1,2)
- 54304.2, 54288.74, 54282.2 — 4d′(0,1,2)
- 48563.522, 48537.623, 48524.093 — $3d^2$(0,1,2)
- 38551.558, 38464.808, 38417.545 — $4p^2$(0,1,2)

np $^3P°$

- 57465. — 5s 4p(1)
- 46260.77, 46258.38 — 8p(1,2)
- 44961.757, 44957.655, 44955.67 — 7p(0,1,2)
- 42526.591, 42518.708, 42514.845 — 6p(0,1,2)
- 39340.080, 39335.322, 39333.382 — 4p′(0,1,2)
- 36575.119, 36554.749, 36547.688 — 5p(0,1,2)
- 15315.943, 15210.063, 15157.901 — 4p(0,1,2)

nl ^3D

- 56494.7, 56469.1, 56444.8 — 5d′(1,2,3)
- 51396.32, 51369.38, 51351.74 — 4d′(1,2,3)
- 48829.78 — 16d(1,2,3)
- 48676.92 — 14d(1,2,3)
- 48570.88 — 13d(1,2,3)
- 48434.29 — 12d(1,2,3)
- 48260.25, 48258.98, 48258.30 — 11d(1,2,3)
- 48033.23, 48031.58 — 10d(1,2,3)
- 47757.9 — 9d(2)
- 47475.915, 47466.014, 47456.452 — 5s(1,2,3)
- 47045.241, 47040.007, 47036.225 — 8d(1,2,3)
- 46306.059, 46303.649, 46301.973 — 7d(1,2,3)
- 45052.374, 45050.419, 45049.073 — 6d(1,2,3)
- 42747.387, 42744.716, 42743.002 — 5d(1,2,3)
- 37757.449, 37751.867, 37748.197 — 4d(1,2,3)
- 20371.000, 20349.260, 20335.360 — 3d(1,2,3)

np $^3D°$

- 38259.124, 38219.118, 38192.392 — 4p′(1,2,3)

nd ^3F

- 56979.5, 56924.1, 56900.7 — 5d′(2,3,4)
- 53260.4, 53247.9, 53214.6 — 4d′(2,3,4)
- 43508.088, 43489.119, 43474.827 — $3d^2$(2,3,4)

nl $^3F°$

- 51318.7, 51259.5, 51235.2 — 5p′(2,3,4)
- 48531.29 — 12f(2,3,4)
- 48382.801, 48382.781, 48382.701 — 11f(2,3,4)
- 48187.118, 48187.075, 48187.045 — 10f(2,3,4)
- 47922.033, 47921.981 — 9f([2,3],4)
- 47550.371, 47550.271, 47550.214 — 8f(2,3,4)
- 47006.400, 47006.280, 47006.194 — 7f(2,3,4)
- 46164.971, 46164.785, 46164.644 — 6f(2,3,4)
- 44763.118, 44762.839, 44762.620 — 5f(2,3,4)
- 42171.026, 42170.558, 42170.214 — 4f(2,3,4)
- 35896.889, 35818.713, 35730.454 — 4p′(2,3,4)

nd ^3G

- 56578.2, 56546.6, 56526.3 — 5d′(3,4,5)
- 51611.5, 51579.0, 51553.6 — 4d′(3,4,5)

KEY

42171.026, 42170.558, 42170.214 — 4f (2,3,4)

- J, lowest to highest
- n
- Energy Level (cm^{-1})

Ionization Limit
49 305.72 cm^{-1}
(6.113 eV)
[Ca I $4s^2\ {}^1S_0$ → Ca II 4s $^2S_{1/2}$]

CORES:
nl = $4s(^2S)$
n′l′ = $3d(^2D)$

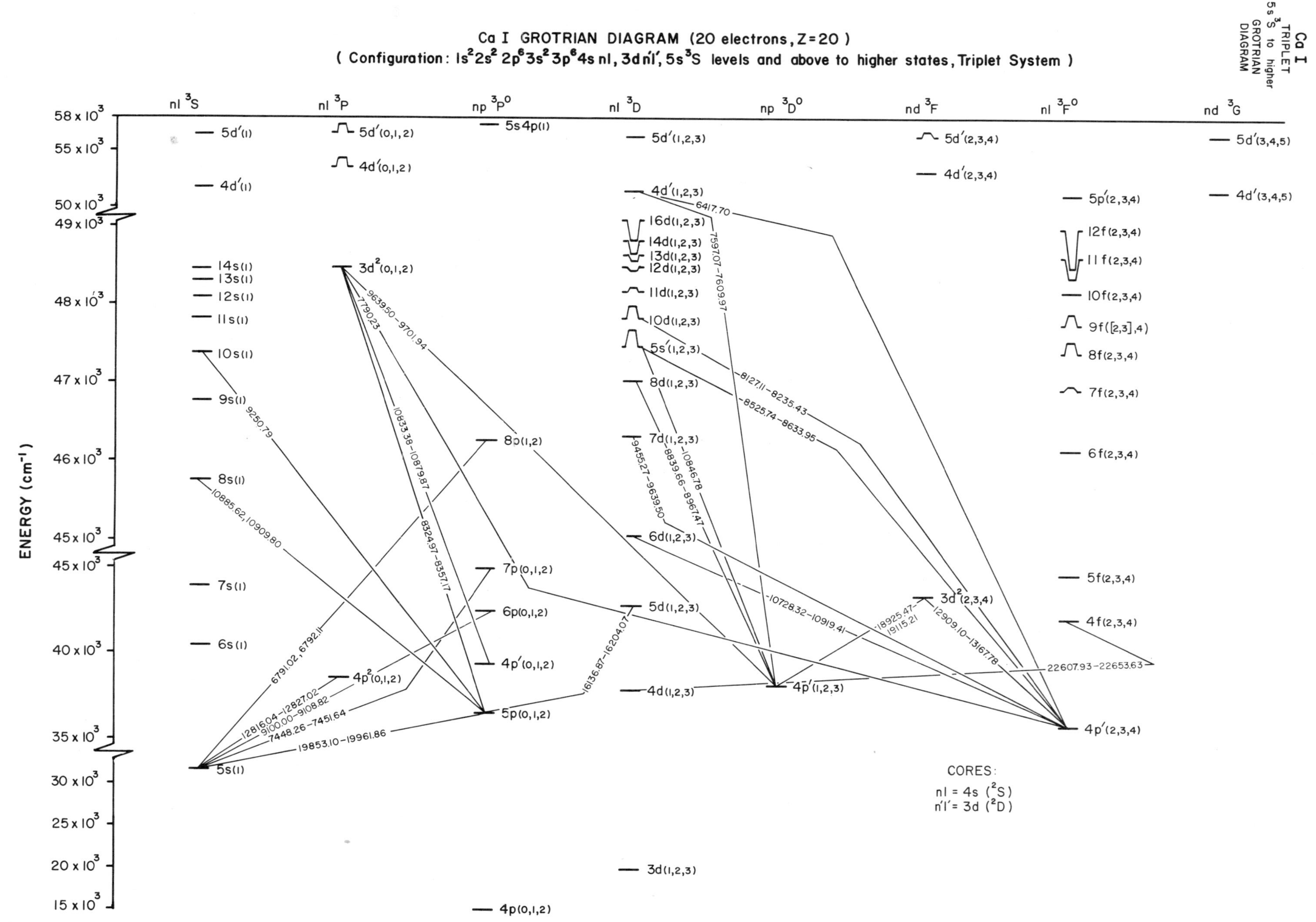

Ca I GROTRIAN DIAGRAM (20 electrons, Z=20)
(Configuration: 1s²2s²2p⁶3s²3p⁶4s nl, 3d n′l′, 5s ³S levels and above to higher states, Triplet System)
Ca I
TRIPLET
5s ³S to higher
GROTRIAN
DIAGRAM
nl ³S
nl ³P
np ³P°
nl ³D
np ³D°
nd ³F
nl ³F°
nd ³G
ENERGY (cm⁻¹)
58 x 10³
55 x 10³
50 x 10³
49 x 10³
48 x 10³
47 x 10³
46 x 10³
45 x 10³
45 x 10³
40 x 10³
35 x 10³
30 x 10³
25 x 10³
20 x 10³
15 x 10³
5d′(1)
4d′(1)
14s(1)
13s(1)
12s(1)
11s(1)
10s(1)
9s(1)
8s(1)
7s(1)
6s(1)
5s(1)
5d′(0,1,2)
4d′(0,1,2)
3d²(0,1,2)
4p²(0,1,2)
5s4p(1)
8p(1,2)
7p(0,1,2)
6p(0,1,2)
4p′(0,1,2)
5p(0,1,2)
4p(0,1,2)
5d′(1,2,3)
4d′(1,2,3)
16d(1,2,3)
14d(1,2,3)
13d(1,2,3)
12d(1,2,3)
11d(1,2,3)
10d(1,2,3)
5s′(1,2,3)
8d(1,2,3)
7d(1,2,3)
6d(1,2,3)
5d(1,2,3)
4d(1,2,3)
3d(1,2,3)
4p′(1,2,3)
5d′(2,3,4)
4d′(2,3,4)
3d²(2,3,4)
5p′(2,3,4)
12f(2,3,4)
11f(2,3,4)
10f(2,3,4)
9f([2,3],4)
8f(2,3,4)
7f(2,3,4)
6f(2,3,4)
5f(2,3,4)
4f(2,3,4)
4p′(2,3,4)
5d′(3,4,5)
4d′(3,4,5)
9250.79
10885.62, 10909.80
6791.02, 6792.11
12816.04 – 12827.02
9100.00 – 9108.82
7448.26 – 7451.64
19853.10 – 19961.86
9639.50 – 9701.94
7790.23
10833.38 – 10879.87
8324.97 – 8357.17
16136.87 – 16204.07
6417.70
7597.07 – 7609.97
8127.11 – 8235.43
8525.74 – 8633.95
10846.78
8839.66 – 8967.47
9455.27 – 9639.50
10728.32 – 10919.41
18925.47
19115.21
12909.10 – 13167.78
22607.93 – 22653.63
CORES:
nl = 4s (²S)
n′l′ = 3d (²D)

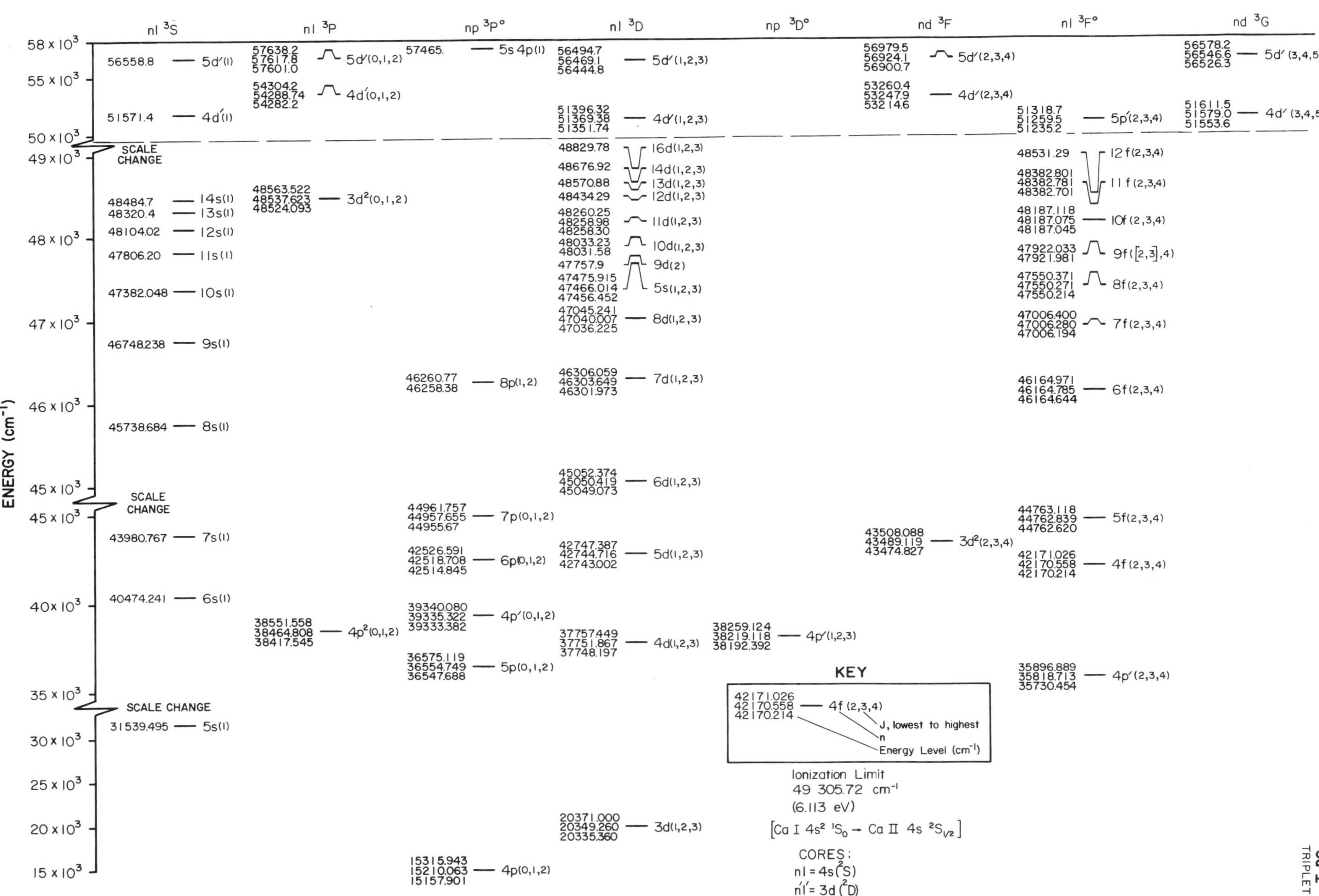
Ca I ENERGY LEVELS (20 electrons, Z=20)
(Configuration: 1s² 2s² 2p⁶ 3s² 3p⁶ 4s nl, 3d n′l′, Triplet System)
ENERGY (cm⁻¹)
58 x 10³
55 x 10³
50 x 10³
49 x 10³
48 x 10³
47 x 10³
46 x 10³
45 x 10³
45 x 10³
40x 10³
35 x 10³
30 x 10³
25 x 10³
20 x 10³
15 x 10³
SCALE CHANGE
SCALE CHANGE
SCALE CHANGE
nl ³S
56558.8 — 5d′(1)
51571.4 — 4d′(1)
48484.7 — 14s(1)
48320.4 — 13s(1)
48104.02 — 12s(1)
47806.20 — 11s(1)
47382.048 — 10s(1)
46748.238 — 9s(1)
45738.684 — 8s(1)
43980.767 — 7s(1)
40474.241 — 6s(1)
31539.495 — 5s(1)
nl ³P
57638.2 57617.8 57601.0 5d′(0,1,2)
54304.2 54288.74 54282.2 4d′(0,1,2)
48563.522 48537.623 48524.093 — 3d²(0,1,2)
38551.558 38464.808 38417.545 — 4p²(0,1,2)
np ³P°
57465. — 5s 4p(1)
46260.77 46258.38 — 8p(1,2)
44961.757 44957.655 44955.67 — 7p(0,1,2)
42526.591 42518.708 42514.845 — 6p(0,1,2)
39340.080 39335.322 39333.382 — 4p′(0,1,2)
36575.119 36554.749 36547.688 — 5p(0,1,2)
15315.943 15210.063 15157.901 — 4p(0,1,2)
nl ³D
56494.7 56469.1 56444.8 — 5d′(1,2,3)
51396.32 51369.38 51351.74 — 4d′(1,2,3)
48829.78 16d(1,2,3)
48676.92 14d(1,2,3)
48570.88 13d(1,2,3)
48434.29 12d(1,2,3)
48260.25 48258.98 48258.30 11d(1,2,3)
48033.23 48031.58 10d(1,2,3)
47757.9 9d(2)
47475.915 47466.014 47456.452 5s(1,2,3)
47045.241 47040.007 47036.225 — 8d(1,2,3)
46306.059 46303.649 46301.973 — 7d(1,2,3)
45052.374 45050.419 45049.073 — 6d(1,2,3)
42747.387 42744.716 42743.002 — 5d(1,2,3)
37757.449 37751.867 37748.197 — 4d(1,2,3)
20371.000 20349.260 20335.360 — 3d(1,2,3)
np ³D°
38259.124 38219.118 38192.392 — 4p′(1,2,3)
nd ³F
56979.5 56924.1 56900.7 5d′(2,3,4)
53260.4 53247.9 53214.6 — 4d′(2,3,4)
43508.088 43489.119 43474.827 — 3d²(2,3,4)
nl ³F°
51318.7 51299.5 51235.2 — 5p′(2,3,4)
48531.29 12f(2,3,4)
48382.801 48382.781 48382.701 11f(2,3,4)
48187.118 48187.075 48187.045 — 10f(2,3,4)
47922.033 47921.981 9f([2,3],4)
47550.371 47550.271 47550.214 8f(2,3,4)
47006.400 47006.280 47006.194 7f(2,3,4)
46164.971 46164.785 46164.644 — 6f(2,3,4)
44763.118 44762.839 44762.620 — 5f(2,3,4)
42171.026 42170.558 42170.214 — 4f(2,3,4)
35896.889 35818.713 35730.454 — 4p′(2,3,4)
nd ³G
56578.2 56546.6 56526.3 — 5d′(3,4,5)
51611.5 51579.0 51553.6 — 4d′(3,4,5)
KEY
42171.026 42170.558 42170.214 — 4f (2,3,4)
J, lowest to highest
n
Energy Level (cm⁻¹)
Ionization Limit
49 305.72 cm⁻¹
(6.113 eV)
[Ca I 4s² ¹S₀ → Ca II 4s ²S₁/₂]
CORES:
nl = 4s(²S)
n′l′ = 3d (²D)

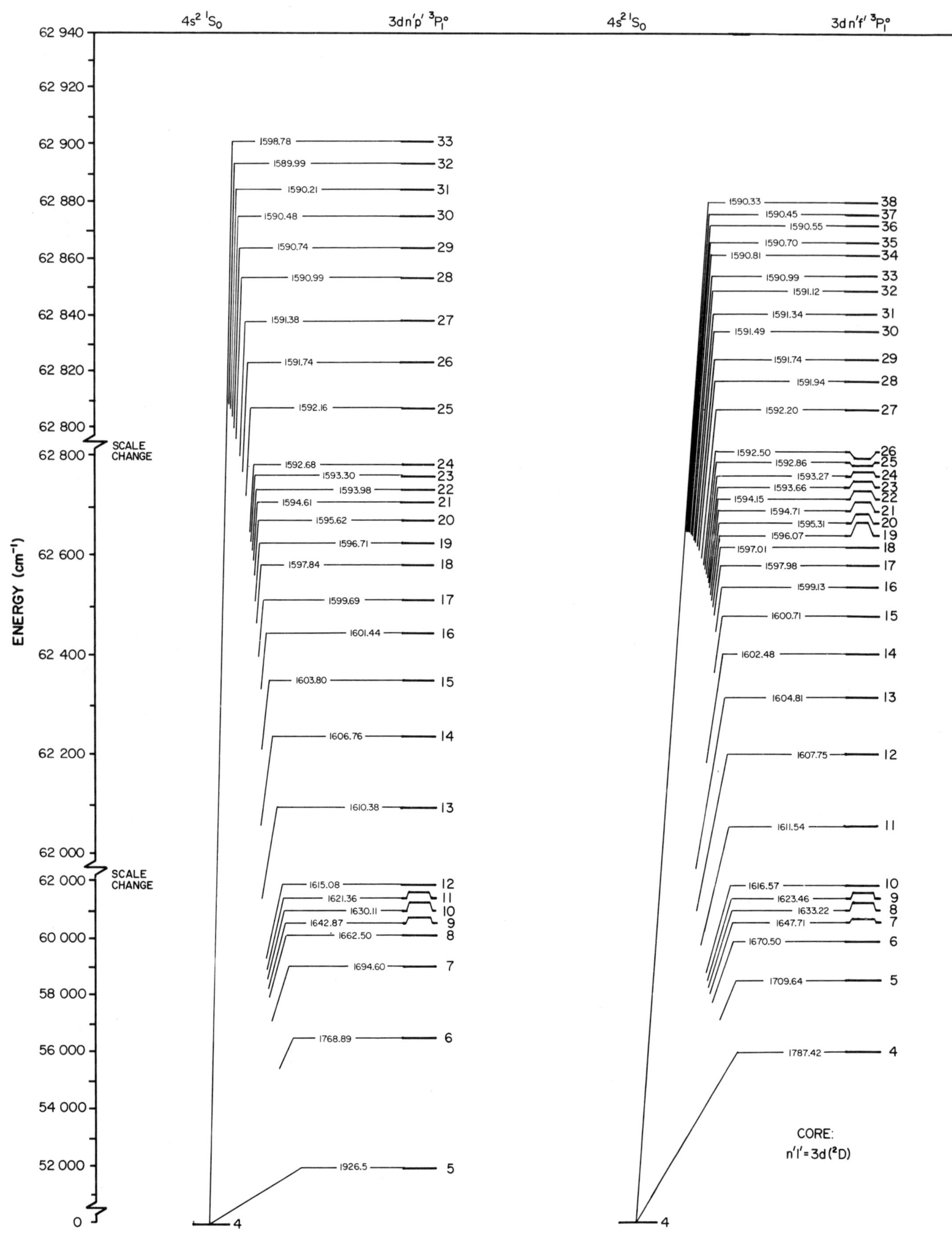
Ca I
GROUND to
AUTOIONIZING
$^3P^o$ TERMS
GROTRIAN DIAGRAM
Ca I GROTRIAN DIAGRAM (20 electrons, Z = 20)
(Configuration: $1s^22s^22p^63s^23p^63d\,n'l'$, Ground to Autoionizing $^3P^o$ terms)
$4s^2\ ^1S_0$
$3dn'p'\ ^3P_1^o$
$4s^2\ ^1S_0$
$3dn'f'\ ^3P_1^o$
ENERGY (cm^{-1})
SCALE CHANGE
SCALE CHANGE
CORE:
$n'l' = 3d(^2D)$

Ca I ENERGY LEVELS (20 electrons, Z = 20)

(Configuration: $1s^2 2s^2 2p^6 3s^2 3p^6 3d\,n'l'$, Triplet System, Autoionizing Levels)

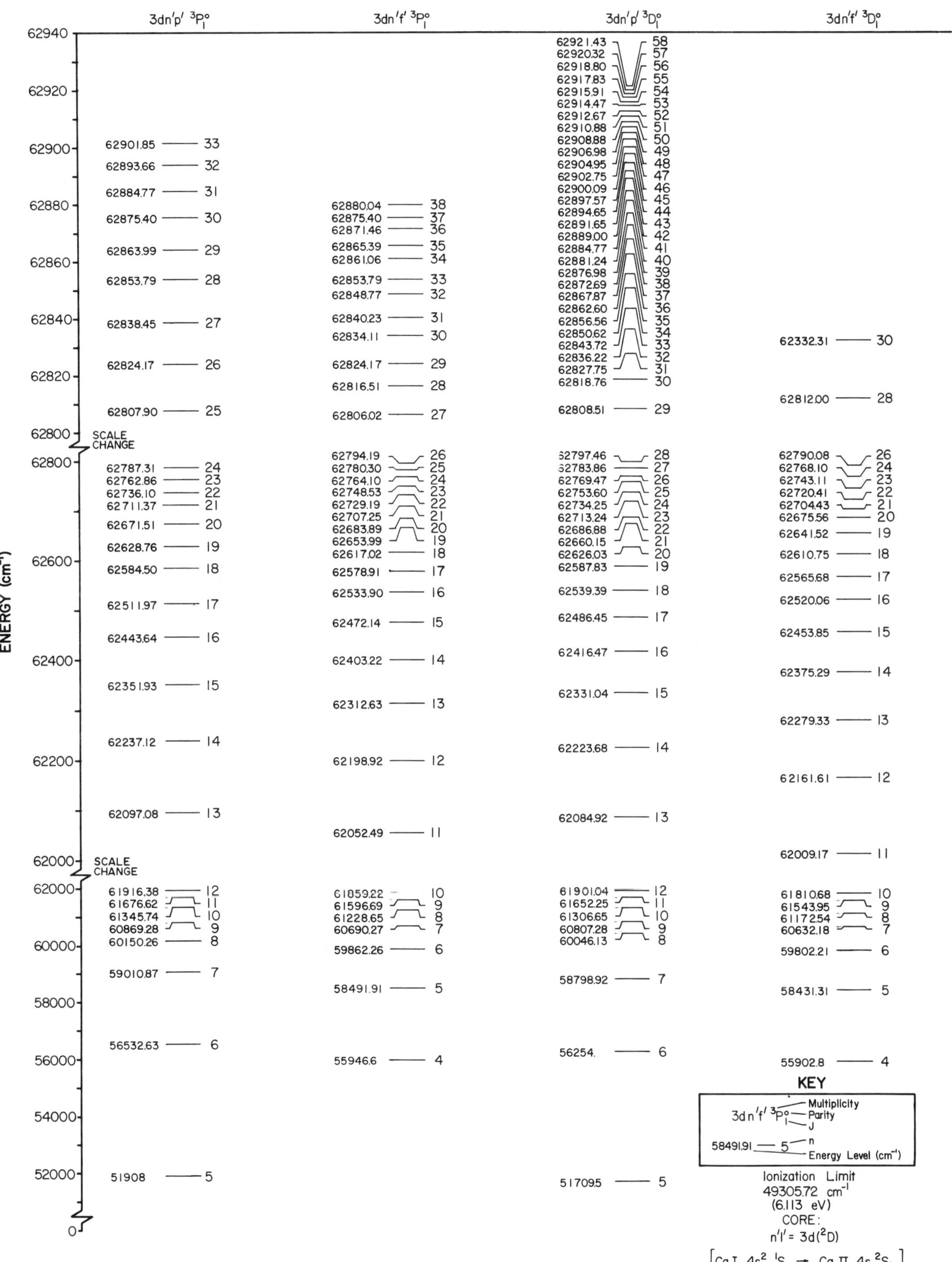

Ca I GROTRIAN DIAGRAM (20 electrons, Z = 20)

(Configuration: $1s^2 2s^2 2p^6 3s^2 3p^6 3d\ n'l'$, Ground to Autoionizing $^3D^o$ terms)

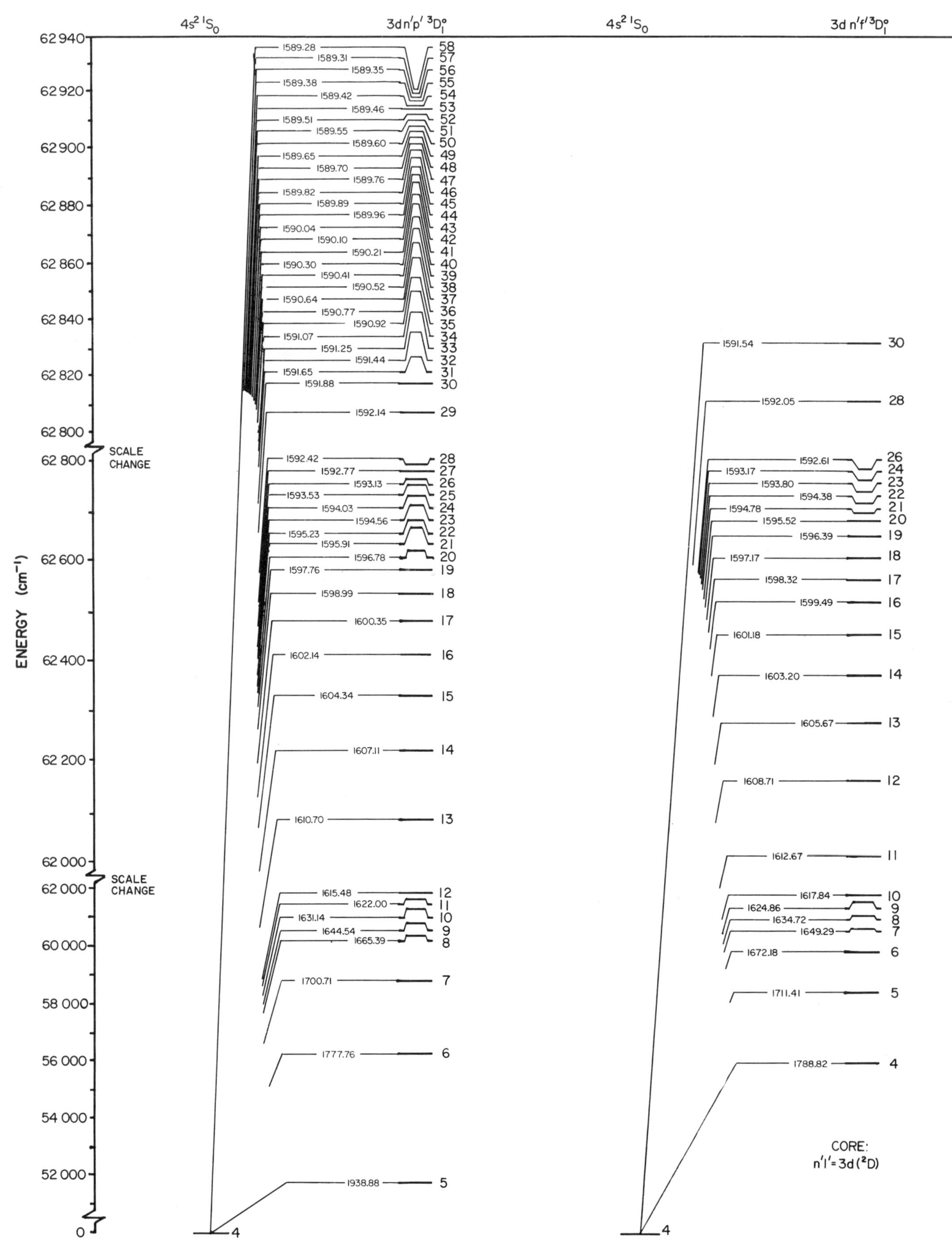

Ca I ENERGY LEVELS (20 electrons, Z = 20)

(Configuration: $1s^2 2s^2 2p^6 3s^2 3p^6 3d\,n'l'$, Triplet System, Autoionizing Levels)

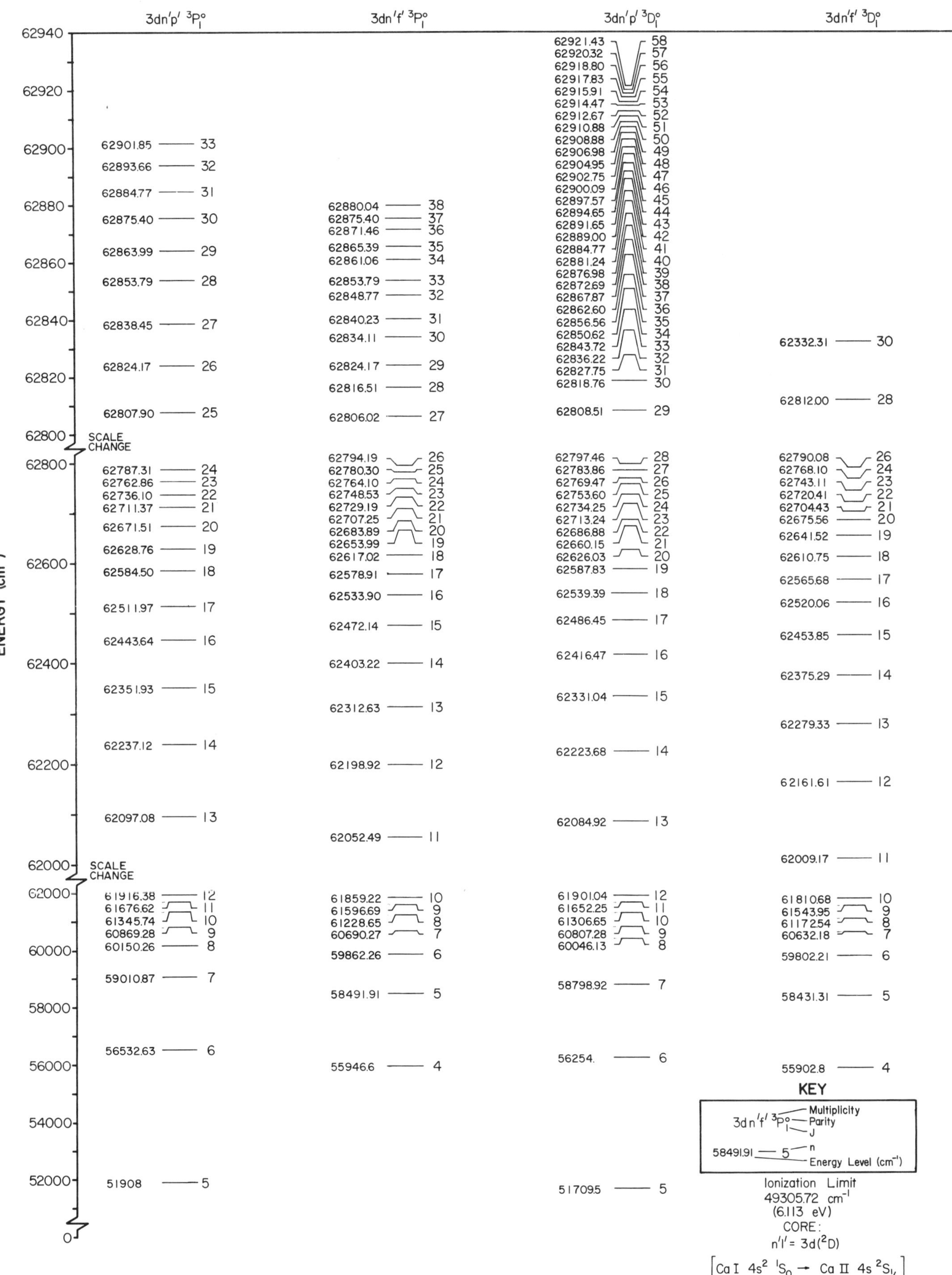

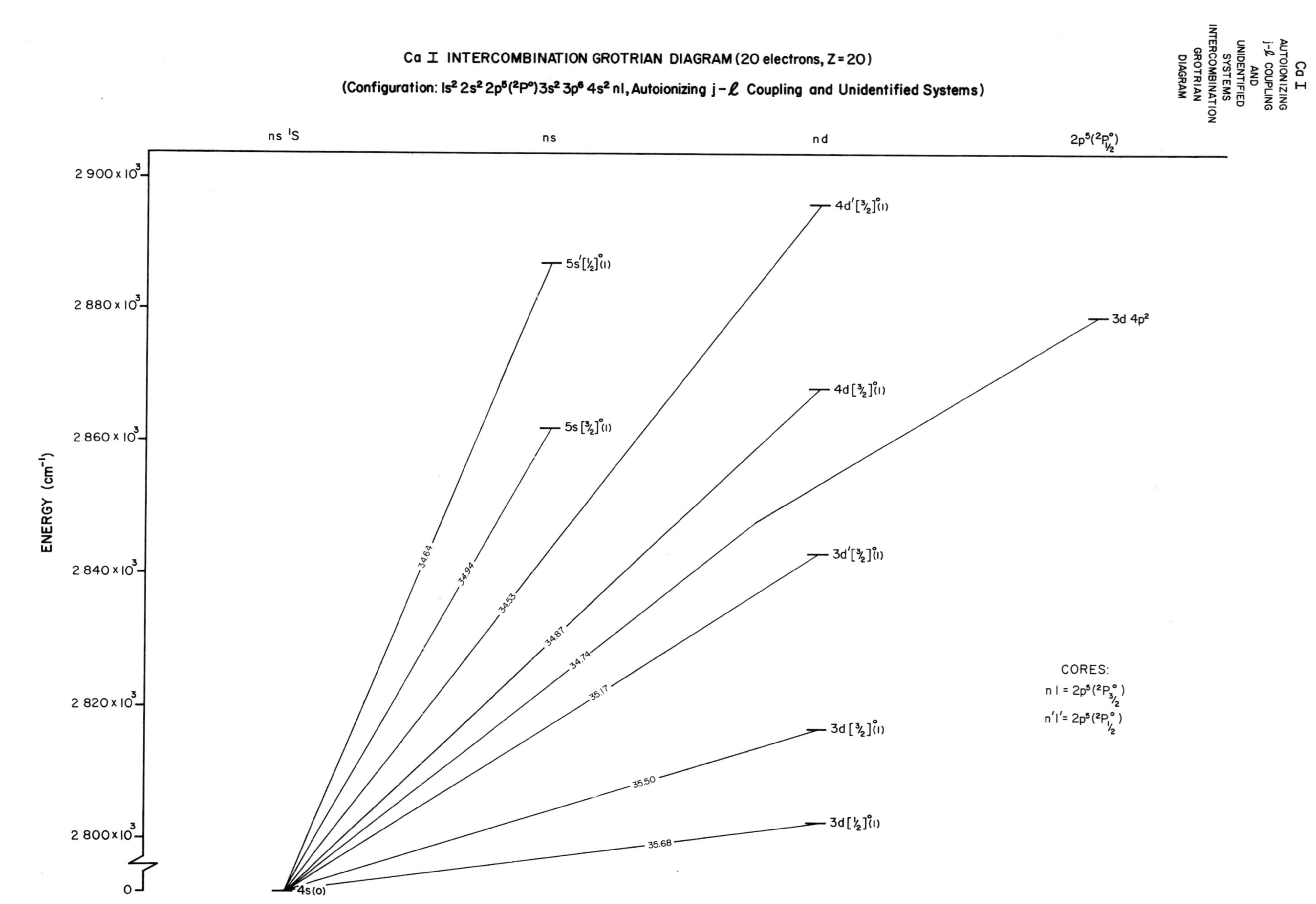
Ca I
AUTOIONIZING
j-ℓ COUPLING
AND
UNIDENTIFIED
SYSTEMS
INTERCOMBINATION
GROTRIAN
DIAGRAM
Ca I INTERCOMBINATION GROTRIAN DIAGRAM (20 electrons, Z=20)
(Configuration: $1s^2 2s^2 2p^5(^2P^o) 3s^2 3p^6 4s^2$ nl, Autoionizing j-ℓ Coupling and Unidentified Systems)
ns ^1S
ns
nd
$2p^5(^2P^o_{1/2})$
$4d'[3/2]^o(1)$
$5s'[1/2]^o(1)$
$3d\,4p^2$
$4d[3/2]^o(1)$
$5s[3/2]^o(1)$
$3d'[3/2]^o(1)$
$3d[3/2]^o(1)$
$3d[1/2]^o(1)$
4s(0)
34.64
34.94
34.53
34.87
34.74
35.17
35.50
35.68
CORES:
$nl = 2p^5(^2P^o_{3/2})$
$n'l' = 2p^5(^2P^o_{1/2})$
ENERGY (cm^{-1})
$2\,900 \times 10^3$
$2\,880 \times 10^3$
$2\,860 \times 10^3$
$2\,840 \times 10^3$
$2\,820 \times 10^3$
$2\,800 \times 10^3$
0

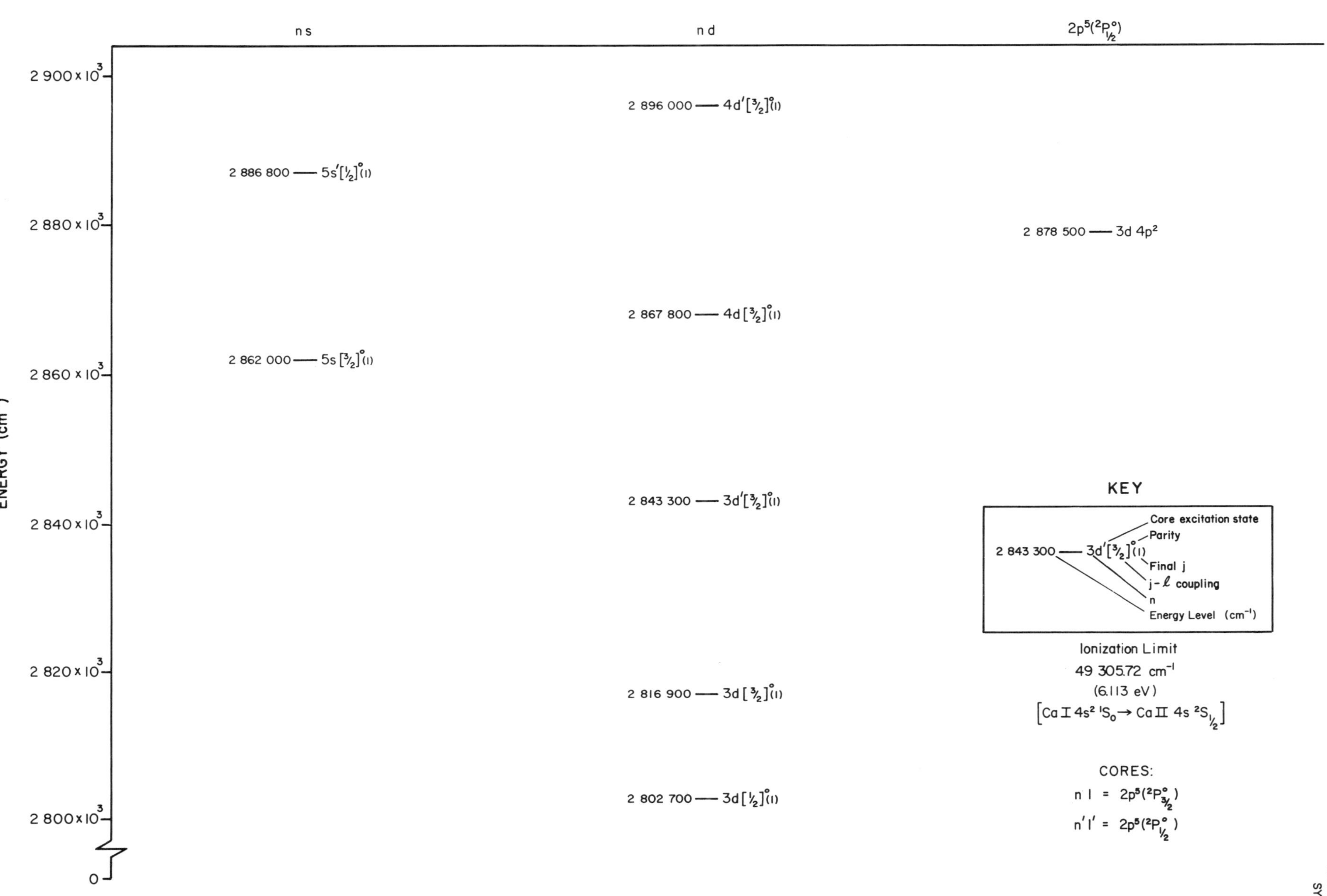
Ca I ENERGY LEVELS (20 electrons, Z = 20)
(Configuration: $1s^2 2s^2 2p^5(^2P^o) 3s^2 3p^6 4s^2$ nl, Autoionizing j - ℓ Coupling and Unidentified Systems)
ns
nd
$2p^5(^2P^o_{1/2})$
2 896 000 —— $4d'[3/2]^o(1)$
2 886 800 —— $5s'[1/2]^o(1)$
2 878 500 —— $3d\,4p^2$
2 867 800 —— $4d[3/2]^o(1)$
2 862 000 —— $5s[3/2]^o(1)$
2 843 300 —— $3d'[3/2]^o(1)$
2 816 900 —— $3d[3/2]^o(1)$
2 802 700 —— $3d[1/2]^o(1)$
KEY
2 843 300 —— $3d'[3/2]^o(1)$
Core excitation state
Parity
Final j
j - ℓ coupling
n
Energy Level (cm^{-1})
Ionization Limit
49 305.72 cm^{-1}
(6.113 eV)
[Ca I $4s^2\ ^1S_0$ → Ca II $4s\ ^2S_{1/2}$]
CORES:
n l = $2p^5(^2P^o_{3/2})$
n' l' = $2p^5(^2P^o_{1/2})$
2 900 x 10^3
2 880 x 10^3
2 860 x 10^3
2 840 x 10^3
2 820 x 10^3
2 800 x 10^3
0
ENERGY (cm^{-1})

Ca I INTERCOMBINATION GROTRIAN DIAGRAM (20 electrons, Z = 20)

(Configuration: $1s^2 2s^2 2p^6 3s^2 3p^6 4s\,nl$, Singlet & Triplet Systems)

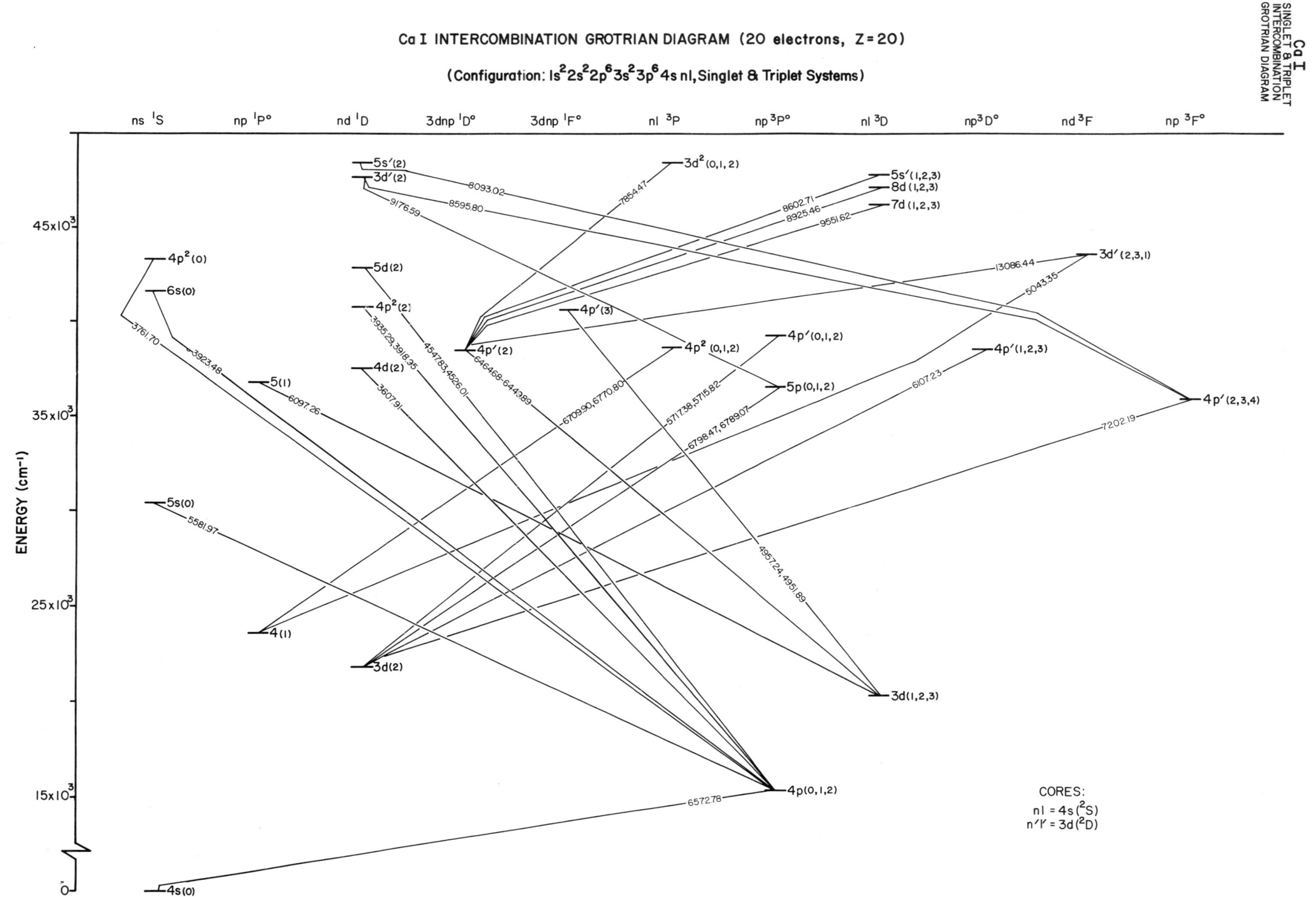

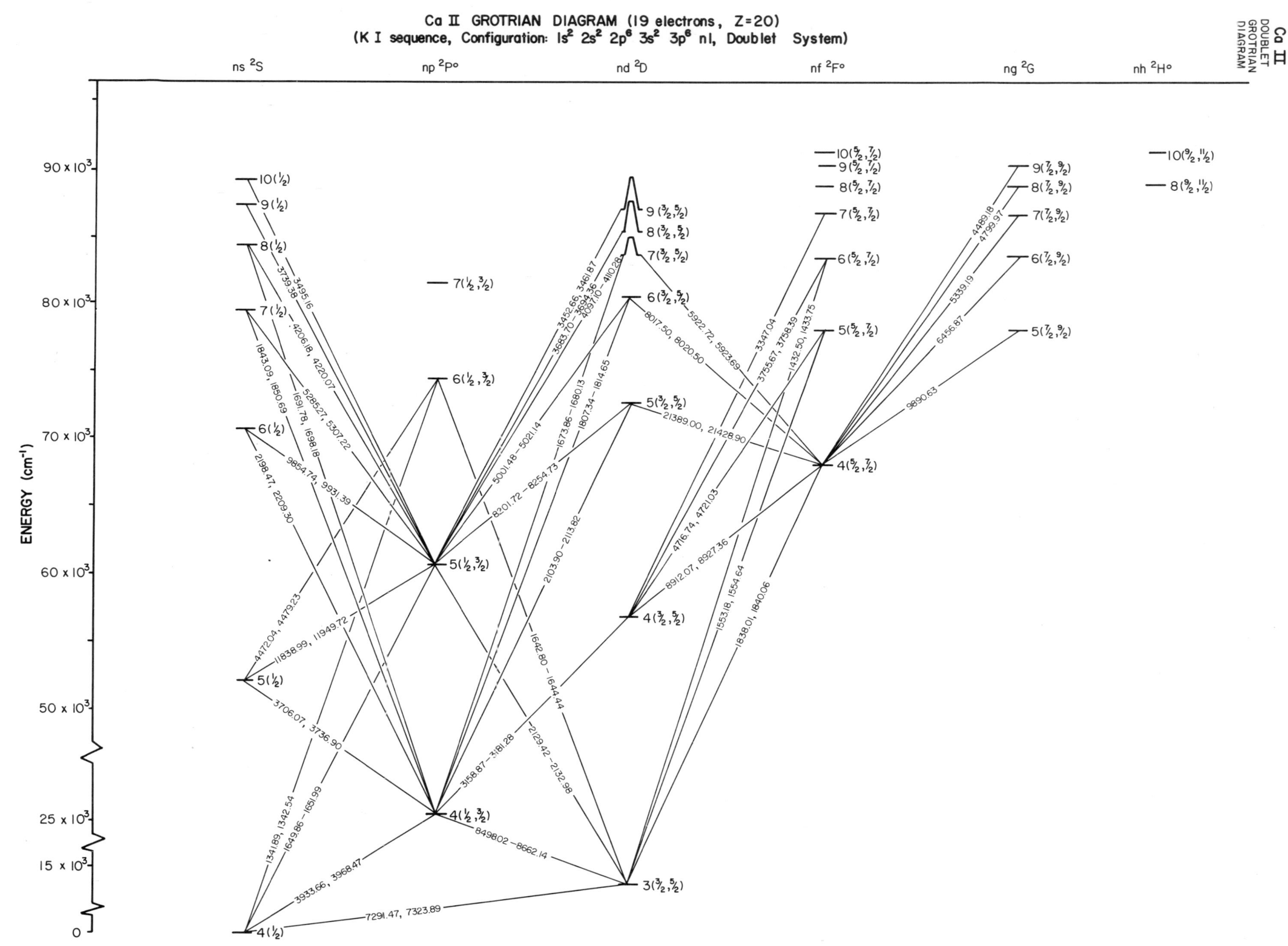

Ca II
DOUBLET
GROTRIAN
DIAGRAM
Ca II GROTRIAN DIAGRAM (19 electrons, Z=20)
(K I sequence, Configuration: $1s^2\ 2s^2\ 2p^6\ 3s^2\ 3p^6\ nl$, Doublet System)
ns ^2S
np $^2P^o$
nd ^2D
nf $^2F^o$
ng ^2G
nh $^2H^o$
ENERGY (cm^{-1})
90 x 10^3
80 x 10^3
70 x 10^3
60 x 10^3
50 x 10^3
25 x 10^3
15 x 10^3
0
10(1/2)
9(1/2)
8(1/2)
7(1/2)
6(1/2)
5(1/2)
4(1/2)
7(1/2, 3/2)
6(1/2, 3/2)
5(1/2, 3/2)
4(1/2, 3/2)
9(3/2, 5/2)
8(3/2, 5/2)
7(3/2, 5/2)
6(3/2, 5/2)
5(3/2, 5/2)
4(3/2, 5/2)
3(3/2, 5/2)
10(5/2, 7/2)
9(5/2, 7/2)
8(5/2, 7/2)
7(5/2, 7/2)
6(5/2, 7/2)
5(5/2, 7/2)
4(5/2, 7/2)
9(7/2, 9/2)
8(7/2, 9/2)
7(7/2, 9/2)
6(7/2, 9/2)
5(7/2, 9/2)
10(9/2, 11/2)
8(9/2, 11/2)
3495.16
3739.38
4206.18, 4220.07
5285.27, 5307.22
1843.09, 1850.69
1691.78, 1698.18
2198.47, 2209.30
9854.74, 9931.39
4472.04, 4479.23
11838.99, 11949.72
3706.07, 3736.90
1341.89, 1342.54
1649.86 – 1651.99
3933.66, 3968.47
7291.47, 7323.89
8498.02 – 8662.14
3158.87 – 3181.28
2129.42 – 2132.98
1642.80 – 1644.44
2103.90 – 2113.82
8201.72 – 8254.73
5001.48 – 5021.14
1673.86 – 1680.13
1807.34 – 1814.65
3452.66, 3461.87
3683.70 – 3694.36
4097.10 – 4110.28
8017.50, 8020.50
5922.72, 5923.69
21389.00, 21428.90
3347.04
3755.67, 3758.39
1432.50, 1433.75
4716.74, 4721.03
8912.07, 8927.36
1553.18, 1554.64
1838.01, 1840.06
9890.63
6456.87
5339.19
4799.97
4489.18

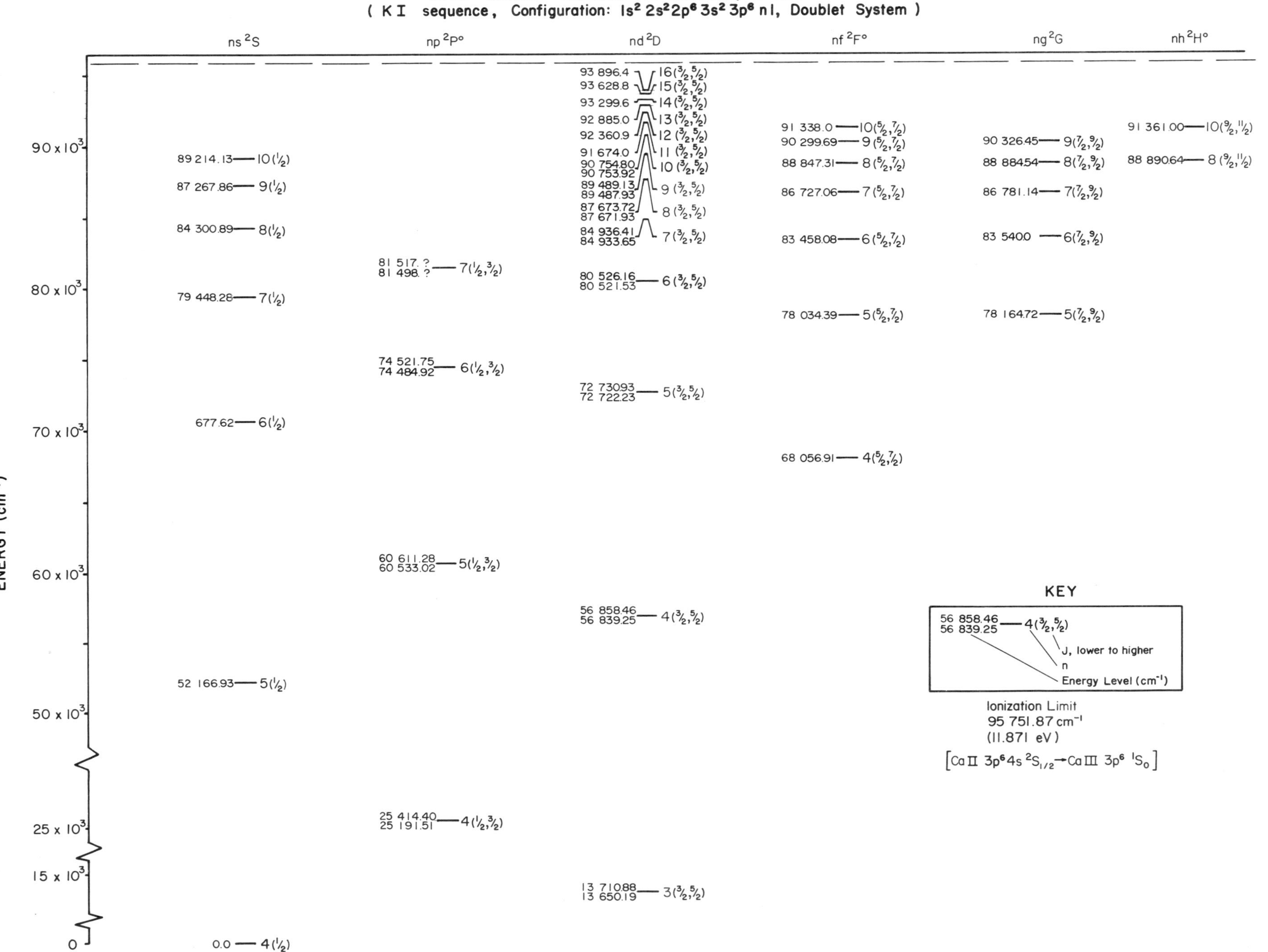
Ca II ENERGY LEVELS (19 electrons, Z=20)
(K I sequence, Configuration: 1s² 2s²2p⁶ 3s² 3p⁶ nl, Doublet System)
ns ²S
np ²P°
nd ²D
nf ²F°
ng ²G
nh ²H°
ENERGY (cm⁻¹)
90 x 10³
80 x 10³
70 x 10³
60 x 10³
50 x 10³
25 x 10³
15 x 10³
0
89 214.13 — 10(½)
87 267.86 — 9(½)
84 300.89 — 8(½)
79 448.28 — 7(½)
677.62 — 6(½)
52 166.93 — 5(½)
0.0 — 4(½)
81 517. ? 81 498. ? — 7(½,3/2)
74 521.75 74 484.92 — 6(½,3/2)
60 611.28 60 533.02 — 5(½,3/2)
25 414.40 25 191.51 — 4(½,3/2)
93 896.4 — 16(3/2,5/2)
93 628.8 — 15(3/2,5/2)
93 299.6 — 14(3/2,5/2)
92 885.0 — 13(3/2,5/2)
92 360.9 — 12(3/2,5/2)
91 674.0 — 11(3/2,5/2)
90 754.80 90 753.92 — 10(3/2,5/2)
89 489.13 89 487.93 — 9(3/2,5/2)
87 673.72 87 671.93 — 8(3/2,5/2)
84 936.41 84 933.65 — 7(3/2,5/2)
80 526.16 80 521.53 — 6(3/2,5/2)
72 730.93 72 722.23 — 5(3/2,5/2)
56 858.46 56 839.25 — 4(3/2,5/2)
13 710.88 13 650.19 — 3(3/2,5/2)
91 338.0 — 10(5/2,7/2)
90 299.69 — 9(5/2,7/2)
88 847.31 — 8(5/2,7/2)
86 727.06 — 7(5/2,7/2)
83 458.08 — 6(5/2,7/2)
78 034.39 — 5(5/2,7/2)
68 056.91 — 4(5/2,7/2)
90 326.45 — 9(7/2,9/2)
88 884.54 — 8(7/2,9/2)
86 781.14 — 7(7/2,9/2)
83 540.0 — 6(7/2,9/2)
78 164.72 — 5(7/2,9/2)
91 361.00 — 10(9/2,11/2)
88 890.64 — 8(9/2,11/2)
KEY
56 858.46 56 839.25 — 4(3/2,5/2)
J, lower to higher
n
Energy Level (cm⁻¹)
Ionization Limit
95 751.87 cm⁻¹
(11.871 eV)
[Ca II 3p⁶4s ²S1/2 → Ca III 3p⁶ ¹S0]

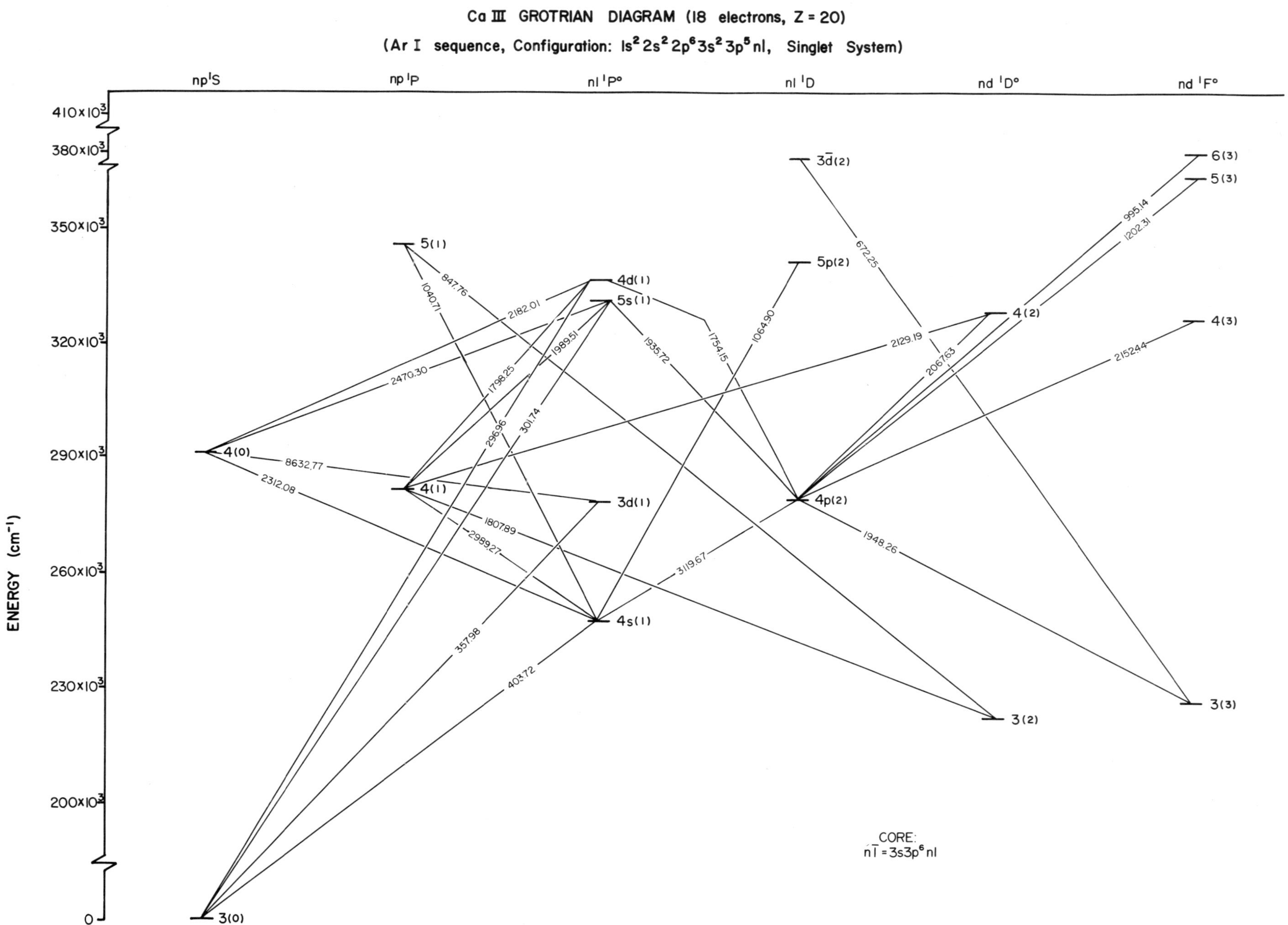
Ca III GROTRIAN DIAGRAM (18 electrons, Z = 20)
(Ar I sequence, Configuration: 1s²2s²2p⁶3s²3p⁵nl, Singlet System)
np¹S
np¹P
nl¹P°
nl¹D
nd¹D°
nd¹F°
410×10³
380×10³
350×10³
320×10³
290×10³
260×10³
230×10³
200×10³
0
ENERGY (cm⁻¹)
3d̄(2)
6(3)
5(3)
5(1)
5p(2)
4d(1)
5s(1)
4(2)
4(3)
4(0)
4(1)
3d(1)
4p(2)
4s(1)
3(3)
3(2)
3(0)
995.14
1202.31
672.25
847.76
1040.71
2182.01
1064.90
1754.15
1935.72
2129.19
2152.44
2067.63
2470.30
1798.25
1989.51
296.96
301.74
8632.77
2312.08
1807.89
2989.27
1948.26
3119.67
357.98
403.72
CORE:
n̄l̄ = 3s3p⁶nl

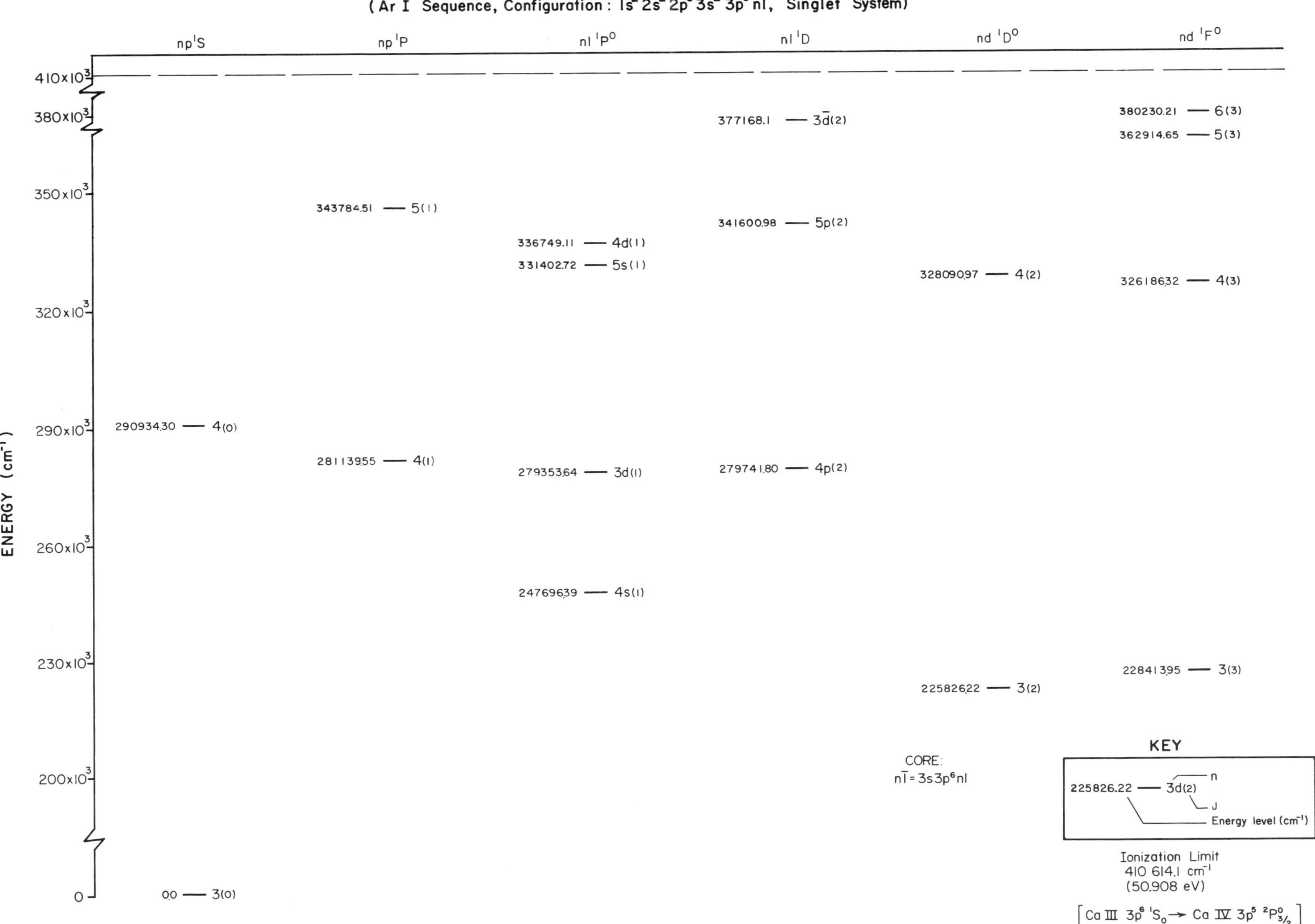
Ca III ENERGY LEVELS (18 electrons, Z=20)
(Ar I Sequence, Configuration: $1s^2 2s^2 2p^6 3s^2 3p^5 nl$, Singlet System)
$np\,^1S$
$np\,^1P$
$nl\,^1P^o$
$nl\,^1D$
$nd\,^1D^o$
$nd\,^1F^o$
ENERGY (cm^{-1})
410×10^3
380×10^3
350×10^3
320×10^3
290×10^3
260×10^3
230×10^3
200×10^3
0
290934.30 — 4(0)
0.0 — 3(0)
343784.51 — 5(1)
281139.55 — 4(1)
336749.11 — 4d(1)
331402.72 — 5s(1)
279353.64 — 3d(1)
247696.39 — 4s(1)
377168.1 — $3\bar{d}(2)$
341600.98 — 5p(2)
279741.80 — 4p(2)
328090.97 — 4(2)
225826.22 — 3(2)
380230.21 — 6(3)
362914.65 — 5(3)
326186.32 — 4(3)
228413.95 — 3(3)
CORE:
$n\bar{l} = 3s3p^6nl$
KEY
225826.22 — 3d(2)
n
J
Energy level (cm^{-1})
Ionization Limit
410 614.1 cm^{-1}
(50.908 eV)
[Ca III $3p^6\,^1S_0$ → Ca IV $3p^5\,^2P^o_{3/2}$]

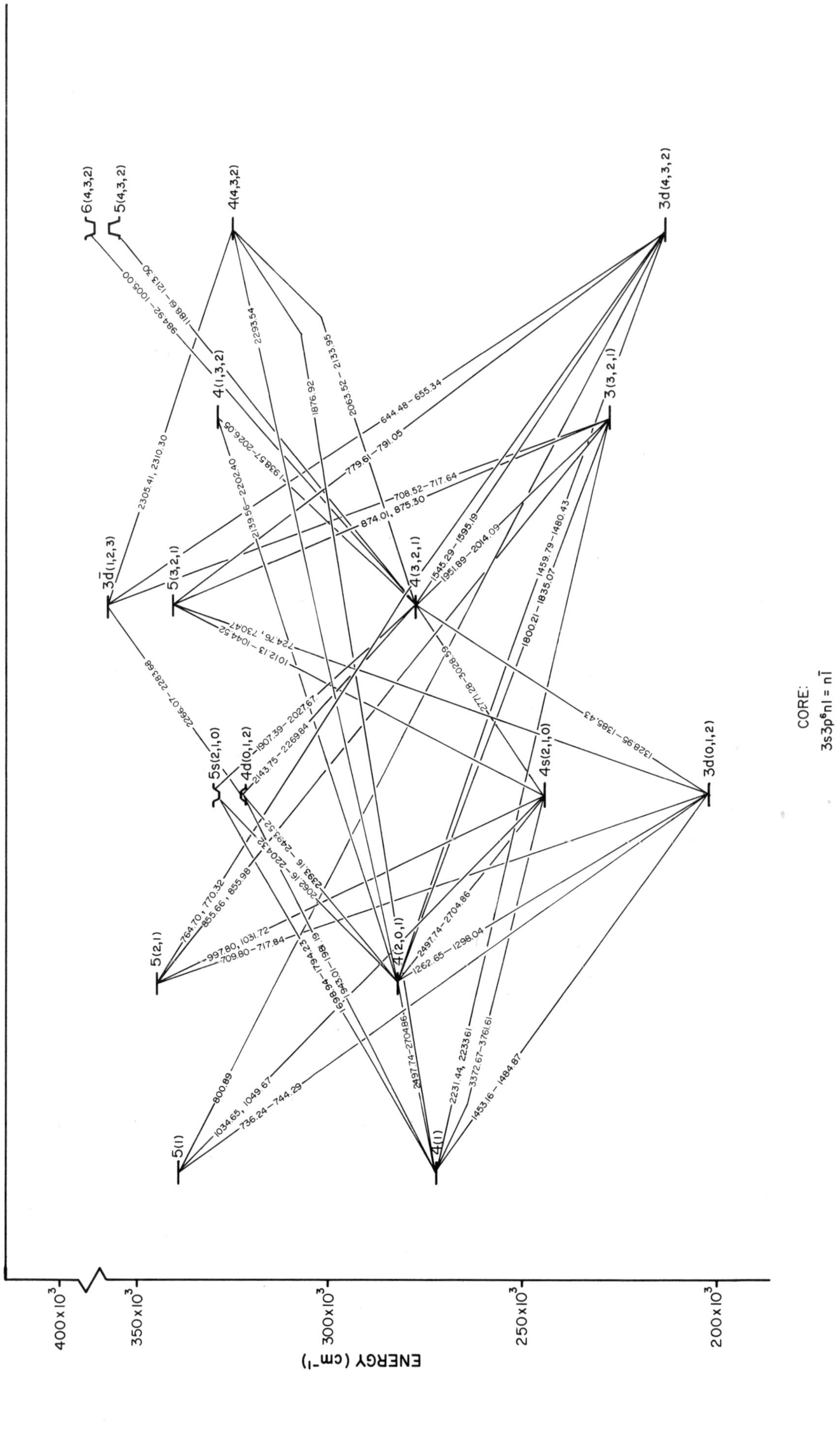

Ca III GROTRIAN DIAGRAM (18 electrons, Z=20)

(Ar I sequence, Configuration: $1s^2 2s^2 2p^6 3s^2 3p^5 nl$, Triplet System)

np ^{3}S | np ^{3}P | nl ^{3}P° | np ^{3}D | nd ^{3}D° | nd ^{3}F°

ENERGY (cm^{-1})

400x10^3
350x10^3
300x10^3
250x10^3
200x10^3

5(1)
4(1)
5(2,1)
4(2,0,1)
5s(2,1,0)
4d(0,1,2)
4s(2,1,0)
3d(0,1,2)
3$\bar{d}$(1,2,3)
5(3,2,1)
4(3,2,1)
4(1,3,2)
3(3,2,1)
6(4,3,2)
5(4,3,2)
4(4,3,2)
3d(4,3,2)

800.89
1034.65, 1049.67
736.24–744.29
2497.74–2704.86
2231.44, 2233.61
3372.67–3761.61
1453.16–1484.87
997.80, 1031.72
709.80–717.84
764.70, 770.32
855.66, 855.98
1698.94–1794.23
1943.01–1981.19
2062.16–2204.32
2393.16–2493.52
2497.74–2704.86
1262.65–1298.04
2266.07–2283.68
1907.39–2027.67
2143.75–2269.84
1012.13–1044.52
724.76, 730.47
2771.28–3028.59
1328.95–1385.43
2305.41, 2310.30
2139.56–2202.40
1938.57–2026.05
874.01, 875.30
708.52–717.64
779.61–791.05
644.48–655.34
1545.29–1595.19
1951.89–2014.09
1800.21–1835.07
1459.79–1480.43
984.92–1005.00
1188.61–1213.30
2293.54
1876.92
2063.52–2133.95

CORE:
$3s3p^6nl = n\bar{l}$

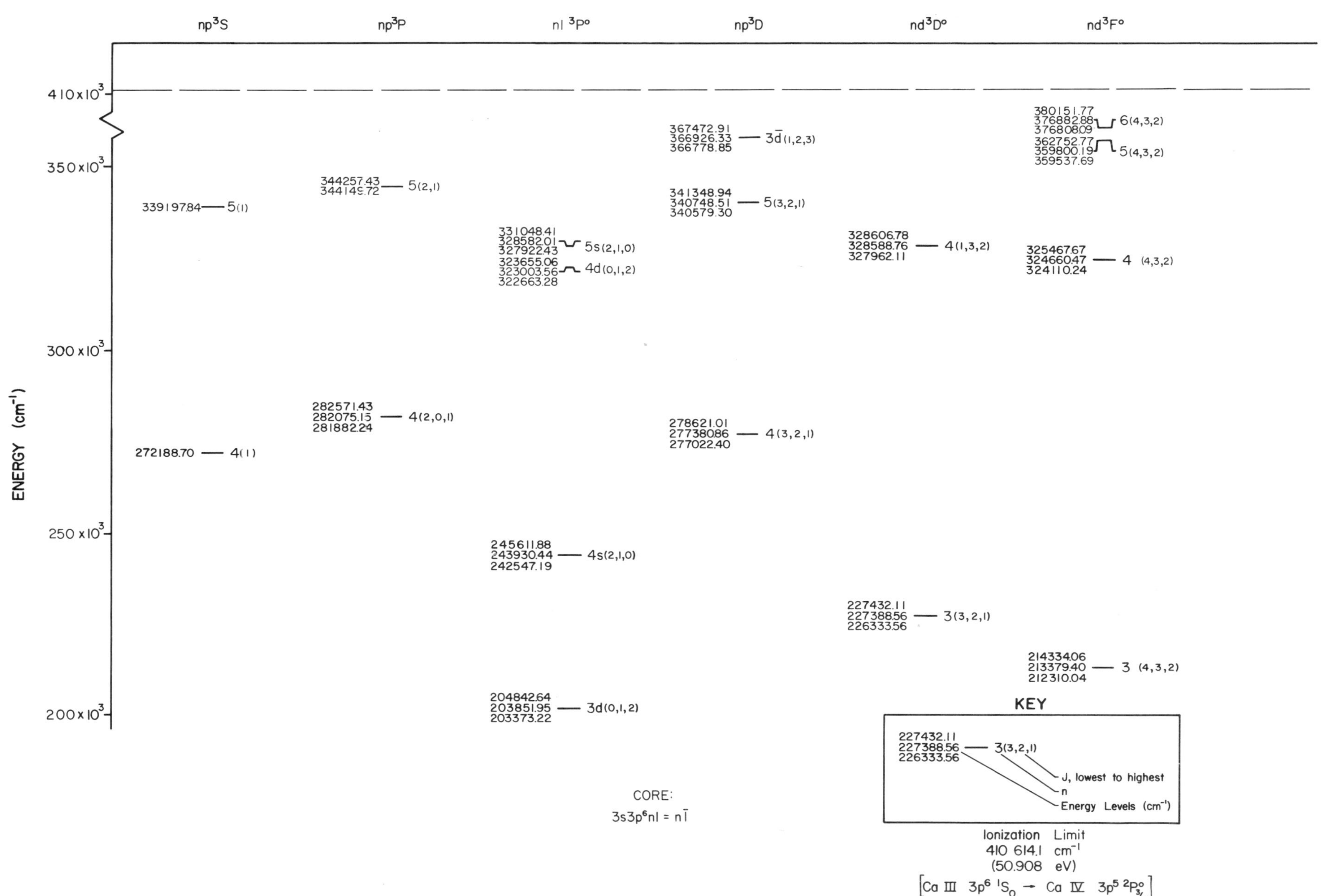
Ca III ENERGY LEVELS (18 electrons, Z = 20)
(Ar I sequence, Configuration: $1s^2 2s^2 2p^6 3s^2 3p^5 nl$, Triplet System)
np³S
np³P
nl ³P°
np³D
nd³D°
nd³F°
ENERGY (cm⁻¹)
410×10³
350×10³
300×10³
250×10³
200×10³
339197.84 — 5(1)
272188.70 — 4(1)
344257.43
344149.72 — 5(2,1)
282571.43
282075.15 — 4(2,0,1)
281882.24
331048.41
328582.01 — 5s(2,1,0)
327922.43
323655.06
323003.56 — 4d(0,1,2)
322663.28
245611.88
243930.44 — 4s(2,1,0)
242547.19
204842.64
203851.95 — 3d(0,1,2)
203373.22
367472.91
366926.33 — 3d̄(1,2,3)
366778.85
341348.94
340748.51 — 5(3,2,1)
340579.30
278621.01
277380.86 — 4(3,2,1)
277022.40
328606.78
328588.76 — 4(1,3,2)
327962.11
227432.11
227388.56 — 3(3,2,1)
226333.56
380151.77
376882.88 — 6(4,3,2)
376808.09
362752.77
359800.19 — 5(4,3,2)
359537.69
325467.67
324660.47 — 4 (4,3,2)
324110.24
214334.06
213379.40 — 3 (4,3,2)
212310.04
CORE:
3s3p⁶nl = nl̄
KEY
227432.11
227388.56 — 3(3,2,1)
226333.56
J, lowest to highest
n
Energy Levels (cm⁻¹)
Ionization Limit
410 614.1 cm⁻¹
(50.908 eV)
[Ca III 3p⁶ ¹S₀ → Ca IV 3p⁵ ²P°3/2]

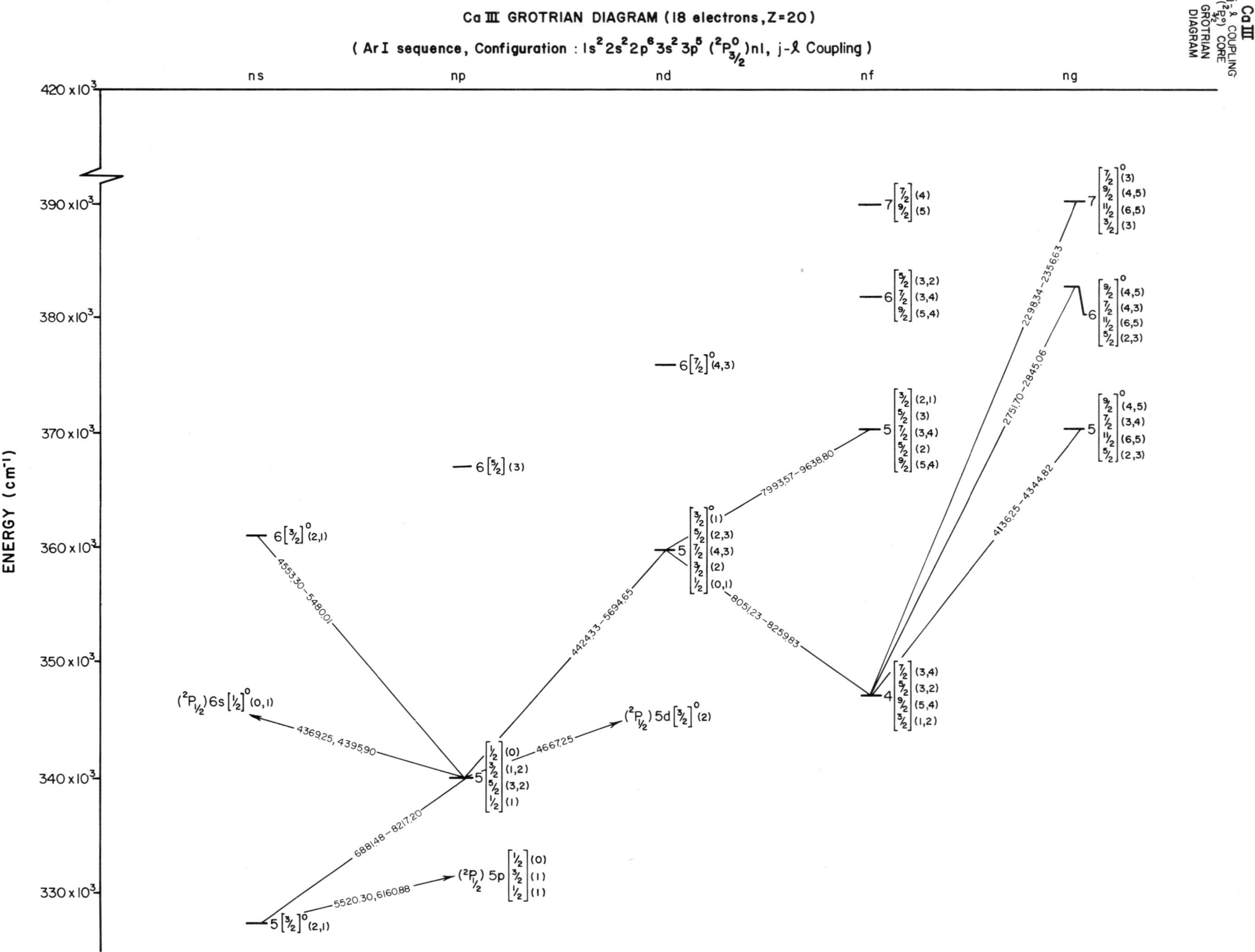

Ca III
j-ℓ COUPLING
($^2P^o_{3/2}$) CORE
GROTRIAN
DIAGRAM
Ca III GROTRIAN DIAGRAM (18 electrons, Z=20)
(Ar I sequence, Configuration : $1s^2 2s^2 2p^6 3s^2 3p^5$ ($^2P^o_{3/2}$)nl, j-ℓ Coupling)
ns
np
nd
nf
ng
ENERGY (cm⁻¹)
420x10³
390x10³
380x10³
370x10³
360x10³
350x10³
340x10³
330x10³
2298.34-2356.63
2751.70-2845.06
4136.25-4344.82
799.357-963.880
805.123-825.983
442.433-569.465
466.725
455.330-548.001
436.925, 439.590
688.148-821.720
5520.30, 6160.88
($^2P_{1/2}$) 5d $[3/2]^o$ (2)
($^2P_{1/2}$) 6s $[1/2]^o$ (0,1)
($^2P_{1/2}$) 5p

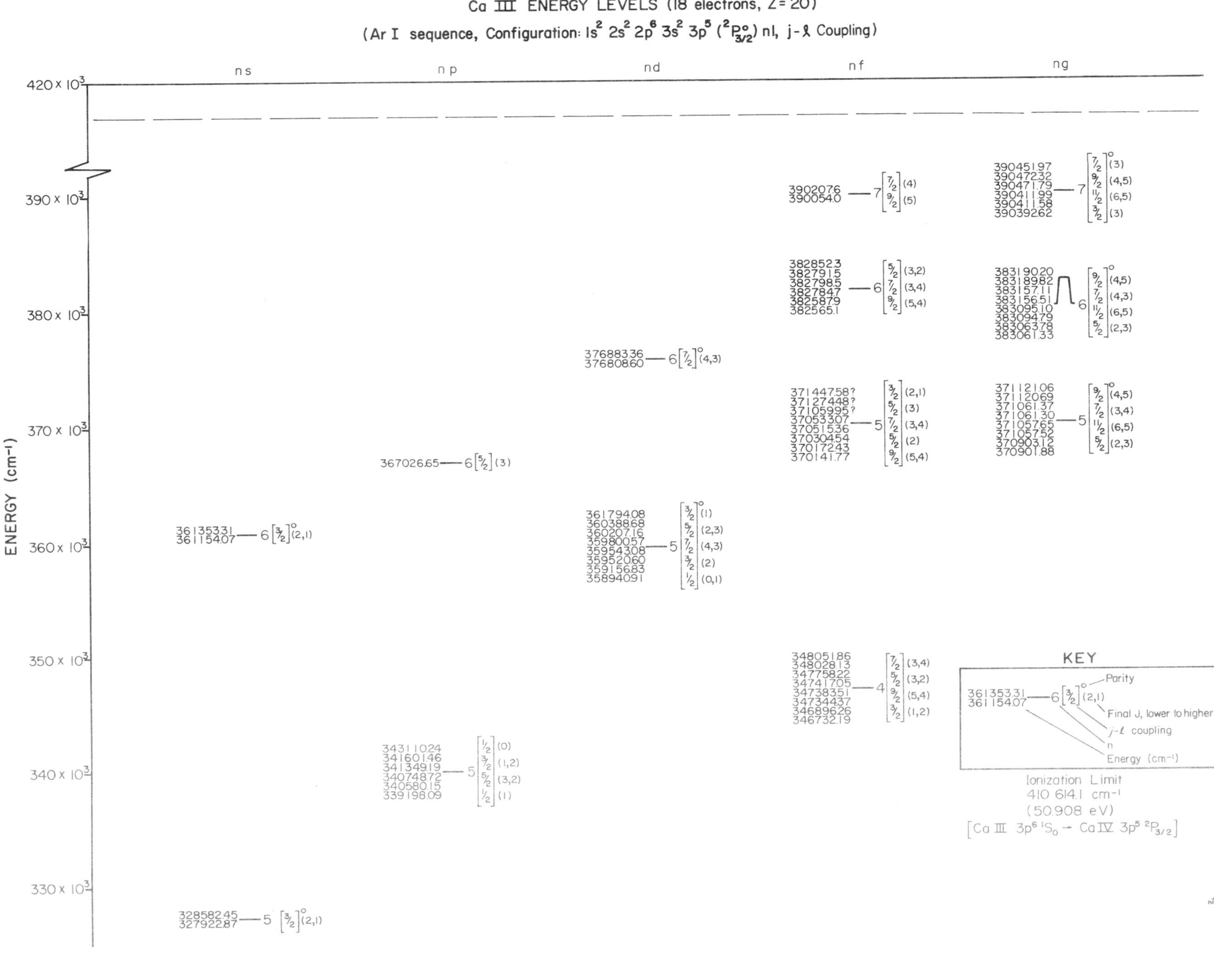

Ca III ENERGY LEVELS (18 electrons, Z = 20)
(Ar I sequence, Configuration: $1s^2\,2s^2\,2p^6\,3s^2\,3p^5\,(^2P^o_{3/2})\,nl$, j-ℓ Coupling)
ns
np
nd
nf
ng
ENERGY (cm⁻¹)
420 x 10³
390 x 10³
380 x 10³
370 x 10³
360 x 10³
350 x 10³
340 x 10³
330 x 10³
390451.97 390472.32 390471.79 390411.99 390411.58 390392.62 — 7 [7/2]° (3) [9/2] (4,5) [11/2] (6,5) [3/2] (3)
390207.6 390054.0 — 7 [7/2] (4) [9/2] (5)
382852.3 382791.5 382798.5 382784.7 382587.9 382565.1 — 6 [5/2] (3,2) [7/2] (3,4) [9/2] (5,4)
383190.20 383189.82 383157.11 383156.51 383095.10 383094.79 383063.78 383061.33 — 6 [9/2]° (4,5) [7/2] (4,3) [11/2] (6,5) [5/2] (2,3)
376883.36 376808.60 — 6 [7/2]° (4,3)
371447.58? 371274.48? 371059.95? 370533.07 370515.36 370304.54 370172.43 370141.77 — 5 [3/2] (2,1) [5/2] (3) [7/2] (3,4) [5/2] (2) [9/2] (5,4)
371121.06 371120.69 371061.37 371061.30 371057.65 371057.52 370903.12 370901.88 — 5 [9/2]° (4,5) [7/2] (3,4) [11/2] (6,5) [5/2] (2,3)
367026.65 — 6 [5/2] (3)
361353.31 361154.07 — 6 [3/2]° (2,1)
361794.08 360388.68 360207.16 359800.57 359543.08 359520.60 359156.83 358940.91 — 5 [3/2]° (1) [5/2] (2,3) [7/2] (4,3) [3/2] (2) [1/2] (0,1)
348051.86 348028.13 347758.22 347417.05 347383.51 347344.37 346896.26 346732.19 — 4 [7/2] (3,4) [5/2] (3,2) [9/2] (5,4) [3/2] (1,2)
343110.24 341601.46 341349.19 340748.72 340580.15 339198.09 — 5 [1/2] (0) [3/2] (1,2) [5/2] (3,2) [1/2] (1)
328582.45 327922.87 — 5 [3/2]° (2,1)
KEY
361353.31 361154.07 — 6 [3/2]° (2,1)
Parity
Final J, lower to higher
j-ℓ coupling
n
Energy (cm⁻¹)
Ionization Limit
410 614.1 cm⁻¹
(50.908 eV)
[Ca III $3p^6\,^1S_0$ → Ca IV $3p^5\,^2P_{3/2}$]
Ca III
$(^2P^o_{3/2})$ j-ℓ COUPLING

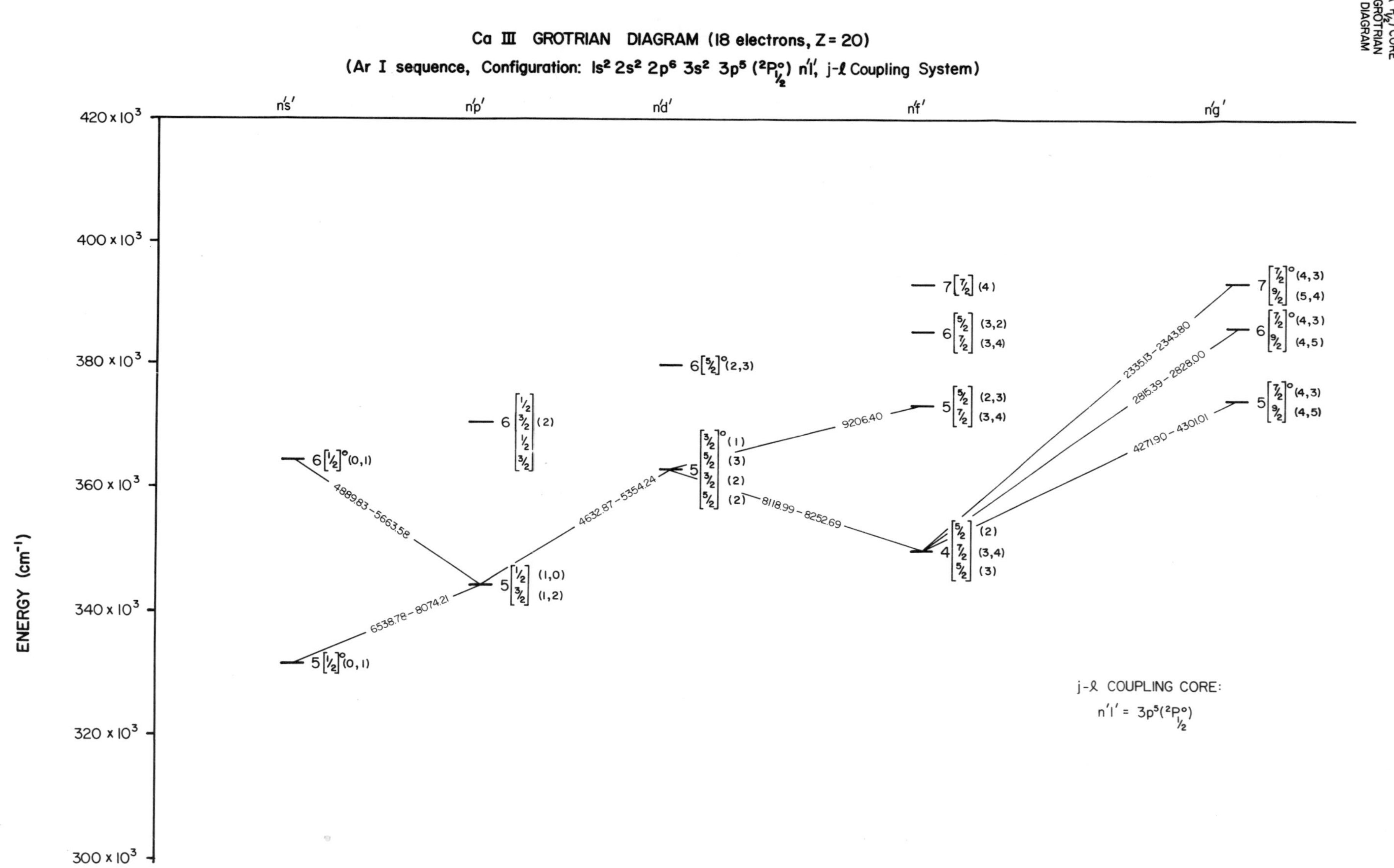
Ca III GROTRIAN DIAGRAM (18 electrons, Z = 20)
(Ar I sequence, Configuration: 1s² 2s² 2p⁶ 3s² 3p⁵ (²P°₁/₂) n′l′, j-ℓ Coupling System)
n′s′
n′p′
n′d′
n′f′
n′g′
ENERGY (cm⁻¹)
420 x 10³
400 x 10³
380 x 10³
360 x 10³
340 x 10³
320 x 10³
300 x 10³
4889.83–5663.58
6538.78–8074.21
4632.87–5354.24
9206.40
8118.99–8252.69
2335.13–2343.80
2815.39–2828.00
4271.90–4301.01
j-ℓ COUPLING CORE:
n′l′ = 3p⁵(²P°₁/₂)

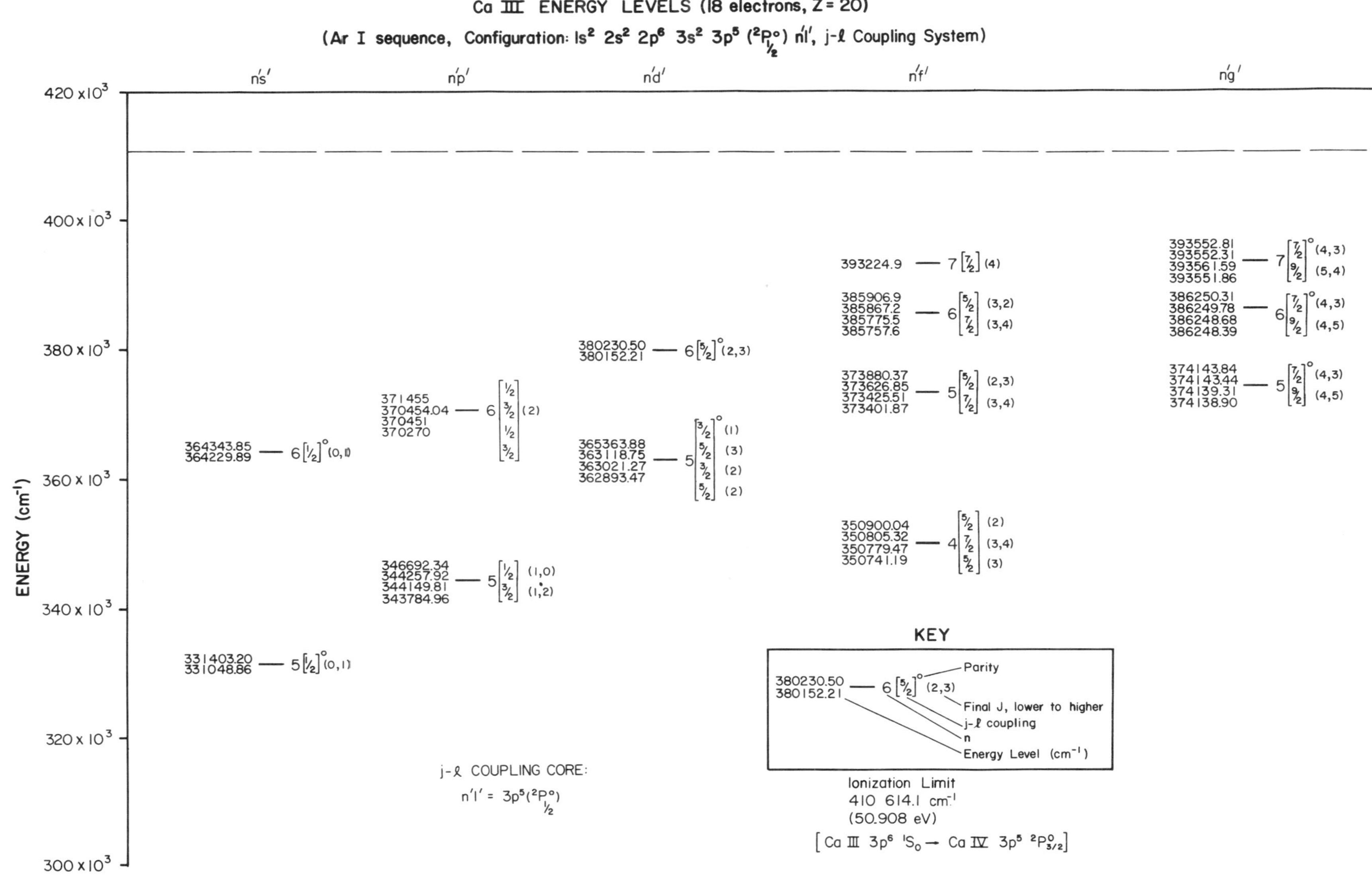
Ca III ENERGY LEVELS (18 electrons, Z = 20)
(Ar I sequence, Configuration: 1s² 2s² 2p⁶ 3s² 3p⁵ (²P°½) n′l′, j-ℓ Coupling System)
ns′
np′
nd′
nf′
ng′
420 x10³
400 x10³
380 x10³
360 x10³
340 x10³
320 x10³
300 x10³
ENERGY (cm⁻¹)
364343.85
364229.89
6 [½]° (0,1)
331403.20
331048.86
5 [½]° (0,1)
371455
370454.04
370451
370270
6
(2)
346692.34
344257.92
344149.81
343784.96
5
(1,0)
(1,2)
380230.50
380152.21
6 [5/2]° (2,3)
365363.88
363118.75
363021.27
362893.47
5
(1)
(3)
(2)
(2)
393224.9
7 [7/2] (4)
385906.9
385867.2
385775.5
385757.6
6
(3,2)
(3,4)
373880.37
373626.85
373425.51
373401.87
5
(2,3)
(3,4)
350900.04
350805.32
350779.47
350741.19
4
(2)
(3,4)
(3)
393552.81
393552.31
393561.59
393551.86
7
(4,3)
(5,4)
386250.31
386249.78
386248.68
386248.39
6
(4,3)
(4,5)
374143.84
374143.44
374139.31
374138.90
5
(4,3)
(4,5)
KEY
380230.50
380152.21
6 [5/2]° (2,3)
Parity
Final J, lower to higher
j-ℓ coupling
n
Energy Level (cm⁻¹)
Ionization Limit
410 614.1 cm⁻¹
(50.908 eV)
[Ca III 3p⁶ ¹S₀ → Ca IV 3p⁵ ²P°3/2]
j-ℓ COUPLING CORE:
n′l′ = 3p⁵(²P°½)

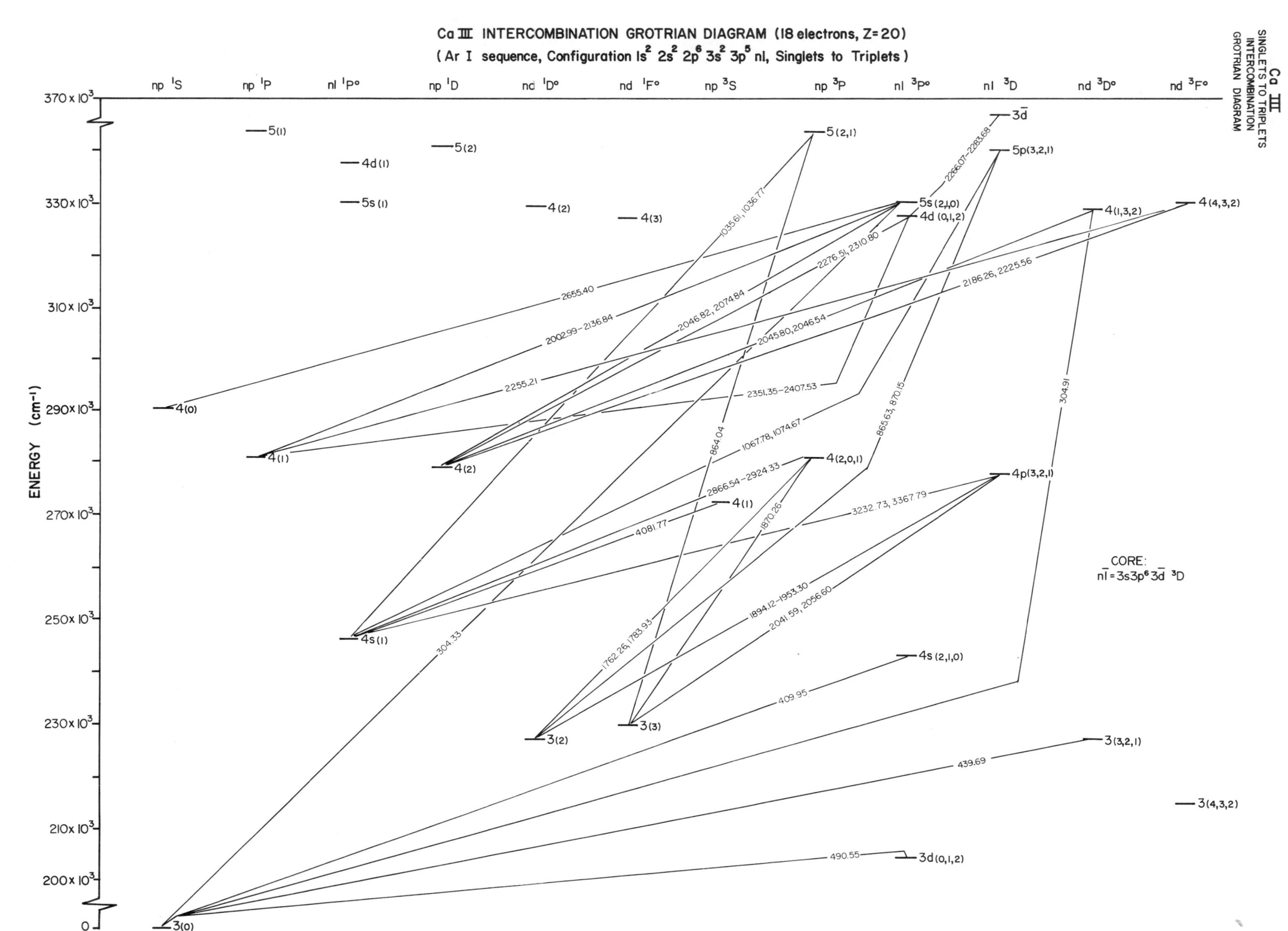
Ca III INTERCOMBINATION GROTRIAN DIAGRAM (18 electrons, Z=20)
(Ar I sequence, Configuration 1s2 2s2 2p6 3s2 3p5 nl, Singlets to Triplets)
Ca III
SINGLETS TO TRIPLETS
INTERCOMBINATION
GROTRIAN DIAGRAM
np 1S
np 1P
nl 1P°
np 1D
nd 1D°
nd 1F°
np 3S
np 3P
nl 3P°
nl 3D
nd 3D°
nd 3F°
ENERGY (cm-1)
370 x 10^3
330 x 10^3
310 x 10^3
290 x 10^3
270 x 10^3
250 x 10^3
230 x 10^3
210 x 10^3
200 x 10^3
0
5(1)
5(2)
4d(1)
5s(1)
4(2)
4(3)
5(2,1)
3d̄
5p(3,2,1)
5s(2,1,0)
4d(0,1,2)
4(1,3,2)
4(4,3,2)
4(0)
4(1)
4(2)
4(2,0,1)
4p(3,2,1)
4(1)
4s(1)
4s(2,1,0)
3(2)
3(3)
3(3,2,1)
3(4,3,2)
3d(0,1,2)
3(0)
CORE:
n̄l̄ = 3s3p6 3d̄ 3D
1035.61, 1036.77
2266.07–2283.68
2276.51, 2310.80
2186.26, 2225.56
2655.40
2002.99–2136.84
2046.82, 2074.84
2045.80, 2046.54
2255.21
2351.35–2407.53
1067.78, 1074.67
865.63, 870.15
864.04
304.91
2866.54–2924.33
3232.73, 3367.79
4081.77
1870.26
1894.12–1953.30
2041.59, 2056.60
1762.26, 1783.93
304.33
409.95
439.69
490.55

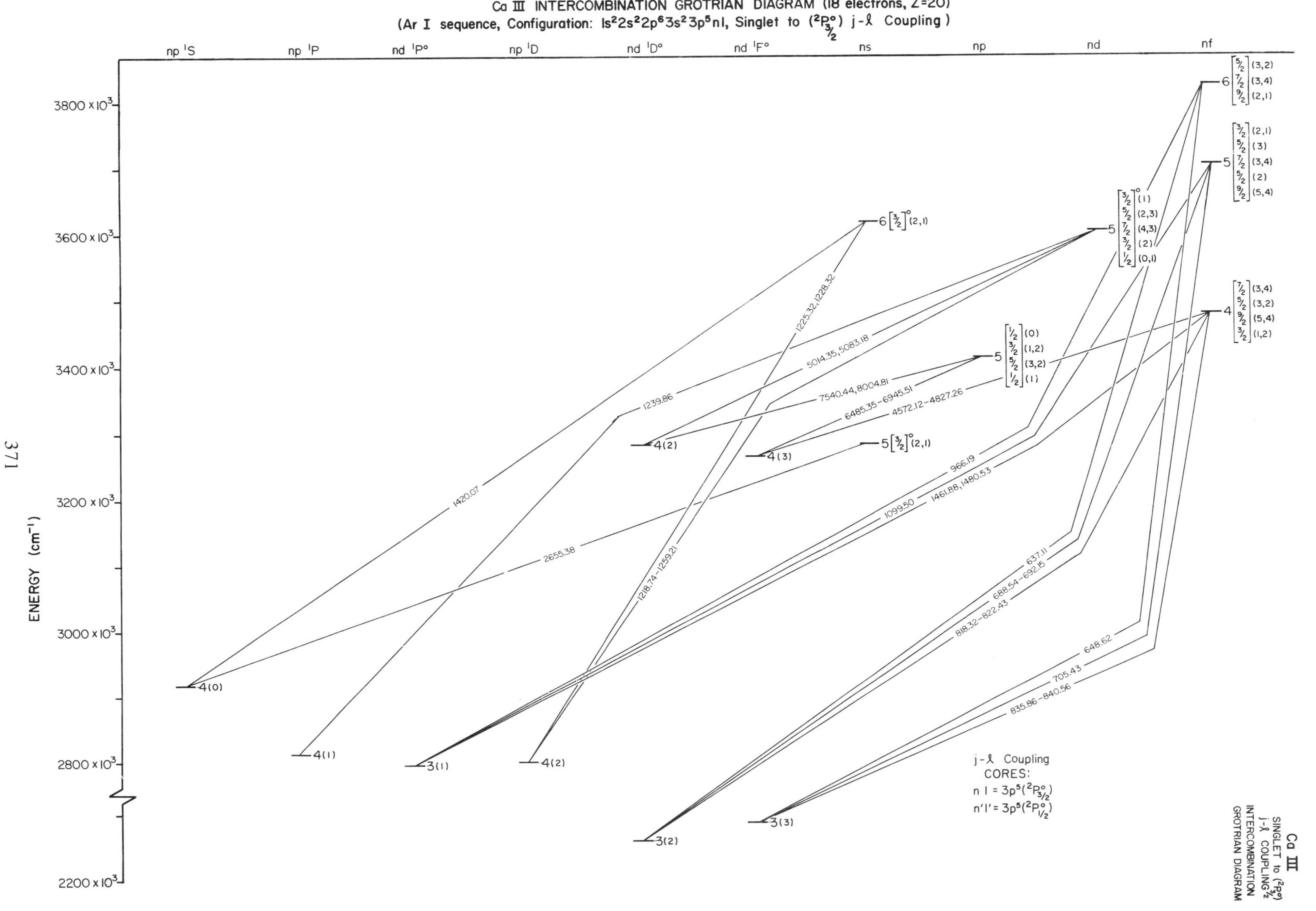

Ca III INTERCOMBINATION GROTRIAN DIAGRAM (18 electrons, Z=20)
(Ar I sequence, Configuration: $1s^2 2s^2 2p^6 3s^2 3p^5 nl$, Singlet to $(^2P^o_{3/2})$ j-ℓ Coupling)
np ^1S
np ^1P
nd $^1P^o$
np ^1D
nd $^1D^o$
nd $^1F^o$
ns
np
nd
nf
ENERGY (cm^{-1})
3800 x10^3
3600 x10^3
3400 x10^3
3200 x10^3
3000 x10^3
2800 x10^3
2200 x10^3
1420.07
2655.38
1239.86
1225.32, 1228.32
1218.74 - 1259.21
5014.35, 5083.18
7540.44, 8004.81
6485.35 - 6945.51
4572.12 - 4827.26
966.19
1099.50
1461.88, 1480.53
637.11
688.54 - 692.15
818.32 - 822.43
648.62
705.43
835.86 - 840.56
j-ℓ Coupling
CORES:
n l = $3p^5(^2P^o_{3/2})$
n′l′ = $3p^5(^2P^o_{1/2})$
Ca III
SINGLET to $(^2P^o_{3/2})$
j-ℓ COUPLING
INTERCOMBINATION
GROTRIAN DIAGRAM

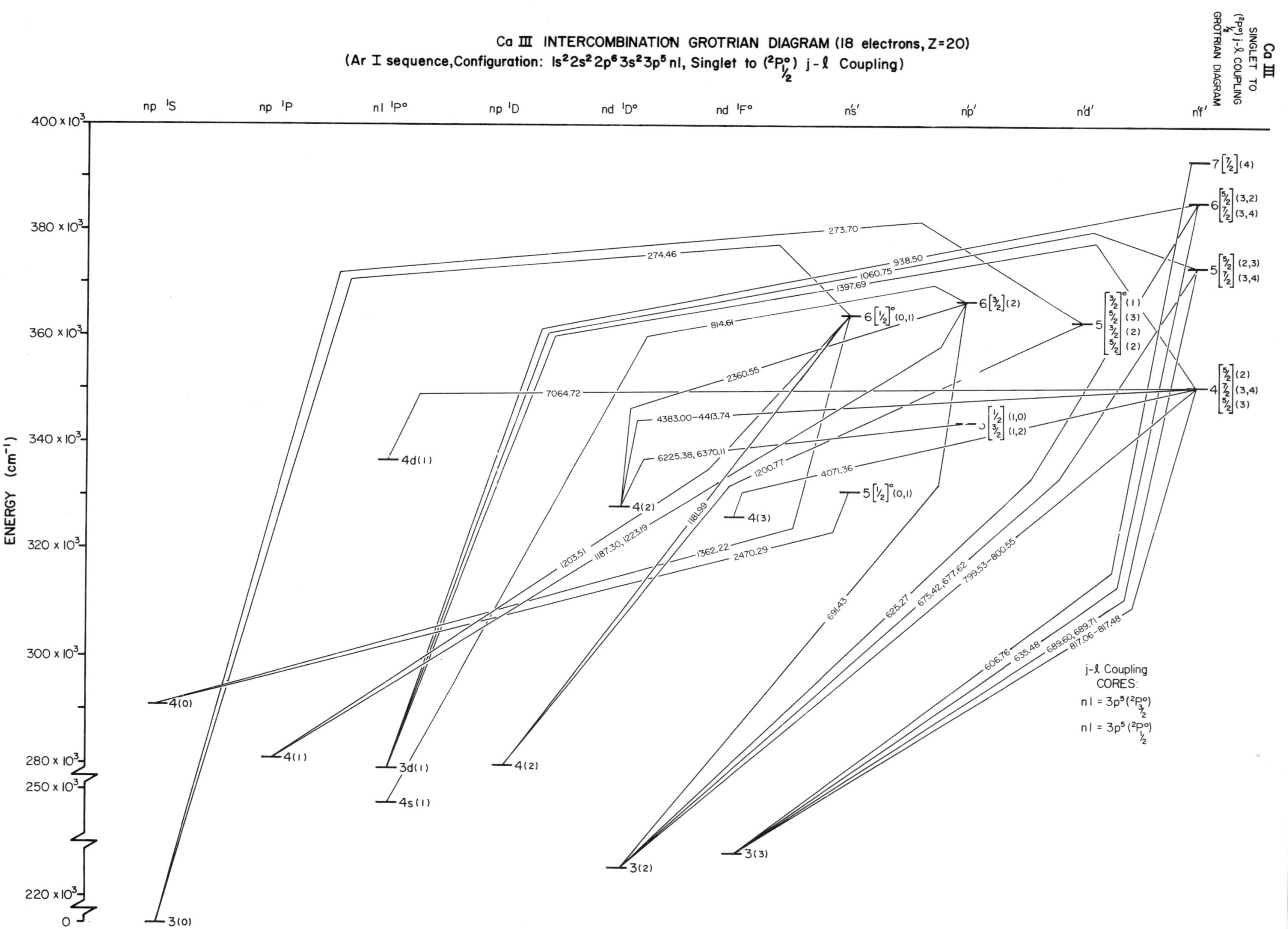
Ca III INTERCOMBINATION GROTRIAN DIAGRAM (18 electrons, Z=20)
(Ar I sequence, Configuration: $1s^2 2s^2 2p^6 3s^2 3p^5 nl$, Singlet to $(^2P^o_{1/2})$ j-ℓ Coupling)
Ca III
SINGLET TO
$(^2P^o_{1/2})$ j-ℓ COUPLING
GROTRIAN DIAGRAM
np ^1S
np ^1P
nl $^1P^o$
np ^1D
nd $^1D^o$
nd $^1F^o$
n's'
n'p'
n'd'
n'f'
ENERGY (cm^{-1})
400 x 10^3
380 x 10^3
360 x 10^3
340 x 10^3
320 x 10^3
300 x 10^3
280 x 10^3
250 x 10^3
220 x 10^3
0
j-ℓ Coupling
CORES:
$nl = 3p^5(^2P^o_{3/2})$
$n'l' = 3p^5(^2P^o_{1/2})$

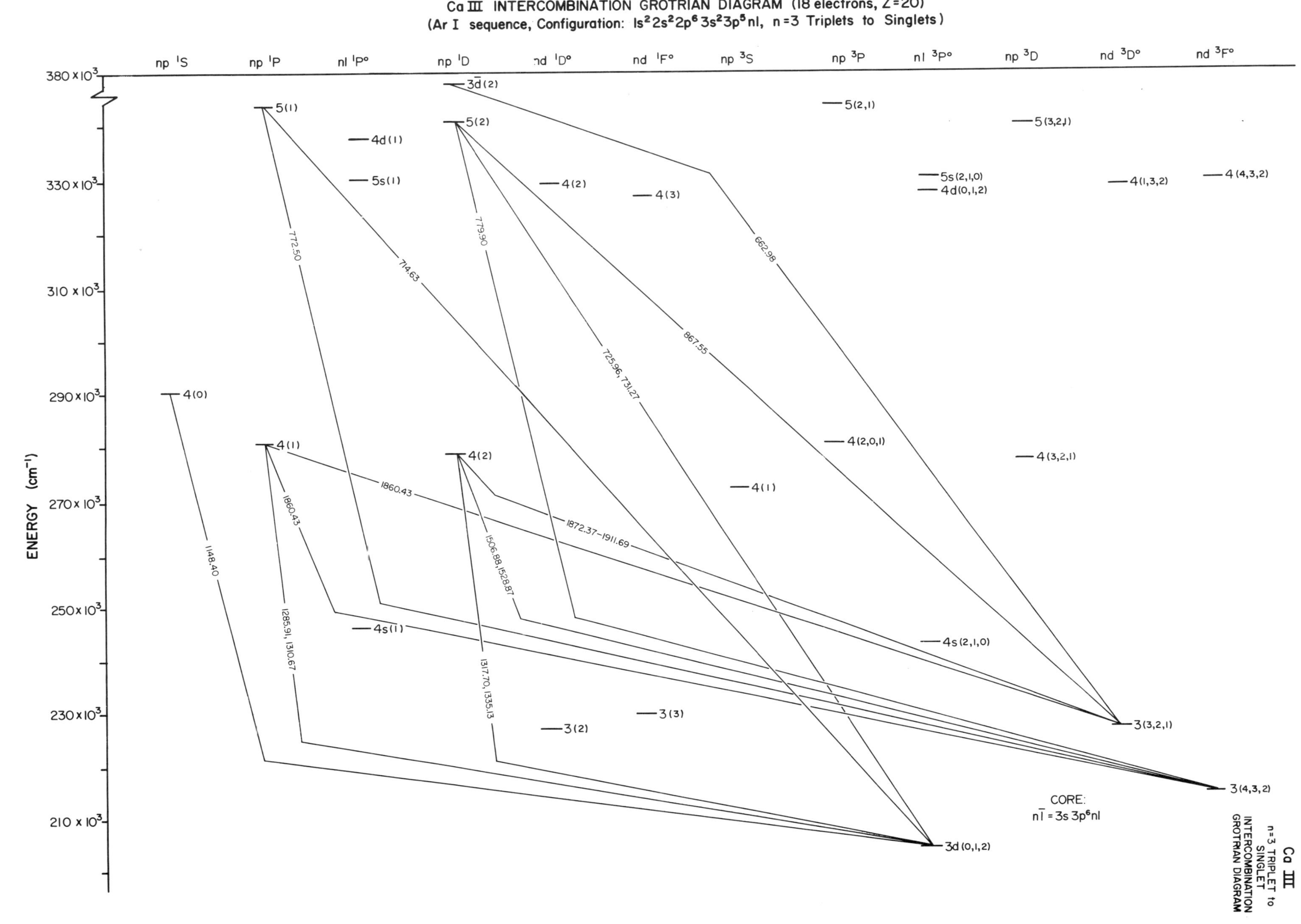
Ca III INTERCOMBINATION GROTRIAN DIAGRAM (18 electrons, Z=20)
(Ar I sequence, Configuration: 1s²2s²2p⁶3s²3p⁵nl, n=3 Triplets to Singlets)
np ¹S
np ¹P
nl ¹P°
np ¹D
nd ¹D°
nd ¹F°
np ³S
np ³P
nl ³P°
np ³D
nd ³D°
nd ³F°
380×10³
330×10³
310×10³
290×10³
270×10³
250×10³
230×10³
210×10³
ENERGY (cm⁻¹)
3d̄(2)
5(1)
5(2)
4d(1)
5s(1)
4(2)
4(3)
5(2,1)
5(3,2,1)
5s(2,1,0)
4d(0,1,2)
4(1,3,2)
4(4,3,2)
4(0)
4(1)
4(2)
4(2,0,1)
4(3,2,1)
4(1)
4s(1)
4s(2,1,0)
3(2)
3(3)
3(3,2,1)
3(4,3,2)
3d(0,1,2)
772.50
714.63
779.90
725.96, 731.27
867.55
662.98
1860.43
1860.43
1872.37–1911.69
1506.88, 1528.87
1148.40
1285.91, 1310.67
1317.70, 1335.13
CORE:
nl̄ = 3s 3p⁶nl
Ca III
n=3 TRIPLET to
SINGLET
INTERCOMBINATION
GROTRIAN DIAGRAM

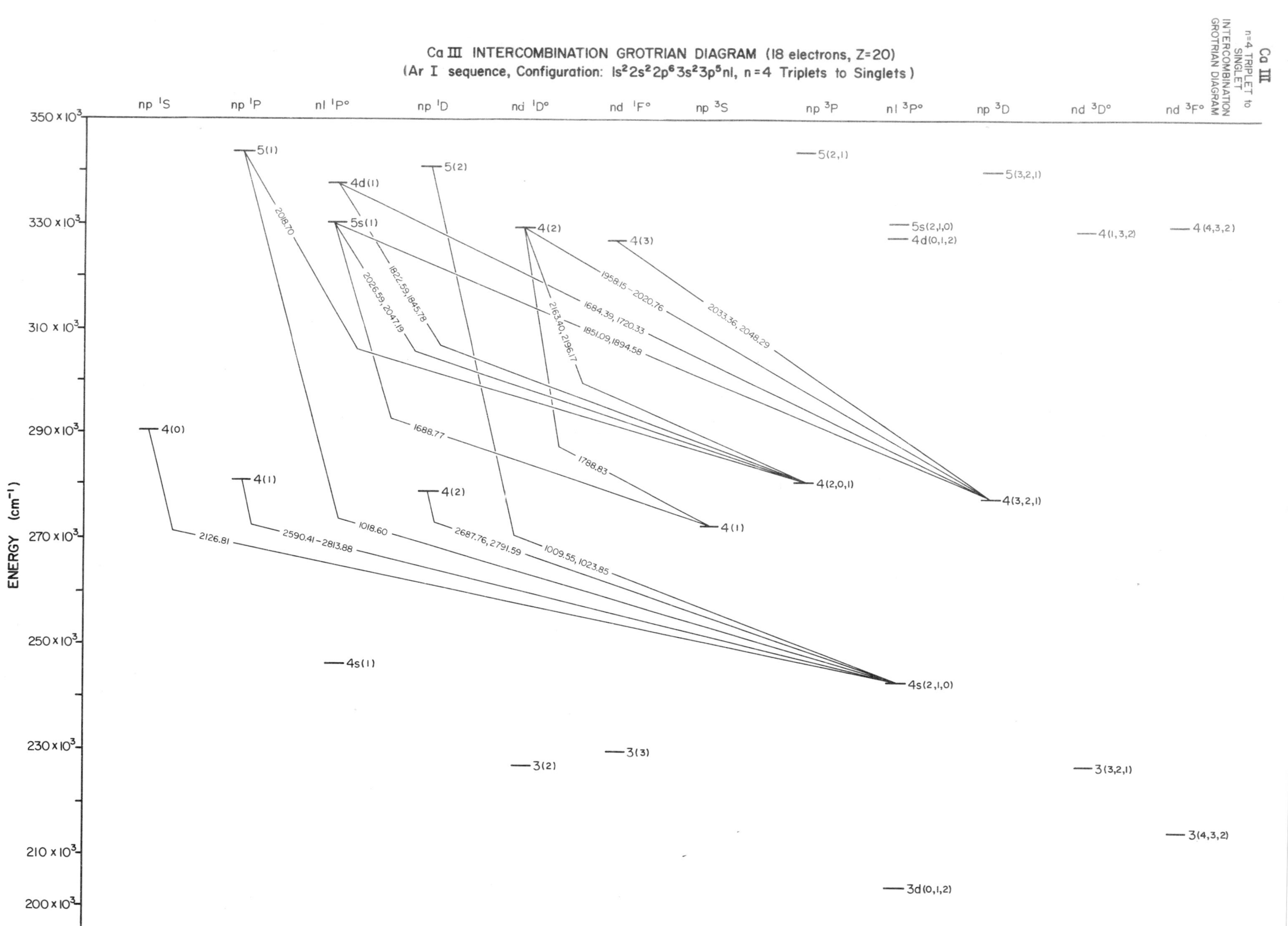
Ca III INTERCOMBINATION GROTRIAN DIAGRAM (18 electrons, Z=20)
(Ar I sequence, Configuration: 1s²2s²2p⁶3s²3p⁵nl, n=4 Triplets to Singlets)
Ca III
n=4 TRIPLET to SINGLET INTERCOMBINATION GROTRIAN DIAGRAM
np 1S
np 1P
nl 1P°
np 1D
nd 1D°
nd 1F°
np 3S
np 3P
nl 3P°
np 3D
nd 3D°
nd 3F°
5(1)
5(2)
4d(1)
5s(1)
4(2)
4(3)
5(2,1)
5(3,2,1)
5s(2,1,0)
4d(0,1,2)
4(1,3,2)
4(4,3,2)
4(0)
4(1)
4(2)
4(2,0,1)
4(3,2,1)
4(1)
4s(1)
4s(2,1,0)
3(2)
3(3)
3(3,2,1)
3(4,3,2)
3d(0,1,2)
2018.70
2026.59, 2047.19
1822.59, 1845.78
1958.15 - 2020.76
1684.39, 1720.33
1851.09, 1894.58
2163.40, 2196.17
2033.36, 2048.29
1688.77
1788.83
2126.81
2590.41 - 2813.88
1018.60
2687.76, 2791.59
1009.55, 1023.85
350 x 10³
330 x 10³
310 x 10³
290 x 10³
270 x 10³
250 x 10³
230 x 10³
210 x 10³
200 x 10³
ENERGY (cm⁻¹)

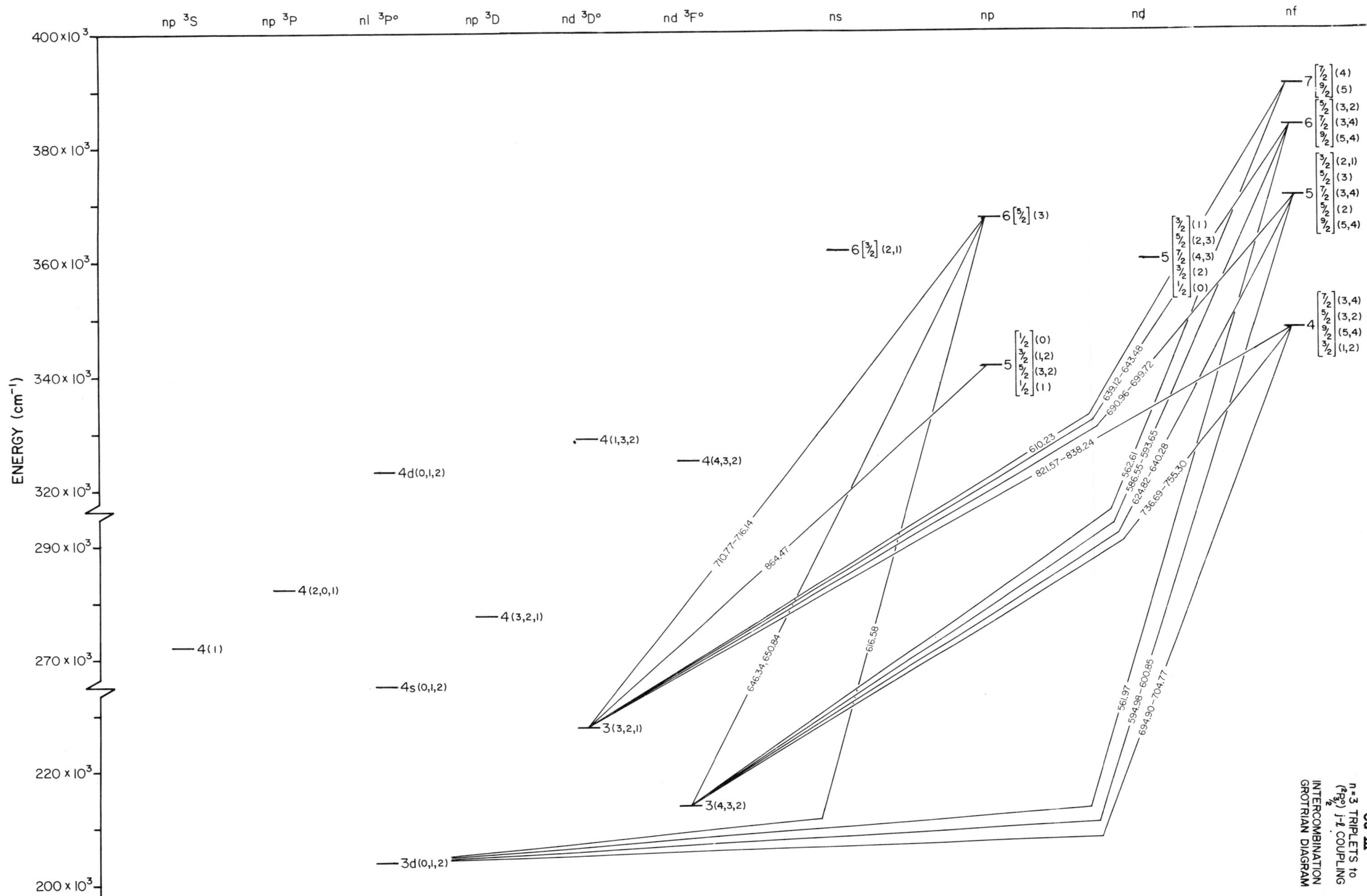
Ca III INTERCOMBINATION GROTRIAN DIAGRAM (18 electrons, Z=20)
(Ar I sequence, Configuration: 1s²2s²2p⁶3s²3p⁵nl, n=3 Triplets to (²P°3/2) Core j-l Coupling)
np ³S
np ³P
nl ³P°
np ³D
nd ³D°
nd ³F°
ns
np
nd
nf
ENERGY (cm⁻¹)
400×10³
380×10³
360×10³
340×10³
320×10³
290×10³
270×10³
220×10³
200×10³
4(1)
4(2,0,1)
4d(0,1,2)
4s(0,1,2)
3d(0,1,2)
4(3,2,1)
4(1,3,2)
3(3,2,1)
4(4,3,2)
3(4,3,2)
6[3/2](2,1)
6[5/2](3)
5
4
5
6
7
710.77–716.14
864.47
646.34, 650.84
616.58
610.23
821.57–838.24
639.12–643.48
690.96–699.72
562.61
586.55–593.65
624.82–640.28
736.69–755.30
561.97
594.98–600.85
694.90–704.77
Ca III
n=3 TRIPLETS to (²P°3/2) j-l COUPLING
INTERCOMBINATION GROTRIAN DIAGRAM

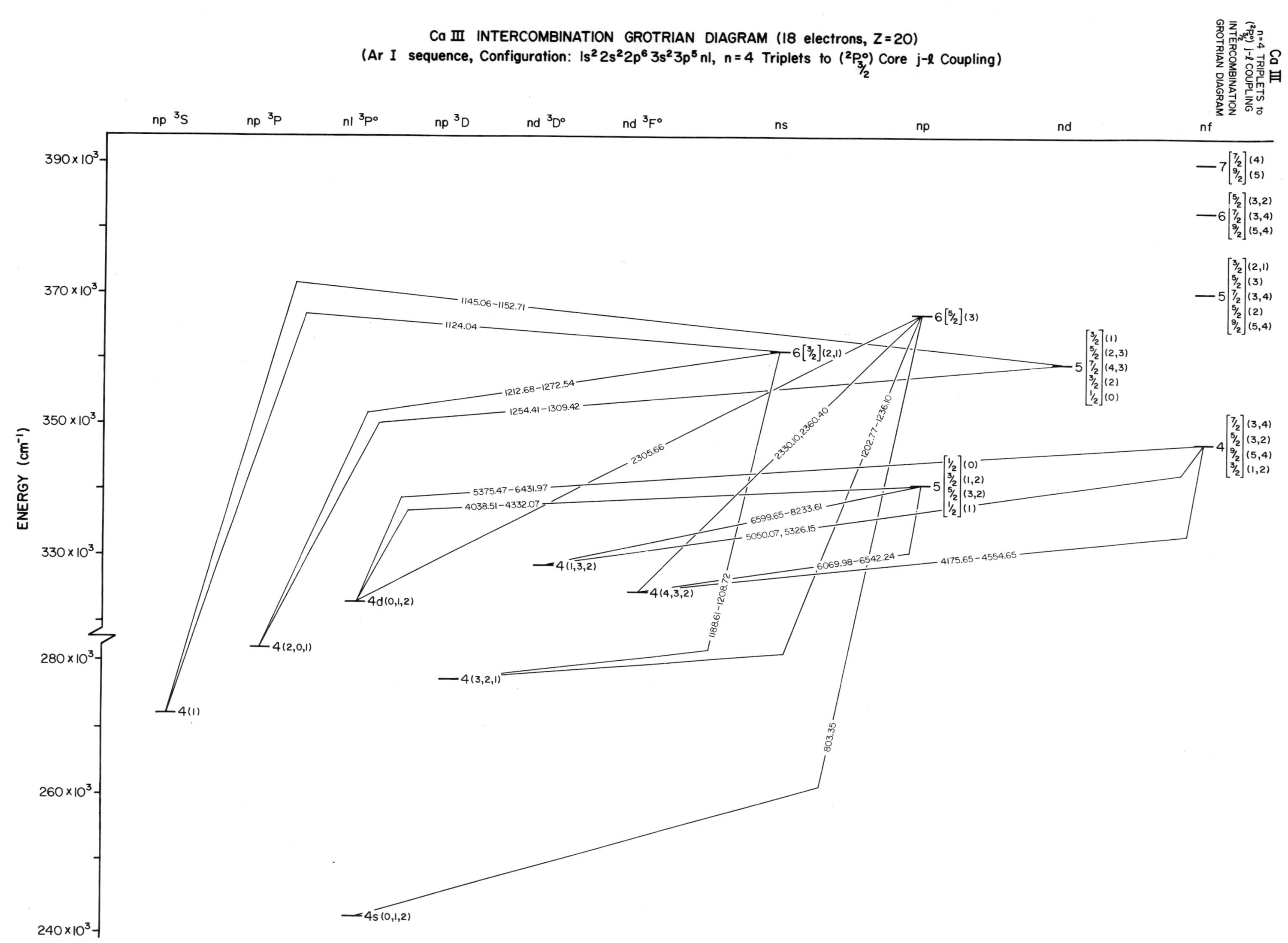
Ca III INTERCOMBINATION GROTRIAN DIAGRAM (18 electrons, Z=20)
(Ar I sequence, Configuration: 1s²2s²2p⁶3s²3p⁵nl, n=4 Triplets to (²P°3/2) Core j-ℓ Coupling)
Ca III
n=4 TRIPLETS to (²P°3/2) j-ℓ COUPLING
INTERCOMBINATION
GROTRIAN DIAGRAM
np ³S
np ³P
nl ³P°
np ³D
nd ³D°
nd ³F°
ns
np
nd
nf
ENERGY (cm⁻¹)
390x10³
370x10³
350x10³
330x10³
280x10³
260x10³
240x10³
4(1)
4(2,0,1)
4d(0,1,2)
4s(0,1,2)
4(3,2,1)
4(1,3,2)
4(4,3,2)
6[3/2](2,1)
6[5/2](3)
1145.06–1152.71
1124.04
1212.68–1272.54
1254.41–1309.42
5375.47–6431.97
4038.51–4332.07
2305.66
2330.10, 2360.40
1202.77–1236.10
1188.61–1208.72
6599.65–8233.61
5050.07, 5326.15
6069.98–6542.24
4175.65–4554.65
803.35

Ca III INTERCOMBINATION GROTRIAN DIAGRAM (18 electrons, Z=20)

(Configuration: $1s^2 2s^2 2p^6 3s^2 3p^5 nl$, Triplet to $(^2P^\circ_{1/2})$ j-ℓ Coupling)

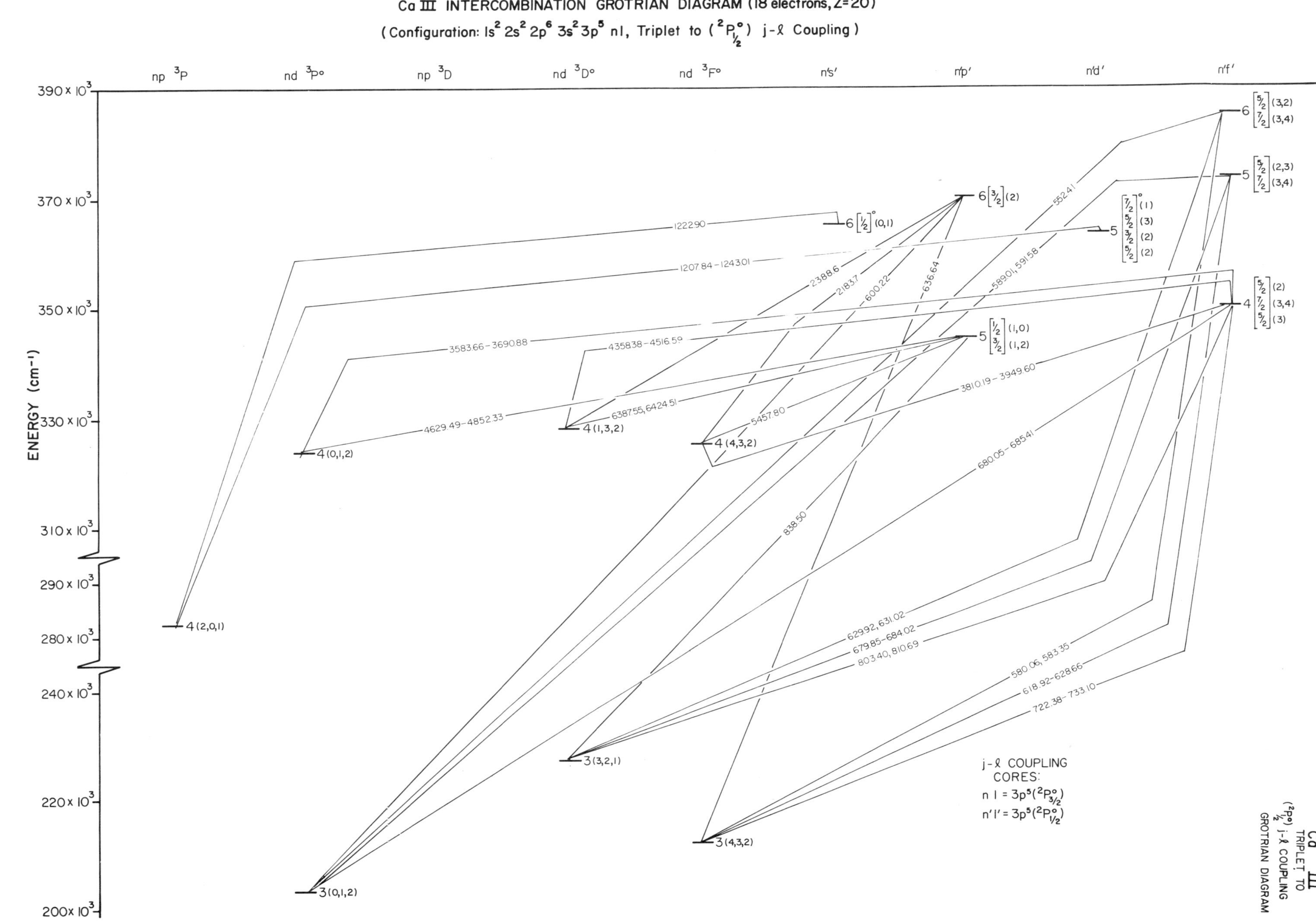

Ca IV GROTRIAN DIAGRAM (17 electrons, Z=20)
(Cl I sequence, Configuration: $1s^2 2s^2 2p^6 3s^2 3p^4 nl$, Doublet and Quartet Systems)

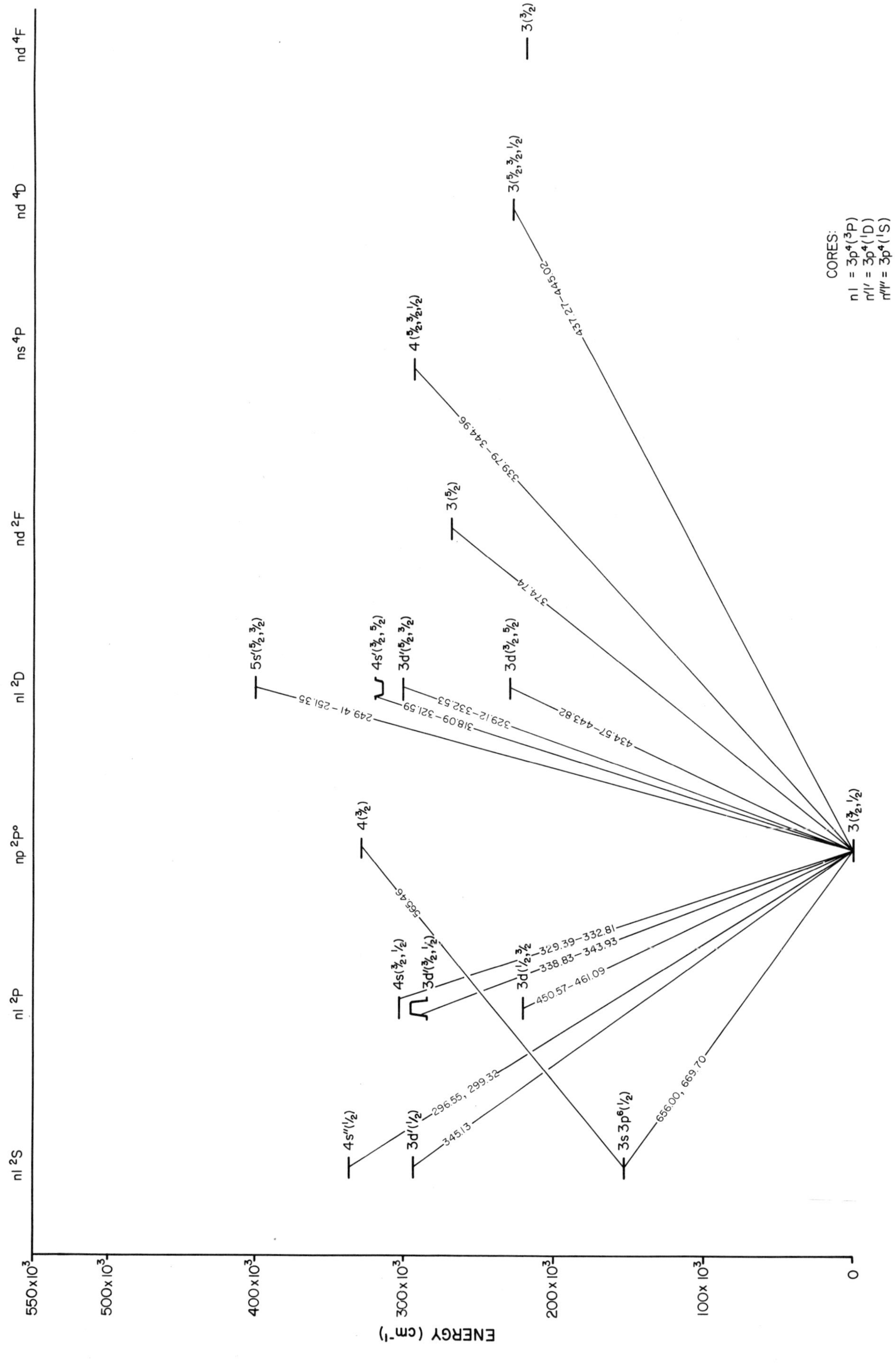

Ca IV ENERGY LEVELS (17 electrons, Z = 20)

(Cl I sequence, Configuration: $1s^2 2s^2 2p^6 3s^2 3p^4 nl$, Doublet & Quartet Systems)

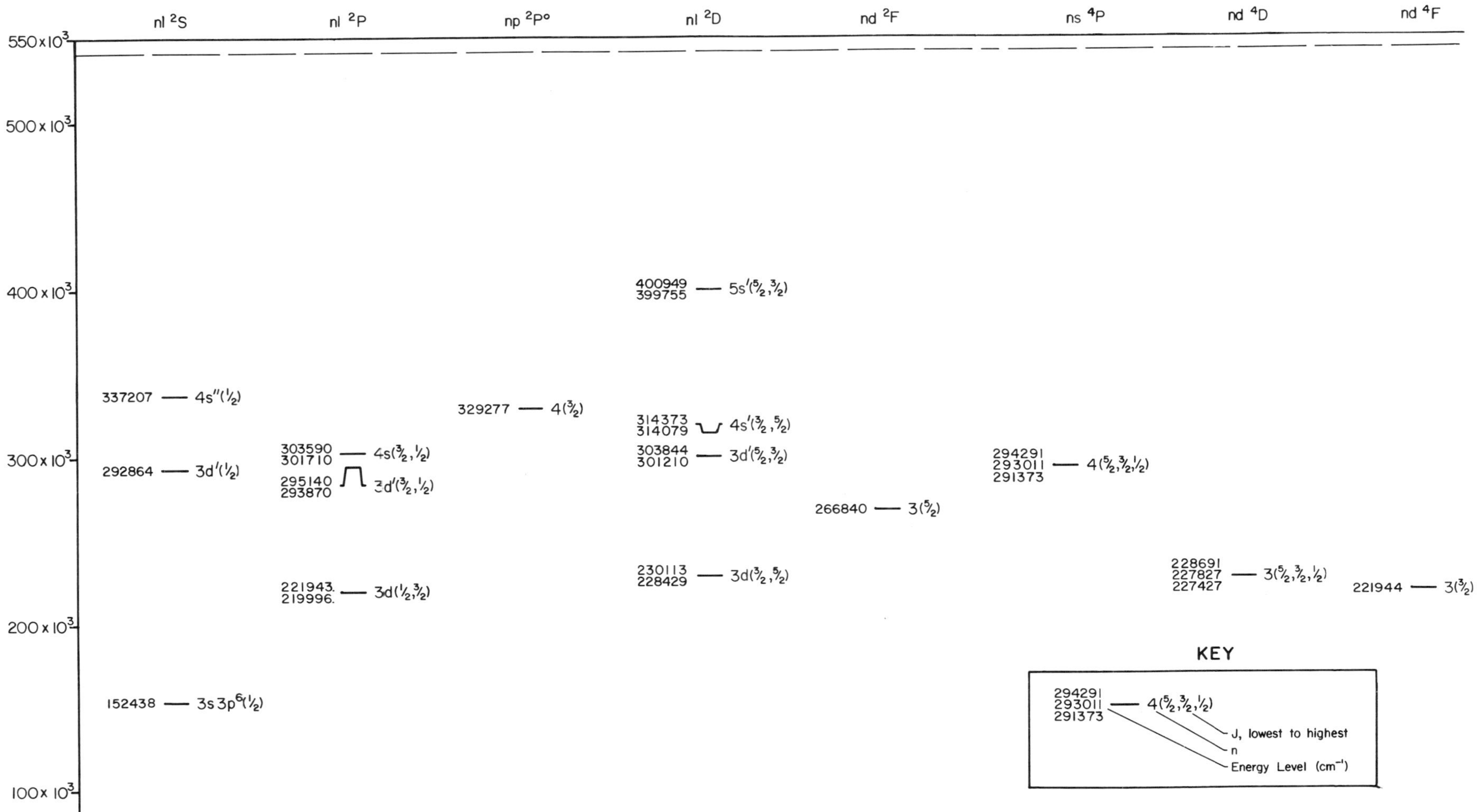

ENERGY (cm^{-1}): 0, 100×10^3, 200×10^3, 300×10^3, 400×10^3, 500×10^3, 550×10^3

nl ^2S
- 337207 — $4s''(\frac{1}{2})$
- 292864 — $3d'(\frac{1}{2})$
- 152438 — $3s3p^6(\frac{1}{2})$

nl ^2P
- 303590, 301710 — $4s(\frac{3}{2},\frac{1}{2})$
- 295140, 293870 — $3d'(\frac{3}{2},\frac{1}{2})$
- 221943., 219996. — $3d(\frac{1}{2},\frac{3}{2})$

np $^2P^o$
- 329277 — $4(\frac{3}{2})$
- 3180, 0 — $3(\frac{3}{2},\frac{1}{2})$

nl ^2D
- 400949, 399755 — $5s'(\frac{5}{2},\frac{3}{2})$
- 314373, 314079 — $4s'(\frac{3}{2},\frac{5}{2})$
- 303844, 301210 — $3d'(\frac{5}{2},\frac{3}{2})$
- 230113, 228429 — $3d(\frac{3}{2},\frac{5}{2})$

nd ^2F
- 266840 — $3(\frac{5}{2})$

ns ^4P
- 294291, 293011, 291373 — $4(\frac{5}{2},\frac{3}{2},\frac{1}{2})$

nd ^4D
- 228691, 227827, 227427 — $3(\frac{5}{2},\frac{3}{2},\frac{1}{2})$

nd ^4F
- 221944 — $3(\frac{3}{2})$

KEY

294291, 293011, 291373 — $4(\frac{5}{2},\frac{3}{2},\frac{1}{2})$
- J, lowest to highest
- n
- Energy Level (cm^{-1})

Ionization Limit
541 200 cm^{-1}
(67.10 eV)

[Ca IV $3p^5\ ^2P^o_{3/2}$ → Ca V $3p^4\ ^3P_2$]

CORES:
- $nl = 3p^4(^3P)$
- $n'l' = 3p^4(^1D)$
- $n''l'' = 3p^4(^1S)$

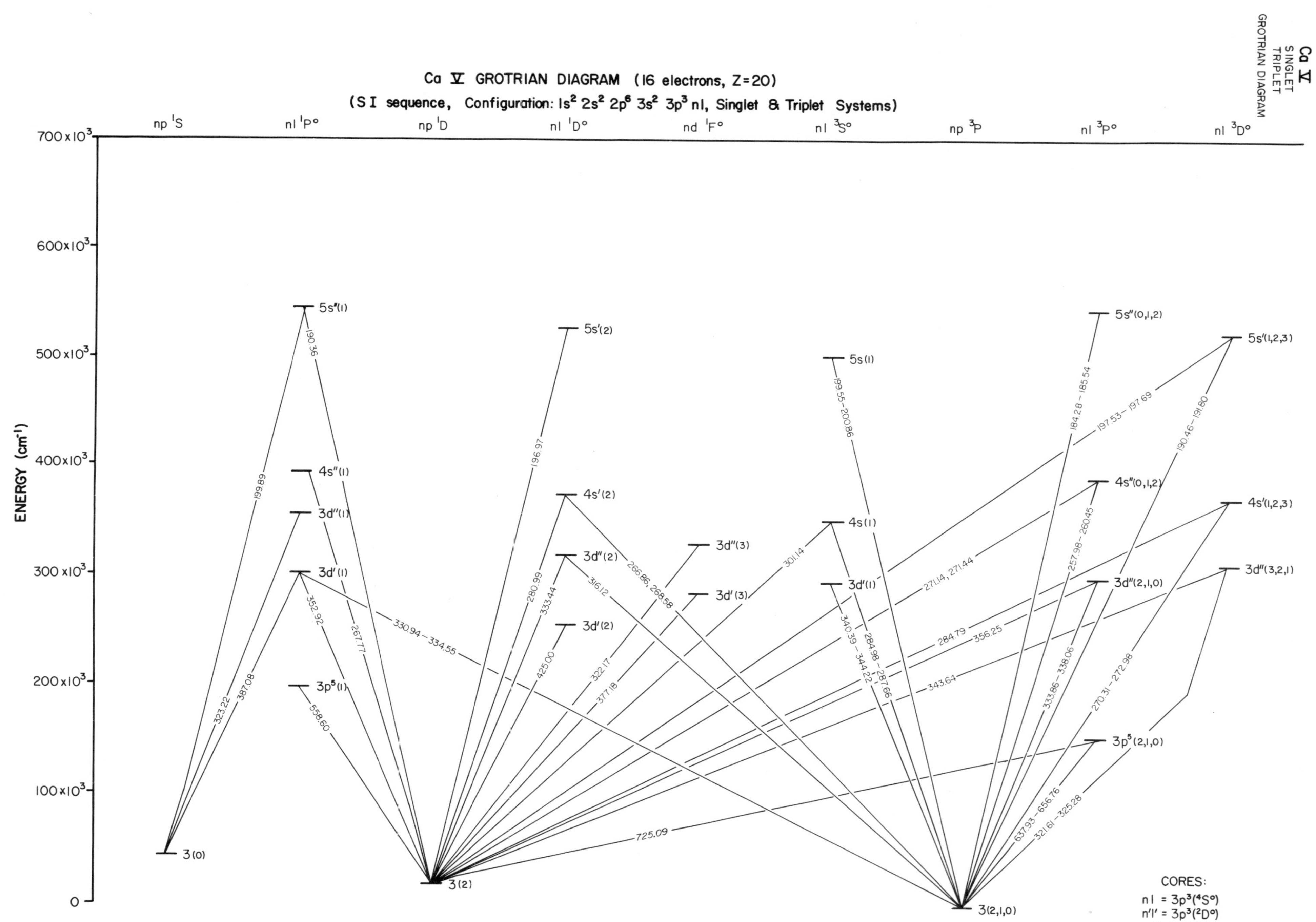

Ca V
SINGLET
TRIPLET
GROTRIAN DIAGRAM
Ca V GROTRIAN DIAGRAM (16 electrons, Z=20)
(S I sequence, Configuration: 1s² 2s² 2p⁶ 3s² 3p³ nl, Singlet & Triplet Systems)
np ¹S
nl ¹P°
np ¹D
nl ¹D°
nd ¹F°
nl ³S°
np ³P
nl ³P°
nl ³D°
ENERGY (cm⁻¹)
700×10³
600×10³
500×10³
400×10³
300×10³
200×10³
100×10³
0
3(0)
3(2)
3(2,1,0)
5s''(1)
4s''(1)
3d''(1)
3d'(1)
3p⁵(1)
5s'(2)
4s'(2)
3d''(2)
3d'(2)
3d''(3)
3d'(3)
5s(1)
4s(1)
3d'(1)
5s''(0,1,2)
4s''(0,1,2)
3d''(2,1,0)
3p⁵(2,1,0)
5s'(1,2,3)
4s'(1,2,3)
3d''(3,2,1)
CORES:
nl = 3p³(⁴S°)
n'l' = 3p³(²D°)
n''l'' = 3p³(²P°)
190.36
199.89
323.22
387.08
267.77
352.92
558.60
330.94 – 334.55
196.97
280.99
333.44
425.00
266.86, 268.58
316.12
322.17
377.18
725.09
301.14
199.55 – 200.86
284.98 – 287.66
340.39 – 344.22
271.14, 271.44
284.79
356.25
343.64
197.53 – 197.69
184.28 – 185.54
190.46 – 191.80
257.98 – 260.45
333.86 – 338.06
270.31 – 272.98
321.61 – 325.28
637.93 – 656.76

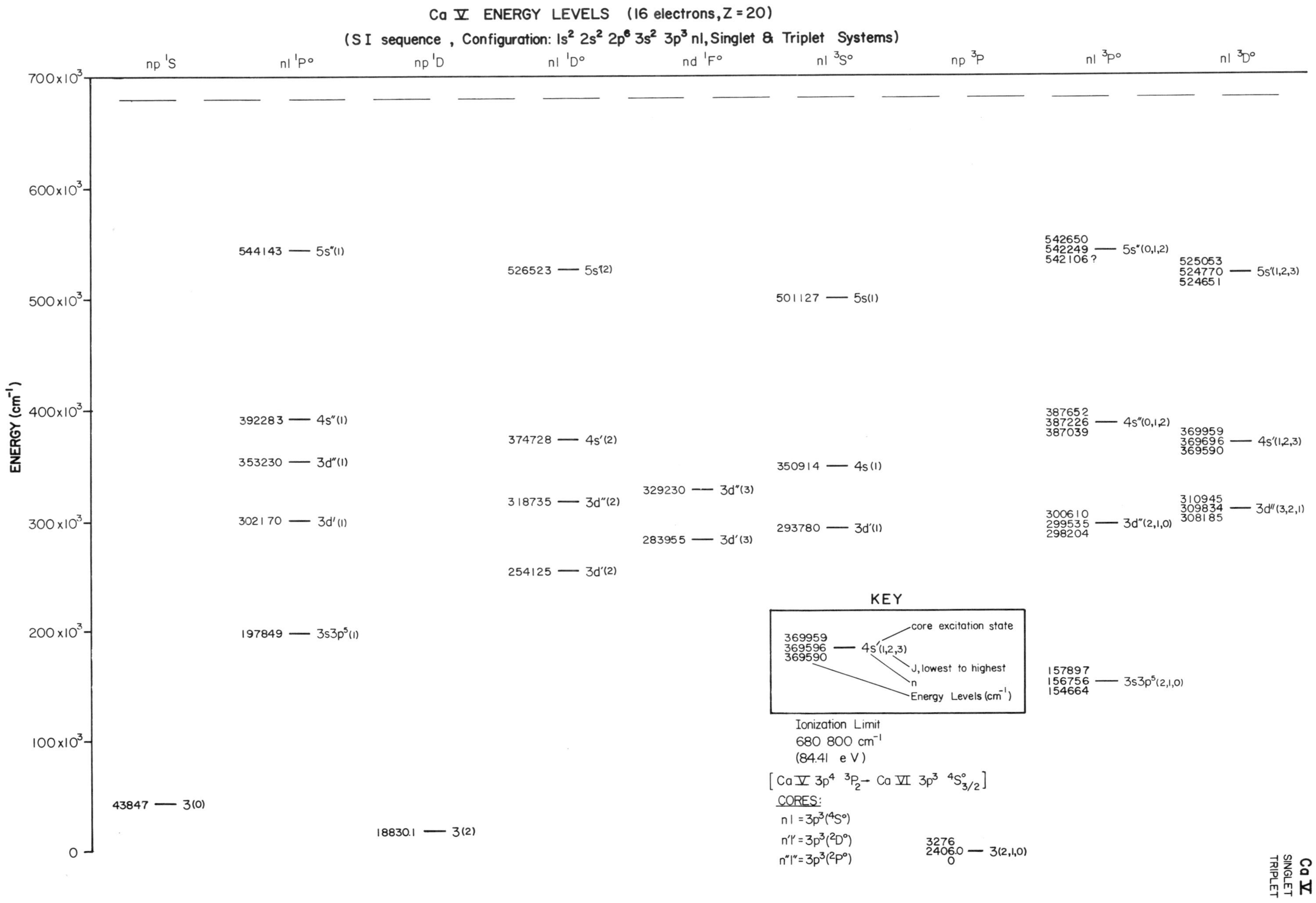

Ca V ENERGY LEVELS (16 electrons, Z = 20)
(S I sequence, Configuration: 1s² 2s² 2p⁶ 3s² 3p³ nl, Singlet & Triplet Systems)
np ¹S
nl ¹P°
np ¹D
nl ¹D°
nd ¹F°
nl ³S°
np ³P
nl ³P°
nl ³D°
ENERGY (cm⁻¹)
700x10³
600x10³
500x10³
400x10³
300x10³
200x10³
100x10³
0
43847 — 3(0)
544143 — 5s″(1)
392283 — 4s″(1)
353230 — 3d″(1)
302170 — 3d′(1)
197849 — 3s3p⁵(1)
18830.1 — 3(2)
526523 — 5s(2)
374728 — 4s′(2)
318735 — 3d″(2)
254125 — 3d′(2)
329230 — 3d″(3)
283955 — 3d′(3)
501127 — 5s(1)
350914 — 4s(1)
293780 — 3d′(1)
3276
2406.0
0 — 3(2,1,0)
542650
542249
542106 ? — 5s″(0,1,2)
387652
387226
387039 — 4s″(0,1,2)
300610
299535
298204 — 3d″(2,1,0)
157897
156756
154664 — 3s3p⁵(2,1,0)
525053
524770
524651 — 5s′(1,2,3)
369959
369696
369590 — 4s′(1,2,3)
310945
309834
308185 — 3d″(3,2,1)
KEY
369959
369596
369590 — 4s′(1,2,3)
core excitation state
J, lowest to highest
n
Energy Levels (cm⁻¹)
Ionization Limit
680 800 cm⁻¹
(84.41 eV)
[Ca V 3p⁴ ³P₂ → Ca VI 3p³ ⁴S°₃/₂]
CORES:
nl = 3p³(⁴S°)
n′l′ = 3p³(²D°)
n″l″ = 3p³(²P°)
Ca V
SINGLET
TRIPLET

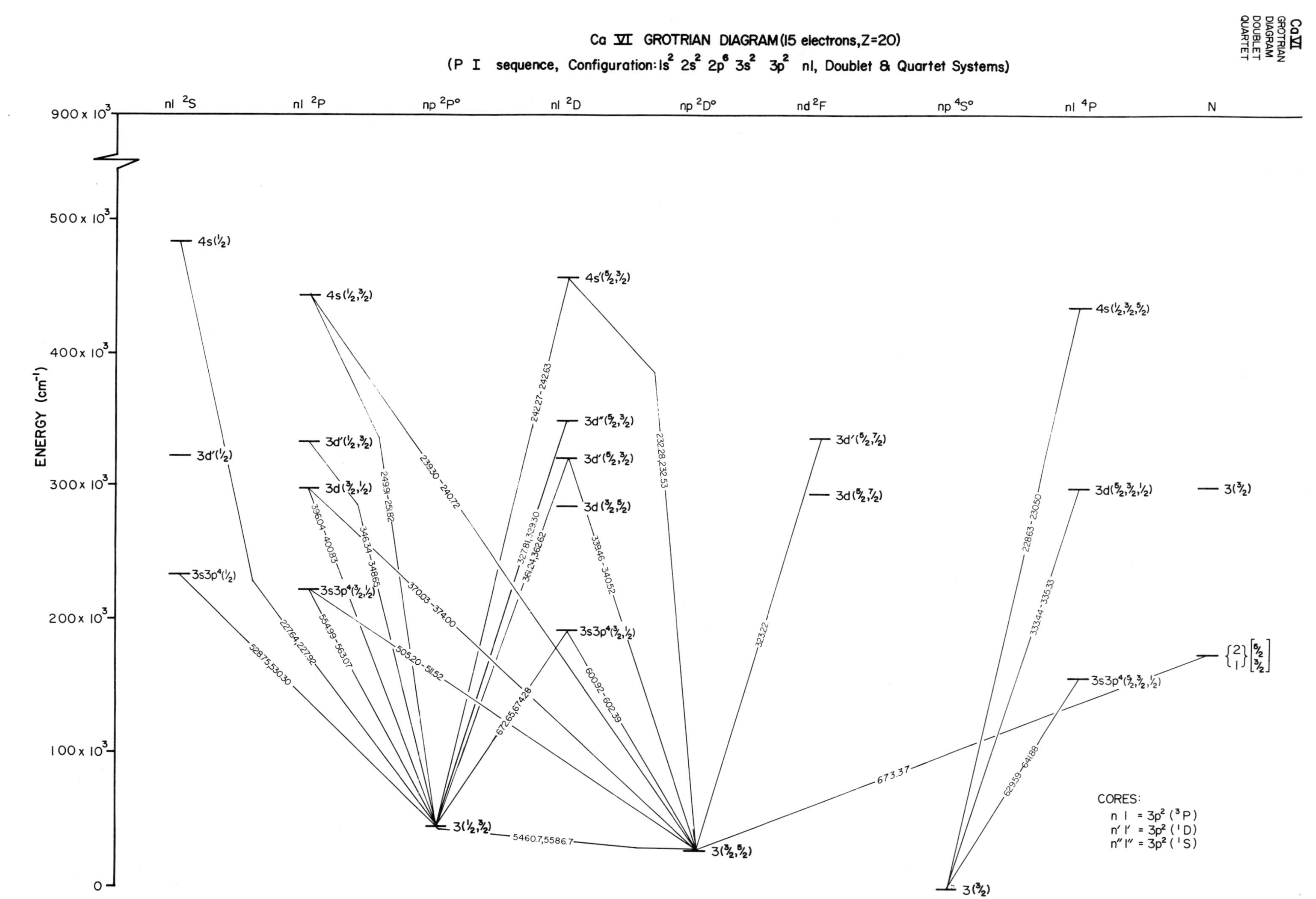
Ca VI
GROTRIAN
DIAGRAM
DOUBLET
QUARTET
Ca VI GROTRIAN DIAGRAM (15 electrons, Z=20)
(P I sequence, Configuration: $1s^2\ 2s^2\ 2p^6\ 3s^2\ 3p^2$ nl, Doublet & Quartet Systems)
nl ^{2}S
nl ^{2}P
np ^{2}P°
nl ^{2}D
np ^{2}D°
nd ^{2}F
np ^{4}S°
nl ^{4}P
N
ENERGY (cm^{-1})
900 x 10^3
500 x 10^3
400 x 10^3
300 x 10^3
200 x 10^3
100 x 10^3
0
4s(1/2)
3d′(1/2)
3s3p^4(1/2)
4s(1/2,3/2)
3d′(1/2,3/2)
3d(3/2,1/2)
3s3p^4(3/2,1/2)
4s′(5/2,3/2)
3d″(5/2,3/2)
3d′(5/2,3/2)
3d(3/2,5/2)
3s3p^4(3/2,1/2)
3d′(5/2,7/2)
3d(5/2,7/2)
4s(1/2,3/2,5/2)
3d(5/2,3/2,1/2)
3s3p^4(5/2,3/2,1/2)
3(3/2)
{2, 1} [5/2, 3/2]
3(1/2,3/2)
3(3/2,5/2)
3(3/2)
52875,53030
22764,22792
554.99-563.07
396.04-400.83
346.34-348.65
249.91-251.82
239.30-240.72
370.03-374.00
505.20-511.52
242.27-242.63
327.81,329.30
361.24,362.62
339.46-340.52
672.65,674.28
600.92-602.39
232.28,232.53
323.22
5460.7,5586.7
673.37
228.63-230.50
333.44-335.33
629.59-641.88
CORES:
n l = 3p^2 (^{3}P)
n′ l′ = 3p^2 (^{1}D)
n″ l″ = 3p^2 (^{1}S)

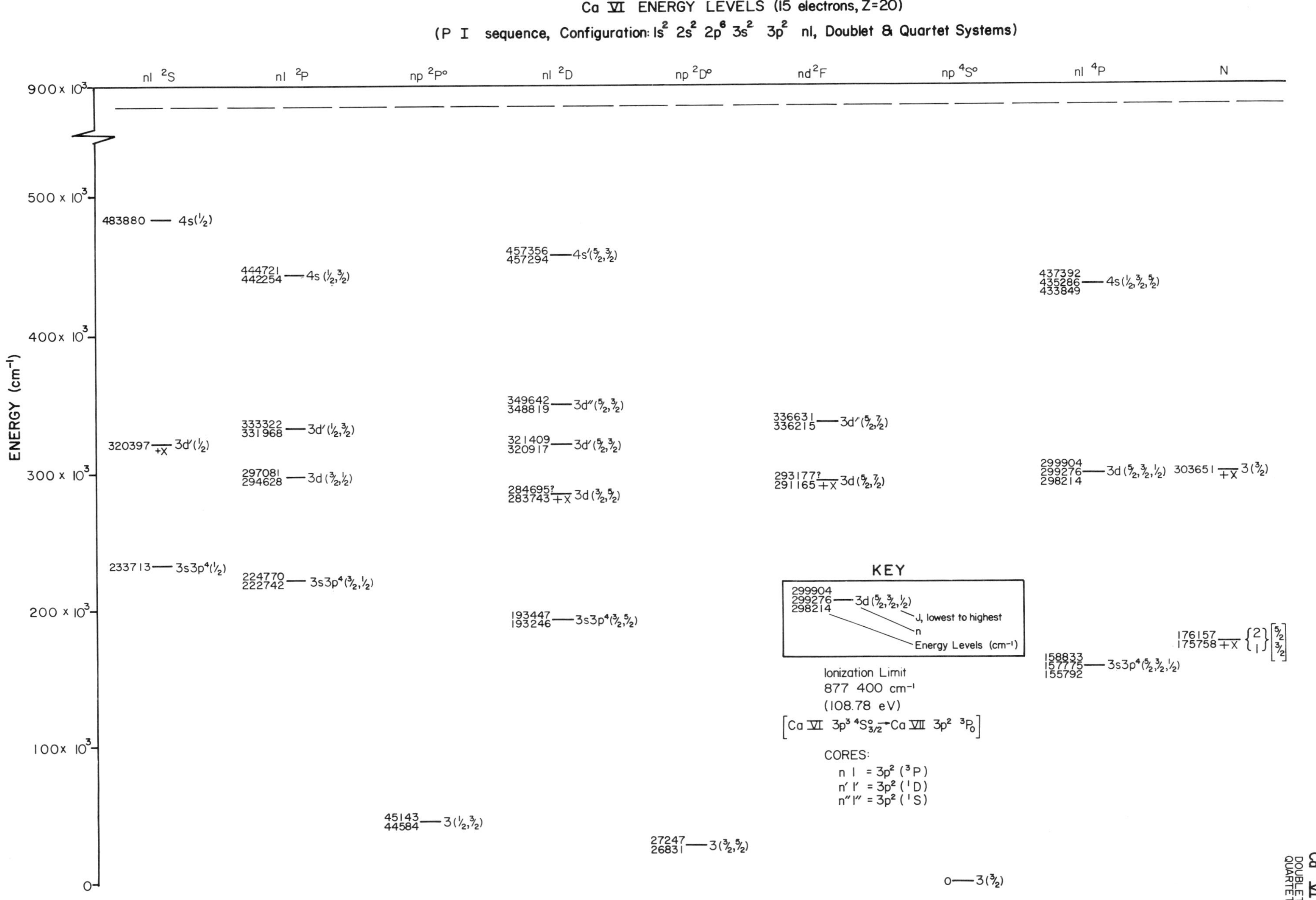
Ca VI ENERGY LEVELS (15 electrons, Z=20)
(P I sequence, Configuration: $1s^2\ 2s^2\ 2p^6\ 3s^2\ 3p^2$ nl, Doublet & Quartet Systems)
nl ^2S
nl ^2P
np $^2P^o$
nl ^2D
np $^2D^o$
nd ^2F
np $^4S^o$
nl ^4P
N
ENERGY (cm^{-1})
900 x 10^3
500 x 10^3
400 x 10^3
300 x 10^3
200 x 10^3
100 x 10^3
0
483880 — 4s(1/2)
320397 +X 3d′(1/2)
233713 — 3s3p⁴(1/2)
444721 442254 — 4s (1/2, 3/2)
333322 331968 — 3d′(1/2, 3/2)
297081 294628 — 3d (3/2, 1/2)
224770 222742 — 3s3p⁴(3/2, 1/2)
45143 44584 — 3(1/2, 3/2)
457356 457294 — 4s′(5/2, 3/2)
349642 348819 — 3d″(5/2, 3/2)
321409 320917 — 3d′(5/2, 3/2)
284695? 283743 +X 3d (3/2, 5/2)
193447 193246 — 3s3p⁴(3/2, 5/2)
27247 26831 — 3(3/2, 5/2)
336631 336215 — 3d′(5/2, 7/2)
293177? 291165 +X 3d (5/2, 7/2)
0 — 3(3/2)
437392 435286 433849 — 4s(1/2, 3/2, 5/2)
299904 299276 298214 — 3d (5/2, 3/2, 1/2)
158833 157775 155792 — 3s3p⁴(5/2, 3/2, 1/2)
303651 +X 3(3/2)
176157 175758 +X {2, 1} [5/2, 3/2]
KEY
299904 299276 298214 — 3d (5/2, 3/2, 1/2)
J, lowest to highest
n
Energy Levels (cm^{-1})
Ionization Limit
877 400 cm^{-1}
(108.78 eV)
[Ca VI $3p^3\ ^4S^o_{3/2}$ → Ca VII $3p^2\ ^3P_0$]
CORES:
n l = $3p^2$ (^3P)
n′ l′ = $3p^2$ (^1D)
n″ l″ = $3p^2$ (^1S)

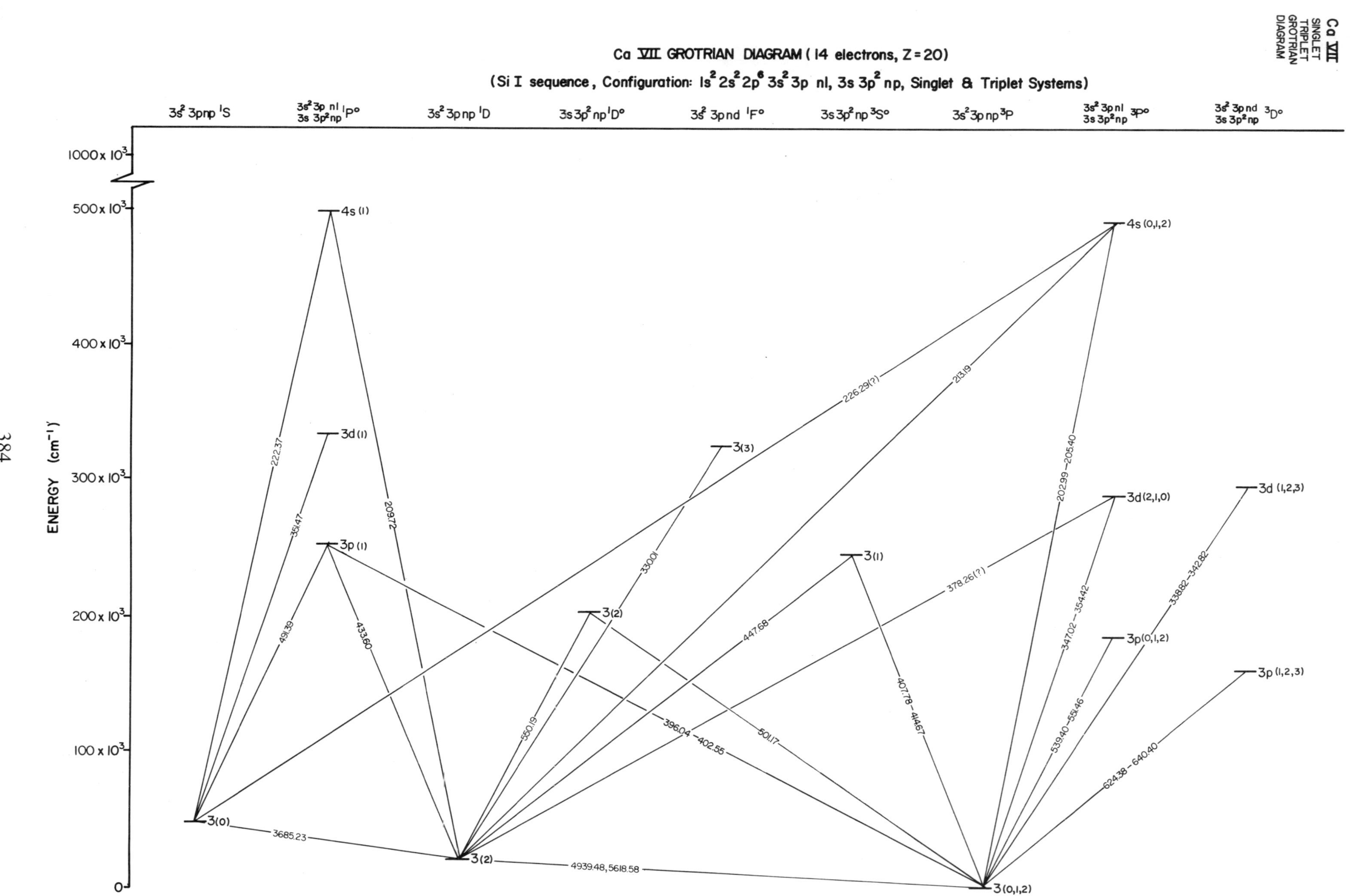

Ca VII
SINGLET
TRIPLET
GROTRIAN
DIAGRAM
Ca VII GROTRIAN DIAGRAM (14 electrons, Z=20)
(Si I sequence, Configuration: 1s² 2s² 2p⁶ 3s² 3p nl, 3s 3p² np, Singlet & Triplet Systems)
3s² 3pnp ¹S
3s² 3p nl ¹P°
3s 3p² np
3s² 3pnp ¹D
3s 3p² np ¹D°
3s² 3pnd ¹F°
3s 3p² np ³S°
3s² 3pnp ³P
3s² 3p nl ³P°
3s 3p² np
3s² 3pnd ³D°
3s 3p² np
1000 x 10³
500 x 10³
400 x 10³
300 x 10³
200 x 10³
100 x 10³
0
ENERGY (cm⁻¹)
4s (1)
3d (1)
3p (1)
3(0)
3(2)
3(2)
3(3)
3(1)
4s (0,1,2)
3d (2,1,0)
3p (0,1,2)
3d (1,2,3)
3p (1,2,3)
3(0,1,2)
222.37
351.47
491.39
433.60
209.72
3685.23
550.19
330.01
226.29(?)
213.19
447.68
396.04 – 402.55
501.17
407.78 – 414.67
378.26(?)
4939.48, 5618.58
202.99 – 205.40
347.02 – 354.42
539.40 – 551.46
338.82 – 342.82
624.38 – 640.40

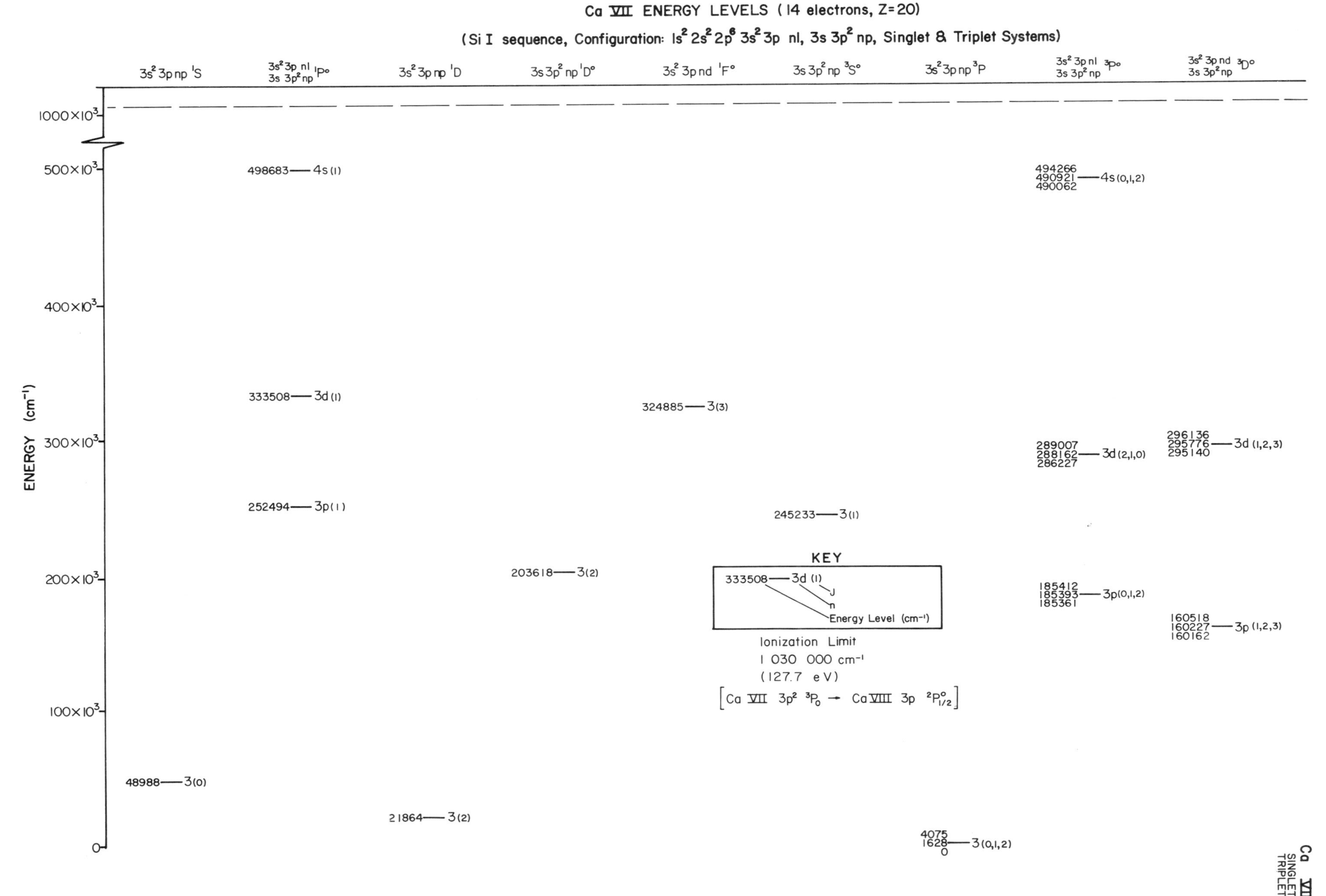
Ca VII ENERGY LEVELS (14 electrons, Z=20)
(Si I sequence, Configuration: $1s^2 2s^2 2p^6 3s^2 3p$ nl, $3s 3p^2$ np, Singlet & Triplet Systems)
$3s^2 3p np\ ^1S$
$3s^2 3p nl$ / $3s 3p^2 np$ $^1P^o$
$3s^2 3p np\ ^1D$
$3s 3p^2 np\ ^1D^o$
$3s^2 3p nd\ ^1F^o$
$3s 3p^2 np\ ^3S^o$
$3s^2 3p np\ ^3P$
$3s^2 3p nl$ / $3s 3p^2 np$ $^3P^o$
$3s^2 3p nd$ / $3s 3p^2 np$ $^3D^o$
1000×10^3
500×10^3
400×10^3
300×10^3
200×10^3
100×10^3
0
ENERGY (cm^{-1})
48988 — 3(0)
498683 — 4s(1)
333508 — 3d(1)
252494 — 3p(1)
21864 — 3(2)
203618 — 3(2)
324885 — 3(3)
245233 — 3(1)
4075 / 1628 / 0 — 3(0,1,2)
494266 / 490921 / 490062 — 4s(0,1,2)
289007 / 288162 / 286227 — 3d(2,1,0)
185412 / 185393 / 185361 — 3p(0,1,2)
296136 / 295776 / 295140 — 3d(1,2,3)
160518 / 160227 / 160162 — 3p(1,2,3)
KEY
333508 — 3d (1)
J
n
Energy Level (cm^{-1})
Ionization Limit
1 030 000 cm^{-1}
(127.7 eV)
[Ca VII $3p^2\ ^3P_0$ → Ca VIII $3p\ ^2P^o_{1/2}$]
Ca VII
SINGLET
TRIPLET

Ca VIII GROTRIAN DIAGRAM (13 electrons, Z=20)

(Al I sequence, Configuration: $1s^2 2s^2 2p^6 3s^2 nl$, $3s\,3p\,nl$, Doublet & Quartet Systems)

$3s^2ns$ / $3s3pnp$ ^2S — $3s3p\,np\,^2P$ — $3s^2\,np\,^2P°$ — $3s^2nd$ / $3s3pnp$ ^2D — $3s^2\,nf\,^2F°$ — $^4S°$ — $3s\,3p\,np\,^4P$ — $3s3p(^3P°)nl\,^4P°$ — $3s3p(^3P°)nd\,^4D°$

ENERGY (cm^{-1})

1200x10³, 1000x10³, 900x10³, 800x10³, 700x10³, 600x10³, 500x10³, 400x10³, 300x10³, 200x10³, 100x10³, 0

4s(1/2); 3p(1/2); 3(1/2,3/2); 6d(3/2,5/2); 5d(3/2,5/2); 4d(3/2,5/2); 3d(3/2,5/2); 3p(3/2,5/2); 7(5/2,7/2); 6(5/2,7/2); 5(5/2,7/2); 4(5/2,7/2); 3(1/2,3/2); $3p^3$(3/2); 3(1/2,3/2,5/2); 4s(1/2,3/2,5/2); 3d(5/2,3/2); 3(1/2–7/2)

182.71, 184.16; 461.71, 471.07; 428.10–441.08; 102.07, 102.48; 112.91, 113.45; 143.22–144.11; 354.17–359.65; 582.83–597.83; 131.42, 131.47; 141.11, 141.15; 160.59, 160.64; 216.96, 217.04; 174.91, 174.98; 462.59–471.18; 177.89–179.51; 356.91–363.39; 353.00–357.98

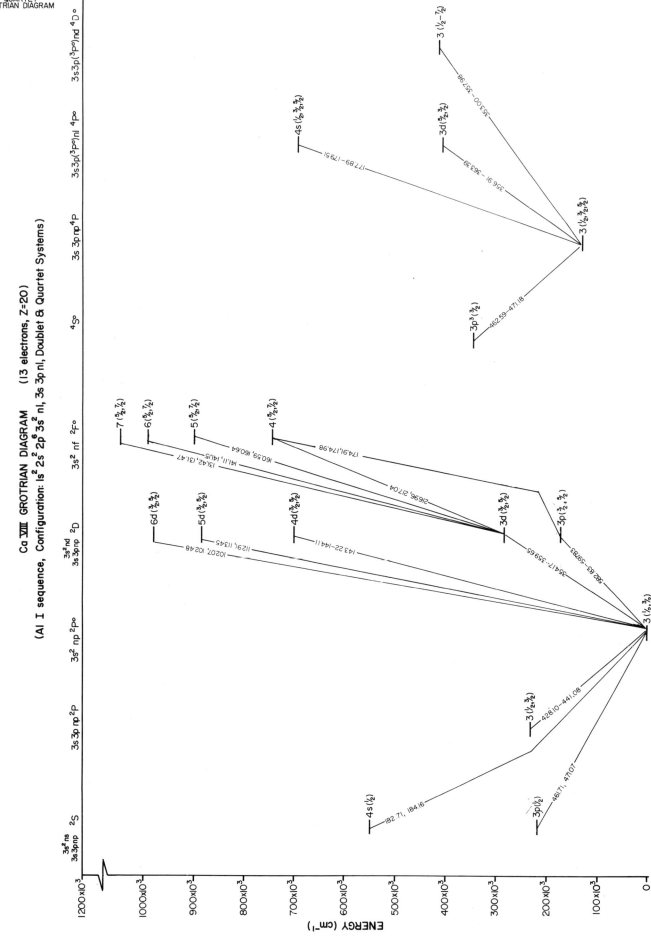

Ca VIII ENERGY LEVELS (13 electrons, Z=20)

(Al I sequence, Configuration: $1s^2 2s^2 2p^6 3s^2 nl$, $3s3p\,nl$, Doublet & Quartet Systems)

Ca VIII
DOUBLET
QUARTET

ENERGY (cm^{-1})

0 – 100×10³ – 200×10³ – 300×10³ – 400×10³ – 500×10³ – 600×10³ – 700×10³ – 800×10³ – 900×10³ – 1000×10³ – 1200×10³

Series	Energy levels (cm^{-1})	n (J)
$3s^2ns$, $3s3pnp$ ^2S	216 588	3p(1/2)
	547 320	4s(1/2)
$3s3p\,np\ ^2P$	233 591 231 018	3(1/2, 3/2)
$3s^2np\ ^2P^o$	4 304 0	3(1/2, 3/2)
$3s^2nd$, $3s3pnp$ ^2D	171 825 171 575	3p(3/2, 5/2)
	282 573 282 354	3d(3/2, 5/2)
	698 416 698 238	4d(3/2, 5/2)
	885 746 885 693	5d(3/2, 5/2)
	980 085 979 749	6d(3/2, 5/2)
$3s^2nf\ ^2F^o$	743 323 743 279	4(5/2, 7/2)
	905 083 905 050	5(5/2, 7/2)
	991 023	6(5/2, 7/2)
	1 043 238	7(5/2, 7/2)
$^4S^o$	345 274	$3p^3$(3/2)
$3s3pnp\ ^4P$	133 042 130 678 129 100	3(1/2, 3/2, 5/2)
$3s3p(^3P^o)nl\ ^4P^o$	409 291 408 227	3d(5/2, 3/2)
	692 833 690 128 688 747	4s(1/2, 3/2, 5/2)
$3s3p(^3P^o)nd\ ^4D^o$	412 881 412 772 412 388 412 816	3(1/2 – 7/2)

KEY

698 416
698 238 — 4d (3/2, 5/2)

J, lower to higher
n
Energy Levels (cm^{-1})

Ionization Limit
1 187 600 cm^{-1}
(147.24 eV)
[Ca VIII $3s^2 3p\ ^2P^o_{1/2}$ → Ca IX $3s^2\ ^1S_0$]

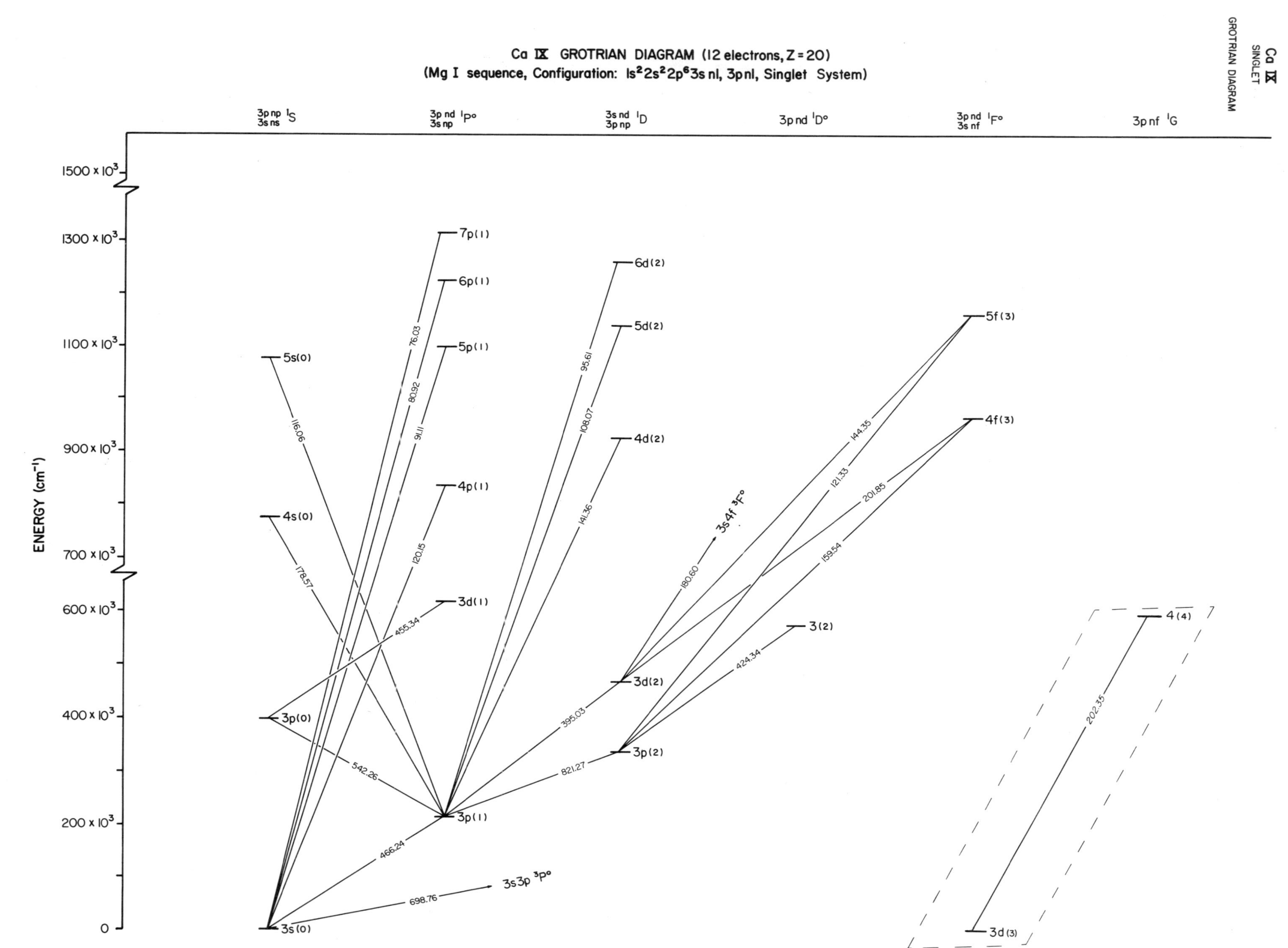

Ca IX
SINGLET
GROTRIAN DIAGRAM
Ca IX GROTRIAN DIAGRAM (12 electrons, Z = 20)
(Mg I sequence, Configuration: $1s^2 2s^2 2p^6 3s\,nl$, 3pnl, Singlet System)
3pnp ^1S
3s ns
3pnd $^1P^o$
3s np
3snd ^1D
3p np
3pnd $^1D^o$
3pnd $^1F^o$
3s nf
3pnf ^1G
ENERGY (cm^{-1})
1500 x 10^3
1300 x 10^3
1100 x 10^3
900 x 10^3
700 x 10^3
600 x 10^3
400 x 10^3
200 x 10^3
0
7p(1)
6p(1)
5p(1)
4p(1)
3d(1)
5s(0)
4s(0)
3p(0)
3s(0)
3p(1)
6d(2)
5d(2)
4d(2)
3d(2)
3p(2)
3(2)
5f(3)
4f(3)
4(4)
3d(3)
3s3p $^3P^o$
3s4f $^3F^o$
76.03
80.92
91.11
120.15
116.06
178.57
455.34
542.26
466.24
698.76
95.61
108.07
141.36
395.03
821.27
180.60
144.35
121.33
201.85
159.54
424.34
202.35

Ca IX ENERGY LEVELS (12 electrons, Z = 20)
(Mg I sequence, Configuration: $1s^2 2s^2 2p^6 3s\,nl$, $3p\,nl$, Singlet System)

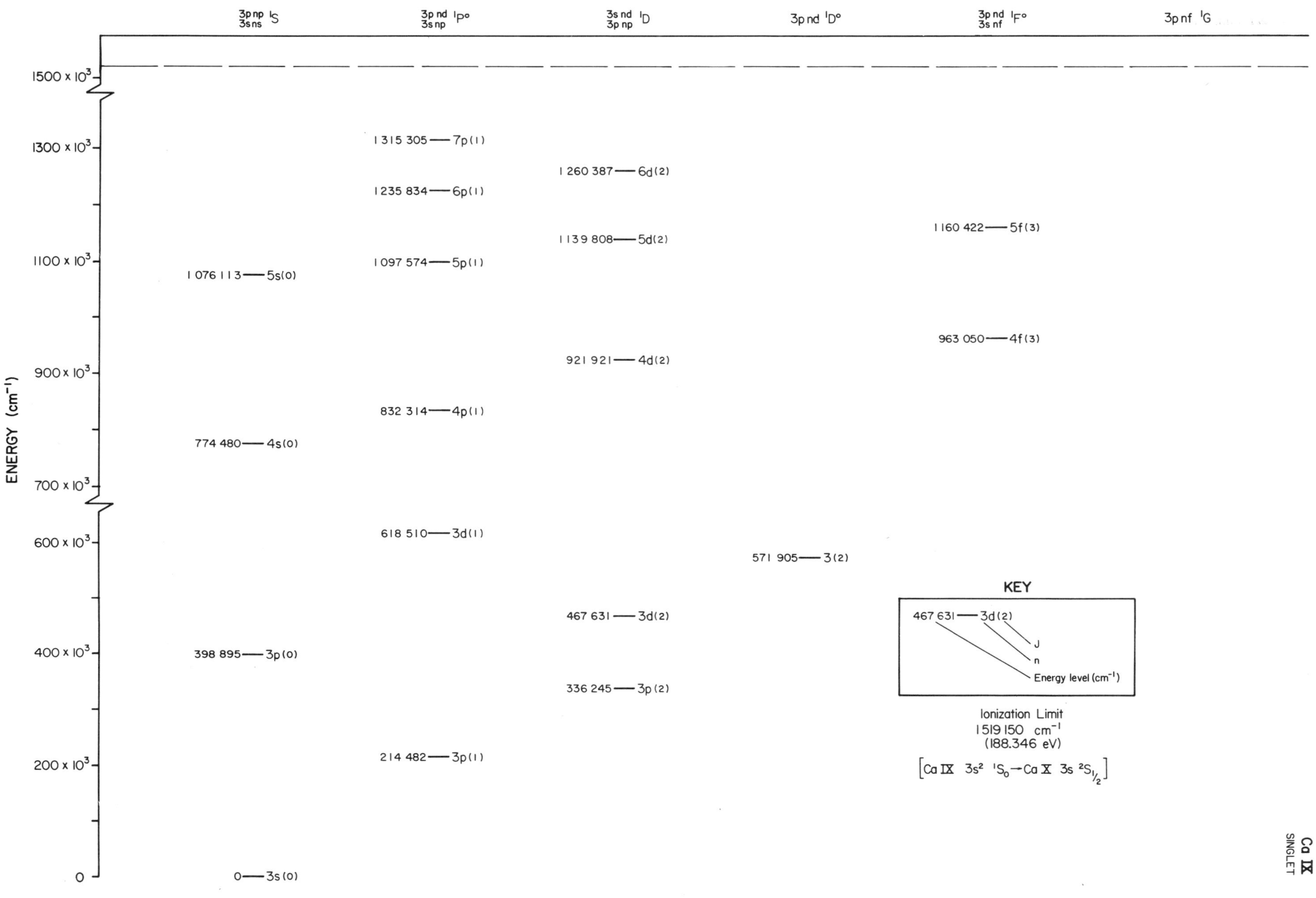

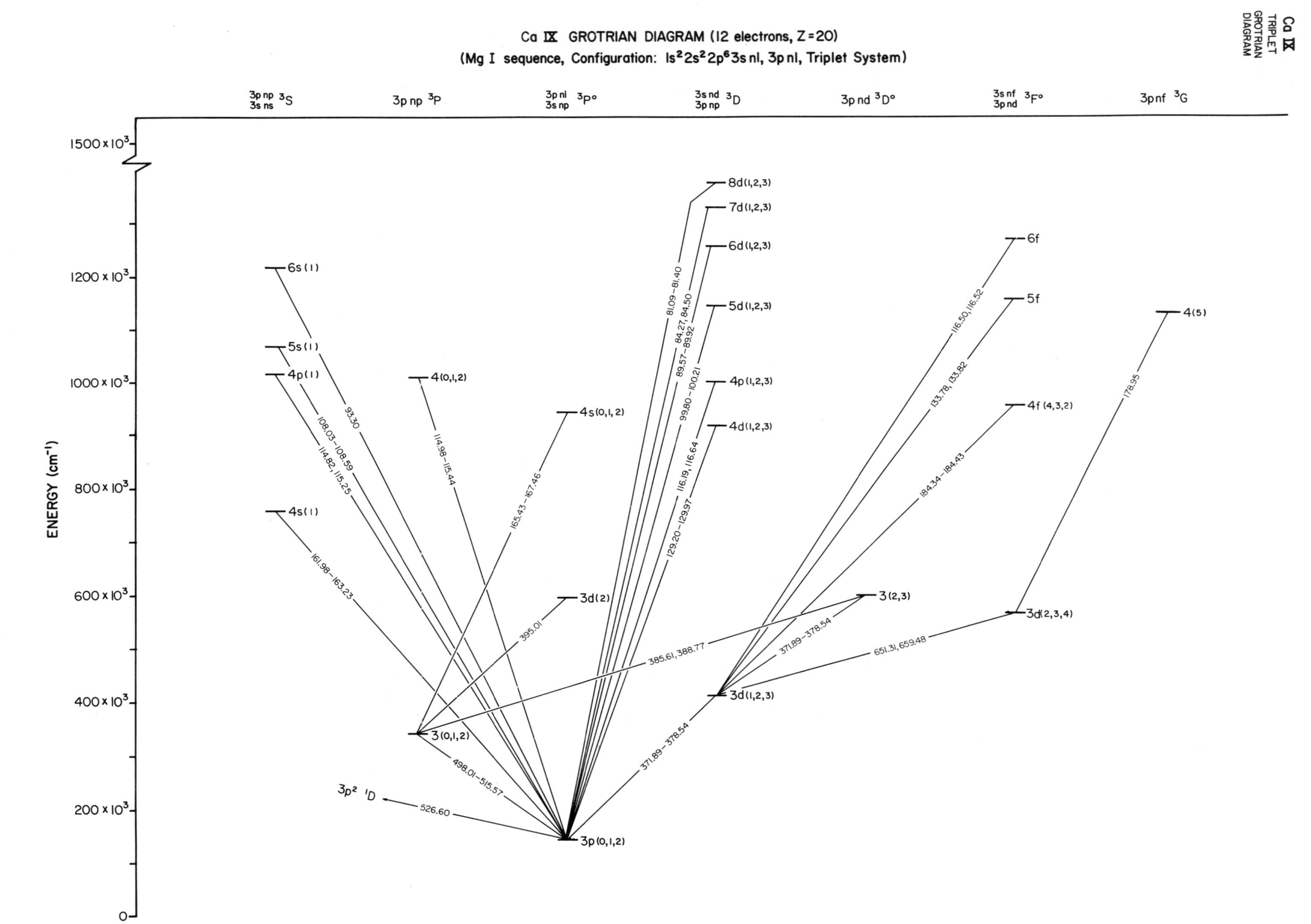

Ca IX
TRIPLET
GROTRIAN
DIAGRAM
Ca IX GROTRIAN DIAGRAM (12 electrons, Z=20)
(Mg I sequence, Configuration: 1s²2s²2p⁶3s nl, 3p nl, Triplet System)
3p np 3s ns ³S
3p np ³P
3p nl 3s np ³P°
3s nd 3p np ³D
3p nd ³D°
3s nf 3p nd ³F°
3p nf ³G
ENERGY (cm⁻¹)
1500 x 10³
1200 x 10³
1000 x 10³
800 x 10³
600 x 10³
400 x 10³
200 x 10³
0
6s(1)
5s(1)
4p(1)
4s(1)
93.30
108.03–108.59
114.82, 115.25
161.98–163.23
4(0,1,2)
114.98–115.44
4s(0,1,2)
165.43–167.46
3d(2)
395.01
3(0,1,2)
498.01–515.57
3p² ¹D
526.60
3p(0,1,2)
8d(1,2,3)
7d(1,2,3)
6d(1,2,3)
5d(1,2,3)
4p(1,2,3)
4d(1,2,3)
81.09–81.40
84.27, 84.50
89.57–89.92
99.80–100.21
116.19, 116.64
129.20–129.97
385.61, 388.77
3(2,3)
371.89–378.54
3d(1,2,3)
371.89–378.54
6f
5f
4f(4,3,2)
116.50, 116.52
133.78, 133.82
184.34–184.43
651.31, 659.48
3d(2,3,4)
4(5)
178.95

Ca IX ENERGY LEVELS (12 electrons, Z = 20)

(Mg I sequence, Configuration: $1s^2 2s^2 2p^6 3snl$, $3pnl$, Triplet System)

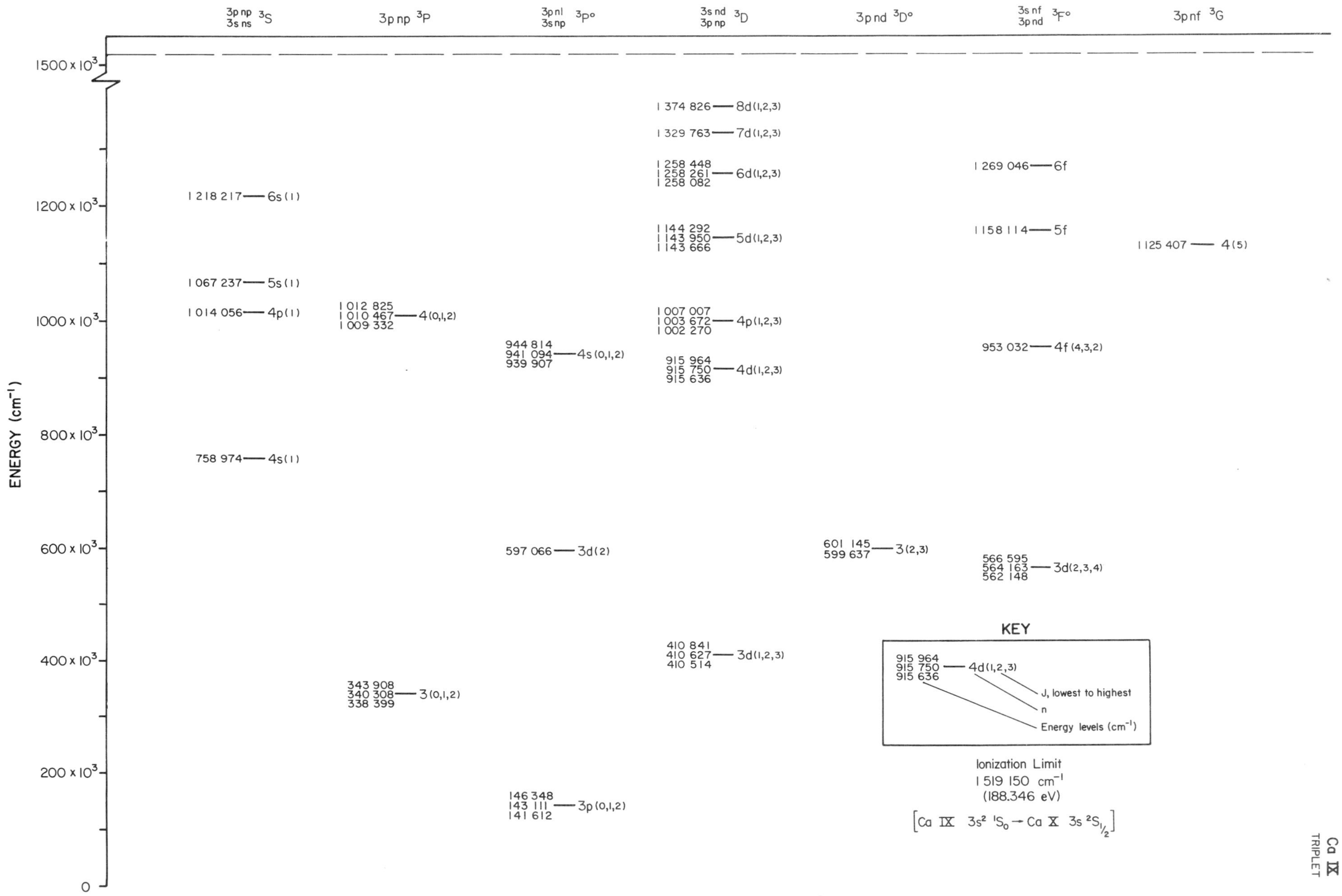

ENERGY (cm^{-1})

0
200 x 10^3
400 x 10^3
600 x 10^3
800 x 10^3
1000 x 10^3
1200 x 10^3
1500 x 10^3

3pnp 3sns ^3S
1 218 217 — 6s(1)
1 067 237 — 5s(1)
1 014 056 — 4p(1)
758 974 — 4s(1)

3pnp ^3P
1 012 825
1 010 467 — 4(0,1,2)
1 009 332
343 908
340 308 — 3(0,1,2)
338 399

3pnl 3snp $^3P°$
944 814
941 094 — 4s(0,1,2)
939 907
597 066 — 3d(2)
146 348
143 111 — 3p(0,1,2)
141 612

3snd 3pnp ^3D
1 374 826 — 8d(1,2,3)
1 329 763 — 7d(1,2,3)
1 258 448
1 258 261 — 6d(1,2,3)
1 258 082
1 144 292
1 143 950 — 5d(1,2,3)
1 143 666
1 007 007
1 003 672 — 4p(1,2,3)
1 002 270
915 964
915 750 — 4d(1,2,3)
915 636
410 841
410 627 — 3d(1,2,3)
410 514

3pnd $^3D°$
601 145
599 637 — 3(2,3)

3snf 3pnd $^3F°$
1 269 046 — 6f
1 158 114 — 5f
953 032 — 4f(4,3,2)
566 595
564 163 — 3d(2,3,4)
562 148

3pnf ^3G
1 125 407 — 4(5)

KEY

915 964
915 750 — 4d(1,2,3)
915 636

J, lowest to highest
n
Energy levels (cm^{-1})

Ionization Limit
1 519 150 cm^{-1}
(188.346 eV)
[Ca IX $3s^2\ ^1S_0$ → Ca X $3s\ ^2S_{1/2}$]

Ca X GROTRIAN DIAGRAM (11 electrons, Z=20)

(Na I sequence, Configuration: $1s^2\,2s^2\,2p^6\,nl$, Doublet System)

ns ^2S | $^2P^o$ | nd ^2D | nf $^2F^o$

$2p^5 3s^2 (3/2, 1/2)$

ENERGY (cm^{-1})

2700 x 10^3 · 1600 x 10^3 · 1400 x 10^3 · 1200 x 10^3 · 1000 x 10^3 · 400 x 10^3 · 0

NOTE SCALE CHANGE

ns ^2S: 8(1/2), 7(1/2), 6(1/2), 5(1/2), 4(1/2), 3(1/2)

$^2P^o$: 10p, 9p(3/2), 8p(1/2, 3/2), 7p(1/2, 3/2), 6p(1/2, 3/2), 5p(1/2, 3/2), 4p(1/2, 3/2), 3p(1/2, 3/2)

nd ^2D: 10(5/2), 9(3/2, 5/2), 8(3/2, 5/2), 7(5/2, 3/2), 6(3/2, 5/2), 5(3/2, 5/2), 4(3/2, 5/2), 3(3/2, 5/2)

nf $^2F^o$: 13, 11(7/2, 5/2), 10(7/2, 5/2), 9(5/2, 7/2), 8(5/2, 7/2), 7(5/2, 7/2), 6(5/2, 7/2), 5(5/2, 7/2), 4(5/2, 7/2)

Transitions: 75.05; 78.46, 78.79; 85.17, 85.53; 371.90; 99.95, 100.46; 62.99; 64.13; 65.83; 68.50; 73.17, 73.20; 82.80, 82.87; 95.90; 105.37; 126.55, 126.63; 110.96, 111.20; 151.84, 153.02; 557.74, 574.01; 70.53, 70.77; 71.86, 72.13; 73.84, 74.12; 76.94, 77.24; 82.27, 82.61; 93.04, 93.48; 122.98–123.75; 206.75, 207.39; 411.71–420.49; 81.82, 81.85; 83.58, 83.62; 84.95, 84.98; 86.86, 86.89; 89.67, 89.70; 94.13, 94.16; 101.91, 101.95; 118.12, 118.17; 166.95, 167.05

For 10f thru 14f $^2F^o$ see Bibliographic note

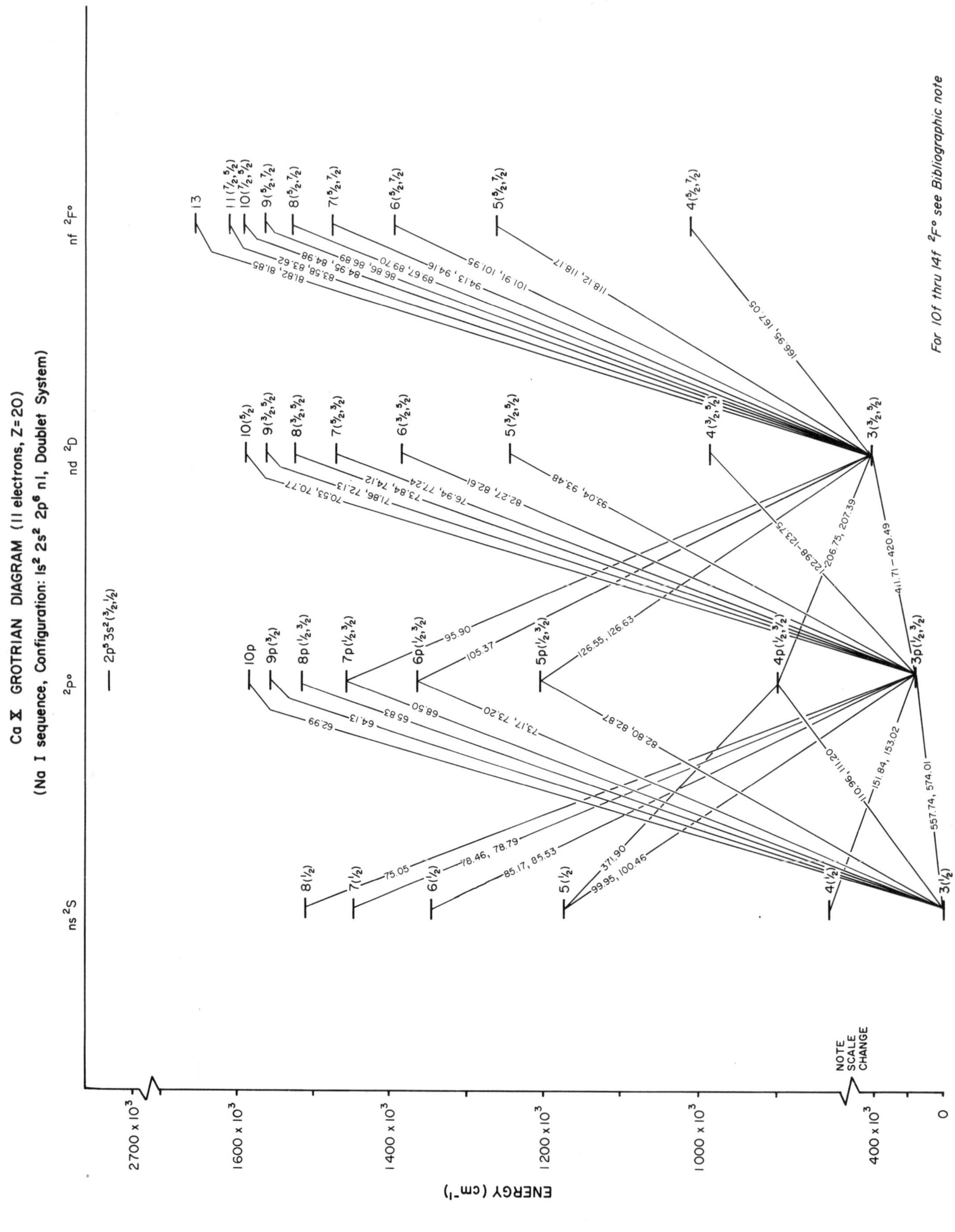

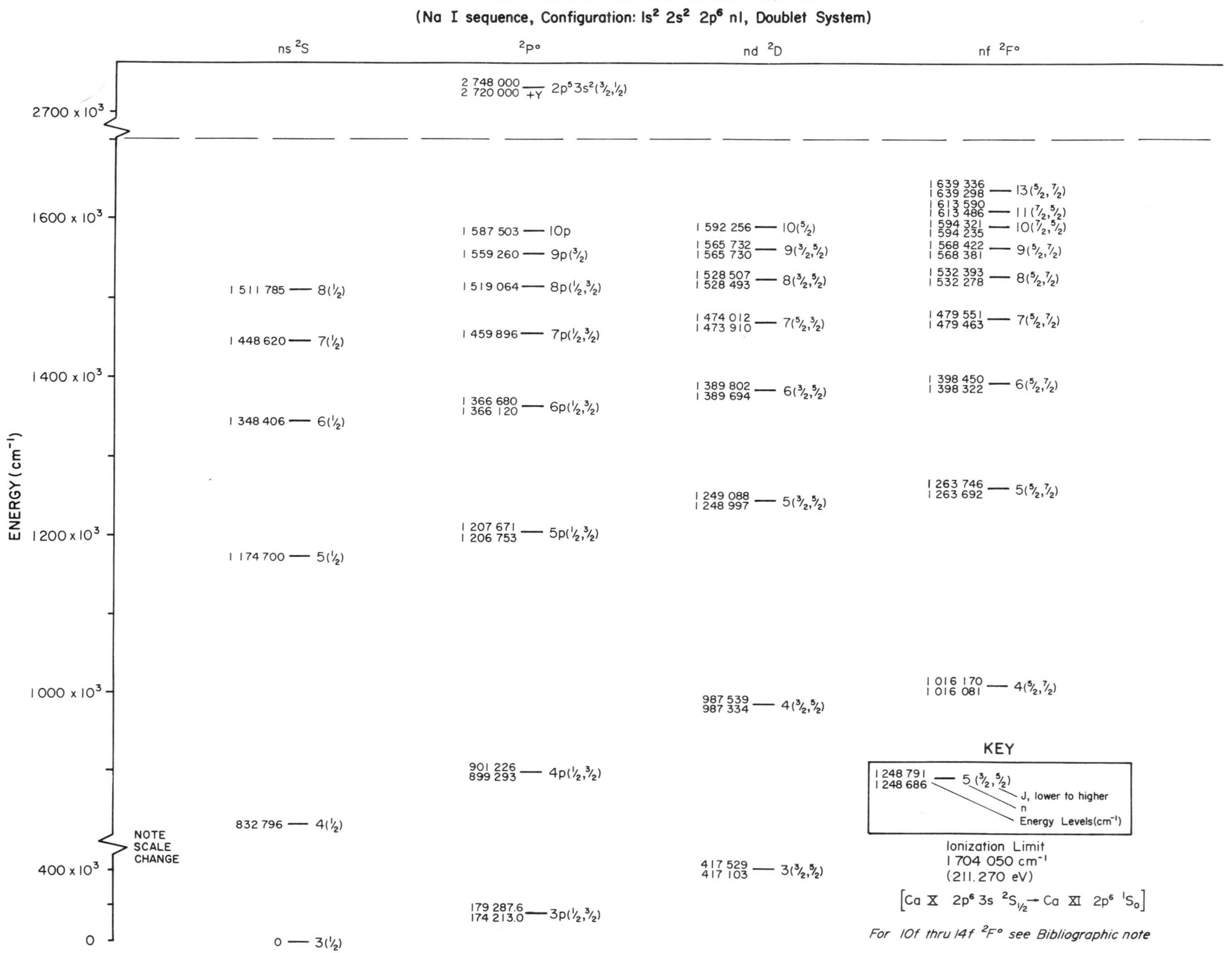
Ca X ENERGY LEVELS (11 electrons, Z=20)
(Na I sequence, Configuration: 1s² 2s² 2p⁶ nl, Doublet System)
ns ²S
²P°
nd ²D
nf ²F°
ENERGY (cm⁻¹)
2700 x 10³
1600 x 10³
1400 x 10³
1200 x 10³
1000 x 10³
400 x 10³
0
NOTE SCALE CHANGE
1 511 785 — 8(1/2)
1 448 620 — 7(1/2)
1 348 406 — 6(1/2)
1 174 700 — 5(1/2)
832 796 — 4(1/2)
0 — 3(1/2)
2 748 000 / 2 720 000 +Y — 2p⁵3s²(3/2,1/2)
1 587 503 — 10p
1 559 260 — 9p(3/2)
1 519 064 — 8p(1/2,3/2)
1 459 896 — 7p(1/2,3/2)
1 366 680 / 1 366 120 — 6p(1/2,3/2)
1 207 671 / 1 206 753 — 5p(1/2,3/2)
901 226 / 899 293 — 4p(1/2,3/2)
179 287.6 / 174 213.0 — 3p(1/2,3/2)
1 592 256 — 10(5/2)
1 565 732 / 1 565 730 — 9(3/2,5/2)
1 528 507 / 1 528 493 — 8(3/2,5/2)
1 474 012 / 1 473 910 — 7(5/2,3/2)
1 389 802 / 1 389 694 — 6(3/2,5/2)
1 249 088 / 1 248 997 — 5(3/2,5/2)
987 539 / 987 334 — 4(3/2,5/2)
417 529 / 417 103 — 3(3/2,5/2)
1 639 336 / 1 639 298 — 13(5/2,7/2)
1 613 590 / 1 613 486 — 11(7/2,5/2)
1 594 321 / 1 594 235 — 10(7/2,5/2)
1 568 422 / 1 568 381 — 9(5/2,7/2)
1 532 393 / 1 532 278 — 8(5/2,7/2)
1 479 551 / 1 479 463 — 7(5/2,7/2)
1 398 450 / 1 398 322 — 6(5/2,7/2)
1 263 746 / 1 263 692 — 5(5/2,7/2)
1 016 170 / 1 016 081 — 4(5/2,7/2)
KEY
1 248 791 / 1 248 686 — 5 (3/2,5/2)
J, lower to higher
n
Energy Levels(cm⁻¹)
Ionization Limit
1 704 050 cm⁻¹
(211.270 eV)
[Ca X 2p⁶ 3s ²S1/2 → Ca XI 2p⁶ ¹S0]
For 10f thru 14f ²F° see Bibliographic note

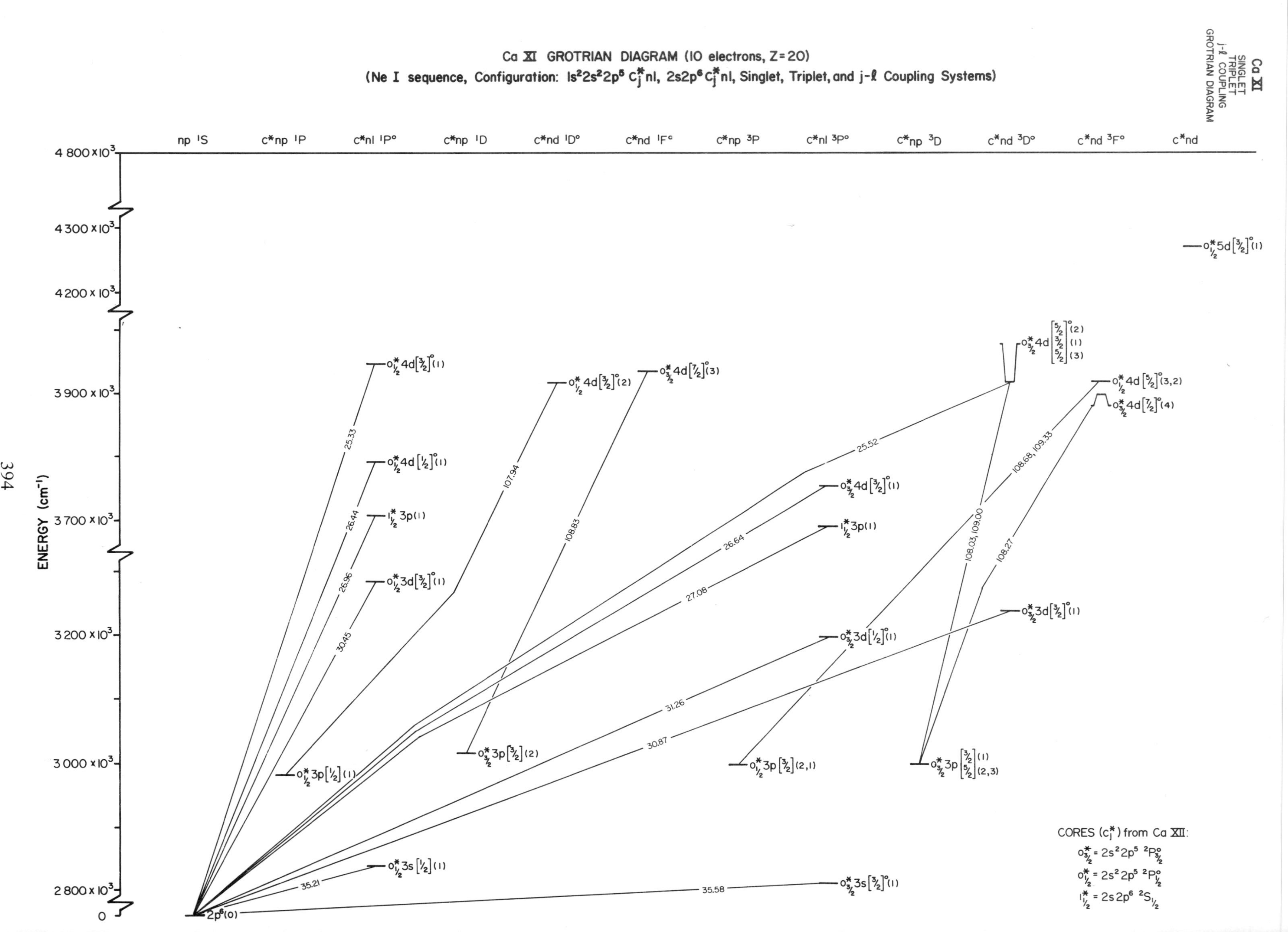
Ca XI
SINGLET
TRIPLET
j-ℓ COUPLING
GROTRIAN DIAGRAM
Ca XI GROTRIAN DIAGRAM (10 electrons, Z=20)
(Ne I sequence, Configuration: 1s²2s²2p⁵ C*j nl, 2s2p⁶ C*j nl, Singlet, Triplet, and j-ℓ Coupling Systems)
np ¹S
c*np ¹P
c*nl ¹P°
c*np ¹D
c*nd ¹D°
c*nd ¹F°
c*np ³P
c*nl ³P°
c*np ³D
c*nd ³D°
c*nd ³F°
c*nd
ENERGY (cm⁻¹)
4 800 x 10³
4 300 x 10³
4 200 x 10³
3 900 x 10³
3 700 x 10³
3 200 x 10³
3 000 x 10³
2 800 x 10³
0
2p⁶(0)
o*1/2 5d[3/2]°(1)
o*1/2 4d[3/2]°(1)
o*1/2 4d[1/2]°(1)
1*1/2 3p(1)
o*1/2 3d[3/2]°(1)
o*1/2 3p[1/2](1)
o*1/2 4d[3/2]°(2)
o*3/2 4d[7/2]°(3)
o*3/2 3p[3/2](2)
o*3/2 4d[3/2]°(1)
1*1/2 3p(1)
o*3/2 3d[1/2]°(1)
o*1/2 3p[3/2](2,1)
o*3/2 4d [5/2]°(2) [3/2](1) [5/2](3)
o*1/2 4d[5/2]°(3,2)
o*3/2 4d[7/2]°(4)
o*3/2 3d[3/2]°(1)
o*3/2 3p [3/2](1) [5/2](2,3)
o*1/2 3s[1/2](1)
o*3/2 3s[3/2]°(1)
25.33
26.44
26.96
30.45
107.94
108.83
25.52
26.64
27.08
31.26
30.87
35.21
35.58
108.68, 109.33
108.03, 109.00
108.27
CORES (c*j) from Ca XII:
o*3/2 = 2s²2p⁵ ²P°3/2
o*1/2 = 2s²2p⁵ ²P°1/2
1*1/2 = 2s2p⁶ ²S1/2

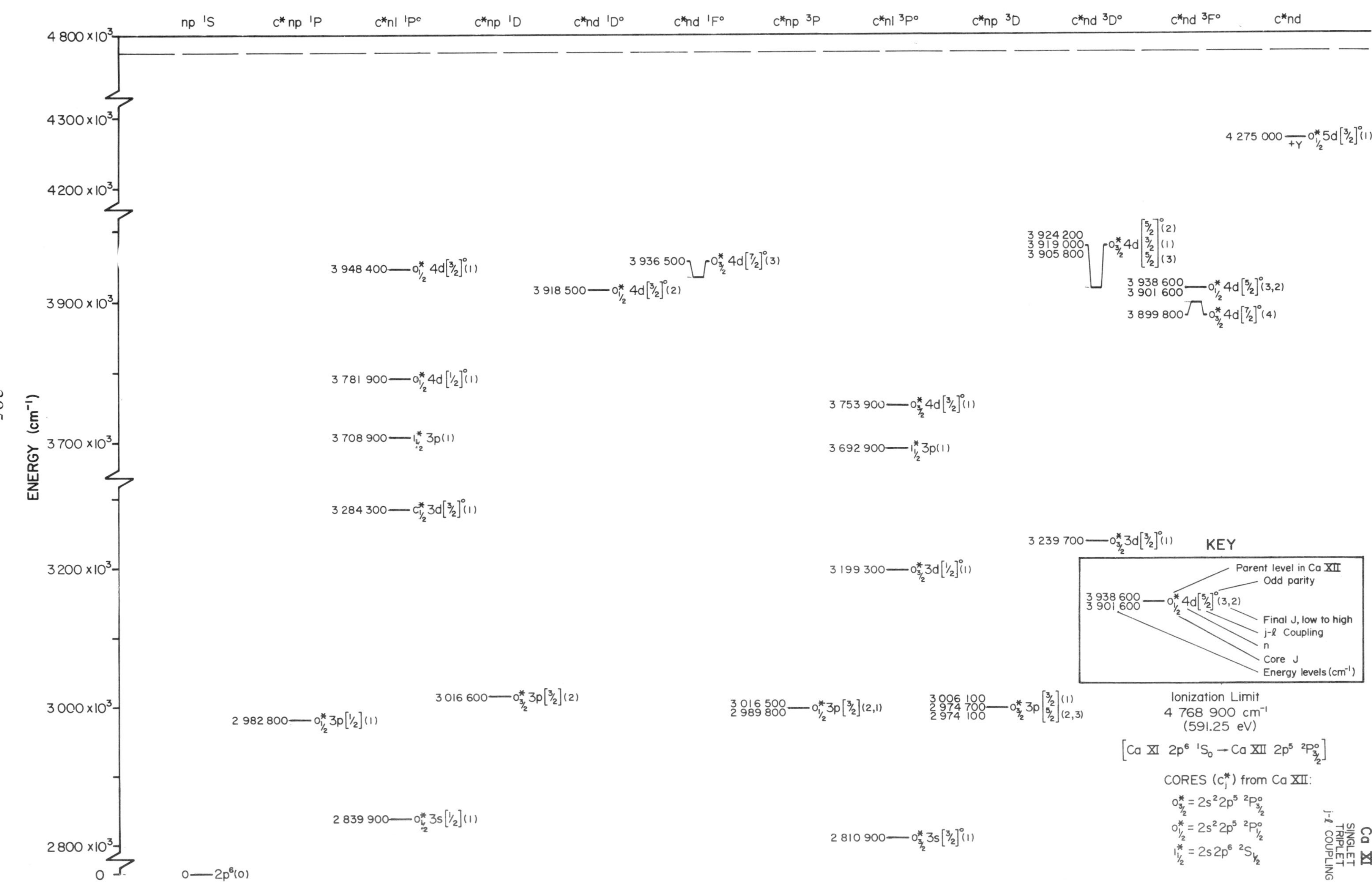

Ca XI ENERGY LEVELS (10 electrons, Z = 20)
(Ne I sequence, Configuration: 1s²2s²2p⁵ c*j nl, 2s2p⁶c*j nl, Singlet, Triplet, and j-ℓ Coupling Systems)
np ¹S
c* np ¹P
c*nl ¹P°
c*np ¹D
c*nd ¹D°
c*nd ¹F°
c*np ³P
c*nl ³P°
c*np ³D
c*nd ³D°
c*nd ³F°
c*nd
4 800 x10³
4 300 x10³
4 200 x10³
3 900 x10³
3 700 x10³
3 200 x10³
3 000 x10³
2 800 x10³
0
ENERGY (cm⁻¹)
4 275 000 +Y — o*1/2 5d[3/2]°(1)
3 948 400 — o*1/2 4d[3/2]°(1)
3 936 500 — o*3/2 4d[7/2]°(3)
3 918 500 — o*1/2 4d[3/2]°(2)
3 924 200
3 919 000
3 905 800 — o*3/2 4d [5/2]°(2) [3/2](1) [5/2](3)
3 938 600
3 901 600 — o*1/2 4d[5/2]°(3,2)
3 899 800 — o*3/2 4d[7/2]°(4)
3 781 900 — o*1/2 4d[1/2]°(1)
3 753 900 — o*3/2 4d[3/2]°(1)
3 708 900 — 1*1/2 3p(1)
3 692 900 — 1*1/2 3p(1)
3 284 300 — c*1/2 3d[3/2]°(1)
3 239 700 — o*3/2 3d[3/2]°(1)
3 199 300 — o*3/2 3d[1/2]°(1)
3 016 600 — o*3/2 3p[3/2](2)
2 982 800 — o*1/2 3p[1/2](1)
3 016 500
2 989 800 — o*1/2 3p[3/2](2,1)
3 006 100
2 974 700
2 974 100 — o*3/2 3p [3/2](1) [5/2](2,3)
2 839 900 — o*1/2 3s[1/2](1)
2 810 900 — o*3/2 3s[3/2]°(1)
0 — 2p⁶(0)
KEY
3 938 600
3 901 600 — o*1/2 4d[5/2]°(3,2)
Parent level in Ca XII
Odd parity
Final J, low to high
j-ℓ Coupling
n
Core J
Energy levels (cm⁻¹)
Ionization Limit
4 768 900 cm⁻¹
(591.25 eV)
[Ca XI 2p⁶ ¹S₀ → Ca XII 2p⁵ ²P°3/2]
CORES (c*j) from Ca XII:
o*3/2 = 2s²2p⁵ ²P°3/2
o*1/2 = 2s²2p⁵ ²P°1/2
1*1/2 = 2s2p⁶ ²S1/2
Ca XI
SINGLET
TRIPLET
j-ℓ COUPLING

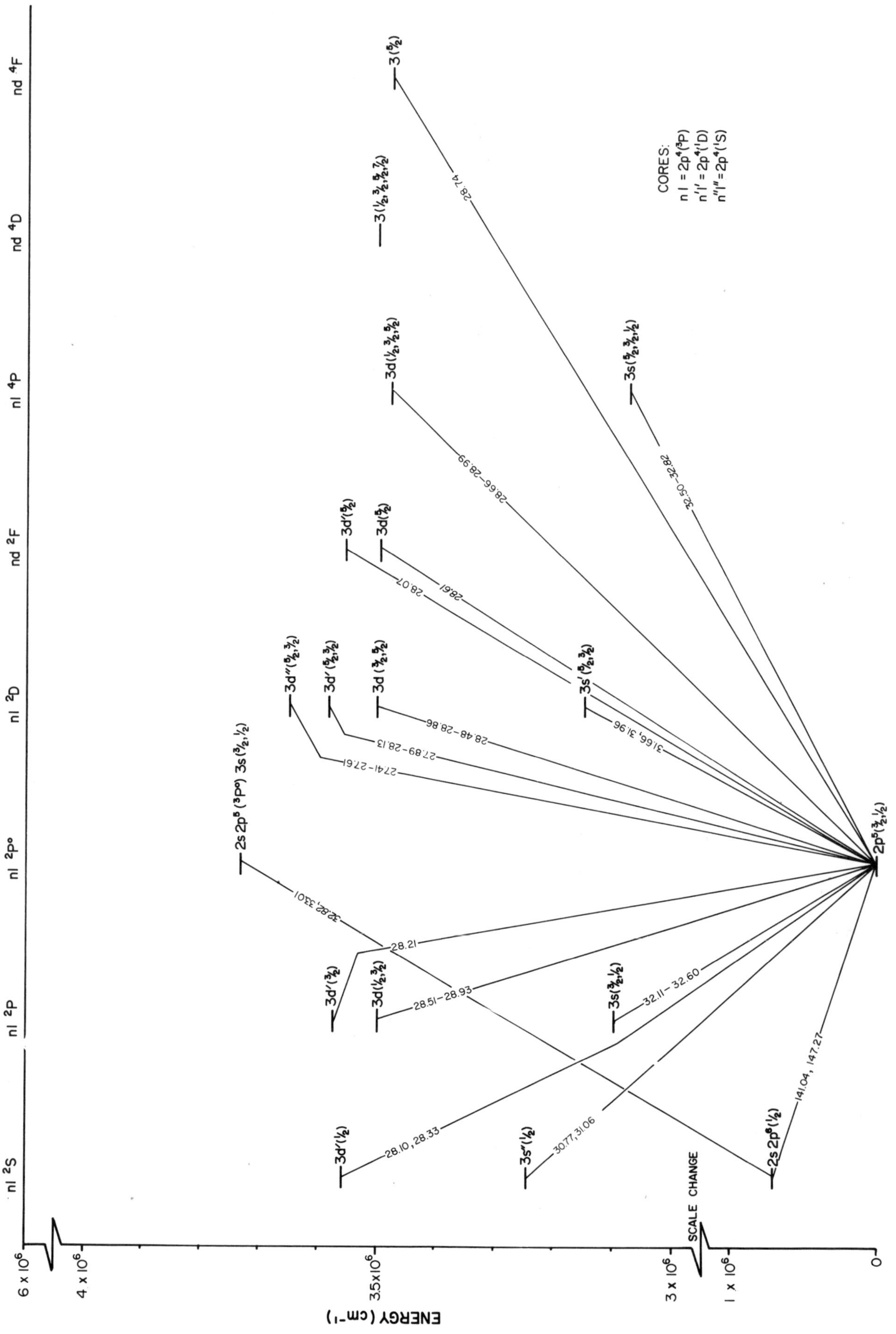
Ca XII GROTRIAN DIAGRAM (9 electrons, Z=20)
(F I sequence, Configuration: 1s² 2s² 2p⁴ nl, Doublet & Quartet Systems)
nl ²S
nl ²P
nl ²P°
nl ²D
nd ²F
nl ⁴P
nd ⁴D
nd ⁴F
CORES:
n l = 2p⁴(³P)
n'l' = 2p⁴(¹D)
n''l'' = 2p⁴(¹S)
ENERGY (cm⁻¹)
6 x 10⁶
4 x 10⁶
3.5x10⁶
3 x 10⁶
1 x 10⁶
0
SCALE CHANGE
2p⁵(3/2,1/2)
2s 2p⁶(1/2)
2s 2p⁵ (³P°) 3s(3/2,1/2)
3d'(1/2)
3s''(1/2)
3d'(3/2)
3d(1/2,3/2)
3s(3/2,1/2)
3d''(5/2,3/2)
3d'(5/2,3/2)
3d (3/2,5/2)
3s'(5/2,3/2)
3d'(5/2)
3d(5/2)
3d(1/2,3/2,5/2)
3s(5/2,3/2,1/2)
3(1/2,3/2,5/2,7/2)
3(5/2)
28.10, 28.33
30.77, 31.06
141.04, 147.27
28.51–28.93
32.11–32.60
28.21
32.82, 33.01
27.41–27.61
27.89–28.13
28.48–28.86
31.66, 31.96
28.07
28.61
28.66–28.99
32.50–32.82
28.74

Ca XII ENERGY LEVELS (9 electrons, Z=20)

(FI sequence, Configuration: $1s^2\,2s^2\,2p^4\,nl$, Doublet & Quartet Systems)

ENERGY (cm⁻¹)

Axis: 0, 1 x 10⁶, SCALE CHANGE, 3 x 10⁶, 3.5x10⁶, 4 x 10⁶, 6 x 10⁶

nl ²S
- 3 559 200 — 3d″(1/2)
- 3 249 500 — 3s″(1/2)
- 709 038.8 — 2s2p⁶(1/2)

nl ²P
- 3 574 900 — 3d′(3/2)
- 3 508 200, 3 486 600 — 3d(1/2, 3/2)
- 3 114 700, 3 097 800 — 3s(3/2, 1/2)

nl ²P°
- 3 755 900, 3 738 000 — 2s2p⁵ (³P°) 3s(3/2, 1/2)
- 30 027.8, 0 — 2p⁵(3/2, 1/2)

nl ²D
- 3 652 400, 3 647 900 — 3d″(5/2, 3/2)
- 3 585 000, 3 574 900 — 3d′(5/2, 3/2)
- 3 511 500, 3 494 900 — 3d(3/2, 5/2)
- 3 159 300, 3 158 500 — 3s′(5/2, 3/2)

nd ²F
- 3 562 800 — 3d′(5/2)
- 3 494 900 — 3d(5/2)

nl ⁴P
- 3 489 400, 3 479 500, 3 475 800 — 3d(1/2, 3/2, 5/2)
- 3 089 200, 3 077 100, 3 062 100 — 3s(5/2, 3/2, 1/2)

nd ⁴D
- 3 500 000 +X — 3(1/2, 3/2, 5/2, 7/2)

nd ⁴F
- 3 479 100 — 3(5/2)

KEY

3 089 200, 3 077 100, 3 062 100 — 3s(5/2, 3/2, 1/2)
- J, lowest to highest
- n
- Energy Levels (cm⁻¹)

Ionization Limit
5 294 300 cm⁻¹
(656.39 eV)

[Ca XII $2p^5\ ^2P^o_{3/2}$ → Ca XIII $2p^4\ ^3P_2$]

CORES:
- n l = 2p⁴(³P)
- n′ l′ = 2p⁴(¹D)
- n″ l″ = 2p⁴(¹S)

Ca XIII
SINGLET
TRIPLET
GROTRIAN
DIAGRAM

Ca XIII GROTRIAN DIAGRAM (8 electrons , Z = 20)

(O I sequence , Configuration : $1s^2 2s^2 2p^3 nl$, $2s 2p^4 nl$, Singlet & Triplet Systems)

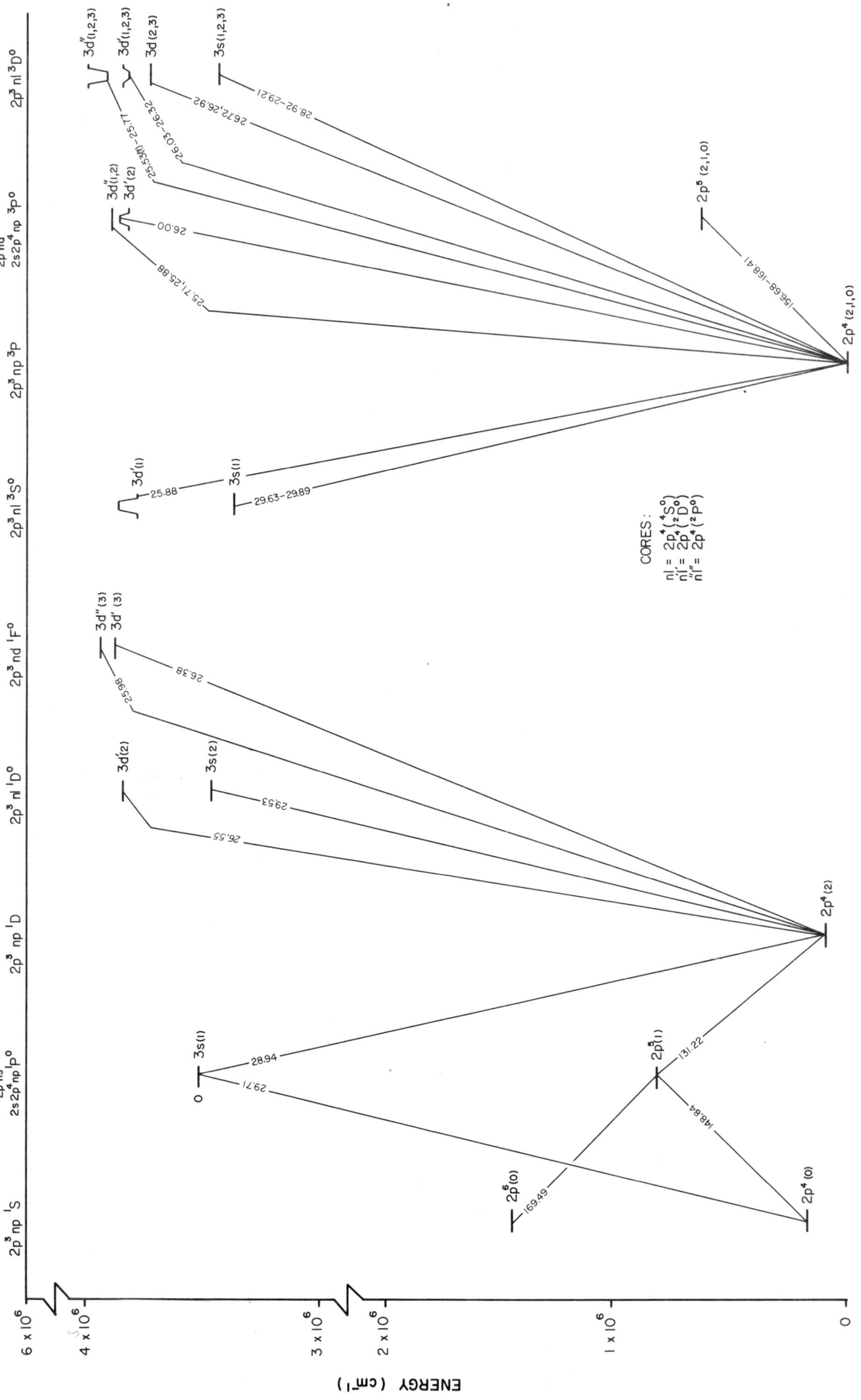

$2p^3 np\ ^1S$ | $2p^3 ns$ / $2s2p^4 np\ ^1P^o$ | $2p^3\ np\ ^1D$ | $2p^3\ nl\ ^1D^o$ | $2p^3 nd\ ^1F^o$ | $2p^3 nl\ ^3S^o$ | $2p^3 np\ ^3P$ | $2p^3 nd$ / $2s2p^4 np\ ^3P^o$ | $2p^3 nl\ ^3D^o$

ENERGY (cm^{-1})

6×10^6
4×10^6
3×10^6
2×10^6
1×10^6
0

$2p^6(0)$
169.49
$2p^5(1)$
148.84
131.22
$2p^4(0)$
0 3s(1)
29.71
28.94
$2p^4(2)$
3d′(2)
26.55
3s(2)
29.53
3d″(3)
3d′ (3)
25.98
26.38
3d′(1)
25.88
3s(1)
29.63–29.89
$2p^4(2,1,0)$
3d″(1,2)
3d′(2)
25.71,25.88
26.00
3d″(1,2,3)
3d′(1,2,3)
3d (2,3)
3s(1,2,3)
25.53(?)–25.77
26.03–26.32
26.72,26.92
28.92–29.21
$2p^5$ (2,1,0)
156.68–168.41

CORES :
$nl = 2p^4(^4S^o)$
$nl' = 2p^4(^2D^o)$
$nl'' = 2p^4(^2P^o)$

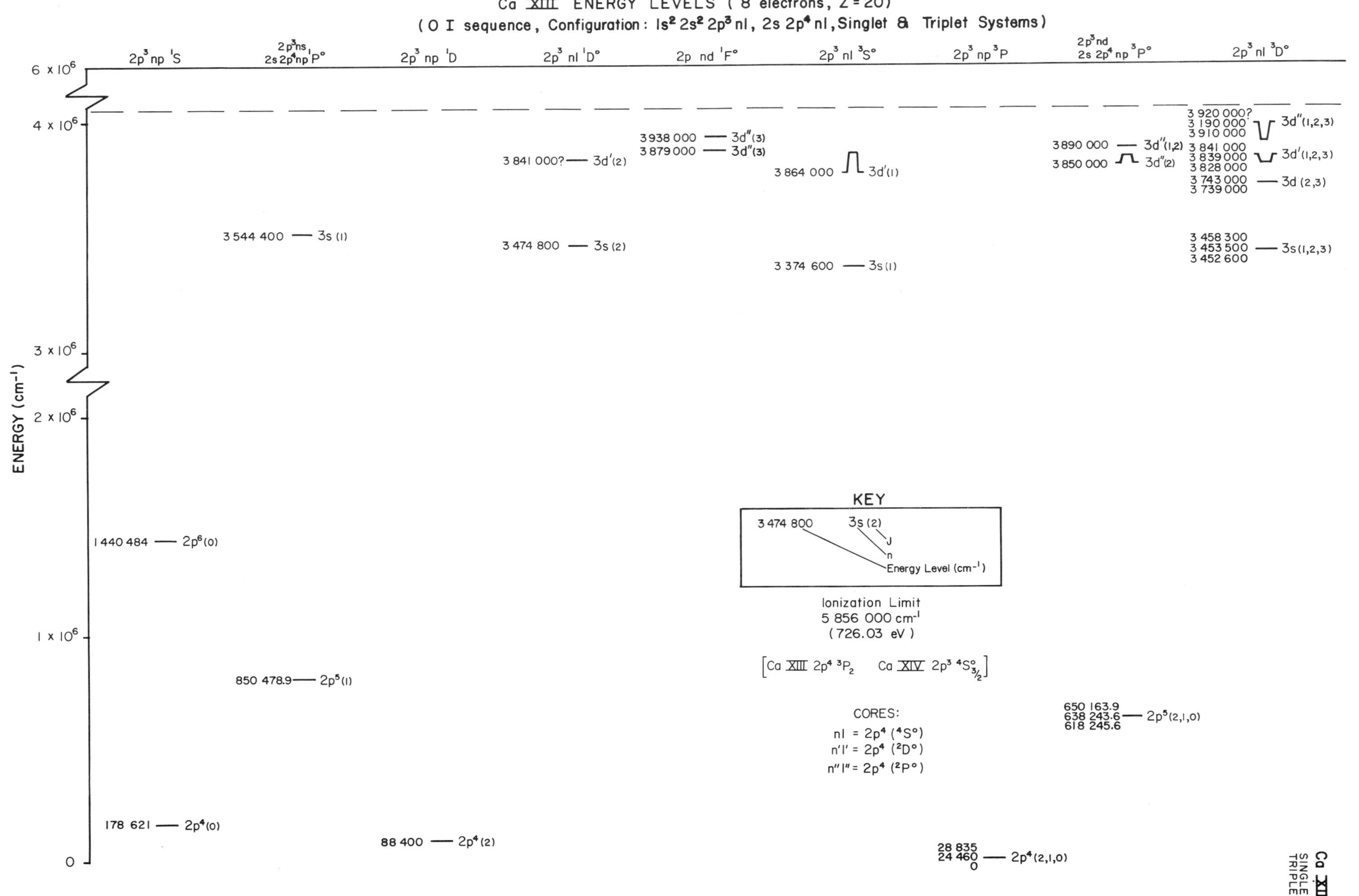

Ca XIII ENERGY LEVELS (8 electrons, Z = 20)
(O I sequence, Configuration: 1s² 2s² 2p³ nl, 2s 2p⁴ nl, Singlet & Triplet Systems)
2p³ np ¹S
2p³ns, 2s 2p⁴ np ¹P°
2p³ np ¹D
2p³ nl ¹D°
2p nd ¹F°
2p³ nl ³S°
2p³ np ³P
2p³nd, 2s 2p⁴ np ³P°
2p³ nl ³D°
ENERGY (cm⁻¹)
6 x 10⁶
4 x 10⁶
3 x 10⁶
2 x 10⁶
1 x 10⁶
0
3 544 400 — 3s (1)
850 478.9 — 2p⁵(1)
1 440 484 — 2p⁶(0)
178 621 — 2p⁴(0)
88 400 — 2p⁴(2)
3 841 000? — 3d′(2)
3 474 800 — 3s (2)
3 938 000 — 3d″(3)
3 879 000 — 3d″(3)
3 864 000 3d′(1)
3 374 600 — 3s(1)
3 890 000 — 3d″(1,2)
3 850 000 3d″(2)
3 920 000?
3 190 000
3 910 000
3d″(1,2,3)
3 841 000
3 839 000
3 828 000
3d′(1,2,3)
3 743 000
3 739 000
— 3d (2,3)
3 458 300
3 453 500
3 452 600
— 3s(1,2,3)
KEY
3 474 800
3s (2)
J
n
Energy Level (cm⁻¹)
Ionization Limit
5 856 000 cm⁻¹
(726.03 eV)
[Ca XIII 2p⁴ ³P₂ Ca XIV 2p³ ⁴S°3/2]
CORES:
nl = 2p⁴ (⁴S°)
n′l′ = 2p⁴ (²D°)
n″l″ = 2p⁴ (²P°)
650 163.9
638 243.6
618 245.6
— 2p⁵(2,1,0)
28 835
24 460
0
— 2p⁴(2,1,0)
Ca XIII
SINGLET
TRIPLET

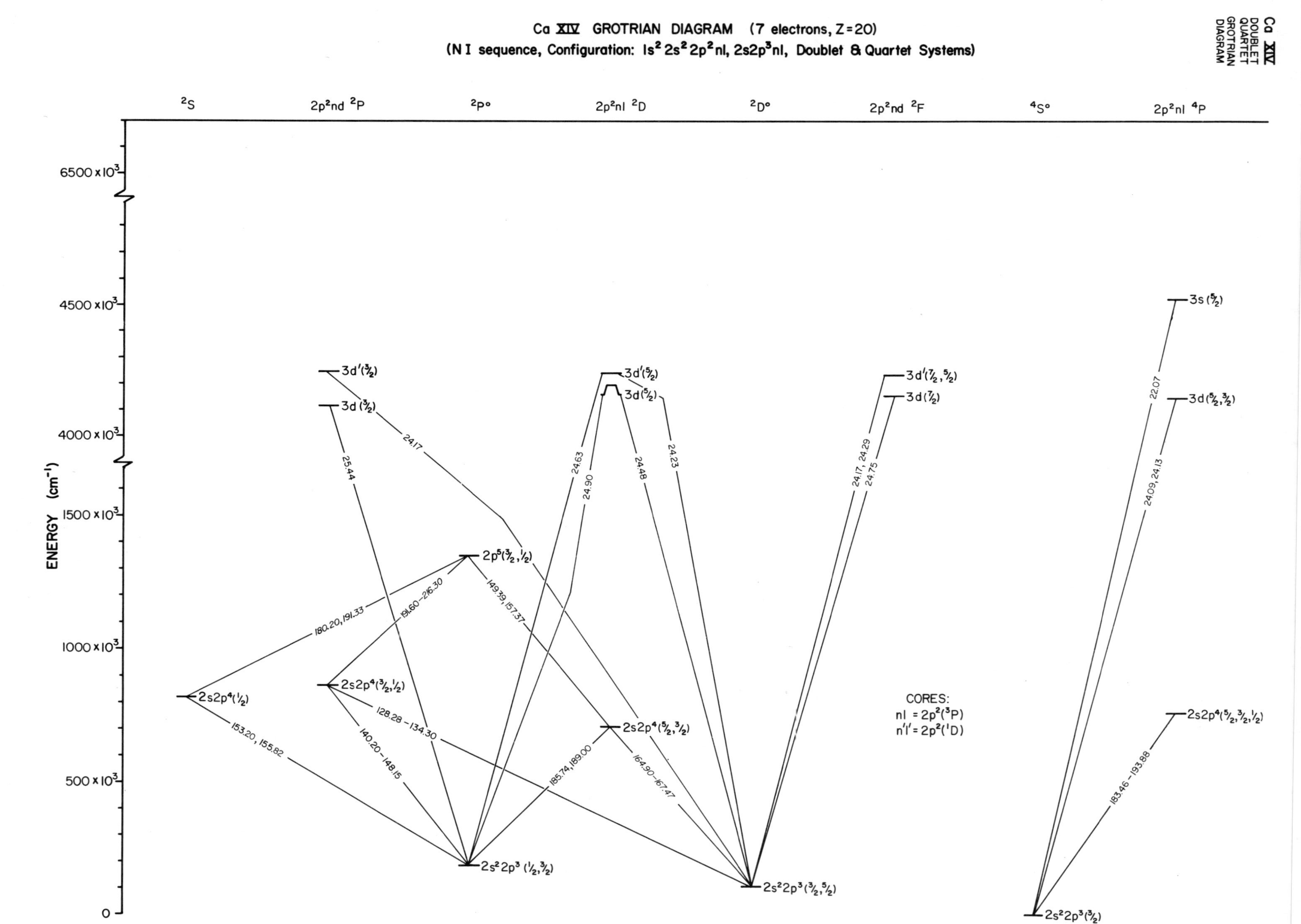

Ca XIV
DOUBLET
QUARTET
GROTRIAN
DIAGRAM
Ca XIV GROTRIAN DIAGRAM (7 electrons, Z=20)
(N I sequence, Configuration: $1s^2 2s^2 2p^2 nl$, $2s2p^3 nl$, Doublet & Quartet Systems)
^2S
$2p^2nd\ ^2P$
$^2P^o$
$2p^2nl\ ^2D$
$^2D^o$
$2p^2nd\ ^2F$
$^4S^o$
$2p^2nl\ ^4P$
ENERGY (cm^{-1})
6500×10^3
4500×10^3
4000×10^3
1500×10^3
1000×10^3
500×10^3
0
$3d'(^3/_2)$
$3d(^3/_2)$
$3d'(^5/_2)$
$3d(^5/_2)$
$3d'(^7/_2,^5/_2)$
$3d(^7/_2)$
$3s(^5/_2)$
$3d(^5/_2,^3/_2)$
$2p^5(^3/_2,^1/_2)$
$2s2p^4(^1/_2)$
$2s2p^4(^3/_2,^1/_2)$
$2s2p^4(^5/_2,^3/_2)$
$2s2p^4(^5/_2,^3/_2,^1/_2)$
$2s^22p^3(^1/_2,^3/_2)$
$2s^22p^3(^3/_2,^5/_2)$
$2s^22p^3(^3/_2)$
CORES:
$nl = 2p^2(^3P)$
$n'l' = 2p^2(^1D)$
25.44
24.17
24.63
24.90
24.48
24.23
24.17, 24.29
24.75
22.07
24.09, 24.13
180.20, 191.33
191.60–216.30
149.39, 157.37
128.28 – 134.30
153.20, 155.82
140.20 – 148.15
185.74, 189.00
164.90–167.47
183.46 – 193.88

Ca XIV ENERGY LEVELS (7 electrons, Z = 20)

(N I sequence, Configuration: $1s^2 2s^2 2p^2 nl, 2s2p^3 nl$, Doublet & Quartet Systems)

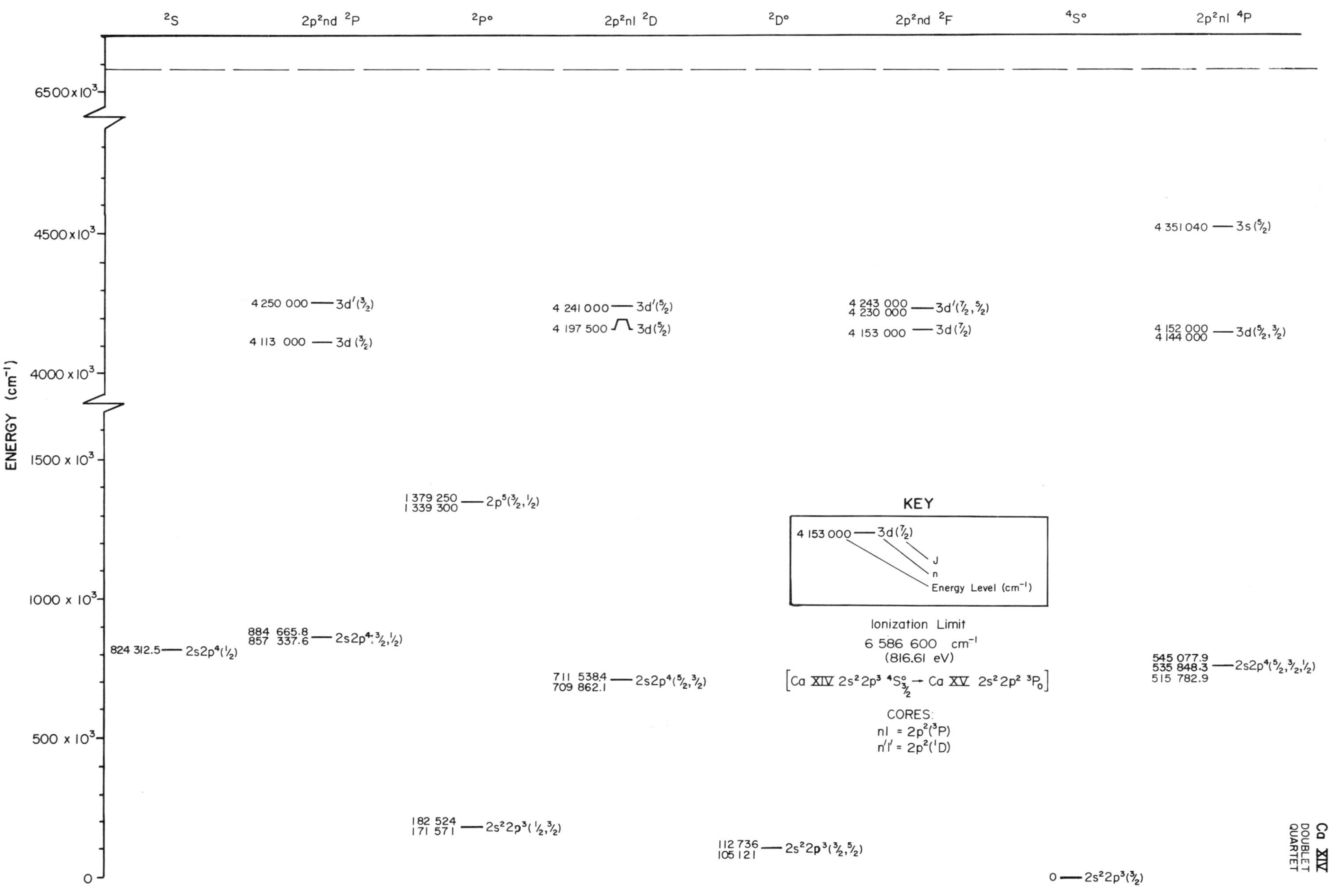

^2S | $2p^2nd\ ^2P$ | $^2P^o$ | $2p^2nl\ ^2D$ | $^2D^o$ | $2p^2nd\ ^2F$ | $^4S^o$ | $2p^2nl\ ^4P$

ENERGY (cm^{-1})

6500 x 10^3
4500 x 10^3
4000 x 10^3
1500 x 10^3
1000 x 10^3
500 x 10^3
0

^2S:
824 312.5 — $2s2p^4(^1/_2)$

$2p^2nd\ ^2P$:
4 250 000 — $3d'(^3/_2)$
4 113 000 — $3d(^3/_2)$
884 665.8, 857 337.6 — $2s2p^4(^3/_2, ^1/_2)$

$^2P^o$:
1 379 250, 1 339 300 — $2p^5(^3/_2, ^1/_2)$
182 524, 171 571 — $2s^22p^3(^1/_2, ^3/_2)$

$2p^2nl\ ^2D$:
4 241 000 — $3d'(^5/_2)$
4 197 500 — $3d(^5/_2)$
711 538.4, 709 862.1 — $2s2p^4(^5/_2, ^3/_2)$

$^2D^o$:
112 736, 105 121 — $2s^22p^3(^3/_2, ^5/_2)$

$2p^2nd\ ^2F$:
4 243 000, 4 230 000 — $3d'(^7/_2, ^5/_2)$
4 153 000 — $3d(^7/_2)$

$^4S^o$:
0 — $2s^22p^3(^3/_2)$

$2p^2nl\ ^4P$:
4 351 040 — $3s(^5/_2)$
4 152 000, 4 144 000 — $3d(^5/_2, ^3/_2)$
545 077.9, 535 848.3, 515 782.9 — $2s2p^4(^5/_2, ^3/_2, ^1/_2)$

KEY

4 153 000 — $3d(^7/_2)$: J; n; Energy Level (cm^{-1})

Ionization Limit
6 586 600 cm^{-1}
(816.61 eV)
[Ca XIV $2s^22p^3\ ^4S^o_{3/2}$ → Ca XV $2s^22p^2\ ^3P_0$]

CORES:
$nl = 2p^2(^3P)$
$n'l' = 2p^2(^1D)$

Ca XIV
DOUBLET
QUARTET

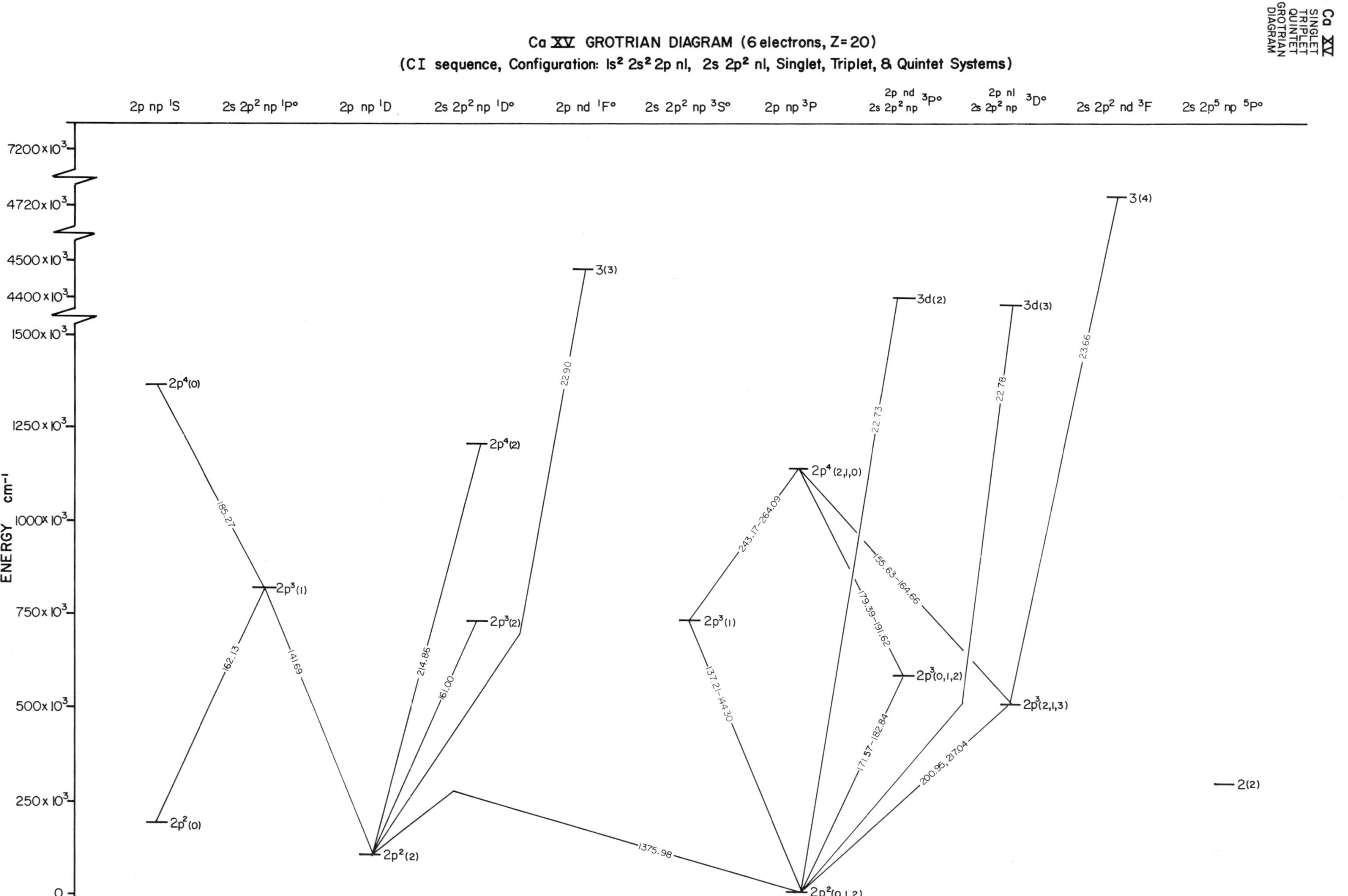
Ca XV
SINGLET
TRIPLET
QUINTET
GROTRIAN
DIAGRAM
Ca XV GROTRIAN DIAGRAM (6 electrons, Z=20)
(C I sequence, Configuration: 1s² 2s² 2p nl, 2s 2p² nl, Singlet, Triplet, & Quintet Systems)
2p np ¹S
2s 2p² np ¹P°
2p np ¹D
2s 2p² np ¹D°
2p nd ¹F°
2s 2p² np ³S°
2p np ³P
2p nd
2s 2p² np ³P°
2p nl
2s 2p² np ³D°
2s 2p² nd ³F
2s 2p⁵ np ⁵P°
ENERGY cm⁻¹
7200x10³
4720x10³
4500x10³
4400x10³
1500x10³
1250x10³
1000x10³
750x10³
500x10³
250x10³
0
2p⁴(0)
185.27
2p³(1)
162.13
141.69
2p²(0)
2p⁴(2)
214.86
2p³(2)
161.00
3(3)
22.90
2p²(2)
1375.98
2p³(1)
243.17–264.09
137.21–144.30
2p⁴(2,1,0)
155.63–164.66
179.39–191.62
2p³(0,1,2)
171.57–182.84
2p²(0,1,2)
3d(2)
22.73
3d(3)
22.78
2p³(2,1,3)
200.95, 217.04
3(4)
23.66
2(2)

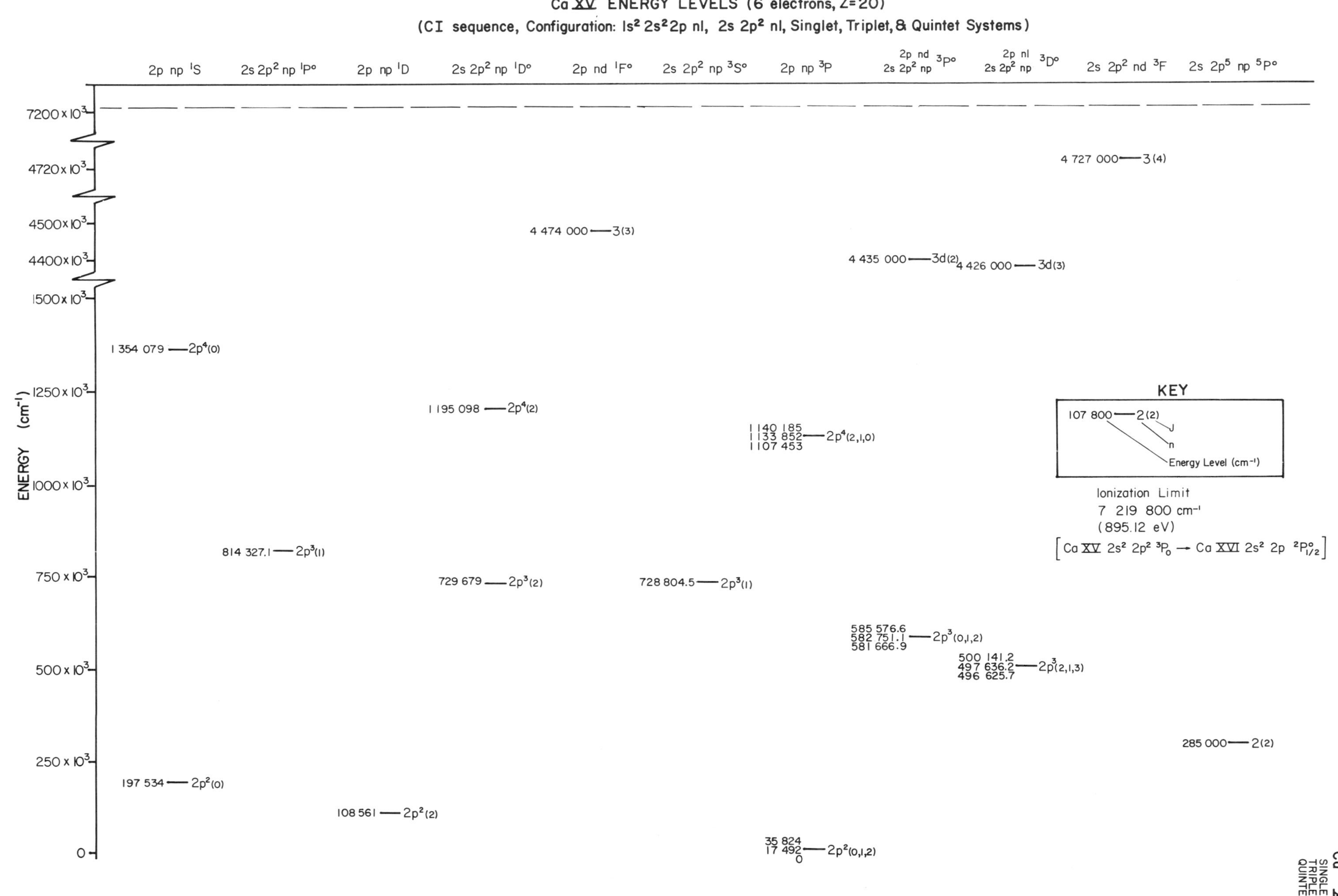
Ca XV ENERGY LEVELS (6 electrons, Z=20)
(CI sequence, Configuration: $1s^2 2s^2 2p\ nl$, $2s\ 2p^2\ nl$, Singlet, Triplet, & Quintet Systems)
2p np ^1S
2s $2p^2$ np $^1P°$
2p np ^1D
2s $2p^2$ np $^1D°$
2p nd $^1F°$
2s $2p^2$ np $^3S°$
2p np ^3P
2p nd / 2s $2p^2$ np $^3P°$
2p nl / 2s $2p^2$ np $^3D°$
2s $2p^2$ nd ^3F
2s $2p^5$ np $^5P°$
ENERGY (cm^{-1})
7200×10^3
4720×10^3
4500×10^3
4400×10^3
1500×10^3
1250×10^3
1000×10^3
750×10^3
500×10^3
250×10^3
0
4 727 000 — 3(4)
4 474 000 — 3(3)
4 435 000 — 3d(2)
4 426 000 — 3d(3)
1 354 079 — $2p^4$(0)
1 195 098 — $2p^4$(2)
1 140 185
1 133 852 — $2p^4$(2,1,0)
1 107 453
814 327.1 — $2p^3$(1)
729 679 — $2p^3$(2)
728 804.5 — $2p^3$(1)
585 576.6
582 751.1 — $2p^3$(0,1,2)
581 666.9
500 141.2
497 636.2 — $2p^3$(2,1,3)
496 625.7
285 000 — 2(2)
197 534 — $2p^2$(0)
108 561 — $2p^2$(2)
35 824
17 492 — $2p^2$(0,1,2)
0
KEY
107 800 — 2(2)
J
n
Energy Level (cm^{-1})
Ionization Limit
7 219 800 cm^{-1}
(895.12 eV)
[Ca XV $2s^2\ 2p^2\ ^3P_0$ → Ca XVI $2s^2\ 2p\ ^2P°_{1/2}$]
Ca XV
SINGLET
TRIPLET
QUINTET

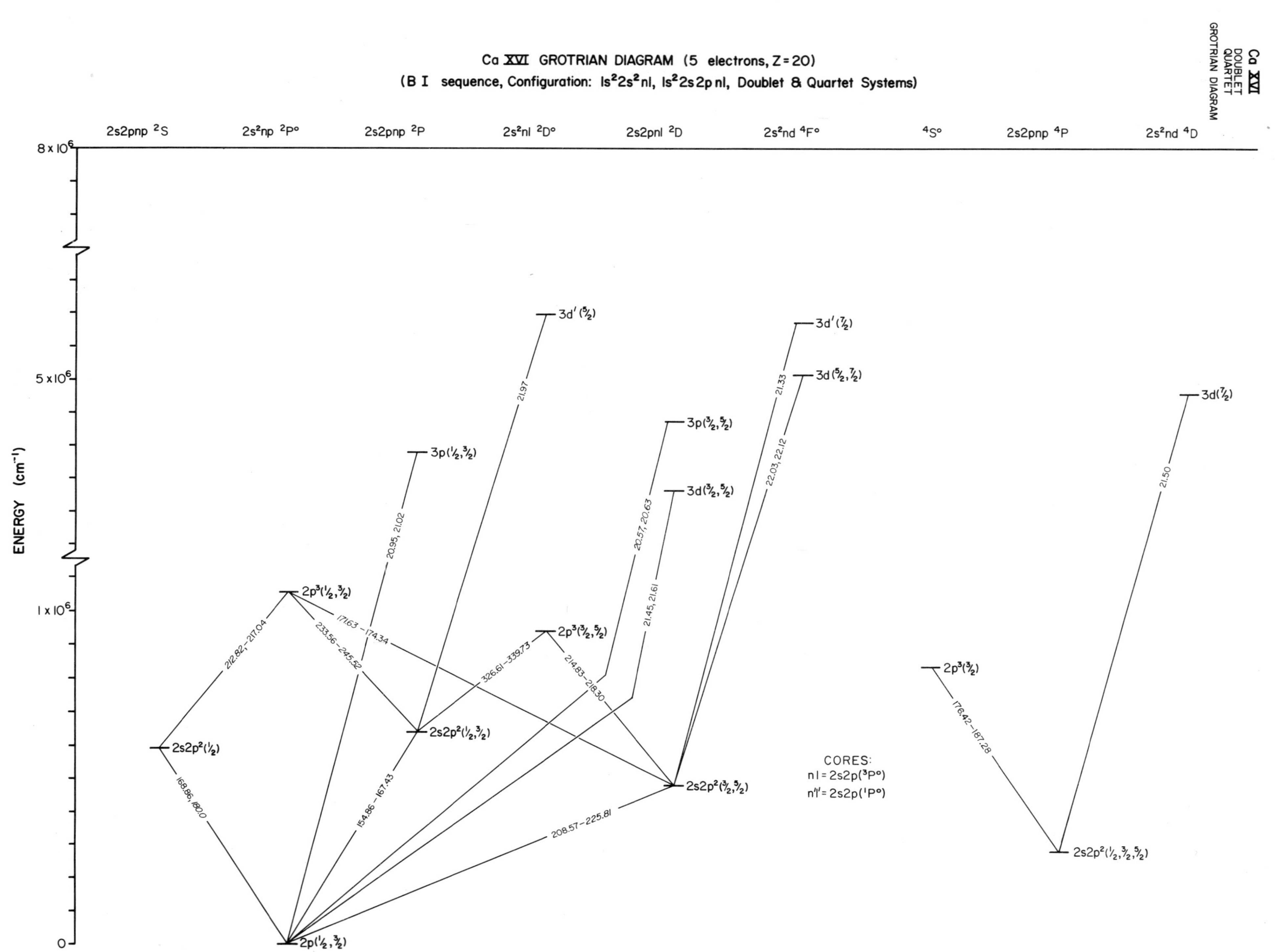
Ca XVI
DOUBLET
QUARTET
GROTRIAN DIAGRAM
Ca XVI GROTRIAN DIAGRAM (5 electrons, Z = 20)
(B I sequence, Configuration: $1s^2 2s^2 nl$, $1s^2 2s 2p nl$, Doublet & Quartet Systems)
2s2pnp ^2S
$2s^2np$ $^2P^o$
2s2pnp ^2P
$2s^2nl$ $^2D^o$
2s2pnl ^2D
$2s^2nd$ $^4F^o$
$^4S^o$
2s2pnp ^4P
$2s^2nd$ ^4D
ENERGY (cm^{-1})
8×10^6
5×10^6
1×10^6
0
$2p(1/2, 3/2)$
$2s2p^2(1/2)$
$2p^3(1/2, 3/2)$
$2s2p^2(1/2, 3/2)$
$3p(1/2, 3/2)$
$3d'(5/2)$
$2p^3(3/2, 5/2)$
$3p(3/2, 5/2)$
$3d(3/2, 5/2)$
$2s2p^2(3/2, 5/2)$
$3d'(7/2)$
$3d(5/2, 7/2)$
$2p^3(3/2)$
$2s2p^2(1/2, 3/2, 5/2)$
$3d(7/2)$
168.86, 180.0
212.82 – 217.04
233.56 – 245.52
171.63 – 174.34
154.86 – 167.43
20.95, 21.02
21.97
326.61 – 339.73
214.83 – 218.30
20.57, 20.63
21.45, 21.61
208.57 – 225.81
21.33
22.03, 22.12
176.42 – 187.28
21.50
CORES:
$nl = 2s2p(^3P^o)$
$n'l' = 2s2p(^1P^o)$

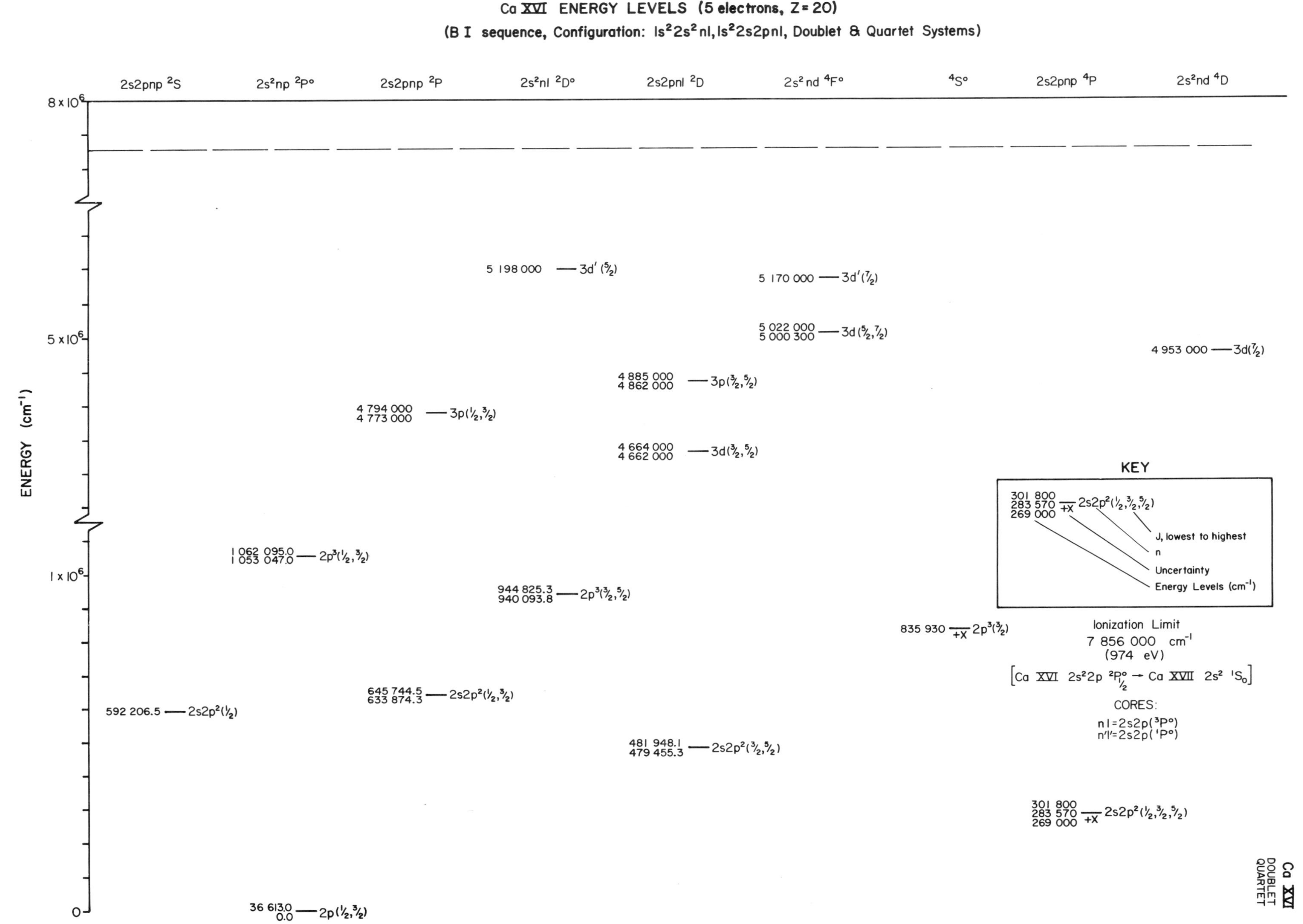
Ca XVI ENERGY LEVELS (5 electrons, Z = 20)
(B I sequence, Configuration: 1s²2s²nl, 1s²2s2pnl, Doublet & Quartet Systems)
2s2pnp ²S
2s²np ²P°
2s2pnp ²P
2s²nl ²D°
2s2pnl ²D
2s² nd ⁴F°
⁴S°
2s2pnp ⁴P
2s²nd ⁴D
8 x 10⁶
5 x 10⁶
1 x 10⁶
0
ENERGY (cm⁻¹)
5 198 000 — 3d′(⁵⁄₂)
5 170 000 — 3d′(⁷⁄₂)
5 022 000
5 000 300
3d(⁵⁄₂,⁷⁄₂)
4 953 000 — 3d(⁷⁄₂)
4 885 000
4 862 000
3p(³⁄₂,⁵⁄₂)
4 794 000
4 773 000
3p(½,³⁄₂)
4 664 000
4 662 000
3d(³⁄₂,⁵⁄₂)
KEY
301 800
283 570
269 000
+X
2s2p²(½,³⁄₂,⁵⁄₂)
J, lowest to highest
n
Uncertainty
Energy Levels (cm⁻¹)
1 062 095.0
1 053 047.0
2p³(½,³⁄₂)
944 825.3
940 093.8
2p³(³⁄₂,⁵⁄₂)
835 930 +X 2p³(³⁄₂)
Ionization Limit
7 856 000 cm⁻¹
(974 eV)
[Ca XVI 2s²2p ²P°½ → Ca XVII 2s² ¹S₀]
CORES:
nl = 2s2p(³P°)
n′l′ = 2s2p(¹P°)
645 744.5
633 874.3
2s2p²(½,³⁄₂)
592 206.5 — 2s2p²(½)
481 948.1
479 455.3
2s2p²(³⁄₂,⁵⁄₂)
301 800
283 570
269 000
+X
2s2p²(½,³⁄₂,⁵⁄₂)
36 613.0
0.0
2p(½,³⁄₂)
Ca XVI
DOUBLET
QUARTET

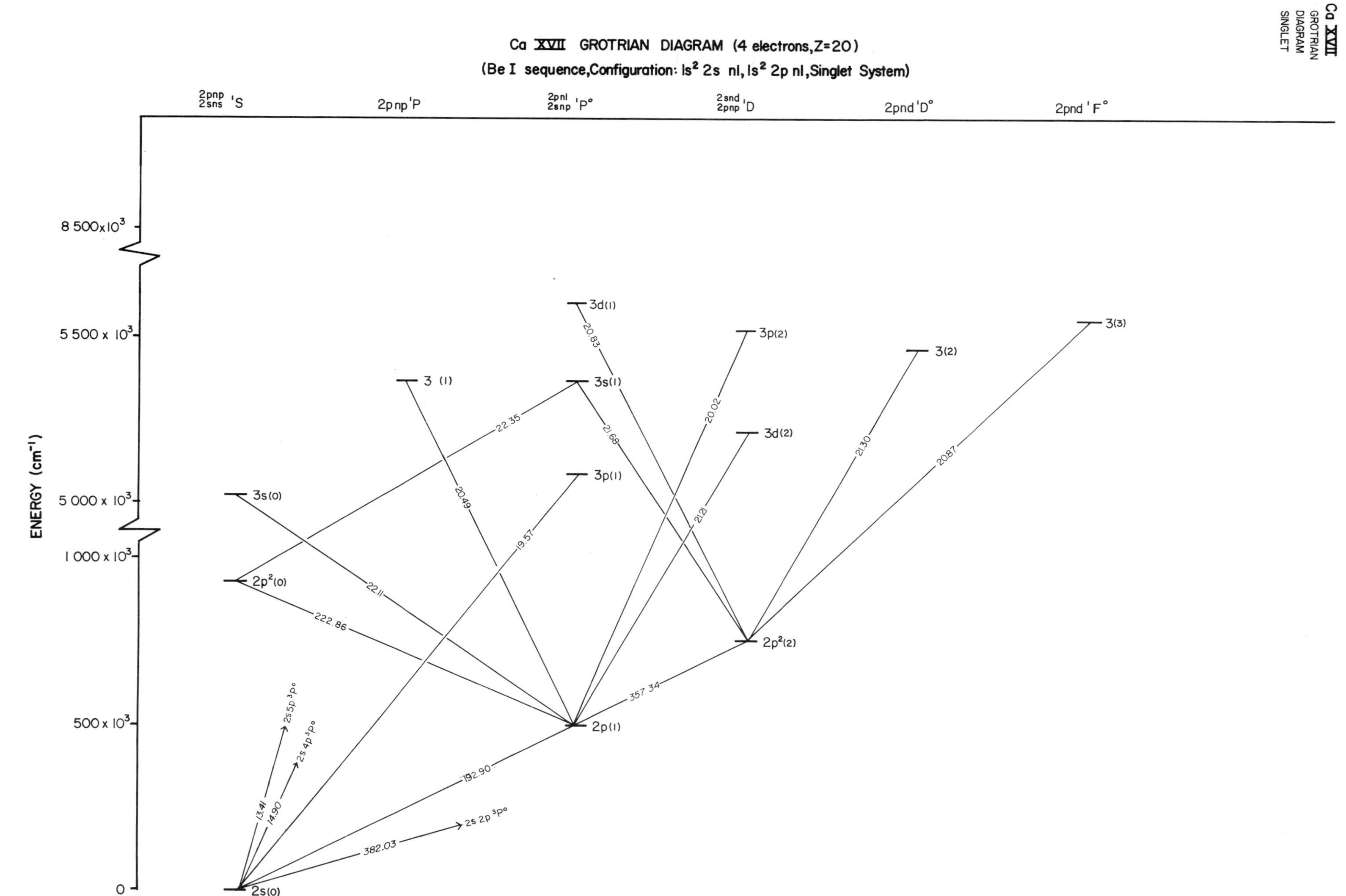
Ca XVII GROTRIAN DIAGRAM (4 electrons, Z=20)
(Be I sequence, Configuration: 1s² 2s nl, 1s² 2p nl, Singlet System)
2pnp 2sns ¹S
2pnp ¹P
2pnl 2snp ¹P°
2snd 2pnp ¹D
2pnd ¹D°
2pnd ¹F°
ENERGY (cm⁻¹)
8 500x10³
5 500 x 10³
5 000 x 10³
1 000 x 10³
500 x 10³
0
3d(1)
3p(2)
3(3)
3(2)
3 (1)
3s(1)
3d(2)
3p(1)
3s(0)
2p²(0)
2p²(2)
2p(1)
2s(0)
20.83
22.35
21.68
20.02
21.30
2087
20.49
19.57
21.21
22.11
222.86
357.34
192.90
382.03
2s 2p ³P°
13.41
14.90
2s 5p ³P°
2s 4p ³P°

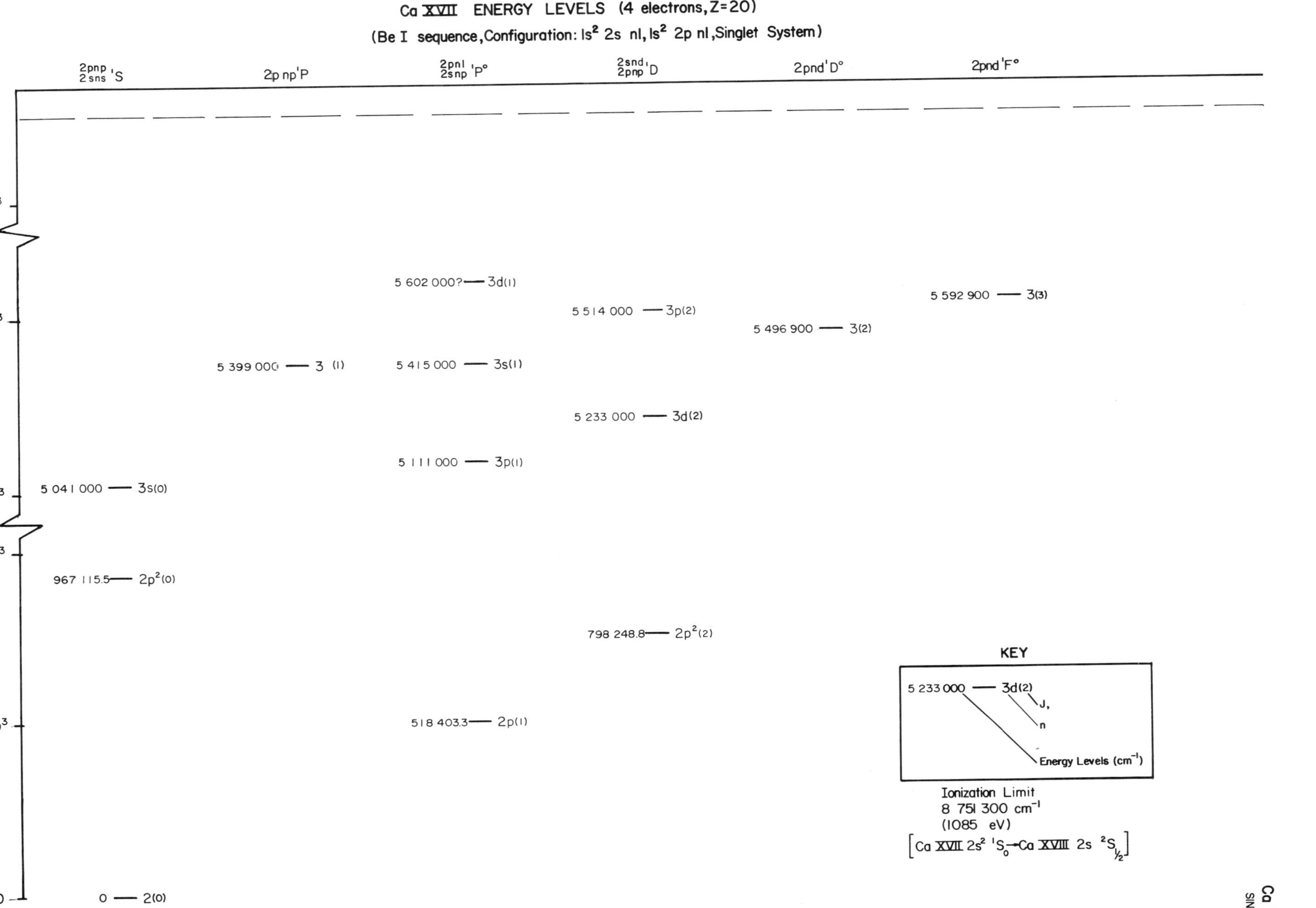
Ca XVII ENERGY LEVELS (4 electrons, Z=20)
(Be I sequence, Configuration: 1s² 2s nl, 1s² 2p nl, Singlet System)
2pnp, 2sns ¹S
2p np ¹P
2pnl, 2snp ¹P°
2snd, 2pnp ¹D
2pnd ¹D°
2pnd ¹F°
ENERGY (cm⁻¹)
8 500×10³
5 500×10³
5 000×10³
1 000×10³
500 ×10³
0
5 602 000? — 3d(1)
5 592 900 — 3(3)
5 514 000 — 3p(2)
5 496 900 — 3(2)
5 399 000 — 3 (1)
5 415 000 — 3s(1)
5 233 000 — 3d(2)
5 111 000 — 3p(1)
5 041 000 — 3s(0)
967 115.5 — 2p²(0)
798 248.8 — 2p²(2)
518 403.3 — 2p(1)
0 — 2(0)
KEY
5 233 000 — 3d(2)
J,
n
Energy Levels (cm⁻¹)
Ionization Limit
8 751 300 cm⁻¹
(1085 eV)
[Ca XVII 2s² ¹S₀ → Ca XVIII 2s ²S½]

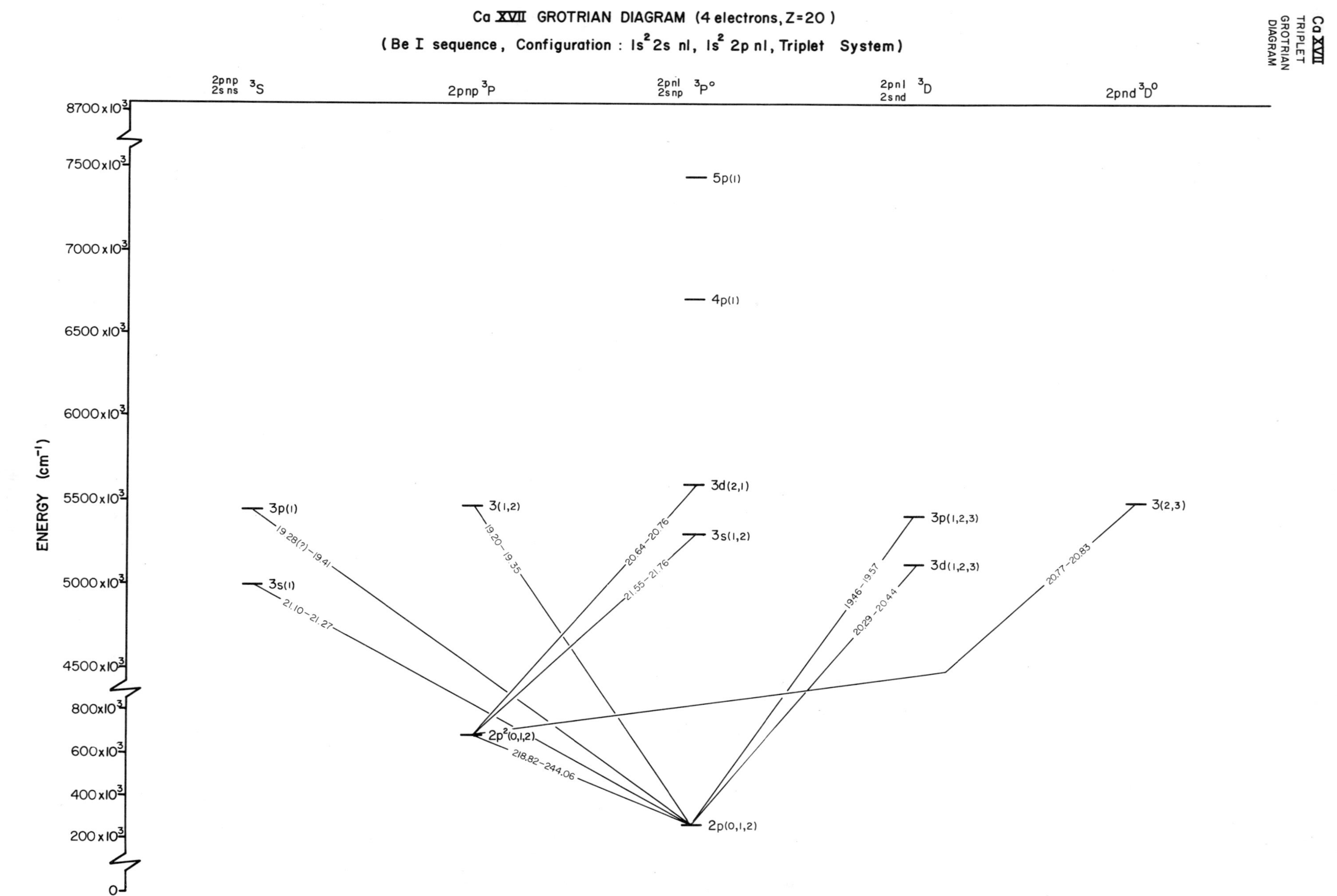
Ca XVII
TRIPLET
GROTRIAN
DIAGRAM
Ca XVII GROTRIAN DIAGRAM (4 electrons, Z=20)
(Be I sequence, Configuration : $1s^2 2s\ nl$, $1s^2 2p\ nl$, Triplet System)
2pnp 2sns ^3S
2pnp ^3P
2pnl 2snp $^3P^o$
2pnl 2snd ^3D
2pnd $^3D^o$
ENERGY (cm^{-1})
8700 x 10^3
7500 x 10^3
7000 x 10^3
6500 x 10^3
6000 x 10^3
5500 x 10^3
5000 x 10^3
4500 x 10^3
800 x 10^3
600 x 10^3
400 x 10^3
200 x 10^3
0
5p(1)
4p(1)
3d(2,1)
3s(1,2)
3p(1)
3s(1)
3(1,2)
3p(1,2,3)
3d(1,2,3)
3(2,3)
$2p^2(0,1,2)$
2p(0,1,2)
19.28(?)–19.41
21.10–21.27
19.20–19.35
20.64–20.76
21.55–21.76
19.46–19.57
20.29–20.44
20.77–20.83
218.82–244.06

Ca XVII ENERGY LEVELS (4 electrons, Z=20)
(Be I sequence, Configuration: $1s^2\,2s\,nl$, $1s^2\,2pnl$, Triplet System)

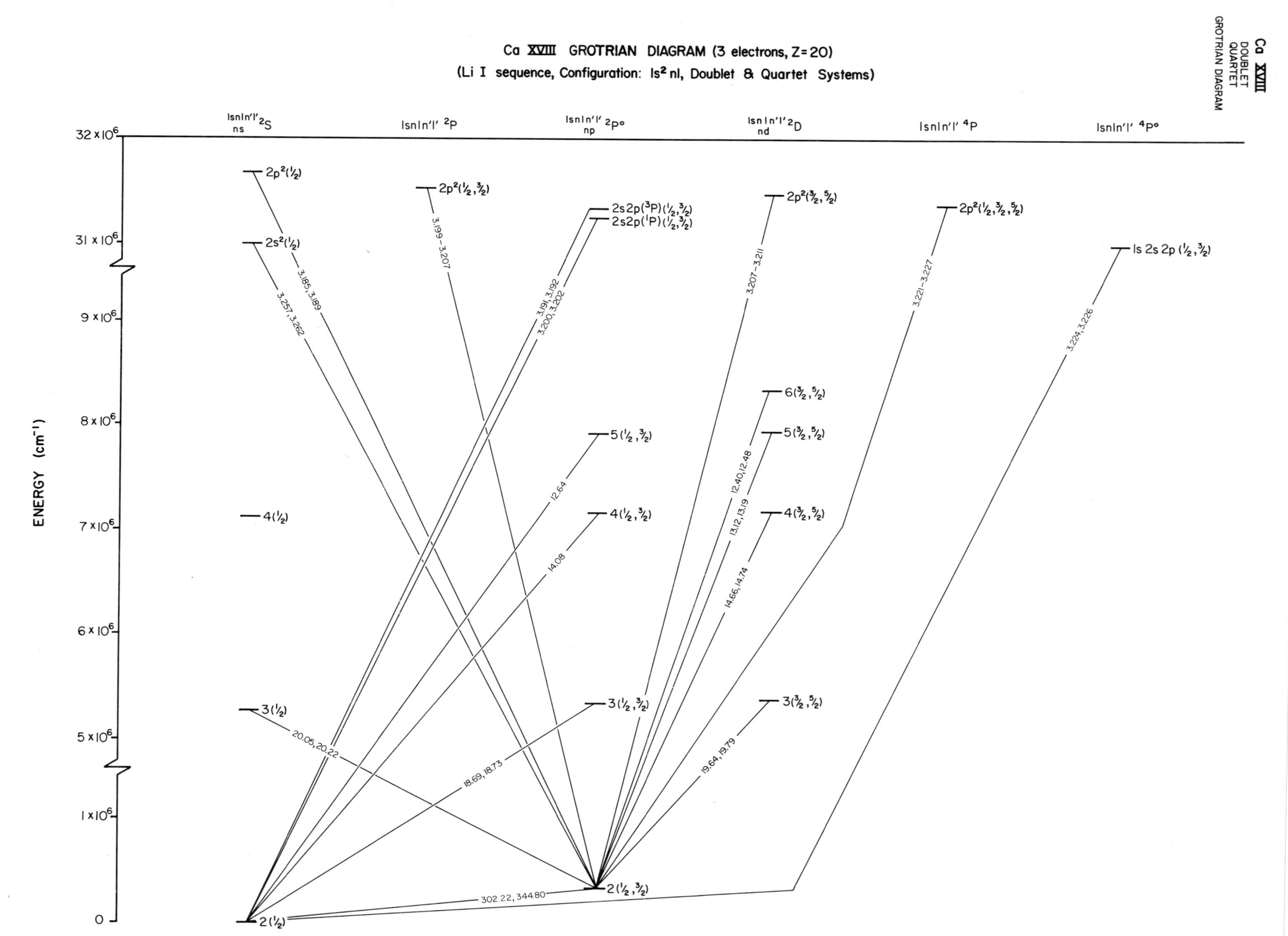
Ca XVIII GROTRIAN DIAGRAM (3 electrons, Z=20)
(Li I sequence, Configuration: 1s² nl, Doublet & Quartet Systems)
1snl'l' ²S ns
1snl'l' ²P
1snl'l' ²P° np
1snl'l' ²D nd
1snl'l' ⁴P
1snl'l' ⁴P°
2p²(1/2)
2s²(1/2)
4(1/2)
3(1/2)
2(1/2)
2p²(1/2,3/2)
2s2p(³P)(1/2,3/2)
2s2p(¹P)(1/2,3/2)
5(1/2,3/2)
4(1/2,3/2)
3(1/2,3/2)
2(1/2,3/2)
2p²(3/2,5/2)
6(3/2,5/2)
5(3/2,5/2)
4(3/2,5/2)
3(3/2,5/2)
2p²(1/2,3/2,5/2)
1s 2s 2p (1/2,3/2)
3.185,3.189
3.257,3.262
3.199–3.207
3.191,3.192
3.200,3.202
3.207–3.211
3.221–3.227
3.224,3.226
12.64
14.08
12.40,12.48
13.12,13.19
14.66,14.74
20.05,20.22
18.69,18.73
19.64,19.79
302.22, 344.80
32×10⁶
31×10⁶
9×10⁶
8×10⁶
7×10⁶
6×10⁶
5×10⁶
1×10⁶
0
ENERGY (cm⁻¹)

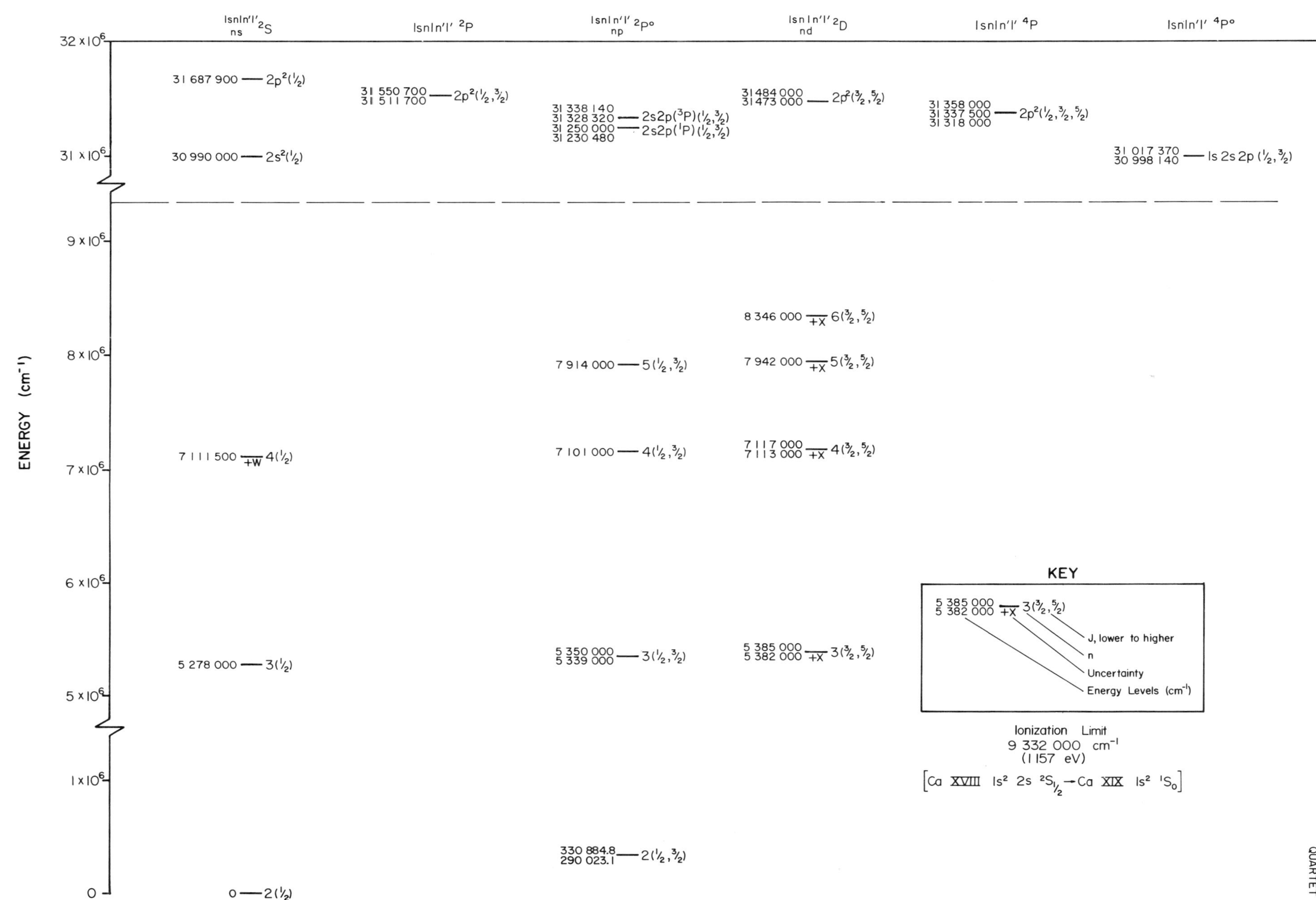
Ca XVIII ENERGY LEVELS (3 electrons, Z = 20)
(Li I sequence, Configuration: 1s² nl, Doublet & Quartet Systems)
1sn1n'1' ²S ns
1sn1n'1' ²P
1sn1n'1' ²P° np
1sn1n'1' ²D nd
1sn1n'1' ⁴P
1sn1n'1' ⁴P°
31 687 900 — 2p²(1/2)
30 990 000 — 2s²(1/2)
7 111 500 +w 4(1/2)
5 278 000 — 3(1/2)
0 — 2(1/2)
31 550 700
31 511 700 — 2p²(1/2, 3/2)
31 338 140
31 328 320 — 2s2p(³P)(1/2, 3/2)
31 250 000
31 230 480 — 2s2p(¹P)(1/2, 3/2)
7 914 000 — 5(1/2, 3/2)
7 101 000 — 4(1/2, 3/2)
5 350 000
5 339 000 — 3(1/2, 3/2)
330 884.8
290 023.1 — 2(1/2, 3/2)
31 484 000
31 473 000 — 2p²(3/2, 5/2)
8 346 000 +x 6(3/2, 5/2)
7 942 000 +x 5(3/2, 5/2)
7 117 000
7 113 000 +x 4(3/2, 5/2)
5 385 000
5 382 000 +x 3(3/2, 5/2)
31 358 000
31 337 500
31 318 000 — 2p²(1/2, 3/2, 5/2)
31 017 370
30 998 140 — 1s 2s 2p (1/2, 3/2)
KEY
5 385 000
5 382 000 +x 3(3/2, 5/2)
J, lower to higher
n
Uncertainty
Energy Levels (cm⁻¹)
Ionization Limit
9 332 000 cm⁻¹
(1157 eV)
[Ca XVIII 1s² 2s ²S1/2 → Ca XIX 1s² ¹S0]
ENERGY (cm⁻¹)
32 x 10⁶
31 x 10⁶
9 x 10⁶
8 x 10⁶
7 x 10⁶
6 x 10⁶
5 x 10⁶
1 x 10⁶
0

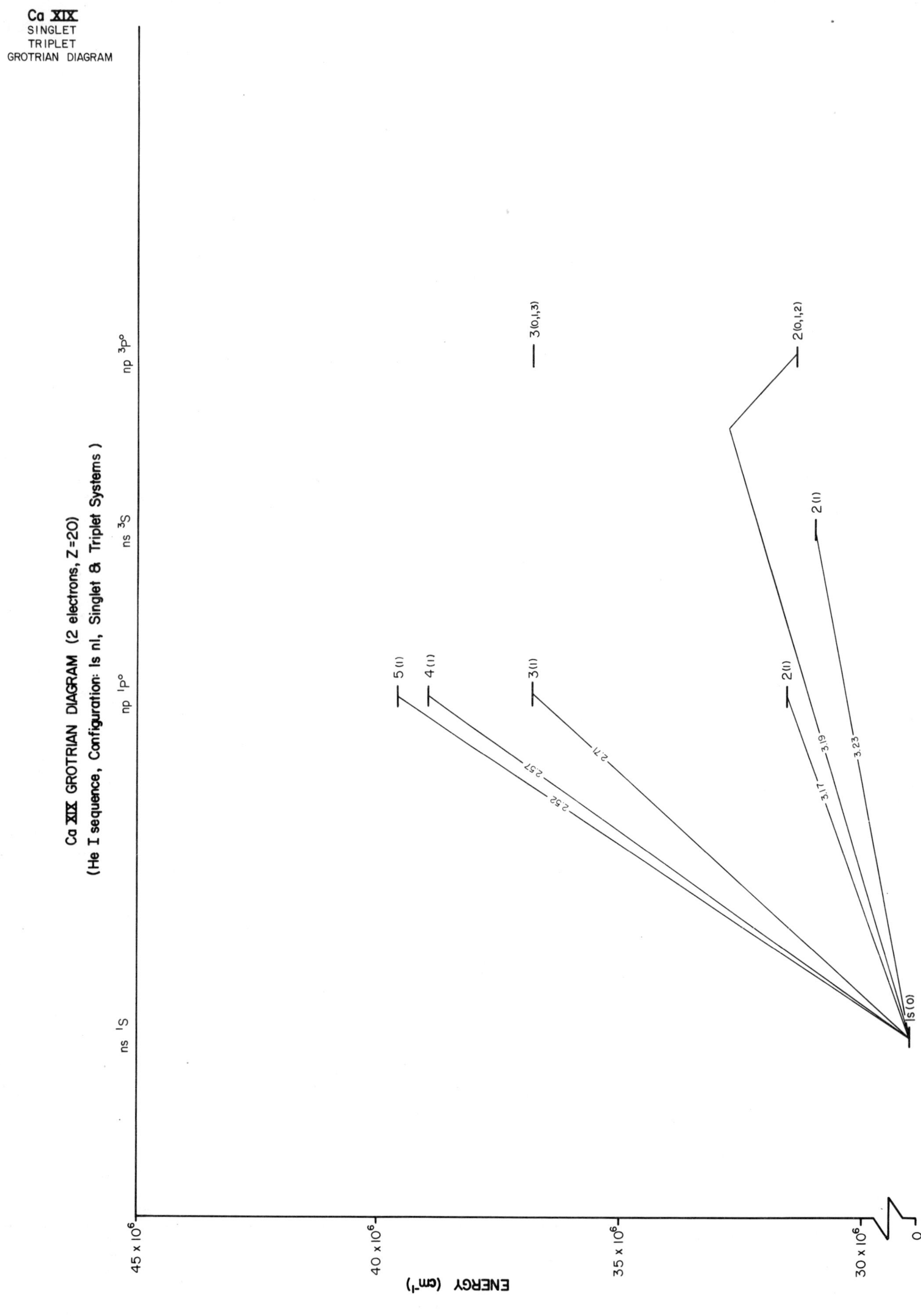
Ca XIX GROTRIAN DIAGRAM (2 electrons, Z=20)
(He I sequence, Configuration: 1s nl, Singlet & Triplet Systems)
ns 1S
np 1P°
ns 3S
np 3P°
ENERGY (cm-1)
45 x 10^6
40 x 10^6
35 x 10^6
30 x 10^6
0
5(1)
4(1)
3(1)
2(1)
2(1)
3(0,1,3)
2(0,1,2)
1s(0)
2.52
2.57
2.71
3.17
3.19
3.23

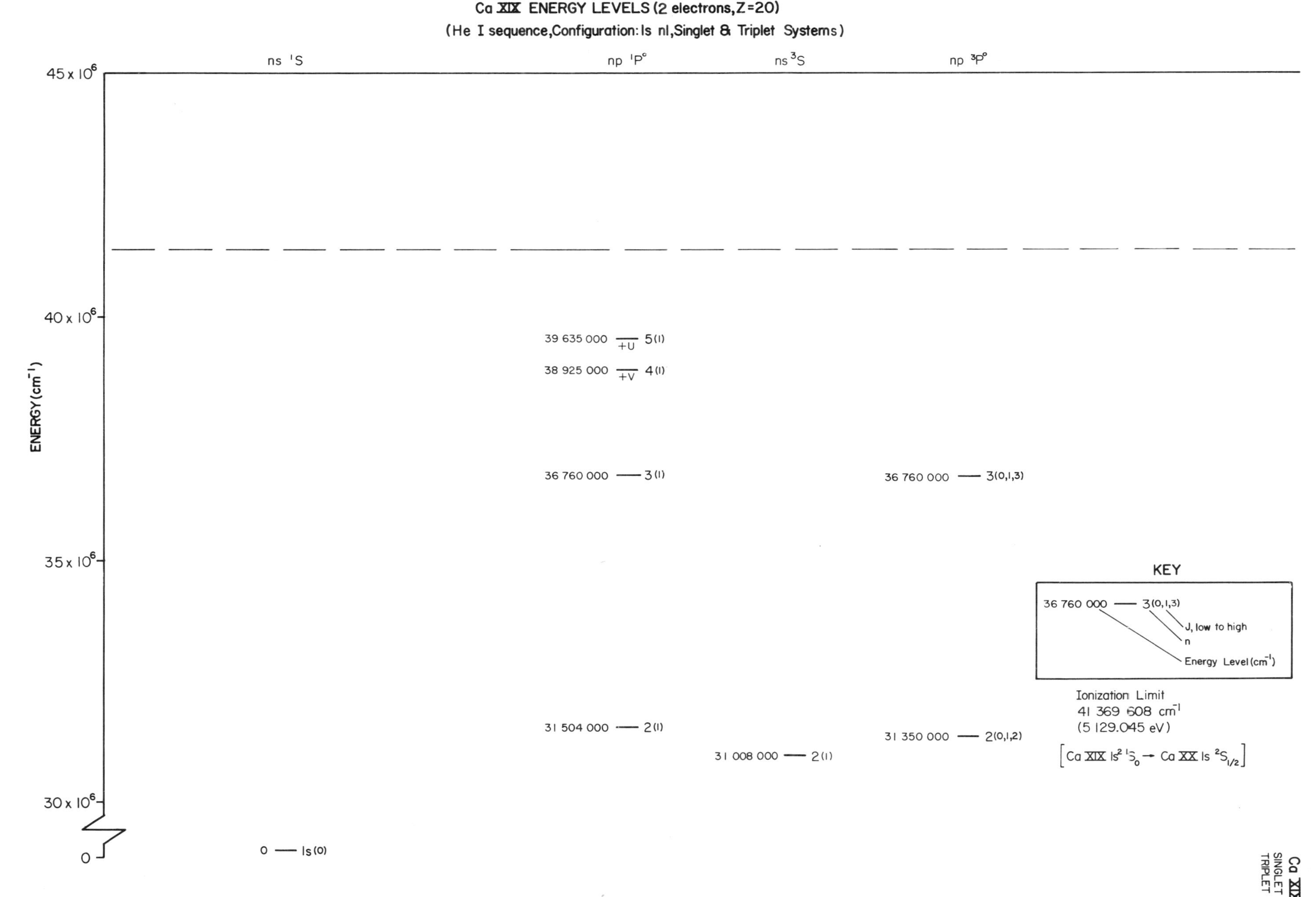
Ca XIX ENERGY LEVELS (2 electrons, Z=20)
(He I sequence, Configuration: 1s nl, Singlet & Triplet Systems)
ns ¹S
np ¹P°
ns ³S
np ³P°
45 x 10⁶
40 x 10⁶
35 x 10⁶
30 x 10⁶
0
ENERGY (cm⁻¹)
39 635 000 +U 5(1)
38 925 000 +V 4(1)
36 760 000 — 3(1)
36 760 000 — 3(0,1,3)
KEY
36 760 000 — 3(0,1,3)
J, low to high
n
Energy Level (cm⁻¹)
Ionization Limit
41 369 608 cm⁻¹
(5 129.045 eV)
[Ca XIX 1s² ¹S₀ → Ca XX 1s ²S₁/₂]
31 504 000 — 2(1)
31 350 000 — 2(0,1,2)
31 008 000 — 2(1)
0 — 1s(0)
Ca XIX
SINGLET
TRIPLET

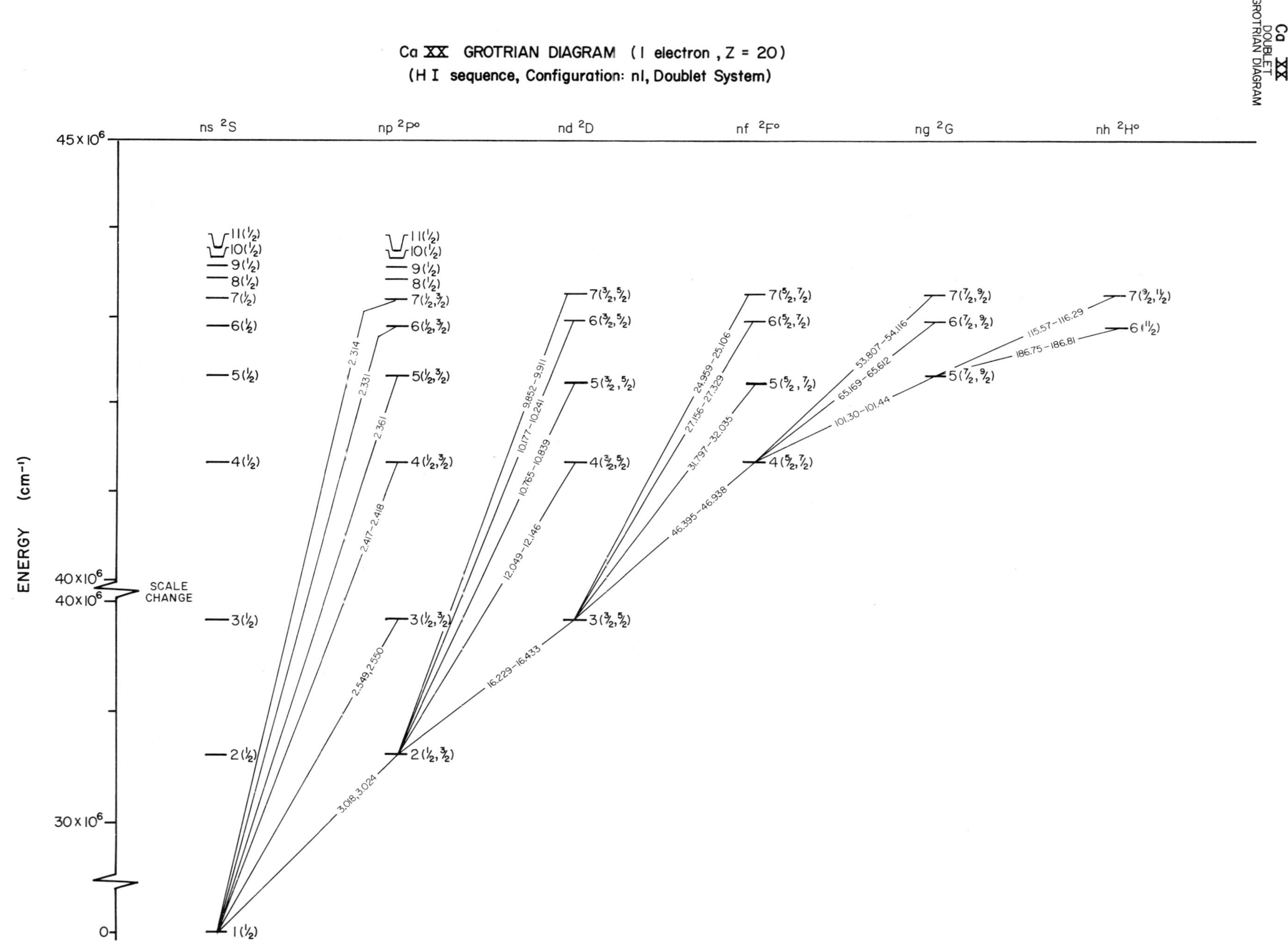
Ca XX GROTRIAN DIAGRAM (I electron , Z = 20)
(H I sequence, Configuration: nl, Doublet System)
ns ^{2}S
np ^{2}P°
nd ^{2}D
nf ^{2}F°
ng ^{2}G
nh ^{2}H°
ENERGY (cm^{-1})
45 x 10^6
40 x 10^6
40 x 10^6
30 x 10^6
0
SCALE CHANGE
3.018, 3.024
2.549, 2.550
2.417-2.418
2.361
2.331
2.314
16.229-16.433
12.049-12.146
10.765-10.839
10.177-10.241
9.852-9.911
46.395-46.938
31.797-32.035
27.156-27.329
24.959-25.106
101.30-101.44
65.169-65.612
53.807-54.116
186.75-186.81
115.57-116.29

Ca XX ENERGY LEVELS (1 electron, Z = 20)
(H I sequence, Configuration: nl, Doublet System)

n	ns ^2S	np $^2P^o$	nd ^2D	nf $^2F^o$	ng ^2G	nh $^2H^o$
11	43 753 799 — 11(½)	43 753 788 — 11(½)				
10	43 677 411 — 10(½)	43 677 395 — 10(½)				
9	43 574 141 — 9(½)	43 574 120 — 9(½)				
8	43 429 723 — 8(½)	43 429 693 — 8(½)				
7	43 218 998 — 7(½)	43 218 954 — 7(½)	— 7(3/2, 5/2)	— 7(5/2, 7/2)	— 7(7/2, 9/2)	— 7(9/2, 11/2)
6	42 894 174 — 6(½)	42 894 103 — 6(½)	— 6(3/2, 5/2)	— 6(5/2, 7/2)	— 6(7/2, 9/2)	42 897 743 — 6(11/2)
5	42 355 144 — 5(½)	42 358 817 42 355 021 — 5(1/2, 3/2)	42 360 066 42 358 811 — 5(3/2, 5/2)	42 360 689 42 360 063 — 5(5/2, 7/2)	42 361 063 42 360 688 — 5(7/2, 9/2)	
4	41 362 068 — 4(½)	41 369 247 41 361 829 — 4(1/2, 3/2)	41 371 685 41 369 234 — 4(3/2, 5/2)	41 372 902 41 371 680 — 4(5/2, 7/2)		
3	39 214 378 — 3(½)	39 231 405 39 213 813 — 3(1/2, 3/2)	39 237 180 39 231 374 — 3(3/2, 5/2)			
2	33 071 589 — 2(½)	33 129 056 33 069 706 — 2(1/2, 3/2)				
1	0 — 1(½)					

ENERGY (cm^{-1})

45×10^6 — 40×10^6 SCALE CHANGE 40×10^6 — 30×10^6 — 0

KEY

42 360 689
42 360 063 — 5(5/2, 7/2)

J, lower to highe-
n
Energy Levels (cm^{-1})

Ionization Limit
44 117 210. ± 7 cm^{-1}
(5 469.70 eV)
[Ca XX 1s $^2S_{1/2}$ → nucleus]

Ca I $Z = 20$ 20 electrons

Although the lowest levels of this atom are easily described and graphed, the higher levels, including many autoionizing levels, are difficult to categorize.
Some changes in level designations have appeared since the publication of AEL. These are partly the result of new experiments and partly the result of calculations of configuration interaction involving 3d3p and 4s4p; because of the nearly simultaneous publication of two important papers, there is some confusion in the literature concerning the proper designations. We tabulate the relevant material below. If an entry reads "AEL", this means that Moore's designation is intended

Level (cm^{-1})	AEL	Risberg (1968)	Roth (1969)	Present designation
15158	$4s(^2S)4p\ ^3P^\circ_0$	AEL	96% AEL	AEL
15210	$4s(^2S)4p\ ^3P^\circ_1$	AEL	96% AEL	AEL
15316	$4s(^2S)4p\ ^3P^\circ_2$	AEL	96% AEL	AEL
23652	$4s(^2S)4p\ ^1P^\circ_1$	AEL	86% AEL+14% $3d(^2D)4p'\ ^1P^\circ_1$	AEL
35730	$3d(^2D)4p'\ ^3F^\circ_2$	AEL	86% AEL+13% $3d(^2D)4p'\ ^1D^\circ_2$	AEL
35819	$3d(^2D)4p'\ ^3F^\circ_3$	AEL	AEL	AEL
35897	$3d(^2D)4p'\ ^3F^\circ_4$	AEL	AEL	AEL
35835	$3d(^2D)4p'\ ^1D^\circ_2$	AEL	87% AEL+13% $3d(^2D)4p'\ ^3F^\circ_2$	AEL
36548	$4s(^2S)5p\ ^3P^\circ_0$	AEL	–	AEL
36555	$4s(^2S)5p\ ^3P^\circ_1$	AEL	–	AEL
36575	$4s(^2S)5p\ ^3P^\circ_2$	AEL	–	AEL
36732	$3d(^2D)4p'\ ^1P^\circ_1$	$4s(^2S)5p\ ^1P^\circ_1$	–	AEL
38192	$3d(^2D)4p'\ ^3D^\circ_1$	AEL	AEL	AEL
38219	$3d(^2D)4p'\ ^3D^\circ_2$	AEL	AEL	AEL
38259	$3d(^2D)4p'\ ^3D^\circ_3$	AEL	AEL	AEL
39333	$3d(^2D)4p'\ ^3P^\circ_0$	AEL	96% AEL	AEL
39335	$3d(^2D)4p'\ ^3P^\circ_1$	AEL	96% AEL	AEL
39340	$3d(^2D)4p'\ ^3P^\circ_2$	AEL	96% AEL	AEL
40538	$3d(^2D)4p'\ ^1F^\circ_3$	AEL	AEL	AEL
41679	$4s(^2S)5p\ ^1P^\circ_1$	$4s(^2S)6p\ ^1P^\circ_1$	86% $3d(^2D)4p'\ ^1P^\circ_1$ +14% $4s(^2S)4p\ ^1P^\circ_1$	$3d(^2D)4p'\ ^1P^\circ_1$
43933	$4s(^2S)6p\ ^1P^\circ_1$	$3d(^2D)4p'\ ^1P^\circ_1$	–	AEL

C.M. Brown, S.G. Tilford, and M.L. Ginter, J. Opt. Soc. Am. **63**, 1454 (1973).

These authors study the absorption spectrum of Ca and identify large numbers of singly and doubly-excited levels.

R.W. Ditchburn and R.D. Hudson, Proc. Roy. Soc. **A256**, 53 (1960).

Authors give term tables from observed absorption spectra.

P. Esherick, J.A. Armstrong, R.W. Dreyfus and J.J. Wynne, Phys. Rev. Letters **36**, 1296 (1976).

These authors locate high ^1S and ^1D levels by double-photon excitation; this reference provides the data for the ^1S and ^1D levels above 46 800 cm^{-1}, and the 4s9d $^3D_2$ level.

W.R.S. Garton and K. Codling, Proc. Phys. Soc. **86**, 1067 (1965).

Authors give tables of absorption lines observed in the region 1590–2400 Å.

C.J. Humpreys, J. Research NBS **47**, 262 (1951).

Author gives line tables and a partial term diagram from lines observed in the region 12 815–22 650 Å.

T.R. Kaiser, Proc. Phys. Soc. **75**, 152 (1960).

Author gives a table of observed absorption lines in the region 1610–1885 Å.

G.H. Newsom, Proc. Phys. Soc. **87**, 975 (1966).

Author gives a table of observed absorption lines in the region 1670–1940 Å.

N.P. Penkin and L.N. Shabanova, Opt. Spectrosc. **18**, 425 (1965).

These authors study absorption spectra in order to get a better ionization potential for Ca I.

N.P. Penkin and L.N. Shabanova, Opt. Spectrosc. **18**, 535 (1965).

These authors study perturbation of series in the spectrum of Ca I and other species.

G. Risberg, Ark. Fys. **37**, 231 (1968).

Author gives line tables, term tables, and a Grotrian diagram based on observations in the region 2115–22 655 Å.

C. Roth, J. Res. NBS, **73A**, 497 (1969).

Analyzes configuration interactions and gives percent core parentage for four terms whose L-quantum numbers are not well-defined.

Ca II $Z = 20$ 19 electrons

B. Edlén and P. Risberg, Ark. Fys. **10**, 553 (1956).

Authors give line tables, term tables, and a term diagram from lines observed in the region 2100–21 430 Å and calculated for the region 1340–2210 Å.

G. Risberg, Ark. Fys. **37**, 231 (1968).

Author gives line and term tables from lines observed in the region 3125–10 225 Å.

E.W.H. Selwyn, Proc. Phys. Soc. **41**, 392 (1929).

Author gives spectrograms and a table of lines observed in the region 1640–2120 Å.

Ca III $Z = 20$ 18 electrons

Kelly and Palumbo (see general references) have interchanged the *K*-values for the transitions taken from their reference B 15 [Borgström (1971)] ; the corrected transitions are shown in our drawings. The level $3s3p^6 3d\ ^1D_2$ is taken from Hansen *et al.* (1975).

A. Borgström, Ark. Fys. **38**, 243 (1968).

Author gives line tables, energy level tables, an energy levels diagram, an energy level array, and spectrograms based on observations in the regions 300–2475 Å and 2490–4720 Å.

A. Borgström, Physica Scripta **3**, 157 (1971).

This author extends considerably the analysis of Ca III.

R.D. Cowan, J. Opt. Soc. Am. **58**, 924 (1968).

Author gives a line table and an energy level table of experimental and calculated values.

A.H. Gabriel, B.C. Fawcett, and C. Jordan, Nature **206**, 390 (1965).

Authors give a wavelength of an observed transition in the vacuum ultraviolet region.

A.H. Gabriel, B.C. Fawcett, and C. Jordan, Proc. Phys. Soc. **87**, 825 (1966).

Authors give a wavelength of an observed transition in the vacuum ultraviolet region.

J.E. Hansen, W. Persson, and A. Borgström, Physica Scripta **11**, 31 (1975).

These authors analyze the interactions between the $3s3p^6 3d$ and $3s^2 3p^5 nf$ configurations.

P.G. Kruger and S.G. Wiessberg, Phys. Rev. **48**, 659 (1935).

Author gives a term table from observations.

P.G. Kruger, S.G. Wiessberg, and L.W. Phillips, Phys. Rev. **51**, 1090 (1937).

Authors give a term table from observations.

G. Risberg, Ark. Fys. **37**, 231 (1968).

Author gives a table of lines observed in the region 2540–4500 Å.

Ca IV | $Z = 20$ | 17 electrons

B.C. Fawcett and A.H. Gabriel, Proc. Phys. Soc. **88**, 262 (1966).

Authors give wavelengths of observed transitions of three multiplets.

A.H. Gabriel, B.C. Fawcett, and Carole Jordan, Nature **206**, 390 (1965).

Authors give a table of observed lines in the vacuum ultraviolet region.

A.H. Gabriel. B.C. Fawcett, and Carole Jordan, Proc. Phys. Soc. **87**, 825 (1966).

Authors give wavelengths of the observed transitions of two multiplets.

P.G. Kruger and L.W. Phillips, Phys. Rev. **51**, 1087 (1937).

Authors give an array of term and observed lines in the vacuum ultraviolet region.

L.Å. Svensson and J.O. Ekberg, Ark. Fys. **37**, 65 (1967).

Authors give a table of lines in the region 430–465 Å.

Ca V | $Z = 20$ | 16 electrons

I.S. Bowen, Ap. J. **121**, 306 (1955).

Author gives the wavelengths of two observed forbidden transitions.

I.S. Bowen, Ap. J. **132**, 1 (1960).

Author gives tables of lines observed and predicted for forbidden transitions.

B. Edlén, Phys. Rev. **61**, 434 (1942).

Author gives a wavelength and term values for the $3s^2 3p^4\ {}^1S$ and $3s3p^5\ {}^1P^\circ_1$ configurations.

A.H. Gabriel, B.C. Fawcett, and C. Jordan, Nature **206**, 390 (1965).

Authors give a wavelength of an observed transition in the vacuum ultraviolet region.

A.H. Gabriel, B.C. Fawcett, and C. Jordan, Proc. Phys. Soc. **87**, 825 (1966).

Authors give the wavelengths of the observed transitions of two multiplets.

L.Å. Svensson and J.O. Ekberg, Ark. Fys. **37**, 65 (1967).

Authors give line and term tables from observed lines.

Ca VI | $Z = 20$ | 15 electrons

Kelly and Palumbo (see general references) list λ 232.275 Å as belonging to $3p^3\ {}^2D^\circ_{3/2}$–$3p^2({}^3P)4s\ {}^2D_{3/2}$; the correct upper level is $3p^2({}^1D)4s'\ {}^2D_{3/2}$. See Ekberg and Svensson (1970).

I.S. Bowen, Ap. J. **132**, 1 (1960).

Author gives a line table from calculations and observations of forbidden transitions.

J.O. Ekberg and L.Å. Svensson, Physica Scripta **2**, 283 (1970).

These authors identify many transitions in this spectrum.

B.C. Fawcett, J. Phys. B**3**, 1732 (1970).

This author identifies several lines in this spectrum.

B.C. Fawcett, A.H. Gabriel, and P.A.H. Saunders, Proc. Phys. Soc. **90**, 863 (1967).

Authors give the wavelengths of the observed transitions of a multiplet.

A.H. Gabriel, B.C. Fawcett, and C. Jordan, Nature **206**, 390 (1965).

Authors give the wavelength of an observed transition in the vacuum ultraviolet region.

A.H. Gabriel, B.C. Fawcett, and C. Jordan, Proc. Phys. Soc. **87**, 825 (1966).

Authors give the wavelength of an observed transition in the vacuum ultraviolet region.

Ca VII — $Z = 20$ — 14 electrons

I.S. Bowen, Ap. J. **132**, 1 (1960).

Author gives wavelengths of observed forbidden transitions.

J.O. Ekberg and L.Å.Svensson, Physica Scripta **2**, 283 (1970).

These authors identify several transitions in this spectrum.

B.C. Fawcett, A.H. Gabriel, and P.A.H. Saunders, Proc. Phys. Soc. **90**, 863 (1967).

Authors give wavelengths of observed transitions in the region 330–340 Å.

L.W. Phillips, Phys. Rev. **55**, 708 (1939).

Author gives line and term tables based on observations in the region 200–640 Å.

H.A. Robinson, Phys. Rev. **52**, 724 (1937).

Author gives line and term tables from lines observed in the region 395–435 Å.

A.D. Thackeray, Mon. Not. Roy. Astron. Soc. **167**, 87 (1974).

This author revises energy levels of Ca VII using astronomical observations.

Ca VIII — $Z = 20$ — 13 electrons

J.O. Ekberg and L.Å. Svensson, Physica Scripta **2**, 283 (1970).

These authors classify many transitions in this spectrum.

B.C. Fawcett, J. Phys. B**3**, 1732 (1970).

This author classifies several transitions.

L.W. Phillips, Phys. Rev. **55**, 708 (1939).

Author gives line and term tables from observations in the region 110–360 Å.

Ca IX — $Z = 20$ — 12 electrons

We have estimated the position of the 3p3d $^3P^\circ_4$ level by assuming that the ratio of the intervals ($J = 3$–$J = 4$)/($J = 2$–$J = 3$) in the 3p3d $^3P^\circ$ term is the mean (1.207) of that ratio for Sc X (1.199) and for K VIII (1.215). We then located the level 3p 4f $^3G_5$ using the known wavelength [178.951 Å, Fawcett (1976)] of the transition 3p3d $^3P^\circ_4$–3p4f $^3G_5$.

J.O. Ekberg, Physica Scripta **4**, 101 (1971).

This author extends considerably the analysis of Ca IX, providing both line and level lists.

B.C. Fawcett, J. Opt. Soc. Am. **66**, 632 (1976).

This author reports a few lines in Ca IX arising from a detailed analysis of Ca X.

B.C. Fawcett, J. Phys. B**3**, 1732 (1970).

This author classifies several transitions in this spectrum.

W.L. Parker and L.W. Phillips, Phys. Rev. **57**, 140 (1940).

Authors give line and term tables from observations in the region 100–515 Å.

Ca X $Z = 20$ 11 electrons

There is a discrepancy between the assignment and/or wavelengths given by Cohen and Behring (1976) and those by Fawcett (1976) for the transitions 3d-*n*f, *n* = 10–14. We have used the values given by Cohen and Behring. There are differences of as much as 100 cm^{-1} between the levels given by Cohen and Behring, and by Edlén and Bodén (1976). Again, the values given by Cohen and Behring have been used except that the highest levels in the three left-hand columns from Edlén and Bodén.

L. Cohen and W.E. Behring, J. Opt. Soc. Am. **66**, 899 (1976).

These authors present energies for many levels in Ca X.

B. Edlén, Z. Physik **100**, 621 (1936).

Author gives line and term tables from lines observed in the region 150–580 Å.

B. Edlén and E. Bodén, Physica Scripta **14**, 31 (1976).

These authors present a classification of lines and levels for Ca X.

B.C. Fawcett, J. Opt. Soc. Am. **66**, 632 (1976).

This author presents a detailed analysis of this spectrum.

U. Feldman, G.A. Doschek, D.K. Prinz, and D.J. Nagel, J. Appl. Phys. **47**, 1341 (1976).

These authors observe this spectrum in a laser-produced plasma.

P.G. Kruger and L.W. Phillips, Phys. Rev. **55**, 352 (1939).

Authors give line and term tables from lines observed in the region 90–580 Å.

Ca XI $Z = 20$ 10 electrons

Kastner *et al.* identify their observed levels in both *LS* and *j–l* coupling and most other levels are also identified in both *LS* (Edlén and Tyrén) and *j–l* (Moore) coupling; we have used both systems simultaneously on the drawings. Where the *LS* coupling in Edlén and Tyrén disagreed with that in Crance, the identification of Crance has been used.

Energies for $2p^5 3p$ and $2p^5 4d$ terms are largely calculated (Crance); although these are observed levels (Kastner *et al.*), they are not connected to the ground state.

M. Crance. Atomic Data **5**, 2 (1973).

Author gives calculated energies for $2p^5 3p$ and $2p^5 4d$ term not listed by other references and corrects *LS*-coupling identifications by Edlén and Tyrén of levels at 3 239 100, 3 284 000, 3 919 000, and 3 948 400 cm^{-1}.

B. Edlén and E. Tyrén, Z. Physik **101**, 206 (1936).

Authors give line and term tables from observations in the region 25–35 Å, in *LS* coupling.

S.O. Kastner, W.E. Behring, and L. Cohen, Ap. J. **199**, 777 (1975).

These authors identify some transitions classified as $2p^5 4d–2p^5 3p$, involving levels not included in the Edlén/Tyrén term system.

R.L. Kelly and L.C. Gapenski, unpublished (1970).

(Identification of level at 4 275 000 cm^{-1}.)

Ca XII $Z = 20$ 9 electrons

Moore gives a level identified as (^{1}D)X at 3 574 200 cm^{-1} from the analysis by Edlén and Tyrén (1936). This level is presumably 3d′ ^{2}D. We have used Feldman *et al.* (1973) for most of the transitions, except that Fawcett (1975) has been used for the level 2s2p^6 $^2S_{1/2}$.

N.G. Basov, V.A. Boiko, Yu.P. Voinov, E.Ya. Kononov, S.L. Mandel'shtam, and G.V. Sklizkov, JETP Letters **5**, 141 (1967).

These authors observe two wavelengths in this spectrum.

B. Edlén and F. Tyrén, Z. Physik **101**, 206 (1936).

These authors present tables of lines and levels for this species.

B.C. Fawcett, Atomic Data and Nuclear Data Tables **16**, 135 (1975).

This author presents tables of spectral lines for transitions $2s^2 2p^n$–$2s2p^{n+1}$ and $2s2p^n$–$2p^{n+1}$.

B.C. Fawcett, D.D. Burgess, and N.J. Peacock, Proc. Phys. Soc. **91**, 970 (1967).

These authors identify two inner-shell transitions in Ca XII.

U. Feldman, G.A. Doschek, R.D. Cowan, and L. Cohen, J. Opt. Soc. Am. **63**, 1445 (1973).

These authors give energies and wavelengths for spectra of ions in the F I isoelectronic sequence K XI–Co XIX.

Ca XIII $Z = 20$ 8 electrons

The levels given by Doschek *et al.* (1973) have been used where possible; other levels are derived from the data of Fawcett, Burgess, and Peacock (1967).

N.G. Basov, V.A. Boiko, Yu.P. Voinov, E.Ya. Kononov, S.L. Mandel'shtam, and G.V. Sklizkov, JETP Letters **5**, 141 (1967).

These authors observe and classify several lines in this spectrum.

G.A. Doschek, U. Feldman, and L. Cohen, J. Opt. Soc. Am. **63**, 1463 (1973).

These authors present levels and wavelengths for transitions of the type $2s^2 2p^4$–$2s^2 2p^3 3s$ for the O I isoelectronic sequence K XII–Mn XVII.

B.C. Fawcett, J. Phys. B**4**, 981 (1971).

This author classifies three lines in this spectrum.

B.C. Fawcett, Atomic Data and Nuclear Data Tables **16**, 135 (1975).

This author presents tables of spectral lines for transitions $2s^2 2p^n$–$2s2p^{n+1}$ and $2s2p^n$–$2p^{n+1}$.

B.C. Fawcett, D.D. Burgess, and N.J. Peacock, Proc. Phys. Soc. **91**, 970 (1967).

These authors classify transitions in this spectrum.

B.C. Fawcett and R.W. Hayes, Mon. Not. Roy. Astron. Soc. **170**, 185 (1975).

These authors compare observed and calculated wavelengths in this spectrum.

Ca XIV $Z = 20$ 7 electrons

Most on the levels have been taken from Kastner (1973); the 3d levels' positions have been calculated from the data of Fawcett and Hayes (1975).

N.G. Basov, V.A. Boiko, Yu.P. Voinov, E.Ya. Kononov, S.L. Mandel'shtam, and G.V. Sklizkov, JETP Letters **5**, 141 (1967).

These authors classify four lines in this spectrum.

B.C. Fawcett, J. Phys. B**3**, 1152 (1970).

This author classifies two wavelengths in this spectrum.

B.C. Fawcett, J. Phys. B**4**, 981 (1971).

This author classifies numerous wavelengths in this spectrum.

B.C. Fawcett, Atomic Data and Nuclear Data Tables **16**, 135 (1975).

The author presents tables of spectral lines for transitions $2s^22p^n$–$2s2p^{n+1}$ and $2s2p^n$–$2p^{n+1}$.

B.C. Fawcett, D.D. Burgess, and N.J. Peacock, Proc. Phys. Soc. **91**, 970 (1967).

These authors classify transitions in this spectrum.

B.C. Fawcett and R.W. Hayes, Mon. Not. Roy. Astron. Soc. **170**, 185 (1975).

These authors compare observed and calculated wavenlengths in this spectrum.

S.O. Kastner, J. Opt. Soc. Am. **63**, 738 (1973).

This author analyzes the levels and intervals for configurations $2s^r2p^k$ in this spectrum.

E.Ya. Kononov, K.N. Koshelev, L.I. Podobedova, and S.S. Churilov, Opt. Spectrosc. **40**, 121 (1976); Opt. Spectrosc. **39**, 458 (1975).

These authors investigate the $n = 2$–$n = 2$ transitions and obtaine values for several wavelengths and energies.

J.D. Purcell and K.G. Widing, Ap. J. **176**, 239 (1972).

These authors observe several wavelengths of Ca XIV in a solar flare.

Ca XV $Z = 20$ 6 electrons

Most of the information has been taken from Fawcett (1975), with some levels taken from Kastner (1973).

B.C. Fawcett, J. Phys. B**4**, 981 (1971).

This author classifies several wavelengths in this spectrum.

B.C. Fawcett, Atomic Data and Nuclear Data Tables **16**, 135 (1975).

This author tabulates levels' energies and wavelengths for transitions $2s^22p^n$–$2s2p^{n+1}$ and $2s2p^n$–$2p^{n+1}$.

B.C. Fawcett and R.W. Hayes, Mon. Not. Astron. Soc. **170**, 185 (1975).

These authors compare observed and calculated wavelengths in this spectrum.

U. Feldman and G.A. Doschek, preprint (April 1976).

These authors observe a forbidden transition (included in our drawings) in the spectrum of the solar corona.

S.O. Kastner, J. Opt. Soc. Am. **63**, 738 (1973).

This author analyzes the levels and intervals for configurations $2s^r2p^k$ in this spectrum.

E.Ya. Kononov, K.N. Koshelev, L.I. Podobedova, and S.S. Churilov, Opt. Spectrosc. **39**, 458 (1975); Opt. Spectrosc. **40**, 121 (1976).

These authors obtain this spectrum from a laser-produced plasma.

J.D. Purcell and K.G. Widing, Ap. J. **176**, 239 (1972).

These authors identify lines of this spectrum of a solar flare.

Ca XVI $Z = 20$ 5 electrons

Kelly and Palumbo list transitions with wavelengths of 22.95 Å, and 16.81 Å, tentatively classified as belonging to the upper levels $3s\,^2S_{1/2}$ and $4s\,^2S_{1/2}$, respectively. Because these assignments are only tentative we have not included them in our drawings.

B.C. Fawcett, Atomic Data and Nuclear Data Tables **16**, 135 (1975).

This author tabulates levels' energies and wavelengths for transitions $2s^2 2p^n - 2s2p^{n+1}$ and $2s2p^n - 2p^{n+1}$.

B.C. Fawcett and R.W. Hayes, Mon. Not. Roy. Astron. Soc. **170**, 185 (1975).

These authors compare calculated and observed wavelengths in the spectrum.

S.O. Kastner, J. Opt. Soc. Am. **61**, 335 (1971).

This author predicts term splittings for this spectrum.

E.Ya. Kononov, K.N. Koshelev, L.I. Podobedova, and S.S. Churilov, Opt. Spectrosc. **39**, 458 (1975); Opt. Spectrosc. **40**, 121 (1976).

These authors present wavelengths and energy levels for this spectrum.

J.D. Purcell and K.G. Widing, Ap. J. **176**, 239 (1972).

These authors identify two transitions of this spectrum of a solar flare.

Ca XVII $Z = 20$ 4 electrons

The recent literature contains several disparate sets of energies for the triplet levels. We have not been able to bring these into agreement by any choice of the uncertainties (X, Y, etc.) in the relative locations of the singlet and triplet levels. We have taken the triplet levels below 6 000 000 cm^{-1} from Goldsmith (1975); other levels are largely from Fawcett (1975).

B.C. Fawcett, Atomic and Nuclear Data Tables **16**, 135 (1975).

This author tabulates levels' energies and wavelengths for the transitions $2s^2 2p^n - 2s2p^{n+1}$ and $2s2p^n - 2p^{n+1}$.

S. Goldsmith, private communication (1975).

S. Goldsmith, L. Oren (Katz), A.M. Crooker, and L. Cohen, Ap. J. **184**, 1021 (1973).

These authors provide observed and predicted wavelengths, and observed energy levels.

E.Ya. Kononov, K.N. Koshelev, L.I. Podobedova, and S.S. Churilov, Opt. Spectrosc. **39**, 458 (1975); Opt. Spectrosc. **40**, 121 (1976).

These authors provide observed and predicted wavelengths, and energy levels.

J.D. Purcell and K.G. Widing, Ap. J. **176**, 239 (1972).

These authors identify a transition in Ca XVII in the spectrum of a solar flare.

Ca XVIII $Z = 20$ 3 electrons

The wavelengths for $2s^2S_{1/2} - 2p^2P^\circ_{1/2,3/2}$ are taken from Purcell and Widing (1972). They give the $2p^2P^\circ_{1/2} - 2p^2P^\circ_{3/2}$ spacing as 40 862 cm^{-1}.

E.V. Aglitskii, V.A. Boiko, S.M. Zakharov, S.A. Pikuz, and A.Ya. Faenov, Sov. J. Quant. Electron. **4**, 500 (1974).

These authors observe and classify several wavelengths in the spectrum.

G.A. Doschek, *Solar Gamma-, X- and EUV Radiation,* IAU Symposium 68, S.R. Kane, Ed., 165 (1975); Space Sci. Rev. **13**, 765 (1972).

This author classifies several satellite lines attributed to Ca XVIII.

B.C. Fawcett, Atomic Data and Nuclear Data Tables **16**, 135 (1975).

This author tabulates levels' energies and wavelengths for the transitions $2s^22p^n-2s2p^{n+1}$ and $2s2p^n-2p^{n+1}$.

U. Feldman, G.A. Doschek, D.J. Nagel, R.D. Cowan, and R.R. Whitlock, Ap. J. **192**, 213 (1974).

These authors study the satellite spectra from laser-produced plasmas and tabulate classified transitions in Ca XVIII. [See also U. Feldman, G.A. Doschek, D.K. Prinz, and D.J. Nagel, J. Appl. Phys. **47**, 1341 (1976).]

A.H. Gabriel, Mon. Not. Roy. Astron. Soc. **160**, 99 (1972).

This authors tabulates many computed wavelengths for this spectrum.

S. Goldsmith, U. Feldman, L. Oren (Katz), and L. Cohen, Ap. J. **174**, 209 (1972).

These authors classify several lines of this spectrum.

J.D. Purcell and K.G. Widing, Ap. J. **176**, 239 (1972).

These authors classify two transitions in this spectrum.

Ca XIX $Z = 20$ 2 electrons

For the transition $1s^2\ ^1S_0-1s2p\ ^3P_1^\circ$, Neupert *et al.* (1971) give the wavelength 3.187 Å; however, Ermolaev *et al.* (1972) calculate this wavelength to be 3.19253 Å and suggest a possible mis-identification by Neupert *et al.*

E.V. Aglitskii, V.A. Boiko, S.M. Zakharov, S.A. Pikuz, and A.Ya. Faenov, Sov. J. Quant. Electron. **4**, 500 (1974).

These authors observe and classify two lines in this spectrum.

L. Cotten, U. Feldman, M. Swartz, and J.H. Underwood, J. Opt. Soc. Am. **58**, 843 (1968).

These authors observe two transitions in this spectrum.

G.A. Doschek, Space Sci. Rev. **13**, 765 (1972).

This author observes several transitions of this spectrum.

A.M. Ermolaev, M. Jones, and K.J.H. Phillips, Ap. J. Letters **12**, 53 (1972).

These authors calculate accurate wavelengths of transitions in this spectrum.

W.M. Neupert, Solar Phys. **18**, 474 (1971); W.M. Neupert, W. Gates, M. Swartz, and R. Young, Ap. J. **149**, L79 (1967).

These authors identify lines observed in the spectrum of a solar flare.

Ca XX $Z = 20$ 1 electron

For hydrogenic systems we show calculated energies and wavelengths. For Ca XX the uncertainty in the QED corrections to the ionization limit is $\pm 7\ \text{cm}^{-1}$. As a result all levels' energies are uncertain by as least this amount. The levels 6d–6g and 7d–7h are taken from Erickson (1976); the others are taken from Garcia and Mack. Erickson also gives calculated energies for 7I, 8K, 9L, 10M, and 11N levels that we do not include on our diagrams.

G.W. Erickson, private communication, (1976) (in press, J. Phys. Chem. Ref. Data).

This author has calculated hydrogenic levels (binding energies) up to $Z = 105$.

J.D. Garcia and J.E. Mack, J. Opt. Soc. Am. **55**, 654 (1965).

Authors give calculated energy-level and line tables for one-electron atomic spectra.

Scandium (Sc)

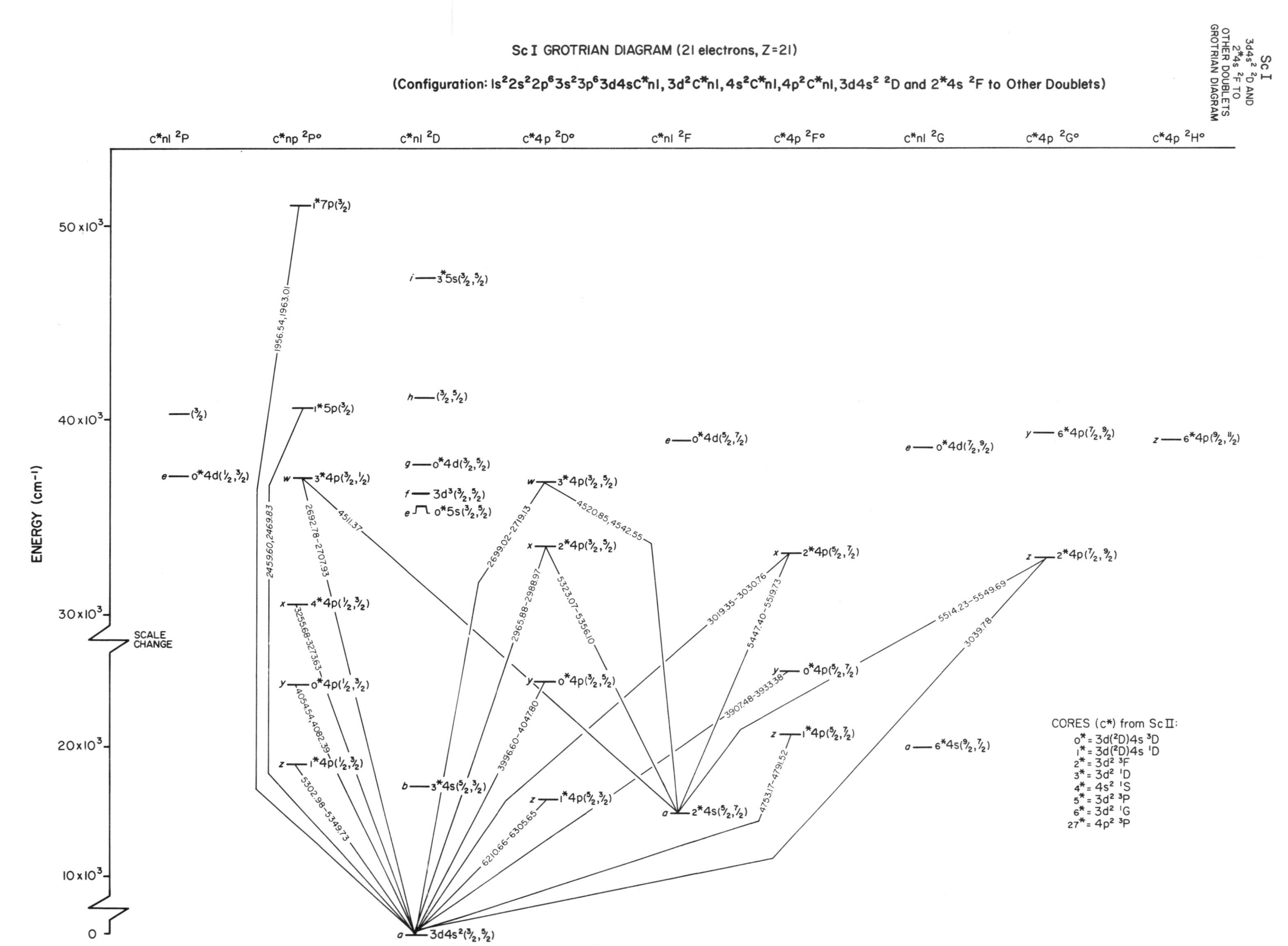
Sc I
3d4s² ²D AND
2*4s ²F TO
OTHER DOUBLETS
GROTRIAN DIAGRAM
Sc I GROTRIAN DIAGRAM (21 electrons, Z=21)
(Configuration: 1s²2s²2p⁶3s²3p⁶3d4sc*nl, 3d²c*nl, 4s²c*nl, 4p²c*nl, 3d4s² ²D and 2*4s ²F to Other Doublets)
c*nl ²P
c*np ²P°
c*nl ²D
c*4p ²D°
c*nl ²F
c*4p ²F°
c*nl ²G
c*4p ²G°
c*4p ²H°
CORES (c*) from Sc II:
o* = 3d(²D)4s ³D
1* = 3d(²D)4s ¹D
2* = 3d² ³F
3* = 3d² ¹D
4* = 4s² ¹S
5* = 3d² ³P
6* = 3d² ¹G
27* = 4p² ³P
ENERGY (cm⁻¹)
50×10³
40×10³
30×10³
20×10³
10×10³
0
SCALE CHANGE
a 3d4s²(³/₂,⁵/₂)
a 2*4s(⁵/₂,⁷/₂)
a 6*4s(⁹/₂,⁷/₂)
b 3*4s(⁵/₂,³/₂)
z 1*4p(¹/₂,³/₂)
y o*4p(¹/₂,³/₂)
x 4*4p(¹/₂,³/₂)
w 3*4p(³/₂,¹/₂)
1*5p(³/₂)
1*7p(³/₂)
z 1*4p(⁵/₂,³/₂)
y o*4p(³/₂,⁵/₂)
x 2*4p(³/₂,⁵/₂)
w 3*4p(³/₂,⁵/₂)
z 1*4p(⁵/₂,⁷/₂)
y o*4p(⁵/₂,⁷/₂)
x 2*4p(⁵/₂,⁷/₂)
z 2*4p(⁷/₂,⁹/₂)
y 6*4p(⁷/₂,⁹/₂)
z 6*4p(⁹/₂,¹¹/₂)
e o*4d(¹/₂,³/₂)
(³/₂)
e o*5s(³/₂,⁵/₂)
f 3d³(³/₂,⁵/₂)
g o*4d(³/₂,⁵/₂)
h (³/₂,⁵/₂)
i 3*5s(³/₂,⁵/₂)
e o*4d(⁵/₂,⁷/₂)
e o*4d(⁷/₂,⁹/₂)
5302.98-5349.73
4054.54,4082.39
3255.68-3273.63
2692.78-2707.93
4511.37
2459.60,2469.83
1956.54,1963.01
6210.66-6305.65
3996.60-4047.80
2965.88-2988.97
2699.02-2719.13
5323.07-5356.10
4520.85,4542.55
3019.35-3030.76
5447.40-5519.73
3907.48-3933.38
4753.17-4791.52
5514.23-5549.69
3039.78

Sc I ENERGY LEVELS (21 electrons, Z=21)

(Configuration: $1s^2 2s^2 2p^6 3s^2 3p^6 3d4sc^*nl$, $3d^2c^*nl$, $4s^2c^*nl$, Doublet System)

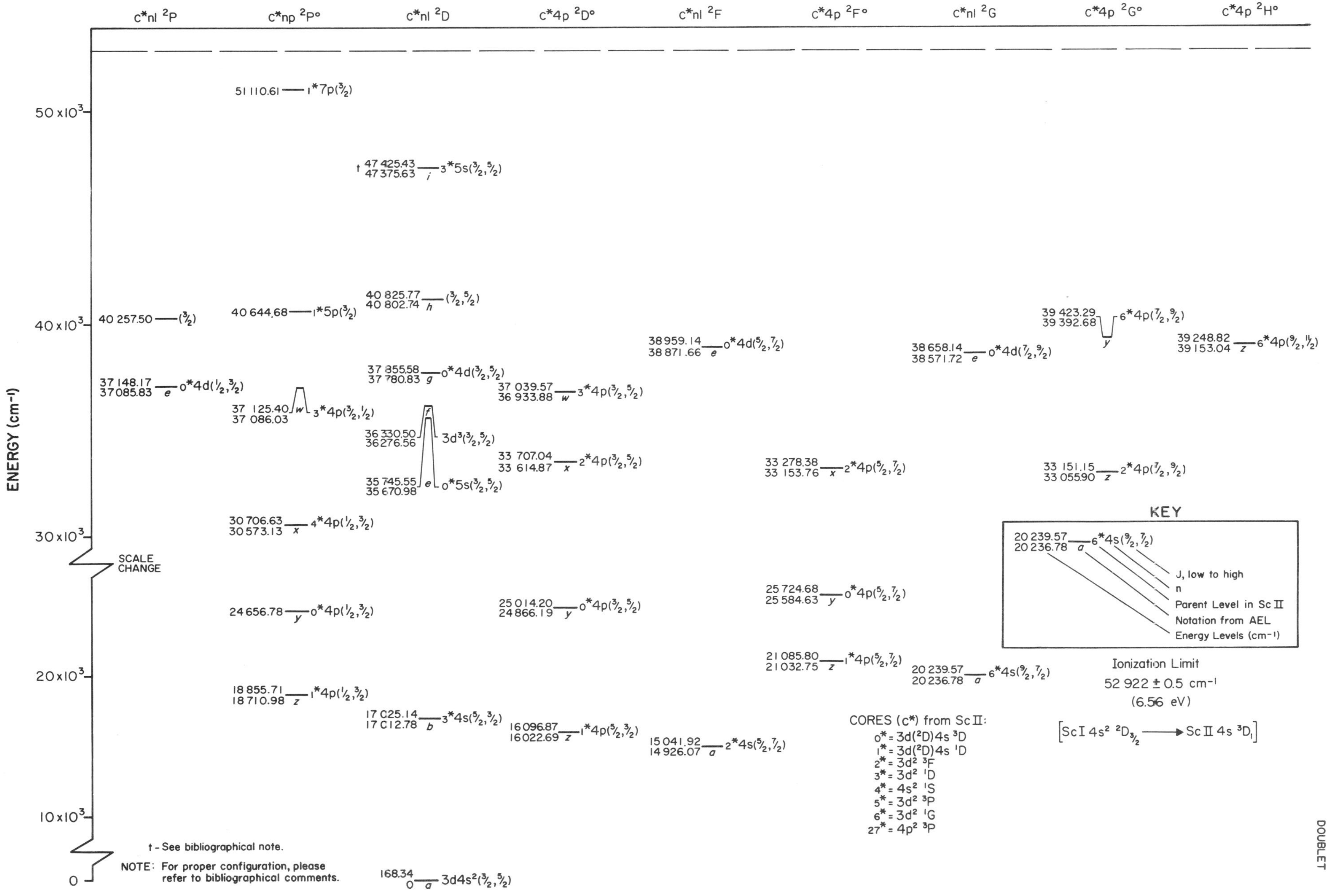

ENERGY (cm⁻¹)

Axis: 0, 10×10^3, 20×10^3, 30×10^3, 40×10^3, 50×10^3 — SCALE CHANGE

$c^*nl\ ^2P$

- 40 257.50 — (3/2)
- 37 148.17, 37 085.83 — e — $0^*4d(1/2, 3/2)$

$c^*np\ ^2P°$

- 51 110.61 — $1^*7p(3/2)$
- 40 644.68 — $1^*5p(3/2)$
- 37 125.40, 37 086.03 — w — $3^*4p(3/2, 1/2)$
- 30 706.63, 30 573.13 — x — $4^*4p(1/2, 3/2)$
- 24 656.78 — y — $0^*4p(1/2, 3/2)$
- 18 855.71, 18 710.98 — z — $1^*4p(1/2, 3/2)$

$c^*nl\ ^2D$

- † 47 425.43, 47 375.63 — i — $3^*5s(3/2, 5/2)$
- 40 825.77, 40 802.74 — h — (3/2, 5/2)
- 37 855.58, 37 780.83 — g — $0^*4d(3/2, 5/2)$
- 36 330.50, 36 276.56 — f — $3d^3(3/2, 5/2)$
- 35 745.55, 35 670.98 — e — $0^*5s(3/2, 5/2)$
- 17 025.14, 17 012.78 — b — $3^*4s(5/2, 3/2)$
- 168.34, 0 — a — $3d4s^2(3/2, 5/2)$

$c^*4p\ ^2D°$

- 37 039.57, 36 933.88 — w — $3^*4p(3/2, 5/2)$
- 33 707.04, 33 614.87 — x — $2^*4p(3/2, 5/2)$
- 25 014.20, 24 866.19 — y — $0^*4p(3/2, 5/2)$
- 16 096.87, 16 022.69 — z — $1^*4p(5/2, 3/2)$

$c^*nl\ ^2F$

- 38 959.14, 38 871.66 — e — $0^*4d(5/2, 7/2)$
- 15 041.92, 14 926.07 — a — $2^*4s(5/2, 7/2)$

$c^*4p\ ^2F°$

- 33 278.38, 33 153.76 — x — $2^*4p(5/2, 7/2)$
- 25 724.68, 25 584.63 — y — $0^*4p(5/2, 7/2)$
- 21 085.80, 21 032.75 — z — $1^*4p(5/2, 7/2)$

$c^*nl\ ^2G$

- 38 658.14, 38 571.72 — e — $0^*4d(7/2, 9/2)$
- 20 239.57, 20 236.78 — a — $6^*4s(9/2, 7/2)$

$c^*4p\ ^2G°$

- 39 423.29, 39 392.68 — y — $6^*4p(7/2, 9/2)$
- 33 151.15, 33 055.90 — z — $2^*4p(7/2, 9/2)$

$c^*4p\ ^2H°$

- 39 248.82, 39 153.04 — z — $6^*4p(9/2, 11/2)$

KEY

20 239.57, 20 236.78 — a — $6^*4s(9/2, 7/2)$

- J, low to high
- n
- Parent Level in Sc II
- Notation from AEL
- Energy Levels (cm⁻¹)

Ionization Limit
52 922 ± 0.5 cm⁻¹
(6.56 eV)

[Sc I $4s^2\ ^2D_{3/2}$ ⟶ Sc II $4s\ ^3D_1$]

CORES (c^*) from Sc II:

- $0^* = 3d(^2D)4s\ ^3D$
- $1^* = 3d(^2D)4s\ ^1D$
- $2^* = 3d^2\ ^3F$
- $3^* = 3d^2\ ^1D$
- $4^* = 4s^2\ ^1S$
- $5^* = 3d^2\ ^3P$
- $6^* = 3d^2\ ^1G$
- $27^* = 4p^2\ ^3P$

† - See bibliographical note.

NOTE: For proper configuration, please refer to bibliographical comments.

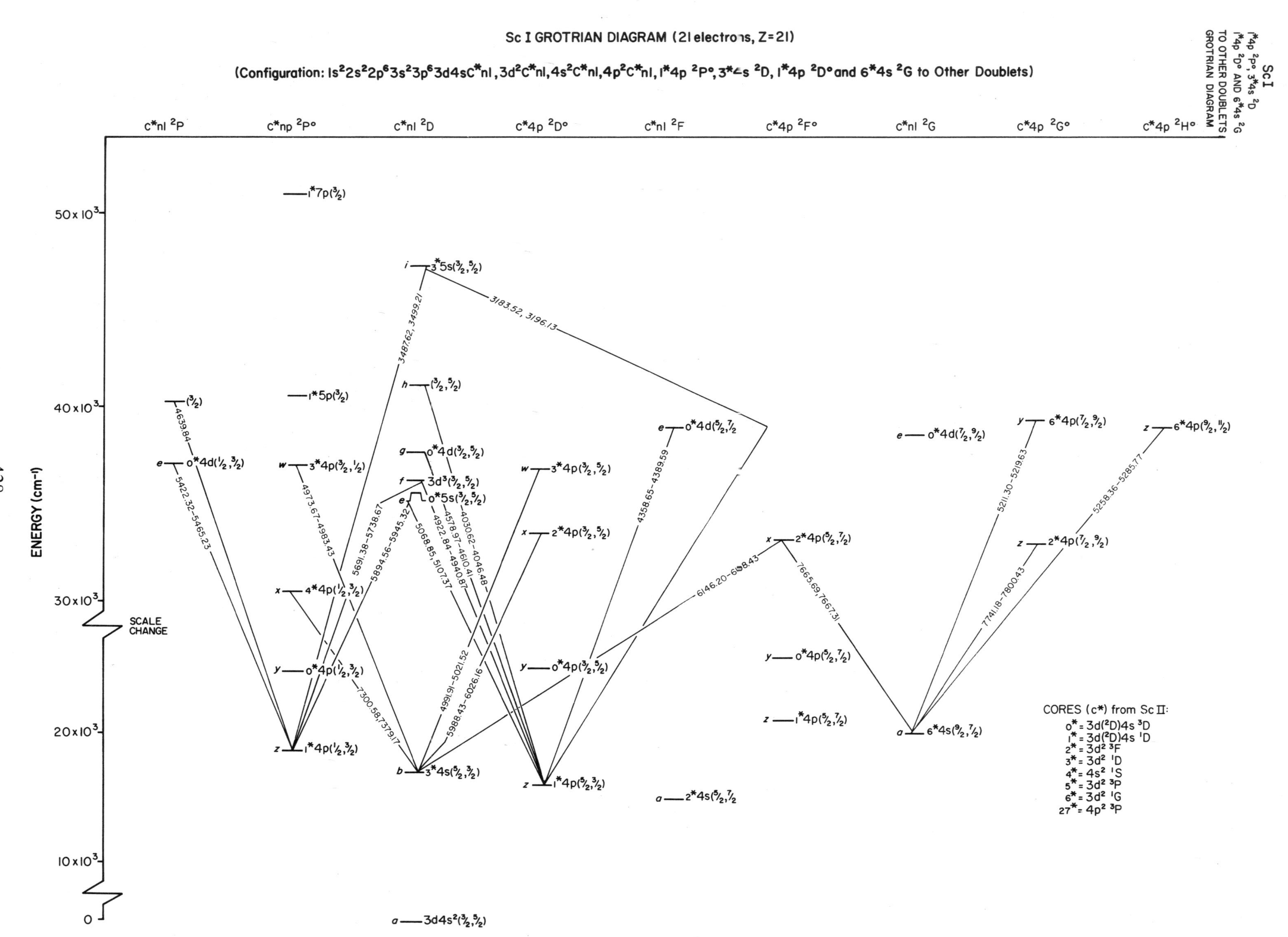
Sc I GROTRIAN DIAGRAM (21 electrons, Z=21)
(Configuration: 1s²2s²2p⁶3s²3p⁶3d4sc*nl, 3d²c*nl, 4s²c*nl, 4p²c*nl, 1*4p ²P°, 3*4s ²D, 1*4p ²D° and 6*4s ²G to Other Doublets)
Sc I
1*4p ²P°, 3*4s ²D
1*4p ²D° AND 6*4s ²G
TO OTHER DOUBLETS
GROTRIAN DIAGRAM
c*nl ²P
c*np ²P°
c*nl ²D
c*4p ²D°
c*nl ²F
c*4p ²F°
c*nl ²G
c*4p ²G°
c*4p ²H°
ENERGY (cm⁻¹)
50x10³
40x10³
30x10³
20x10³
10x10³
0
SCALE CHANGE
1*7p(3/2)
1*5p(3/2)
i 3*5s(3/2,5/2)
h (3/2,5/2)
g o*4d(3/2,5/2)
f 3d³(3/2,5/2)
e o*5s(3/2,5/2)
(3/2)
e o*4d(1/2,3/2)
w 3*4p(3/2,1/2)
x 4*4p(1/2,3/2)
y o*4p(1/2,3/2)
z 1*4p(1/2,3/2)
b 3*4s(5/2,3/2)
a 3d4s²(3/2,5/2)
w 3*4p(3/2,5/2)
x 2*4p(3/2,5/2)
y o*4p(3/2,5/2)
z 1*4p(5/2,3/2)
e o*4d(5/2,7/2)
a 2*4s(5/2,7/2)
x 2*4p(5/2,7/2)
y o*4p(5/2,7/2)
z 1*4p(5/2,7/2)
e o*4d(7/2,9/2)
a 6*4s(9/2,7/2)
y 6*4p(7/2,9/2)
z 2*4p(7/2,9/2)
z 6*4p(9/2,11/2)
4639.84
5422.32–5465.23
4973.67–4983.43
3487.62, 3499.21
3183.52, 3196.13
5691.38–5738.67
5894.56–5945.32
7300.58, 7379.17
5068.85, 5107.37
4922.84–4940.87
4578.97–4610.41
4030.62–4046.48
4991.91–5021.52
5988.43–6026.16
4358.65–4389.59
6146.20–6198.43
7665.69, 7667.31
5211.30–5219.63
7741.18–7800.43
5258.36–5285.77
CORES (c*) from Sc II:
o* = 3d(²D)4s ³D
1* = 3d(²D)4s ¹D
2* = 3d² ³F
3* = 3d² ¹D
4* = 4s² ¹S
5* = 3d² ³P
6* = 3d² ¹G
27* = 4p² ³P

Sc I ENERGY LEVELS (21 electrons, Z=21)

(Configuration: $1s^2 2s^2 2p^6 3s^2 3p^6 3d4sc^* nl$, $3d^2 c^* nl$, $4s^2 c^* nl$, Doublet System)

ENERGY (cm⁻¹): 50 x10³, 40 x10³, 30 x10³, SCALE CHANGE, 20 x10³, 10 x10³, 0

$c^*nl\ ^2P$
- 40 257.50 — ($^3/_2$)
- 37 148.17, 37 085.83 — e — $0^*4d(^1/_2, ^3/_2)$

$c^*np\ ^2P°$
- 51 110.61 — $1^*7p(^3/_2)$
- 40 644.68 — $1^*5p(^3/_2)$
- 37 125.40, 37 086.03 — w — $3^*4p(^3/_2, ^1/_2)$
- 30 706.63, 30 573.13 — x — $4^*4p(^1/_2, ^3/_2)$
- 24 656.78 — y — $0^*4p(^1/_2, ^3/_2)$
- 18 855.71, 18 710.98 — z — $1^*4p(^1/_2, ^3/_2)$

$c^*nl\ ^2D$
- † 47 425.43, 47 375.63 — i — $3^*5s(^3/_2, ^5/_2)$
- 40 825.77, 40 802.74 — h — ($^3/_2, ^5/_2$)
- 37 855.58, 37 780.83 — g — $0^*4d(^3/_2, ^5/_2)$
- 36 330.50, 36 276.56 — f — $3d^3(^3/_2, ^5/_2)$
- 35 745.55, 35 670.98 — e — $0^*5s(^3/_2, ^5/_2)$
- 17 025.14, 17 012.78 — b — $3^*4s(^5/_2, ^3/_2)$
- 168.34, 0 — a — $3d4s^2(^3/_2, ^5/_2)$

$c^*4p\ ^2D°$
- 37 039.57, 36 933.88 — w — $3^*4p(^3/_2, ^5/_2)$
- 33 707.04, 33 614.87 — x — $2^*4p(^3/_2, ^5/_2)$
- 25 014.20, 24 866.19 — y — $0^*4p(^3/_2, ^5/_2)$
- 16 096.87, 16 022.69 — z — $1^*4p(^5/_2, ^3/_2)$

$c^*nl\ ^2F$
- 38 959.14, 38 871.66 — e — $0^*4d(^5/_2, ^7/_2)$
- 15 041.92, 14 926.07 — a — $2^*4s(^5/_2, ^7/_2)$

$c^*4p\ ^2F°$
- 33 278.38, 33 153.76 — x — $2^*4p(^5/_2, ^7/_2)$
- 25 724.68, 25 584.63 — y — $0^*4p(^5/_2, ^7/_2)$
- 21 085.80, 21 032.75 — z — $1^*4p(^5/_2, ^7/_2)$

$c^*nl\ ^2G$
- 38 658.14, 38 571.72 — e — $0^*4d(^7/_2, ^9/_2)$
- 20 239.57, 20 236.78 — a — $6^*4s(^9/_2, ^7/_2)$

$c^*4p\ ^2G°$
- 39 423.29, 39 392.68 — y — $6^*4p(^7/_2, ^9/_2)$
- 33 151.15, 33 055.90 — z — $2^*4p(^7/_2, ^9/_2)$

$c^*4p\ ^2H°$
- 39 248.82, 39 153.04 — z — $6^*4p(^9/_2, ^{11}/_2)$

KEY

20 239.57, 20 236.78 — a — $6^*4s(^9/_2, ^7/_2)$
- J, low to high
- n
- Parent Level in Sc II
- Notation from AEL
- Energy Levels (cm⁻¹)

Ionization Limit
52 922 ± 0.5 cm⁻¹
(6.56 eV)

[Sc I $4s^2\ ^2D_{3/2}$ ⟶ Sc II $4s\ ^3D_1$]

CORES (c^*) from Sc II:
- $0^* = 3d(^2D)4s\ ^3D$
- $1^* = 3d(^2D)4s\ ^1D$
- $2^* = 3d^2\ ^3F$
- $3^* = 3d^2\ ^1D$
- $4^* = 4s^2\ ^1S$
- $5^* = 3d^2\ ^3P$
- $6^* = 3d^2\ ^1G$
- $27^* = 4p^2\ ^3P$

† - See bibliographical note.

NOTE: For proper configuration, please refer to bibliographical comments.

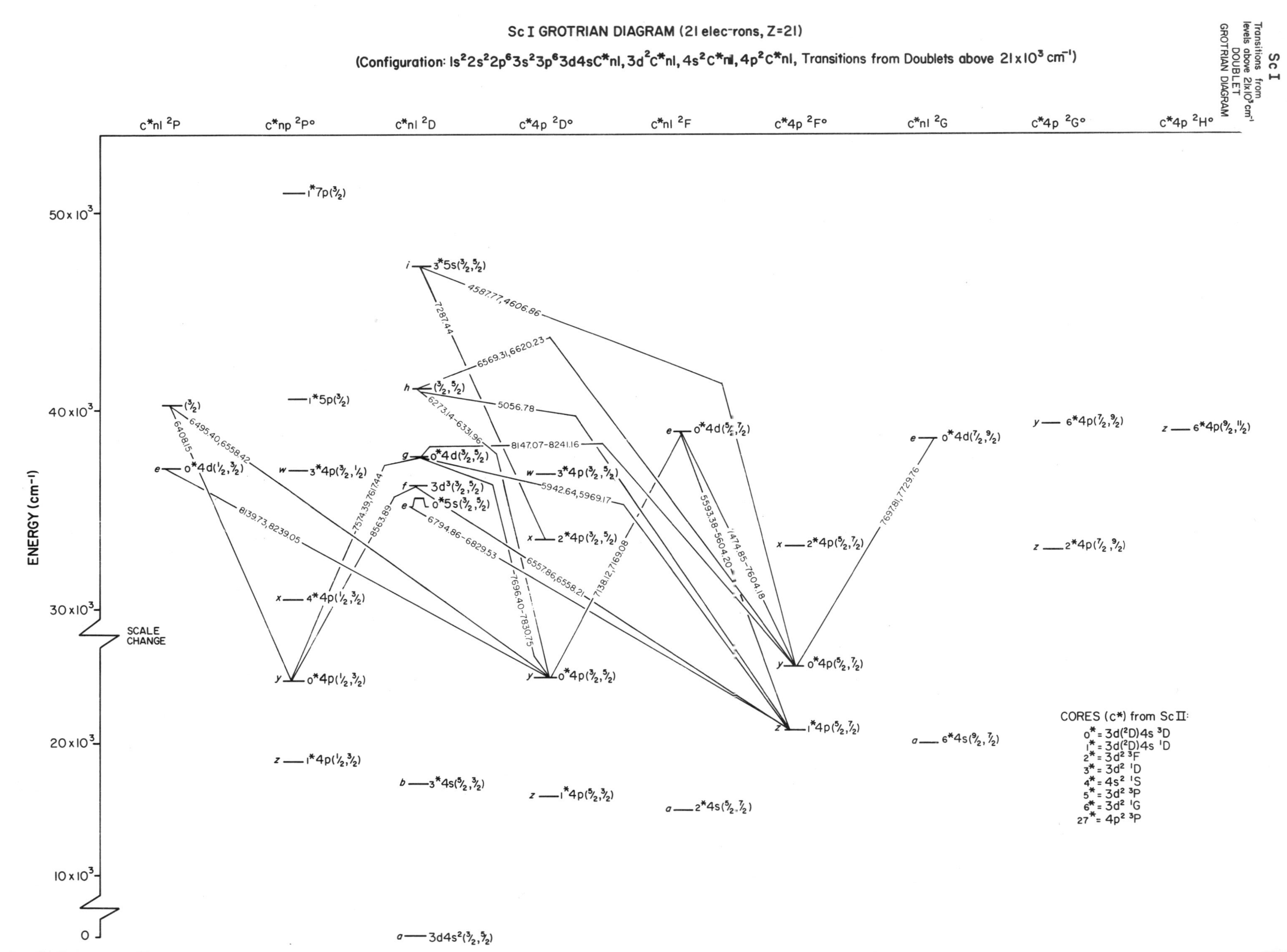

Sc I GROTRIAN DIAGRAM (21 electrons, Z=21)
(Configuration: 1s2 2s2 2p6 3s2 3p6 3d4sC*nl, 3d2C*nl, 4s2C*nl, 4p2C*nl, Transitions from Doublets above 21x10^3 cm^-1)
Sc I
Transitions from levels above 21x10^3 cm^-1
DOUBLET
GROTRIAN DIAGRAM
c*nl 2P
c*np 2P°
c*nl 2D
c*4p 2D°
c*nl 2F
c*4p 2F°
c*nl 2G
c*4p 2G°
c*4p 2H°
ENERGY (cm-1)
50x10^3
40x10^3
30x10^3
20x10^3
10x10^3
0
SCALE CHANGE
CORES (c*) from Sc II:
0* = 3d(2D)4s 3D
1* = 3d(2D)4s 1D
2* = 3d2 3F
3* = 3d2 1D
4* = 4s2 1S
5* = 3d2 3P
6* = 3d2 1G
27* = 4p2 3P
6408.15
6495.40,6558.42
8139.73,8239.05
7287.44
4587.77,4606.86
6569.31,6620.23
5056.78
6273.14-6331.96
8147.07-8241.16
5942.64,5969.17
6557.86,6558.21
6794.86-6829.53
7696.40-7830.75
7574.39,7617.44
8563.89
7138.12,7169.08
5593.38-5604.20
7474.85-7604.18
7697.81,7729.76

Sc I ENERGY LEVELS (21 electrons, Z=21)

(Configuration: $1s^2 2s^2 2p^6 3s^2 3p^6 3d4sc^*nl$, $3d^2c^*nl$, $4s^2c^*nl$, Doublet System)

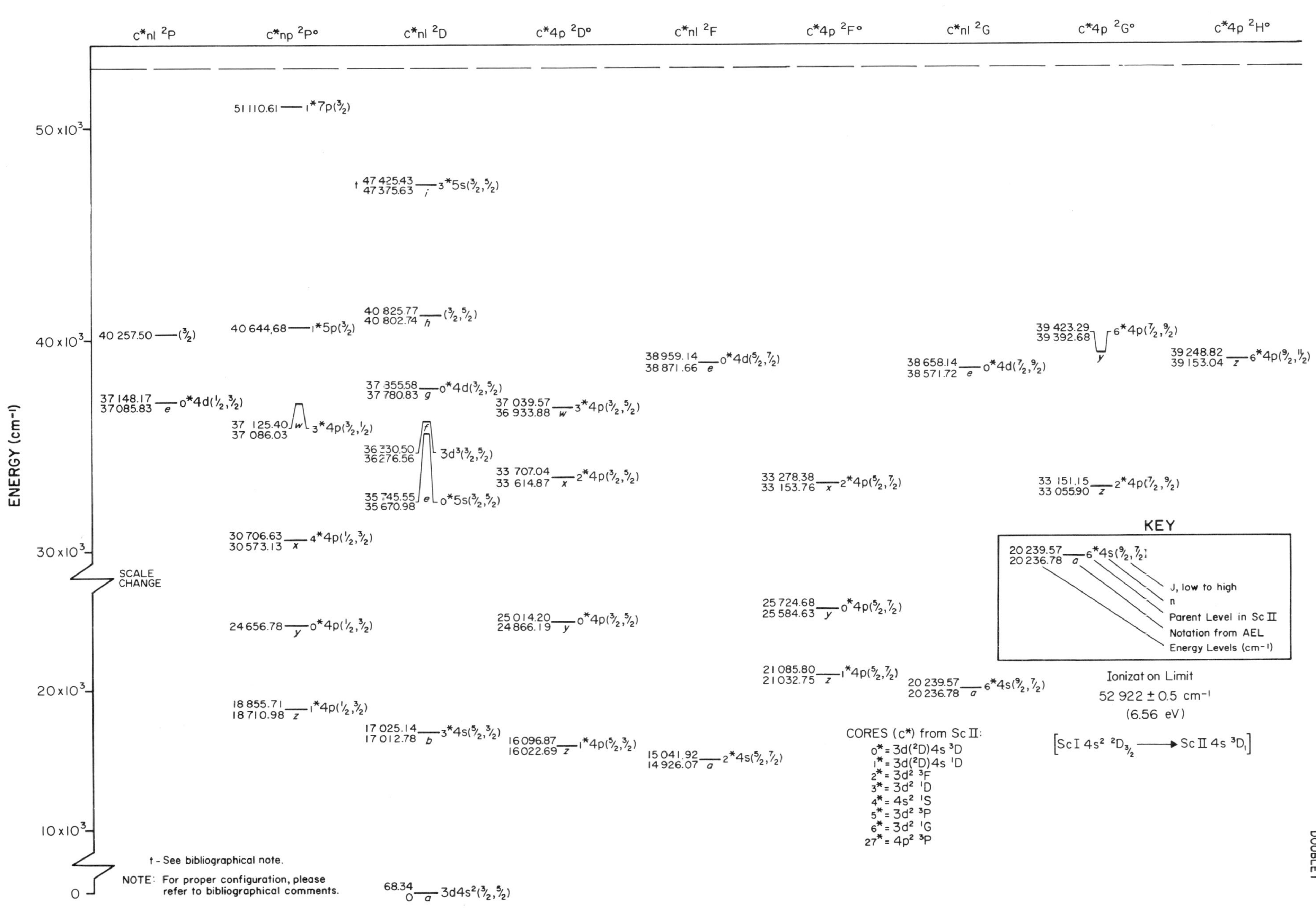

Sc I GROTRIAN DIAGRAM (21 electrons, Z = 21)
(Configuration: $1s^2 2s^2 2p^6 3s^2 3p^6 3d\,4s\,c^*nl$, Ground to $^2P^o$ and $^2D^o$ Levels, n ≥ 9)

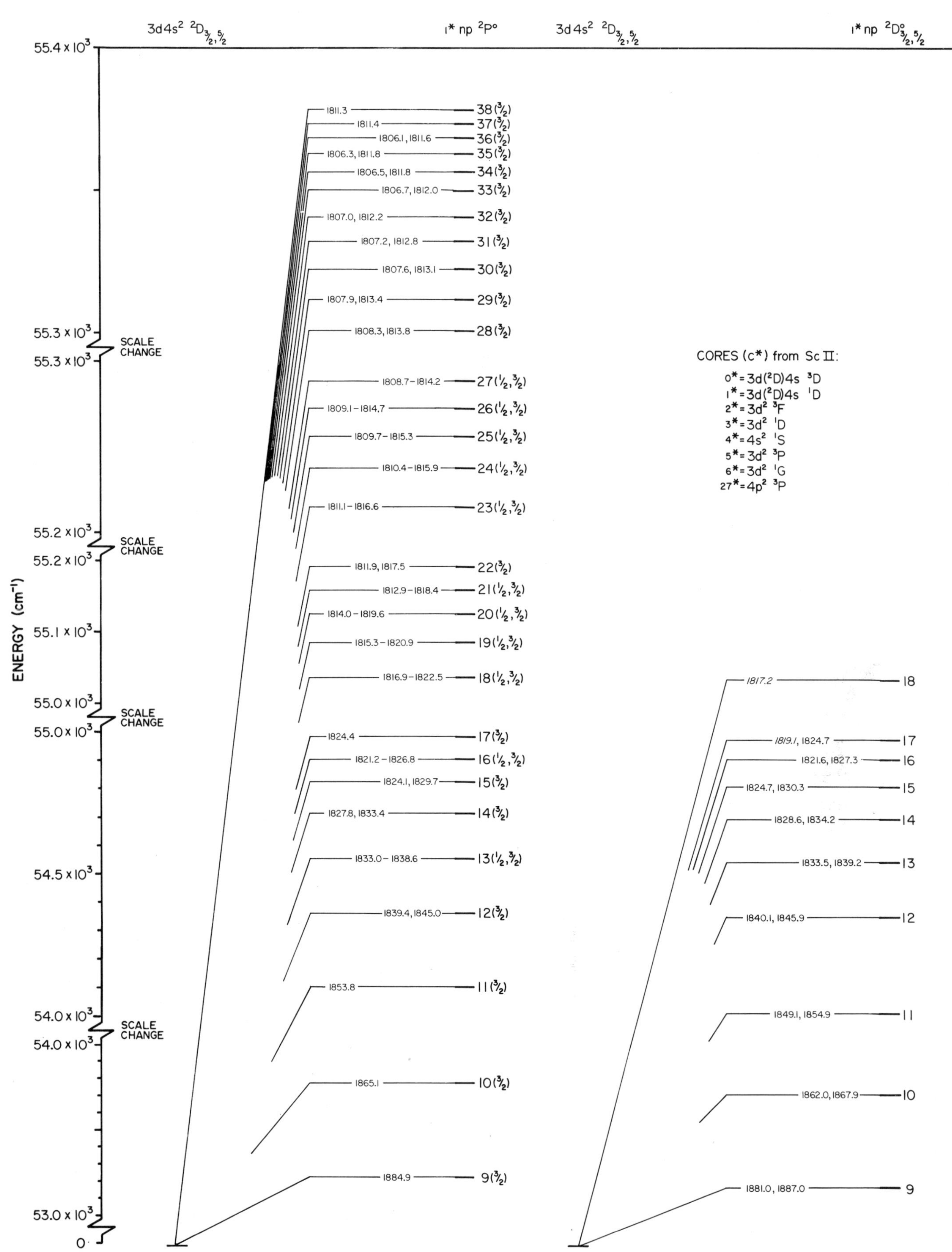

Sc I ENERGY LEVELS (21 electrons, Z = 21)
(Configuration: $1s^2 2s^2 2p^6 3s^2 3p^6 3d\,4s\,c^* nl$, $^2P^o$ and $^2D^o$ Levels, n ≥ 9)

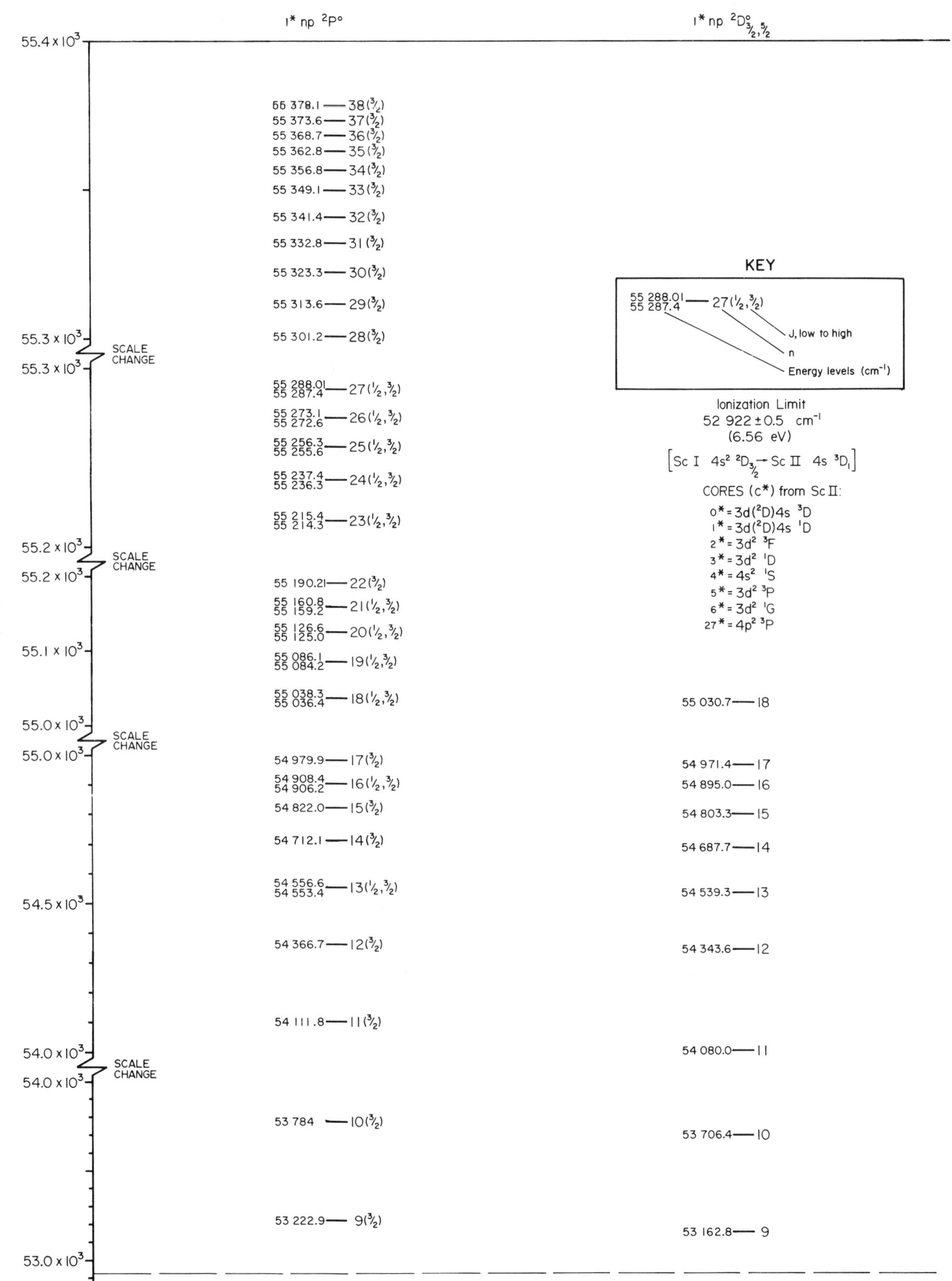

Sc I
Ground to 5*nf $^2F^o$
2*np $^2D^o$, $^2F^o$, $^2G^o$, n ≥ 4
GROTRIAN DIAGRAM

Sc I GROTRIAN DIAGRAM (21 electrons, Z=21)
(Configuration: $1s^2 2s^2 2p^6 3s^2 3p^6 3d^6$ C*nl, 5*nf $^2F^o$ and 2*np $^2D^o$, $^2F^o$, $^2G^o$ Levels, n≥4)

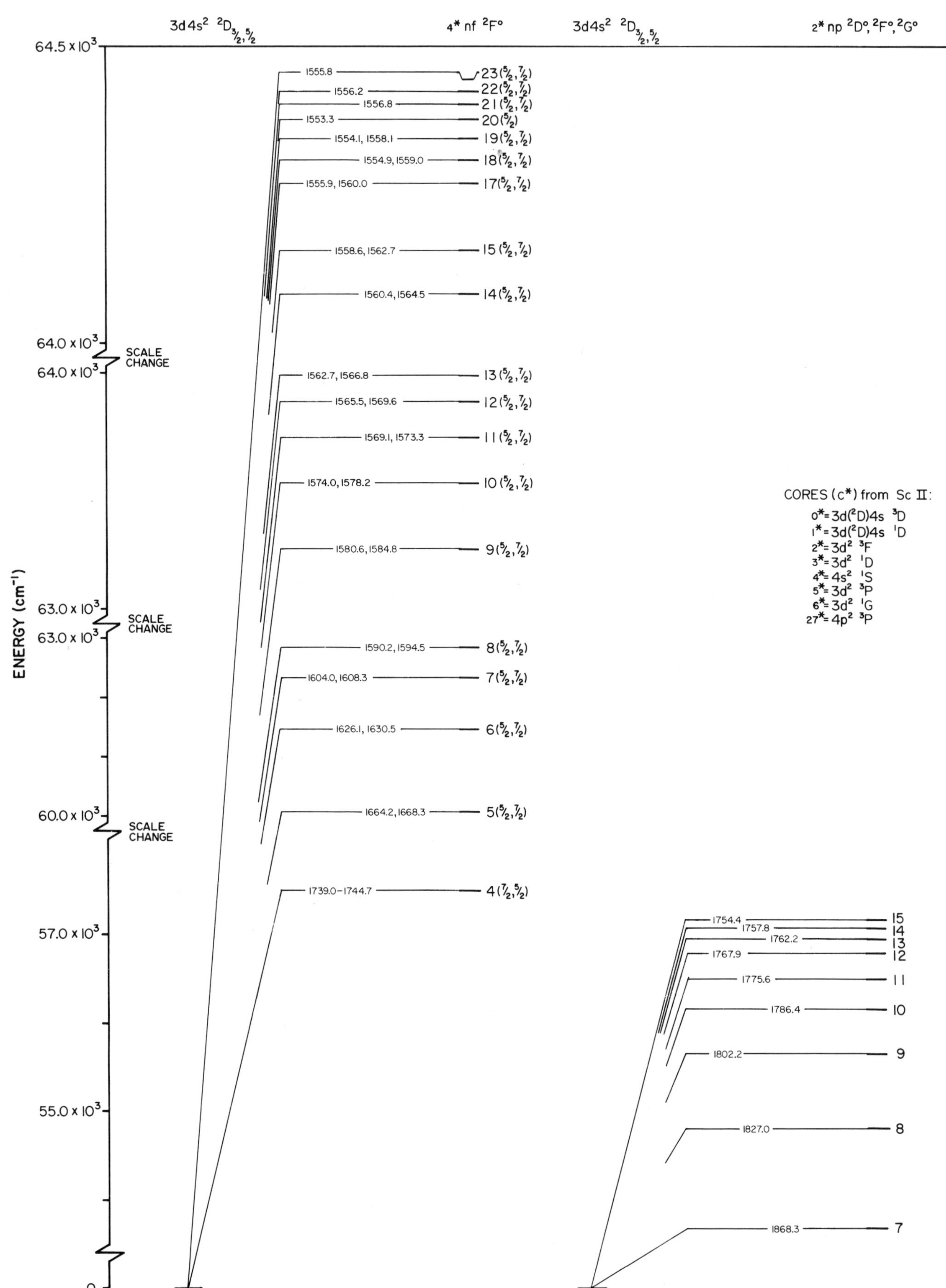

Sc I ENERGY LEVELS (21 electrons, Z = 21)

(Configuration: $1s^2 2s^2 2p^6 3s^2 3p^6 3d^2$ c^*nl, $5^*nf\ ^2F^o$ and $2^*np\ ^2D^o, ^2F^o, ^2G^o$ Levels, $n \geq 4$)

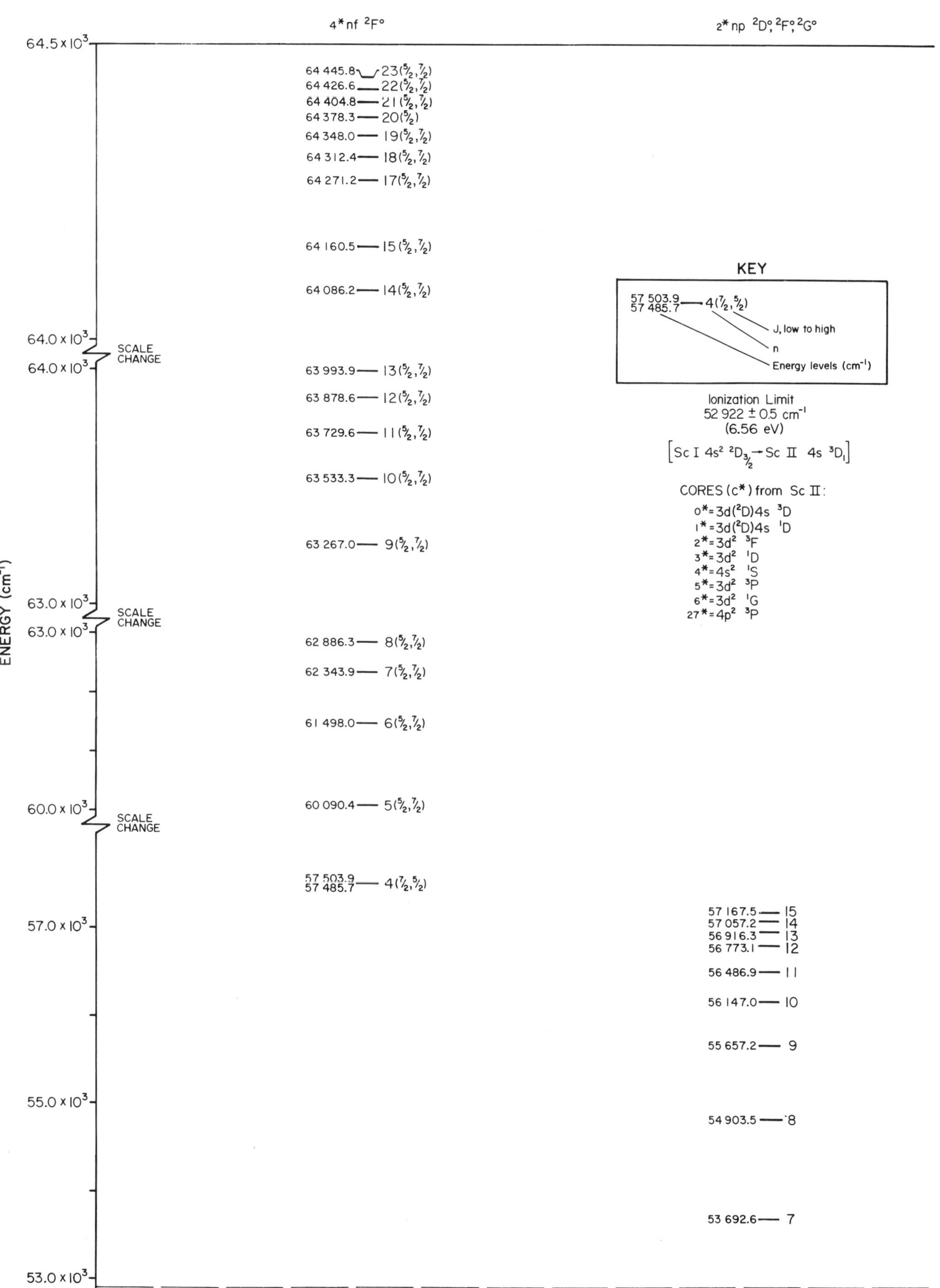

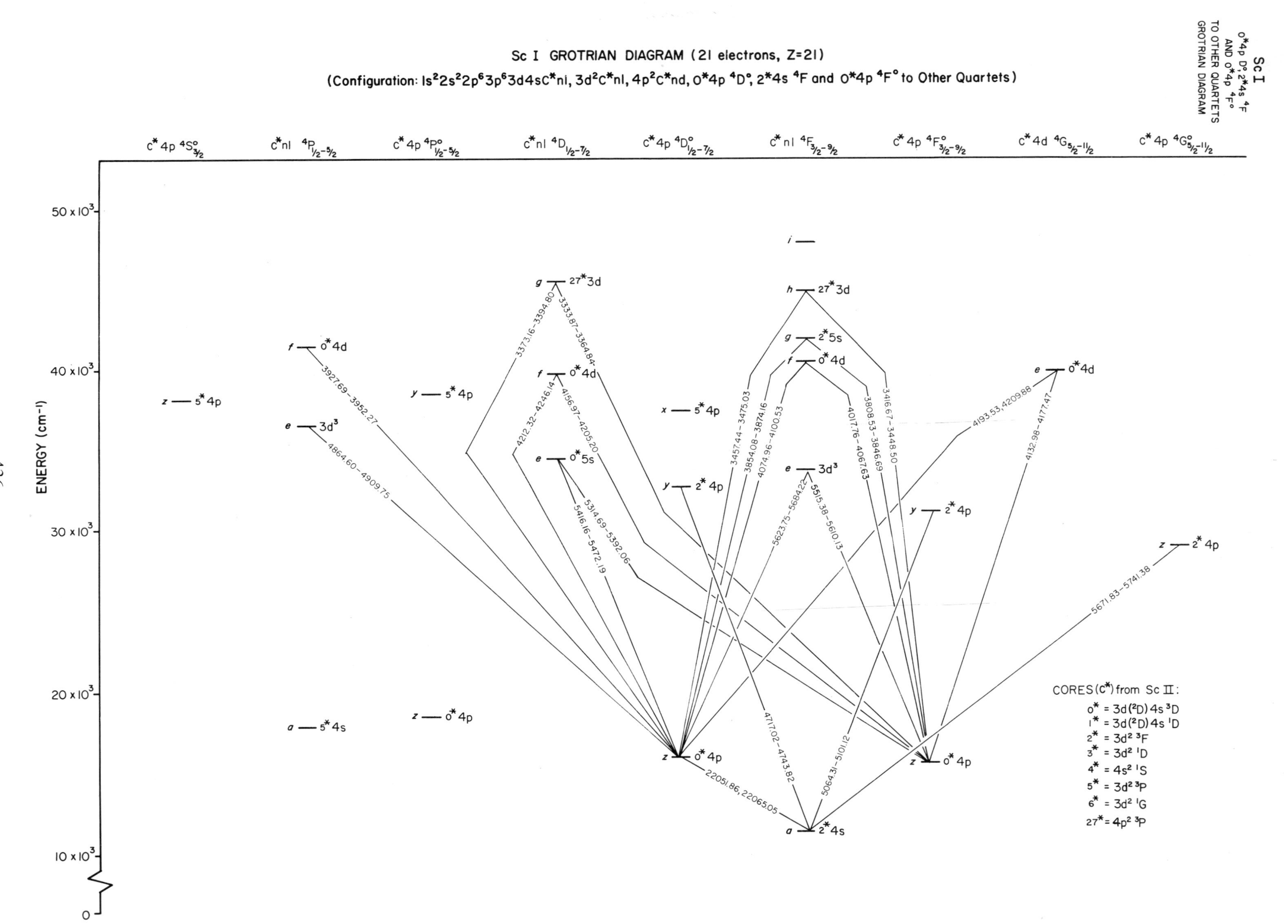
Sc I
o*4p D°, 2*4s ⁴F
AND o*4p ⁴F°
TO OTHER QUARTETS
GROTRIAN DIAGRAM
Sc I GROTRIAN DIAGRAM (21 electrons, Z=21)
(Configuration: 1s²2s²2p⁶3p⁶3d4sc*nl, 3d²c*nl, 4p²c*nd, o*4p ⁴D°, 2*4s ⁴F and o*4p ⁴F° to Other Quartets)
c*4p ⁴S°3/2
c*nl ⁴P1/2–5/2
c*4p ⁴P°1/2–5/2
c*nl ⁴D1/2–7/2
c*4p ⁴D°1/2–7/2
c*nl ⁴F3/2–9/2
c*4p ⁴F°3/2–9/2
c*4d ⁴G5/2–11/2
c*4p ⁴G°5/2–11/2
ENERGY (cm⁻¹)
50 x 10³
40 x 10³
30 x 10³
20 x 10³
10 x 10³
0
CORES (c*) from Sc II:
o* = 3d(²D) 4s ³D
1* = 3d(²D) 4s ¹D
2* = 3d² ³F
3* = 3d² ¹D
4* = 4s² ¹S
5* = 3d² ³P
6* = 3d² ¹G
27* = 4p² ³P

Sc I ENERGY LEVELS (21 electrons, Z = 21)

(Configuration: $1s^2 2s^2 2p^6 3s^2 3p^6 3d 4s nl$, $3d^2 nl$, $4p^2 nd$, Quartet System)

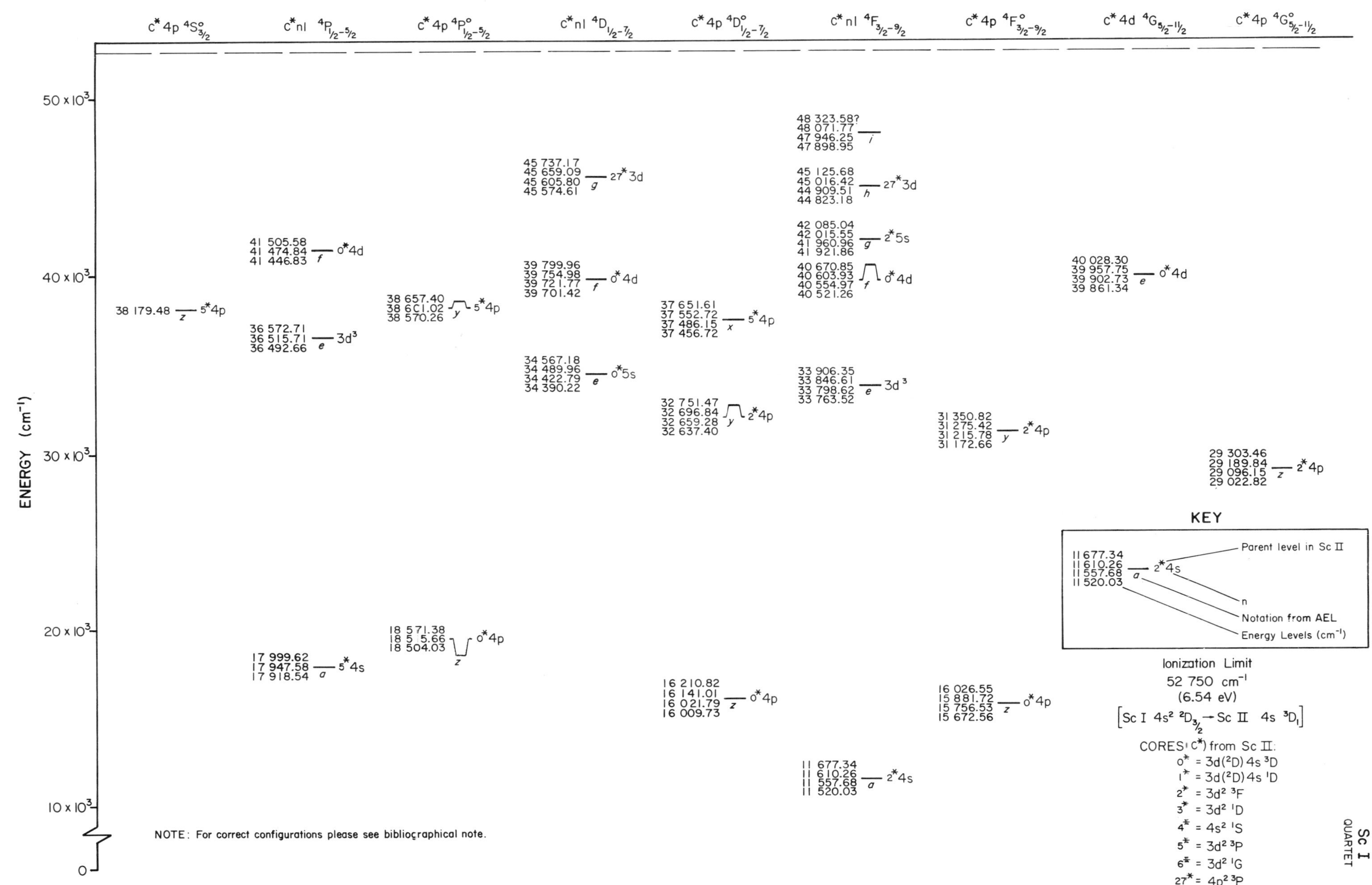

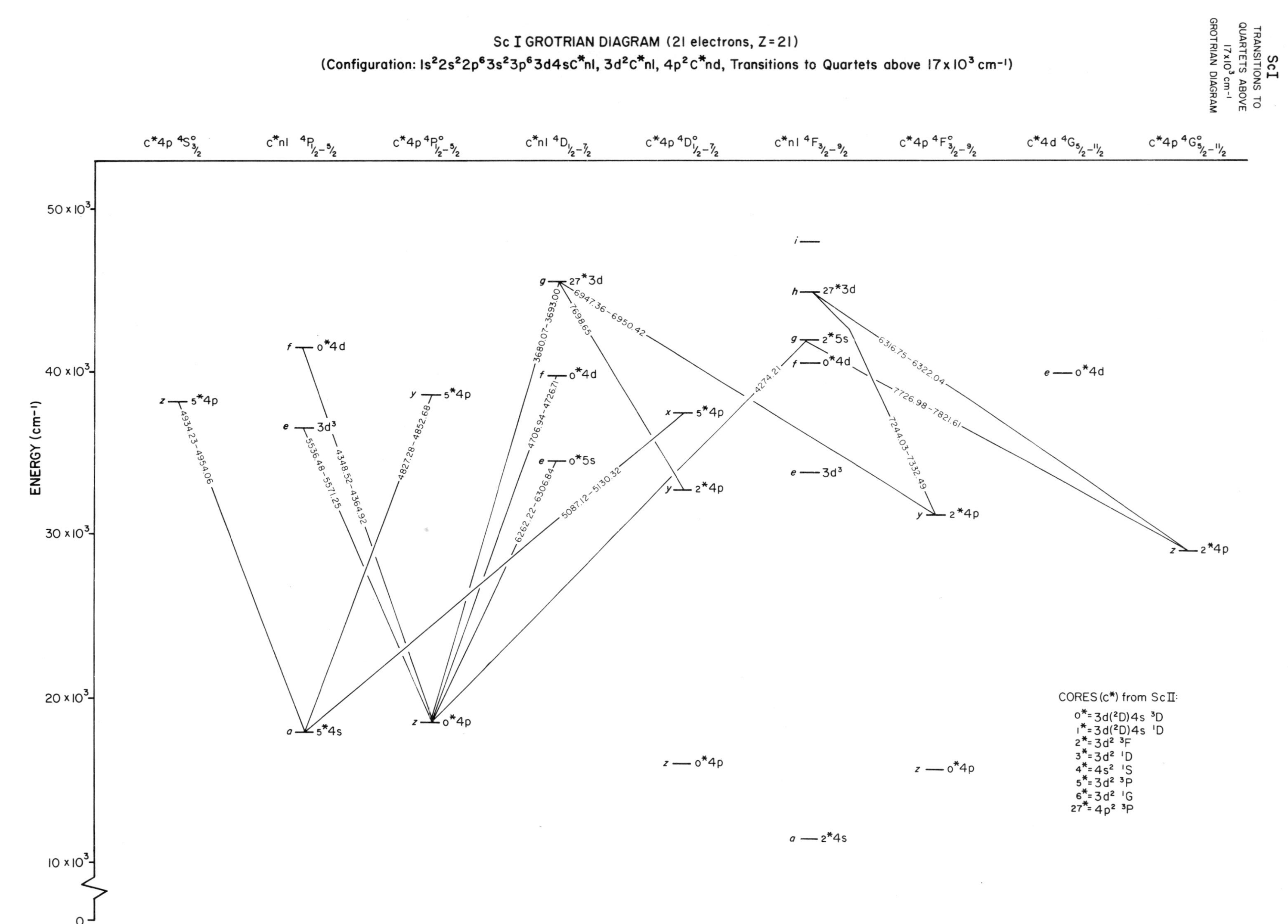

ScI
TRANSITIONS TO QUARTETS ABOVE 17x10³ cm⁻¹
GROTRIAN DIAGRAM
Sc I GROTRIAN DIAGRAM (21 electrons, Z=21)
(Configuration: 1s²2s²2p⁶3s²3p⁶3d4sc*nl, 3d²c*nl, 4p²c*nd, Transitions to Quartets above 17x10³ cm⁻¹)
c*4p ⁴S°3/2
c*nl ⁴P1/2–5/2
c*4p ⁴P°1/2–5/2
c*nl ⁴D1/2–7/2
c*4p ⁴D°1/2–7/2
c*nl ⁴F3/2–9/2
c*4p ⁴F°3/2–9/2
c*4d ⁴G5/2–11/2
c*4p ⁴G°5/2–11/2
ENERGY (cm⁻¹)
50x10³
40x10³
30x10³
20x10³
10x10³
0
z 5*4p
f o*4d
e 3d³
y 5*4p
g 27*3d
f o*4d
e o*5s
x 5*4p
y 2*4p
i
h 27*3d
g 2*5s
f o*4d
e 3d³
y 2*4p
e o*4d
z 2*4p
a 5*4s
z o*4p
z o*4p
z o*4p
a 2*4s
4934.23–4954.06
5536.48–5571.25
4348.52–4364.92
4827.28–4852.68
3680.07–3693.00
4706.94–4726.71
6262.22–6306.84
5087.12–5130.32
6947.36–6950.42
7698.65
4274.21
6316.75–6322.04
7726.98–7821.61
7244.03–7332.49
CORES (c*) from ScII:
o* = 3d(²D)4s ³D
1* = 3d(²D)4s ¹D
2* = 3d² ³F
3* = 3d² ¹D
4* = 4s² ¹S
5* = 3d² ³P
6* = 3d² ¹G
27* = 4p² ³P

Sc I ENERGY LEVELS (21 electrons, Z = 21)

(Configuration: $1s^2 2s^2 2p^6 3s^2 3p^6 3d4snl, 3d^2nl, 4p^2nd$, Quartet System)

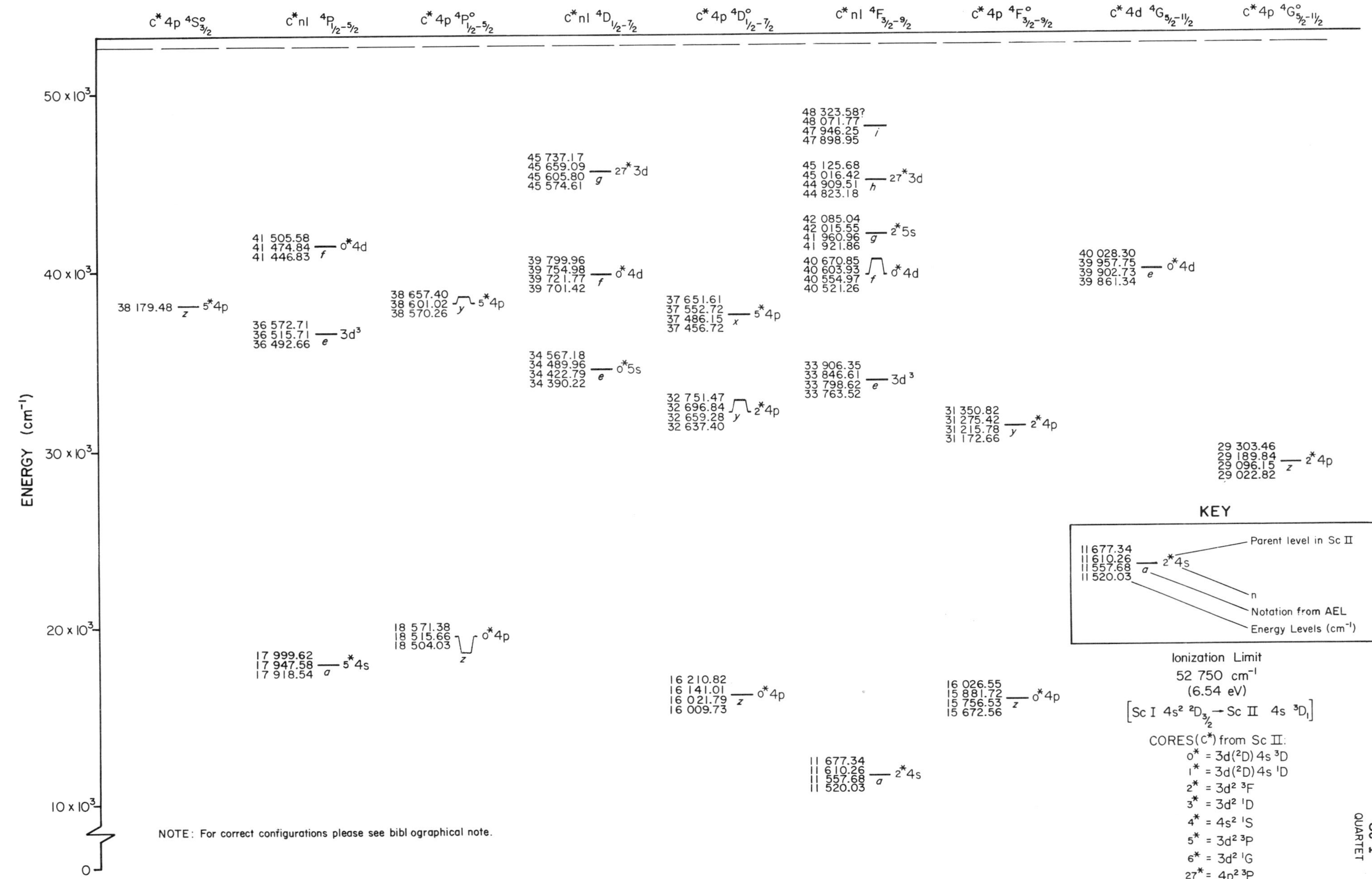

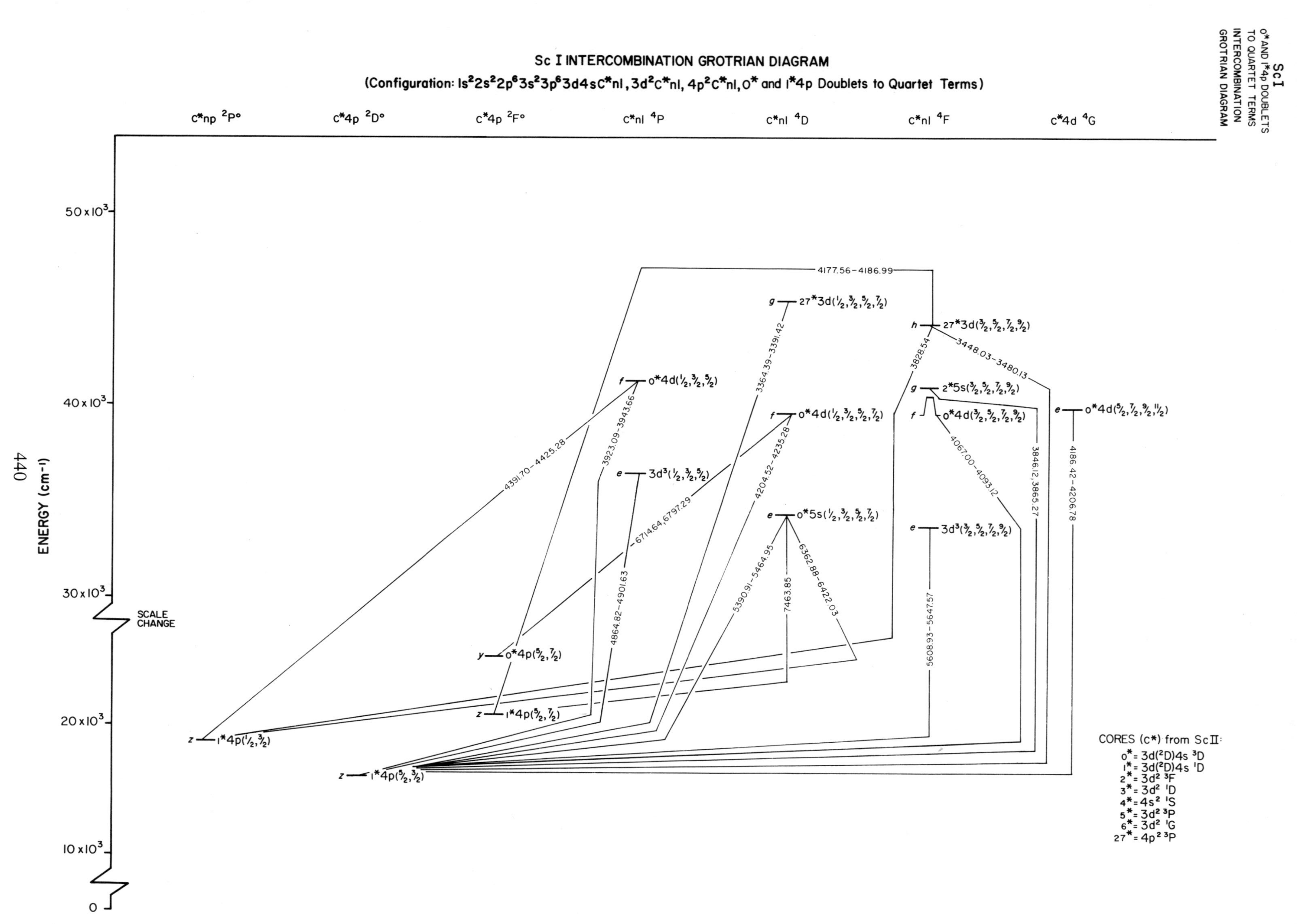

Sc I
o* AND 1*4p DOUBLETS TO QUARTET TERMS
INTERCOMBINATION GROTRIAN DIAGRAM
Sc I INTERCOMBINATION GROTRIAN DIAGRAM
(Configuration: 1s²2s²2p⁶3s²3p⁶3d4sc*nl, 3d²c*nl, 4p²c*nl, o* and 1*4p Doublets to Quartet Terms)
c*np ²P°
c*4p ²D°
c*4p ²F°
c*nl ⁴P
c*nl ⁴D
c*nl ⁴F
c*4d ⁴G
ENERGY (cm-1)
50x10³
40x10³
30x10³
20x10³
10x10³
0
SCALE CHANGE
4177.56-4186.99
g 27*3d(1/2,3/2,5/2,7/2)
h 27*3d(3/2,5/2,7/2,9/2)
3448.03-3480.13
3828.54
3364.39-3391.42
f o*4d(1/2,3/2,5/2)
g 2*5s(3/2,5/2,7/2,9/2)
f o*4d(1/2,3/2,5/2,7/2)
f o*4d(3/2,5/2,7/2,9/2)
e o*4d(5/2,7/2,9/2,11/2)
3923.09-3943.66
4391.70-4425.28
4204.52-4235.28
4067.00-4093.12
3846.12, 3865.27
4186.42-4206.78
e 3d³(1/2,3/2,5/2)
6714.64, 6797.29
e o*5s(1/2,3/2,5/2,7/2)
e 3d³(3/2,5/2,7/2,9/2)
4864.82-4901.63
5390.91-5464.95
7463.85
6362.88-6422.03
5608.93-5647.57
y o*4p(5/2,7/2)
z 1*4p(5/2,7/2)
z 1*4p(1/2,3/2)
z 1*4p(5/2,3/2)
CORES (c*) from Sc II:
o* = 3d(²D)4s ³D
1* = 3d(²D)4s ¹D
2* = 3d² ³F
3* = 3d² ¹D
4* = 4s² ¹S
5* = 3d² ³P
6* = 3d² ¹G
27* = 4p² ³P

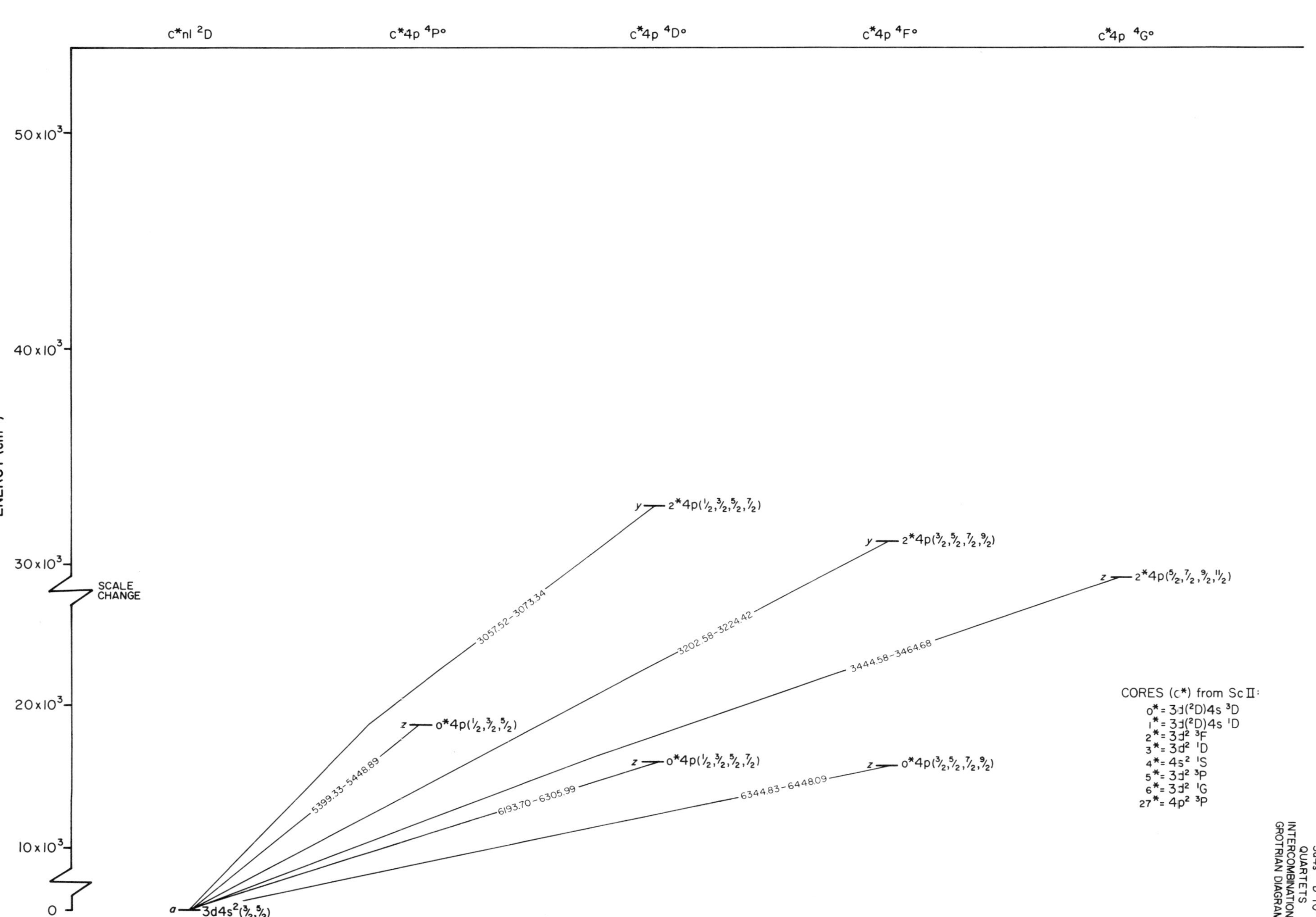

ScI INTERCOMBINATION GROTRIAN DIAGRAM
(Configuration: 1s²2s²2p⁶3s²3p⁶3d4sC*nl, 3d²C*nl, 3d4s² ²D to Quartet Terms)
c*nl ²D
c*4p ⁴P°
c*4p ⁴D°
c*4p ⁴F°
c*4p ⁴G°
ENERGY (cm⁻¹)
50x10³
40x10³
30x10³
20x10³
10x10³
0
SCALE CHANGE
y 2*4p(1/2,3/2,5/2,7/2)
y 2*4p(3/2,5/2,7/2,9/2)
z 2*4p(5/2,7/2,9/2,11/2)
z o*4p(1/2,3/2,5/2)
z o*4p(1/2,3/2,5/2,7/2)
z o*4p(3/2,5/2,7/2,9/2)
a 3d4s²(3/2,5/2)
3057.52-3073.34
3202.58-3224.42
3444.58-3464.68
5399.33-5448.89
6193.70-6305.99
6344.83-6448.09
CORES (c*) from ScII:
o* = 3d(²D)4s ³D
1* = 3d(²D)4s ¹D
2* = 3d² ³F
3* = 3d² ¹D
4* = 4s² ¹S
5* = 3d² ³P
6* = 3d² ¹G
27* = 4p² ³P
ScI
3d4s² ²D TO
QUARTETS
INTERCOMBINATION
GROTRIAN DIAGRAM

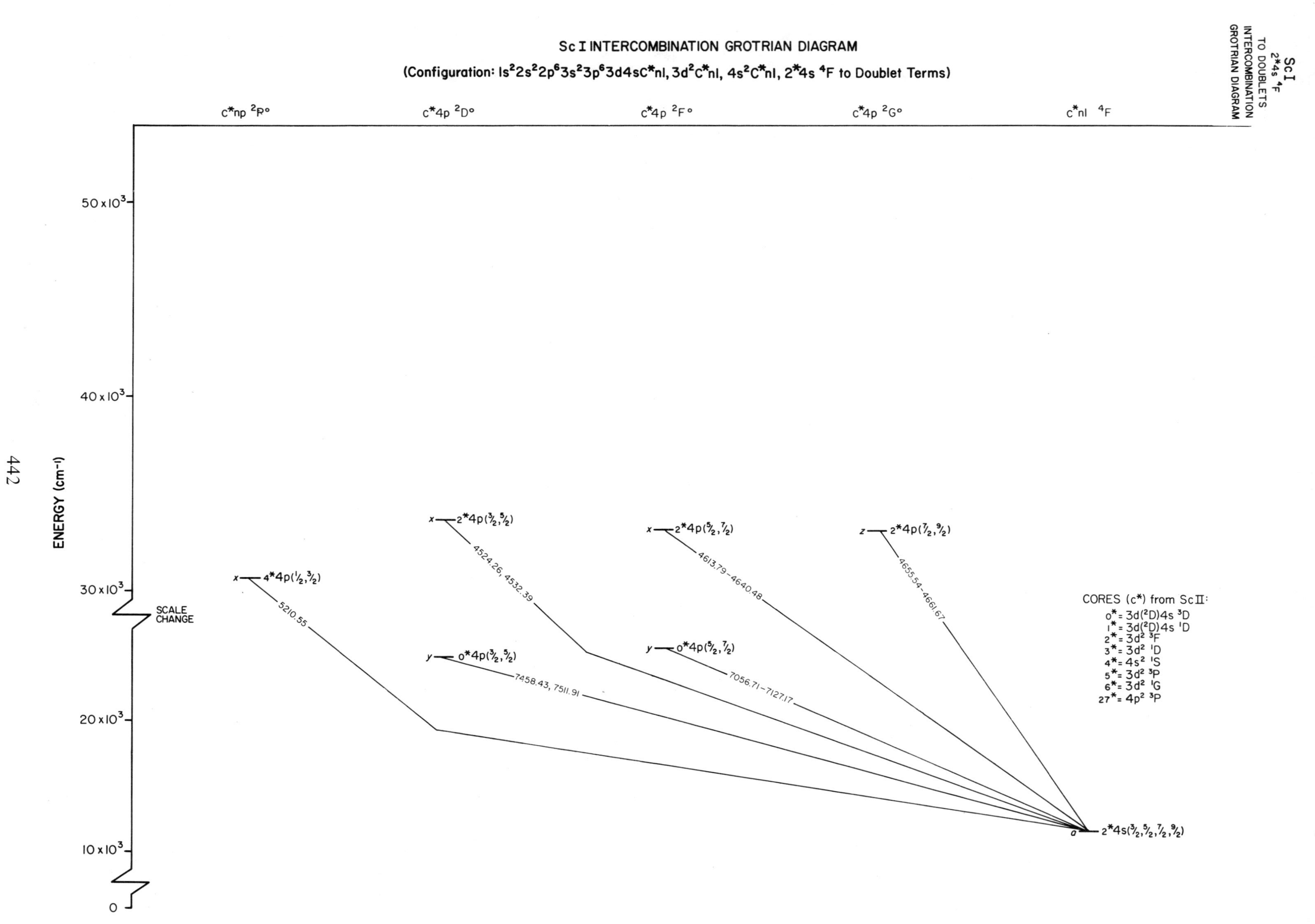

Sc I INTERCOMBINATION GROTRIAN DIAGRAM
(Configuration: 1s²2s²2p⁶3s²3p⁶3d4sC*nl, 3d²C*nl, 4s²C*nl, 2*4s ⁴F to Doublet Terms)
Sc I
2*4s ⁴F
TO DOUBLETS
INTERCOMBINATION
GROTRIAN DIAGRAM
c*np ²P°
c*4p ²D°
c*4p ²F°
c*4p ²G°
c*nl ⁴F
x 4*4p(1/2, 3/2)
x 2*4p(3/2, 5/2)
x 2*4p(5/2, 7/2)
z 2*4p(7/2, 9/2)
y 0*4p(3/2, 5/2)
y 0*4p(5/2, 7/2)
a 2*4s(3/2, 5/2, 7/2, 9/2)
5210.55
4524.26, 4532.39
4613.79–4640.48
4655.54–4661.67
7458.43, 7511.91
7056.71–7127.17
CORES (c*) from Sc II:
0* = 3d(²D)4s ³D
1* = 3d(²D)4s ¹D
2* = 3d² ³F
3* = 3d² ¹D
4* = 4s² ¹S
5* = 3d² ³P
6* = 3d² ¹G
27* = 4p² ³P
SCALE CHANGE
50×10³
40×10³
30×10³
20×10³
10×10³
0
ENERGY (cm⁻¹)

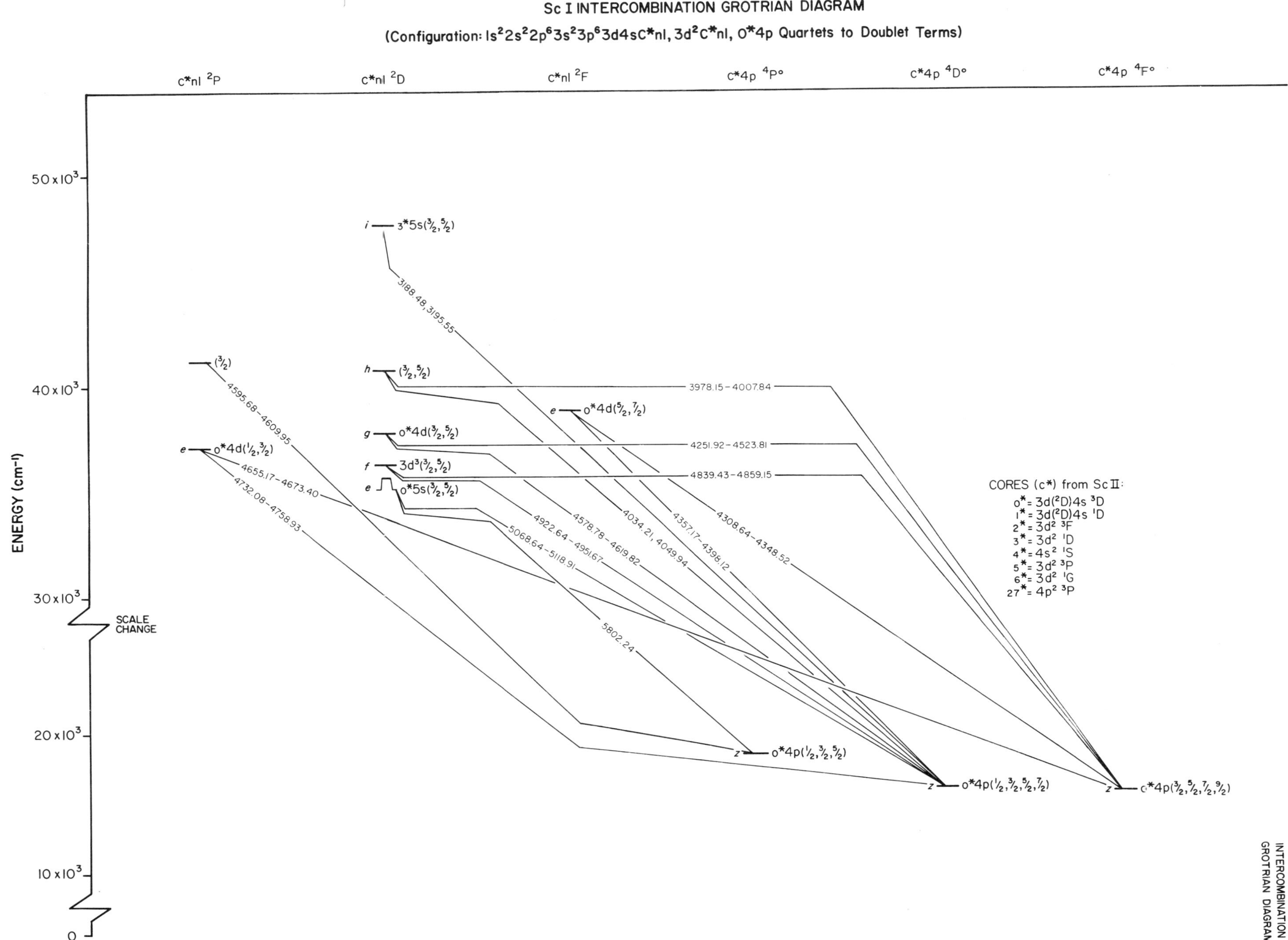
Sc I INTERCOMBINATION GROTRIAN DIAGRAM
(Configuration: 1s²2s²2p⁶3s²3p⁶3d4sC*nl, 3d²C*nl, 0*4p Quartets to Doublet Terms)
c*nl ²P
c*nl ²D
c*nl ²F
c*4p ⁴P°
c*4p ⁴D°
c*4p ⁴F°
ENERGY (cm⁻¹)
50 x 10³
40 x 10³
30 x 10³
20 x 10³
10 x 10³
0
SCALE CHANGE
i 3*5s(3/2,5/2)
(3/2)
h (3/2,5/2)
e 0*4d(5/2,7/2)
g 0*4d(3/2,5/2)
e 0*4d(1/2,3/2)
f 3d³(3/2,5/2)
e 0*5s(3/2,5/2)
3188.48, 3195.55
3978.15–4007.84
4595.68–4609.95
4251.92–4523.81
4839.43–4859.15
4655.17–4673.40
4732.08–4758.93
4578.78–4619.82
4034.21, 4049.94
4357.17–4398.12
4308.64–4348.52
4922.64–4951.67
5068.64–5118.91
5802.24
CORES (c*) from Sc II:
0* = 3d(²D)4s ³D
1* = 3d(²D)4s ¹D
2* = 3d² ³F
3* = 3d² ¹D
4* = 4s² ¹S
5* = 3d² ³P
6* = 3d² ¹G
27* = 4p² ³P
z 0*4p(1/2,3/2,5/2)
z 0*4p(1/2,3/2,5/2,7/2)
z c*4p(3/2,5/2,7/2,9/2)
Sc I
0*4p QUARTETS
TO DOUBLETS
INTERCOMBINATION
GROTRIAN DIAGRAM

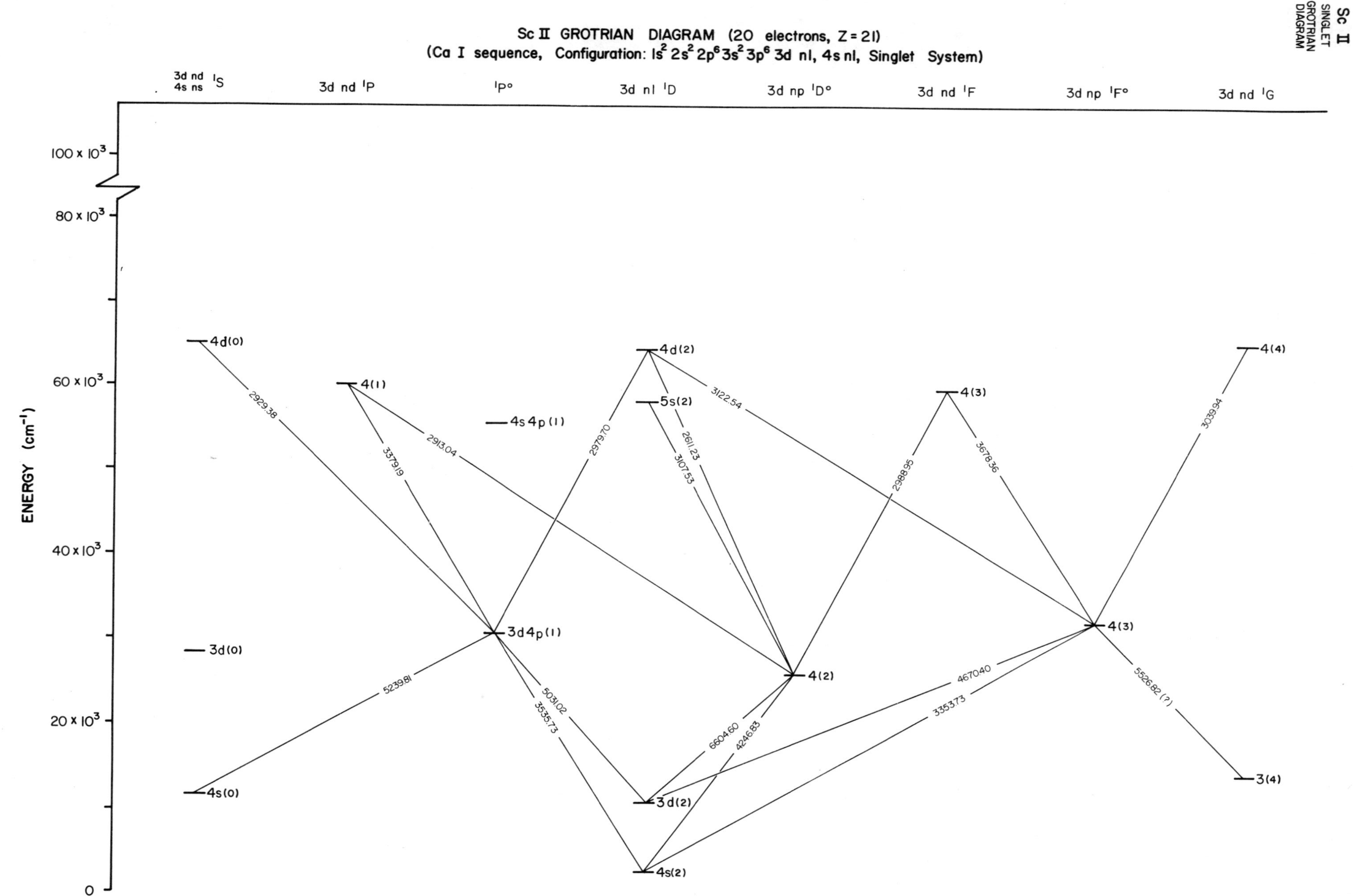
Sc II
SINGLET
GROTRIAN
DIAGRAM
Sc II GROTRIAN DIAGRAM (20 electrons, Z = 21)
(Ca I sequence, Configuration: $1s^2 2s^2 2p^6 3s^2 3p^6$ 3d nl, 4s nl, Singlet System)
3d nd
4s ns ^{1}S
3d nd ^{1}P
^{1}P°
3d nl ^{1}D
3d np ^{1}D°
3d nd ^{1}F
3d np ^{1}F°
3d nd ^{1}G
ENERGY (cm^{-1})
100 x 10^3
80 x 10^3
60 x 10^3
40 x 10^3
20 x 10^3
0
4d(0)
3d(0)
4s(0)
4(1)
4s 4p (1)
3d 4p (1)
4d(2)
5s(2)
3d(2)
4s(2)
4(2)
4(3)
4(3)
4(4)
3(4)
2929.38
3379.19
2913.04
5239.81
2979.70
5031.02
3535.73
2611.23
3107.53
3122.54
6604.60
4246.83
2988.95
3678.36
4670.40
3353.73
3039.94
5526.82 (?)

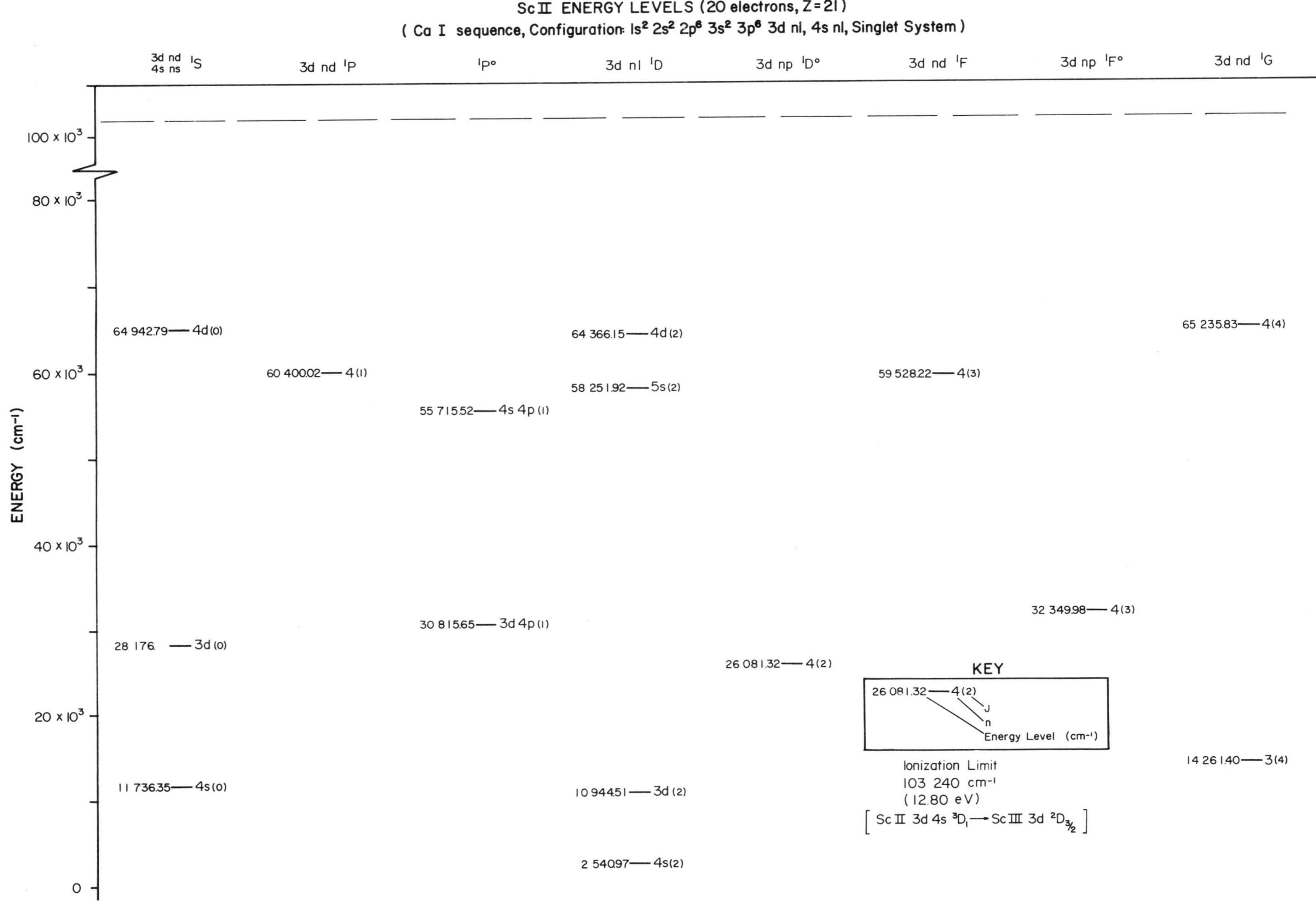
Sc II ENERGY LEVELS (20 electrons, Z=21)
(Ca I sequence, Configuration: 1s² 2s² 2p⁶ 3s² 3p⁶ 3d nl, 4s nl, Singlet System)
3d nd
4s ns ¹S
3d nd ¹P
¹P°
3d nl ¹D
3d np ¹D°
3d nd ¹F
3d np ¹F°
3d nd ¹G
100 x 10³
80 x 10³
60 x 10³
40 x 10³
20 x 10³
0
ENERGY (cm⁻¹)
64 942.79—4d(0)
28 176. —3d(0)
11 736.35—4s(0)
60 400.02—4(1)
55 715.52—4s 4p(1)
30 815.65—3d 4p(1)
64 366.15—4d(2)
58 251.92—5s(2)
10 944.51—3d(2)
2 540.97—4s(2)
26 081.32—4(2)
59 528.22—4(3)
32 349.98—4(3)
65 235.83—4(4)
14 261.40—3(4)
KEY
26 081.32—4(2)
J
n
Energy Level (cm⁻¹)
Ionization Limit
103 240 cm⁻¹
(12.80 eV)
[Sc II 3d 4s ³D₁ → Sc III 3d ²D3/2]
Sc II
SINGLET

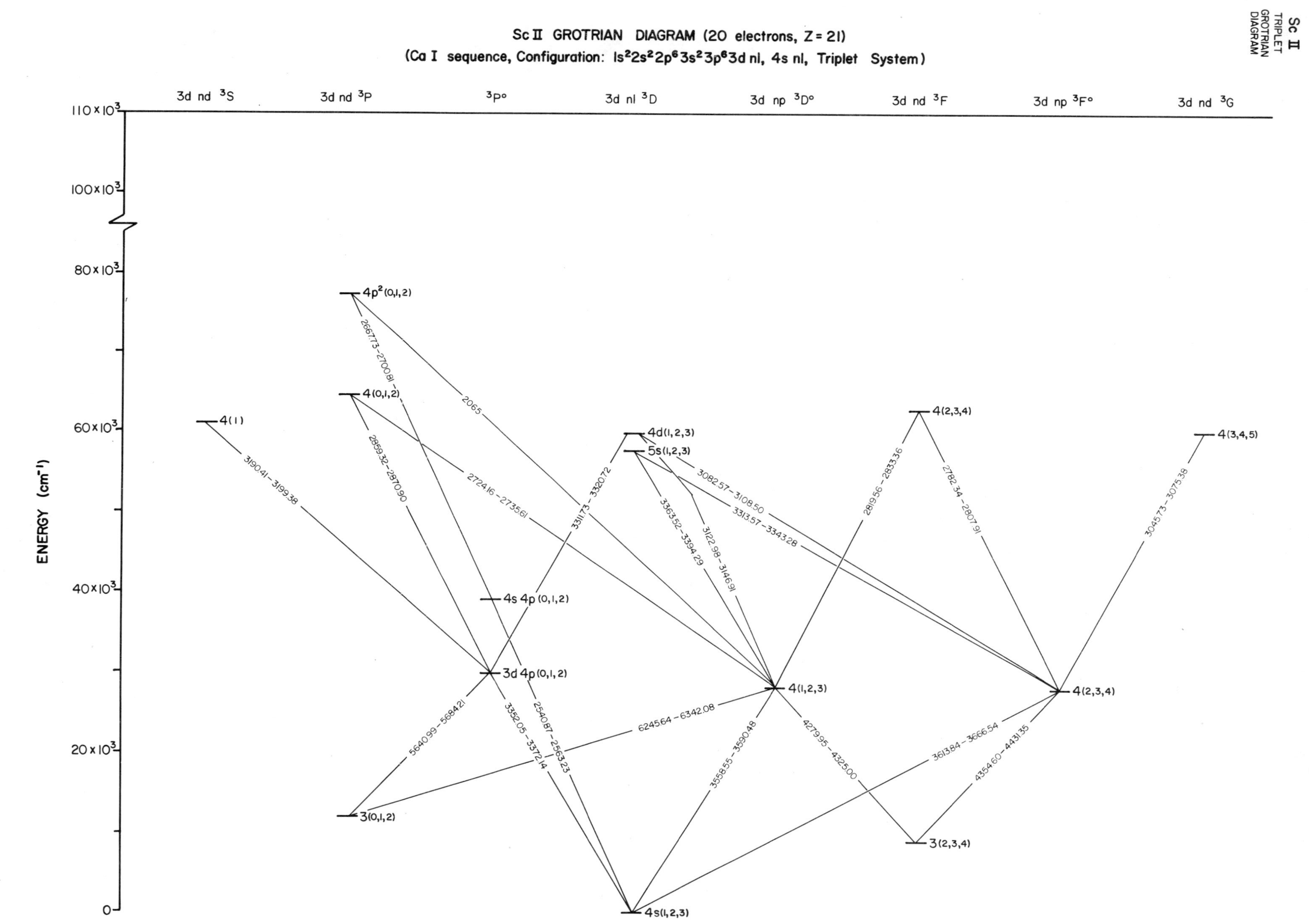
Sc II
TRIPLET
GROTRIAN
DIAGRAM
Sc II GROTRIAN DIAGRAM (20 electrons, Z = 21)
(Ca I sequence, Configuration: $1s^2 2s^2 2p^6 3s^2 3p^6 3d\,nl$, 4s nl, Triplet System)
3d nd ^{3}S
3d nd ^{3}P
^{3}P°
3d nl ^{3}D
3d np ^{3}D°
3d nd ^{3}F
3d np ^{3}F°
3d nd ^{3}G
ENERGY (cm^{-1})
110×10^3
100×10^3
80×10^3
60×10^3
40×10^3
20×10^3
0
4p^2(0,1,2)
2667.73–2700.81
2065
4(0,1,2)
2859.32–2870.90
2724.16–2735.61
4(1)
3190.41–3199.38
4d(1,2,3)
5s(1,2,3)
3311.73–3320.72
3363.52–3394.29
3122.98–3146.91
3082.57–3108.50
3313.57–3343.28
4(2,3,4)
2819.56–2833.36
2782.34–2807.91
4(3,4,5)
3045.73–3075.38
4s 4p (0,1,2)
3d 4p (0,1,2)
4(1,2,3)
4(2,3,4)
5640.99–5684.21
3352.05–3372.14
2540.87–2563.23
6245.64–6342.08
3558.55–3590.48
4279.95–4325.00
3613.84–3666.54
4354.60–4431.35
3(0,1,2)
3(2,3,4)
4s(1,2,3)

Sc II ENERGY LEVELS (20 electrons, Z=21)

(Ca I sequence, Configuration: $1s^2 2s^2 2p^6 3s^2 3p^6$ 3d nl, 4s nl, Triplet System)

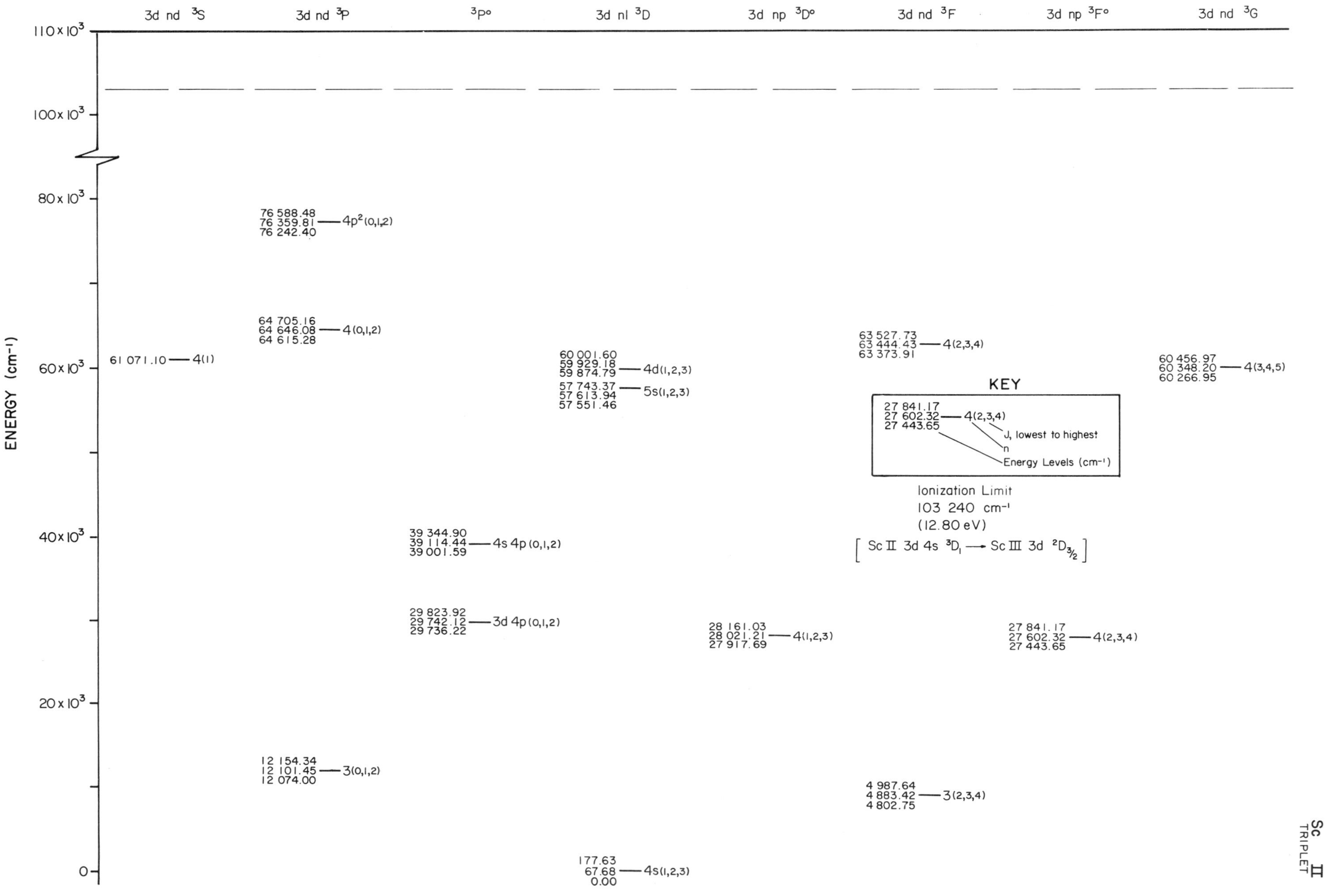

ENERGY (cm^{-1})

110 x 10^3
100 x 10^3
80 x 10^3
60 x 10^3
40 x 10^3
20 x 10^3
0

3d nd ^3S
61 071.10 — 4(1)

3d nd ^3P
76 588.48
76 359.81 — $4p^2$(0,1,2)
76 242.40
64 705.16
64 646.08 — 4(0,1,2)
64 615.28
12 154.34
12 101.45 — 3(0,1,2)
12 074.00

$^3P^o$
39 344.90
39 114.44 — 4s 4p (0,1,2)
39 001.59
29 823.92
29 742.12 — 3d 4p (0,1,2)
29 736.22

3d nl ^3D
60 001.60
59 929.18 — 4d(1,2,3)
59 874.79
57 743.37 — 5s(1,2,3)
57 613.94
57 551.46
177.63
67.68 — 4s(1,2,3)
0.00

3d np $^3D^o$
28 161.03
28 021.21 — 4(1,2,3)
27 917.69

3d nd ^3F
63 527.73
63 444.43 — 4(2,3,4)
63 373.91
4 987.64
4 883.42 — 3(2,3,4)
4 802.75

3d np $^3F^o$
27 841.17
27 602.32 — 4(2,3,4)
27 443.65

3d nd ^3G
60 456.97
60 348.20 — 4(3,4,5)
60 266.95

KEY
27 841.17
27 602.32 — 4(2,3,4)
27 443.65
J, lowest to highest
n
Energy Levels (cm^{-1})

Ionization Limit
103 240 cm^{-1}
(12.80 eV)
[Sc II 3d 4s $^3D_1$ → Sc III 3d $^2D_{3/2}$]

Sc II
SINGLET
TRIPLET
INTERCOMBINATION
GROTRIAN DIAGRAM

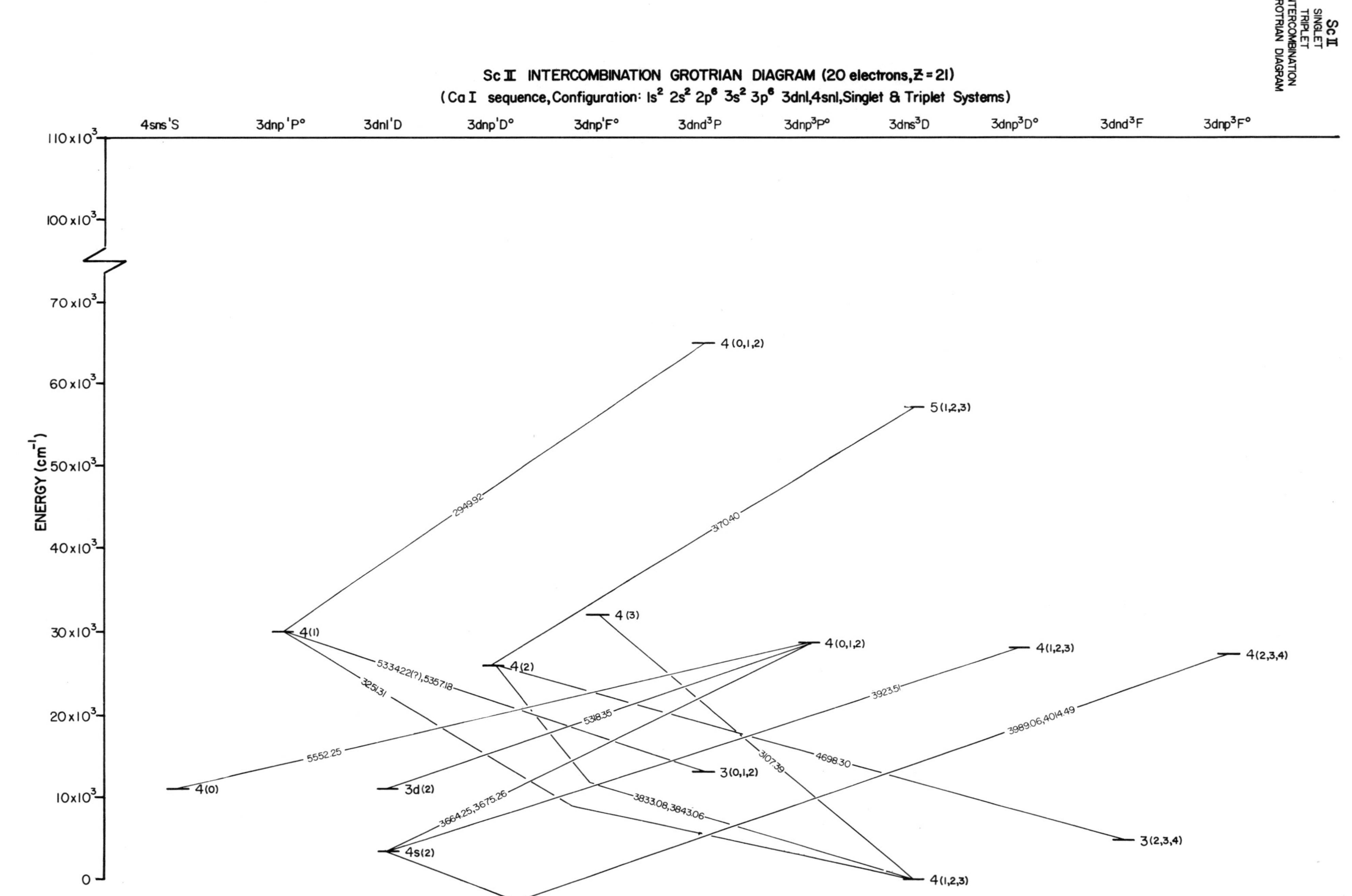

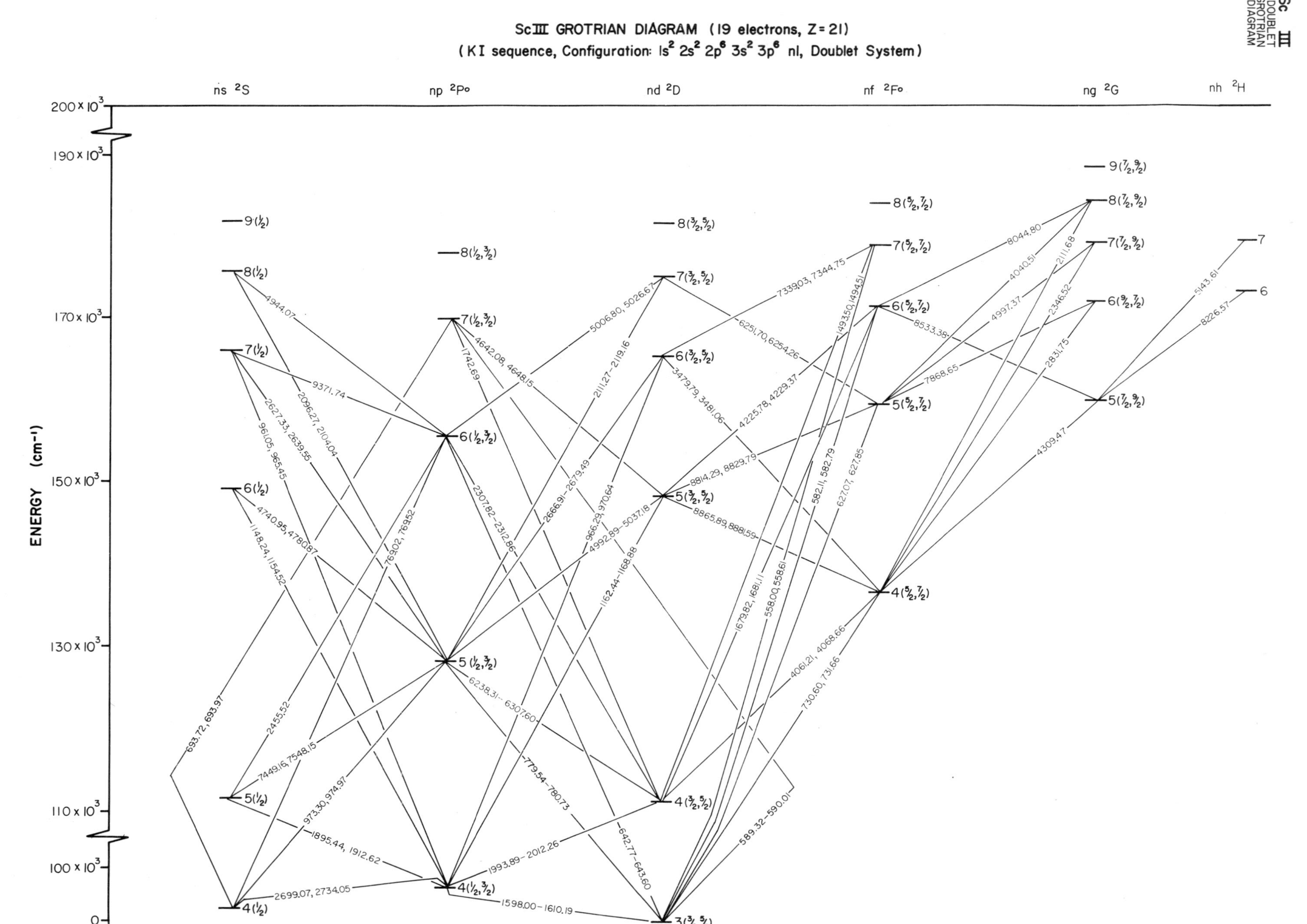
ScIII GROTRIAN DIAGRAM (19 electrons, Z = 21)
(KI sequence, Configuration: $1s^2\,2s^2\,2p^6\,3s^2\,3p^6$ nl, Doublet System)
ns ²S
np ²P°
nd ²D
nf ²F°
ng ²G
nh ²H
ENERGY (cm⁻¹)
200 x 10³
190 x 10³
170 x 10³
150 x 10³
130 x 10³
110 x 10³
100 x 10³
0
9(½)
8(½)
7(½)
6(½)
5(½)
4(½)
8(½,3/2)
7(½,3/2)
6(½,3/2)
5(½,3/2)
4(½,3/2)
8(3/2,5/2)
7(3/2,5/2)
6(3/2,5/2)
5(3/2,5/2)
4(3/2,5/2)
3(3/2,5/2)
8(5/2,7/2)
7(5/2,7/2)
6(5/2,7/2)
5(5/2,7/2)
4(5/2,7/2)
9(7/2,9/2)
8(7/2,9/2)
7(7/2,9/2)
6(9/2,7/2)
5(7/2,9/2)
7
6
693.72, 693.97
4944.07
9371.74
2627.33, 2639.55
961.05, 965.45
2096.27, 2104.04
4740.95, 4780.87
1148.24, 1154.52
2455.52
7449.16, 7548.15
973.30, 974.97
1895.44, 1912.62
2699.07, 2734.05
769.02, 769.52
1742.69
4642.08, 4648.15
2307.82–2312.86
6238.31–6307.60
779.54–780.73
1993.89–2012.26
1598.00–1610.19
5006.80, 5026.67
2111.27–2119.16
2666.91–2679.49
966.29, 970.64
4992.89–5037.18
1162.44–1168.88
642.77–643.60
3479.79, 3481.06
6251.70, 6254.26
7339.03, 7344.75
4225.78, 4229.37
8814.29, 8829.79
8865.89, 8881.59
1493.50, 1494.51
1679.82, 1681.11
558.00, 558.61
582.11, 582.79
627.07, 627.85
589.32–590.01
4061.21, 4068.66
730.60, 731.66
8044.80
4040.51
2111.68
4997.37
2346.52
8533.38
7868.65
2831.75
4309.47
5143.61
8226.57

Sc III ENERGY LEVELS (19 electrons, Z=21)

(KI sequence, Configuration: $1s^2 2s^2 2p^6 3s^2 3p^6 nl$, Doublet System)

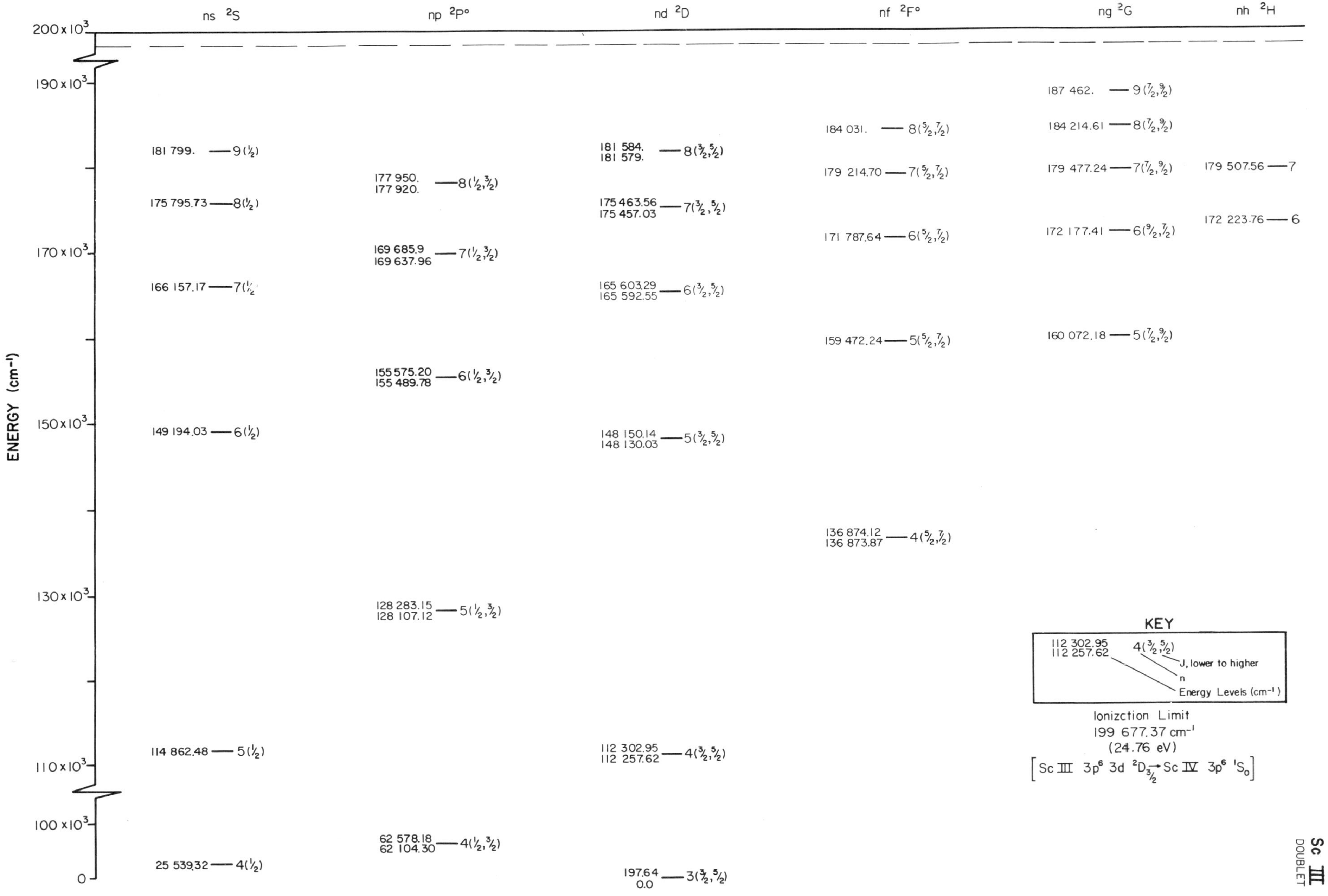

ENERGY (cm^{-1})

Axis: 0, 100×10^3, 110×10^3, 130×10^3, 150×10^3, 170×10^3, 190×10^3, 200×10^3

ns ^2S

- 181 799. — 9(½)
- 175 795.73 — 8(½)
- 166 157.17 — 7(½
- 149 194.03 — 6(½)
- 114 862.48 — 5(½)
- 25 539.32 — 4(½)

np $^2P^o$

- 177 950., 177 920. — 8(½,3/2)
- 169 685.9, 169 637.96 — 7(½,3/2)
- 155 575.20, 155 489.78 — 6(½,3/2)
- 128 283.15, 128 107.12 — 5(½,3/2)
- 62 578.18, 62 104.30 — 4(½,3/2)

nd ^2D

- 181 584., 181 579. — 8(3/2,5/2)
- 175 463.56, 175 457.03 — 7(3/2,5/2)
- 165 603.29, 165 592.55 — 6(3/2,5/2)
- 148 150.14, 148 130.03 — 5(3/2,5/2)
- 112 302.95, 112 257.62 — 4(3/2,5/2)
- 197.64, 0.0 — 3(3/2,5/2)

nf $^2F^o$

- 184 031. — 8(5/2,7/2)
- 179 214.70 — 7(5/2,7/2)
- 171 787.64 — 6(5/2,7/2)
- 159 472.24 — 5(5/2,7/2)
- 136 874.12, 136 873.87 — 4(5/2,7/2)

ng ^2G

- 187 462. — 9(7/2,9/2)
- 184 214.61 — 8(7/2,9/2)
- 179 477.24 — 7(7/2,9/2)
- 172 177.41 — 6(9/2,7/2)
- 160 072.18 — 5(7/2,9/2)

nh ^2H

- 179 507.56 — 7
- 172 223.76 — 6

KEY

112 302.95, 112 257.62 — 4(3/2,5/2): Energy Levels (cm^{-1}); n; J, lower to higher

Ionizction Limit
199 677.37 cm^{-1}
(24.76 eV)
[Sc III $3p^6 3d\ ^2D_{3/2}$ → Sc IV $3p^6\ ^1S_0$]

Sc IV GROTRIAN DIAGRAM (18 electrons, Z=21)
(Ar I sequence, Configuration: $1s^2 2s^2 2p^6 3s^2 3p^5 nl$, Singlet System)

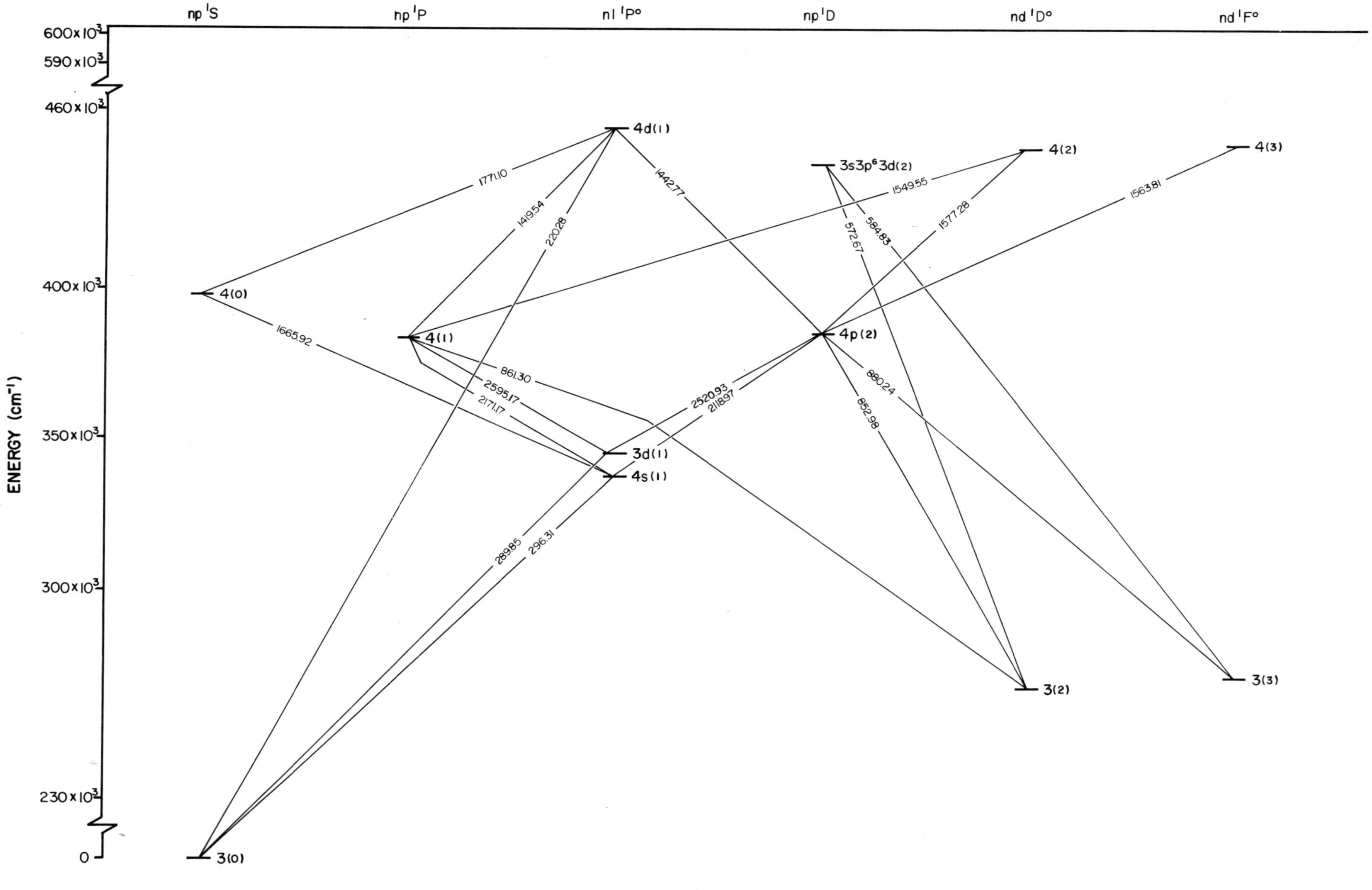

np ^{1}S | np ^{1}P | nl ^{1}P° | np ^{1}D | nd ^{1}D° | nd ^{1}F°

ENERGY (cm^{-1})

600x10^3
590x10^3
460x10^3
400x10^3
350x10^3
300x10^3
230x10^3
0

4d(1)
3s3p^{6}3d(2)
4(2)
4(3)
4(0)
4(1)
4p(2)
3d(1)
4s(1)
3(2)
3(3)
3(0)

1771.10
1419.54
220.28
1442.77
1549.55
572.67
584.83
1577.28
1563.81
1665.92
861.30
2595.17
2171.17
2520.93
2118.97
880.24
852.98
289.85
296.31

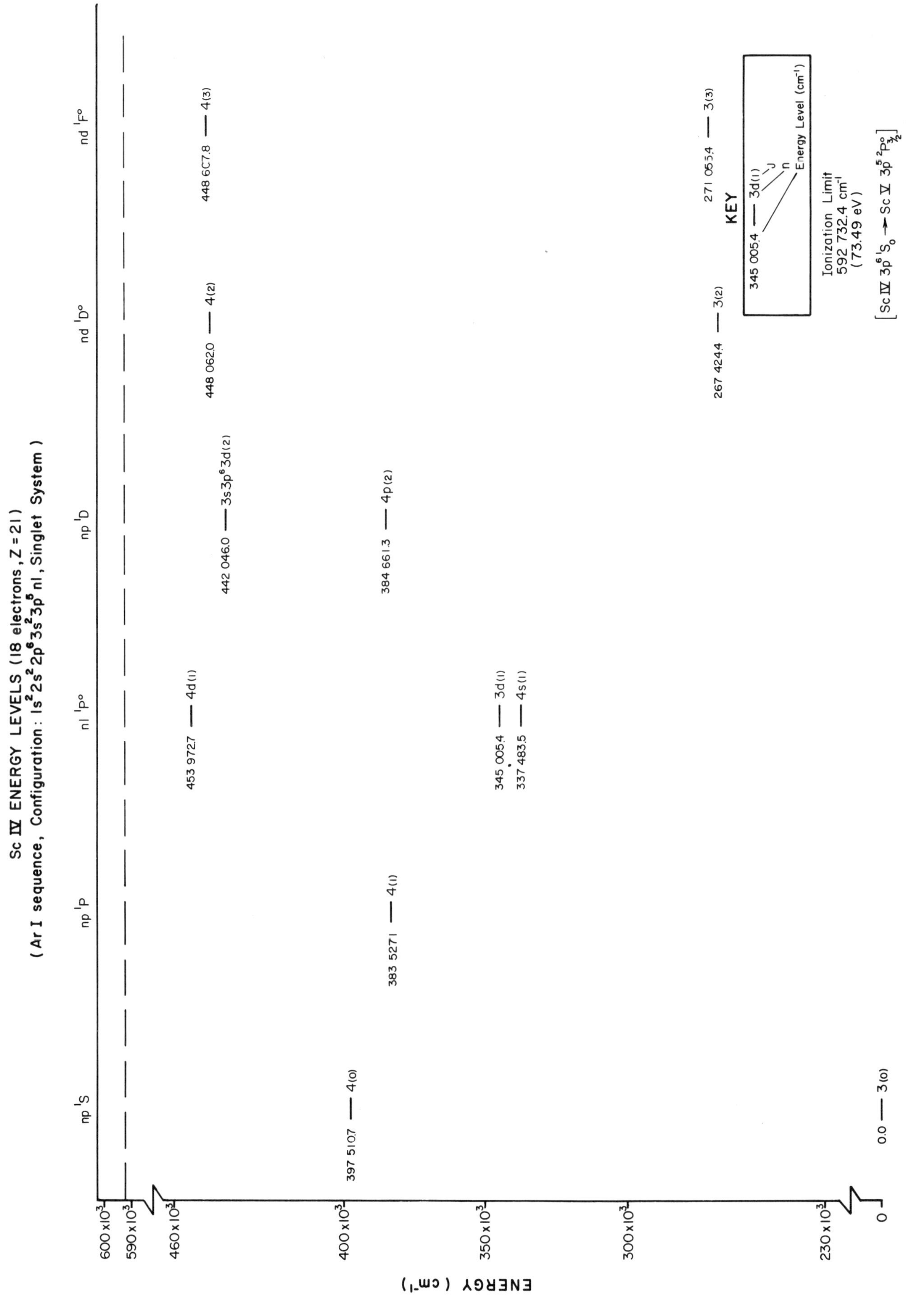

Sc IV ENERGY LEVELS (18 electrons, Z = 21)

(Ar I sequence, Configuration: $1s^2 2s^2 2p^6 3s^2 3p^5 nl$, Singlet System)

ENERGY (cm^{-1})

600 x 10^3
590 x 10^3
460 x 10^3
400 x 10^3
350 x 10^3
300 x 10^3
230 x 10^3
0

np ^1S
397 510.7 — 4(0)
0.0 — 3(0)

np ^1P
383 527.1 — 4(1)

nl $^1P^o$
453 972.7 — 4d(1)
345 005.4 — 3d(1)
337 483.5 — 4s(1)

np ^1D
442 046.0 — $3s3p^63d$(2)
384 661.3 — 4p(2)

nd $^1D^o$
448 062.0 — 4(2)
267 424.4 — 3(2)

nd $^1F^o$
448 6C7.8 — 4(3)
271 055.4 — 3(3)

KEY

345 005.4 — 3d(1)
J
n
Energy Level (cm^{-1})

Ionization Limit
592 732.4 cm^{-1}
(73.49 eV)

[Sc IV $3p^6\ ^1S_0$ → Sc V $3p^5\ ^2P^o_{3/2}$]

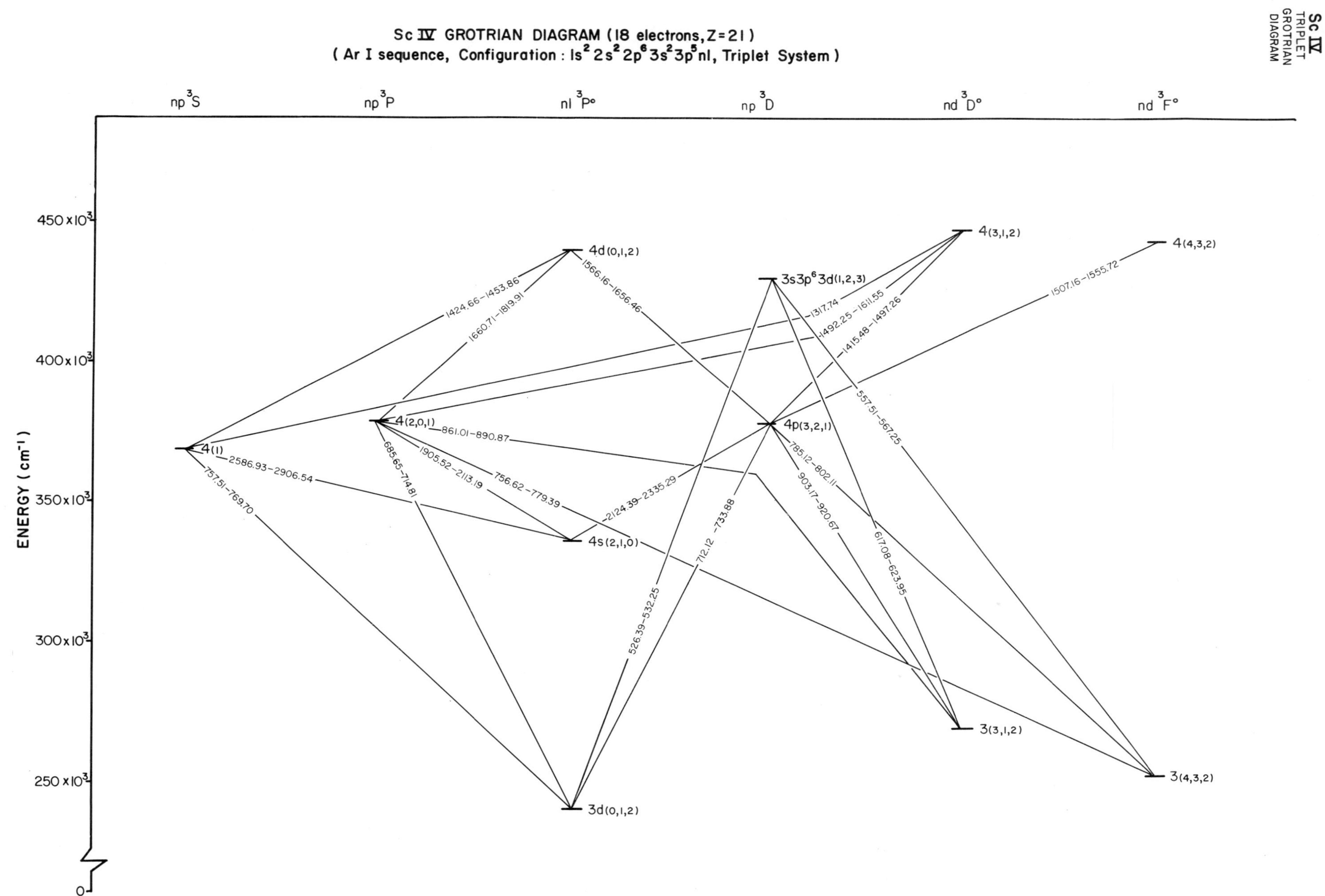
Sc IV GROTRIAN DIAGRAM (18 electrons, Z=21)
(Ar I sequence, Configuration : 1s2 2s2 2p6 3s2 3p5 nl, Triplet System)
np 3S
np 3P
nl 3P°
np 3D
nd 3D°
nd 3F°
ENERGY (cm-1)
450x103
400x103
350x103
300x103
250x103
0
4(1)
4(2,0,1)
4d(0,1,2)
4s(2,1,0)
3d(0,1,2)
3s3p6 3d(1,2,3)
4p(3,2,1)
4(3,1,2)
3(3,1,2)
4(4,3,2)
3(4,3,2)
1424.66-1453.86
1660.71-1819.91
1566.16-1656.46
2586.93-2906.54
757.51-769.70
861.01-890.87
685.65-714.81
1905.52-2113.19
756.62-779.39
2124.39-2335.29
1317.74
1492.25-1611.55
1415.48-1497.26
1507.16-1555.72
557.51-567.25
785.12-802.11
903.17-920.67
617.08-623.95
712.12 -733.88
526.39-532.25

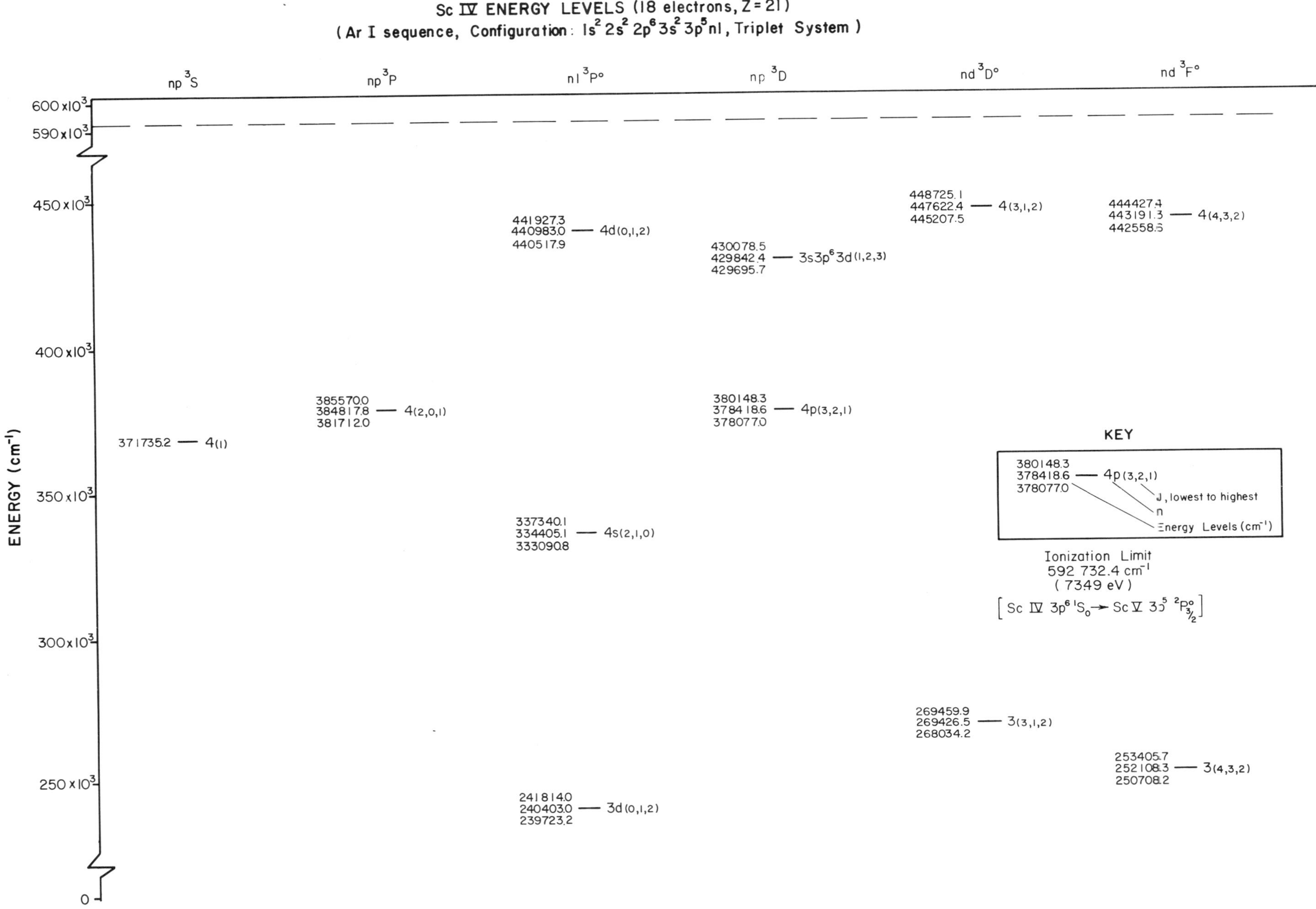
Sc IV ENERGY LEVELS (18 electrons, Z = 21)
(Ar I sequence, Configuration: $1s^2 2s^2 2p^6 3s^2 3p^5 nl$, Triplet System)
$np\ ^3S$
$np\ ^3P$
$nl\ ^3P^o$
$np\ ^3D$
$nd\ ^3D^o$
$nd\ ^3F^o$
ENERGY (cm^{-1})
600×10^3
590×10^3
450×10^3
400×10^3
350×10^3
300×10^3
250×10^3
0
371735.2 — 4(1)
385570.0
384817.8 — 4(2,0,1)
381712.0
441927.3
440983.0 — 4d(0,1,2)
440517.9
337340.1
334405.1 — 4s(2,1,0)
333090.8
241814.0
240403.0 — 3d(0,1,2)
239723.2
430078.5
429842.4 — $3s3p^63d$(1,2,3)
429695.7
380148.3
378418.6 — 4p(3,2,1)
378077.0
448725.1
447622.4 — 4(3,1,2)
445207.5
269459.9
269426.5 — 3(3,1,2)
268034.2
444427.4
443191.3 — 4(4,3,2)
442558.6
253405.7
252108.3 — 3(4,3,2)
250708.2
KEY
380148.3
378418.6 — 4p(3,2,1)
378077.0
J, lowest to highest
n
Energy Levels (cm^{-1})
Ionization Limit
592 732.4 cm^{-1}
(73.49 eV)
[Sc IV $3p^6\ ^1S_0$ → Sc V $3p^5\ ^2P^o_{3/2}$]

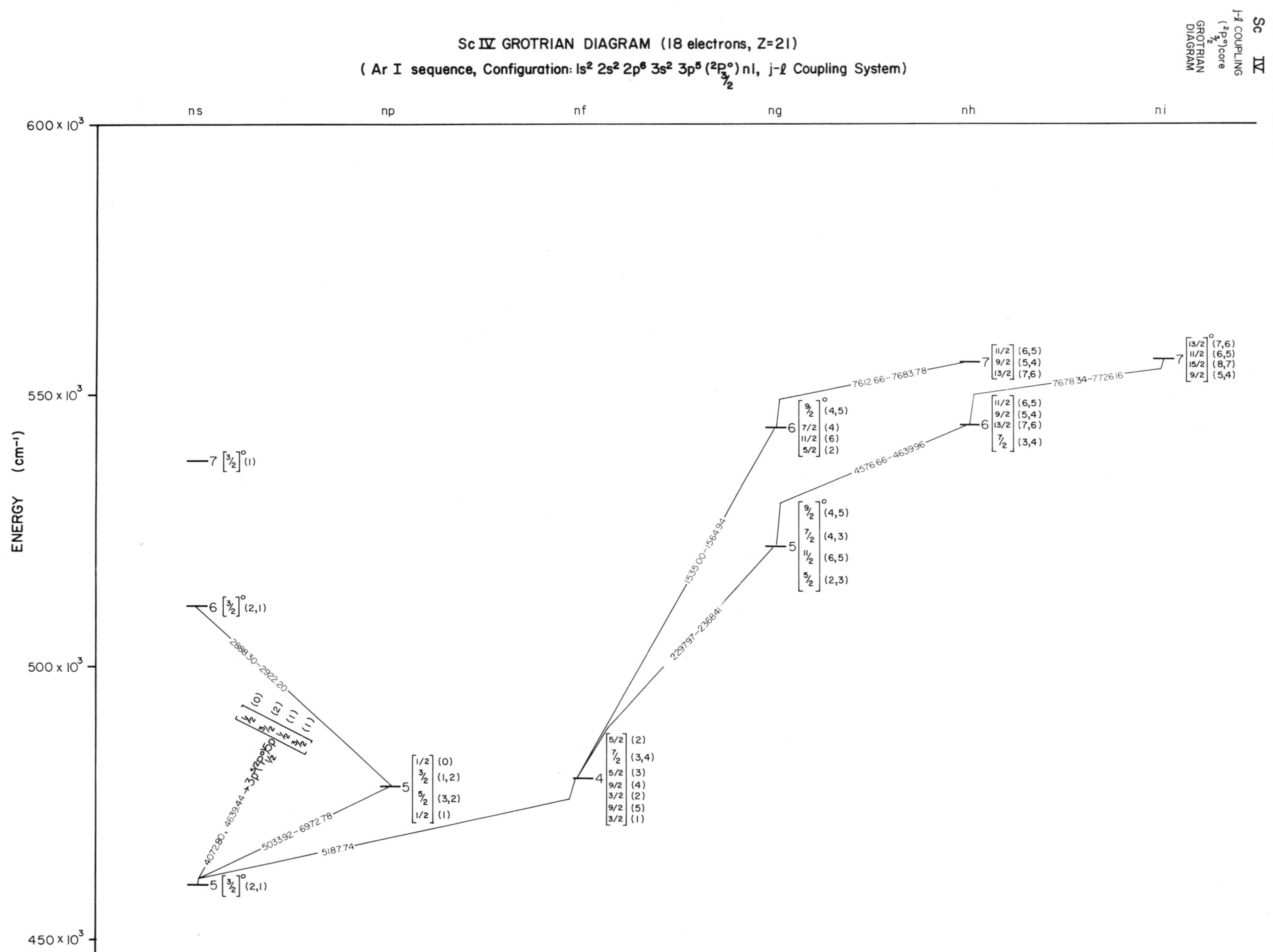

Sc IV
j-ℓ COUPLING
($^2P^o_{3/2}$)core
GROTRIAN
DIAGRAM
Sc IV GROTRIAN DIAGRAM (18 electrons, Z=21)
(Ar I sequence, Configuration: $1s^2 2s^2 2p^6 3s^2 3p^5 (^2P^o_{3/2})$ nl, j-ℓ Coupling System)
ns
np
nf
ng
nh
ni
600 x 10^3
550 x 10^3
500 x 10^3
450 x 10^3
ENERGY (cm⁻¹)
7 $[3/2]^o$ (1)
6 $[3/2]^o$ (2,1)
5 $[3/2]^o$ (2,1)
2888.30–2922.20
4072.80, 4639.44 → $3p^5(^2P^o_{1/2})$5p
1/2 (0)
3/2 (2)
1/2 (1)
3/2 (1)
5033.92–6972.78
5187.74
5 1/2 (0)
3/2 (1,2)
5/2 (3,2)
1/2 (1)
4 5/2 (2)
7/2 (3,4)
5/2 (3)
9/2 (4)
3/2 (2)
9/2 (5)
3/2 (1)
1535.00–1564.94
2297.97–2368.41
6 $[9/2]^o$ (4,5)
7/2 (4)
11/2 (6)
5/2 (2)
5 $[9/2]^o$ (4,5)
7/2 (4,3)
11/2 (6,5)
5/2 (2,3)
7612.66–7683.78
4576.66–4639.96
7 11/2 (6,5)
9/2 (5,4)
13/2 (7,6)
6 11/2 (6,5)
9/2 (5,4)
13/2 (7,6)
7/2 (3,4)
7678.34–7726.16
7 $[13/2]^o$ (7,6)
11/2 (6,5)
15/2 (8,7)
9/2 (5,4)

Sc IV ENERGY LEVELS (18 electrons, Z = 21)

(Ar I sequence, Configuration: $1s^2\ 2s^2\ 2p^6\ 3s^2\ 3p^5\ (^2P^o_{3/2})\ nl$, j-ℓ Coupling System)

ENERGY (cm⁻¹): 600 x 10³, 550 x 10³, 500 x 10³, 450 x 10³

ns

537 845 — 7 [3/2]° (1)

511 630.3, 511 228.2 — 6 [3/2]° (2,1)

460 426.9, 459 496.9 — 5 [3/2]° (2,1)

np

480 286.6, 478 213.2, 477 747.6, 477 018.0, 476 818.5, 474 764.4 — 5 [1/2] (0), [3/2] (1,2), [5/2] (3,2), [1/2] (1)

nf

480 097.9, 479 993.2, 479 908.9, 479 452.9, 478 768.4, 478 767.7, 478 609.5, 478 495.9 — 4 [5/2] (2), [7/2] (3,4), [5/2] (3), [9/2] (4), [3/2] (2), [9/2] (5), [3/2] (1)

ng

543 893.3, 543 892.1, 543 824.9, 543 708.3, 543 642.4 — 6 [9/2]° (4,5), [7/2] (4), [11/2] (6), [5/2] (2)

522 425.7, 522 424.8, 522 307.3, 522 304.5, 522 110.7, 522 110.2, 522 004.8, 521 999.2 — 5 [9/2]° (4,5), [7/2] (4,3), [11/2] (6,5), [5/2] (2,3)

nh

556 903.5, 556 886.9, 556 840.9 — 7 [11/2] (6,5), [9/2] (5,4), [13/2] (7,6)

543 971.2, 543 942.1, 543 869.8, 543 843.6, 543 843.1 — 6 [11/2] (6,5), [9/2] (5,4), [13/2] (7,6), [7/2] (3,4)

ni

556 910.7, 556 900.8, 556 871.3, 556 863.5 — 7 [13/2]° (7,6), [11/2] (6,5), [15/2] (8,7), [9/2] (5,4)

KEY

543 893.3, 543 892.1, 543 824.9, 543 708.3, 543 642.4 — 6 [9/2]° (4,5), [7/2] (4), [11/2] (6), [5/2] (2)

° — Parity
( ) — Final J (lower to higher)
[ ] — j-ℓ coupling
6 — n
Numbers — Energy Levels (cm⁻¹)

Ionization Limit
592 732.4 cm⁻¹
(73.49 eV)
[Sc IV $3p^6\ ^1S_0$ → Sc V $3p^5\ ^3P^o_{3/2}$]

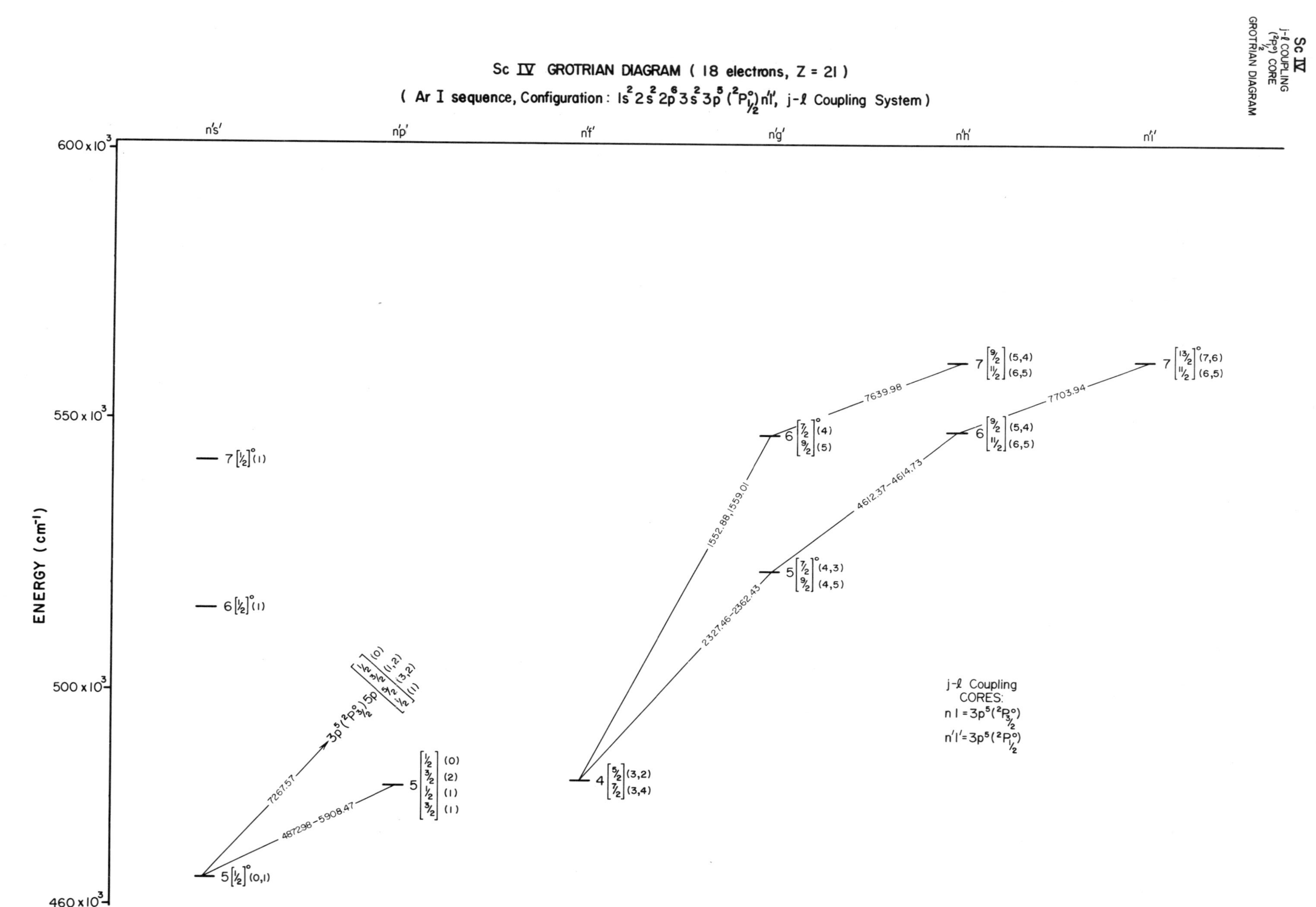
Sc IV GROTRIAN DIAGRAM (18 electrons, Z = 21)
(Ar I sequence, Configuration : 1s² 2s² 2p⁶ 3s² 3p⁵ (²P°1/2) n′l′, j-ℓ Coupling System)
Sc IV
j-ℓ COUPLING
(²P°1/2) CORE
GROTRIAN DIAGRAM
n′s′
n′p′
n′f′
n′g′
n′h′
n′i′
600 x 10³
550 x 10³
500 x 10³
460 x 10³
ENERGY (cm⁻¹)
7 [1/2]° (1)
6 [1/2]° (1)
5 [1/2]° (0,1)
7267.57
487298-5908.47
3p⁵(²P°3/2)5p [1/2] (0) [3/2] (1,2) [5/2] (3,2) [1/2] (1)
5 [1/2] (0) [3/2] (2) [1/2] (1) [3/2] (1)
4 [5/2] (3,2) [7/2] (3,4)
1552.88, 1559.01
2327.46-2362.43
6 [7/2]° (4) [9/2] (5)
5 [7/2]° (4,3) [9/2] (4,5)
7639.98
4612.37-4614.73
7 [9/2] (5,4) [11/2] (6,5)
6 [9/2] (5,4) [11/2] (6,5)
7703.94
7 [13/2]° (7,6) [11/2] (6,5)
j-ℓ Coupling
CORES:
n l = 3p⁵(²P°3/2)
n′l′ = 3p⁵(²P°1/2)

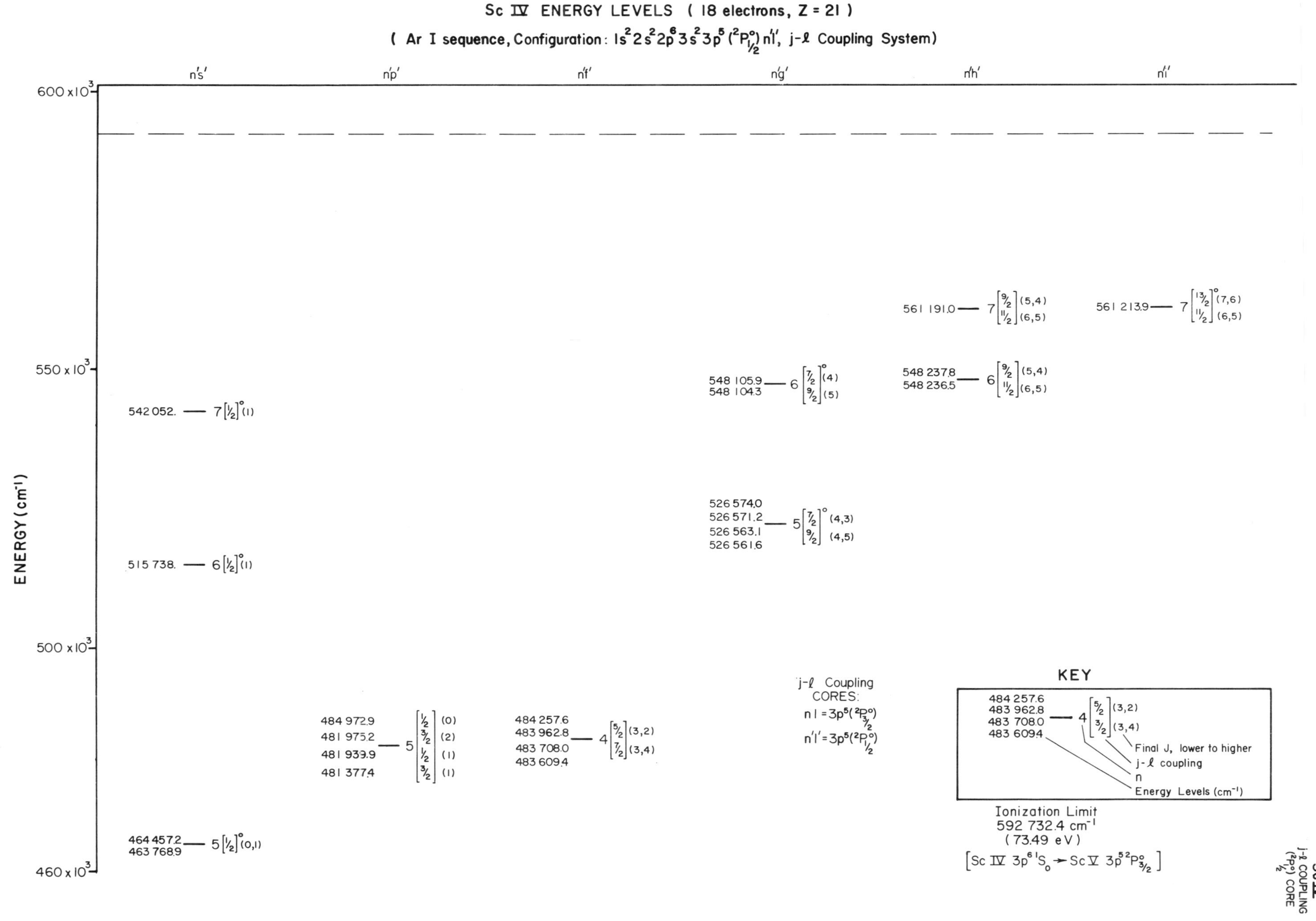

Sc IV ENERGY LEVELS (18 electrons, Z = 21)
(Ar I sequence, Configuration: 1s²2s²2p⁶3s²3p⁵(²P°1/2)n'l', j-l Coupling System)
n's'
n'p'
n'f'
n'g'
n'h'
n'i'
600 x 10³
550 x 10³
500 x 10³
460 x 10³
ENERGY (cm⁻¹)
561 191.0 — 7 [9/2 (5,4) 11/2 (6,5)]
561 213.9 — 7 [13/2 (7,6) 11/2 (6,5)]°
548 105.9
548 104.3 — 6 [7/2 (4) 9/2 (5)]°
548 237.8
548 236.5 — 6 [9/2 (5,4) 11/2 (6,5)]
542 052. — 7[1/2]° (1)
526 574.0
526 571.2
526 563.1
526 561.6 — 5 [7/2 (4,3) 9/2 (4,5)]°
515 738. — 6[1/2]° (1)
j-l Coupling
CORES:
n l = 3p⁵(²P°3/2)
n'l' = 3p⁵(²P°1/2)
KEY
484 257.6
483 962.8
483 708.0
483 609.4 — 4 [5/2 (3,2) 3/2 (3,4)]
Final J, lower to higher
j-l coupling
n
Energy Levels (cm⁻¹)
484 972.9
481 975.2
481 939.9
481 377.4 — 5 [1/2 (0) 3/2 (2) 1/2 (1) 3/2 (1)]
484 257.6
483 962.8
483 708.0
483 609.4 — 4 [5/2 (3,2) 7/2 (3,4)]
Ionization Limit
592 732.4 cm⁻¹
(73.49 eV)
[Sc IV 3p⁶ ¹S₀ → Sc V 3p⁵ ²P°3/2]
464 457.2
463 768.9 — 5[1/2]° (0,1)
Sc IV
j-l COUPLING
(²P°1/2) CORE

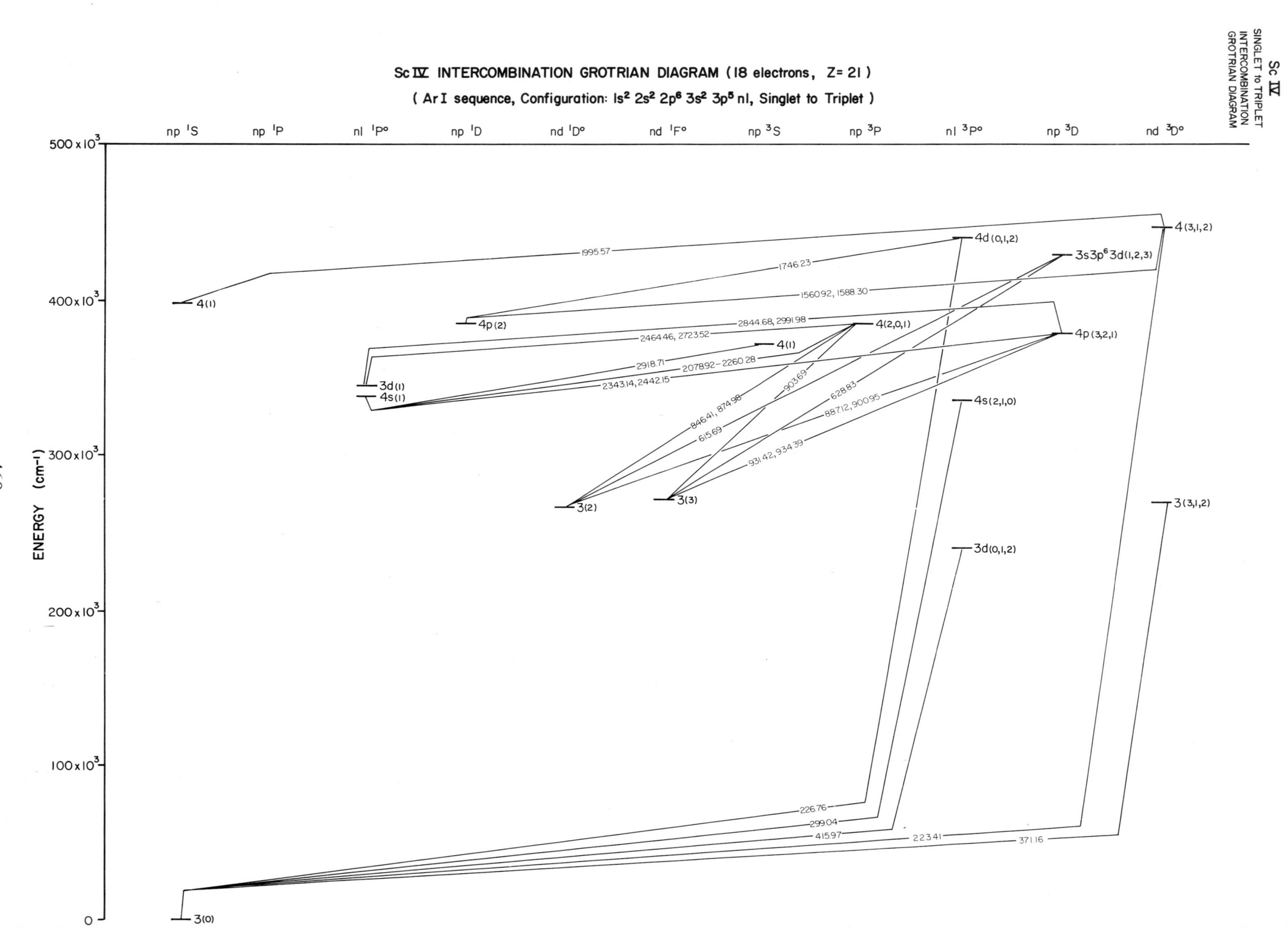
Sc IV
SINGLET to TRIPLET
INTERCOMBINATION
GROTRIAN DIAGRAM
Sc IV INTERCOMBINATION GROTRIAN DIAGRAM (18 electrons, Z= 21)
(Ar I sequence, Configuration: $1s^2\ 2s^2\ 2p^6\ 3s^2\ 3p^5$ nl, Singlet to Triplet)
np ¹S
np ¹P
nl ¹P°
np ¹D
nd ¹D°
nd ¹F°
np ³S
np ³P
nl ³P°
np ³D
nd ³D°
500 x 10³
400 x 10³
300 x 10³
200 x 10³
100 x 10³
0
ENERGY (cm⁻¹)
4(3,1,2)
4d(0,1,2)
3s3p⁶3d(1,2,3)
4(1)
4p(2)
4(2,0,1)
4p(3,2,1)
4(1)
3d(1)
4s(1)
4s(2,1,0)
3(2)
3(3)
3(3,1,2)
3d(0,1,2)
3(0)
1995.57
1746.23
1560.92, 1588.30
2844.68, 2991.98
2464.46, 2723.52
2918.71
2078.92–2260.28
2343.14, 2442.15
903.69
628.83
846.41, 874.98
615.69
887.12, 900.95
931.42, 934.39
226.76
299.04
415.97
223.41
371.16

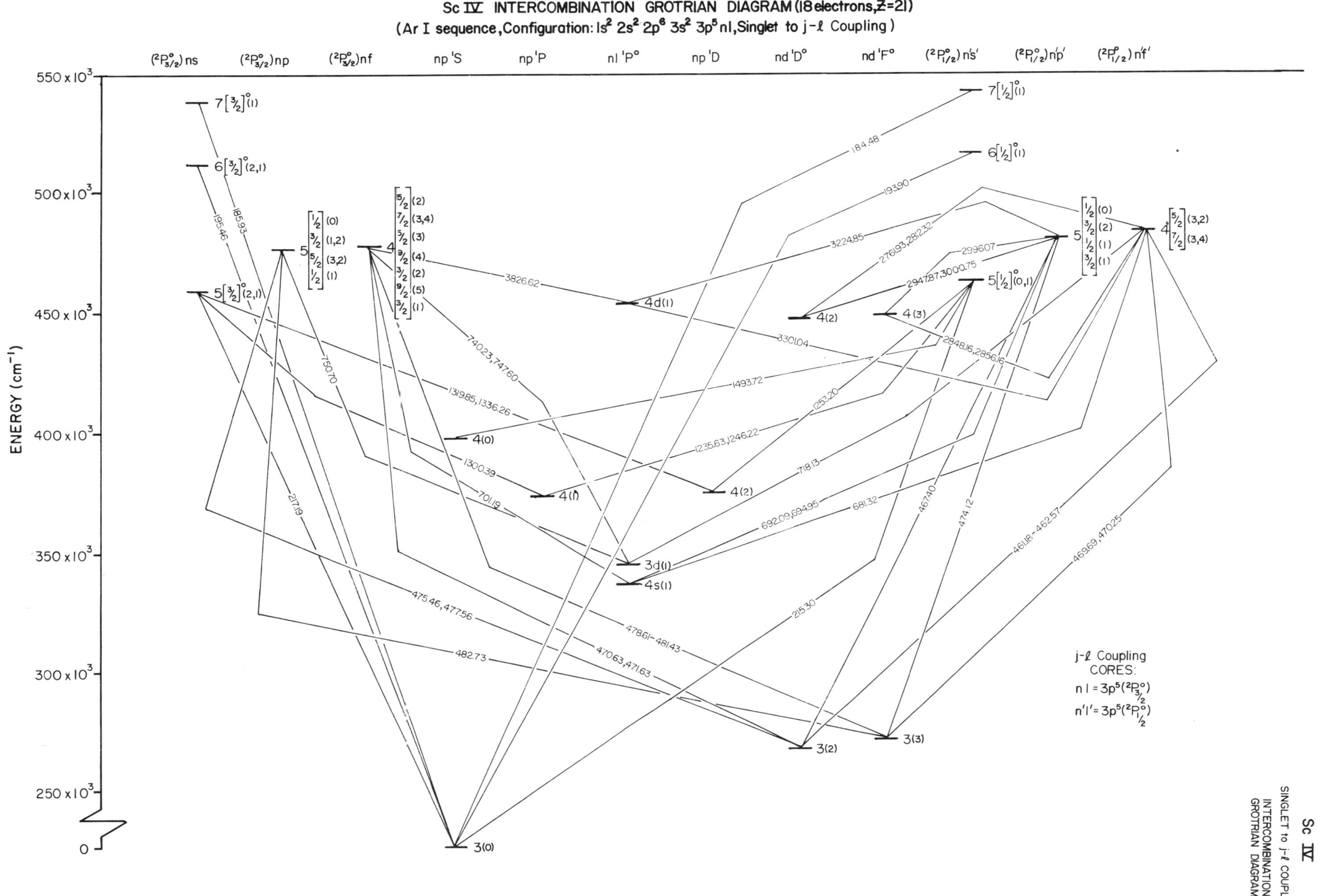
Sc IV INTERCOMBINATION GROTRIAN DIAGRAM (18 electrons, Z=21)
(Ar I sequence, Configuration: 1s² 2s² 2p⁶ 3s² 3p⁵ nl, Singlet to j-ℓ Coupling)
(²P°3/2)ns
(²P°3/2)np
(²P°3/2)nf
np ¹S
np ¹P
nl ¹P°
np ¹D
nd ¹D°
nd ¹F°
(²P°1/2)n's'
(²P°1/2)n'p'
(²P°1/2)n'f'
ENERGY (cm⁻¹)
550 x 10³
500 x 10³
450 x 10³
400 x 10³
350 x 10³
300 x 10³
250 x 10³
0
j-ℓ Coupling
CORES:
nl = 3p⁵(²P°3/2)
n'l' = 3p⁵(²P°1/2)
Sc IV
SINGLET to j-ℓ COUPLING
INTERCOMBINATION
GROTRIAN DIAGRAM

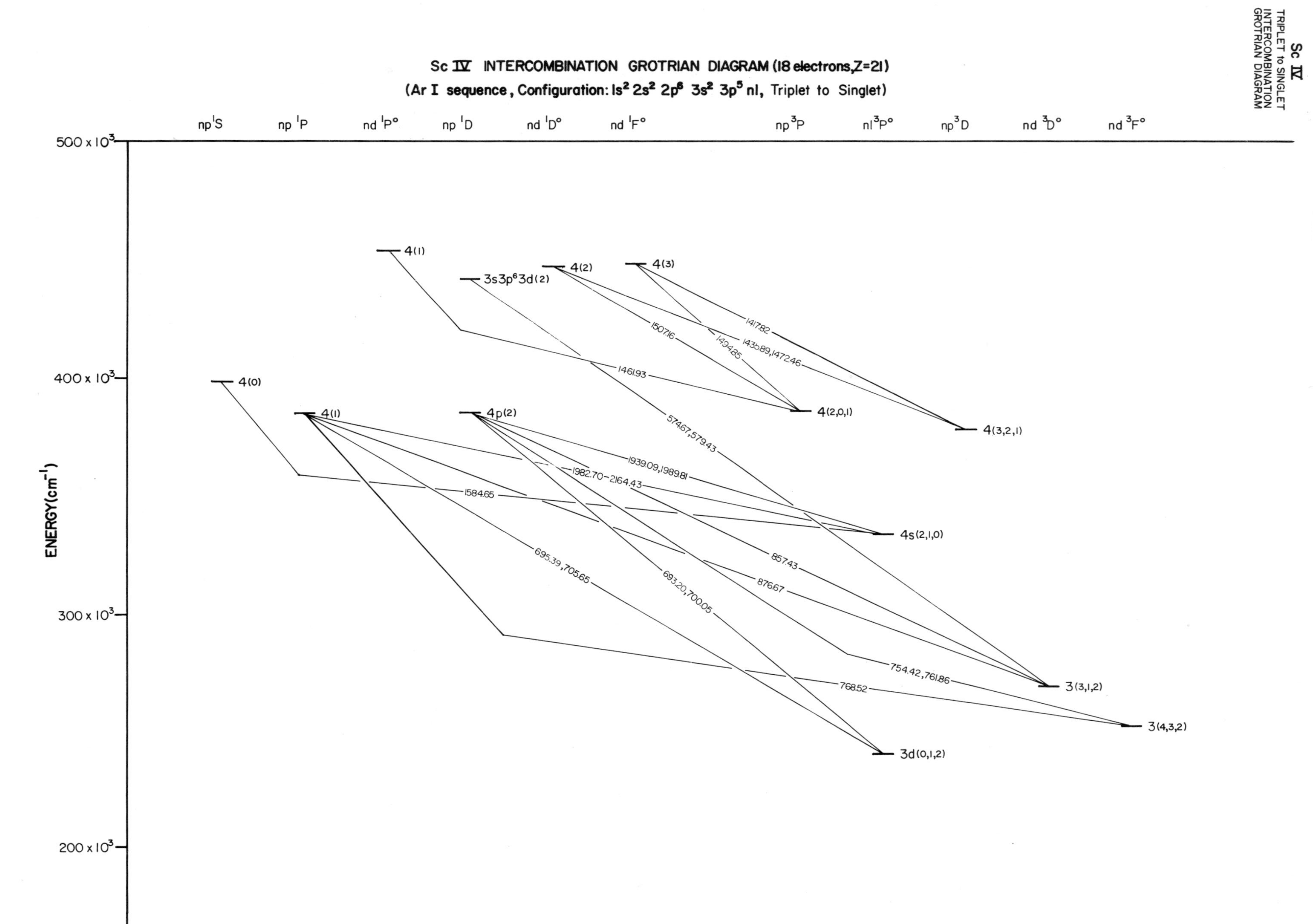
Sc IV INTERCOMBINATION GROTRIAN DIAGRAM (18 electrons, Z=21)
(Ar I sequence, Configuration: $1s^2\ 2s^2\ 2p^6\ 3s^2\ 3p^5$ nl, Triplet to Singlet)
np ^{1}S
np ^{1}P
nd ^{1}P°
np ^{1}D
nd ^{1}D°
nd ^{1}F°
np ^{3}P
nl ^{3}P°
np ^{3}D
nd ^{3}D°
nd ^{3}F°
4(1)
3s3p^{6}3d(2)
4(2)
4(3)
1417.82
1435.89,1472.46
1434.85
1507.16
1461.93
4(0)
4(1)
4p(2)
4(2,0,1)
4(3,2,1)
574.67,579.43
1939.09,1989.81
1982.70–2164.43
1584.65
4s(2,1,0)
857.43
876.67
695.39,705.65
693.20,700.05
754.42,761.86
768.52
3(3,1,2)
3(4,3,2)
3d(0,1,2)
ENERGY(cm^{-1})
500 x 10^3
400 x 10^3
300 x 10^3
200 x 10^3

Sc IV INTERCOMBINATION GROTRIAN DIAGRAM (18 electrons, Z = 21)
(Ar I sequence, Configuration: $1s^2 2s^2 2p^6 3s^2 3p^5 nl$, Triplet to $(^2P^o_{3/2})$ j-ℓ Coupling)

Sc IV
TRIPLET to $(^2P^o_{3/2})$
j-ℓ COUPLING
GROTRIAN DIAGRAM

np³S np³P nl³P° np³D nd³D° nd³F° ns np nf

ENERGY (cm⁻¹)

500 x 10³
450 x 10³
400 x 10³
350 x 10³
300 x 10³
250 x 10³

6[3/2]° (2,1)
5 [1/2] (0) [3/2] (1,2) [5/2] (3,2) [1/2] (1)
4 [5/2] (2) [7/2] (3,4) [5/2] (3) [9/2] (4) [3/2] (2) [9/2] (5) [3/2] (1)
5[3/2]° (2,1)
4d (0,1,2)
4 (3,1,2)
4 (4,3,2)
4 (2,0,1)
4 (3,2,1)
4 (1)
4s (2,1,0)
3 (3,1,2)
3 (4,3,2)
3d (0,1,2)

716.88
750.70, 751.03
2755.07, 2959.34
3028.90, 3390.25
2918.02 – 3067.48
2632.33 – 2713.61
2873.90 – 3078.36
2670.53 – 2934.41
686.38
1139.45
1219.40 – 1260.25
689.06 – 712.45
1270.41 – 1322.59
417.20 – 422.51
471.79 – 478.30
436.14 – 444.27
416.86 – 429.28
474.25 – 487.09
442.26 – 447.19

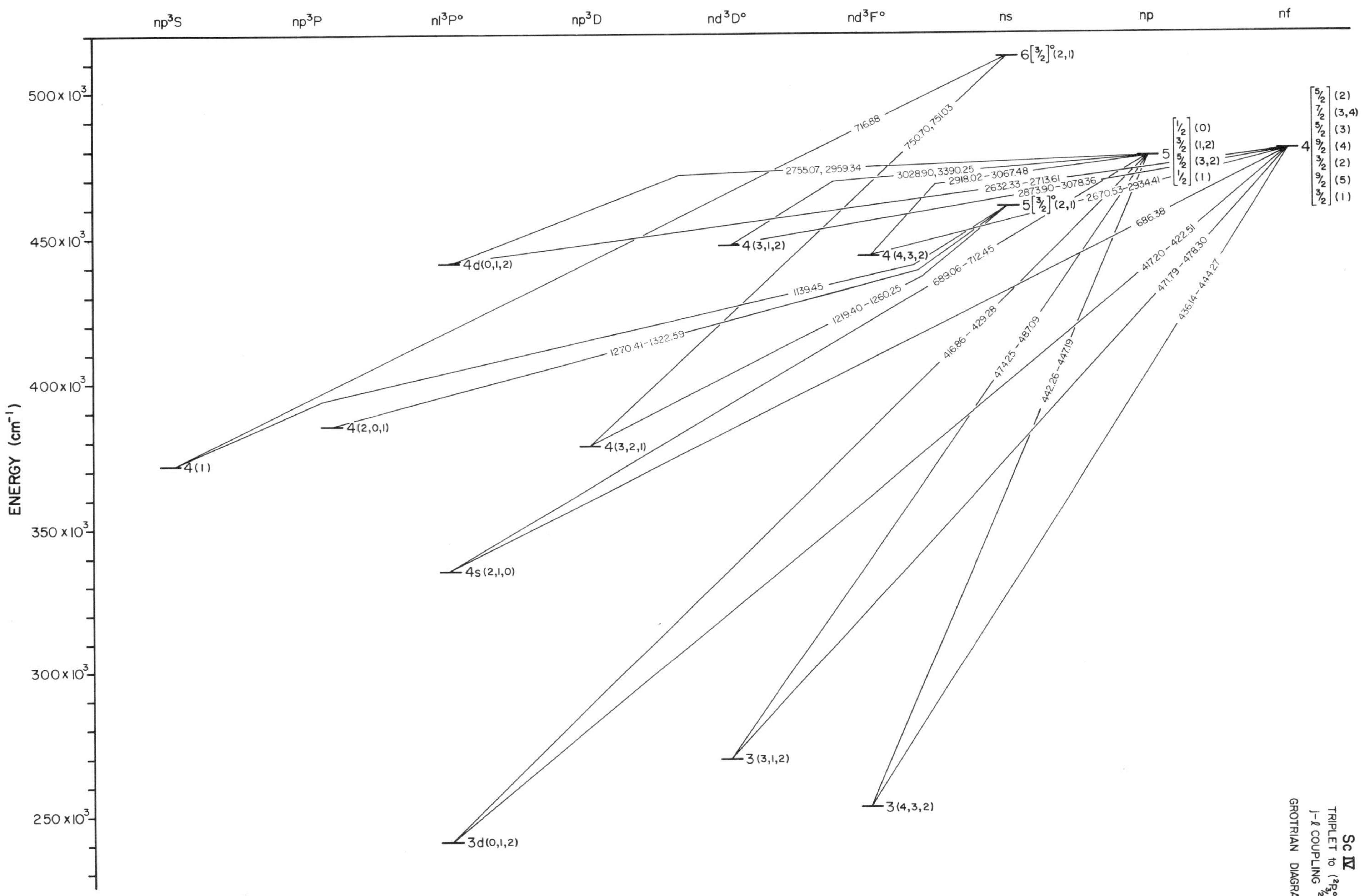

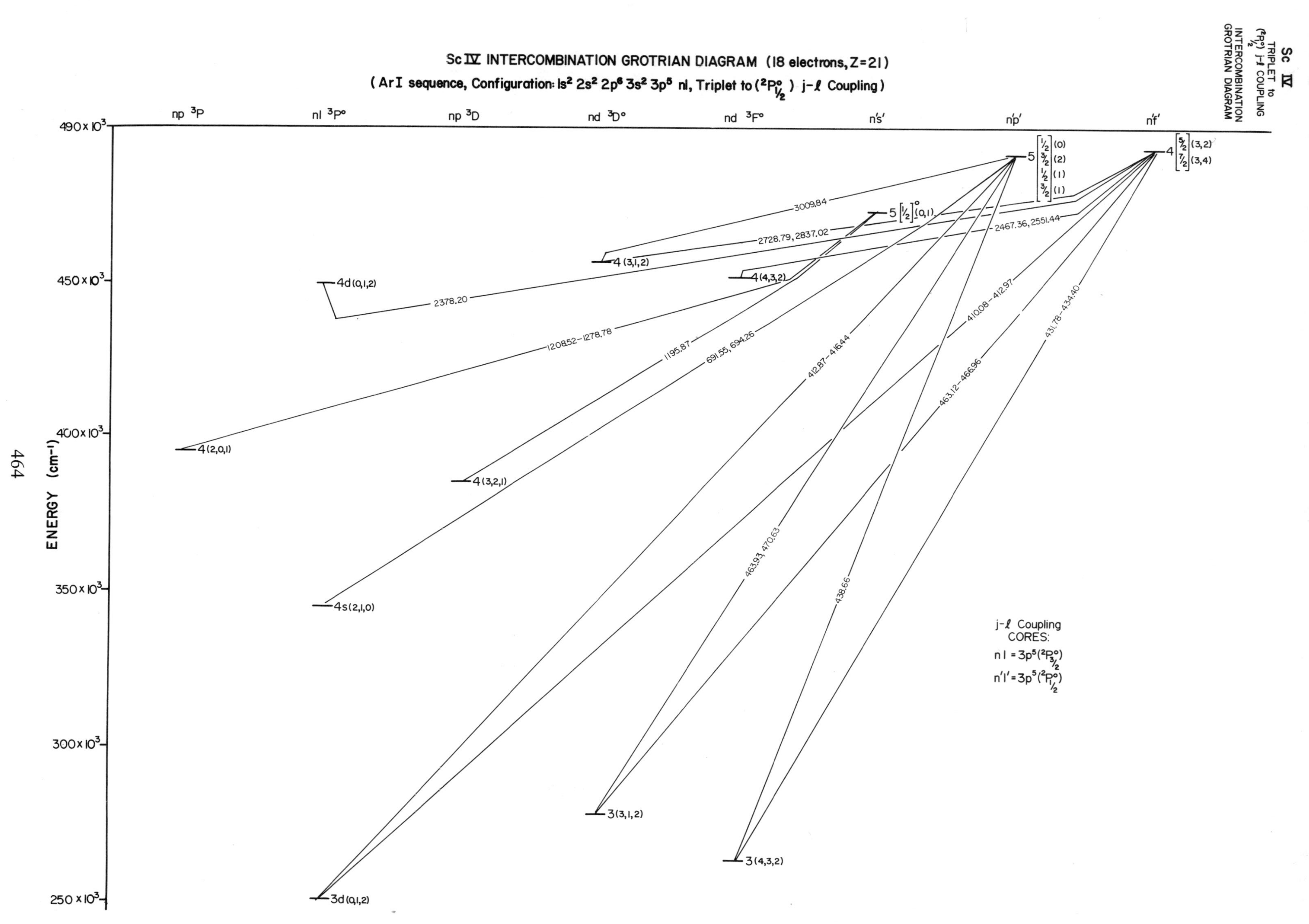

Sc IV INTERCOMBINATION GROTRIAN DIAGRAM (18 electrons, Z=21)
(Ar I sequence, Configuration: $1s^2\ 2s^2\ 2p^6\ 3s^2\ 3p^5$ nl, Triplet to ($^2P^o_{1/2}$) j-l Coupling)
Sc IV
TRIPLET to
($^2P^o_{1/2}$) j-l COUPLING
INTERCOMBINATION
GROTRIAN DIAGRAM
np ^3P
nl $^3P^o$
np ^3D
nd $^3D^o$
nd $^3F^o$
n's'
n'p'
n'f'
ENERGY (cm^{-1})
490 x 10^3
450 x 10^3
400 x 10^3
350 x 10^3
300 x 10^3
250 x 10^3
5 [1/2] (0) [3/2] (2) [1/2] (1) [3/2] (1)
4 [5/2] (3,2) [7/2] (3,4)
5 [1/2]° (0,1)
4 (3,1,2)
4 (4,3,2)
4d (0,1,2)
4 (2,0,1)
4 (3,2,1)
4s (2,1,0)
3 (3,1,2)
3 (4,3,2)
3d (0,1,2)
3009.84
2728.79, 2837.02
2467.36, 2551.44
2378.20
1208.52–1278.78
1195.87
691.55, 694.26
412.87–416.44
410.08–412.97
431.78–434.40
463.12–466.96
463.93, 470.63
438.66
j-l Coupling
CORES:
nl = $3p^5(^2P^o_{3/2})$
n'l' = $3p^5(^2P^o_{1/2})$

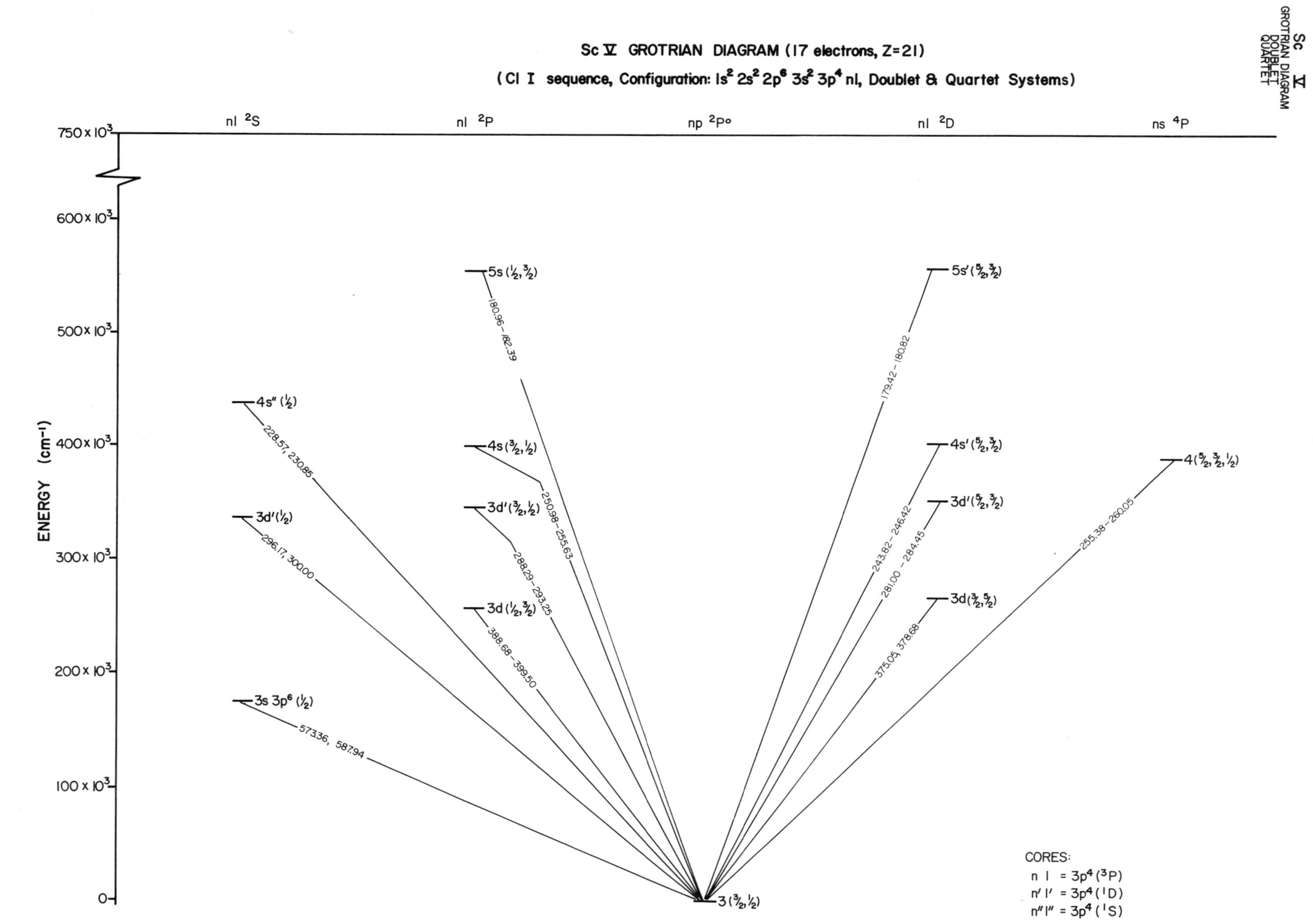
Sc V
GROTRIAN DIAGRAM
DOUBLET
QUARTET
Sc V GROTRIAN DIAGRAM (17 electrons, Z=21)
(Cl I sequence, Configuration: $1s^2 2s^2 2p^6 3s^2 3p^4$ nl, Doublet & Quartet Systems)
nl ^{2}S
nl ^{2}P
np ^{2}P°
nl ^{2}D
ns ^{4}P
ENERGY (cm-1)
750 x 10^3
600 x 10^3
500 x 10^3
400 x 10^3
300 x 10^3
200 x 10^3
100 x 10^3
0
4s″(1/2)
228.57, 230.85
3d′(1/2)
296.17, 300.00
3s 3p^6 (1/2)
573.36, 587.94
5s(1/2, 3/2)
180.96 - 182.39
4s(3/2, 1/2)
250.98 - 255.63
3d′(3/2, 1/2)
288.29 - 293.25
3d(1/2, 3/2)
388.68 - 399.50
3(3/2, 1/2)
5s′(5/2, 3/2)
179.42 - 180.82
4s′(5/2, 3/2)
243.82 - 246.42
3d′(5/2, 3/2)
281.00 - 284.45
3d(3/2, 5/2)
375.05, 378.68
4(5/2, 3/2, 1/2)
255.38 - 260.05
CORES:
n l = 3p^4 (^{3}P)
n′ l′ = 3p^4 (^{1}D)
n″ l″ = 3p^4 (^{1}S)

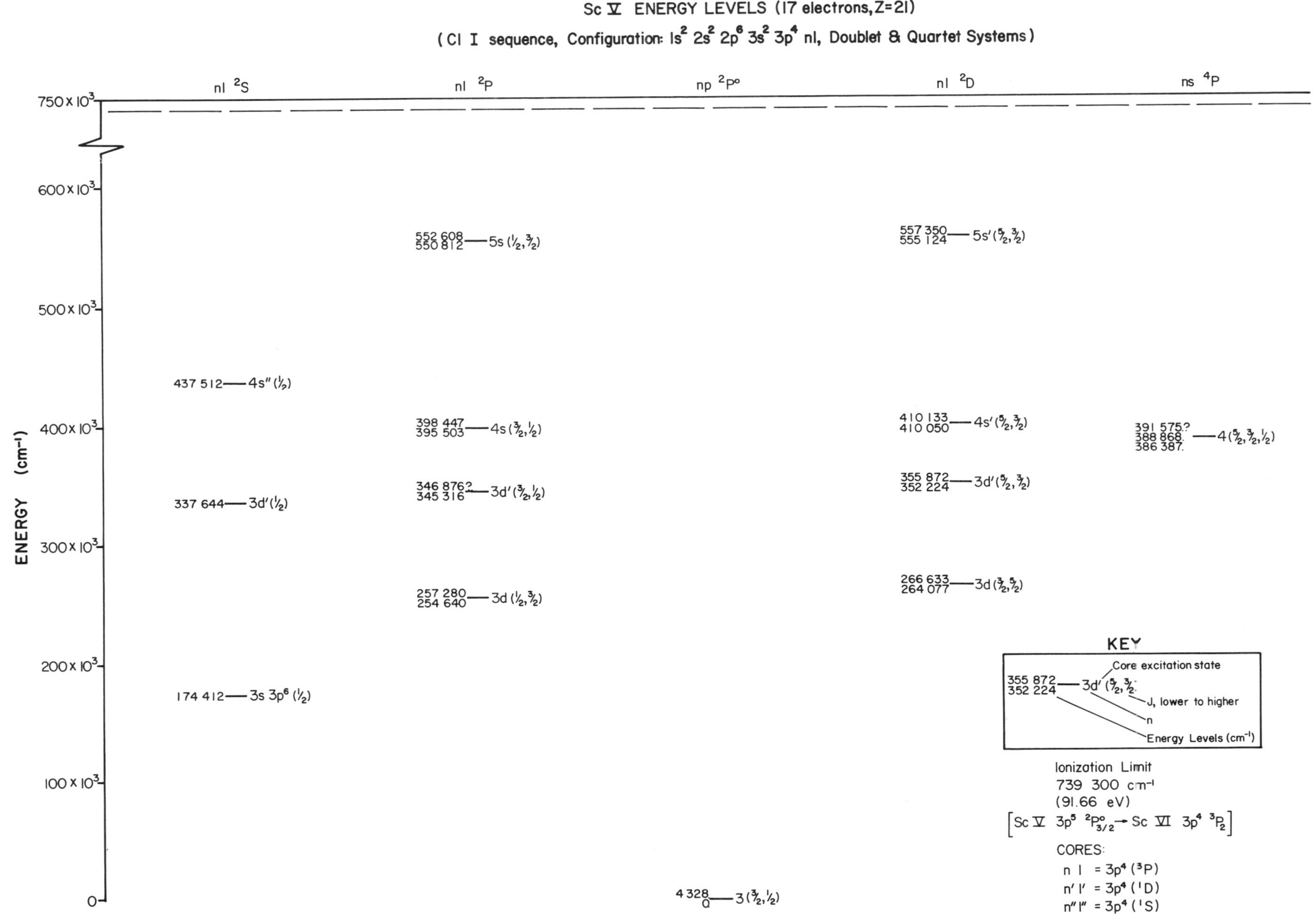
Sc V ENERGY LEVELS (17 electrons, Z=21)
(Cl I sequence, Configuration: 1s² 2s² 2p⁶ 3s² 3p⁴ nl, Doublet & Quartet Systems)
nl ²S
nl ²P
np ²P°
nl ²D
ns ⁴P
750 x 10³
600 x 10³
500 x 10³
400 x 10³
300 x 10³
200 x 10³
100 x 10³
0
ENERGY (cm⁻¹)
437 512 — 4s″(½)
337 644 — 3d′(½)
174 412 — 3s 3p⁶ (½)
552 608 / 550 812 — 5s (½, 3/2)
398 447 / 395 503 — 4s (3/2, ½)
346 876? / 345 316 — 3d′(3/2, ½)
257 280 / 254 640 — 3d (½, 3/2)
4 328 / 0 — 3(3/2, ½)
557 350 / 555 124 — 5s′(5/2, 3/2)
410 133 / 410 050 — 4s′(5/2, 3/2)
355 872 / 352 224 — 3d′(5/2, 3/2)
266 633 / 264 077 — 3d (3/2, 5/2)
391 575? / 388 868. / 386 387. — 4(5/2, 3/2, ½)
KEY
355 872 / 352 224 — 3d′(5/2, 3/2)
Core excitation state
J, lower to higher
n
Energy Levels (cm⁻¹)
Ionization Limit
739 300 cm⁻¹
(91.66 eV)
[Sc V 3p⁵ ²P°3/2 → Sc VI 3p⁴ ³P2]
CORES:
n l = 3p⁴ (³P)
n′ l′ = 3p⁴ (¹D)
n″ l″ = 3p⁴ (¹S)

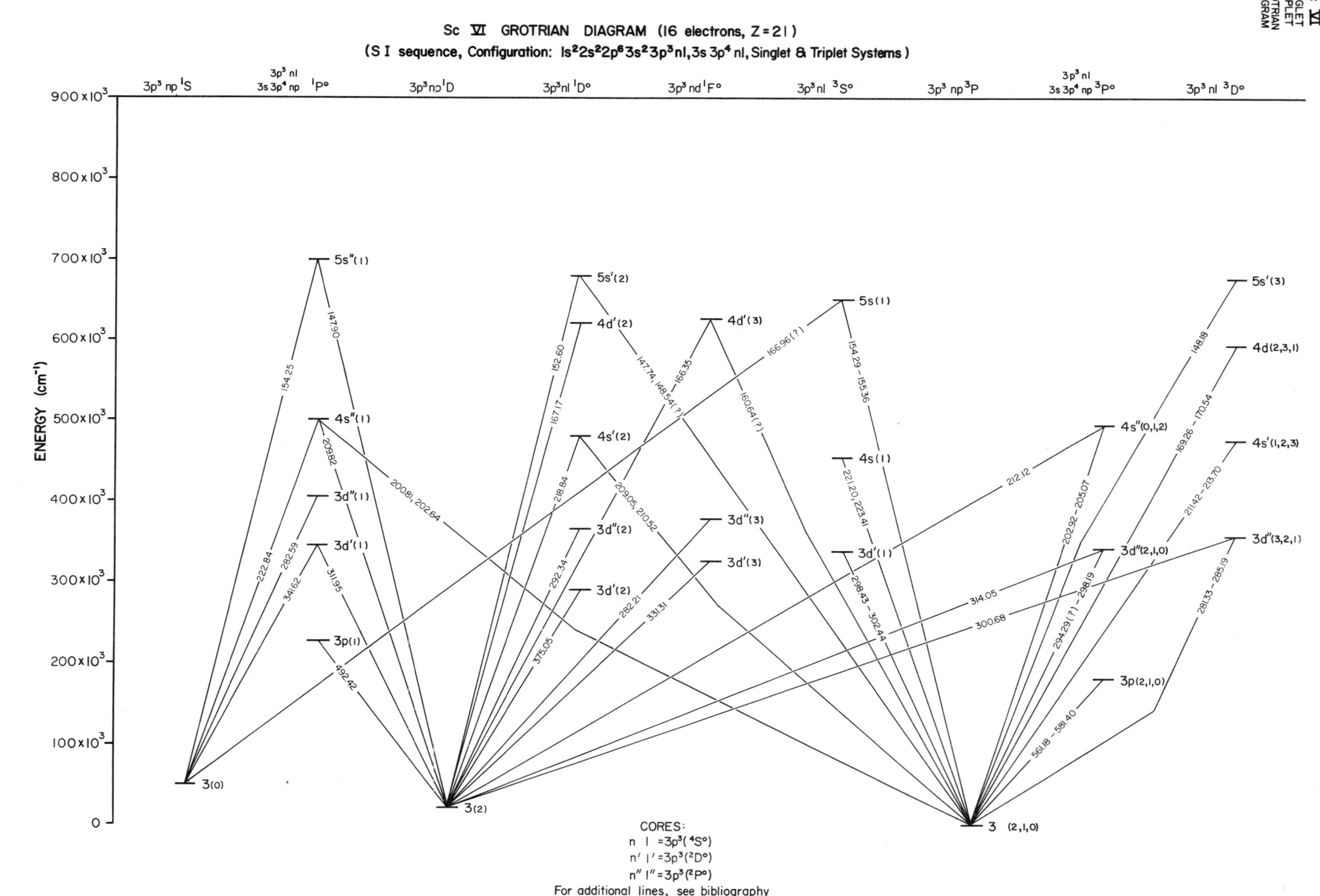
Sc VI GROTRIAN DIAGRAM (16 electrons, Z = 21)
(S I sequence, Configuration: 1s²2s²2p⁶3s²3p³nl, 3s 3p⁴ nl, Singlet & Triplet Systems)
3p³ np ¹S
3p³ nl 3s 3p⁴ np ¹P°
3p³ np ¹D
3p³ nl ¹D°
3p³ nd ¹F°
3p³ nl ³S°
3p³ np ³P
3p³ nl 3s 3p⁴ np ³P°
3p³ nl ³D°
ENERGY (cm⁻¹)
900 x 10³
800 x 10³
700 x 10³
600 x 10³
500 x 10³
400 x 10³
300 x 10³
200 x 10³
100 x 10³
0
3(0)
3(2)
3 (2,1,0)
5s″(1)
4s″(1)
3d″(1)
3d′(1)
3p(1)
5s′(2)
4d′(2)
4s′(2)
3d″(2)
3d′(2)
4d′(3)
3d″(3)
3d′(3)
5s(1)
4s(1)
3d′(1)
4s″(0,1,2)
3d″(2,1,0)
3p(2,1,0)
5s′(3)
4d(2,3,1)
4s′(1,2,3)
3d″(3,2,1)
154.25
147.90
222.84
282.59
341.62
209.82
311.95
492.42
200.81, 202.64
152.60
167.17
218.84
292.34
375.05
147.74, 148.54(?)
166.35
209.05, 210.52
282.21
333.31
166.96(?)
160.64(?)
154.29 – 155.36
221.20, 223.41
298.43 – 302.44
212.12
314.05
300.68
202.92 – 205.07
294.29(?) – 298.19
561.18 – 581.40
148.18
169.26 – 170.54
211.42 – 213.70
281.33 – 285.19
CORES:
n l = 3p³(⁴S°)
n′ l′ = 3p³(²D°)
n″ l″ = 3p³(²P°)
For additional lines, see bibliography

Sc VI ENERGY LEVELS (16 electrons, Z = 21)

(S I sequence, Configuration: $1s^2 2s^2 2p^6 3s^2 3p^3 nl$, $3s3p^4 nl$, Singlet & Triplet Systems)

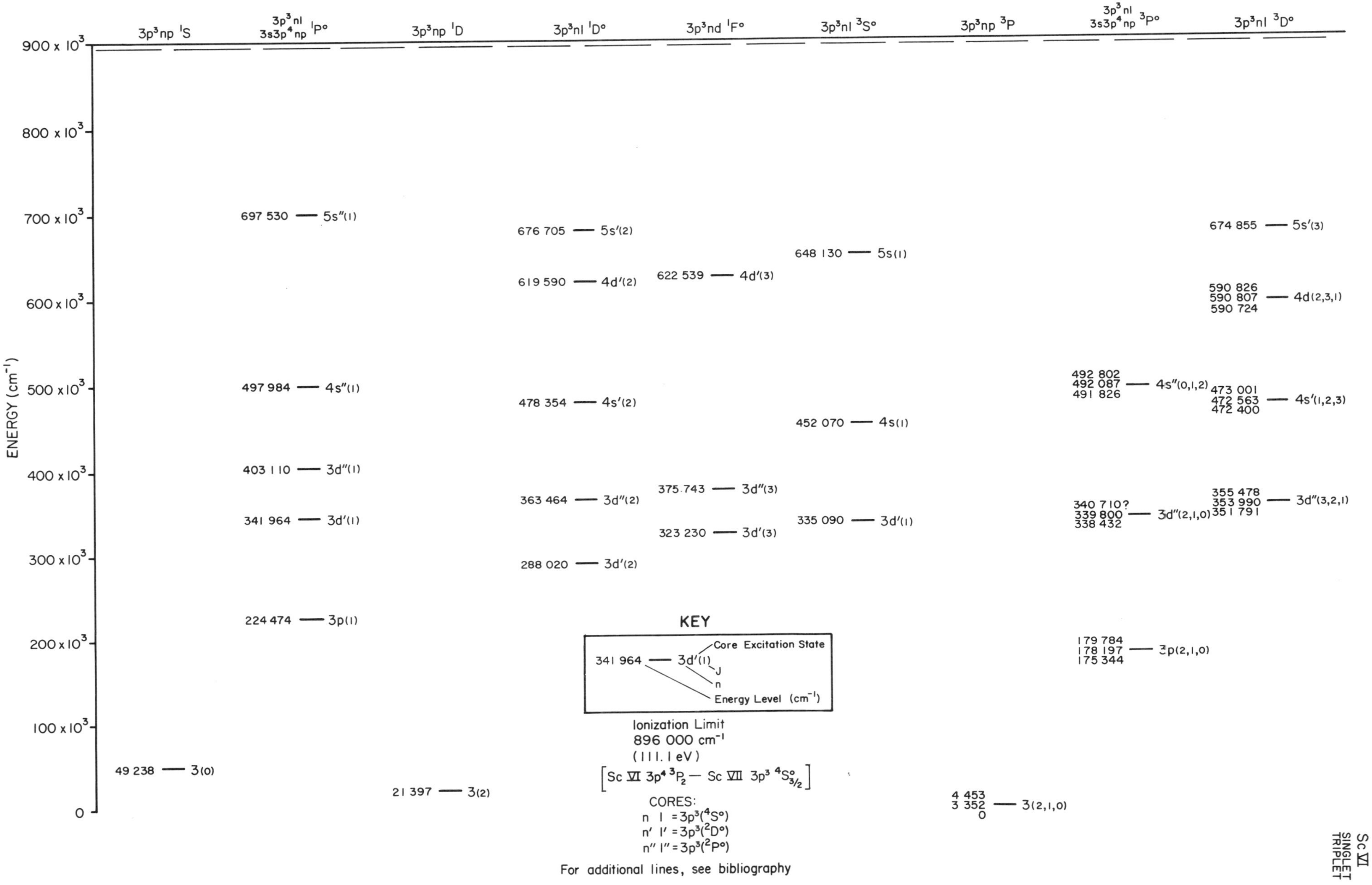

ENERGY (cm^{-1}): 900 x 10^3, 800 x 10^3, 700 x 10^3, 600 x 10^3, 500 x 10^3, 400 x 10^3, 300 x 10^3, 200 x 10^3, 100 x 10^3, 0

$3p^3np\ ^1S$
- 49 238 — 3(0)

$3p^3nl$ / $3s3p^4np$ $^1P^o$
- 697 530 — 5s″(1)
- 497 984 — 4s″(1)
- 403 110 — 3d″(1)
- 341 964 — 3d′(1)
- 224 474 — 3p(1)

$3p^3np\ ^1D$
- 21 397 — 3(2)

$3p^3nl\ ^1D^o$
- 676 705 — 5s′(2)
- 619 590 — 4d′(2)
- 478 354 — 4s′(2)
- 363 464 — 3d″(2)
- 288 020 — 3d′(2)

$3p^3nd\ ^1F^o$
- 622 539 — 4d′(3)
- 375.743 — 3d″(3)
- 323 230 — 3d′(3)

$3p^3nl\ ^3S^o$
- 648 130 — 5s(1)
- 452 070 — 4s(1)
- 335 090 — 3d′(1)

$3p^3np\ ^3P$
- 4 453, 3 352, 0 — 3(2,1,0)

$3p^3nl$ / $3s3p^4np$ $^3P^o$
- 492 802, 492 087, 491 826 — 4s″(0,1,2)
- 340 710?, 339 800, 338 432 — 3d″(2,1,0)
- 179 784, 178 197, 175 344 — 3p(2,1,0)

$3p^3nl\ ^3D^o$
- 674 855 — 5s′(3)
- 590 826, 590 807, 590 724 — 4d(2,3,1)
- 473 001, 472 563, 472 400 — 4s′(1,2,3)
- 355 478, 353 990, 351 791 — 3d″(3,2,1)

KEY

341 964 — 3d′(1): 341 964 = Energy Level (cm^{-1}); 3 = n; ′ = Core Excitation State; (1) = J

Ionization Limit
896 000 cm^{-1}
(111.1 eV)
[Sc VI $3p^4\ ^3P_2$ — Sc VII $3p^3\ ^4S^o_{3/2}$]

CORES:
n l = $3p^3(^4S^o)$
n′ l′ = $3p^3(^2D^o)$
n″ l″ = $3p^3(^2P^o)$

For additional lines, see bibliography

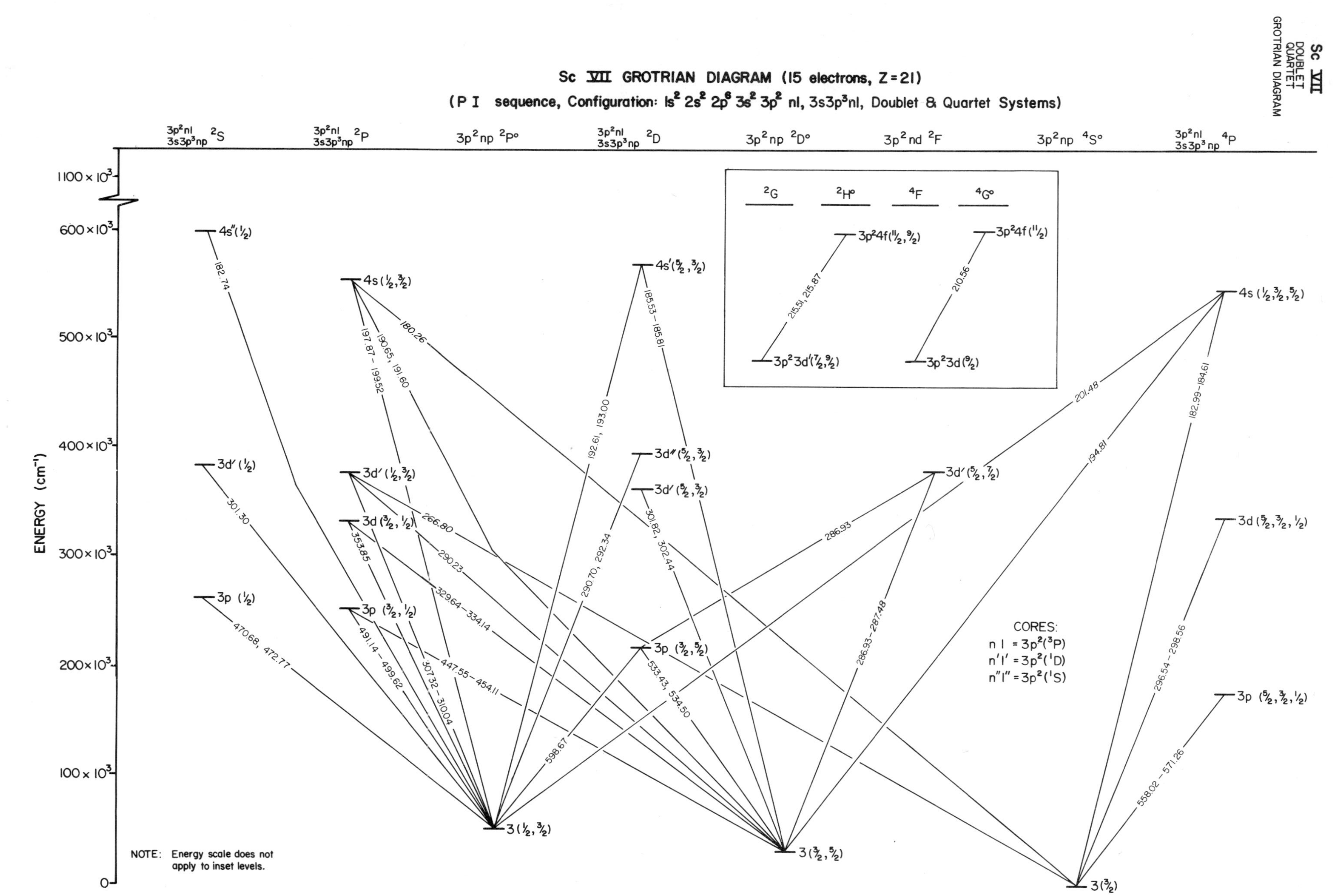
Sc VII
DOUBLET
QUARTET
GROTRIAN DIAGRAM
Sc VII GROTRIAN DIAGRAM (15 electrons, Z=21)
(P I sequence, Configuration: $1s^2 2s^2 2p^6 3s^2 3p^2$ nl, $3s3p^3$nl, Doublet & Quartet Systems)
$3p^2nl$ $3s3p^3np$ ^2S
$3p^2nl$ $3s3p^3np$ ^2P
$3p^2np$ $^2P^o$
$3p^2nl$ $3s3p^3np$ ^2D
$3p^2np$ $^2D^o$
$3p^2nd$ ^2F
$3p^2np$ $^4S^o$
$3p^2nl$ $3s3p^3np$ ^4P
ENERGY (cm^{-1})
1100×10^3
600×10^3
500×10^3
400×10^3
300×10^3
200×10^3
100×10^3
0
^2G
$^2H^o$
^4F
$^4G^o$
$3p^24f(^{11}/_2, ^9/_2)$
$3p^24f(^{11}/_2)$
215.51, 215.87
210.56
$3p^23d'(^7/_2, ^9/_2)$
$3p^23d(^9/_2)$
$4s''(^1/_2)$
182.74
$4s(^1/_2, ^3/_2)$
180.26
197.87 – 199.52
190.65, 191.60
$4s'(^5/_2, ^3/_2)$
185.53 – 185.81
192.61, 193.00
$4s(^1/_2, ^3/_2, ^5/_2)$
201.48
182.99 – 184.61
194.81
$3d'(^1/_2)$
301.30
$3d'(^1/_2, ^3/_2)$
$3d(^3/_2, ^1/_2)$
353.85
266.80
290.23
329.64 – 334.14
$3d''(^5/_2, ^3/_2)$
$3d'(^5/_2, ^3/_2)$
301.82, 302.44
290.70, 292.34
$3d'(^5/_2, ^7/_2)$
286.93
286.93 – 287.48
$3d(^5/_2, ^3/_2, ^1/_2)$
296.54 – 298.56
$3p(^1/_2)$
470.68, 472.77
$3p(^3/_2, ^1/_2)$
491.14 – 499.62
307.32 – 310.04
447.55 – 454.11
$3p(^3/_2, ^5/_2)$
533.43, 534.50
598.67
$3p(^5/_2, ^3/_2, ^1/_2)$
558.02 – 571.26
CORES:
n l = $3p^2(^3P)$
n'l' = $3p^2(^1D)$
n"l" = $3p^2(^1S)$
$3(^1/_2, ^3/_2)$
$3(^3/_2, ^5/_2)$
$3(^3/_2)$
NOTE: Energy scale does not apply to inset levels.

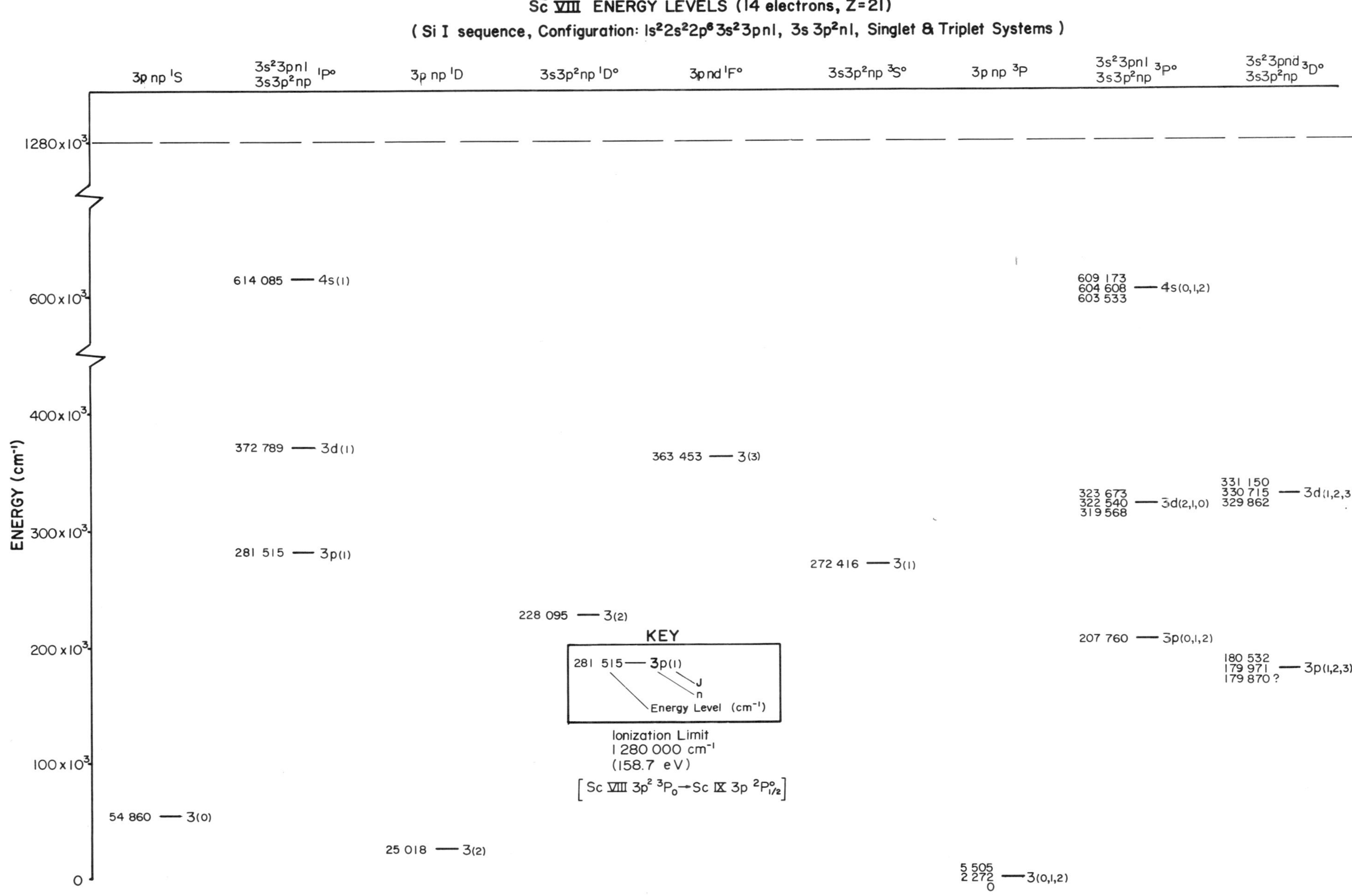
Sc VIII ENERGY LEVELS (14 electrons, Z=21)
(Si I sequence, Configuration: 1s²2s²2p⁶3s²3pnl, 3s 3p²nl, Singlet & Triplet Systems)
3p np ¹S
3s²3pnl ¹P°
3s3p²np
3p np ¹D
3s3p²np ¹D°
3pnd ¹F°
3s3p²np ³S°
3p np ³P
3s²3pnl ³P°
3s3p²np
3s²3pnd ³D°
3s3p²np
1280x10³
600x10³
400x10³
300x10³
200x10³
100x10³
0
ENERGY (cm⁻¹)
614 085 — 4s(1)
372 789 — 3d(1)
281 515 — 3p(1)
54 860 — 3(0)
25 018 — 3(2)
228 095 — 3(2)
363 453 — 3(3)
272 416 — 3(1)
5 505
2 272 — 3(0,1,2)
0
609 173
604 608 — 4s(0,1,2)
603 533
323 673
322 540 — 3d(2,1,0)
319 568
207 760 — 3p(0,1,2)
331 150
330 715 — 3d(1,2,3)
329 862
180 532
179 971 — 3p(1,2,3)
179 870 ?
KEY
281 515 — 3p(1)
J
n
Energy Level (cm⁻¹)
Ionization Limit
1 280 000 cm⁻¹
(158.7 eV)
[Sc VIII 3p² ³P₀ → Sc IX 3p ²P°₁/₂]

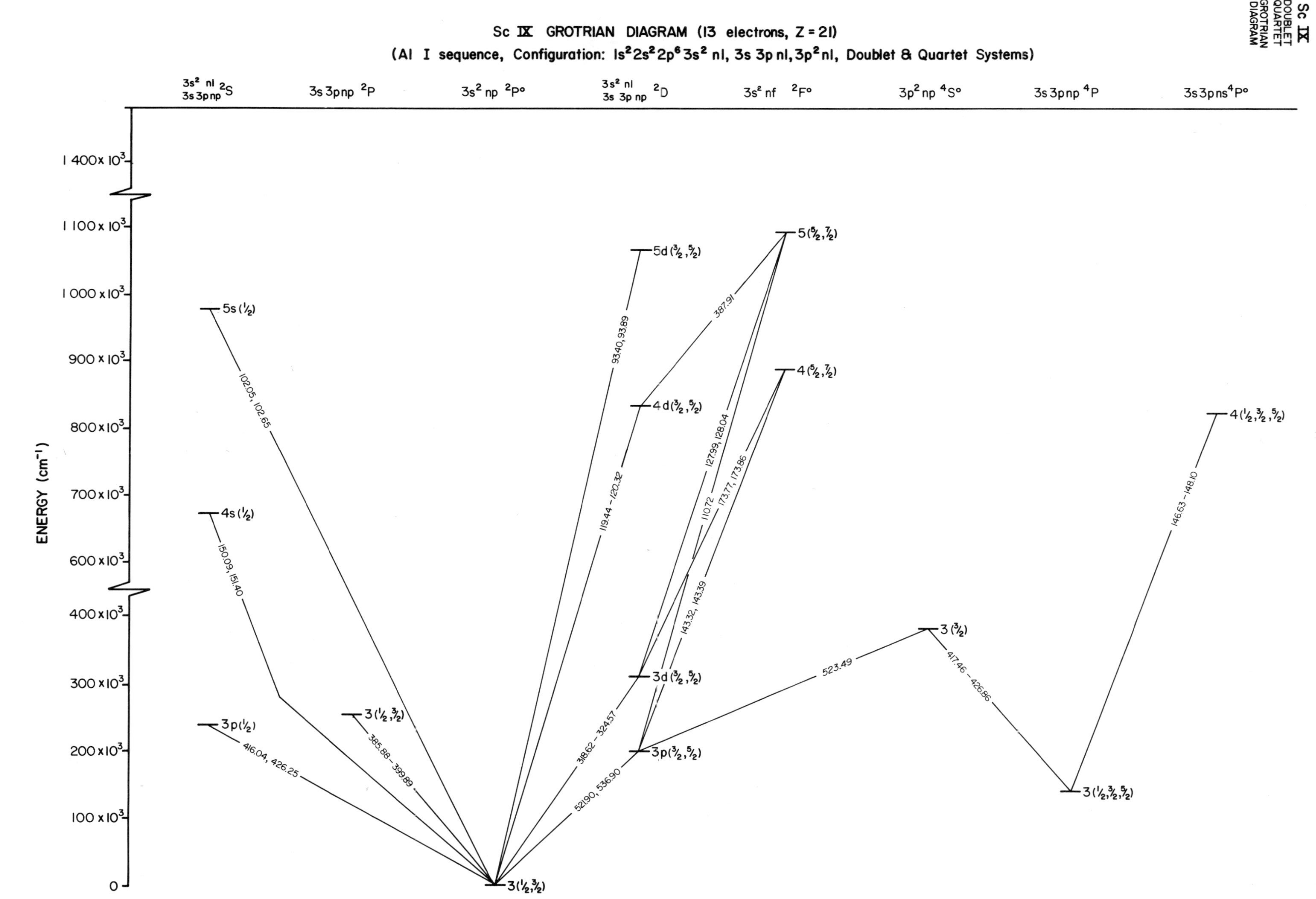
Sc IX
DOUBLET
QUARTET
GROTRIAN
DIAGRAM
Sc IX GROTRIAN DIAGRAM (13 electrons, Z = 21)
(Al I sequence, Configuration: 1s²2s²2p⁶3s² nl, 3s 3p nl, 3p² nl, Doublet & Quartet Systems)
3s² nl 3s 3pnp ²S
3s 3pnp ²P
3s² np ²P°
3s² nl 3s 3p np ²D
3s² nf ²F°
3p² np ⁴S°
3s 3pnp ⁴P
3s 3pns ⁴P°
1 400 x 10³
1 100 x 10³
1 000 x 10³
900 x 10³
800 x 10³
700 x 10³
600 x 10³
400 x 10³
300 x 10³
200 x 10³
100 x 10³
0
ENERGY (cm⁻¹)
5s(½)
4s(½)
3p(½)
3(½,3/2)
3(½,3/2)
5d(3/2,5/2)
4d(3/2,5/2)
3d(3/2,5/2)
3p(3/2,5/2)
5(5/2,7/2)
4(5/2,7/2)
3(3/2)
3(½,3/2,5/2)
4(½,3/2,5/2)
102.05, 102.65
150.09, 151.40
416.04, 426.25
385.88 - 399.89
93.40, 93.89
119.44 - 120.32
318.62 - 324.57
521.90, 536.90
387.91
127.99, 128.04
110.72
173.77, 173.86
143.32, 143.39
523.49
417.46 - 426.86
146.63 - 148.10

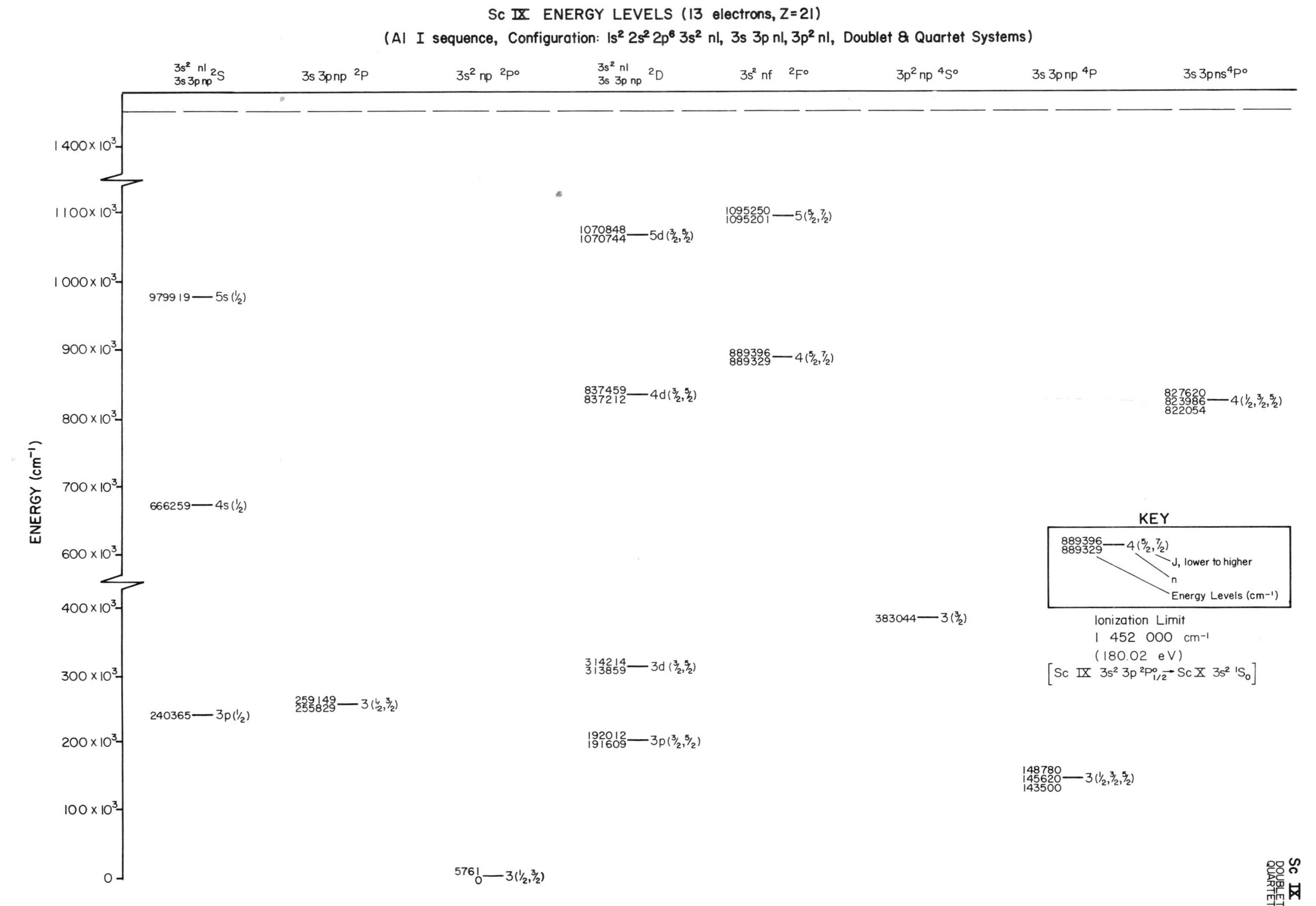

Sc IX ENERGY LEVELS (13 electrons, Z=21)
(Al I sequence, Configuration: 1s² 2s² 2p⁶ 3s² nl, 3s 3p nl, 3p² nl, Doublet & Quartet Systems)
3s² nl
3s 3p np ²S
3s 3p np ²P
3s² np ²P°
3s² nl
3s 3p np ²D
3s² nf ²F°
3p² np ⁴S°
3s 3p np ⁴P
3s 3p ns ⁴P°
979919 — 5s(½)
666259 — 4s(½)
240365 — 3p(½)
259149
255829 — 3(½,³⁄₂)
5761
0 — 3(½,³⁄₂)
1070848
1070744 — 5d(³⁄₂,⁵⁄₂)
837459
837212 — 4d(³⁄₂,⁵⁄₂)
314214
313859 — 3d(³⁄₂,⁵⁄₂)
192012
191609 — 3p(³⁄₂,⁵⁄₂)
1095250
1095201 — 5(⁵⁄₂,⁷⁄₂)
889396
889329 — 4(⁵⁄₂,⁷⁄₂)
383044 — 3(³⁄₂)
148780
145620
143500 — 3(½,³⁄₂,⁵⁄₂)
827620
823986
822054 — 4(½,³⁄₂,⁵⁄₂)
KEY
889396
889329 — 4(⁵⁄₂,⁷⁄₂)
J, lower to higher
n
Energy Levels (cm⁻¹)
Ionization Limit
1 452 000 cm⁻¹
(180.02 eV)
[Sc IX 3s² 3p ²P°½ → Sc X 3s² ¹S₀]
ENERGY (cm⁻¹)
1 400 x 10³
1 100 x 10³
1 000 x 10³
900 x 10³
800 x 10³
700 x 10³
600 x 10³
400 x 10³
300 x 10³
200 x 10³
100 x 10³
0
Sc IX
DOUBLET
QUARTET

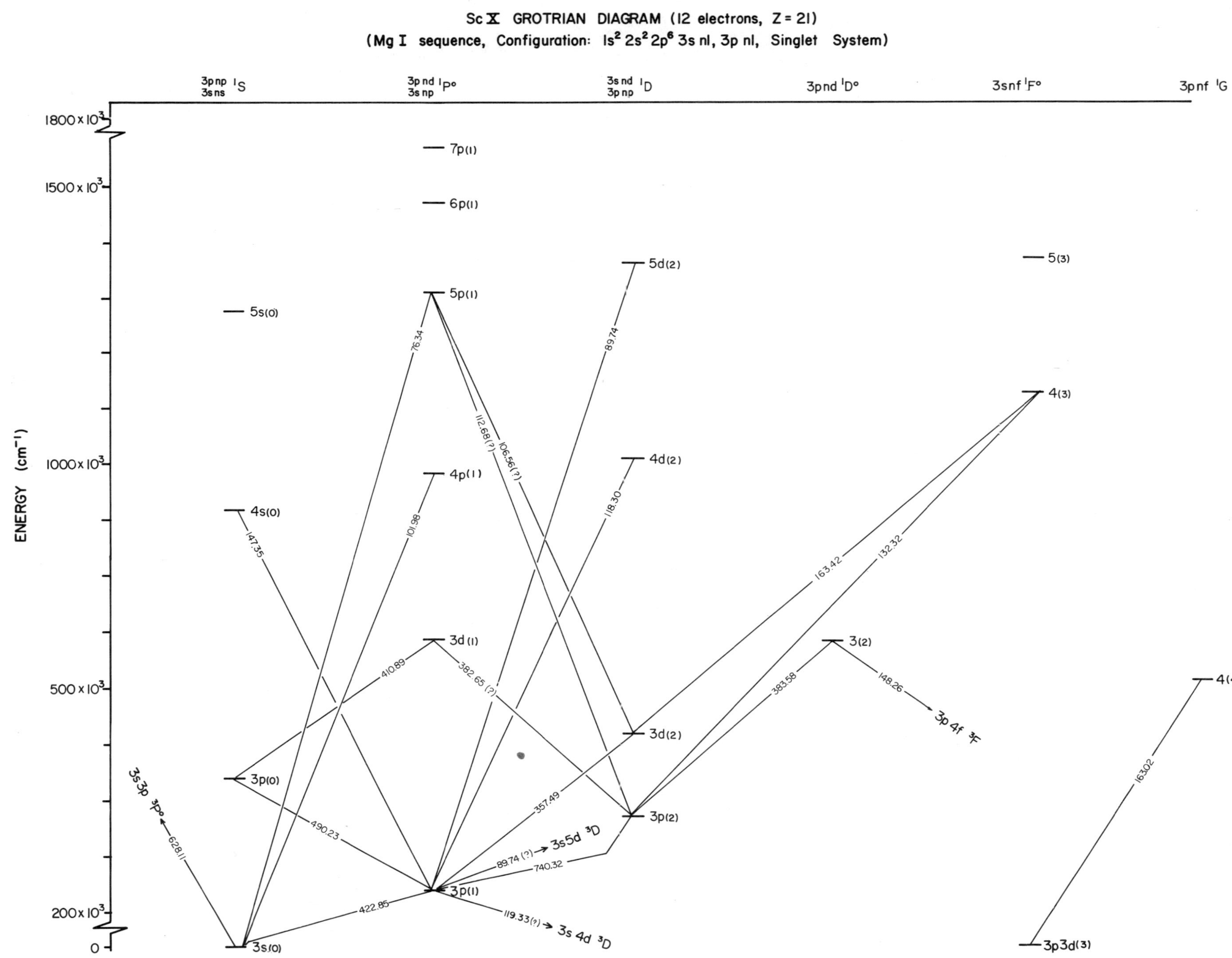
Sc X GROTRIAN DIAGRAM (12 electrons, Z = 21)
(Mg I sequence, Configuration: 1s² 2s² 2p⁶ 3s nl, 3p nl, Singlet System)
Sc X
SINGLET
GROTRIAN DIAGRAM
3pnp ¹S
3sns
3pnd ¹P°
3snp
3snd ¹D
3pnp
3pnd ¹D°
3snf ¹F°
3pnf ¹G
ENERGY (cm⁻¹)
1800 x 10³
1500 x 10³
1000 x 10³
500 x 10³
200 x 10³
0
7p(1)
6p(1)
5p(1)
4p(1)
3d(1)
3p(1)
5s(0)
4s(0)
3p(0)
3s(0)
5d(2)
4d(2)
3d(2)
3p(2)
5(3)
4(3)
3(2)
4(4)
3p3d(3)
76.34
101.98
147.35
410.89
382.65 (?)
112.68(?)
106.56(?)
89.74
118.30
163.42
132.32
383.58
148.26
3p 4f ³F
163.02
357.49
490.23
89.74 (?)
3s5d ³D
740.32
422.85
119.33(?)
3s 4d ³D
628.11
3s3p ³P°

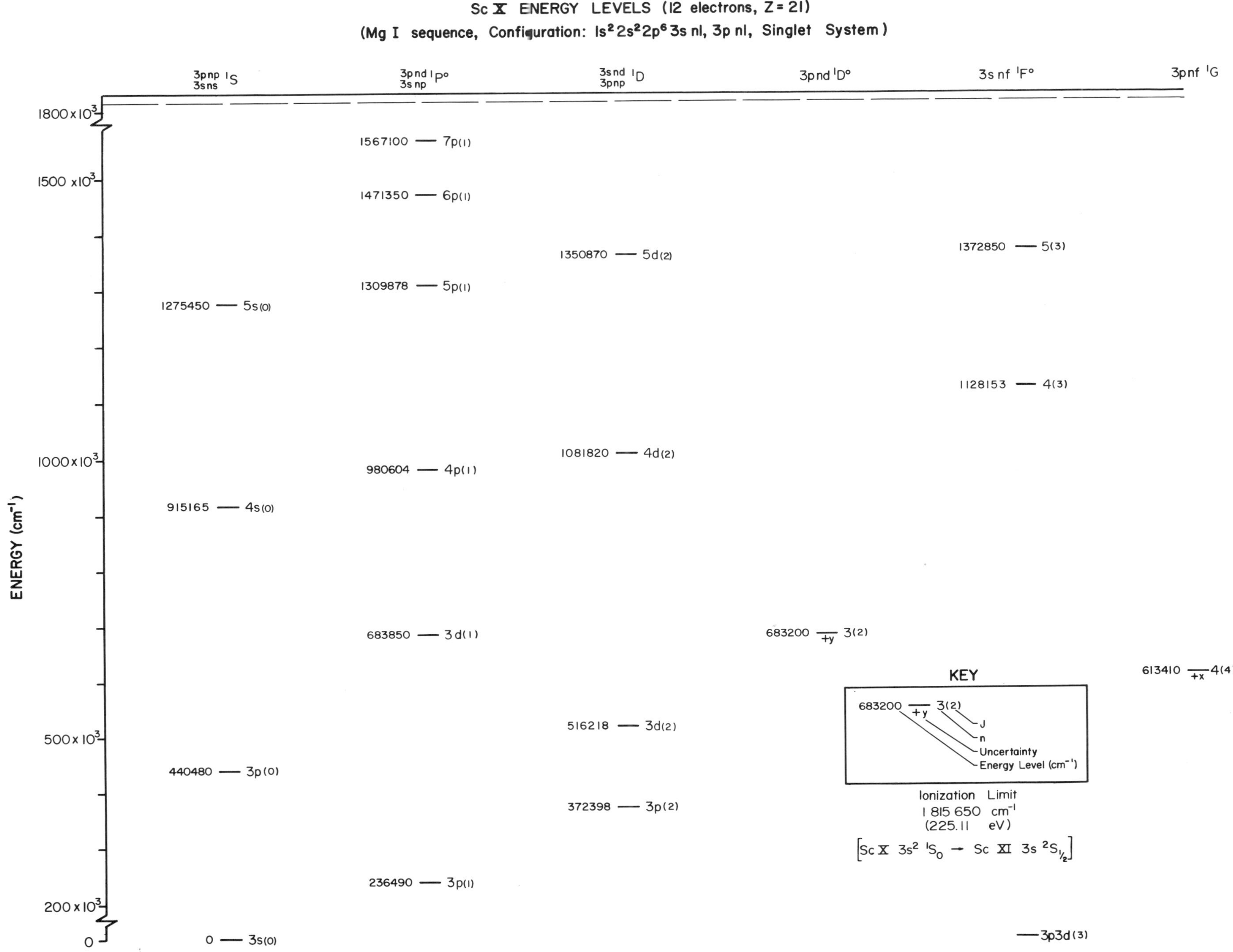

Sc X ENERGY LEVELS (12 electrons, Z = 21)
(Mg I sequence, Configuration: 1s²2s²2p⁶3s nl, 3p nl, Singlet System)
3pnp ¹S
3sns
3pnd ¹P°
3snp
3snd ¹D
3pnp
3pnd ¹D°
3snf ¹F°
3pnf ¹G
1800×10³
1500×10³
1000×10³
500×10³
200×10³
0
ENERGY (cm⁻¹)
1567100 — 7p(1)
1471350 — 6p(1)
1350870 — 5d(2)
1372850 — 5(3)
1309878 — 5p(1)
1275450 — 5s(0)
1128153 — 4(3)
1081820 — 4d(2)
980604 — 4p(1)
915165 — 4s(0)
683850 — 3d(1)
683200 +y 3(2)
613410 +x 4(4)
516218 — 3d(2)
440480 — 3p(0)
372398 — 3p(2)
236490 — 3p(1)
0 — 3s(0)
—3p3d(3)
KEY
683200 +y 3(2)
J
n
Uncertainty
Energy Level (cm⁻¹)
Ionization Limit
1 815 650 cm⁻¹
(225.11 eV)
[Sc X 3s² ¹S₀ → Sc XI 3s ²S½]
Sc X
SINGLET

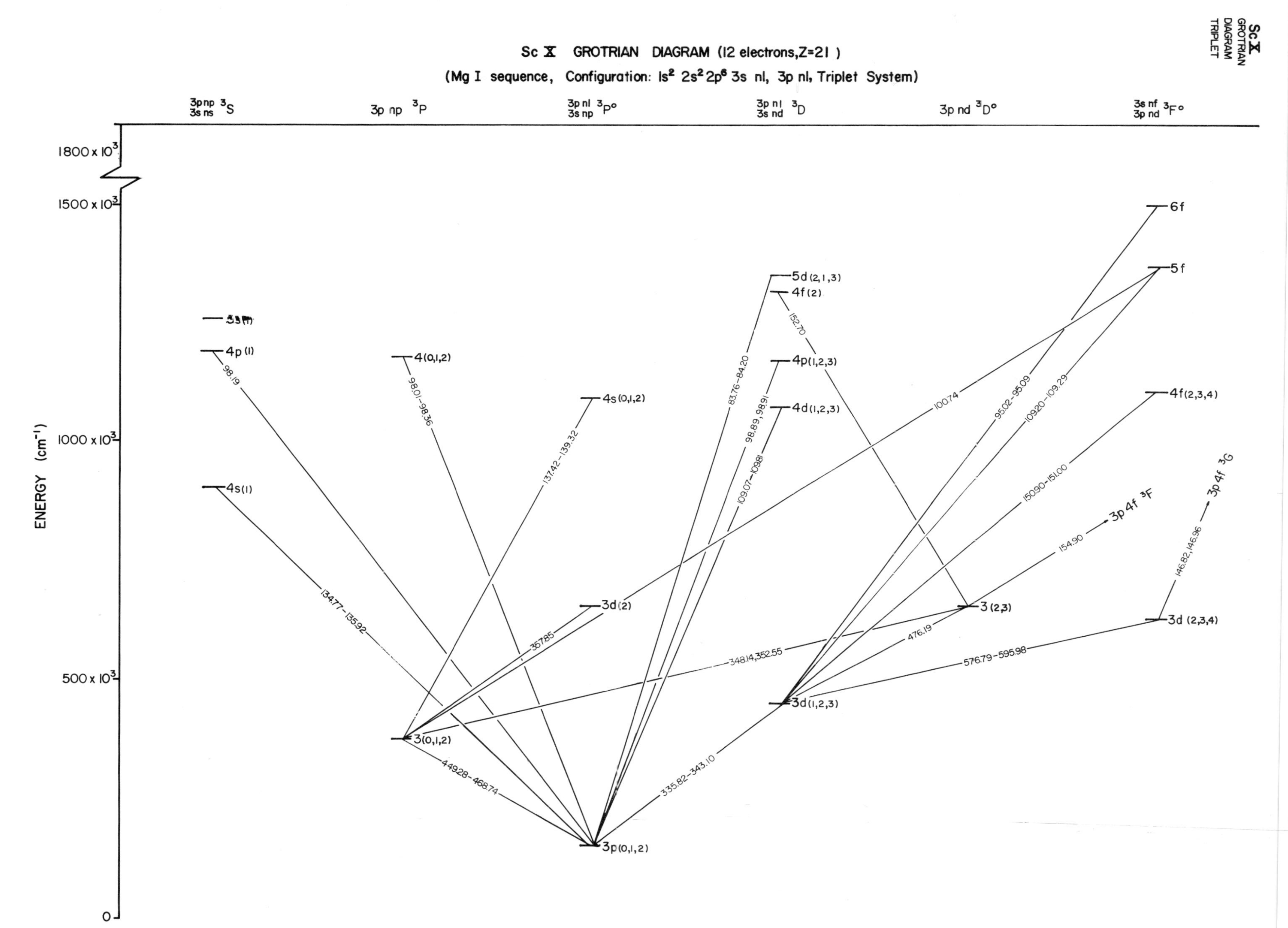

Sc X
GROTRIAN
DIAGRAM
TRIPLET
Sc X GROTRIAN DIAGRAM (12 electrons, Z=21)
(Mg I sequence, Configuration: 1s² 2s² 2p⁶ 3s nl, 3p nl, Triplet System)
3p np ³S
3s ns
3p np ³P
3p nl ³P°
3s np
3p nl ³D
3s nd
3p nd ³D°
3s nf ³F°
3p nd
ENERGY (cm⁻¹)
1800 x 10³
1500 x 10³
1000 x 10³
500 x 10³
0
4p(1)
4s(1)
4(0,1,2)
3(0,1,2)
4s(0,1,2)
3d(2)
3p(0,1,2)
5d(2,1,3)
4f(2)
4p(1,2,3)
4d(1,2,3)
3d(1,2,3)
3(2,3)
6f
5f
4f(2,3,4)
3d(2,3,4)
3p 4f ³F
3p 4f ³G
98.19
98.01-98.36
137.42-139.32
134.77-135.92
357.85
449.28-468.74
83.76-84.20
98.89, 98.91
109.07-109.81
152.70
100.74
95.02-95.09
109.20-109.29
150.90-151.00
154.90
146.82, 146.96
348.14, 352.55
476.19
576.79-595.98
335.82-343.10

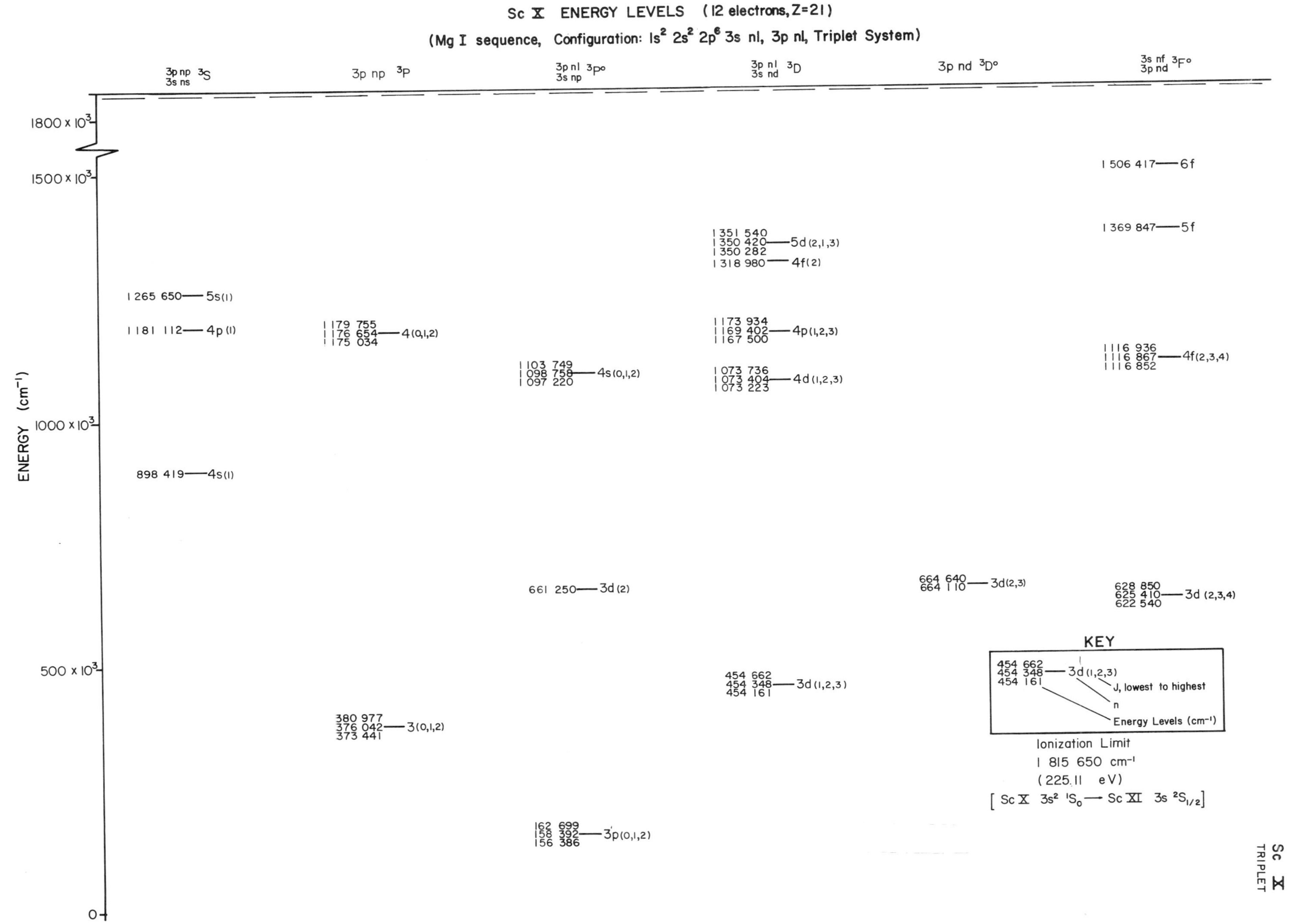
Sc X ENERGY LEVELS (12 electrons, Z=21)
(Mg I sequence, Configuration: $1s^2 2s^2 2p^6$ 3s nl, 3p nl, Triplet System)
3p np ^3S / 3s ns
3p np ^3P
3p nl $^3P^o$ / 3s np
3p nl ^3D / 3s nd
3p nd $^3D^o$
3s nf $^3F^o$ / 3p nd
ENERGY (cm^{-1})
1800 x 10^3
1500 x 10^3
1000 x 10^3
500 x 10^3
0
1 265 650 — 5s(1)
1 181 112 — 4p(1)
898 419 — 4s(1)
1 179 755
1 176 654 — 4(0,1,2)
1 175 034
380 977
376 042 — 3(0,1,2)
373 441
1 103 749
1 098 758 — 4s(0,1,2)
1 097 220
661 250 — 3d(2)
162 699
158 392 — 3p(0,1,2)
156 386
1 351 540
1 350 420 — 5d(2,1,3)
1 350 282
1 318 980 — 4f(2)
1 173 934
1 169 402 — 4p(1,2,3)
1 167 500
1 073 736
1 073 404 — 4d(1,2,3)
1 073 223
454 662
454 348 — 3d(1,2,3)
454 161
664 640
664 110 — 3d(2,3)
1 506 417 — 6f
1 369 847 — 5f
1 116 936
1 116 867 — 4f(2,3,4)
1 116 852
628 850
625 410 — 3d(2,3,4)
622 540
KEY
454 662
454 348 — 3d(1,2,3)
454 161
J, lowest to highest
n
Energy Levels (cm^{-1})
Ionization Limit
1 815 650 cm^{-1}
(225.11 eV)
[Sc X $3s^2$ $^1S_0$ → Sc XI 3s $^2S_{1/2}$]
Sc X
TRIPLET

Sc XI
DOUBLET
GROTRIAN DIAGRAM

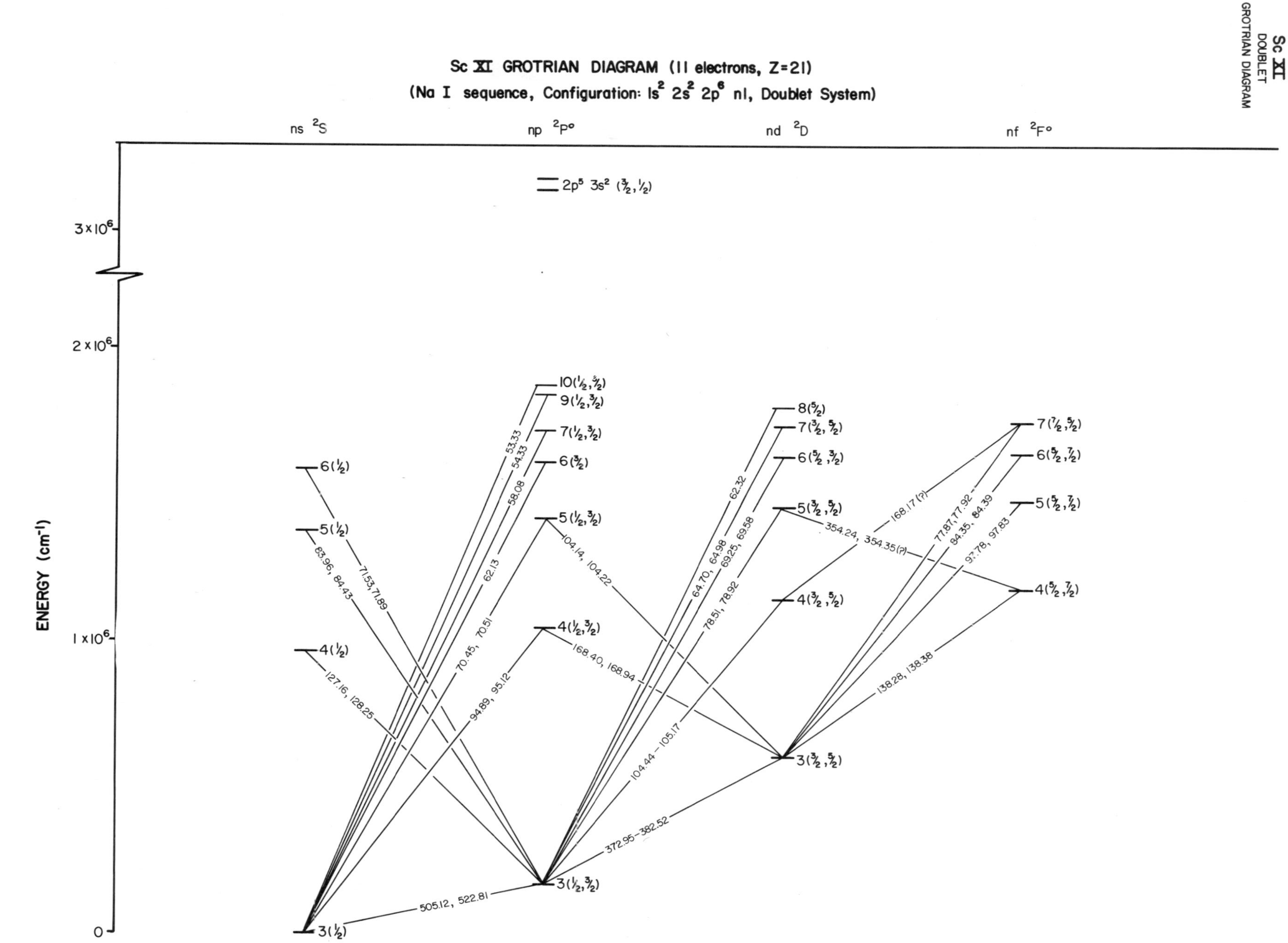

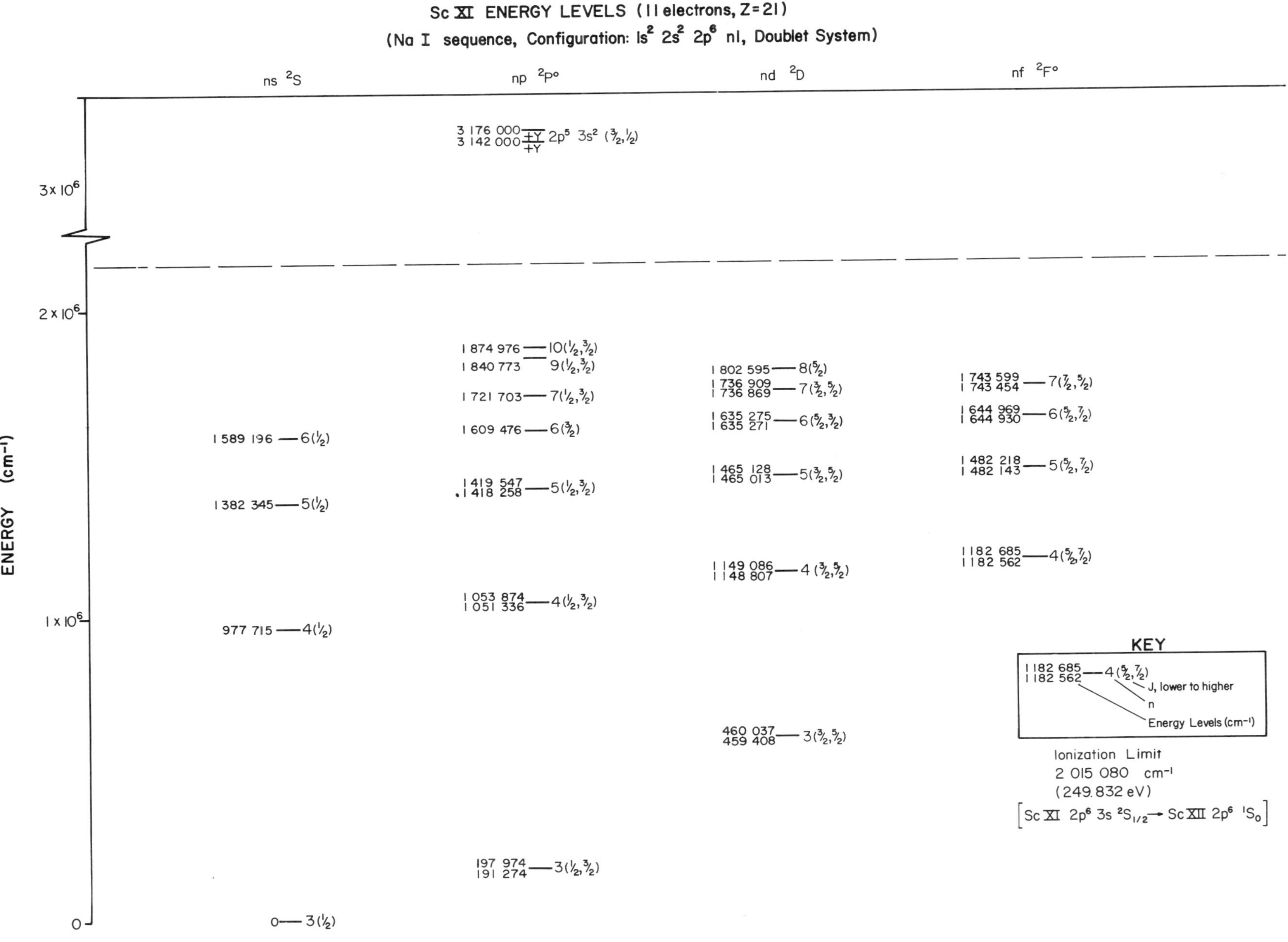
Sc XI ENERGY LEVELS (11 electrons, Z=21)
(Na I sequence, Configuration: $1s^2\ 2s^2\ 2p^6\ nl$, Doublet System)
ns ^2S
np $^2P^o$
nd ^2D
nf $^2F^o$
ENERGY (cm^{-1})
3 x 10^6
2 x 10^6
1 x 10^6
0
3 176 000 +Y, 3 142 000 +Y — $2p^5\ 3s^2$ (3/2, 1/2)
1 589 196 — 6(1/2)
1 382 345 — 5(1/2)
977 715 — 4(1/2)
0 — 3(1/2)
1 874 976 — 10(1/2, 3/2)
1 840 773 — 9(1/2, 3/2)
1 721 703 — 7(1/2, 3/2)
1 609 476 — 6(3/2)
1 419 547, 1 418 258 — 5(1/2, 3/2)
1 053 874, 1 051 336 — 4(1/2, 3/2)
197 974, 191 274 — 3(1/2, 3/2)
1 802 595 — 8(5/2)
1 736 909, 1 736 869 — 7(3/2, 5/2)
1 635 275, 1 635 271 — 6(5/2, 3/2)
1 465 128, 1 465 013 — 5(3/2, 5/2)
1 149 086, 1 148 807 — 4(3/2, 5/2)
460 037, 459 408 — 3(3/2, 5/2)
1 743 599, 1 743 454 — 7(7/2, 5/2)
1 644 969, 1 644 930 — 6(5/2, 7/2)
1 482 218, 1 482 143 — 5(5/2, 7/2)
1 182 685, 1 182 562 — 4(5/2, 7/2)
KEY
1 182 685, 1 182 562 — 4(5/2, 7/2)
J, lower to higher
n
Energy Levels (cm^{-1})
Ionization Limit
2 015 080 cm^{-1}
(249.832 eV)
[Sc XI $2p^6\ 3s\ ^2S_{1/2}$ → Sc XII $2p^6\ ^1S_0$]

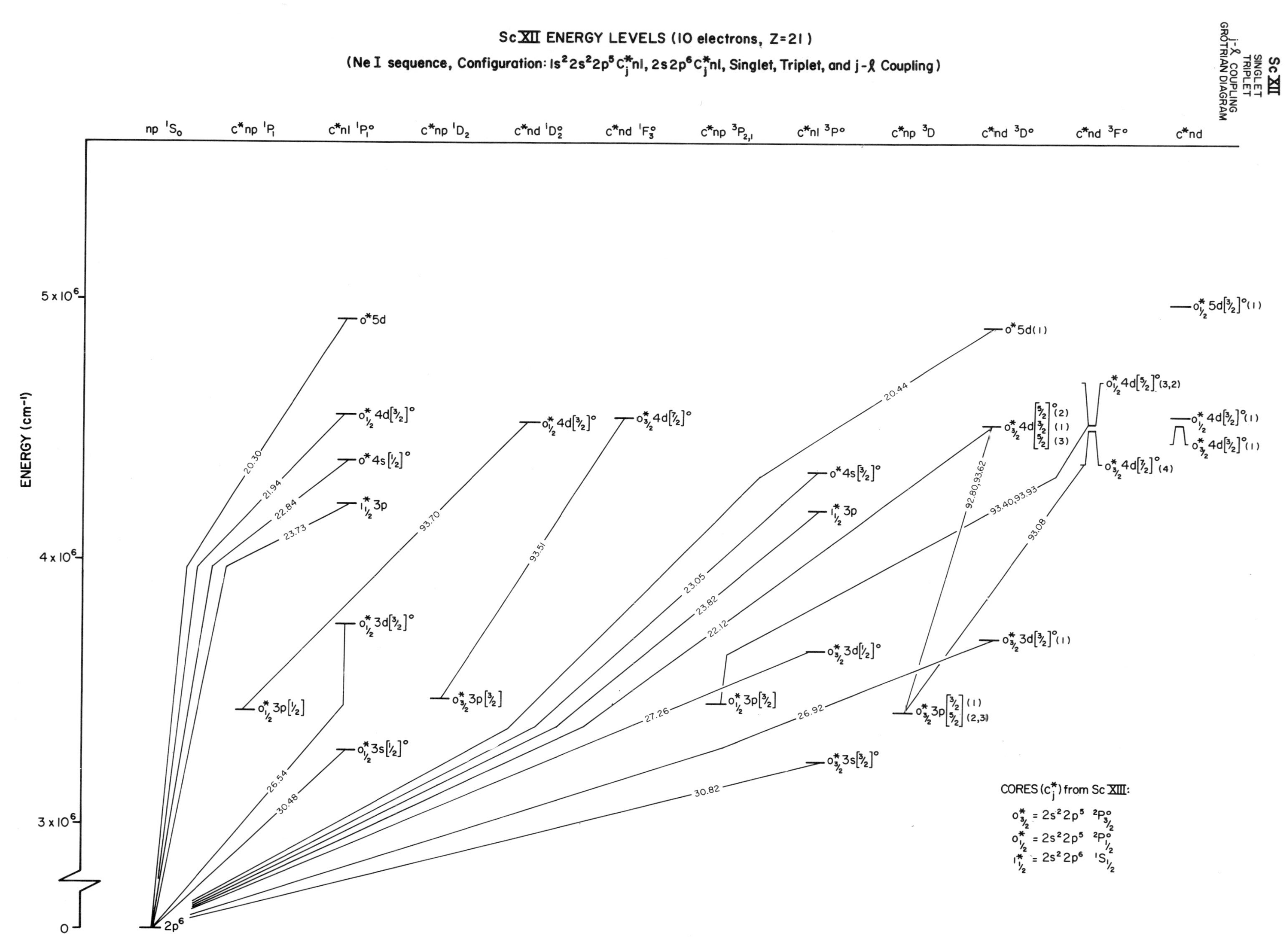

Sc XII
SINGLET
TRIPLET
j-ℓ COUPLING
GROTRIAN DIAGRAM
Sc XII ENERGY LEVELS (10 electrons, Z=21)
(Ne I sequence, Configuration: 1s²2s²2p⁵c*ⱼnl, 2s2p⁶c*ⱼnl, Singlet, Triplet, and j-ℓ Coupling)
np ¹S₀
c*np ¹P₁
c*nl ¹P°₁
c*np ¹D₂
c*nd ¹D°₂
c*nd ¹F°₃
c*np ³P₂,₁
c*nl ³P°
c*np ³D
c*nd ³D°
c*nd ³F°
c*nd
ENERGY (cm⁻¹)
5 x 10⁶
4 x 10⁶
3 x 10⁶
0
2p⁶
o*5d
o*1/2 4d[3/2]°
o*4s[1/2]°
1*1/2 3p
o*1/2 3d[3/2]°
o*1/2 3p[1/2]
o*1/2 3s[1/2]°
o*1/2 4d[3/2]°
o*3/2 4d[7/2]°
o*3/2 3p[3/2]
o*5d(1)
o*4s[3/2]°
1*1/2 3p
o*3/2 3d[1/2]°
o*1/2 3p[3/2]
o*3/2 3s[3/2]°
o*3/2 4d[5/2]°(2) [3/2](1) [5/2](3)
o*3/2 3d[3/2]°(1)
o*3/2 3p[3/2](1) [5/2](2,3)
o*1/2 4d[5/2]°(3,2)
o*3/2 4d[7/2]°(4)
o*1/2 5d[3/2]°(1)
o*1/2 4d[3/2]°(1)
o*3/2 4d[3/2]°(1)
20.30
21.94
22.84
23.73
26.54
30.48
93.70
93.51
20.44
23.05
23.82
22.12
27.26
26.92
30.82
92.80,93.62
93.40,93.93
93.08
CORES (c*ⱼ) from Sc XIII:
o*3/2 = 2s²2p⁵ ²P°3/2
o*1/2 = 2s²2p⁵ ²P°1/2
1*1/2 = 2s²2p⁶ ¹S1/2

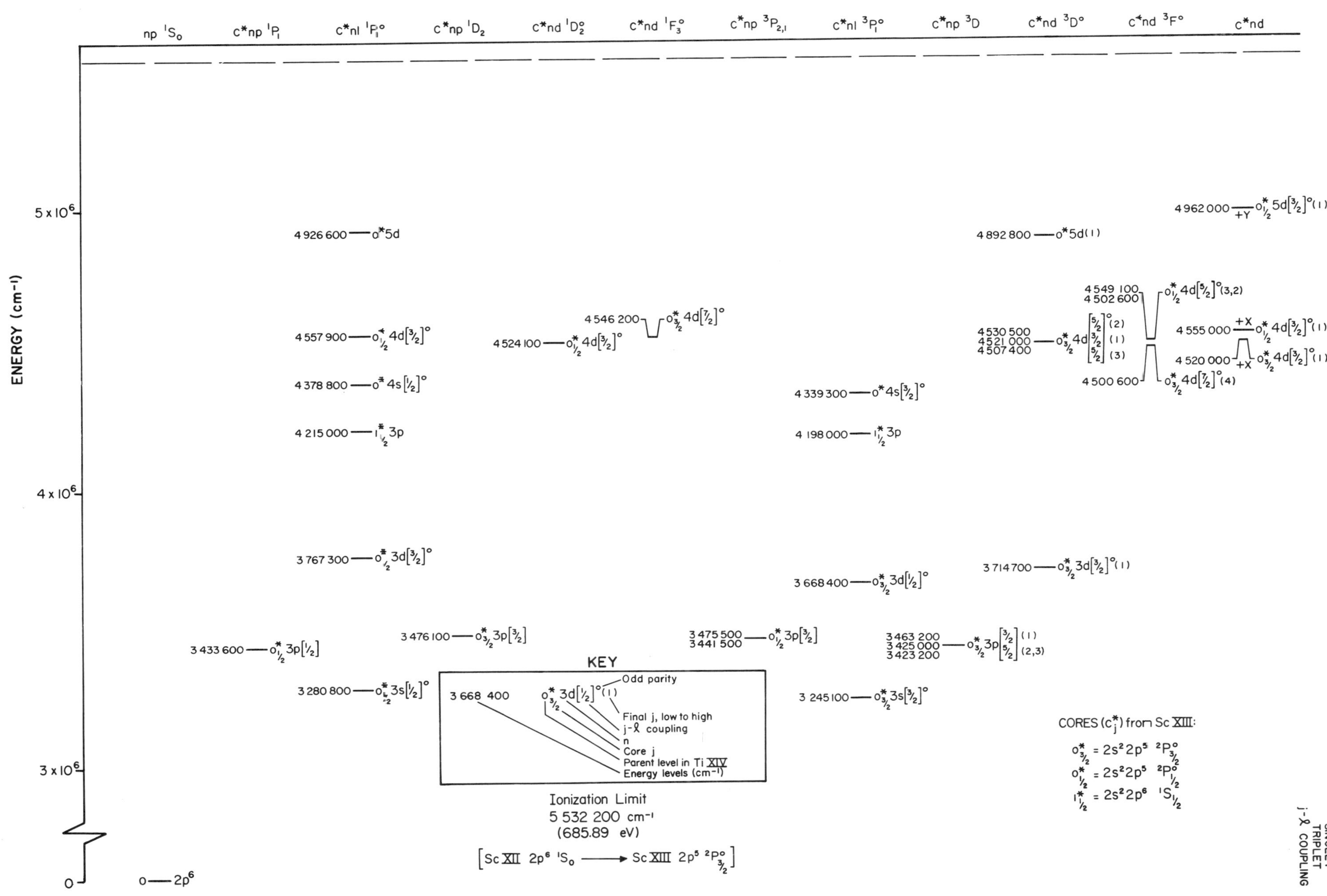
Sc XII ENERGY LEVELS (10 electrons, Z=21)
(Ne I sequence, Configuration: 1s²2s²2p⁵c*ⱼnl, 2s2p⁶c*ⱼnl, Singlet, Triplet, and j-l Coupling)
np ¹S₀
c*np ¹P₁
c*nl ¹F°₁
c*np ¹D₂
c*nd ¹D°₂
c*nd ¹F°₃
c*np ³P₂,₁
c*nl ³P°₁
c*np ³D
c*nd ³D°
c*nd ³F°
c*nd
ENERGY (cm⁻¹)
5 x 10⁶
4 x 10⁶
3 x 10⁶
0
4 962 000 +Y o*₁/₂ 5d[3/2]° (1)
4 926 600 — o*5d
4 892 800 — o*5d (1)
4 549 100 o*₁/₂ 4d[5/2]° (3,2)
4 502 600
4 546 200 o*₃/₂ 4d[7/2]°
4 557 900 — o*₁/₂ 4d[3/2]°
4 524 100 — o*₁/₂ 4d[3/2]°
4 530 500
4 521 000
4 507 400 — o*₃/₂ 4d [5/2]° (2) [3/2] (1) [5/2] (3)
4 555 000 +X o*₁/₂ 4d[3/2]° (1)
4 520 000 +X o*₃/₂ 4d[3/2]° (1)
4 500 600 o*₃/₂ 4d[7/2]° (4)
4 378 800 — o* 4s[1/2]°
4 339 300 — o* 4s[3/2]°
4 215 000 — 1*₁/₂ 3p
4 198 000 — 1*₁/₂ 3p
3 767 300 — o*₁/₂ 3d[3/2]°
3 714 700 — o*₃/₂ 3d[3/2]° (1)
3 668 400 — o*₃/₂ 3d[1/2]°
3 476 100 — o*₃/₂ 3p[3/2]
3 475 500
3 441 500 — o*₁/₂ 3p[3/2]
3 433 600 — o*₁/₂ 3p[1/2]
3 463 200
3 425 000
3 423 200 — o*₃/₂ 3p [3/2] (1) [5/2] (2,3)
3 280 800 — o*₁/₂ 3s[1/2]°
3 245 100 — o*₃/₂ 3s[3/2]°
0 — 2p⁶
KEY
3 668 400 o*₃/₂ 3d[1/2]° (1)
Odd parity
Final j, low to high
j-l coupling
n
Core j
Parent level in Ti XIV
Energy levels (cm⁻¹)
Ionization Limit
5 532 200 cm⁻¹
(685.89 eV)
[Sc XII 2p⁶ ¹S₀ ⟶ Sc XIII 2p⁵ ²P°₃/₂]
CORES (c*ⱼ) from Sc XIII:
o*₃/₂ = 2s²2p⁵ ²P°₃/₂
o*₁/₂ = 2s²2p⁵ ²P°₁/₂
1*₁/₂ = 2s2p⁶ ²S₁/₂
Sc XII
SINGLET
TRIPLET
j-l COUPLING

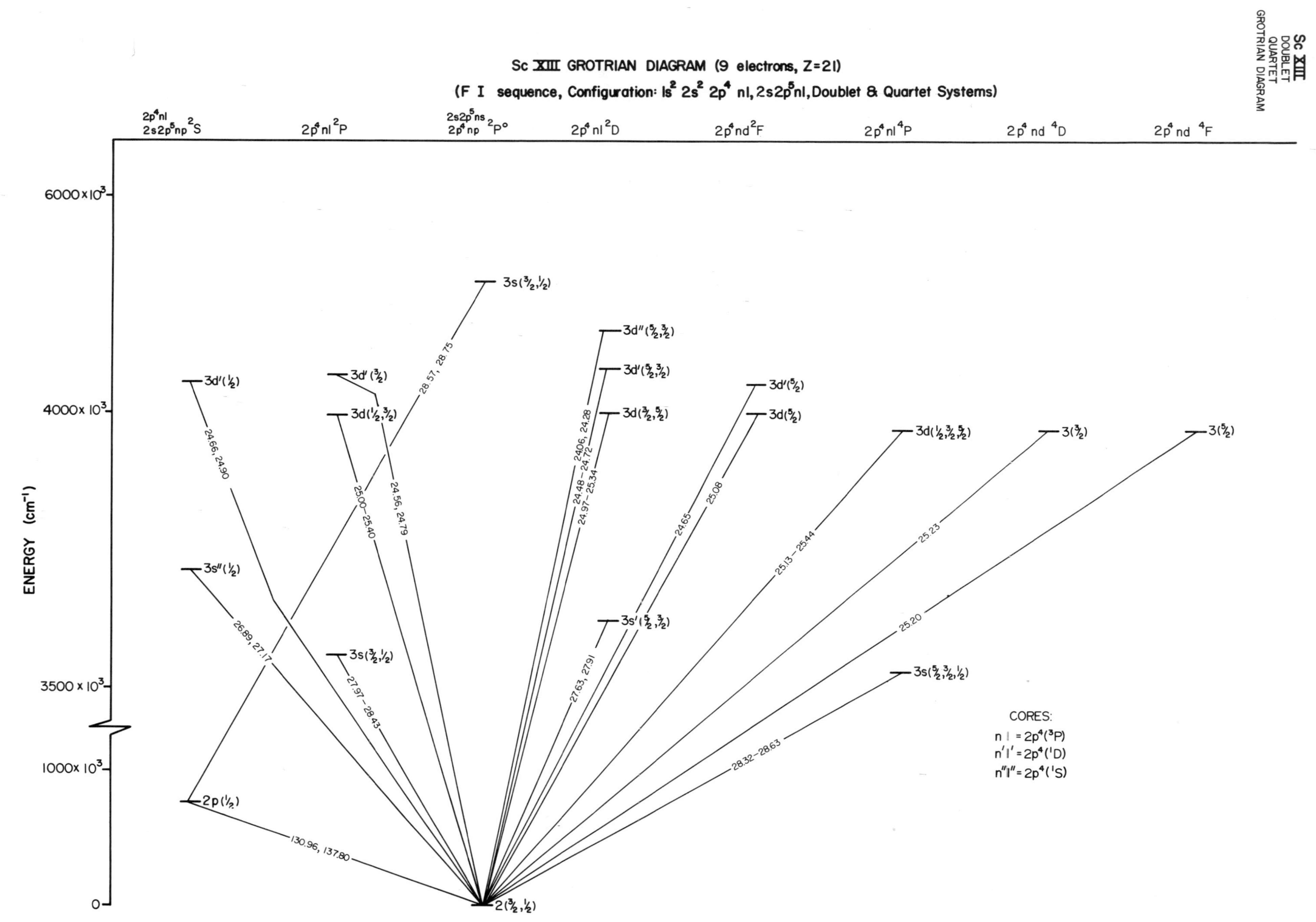

Sc XIII
DOUBLET
QUARTET
GROTRIAN DIAGRAM
Sc XIII GROTRIAN DIAGRAM (9 electrons, Z=21)
(F I sequence, Configuration: ls² 2s² 2p⁴ nl, 2s2p⁵nl, Doublet & Quartet Systems)
2p⁴nl 2s2p⁵np ²S
2p⁴nl ²P
2s2p⁵ns 2p⁴np ²P°
2p⁴nl ²D
2p⁴nd ²F
2p⁴nl ⁴P
2p⁴nd ⁴D
2p⁴nd ⁴F
ENERGY (cm⁻¹)
6000×10³
4000×10³
3500×10³
1000×10³
0
3s(3/2,1/2)
28.57, 28.75
3d″(5/2,3/2)
3d′(5/2,3/2)
3d(3/2,5/2)
2406, 24.28
24.48 – 24.72
24.97 – 25.34
3d′(1/2)
24.66, 24.90
3d′(3/2)
3d(1/2,3/2)
25.00 – 25.40
24.56, 24.79
3d′(5/2)
3d(5/2)
24.65
25.08
3d(1/2,3/2,5/2)
25.13 – 25.44
3(3/2)
25.23
3(5/2)
25.20
3s″(1/2)
26.89, 27.17
3s′(5/2,3/2)
27.63, 27.91
3s(3/2,1/2)
27.97 – 28.43
3s(5/2,3/2,1/2)
28.32 – 28.63
CORES:
n l = 2p⁴(³P)
n′l′ = 2p⁴(¹D)
n″l″ = 2p⁴(¹S)
2p(1/2)
130.96, 137.80
2(3/2,1/2)

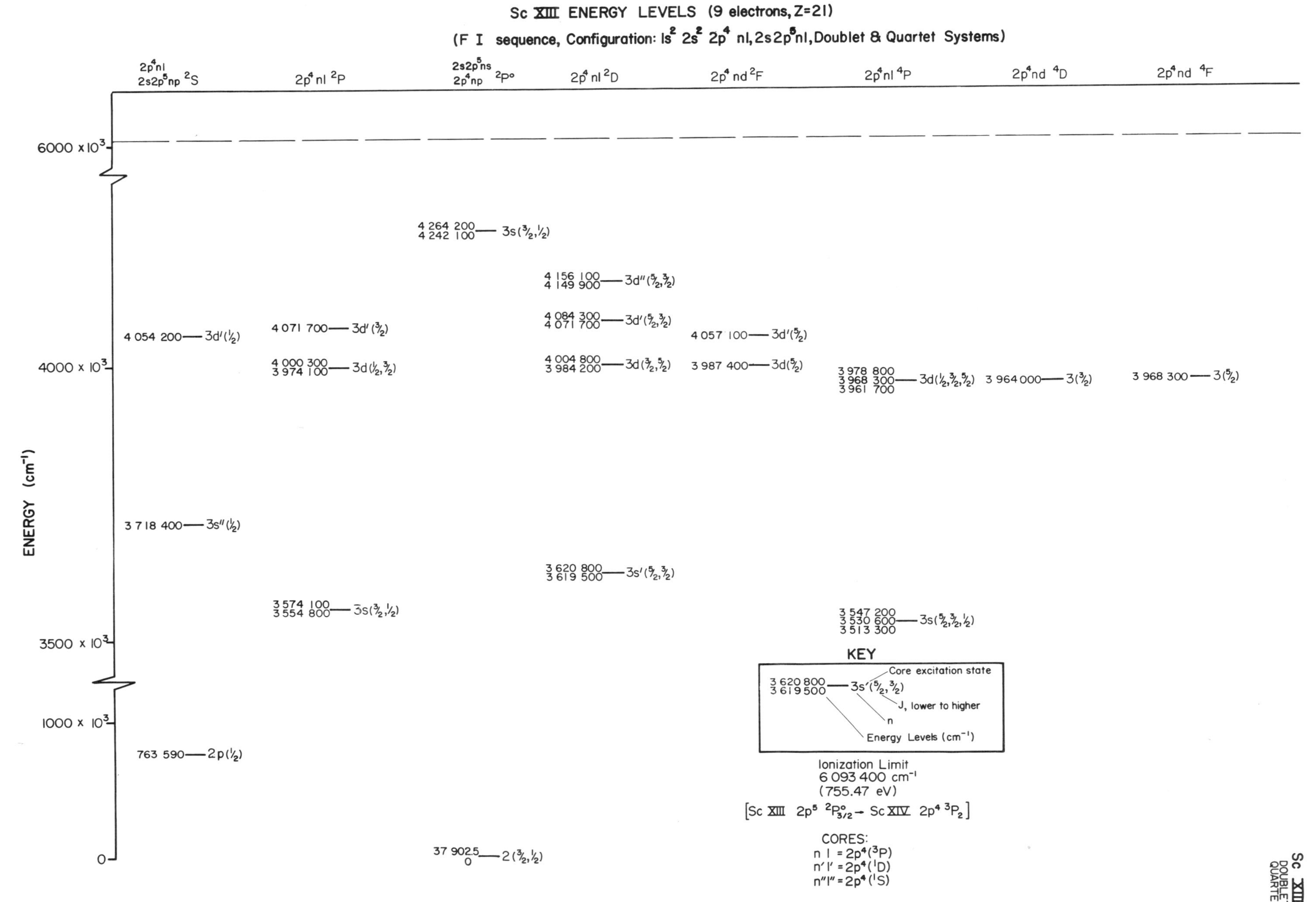

Sc XIII ENERGY LEVELS (9 electrons, Z=21)
(F I sequence, Configuration: $1s^2 2s^2 2p^4$ nl, $2s2p^5$nl, Doublet & Quartet Systems)
$2p^4$nl $2s2p^5$np ^{2}S
$2p^4$nl ^{2}P
$2s2p^5$ns $2p^4$np ^{2}P°
$2p^4$nl ^{2}D
$2p^4$nd ^{2}F
$2p^4$nl ^{4}P
$2p^4$nd ^{4}D
$2p^4$nd ^{4}F
6000 x 10³
4000 x 10³
3500 x 10³
1000 x 10³
0
ENERGY (cm⁻¹)
4 264 200
4 242 100 — 3s(3/2, 1/2)
4 156 100
4 149 900 — 3d''(5/2, 3/2)
4 084 300
4 071 700 — 3d'(5/2, 3/2)
4 054 200 — 3d'(1/2)
4 071 700 — 3d'(3/2)
4 057 100 — 3d'(5/2)
4 000 300
3 974 100 — 3d(1/2, 3/2)
4 004 800
3 984 200 — 3d(3/2, 5/2)
3 987 400 — 3d(5/2)
3 978 800
3 968 300
3 961 700 — 3d(1/2, 3/2, 5/2)
3 964 000 — 3(7/2)
3 968 300 — 3(9/2)
3 718 400 — 3s''(1/2)
3 620 800
3 619 500 — 3s'(5/2, 3/2)
3 574 100
3 554 800 — 3s(3/2, 1/2)
3 547 200
3 530 600
3 513 300 — 3s(5/2, 3/2, 1/2)
763 590 — 2p(1/2)
37 902.5
0 — 2(3/2, 1/2)
KEY
3 620 800
3 619 500 — 3s'(5/2, 3/2)
Core excitation state
J, lower to higher
n
Energy Levels (cm⁻¹)
Ionization Limit
6 093 400 cm⁻¹
(755.47 eV)
[Sc XIII $2p^5$ $^2P^o_{3/2}$ → Sc XIV $2p^4$ $^3P_2$]
CORES:
n l = $2p^4$(^{3}P)
n'l' = $2p^4$(^{1}D)
n''l'' = $2p^4$(^{1}S)
Sc XIII
DOUBLET
QUARTET

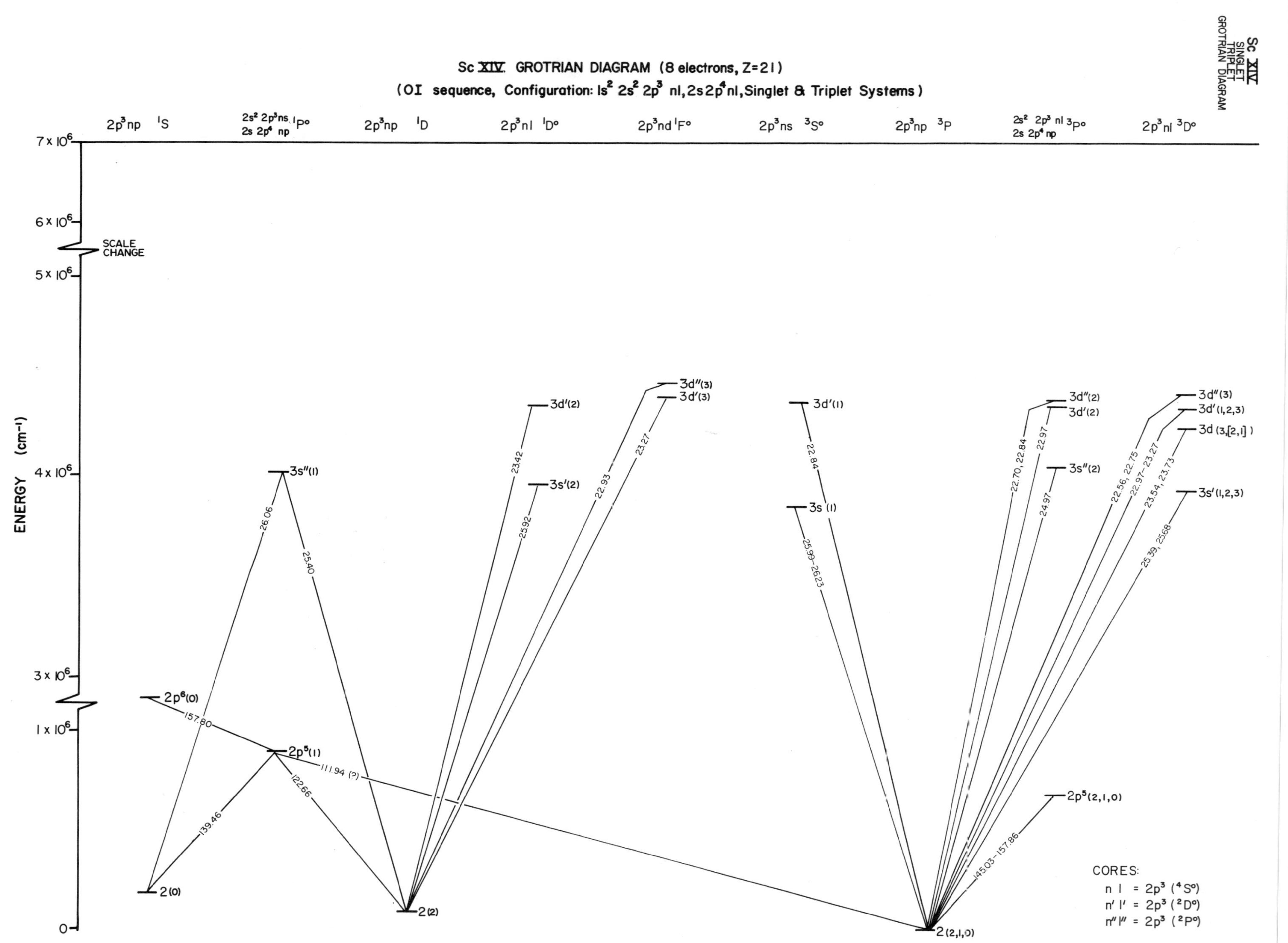
Sc XIV
SINGLET
TRIPLET
GROTRIAN DIAGRAM
Sc XIV GROTRIAN DIAGRAM (8 electrons, Z=21)
(OI sequence, Configuration: $1s^2 2s^2 2p^3$ nl, $2s2p^4$nl, Singlet & Triplet Systems)
$2p^3np$ ^1S
$2s^2 2p^3ns$ $^1P^o$
$2s\,2p^4$ np
$2p^3np$ ^1D
$2p^3nl$ $^1D^o$
$2p^3nd$ $^1F^o$
$2p^3ns$ $^3S^o$
$2p^3np$ ^3P
$2s^2$ $2p^3$ nl $^3P^o$
$2s\,2p^4$ np
$2p^3nl$ $^3D^o$
7×10^6
6×10^6
SCALE CHANGE
5×10^6
4×10^6
3×10^6
1×10^6
0
ENERGY (cm⁻¹)
$2p^6(0)$
157.80
$2p^5(1)$
139.46
2(0)
$3s''(1)$
26.06
25.40
122.66
111.94 (?)
2(2)
$3d'(2)$
23.42
$3s'(2)$
25.92
$3d''(3)$
$3d'(3)$
22.93
23.27
$3d'(1)$
22.84
3s (1)
25.99-26.23
$3d''(2)$
$3d'(2)$
22.70, 22.84
22.97
$3s''(2)$
24.97
$3d''(3)$
$3d'(1,2,3)$
3d (3,[2,1])
$3s'(1,2,3)$
22.56, 22.75
22.97-23.27
23.54, 23.73
25.39, 25.68
$2p^5(2,1,0)$
145.03-157.86
2(2,1,0)
CORES:
n l = $2p^3$ ($^4S^o$)
n' l' = $2p^3$ ($^2D^o$)
n'' l'' = $2p^3$ ($^2P^o$)

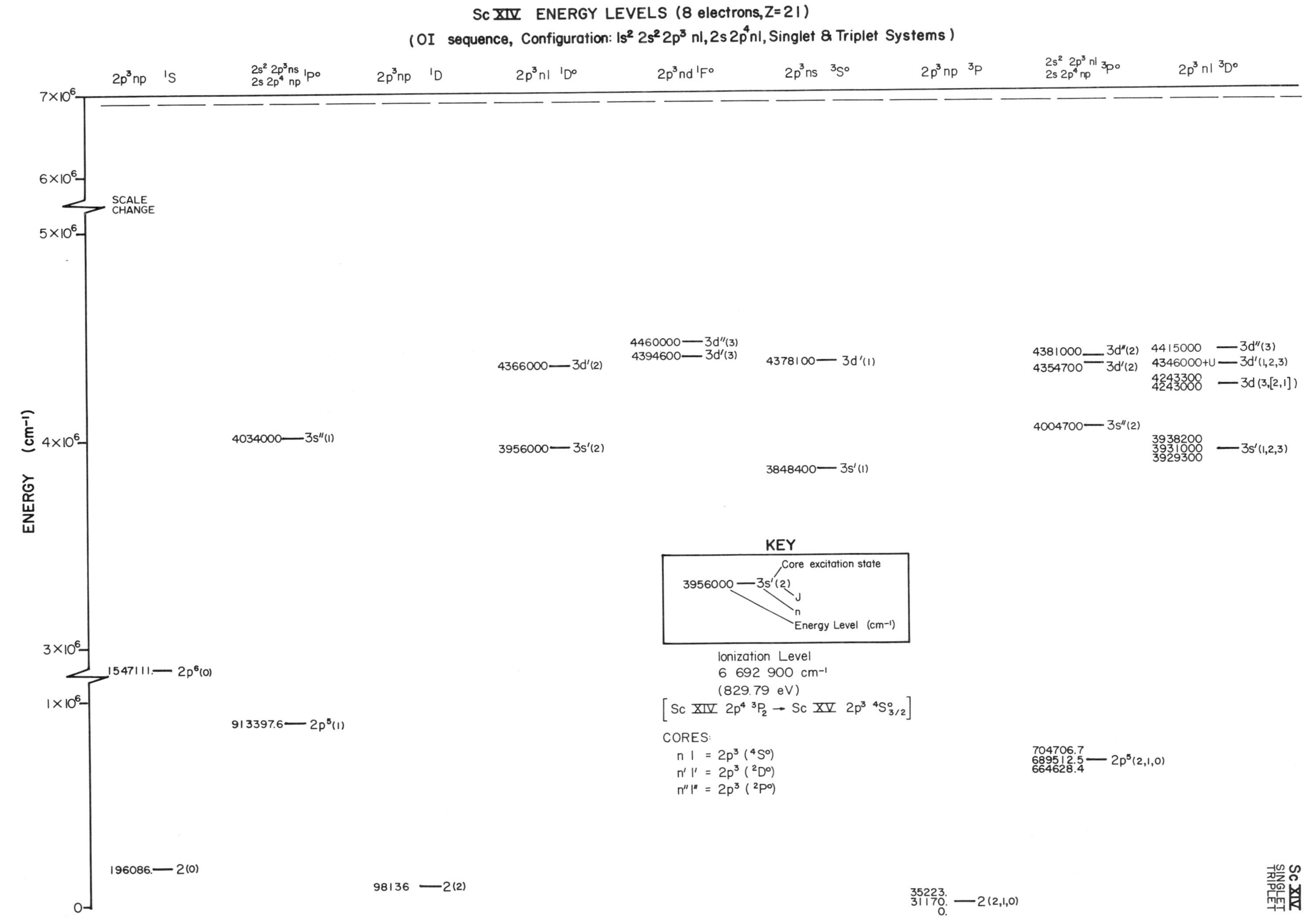

Sc XIV ENERGY LEVELS (8 electrons, Z=21)
(OI sequence, Configuration: 1s² 2s² 2p³ nl, 2s 2p⁴ nl, Singlet & Triplet Systems)
2p³np ¹S
2s² 2p³ns ¹P°
2s 2p⁴ np
2p³np ¹D
2p³nl ¹D°
2p³nd ¹F°
2p³ns ³S°
2p³np ³P
2s² 2p³ nl ³P°
2s 2p⁴ np
2p³ nl ³D°
7×10⁶
6×10⁶
SCALE CHANGE
5×10⁶
4×10⁶
3×10⁶
1×10⁶
0
ENERGY (cm⁻¹)
1547111. — 2p⁶(0)
196086. — 2(0)
4034000 — 3s″(1)
913397.6 — 2p⁵(1)
98136 — 2(2)
4366000 — 3d′(2)
3956000 — 3s′(2)
4460000 — 3d″(3)
4394600 — 3d′(3)
4378100 — 3d′(1)
3848400 — 3s′(1)
35223.
31170. — 2(2,1,0)
0.
4381000 — 3d″(2)
4354700 — 3d′(2)
4004700 — 3s″(2)
704706.7
689512.5 — 2p⁵(2,1,0)
664628.4
4415000 — 3d″(3)
4346000+U — 3d′(1,2,3)
4243300
4243000 — 3d (3,[2,1])
3938200
3931000 — 3s′(1,2,3)
3929300
KEY
3956000 — 3s′(2)
Core excitation state
J
n
Energy Level (cm⁻¹)
Ionization Level
6 692 900 cm⁻¹
(829.79 eV)
[Sc XIV 2p⁴ ³P₂ → Sc XV 2p³ ⁴S°3/2]
CORES:
n l = 2p³ (⁴S°)
n′ l′ = 2p³ (²D°)
n″ l″ = 2p³ (²P°)
Sc XIV
SINGLET
TRIPLET

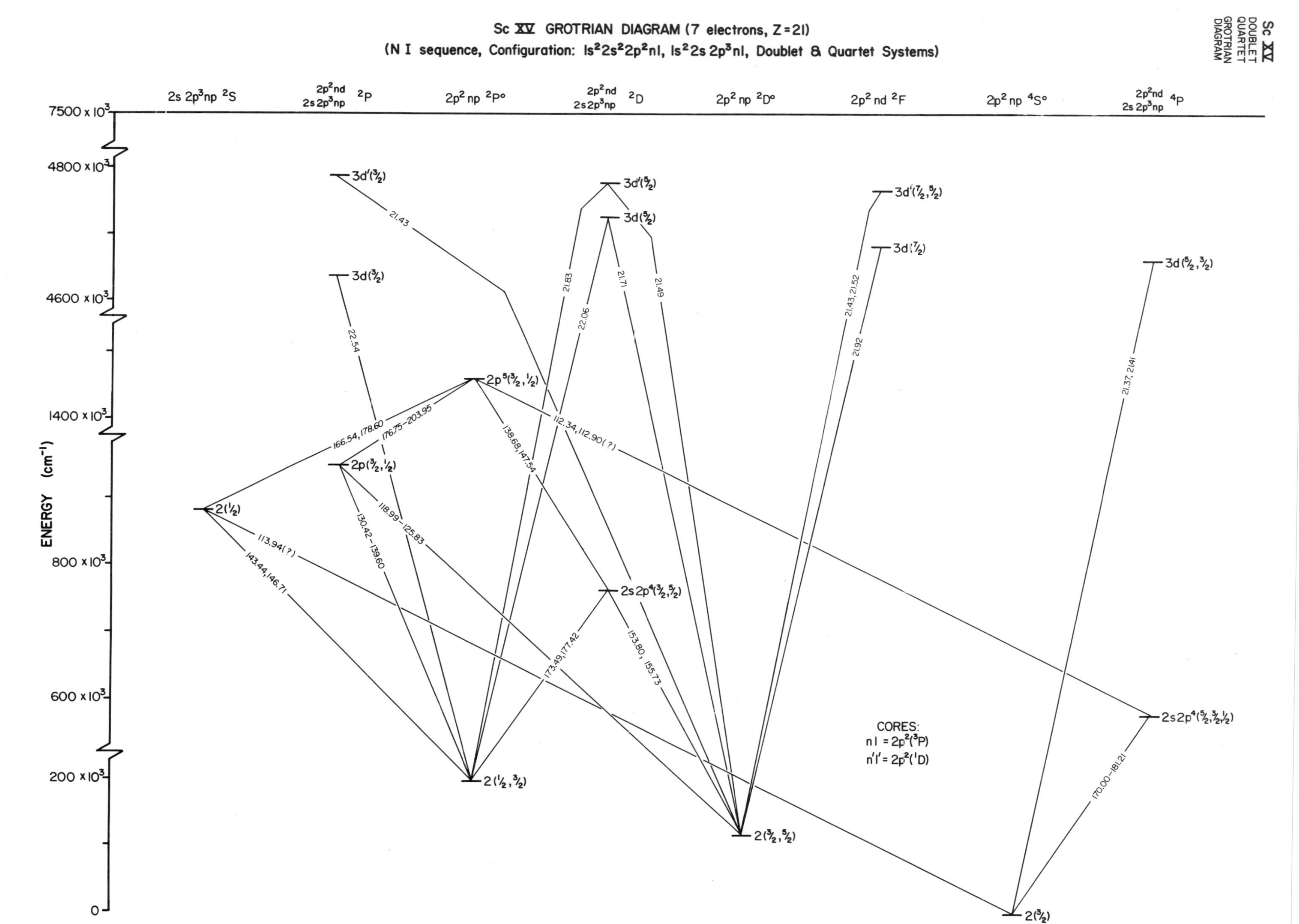

Sc XV
DOUBLET
QUARTET
GROTRIAN
DIAGRAM
Sc XV GROTRIAN DIAGRAM (7 electrons, Z = 21)
(N I sequence, Configuration: $1s^2 2s^2 2p^2 nl$, $1s^2 2s\, 2p^3 nl$, Doublet & Quartet Systems)
$2s\,2p^3np\ ^2S$
$2p^2nd$ / $2s\,2p^3np$ ^2P
$2p^2\,np\ ^2P^o$
$2p^2nd$ / $2s\,2p^3np$ ^2D
$2p^2\,np\ ^2D^o$
$2p^2\,nd\ ^2F$
$2p^2\,np\ ^4S^o$
$2p^2nd$ / $2s\,2p^3np$ ^4P
ENERGY (cm^{-1})
7500×10^3
4800×10^3
4600×10^3
1400×10^3
800×10^3
600×10^3
200×10^3
0
$3d'(3/2)$
$3d(3/2)$
$2p^5(3/2, 1/2)$
$2p(3/2, 1/2)$
$2(1/2)$
$2(1/2, 3/2)$
$3d'(5/2)$
$3d(5/2)$
$2s2p^4(3/2, 5/2)$
$2(3/2, 5/2)$
$3d'(7/2, 5/2)$
$3d(7/2)$
$3d(5/2, 3/2)$
$2s2p^4(5/2, 3/2, 1/2)$
$2(3/2)$
21.43
22.54
166.54, 178.60
176.75–203.95
130.42–139.60
118.99–125.83
113.94(?)
143.44, 146.71
21.83
22.06
21.71
21.49
138.68, 147.54
112.34, 112.90(?)
173.49, 177.42
153.80, 155.73
21.43, 21.52
21.92
21.37, 21.41
170.00–181.21
CORES:
$nl = 2p^2(^3P)$
$n'l' = 2p^2(^1D)$

Sc XV ENERGY LEVELS (7 electrons, Z = 21)

(N I sequence, Configuration: $1s^2 2s^2 2p^2 nl$, $1s^2 2s\, 2p^3 nl$, Doublet & Quartet Systems)

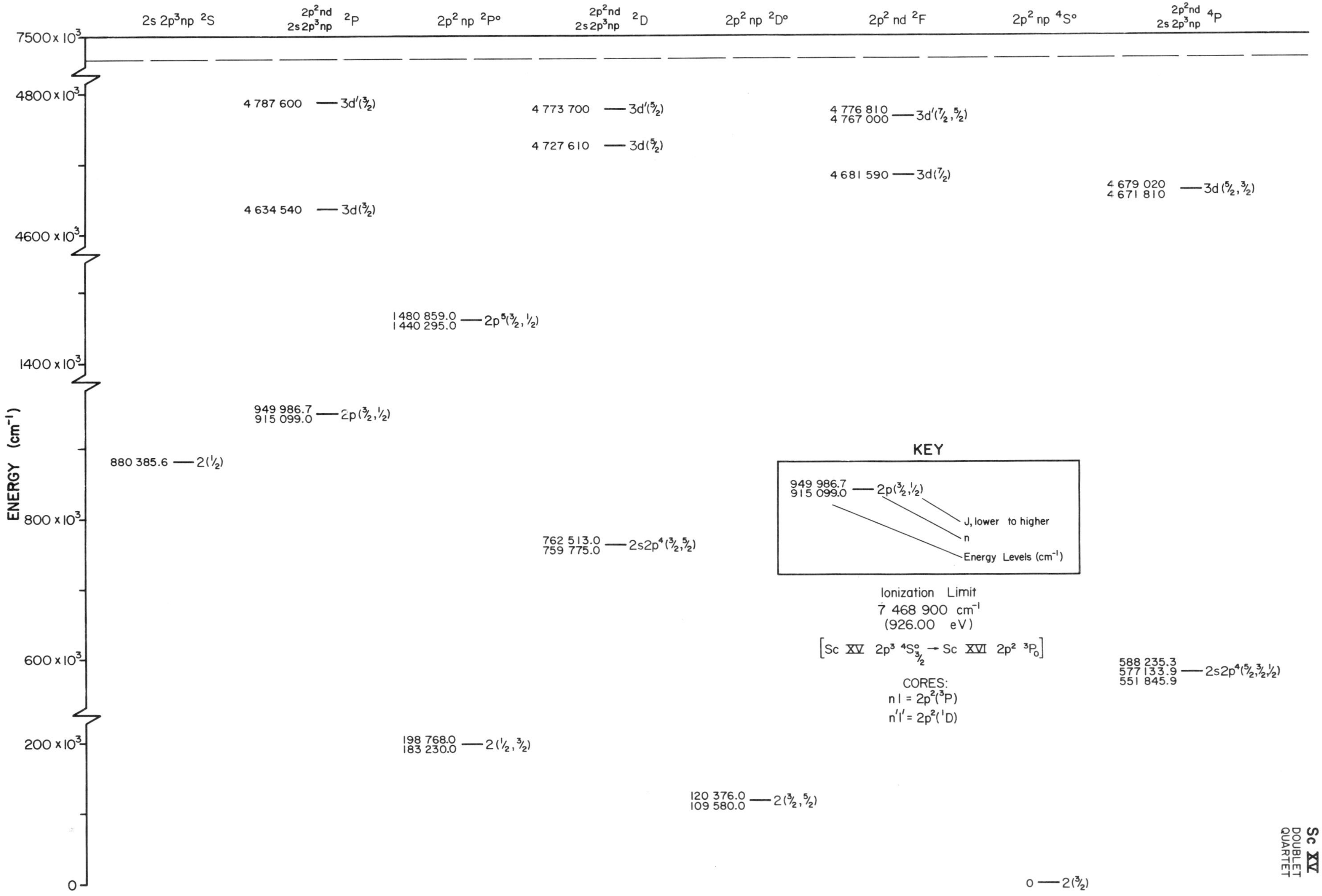

ENERGY (cm^{-1})

Axis: 7500 x 10^3; 4800 x 10^3; 4600 x 10^3; 1400 x 10^3; 800 x 10^3; 600 x 10^3; 200 x 10^3; 0

$2s\,2p^3np$ ^2S

880 385.6 — $2(\frac{1}{2})$

$2p^2nd$ / $2s\,2p^3np$ ^2P

4 787 600 — $3d'(\frac{3}{2})$

4 634 540 — $3d(\frac{3}{2})$

949 986.7, 915 099.0 — $2p(\frac{3}{2},\frac{1}{2})$

$2p^2\,np$ $^2P^o$

1 480 859.0, 1 440 295.0 — $2p^5(\frac{3}{2},\frac{1}{2})$

198 768.0, 183 230.0 — $2(\frac{1}{2},\frac{3}{2})$

$2p^2nd$ / $2s\,2p^3np$ ^2D

4 773 700 — $3d'(\frac{5}{2})$

4 727 610 — $3d(\frac{5}{2})$

762 513.0, 759 775.0 — $2s2p^4(\frac{3}{2},\frac{5}{2})$

$2p^2\,np$ $^2D^o$

120 376.0, 109 580.0 — $2(\frac{3}{2},\frac{5}{2})$

$2p^2\,nd$ ^2F

4 776 810, 4 767 000 — $3d'(\frac{7}{2},\frac{5}{2})$

4 681 590 — $3d(\frac{7}{2})$

$2p^2\,np$ $^4S^o$

0 — $2(\frac{3}{2})$

$2p^2nd$ / $2s\,2p^3np$ ^4P

4 679 020, 4 671 810 — $3d(\frac{5}{2},\frac{3}{2})$

588 235.3, 577 133.9, 551 845.9 — $2s2p^4(\frac{5}{2},\frac{3}{2},\frac{1}{2})$

KEY

949 986.7, 915 099.0 — $2p(\frac{3}{2},\frac{1}{2})$: J, lower to higher; n; Energy Levels (cm^{-1})

Ionization Limit
7 468 900 cm^{-1}
(926.00 eV)

[Sc XV $2p^3\ ^4S^o_{3/2}$ → Sc XVI $2p^2\ ^3P_0$]

CORES:
$nl = 2p^2(^3P)$
$n'l' = 2p^2(^1D)$

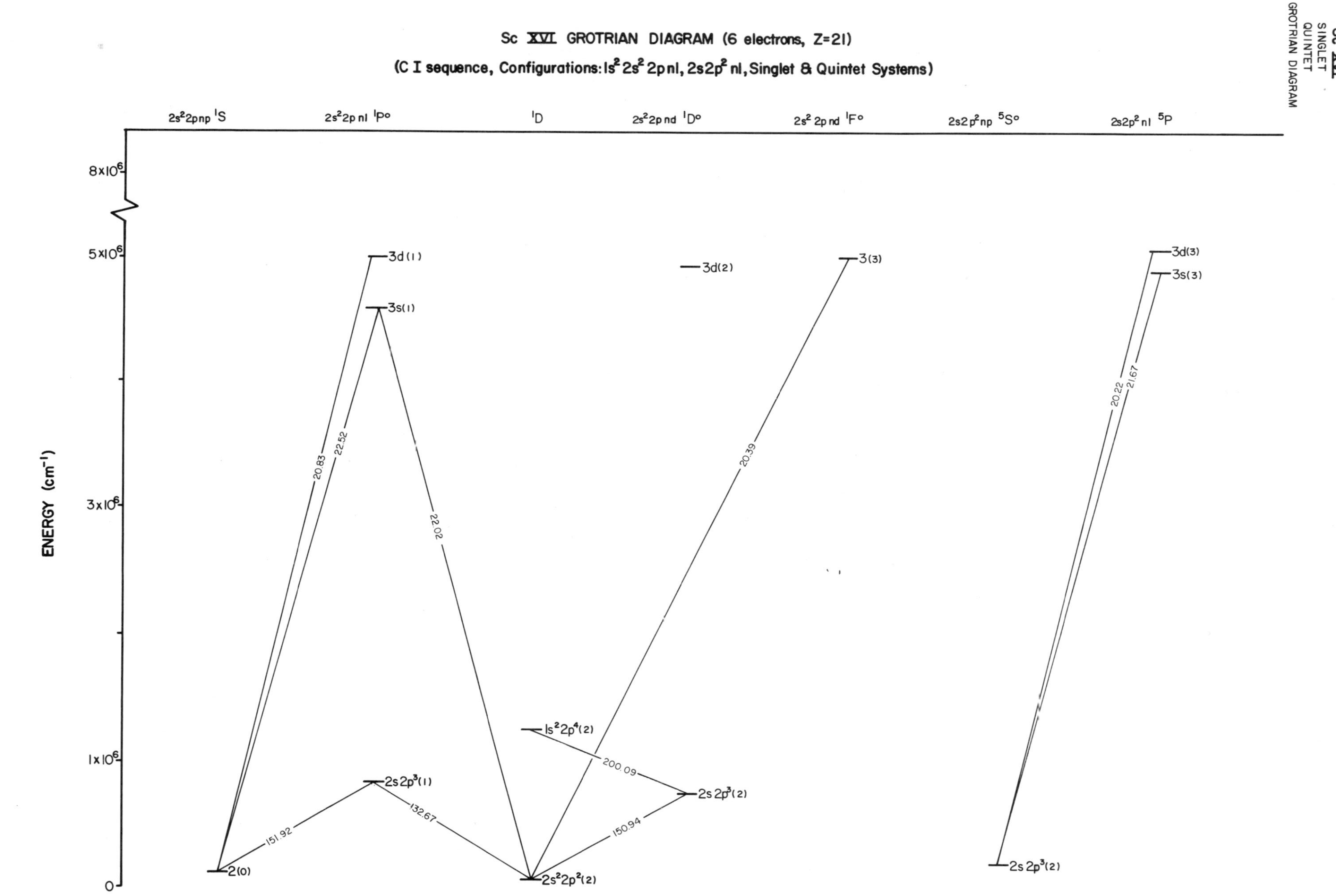
Sc XVI GROTRIAN DIAGRAM (6 electrons, Z=21)
(C I sequence, Configurations: $1s^22s^22pnl$, $2s2p^2nl$, Singlet & Quintet Systems)
$2s^22pnp$ ^1S
$2s^22pnl$ $^1P^o$
^1D
$2s^22pnd$ $^1D^o$
$2s^22pnd$ $^1F^o$
$2s2p^2np$ $^5S^o$
$2s2p^2nl$ ^5P
ENERGY (cm^{-1})
8×10^6
5×10^6
3×10^6
1×10^6
0
3d(1)
3s(1)
3d(2)
3(3)
3d(3)
3s(3)
20.83
22.52
22.02
20.39
20.22
21.67
$1s^22p^4(2)$
200.09
$2s2p^3(1)$
$2s2p^3(2)$
151.92
132.67
150.94
2(0)
$2s^22p^2(2)$
$2s2p^3(2)$

Sc XVI ENERGY LEVELS (6 electrons, Z=21)

(C I sequence, Configurations: $1s^2 2s^2 2p\,nl$, $2s2p^2 nl$, Singlet & Quintet Systems)

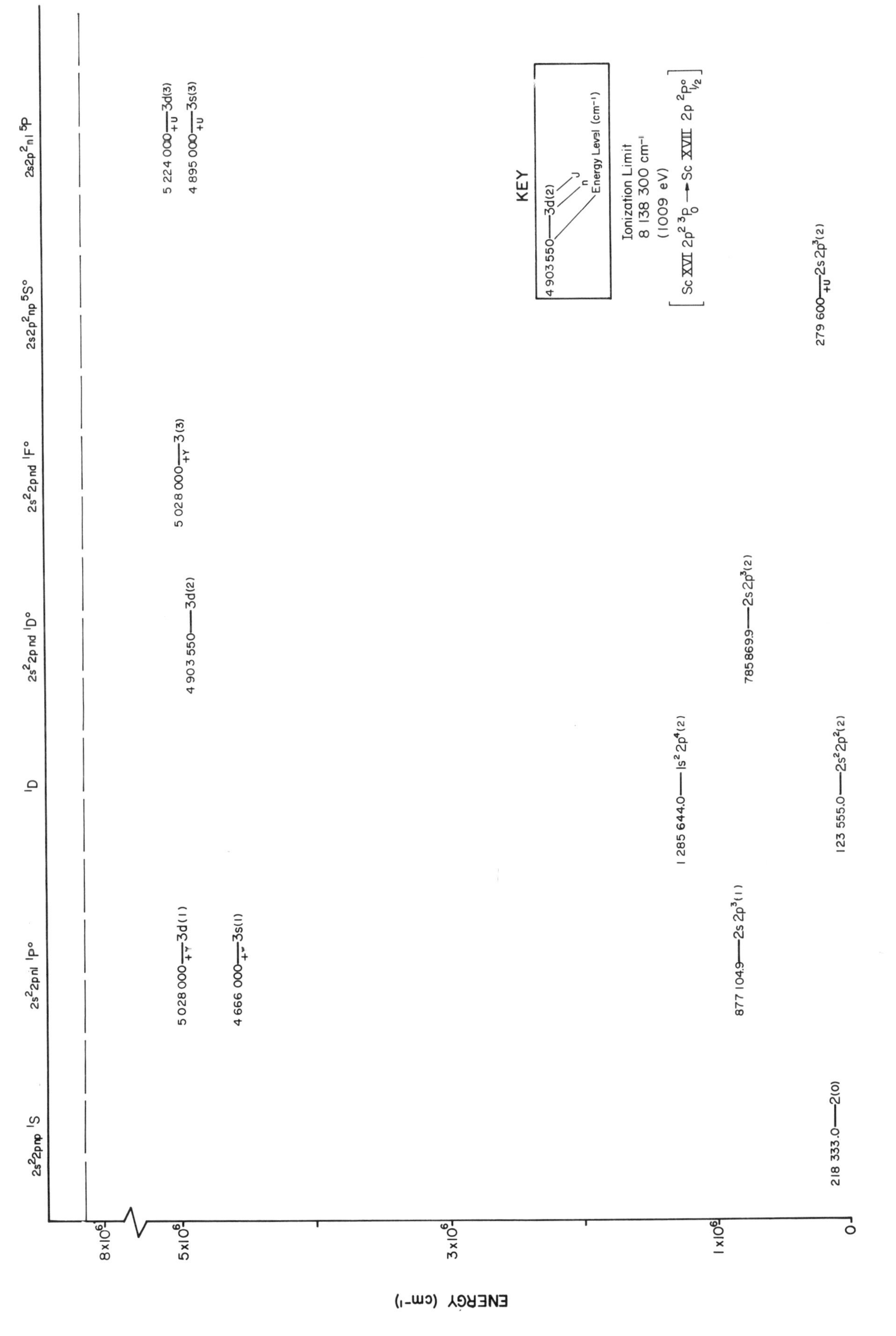

$2s^2 2pnp\ ^1S$

218 333.0 — 2(0)

$2s^2 2pnl\ ^1P°$

5 028 000 +Y — 3d(1)

4 666 000 +ν — 3s(1)

877 104.9 — $2s2p^3$(1)

^1D

1 285 644.0 — $1s^2 2p^4$(2)

123 555.0 — $2s^2 2p^2$(2)

$2s^2 2p\,nd\ ^1D°$

4 903 550 — 3d(2)

785 869.9 — $2s2p^3$(2)

$2s^2 2p\,nd\ ^1F°$

5 028 000 +Y — 3(3)

$2s2p^2 np\ ^5S°$

279 600 +U — $2s2p^3$(2)

$2s2p^2 nl\ ^5P$

5 224 000 +U — 3d(3)

4 895 000 +U — 3s(3)

ENERGY (cm⁻¹): 0, 1x10⁶, 3x10⁶, 5x10⁶, 8x10⁶

KEY

4 903 550 — 3d(2): Energy Level (cm⁻¹), n, J

Ionization Limit
8 138 300 cm⁻¹
(1009 eV)
[Sc XVI $2p^2\ ^3P_0$ → Sc XVII $2p\ ^2P°_{1/2}$]

Sc XVI
TRIPLET
GROTRIAN
DIAGRAM

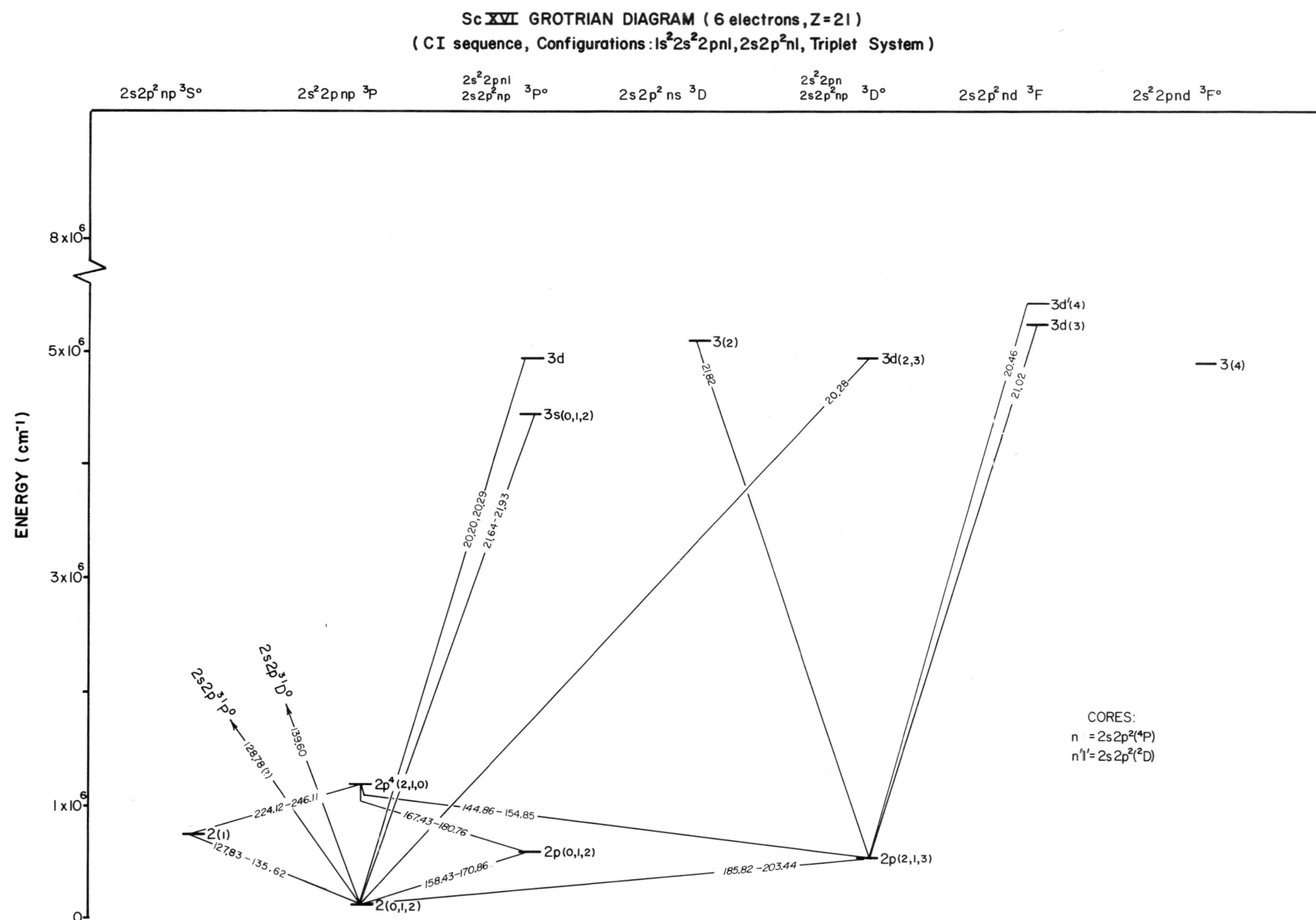

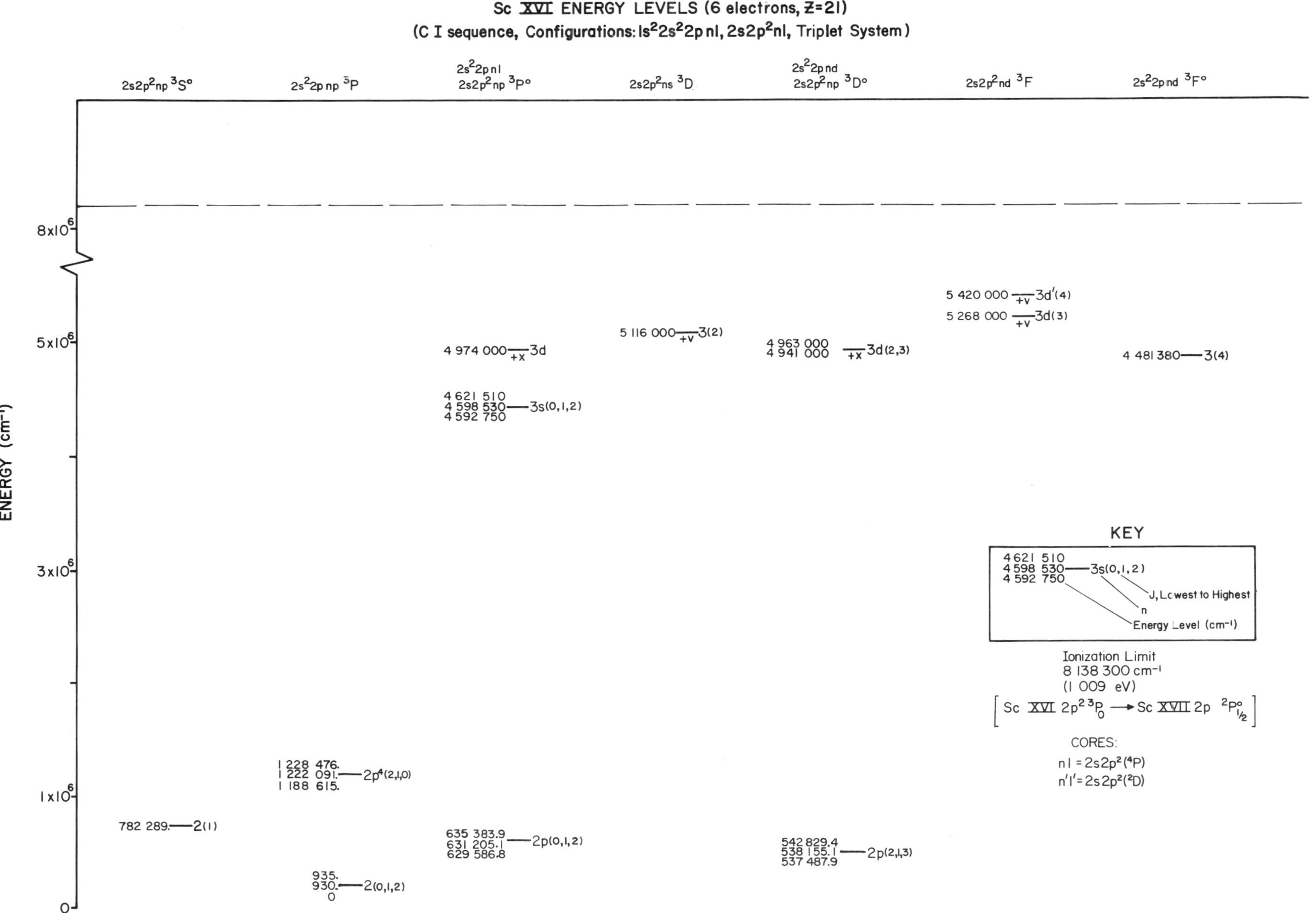
Sc XVI ENERGY LEVELS (6 electrons, Z=21)
(C I sequence, Configurations: 1s²2s²2pnl, 2s2p²nl, Triplet System)
2s2p²np ³S°
2s²2pnp ³P
2s²2pnl
2s2p²np ³P°
2s2p²ns ³D
2s²2pnd
2s2p²np ³D°
2s2p²nd ³F
2s²2pnd ³F°
8x10⁶
5x10⁶
3x10⁶
1x10⁶
0
ENERGY (cm⁻¹)
5 420 000 —3d′(4)
+v
5 268 000 —3d(3)
+v
5 116 000—3(2)
+v
4 974 000—3d
+x
4 963 000
4 941 000
—3d(2,3)
+x
4 481 380—3(4)
4 621 510
4 598 530—3s(0,1,2)
4 592 750
KEY
4 621 510
4 598 530—3s(0,1,2)
4 592 750
J, Lowest to Highest
n
Energy Level (cm⁻¹)
Ionization Limit
8 138 300 cm⁻¹
(1 009 eV)
[Sc XVI 2p² ³P₀ → Sc XVII 2p ²P°½]
CORES:
nl = 2s2p²(⁴P)
n′l′ = 2s2p²(²D)
1 228 476.
1 222 091.—2p⁴(2,1,0)
1 188 615.
782 289.—2(1)
635 383.9
631 205.1—2p(0,1,2)
629 586.8
542 829.4
538 155.1—2p(2,1,3)
537 487.9
935.
930.—2(0,1,2)
0

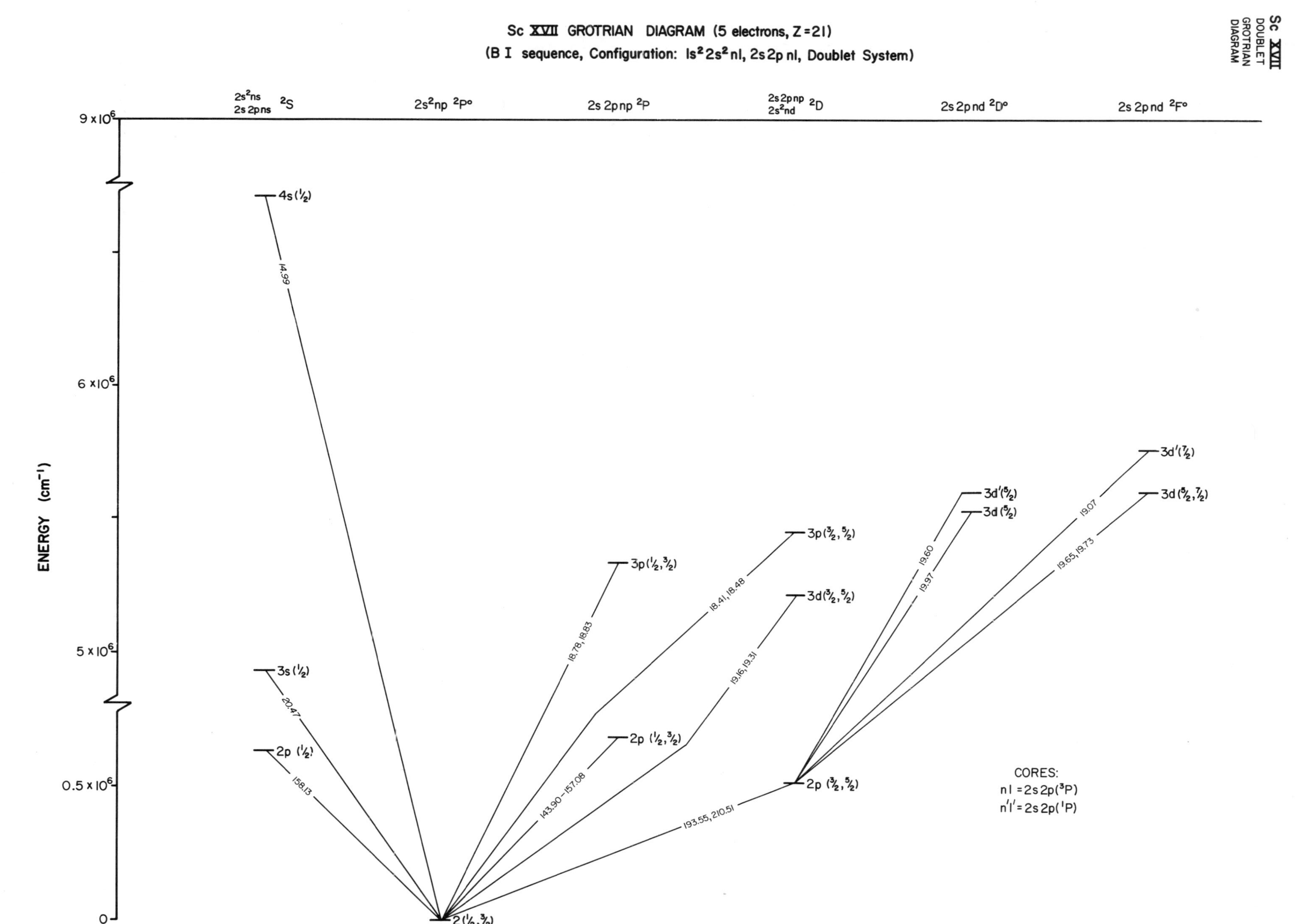
Sc XVII GROTRIAN DIAGRAM (5 electrons, Z = 21)
(B I sequence, Configuration: 1s²2s²nl, 2s2p nl, Doublet System)
2s²ns 2s2pns ²S
2s²np ²P°
2s2pnp ²P
2s2pnp 2s²nd ²D
2s2pnd ²D°
2s2pnd ²F°
4s (1/2)
3s (1/2)
2p (1/2)
2 (1/2, 3/2)
3p (1/2, 3/2)
2p (1/2, 3/2)
3p (3/2, 5/2)
3d (3/2, 5/2)
2p (3/2, 5/2)
3d' (5/2)
3d (5/2)
3d' (7/2)
3d (5/2, 7/2)
14.99
20.47
158.13
18.78, 18.83
143.90–157.08
18.41, 18.48
19.16, 19.31
193.55, 210.51
19.60
19.97
19.07
19.65, 19.73
CORES:
nl = 2s2p(³P)
n'l' = 2s2p(¹P)
ENERGY (cm⁻¹)
9 x 10⁶
6 x 10⁶
5 x 10⁶
0.5 x 10⁶
0

Sc XVII ENERGY LEVELS (5 electrons, Z = 21)

(B I sequence, Configuration: $1s^2 2s^2 nl$, $2s2p\,nl$, Doublet System)

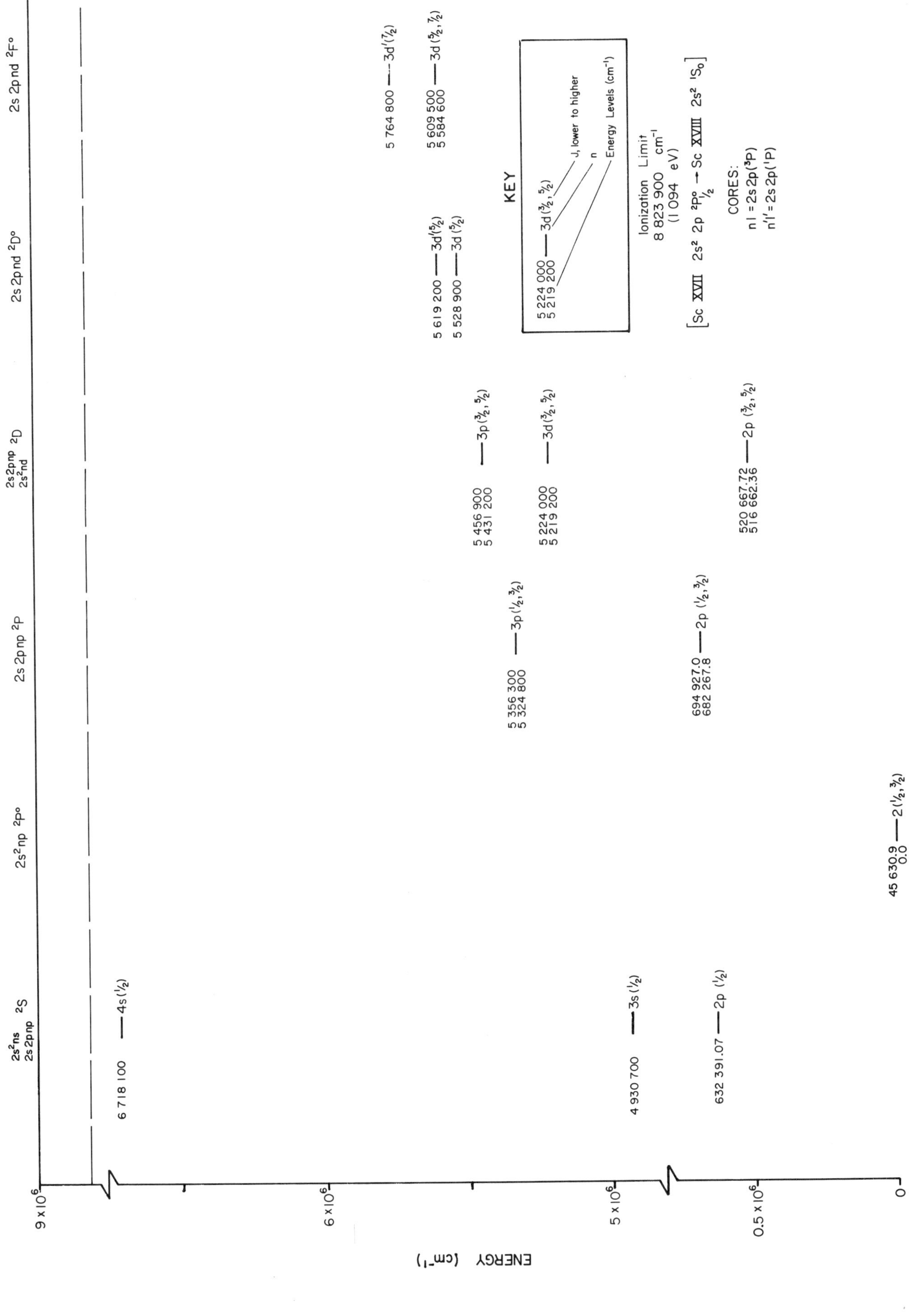

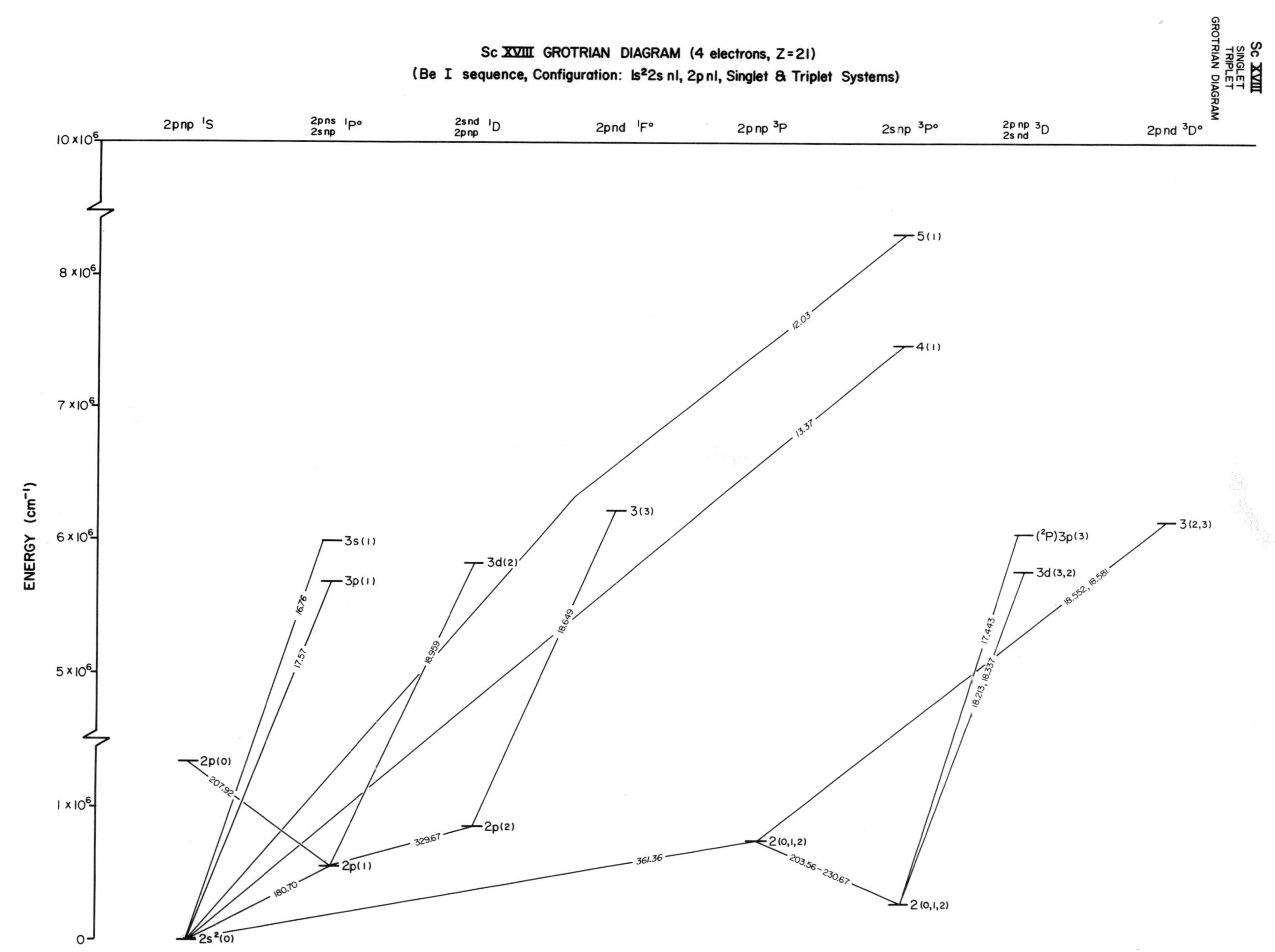
Sc XVIII GROTRIAN DIAGRAM (4 electrons, Z=21)
(Be I sequence, Configuration: ls²2s nl, 2p nl, Singlet & Triplet Systems)
Sc XVIII
SINGLET
TRIPLET
GROTRIAN DIAGRAM
2pnp ¹S
2pns ¹P°
2snp
2snd ¹D
2pnp
2pnd ¹F°
2pnp ³P
2snp ³P°
2pnp ³D
2snd
2pnd ³D°
ENERGY (cm⁻¹)
10 x 10⁶
8 x 10⁶
7 x 10⁶
6 x 10⁶
5 x 10⁶
1 x 10⁶
0
5(1)
4(1)
3(3)
3s(1)
3p(1)
3d(2)
(²P)3p(3)
3d(3,2)
3(2,3)
2p(0)
2p(2)
2p(1)
2(0,1,2)
2(0,1,2)
2s²(0)
12.03
13.37
16.76
17.57
18.959
18.649
17.443
18.213, 18.337
18.552, 18.581
207.92
329.67
180.70
361.36
203.56-230.67

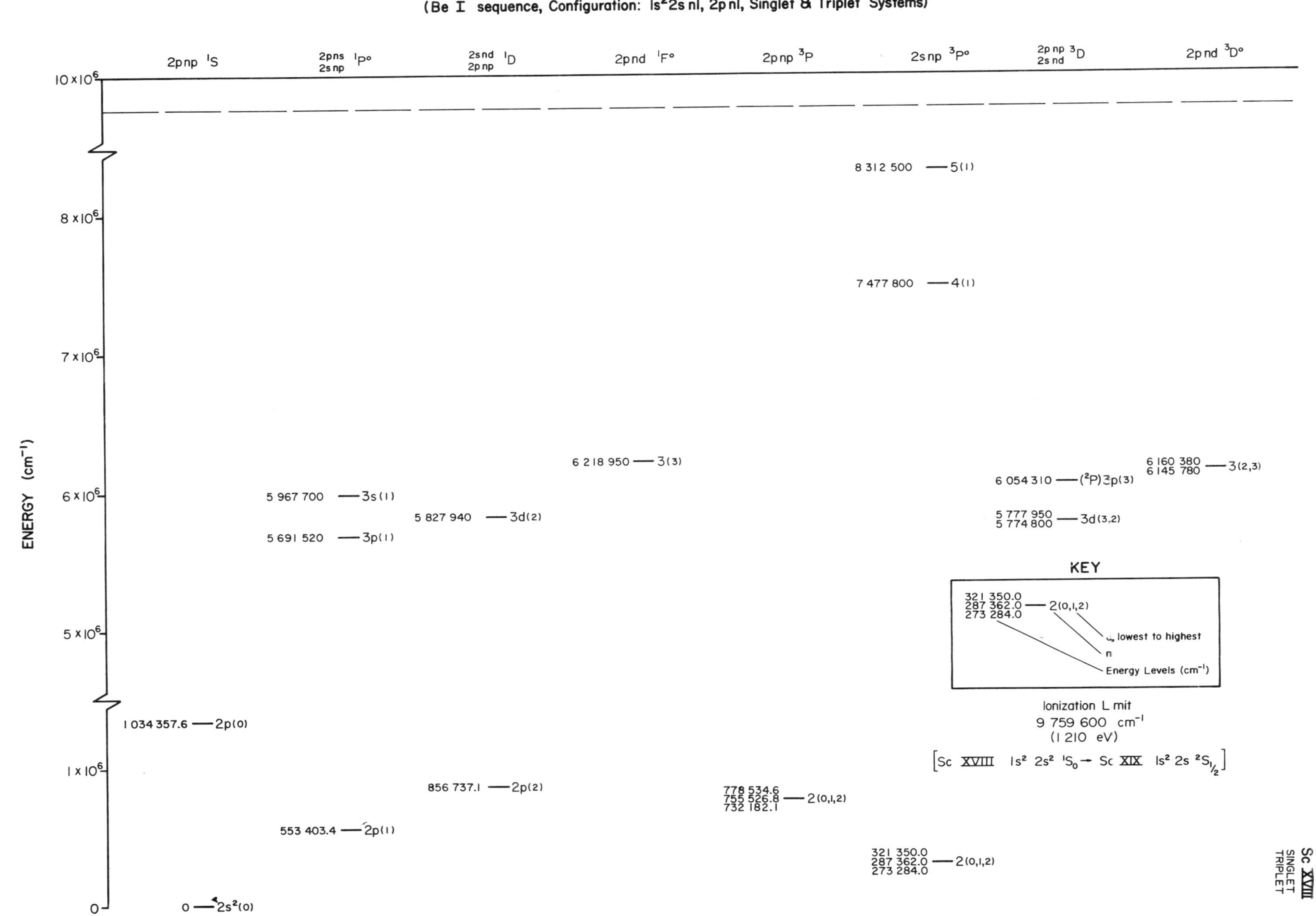
Sc XVIII ENERGY LEVELS (4 electrons, Z=21)
(Be I sequence, Configuration: 1s²2s nl, 2p nl, Singlet & Triplet Systems)
2pnp ¹S
2pns 2snp ¹P°
2snd 2pnp ¹D
2pnd ¹F°
2pnp ³P
2snp ³P°
2pnp 2snd ³D
2pnd ³D°
ENERGY (cm⁻¹)
10×10⁶
8×10⁶
7×10⁶
6×10⁶
5×10⁶
1×10⁶
0
8 312 500 — 5(1)
7 477 800 — 4(1)
6 218 950 — 3(3)
6 160 380
6 145 780 — 3(2,3)
6 054 310 — (²P)3p(3)
5 967 700 — 3s(1)
5 827 940 — 3d(2)
5 691 520 — 3p(1)
5 777 950
5 774 800 — 3d(3,2)
KEY
321 350.0
287 362.0 — 2(0,1,2)
273 284.0
J, lowest to highest
n
Energy Levels (cm⁻¹)
Ionization Limit
9 759 600 cm⁻¹
(1 210 eV)
[Sc XVIII 1s² 2s² ¹S₀ → Sc XIX 1s² 2s ²S½]
1 034 357.6 — 2p(0)
856 737.1 — 2p(2)
778 534.6
755 526.8 — 2(0,1,2)
732 182.1
553 403.4 — 2p(1)
321 350.0
287 362.0 — 2(0,1,2)
273 284.0
0 — 2s²(0)

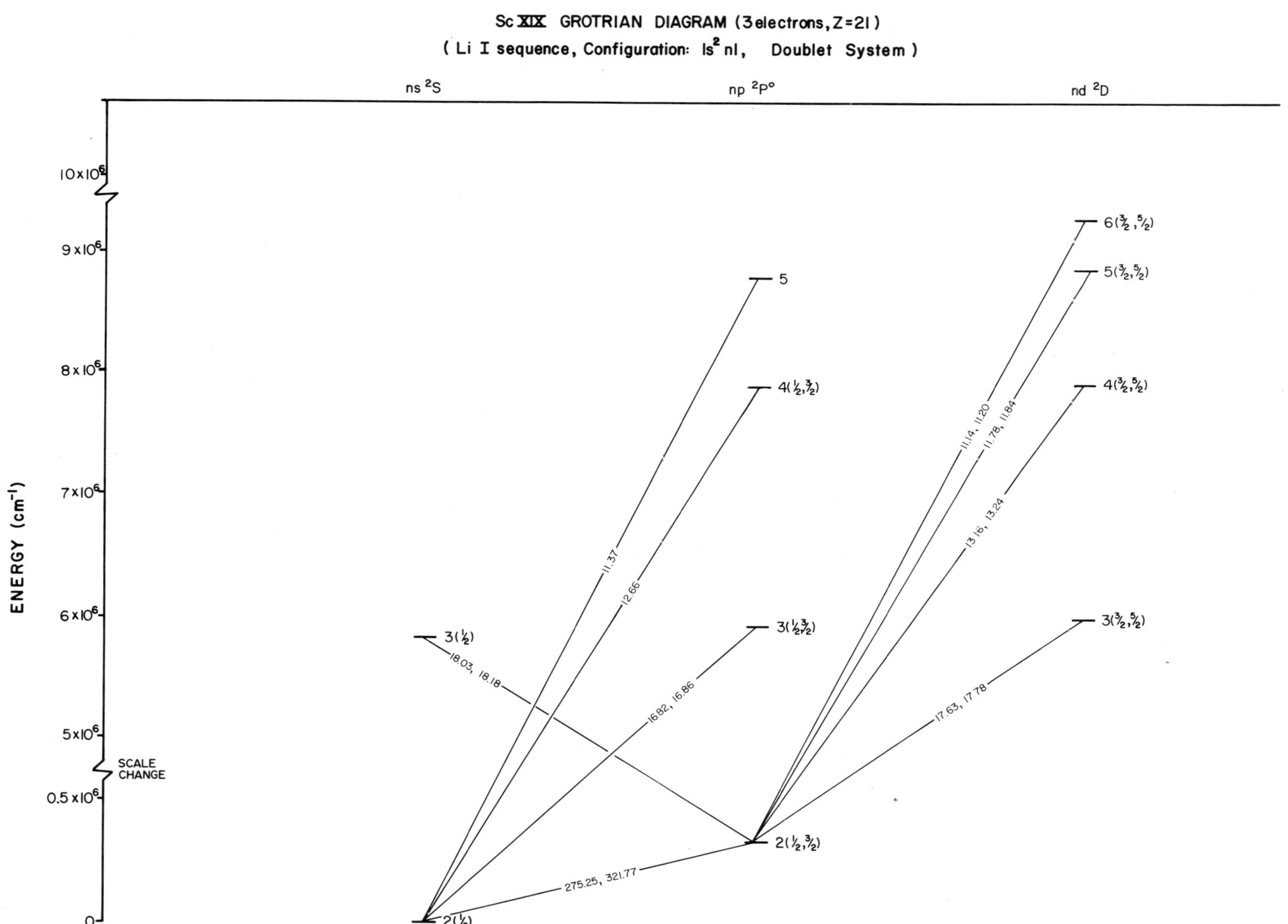
Sc XIX GROTRIAN DIAGRAM (3electrons, Z=21)
(Li I sequence, Configuration: 1s²nl, Doublet System)
ns ²S
np ²P°
nd ²D
ENERGY (cm⁻¹)
10x10⁶
9x10⁶
8x10⁶
7x10⁶
6x10⁶
5x10⁶
0.5x10⁶
0
SCALE CHANGE
2(½)
3(½)
2(½,3/2)
3(½,3/2)
4(½,3/2)
5
3(3/2,5/2)
4(3/2,5/2)
5(3/2,5/2)
6(3/2,5/2)
18.03, 18.18
275.25, 321.77
16.82, 16.86
12.66
11.37
17.63, 17.78
13.16, 13.24
11.78, 11.84
11.14, 11.20

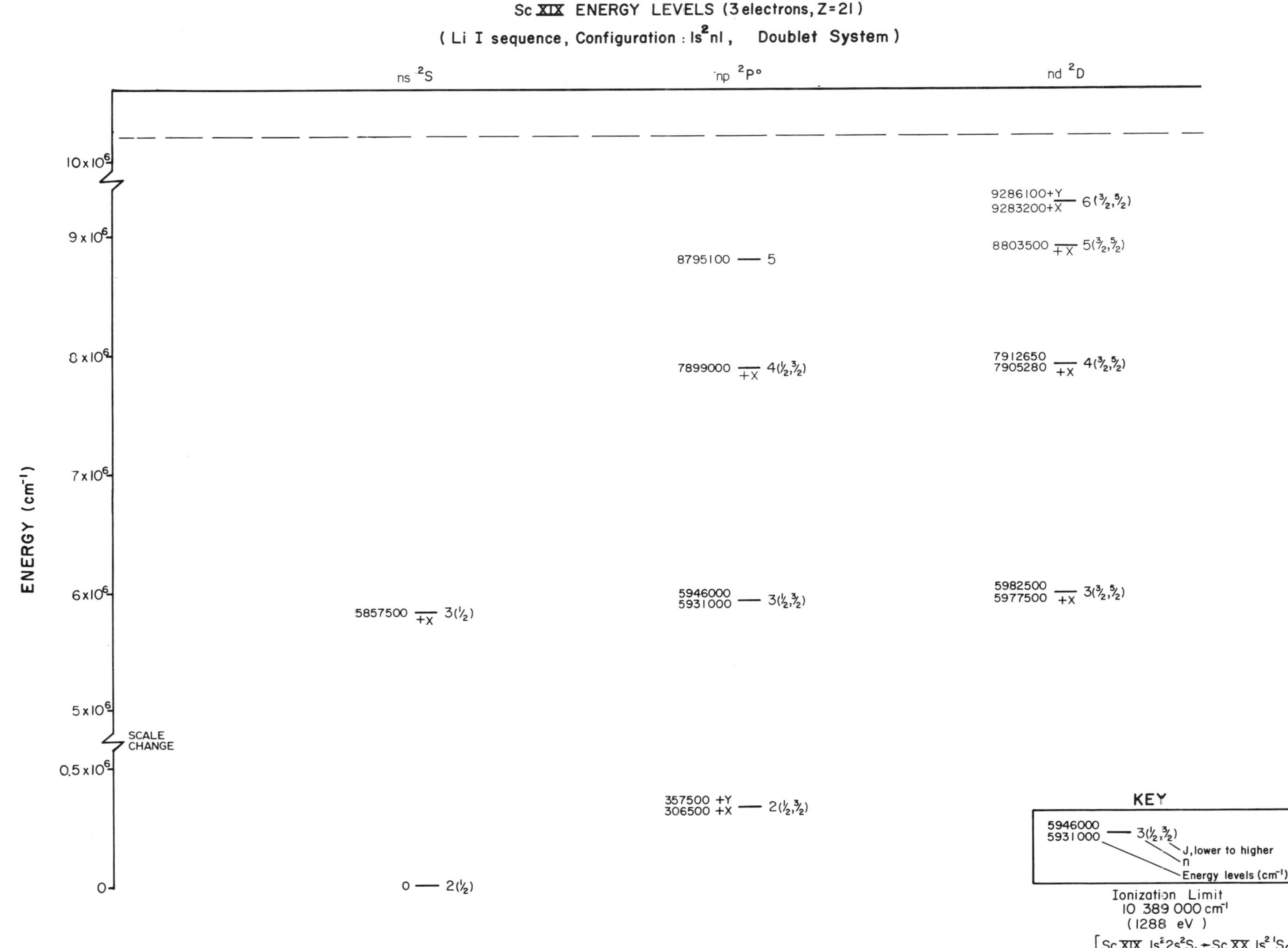
Sc XIX ENERGY LEVELS (3 electrons, Z=21)
(Li I sequence, Configuration : 1s²nl , Doublet System)
ns ²S
np ²P°
nd ²D
ENERGY (cm⁻¹)
10x10⁶
9x10⁶
8x10⁶
7x10⁶
6x10⁶
5x10⁶
SCALE CHANGE
0.5x10⁶
0
9286100+Y
9283200+X 6(3/2,5/2)
8803500 +X 5(3/2,5/2)
8795100 — 5
7899000 +X 4(1/2,3/2)
7912650
7905280 +X 4(3/2,5/2)
5857500 +X 3(1/2)
5946000
5931000 — 3(1/2,3/2)
5982500
5977500 +X 3(3/2,5/2)
357500 +Y
306500 +X — 2(1/2,3/2)
0 — 2(1/2)
KEY
5946000
5931000 — 3(1/2,3/2)
J, lower to higher
n
Energy levels (cm⁻¹)
Ionization Limit
10 389 000 cm⁻¹
(1288 eV)
[Sc XIX 1s²2s²S1/2 → Sc XX 1s² ¹S0]

Sc XX GROTRIAN DIAGRAM (2 electrons, Z=21)
(He I sequence, Configuration: 1s nl, Singlet and Triplet Systems)

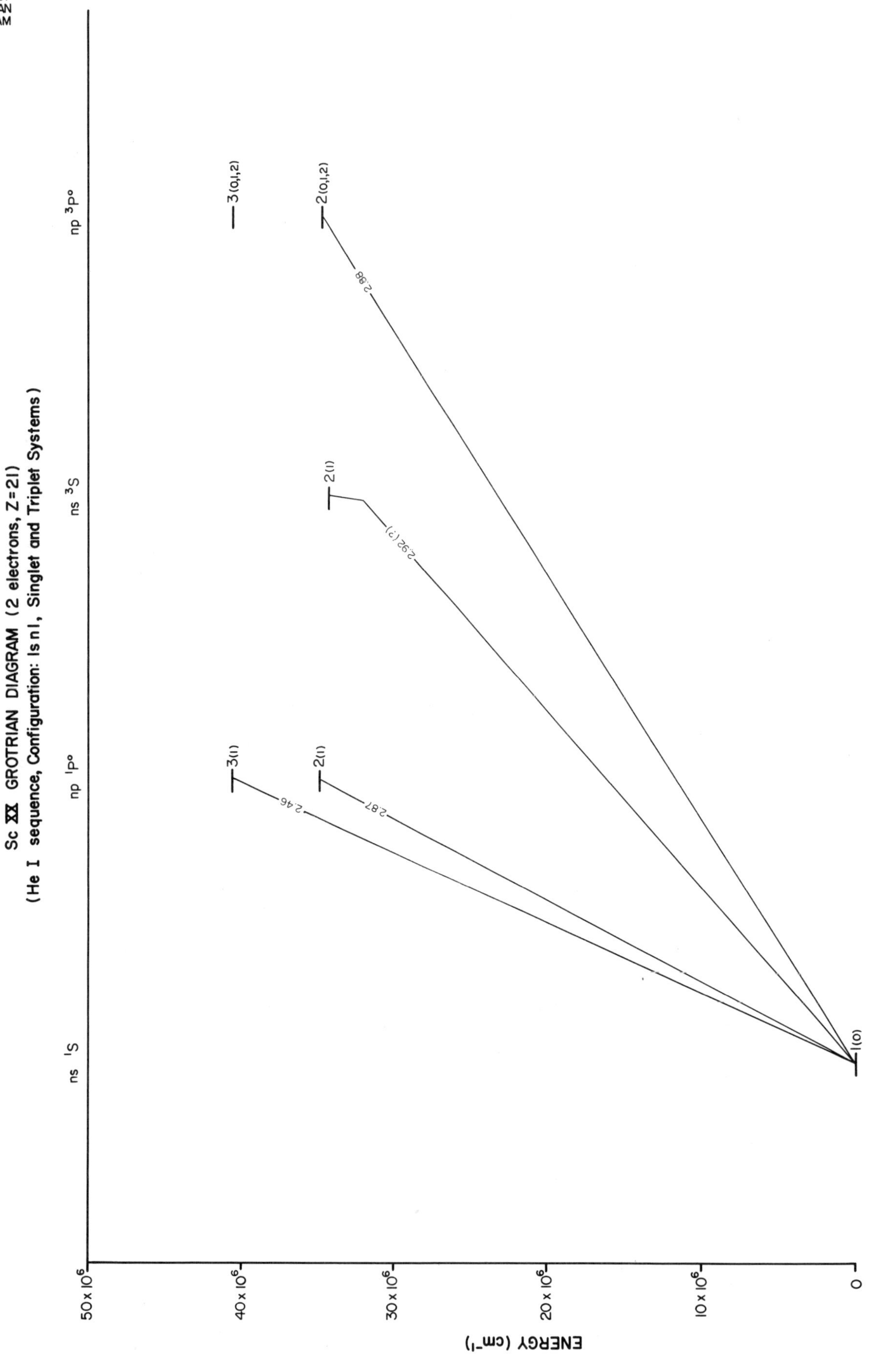

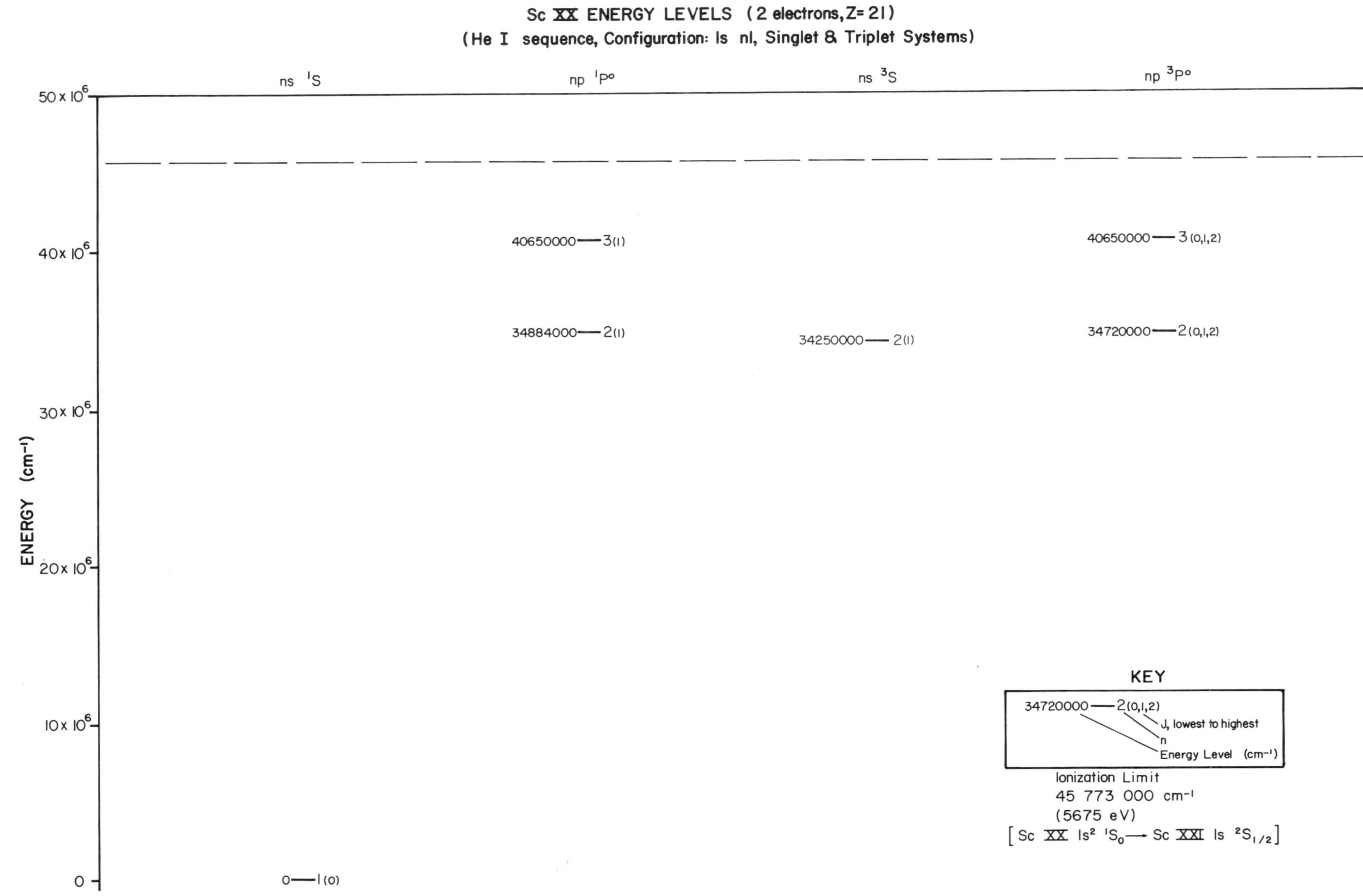
Sc XX ENERGY LEVELS (2 electrons, Z= 21)
(He I sequence, Configuration: 1s nl, Singlet & Triplet Systems)
ns ¹S
np ¹P°
ns ³S
np ³P°
50 x 10⁶
40 x 10⁶
30 x 10⁶
20 x 10⁶
10 x 10⁶
0
ENERGY (cm⁻¹)
40650000 — 3(1)
34884000 — 2(1)
34250000 — 2(1)
40650000 — 3(0,1,2)
34720000 — 2(0,1,2)
0 — 1(0)
KEY
34720000 — 2(0,1,2)
J, lowest to highest
n
Energy Level (cm⁻¹)
Ionization Limit
45 773 000 cm⁻¹
(5675 eV)
[Sc XX 1s² ¹S₀ → Sc XXI 1s ²S₁/₂]

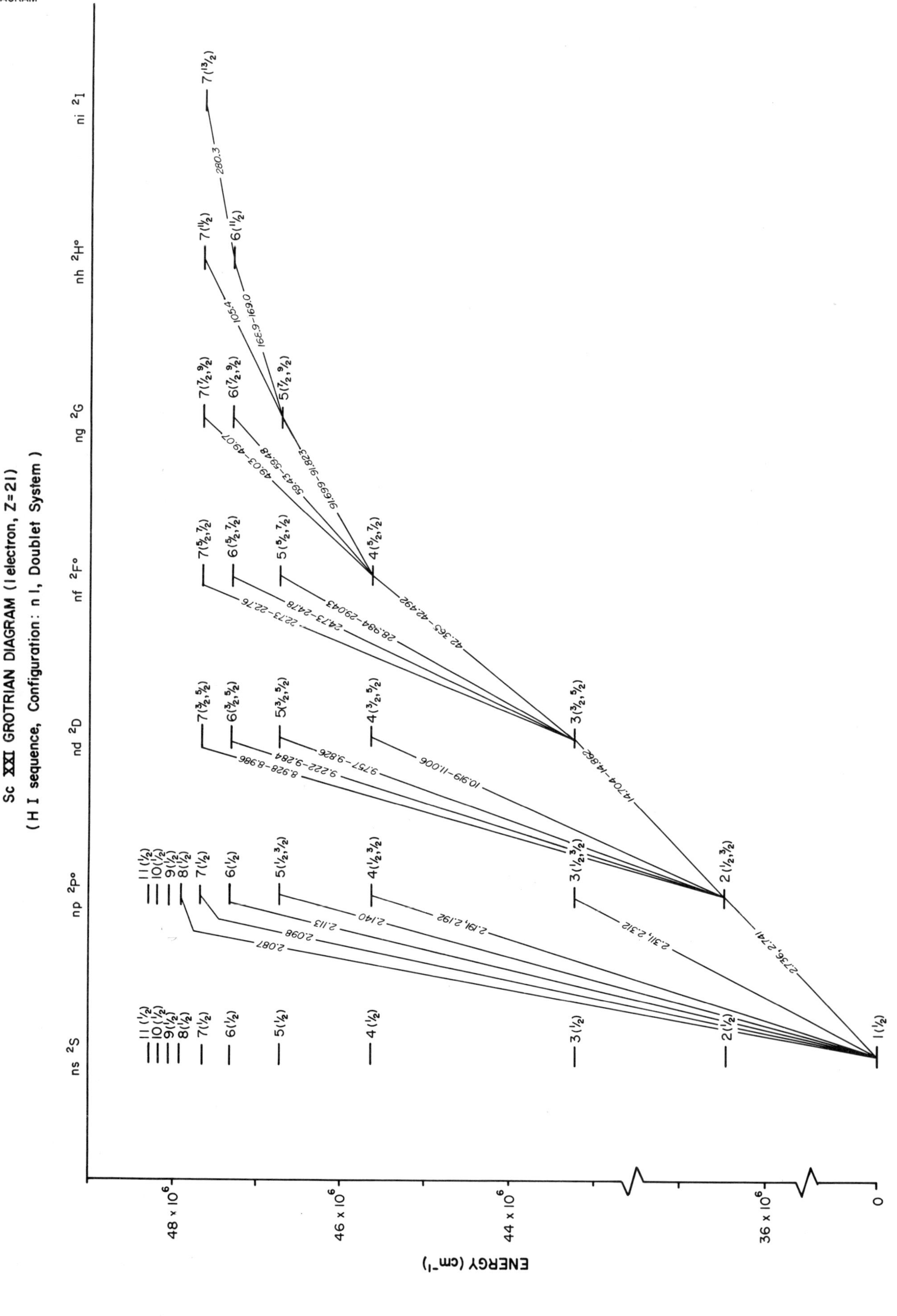

Sc XXI GROTRIAN DIAGRAM (1 electron, Z=21)

(H I sequence, Configuration: n l, Doublet System)

ns ^{2}S | np ^{2}P° | nd ^{2}D | nf ^{2}F° | ng ^{2}G | nh ^{2}H° | ni ^{2}I

ns ^{2}S: 1(1/2), 2(1/2), 3(1/2), 4(1/2), 5(1/2), 6(1/2), 7(1/2), 8(1/2), 9(1/2), 10(1/2), 11(1/2)

np ^{2}P°: 2(1/2,3/2), 3(1/2,3/2), 4(1/2,3/2), 5(1/2,3/2), 6(1/2), 7(1/2), 8(1/2), 9(1/2), 10(1/2), 11(1/2)

nd ^{2}D: 3(3/2,5/2), 4(3/2,5/2), 5(3/2,5/2), 6(3/2,5/2), 7(3/2,5/2)

nf ^{2}F°: 4(5/2,7/2), 5(5/2,7/2), 6(5/2,7/2), 7(5/2,7/2)

ng ^{2}G: 5(7/2,9/2), 6(7/2,9/2), 7(7/2,9/2)

nh ^{2}H°: 6(11/2), 7(11/2)

ni ^{2}I: 7(13/2)

2.736, 2.741
2.311, 2.312
2.191, 2.192
2.140
2.113
2.098
2.087
14.704–14.862
10.919–11.006
9.757–9.826
9.222–9.284
8.928–8.986
42.363–42.492
28.984–29.043
24.73–24.78
22.73–22.76
91.699–91.823
59.43–59.48
49.03–49.07
168.9–169.0
105.4
280.3

ENERGY (cm^{-1})

0 | 36 x 10^6 | 44 x 10^6 | 46 x 10^6 | 48 x 10^6

Sc XXI ENERGY LEVELS (I electron, Z= 21)

(H I sequence, Configuration: n l, Doublet System)

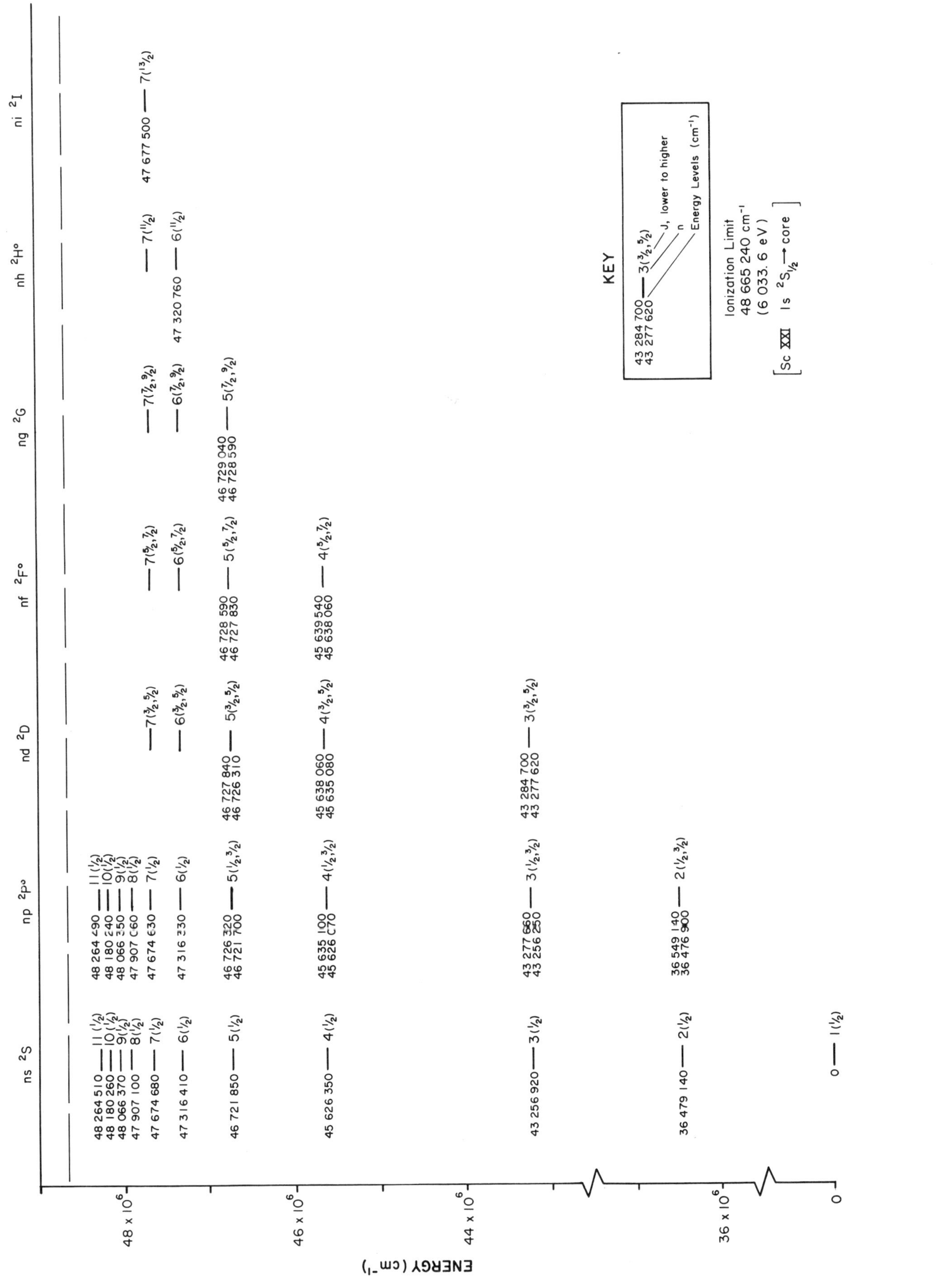

ENERGY (cm^{-1})

Axis: 0; 36×10^6; 44×10^6; 46×10^6; 48×10^6

ns ^2S

48 264 510 — 11(1/2)
48 180 260 — 10(1/2)
48 066 370 — 9(1/2)
47 907 100 — 8(1/2)
47 674 680 — 7(1/2)
47 316 410 — 6(1/2)
46 721 850 — 5(1/2)
45 626 350 — 4(1/2)
43 256 920 — 3(1/2)
36 479 140 — 2(1/2)
0 — 1(1/2)

np $^2P^o$

48 264 490 — 11(1/2)
48 180 240 — 10(1/2)
48 066 350 — 9(1/2)
47 907 060 — 8(1/2)
47 674 630 — 7(1/2)
47 316 330 — 6(1/2)
46 726 320, 46 721 700 — 5(1/2, 3/2)
45 635 100, 45 626 070 — 4(1/2, 3/2)
43 277 660, 43 256 250 — 3(1/2, 3/2)
36 549 140, 36 476 900 — 2(1/2, 3/2)

nd ^2D

— 7(3/2, 5/2)
— 6(3/2, 5/2)
46 727 840, 46 726 310 — 5(3/2, 5/2)
45 638 060, 45 635 080 — 4(3/2, 5/2)
43 284 700, 43 277 620 — 3(3/2, 5/2)

nf $^2F^o$

— 7(5/2, 7/2)
— 6(5/2, 7/2)
46 728 590, 46 727 830 — 5(5/2, 7/2)
45 639 540, 45 638 060 — 4(5/2, 7/2)

ng ^2G

— 7(7/2, 9/2)
— 6(7/2, 9/2)
46 729 040, 46 728 590 — 5(7/2, 9/2)

nh $^2H^o$

— 7(11/2)
47 320 760 — 6(11/2)

ni ^2I

47 677 500 — 7(13/2)

KEY

43 284 700, 43 277 620 — 3(3/2, 5/2)
J, lower to higher
n
Energy Levels (cm^{-1})

Ionization Limit
48 665 240 cm^{-1}
(6 033.6 eV)
[Sc XXI 1s $^2S_{1/2}$ → core]

Sc I $Z = 21$ 21 electrons

The experimental data are taken from Beufeld and Schrenk (1975), and Garton *et al.* (1973). The analysis of the configurations of the odd-parity levels founded on $(3d+4s)^2 4p$ has been published by C. Roth, J. Res. N.B.S. **73A**, 497 (1968). Several points should be borne in mind:

1. According to Roth, there is considerable configuration interaction, and spin is not a good quantum number for many of the levels. The combinations he presents are too complicated to be described in simple diagrams, and we fall back on the configurations suggested in AEL as modified by the other above references.

2. Some levels have a tentative classification in Neufeld's thesis, but not in the paper by Neufeld and Schrenk. We have designated such levels by T.

3. Garton *et al.* report 800 new lines of Sc I, of which they classify 236, locating 118 new odd levels, mostly autoionizing.

According to Neufeld and Schrenk (1975), the term $w\ ^2F^\circ$ and the four terms listed with uncertainty $+y$ ($a\ ^2P$, $v\ ^2D^\circ$, $z\ ^2S^\circ$, and $u\ ^2D^\circ$) are now assumed false on the basis of Zeeman-effect measurements. The $2{*}np$ doublet series, $7 \leqslant n \leqslant 15$ is taken from data by Garton *et al.* (1973). For these, L is not reported, and the series is labeled $^2D^\circ$, $^2F^\circ$, $^2G^\circ$ on the diagrams. Level compositions are calculated by Roth (1969), for which, see his table.

R.A. Fisher, W.C. Knopf, Jr., and F.E. Kinney, Ap. J. **130**, 683 (1959).

These authors list observed wavelengths in the range 11 000–23 000 Å (with some classifications) for this system.

W.R.S. Garton, E.M. Reeves, F.S. Tompkins, and B. Ercoli, Proc. Roy. Soc. (London) **A333**, 1 (1973).

These authors study in detail the absorption spectrum in the range 1200–3200 Å, presenting many new lines and levels.

W.F. Meggers, Sci. Papers. NBS **22**, 61 (1927).

This author presents the spectrum observed in the range 2540–8645 Å.

L.W. Neufeld and W.G. Schrenk, Spectrochim. Acta. **30B**, 45 (1975); see also L.W. Neufeld, Ph. D. Dissertation, Kansas State University (1970), University Microfilms #71–17, 365; also, private communication from B. Curnutte.

These authors reanalyze this spectrum, 2300–9000 Å, and present new energy levels and wavelengths.

C. Roth, J. Res. N.B.S. **73A**, 497 (1969).

This author calculates wavefunctions for some levels in this spectrum.

H.N. Russell and W.F. Meggers, Sci. Papers NBS **22**, 329 (1927).

These writers analyze this spectrum in detail, giving a line table, term tables, and energy-level diagram, and spectrograms in the region 2690–8240 Å.

Sc II $Z = 21$ 20 electrons

Information for the diagrams has been taken from Russell and Meggers (1927).

H.N. Russell and W.F. Meggers, Sci. Papers NBS **22**, 329 (1927).

Authors give a line table, term tables, an energy level diagram, multiplet arrays, and spectrograms from lines observed in the region 2540–6610 Å.

R.E. Trees, Phys. Rev. **97**, 686 (1955).

Author gives a table of average energies.

Sc III $Z = 21$ 19 electrons

We have used the data of Van Deurzen *et al.* (1973) in most cases, and the data of Holmström (1972) for the ^{2}H levels and transitions not indicated by Van Deurzen *et al.*

J.-E. Holmström, Physica Scripta **5**, 249 (1949).

This author classifies spectral lines in the range 550–9000 Å and presents levels and a Grotrian diagram.

H.N. Russell and R.J. Lang, Ap. J. **66**, 13 (1927).

Authors give a term table of observed and predicted values.

C. Roth, J. Research NBS-A. Phys. and Chem. **72A**, 505 (1968).

Author gives observed and calculated energy levels.

C.H.H. Van Deurzen, J.G. Conway, and S.P. Davis, J. Opt. Soc. Am. **63**, 158 (1973).

These authors present a line list, energy levels derived from them, and an improved value for the ionization limit.

Sc IV $Z = 21$ 18 electrons

We have used the data of Smitt (1973). Terms with configuration 3s3p^{6}3d are designated 3d* in his paper.

B.C. Fawcett, N.J. Peacock, and R.D. Cowan, J. Phys. **B1**, 295 (1968).

Authors give wavelengths of two observed transitions in the vacuum ultraviolet region.

A.H. Gabriel, B.C. Fawcett, and Carole Jordan, Nature **206**, 390 (1965).

Authors give a wavelength of an observed transition in the vacuum ultraviolet region.

A.H. Gabriel, B.C. Fawcett, and Carole Jordan, Proc. Phys. Soc. **87**, 825 (1966).

Authors give a wavelength of an observed transition in the vacuum ultraviolet region.

P.G. Kruger and S.G. Weissberg, Phys. Rev. **48**, 659 (1935).

Authors give calculated term values from Moseley diagrams.

P.G. Kruger, S.G. Weissberg, and L.W. Phillips, Phys. Rev. **51**, 1090 (1937).

Authors give line and term tables from lines observed in the region 215–300 Å.

R. Smitt, Physica Scripta **8**, 292 (1973).

This author presents an extended analysis of this spectrum, with lists of lines and levels, and a Grotrian diagram.

L.Å. Svensson and J.O. Ekberg, Ark. Fys. **37**, 65 (1967).

Authors identify a wavelength in the vacuum ultraviolet region.

Sc V $Z = 21$ 17 electrons

B.C. Fawcett and A.H. Gabriel, Proc. Phys. Soc. **88**, 262 (1966).

Authors give a line table for observed transitions of three multiplets in the region 280–300 Å.

B.C. Fawcett, N.J. Peacock, and R.D. Cowan, J. Phys. **B1**, 295 (1968).

Authors give a table of lines observed in the region 175–185 Å.

A.H. Gabriel, B.C. Fawcett, and Carole Jordan, Nature **206**, 390 (1965).

Authors give wavelengths of transitions observed in the region 280–290 Å.

A.H. Gabriel, B.C. Fawcett, and Carole Jordan, Proc. Phys. Soc. **87**, 825 (1966).

Authors give wavelengths of observed transitions of two multiplets.

P.G. Kruger and L.W. Phillips, Phys. Rev. **51**, 1087 (1937).

Authors give an array of terms and observed lines in the vacuum ultraviolet region.

L.Å. Svensson and J.O. Ekberg, Ark. Fys. **37**, 65 (1967).

Authors give a table of lines identified in the region 375–400 Å.

Sc VI $Z = 21$ 16 electrons

The levels are taken from Kelly's unpublished tabulation. Fawcett *et al.* (1972) present some transitions in the quintet and triplet systems whose connections to lower levels are uncertain, and these have not been included in the present drawings.

I.S. Bowen, Ap. J. **132**, 1 (1960).

Author gives a line table of observed forbidden transitions.

B. Edlén, Phys. Rev. **62**, 434 (1942).

Author gives wavelength and term values for the $3s^2 3p^4\ {}^1S$ and $3s3p^5\ {}^1P_1$ configurations.

B.C. Fawcett, R.D. Cowan, and R.W. Hayes, J. Phys. **B5**, 2143 (1972).

These authors present wavelengths.

B.C. Fawcett and A.H. Gabriel, Proc. Phys. Soc. **88**, 262 (1966).

Authors give wavelengths of the observed transitions of two multiplets.

B.C. Fawcett, N.J. Peacock, and R.D. Cowan, J. Phys. **B1**, 295 (1968).

Authors give a table of lines observed in the region 145–170 Å.

A.H. Gabriel, B.C. Fawcett, and Carole Jordan, Nature **206**, 890 (1965).

Authors give a wavelength of an observed transition in the vacuum ultraviolet region.

A.H. Gabriel, B.C. Fawcett, and Carole Jordan, Proc. Phys. Soc. **87**, 825 (1966).

Authors give a table of lines observed in the region 280–290 Å.

P.G. Kruger and H.S. Pattin, Phys. Rev. **52**, 621 (1937).

Authors give a multiplet array giving wavelengths and term values from observations in the extreme ultraviolet region.

L.Å. Svensson and J.O. Ekberg, Ark. Fys. **37**, 65 (1967).

Authors give a term table from observations.

Sc VII $Z = 21$ 15 electrons

Fawcett *et al.* (1972) give wavelengths for several transitions involving levels whose locations cannot be given on our drawings, since there are no identified transitions connecting them to known levels.

J.O. Ekberg and L.Å. Svensson, Physica Scripta **2**, 283 (1970).

These authors identify many lines in this spectrum and give lists of wavelengths and levels' energies.

B.C. Fawcett, J. Phys. B3, 1732 (1970).

This author lists many unidentified Sc lines in addition to classifying several transitions in this spectrum.

B.C. Fawcett, R.D. Cowan, and R.W. Hayes, J. Phys. **B5**, 2143 (1972).

These authors classify a few new lines in this spectrum.

B.C. Fawcett, A.H. Gabriel, and P.A.H. Saunders, Proc. Phys. Soc. **90**, 863 (1967).

Authors give the wavelengths of the transitions of one multiplet.

A.H. Gabriel, B.C. Fawcett, and Carole Jordan, Nature **206**, 390 (1965).

Authors give a wavelength of an observed transition.

A.H. Gabriel, B.C. Fawcett, and Carole Jordan, Proc. Phys. Soc. **87**, 825 (1966).

Authors give a wavelength of an observed transition.

P.G. Kruger and H.S. Pattin, Phys. Rev. **52**, 621 (1937).

Authors give an array of observed multiplets.

SC VIII $Z = 21$ 14 electrons

Fawcett *et al.* (1972) give wavelengths for $3p3d^3F^\circ$–$3p4f^3G$ transitions, but the energies of these levels relative to others are unknown and have not been included.

J.O. Ekberg and L.Å. Svensson, Physica Scripta **2**, 283 (1970).

These authors identify many levels in this spectrum and give lists of wavelengths and levels' energies.

B.C. Fawcett, J. Phys. **B3**, 1732 (1970).

This author lists many unidentified Sc lines in addition to classifying several transitions in this spectrum.

B.C. Fawcett, R.D. Cowan, and R.W. Hayes, J. Phys. **B5**, 2143 (1972).

These authors classify two lines in this spectrum.

B.C. Fawcett, A.H. Gabriel, and P.A.H. Saunders, Proc. Phys. Soc. **90**, 863 (1967).

Authors give wavelengths of observed transitions in the region 295–310 Å.

P.G. Kruger and L.W. Phillips, Phys. Rev. **52**, 97 (1937).

Authors give an array of observed wavelengths and term values.

L.W. Phillips, Phys. Rev. **55**, 708 (1939).

Author gives line and term tables from lines observed in the region 350–500 Å.

Sc IX $Z = 21$ 13 electrons

J.O. Ekberg and L.Å. Svensson, Physica Scripta **2**, 283 (1970).

These authors identify many levels in this spectrum and give lists of wavelengths and levels' energies.

B.C. Fawcett, J. Phys. **B3**, 1732 (1970).

This author lists many unidentified Sc lines in addition to classifying several transitions in this spectrum.

Sc X $Z = 21$ 12 electrons

J.O. Ekberg, Physica Scripta **4**, 101 (1971).

This author classifies many lines in this spectrum and gives lists of wavelengths and levels' energies.

B.C. Fawcett, J. Opt. Soc. Am. **66**, 632 (1976).

This author presents a detailed analysis of Ca X with some additional identifications in Sc X.

W.L. Parker and L.W. Phillips, Phys. Rev. **57**, 140 (1940).

These authors present classifications for about twelve lines in this spectrum.

Sc XI | $Z = 21$ | 11 electrons

L. Cohen and W.E. Behring, J. Opt. Soc. Am. **66**, 899 (1976).

These authors present many wavelengths and energy levels for this system.

B. Edlén, Z. Physik **100**, 621 (1936).

This author presents several classified lines in the spectrum.

U. Feldman, G.A. Doschek, D.K. Prinz, and D.J. Nagel, J. Appl. Phys. **47**, 1341 (1976).

These authors observe several lines in this spectrum.

P.G. Kruger and L.W. Phillips, Phys. Rev. **55**, 352 (1939).

These authors classify several lines in this spectrum.

B.C. Fawcett, J. Phys. B3, 1732 (1970).

This author identifies the transition $4f\,^2F^{\circ}_{5/2}$–$5d\,^2D_{5/2}$.

Sc XII | $Z = 21$ | 10 electrons

Kastner *et al.* identify their observed levels in both *LS* and *j–l* coupling, and most other levels are also identified in both *LS* (Edlén and Tyrén, Feldman and Cohen) and *j–l* (Moore, Fawcett) coupling; we have used both systems simultaneously on the drawings. Where the *LS* coupling in Edlén and Tyrén disagreed with that in Crance, the identification of Crance has been used.

Energies for $2p^5 3p$ and $2p^5 4d$ terms are largely calculated (Crance); although these are observed levels (Kastner *et al.*), they are not connected to the ground state.

M. Crance, Atomic Data **5**, 2 (1973).

Author gives calculated energies for $2p^5 3p$ and $2p^5 4d$ terms not listed by other references. Also corrects *LS*-coupling identifications by Edlén and Tyrén of levels at 3 714 700 and 3 767 300 cm^{-1}.

B. Edlén and F. Tyrén, Z. Physik **101**, 206 (1936).

Authors give line and term tables from observations in the region 25–30 Å, in *LS* coupling.

B.C. Fawcett, Proc. Phys. Soc. **86**, 1087 (1965).

Author gives *j–l* coupling information on levels at 4 521 000 and 4 557 900 cm^{-1}.

U. Feldman and L. Cohen, Ap. J. **149**, 265 (1967).

Authors give line and term tables from observations in the region 20–25 Å, in *LS* coupling.

S.O. Kastner, W.E. Behring, and L. Cohen, Ap. J. **199**, 777 (1975).

These authors identify some transitions classified as $2p^5 3p$–$2p^5 4d$, involving levels not observed by other references (no observed transitions to the ground state).

R.L. Kelly and L.C. Gapenski, unpublished (1970).

(Identification of three levels in *j–l* coupling, with uncertainties.)

Sc XIII | $Z = 21$ | 9 electrons

R.D. Chapman and Y. Shadmi, J. Opt. Soc. Am. **63**, 1440 (1973).

These authors present calculations of the energies of levels in this spectrum.

L. Cohen, U. Feldman, and S.O. Kastner, J. Opt. Soc. Am. **58**, 331 (1968).

Authors give wavelengths of observed transitions in the region 24–28 Å.

B.C. Fawcett, Proc. Phys. Soc. **86**, 1087 (1965).

Author gives a table of lines observed in the region 20–30 Å.

B.C. Fawcett, J. Phys. B**4**, 981 (1971).

This author classifies four lines in this spectrum.

B.C. Fawcett, Atomic Data and Nuclear Data Tables **16**, 135 (1975).

This author tabulates energies and spectral lines arising from transitions of the sort $2s^22p^n$–$2s2p^{n+1}$ and $2s2p^n$–$2p^{n+1}$.

B.C. Fawcett, D.D. Burgess, and N.J. Peacock, Proc. Phys. Soc. **91**, 970 (1967).

Author gives wavelengths of the transitions in a multiplet.

U. Feldman, G.A. Doschek, R.D. Cowan, and L. Cohen, J. Opt. Soc. Am. **63**, 1445 (1973).

These authors classify and tabulate many wavelengths and energy levels in this system.

Sc XIV $Z = 21$ 8 electrons

G.A. Doschek, U. Feldman and L. Cohen, J. Opt. Soc. Am. **63**, 1463 (1973).

These authors present several wavelengths and energy levels for this system.

B.C. Fawcett, J. Phys. B**4**, 981 (1971).

This author classifies several wavelengths in this spectrum.

B.C. Fawcett, Atomic Data and Nuclear Data Tables **16**, 135 (1975).

This author tabulates energies and wavelengths for transition of the sort $2s^22p^n$–$2s2p^{n+1}$ and $2s2p^n$–$2p^{n+1}$.

B.C. Fawcett and R.W. Hayes, Mon. Not. Roy. Astron. Soc. **170**, 185 (1975).

These authors present numerous comparisons between observed and calculated wavelengths in this system.

S. Goldsmith, U. Feldman, and L. Cohen, J. Opt. Soc. Am. **61**, 615 (1971).

These authors present several classifications of lines in this spectrum.

Sc XV $Z = 21$ 7 electrons

The information given by Fawcett (1971) has been used, except that the positions of the ^2F and ^4P levels have been calculated from the transitions in Kelly and Palumbo.

B.C. Fawcett, J. Phys. B**4**, 981 (1971).

This author classifies numerous lines in the spectrum.

B.C. Fawcett, Atomic Data and Nuclear Tables **16**, 135 (1975).

This author tabulates energies and wavelengths for transitions of the sort $2s^22p^n$–$2s2p^{n+1}$ and $2s2p^n$–$2p^{n+1}$.

B.C. Fawcett and R.W. Hayes, Mon. Not. Roy. Astron. Soc. **170**, 185 (1975).

These authors compare calculated and observed wavelengths in this spectrum.

S.O. Kastner, J. Opt. Soc. Am. **63**, 738 (1973).

This author calculates energy levels in this system.

Sc XVI $Z = 21$ 6 electrons

B.C. Fawcett, J. Phys. B4, 981 (1971).

This author classifies several transitions in this spectrum.

B.C. Fawcett, Atomic Data and Nuclear Data Tables **16**, 135 (1975).

This author tabulates energy levels and wavelengths for transitions of the sort $2s^2 2p^n - 2s2p^{n+1}$ and $2s2p^n - 2p^{n+1}$.

B.C. Fawcett and R.W. Hayes, Mon. Not. Roy. Astron. Soc. **170**, 190 (1975).

These authors compare several observed and calculated wavelengths for this spectrum.

S. Goldsmith, U. Feldman, A. Crooker, and L. Cohen, J. Opt. Soc. Am. **62**, 260 (1972).

These authors present new spectra of Sc XVI in the range 10–22 Å, with derived energy levels.

Sc XVII $Z = 21$ 5 electrons

B.C. Fawcett, Atomic Data and Nuclear Data Tables **16**, 135 (1975).

This author tabulates energies and wavelengths for transitions of the sort $2s^2 2p^n - 2s2p^{n+1}$ and $2s2p^n - 2p^{n+1}$.

B.C. Fawcett and R.D. Cowan, Mon. Not. Roy. Astron. Soc. **171**, 1 (1975).

These authors list calculated and observed values for the wavelengths of the transitions $2s^2 2p^n - 2s2p^{n+1}$ and $2s2p^n - 2p^{n+1}$.

B.C. Fawcett and R.W. Hayes, Mon. Not. Roy. Astron. Soc. **170**, 185 (1975).

These authors compare several observed and calculated wavelengths for this spectrum.

Sc XVIII $Z = 21$ 4 electrons

B.C. Fawcett, Atomic Data and Nuclear Data Tables **16**, 135 (1975).

This author tabulates energies and wavelengths of transitions of the sort $2s^2 2p^n - 2s2p^{n+1}$ and $2s2p^n - 2p^{n+1}$.

B.C. Fawcett and R.W. Hayes, Mon. Not. Roy. Astron. Soc. **170**, 185 (1975).

These authors compare several observed and calculated wavelengths for this spectrum.

S. Goldsmith, L. Oren (Katz), A.M. Crooker, and L. Cohen, Ap. J. **184**, 1021 (1973).

These authors tabulate three calculated intervals in this system.

Sc XIX $Z = 21$ 3 electrons

U. Feldman, G.A. Doschek, D.K. Prinz, and D.J. Nagel, J. Appl. Phys. **4**, 1341 (1976).

These authors list the wavelengths of a doublet in this spectrum.

S. Goldsmith, U. Feldman, L. Oren (Katz), and L. Cohen, Ap. J. **174**, 209 (1972).

These authors tabulate extrapolated and observed wavelengths and derived energy levels for this system.

Sc XX $Z = 21$ 2 electrons

L. Cohen, U. Feldman. M. Swartz, and J.H. Underwood, J. Opt. Soc. Am. **58**, 843 (1968).

These authors observe two wavelengths in this spectrum.

For hydrogenic systems we show calculated energies and wavelengths. Since all transitions in a set nl–$n'l'$ have nearly the same wavelength if n, n' are fixed, we have indicated only the range of wavelengths and have not tried to show each transition l–l' in a set.

G.W. Erickson, private communication (1977) and in press, J. Phys. Chem. Ref. Data.

This author provides relativistic calculations for one-electron spectra.

Titanium (Ti)

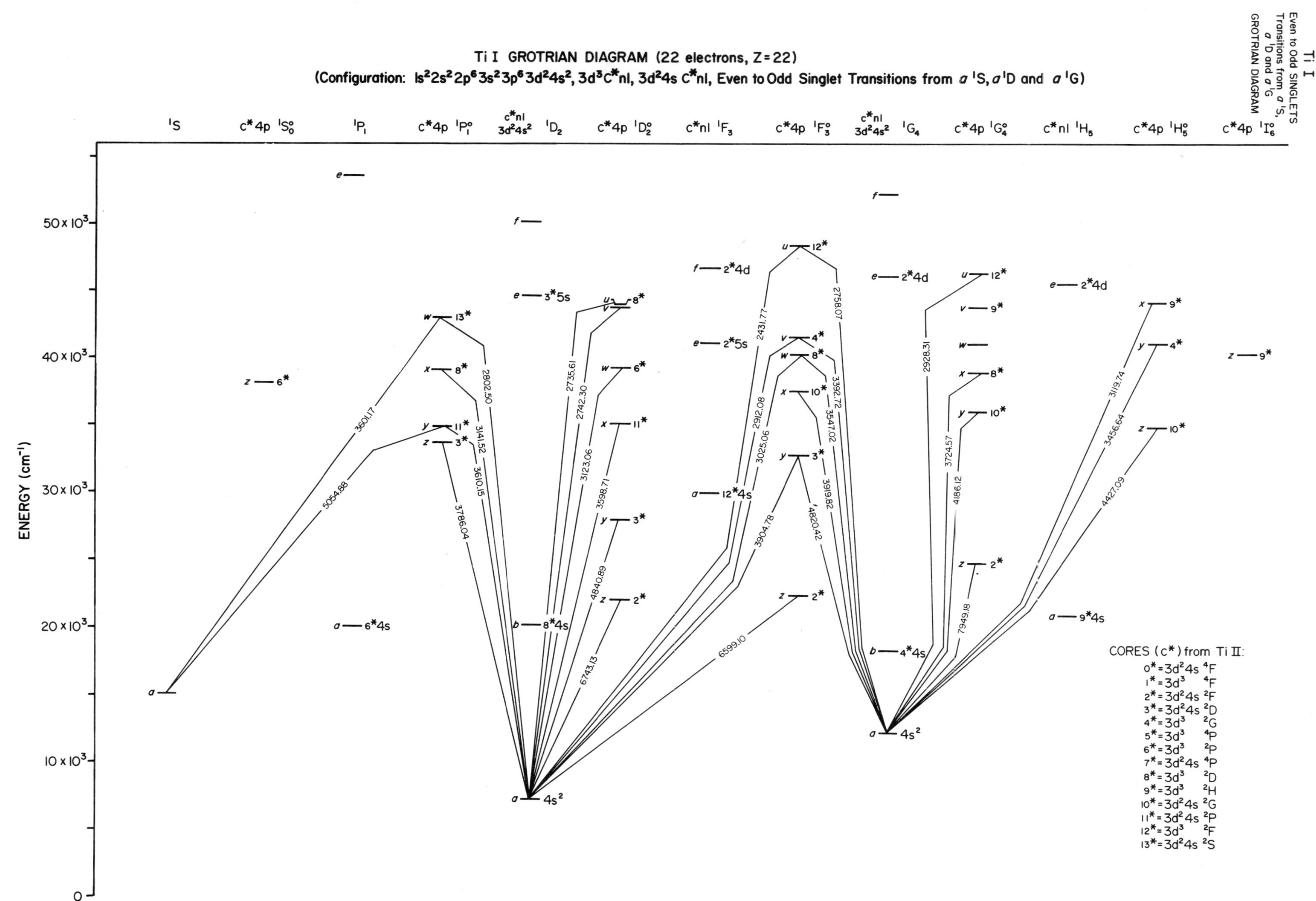
Ti I
Even to Odd SINGLETS
Transitions from a ¹S, a ¹D and a ¹G
GROTRIAN DIAGRAM
Ti I GROTRIAN DIAGRAM (22 electrons, Z=22)
(Configuration: 1s²2s²2p⁶3s²3p⁶3d²4s², 3d³c*nl, 3d²4s c*nl, Even to Odd Singlet Transitions from a ¹S, a ¹D and a ¹G)
¹S
c*4p ¹S°₀
¹P₁
c*4p ¹P°₁
c*nl 3d²4s²
¹D₂
c*4p ¹D°₂
c*nl ¹F₃
c*4p ¹F°₃
c*nl 3d²4s²
¹G₄
c*4p ¹G°₄
c*nl ¹H₅
c*4p ¹H°₅
c*4p ¹I°₆
CORES (c*) from Ti II:
0* = 3d²4s ⁴F
1* = 3d³ ⁴F
2* = 3d²4s ²F
3* = 3d²4s ²D
4* = 3d³ ²G
5* = 3d³ ⁴P
6* = 3d³ ²P
7* = 3d²4s ⁴P
8* = 3d³ ²D
9* = 3d³ ²H
10* = 3d²4s ²G
11* = 3d²4s ²P
12* = 3d³ ²F
13* = 3d²4s ²S
3601.17
5054.88
2802.50
3141.52
3610.15
3786.04
2735.61
2742.30
3123.06
3598.71
4840.89
6743.13
2431.77
2912.08
3025.06
3904.78
6599.10
2758.07
3392.72
3547.02
3919.82
4820.42
2928.31
3724.57
4186.12
7949.18
3119.74
3456.64
4427.09
ENERGY (cm⁻¹)
50×10³
40×10³
30×10³
20×10³
10×10³
0

Ti I ENERGY LEVELS (22 electrons, Z = 22)

(Configuration: $1s^2 2s^2 2p^6 3s^2 3p^6 3d^2 4s^2$, $3d^3$ c^*nl, $3d^2 4s$ c^*nl, Singlet System)

Term	Energy level (cm^{-1})	Notation from A.E.L.	Parent level in Ti II, n
^1S	15 166.55	a	
c^*4p $^1S^\circ_0$	38 200.94	z	6*
$^1P_1$	53 663.62	e	
$^1P_1$	20 062.98	a	6*4s
c^*4p $^1P^\circ_1$	42 927.55	w	13*
c^*4p $^1P^\circ_1$	39 077.713	x	8*
c^*4p $^1P^\circ_1$	34 947.120	y	11*
c^*4p $^1P^\circ_1$	33 650.671	z	3*
c^*nl, $3d^2 4s^2$ $^1D_2$	50 128.08	f	
c^*nl, $3d^2 4s^2$ $^1D_2$	44 581.16	e	3*5s
c^*nl, $3d^2 4s^2$ $^1D_2$	20 209.444	b	8*4s
c^*nl, $3d^2 4s^2$ $^1D_2$	7 255.369	a	$4s^2$
c^*4p $^1D^\circ_2$	43 799.455	u	8*
c^*4p $^1D^\circ_2$	43 710.28	v	8*
c^*4p $^1D^\circ_2$	39 265.80	w	6*
c^*4p $^1D^\circ_2$	35 035.147	x	11*
c^*4p $^1D^\circ_2$	27 907.026	y	3*
c^*4p $^1D^\circ_2$	22 081.198	z	2*
c^*nl $^1F_3$	46 650.26	f	2*4d
c^*nl $^1F_3$	41 087.31	e	2*5s
c^*nl $^1F_3$	29 818.31	a	12*4s
c^*4p $^1F^\circ_3$	48 365.09	u	12*
c^*4p $^1F^\circ_3$	41 585.24	v	4*
c^*4p $^1F^\circ_3$	40 302.950	w	8*
c^*4p $^1F^\circ_3$	37 622.573	x	10*
c^*4p $^1F^\circ_3$	32 857.721	y	3*
c^*4p $^1F^\circ_3$	22 404.69	z	2*
c^*nl, $3d^2 4s^2$ $^1G_4$	52 125.98	f	
c^*nl, $3d^2 4s^2$ $^1G_4$	46 068.04	e	2*4d
c^*nl, $3d^2 4s^2$ $^1G_4$	18 287.560	b	4*4s
c^*nl, $3d^2 4s^2$ $^1G_4$	12 118.394	a	$4s^2$
c^*4p $^1G^\circ_4$	46 257.67	u	12*
c^*4p $^1G^\circ_4$	43 674.130	v	9*
c^*4p $^1G^\circ_4$	40 883.30	w	
c^*4p $^1G^\circ_4$	38 959.499	x	8*
c^*4p $^1G^\circ_4$	36 000.144	y	10*
c^*4p $^1G^\circ_4$	24 694.895	z	2*
c^*nl $^1H_5$	45 485.35	e	2*4d
c^*nl $^1H_5$	20 795.599	a	9*4s
c^*4p $^1H^\circ_5$	44 163.24	x	9*
c^*4p $^1H^\circ_5$	41 039.874	y	4*
c^*4p $^1H^\circ_5$	34 700.212	z	10*
c^*4p $^1I^\circ_6$	40 319.80	z	9*

ENERGY (cm^{-1}): 0, 10×10^3, 20×10^3, 30×10^3, 40×10^3, 50×10^3

KEY

20 209.444 — b — 8*4s
- n
- Parent level in Ti II
- Notation from A.E.L.
- Energy level (cm^{-1})

Ionization Limit
55 010 cm^{-1}
(6.821 eV)
[Ti I $4s^2$ $^3F_2$ → Ti II $4s$ $^4F_{3/2}$]

CORES (c^*) from Ti II:
- 0* = $3d^2 4s$ ^4F
- 1* = $3d^3$ ^4F
- 2* = $3d^2 4s$ ^2F
- 3* = $3d^2 4s$ ^2D
- 4* = $3d^3$ ^2G
- 5* = $3d^3$ ^4P
- 6* = $3d^3$ ^2P
- 7* = $3d^2 4s$ ^4P
- 8* = $3d^3$ ^2D
- 9* = $3d^3$ ^2H
- 10* = $3d^2 4s$ ^2G
- 11* = $3d^2 4s$ ^2P
- 12* = $3d^3$ ^2F
- 13* = $3d^2 4s$ ^2S

NOTE: For proper configuration, please refer to bibliographical comments.

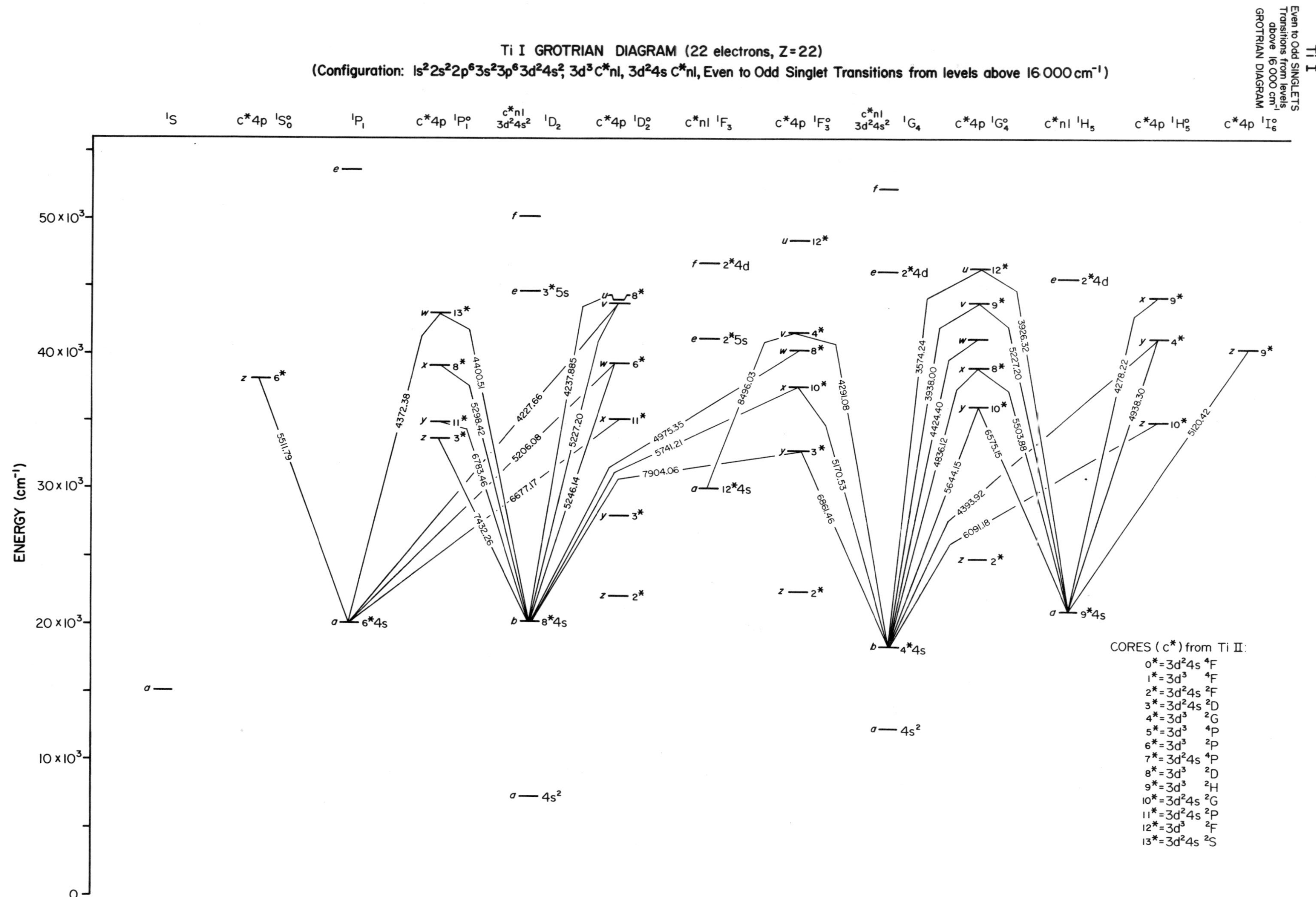

Ti I
Even to Odd SINGLETS
Transitions from levels above 16 000 cm⁻¹
GROTRIAN DIAGRAM
Ti I GROTRIAN DIAGRAM (22 electrons, Z=22)
(Configuration: 1s²2s²2p⁶3s²3p⁶3d²4s², 3d³c*nl, 3d²4s c*nl, Even to Odd Singlet Transitions from levels above 16 000 cm⁻¹)
¹S
c*4p ¹S°₀
¹P₁
c*4p ¹P°₁
c*nl 3d²4s²
¹D₂
c*4p ¹D°₂
c*nl ¹F₃
c*4p ¹F°₃
c*nl 3d²4s²
¹G₄
c*4p ¹G°₄
c*nl ¹H₅
c*4p ¹H°₅
c*4p ¹I°₆
ENERGY (cm⁻¹)
50×10³
40×10³
30×10³
20×10³
10×10³
0
5511.79
4372.38
4400.51
5298.42
6783.46
7432.26
4227.66
5206.08
6677.17
4237.885
5227.20
5246.14
4975.35
5741.21
7904.06
8496.03
4291.08
5170.53
6861.46
3574.24
3938.00
4424.40
4836.12
5644.15
4393.92
6091.18
3926.32
5227.20
5503.88
6575.15
4278.22
4938.30
5120.42
CORES (c*) from Ti II:
0*=3d²4s ⁴F
1*=3d³ ⁴F
2*=3d²4s ²F
3*=3d²4s ²D
4*=3d³ ²G
5*=3d³ ⁴P
6*=3d³ ²P
7*=3d²4s ⁴P
8*=3d³ ²D
9*=3d³ ²H
10*=3d²4s ²G
11*=3d²4s ²P
12*=3d³ ²F
13*=3d²4s ²S

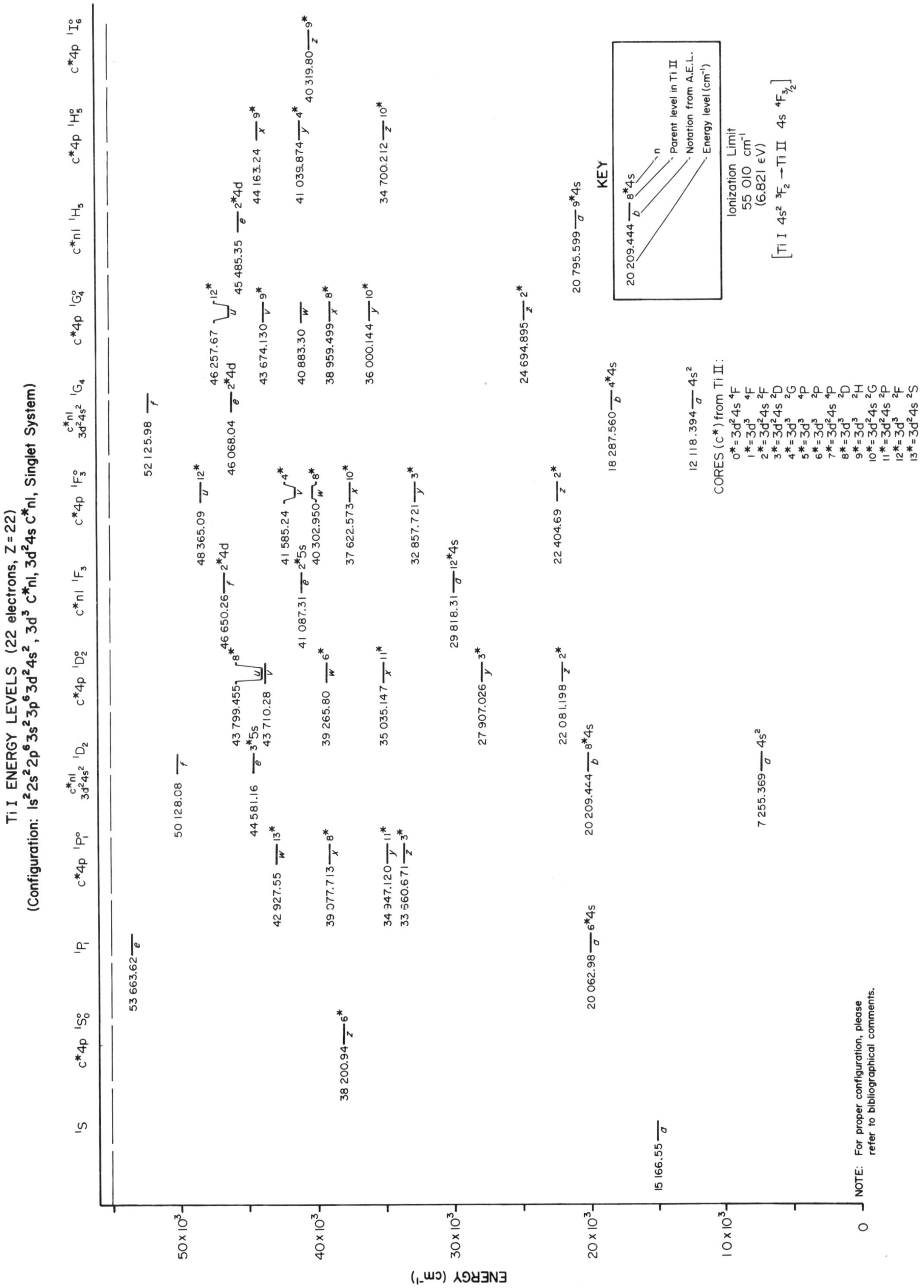

Ti I ENERGY LEVELS (22 electrons, Z = 22)

(Configuration: $1s^2 2s^2 2p^6 3s^2 3p^6 3d^2 4s^2$, $3d^3$ c*nl, $3d^2 4s$ c*nl, Singlet System)

ENERGY (cm^{-1}): 0, 10 x 10³, 20 x 10³, 30 x 10³, 40 x 10³, 50 x 10³

^1S
- 15 166.55 — a

c*4p $^1S^o_0$
- 38 200.94 — z 6*

$^1P_1$
- 53 663.62 — e
- 20 062.98 — a 6*4s

c*4p $^1P^o_1$
- 42 927.55 — w 13*
- 39 077.713 — x 8*
- 34 947.120 — y 11*
- 33 560.671 — z 3*

c*nl $3d^2 4s^2$ $^1D_2$
- 50 128.08 — f
- 44 581.16 — e 3*5s
- 20 209.444 — b 8*4s
- 7 255.369 — a $4s^2$

c*4p $^1D^o_2$
- 43 799.455 — u 8*
- 43 710.28 — v
- 39 265.80 — w 6*
- 35 035.147 — x 11*
- 27 907.026 — y 3*
- 22 081.198 — z 2*

c*nl $^1F_3$
- 46 650.26 — f 2*4d
- 41 087.31 — e 2*5s
- 29 818.31 — a 12*4s

c*4p $^1F^o_3$
- 48 365.09 — u 12*
- 41 585.24 — v 4*
- 40 302.950 — w 8*
- 37 622.573 — x 10*
- 32 857.721 — y 3*
- 22 404.69 — z 2*

c*nl $3d^2 4s^2$ $^1G_4$
- 52 125.98 — f
- 46 068.04 — e 2*4d
- 18 287.560 — b 4*4s
- 12 118.394 — a $4s^2$

c*4p $^1G^o_4$
- 46 257.67 — u 12*
- 43 674.130 — v 9*
- 40 883.30 — w
- 38 959.499 — x 8*
- 36 000.144 — y 10*
- 24 694.895 — z 2*

c*nl $^1H_5$
- 45 485.35 — e 2*4d
- 20 795.599 — a 9*4s

c*4p $^1H^o_5$
- 44 163.24 — x 9*
- 41 039.874 — y 4*
- 34 700.212 — z 10*

c*4p $^1I^o_6$
- 40 319.80 — z 9*

KEY

20 209.444 — b 8*4s
- n
- Parent level in Ti II
- Notation from A.E.L.
- Energy level (cm^{-1})

Ionization Limit
55 010 cm^{-1}
(6.821 eV)
[Ti I $4s^2$ $^3F_2$ → Ti II $4s$ $^4F_{3/2}$]

CORES (c*) from Ti II:
- 0* = $3d^2 4s$ ^4F
- 1* = $3d^3$ ^4F
- 2* = $3d^2 4s$ ^2F
- 3* = $3d^2 4s$ ^2D
- 4* = $3d^3$ ^2G
- 5* = $3d^3$ ^4P
- 6* = $3d^3$ ^2P
- 7* = $3d^2 4s$ ^4P
- 8* = $3d^3$ ^2D
- 9* = $3d^3$ ^2H
- 10* = $3d^2 4s$ ^2G
- 11* = $3d^2 4s$ ^2P
- 12* = $3d^3$ ^2F
- 13* = $3d^2 4s$ ^2S

NOTE: For proper configuration, please refer to bibliographical comments.

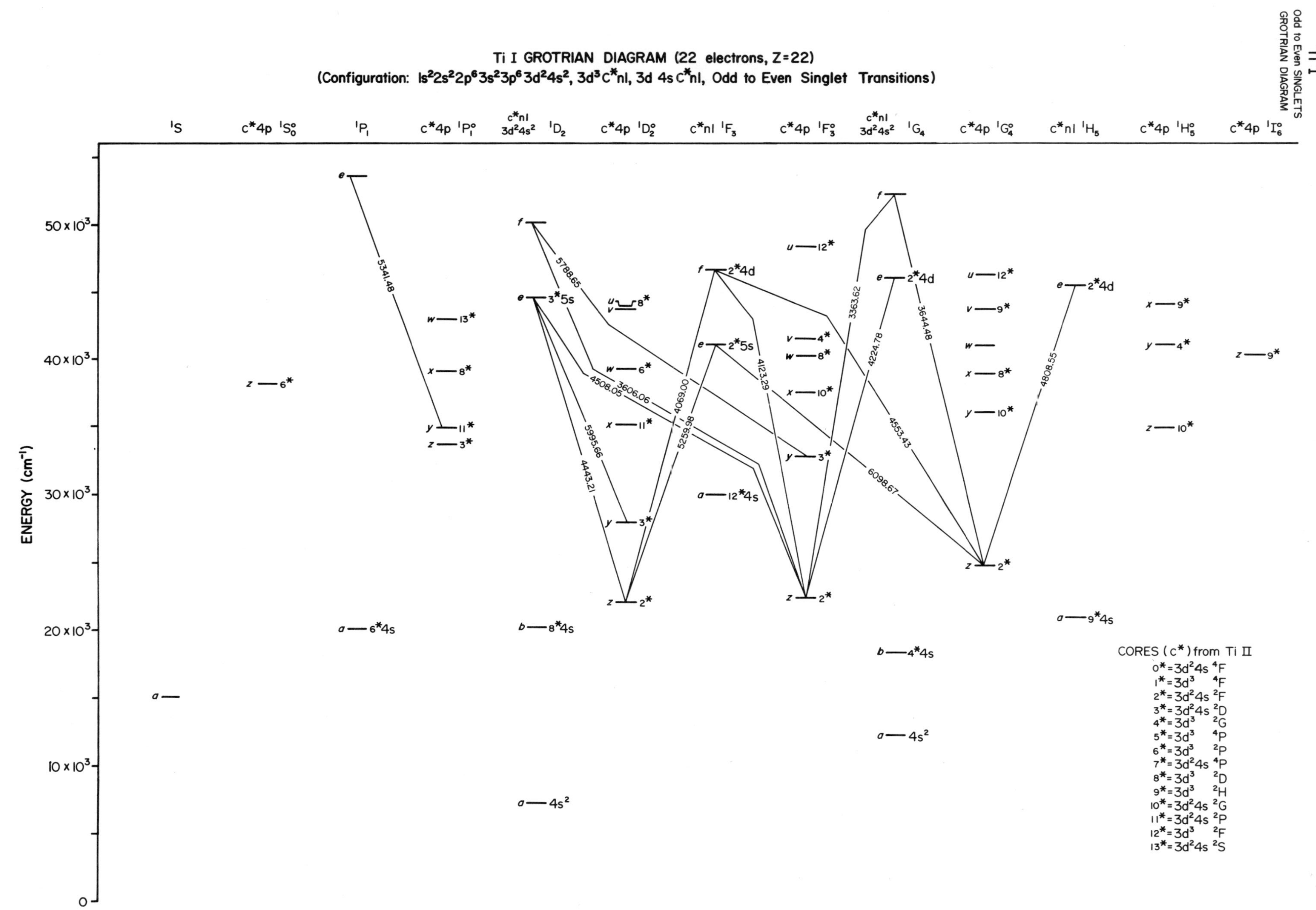
Ti I GROTRIAN DIAGRAM (22 electrons, Z=22)
(Configuration: 1s²2s²2p⁶3s²3p⁶3d²4s², 3d³c*nl, 3d 4sc*nl, Odd to Even Singlet Transitions)
¹S
c*4p ¹S°₀
¹P₁
c*4p ¹P°₁
c*nl 3d²4s² ¹D₂
c*4p ¹D°₂
c*nl ¹F₃
c*4p ¹F°₃
c*nl 3d²4s² ¹G₄
c*4p ¹G°₄
c*nl ¹H₅
c*4p ¹H°₅
c*4p ¹I°₆
ENERGY (cm⁻¹)
50 x 10³
40 x 10³
30 x 10³
20 x 10³
10 x 10³
0
5341.48
5788.65
4508.05
3606.06
5995.66
4443.21
4069.00
5259.98
4123.29
3363.62
4224.78
3644.48
4553.43
6098.67
4808.55
CORES (c*) from Ti II
0* = 3d²4s ⁴F
1* = 3d³ ⁴F
2* = 3d²4s ²F
3* = 3d²4s ²D
4* = 3d³ ²G
5* = 3d³ ⁴P
6* = 3d³ ²P
7* = 3d²4s ⁴P
8* = 3d³ ²D
9* = 3d³ ²H
10* = 3d²4s ²G
11* = 3d²4s ²P
12* = 3d³ ²F
13* = 3d²4s ²S

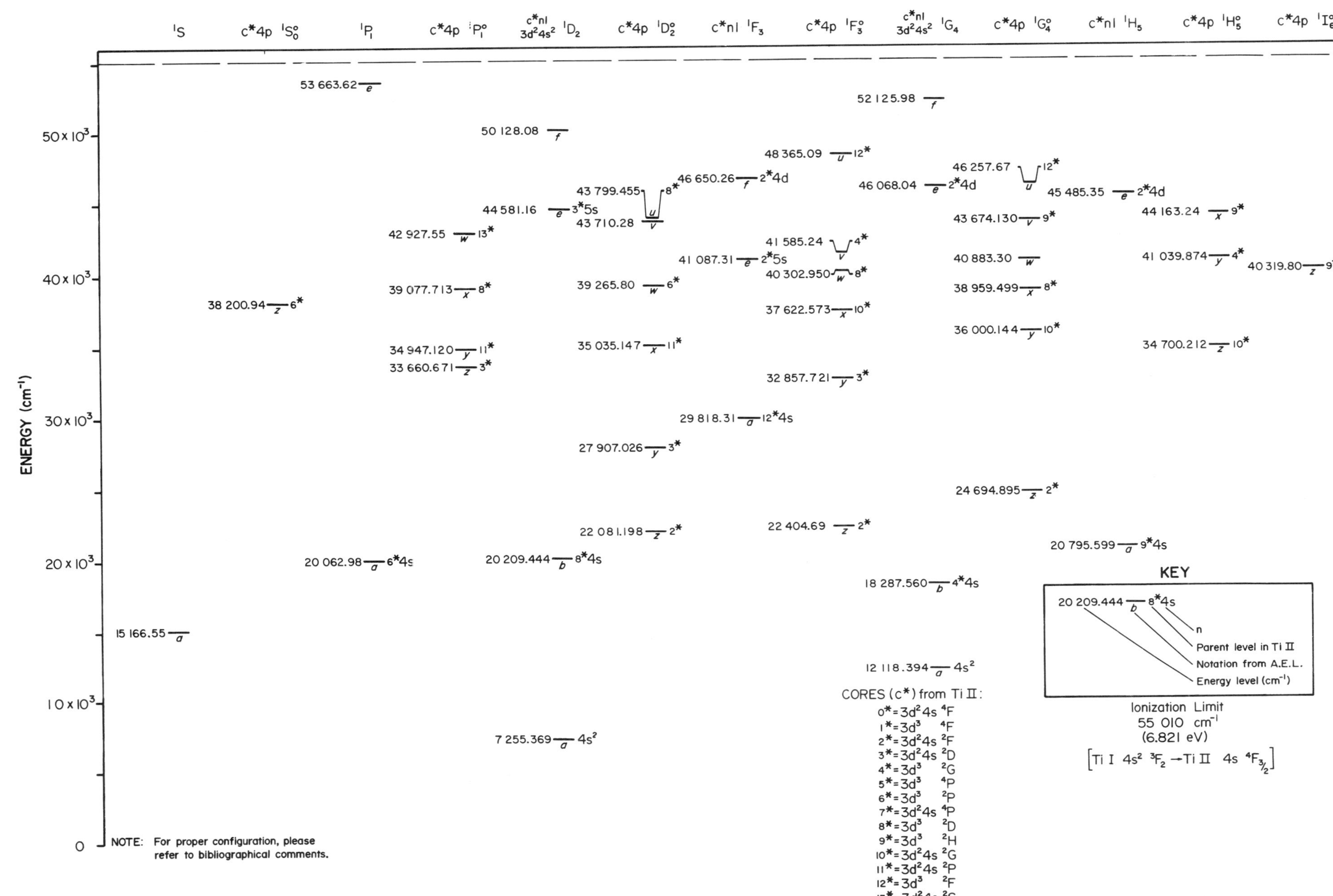
Ti I ENERGY LEVELS (22 electrons, Z=22)
(Configuration: 1s²2s²2p⁶3s²3p⁶3d²4s², 3d³ c*nl, 3d²4s c*nl, Singlet System)
¹S
c*4p ¹S°₀
¹P₁
c*4p ¹P°₁
c*nl 3d²4s² ¹D₂
c*4p ¹D°₂
c*nl ¹F₃
c*4p ¹F°₃
c*nl 3d²4s² ¹G₄
c*4p ¹G°₄
c*nl ¹H₅
c*4p ¹H°₅
c*4p ¹I°₆
ENERGY (cm⁻¹)
50 x 10³
40 x 10³
30 x 10³
20 x 10³
10 x 10³
0
15 166.55 a
38 200.94 z 6*
53 663.62 e
20 062.98 a 6*4s
42 927.55 w 13*
39 077.713 x 8*
34 947.120 y 11*
33 660.671 z 3*
50 128.08 f
44 581.16 e 3*5s
20 209.444 b 8*4s
7 255.369 a 4s²
43 799.455 u 8*
43 710.28 v
39 265.80 w 6*
35 035.147 x 11*
27 907.026 y 3*
22 081.198 z 2*
46 650.26 f 2*4d
41 087.31 e 2*5s
29 818.31 a 12*4s
48 365.09 u 12*
41 585.24 v 4*
40 302.950 w 8*
37 622.573 x 10*
32 857.721 y 3*
22 404.69 z 2*
52 125.98 f
46 068.04 e 2*4d
18 287.560 b 4*4s
12 118.394 a 4s²
46 257.67 u 12*
43 674.130 v 9*
40 883.30 w
38 959.499 x 8*
36 000.144 y 10*
24 694.895 z 2*
45 485.35 e 2*4d
20 795.599 a 9*4s
44 163.24 x 9*
41 039.874 y 4*
34 700.212 z 10*
40 319.80 z 9*
KEY
20 209.444 b 8*4s
n
Parent level in Ti II
Notation from A.E.L.
Energy level (cm⁻¹)
Ionization Limit
55 010 cm⁻¹
(6.821 eV)
[Ti I 4s² ³F₂ → Ti II 4s ⁴F₃/₂]
CORES (c*) from Ti II:
0* = 3d²4s ⁴F
1* = 3d³ ⁴F
2* = 3d²4s ²F
3* = 3d²4s ²D
4* = 3d³ ²G
5* = 3d³ ⁴P
6* = 3d³ ²P
7* = 3d²4s ⁴P
8* = 3d³ ²D
9* = 3d³ ²H
10* = 3d²4s ²G
11* = 3d²4s ²P
12* = 3d³ ²F
13* = 3d²4s ²S
NOTE: For proper configuration, please refer to bibliographical comments.

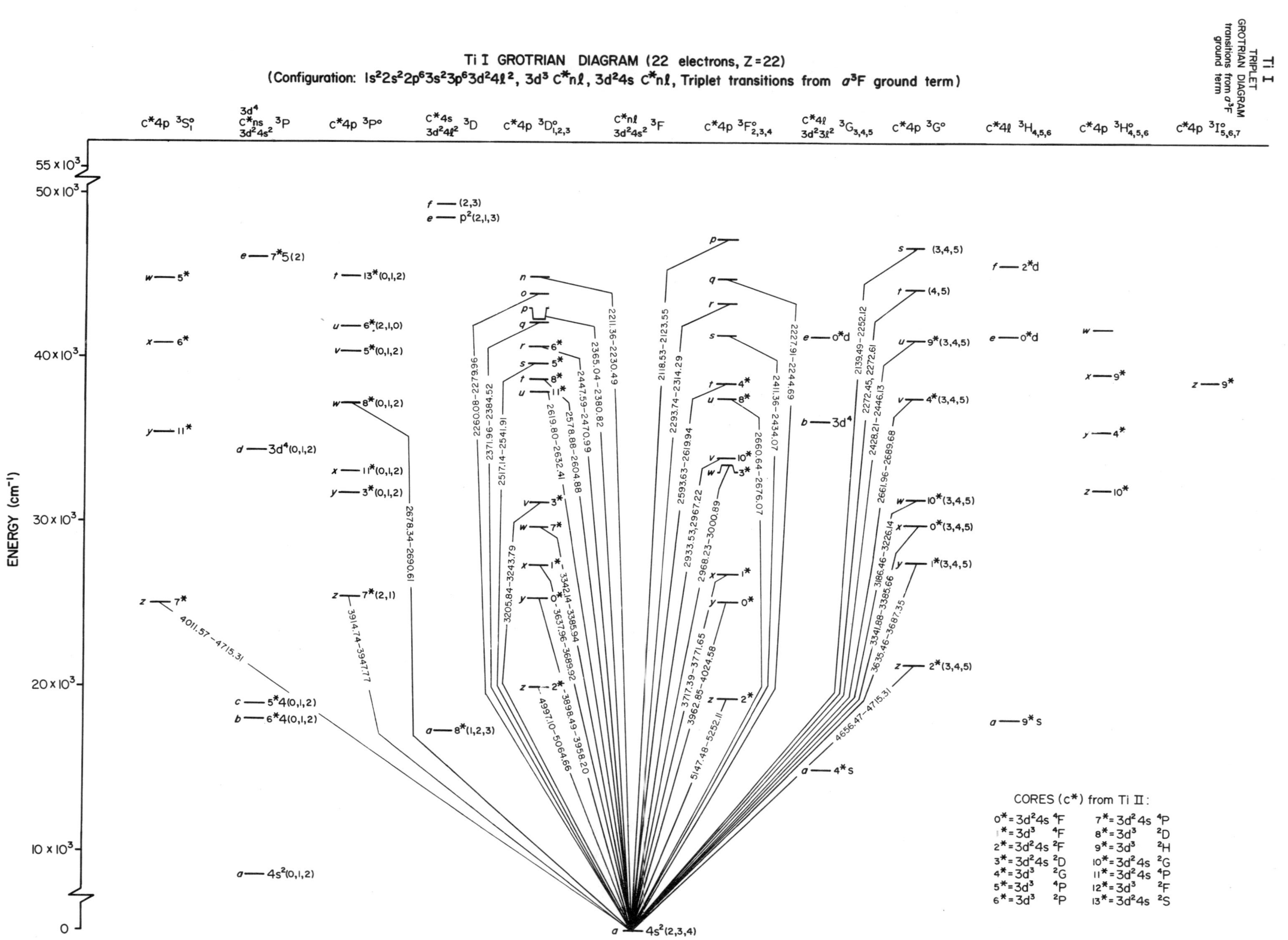
Ti I
TRIPLET
GROTRIAN DIAGRAM
transitions from a³F ground term
Ti I GROTRIAN DIAGRAM (22 electrons, Z=22)
(Configuration: 1s²2s²2p⁶3s²3p⁶3d²4ℓ², 3d³ c*nℓ, 3d²4s c*nℓ, Triplet transitions from a³F ground term)
c*4p ³S°₁
3d⁴ c*ns 3d²4s² ³P
c*4p ³P°
c*4s 3d²4ℓ² ³D
c*4p ³D°₁,₂,₃
c*nℓ 3d²4s² ³F
c*4p ³F°₂,₃,₄
c*4ℓ 3d²3ℓ² ³G₃,₄,₅
c*4p ³G°
c*4ℓ ³H₄,₅,₆
c*4p ³H°₄,₅,₆
c*4p ³I°₅,₆,₇
ENERGY (cm⁻¹)
55 x 10³
50 x 10³
40 x 10³
30 x 10³
20 x 10³
10 x 10³
0
w — 5*
x — 6*
y — 11*
z — 7*
4011.57 – 4715.31
e — 7*5(2)
d — 3d⁴(0,1,2)
c — 5*4(0,1,2)
b — 6*4(0,1,2)
a — 4s²(0,1,2)
t — 13*(0,1,2)
u — 6*(2,1,0)
v — 5*(0,1,2)
w — 8*(0,1,2)
x — 11*(0,1,2)
y — 3*(0,1,2)
z — 7*(2,1)
3914.74–3947.77
2678.34–2690.61
f — (2,3)
e — p²(2,1,3)
a — 8*(1,2,3)
n
o
p
q
r — 6*
s — 5*
t — 8*
u — 11*
v — 3*
w — 7*
x — 1*
y — 0*
z — 2*
2260.08–2279.96
2371.96–2384.52
2517.14–2541.91
2619.80–2632.41
2578.88–2604.88
2447.59–2470.99
2365.04–2380.82
2211.36–2230.49
3205.84–3243.79
3342.14–3385.94
3637.96–3689.92
3898.49–3958.20
4997.10–5064.66
a — 4s²(2,3,4)
p
q
r
s
t — 4*
u — 8*
v — 10*
w — 3*
x — 1*
y — 0*
z — 2*
2118.53–2123.55
2293.74–2314.29
2593.63–2619.94
2933.53, 2967.22
2968.23–3000.89
3717.39–3771.65
3962.85–4024.58
5147.48–5252.11
2411.36–2434.07
2660.64–2676.07
2227.91–2244.69
e — 0*d
b — 3d⁴
a — 4*s
s — (3,4,5)
t — (4,5)
u — 9*(3,4,5)
v — 4*(3,4,5)
w — 10*(3,4,5)
x — 0*(3,4,5)
y — 1*(3,4,5)
z — 2*(3,4,5)
2139.49–2252.12
2272.45, 2272.61
2428.21–2446.13
2661.96–2689.68
3186.46–3226.14
3341.88–3385.66
3635.46–3687.35
4656.47–4715.31
f — 2*d
e — 0*d
a — 9*s
w —
x — 9*
y — 4*
z — 10*
z — 9*
CORES (c*) from Ti II:
0* = 3d²4s ⁴F
1* = 3d³ ⁴F
2* = 3d²4s ²F
3* = 3d²4s ²D
4* = 3d³ ²G
5* = 3d³ ⁴P
6* = 3d³ ²P
7* = 3d²4s ⁴P
8* = 3d³ ²D
9* = 3d³ ²H
10* = 3d²4s ²G
11* = 3d²4s ⁴P
12* = 3d³ ²F
13* = 3d²4s ²S

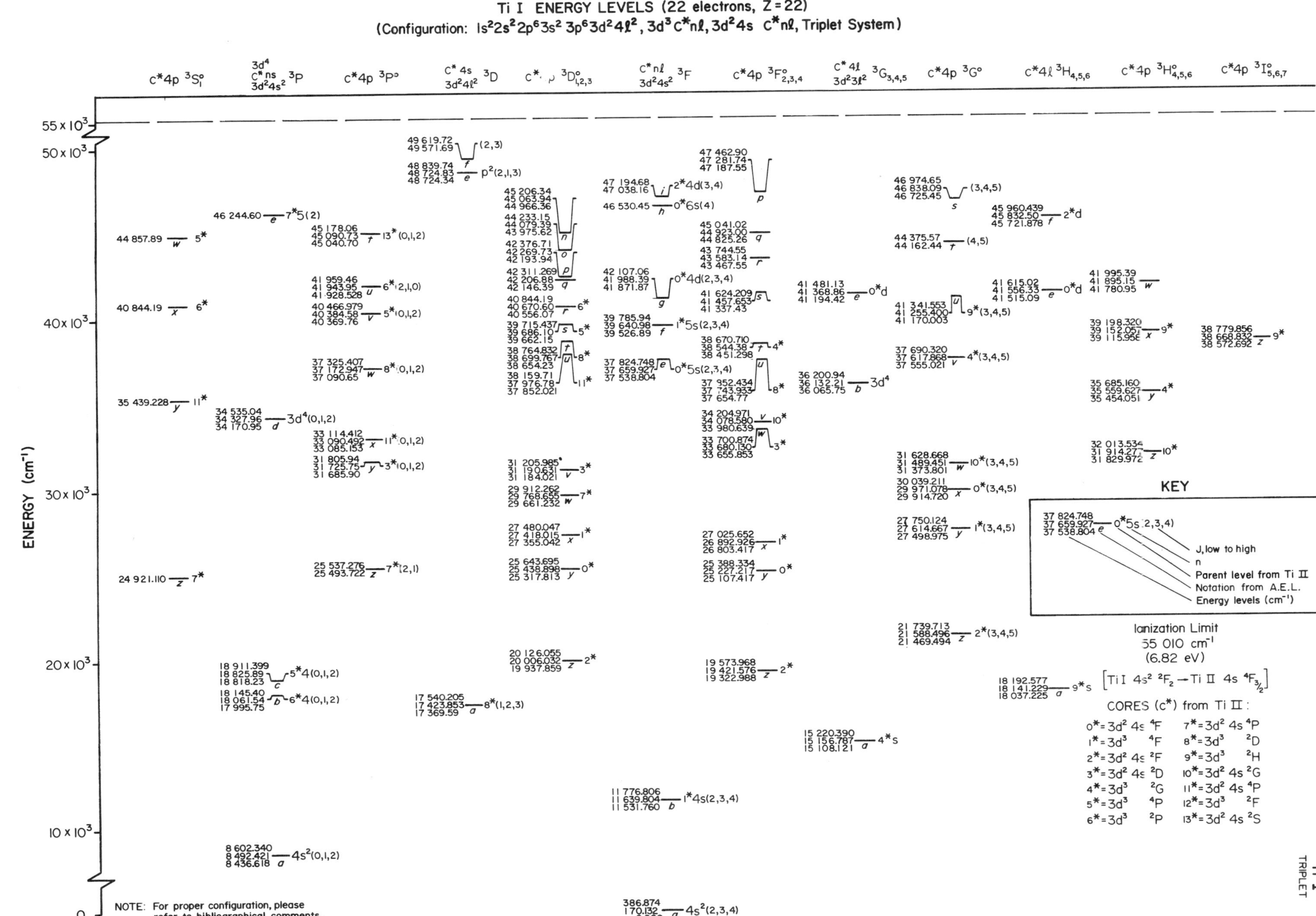

Ti I ENERGY LEVELS (22 electrons, Z=22)
(Configuration: 1s²2s²2p⁶3s² 3p⁶3d²4ℓ², 3d³ c*nℓ, 3d²4s c*nℓ, Triplet System)
c*4p ³S°₁
3d⁴ c*ns 3d²4s² ³P
c*4p ³P°
c*4s 3d²4ℓ² ³D
c*, p ³D°₁,₂,₃
c*nℓ 3d²4s² ³F
c*4p ³F°₂,₃,₄
c*4ℓ 3d²3ℓ² ³G₃,₄,₅
c*4p ³G°
c*4ℓ ³H₄,₅,₆
c*4p ³H°₄,₅,₆
c*4p ³I°₅,₆,₇
ENERGY (cm⁻¹)
55 x 10³
50 x 10³
40 x 10³
30 x 10³
20 x 10³
10 x 10³
0
44 857.89 w 5*
40 844.19 x 6*
35 439.228 y 11*
24 921.110 z 7*
46 244.60 e 7*5(2)
34 535.04 34 327.96 34 170.95 d 3d⁴(0,1,2)
18 911.399 18 825.89 18 818.23 c 5*4(0,1,2)
18 145.40 18 061.54 17 995.75 b 6*4(0,1,2)
8 602.340 8 492.421 8 436.618 a 4s²(0,1,2)
45 178.06 45 090.73 45 040.70 t 13*(0,1,2)
41 959.46 41 943.95 41 928.528 u 6*(2,1,0)
40 466.979 40 384.58 40 369.76 v 5*(0,1,2)
37 325.407 37 172.947 37 090.65 w 8*(0,1,2)
33 114.412 33 090.492 33 085.153 x 11*(0,1,2)
31 805.94 31 725.75 31 685.90 y 3*(0,1,2)
25 537.276 25 493.722 z 7*(2,1)
49 619.72 49 571.69 f (2,3)
48 839.74 48 724.83 48 724.34 e p²(2,1,3)
17 540.205 17 423.853 17 369.59 a 8*(1,2,3)
45 206.34 45 063.94 44 966.36
44 233.15 44 079.39 43 975.62 n
42 376.71 42 269.73 42 193.94 o
42 311.269 42 206.88 42 146.39 p q
40 844.19 40 670.60 40 556.07 r 6*
39 715.437 39 686.10 39 662.15 s 5*
38 764.832 38 699.767 38 654.23 t u 8*
38 159.71 37 976.78 37 852.021 11*
31 205.985 31 190.631 31 184.021 v 3*
29 912.262 29 768.655 29 661.232 w 7*
27 480.047 27 418.015 27 355.042 x 1*
25 643.695 25 438.898 25 317.813 y 0*
20 126.055 20 006.032 19 937.859 z 2*
47 194.68 47 038.16 i 2*4d(3,4)
46 530.45 h 0*6s(4)
42 107.06 41 988.39 41 871.87 g 0*4d(2,3,4)
39 785.94 39 640.98 39 526.89 f 1*5s(2,3,4)
37 824.748 37 659.927 37 538.804 e 0*5s(2,3,4)
11 776.806 11 639.804 11 531.760 b 1*4s(2,3,4)
386.874 170.132 0.000 a 4s²(2,3,4)
47 462.90 47 281.74 47 187.55 p
45 041.02 44 923.00 44 825.26 q
43 744.55 43 583.14 43 467.55 r
41 624.209 41 457.653 41 337.43 s
38 670.710 38 544.38 38 451.298 t 4*
37 952.434 37 743.933 37 654.77 u 8*
34 204.971 34 078.580 33 980.639 v 10*
33 700.874 33 680.130 33 655.853 w 3*
27 025.652 26 892.926 26 803.417 x 1*
25 388.334 25 227.217 25 107.417 y 0*
19 573.968 19 421.576 19 322.988 z 2*
41 481.13 41 368.86 41 194.42 e 0*d
36 200.94 36 132.21 36 065.75 b 3d⁴
15 220.390 15 156.787 15 108.121 a 4*s
46 974.65 46 838.09 46 725.45 s (3,4,5)
44 375.57 44 162.44 t (4,5)
41 341.553 41 255.400 41 170.003 u 9*(3,4,5)
37 690.320 37 617.868 37 555.021 v 4*(3,4,5)
31 628.668 31 489.451 31 373.801 w 10*(3,4,5)
30 039.211 29 971.078 29 914.720 x 0*(3,4,5)
27 750.124 27 614.667 27 498.975 y 1*(3,4,5)
21 739.713 21 588.496 21 469.494 z 2*(3,4,5)
45 960.439 45 832.50 45 721.878 f 2*d
41 615.02 41 556.33 41 515.09 e 0*d
18 192.577 18 141.229 18 037.225 a 9*s
41 995.39 41 895.15 41 780.95 w
39 198.320 39 152.057 39 115.958 x 9*
35 685.160 35 559.627 35 454.051 y 4*
32 013.534 31 914.277 31 829.972 z 10*
38 779.856 38 668.832 38 572.692 z 9*
KEY
37 824.748 37 659.927 37 538.804 e 0*5s(2,3,4)
J, low to high
n
Parent level from Ti II
Notation from A.E.L.
Energy levels (cm⁻¹)
Ionization Limit
55 010 cm⁻¹
(6.82 eV)
[Ti I 4s² ²F₂ → Ti II 4s ⁴F₃/₂]
CORES (c*) from Ti II:
0*=3d² 4s ⁴F
1*=3d³ ⁴F
2*=3d² 4s ²F
3*=3d² 4s ²D
4*=3d³ ²G
5*=3d³ ⁴P
6*=3d³ ²P
7*=3d² 4s ⁴P
8*=3d³ ²D
9*=3d³ ²H
10*=3d² 4s ²G
11*=3d² 4s ⁴P
12*=3d³ ²F
13*=3d² 4s ²S
NOTE: For proper configuration, please refer to bibliographical comments.
Ti I TRIPLET

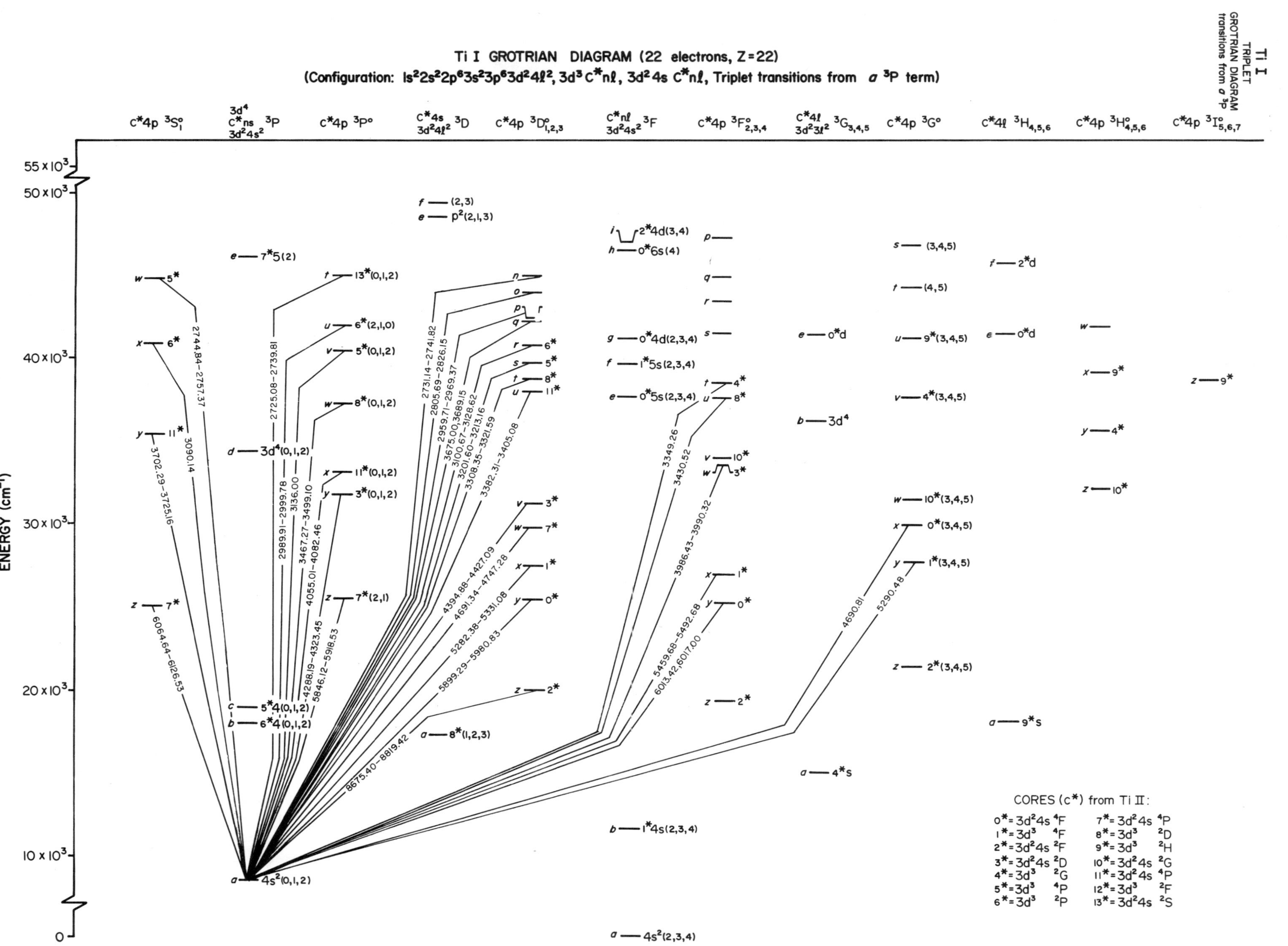
Ti I
TRIPLET
GROTRIAN DIAGRAM
transitions from a ³P
Ti I GROTRIAN DIAGRAM (22 electrons, Z=22)
(Configuration: 1s²2s²2p⁶3s²3p⁶3d²4ℓ², 3d³ c*nℓ, 3d² 4s c*nℓ, Triplet transitions from a ³P term)
c*4p ³S°₁
3d⁴ c*ns 3d²4s² ³P
c*4p ³P°
c*4s 3d²4ℓ² ³D
c*4p ³D°₁,₂,₃
c*nℓ 3d²4s² ³F
c*4p ³F°₂,₃,₄
c*4ℓ 3d²3ℓ² ³G₃,₄,₅
c*4p ³G°
c*4ℓ ³H₄,₅,₆
c*4p ³H°₄,₅,₆
c*4p ³I°₅,₆,₇
ENERGY (cm⁻¹)
55 x 10³
50 x 10³
40 x 10³
30 x 10³
20 x 10³
10 x 10³
0
2744.84–2757.37
3090.14
3702.29–3725.16
6064.64–6126.53
2725.08–2739.81
2989.91–2999.78
3136.00
3467.27–3499.10
4055.01–4082.46
4288.19–4323.45
5846.12–5918.53
2731.14–2741.82
2805.69–2826.15
2959.71–2969.37
3675.00, 3689.15
3100.67–3128.62
3201.60–3213.16
3308.35–3321.59
3382.31–3405.08
4394.88–4427.09
4691.34–4747.28
5282.38–5331.08
5899.29–5980.83
8675.40–8819.42
3349.26
3430.52
3986.43–3990.32
5459.68–5492.68
6013.42, 6017.00
4690.81
5290.48
a — 4s²(0,1,2)
b — 6*4(0,1,2)
c — 5*4(0,1,2)
d — 3d⁴(0,1,2)
e — 7*5(2)
f — (2,3)
e — p²(2,1,3)
a — 8*(1,2,3)
a — 4s²(2,3,4)
b — 1*4s(2,3,4)
e — 0*5s(2,3,4)
f — 1*5s(2,3,4)
g — 0*4d(2,3,4)
h — 0*6s(4)
i — 2*4d(3,4)
b — 3d⁴
a — 4*s
e — 0*d
a — 9*s
f — 2*d
z — 2*(3,4,5)
y — 1*(3,4,5)
x — 0*(3,4,5)
w — 10*(3,4,5)
v — 4*(3,4,5)
u — 9*(3,4,5)
t — (4,5)
s — (3,4,5)
CORES (c*) from Ti II:
0* = 3d²4s ⁴F
1* = 3d³ ⁴F
2* = 3d²4s ²F
3* = 3d²4s ²D
4* = 3d³ ²G
5* = 3d³ ⁴P
6* = 3d³ ²P
7* = 3d²4s ⁴P
8* = 3d³ ²D
9* = 3d³ ²H
10* = 3d²4s ²G
11* = 3d²4s ⁴P
12* = 3d³ ²F
13* = 3d²4s ²S

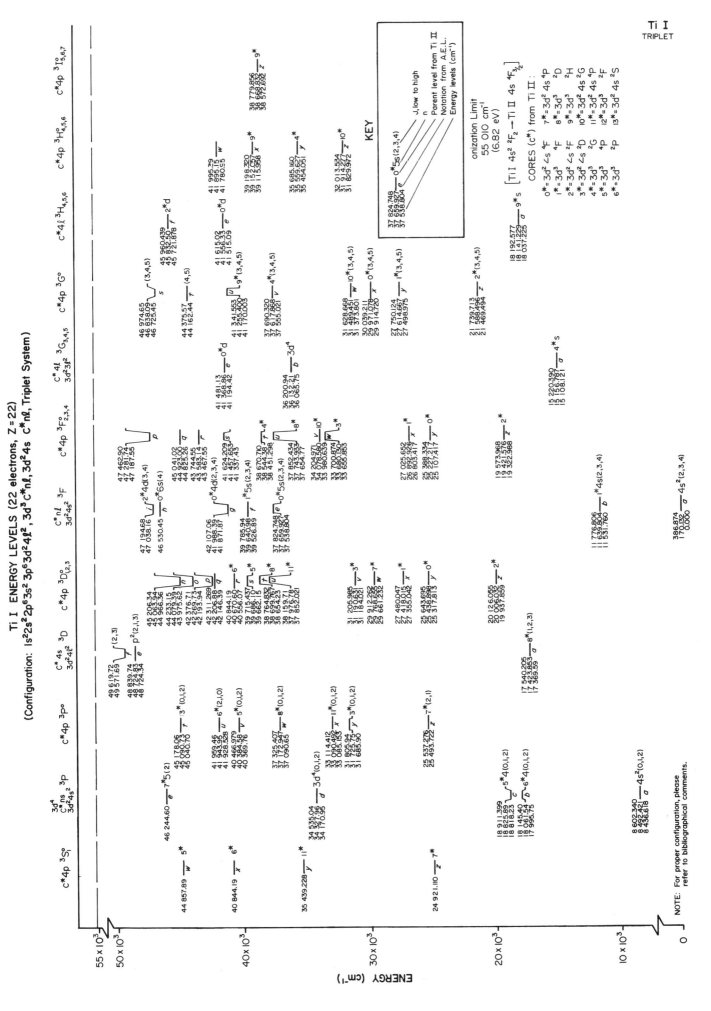

Ti I
TRIPLET

Ti I ENERGY LEVELS (22 electrons, Z = 22)

(Configuration: $1s^2 2s^2 2p^6 3s^2 3p^6 3d^2 4\ell^2$, $3d^3 c^* n\ell$, $3d^2 4s$ $c^* n\ell$, Triplet System)

ENERGY (cm^{-1}): 0, 10×10^3, 20×10^3, 30×10^3, 40×10^3, 50×10^3, 55×10^3

c^*4p $^3S^o_1$

- 44 857.89 w 5*
- 40 844.19 x 6*
- 35 439.228 y 11*
- 24 921.110 z 7*

$3d^4$, c^*ns, $3d^2 4s^2$ ^3P

- 46 244.60 e $7^*5(2)$
- 34 535.04 / 34 327.96 / 34 170.95 d $3d^4(0,1,2)$
- 18 911.399 / 18 825.89 / 18 818.23 c $5^*4(0,1,2)$
- 18 145.40 / 18 061.54 / 17 995.75 b $6^*4(0,1,2)$
- 8 602.340 / 8 492.421 / 8 436.618 a $4s^2(0,1,2)$

c^*4p $^3P^o$

- 45 178.06 / 45 090.73 / 45 040.70 t 13*(0,1,2)
- 41 959.46 / 41 943.95 / 41 928.528 u 6*(2,1,0)
- 40 466.979 / 40 384.58 / 40 369.76 v 5*(0,1,2)
- 37 325.407 / 37 172.947 / 37 090.65 w 8*(0,1,2)
- 33 114.412 / 33 090.492 / 33 085.153 x 11*(0,1,2)
- 31 805.94 / 31 725.75 / 31 685.90 y 3*(0,1,2)
- 25 537.276 / 25 493.722 z 1*(2,1)

c^*4s $3d^2 4\ell^2$ ^3D

- 49 619.72 / 49 571.69 f (2,3)
- 48 839.74 / 48 724.83 / 48 724.34 e $p^2(2,1,3)$
- 17 540.205 / 17 423.853 / 17 369.59 a 8*(1,2,3)

c^*4p $^3D^o_{1,2,3}$

- 45 206.34 / 45 063.94 / 44 966.36 n
- 44 233.15 / 44 079.39 / 43 975.62 o
- 42 376.71 / 42 269.73 / 42 193.94 p
- 42 311.269 / 42 206.88 / 42 146.39 q
- 40 844.19 / 40 670.60 / 40 556.07 r 6*
- 39 715.437 / 39 686.10 / 39 662.15 s 5*
- 38 764.832 / 38 699.767 / 38 654.23 t u 8*
- 38 159.71 / 37 976.78 / 37 852.021 11*
- 31 205.985 / 31 190.631 / 31 184.021 v 3*
- 29 912.262 / 29 768.655 / 29 661.232 w 7*
- 27 480.047 / 27 418.015 / 27 355.042 x 1*
- 25 643.695 / 25 438.898 / 25 317.813 y 0*
- 20 126.055 / 20 006.032 / 19 937.859 z 2*

$c^*n\ell$ $3d^2 4s^2$ ^3F

- 47 194.68 / 47 038.16 i $2^*4d(3,4)$
- 46 530.45 h $0^*6s(4)$
- 42 107.06 / 41 988.39 / 41 871.87 g $0^*4d(2,3,4)$
- 39 785.94 / 39 640.98 / 39 526.89 f $1^*5s(2,3,4)$
- 37 824.748 / 37 659.927 / 37 538.804 e $0^*5s(2,3,4)$
- 11 776.806 / 11 639.804 / 11 531.760 b $1^*4s(2,3,4)$
- 386.874 / 170.132 / 0.000 a $4s^2(2,3,4)$

c^*4p $^3F^o_{2,3,4}$

- 47 462.90 / 47 281.74 / 47 187.55 p
- 45 041.02 / 44 923.00 / 44 825.26 q
- 43 744.55 / 43 583.14 / 43 467.55 r
- 41 624.209 / 41 457.653 / 41 337.43 s
- 38 670.710 / 38 544.38 / 38 451.298 t 4*
- 37 852.434 / 37 743.933 / 37 654.77 u 8*
- 34 204.971 / 34 078.580 / 33 980.639 v 10*
- 33 700.874 / 33 680.130 / 33 655.853 w 3*
- 27 025.652 / 26 892.926 / 26 803.417 x 1*
- 25 388.334 / 25 227.217 / 25 107.417 y 0*
- 19 573.968 / 19 421.576 / 19 322.988 z 2*

$c^*4\ell$ $3d^2 3\ell^2$ $^3G_{3,4,5}$

- 41 481.13 / 41 368.86 / 41 194.42 e 0*d
- 36 200.94 / 36 132.21 / 36 065.75 b $3d^4$
- 15 220.390 / 15 156.787 / 15 108.121 a 4*s

c^*4p $^3G^o$

- 46 974.65 / 46 838.09 / 46 725.45 s (3,4,5)
- 44 375.57 / 44 162.44 t (4,5)
- 41 341.553 / 41 255.400 / 41 170.003 u 9*(3,4,5)
- 37 690.320 / 37 617.868 / 37 555.021 v 4*(3,4,5)
- 31 628.668 / 31 489.451 / 31 373.801 w 10*(3,4,5)
- 30 039.211 / 29 971.078 / 29 914.720 x 0*(3,4,5)
- 27 750.124 / 27 614.667 / 27 498.975 y 1*(3,4,5)
- 21 739.713 / 21 588.496 / 21 469.494 z 2*(3,4,5)

$c^*4\ell$ $^3H_{4,5,6}$

- 45 960.439 / 45 832.50 / 45 721.878 f 2*d
- 41 615.02 / 41 556.33 / 41 515.09 e 0*d
- 18 192.577 / 18 141.229 / 18 037.225 a 9*s

c^*4p $^3H^o_{4,5,6}$

- 41 995.29 / 41 895.15 / 41 780.55 w
- 39 198.320 / 39 152.057 / 39 115.558 x 9*
- 35 685.160 / 35 559.627 / 35 454.051 y 4*
- 32 013.534 / 31 914.277 / 31 829.972 z 10*

c^*4p $^3I^o_{5,6,7}$

- 38 779.856 / 38 668.832 / 38 572.692 z 9*

KEY

37 824.748 / 37 659.927 / 37 538.804 e $0^*5s(2,3,4)$

- J, low to high
- n
- Parent level from Ti II
- Notation from A.E.L.
- Energy levels (cm^{-1})

onization Limit
55 010 cm^{-1}
(6.82 eV)

[Ti I $4s^2$ $^2F_2$ → Ti II $4s$ $^4F_{3/2}$]

CORES (c^*) from Ti II:

$0^* = 3d^2 4s$ ^4F	$7^* = 3d^2 4s$ ^4P
$1^* = 3d^3$ ^4F	$8^* = 3d^3$ ^2D
$2^* = 3d^2 4s$ ^2F	$9^* = 3d^3$ ^2H
$3^* = 3d^2 4s$ ^2D	$10^* = 3d^2 4s$ ^2G
$4^* = 3d^3$ ^2G	$11^* = 3d^2 4s$ ^4P
$5^* = 3d^3$ ^4P	$12^* = 3d^3$ ^2F
$6^* = 3d^3$ ^2P	$13^* = 3d^2 4s$ ^2S

NOTE: For proper configuration, please refer to bibliographical comments.

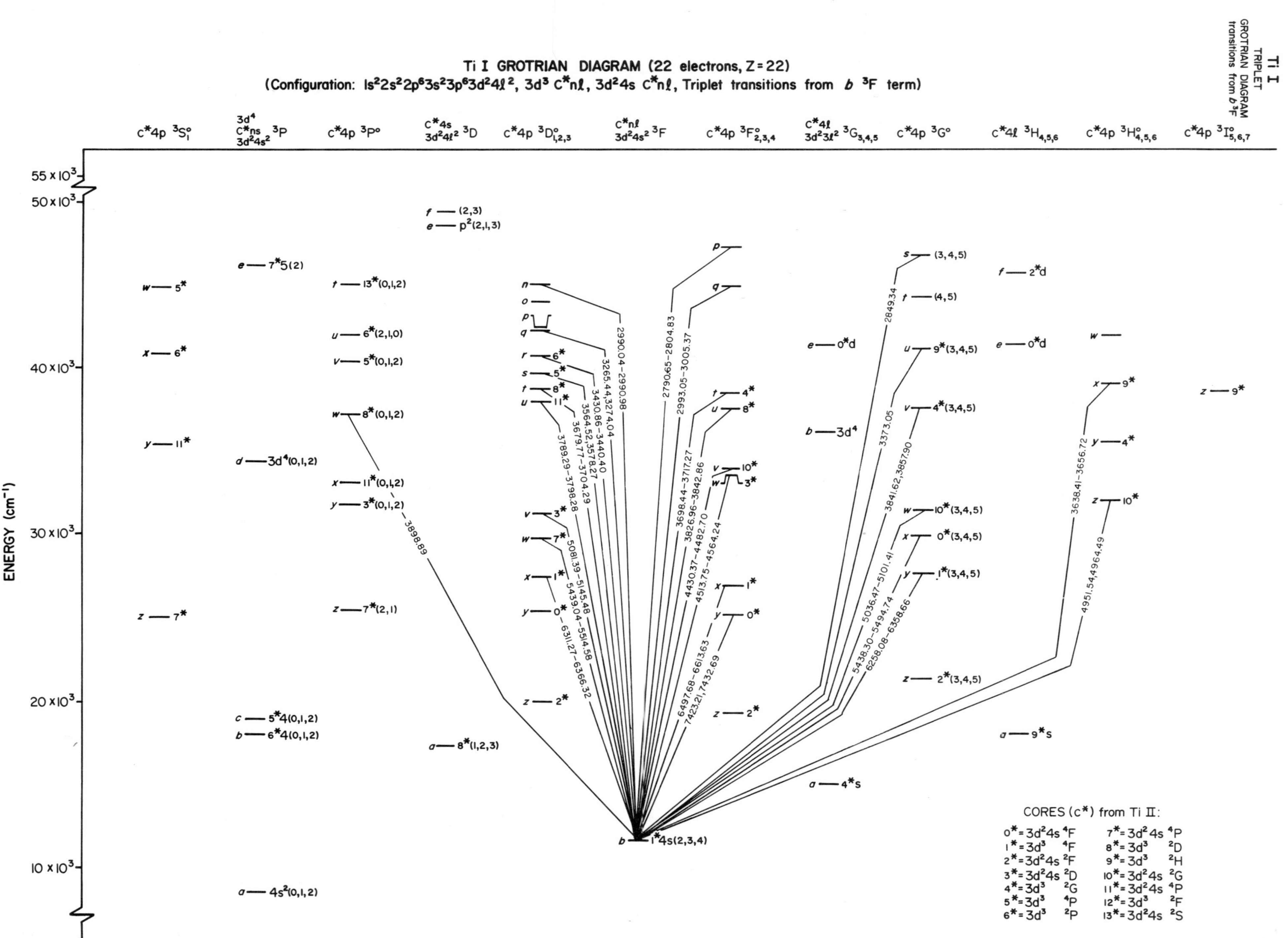

Ti I
TRIPLET
GROTRIAN DIAGRAM
transitions from $b\,^3F$
Ti I GROTRIAN DIAGRAM (22 electrons, Z = 22)
(Configuration: $1s^22s^22p^63s^23p^63d^24\ell^2$, $3d^3\,c^*n\ell$, $3d^24s\,c^*n\ell$, Triplet transitions from $b\,^3F$ term)
ENERGY (cm^{-1})
CORES (c*) from Ti II:
$0^* = 3d^24s\ ^4F$
$1^* = 3d^3\ ^4F$
$2^* = 3d^24s\ ^2F$
$3^* = 3d^24s\ ^2D$
$4^* = 3d^3\ ^2G$
$5^* = 3d^3\ ^4P$
$6^* = 3d^3\ ^2P$
$7^* = 3d^24s\ ^4P$
$8^* = 3d^3\ ^2D$
$9^* = 3d^3\ ^2H$
$10^* = 3d^24s\ ^2G$
$11^* = 3d^24s\ ^4P$
$12^* = 3d^3\ ^2F$
$13^* = 3d^24s\ ^2S$

Ti I
TRIPLET

Ti I ENERGY LEVELS (22 electrons, Z = 22)

(Configuration: $1s^2 2s^2 2p^6 3s^2 3p^6 3d^2 4\ell^2$, $3d^3 c^* n\ell$, $3d^2 4s\ c^* n\ell$, Triplet System)

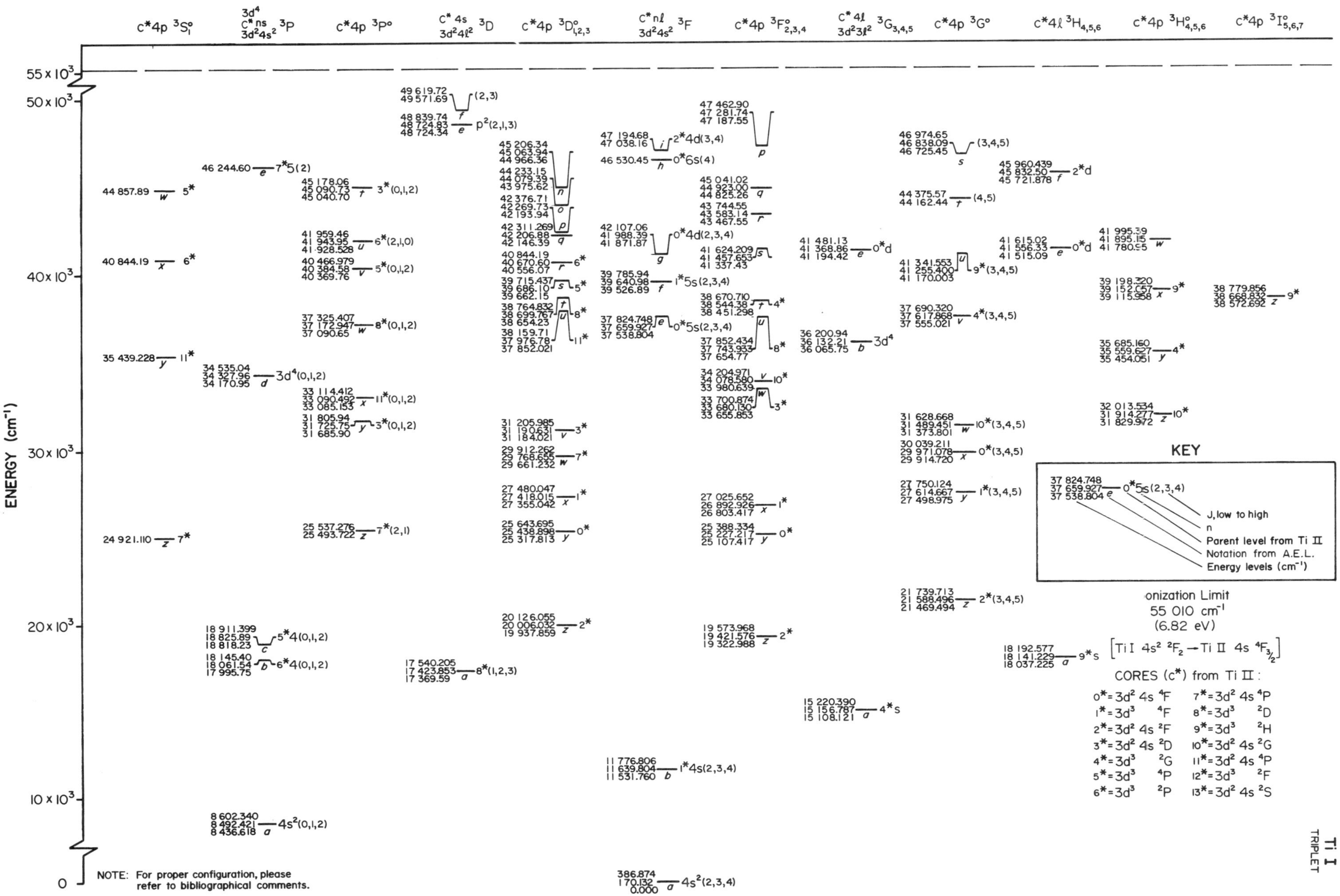

ENERGY (cm^{-1}): 0, 10×10^3, 20×10^3, 30×10^3, 40×10^3, 50×10^3, 55×10^3

$c^*4p\ ^3S^o_1$

- 44 857.89 w 5*
- 40 844.19 x 6*
- 35 439.228 y 11*
- 24 921.110 z 7*

$3d^4$, c^*ns, $3d^2 4s^2$ ^3P

- 46 244.60 e 7*5(2)
- 34 535.04 / 34 327.96 / 34 170.95 d $3d^4$(0,1,2)
- 18 911.399 / 18 825.89 / 18 818.23 c 5*4(0,1,2)
- 18 145.40 / 18 061.54 / 17 995.75 b 6*4(0,1,2)
- 8 602.340 / 8 492.421 / 8 436.618 a $4s^2$(0,1,2)

$c^*4p\ ^3P^o$

- 45 178.06 / 45 090.73 / 45 040.70 t 3*(0,1,2)
- 41 959.46 / 41 943.95 / 41 928.528 u 6*(2,1,0)
- 40 466.979 / 40 384.58 / 40 369.76 v 5*(0,1,2)
- 37 325.407 / 37 172.947 / 37 090.65 w 8*(0,1,2)
- 33 114.412 / 33 090.492 / 33 085.153 x 11*(0,1,2)
- 31 805.94 / 31 725.75 / 31 685.90 y 3*(0,1,2)
- 25 537.276 / 25 493.722 z 7*(2,1)

c^*4s, $3d^2 4\ell^2$ ^3D

- 49 619.72 / 49 571.69 f (2,3)
- 48 839.74 / 48 724.83 / 48 724.34 e p^2(2,1,3)
- 17 540.205 / 17 423.853 / 17 369.59 a 8*(1,2,3)

$c^*4p\ ^3D^o_{1,2,3}$

- 45 206.34 / 45 063.94 / 44 966.36 n
- 44 233.15 / 44 079.39 / 43 975.62 o
- 42 376.71 / 42 269.73 / 42 193.94 p
- 42 311.269 / 42 206.88 / 42 146.39 q
- 40 844.19 / 40 670.60 / 40 556.07 r 6*
- 39 715.437 / 39 686.10 / 39 662.15 s 5*
- 38 764.832 / 38 699.767 / 38 654.23 t, u 8*
- 38 159.71 / 37 976.78 / 37 852.021 11*
- 31 205.985 / 31 190.631 / 31 184.021 v 3*
- 29 912.262 / 29 768.655 / 29 661.232 w 7*
- 27 480.047 / 27 418.015 / 27 355.042 x 1*
- 25 643.695 / 25 438.898 / 25 317.813 y 0*
- 20 126.055 / 20 006.032 / 19 937.859 z 2*

$c^*n\ell$, $3d^2 4s^2$ ^3F

- 47 194.68 / 47 038.16 i 2*4d(3,4)
- 46 530.45 h 0*6s(4)
- 42 107.06 / 41 988.39 / 41 871.87 g 0*4d(2,3,4)
- 39 785.94 / 39 640.98 / 39 526.89 f 1*5s(2,3,4)
- 37 824.748 / 37 659.927 / 37 538.804 e 0*5s(2,3,4)
- 11 776.806 / 11 639.804 / 11 531.760 b 1*4s(2,3,4)
- 386.874 / 170.132 / 0.000 a $4s^2$(2,3,4)

$c^*4p\ ^3F^o_{2,3,4}$

- 47 462.90 / 47 281.74 / 47 187.55 p
- 45 041.02 / 44 923.00 / 44 825.26 q
- 43 744.55 / 43 583.14 / 43 467.55 r
- 41 624.209 / 41 457.653 / 41 337.43 s
- 38 670.710 / 38 544.38 / 38 451.298 t 4*
- 37 852.434 / 37 743.933 / 37 654.77 u 8*
- 34 204.971 / 34 078.580 / 33 980.639 v 10*
- 33 700.874 / 33 680.130 / 33 655.853 w 3*
- 27 025.652 / 26 892.926 / 26 803.417 x 1*
- 25 388.334 / 25 227.217 / 25 107.417 y 0*
- 19 573.968 / 19 421.576 / 19 322.988 z 2*

$c^*4\ell$, $3d^2 3\ell^2$ $^3G_{3,4,5}$

- 41 481.13 / 41 368.86 / 41 194.42 e 0*d
- 36 200.94 / 36 132.21 / 36 065.75 b $3d^4$
- 15 220.390 / 15 156.787 / 15 108.121 a 4*s

$c^*4p\ ^3G^o$

- 46 974.65 / 46 838.09 / 46 725.45 s (3,4,5)
- 44 375.57 / 44 162.44 r (4,5)
- 41 341.553 / 41 255.400 / 41 170.003 u 9*(3,4,5)
- 37 690.320 / 37 617.868 / 37 555.021 v 4*(3,4,5)
- 31 628.668 / 31 489.451 / 31 373.801 w 10*(3,4,5)
- 30 039.211 / 29 971.078 / 29 914.720 x 0*(3,4,5)
- 27 750.124 / 27 614.667 / 27 498.975 y 1*(3,4,5)
- 21 739.713 / 21 588.496 / 21 469.494 z 2*(3,4,5)

$c^*4\ell\ ^3H_{4,5,6}$

- 45 960.439 / 45 832.50 / 45 721.878 f 2*d
- 41 615.02 / 41 556.33 / 41 515.09 e 0*d
- 18 192.577 / 18 141.229 / 18 037.225 a 9*s

$c^*4p\ ^3H^o_{4,5,6}$

- 41 995.39 / 41 895.15 / 41 780.55 w
- 39 198.320 / 39 152.057 / 39 115.958 x 9*
- 35 685.160 / 35 559.627 / 35 454.051 y 4*
- 32 013.534 / 31 914.277 / 31 829.972 z 10*

$c^*4p\ ^3I^o_{5,6,7}$

- 38 779.856 / 38 668.832 / 38 572.692 z 9*

KEY

37 824.748 / 37 659.927 / 37 538.804 e 0*5s(2,3,4)

- J, low to high
- n
- Parent level from Ti II
- Notation from A.E.L.
- Energy levels (cm^{-1})

onization Limit
55 010 cm^{-1}
(6.82 eV)

[Ti I $4s^2\ ^2F_2$ → Ti II $4s\ ^4F_{3/2}$]

CORES (c^*) from Ti II:

$0^* = 3d^2 4s\ ^4F$	$7^* = 3d^2 4s\ ^4P$
$1^* = 3d^3\ ^4F$	$8^* = 3d^3\ ^2D$
$2^* = 3d^2 4s\ ^2F$	$9^* = 3d^3\ ^2H$
$3^* = 3d^2 4s\ ^2D$	$10^* = 3d^2 4s\ ^2G$
$4^* = 3d^3\ ^2G$	$11^* = 3d^2 4s\ ^4P$
$5^* = 3d^3\ ^4P$	$12^* = 3d^3\ ^2F$
$6^* = 3d^3\ ^2P$	$13^* = 3d^2 4s\ ^2S$

NOTE: For proper configuration, please refer to bibliographical comments.

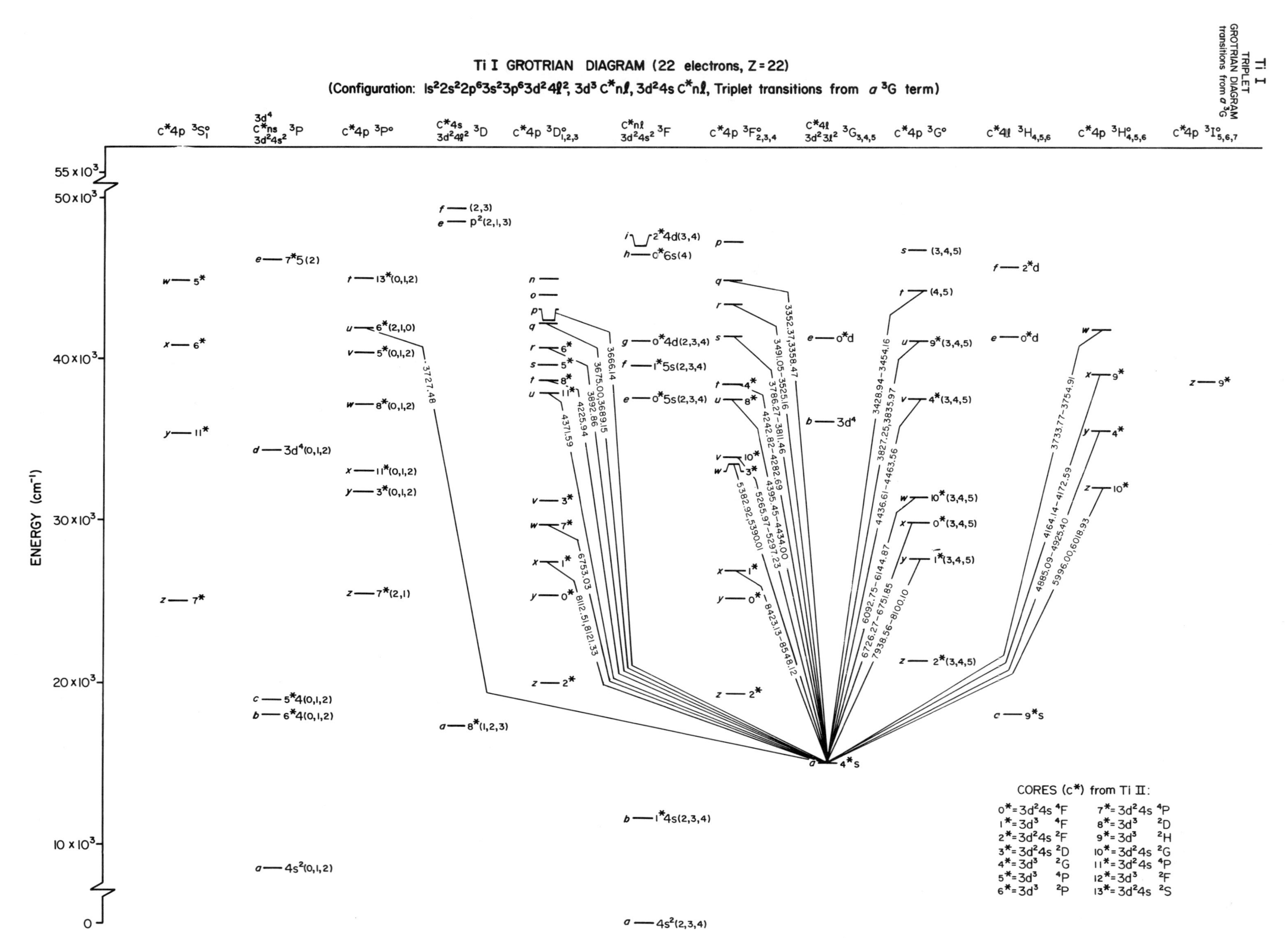

Ti I
TRIPLET
GROTRIAN DIAGRAM
transitions from a 3G
Ti I GROTRIAN DIAGRAM (22 electrons, Z=22)
(Configuration: 1s22s22p63s23p63d24l2, 3d3 c*nl, 3d24s c*nl, Triplet transitions from a 3G term)
c*4p 3S°1
3d4 c*ns 3d24s2 3P
c*4p 3P°
c*4s 3d24l2 3D
c*4p 3D°1,2,3
c*nl 3d24s2 3F
c*4p 3F°2,3,4
c*4l 3d23l2 3G3,4,5
c*4p 3G°
c*4l 3H4,5,6
c*4p 3H°4,5,6
c*4p 3I°5,6,7
3727.48
3666.14
3675.00,3689.15
3892.86
4225.94
4371.59
6753.03
8112.51,8121.33
3352.37,3358.47
3491.05–3525.16
3786.27–3811.46
4242.82–4282.69
4395.45–4434.00
5265.97–5297.23
5382.92,5390.01
8423.13–8548.12
3428.94–3454.16
3827.25,3835.97
4436.61–4463.56
6092.75–6144.87
6726.27–6751.85
7938.56–8100.10
3733.77–3754.91
4164.14–4172.59
4885.09–4925.40
5996.00,6018.93
a—4*s
CORES (c*) from Ti II:
0*=3d24s 4F
1*=3d3 4F
2*=3d24s 2F
3*=3d24s 2D
4*=3d3 2G
5*=3d3 4P
6*=3d3 2P
7*=3d24s 4P
8*=3d3 2D
9*=3d3 2H
10*=3d24s 2G
11*=3d24s 4P
12*=3d3 2F
13*=3d24s 2S
ENERGY (cm-1)
55×103
50×103
40×103
30×103
20×103
10×103
0

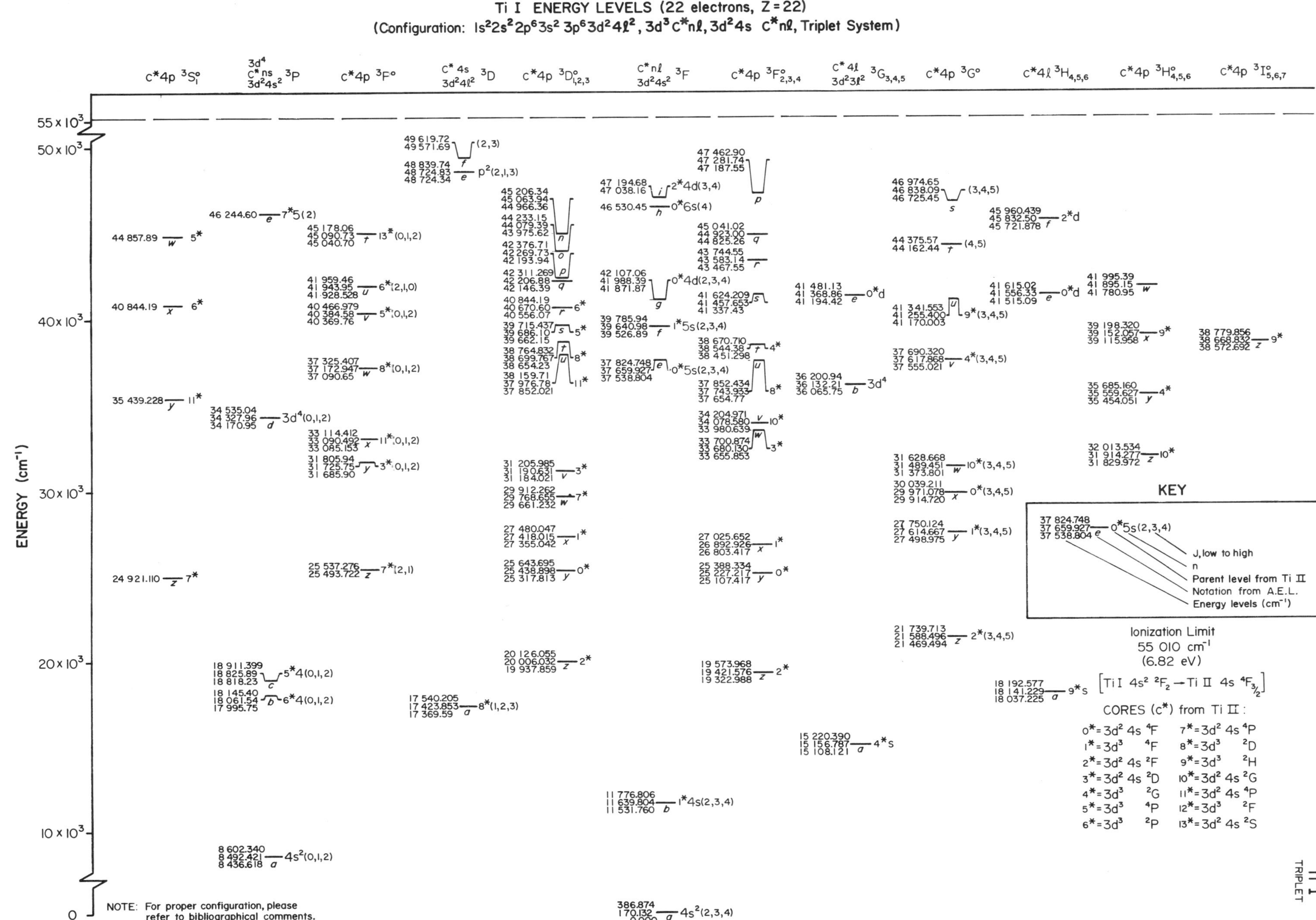
Ti I ENERGY LEVELS (22 electrons, Z = 22)
(Configuration: 1s²2s²2p⁶3s² 3p⁶3d²4ℓ², 3d³c*nℓ, 3d²4s c*nℓ, Triplet System)
c*4p ³S°₁
3d⁴ c*ns 3d²4s² ³P
c*4p ³F°
c*4s 3d²4ℓ² ³D
c*4p ³D°1,2,3
c*nℓ 3d²4s² ³F
c*4p ³F°2,3,4
c*4ℓ 3d²3ℓ² ³G3,4,5
c*4p ³G°
c*4ℓ ³H4,5,6
c*4p ³H°4,5,6
c*4p ³I°5,6,7
ENERGY (cm⁻¹)
KEY
J, low to high
n
Parent level from Ti II
Notation from A.E.L.
Energy levels (cm⁻¹)
Ionization Limit
55 010 cm⁻¹
(6.82 eV)
[Ti I 4s² ²F₂ → Ti II 4s ⁴F3/2]
CORES (c*) from Ti II:
0*=3d² 4s ⁴F
1*=3d³ ⁴F
2*=3d² 4s ²F
3*=3d² 4s ²D
4*=3d³ ²G
5*=3d³ ⁴P
6*=3d³ ²P
7*=3d² 4s ⁴P
8*=3d³ ²D
9*=3d³ ²H
10*=3d² 4s ²G
11*=3d² 4s ⁴P
12*=3d³ ²F
13*=3d² 4s ²S
NOTE: For proper configuration, please refer to bibliographical comments.
Ti I
TRIPLET

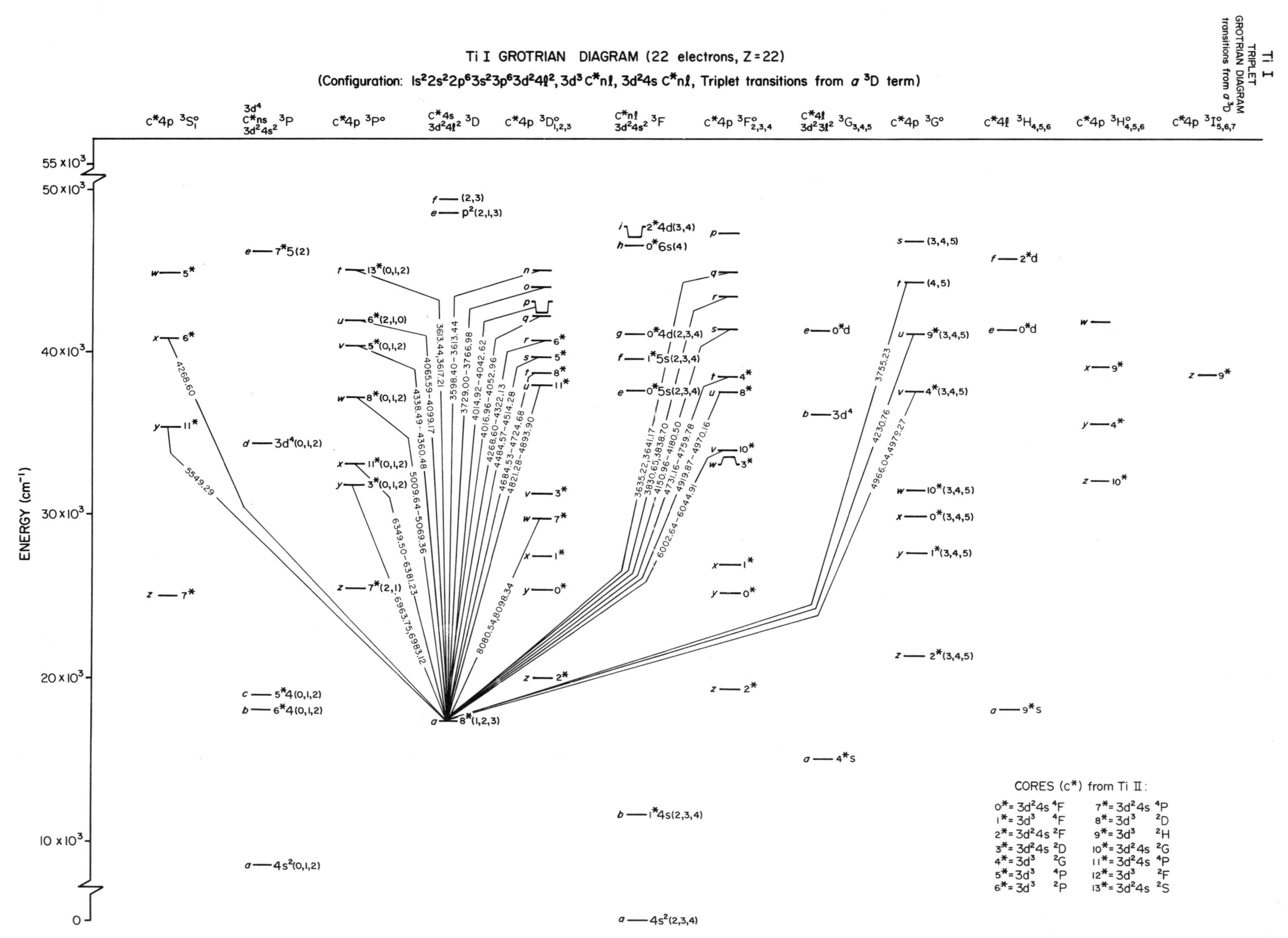
Ti I GROTRIAN DIAGRAM (22 electrons, Z = 22)
(Configuration: $1s^22s^22p^63s^23p^63d^24l^2$, $3d^3c^*nl$, $3d^24s\,c^*nl$, Triplet transitions from $a\,^3D$ term)
Ti I
TRIPLET
GROTRIAN DIAGRAM
transitions from $a\,^3D$
$c^*4p\ ^3S^o_1$
$3d^4$ c^*ns $3d^24s^2$ ^3P
$c^*4p\ ^3P^o$
c^*4s $3d^24l^2$ ^3D
$c^*4p\ ^3D^o_{1,2,3}$
c^*nl $3d^24s^2$ ^3F
$c^*4p\ ^3F^o_{2,3,4}$
c^*4l $3d^23l^2$ $^3G_{3,4,5}$
$c^*4p\ ^3G^o$
$c^*4l\ ^3H_{4,5,6}$
$c^*4p\ ^3H^o_{4,5,6}$
$c^*4p\ ^3I^o_{5,6,7}$
ENERGY (cm^{-1})
55×10^3
50×10^3
40×10^3
30×10^3
20×10^3
10×10^3
0
CORES (c^*) from Ti II:
$0^* = 3d^24s\ ^4F$
$1^* = 3d^3\ ^4F$
$2^* = 3d^24s\ ^2F$
$3^* = 3d^24s\ ^2D$
$4^* = 3d^3\ ^2G$
$5^* = 3d^3\ ^4P$
$6^* = 3d^3\ ^2P$
$7^* = 3d^24s\ ^4P$
$8^* = 3d^3\ ^2D$
$9^* = 3d^3\ ^2H$
$10^* = 3d^24s\ ^2G$
$11^* = 3d^24s\ ^4P$
$12^* = 3d^3\ ^2F$
$13^* = 3d^24s\ ^2S$

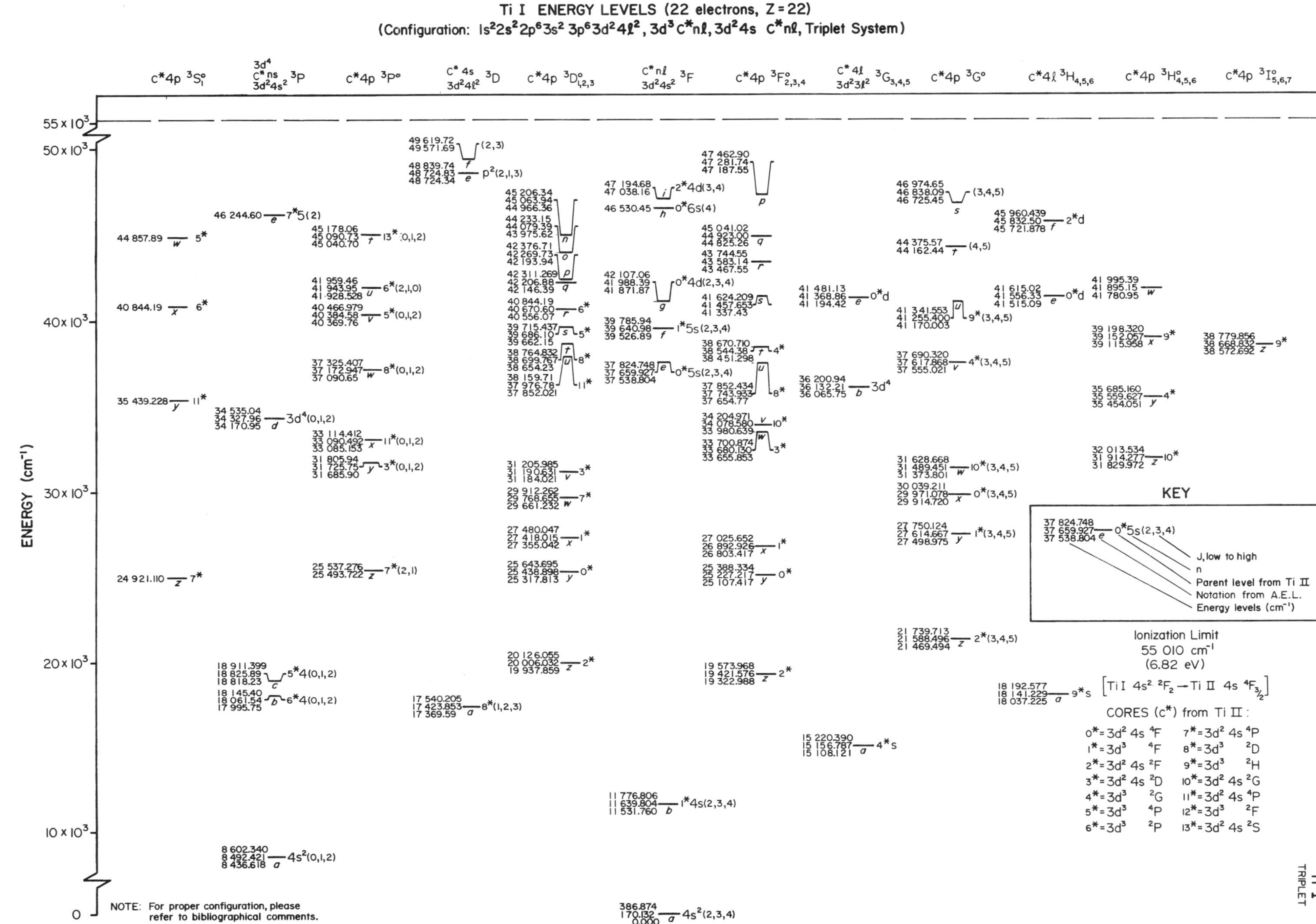

Ti I ENERGY LEVELS (22 electrons, Z=22)
(Configuration: 1s²2s²2p⁶3s² 3p⁶3d²4ℓ², 3d³ c*nℓ, 3d²4s c*nℓ, Triplet System)
c*4p ³S°₁
3d⁴ c*ns 3d²4s² ³P
c*4p ³P°
c*4s 3d²4ℓ² ³D
c*4p ³D°₁,₂,₃
c*nℓ 3d²4s² ³F
c*4p ³F°₂,₃,₄
c*4ℓ 3d²3ℓ² ³G₃,₄,₅
c*4p ³G°
c*4ℓ ³H₄,₅,₆
c*4p ³H°₄,₅,₆
c*4p ³I°₅,₆,₇
55 x 10³
50 x 10³
40 x 10³
30 x 10³
20 x 10³
10 x 10³
0
ENERGY (cm⁻¹)
44 857.89 w 5*
40 844.19 x 6*
35 439.228 y 11*
24 921.110 z 7*
46 244.60 e 7*5(2)
34 535.04 34 327.96 34 170.95 d 3d⁴(0,1,2)
18 911.399 18 825.89 18 818.23 c 5*4(0,1,2)
18 145.40 18 061.54 17 995.75 b 6*4(0,1,2)
8 602.340 8 492.421 8 436.618 a 4s²(0,1,2)
45 178.06 45 090.73 45 040.70 t 13*(0,1,2)
41 959.46 41 943.95 41 928.528 u 6*(2,1,0)
40 466.979 40 384.58 40 369.76 v 5*(0,1,2)
37 325.407 37 172.947 37 090.65 w 8*(0,1,2)
33 114.412 33 090.492 33 085.153 x 11*(0,1,2)
31 805.94 31 725.75 31 685.90 y 3*(0,1,2)
25 537.276 25 493.722 z 7*(2,1)
49 619.72 49 571.69 f (2,3)
48 839.74 48 724.83 48 724.34 e p²(2,1,3)
17 540.205 17 423.853 17 369.59 a 8*(1,2,3)
45 206.34 45 063.94 44 966.36 n
44 233.15 44 079.39 43 975.62 o
42 376.71 42 269.73 42 193.94 p
42 311.269 42 206.88 42 146.39 q
40 844.19 40 670.60 40 556.07 r 6*
39 715.437 39 686.10 39 662.15 s 5*
38 764.832 38 699.767 38 654.23 t 8*
38 159.71 37 976.78 37 852.021 u 11*
31 205.985 31 190.631 31 184.021 v 3*
29 912.262 29 768.655 29 661.232 w 7*
27 480.047 27 418.015 27 355.042 x 1*
25 643.695 25 438.898 25 317.813 y 0*
20 126.055 20 006.032 19 937.859 z 2*
47 194.68 47 038.16 i 2*4d(3,4)
46 530.45 h 0*6s(4)
42 107.06 41 988.39 41 871.87 g 0*4d(2,3,4)
39 785.94 39 640.98 39 526.89 f 1*5s(2,3,4)
37 824.748 37 659.927 37 538.804 e 0*5s(2,3,4)
11 776.806 11 639.804 11 531.760 b 1*4s(2,3,4)
386.874 170.132 0.000 a 4s²(2,3,4)
47 462.90 47 281.74 47 187.55 p
45 041.02 44 923.00 44 825.26 q
43 744.55 43 583.14 43 467.55 r
41 624.209 41 457.653 41 337.43 s
38 670.710 38 544.38 38 451.298 t 4*
37 852.434 37 743.933 37 654.77 u 8*
34 204.971 34 078.580 33 980.639 v 10*
33 700.874 33 680.130 33 655.853 w 3*
27 025.652 26 892.926 26 803.417 x 1*
25 388.334 25 227.217 25 107.417 y 0*
19 573.968 19 421.576 19 322.988 z 2*
41 481.13 41 368.86 41 194.42 e 0*d
36 200.94 36 132.21 36 065.75 b 3d⁴
15 220.390 15 156.787 15 108.121 a 4*s
46 974.65 46 838.09 46 725.45 s (3,4,5)
44 375.57 44 162.44 t (4,5)
41 341.553 41 255.400 41 170.003 u 9*(3,4,5)
37 690.320 37 617.868 37 555.021 v 4*(3,4,5)
31 628.668 31 489.451 31 373.801 w 10*(3,4,5)
30 039.211 29 971.078 29 914.720 x 0*(3,4,5)
27 750.124 27 614.667 27 498.975 y 1*(3,4,5)
21 739.713 21 588.496 21 469.494 z 2*(3,4,5)
45 960.439 45 832.50 45 721.878 f 2*d
41 615.02 41 556.33 41 515.09 e 0*d
18 192.577 18 141.229 18 037.225 a 9*s
41 995.39 41 895.15 41 780.95 w
39 198.320 39 152.057 39 115.958 x 9*
35 685.160 35 559.627 35 454.051 y 4*
32 013.534 31 914.277 31 829.972 z 10*
38 779.856 38 668.832 38 572.692 z 9*
KEY
37 824.748 37 659.927 37 538.804 e 0*5s(2,3,4)
J, low to high
n
Parent level from Ti II
Notation from A.E.L.
Energy levels (cm⁻¹)
Ionization Limit
55 010 cm⁻¹
(6.82 eV)
[Ti I 4s² ²F₂ → Ti II 4s ⁴F₃/₂]
CORES (c*) from Ti II:
0* = 3d² 4s ⁴F
1* = 3d³ ⁴F
2* = 3d² 4s ²F
3* = 3d² 4s ²D
4* = 3d³ ²G
5* = 3d³ ⁴P
6* = 3d³ ²P
7* = 3d² 4s ⁴P
8* = 3d³ ²D
9* = 3d³ ²H
10* = 3d² 4s ²G
11* = 3d² 4s ⁴P
12* = 3d³ ²F
13* = 3d² 4s ²S
NOTE: For proper configuration, please refer to bibliographical comments.
Ti I
TRIPLET

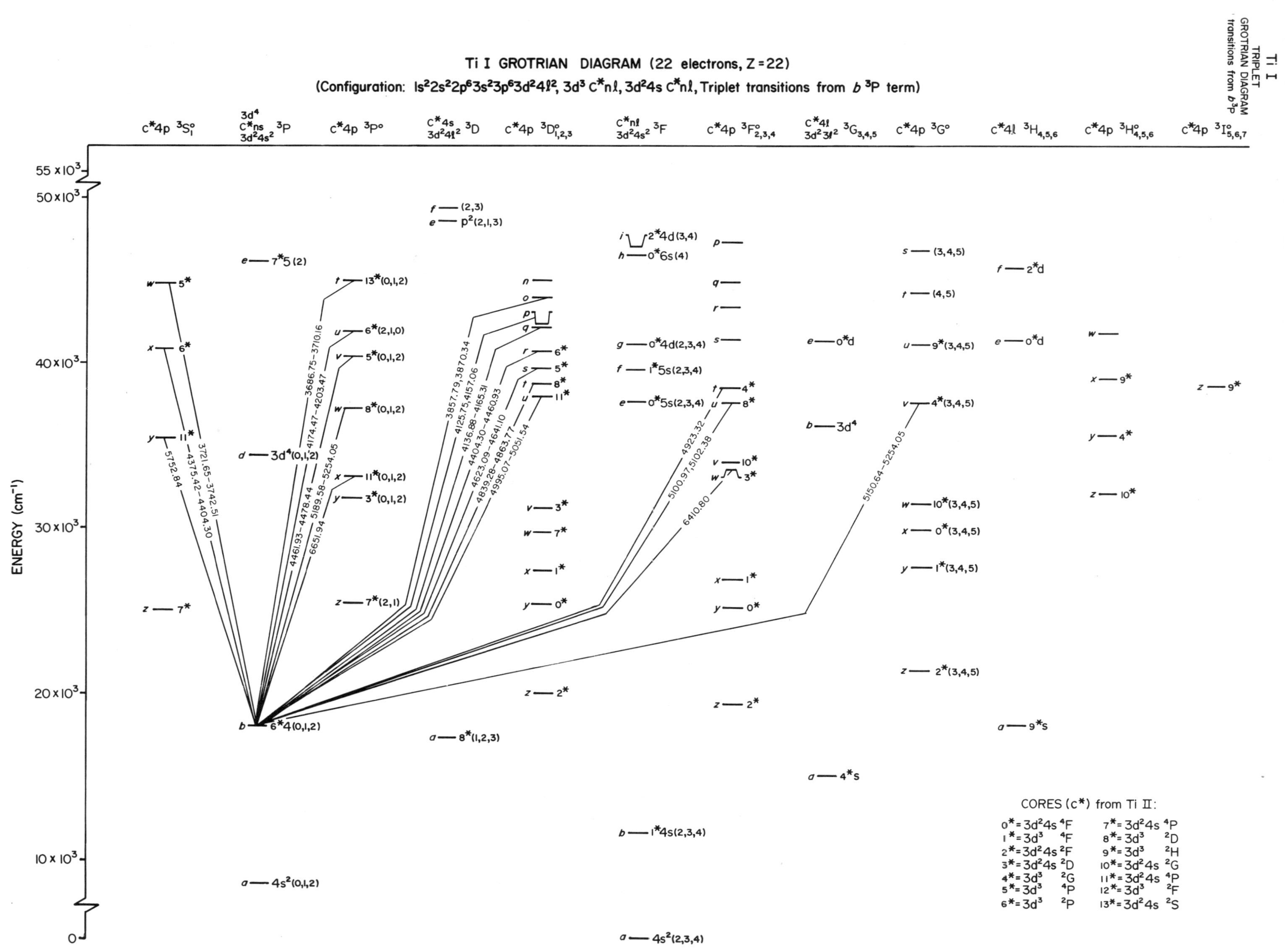

Ti I
TRIPLET
GROTRIAN DIAGRAM
transitions from b^3P
Ti I GROTRIAN DIAGRAM (22 electrons, Z=22)
(Configuration: $1s^22s^22p^63s^23p^63d^24l^2$, $3d^3c^*nl$, $3d^24s\,c^*nl$, Triplet transitions from $b\,^3P$ term)
ENERGY (cm^{-1})
55×10^3
50×10^3
40×10^3
30×10^3
20×10^3
10×10^3
0
$c^*4p\ ^3S^o_1$
$3d^4$ c^*ns $3d^24s^2$ ^3P
$c^*4p\ ^3P^o$
c^*4s $3d^24l^2$ ^3D
$c^*4p\ ^3D^o_{1,2,3}$
c^*nl $3d^24s^2$ ^3F
$c^*4p\ ^3F^o_{2,3,4}$
c^*4l $3d^23l^2$ $^3G_{3,4,5}$
$c^*4p\ ^3G^o$
$c^*4l\ ^3H_{4,5,6}$
$c^*4p\ ^3H^o_{4,5,6}$
$c^*4p\ ^3I^o_{5,6,7}$
3721.65–3742.51
4375.42–4404.30
5752.84
3686.75–3710.16
4174.47–4203.47
4461.93–4478.44
5189.58–5254.05
6651.94
3857.79,3870.34
4125.75,4157.06
4136.88–4165.31
4404.30–4460.93
4623.09–4641.10
4839.28–4863.77
4995.07–5051.54
4923.32
5100.97,5102.38
6410.80
5150.64–5254.05
CORES (c^*) from Ti II:
$0^*=3d^24s\ ^4F$
$1^*=3d^3\ ^4F$
$2^*=3d^24s\ ^2F$
$3^*=3d^24s\ ^2D$
$4^*=3d^3\ ^2G$
$5^*=3d^3\ ^4P$
$6^*=3d^3\ ^2P$
$7^*=3d^24s\ ^4P$
$8^*=3d^3\ ^2D$
$9^*=3d^3\ ^2H$
$10^*=3d^24s\ ^2G$
$11^*=3d^24s\ ^4P$
$12^*=3d^3\ ^2F$
$13^*=3d^24s\ ^2S$

Ti I ENERGY LEVELS (22 electrons, Z = 22)

(Configuration: $1s^2 2s^2 2p^6 3s^2 3p^6 3d^2 4\ell^2$, $3d^3 c^* n\ell$, $3d^2 4s\ c^* n\ell$, Triplet System)

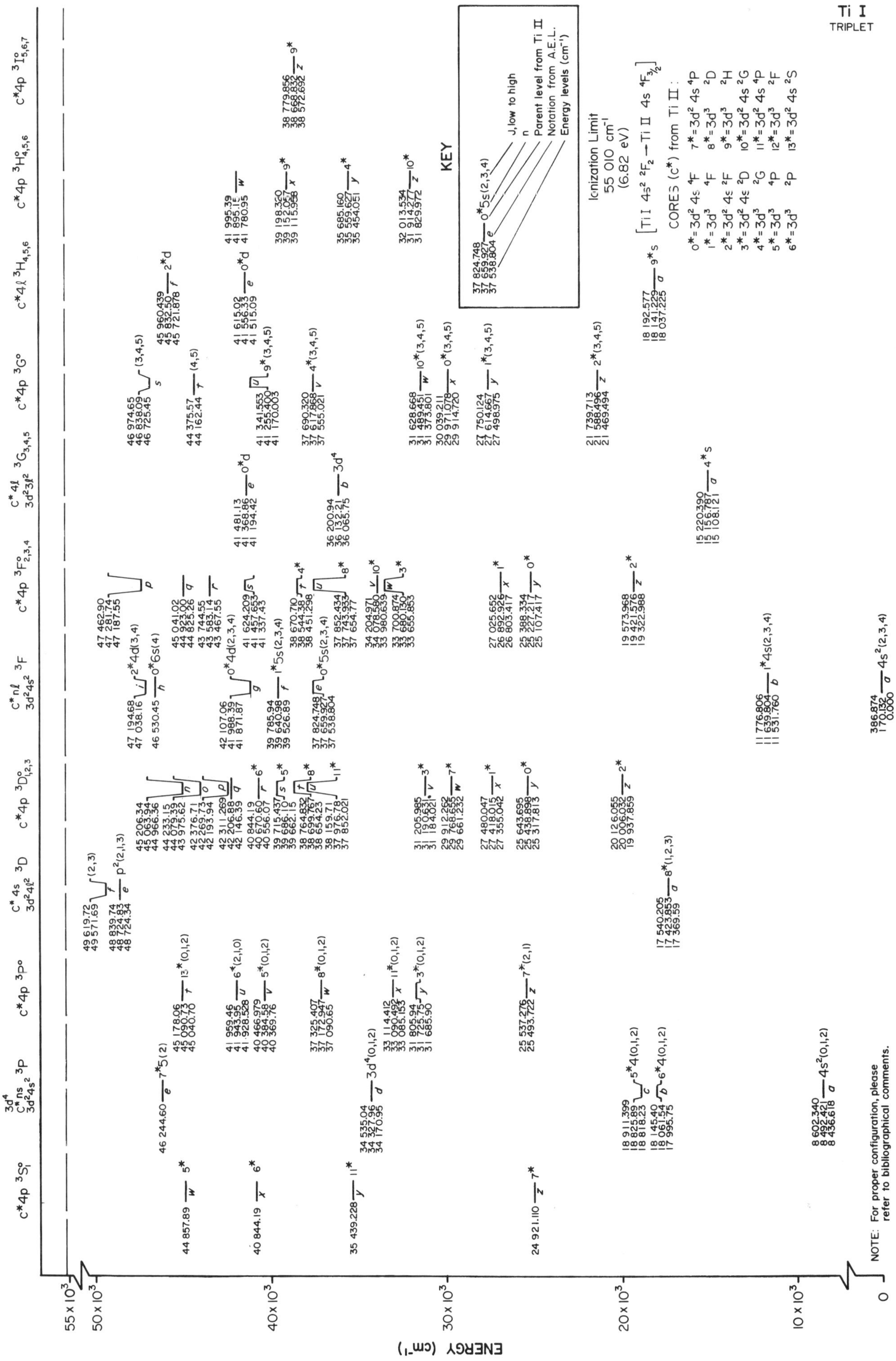

ENERGY (cm^{-1}): 0, 10×10^3, 20×10^3, 30×10^3, 40×10^3, 50×10^3, 55×10^3

$c^*4p\ ^3S^o_1$

- 44 857.89 — w — 5*
- 40 844.19 — x — 6*
- 35 439.228 — y — 11*
- 24 921.110 — z — 7*

$3d^4$, $c^* ns$, $3d^2 4s^2\ ^3P$

- 46 244.60 — e — 7*5(2)
- 34 535.04, 34 327.96, 34 170.95 — d — $3d^4$(0,1,2)
- 18 911.399, 18 825.89, 18 818.23 — c — 5*4(0,1,2)
- 18 145.40, 18 061.54, 17 995.75 — b — 6*4(0,1,2)
- 8 602.340, 8 492.421, 8 436.618 — a — $4s^2$(0,1,2)

$c^*4p\ ^3P^o$

- 45 178.06, 45 090.73, 45 040.70 — t — 13*(0,1,2)
- 41 959.46, 41 943.95, 41 928.528 — u — 6*(2,1,0)
- 40 466.979, 40 384.58, 40 369.76 — v — 5*(0,1,2)
- 37 325.407, 37 172.947, 37 090.65 — w — 8*(0,1,2)
- 33 114.412, 33 090.492, 33 085.153 — x — 11*(0,1,2)
- 31 805.94, 31 725.75, 31 685.90 — y — 3*(0,1,2)
- 25 537.276, 25 493.722 — z — 7*(2,1)

c^*4s, $3d^2 4\ell^2\ ^3D$

- 49 619.72, 49 571.69 — f — (2,3)
- 48 839.74, 48 724.83, 48 724.34 — e — p^2(2,1,3)
- 17 540.205, 17 423.853, 17 369.59 — a — 8*(1,2,3)

$c^*4p\ ^3D^o_{1,2,3}$

- 45 206.34, 45 063.94, 44 966.36 — n
- 44 233.15, 44 079.39, 43 975.62 — o
- 42 376.71, 42 269.73, 42 193.94 — p
- 42 311.269, 42 206.88, 42 146.39 — q
- 40 844.19, 40 670.60, 40 556.07 — r — 6*
- 39 715.437, 39 686.10, 39 662.15 — s — 5*
- 38 764.832, 38 699.767, 38 654.23 — t — 8*
- 38 159.71, 37 976.78, 37 852.021 — u — 11*
- 31 205.985, 31 190.631, 31 184.021 — v — 3*
- 29 912.262, 29 768.655, 29 661.232 — w — 7*
- 27 480.047, 27 418.015, 27 355.042 — x — 1*
- 25 643.695, 25 438.898, 25 317.813 — y — 0*
- 20 126.055, 20 006.032, 19 937.859 — z — 2*

$c^* n\ell$, $3d^2 4s^2\ ^3F$

- 47 194.68, 47 038.16 — j — 2*4d(3,4)
- 46 530.45 — h — 0*6s(4)
- 42 107.06, 41 988.39, 41 871.87 — g — 0*4d(2,3,4)
- 39 785.94, 39 640.98, 39 526.89 — f — 1*5s(2,3,4)
- 37 824.748, 37 659.927, 37 538.804 — e — 0*5s(2,3,4)
- 11 776.806, 11 639.804, 11 531.760 — b — 1*4s(2,3,4)
- 386.874, 170.132, 0.000 — a — $4s^2$(2,3,4)

$c^*4p\ ^3F^o_{2,3,4}$

- 47 462.90, 47 281.74, 47 187.55 — p
- 45 041.02, 44 923.00, 44 825.26 — q
- 43 744.55, 43 583.14, 43 467.55 — r
- 41 624.209, 41 457.653, 41 337.43 — s
- 38 670.710, 38 544.38, 38 451.298 — t — 4*
- 37 852.434, 37 743.933, 37 654.77 — u — 8*
- 34 204.971, 34 078.580, 33 980.639 — v — 10*
- 33 700.874, 33 680.130, 33 655.853 — w — 3*
- 27 025.652, 26 892.926, 26 803.417 — x — 1*
- 25 388.334, 25 227.217, 25 107.417 — y — 0*
- 19 573.968, 19 421.576, 19 322.988 — z — 2*

$c^*4\ell$, $3d^2 3\ell^2\ ^3G_{3,4,5}$

- 41 481.13, 41 368.86, 41 194.42 — e — 0*d
- 36 200.94, 36 132.21, 36 065.75 — b — $3d^4$
- 15 220.390, 15 156.787, 15 108.121 — a — 4*s

$c^*4p\ ^3G^o$

- 46 974.65, 46 838.09, 46 725.45 — s — (3,4,5)
- 44 375.57, 44 162.44 — t — (4,5)
- 41 341.553, 41 255.400, 41 170.003 — u — 9*(3,4,5)
- 37 690.320, 37 617.868, 37 555.021 — v — 4*(3,4,5)
- 31 628.668, 31 489.451, 31 373.801 — w — 10*(3,4,5)
- 30 039.211, 29 971.078, 29 914.720 — x — 0*(3,4,5)
- 27 750.124, 27 614.667, 27 498.975 — y — 1*(3,4,5)
- 21 739.713, 21 588.496, 21 469.494 — z — 2*(3,4,5)

$c^*4\ell\ ^3H_{4,5,6}$

- 45 960.439, 45 832.50, 45 721.878 — f — 2*d
- 41 615.02, 41 556.33, 41 515.09 — e — 0*d
- 18 192.577, 18 141.229, 18 037.225 — a — 9*s

$c^*4p\ ^3H^o_{4,5,6}$

- 41 995.39, 41 895.15, 41 780.95 — w
- 39 198.320, 39 152.057, 39 115.958 — x — 9*
- 35 685.160, 35 559.627, 35 454.051 — y — 4*
- 32 013.534, 31 914.277, 31 829.972 — z — 10*

$c^*4p\ ^3I^o_{5,6,7}$

- 38 779.856, 38 668.832, 38 572.692 — z — 9*

KEY

37 824.748, 37 659.927, 37 538.804 — e — 0*5s(2,3,4)

- J, low to high
- n
- Parent level from Ti II
- Notation from A.E.L.
- Energy levels (cm^{-1})

Ionization Limit
55 010 cm^{-1}
(6.82 eV)
[Ti I $4s^2\ ^2F_2$ → Ti II $4s\ ^4F_{3/2}$]

CORES (c^*) from Ti II:

0* = $3d^2 4s\ ^4F$	7* = $3d^2 4s\ ^4P$
1* = $3d^3\ ^4F$	8* = $3d^3\ ^2D$
2* = $3d^2 4s\ ^2F$	9* = $3d^3\ ^2H$
3* = $3d^2 4s\ ^2D$	10* = $3d^2 4s\ ^2G$
4* = $3d^3\ ^2G$	11* = $3d^2 4s\ ^4P$
5* = $3d^3\ ^4P$	12* = $3d^3\ ^2F$
6* = $3d^3\ ^2P$	13* = $3d^2 4s\ ^2S$

NOTE: For proper configuration, please refer to bibliographical comments.

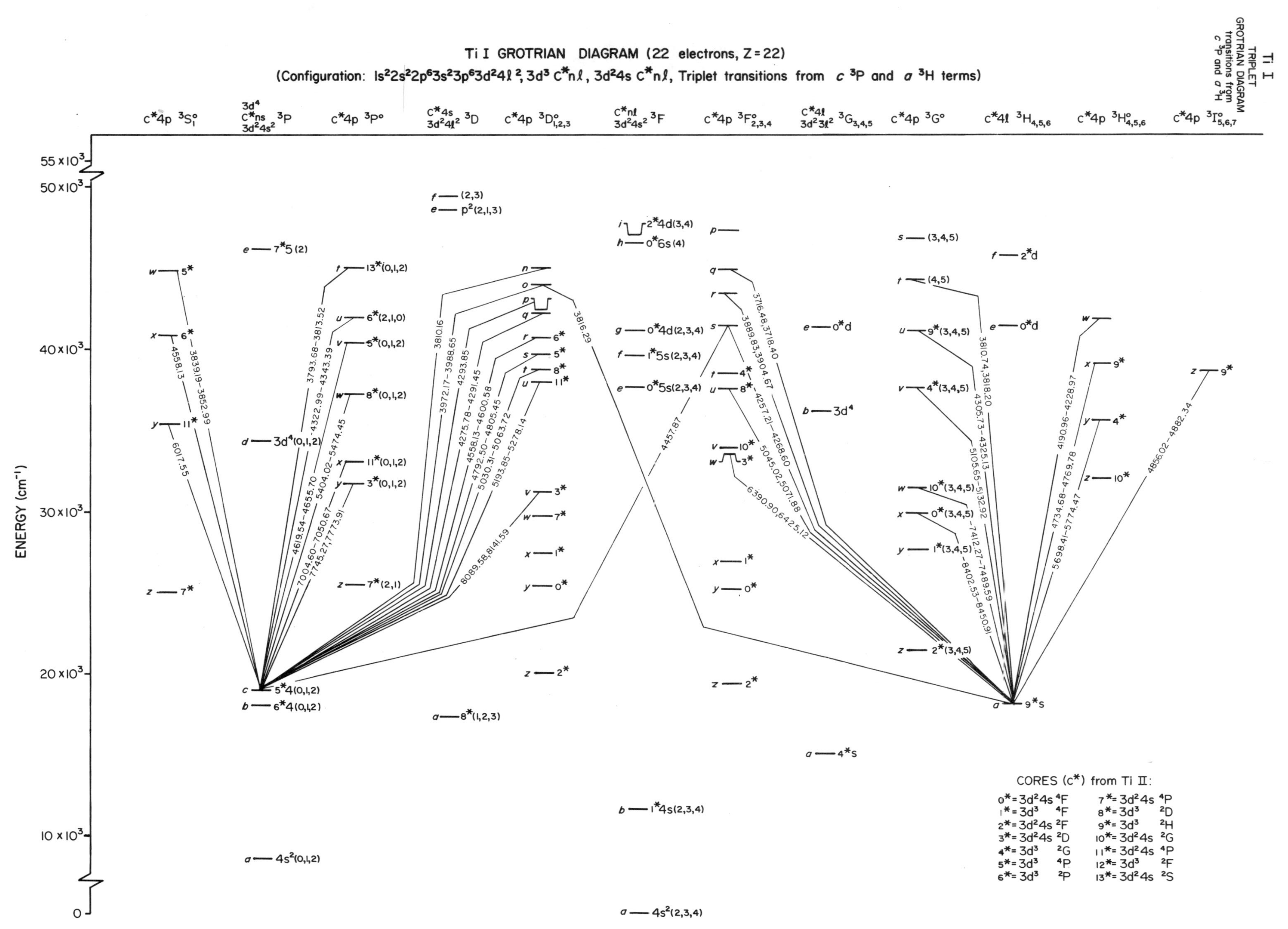
Ti I
TRIPLET
GROTRIAN DIAGRAM
transitions from
$c\,^3P$ and $a\,^3H$
Ti I GROTRIAN DIAGRAM (22 electrons, Z = 22)
(Configuration: $1s^22s^22p^63s^23p^63d^24\ell^2$, $3d^3c^*n\ell$, $3d^24s\,c^*n\ell$, Triplet transitions from $c\,^3P$ and $a\,^3H$ terms)
ENERGY (cm⁻¹)
55 x 10³
50 x 10³
40 x 10³
30 x 10³
20 x 10³
10 x 10³
0
CORES (c*) from Ti II:
0* = 3d²4s ⁴F
1* = 3d³ ⁴F
2* = 3d²4s ²F
3* = 3d²4s ²D
4* = 3d³ ²G
5* = 3d³ ⁴P
6* = 3d³ ²P
7* = 3d²4s ⁴P
8* = 3d³ ²D
9* = 3d³ ²H
10* = 3d²4s ²G
11* = 3d²4s ⁴P
12* = 3d³ ²F
13* = 3d²4s ²S

Ti I ENERGY LEVELS (22 electrons, Z = 22)

(Configuration: $1s^2 2s^2 2p^6 3s^2 3p^6 3d^2 4\ell^2$, $3d^3 c^* n\ell$, $3d^2 4s$ $c^* n\ell$, Triplet System)

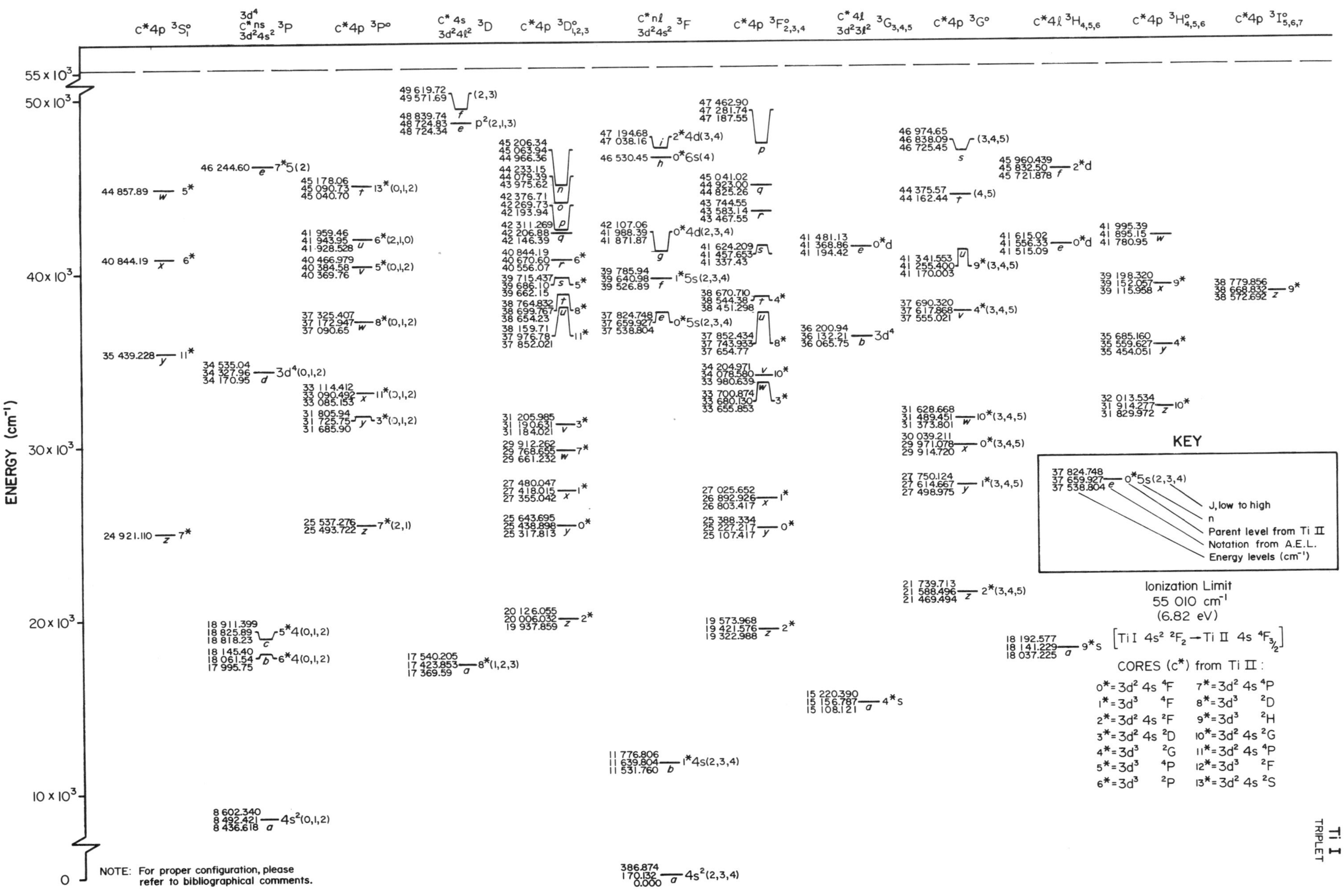

Ti I
TRIPLET

ENERGY (cm^{-1}): 0, 10×10^3, 20×10^3, 30×10^3, 40×10^3, 50×10^3, 55×10^3

c^*4p $^3S^o_1$
- 44 857.89 w 5*
- 40 844.19 x 6*
- 35 439.228 y 11*
- 24 921.110 z 7*

$3d^4$, c^*ns, $3d^2 4s^2$ ^3P
- 46 244.60 e 7*5(2)
- 34 535.04, 34 327.96, 34 170.95 d $3d^4$(0,1,2)
- 18 911.399, 18 825.89, 18 818.23 c 5*4(0,1,2)
- 18 145.40, 18 061.54, 17 995.75 b 6*4(0,1,2)
- 8 602.340, 8 492.421, 8 436.618 a $4s^2$(0,1,2)

c^*4p $^3P^o$
- 45 178.06, 45 090.73, 45 040.70 r 13*(0,1,2)
- 41 959.46, 41 943.95, 41 928.528 u 6*(2,1,0)
- 40 466.979, 40 384.58, 40 369.76 v 5*(0,1,2)
- 37 325.407, 37 172.947, 37 090.65 w 8*(0,1,2)
- 33 114.412, 33 090.492, 33 085.153 x 11*(0,1,2)
- 31 805.94, 31 725.75, 31 685.90 y 3*(0,1,2)
- 25 537.276, 25 493.722 z 7*(2,1)

c^*4s, $3d^2 4\ell^2$ ^3D
- 49 619.72, 49 571.69 f (2,3)
- 48 839.74, 48 724.83, 48 724.34 e p^2(2,1,3)
- 17 540.205, 17 423.853, 17 369.59 a 8*(1,2,3)

c^*4p $^3D^o_{1,2,3}$
- 45 206.34, 45 063.94, 44 966.36 n
- 44 233.15, 44 079.39, 43 975.62 o
- 42 376.71, 42 269.73, 42 193.94 p
- 42 311.269, 42 206.88, 42 146.39 q
- 40 844.19, 40 670.60, 40 556.07 r 6*
- 39 715.437, 39 686.10, 39 662.15 s 5*
- 38 764.832, 38 699.767, 38 654.23 t, u 8*
- 38 159.71, 37 976.78, 37 852.021 11*
- 31 205.985, 31 190.631, 31 184.021 v 3*
- 29 912.262, 29 768.655, 29 661.232 w 7*
- 27 480.047, 27 418.015, 27 355.042 x 1*
- 25 643.695, 25 438.898, 25 317.813 y 0*
- 20 126.055, 20 006.032, 19 937.859 z 2*

$c^*n\ell$, $3d^2 4s^2$ ^3F
- 47 194.68, 47 038.16 j, i 2*4d(3,4)
- 46 530.45 h 0*6s(4)
- 42 107.06, 41 988.39, 41 871.87 g 0*4d(2,3,4)
- 39 785.94, 39 640.98, 39 526.89 f 1*5s(2,3,4)
- 37 824.748, 37 659.927, 37 538.804 e 0*5s(2,3,4)
- 11 776.806, 11 639.804, 11 531.760 b 1*4s(2,3,4)
- 386.874, 170.132, 0.000 a $4s^2$(2,3,4)

c^*4p $^3F^o_{2,3,4}$
- 47 462.90, 47 281.74, 47 187.55 p
- 45 041.02, 44 923.00, 44 825.26 q
- 43 744.55, 43 583.14, 43 467.55 r
- 41 624.209, 41 457.653, 41 337.43 s
- 38 670.710, 38 544.38, 38 451.298 t 4*
- 37 852.434, 37 743.933, 37 654.77 u 8*
- 34 204.971, 34 078.580, 33 980.639 v 10*
- 33 700.874, 33 680.130, 33 655.853 w 3*
- 27 025.652, 26 892.926, 26 803.417 x 1*
- 25 388.334, 25 227.217, 25 107.417 y 0*
- 19 573.968, 19 421.576, 19 322.988 z 2*

$c^*4\ell$, $3d^2 3\ell^2$ $^3G_{3,4,5}$
- 41 481.13, 41 368.86, 41 194.42 e 0*d
- 36 200.94, 36 132.21, 36 065.75 b $3d^4$
- 15 220.390, 15 156.787, 15 108.121 a 4*s

c^*4p $^3G^o$
- 46 974.65, 46 838.09, 46 725.45 s (3,4,5)
- 44 375.57, 44 162.44 r (4,5)
- 41 341.553, 41 255.400, 41 170.003 u 9*(3,4,5)
- 37 690.320, 37 617.868, 37 555.021 v 4*(3,4,5)
- 31 628.668, 31 489.451, 31 373.801 w 10*(3,4,5)
- 30 039.211, 29 971.078, 29 914.720 x 0*(3,4,5)
- 27 750.124, 27 614.667, 27 498.975 y 1*(3,4,5)
- 21 739.713, 21 588.496, 21 469.494 z 2*(3,4,5)

$c^*4\ell$ $^3H_{4,5,6}$
- 45 960.439, 45 832.50, 45 721.878 f 2*d
- 41 615.02, 41 556.33, 41 515.09 e 0*d
- 18 192.577, 18 141.229, 18 037.225 a 9*s

c^*4p $^3H^o_{4,5,6}$
- 41 995.39, 41 895.15, 41 780.95 w
- 39 198.320, 39 152.057, 39 115.958 x 9*
- 35 685.160, 35 559.627, 35 454.051 y 4*
- 32 013.534, 31 914.277, 31 829.972 z 10*

c^*4p $^3I^o_{5,6,7}$
- 38 779.856, 38 668.832, 38 572.692 z 9*

KEY

37 824.748, 37 659.927, 37 538.804 e 0*5s(2,3,4)

- J, low to high
- n
- Parent level from Ti II
- Notation from A.E.L.
- Energy levels (cm^{-1})

Ionization Limit
55 010 cm^{-1}
(6.82 eV)
[Ti I $4s^2$ $^2F_2$ → Ti II 4s $^4F_{3/2}$]

CORES (c^*) from Ti II:

0* = $3d^2$ 4s	^4F	7* = $3d^2$ 4s	^4P
1* = $3d^3$	^4F	8* = $3d^3$	^2D
2* = $3d^2$ 4s	^2F	9* = $3d^3$	^2H
3* = $3d^2$ 4s	^2D	10* = $3d^2$ 4s	^2G
4* = $3d^3$	^2G	11* = $3d^2$ 4s	^4P
5* = $3d^3$	^4P	12* = $3d^3$	^2F
6* = $3d^3$	^2P	13* = $3d^2$ 4s	^2S

NOTE: For proper configuration, please refer to bibliographical comments.

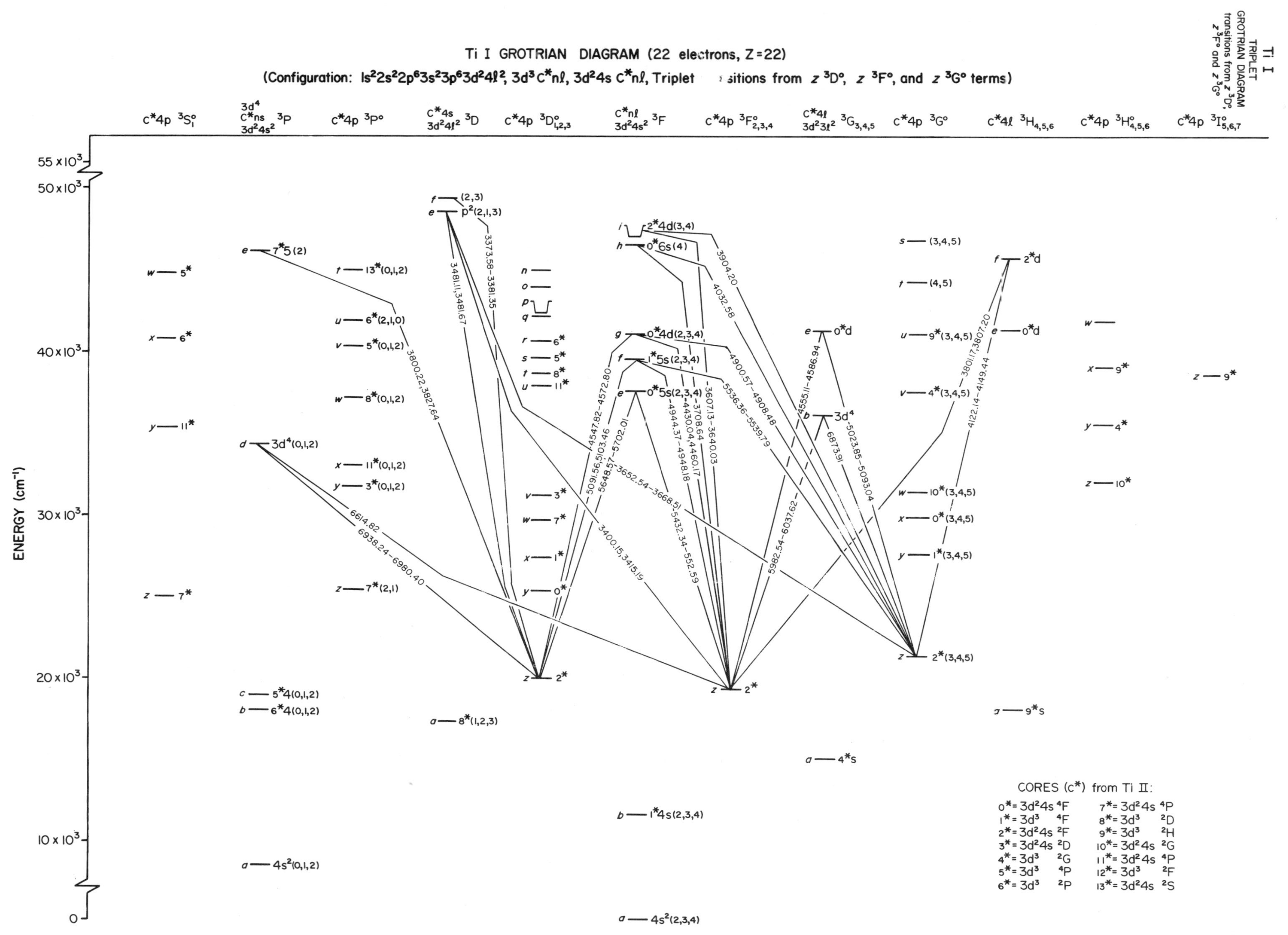
Ti I GROTRIAN DIAGRAM (22 electrons, Z=22)
(Configuration: 1s²2s²2p⁶3s²3p⁶3d²4ℓ², 3d³c*nℓ, 3d²4s c*nℓ, Triplet transitions from z ³D°, z ³F°, and z ³G° terms)
Ti I
TRIPLET
GROTRIAN DIAGRAM
transitions from z ³D°, z ³F° and z ³G°
c*4p ³S°₁
3d⁴ c*ns 3d²4s² ³P
c*4p ³P°
c*4s 3d²4ℓ² ³D
c*4p ³D°₁,₂,₃
c*nℓ 3d²4s² ³F
c*4p ³F°₂,₃,₄
c*4ℓ 3d²3ℓ² ³G₃,₄,₅
c*4p ³G°
c*4ℓ ³H₄,₅,₆
c*4p ³H°₄,₅,₆
c*4p ³I°₅,₆,₇
ENERGY (cm⁻¹)
55 x 10³
50 x 10³
40 x 10³
30 x 10³
20 x 10³
10 x 10³
0
CORES (c*) from Ti II:
0* = 3d²4s ⁴F
1* = 3d³ ⁴F
2* = 3d²4s ²F
3* = 3d²4s ²D
4* = 3d³ ²G
5* = 3d³ ⁴P
6* = 3d³ ²P
7* = 3d²4s ⁴P
8* = 3d³ ²D
9* = 3d³ ²H
10* = 3d²4s ²G
11* = 3d²4s ⁴P
12* = 3d³ ²F
13* = 3d²4s ²S
3800.22,3827.64
3481.11,3481.67
3373.58–3381.35
6614.82
6938.24–6980.40
4547.82–4572.80
5091.56,5103.46
5648.57–5702.01
3652.54–3668.51
3400.15,3415.15
4944.37–4948.18
4430.04,4460.17
3708.64
3607.13–3640.03
5432.34–552.59
3904.20
4032.58
4900.57–4908.48
5536.36–5539.79
4555.11–4586.94
5982.54–6037.62
6873.91
5023.85–5093.04
3801.17,3807.20
4122.14–4149.44

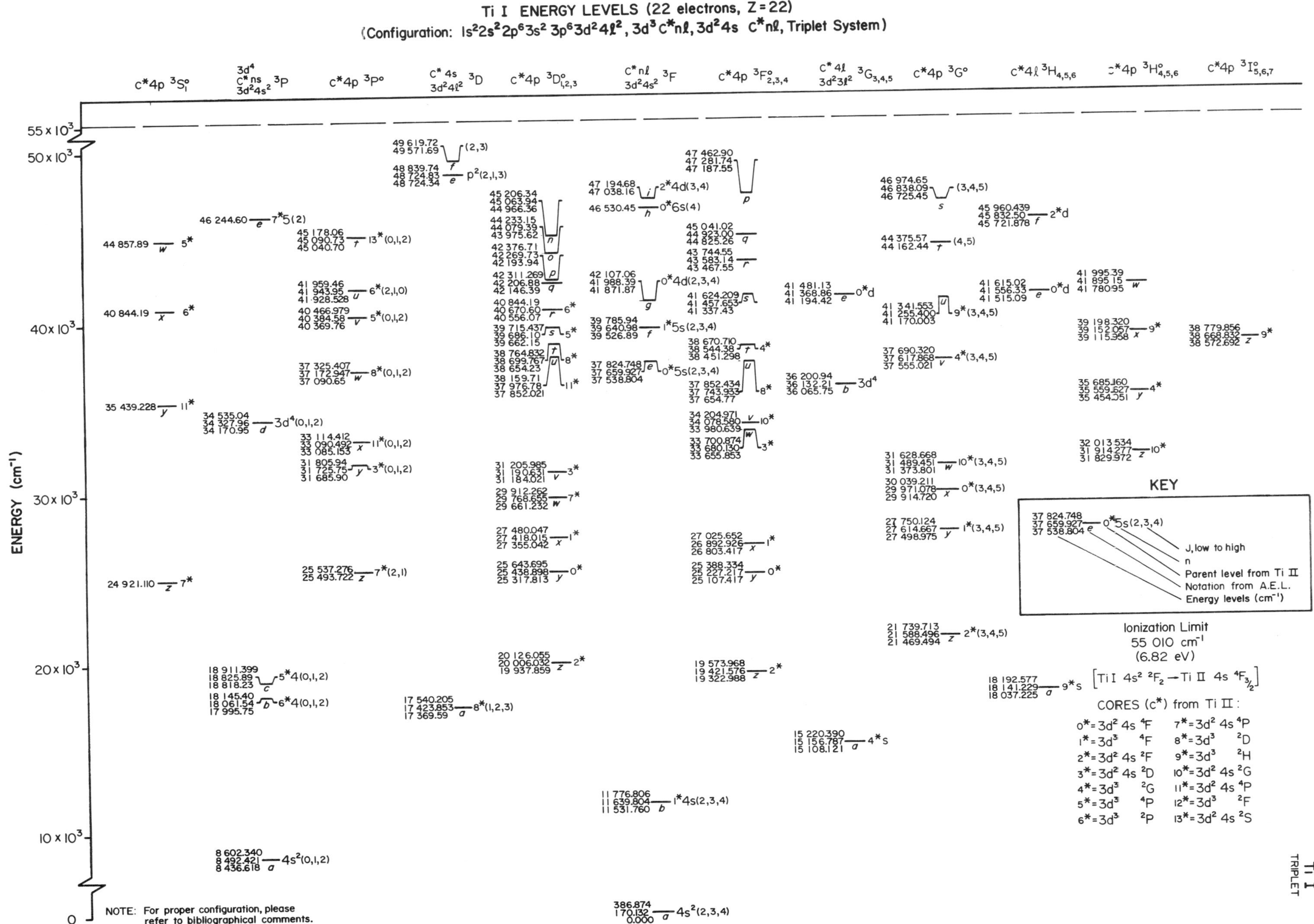

Ti I ENERGY LEVELS (22 electrons, Z = 22)
(Configuration: $1s^2 2s^2 2p^6 3s^2 3p^6 3d^2 4\ell^2$, $3d^3 c^* n\ell$, $3d^2 4s\ c^* n\ell$, Triplet System)
ENERGY (cm⁻¹)
0
10 x 10³
20 x 10³
30 x 10³
40 x 10³
50 x 10³
55 x 10³
$c^*4p\ ^3S^o_1$
44 857.89 w 5*
40 844.19 x 6*
35 439.228 y 11*
24 921.110 z 7*
$3d^4$, c^*ns, $3d^2 4s^2$ ^3P
46 244.60 e 7*5(2)
34 535.04, 34 327.96, 34 170.95 d 3d⁴(0,1,2)
18 911.399, 18 825.89, 18 818.23 c 5*4(0,1,2)
18 145.40, 18 061.54, 17 995.75 b 6*4(0,1,2)
8 602.340, 8 492.421, 8 436.618 a 4s²(0,1,2)
$c^*4p\ ^3P^o$
45 178.06, 45 090.73, 45 040.70 t 13*(0,1,2)
41 959.46, 41 943.95, 41 928.528 u 6*(2,1,0)
40 466.979, 40 384.58, 40 369.76 v 5*(0,1,2)
37 325.407, 37 172.947, 37 090.65 w 8*(0,1,2)
33 114.412, 33 090.492, 33 085.153 x 11*(0,1,2)
31 805.94, 31 725.75, 31 685.90 y 3*(0,1,2)
25 537.276, 25 493.722 z 7*(2,1)
c^*4s, $3d^2 4\ell^2$ ^3D
49 619.72, 49 571.69 f (2,3)
48 839.74, 48 724.83, 48 724.34 e p²(2,1,3)
17 540.205, 17 423.853, 17 369.59 a 8*(1,2,3)
$c^*4p\ ^3D^o_{1,2,3}$
45 206.34, 45 063.94, 44 966.36
44 233.15, 44 079.39, 43 975.62 n
42 376.71, 42 269.73, 42 193.94 o
42 311.269, 42 206.88, 42 146.39 p q
40 844.19, 40 670.60, 40 556.07 r 6*
39 715.437, 39 686.10, 39 662.15 s 5*
38 764.832, 38 699.767, 38 654.23 t u 8*
38 159.71, 37 976.78, 37 852.021 11*
31 205.985, 31 190.631, 31 184.021 v 3*
29 912.262, 29 768.655, 29 661.232 w 7*
27 480.047, 27 418.015, 27 355.042 x 1*
25 643.695, 25 438.898, 25 317.813 y 0*
20 126.055, 20 006.032, 19 937.859 z 2*
$c^*n\ell$, $3d^2 4s^2$ ^3F
47 194.68, 47 038.16 i 2*4d(3,4)
46 530.45 h 0*6s(4)
42 107.06, 41 988.39, 41 871.87 g 0*4d(2,3,4)
39 785.94, 39 640.98, 39 526.89 f 1*5s(2,3,4)
37 824.748, 37 659.927, 37 538.804 e 0*5s(2,3,4)
11 776.806, 11 639.804, 11 531.760 b 1*4s(2,3,4)
386.874, 170.132, 0.000 a 4s²(2,3,4)
$c^*4p\ ^3F^o_{2,3,4}$
47 462.90, 47 281.74, 47 187.55 p
45 041.02, 44 923.00, 44 825.26 q
43 744.55, 43 583.14, 43 467.55 r
41 624.209, 41 457.653, 41 337.43 s
38 670.710, 38 544.38, 38 451.298 t 4*
37 852.434, 37 743.933, 37 654.77 u 8*
34 204.971, 34 078.580, 33 980.639 v 10*
33 700.874, 33 680.130, 33 655.853 w 3*
27 025.652, 26 892.926, 26 803.417 x 1*
25 388.334, 25 227.217, 25 107.417 y 0*
19 573.968, 19 421.576, 19 322.988 z 2*
$c^*4\ell$, $3d^2 3\ell^2$ $^3G_{3,4,5}$
41 481.13, 41 368.86, 41 194.42 e 0*d
36 200.94, 36 132.21, 36 065.75 b 3d⁴
15 220.390, 15 156.787, 15 108.121 a 4*s
$c^*4p\ ^3G^o$
46 974.65, 46 838.09, 46 725.45 s (3,4,5)
44 375.57, 44 162.44 t (4,5)
41 341.553, 41 255.400, 41 170.003 u 9*(3,4,5)
37 690.320, 37 617.868, 37 555.021 v 4*(3,4,5)
31 628.668, 31 489.451, 31 373.801 w 10*(3,4,5)
30 039.211, 29 971.078, 29 914.720 x 0*(3,4,5)
27 750.124, 27 614.667, 27 498.975 y 1*(3,4,5)
21 739.713, 21 588.496, 21 469.494 z 2*(3,4,5)
$c^*4\ell\ ^3H_{4,5,6}$
45 960.439, 45 832.50, 45 721.878 f 2*d
41 615.02, 41 556.33, 41 515.09 e 0*d
18 192.577, 18 141.229, 18 037.225 a 9*s
$c^*4p\ ^3H^o_{4,5,6}$
41 995.39, 41 895.15, 41 780.95 w
39 198.320, 39 152.057, 39 115.358 x 9*
35 685.160, 35 559.527, 35 454.351 y 4*
32 013.534, 31 914.277, 31 829.572 z 10*
$c^*4p\ ^3I^o_{5,6,7}$
38 779.856, 38 668.832, 38 572.692 z 9*
KEY
37 824.748, 37 659.927, 37 538.804 e 0*5s(2,3,4)
J, low to high
n
Parent level from Ti II
Notation from A.E.L.
Energy levels (cm⁻¹)
Ionization Limit
55 010 cm⁻¹
(6.82 eV)
[Ti I 4s² ²F₂ → Ti II 4s ⁴F₃/₂]
CORES (c*) from Ti II:
0* = 3d² 4s ⁴F
1* = 3d³ ⁴F
2* = 3d² 4s ²F
3* = 3d² 4s ²D
4* = 3d³ ²G
5* = 3d³ ⁴P
6* = 3d³ ²P
7* = 3d² 4s ⁴P
8* = 3d³ ²D
9* = 3d³ ²H
10* = 3d² 4s ²G
11* = 3d² 4s ⁴P
12* = 3d³ ²F
13* = 3d² 4s ²S
NOTE: For proper configuration, please refer to bibliographical comments.
Ti I
TRIPLET

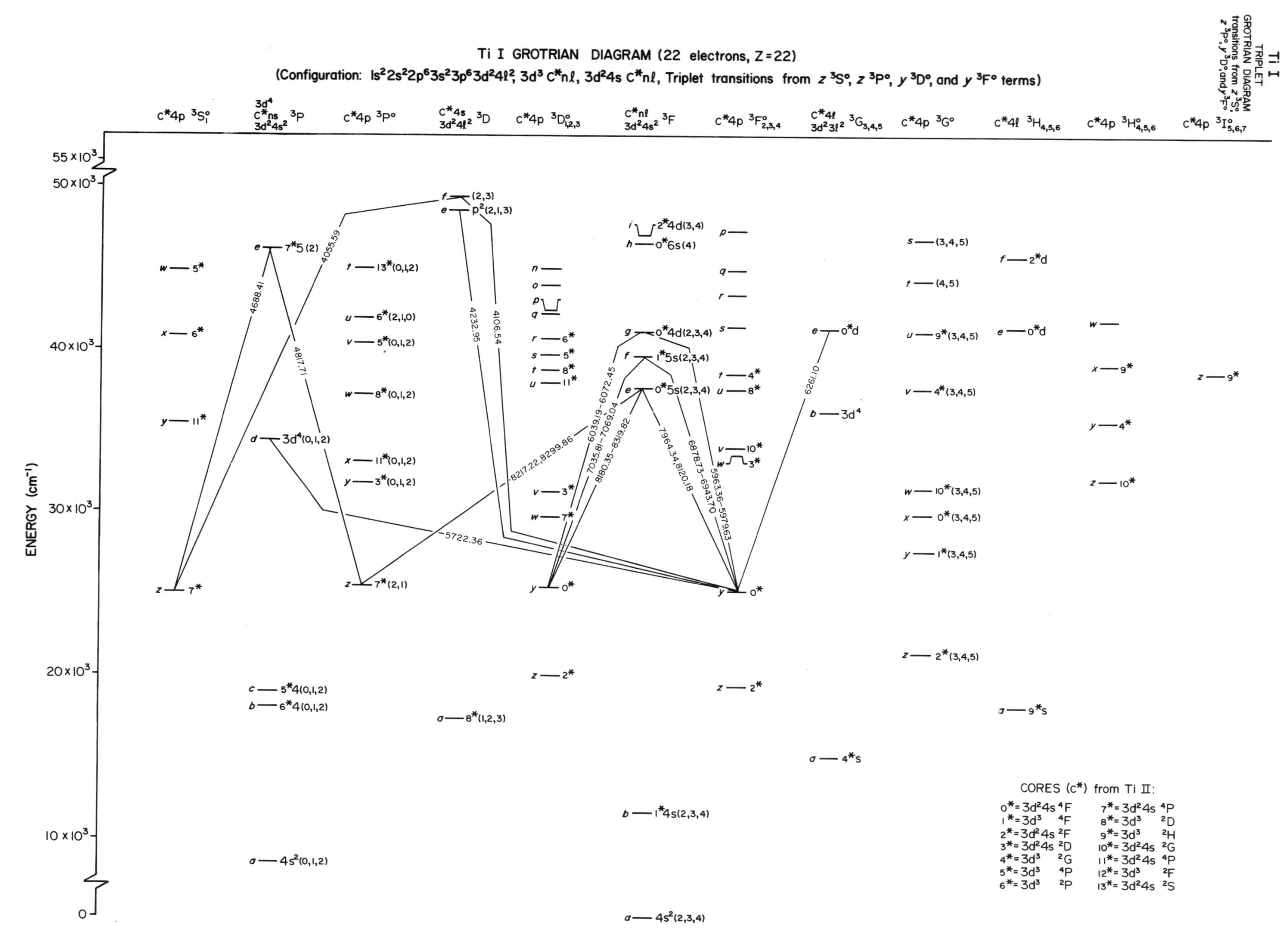
Ti I
TRIPLET
GROTRIAN DIAGRAM
transitions from z 3S°, z 3P°, y 3D°, and y 3F°
Ti I GROTRIAN DIAGRAM (22 electrons, Z=22)
(Configuration: 1s²2s²2p⁶3s²3p⁶3d²4l²; 3d³ c*nl, 3d²4s c*nl, Triplet transitions from z 3S°, z 3P°, y 3D°, and y 3F° terms)
c*4p 3S°1
3d4 c*ns 3d24s2 3P
c*4p 3P°
c*4s 3d24l2 3D
c*4p 3D°1,2,3
c*nl 3d24s2 3F
c*4p 3F°2,3,4
c*4l 3d23l2 3G3,4,5
c*4p 3G°
c*4l 3H4,5,6
c*4p 3H°4,5,6
c*4p 3I°5,6,7
ENERGY (cm-1)
55 x 10³
50 x 10³
40 x 10³
30 x 10³
20 x 10³
10 x 10³
0
4688.41
4055.59
4817.71
4232.95
4106.54
5722.36
8217.22, 8299.86
6039.19–6072.45
7035.81–7069.04
8180.35–8319.82
7964.34, 8120.18
6878.73–6943.70
5963.36–5979.63
6261.10
CORES (c*) from Ti II:
0* = 3d²4s 4F
1* = 3d³ 4F
2* = 3d²4s 2F
3* = 3d²4s 2D
4* = 3d³ 2G
5* = 3d³ 4P
6* = 3d³ 2P
7* = 3d²4s 4P
8* = 3d³ 2D
9* = 3d³ 2H
10* = 3d²4s 2G
11* = 3d²4s 4P
12* = 3d³ 2F
13* = 3d²4s 2S

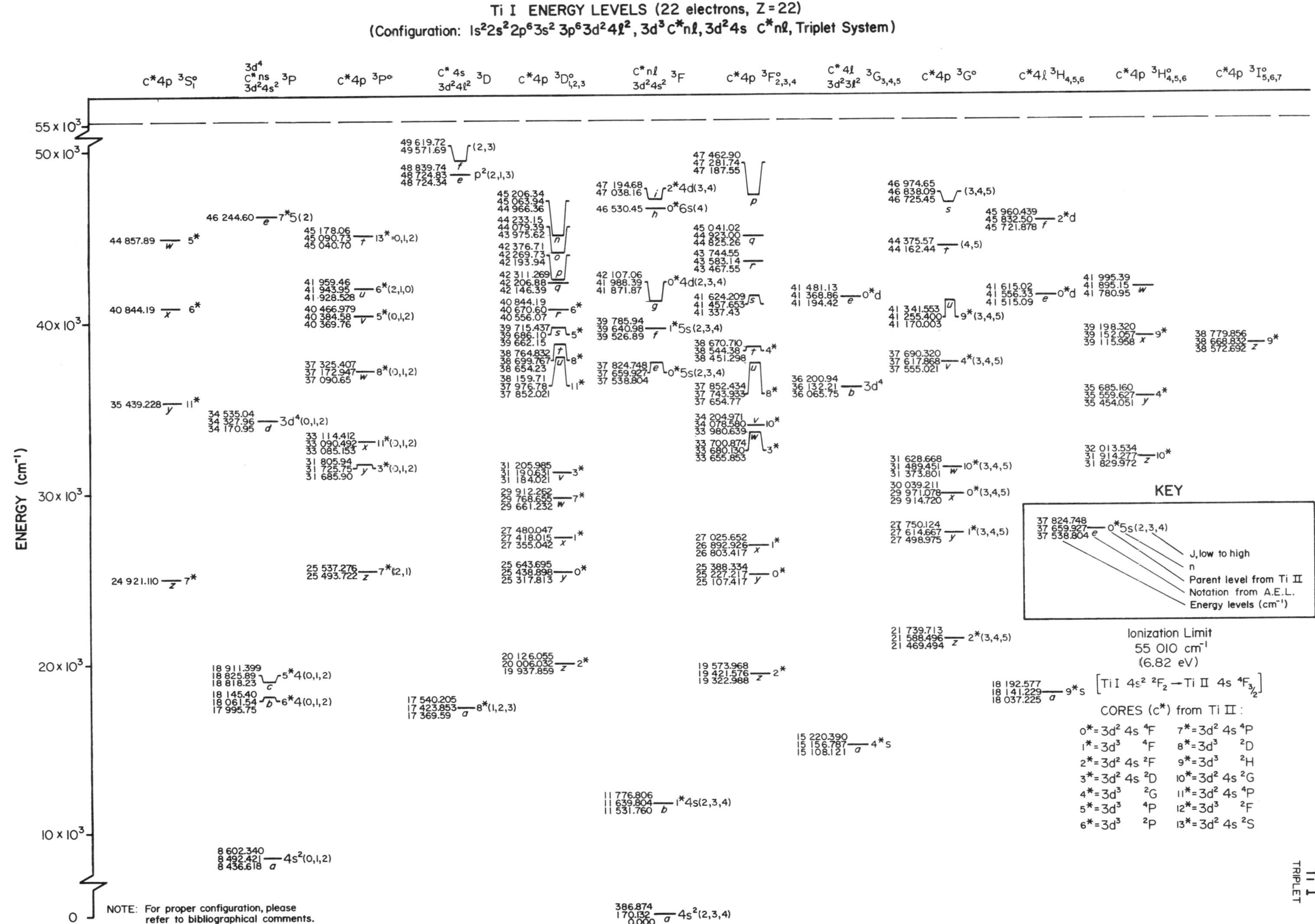
Ti I ENERGY LEVELS (22 electrons, Z = 22)
(Configuration: $1s^2 2s^2 2p^6 3s^2 3p^6 3d^2 4\ell^2$, $3d^3 c^* n\ell$, $3d^2 4s\ c^* n\ell$, Triplet System)
$c^*4p\ ^3S^o_1$
$3d^4$ / c^*ns / $3d^2 4s^2$ ^3P
$c^*4p\ ^3P^o$
c^*4s / $3d^2 4\ell^2$ ^3D
$c^*4p\ ^3D^o_{1,2,3}$
$c^*n\ell$ / $3d^2 4s^2$ ^3F
$c^*4p\ ^3F^o_{2,3,4}$
$c^*4\ell$ / $3d^2 3\ell^2$ $^3G_{3,4,5}$
$c^*4p\ ^3G^o$
$c^*4\ell\ ^3H_{4,5,6}$
$c^*4p\ ^3H^o_{4,5,6}$
$c^*4p\ ^3I^o_{5,6,7}$
55×10^3
50×10^3
40×10^3
30×10^3
20×10^3
10×10^3
0
ENERGY (cm^{-1})
44 857.89 w 5*
40 844.19 x 6*
35 439.228 y 11*
24 921.110 z 7*
46 244.60 e 7*5(2)
34 535.04 / 34 327.96 / 34 170.95 d 3d⁴(0,1,2)
18 911.399 / 18 825.89 / 18 818.23 c 5*4(0,1,2)
18 145.40 / 18 061.54 / 17 995.75 b 6*4(0,1,2)
8 602.340 / 8 492.421 / 8 436.618 a 4s²(0,1,2)
45 178.06 / 45 090.73 / 45 040.70 t 13*(0,1,2)
41 959.46 / 41 943.95 / 41 928.528 u 6*(2,1,0)
40 466.979 / 40 384.58 / 40 369.76 v 5*(0,1,2)
37 325.407 / 37 172.947 / 37 090.65 w 8*(0,1,2)
33 114.412 / 33 090.492 / 33 085.153 x 11*(0,1,2)
31 805.94 / 31 725.75 / 31 685.90 y 3*(0,1,2)
25 537.276 / 25 493.722 z 7*(2,1)
49 619.72 / 49 571.69 f (2,3)
48 839.74 / 48 724.83 / 48 724.34 e p²(2,1,3)
17 540.205 / 17 423.853 / 17 369.59 a 8*(1,2,3)
45 206.34 / 45 063.94 / 44 966.36 n
44 233.15 / 44 079.39 / 43 975.62 o
42 376.71 / 42 269.73 / 42 193.94 p
42 311.269 / 42 206.88 / 42 146.39 q
40 844.19 / 40 670.60 / 40 556.07 r 6*
39 715.437 / 39 686.10 / 39 662.15 s 5*
38 764.832 / 38 699.767 / 38 654.23 t 8*
38 159.71 / 37 976.78 / 37 852.021 u 11*
31 205.985 / 31 190.631 / 31 184.021 v 3*
29 912.262 / 29 768.655 / 29 661.232 w 7*
27 480.047 / 27 418.015 / 27 355.042 x 1*
25 643.695 / 25 438.898 / 25 317.813 y 0*
20 126.055 / 20 006.032 / 19 937.859 z 2*
47 194.68 / 47 038.16 i 2*4d(3,4)
46 530.45 h 0*6s(4)
42 107.06 / 41 988.39 / 41 871.87 g 0*4d(2,3,4)
39 785.94 / 39 640.98 / 39 526.89 f 1*5s(2,3,4)
37 824.748 / 37 659.927 / 37 538.804 e 0*5s(2,3,4)
11 776.806 / 11 639.804 / 11 531.760 b 1*4s(2,3,4)
386.874 / 170.132 / 0.000 a 4s²(2,3,4)
47 462.90 / 47 281.74 / 47 187.55 p
45 041.02 / 44 923.00 / 44 825.26 q
43 744.55 / 43 583.14 / 43 467.55 r
41 624.209 / 41 457.653 / 41 337.43 s
38 670.710 / 38 544.38 / 38 451.298 t 4*
37 852.434 / 37 743.933 / 37 654.77 u 8*
34 204.971 / 34 078.580 / 33 980.639 v 10*
33 700.874 / 33 680.130 / 33 655.853 w 3*
27 025.652 / 26 892.926 / 26 803.417 x 1*
25 388.334 / 25 227.217 / 25 107.417 y 0*
19 573.968 / 19 421.576 / 19 322.988 z 2*
41 481.13 / 41 368.86 / 41 194.42 e 0*d
36 200.94 / 36 132.21 / 36 065.75 b 3d⁴
15 220.390 / 15 156.787 / 15 108.121 a 4*s
46 974.65 / 46 838.09 / 46 725.45 s (3,4,5)
44 375.57 / 44 162.44 T (4,5)
41 341.553 / 41 255.400 / 41 170.003 u 9*(3,4,5)
37 690.320 / 37 617.868 / 37 555.021 v 4*(3,4,5)
31 628.668 / 31 489.451 / 31 373.801 w 10*(3,4,5)
30 039.211 / 29 971.078 / 29 914.720 x 0*(3,4,5)
27 750.124 / 27 614.667 / 27 498.975 y 1*(3,4,5)
21 739.713 / 21 588.496 / 21 469.494 z 2*(3,4,5)
45 960.439 / 45 832.50 / 45 721.878 f 2*d
41 615.02 / 41 556.33 / 41 515.09 e 0*d
18 192.577 / 18 141.229 / 18 037.225 a 9*s
41 995.39 / 41 895.15 / 41 780.95 w
39 198.320 / 39 152.057 / 39 115.958 x 9*
35 685.160 / 35 559.627 / 35 454.051 y 4*
32 013.534 / 31 914.277 / 31 829.972 z 10*
38 779.856 / 38 668.832 / 38 572.692 z 9*
KEY
37 824.748 / 37 659.927 / 37 538.804 e 0*5s(2,3,4)
J, low to high
n
Parent level from Ti II
Notation from A.E.L.
Energy levels (cm^{-1})
Ionization Limit
55 010 cm^{-1}
(6.82 eV)
[Ti I $4s^2\ ^2F_2$ → Ti II $4s\ ^4F_{3/2}$]
CORES (c*) from Ti II:
0* = $3d^2 4s\ ^4F$
1* = $3d^3\ ^4F$
2* = $3d^2 4s\ ^2F$
3* = $3d^2 4s\ ^2D$
4* = $3d^3\ ^2G$
5* = $3d^3\ ^4P$
6* = $3d^3\ ^2P$
7* = $3d^2 4s\ ^4P$
8* = $3d^3\ ^2D$
9* = $3d^3\ ^2H$
10* = $3d^2 4s\ ^2G$
11* = $3d^2 4s\ ^4P$
12* = $3d^3\ ^2F$
13* = $3d^2 4s\ ^2S$
NOTE: For proper configuration, please refer to bibliographical comments.
Ti I
TRIPLET

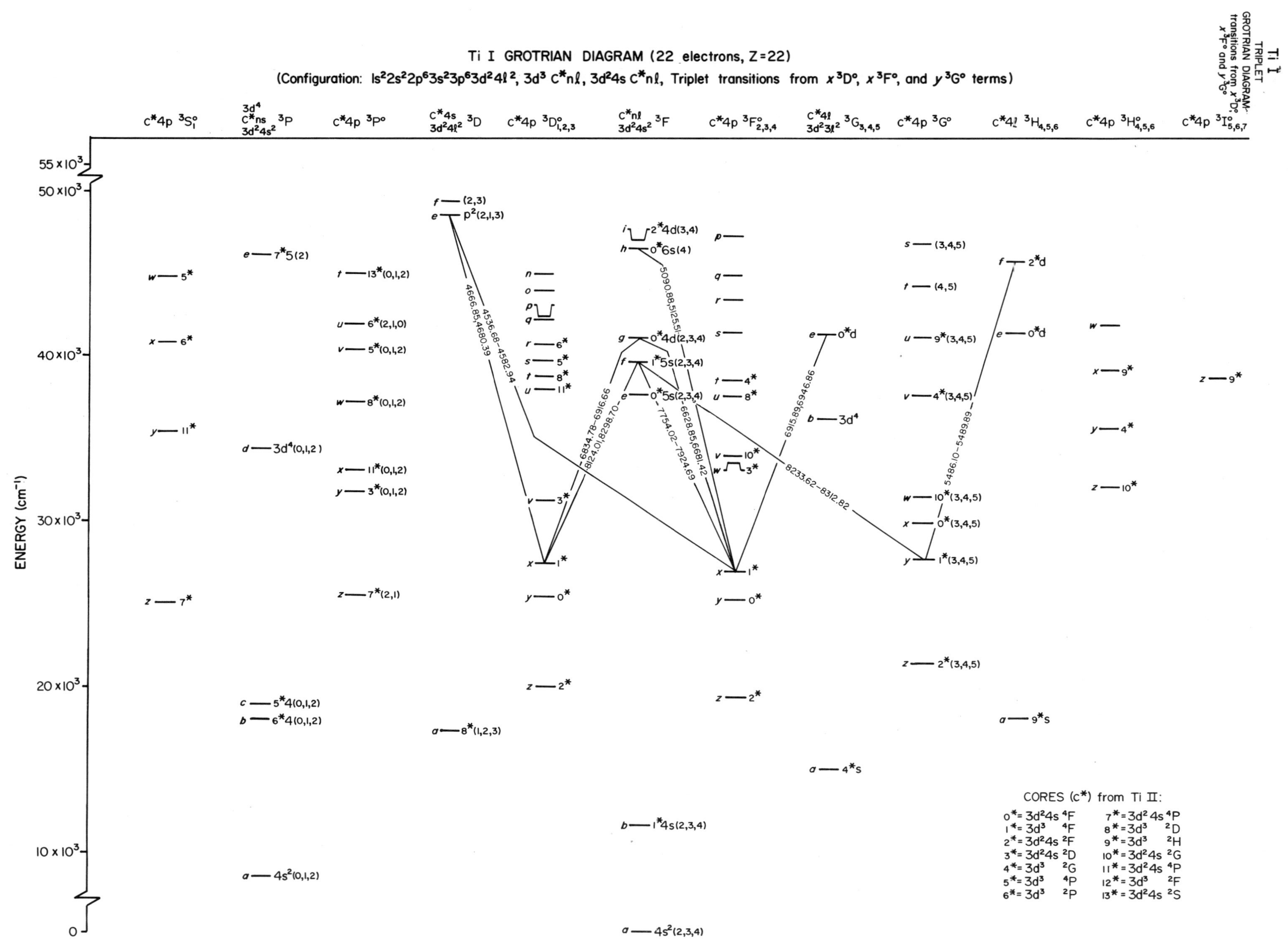
Ti I
TRIPLET
GROTRIAN DIAGRAM
transitions from $x^3D°$, $x^3F°$ and $y^3G°$
Ti I GROTRIAN DIAGRAM (22 electrons, Z=22)
(Configuration: $1s^22s^22p^63s^23p^63d^24l^2$, $3d^3\,c^*nl$, $3d^24s\,c^*nl$, Triplet transitions from $x^3D°$, $x^3F°$, and $y^3G°$ terms)
$c^*4p\ ^3S°_1$
$3d^4$ c^*ns $3d^24s^2$ ^3P
$c^*4p\ ^3P°$
c^*4s $3d^24l^2$ ^3D
$c^*4p\ ^3D°_{1,2,3}$
c^*nl $3d^24s^2$ ^3F
$c^*4p\ ^3F°_{2,3,4}$
c^*4l $3d^23l^2$ $^3G_{3,4,5}$
$c^*4p\ ^3G°$
$c^*4l\ ^3H_{4,5,6}$
$c^*4p\ ^3H°_{4,5,6}$
$c^*4p\ ^3I°_{5,6,7}$
ENERGY (cm^{-1})
55×10^3
50×10^3
40×10^3
30×10^3
20×10^3
10×10^3
0
4666.85,4680.39
4536.68–4582.94
6834.78–6916.66
8124.01,8298.70
5090.88,5125.51
7754.02–7924.69
6628.85,6681.42
6915.89,6946.86
8233.62–8312.82
5486.10–5489.89
CORES (c^*) from Ti II:
$0^* = 3d^24s\ ^4F$
$1^* = 3d^3\ ^4F$
$2^* = 3d^24s\ ^2F$
$3^* = 3d^24s\ ^2D$
$4^* = 3d^3\ ^2G$
$5^* = 3d^3\ ^4P$
$6^* = 3d^3\ ^2P$
$7^* = 3d^24s\ ^4P$
$8^* = 3d^3\ ^2D$
$9^* = 3d^3\ ^2H$
$10^* = 3d^24s\ ^2G$
$11^* = 3d^24s\ ^4P$
$12^* = 3d^3\ ^2F$
$13^* = 3d^24s\ ^2S$

Ti I ENERGY LEVELS (22 electrons, Z = 22)

(Configuration: $1s^2 2s^2 2p^6 3s^2 3p^6 3d^2 4\ell^2$, $3d^3 c^* n\ell$, $3d^2 4s$ $c^* n\ell$, Triplet System)

Ti I
TRIPLET

ENERGY (cm⁻¹): 0; 10 x 10³; 20 x 10³; 30 x 10³; 40 x 10³; 50 x 10³; 55 x 10³

c^*4p $^3S^o_1$

- 44 857.89 w 5*
- 40 844.19 x 6*
- 35 439.228 y 11*
- 24 921.110 z 7*

$3d^4$ / c^*ns / $3d^2 4s^2$ ^3P

- 46 244.60 e 7*5(2)
- 34 535.04, 34 327.96, 34 170.95 d $3d^4$(0,1,2)
- 18 911.399, 18 825.89, 18 818.23 c 5*4(0,1,2)
- 18 145.40, 18 061.54, 17 995.75 b 6*4(0,1,2)
- 8 602.340, 8 492.421, 8 436.618 a $4s^2$(0,1,2)

c^*4p $^3P^o$

- 45 178.06, 45 090.73, 45 040.70 t 13*(0,1,2)
- 41 959.46, 41 943.95, 41 928.528 u 6*(2,1,0)
- 40 466.979, 40 384.58, 40 369.76 v 5*(0,1,2)
- 37 325.407, 37 172.947, 37 090.65 w 8*(0,1,2)
- 33 114.412, 33 090.492, 33 085.153 x 11*(0,1,2)
- 31 805.94, 31 725.75, 31 685.90 y 3*(0,1,2)
- 25 537.276, 25 493.722 z 7*(2,1)

c^*4s / $3d^2 4\ell^2$ ^3D

- 49 619.72, 49 571.69 f (2,3)
- 48 839.74, 48 724.83, 48 724.34 e p^2(2,1,3)
- 17 540.205, 17 423.853, 17 369.59 a 8*(1,2,3)

c^*4p $^3D^o_{1,2,3}$

- 45 206.34, 45 063.94, 44 966.36 n
- 44 233.15, 44 079.39, 43 975.62 o
- 42 376.71, 42 269.73, 42 193.94 p
- 42 311.269, 42 206.88, 42 146.39 q
- 40 844.19, 40 670.60, 40 556.07 r 6*
- 39 715.437, 39 686.10, 39 662.15 s 5*
- 38 764.832, 38 699.767, 38 654.23 t, u 8*
- 38 159.71, 37 976.78, 37 852.021 11*
- 31 205.985, 31 190.631, 31 184.021 v 3*
- 29 912.262, 29 768.655, 29 661.232 w 7*
- 27 480.047, 27 418.015, 27 355.042 x 1*
- 25 643.695, 25 438.898, 25 317.813 y 0*
- 20 126.055, 20 006.032, 19 937.859 z 2*

$c^*n\ell$ / $3d^2 4s^2$ ^3F

- 47 194.68, 47 038.16 i 2*4d(3,4)
- 46 530.45 h 0*6s(4)
- 42 107.06, 41 988.39, 41 871.87 g 0*4d(2,3,4)
- 39 785.94, 39 640.98, 39 526.89 f 1*5s(2,3,4)
- 37 824.748, 37 659.927, 37 538.804 e 0*5s(2,3,4)
- 11 776.806, 11 639.804, 11 531.760 b 1*4s(2,3,4)
- 386.874, 170.132, 0.000 a $4s^2$(2,3,4)

c^*4p $^3F^o_{2,3,4}$

- 47 462.90, 47 281.74, 47 187.55 p
- 45 041.02, 44 923.00, 44 825.26 q
- 43 744.55, 43 583.14, 43 467.55 r
- 41 624.209, 41 457.653, 41 337.43 s
- 38 670.710, 38 544.38, 38 451.298 t 4*
- 37 852.434, 37 743.933, 37 654.77 u 8*
- 34 204.971, 34 078.580, 33 980.639 v 10*
- 33 700.874, 33 680.130, 33 655.853 w 3*
- 27 025.652, 26 892.926, 26 803.417 x 1*
- 25 388.334, 25 227.217, 25 107.417 y 0*
- 19 573.968, 19 421.576, 19 322.988 z 2*

$c^*4\ell$ / $3d^2 3\ell^2$ $^3G_{3,4,5}$

- 41 481.13, 41 368.86, 41 194.42 e 0*d
- 36 200.94, 36 132.21, 36 065.75 b $3d^4$
- 15 220.390, 15 156.787, 15 108.121 a 4*s

c^*4p $^3G^o$

- 46 974.65, 46 838.09, 46 725.45 s (3,4,5)
- 44 375.57, 44 162.44 t (4,5)
- 41 341.553, 41 255.400, 41 170.003 u 9*(3,4,5)
- 37 690.320, 37 617.868, 37 555.021 v 4*(3,4,5)
- 31 628.668, 31 489.451, 31 373.801 w 10*(3,4,5)
- 30 039.211, 29 971.078, 29 914.720 x 0*(3,4,5)
- 27 750.124, 27 614.667, 27 498.975 y 1*(3,4,5)
- 21 739.713, 21 588.496, 21 469.494 z 2*(3,4,5)

$c^*4\ell$ $^3H_{4,5,6}$

- 45 960.439, 45 832.50, 45 721.878 f 2*d
- 41 615.02, 41 556.33, 41 515.09 e 0*d
- 18 192.577, 18 141.229, 18 037.225 a 9*s

c^*4p $^3H^o_{4,5,6}$

- 41 995.39, 41 895.15, 41 780.95 w
- 39 198.320, 39 152.057, 39 115.958 x 9*
- 35 685.160, 35 559.627, 35 454.051 y 4*
- 32 013.534, 31 914.277, 31 829.972 z 10*

c^*4p $^3I^o_{5,6,7}$

- 38 779.856, 38 668.832, 38 572.692 z 9*

KEY

37 824.748, 37 659.927, 37 538.804 e 0*5s(2,3,4)

- J, low to high
- n
- Parent level from Ti II
- Notation from A.E.L.
- Energy levels (cm⁻¹)

Ionization Limit
55 010 cm⁻¹
(6.82 eV)
[Ti I $4s^2$ $^2F_2$ → Ti II 4s $^4F_{3/2}$]

CORES (c^*) from Ti II:

0* = $3d^2 4s$ ^4F	7* = $3d^2 4s$ ^4P
1* = $3d^3$ ^4F	8* = $3d^3$ ^2D
2* = $3d^2 4s$ ^2F	9* = $3d^3$ ^2H
3* = $3d^2 4s$ ^2D	10* = $3d^2 4s$ ^2G
4* = $3d^3$ ^2G	11* = $3d^2 4s$ ^4P
5* = $3d^3$ ^4P	12* = $3d^3$ ^2F
6* = $3d^3$ ^2P	13* = $3d^2 4s$ ^2S

NOTE: For proper configuration, please refer to bibliographical comments.

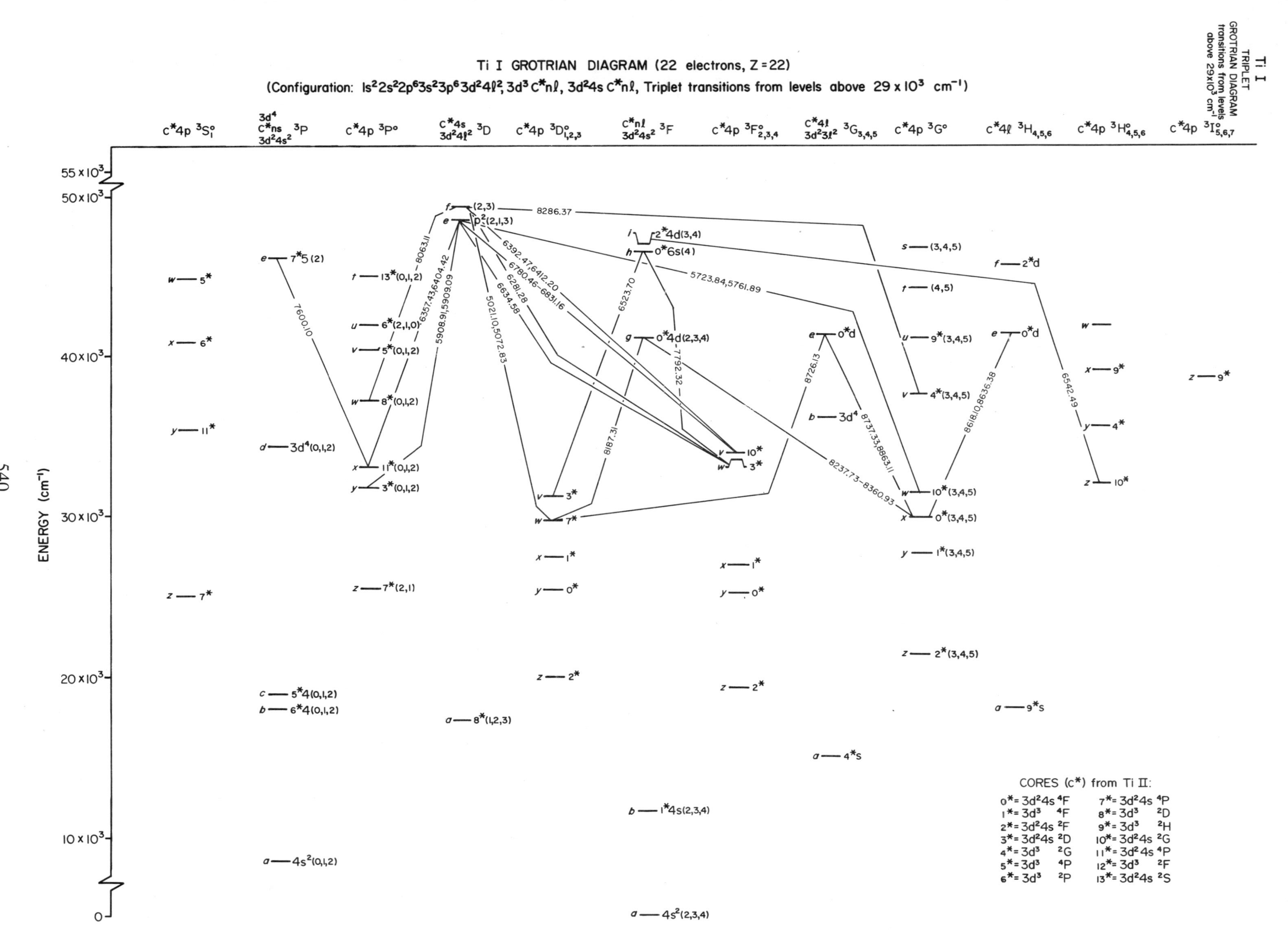
Ti I
TRIPLET
GROTRIAN DIAGRAM
transitions from levels above 29×10^3 cm^{-1}
Ti I GROTRIAN DIAGRAM (22 electrons, Z = 22)
(Configuration: $1s^22s^22p^63s^23p^63d^24l^2$, $3d^3c^*nl$, $3d^24s\,c^*nl$, Triplet transitions from levels above 29×10^3 cm^{-1})
ENERGY (cm^{-1})
55×10^3
50×10^3
40×10^3
30×10^3
20×10^3
10×10^3
0
$c^*4p\ ^3S^o_1$
$3d^4\ c^*ns\ 3d^24s^2\ ^3P$
$c^*4p\ ^3P^o$
$c^*4s\ 3d^24l^2\ ^3D$
$c^*4p\ ^3D^o_{1,2,3}$
$c^*nl\ 3d^24s^2\ ^3F$
$c^*4p\ ^3F^o_{2,3,4}$
$c^*4l\ 3d^23l^2\ ^3G_{3,4,5}$
$c^*4p\ ^3G^o$
$c^*4l\ ^3H_{4,5,6}$
$c^*4p\ ^3H^o_{4,5,6}$
$c^*4p\ ^3I^o_{5,6,7}$
7600.10
8063.11
6357.43,6404.42
5908.91,5909.09
5021.10,5072.83
6634.58
6281.28
6780.46-6831.16
6392.47,6412.20
8286.37
6523.70
8187.31
5723.84,5761.89
7792.32
8726.13
8737.33,8863.11
8237.73-8360.93
8618.10,8636.38
6542.49
CORES (c^*) from Ti II:
$0^* = 3d^24s\ ^4F$
$1^* = 3d^3\ ^4F$
$2^* = 3d^24s\ ^2F$
$3^* = 3d^24s\ ^2D$
$4^* = 3d^3\ ^2G$
$5^* = 3d^3\ ^4P$
$6^* = 3d^3\ ^2P$
$7^* = 3d^24s\ ^4P$
$8^* = 3d^3\ ^2D$
$9^* = 3d^3\ ^2H$
$10^* = 3d^24s\ ^2G$
$11^* = 3d^24s\ ^4P$
$12^* = 3d^3\ ^2F$
$13^* = 3d^24s\ ^2S$

Ti I TRIPLET

Ti I ENERGY LEVELS (22 electrons, Z = 22)

(Configuration: $1s^2 2s^2 2p^6 3s^2 3p^6 3d^2 4\ell^2$, $3d^3 c^* n\ell$, $3d^2 4s\ c^* n\ell$, Triplet System)

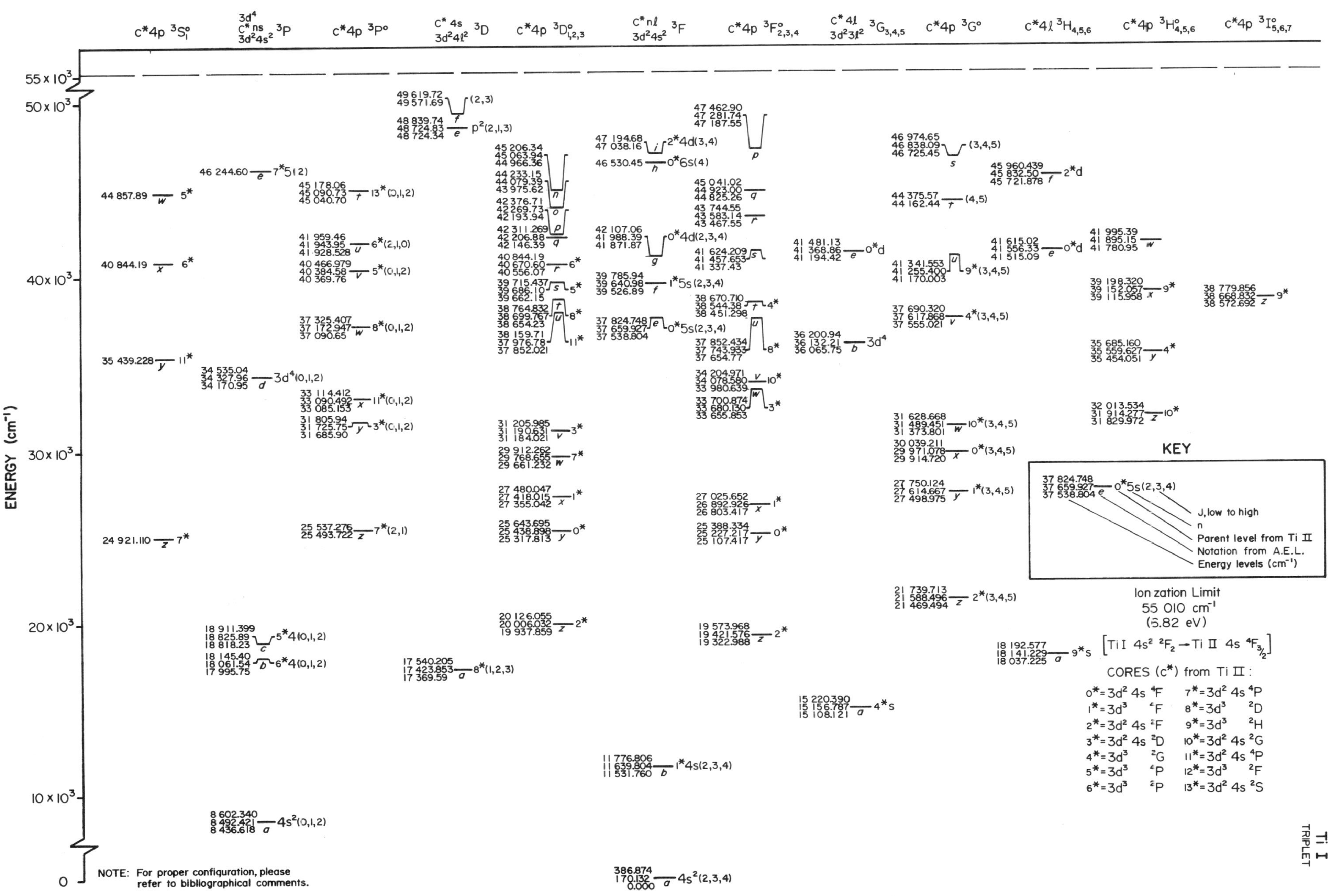

ENERGY (cm⁻¹): 0, 10 x 10³, 20 x 10³, 30 x 10³, 40 x 10³, 50 x 10³, 55 x 10³

$c^*4p\ ^3S^o_1$
- 44 857.89 w 5*
- 40 844.19 x 6*
- 35 439.228 y 11*
- 24 921.110 z 7*

$3d^4$, c^*ns, $3d^2 4s^2$ ^3P
- 46 244.60 e 7*5(2)
- 34 535.04, 34 327.96, 34 170.95 d 3d⁴(0,1,2)
- 18 911.399, 18 825.89, 18 818.23 c 5*4(0,1,2)
- 18 145.40, 18 061.54, 17 995.75 b 6*4(0,1,2)
- 8 602.340, 8 492.421, 8 436.618 a 4s²(0,1,2)

$c^*4p\ ^3P^o$
- 45 178.06, 45 090.73, 45 040.70 t 13*(0,1,2)
- 41 959.46, 41 943.95, 41 928.528 u 6*(2,1,0)
- 40 466.979, 40 384.58, 40 369.76 v 5*(0,1,2)
- 37 325.407, 37 172.947, 37 090.65 w 8*(0,1,2)
- 33 114.412, 33 090.492, 33 085.153 x 11*(0,1,2)
- 31 805.94, 31 725.75, 31 685.90 y 3*(0,1,2)
- 25 537.276, 25 493.722 z 7*(2,1)

c^*4s $3d^2 4\ell^2$ ^3D
- 49 619.72, 49 571.69 f (2,3)
- 48 839.74, 48 724.83, 48 724.34 e p²(2,1,3)
- 17 540.205, 17 423.853, 17 369.59 a 8*(1,2,3)

$c^*4p\ ^3D^o_{1,2,3}$
- 45 206.34, 45 063.94, 44 966.36 n
- 44 233.15, 44 079.39, 43 975.62 o
- 42 376.71, 42 269.73, 42 193.94 p
- 42 311.269, 42 206.88, 42 146.39 q
- 40 844.19, 40 670.60, 40 556.07 r 6*
- 39 715.437, 39 686.10, 39 662.15 s 5*
- 38 764.832, 38 699.767, 38 654.23 t, u 8*
- 38 159.71, 37 976.78, 37 852.021 11*
- 31 205.985, 31 190.631, 31 184.021 v 3*
- 29 912.262, 29 768.655, 29 661.232 w 7*
- 27 480.047, 27 418.015, 27 355.042 x 1*
- 25 643.695, 25 438.898, 25 317.813 y 0*
- 20 126.055, 20 006.032, 19 937.859 z 2*

$c^*n\ell$ $3d^2 4s^2$ ^3F
- 47 194.68, 47 038.16 i 2*4d(3,4)
- 46 530.45 h 0*6s(4)
- 42 107.06, 41 988.39, 41 871.87 g 0*4d(2,3,4)
- 39 785.94, 39 640.98, 39 526.89 f 1*5s(2,3,4)
- 37 824.748, 37 659.927, 37 538.804 e 0*5s(2,3,4)
- 11 776.806, 11 639.804, 11 531.760 b 1*4s(2,3,4)
- 386.874, 170.132, 0.000 a 4s²(2,3,4)

$c^*4p\ ^3F^o_{2,3,4}$
- 47 462.90, 47 281.74, 47 187.55 p
- 45 041.02, 44 923.00, 44 825.26 q
- 43 744.55, 43 583.14, 43 467.55 r
- 41 624.209, 41 457.653, 41 337.43 s
- 38 670.710, 38 544.38, 38 451.298 t 4*
- 37 852.434, 37 743.933, 37 654.77 u 8*
- 34 204.971, 34 078.580, 33 980.639 v 10*
- 33 700.874, 33 680.130, 33 655.853 w 3*
- 27 025.652, 26 892.926, 26 803.417 x 1*
- 25 388.334, 25 227.217, 25 107.417 y 0*
- 19 573.968, 19 421.576, 19 322.988 z 2*

$c^*4\ell$ $3d^2 3\ell^2$ $^3G_{3,4,5}$
- 41 481.13, 41 368.86, 41 194.42 e 0*d
- 36 200.94, 36 132.21, 36 065.75 b 3d⁴
- 15 220.390, 15 156.787, 15 108.121 a 4*s

$c^*4p\ ^3G^o$
- 46 974.65, 46 838.09, 46 725.45 s (3,4,5)
- 44 375.57, 44 162.44 t (4,5)
- 41 341.553, 41 255.400, 41 170.003 u 9*(3,4,5)
- 37 690.320, 37 617.868, 37 555.021 v 4*(3,4,5)
- 31 628.668, 31 489.451, 31 373.801 w 10*(3,4,5)
- 30 039.211, 29 971.078, 29 914.720 x 0*(3,4,5)
- 27 750.124, 27 614.667, 27 498.975 y 1*(3,4,5)
- 21 739.713, 21 588.496, 21 469.494 z 2*(3,4,5)

$c^*4\ell\ ^3H_{4,5,6}$
- 45 960.439, 45 832.50, 45 721.878 f 2*d
- 41 615.02, 41 556.33, 41 515.09 e 0*d
- 18 192.577, 18 141.229, 18 037.225 a 9*s

$c^*4p\ ^3H^o_{4,5,6}$
- 41 995.39, 41 895.15, 41 780.95 w
- 39 198.320, 39 152.057, 39 115.958 x 9*
- 35 685.160, 35 559.627, 35 454.051 y 4*
- 32 013.534, 31 914.277, 31 829.972 z 10*

$c^*4p\ ^3I^o_{5,6,7}$
- 38 779.856, 38 668.832, 38 572.692 z 9*

KEY

37 824.748, 37 659.927, 37 538.804 e 0*5s(2,3,4)

- J, low to high
- n
- Parent level from Ti II
- Notation from A.E.L.
- Energy levels (cm⁻¹)

Ion zation Limit
55 010 cm⁻¹
(6.82 eV)

[Ti I $4s^2\ ^2F_2$ → Ti II $4s\ ^4F_{3/2}$]

CORES (c^*) from Ti II:

0* = 3d² 4s ⁴F	7* = 3d² 4s ⁴P
1* = 3d³ ⁴F	8* = 3d³ ²D
2* = 3d² 4s ²F	9* = 3d³ ²H
3* = 3d² 4s ²D	10* = 3d² 4s ²G
4* = 3d³ ²G	11* = 3d² 4s ⁴P
5* = 3d³ ⁴P	12* = 3d³ ²F
6* = 3d³ ²P	13* = 3d² 4s ²S

NOTE: For proper configuration, please refer to bibliographical comments.

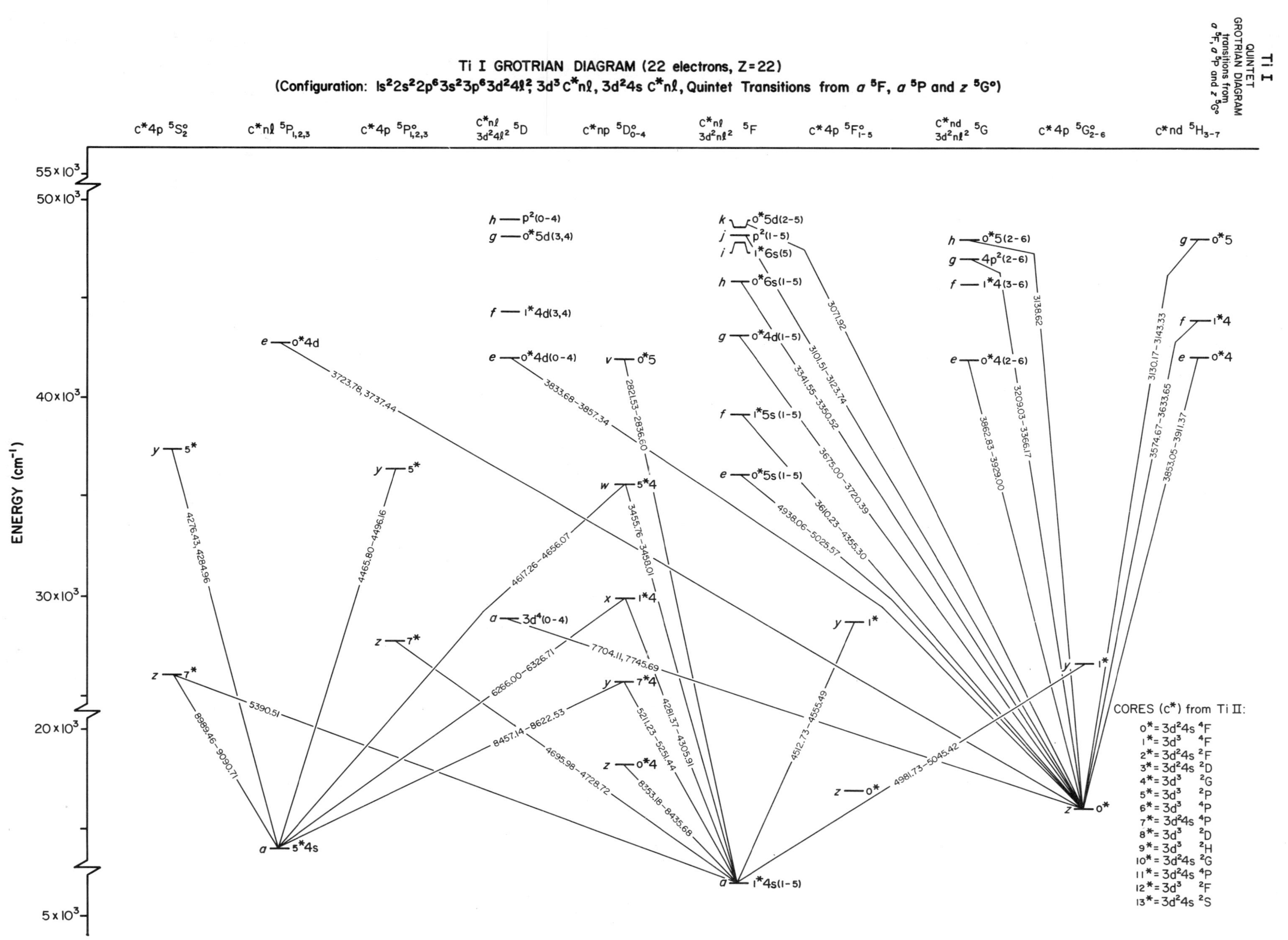
Ti I
QUINTET
GROTRIAN DIAGRAM
transitions from
a ⁵F, a ⁵P and z ⁵G°
Ti I GROTRIAN DIAGRAM (22 electrons, Z=22)
(Configuration: $1s^2 2s^2 2p^6 3s^2 3p^6 3d^2 4l^2$, $3d^3 c^* nl$, $3d^2 4s\, c^* nl$, Quintet Transitions from $a\ ^5F$, $a\ ^5P$ and $z\ ^5G°$)
$c^*4p\ ^5S°_2$
$c^*nl\ ^5P_{1,2,3}$
$c^*4p\ ^5P°_{1,2,3}$
c^*nl $3d^2 4l^2$ ^5D
$c^*np\ ^5D°_{0-4}$
c^*nl $3d^2 nl^2$ ^5F
$c^*4p\ ^5F°_{1-5}$
c^*nd $3d^2 nl^2$ ^5G
$c^*4p\ ^5G°_{2-6}$
$c^*nd\ ^5H_{3-7}$
ENERGY (cm^{-1})
55×10^3
50×10^3
40×10^3
30×10^3
20×10^3
5×10^3
3723.78, 3737.44
3833.68–3857.34
2821.53–2836.60
3455.76–3458.01
4617.26–4656.07
6266.00–6326.71
7704.11, 7745.69
4281.37–4305.91
5211.23–5251.44
8457.14–8622.53
4695.98–4728.72
8353.18–8435.68
4276.43, 4284.96
4465.80–4496.16
5390.51
8989.46–9090.71
4512.73–4555.49
4981.73–5045.42
3071.92
3101.51–3123.74
3341.55–3350.52
3675.00–3720.39
3610.23–4355.30
4938.06–5025.57
3138.62
3209.03–3366.17
3862.83–3929.00
3130.17–3143.33
3574.67–3633.65
3853.05–3911.37
CORES (c*) from Ti II:
$0^* = 3d^2 4s\ ^4F$
$1^* = 3d^3\ ^4F$
$2^* = 3d^2 4s\ ^2F$
$3^* = 3d^2 4s\ ^2D$
$4^* = 3d^3\ ^2G$
$5^* = 3d^3\ ^2P$
$6^* = 3d^3\ ^4P$
$7^* = 3d^2 4s\ ^4P$
$8^* = 3d^3\ ^2D$
$9^* = 3d^3\ ^2H$
$10^* = 3d^2 4s\ ^2G$
$11^* = 3d^2 4s\ ^4P$
$12^* = 3d^3\ ^2F$
$13^* = 3d^2 4s\ ^2S$

Ti I ENERGY LEVELS (22 electrons, Z = 22)

(Configuration: $1s^2 2s^2 2p^6 3s^2 3p^6 3d^2 4\ell^2, 3d^3 c^* n\ell, 3d^2 4s\ c^* n\ell$, Quintet System)

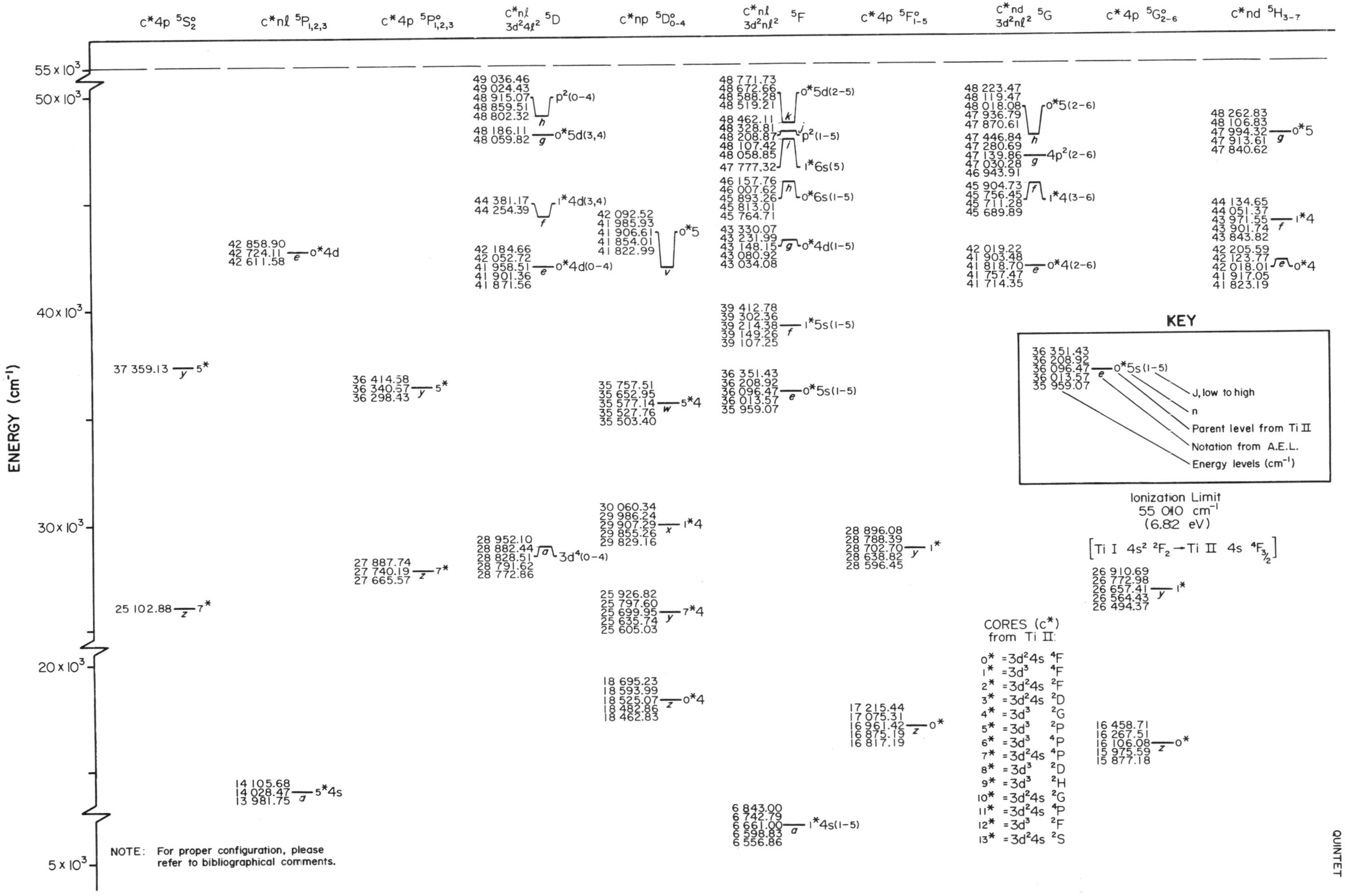

ENERGY (cm^{-1}): 55×10^3, 50×10^3, 40×10^3, 30×10^3, 20×10^3, 5×10^3

$c^*4p\ ^5S^o_2$

37 359.13 — y — 5*

25 102.88 — z — 7*

$c^*n\ell\ ^5P_{1,2,3}$

42 858.90
42 724.11
42 611.58 — e — o*4d

14 105.68
14 028.47
13 981.75 — a — 5*4s

$c^*4p\ ^5P^o_{1,2,3}$

36 414.58
36 340.57
36 298.43 — y — 5*

27 887.74
27 740.19
27 665.57 — z — 7*

$c^*n\ell\ 3d^2 4\ell^2\ ^5D$

49 036.46
49 024.43
48 915.07
48 859.51
48 802.32 — h — p^2(0–4)

48 186.11
48 059.82 — g — o*5d(3,4)

44 381.17
44 254.39 — f — 1*4d(3,4)

42 184.66
42 052.72
41 958.51
41 901.36
41 871.56 — e — o*4d(0–4)

28 952.10
28 882.44
28 828.51
28 791.62
28 772.86 — a — $3d^4$(0–4)

$c^*np\ ^5D^o_{0-4}$

42 092.52
41 985.93
41 906.61
41 854.01
41 822.99 — v — o*5

35 757.51
35 652.95
35 577.14
35 527.76
35 503.40 — w — 5*4

30 060.34
29 986.24
29 907.29
29 855.26
29 829.16 — x — 1*4

25 926.82
25 797.60
25 699.95
25 635.74
25 605.03 — y — 7*4

18 695.23
18 593.99
18 525.07
18 482.86
18 462.83 — z — o*4

$c^*n\ell\ 3d^2 n\ell^2\ ^5F$

48 771.73
48 672.66
48 588.28
48 519.21 — k — o*5d(2–5)

48 462.11
48 328.81
48 208.87
48 107.42
48 058.85 — j — p^2(1–5)

47 777.32 — l — 1*6s(5)

46 157.76
46 007.62
45 893.26
45 813.01
45 764.71 — h — o*6s(1–5)

43 330.07
43 231.99
43 148.15
43 080.92
43 034.08 — g — o*4d(1–5)

39 412.78
39 302.36
39 214.38
39 149.26
39 107.25 — f — 1*5s(1–5)

36 351.43
36 208.92
36 096.47
36 013.57
35 959.07 — e — o*5s(1–5)

6 843.00
6 742.79
6 661.00
6 598.83
6 556.86 — a — 1*4s(1–5)

$c^*4p\ ^5F^o_{1-5}$

28 896.08
28 788.39
28 702.70
28 638.82
28 596.45 — y — 1*

17 215.44
17 075.31
16 961.42
16 875.19
16 817.19 — z — o*

$c^*nd\ 3d^2 n\ell^2\ ^5G$

48 223.47
48 119.47
48 018.08
47 936.79
47 870.61 — h — o*5(2–6)

47 446.84
47 280.69
47 139.86
47 030.28
46 943.91 — g — $4p^2$(2–6)

45 904.73
45 756.45
45 711.28
45 689.89 — f — 1*4(3–6)

42 019.22
41 903.48
41 818.70
41 757.47
41 714.35 — e — o*4(2–6)

$c^*4p\ ^5G^o_{2-6}$

26 910.69
26 772.98
26 657.41
26 564.43
26 494.37 — y — 1*

16 458.71
16 267.51
16 106.08
15 975.59
15 877.18 — z — o*

$c^*nd\ ^5H_{3-7}$

48 262.83
48 106.83
47 994.32
47 913.61
47 840.62 — g — o*5

44 134.65
44 051.37
43 971.55
43 901.74
43 843.82 — f — 1*4

42 205.59
42 123.77
42 018.01
41 917.05
41 823.19 — e — o*4

KEY

36 351.43
36 208.92
36 096.47 — e — o*5s(1–5)
36 013.57
35 959.07

- J, low to high
- n
- Parent level from Ti II
- Notation from A.E.L.
- Energy levels (cm^{-1})

Ionization Limit
55 010 cm^{-1}
(6.82 eV)

[Ti I $4s^2\ ^2F_2$ → Ti II $4s\ ^4F_{3/2}$]

CORES (c*) from Ti II:

o*	$= 3d^2 4s\ ^4F$
1*	$= 3d^3\ ^4F$
2*	$= 3d^2 4s\ ^2F$
3*	$= 3d^2 4s\ ^2D$
4*	$= 3d^3\ ^2G$
5*	$= 3d^3\ ^2P$
6*	$= 3d^3\ ^4P$
7*	$= 3d^2 4s\ ^4P$
8*	$= 3d^3\ ^2D$
9*	$= 3d^3\ ^2H$
10*	$= 3d^2 4s\ ^2G$
11*	$= 3d^2 4s\ ^4P$
12*	$= 3d^3\ ^2F$
13*	$= 3d^2 4s\ ^2S$

NOTE: For proper configuration, please refer to bibliographical comments.

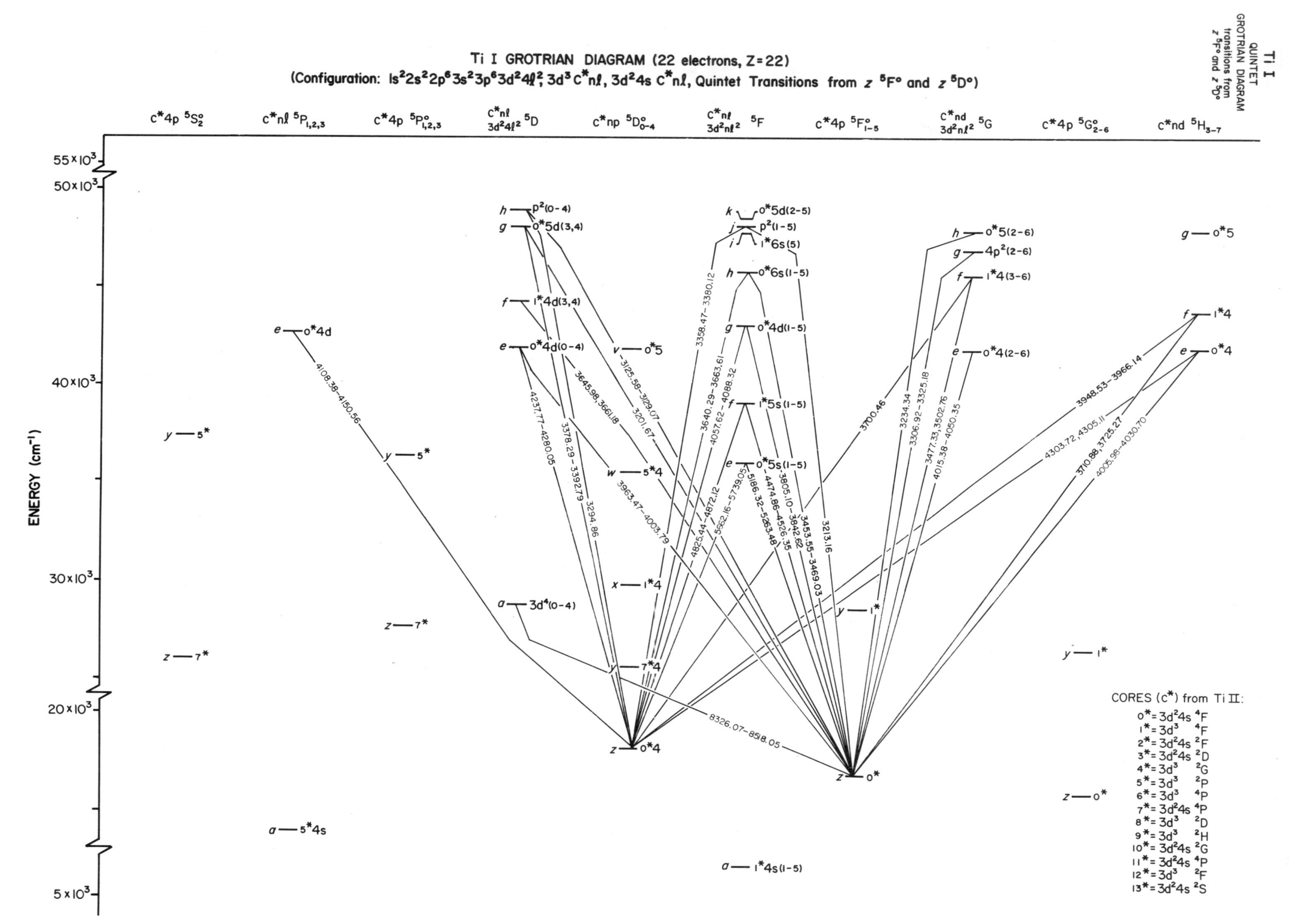

Ti I
QUINTET
GROTRIAN DIAGRAM
transitions from
z ⁵F° and z ⁵D°
Ti I GROTRIAN DIAGRAM (22 electrons, Z=22)
(Configuration: 1s²2s²2p⁶3s²3p⁶3d²4ℓ², 3d³c*nℓ, 3d²4s c*nℓ, Quintet Transitions from z ⁵F° and z ⁵D°)
c*4p ⁵S°₂
c*nℓ ⁵P₁,₂,₃
c*4p ⁵P°₁,₂,₃
c*nℓ 3d²4ℓ² ⁵D
c*np ⁵D°₀₋₄
c*nℓ 3d²nℓ² ⁵F
c*4p ⁵F°₁₋₅
c*nd 3d²nℓ² ⁵G
c*4p ⁵G°₂₋₆
c*nd ⁵H₃₋₇
CORES (c*) from Ti II:
o* = 3d²4s ⁴F
1* = 3d³ ⁴F
2* = 3d²4s ²F
3* = 3d²4s ²D
4* = 3d³ ²G
5* = 3d³ ²P
6* = 3d³ ⁴P
7* = 3d²4s ⁴P
8* = 3d³ ²D
9* = 3d³ ²H
10* = 3d²4s ²G
11* = 3d²4s ⁴P
12* = 3d³ ²F
13* = 3d²4s ²S
4108.38–4150.56
4237.77–4280.05
3378.29–3392.79
3294.86
3645.98,3661.18
3201.67
3125.58–3129.07
3963.47–4003.79
3358.47–3380.12
3640.29–3663.61
4057.62–4088.32
4825.44–4872.12
5662.16–5739.05
5186.32–5263.48
4474.86–4526.35
3805.10–3842.62
3453.55–3469.03
3213.16
8326.07–8518.05
3700.46
3234.34
3306.92–3325.18
3477.33,3502.76
4015.38–4050.35
4303.72,4305.11
3948.53–3966.14
3710.88,3725.27
4005.98–4030.70
ENERGY (cm⁻¹)
55×10³
50×10³
40×10³
30×10³
20×10³
5×10³

Ti I ENERGY LEVELS (22 electrons, Z = 22)

(Configuration: $1s^2 2s^2 2p^6 3s^2 3p^6 3d^2 4\ell^2$, $3d^3 c^* n\ell$, $3d^2 4s\ c^* n\ell$, Quintet System)

ENERGY (cm^{-1})

Axis: 55×10^3, 50×10^3, 40×10^3, 30×10^3, 20×10^3, 5×10^3

$c^*4p\ ^5S^\circ_2$

- 37 359.13 — y — 5^*
- 25 102.88 — z — 7^*

$c^*n\ell\ ^5P_{1,2,3}$

- 42 858.90, 42 724.11, 42 611.58 — e — o^*4d
- 14 105.68, 14 028.47, 13 981.75 — a — 5^*4s

$c^*4p\ ^5P^\circ_{1,2,3}$

- 36 414.58, 36 340.67, 36 298.43 — y — 5^*
- 27 887.74, 27 740.19, 27 665.57 — z — 7^*

$c^*n\ell$ $3d^2 4\ell^2$ ^5D

- 49 036.46, 49 024.43, 48 915.07, 48 859.51, 48 802.32 — h — $p^2(0-4)$
- 48 186.11, 48 059.82 — g — $o^*5d(3,4)$
- 44 381.17, 44 254.39 — f — $1^*4d(3,4)$
- 42 184.66, 42 052.72, 41 958.51, 41 901.36, 41 871.56 — e — $o^*4d(0-4)$
- 28 952.10, 28 882.44, 28 828.51, 28 791.62, 28 772.86 — a — $3d^4(0-4)$

$c^*np\ ^5D^\circ_{0-4}$

- 42 092.52, 41 985.93, 41 906.61, 41 854.01, 41 822.99 — v — o^*5
- 35 757.51, 35 652.95, 35 577.14, 35 527.76, 35 503.40 — w — 5^*4
- 30 060.34, 29 986.24, 29 907.29, 29 855.26, 29 829.16 — x — 1^*4
- 25 926.82, 25 797.60, 25 699.95, 25 635.74, 25 605.03 — y — 7^*4
- 18 695.23, 18 593.99, 18 525.07, 18 482.86, 18 462.83 — z — o^*4

$c^*n\ell$ $3d^2 n\ell^2$ ^5F

- 48 771.73, 48 672.66, 48 588.28, 48 519.21 — k — $o^*5d(2-5)$
- 48 462.11, 48 328.81, 48 208.87 — j — $p^2(1-5)$
- 48 107.42, 48 058.85, 47 777.32 — i — $1^*6s(5)$
- 46 157.76, 46 007.62, 45 893.26, 45 813.01, 45 764.71 — h — $o^*6s(1-5)$
- 43 330.07, 43 231.99, 43 148.15, 43 080.92, 43 034.08 — g — $o^*4d(1-5)$
- 39 412.78, 39 302.36, 39 214.38, 39 149.26, 39 107.25 — f — $1^*5s(1-5)$
- 36 351.43, 36 208.92, 36 096.47, 36 013.57, 35 959.07 — e — $o^*5s(1-5)$
- 6 843.00, 6 742.79, 6 661.00, 6 598.83, 6 556.86 — a — $1^*4s(1-5)$

$c^*4p\ ^5F^\circ_{1-5}$

- 28 896.08, 28 788.39, 28 702.70, 28 638.82, 28 596.45 — y — 1^*
- 17 215.44, 17 075.31, 16 961.42, 16 875.19, 16 817.19 — z — o^*

c^*nd $3d^2 n\ell^2$ ^5G

- 48 223.47, 48 119.47, 48 018.08, 47 936.79, 47 870.61 — h — $o^*5(2-6)$
- 47 446.84, 47 280.69, 47 139.86, 47 030.28, 46 943.91 — g — $4p^2(2-6)$
- 45 904.73, 45 756.45, 45 711.28, 45 689.89 — f — $1^*4(3-6)$
- 42 019.22, 41 903.48, 41 818.70, 41 757.47, 41 714.35 — e — $o^*4(2-6)$

$c^*4p\ ^5G^\circ_{2-6}$

- 26 910.69, 26 772.98, 26 657.41, 26 564.43, 26 494.37 — y — 1^*
- 16 458.71, 16 267.51, 16 106.08, 15 975.59, 15 877.18 — z — o^*

$c^*nd\ ^5H_{3-7}$

- 48 262.83, 48 106.83, 47 994.32, 47 913.61, 47 840.62 — g — o^*5
- 44 134.65, 44 051.37, 43 971.55, 43 901.74, 43 843.82 — f — 1^*4
- 42 205.59, 42 123.77, 42 018.01, 41 917.05, 41 823.19 — e — o^*4

KEY

36 351.43, 36 208.92, 36 096.47, 36 013.57, 35 959.07 — e — $o^*5s(1-5)$

- J, low to high
- n
- Parent level from Ti II
- Notation from A.E.L.
- Energy levels (cm^{-1})

Ionization Limit
55 010 cm^{-1}
(6.82 eV)

[Ti I $4s^2\ ^2F_2 \rightarrow$ Ti II $4s\ ^4F_{3/2}$]

CORES (c^*) from Ti II:

- $0^* = 3d^2 4s\ ^4F$
- $1^* = 3d^3\ ^4F$
- $2^* = 3d^2 4s\ ^2F$
- $3^* = 3d^2 4s\ ^2D$
- $4^* = 3d^3\ ^2G$
- $5^* = 3d^3\ ^2P$
- $6^* = 3d^3\ ^4P$
- $7^* = 3d^2 4s\ ^4P$
- $8^* = 3d^3\ ^2D$
- $9^* = 3d^3\ ^2H$
- $10^* = 3d^2 4s\ ^2G$
- $11^* = 3d^2 4s\ ^4P$
- $12^* = 3d^3\ ^2F$
- $13^* = 3d^2 4s\ ^2S$

NOTE: For proper configuration, please refer to bibliographical comments.

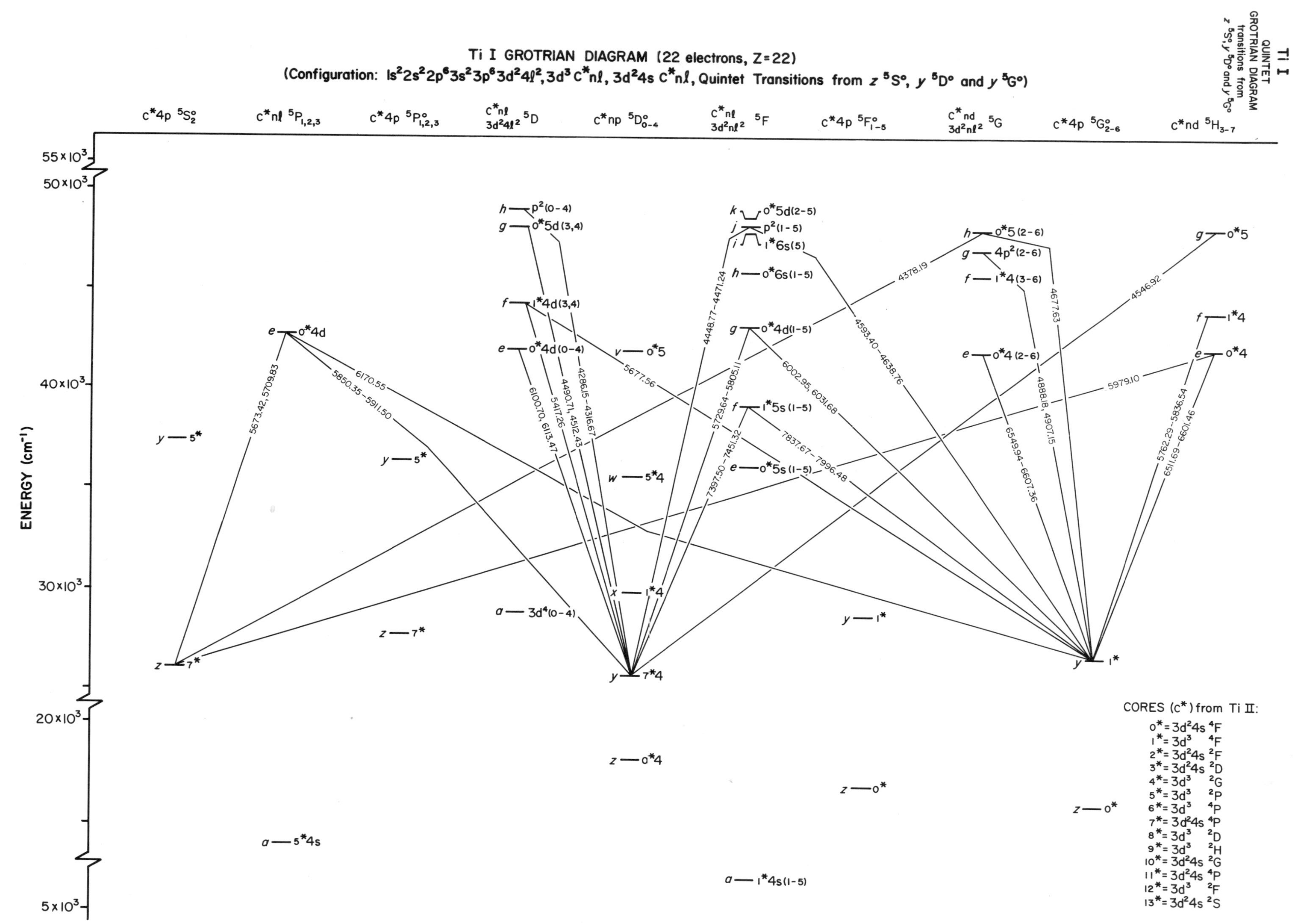

Ti I
QUINTET
GROTRIAN DIAGRAM
transitions from
z 5S°, y 5D° and y 5G°
Ti I GROTRIAN DIAGRAM (22 electrons, Z=22)
(Configuration: 1s²2s²2p⁶3s²3p⁶3d²4l², 3d³ c*nl, 3d²4s c*nl, Quintet Transitions from z 5S°, y 5D° and y 5G°)
c*4p 5S°2
c*nl 5P1,2,3
c*4p 5P°1,2,3
c*nl 3d²4l² 5D
c*np 5D°0-4
c*nl 3d²nl² 5F
c*4p 5F°1-5
c*nd 3d²nl² 5G
c*4p 5G°2-6
c*nd 5H3-7
ENERGY (cm⁻¹)
55×10³
50×10³
40×10³
30×10³
20×10³
5×10³
5673.42, 5709.83
6170.55
5850.35–5911.50
6100.70, 6113.47
5417.26
4490.71, 4512.43
4286.15–4316.67
5677.56
4448.77–4471.24
5729.64–5805.11
7397.50–7451.32
6002.95, 6031.68
7837.67–7996.48
4593.40–4638.76
4378.19
6549.94–6607.36
4888.18, 4907.15
4677.63
4546.92
5979.10
5762.29–5836.54
6511.69–6601.46
CORES (c*) from Ti II:
0* = 3d²4s 4F
1* = 3d³ 4F
2* = 3d²4s 2F
3* = 3d²4s 2D
4* = 3d³ 2G
5* = 3d³ 2P
6* = 3d³ 4P
7* = 3d²4s 4P
8* = 3d³ 2D
9* = 3d³ 2H
10* = 3d²4s 2G
11* = 3d²4s 4P
12* = 3d³ 2F
13* = 3d²4s 2S

Ti I ENERGY LEVELS (22 electrons, Z = 22)

(Configuration: $1s^2 2s^2 2p^6 3s^2 3p^6 3d^2 4l^2$, $3d^3 c^* nl$, $3d^2 4s\ c^* nl$, Quintet System)

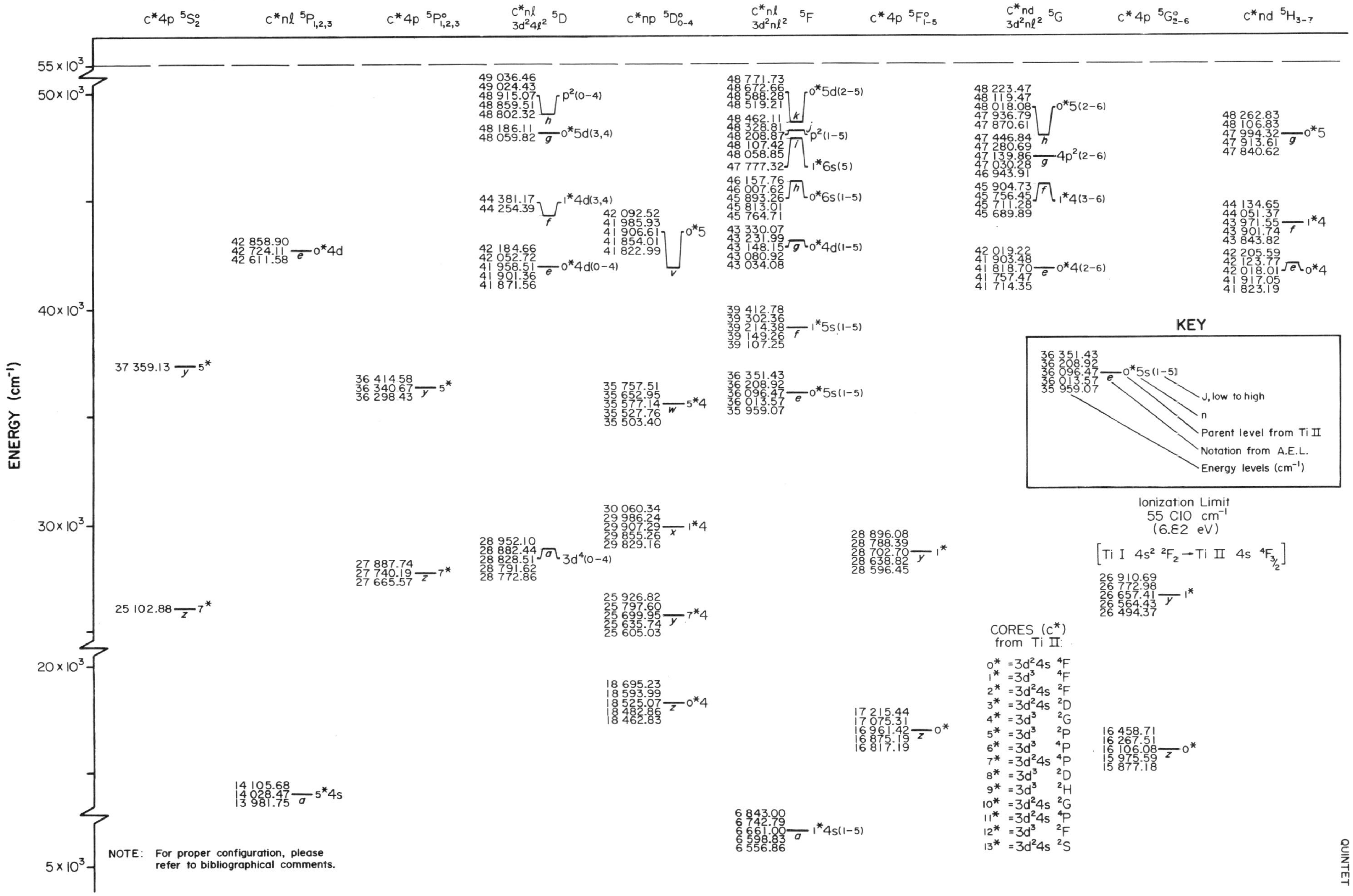

ENERGY (cm^{-1})

Axis: 55 x 10^3, 50 x 10^3, 40 x 10^3, 30 x 10^3, 20 x 10^3, 5 x 10^3

$c^*4p\ ^5S^o_2$

- 37 359.13 — y — 5*
- 25 102.88 — z — 7*

$c^*nl\ ^5P_{1,2,3}$

- 42 858.90, 42 724.11, 42 611.58 — e — o*4d
- 14 105.68, 14 028.47, 13 981.75 — a — 5*4s

$c^*4p\ ^5P^o_{1,2,3}$

- 36 414 58, 36 340 67, 36 298 43 — y — 5*
- 27 887.74, 27 740.19, 27 665.57 — z — 7*

c^*nl $3d^2 4l^2$ ^5D

- 49 036.46, 49 024.43, 48 915.07, 48 859.51, 48 802.32 — h — p^2(0–4)
- 48 186.11, 48 059.82 — g — o*5d(3,4)
- 44 381.17, 44 254.39 — f — 1*4d(3,4)
- 42 184.66, 42 052.72, 41 958.51, 41 901.36, 41 871.56 — e — o*4d(0–4)
- 28 952.10, 28 882.44, 28 828.51, 28 791.62, 28 772.86 — a — $3d^4$(0–4)

$c^*np\ ^5D^o_{0-4}$

- 42 092.52, 41 985.93, 41 906.61, 41 854.01, 41 822.99 — v — o*5
- 35 757.51, 35 652.95, 35 577.14, 35 527.76, 35 503.40 — w — 5*4
- 30 060.34, 29 986.24, 29 907.29, 29 855.26, 29 829.16 — x — 1*4
- 25 926.82, 25 797.60, 25 699.95, 25 635.74, 25 605.03 — y — 7*4
- 18 695.23, 18 593.99, 18 525.07, 18 482.86, 18 462.83 — z — o*4

c^*nl $3d^2 nl^2$ ^5F

- 48 771.73, 48 672.66, 48 588.28, 48 519.21 — k — o*5d(2–5)
- 48 462.11, 48 328.81, 48 208.87 — j — p^2(1–5)
- 48 107.42, 48 058.85, 47 777.32 — i — 1*6s(5)
- 46 157.76, 46 007.62, 45 893.26, 45 813.01, 45 764.71 — h — o*6s(1–5)
- 43 330.07, 43 231.99, 43 148.15, 43 080.92, 43 034.08 — g — o*4d(1–5)
- 39 412.78, 39 302.36, 39 214.38, 39 149.26, 39 107.25 — f — 1*5s(1–5)
- 36 351.43, 36 208.92, 36 096.47, 36 013.57, 35 959.07 — e — o*5s(1–5)
- 6 843.00, 6 742.79, 6 661.00, 6 598.83, 6 556.86 — a — 1*4s(1–5)

$c^*4p\ ^5F^o_{1-5}$

- 28 896.08, 28 788.39, 28 702.70, 28 638.82, 28 596.45 — y — 1*
- 17 215.44, 17 075.31, 16 961.42, 16 875.19, 16 817.19 — z — o*

c^*nd $3d^2 nl^2$ ^5G

- 48 223.47, 48 119.47, 48 018.08, 47 936.79, 47 870.61 — h — o*5(2–6)
- 47 446.84, 47 280.69, 47 139.86, 47 030.28, 46 943.91 — g — $4p^2$(2–6)
- 45 904.73, 45 756.45, 45 711.28, 45 689.89 — f — 1*4(3–6)
- 42 019.22, 41 903.48, 41 818.70, 41 757.47, 41 714.35 — e — o*4(2–6)

$c^*4p\ ^5G^o_{2-6}$

- 26 910.69, 26 772.98, 26 657.41, 26 564.43, 26 494.37 — y — 1*
- 16 458.71, 16 267.51, 16 106.08, 15 975.59, 15 877.18 — z — o*

$c^*nd\ ^5H_{3-7}$

- 48 262.83, 48 106.83, 47 994.32, 47 913.61, 47 840.62 — g — o*5
- 44 134.65, 44 051.37, 43 971.55, 43 901.74, 43 843.82 — f — 1*4
- 42 205.59, 42 123.77, 42 018.01, 41 917.05, 41 823.19 — e — o*4

KEY

36 351.43, 36 208.92, 36 096.47, 36 013.57, 35 959.07 — e — o*5s(1–5)

- J, low to high
- n
- Parent level from Ti II
- Notation from A.E.L.
- Energy levels (cm^{-1})

Ionization Limit
55 010 cm^{-1}
(6.82 eV)

[Ti I $4s^2\ ^2F_2$ → Ti II $4s\ ^4F_{3/2}$]

CORES (c*) from Ti II:

- 0* = $3d^2 4s\ ^4F$
- 1* = $3d^3\ ^4F$
- 2* = $3d^2 4s\ ^2F$
- 3* = $3d^2 4s\ ^2D$
- 4* = $3d^3\ ^2G$
- 5* = $3d^3\ ^2P$
- 6* = $3d^3\ ^4P$
- 7* = $3d^2 4s\ ^4P$
- 8* = $3d^3\ ^2D$
- 9* = $3d^3\ ^2H$
- 10* = $3d^2 4s\ ^2G$
- 11* = $3d^2 4s\ ^4P$
- 12* = $3d^3\ ^2F$
- 13* = $3d^2 4s\ ^2S$

NOTE: For proper configuration, please refer to bibliographical comments.

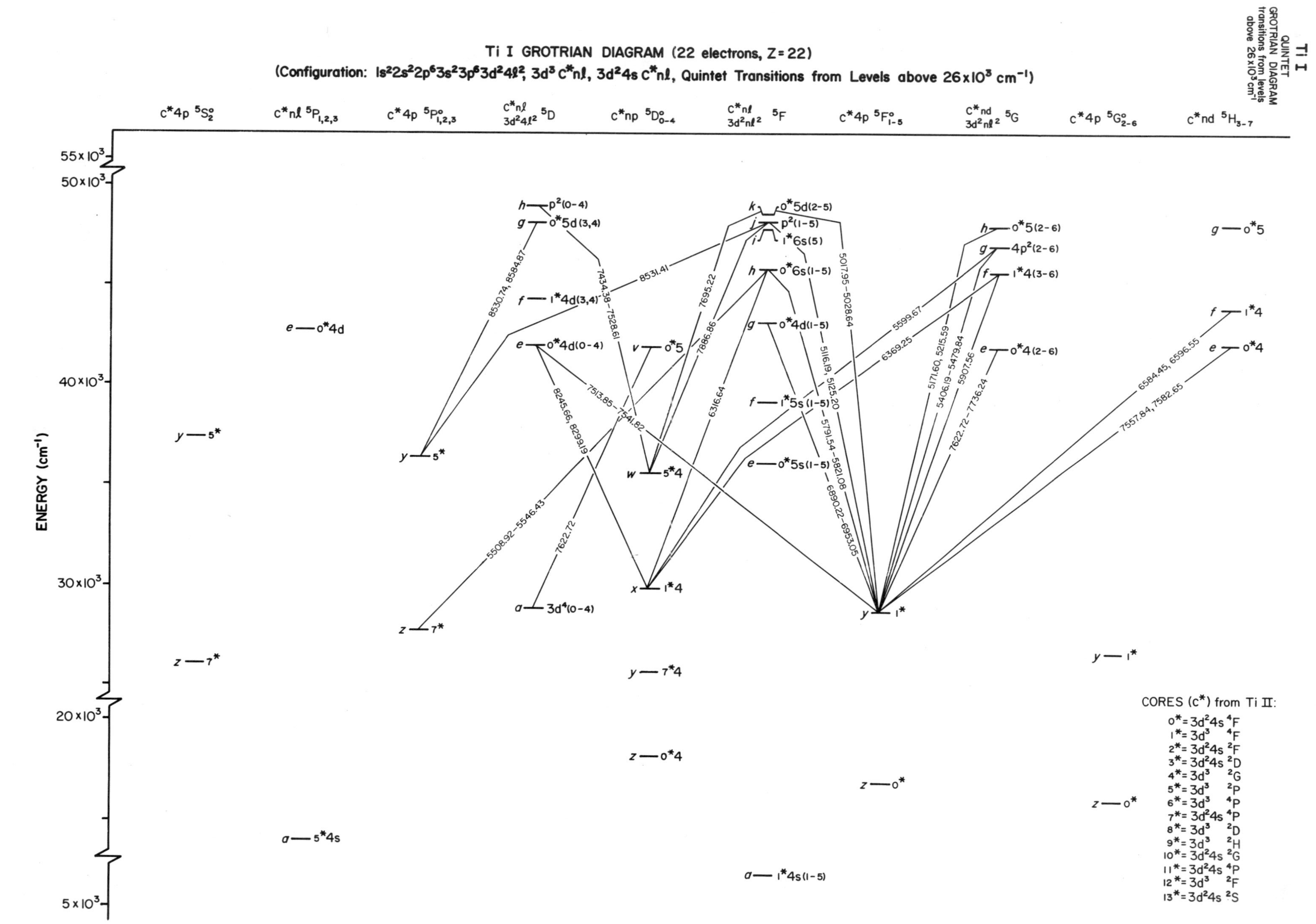
Ti I
QUINTET
GROTRIAN DIAGRAM
transitions from levels above 26x10³ cm⁻¹
Ti I GROTRIAN DIAGRAM (22 electrons, Z=22)
(Configuration: 1s²2s²2p⁶3s²3p⁶3d²4ℓ², 3d³ c*nℓ, 3d²4s c*nℓ, Quintet Transitions from Levels above 26x10³ cm⁻¹)
c*4p ⁵S°₂
c*nℓ ⁵P₁,₂,₃
c*4p ⁵P°₁,₂,₃
c*nℓ 3d²4ℓ² ⁵D
c*np ⁵D°₀₋₄
c*nℓ 3d²nℓ² ⁵F
c*4p ⁵F°₁₋₅
c*nd 3d²nℓ² ⁵G
c*4p ⁵G°₂₋₆
c*nd ⁵H₃₋₇
ENERGY (cm⁻¹)
55x10³
50x10³
40x10³
30x10³
20x10³
5x10³
y — 5*
z — 7*
e — o*4d
a — 5*4s
y — 5*
z — 7*
h — p²(0-4)
g — o*5d(3,4)
f — 1*4d(3,4)
e — o*4d(0-4)
a — 3d⁴(0-4)
v — o*5
w — 5*4
x — 1*4
y — 7*4
z — o*4
k — o*5d(2-5)
j — p²(1-5)
i — 1*6s(5)
h — o*6s(1-5)
g — o*4d(1-5)
f — 1*5s(1-5)
e — o*5s(1-5)
a — 1*4s(1-5)
y — 1*
z — o*
h — o*5(2-6)
g — 4p²(2-6)
f — 1*4(3-6)
e — o*4(2-6)
y — 1*
z — o*
g — o*5
f — 1*4
e — o*4
8530.74, 8584.87
7434.38–7528.61
8531.41
7695.22
7886.86
6316.64
8245.66, 8299.19
7513.85–7541.82
5508.92–5546.43
7622.72
5017.95–5028.64
5116.19, 5125.20
5791.54–5821.08
6890.22–6953.05
5599.67
6369.25
5171.60, 5215.59
5406.19–5479.84
5907.56
7622.72–7736.24
6584.45, 6596.55
7557.84, 7582.65
CORES (c*) from Ti II:
o* = 3d²4s ⁴F
1* = 3d³ ⁴F
2* = 3d²4s ²F
3* = 3d²4s ²D
4* = 3d³ ²G
5* = 3d³ ²P
6* = 3d³ ⁴P
7* = 3d²4s ⁴P
8* = 3d³ ²D
9* = 3d³ ²H
10* = 3d²4s ²G
11* = 3d²4s ⁴P
12* = 3d³ ²F
13* = 3d²4s ²S

Ti I ENERGY LEVELS (22 electrons, Z = 22)

(Configuration: $1s^2 2s^2 2p^6 3s^2 3p^6 3d^2 4l^2$, $3d^3 c^* nl$, $3d^2 4s\ c^* nl$, Quintet System)

ENERGY (cm^{-1}) axis: 55 x 10³, 50 x 10³, 40 x 10³, 30 x 10³, 20 x 10³, 5 x 10³

$c^*4p\ ^5S^o_2$

- 37 359.13 — y — 5*
- 25 102.88 — z — 7*

$c^*nl\ ^5P_{1,2,3}$

- 42 858.90, 42 724.11, 42 611.58 — e — o*4d
- 14 105.68, 14 028.47, 13 981.75 — a — 5*4s

$c^*4p\ ^5P^o_{1,2,3}$

- 36 41<.58, 36 340.67, 36 29E.43 — y — 5*
- 27 887.74, 27 740.19, 27 665.57 — z — 7*

c^*nl $3d^2 4l^2$ ^5D

- 49 036.46, 49 024.43, 48 915.07, 48 859.51, 48 802.32 — h — p^2(0–4)
- 48 186.11, 48 059.82 — g — o*5d(3,4)
- 44 381.17, 44 254.39 — f — 1*4d(3,4)
- 42 184.66, 42 052.72, 41 958.51, 41 901.36, 41 871.56 — e — o*4d(0–4)
- 28 952.10, 28 882.44, 28 828.51, 28 791.62, 28 772.86 — a — $3d^4$(0–4)

$c^*np\ ^5D^o_{0-4}$

- 42 092.52, 41 985.93, 41 906.61, 41 854.01, 41 822.99 — v — o*5
- 35 757.51, 35 652.95, 35 577.14, 35 527.76, 35 503.40 — w — 5*4
- 30 060.34, 29 986.24, 29 907.29, 29 855.26, 29 829.16 — x — 1*4
- 25 926.82, 25 797.60, 25 699.95, 25 635.74, 25 605.03 — y — 7*4
- 18 695.23, 18 593.99, 18 525.07, 18 482.86, 18 462.83 — z — o*4

c^*nl $3d^2 nl^2$ ^5F

- 48 771.73, 48 672.66, 48 588.28, 48 519.21 — k — o*5d(2–5)
- 48 462.11, 48 328.81, 48 208.87 — j — p^2(1–5)
- 48 107.42, 48 058.85, 47 777.32 — i — 1*6s(5)
- 46 157.76, 46 007.62, 45 893.26, 45 813.01, 45 764.71 — h — o*6s(1–5)
- 43 330.07, 43 231.99, 43 148.15, 43 080.92, 43 034.08 — g — o*4d(1–5)
- 39 412.78, 39 302.36, 39 214.38, 39 149.26, 39 107.25 — f — 1*5s(1–5)
- 36 351.43, 36 208.92, 36 096.47, 36 013.57, 35 959.07 — e — o*5s(1–5)
- 6 843.00, 6 742.79, 6 661.00, 6 598.83, 6 556.86 — a — 1*4s(1–5)

$c^*4p\ ^5F^o_{1-5}$

- 28 896.08, 28 788.39, 28 702.70, 28 638.82, 28 596.45 — y — 1*
- 17 215.44, 17 075.31, 16 961.42, 16 875.19, 16 817.19 — z — o*

c^*nd $3d^2 nl^2$ ^5G

- 48 223.47, 48 119.47, 48 018.08, 47 936.79, 47 870.61 — h — o*5(2–6)
- 47 446.84, 47 280.69, 47 139.86, 47 030.28, 46 943.91 — g — $4p^2$(2–6)
- 45 904.73, 45 756.45, 45 711.28, 45 689.89 — f — 1*4(3–6)
- 42 019.22, 41 903.48, 41 818.70, 41 757.47, 41 714.35 — e — o*4(2–6)

$c^*4p\ ^5G^o_{2-6}$

- 26 910.69, 26 772.98, 26 657.41, 26 564.43, 26 494.37 — y — 1*
- 16 458.71, 16 267.51, 16 106.08, 15 975.59, 15 877.18 — z — o*

$c^*nd\ ^5H_{3-7}$

- 48 262.83, 48 106.83, 47 994.32, 47 913.61, 47 840.62 — g — o*5
- 44 134.65, 44 051.37, 43 971.55, 43 901.74, 43 843.82 — f — 1*4
- 42 205.59, 42 123.77, 42 018.01, 41 917.05, 41 823.19 — e — o*4

KEY

36 351.43, 36 208.92, 36 096.47, 36 013.57, 35 959.07 — e — o*5s(1–5)

- J, low to high
- n
- Parent level from Ti II
- Notation from A.E.L.
- Energy levels (cm^{-1})

Ioniza-ion Limit
55 010 cm^{-1}
(6.82 eV)

[Ti I $4s^2\ ^2F_2$ → Ti II $4s\ ^4F_{3/2}$]

CORES (c*) from Ti II:

- o* = $3d^2 4s\ ^4F$
- 1* = $3d^3\ ^4F$
- 2* = $3d^2 4s\ ^2F$
- 3* = $3d^2 4s\ ^2D$
- 4* = $3d^3\ ^2G$
- 5* = $3d^3\ ^2P$
- 6* = $3d^3\ ^4P$
- 7* = $3d^2 4s\ ^4P$
- 8* = $3d^3\ ^2D$
- 9* = $3d^3\ ^2H$
- 10* = $3d^2 4s\ ^2G$
- 11* = $3d^2 4s\ ^4P$
- 12* = $3d^3\ ^2F$
- 13* = $3d^2 4s\ ^2S$

NOTE: For proper configuration, please refer to bibliographical comments.

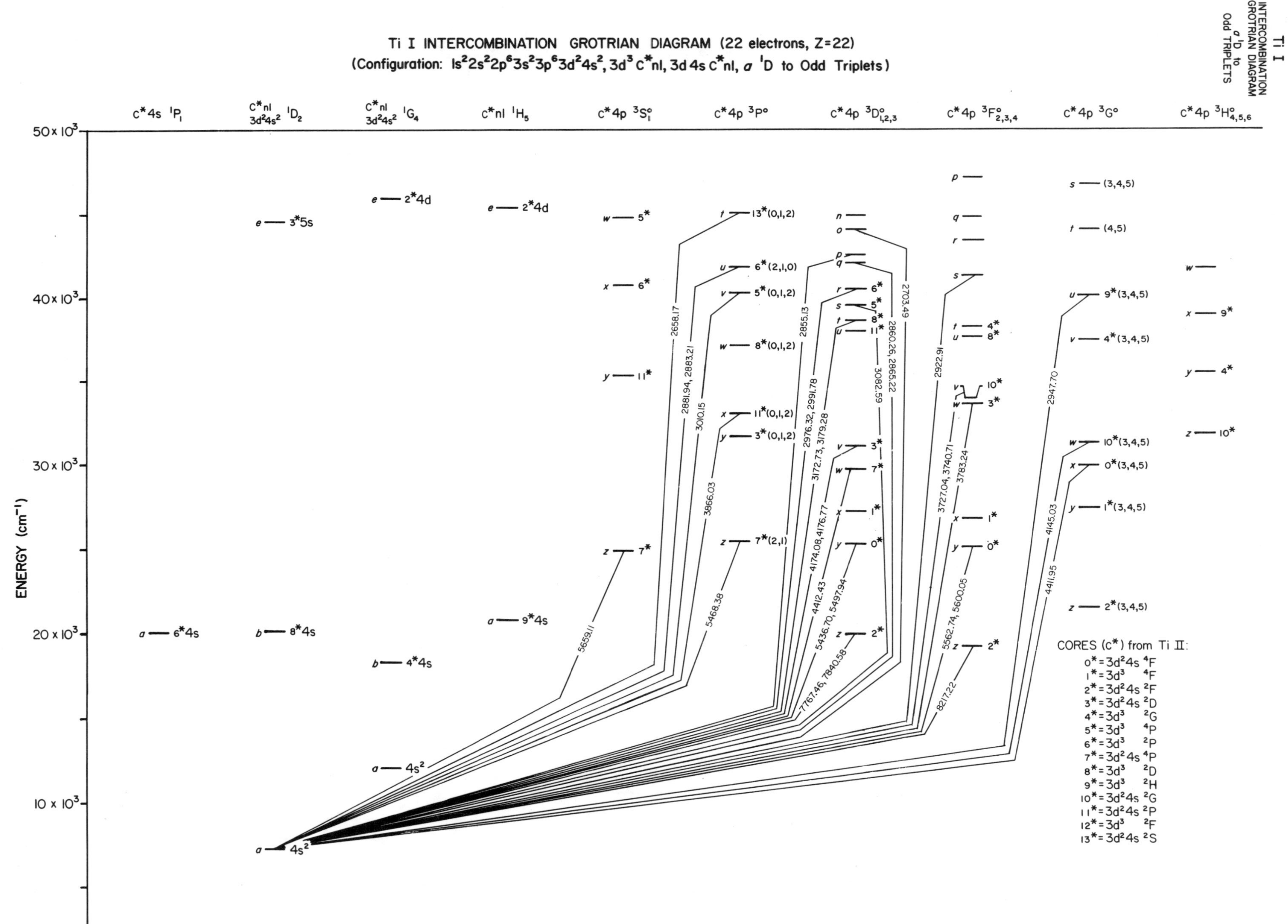

Ti I
INTERCOMBINATION GROTRIAN DIAGRAM
a 1D to Odd TRIPLETS
Ti I INTERCOMBINATION GROTRIAN DIAGRAM (22 electrons, Z=22)
(Configuration: 1s²2s²2p⁶3s²3p⁶3d²4s², 3d³C*nl, 3d4sC*nl, a 1D to Odd Triplets)
C*4s 1P1
C*nl 3d²4s² 1D2
C*nl 3d²4s² 1G4
C*nl 1H5
C*4p 3S°1
C*4p 3P°
C*4p 3D°1,2,3
C*4p 3F°2,3,4
C*4p 3G°
C*4p 3H°4,5,6
ENERGY (cm⁻¹)
50 x 10³
40 x 10³
30 x 10³
20 x 10³
10 x 10³
2658.17
2881.94, 2883.21
3010.15
3866.03
5468.38
5659.11
2703.49
2860.26, 2865.22
3082.59
2855.13
2976.32, 2991.78
3172.73, 3179.28
4174.08, 4176.77
4412.43
5436.70, 5497.94
7767.46, 7840.58
2922.91
3727.04, 3740.71
3783.24
5562.74, 5600.05
8217.22
2947.70
4145.03
4411.95
CORES (C*) from Ti II:
0* = 3d²4s 4F
1* = 3d³ 4F
2* = 3d²4s 2F
3* = 3d²4s 2D
4* = 3d³ 2G
5* = 3d³ 4P
6* = 3d³ 2P
7* = 3d²4s 4P
8* = 3d³ 2D
9* = 3d³ 2H
10* = 3d²4s 2G
11* = 3d²4s 2P
12* = 3d³ 2F
13* = 3d²4s 2S

Ti I INTERCOMBINATION GROTRIAN DIAGRAM (22 electrons, Z=22)

(Configuration: $1s^2 2s^2 2p^6 3s^2 3p^6 3d^2 4s^2$, $3d^3\ c^* nl$, $3d\,4s\ c^* nl$, $a\ ^1G$ to Odd Triplets)

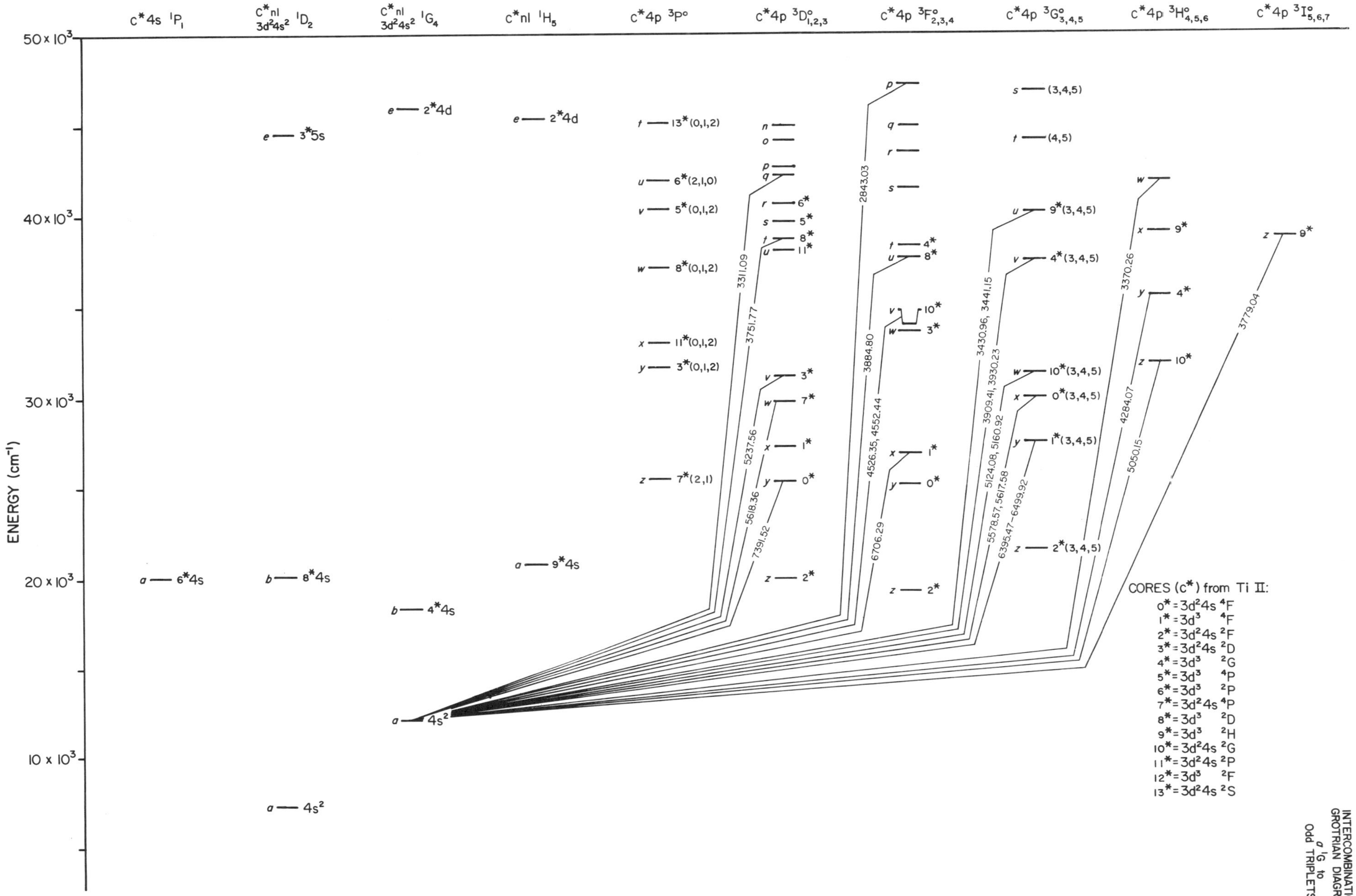

$c^*4s\ ^1P_1$ | $c^*nl\ 3d^24s^2\ ^1D_2$ | $c^*nl\ 3d^24s^2\ ^1G_4$ | $c^*nl\ ^1H_5$ | $c^*4p\ ^3P^o$ | $c^*4p\ ^3D^o_{1,2,3}$ | $c^*4p\ ^3F^o_{2,3,4}$ | $c^*4p\ ^3G^o_{3,4,5}$ | $c^*4p\ ^3H^o_{4,5,6}$ | $c^*4p\ ^3I^o_{5,6,7}$

ENERGY (cm^{-1}): 50 x 10^3, 40 x 10^3, 30 x 10^3, 20 x 10^3, 10 x 10^3

$^1P_1$: a — 6*4s

$^1D_2$: e — 3*5s; b — 8*4s; a — 4s^2

$^1G_4$: e — 2*4d; b — 4*4s; a — 4s^2

$^1H_5$: e — 2*4d; a — 9*4s

$^3P^o$: t — 13*(0,1,2); u — 6*(2,1,0); v — 5*(0,1,2); w — 8*(0,1,2); x — 11*(0,1,2); y — 3*(0,1,2); z — 7*(2,1)

$^3D^o$: n; o; p; q; r — 6*; s — 5*; t — 8*; u — 11*; v — 3*; w — 7*; x — 1*; y — 0*; z — 2*

Lines: 3311.09, 3751.77, 5237.56, 5618.36, 7391.52

$^3F^o$: p; q; r; s; t — 4*; u — 8*; v — 10*; w — 3*; x — 1*; y — 0*; z — 2*

Lines: 2843.03, 3884.80, 4526.35, 4552.44, 6706.29

$^3G^o$: s — (3,4,5); t — (4,5); u — 9*(3,4,5); v — 4*(3,4,5); w — 10*(3,4,5); x — 0*(3,4,5); y — 1*(3,4,5); z — 2*(3,4,5)

Lines: 3430.96, 3441.15; 3909.41, 3930.23; 5124.08, 5160.92; 5578.57, 5617.58; 6395.47–6499.92

$^3H^o$: w; x — 9*; y — 4*; z — 10*

Lines: 3370.26, 4284.07, 5050.15

$^3I^o$: z — 9*

Line: 3779.04

CORES (c*) from Ti II:

0* = 3d^{2}4s ^{4}F
1* = 3d^3 ^{4}F
2* = 3d^{2}4s ^{2}F
3* = 3d^{2}4s ^{2}D
4* = 3d^3 ^{2}G
5* = 3d^3 ^{4}P
6* = 3d^3 ^{2}P
7* = 3d^{2}4s ^{4}P
8* = 3d^3 ^{2}D
9* = 3d^3 ^{2}H
10* = 3d^{2}4s ^{2}G
11* = 3d^{2}4s ^{2}P
12* = 3d^3 ^{2}F
13* = 3d^{2}4s ^{2}S

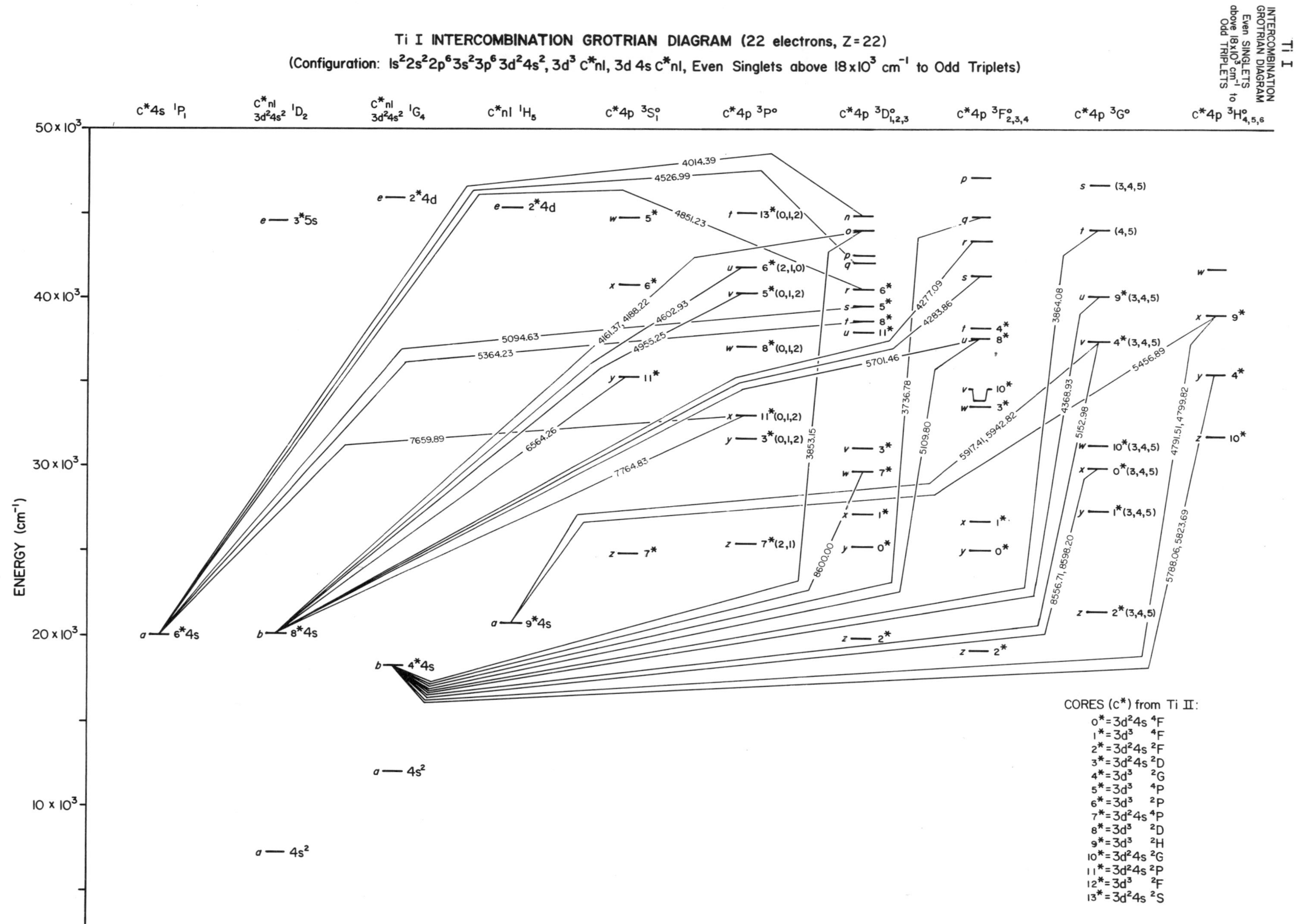
Ti I
INTERCOMBINATION
GROTRIAN DIAGRAM
Even SINGLETS
above 18x10^3 cm^{-1} to
Odd TRIPLETS
Ti I INTERCOMBINATION GROTRIAN DIAGRAM (22 electrons, Z=22)
(Configuration: $1s^2 2s^2 2p^6 3s^2 3p^6 3d^2 4s^2$, $3d^3$ c*nl, 3d 4s c*nl, Even Singlets above 18×10^3 cm^{-1} to Odd Triplets)
c*4s $^1P_1$
c*nl $3d^2 4s^2$ $^1D_2$
c*nl $3d^2 4s^2$ $^1G_4$
c*nl $^1H_5$
c*4p $^3S^o_1$
c*4p $^3P^o$
c*4p $^3D^o_{1,2,3}$
c*4p $^3F^o_{2,3,4}$
c*4p $^3G^o$
c*4p $^3H^o_{4,5,6}$
50 x 10^3
40 x 10^3
30 x 10^3
20 x 10^3
10 x 10^3
ENERGY (cm^{-1})
a — 6*4s
b — 8*4s
e — 3*5s
a — 4s^2
b — 4*4s
e — 2*4d
a — 4s^2
e — 2*4d
a — 9*4s
w — 5*
x — 6*
y — 11*
z — 7*
t — 13*(0,1,2)
u — 6*(2,1,0)
v — 5*(0,1,2)
w — 8*(0,1,2)
x — 11*(0,1,2)
y — 3*(0,1,2)
z — 7*(2,1)
n
o
p
q
r — 6*
s — 5*
t — 8*
u — 11*
v — 3*
w — 7*
x — 1*
y — 0*
z — 2*
p
q
r
s
t — 4*
u — 8*
v — 10*
w — 3*
x — 1*
y — 0*
z — 2*
s — (3,4,5)
t — (4,5)
u — 9*(3,4,5)
v — 4*(3,4,5)
w — 10*(3,4,5)
x — 0*(3,4,5)
y — 1*(3,4,5)
z — 2*(3,4,5)
w
x — 9*
y — 4*
z — 10*
4014.39
4526.99
4851.23
5094.63
5364.23
4161.37, 4188.22
4602.93
4955.25
7659.89
6564.26
7764.83
3853.15
8600.00
3736.78
5109.80
4277.09
4283.86
5701.46
5917.41, 5942.82
3864.08
4368.93
5152.98
8556.71, 8598.20
5456.89
4791.51, 4799.82
5788.06, 5823.69
CORES (c*) from Ti II:
0* = $3d^2 4s$ ^4F
1* = $3d^3$ ^4F
2* = $3d^2 4s$ ^2F
3* = $3d^2 4s$ ^2D
4* = $3d^3$ ^2G
5* = $3d^3$ ^4P
6* = $3d^3$ ^2P
7* = $3d^2 4s$ ^4P
8* = $3d^3$ ^2D
9* = $3d^3$ ^2H
10* = $3d^2 4s$ ^2G
11* = $3d^2 4s$ ^2P
12* = $3d^3$ ^2F
13* = $3d^2 4s$ ^2S

Ti I INTERCOMBINATION GROTRIAN DIAGRAM (22 electrons, Z=22)
(Configuration: $1s^2 2s^2 2p^6 3s^2 3p^6 3d^2 4s^2$, $3d^3 c^* nl$, $3d\,4s\,c^* nl$, Odd Singlets to Even Triplets)

$c^*4p\ ^1P^o_1$ | $c^*4p\ ^1D^o_2$ | $c^*4p\ ^1F^o_3$ | $c^*4p\ ^1G^o_4$ | $c^*4p\ ^1H^o_5$ | c^*ns, $3d^4$ ^3P | c^*4s, $3d^2 4l^2$ ^3D | $c^*nl\ ^3F$ | c^*4l, $3d^4$ $^3G_{3\,4\,5}$

ENERGY (cm^{-1})

50×10^3

40×10^3

30×10^3

20×10^3

$^1P^o_1$: w 13*, x 8*, y 11*, z 3*

$^1D^o_2$: u 8*, v, w 6*, x 11*, y 3*, z 2*

$^1F^o_3$: u 12*, v 4*, w 8*, y 10*, y 3*, z 2*

$^1G^o_4$: u 12*, v 9*, w, x 8*, y 10*, z 2*

$^1H^o_5$: x 9*, y 4*, z 10*

^3P: e 7*5(2), d $3d^4$(0,1,2)

^3D: f (2,3), e p^2(2,1,3)

^3F: i 2*4d(3,4), h 0*6s(4), g 0*4d(2,3,4), f *5s(2,3,4), e 0*5s(2,3,4)

^3G: e 0*d, b $3d^4$

6283.20

4010.93

3752.18

8980.44

6624.63

8241.55

8740.87

5751.75, 5800.07

5271.62

6417.19

5230.57

CORES (c^*) from Ti II:

0* = $3d^2 4s\ ^4F$
1* = $3d^3\ ^4F$
2* = $3d^2 4s\ ^2F$
3* = $3d^2 4s\ ^2D$
4* = $3d^3\ ^2G$
5* = $3d^3\ ^4P$
6* = $3d^3\ ^2P$
7* = $3d^2 4s\ ^4P$
8* = $3d^3\ ^2D$
9* = $3d^3\ ^2H$
10* = $3d^2 4s\ ^2G$
11* = $3d^2 4s\ ^2P$
12* = $3d^3\ ^2F$
13* = $3d^2 4s\ ^2S$

Ti I
SINGLET to TRIPLET
INTERCOMBINATION
GROTRIAN DIAGRAM
Odd to Even

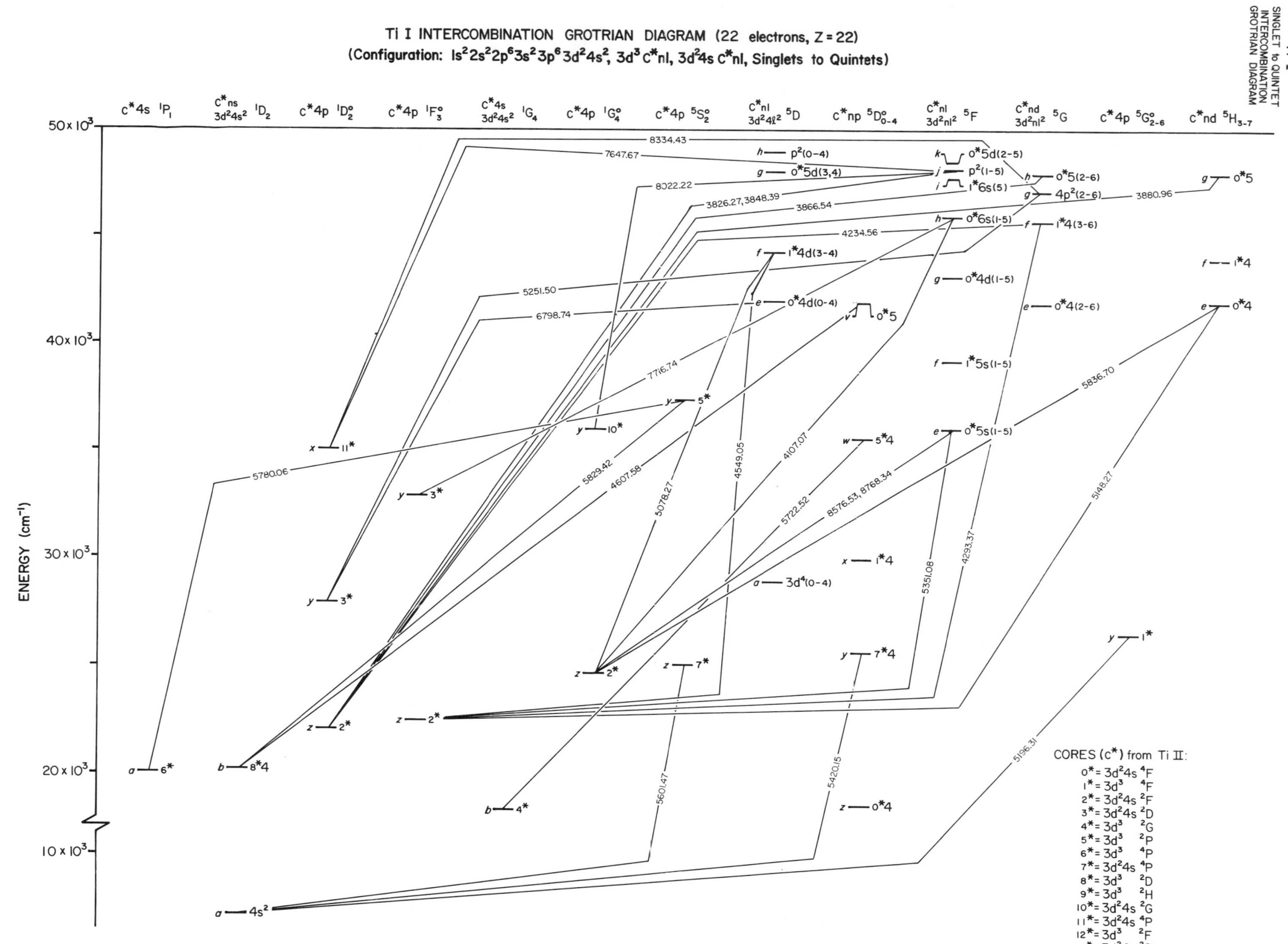
Ti I
SINGLET to QUINTET
INTERCOMBINATION
GROTRIAN DIAGRAM
Ti I INTERCOMBINATION GROTRIAN DIAGRAM (22 electrons, Z = 22)
(Configuration: 1s²2s²2p⁶3s²3p⁶3d²4s², 3d³c*nl, 3d²4s c*nl, Singlets to Quintets)
c*4s ¹P₁
c*ns 3d²4s² ¹D₂
c*4p ¹D°₂
c*4p ¹F°₃
c*4s 3d²4s² ¹G₄
c*4p ¹G°₄
c*4p ⁵S°₂
c*nl 3d²4l² ⁵D
c*np ⁵D°₀₋₄
c*nl 3d²nl² ⁵F
c*nd 3d²nl² ⁵G
c*4p ⁵G°₂₋₆
c*nd ⁵H₃₋₇
ENERGY (cm⁻¹)
50 x 10³
40 x 10³
30 x 10³
20 x 10³
10 x 10³
CORES (c*) from Ti II:
o* = 3d²4s ⁴F
1* = 3d³ ⁴F
2* = 3d²4s ²F
3* = 3d²4s ²D
4* = 3d³ ²G
5* = 3d³ ²P
6* = 3d³ ⁴P
7* = 3d²4s ⁴P
8* = 3d³ ²D
9* = 3d³ ²H
10* = 3d²4s ²G
11* = 3d²4s ⁴P
12* = 3d³ ²F
13* = 3d²4s ²S
8334.43
7647.67
8022.22
3826.27, 3848.39
3866.54
3880.96
4234.56
5251.50
6798.74
7716.74
5780.06
5829.42
4607.58
5078.27
4549.05
4107.07
5722.52
8576.53, 8768.34
5351.08
4293.37
5836.70
5148.27
5196.31
5601.47
5420.15

Ti I INTERCOMBINATION GROTRIAN DIAGRAM (22 electrons, Z = 22)

(Configuration: $1s^2 2s^2 2p^6 3s^2 3p^6 3d^2 4s^2$, $3d^3 c^* nl$, $3d^2 4s\ c^* nl$, Even Triplets below $10 \times 10^3\ cm^{-1}$ to Odd Singlets)

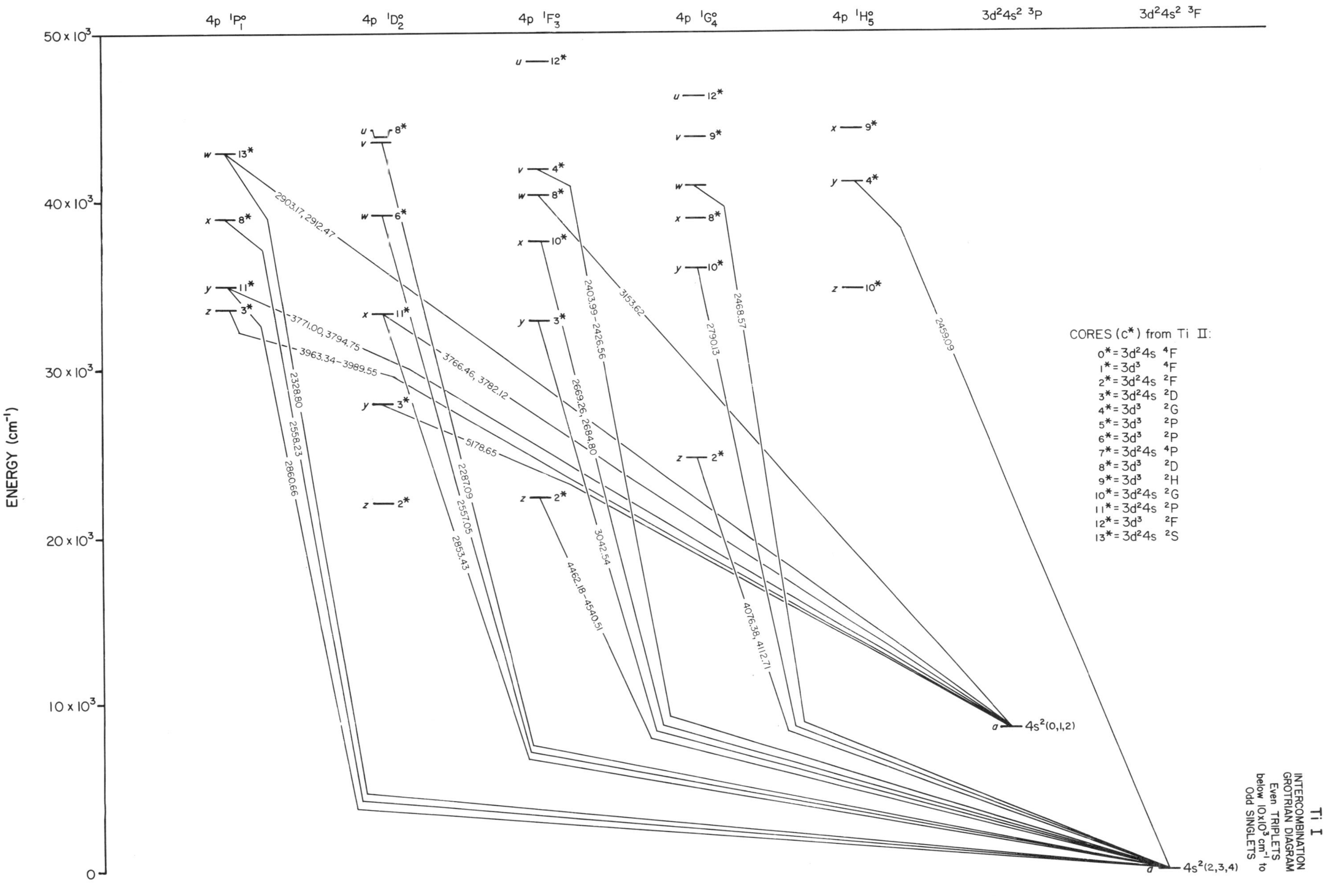

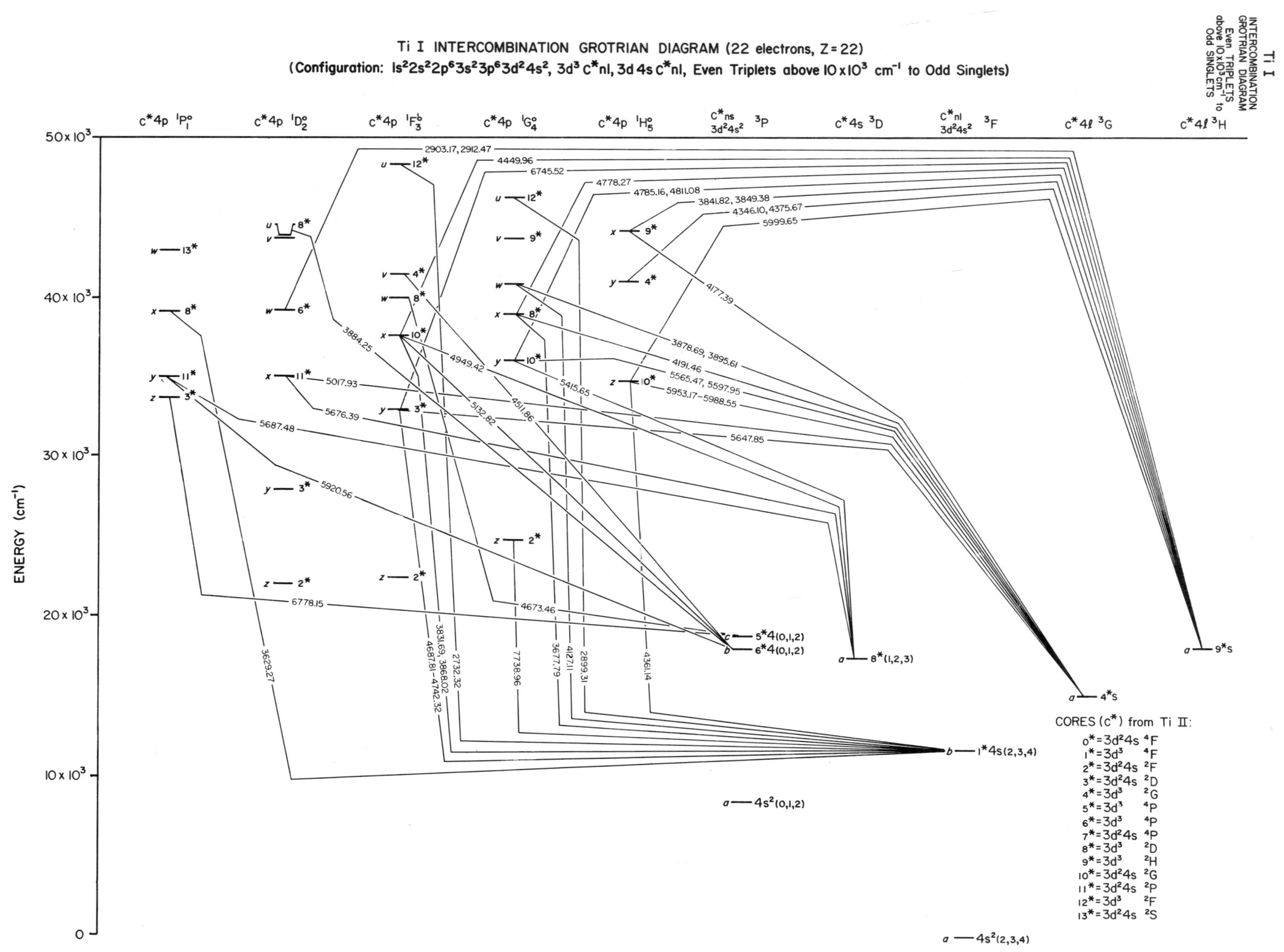

Ti I
INTERCOMBINATION GROTRIAN DIAGRAM
Even TRIPLETS above 10 x 10³ cm⁻¹ to Odd SINGLETS
Ti I INTERCOMBINATION GROTRIAN DIAGRAM (22 electrons, Z = 22)
(Configuration: 1s²2s²2p⁶3s²3p⁶3d²4s², 3d³ c*nl, 3d 4s c*nl, Even Triplets above 10 x 10³ cm⁻¹ to Odd Singlets)
c*4p ¹P°₁
c*4p ¹D°₂
c*4p ¹F°₃
c*4p ¹G°₄
c*4p ¹H°₅
c*ns 3d²4s² ³P
c*4s ³D
c*nl 3d²4s² ³F
c*4l ³G
c*4l ³H
ENERGY (cm⁻¹)
50 x 10³
40 x 10³
30 x 10³
20 x 10³
10 x 10³
0
2903.17, 2912.47
4449.96
6745.52
4778.27
4785.16, 4811.08
3841.82, 3849.38
4346.10, 4375.67
5999.65
4177.39
3884.25
3878.69, 3895.61
4191.46
5565.47, 5597.95
5953.17–5988.55
4949.42
5415.65
5017.93
5132.82
4511.86
5676.39
5647.85
5687.48
5920.56
6778.15
4673.46
3629.27
3831.69, 3868.02
4687.81–4742.32
2732.32
7738.96
3677.79
4127.11
2899.31
4361.14
5*4(0,1,2)
6*4(0,1,2)
8*(1,2,3)
4*s
9*s
1*4s(2,3,4)
4s²(0,1,2)
4s²(2,3,4)
CORES (c*) from Ti II:
0* = 3d²4s ⁴F
1* = 3d³ ⁴F
2* = 3d²4s ²F
3* = 3d²4s ²D
4* = 3d³ ²G
5* = 3d³ ⁴P
6* = 3d³ ⁴P
7* = 3d²4s ⁴P
8* = 3d³ ²D
9* = 3d³ ²H
10* = 3d²4s ²G
11* = 3d²4s ²P
12* = 3d³ ²F
13* = 3d²4s ²S

Ti I INTERCOMBINATION GROTRIAN DIAGRAM (22 electrons, Z = 22)
(Configuration: $1s^2 2s^2 2p^6 3s^2 3p^6 3d^2 4s^2$, $3d^3 c^* nl$, $3d\,4s\, c^* nl$, Odd Triplets to Even Singlets)

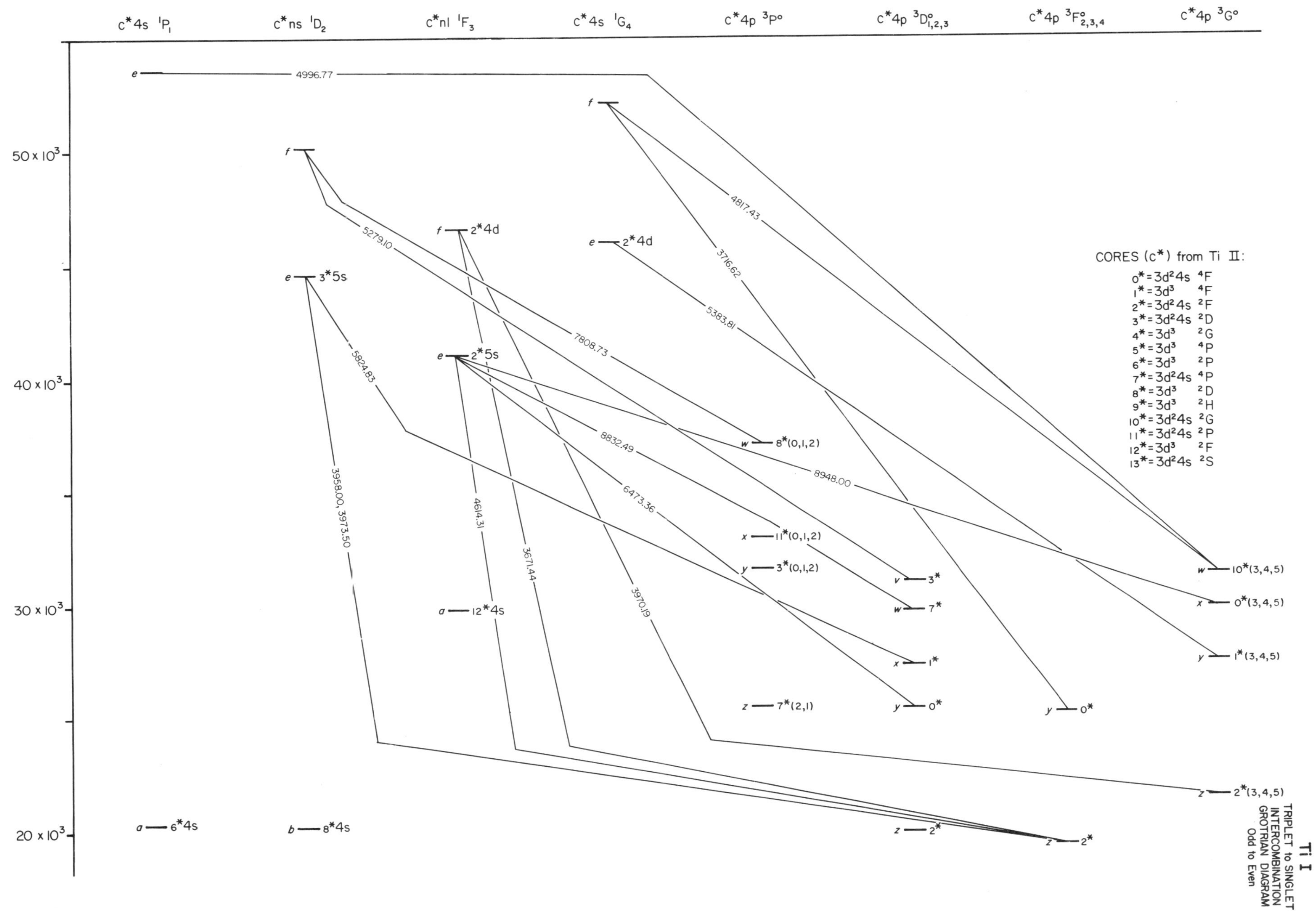

Ti I
TRIPLET to SINGLET
INTERCOMBINATION
GROTRIAN DIAGRAM
Odd to Even

Columns: $c^*4s\ ^1P_1$ | $c^*ns\ ^1D_2$ | $c^*nl\ ^1F_3$ | $c^*4s\ ^1G_4$ | $c^*4p\ ^3P^o$ | $c^*4p\ ^3D^o_{1,2,3}$ | $c^*4p\ ^3F^o_{2,3,4}$ | $c^*4p\ ^3G^o$

ENERGY (cm^{-1}): 20×10^3, 30×10^3, 40×10^3, 50×10^3

Levels: e; a — 6^*4s; f; e — 3^*5s; b — 8^*4s; f — 2^*4d; e — 2^*5s; a — 12^*4s; f; e — 2^*4d; w — $8^*(0,1,2)$; x — $11^*(0,1,2)$; y — $3^*(0,1,2)$; z — $7^*(2,1)$; v — 3^*; w — 7^*; x — 1^*; y — 0^*; z — 2^*; y — 0^*; z — 2^*; w — $10^*(3,4,5)$; x — $0^*(3,4,5)$; y — $1^*(3,4,5)$; z — $2^*(3,4,5)$

Transitions (Å): 4996.77; 5279.10; 5824.83; 3958.00, 3973.50; 7808.73; 4614.31; 3671.44; 8832.49; 6473.36; 3970.19; 4817.43; 3716.62; 5383.81; 8948.00

CORES (c^*) from Ti II:
$0^* = 3d^2 4s\ ^4F$
$1^* = 3d^3\ ^4F$
$2^* = 3d^2 4s\ ^2F$
$3^* = 3d^2 4s\ ^2D$
$4^* = 3d^3\ ^2G$
$5^* = 3d^3\ ^4P$
$6^* = 3d^3\ ^2P$
$7^* = 3d^2 4s\ ^4P$
$8^* = 3d^3\ ^2D$
$9^* = 3d^3\ ^2H$
$10^* = 3d^2 4s\ ^2G$
$11^* = 3d^2 4s\ ^2P$
$12^* = 3d^3\ ^2F$
$13^* = 3d^2 4s\ ^2S$

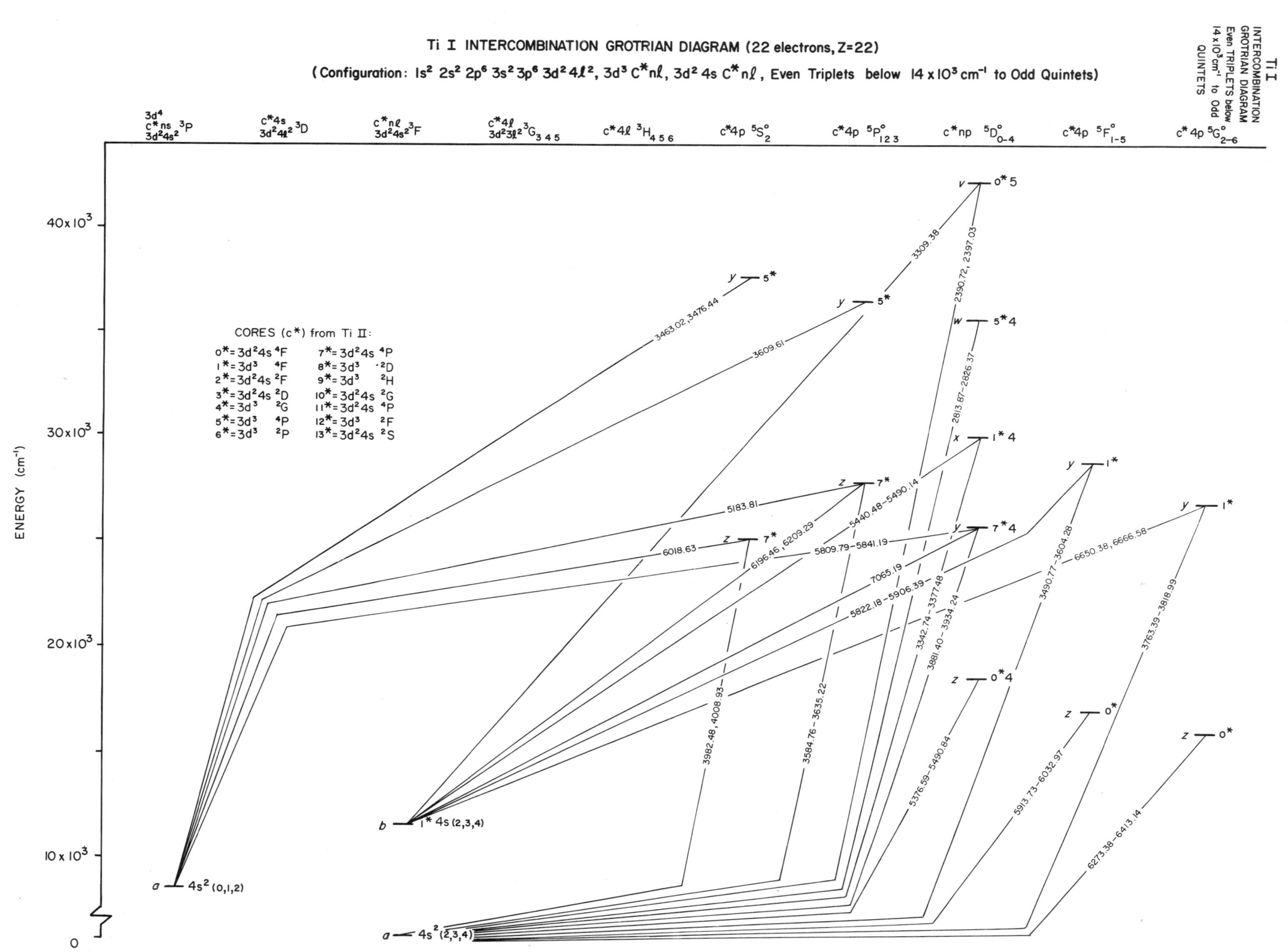
Ti I INTERCOMBINATION GROTRIAN DIAGRAM (22 electrons, Z=22)
(Configuration: $1s^2\ 2s^2\ 2p^6\ 3s^2\ 3p^6\ 3d^2 4\ell^2$, $3d^3\ c^* n\ell$, $3d^2\ 4s\ c^* n\ell$, Even Triplets below $14 \times 10^3\ cm^{-1}$ to Odd Quintets)
Ti I
INTERCOMBINATION
GROTRIAN DIAGRAM
Even TRIPLETS below
$14 \times 10^3 cm^{-1}$ to Odd
QUINTETS
$3d^4$ $c^* ns\ ^3P$ $3d^2 4s^2$
$c^* 4s$ $3d^2 4\ell^2\ ^3D$
$c^* n\ell$ $3d^2 4s^2\ ^3F$
$c^* 4\ell$ $3d^2 3\ell^2\ ^3G_{3\,4\,5}$
$c^* 4\ell\ ^3H_{4\,5\,6}$
$c^* 4p\ ^5S^o_2$
$c^* 4p\ ^5P^o_{1\,2\,3}$
$c^* np\ ^5D^o_{0-4}$
$c^* 4p\ ^5F^o_{1-5}$
$c^* 4p\ ^5G^o_{2-6}$
ENERGY (cm^{-1})
40×10^3
30×10^3
20×10^3
10×10^3
0
CORES (c^*) from Ti II:
$0^* = 3d^2 4s\ ^4F$
$1^* = 3d^3\ ^4F$
$2^* = 3d^2 4s\ ^2F$
$3^* = 3d^2 4s\ ^2D$
$4^* = 3d^3\ ^2G$
$5^* = 3d^3\ ^4P$
$6^* = 3d^3\ ^2P$
$7^* = 3d^2 4s\ ^4P$
$8^* = 3d^3\ ^2D$
$9^* = 3d^3\ ^2H$
$10^* = 3d^2 4s\ ^2G$
$11^* = 3d^2 4s\ ^4P$
$12^* = 3d^3\ ^2F$
$13^* = 3d^2 4s\ ^2S$
a $4s^2$ (0,1,2)
b 1^* 4s (2,3,4)
a $4s^2$ (2,3,4)
y 5^*
y 5^*
v 0^*5
w 5^*4
x 1^*4
z 7^*
z 7^*
y 7^*4
y 1^*
y 1^*
z 0^*4
z 0^*
z 0^*
3463.02, 3476.44
3609.61
3309.38
2390.72, 2397.03
2813.87–2826.37
5183.81
6018.63
5440.48–5490.14
6196.46, 6209.29
5809.79–5841.19
7065.19
5822.18–5906.39
3490.77–3604.28
6650.38, 6666.58
3342.74–3377.48
3881.40–3934.24
3763.39–3818.99
3982.48, 4008.93
3584.76–3635.22
5376.59–5490.84
5913.73–6032.97
6273.38–6413.14

Ti I INTERCOMBINATION GROTRIAN DIAGRAM (22 electrons, Z = 22)

(Configuration: $1s^2\,2s^2\,2p^6\,3s^2\,3p^6\,3d^2\,4\ell^2$, $3d^3\,c^*n\ell$, $3d^2\,4s\,c^*n\ell$, Even Triplets above 14×10^3 cm^{-1} to Odd Quintets)

Ti I
INTERCOMBINATION GROTRIAN DIAGRAM
Even TRIPLETS above 14×10 cm^{-1} to Odd QUINTETS

$3d^4$, c^*ns, $3d^24s^2$ ^3P | c^*4s, $3d^24\ell^2$ ^3D | $c^*n\ell$, $3d^24s^2$ ^3F | $c^*4\ell$, $3d^23\ell^2$ $^3G_{3,4,5}$ | $c^*4\ell$ $^3H_{4,5,6}$ | c^*4p $^5S^o_2$ | c^*4p $^5P^o_{1,2,3}$ | c^*np $^5D^o_{0-4}$ | c^*4p $^5F^o_{1-5}$ | c^*4p $^5G^o_{2-6}$

40×10^3 — 30×10^3 — 20×10^3 — 10×10^3 — 0

d — $3d^4$ (0,1,2)
c — 5^*4 (0,1,2)
b — 6^*4 (0,1,2)
a — $4s^2$ (0,1,2)
a — 8^* (1,2,3)
g — o^*4d (2,3,4)
f — 1^*5s (2,3,4)
e — o^*5s (2,3,4)
b — 1^*4s (2,3,4)
a — $4s^2$ (2,3,4)
e — o^*d
b — $3d^4$
a — 4^*s
e — o^*d
a — 9^*s
y — 5^*
z — 7^*
y — 5^*
z — 7^*
v — o^*5
w — 5^*4
x — 1^*4
y — 7^*4
z — o^*4
y — 1^*
z — o^*
y — 1^*
z — o^*

5968.05
5419.18
5180.51, 5203.14
5289.52, 5317.57
4088.32, 4089.54
5702.25, 5735.03
5490.70
5641.67, 5675.21
3726.23
8595.57
8032.50
7852.70
8741.15, 8763.69

CORES ( ) from Ti II:

$o^* = 3d^24s\ ^4F$	$7^* = 3d^24s\ ^4P$
$1^* = 3d^3\ ^4F$	$8^* = 3d^3\ ^2D$
$2^* = 3d^24s\ ^2F$	$9^* = 3d^3\ ^2H$
$3^* = 3d^24s\ ^2D$	$10^* = 3d^24s\ ^2G$
$4^* = 3d^3\ ^2G$	$11^* = 3d^24s\ ^4P$
$5^* = 3d^3\ ^4P$	$12^* = 3d^3\ ^2F$
$6^* = 3d^3\ ^2P$	$13^* = 3d^24s\ ^2S$

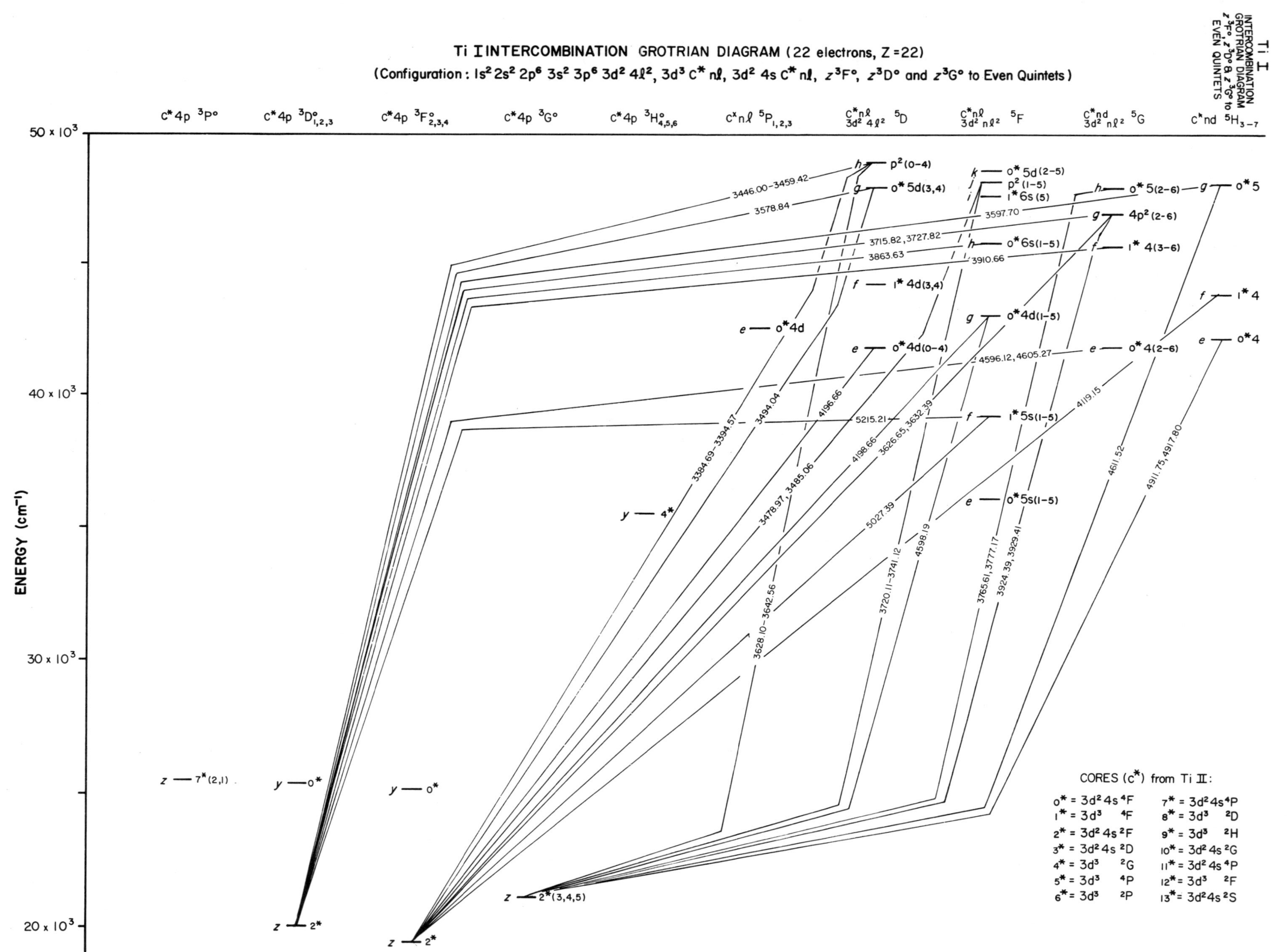

Ti I INTERCOMBINATION GROTRIAN DIAGRAM (22 electrons, Z=22)
(Configuration: $1s^2 2s^2 2p^6 3s^2 3p^6 3d^2 4\ell^2$, $3d^3 C^* n\ell$, $3d^2 4s C^* n\ell$, z^3F^o, z^3D^o and z^3G^o to Even Quintets)
Ti I
INTERCOMBINATION GROTRIAN DIAGRAM
z^3F^o, z^3D^o & z^3G^o to EVEN QUINTETS
$c^*4p\ ^3P^o$
$c^*4p\ ^3D^o_{1,2,3}$
$c^*4p\ ^3F^o_{2,3,4}$
$c^*4p\ ^3G^o$
$c^*4p\ ^3H^o_{4,5,6}$
$c^*n\ell\ ^5P_{1,2,3}$
$C^*n\ell\ 3d^2 4\ell^2\ ^5D$
$C^*n\ell\ 3d^2 n\ell^2\ ^5F$
$C^*nd\ 3d^2 n\ell^2\ ^5G$
$c^*nd\ ^5H_{3-7}$
ENERGY (cm^{-1})
50×10^3
40×10^3
30×10^3
20×10^3
CORES (c^*) from Ti II:
$o^* = 3d^2 4s\ ^4F$
$1^* = 3d^3\ ^4F$
$2^* = 3d^2 4s\ ^2F$
$3^* = 3d^2 4s\ ^2D$
$4^* = 3d^3\ ^2G$
$5^* = 3d^3\ ^4P$
$6^* = 3d^3\ ^2P$
$7^* = 3d^2 4s\ ^4P$
$8^* = 3d^3\ ^2D$
$9^* = 3d^3\ ^2H$
$10^* = 3d^2 4s\ ^2G$
$11^* = 3d^2 4s\ ^4P$
$12^* = 3d^3\ ^2F$
$13^* = 3d^2 4s\ ^2S$
3446.00–3459.42
3578.84
3597.70
3715.82, 3727.82
3863.63
3910.66
4596.12, 4605.27
5215.21
3384.69–3394.57
3494.04
4196.66
4198.66
3626.65, 3632.39
4119.15
4611.52
4911.75, 4917.80
3478.97, 3485.06
5027.39
4598.19
3628.10–3642.56
3720.11–3741.12
3765.61, 3777.17
3924.39, 3929.41

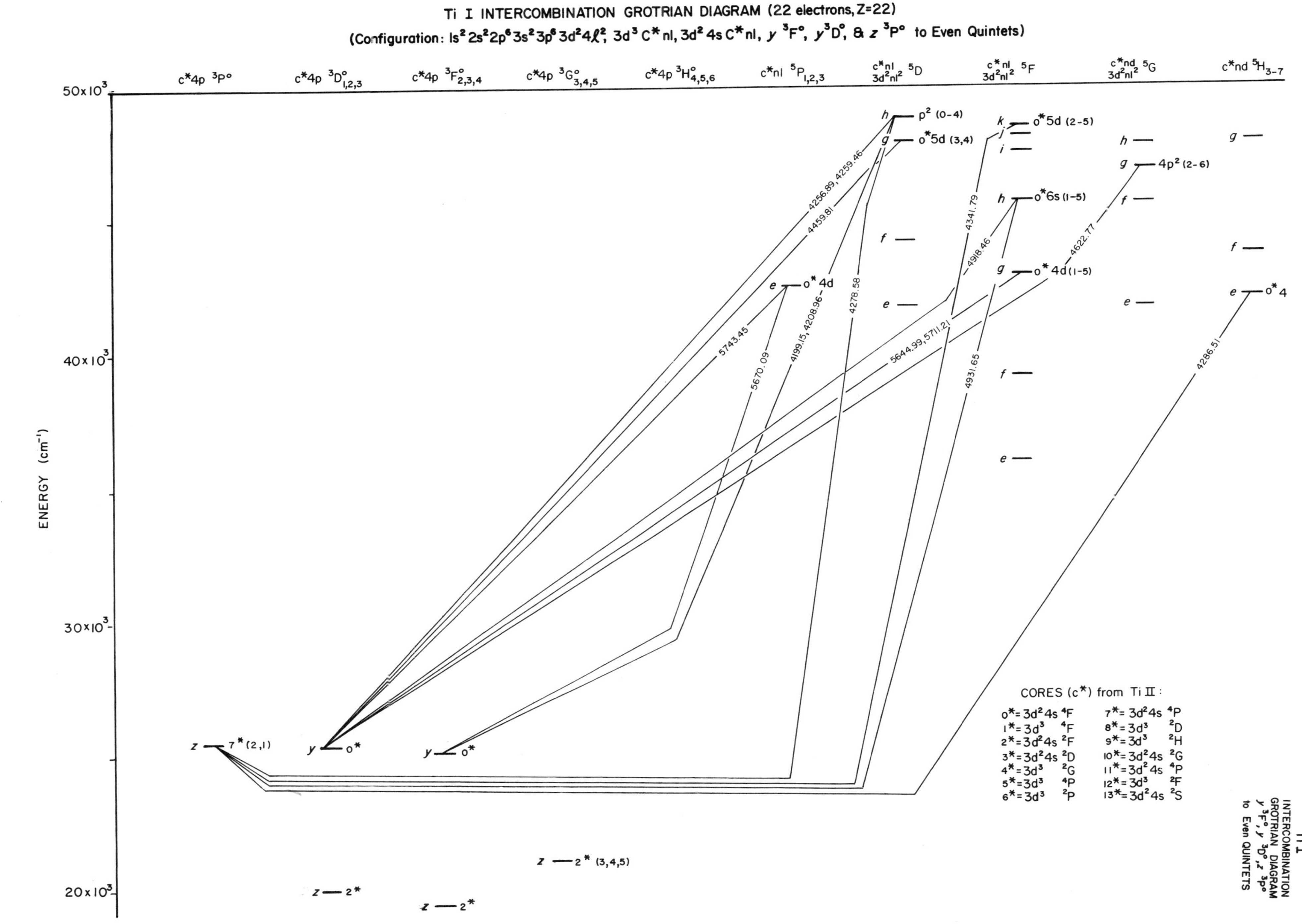
Ti I INTERCOMBINATION GROTRIAN DIAGRAM (22 electrons, Z=22)
(Configuration: $1s^2 2s^2 2p^6 3s^2 3p^6 3d^2 4\ell^2$, $3d^3$ C* nl, $3d^2$ 4s C* nl, y $^3F^o$, y $^3D^o$, & z $^3P^o$ to Even Quintets)
c*4p $^3P^o$
c*4p $^3D^o_{1,2,3}$
c*4p $^3F^o_{2,3,4}$
c*4p $^3G^o_{3,4,5}$
c*4p $^3H^o_{4,5,6}$
c*nl $^5P_{1,2,3}$
c*nl $3d^2nl^2$ ^5D
c*nl $3d^2nl^2$ ^5F
c*nd $3d^2nl^2$ ^5G
c*nd $^5H_{3-7}$
50x10³
40x10³
30x10³
20x10³
ENERGY (cm⁻¹)
z 7* (2,1)
y o*
y o*
z 2* (3,4,5)
z 2*
z 2*
h p² (0-4)
g o*5d (3,4)
f
e
e o*4d
k o*5d (2-5)
j
i
h o*6s (1-5)
g o*4d (1-5)
f
e
h
g 4p² (2-6)
f
e
g
f
e o*4
4256.89, 4259.46
4459.81
5743.45
5670.09
4199.15, 4208.96
4278.58
4341.79
4918.46
5644.99, 5711.21
4931.65
4622.77
4286.51
CORES (c*) from Ti II:
o* = 3d²4s ⁴F
1* = 3d³ ⁴F
2* = 3d²4s ²F
3* = 3d²4s ²D
4* = 3d³ ²G
5* = 3d³ ⁴P
6* = 3d³ ²P
7* = 3d²4s ⁴P
8* = 3d³ ²D
9* = 3d³ ²H
10* = 3d²4s ²G
11* = 3d²4s ⁴P
12* = 3d³ ²F
13* = 3d²4s ²S
Ti I
INTERCOMBINATION
GROTRIAN DIAGRAM
y ³F°, y ³D°, z ³P°
to Even QUINTETS

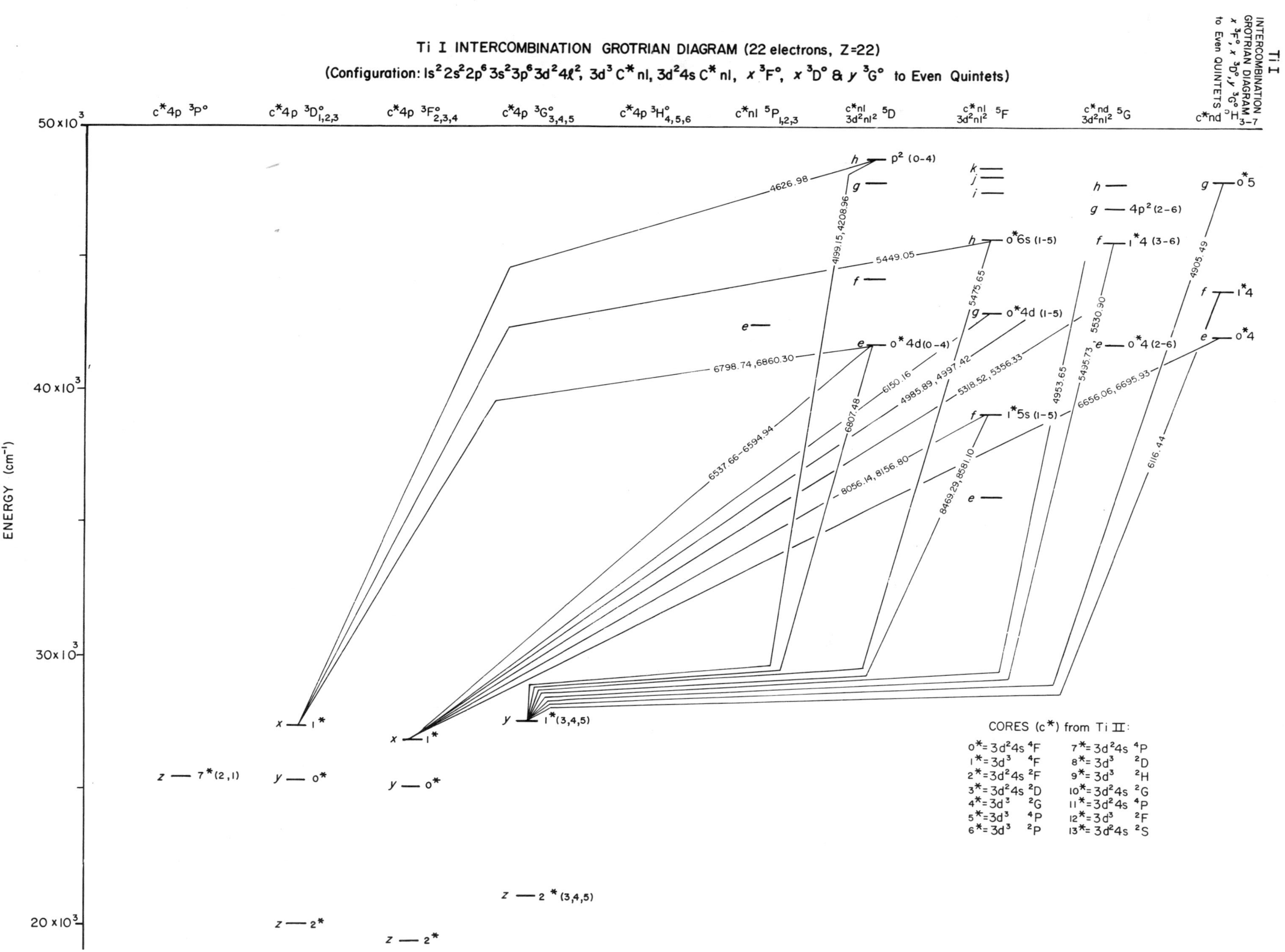

Ti I
INTERCOMBINATION GROTRIAN DIAGRAM
$x\ ^3F^o$, $x\ ^3D^o$, $y\ ^3G^o$
to Even QUINTETS
Ti I INTERCOMBINATION GROTRIAN DIAGRAM (22 electrons, Z=22)
(Configuration: $1s^22s^22p^63s^23p^63d^24\ell^2$, $3d^3 C^*nl$, $3d^24s\, C^*nl$, $x\ ^3F^o$, $x\ ^3D^o$ & $y\ ^3G^o$ to Even Quintets)
$c^*4p\ ^3P^o$
$c^*4p\ ^3D^o_{1,2,3}$
$c^*4p\ ^3F^o_{2,3,4}$
$c^*4p\ ^3G^o_{3,4,5}$
$c^*4p\ ^3H^o_{4,5,6}$
$c^*nl\ ^5P_{1,2,3}$
$c^*nl\ 3d^2nl^2\ ^5D$
$c^*nl\ 3d^2nl^2\ ^5F$
$c^*nd\ 3d^2nl^2\ ^5G$
$c^*nd\ ^5H_{3-7}$
50×10^3
40×10^3
30×10^3
20×10^3
ENERGY (cm^{-1})
h — p^2 (0-4)
g
f
e
e — 0^*4d (0-4)
k
j
i
h — 0^*6s (1-5)
g — 0^*4d (1-5)
f — 1^*5s (1-5)
e
h
g — $4p^2$ (2-6)
f — 1^*4 (3-6)
e — 0^*4 (2-6)
g — 0^*5
f — 1^*4
e — 0^*4
x — 1^*
x — 1^*
y — $1^*(3,4,5)$
z — $7^*(2,1)$
y — 0^*
y — 0^*
z — $2^*(3,4,5)$
z — 2^*
z — 2^*
4626.98
4199.15, 4208.96
5449.05
5475.65
6798.74, 6860.30
6150.16
4985.89, 4997.42
5318.52, 5356.33
6807.48
6537.66–6594.94
8056.14, 8156.80
8469.29, 8581.10
4953.65
5435.73
5530.90
6656.06, 6695.93
4905.49
6116.44
CORES (c^*) from Ti II:
$0^* = 3d^24s\ ^4F$
$1^* = 3d^3\ ^4F$
$2^* = 3d^24s\ ^2F$
$3^* = 3d^24s\ ^2D$
$4^* = 3d^3\ ^2G$
$5^* = 3d^3\ ^4P$
$6^* = 3d^3\ ^2P$
$7^* = 3d^24s\ ^4P$
$8^* = 3d^3\ ^2D$
$9^* = 3d^3\ ^2H$
$10^* = 3d^24s\ ^2G$
$11^* = 3d^24s\ ^4P$
$12^* = 3d^3\ ^2F$
$13^* = 3d^24s\ ^2S$

Ti I INTERCOMBINATION GROTRIAN DIAGRAM (22 electrons, Z=22)

(Configuration: $1s^2 2s^2 2p^6 3s^2 3p^6 3d^2 4\ell^2$, $3d^3$ C*nl, 3d 4s C*nl, $w\,^3D^\circ$, $x\,^3G^\circ$, $v\,^3D^\circ$, $w\,^3G^\circ$, $y\,^3P^\circ$, to Even Quintets)

CORES (c*) from Ti II:

o* = $3d^2 4s\ ^4F$	7* = $3d^2 4s\ ^4P$
1* = $3d^3\ ^4F$	8* = $3d^3\ ^2D$
2* = $3d^2 4s\ ^2F$	9* = $3d^3\ ^2H$
3* = $3d^2 4s\ ^2D$	10* = $3d^2 4s\ ^2G$
4* = $3d^3\ ^2G$	11* = $3d^2 4s\ ^4P$
5* = $3d^3\ ^4P$	12* = $3d^3\ ^2F$
6* = $3d^3\ ^2P$	13* = $3d^2 4s\ ^2S$

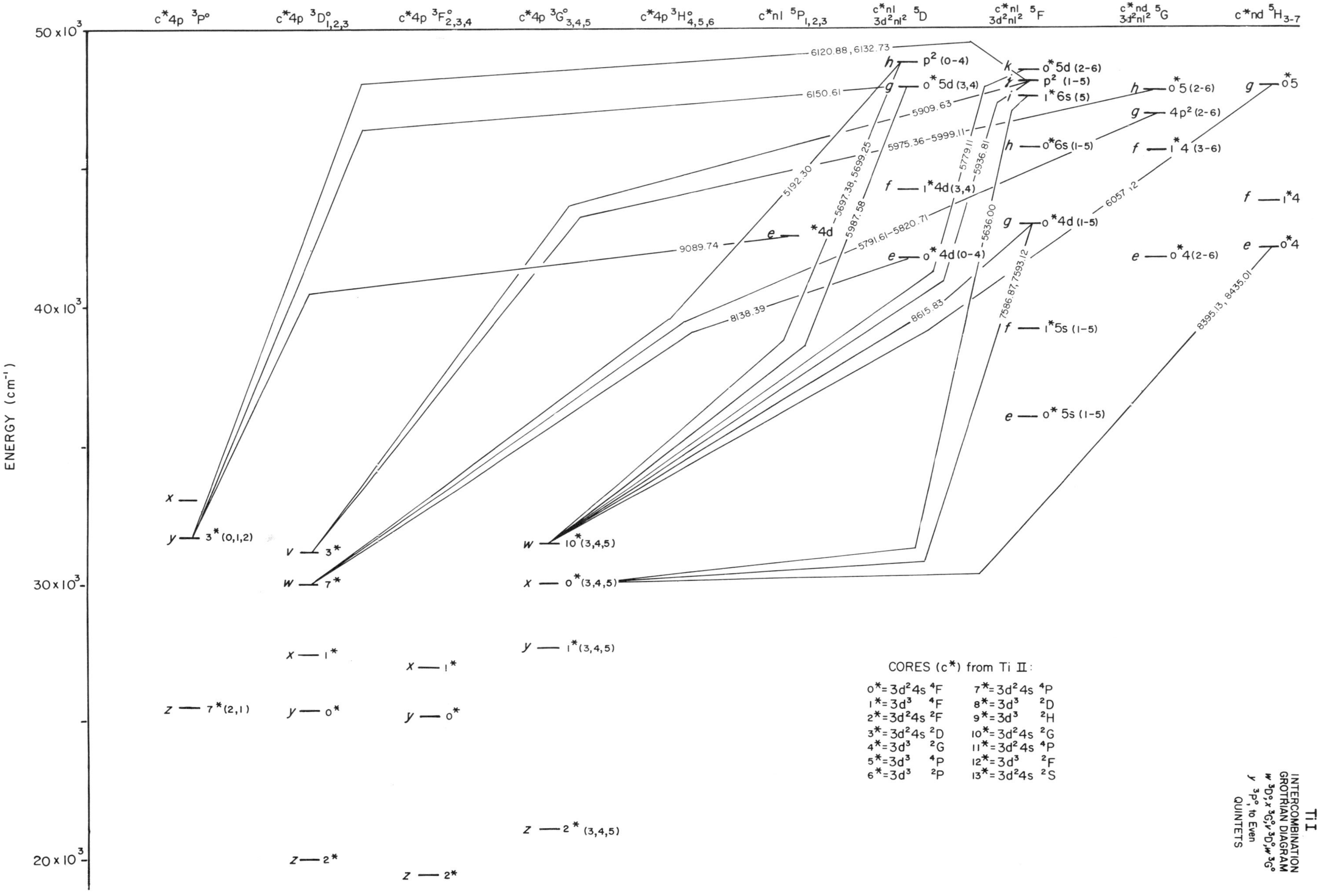

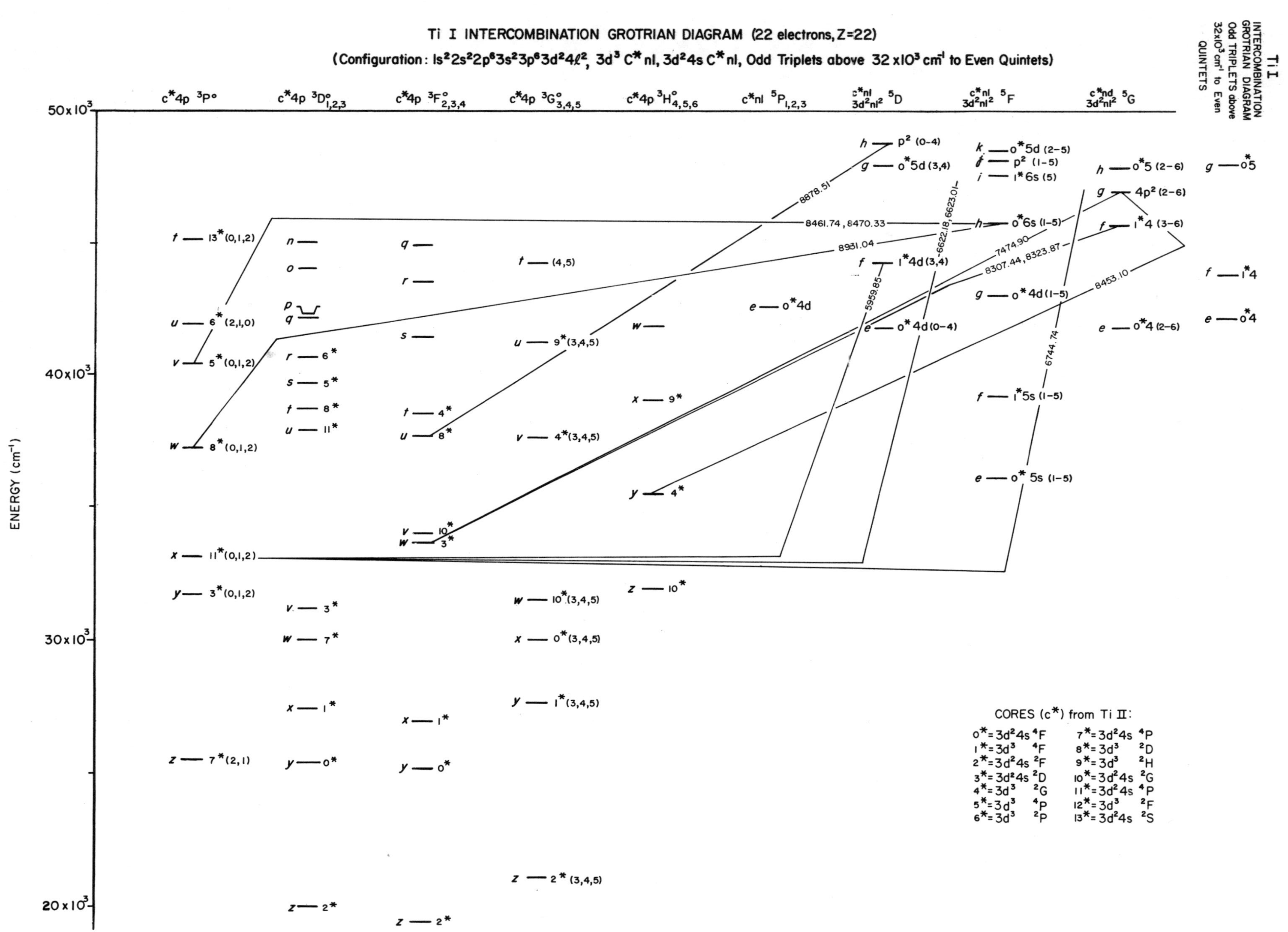

Ti I INTERCOMBINATION GROTRIAN DIAGRAM (22 electrons, Z=22)
(Configuration: $1s^22s^22p^63s^23p^63d^24\ell^2$, $3d^3$ c^*nl, $3d^24s$ c^*nl, Odd Triplets above 32×10^3 cm^{-1} to Even Quintets)
Ti I
INTERCOMBINATION GROTRIAN DIAGRAM
Odd TRIPLETS above 32×10^3 cm^{-1} to Even QUINTETS
c^*4p $^3P^o$
c^*4p $^3D^o_{1,2,3}$
c^*4p $^3F^o_{2,3,4}$
c^*4p $^3G^o_{3,4,5}$
c^*4p $^3H^o_{4,5,6}$
c^*nl $^5P_{1,2,3}$
c^*nl $3d^2nl^2$ ^5D
c^*nl $3d^2nl^2$ ^5F
c^*nd $3d^2nl^2$ ^5G
8878.51
8461.74, 8470.33
8931.04
5959.85
6622.18, 6623.01
7474.90
8307.44, 8323.87
8453.10
6744.74
ENERGY (cm^{-1})
50×10^3
40×10^3
30×10^3
20×10^3
CORES (c^*) from Ti II:
$0^*=3d^24s$ ^4F
$1^*=3d^3$ ^4F
$2^*=3d^24s$ ^2F
$3^*=3d^24s$ ^2D
$4^*=3d^3$ ^2G
$5^*=3d^3$ ^4P
$6^*=3d^3$ ^2P
$7^*=3d^24s$ ^4P
$8^*=3d^3$ ^2D
$9^*=3d^3$ ^2H
$10^*=3d^24s$ ^2G
$11^*=3d^24s$ ^4P
$12^*=3d^3$ ^2F
$13^*=3d^24s$ ^2S

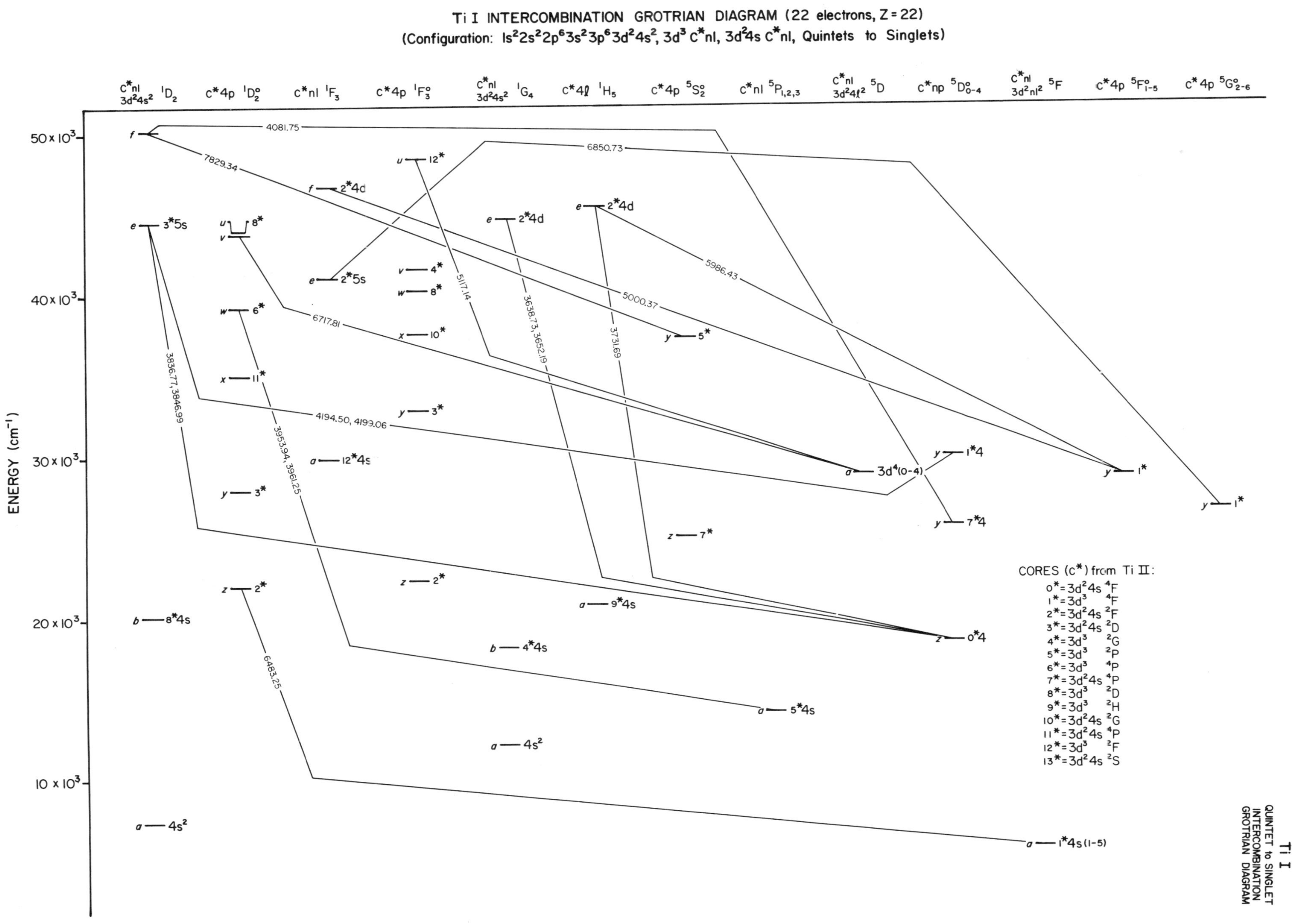
Ti I INTERCOMBINATION GROTRIAN DIAGRAM (22 electrons, Z = 22)
(Configuration: 1s²2s²2p⁶3s²3p⁶3d²4s², 3d³ c*nl, 3d²4s c*nl, Quintets to Singlets)
c*nl 3d²4s² ¹D₂
c*4p ¹D°₂
c*nl ¹F₃
c*4p ¹F°₃
c*nl 3d²4s² ¹G₄
c*4l ¹H₅
c*4p ⁵S°₂
c*nl ⁵P₁,₂,₃
c*nl 3d²4l² ⁵D
c*np ⁵D°₀₋₄
c*nl 3d²nl² ⁵F
c*4p ⁵F°₁₋₅
c*4p ⁵G°₂₋₆
ENERGY (cm⁻¹)
50×10³
40×10³
30×10³
20×10³
10×10³
4081.75
7829.34
6850.73
5117.14
3638.73, 3652.19
3731.69
5986.43
5000.37
6717.81
3836.77, 3846.99
4194.50, 4193.06
3953.94, 3961.25
6483.25
CORES (c*) from Ti II:
0* = 3d²4s ⁴F
1* = 3d³ ⁴F
2* = 3d²4s ²F
3* = 3d²4s ²D
4* = 3d³ ²G
5* = 3d³ ²P
6* = 3d³ ⁴P
7* = 3d²4s ⁴P
8* = 3d³ ²D
9* = 3d³ ²H
10* = 3d²4s ²G
11* = 3d²4s ⁴P
12* = 3d³ ²F
13* = 3d²4s ²S
Ti I
QUINTET to SINGLET
INTERCOMBINATION
GROTRIAN DIAGRAM

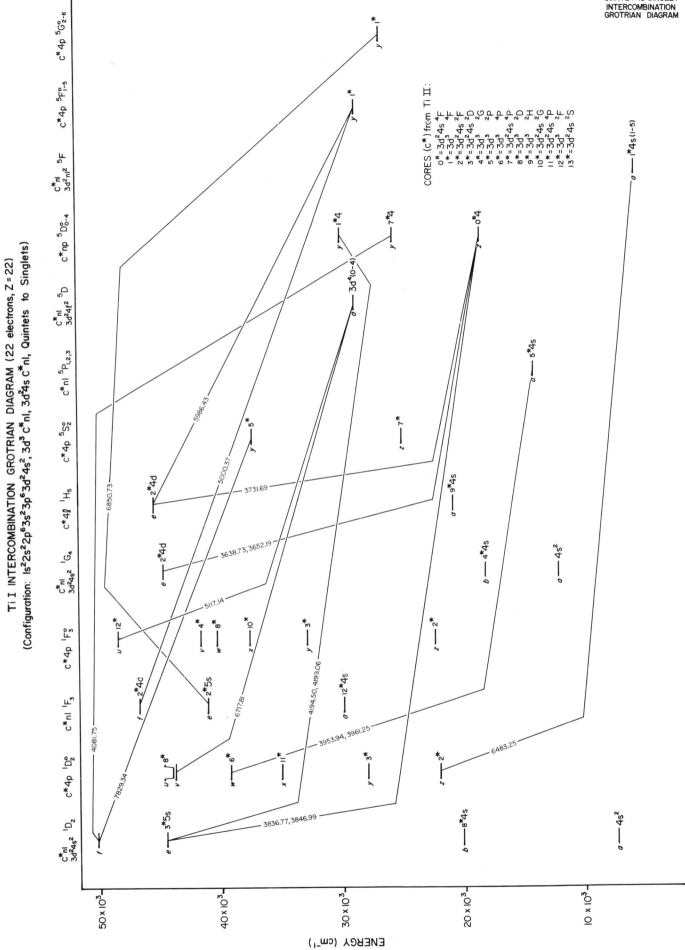

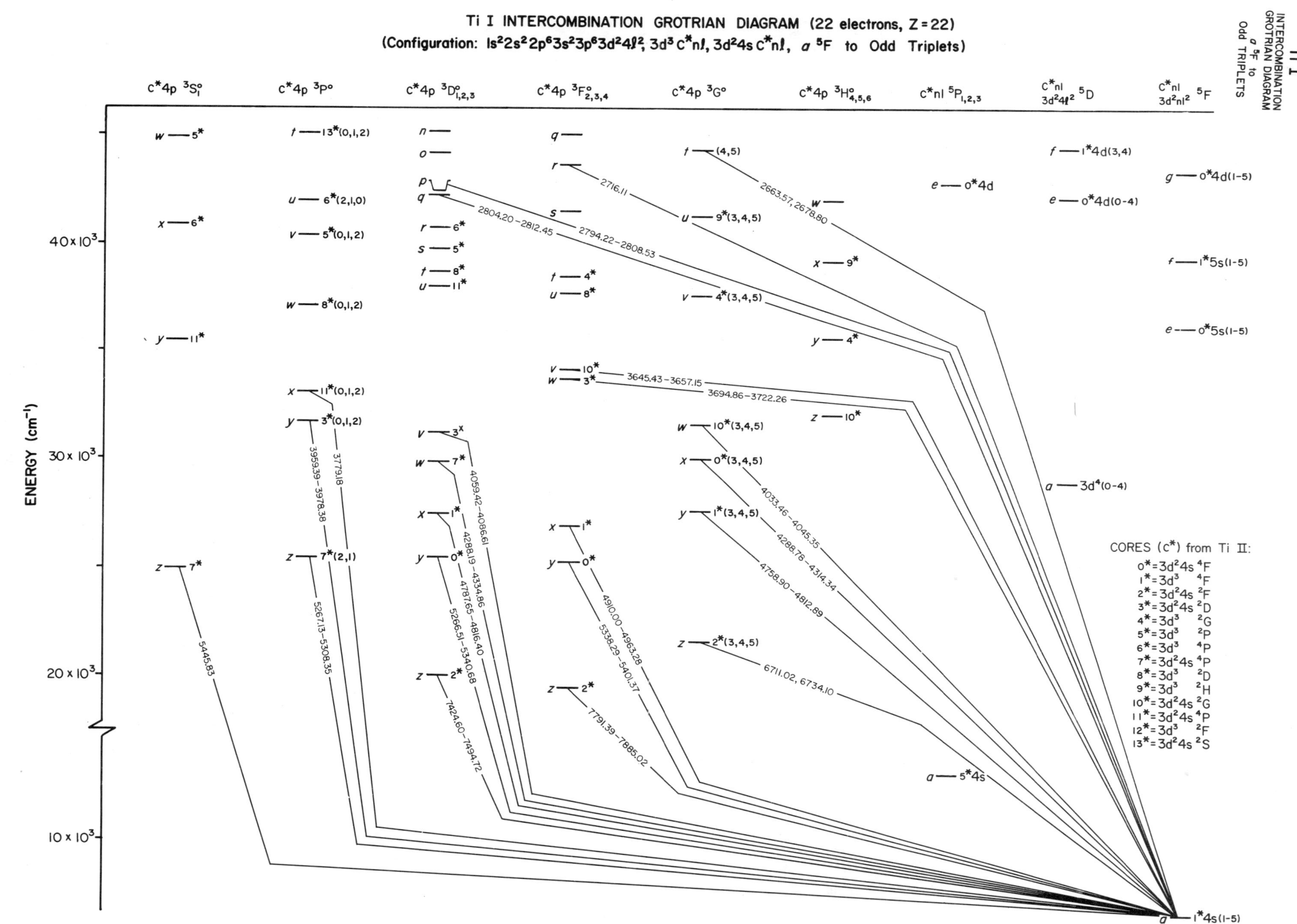

Ti I INTERCOMBINATION GROTRIAN DIAGRAM (22 electrons, Z=22)
(Configuration: 1s²2s²2p⁶3s²3p⁶3d²4l², 3d³ c*nl, 3d²4s c*nl, a ⁵F to Odd Triplets)
Ti I
INTERCOMBINATION
GROTRIAN DIAGRAM
a ⁵F to
Odd TRIPLETS
c*4p ³S°₁
c*4p ³P°
c*4p ³D°₁,₂,₃
c*4p ³F°₂,₃,₄
c*4p ³G°
c*4p ³H°₄,₅,₆
c*nl ⁵P₁,₂,₃
c*nl 3d²4l² ⁵D
c*nl 3d²nl² ⁵F
ENERGY (cm⁻¹)
40×10³
30×10³
20×10³
10×10³
5445.83
3779.18
3959.39–3978.38
5267.13–5308.35
2804.20–2812.45
4059.42–4086.61
4288.19–4334.86
4787.65–4816.40
5266.51–5340.68
7424.60–7494.72
2716.11
2794.22–2808.53
3645.43–3657.15
3694.86–3722.26
4910.00–4963.28
5338.29–5401.37
7791.39–7885.02
2663.57, 2678.80
4033.46–4045.35
4288.78–4314.34
4758.90–4812.89
6711.02, 6734.10
a — 1*4s(1-5)
a — 5*4s
a — 3d⁴(0-4)
e — o*4d
f — 1*4d(3,4)
e — o*4d(0-4)
g — o*4d(1-5)
f — 1*5s(1-5)
e — o*5s(1-5)
CORES (c*) from Ti II:
o*=3d²4s ⁴F
1*=3d³ ⁴F
2*=3d²4s ²F
3*=3d²4s ²D
4*=3d³ ²G
5*=3d³ ²P
6*=3d³ ⁴P
7*=3d²4s ⁴P
8*=3d³ ²D
9*=3d³ ²H
10*=3d²4s ²G
11*=3d²4s ⁴P
12*=3d³ ²F
13*=3d²4s ²S

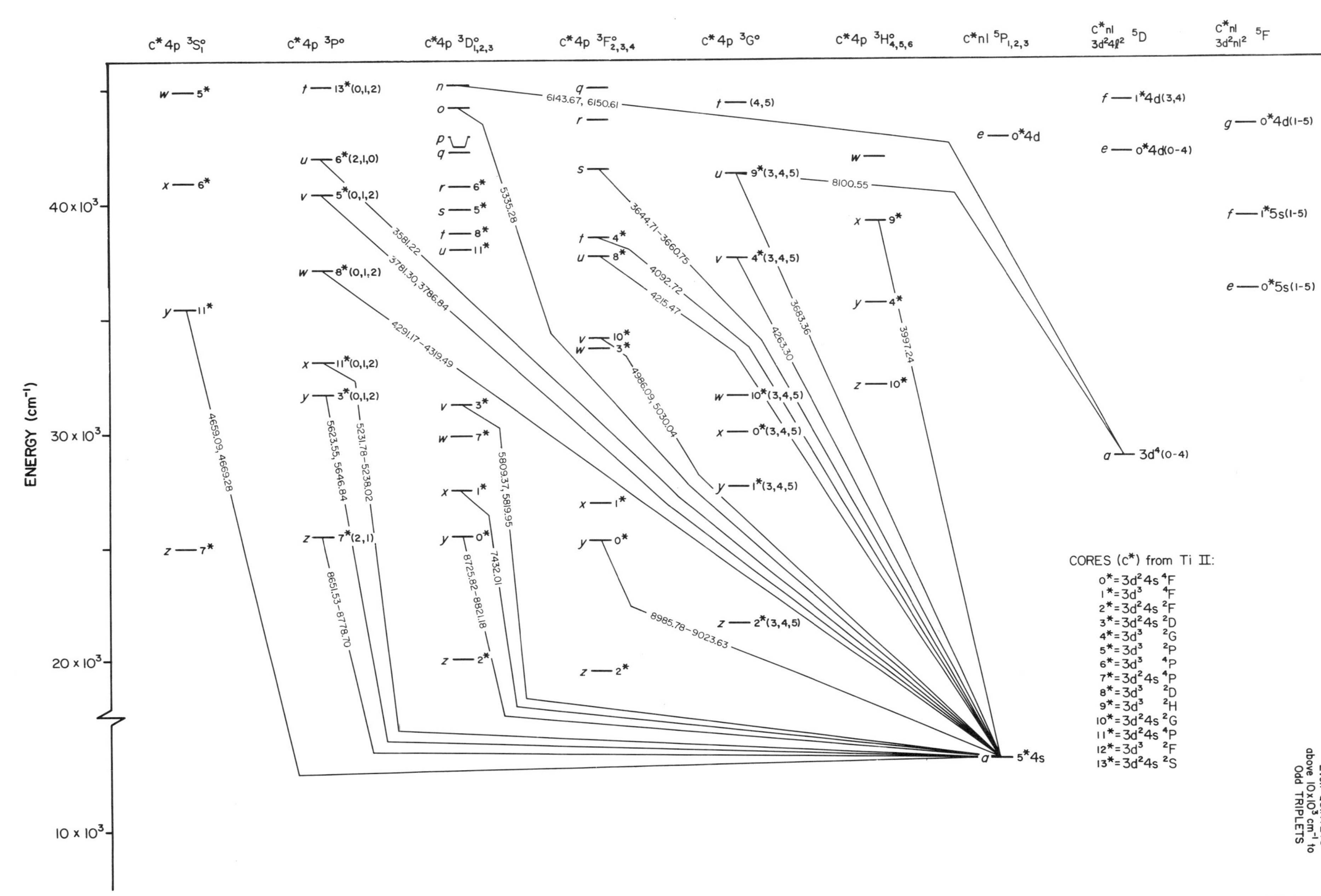
Ti I INTERCOMBINATION GROTRIAN DIAGRAM (22 electrons, Z = 22)
(Configuration: 1s²2s²2p⁶3s²3p⁶3d²4ℓ², 3d³c*nℓ, 3d²4s c*nℓ, Even Quintets above 10×10³ cm⁻¹ to Odd Triplets)
c*4p ³S°₁
c*4p ³P°
c*4p ³D°₁,₂,₃
c*4p ³F°₂,₃,₄
c*4p ³G°
c*4p ³H°₄,₅,₆
c*nl ⁵P₁,₂,₃
c*nl 3d²4ℓ² ⁵D
c*nl 3d²nl² ⁵F
ENERGY (cm⁻¹)
40×10³
30×10³
20×10³
10×10³
w — 5*
x — 6*
y — 11*
z — 7*
t — 13*(0,1,2)
u — 6*(2,1,0)
v — 5*(0,1,2)
w — 8*(0,1,2)
x — 11*(0,1,2)
y — 3*(0,1,2)
z — 7*(2,1)
n
o
p
q
r — 6*
s — 5*
t — 8*
u — 11*
v — 3*
w — 7*
x — 1*
y — 0*
z — 2*
q
r
s
t — 4*
u — 8*
v — 10*
w — 3*
x — 1*
y — 0*
z — 2*
t — (4,5)
u — 9*(3,4,5)
v — 4*(3,4,5)
w — 10*(3,4,5)
x — 0*(3,4,5)
y — 1*(3,4,5)
z — 2*(3,4,5)
w
x — 9*
y — 4*
z — 10*
e — 0*4d
f — 1*4d(3,4)
e — 0*4d(0-4)
a — 3d⁴(0-4)
g — 0*4d(1-5)
f — 1*5s(1-5)
e — 0*5s(1-5)
a — 5*4s
a — 1*4s(1-5)
4659.09, 4669.28
5623.55, 5646.84
5231.78–5238.02
8651.53–8778.70
3581.22
3781.30, 3786.84
4291.17–4319.49
5335.28
5809.37, 5819.95
7432.01
8725.82–8821.18
6143.67, 6150.61
3644.71–3660.75
4092.72
4215.47
4986.09, 5030.04
8985.78–9023.63
3683.36
4263.30
8100.55
3997.24
CORES (c*) from Ti II:
0* = 3d²4s ⁴F
1* = 3d³ ⁴F
2* = 3d²4s ²F
3* = 3d²4s ²D
4* = 3d³ ²G
5* = 3d³ ²P
6* = 3d³ ⁴P
7* = 3d²4s ⁴P
8* = 3d³ ²D
9* = 3d³ ²H
10* = 3d²4s ²G
11* = 3d²4s ⁴P
12* = 3d³ ²F
13* = 3d²4s ²S
Ti I
INTERCOMBINATION
GROTRIAN DIAGRAM
Even QUINTETS
above 10×10³ cm⁻¹ to
Odd TRIPLETS

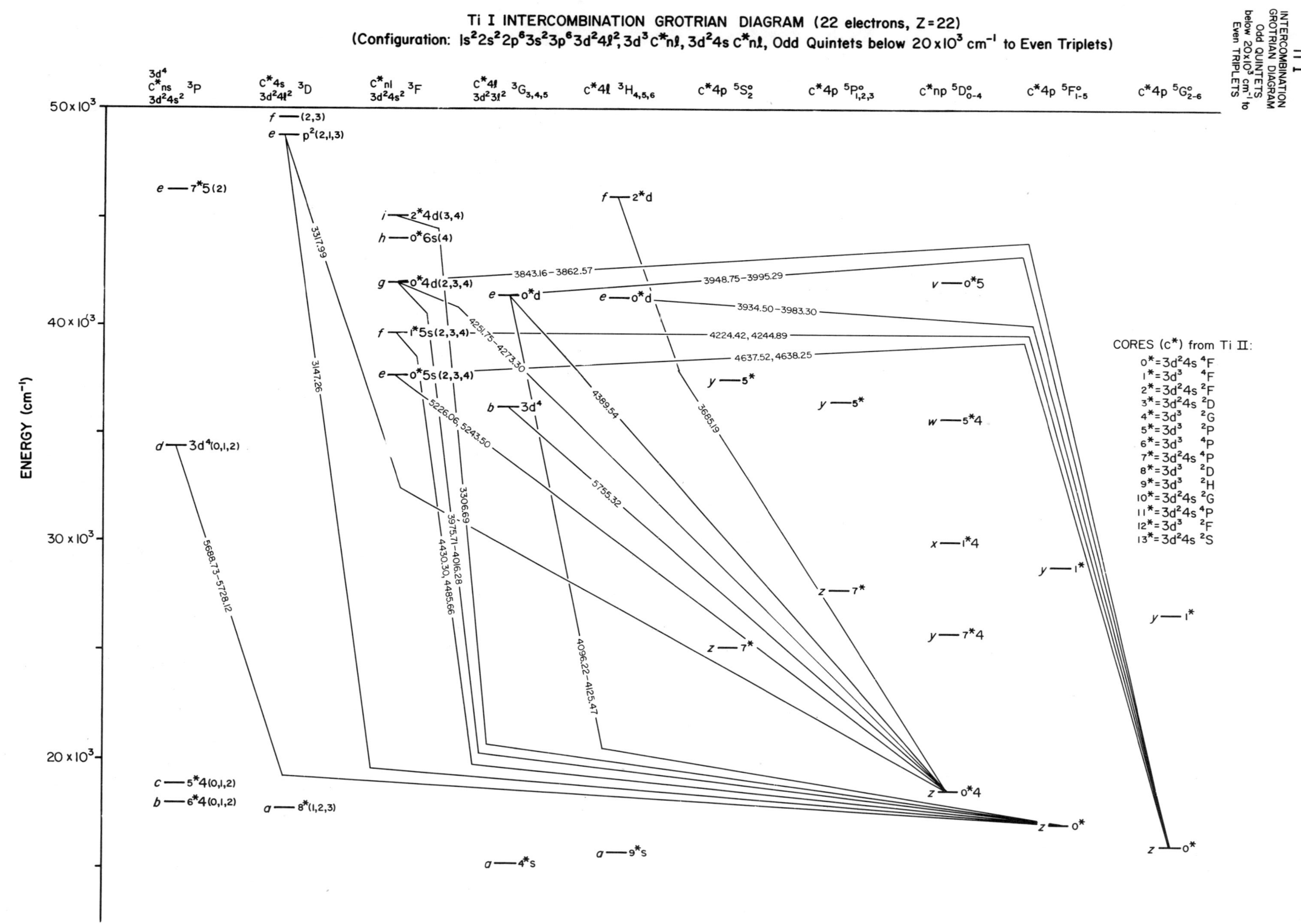

Ti I INTERCOMBINATION GROTRIAN DIAGRAM (22 electrons, Z=22)
(Configuration: $1s^2 2s^2 2p^6 3s^2 3p^6 3d^2 4\ell^2, 3d^3 c^* n\ell, 3d^2 4s\, c^* n\ell$, Odd Quintets below 20×10^3 cm^{-1} to Even Triplets)
Ti I
INTERCOMBINATION
GROTRIAN DIAGRAM
Odd QUINTETS
below 20×10^3 cm^{-1} to
Even TRIPLETS
$3d^4$ c^*ns $3d^2 4s^2$ ^3P
c^*4s $3d^2 4\ell^2$ ^3D
c^*nl $3d^2 4s^2$ ^3F
$c^*4\ell$ $3d^2 3\ell^2$ $^3G_{3,4,5}$
$c^*4\ell$ $^3H_{4,5,6}$
c^*4p $^5S^o_2$
c^*4p $^5P^o_{1,2,3}$
c^*np $^5D^o_{0-4}$
c^*4p $^5F^o_{1-5}$
c^*4p $^5G^o_{2-6}$
ENERGY (cm^{-1})
50×10^3
40×10^3
30×10^3
20×10^3
CORES (c^*) from Ti II:
$0^* = 3d^2 4s$ ^4F
$1^* = 3d^3$ ^4F
$2^* = 3d^2 4s$ ^2F
$3^* = 3d^2 4s$ ^2D
$4^* = 3d^3$ ^2G
$5^* = 3d^3$ ^2P
$6^* = 3d^3$ ^4P
$7^* = 3d^2 4s$ ^4P
$8^* = 3d^3$ ^2D
$9^* = 3d^3$ ^2H
$10^* = 3d^2 4s$ ^2G
$11^* = 3d^2 4s$ ^4P
$12^* = 3d^3$ ^2F
$13^* = 3d^2 4s$ ^2S
3317.99
3147.26
5688.73–5728.12
3843.16–3862.57
3948.75–3995.29
3934.50–3983.30
4224.42, 4244.89
4637.52, 4638.25
4251.75–4273.30
5226.06, 5243.50
4389.54
3685.19
5755.32
3306.69
3975.71–4016.28
4430.30, 4485.66
4096.22–4125.47

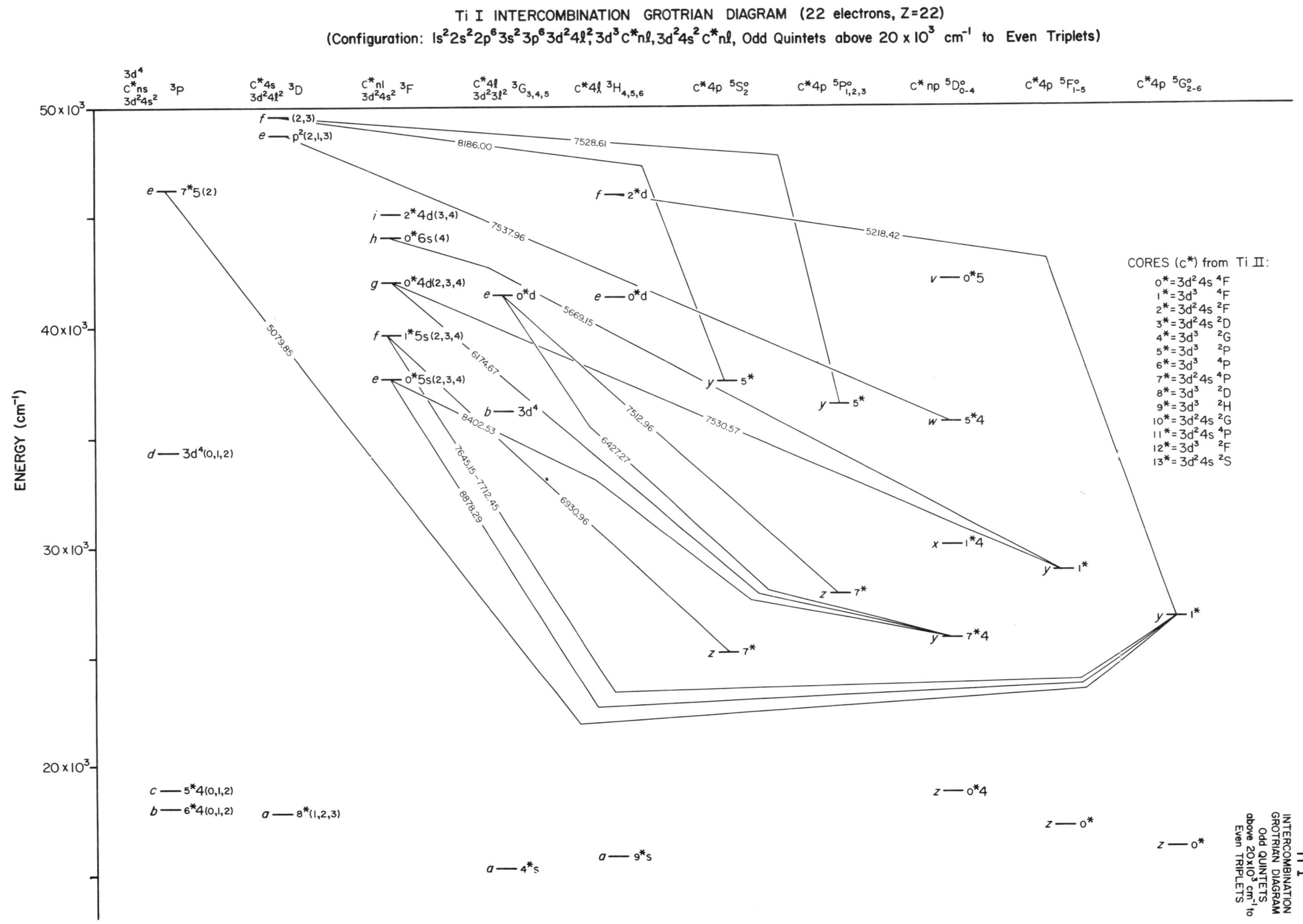
Ti I INTERCOMBINATION GROTRIAN DIAGRAM (22 electrons, Z=22)
(Configuration: 1s²2s²2p⁶3s²3p⁶3d²4l²,3d³c*nl,3d²4s²c*nl, Odd Quintets above 20 x 10³ cm⁻¹ to Even Triplets)
3d⁴ c*ns 3d²4s² ³P
c*4s 3d²4l² ³D
c*nl 3d²4s² ³F
c*4l 3d²3l² ³G3,4,5
c*4l ³H4,5,6
c*4p ⁵S°2
c*4p ⁵P°1,2,3
c* np ⁵D°0-4
c*4p ⁵F°1-5
c*4p ⁵G°2-6
50x10³
40x10³
30x10³
20x10³
ENERGY (cm⁻¹)
f — (2,3)
e — p²(2,1,3)
e — 7*5(2)
f — 2*d
i — 2*4d(3,4)
h — o*6s(4)
g — o*4d(2,3,4)
e — o*d
e — o*d
f — 1*5s(2,3,4)
e — o*5s(2,3,4)
b — 3d⁴
d — 3d⁴(0,1,2)
c — 5*4(0,1,2)
b — 6*4(0,1,2)
a — 8*(1,2,3)
a — 4*s
a — 9*s
y — 5*
y — 5*
v — o*5
w — 5*4
x — 1*4
y — 1*
z — 7*
y — 7*4
z — 7*
y — 1*
z — o*4
z — o*
z — o*
7528.61
8186.00
5218.42
7537.96
5669.15
5079.85
6174.67
8402.53
7512.96
7530.57
6427.27
7645.15 – 7712.45
8878.29
6930.96
CORES (c*) from Ti II:
o*=3d²4s ⁴F
1*=3d³ ⁴F
2*=3d²4s ²F
3*=3d²4s ²D
4*=3d³ ²G
5*=3d³ ²P
6*=3d³ ⁴P
7*=3d²4s ⁴P
8*=3d³ ²D
9*=3d³ ²H
10*=3d²4s ²G
11*=3d²4s ⁴P
12*=3d³ ²F
13*=3d²4s ²S
Ti I
INTERCOMBINATION GROTRIAN DIAGRAM
Odd QUINTETS above 20x10³ cm⁻¹ to Even TRIPLETS

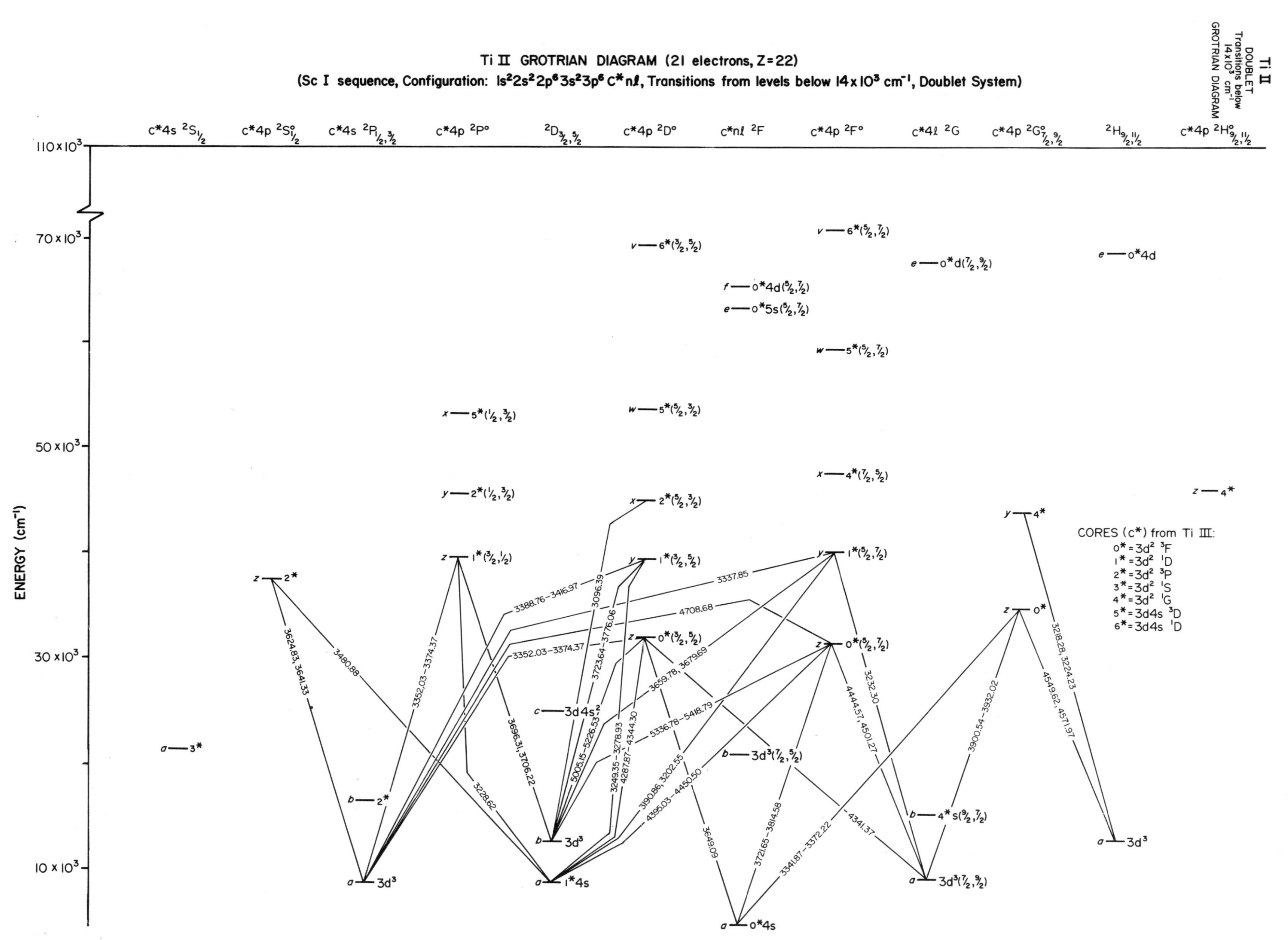
Ti II GROTRIAN DIAGRAM (21 electrons, Z=22)
(Sc I sequence, Configuration: 1s²2s²2p⁶3s²3p⁶ C*nℓ, Transitions from levels below 14×10³ cm⁻¹, Doublet System)
Ti II
DOUBLET
Transitions below
14×10³ cm⁻¹
GROTRIAN DIAGRAM
c*4s ²S½
c*4p ²S°½
c*4s ²P½,3/2
c*4p ²P°
²D3/2,5/2
c*4p ²D°
c*nℓ ²F
c*4p ²F°
c*4ℓ ²G
c*4p ²G°7/2,9/2
²H9/2,11/2
c*4p ²H°9/2,11/2
ENERGY (cm⁻¹)
110×10³
70×10³
50×10³
30×10³
10×10³
CORES (c*) from Ti III:
0* = 3d² ³F
1* = 3d² ¹D
2* = 3d² ³P
3* = 3d² ¹S
4* = 3d² ¹G
5* = 3d4s ³D
6* = 3d4s ¹D
3624.83, 3641.33
3480.88
3352.03–3374.37
3228.62
3696.31, 3706.22
3388.76–3416.97
3352.03–3374.37
3096.39
3723.64–3776.06
5005.15–5226.53
3249.35–3278.93
4287.87–4344.30
3659.78, 3679.69
5336.78–5418.79
3190.86, 3202.55
4395.03–4450.50
3649.09
3721.65–3814.58
3341.87–3372.22
4708.68
3337.85
3232.30
4444.57, 4501.27
4341.37
3900.54–3932.02
4549.62, 4571.97
3218.28, 3224.23

Ti II ENERGY LEVELS (21 electrons, Z = 22)

(Sc I sequence, Configuration: $1s^2 2s^2 2p^6 3s^2 3p^6$ C*nℓ, Doublet System)

ENERGY (cm^{-1}): 110 x 10³, 70 x 10³, 50 x 10³, 30 x 10³, 10 x 10³

c*4s $^2S_{1/2}$
- 21 338.00 — a — 3*

c*4p $^2S^o_{1/2}$
- 37 430.55 — z — 2*

c*4s $^2P_{1/2,3/2}$
- 16 625.25 / 16 515.79 — b — 2*
- 9 975.92 / 9 850.90 — a — 3d³

c*4p $^2P^o$
- 53 128.17 / 53 121.48 — x — 5*(1/2, 3/2)
- 45 548.90 / 45 472.89 — y — 2*(1/2, 3/2)
- 39 674.64 / 39 602.90 — z — 1*(3/2, 1/2)

$^2D_{3/2,5/2}$
- 25 193.04 / 24 961.34 — c — 3d4s²
- 12 758.15 / 12 628.77 — b — 3d³
- 8 744.27 / 8 710.47 — a — 1*4s

c*4p $^2D^o$
- 69 622.15 / 69 327.32 — v — 6*(3/2, 5/2)
- 53 596.70 / 53 554.90 — w — 5*(5/2, 3/2)
- 44 914.80 / 44 902.42 — x — 2*(5/2, 3/2)
- 39 476.87 / 39 233.44 — y — 1*(3/2, 5/2)
- 32 025.50 / 31 756.50 — z — 0*(3/2, 5/2)

c*nℓ ^2F
- 65 458.65 / 65 312.71 — f — 0*4d(5/2, 7/2)
- 63 444.76 / 63 168.23 — e — 0*5s(5/2, 7/2)
- 20 951.77 / 20 891.88 — b — 3d³(7/2, 5/2)
- 4 897.24 / 4 628.61 — a — 0*4s

c*4p $^2F^o$
- 70 893.00 / 70 606.35 — v — 6*(5/2, 7/2)
- 59 467.81 / 59 321.79 — w — 5*(5/2, 7/2)
- 47 625.17 / 47 466.80 — x — 4*(7/2, 5/2)
- 40 074.71 / 39 926.83 — y — 1*(5/2, 7/2)
- 31 490.82 / 31 207.44 — z — 0*(5/2, 7/2)

c*4ℓ ^2G
- 67 820.87 / 67 604.20 — e — 0*d(7/2, 9/2)
- 15 265.60 / 15 257.53 — b — 4*s(9/2, 7/2)
- 9 118.15 / 8 997.69 — a — 3d³(7/2, 9/2)

c*4p $^2G^o_{7/2,9/2}$
- 43 780.99 / 43 740.77 — y — 4*
- 34 748.50 / 34 543.36 — z — 0*

$^2H_{9/2,11/2}$
- 68 582.34 / 68 323.95 — e — 0*4d
- 12 774.81 / 12 676.99 — a — 3d³

c*4p $^2H^o_{9/2,11/2}$
- 45 908.56 / 45 673.75 — z — 4*

KEY

65 458.65 / 65 312.71 — f — 0*4d(5/2, 7/2)
- J, low to high
- n
- Parent level from Ti III
- Notation from A.E.L.
- Energy levels (cm^{-1})

Ionization Limit
109 506 cm^{-1}
(13.58 eV)
[Ti II $3d^2 4s$ $^4F_{3/2}$ → Ti III $3d^2$ $^3F_2$]

CORES (c*) from Ti III:

0* = 3d² ³F	4* = 3d² ¹G
1* = 3d² ¹D	5* = 3d4s ³D
2* = 3d² ³P	6* = 3d4s ¹D
3* = 3d² ¹S	

NOTE: For proper configuration, please refer to bibliographical comments.

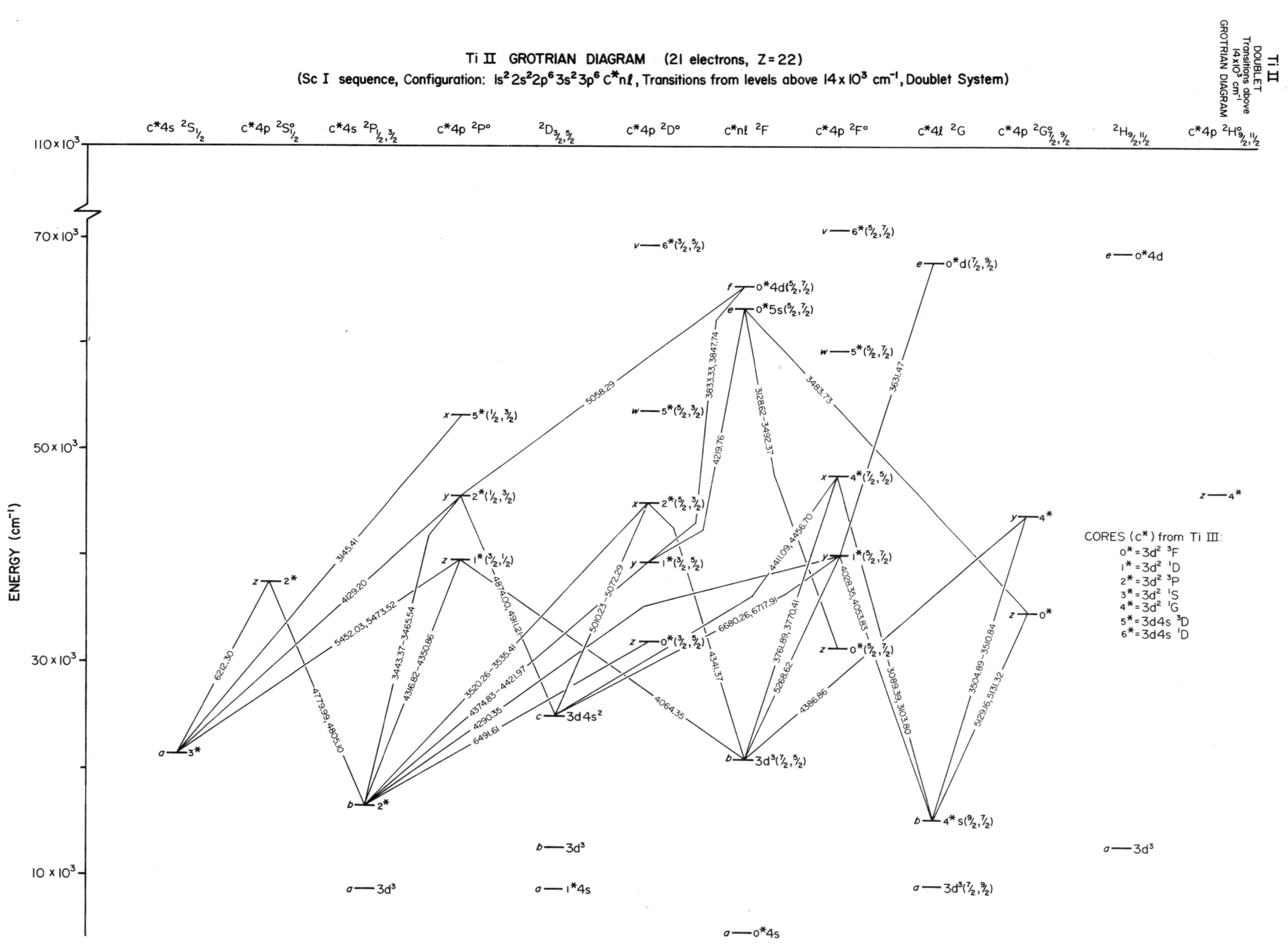
Ti II GROTRIAN DIAGRAM (21 electrons, Z = 22)
(Sc I sequence, Configuration: 1s²2s²2p⁶3s²3p⁶ c*nℓ, Transitions from levels above 14 x 10³ cm⁻¹, Doublet System)
c*4s ²S1/2
c*4p ²S°1/2
c*4s ²P1/2,3/2
c*4p ²P°
²D3/2,5/2
c*4p ²D°
c*nℓ ²F
c*4p ²F°
c*4ℓ ²G
c*4p ²G°7/2,9/2
²H9/2,11/2
c*4p ²H°9/2,11/2
110 x 10³
70 x 10³
50 x 10³
30 x 10³
10 x 10³
ENERGY (cm⁻¹)
CORES (c*) from Ti III:
0* = 3d² ³F
1* = 3d² ¹D
2* = 3d² ³P
3* = 3d² ¹S
4* = 3d² ¹G
5* = 3d4s ³D
6* = 3d4s ¹D
a — 3*
z — 2*
b — 2*
x — 5*(1/2, 3/2)
y — 2*(1/2, 3/2)
z — 1*(3/2, 1/2)
a — 3d³
b — 3d³
a — 1*4s
c — 3d4s²
v — 6*(3/2, 5/2)
w — 5*(5/2, 3/2)
x — 2*(5/2, 3/2)
y — 1*(3/2, 5/2)
z — 0*(3/2, 5/2)
f — 0*4d(5/2, 7/2)
e — 0*5s(5/2, 7/2)
b — 3d³(7/2, 5/2)
a — 0*4s
v — 6*(5/2, 7/2)
w — 5*(5/2, 7/2)
x — 4*(7/2, 5/2)
y — 1*(5/2, 7/2)
z — 0*(5/2, 7/2)
e — 0*d(7/2, 9/2)
b — 4*s(9/2, 7/2)
a — 3d³(7/2, 9/2)
y — 4*
z — 0*
e — 0*4d
a — 3d³
z — 4*
6212.30
4779.99, 4805.10
3145.41
4129.20
5452.03, 5473.52
3443.37–3465.54
4316.82–4350.86
4874.00, 4911.21
5058.29
3520.26–3535.41
4374.83–4421.97
4290.35
6491.61
5010.23–5072.29
4064.35
3833.33, 3847.74
4219.76
3128.62–3492.37
3483.73
4341.37
6680.26, 6717.91
4411.09, 4456.70
3761.89, 3770.41
5268.62
4028.35, 4053.83
4386.86
3631.47
3089.39, 3103.80
3504.89–3510.84
5129.16, 5131.32

Ti II ENERGY LEVELS (21 electrons, $\bar{Z}$ = 22)

(Sc I sequence, Configuration: $1s^2 2s^2 2p^6 3s^2 3p^6\ c^* n\ell$, Doublet System)

Term	Energy levels (cm^{-1})	Notation from A.E.L.	Parent level from Ti III, n, J
$c^*4s\ ^2S_{1/2}$	21 338.00	a	3*
$c^*4p\ ^2S^\circ_{1/2}$	37 430.55	z	2*
$c^*4s\ ^2P_{1/2,3/2}$	16 625.25, 16 515.79	b	2*
	9 975.92, 9 850.90	a	$3d^3$
$c^*4p\ ^2P^\circ$	53 128.17, 53 121.48	x	5*(1/2, 3/2)
	45 548.90, 45 472.89	y	2*(1/2, 3/2)
	39 674.64, 39 602.90	z	1*(3/2, 1/2)
$^2D_{3/2,5/2}$	25 193.04, 24 961.34	c	$3d4s^2$
	12 758.15, 12 628.77	b	$3d^3$
	8 744.27, 8 710.47	a	1*4s
$c^*4p\ ^2D^\circ$	69 622.15, 69 327.32	v	6*(3/2, 5/2)
	53 596.70, 53 554.90	w	5*(5/2, 3/2)
	44 914.80, 44 902.42	x	2*(5/2, 3/2)
	39 476.87, 39 233.44	y	1*(3/2, 5/2)
	32 025.50, 31 756.50	z	0*(3/2, 5/2)
$c^*n\ell\ ^2F$	65 458.65, 65 312.71	f	0*4d(5/2, 7/2)
	63 444.76, 63 168.23	e	0*5s(5/2, 7/2)
	20 951.77, 20 891.88	b	$3d^3$(7/2, 5/2)
	4 897.24, 4 628.61	a	0*4s
$c^*4p\ ^2F^\circ$	70 893.00, 70 606.35	v	6*(5/2, 7/2)
	59 467.81, 59 321.79	w	5*(5/2, 7/2)
	47 625.17, 47 466.80	x	4*(7/2, 5/2)
	40 074.71, 39 926.83	y	1*(5/2, 7/2)
	31 490.82, 31 207.44	z	0*(5/2, 7/2)
$c^*4\ell\ ^2G$	67 820.87, 67 604.20	e	0*d(7/2, 9/2)
	15 265.60, 15 257.53	b	4*s(9/2, 7/2)
	9 118.15, 8 997.69	a	$3d^3$(7/2, 9/2)
$c^*4p\ ^2G^\circ_{7/2,9/2}$	43 780.99, 43 740.77	y	4*
	34 748.50, 34 543.36	z	0*
$^2H_{9/2,11/2}$	68 582.34, 68 328.95	e	0*4d
	12 774.81, 12 676.99	a	$3d^3$
$c^*4p\ ^2H^\circ_{9/2,11/2}$	45 908.56, 45 673.75	z	4*

ENERGY (cm^{-1}): 110×10^3, 70×10^3, 50×10^3, 30×10^3, 10×10^3

KEY

65 458.65, 65 312.71 — f — 0*4d(5/2, 7/2)

- J, low to high
- n
- Parent level from Ti III
- Notation from A.E.L.
- Energy levels (cm^{-1})

Ionization Limit
109 506 cm^{-1}
(13.58 eV)
[Ti II $3d^2 4s\ ^4F_{3/2}$ → Ti III $3d^2\ ^3F_2$]

CORES (c*) from Ti III:

0* = $3d^2\ ^3F$	4* = $3d^2\ ^1G$
1* = $3d^2\ ^1D$	5* = $3d4s\ ^3D$
2* = $3d^2\ ^3P$	6* = $3d4s\ ^1D$
3* = $3d^2\ ^1S$	

NOTE: For proper configuration, please refer to bibliographical comments.

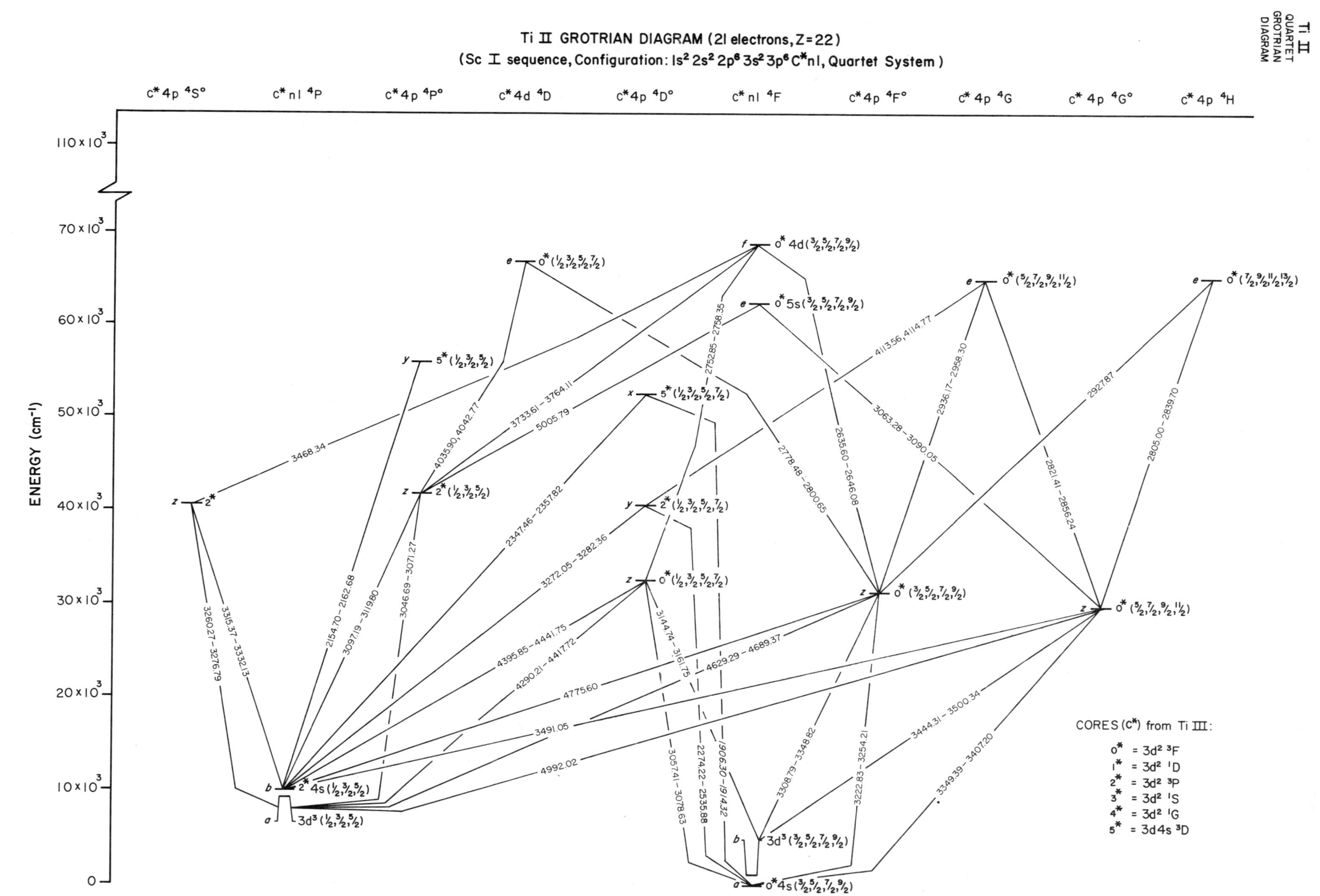
Ti II GROTRIAN DIAGRAM (21 electrons, Z=22)
(Sc I sequence, Configuration: 1s² 2s² 2p⁶ 3s² 3p⁶ C*nl, Quartet System)
c* 4p ⁴S°
c* nl ⁴P
c* 4p ⁴P°
c* 4d ⁴D
c* 4p ⁴D°
c* nl ⁴F
c* 4p ⁴F°
c* 4p ⁴G
c* 4p ⁴G°
c* 4p ⁴H
ENERGY (cm⁻¹)
110×10³
70×10³
60×10³
50×10³
40×10³
30×10³
20×10³
10×10³
0
z 2*
b 2* 4s (1/2, 3/2, 5/2)
a 3d³ (1/2, 3/2, 5/2)
y 5* (1/2, 3/2, 5/2)
z 2* (1/2, 3/2, 5/2)
e o* (1/2, 3/2, 5/2, 7/2)
x 5* (1/2, 3/2, 5/2, 7/2)
y 2* (1/2, 3/2, 5/2, 7/2)
z o* (1/2, 3/2, 5/2, 7/2)
f o* 4d (3/2, 5/2, 7/2, 9/2)
e o* 5s (3/2, 5/2, 7/2, 9/2)
b 3d³ (3/2, 5/2, 7/2, 9/2)
a o* 4s (3/2, 5/2, 7/2, 9/2)
z o* (3/2, 5/2, 7/2, 9/2)
e o* (5/2, 7/2, 9/2, 11/2)
z o* (5/2, 7/2, 9/2, 11/2)
e o* (7/2, 9/2, 11/2, 13/2)
3468.34
3260.27 – 3276.79
3315.37 – 3332.13
2154.70 – 2162.68
3097.19 – 3119.80
3046.69 – 3071.27
4035.90, 4042.77
3733.61 – 3764.11
5005.79
2347.46 – 2357.82
3272.05 – 3282.36
4395.85 – 4441.75
4290.21 – 4417.72
4775.60
3491.05
4992.02
3144.74 – 3161.75
3057.41 – 3078.63
2274.22 – 2535.88
1906.30 – 1914.32
2752.85 – 2758.35
4629.29 – 4689.37
2778.48 – 2800.65
2635.60 – 2646.08
4113.56, 4114.77
3063.28 – 3090.05
2936.17 – 2958.30
3308.79 – 3348.82
3222.83 – 3254.21
2927.87
2821.41 – 2856.24
2805.00 – 2839.70
3444.31 – 3500.34
3349.39 – 3407.20
CORES (C*) from Ti III:
o* = 3d² ³F
1* = 3d² ¹D
2* = 3d² ³P
3* = 3d² ¹S
4* = 3d² ¹G
5* = 3d4s ³D

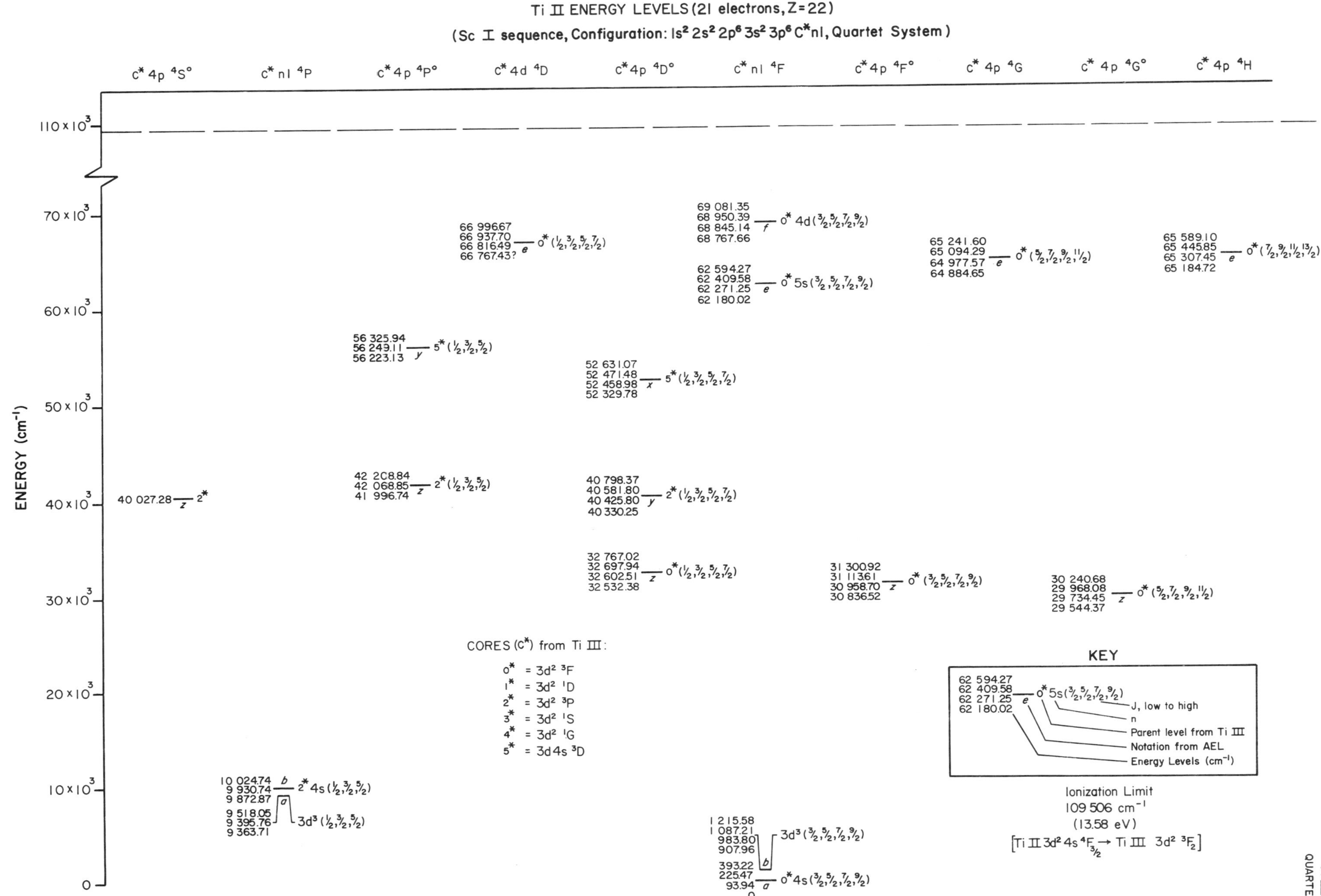
Ti II ENERGY LEVELS (21 electrons, Z=22)
(Sc I sequence, Configuration: 1s² 2s² 2p⁶ 3s² 3p⁶ C*nl, Quartet System)
c* 4p ⁴S°
c* nl ⁴P
c* 4p ⁴P°
c* 4d ⁴D
c* 4p ⁴D°
c* nl ⁴F
c* 4p ⁴F°
c* 4p ⁴G
c* 4p ⁴G°
c* 4p ⁴H
ENERGY (cm⁻¹)
110×10³
70×10³
60×10³
50×10³
40×10³
30×10³
20×10³
10×10³
0
40 027.28 z 2*
10 024.74 9 930.74 9 872.87 b 2* 4s (1/2, 3/2, 5/2)
9 518.05 9 395.76 9 363.71 a 3d³ (1/2, 3/2, 5/2)
56 325.94 56 249.11 56 223.13 y 5* (1/2, 3/2, 5/2)
42 208.84 42 068.85 41 996.74 z 2* (1/2, 3/2, 5/2)
66 996.67 66 937.70 66 816.49 66 767.43? e o* (1/2, 3/2, 5/2, 7/2)
52 631.07 52 471.48 52 458.98 52 329.78 x 5* (1/2, 3/2, 5/2, 7/2)
40 798.37 40 581.80 40 425.80 40 330.25 y 2* (1/2, 3/2, 5/2, 7/2)
32 767.02 32 697.94 32 602.51 32 532.38 z o* (1/2, 3/2, 5/2, 7/2)
69 081.35 68 950.39 68 845.14 68 767.66 f o* 4d (3/2, 5/2, 7/2, 9/2)
62 594.27 62 409.58 62 271.25 62 180.02 e o* 5s (3/2, 5/2, 7/2, 9/2)
1 215.58 1 087.21 983.80 907.96 b 3d³ (3/2, 5/2, 7/2, 9/2)
393.22 225.47 93.94 0 a o* 4s (3/2, 5/2, 7/2, 9/2)
31 300.92 31 113.61 30 958.70 30 836.52 z o* (3/2, 5/2, 7/2, 9/2)
65 241.60 65 094.29 64 977.57 64 884.65 e o* (5/2, 7/2, 9/2, 11/2)
30 240.68 29 968.08 29 734.45 29 544.37 z o* (5/2, 7/2, 9/2, 11/2)
65 589.10 65 445.85 65 307.45 65 184.72 e o* (7/2, 9/2, 11/2, 13/2)
CORES (c*) from Ti III:
o* = 3d² ³F
1* = 3d² ¹D
2* = 3d² ³P
3* = 3d² ¹S
4* = 3d² ¹G
5* = 3d4s ³D
KEY
62 594.27 62 409.58 62 271.25 62 180.02 e o* 5s (3/2, 5/2, 7/2, 9/2)
J, low to high
n
Parent level from Ti III
Notation from AEL
Energy Levels (cm⁻¹)
Ionization Limit
109 506 cm⁻¹
(13.58 eV)
[Ti II 3d² 4s ⁴F3/2 → Ti III 3d² ³F2]

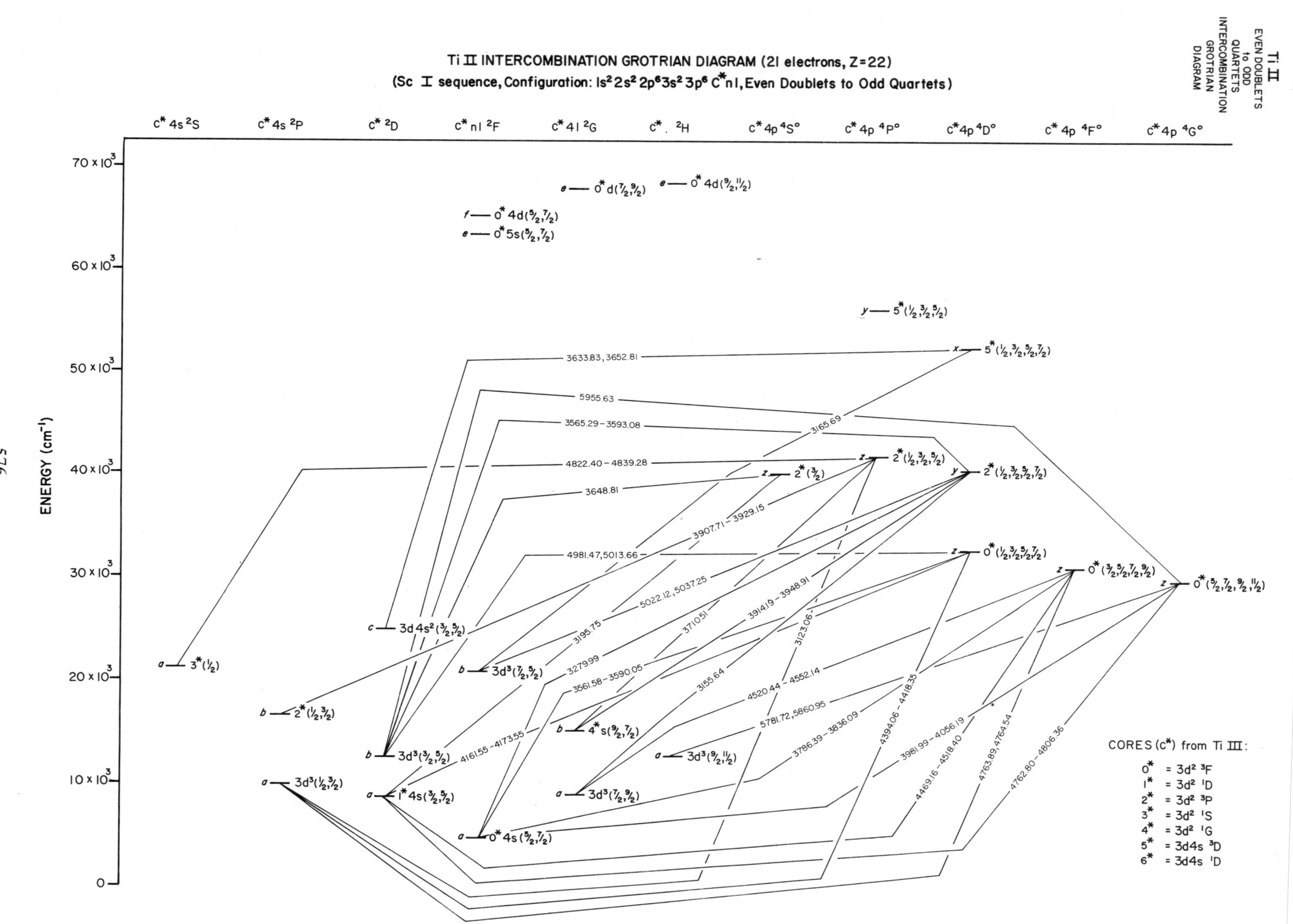

Ti II
EVEN DOUBLETS to ODD QUARTETS INTERCOMBINATION GROTRIAN DIAGRAM
Ti II INTERCOMBINATION GROTRIAN DIAGRAM (21 electrons, Z=22)
(Sc I sequence, Configuration: 1s² 2s² 2p⁶ 3s² 3p⁶ C* nl, Even Doublets to Odd Quartets)
c* 4s ²S
c* 4s ²P
c* ²D
c* nl ²F
c* 4l ²G
c* . ²H
c* 4p ⁴S°
c* 4p ⁴P°
c* 4p ⁴D°
c* 4p ⁴F°
c* 4p ⁴G°
ENERGY (cm⁻¹)
70 x 10³
60 x 10³
50 x 10³
40 x 10³
30 x 10³
20 x 10³
10 x 10³
0
e — 0* d(7/2, 9/2)
e — 0* 4d(9/2, 11/2)
f — 0* 4d(5/2, 7/2)
e — 0* 5s(5/2, 7/2)
y — 5*(1/2, 3/2, 5/2)
x — 5*(1/2, 3/2, 5/2, 7/2)
z — 2*(1/2, 3/2, 5/2)
z — 2*(3/2)
y — 2*(1/2, 3/2, 5/2, 7/2)
z — 0*(1/2, 3/2, 5/2, 7/2)
z — 0*(3/2, 5/2, 7/2, 9/2)
z — 0*(5/2, 7/2, 9/2, 11/2)
c — 3d4s²(3/2, 5/2)
a — 3*(1/2)
b — 3d³(7/2, 5/2)
b — 2*(1/2, 3/2)
b — 4*s(9/2, 7/2)
a — 3d³(9/2, 11/2)
b — 3d³(3/2, 5/2)
a — 3d³(1/2, 3/2)
a — 1* 4s(3/2, 5/2)
a — 3d³(7/2, 9/2)
a — 0* 4s(5/2, 7/2)
3633.83, 3652.81
5955.63
3565.29 – 3593.08
4822.40 – 4839.28
3648.81
3165.69
3907.71 – 3929.15
4981.47, 5013.66
5022.12, 5037.25
3195.75
3710.51
3914.19 – 3948.91
3123.06
3279.99
3561.58 – 3590.05
3155.64
4520.44 – 4552.14
4161.55 – 4173.55
5781.72, 5860.95
3786.39 – 3836.09
4394.06 – 4418.35
3981.99 – 4056.19
4469.16 – 4518.40
4763.89, 4764.54
4762.80 – 4806.36
CORES (c*) from Ti III:
0* = 3d² ³F
1* = 3d² ¹D
2* = 3d² ³P
3* = 3d² ¹S
4* = 3d² ¹G
5* = 3d4s ³D
6* = 3d4s ¹D

Ti II INTERCOMBINATION GROTRIAN DIAGRAM (21 electrons, Z=22)

(Sc I sequence, Configuration: $1s^2 2s^2 2p^6 3s^2 3p^6 c^* nl$, Odd Doublets to Even Quartets)

Ti II
ODD DOUBLETS
to EVEN
QUARTETS
INTERCOMBINATION
GROTRIAN
DIAGRAM

$c^* 4p\ ^2S^o$ | $c^* 4p\ ^2P^o$ | $c^* 4p\ ^2D^o$ | $c^* 4p\ ^2F^o$ | $c^* 4p\ ^2G^o$ | $c^* 4p\ ^2H^o$ | $c^* nl\ ^4P$ | $c^* 4d\ ^4D$ | $c^* nl\ ^4F$ | $c^* 4p\ ^4G$ | $c^* 4p\ ^4H$

ENERGY (cm^{-1})

70×10^3, 60×10^3, 50×10^3, 40×10^3, 30×10^3, 20×10^3, 10×10^3, 0

z — $2^*(1/2)$

x — $5^*(1/2, 3/2)$
y — $2^*(1/2, 3/2)$
z — $1^*(3/2, 1/2)$

v — $6^*(3/2, 5/2)$
w — $5^*(5/2, 3/2)$
x — $2^*(5/2, 3/2)$
y — $1^*(3/2, 5/2)$
z — $0^*(3/2, 5/2)$

v — $6^*(5/2, 7/2)$
w — $5^*(5/2, 7/2)$
x — $4^*(7/2, 5/2)$
y — $1^*(5/2, 7/2)$
z — $0^*(5/2, 7/2)$

y — $4^*(7/2, 9/2)$
z — $0^*(7/2, 9/2)$

z — $4^*(9/2, 11/2)$

e — $0^*(1/2, 3/2, 5/2, 7/2)$

f — $0^* 4d(3/2, 5/2, 7/2, 9/2)$
e — $0^* 5s(3/2, 5/2, 7/2, 9/2)$

e — $0^*(5/2, 7/2, 9/2, 11/2)$

e — $0^*(7/2, 9/2, 11/2, 13/2)$

4700.67
5003.06
5755.89
3657.32, 3690.00
5792.26
3897.36
3693.07 – 3721.50
4298.81, 4309.68
3951.35 – 3982.26
5908.33, 5973.62
5178.91
3995.74, 4029.48
3249.65
3958.07 – 3981.31
3614.19
3214.18 – 3233.31
3888.74
4442.13

CORES (c^*) from Ti III:

$0^* = 3d^2\ ^3F$
$1^* = 3d^2\ ^1D$
$2^* = 3d^2\ ^3P$
$3^* = 3d^2\ ^1S$
$4^* = 3d^2\ ^1G$
$5^* = 3d4s\ ^3D$
$6^* = 3d4s\ ^1D$

b — $2^* 4s(1/2, 3/2, 5/2)$
a — $3d^3(1/2, 3/2, 5/2)$

b — $3d^3(3/2, 5/2, 7/2, 9/2)$
a — $0^* 4s(3/2, 5/2, 7/2, 9/2)$

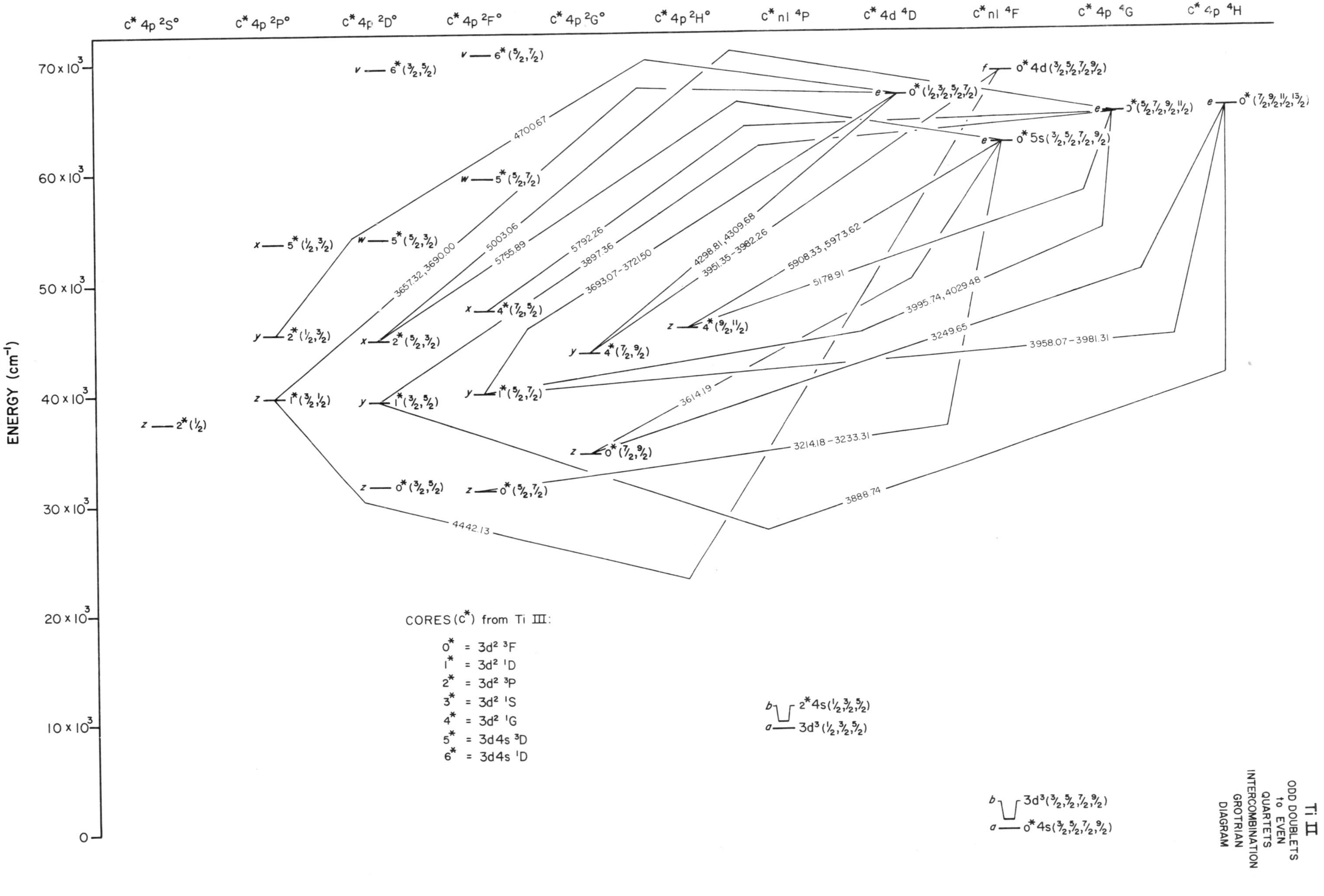

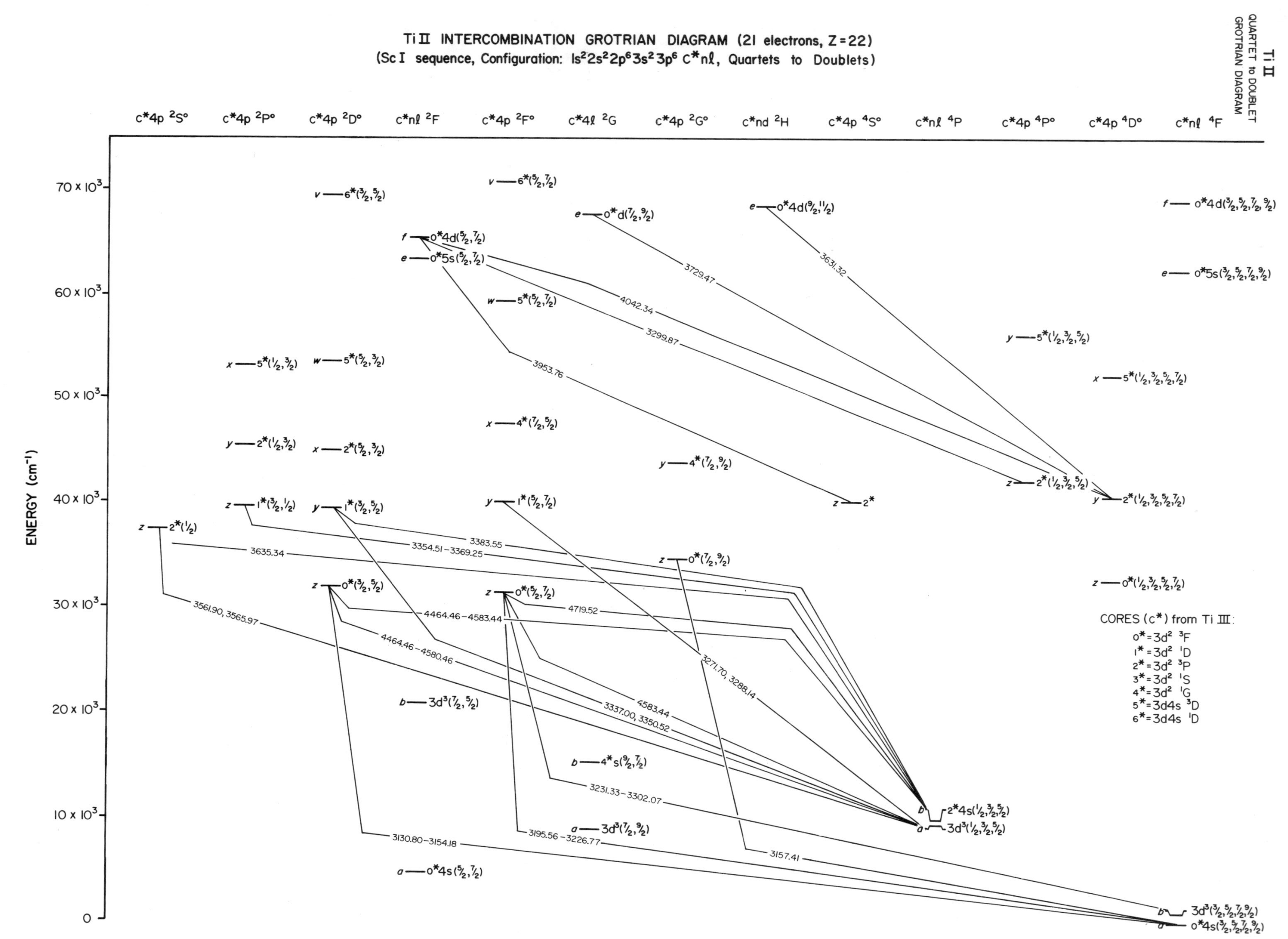
Ti II INTERCOMBINATION GROTRIAN DIAGRAM (21 electrons, Z=22)
(Sc I sequence, Configuration: 1s²2s²2p⁶3s²3p⁶ c*nℓ, Quartets to Doublets)
c*4p ²S°
c*4p ²P°
c*4p ²D°
c*nℓ ²F
c*4p ²F°
c*4ℓ ²G
c*4p ²G°
c*nd ²H
c*4p ⁴S°
c*nℓ ⁴P
c*4p ⁴P°
c*4p ⁴D°
c*nℓ ⁴F
ENERGY (cm⁻¹)
70 x 10³
60 x 10³
50 x 10³
40 x 10³
30 x 10³
20 x 10³
10 x 10³
0
v — 6*(3/2,5/2)
v — 6*(5/2,7/2)
e — o*d(7/2,9/2)
e — o*4d(9/2,11/2)
f — o*4d(3/2,5/2,7/2,9/2)
f — o*4d(5/2,7/2)
e — o*5s(5/2,7/2)
e — o*5s(3/2,5/2,7/2,9/2)
w — 5*(5/2,7/2)
y — 5*(1/2,3/2,5/2)
x — 5*(1/2,3/2)
w — 5*(5/2,3/2)
x — 5*(1/2,3/2,5/2,7/2)
x — 4*(7/2,5/2)
y — 2*(1/2,3/2)
x — 2*(5/2,3/2)
y — 4*(7/2,9/2)
z — 2*(1/2,3/2,5/2)
y — 2*(1/2,3/2,5/2,7/2)
z — 2*
z — 1*(3/2,1/2)
y — 1*(3/2,5/2)
y — 1*(5/2,7/2)
z — 2*(1/2)
z — o*(7/2,9/2)
z — o*(3/2,5/2)
z — o*(5/2,7/2)
z — o*(1/2,3/2,5/2,7/2)
b — 3d³(7/2,5/2)
b — 4*s(9/2,7/2)
a — 3d³(7/2,9/2)
a — o*4s(5/2,7/2)
b 2*4s(1/2,3/2,5/2)
a 3d³(1/2,3/2,5/2)
b 3d³(3/2,5/2,7/2,9/2)
a o*4s(3/2,5/2,7/2,9/2)
3729.47
3631.32
4042.34
3299.87
3953.76
3383.55
3354.51 – 3369.25
3635.34
4719.52
4464.46 – 4583.44
4464.46 – 4580.46
3561.90, 3565.97
3271.70, 3288.14
4583.44
3337.00, 3350.52
3231.33 – 3302.07
3195.56 – 3226.77
3130.80 – 3154.18
3157.41
CORES (c*) from Ti III:
o* = 3d² ³F
1* = 3d² ¹D
2* = 3d² ³P
3* = 3d² ¹S
4* = 3d² ¹G
5* = 3d4s ³D
6* = 3d4s ¹D

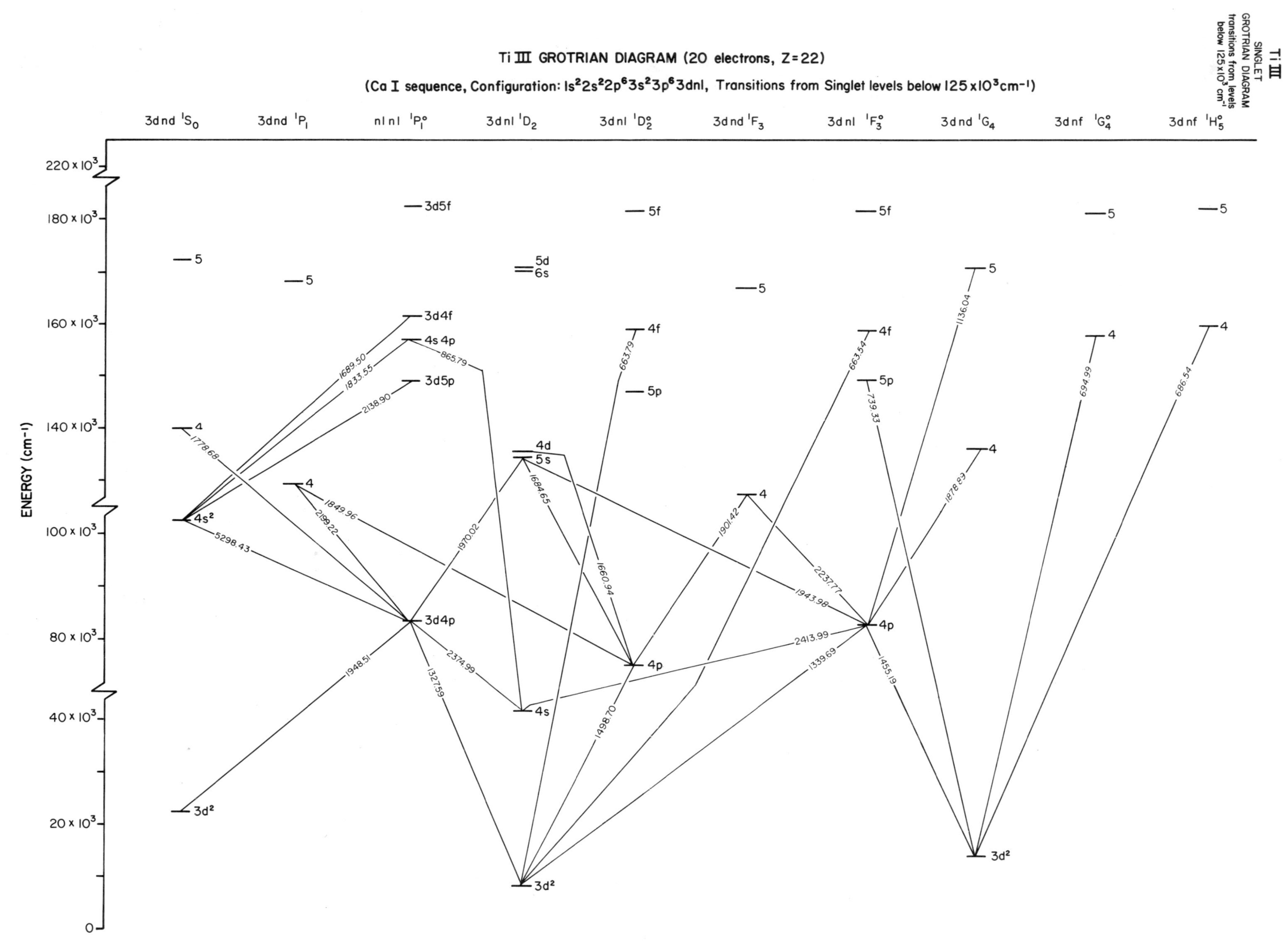

Ti III
SINGLET
GROTRIAN DIAGRAM
transitions from levels below 125×10^3 cm^{-1}
Ti III GROTRIAN DIAGRAM (20 electrons, Z=22)
(Ca I sequence, Configuration: $1s^2 2s^2 2p^6 3s^2 3p^6 3dnl$, Transitions from Singlet levels below 125×10^3 cm^{-1})
3dnd $^1S_0$
3dnd $^1P_1$
nlnl $^1P^o_1$
3dnl $^1D_2$
3dnl $^1D^o_2$
3dnd $^1F_3$
3dnl $^1F^o_3$
3dnd $^1G_4$
3dnf $^1G^o_4$
3dnf $^1H^o_5$
ENERGY (cm^{-1})
220×10^3
180×10^3
160×10^3
140×10^3
100×10^3
80×10^3
40×10^3
20×10^3
0
$3d^2$
$4s^2$
3d4p
3d4f
4s 4p
3d5p
3d5f
4s
4p
4d
5s
5d
6s
4f
5f
5p
686.54
694.99
1136.04
1878.83
739.33
1455.19
663.54
2237.77
1943.98
2413.99
1339.69
1901.42
663.79
1660.94
1498.70
1684.65
865.79
1970.02
2374.99
1327.59
1948.51
1689.50
1833.55
2138.90
1849.96
2199.22
1778.68
5298.43

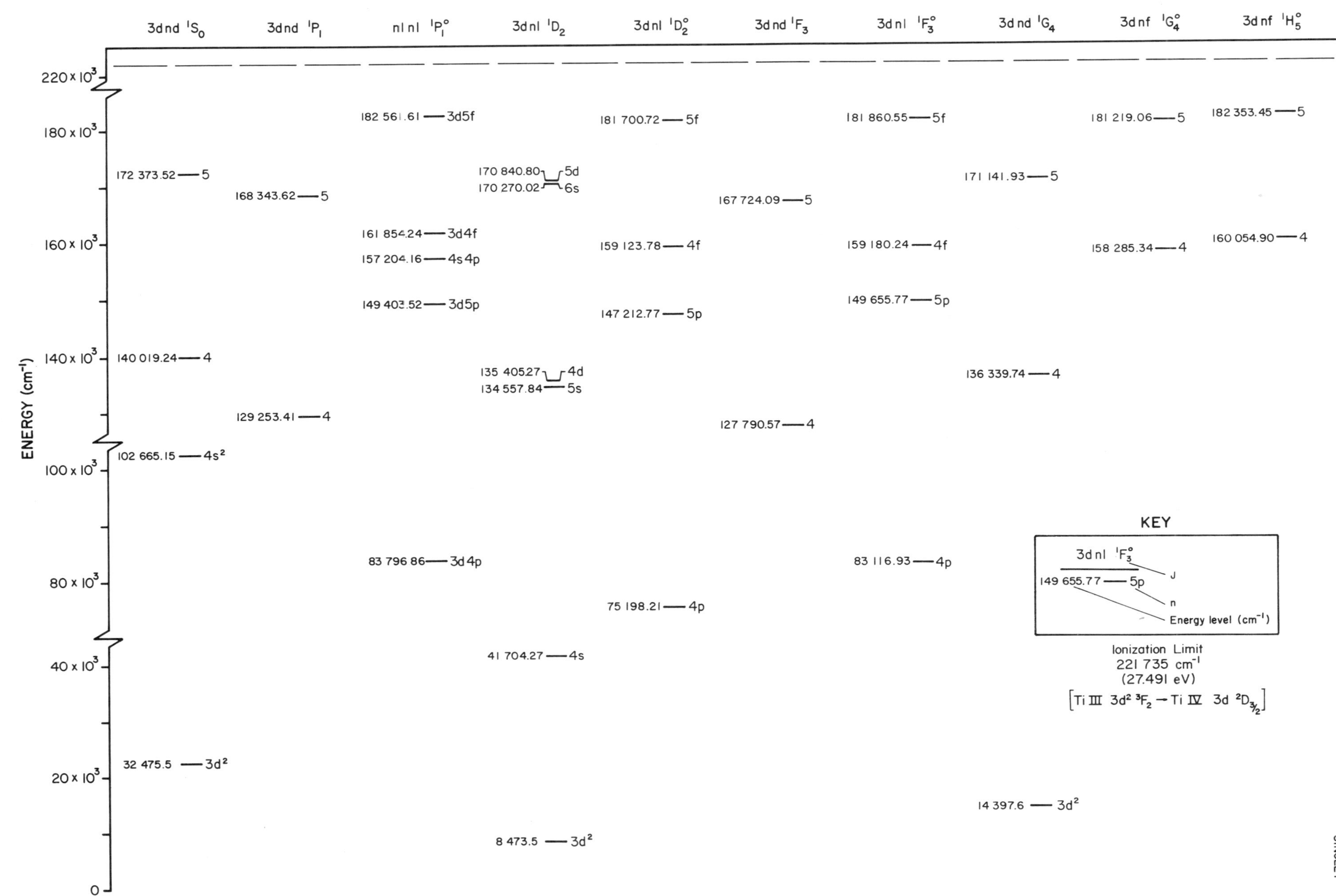
Ti III ENERGY LEVELS (20 electrons, Z=22)
(Ca I sequence, Configuration: $1s^2 2s^2 2p^6 3s^2 3p^6 3dnl$, Singlet System)
Ti III
SINGLET
3dnd $^1S_0$
3dnd $^1P_1$
nlnl $^1P^o_1$
3dnl $^1D_2$
3dnl $^1D^o_2$
3dnd $^1F_3$
3dnl $^1F^o_3$
3dnd $^1G_4$
3dnf $^1G^o_4$
3dnf $^1H^o_5$
ENERGY (cm^{-1})
220 x 10^3
180 x 10^3
160 x 10^3
140 x 10^3
100 x 10^3
80 x 10^3
40 x 10^3
20 x 10^3
0
172 373.52 — 5
140 019.24 — 4
102 665.15 — $4s^2$
32 475.5 — $3d^2$
168 343.62 — 5
129 253.41 — 4
182 561.61 — 3d5f
161 854.24 — 3d4f
157 204.16 — 4s4p
149 403.52 — 3d5p
83 796 86 — 3d4p
170 840.80 — 5d
170 270.02 — 6s
135 405.27 — 4d
134 557.84 — 5s
41 704.27 — 4s
8 473.5 — $3d^2$
181 700.72 — 5f
159 123.78 — 4f
147 212.77 — 5p
75 198.21 — 4p
167 724.09 — 5
127 790.57 — 4
181 860.55 — 5f
159 180.24 — 4f
149 655.77 — 5p
83 116.93 — 4p
171 141.93 — 5
136 339.74 — 4
14 397.6 — $3d^2$
181 219.06 — 5
158 285.34 — 4
182 353.45 — 5
160 054.90 — 4
KEY
3dnl $^1F^o_3$
149 655.77 — 5p
J
n
Energy level (cm^{-1})
Ionization Limit
221 735 cm^{-1}
(27.491 eV)
[Ti III $3d^2\ ^3F_2$ → Ti IV $3d\ ^2D_{3/2}$]

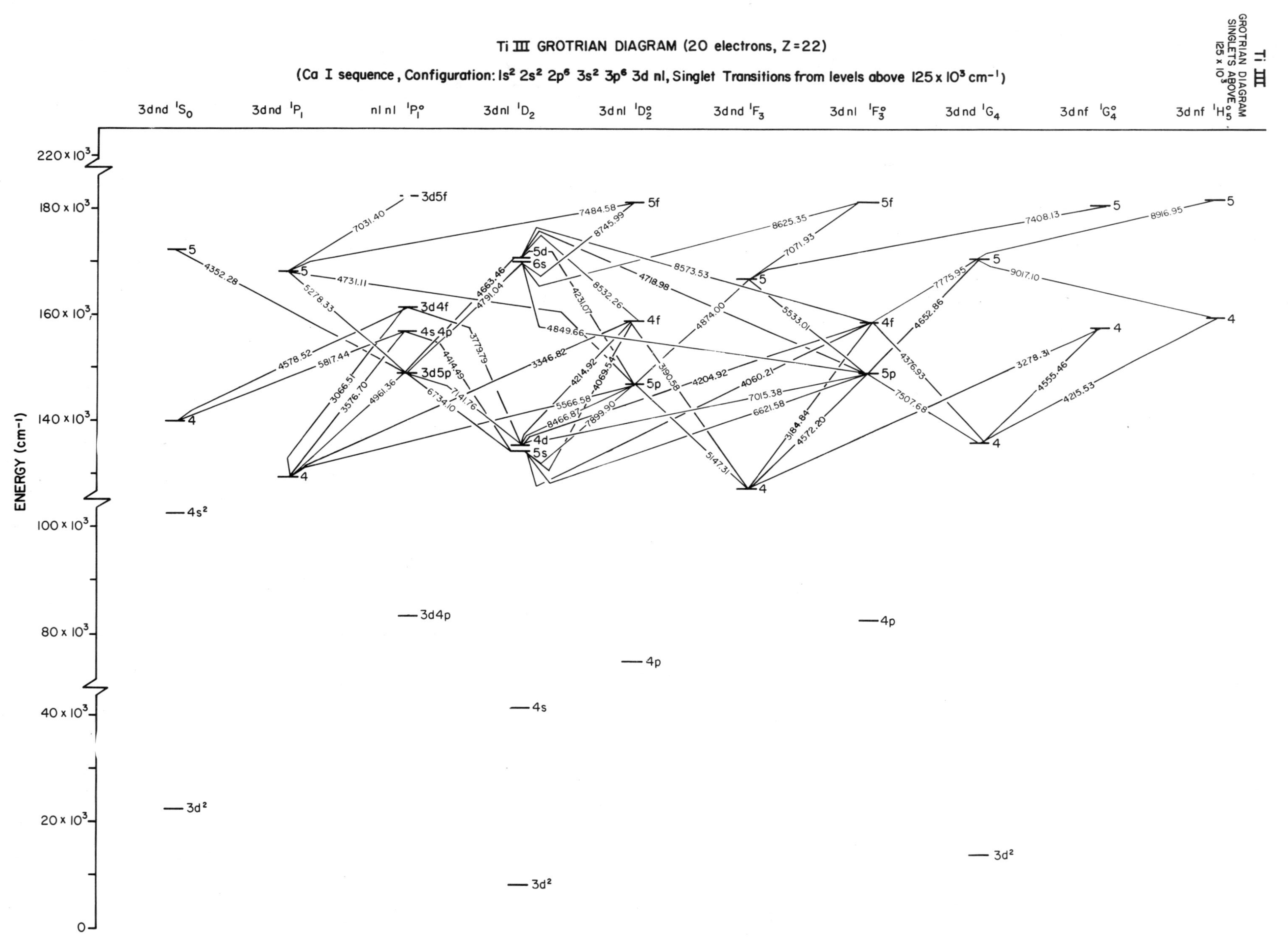
Ti III
GROTRIAN DIAGRAM
SINGLETS ABOVE
125 x 10³
Ti III GROTRIAN DIAGRAM (20 electrons, Z=22)
(Ca I sequence, Configuration: 1s² 2s² 2p⁶ 3s² 3p⁶ 3d nl, Singlet Transitions from levels above 125 x 10³ cm⁻¹)
3dnd ¹S₀
3dnd ¹P₁
nlnl ¹P°₁
3dnl ¹D₂
3dnl ¹D°₂
3dnd ¹F₃
3dnl ¹F°₃
3dnd ¹G₄
3dnf ¹G°₄
3dnf ¹H°₅
ENERGY (cm⁻¹)
220 x 10³
180 x 10³
160 x 10³
140 x 10³
100 x 10³
80 x 10³
40 x 10³
20 x 10³
0
4s²
3d²
3d5f
3d4f
4s4p
3d5p
3d4p
5d
6s
4d
5s
4s
5f
4f
5p
4p
4352.28
7031.40
4731.11
5278.33
4578.52
5817.44
3066.51
3576.70
4961.36
3779.79
4414.49
6734.10
7141.76
4663.46
4791.04
7484.58
8745.99
4231.07
8532.26
4849.66
3346.82
4214.92
4069.54
5566.58
8466.87
7899.90
3190.58
5147.31
4718.98
4874.00
4204.92
8625.35
7071.93
8573.53
5533.01
4060.21
7015.38
6621.58
3184.84
4572.20
7775.95
4652.86
4376.93
7507.68
9017.10
7408.13
8916.95
3278.31
4555.46
4215.53

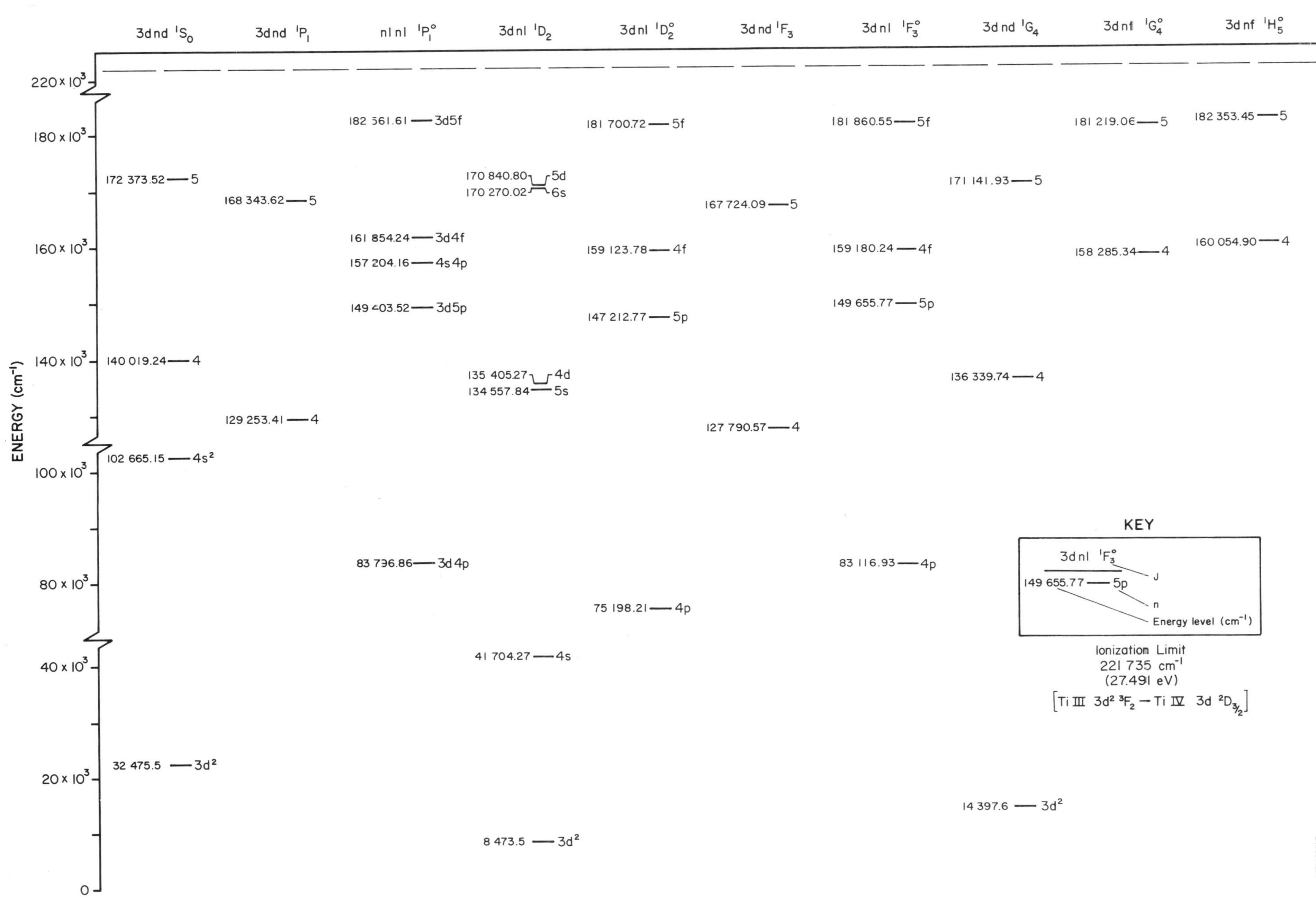
Ti III ENERGY LEVELS (20 electrons, Z=22)
(Ca I sequence, Configuration: $1s^2 2s^2 2p^6 3s^2 3p^6 3dnl$, Singlet System)
3dnd $^1S_0$
3dnd $^1P_1$
nlnl $^1P^o_1$
3dnl $^1D_2$
3dnl $^1D^o_2$
3dnd $^1F_3$
3dnl $^1F^o_3$
3dnd $^1G_4$
3dnf $^1G^o_4$
3dnf $^1H^o_5$
172 373.52 — 5
140 019.24 — 4
102 665.15 — $4s^2$
32 475.5 — $3d^2$
168 343.62 — 5
129 253.41 — 4
182 561.61 — 3d5f
161 854.24 — 3d4f
157 204.16 — 4s4p
149 403.52 — 3d5p
83 736.86 — 3d4p
170 840.80 — 5d
170 270.02 — 6s
135 405.27 — 4d
134 557.84 — 5s
41 704.27 — 4s
8 473.5 — $3d^2$
181 700.72 — 5f
159 123.78 — 4f
147 212.77 — 5p
75 198.21 — 4p
167 724.09 — 5
127 790.57 — 4
181 860.55 — 5f
159 180.24 — 4f
149 655.77 — 5p
83 116.93 — 4p
171 141.93 — 5
136 339.74 — 4
14 397.6 — $3d^2$
181 219.06 — 5
158 285.34 — 4
182 353.45 — 5
160 054.90 — 4
KEY
3dnl $^1F^o_3$
149 655.77 — 5p
J
n
Energy level (cm^{-1})
Ionization Limit
221 735 cm^{-1}
(27.491 eV)
[Ti III $3d^2\ ^3F_2$ — Ti IV $3d\ ^2D_{3/2}$]
ENERGY (cm^{-1})
220×10^3
180×10^3
160×10^3
140×10^3
100×10^3
80×10^3
40×10^3
20×10^3
0
Ti III
SINGLET

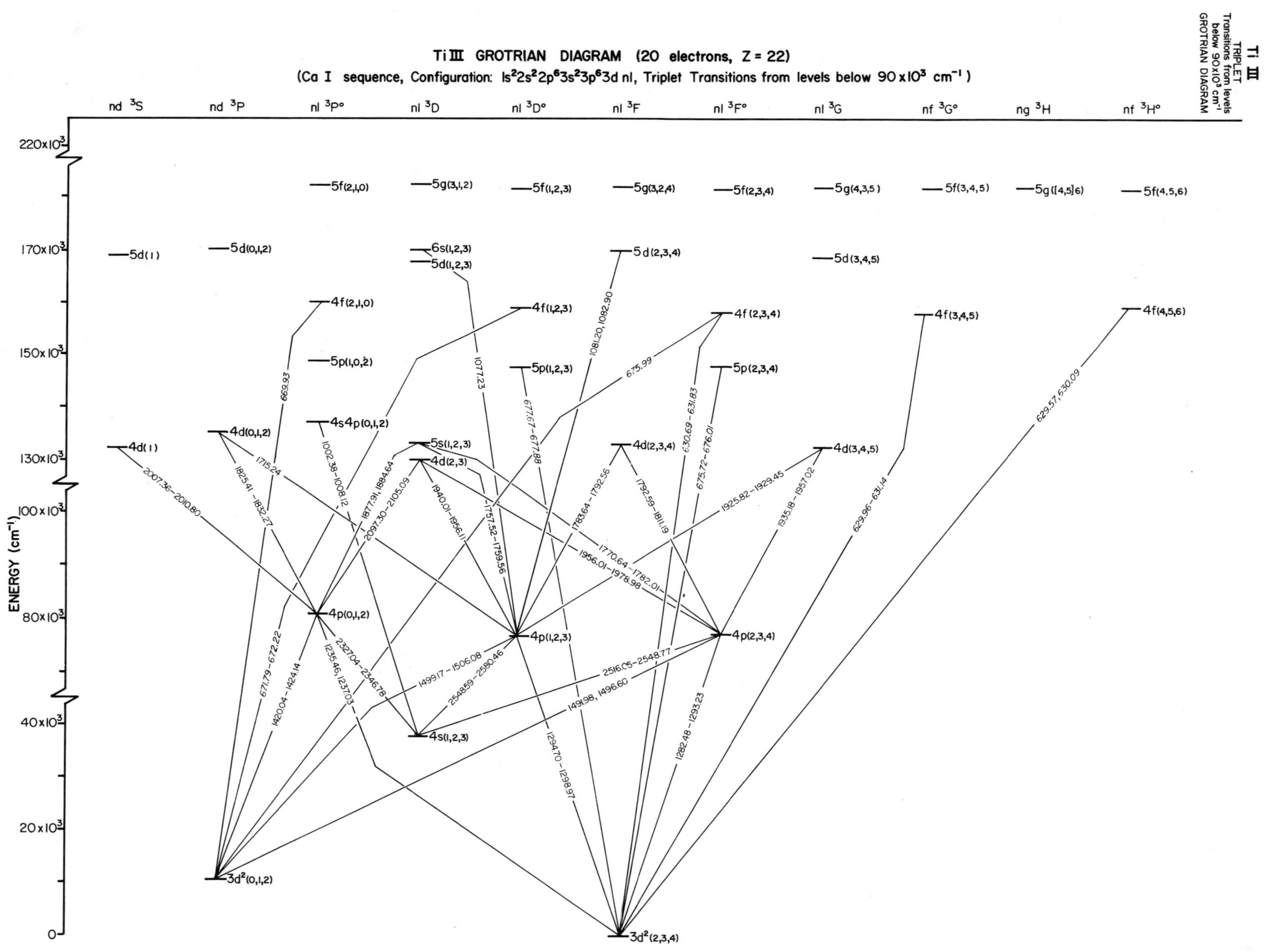

Ti III
TRIPLET
Transitions from levels below 90x10³ cm⁻¹
GROTRIAN DIAGRAM
Ti III GROTRIAN DIAGRAM (20 electrons, Z = 22)
(Ca I sequence, Configuration: 1s²2s²2p⁶3s²3p⁶3d nl, Triplet Transitions from levels below 90x10³ cm⁻¹)
nd ³S
nd ³P
nl ³P°
nl ³D
nl ³D°
nl ³F
nl ³F°
nl ³G
nf ³G°
ng ³H
nf ³H°
5f(2,1,0)
5g(3,1,2)
5f(1,2,3)
5g(3,2,4)
5f(2,3,4)
5g(4,3,5)
5f(3,4,5)
5g([4,5]6)
5f(4,5,6)
5d(1)
5d(0,1,2)
6s(1,2,3)
5d(1,2,3)
5d(2,3,4)
5d(3,4,5)
4f(2,1,0)
4f(1,2,3)
4f(2,3,4)
4f(3,4,5)
4f(4,5,6)
5p(1,0,2)
5p(1,2,3)
5p(2,3,4)
4s4p(0,1,2)
4d(1)
4d(0,1,2)
5s(1,2,3)
4d(2,3)
4d(2,3,4)
4d(3,4,5)
4p(0,1,2)
4p(1,2,3)
4p(2,3,4)
4s(1,2,3)
3d²(0,1,2)
3d²(2,3,4)
669.93
1077.23
1081.20, 1082.90
675.99
630.69–631.83
677.67–677.88
675.72–676.01
629.57, 630.09
1715.24
1002.38–1008.12
1877.91, 1884.64
2097.30–2105.09
1940.01–1956.11
1757.52–1759.56
1783.64–1792.56
1792.59–1811.19
1925.82–1929.45
1935.18–1957.02
629.96–631.14
2007.36–2010.80
1825.41–1832.27
1770.64–1782.01
1956.01–1978.98
671.79–672.22
1420.04–1424.14
2327.04–2346.78
1235.46, 1237.03
1499.17–1506.08
2548.59–2580.46
2516.05–2548.77
1491.98, 1496.60
1294.70–1298.97
1282.48–1293.23
ENERGY (cm⁻¹)
220x10³
170x10³
150x10³
130x10³
100x10³
80x10³
40x10³
20x10³
0

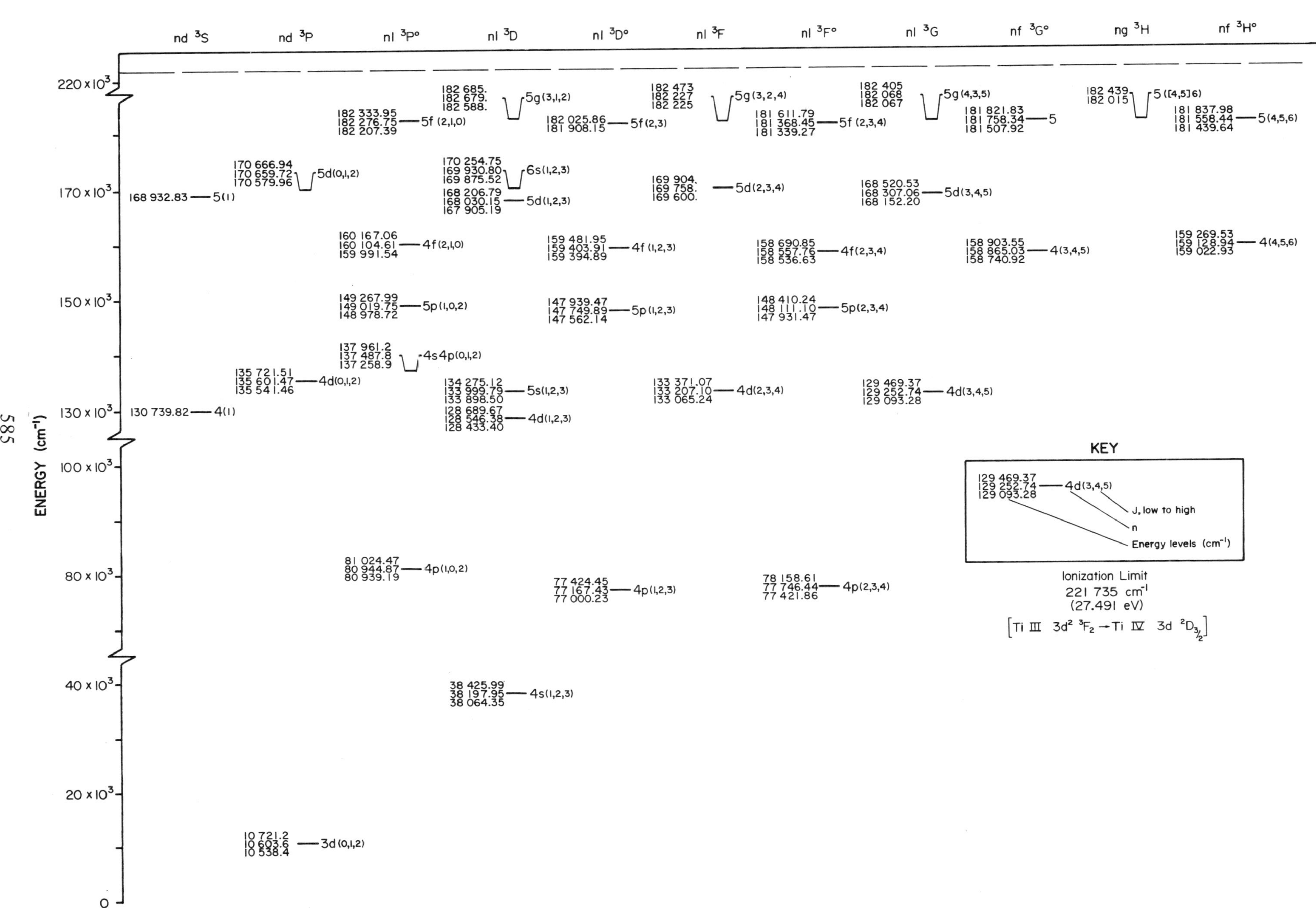
Ti III ENERGY LEVELS (20 electrons, Z=22)
(Ca I sequence, Configuration: $1s^2 2s^2 2p^6 3s^2 3p^6 3d\,nl$, Triplet System)
nd ^{3}S
nd ^{3}P
nl ^{3}P°
nl ^{3}D
nl ^{3}D°
nl ^{3}F
nl ^{3}F°
nl ^{3}G
nf ^{3}G°
ng ^{3}H
nf ^{3}H°
ENERGY (cm^{-1})
220 x 10^3
170 x 10^3
150 x 10^3
130 x 10^3
100 x 10^3
80 x 10^3
40 x 10^3
20 x 10^3
0
168 932.83 — 5(1)
130 739.82 — 4(1)
170 666.94
170 659.72
170 579.96 — 5d(0,1,2)
135 721.51
135 601.47
135 541.46 — 4d(0,1,2)
10 721.2
10 603.6
10 538.4 — 3d(0,1,2)
182 333.95
182 276.75
182 207.39 — 5f (2,1,0)
160 167.06
160 104.61
159 991.54 — 4f (2,1,0)
149 267.99
149 019.75
148 978.72 — 5p (1,0,2)
137 961.2
137 487.8
137 258.9 — 4s4p(0,1,2)
81 024.47
80 944.87
80 939.19 — 4p (1,0,2)
182 685.
182 679.
182 588. — 5g (3,1,2)
170 254.75
169 930.80
169 875.52 — 6s (1,2,3)
168 206.79
168 030.15
167 905.19 — 5d (1,2,3)
134 275.12
133 999.79
133 898.50 — 5s (1,2,3)
128 689.67
128 546.38
128 433.40 — 4d (1,2,3)
38 425.99
38 197.95
38 064.35 — 4s (1,2,3)
182 025.86
181 908.15 — 5f (2,3)
159 481.95
159 403.91
159 394.89 — 4f (1,2,3)
147 939.47
147 749.89
147 562.14 — 5p (1,2,3)
77 424.45
77 167.43
77 000.23 — 4p (1,2,3)
182 473
182 227
182 225 — 5g (3,2,4)
169 904.
169 758.
169 600. — 5d (2,3,4)
133 371.07
133 207.10
133 065.24 — 4d (2,3,4)
181 611.79
181 368.45
181 339.27 — 5f (2,3,4)
158 690.85
158 557.76
158 536.63 — 4f (2,3,4)
148 410.24
148 111.10
147 931.47 — 5p (2,3,4)
78 158.61
77 746.44
77 421.86 — 4p (2,3,4)
182 405
182 068
182 067 — 5g (4,3,5)
168 520.53
168 307.06
168 152.20 — 5d (3,4,5)
129 469.37
129 252.74
129 093.28 — 4d (3,4,5)
181 821.83
181 758.34
181 507.92 — 5
158 903.55
158 865.03
158 740.92 — 4 (3,4,5)
182 439
182 015 — 5 ([4,5]6)
181 837.98
181 558.44
181 439.64 — 5 (4,5,6)
159 269.53
159 128.94
159 022.93 — 4 (4,5,6)
KEY
129 469.37
129 252.74
129 093.28 — 4d (3,4,5)
J, low to high
n
Energy levels (cm^{-1})
Ionization Limit
221 735 cm^{-1}
(27.491 eV)
[Ti III $3d^2\ ^3F_2$ → Ti IV $3d\ ^2D_{3/2}$]

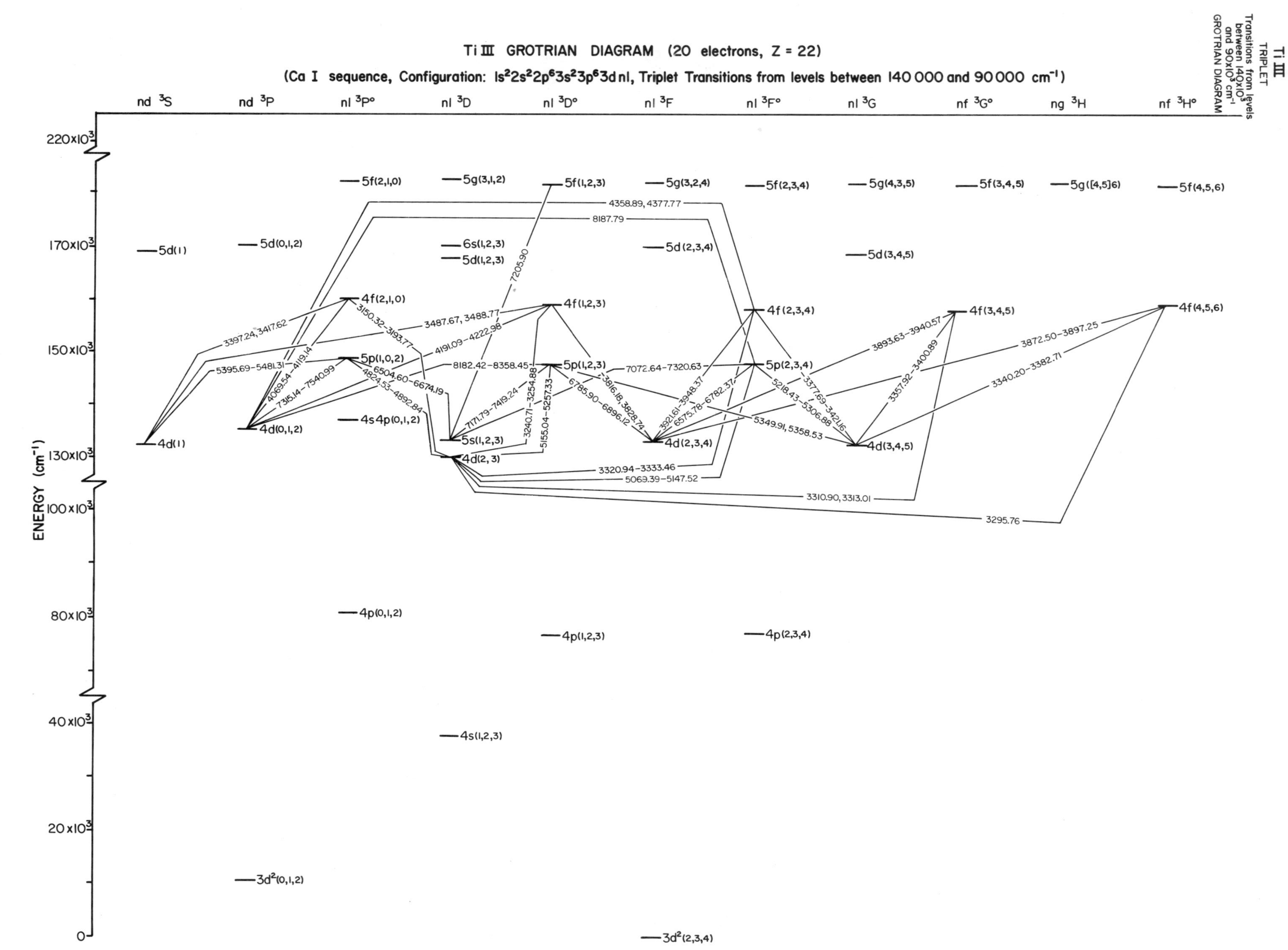
Ti III GROTRIAN DIAGRAM (20 electrons, Z = 22)
(Ca I sequence, Configuration: $1s^2 2s^2 2p^6 3s^2 3p^6 3d\,nl$, Triplet Transitions from levels between 140 000 and 90 000 cm^{-1})
Ti III
TRIPLET
Transitions from levels between 140×10^3 and 90×10^3 cm^{-1}
GROTRIAN DIAGRAM
nd ^{3}S
nd ^{3}P
nl ^{3}P°
nl ^{3}D
nl ^{3}D°
nl ^{3}F
nl ^{3}F°
nl ^{3}G
nf ^{3}G°
ng ^{3}H
nf ^{3}H°
ENERGY (cm^{-1})
220×10^3
170×10^3
150×10^3
130×10^3
100×10^3
80×10^3
40×10^3
20×10^3
0
5f(2,1,0)
5g(3,1,2)
5f(1,2,3)
5g(3,2,4)
5f(2,3,4)
5g(4,3,5)
5f(3,4,5)
5g([4,5]6)
5f(4,5,6)
5d(1)
5d(0,1,2)
6s(1,2,3)
5d(1,2,3)
5d(2,3,4)
5d(3,4,5)
4f(2,1,0)
4f(1,2,3)
4f(2,3,4)
4f(3,4,5)
4f(4,5,6)
5p(1,0,2)
5p(1,2,3)
5p(2,3,4)
4s4p(0,1,2)
4d(1)
4d(0,1,2)
5s(1,2,3)
4d(2,3)
4d(2,3,4)
4d(3,4,5)
4p(0,1,2)
4p(1,2,3)
4p(2,3,4)
4s(1,2,3)
$3d^2$(0,1,2)
$3d^2$(2,3,4)
4358.89, 4377.77
8187.79
7205.90
3397.24, 3417.62
3150.32–3193.77
3487.67, 3488.77
4191.09–4222.98
5395.69–5481.31
4069.54–4119.14
7315.14–7540.99
6504.60–6674.19
4824.53–4892.84
8182.42–8358.45
7171.79–7419.24
3240.71–3254.88
5155.04–5257.33
7072.64–7320.63
3816.18, 3828.74
6785.90–6896.12
3921.61–3948.37
6575.78–6782.37
5218.43–5306.88
3377.69–3421.16
5349.91, 5358.53
3893.63–3940.57
3357.92–3400.89
3872.50–3897.25
3340.20–3382.71
3320.94–3333.46
5069.39–5147.52
3310.90, 3313.01
3295.76

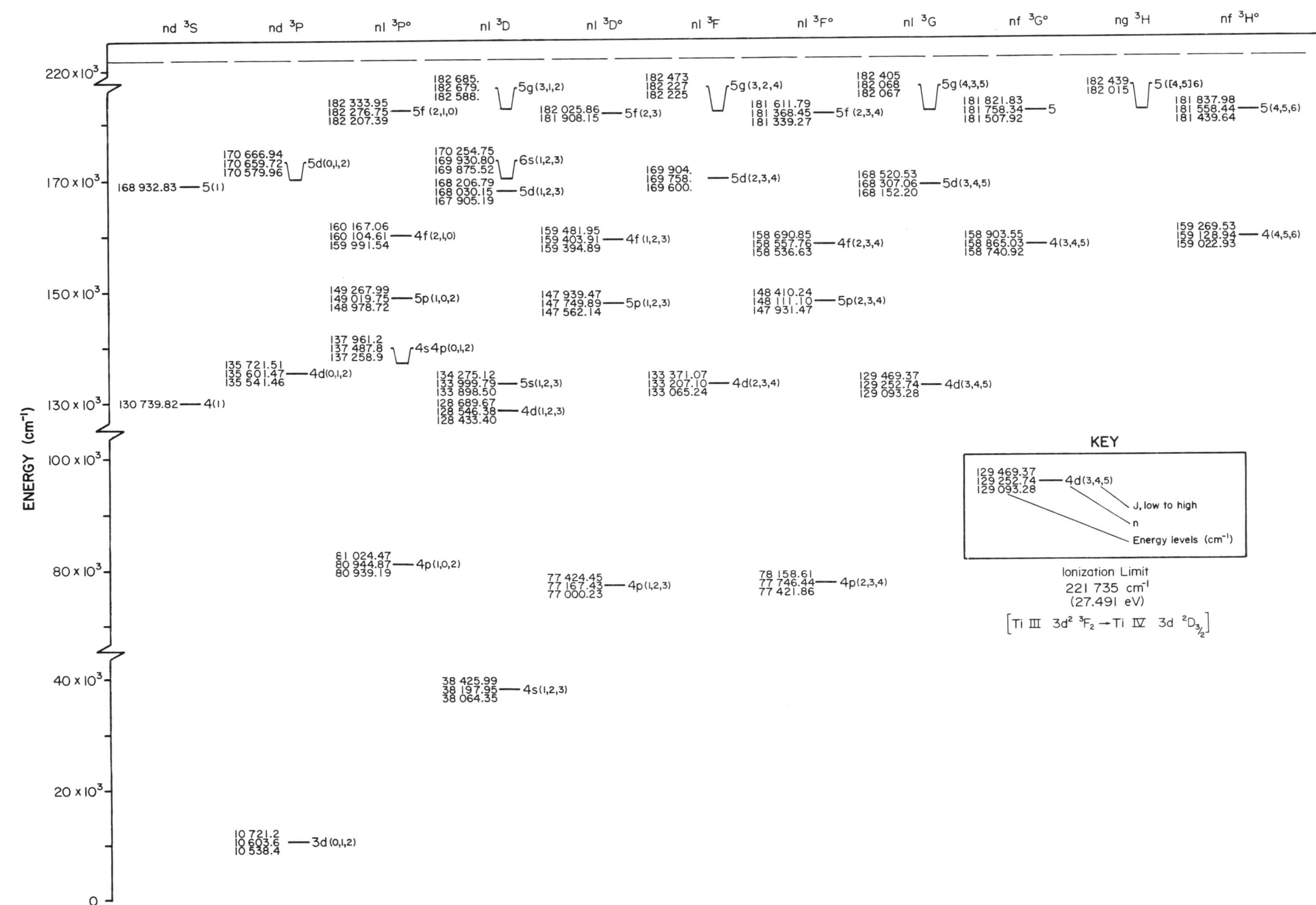
Ti III ENERGY LEVELS (20 electrons, Z=22)
(Ca I sequence, Configuration: $1s^2 2s^2 2p^6 3s^2 3p^6 3d\,nl$, Triplet System)
nd ^3S
nd ^3P
nl $^3P^o$
nl ^3D
nl $^3D^o$
nl ^3F
nl $^3F^o$
nl ^3G
nf $^3G^o$
ng ^3H
nf $^3H^o$
ENERGY (cm^{-1})
220 x 10^3
170 x 10^3
150 x 10^3
130 x 10^3
100 x 10^3
80 x 10^3
40 x 10^3
20 x 10^3
0
168 932.83 — 5(1)
130 739.82 — 4(1)
170 666.94 170 659.72 170 579.96 5d(0,1,2)
135 721.51 135 601.47 135 541.46 — 4d(0,1,2)
10 721.2 10 603.6 10 538.4 — 3d(0,1,2)
182 333.95 182 276.75 182 207.39 — 5f (2,1,0)
160 167.06 160 104.61 159 991.54 — 4f (2,1,0)
149 267.99 149 019.75 148 978.72 — 5p(1,0,2)
137 961.2 137 487.8 137 258.9 4s4p(0,1,2)
81 024.47 80 944.87 80 939.19 — 4p(1,0,2)
182 685. 182 679. 182 588. 5g(3,1,2)
170 254.75 169 930.80 169 875.52 6s(1,2,3)
168 206.79 168 030.15 167 905.19 — 5d(1,2,3)
134 275.12 133 999.79 133 898.50 — 5s(1,2,3)
128 689.67 128 546.38 128 433.40 — 4d(1,2,3)
38 425.99 38 197.95 38 064.35 — 4s(1,2,3)
182 025.86 181 908.15 — 5f(2,3)
159 481.95 159 403.91 159 394.89 — 4f (1,2,3)
147 939.47 147 749.89 147 562.14 — 5p(1,2,3)
77 424.45 77 167.43 77 000.23 — 4p(1,2,3)
182 473 182 227 182 225 5g(3,2,4)
169 904. 169 758. 169 600. — 5d(2,3,4)
133 371.07 133 207.10 133 065.24 — 4d(2,3,4)
181 611.79 181 368.45 181 339.27 — 5f (2,3,4)
158 690.85 158 557.76 158 536.63 — 4f(2,3,4)
148 410.24 148 111.10 147 931.47 — 5p(2,3,4)
78 158.61 77 746.44 77 421.86 — 4p(2,3,4)
182 405 182 068 182 067 5g(4,3,5)
168 520.53 168 307.06 168 152.20 — 5d(3,4,5)
129 469.37 129 252.74 129 093.28 — 4d(3,4,5)
181 821.83 181 758.34 181 507.92 — 5
158 903.55 158 865.03 158 740.92 — 4(3,4,5)
182 439 182 015 5([4,5]6)
181 837.98 181 558.44 181 439.64 — 5(4,5,6)
159 269.53 159 128.94 159 022.93 — 4(4,5,6)
KEY
129 469.37 129 252.74 129 093.28 — 4d(3,4,5)
J, low to high
n
Energy levels (cm^{-1})
Ionization Limit
221 735 cm^{-1}
(27.491 eV)
[Ti III $3d^2\ ^3F_2$ → Ti IV $3d\ ^2D_{3/2}$]

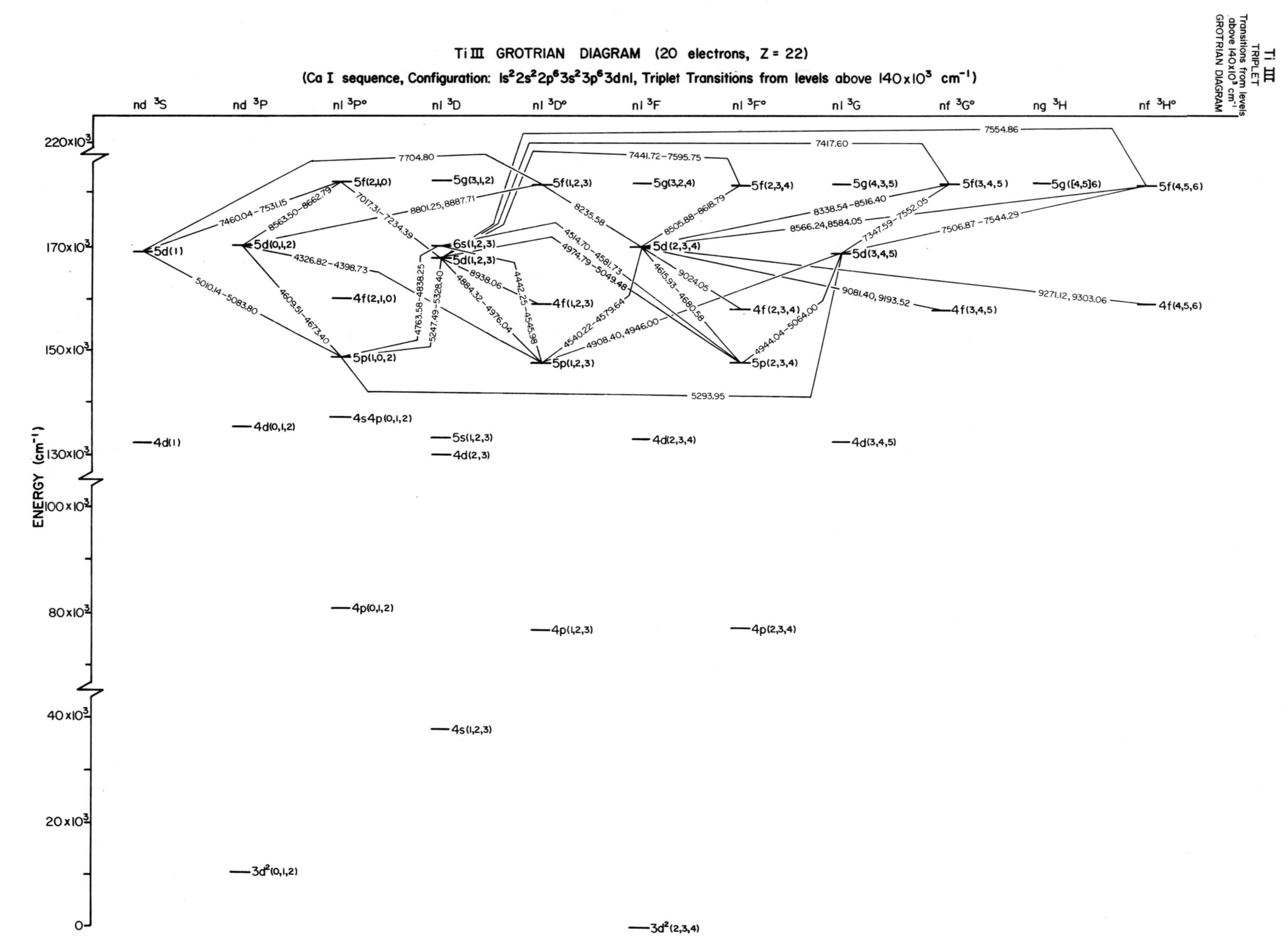
Ti III
TRIPLET
Transitions from levels above 140x10³ cm⁻¹
GROTRIAN DIAGRAM
Ti III GROTRIAN DIAGRAM (20 electrons, Z = 22)
(Ca I sequence, Configuration: 1s²2s²2p⁶3s²3p⁶3dnl, Triplet Transitions from levels above 140x10³ cm⁻¹)
nd ³S
nd ³P
nl ³P°
nl ³D
nl ³D°
nl ³F
nl ³F°
nl ³G
nf ³G°
ng ³H
nf ³H°
5d(1)
4d(1)
5d(0,1,2)
4d(0,1,2)
3d²(0,1,2)
5f(2,1,0)
4f(2,1,0)
5p(1,0,2)
4s4p(0,1,2)
4p(0,1,2)
5g(3,1,2)
6s(1,2,3)
5d(1,2,3)
5s(1,2,3)
4d(2,3)
4s(1,2,3)
5f(1,2,3)
4f(1,2,3)
5p(1,2,3)
4p(1,2,3)
5g(3,2,4)
5d(2,3,4)
4d(2,3,4)
3d²(2,3,4)
5f(2,3,4)
4f(2,3,4)
5p(2,3,4)
4p(2,3,4)
5g(4,3,5)
5d(3,4,5)
4d(3,4,5)
5f(3,4,5)
4f(3,4,5)
5g([4,5]6)
5f(4,5,6)
4f(4,5,6)
7554.86
7417.60
7441.72–7595.75
7704.80
7460.04–7531.15
8563.50–8662.79
7017.31–7234.39
8801.25, 8887.71
8235.58
8505.88–8618.79
8338.54–8516.40
8566.24, 8584.05
7347.59–7552.05
7506.87–7544.29
4326.82–4398.73
5010.14–5083.80
4609.51–4673.40
4763.58–4838.25
5247.49–5328.40
8938.06
4884.32–4976.04
4442.25–4545.98
4514.70–4581.73
4974.79–5049.48
4540.22–4579.64
4908.40, 4946.00
4615.93–4680.58
9024.05
4944.04–5064.00
9081.40, 9193.52
9271.12, 9303.06
5293.95
ENERGY (cm⁻¹)
220x10³
170x10³
150x10³
130x10³
100x10³
80x10³
40x10³
20x10³
0

Ti III ENERGY LEVELS (20 electrons, Z=22)
(Ca I sequence, Configuration: $1s^2 2s^2 2p^6 3s^2 3p^6 3d\,nl$, Triplet System)

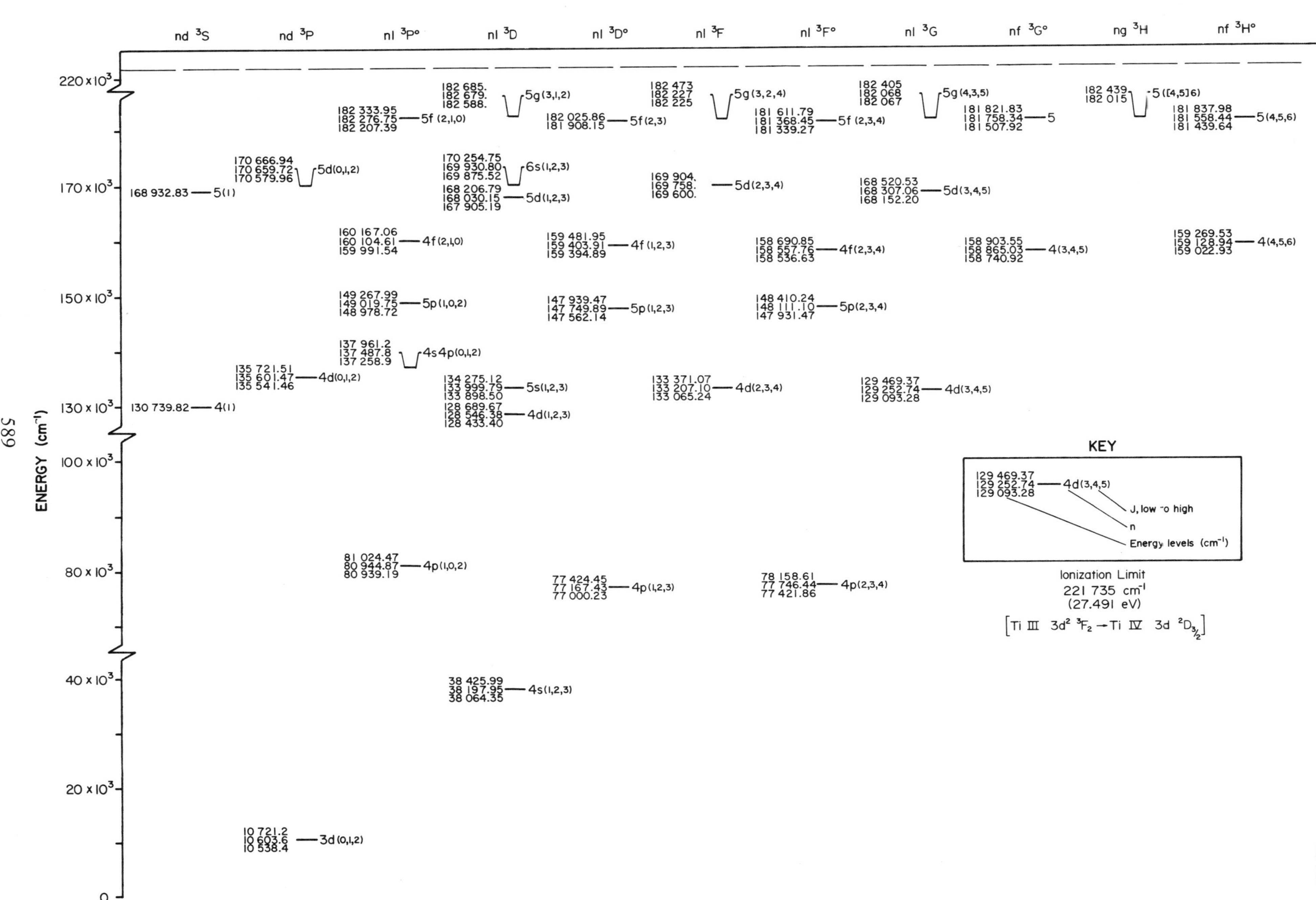

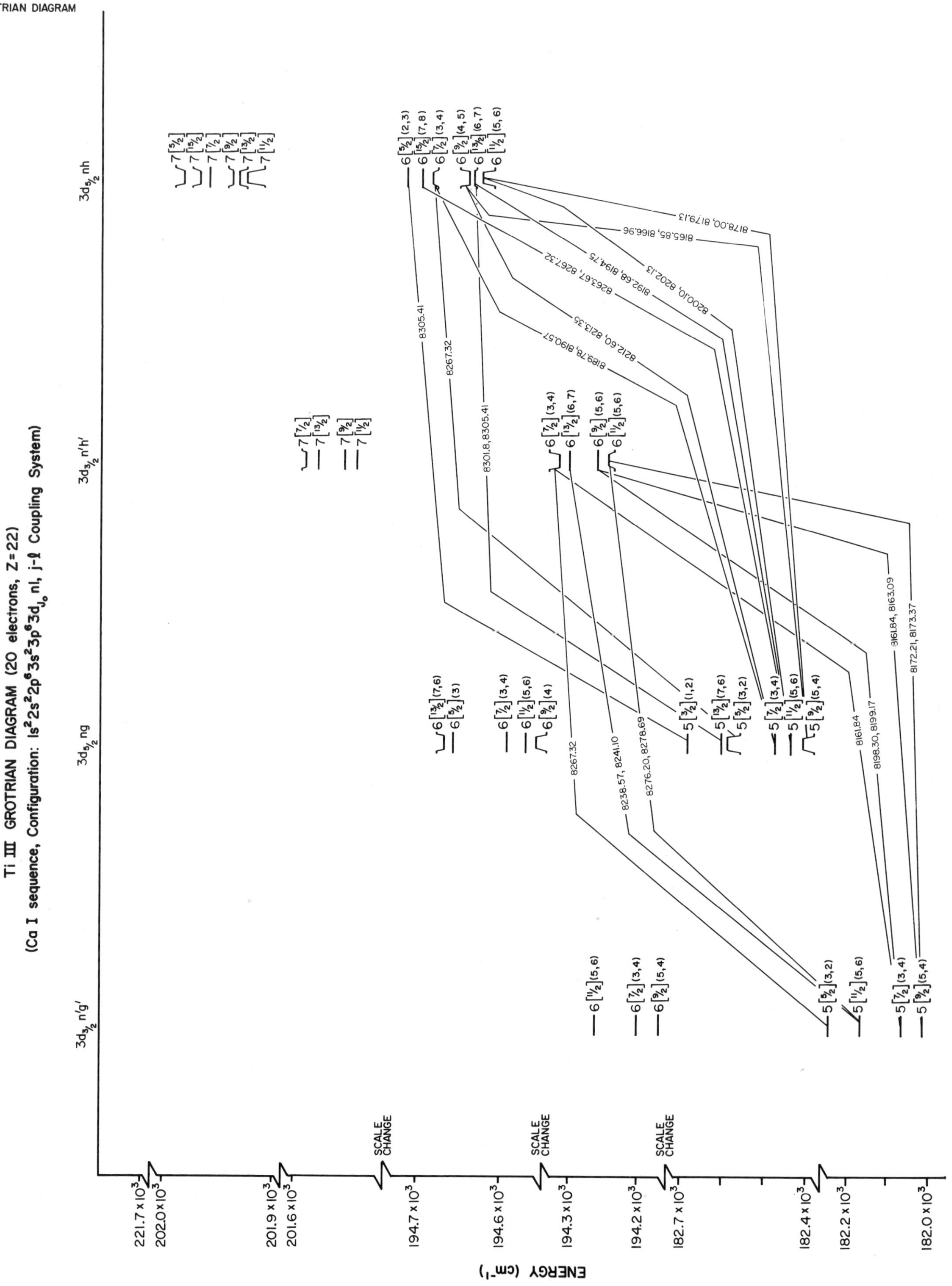

Ti III GROTRIAN DIAGRAM (20 electrons, Z = 22)

(Ca I sequence, Configuration: $1s^2 2s^2 2p^6 3s^2 3p^6 3d_{J_0}\, nl$, j-ℓ Coupling System)

$3d_{3/2}\, n'g'$ — $3d_{5/2}\, ng$ — $3d_{3/2}\, n'h'$ — $3d_{5/2}\, nh$

ENERGY (cm^{-1})

221.7×10^3, 202.0×10^3, 201.9×10^3, 201.6×10^3, SCALE CHANGE, 194.7×10^3, 194.6×10^3, SCALE CHANGE, 194.3×10^3, 194.2×10^3, SCALE CHANGE, 182.7×10^3, 182.4×10^3, 182.2×10^3, 182.0×10^3

$3d_{3/2}\, n'g'$:
$6[11/2](5,6)$
$6[7/2](3,4)$
$6[9/2](5,4)$
$5[5/2](3,2)$
$5[11/2](5,6)$
$5[7/2](3,4)$
$5[9/2](5,4)$

$3d_{5/2}\, ng$:
$6[13/2](7,6)$
$6[5/2](3)$
$6[7/2](3,4)$
$6[11/2](5,6)$
$6[9/2](4)$
$5[3/2](1,2)$
$5[13/2](7,6)$
$5[5/2](3,2)$
$5[7/2](3,4)$
$5[11/2](5,6)$
$5[9/2](5,4)$

$3d_{3/2}\, n'h'$:
$7[7/2]$
$7[13/2]$
$7[9/2]$
$7[11/2]$
$6[7/2](3,4)$
$6[13/2](6,7)$
$6[9/2](5,6)$
$6[11/2](5,6)$

$3d_{5/2}\, nh$:
$7[5/2]$
$7[15/2]$
$7[7/2]$
$7[9/2]$
$7[13/2]$
$7[11/2]$
$6[5/2](2,3)$
$6[15/2](7,8)$
$6[7/2](3,4)$
$6[9/2](4,5)$
$6[13/2](6,7)$
$6[11/2](5,6)$

Transitions:
8305.41
8267.32
8301.8, 8305.41
8189.78, 8190.57
8212.60, 8213.35
8263.67, 8267.32
8192.68, 8194.75
8200.10, 8202.13
8165.85, 8166.96
8178.00, 8179.13
8267.32
8238.57, 8241.10
8276.20, 8278.69
8161.84
8198.30, 8199.17
8161.84, 8163.09
8172.21, 8173.37

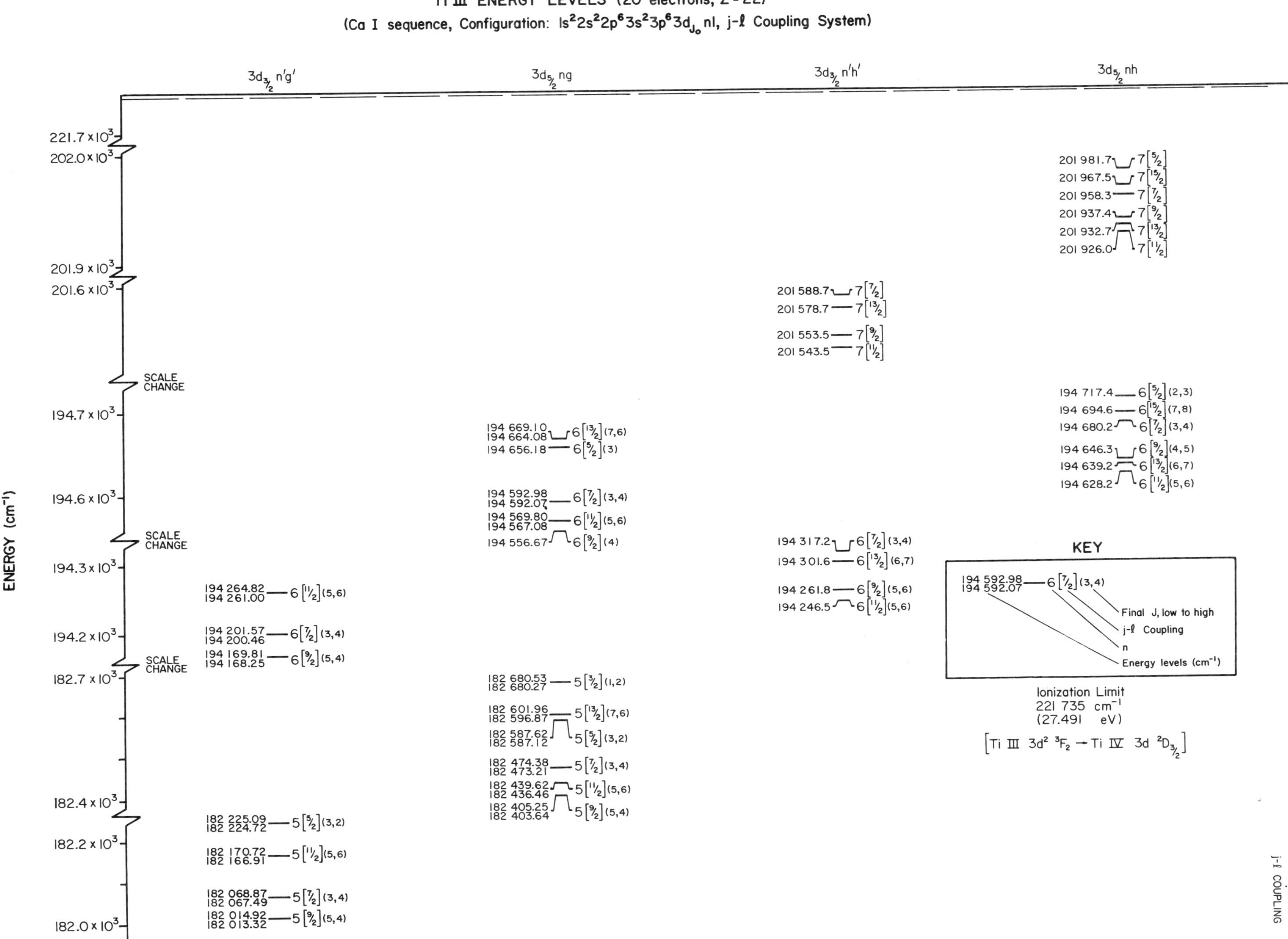
Ti III ENERGY LEVELS (20 electrons, Z = 22)
(Ca I sequence, Configuration: $1s^2 2s^2 2p^6 3s^2 3p^6 3d_{J_0} nl$, j-ℓ Coupling System)
$3d_{3/2}\,n'g'$
$3d_{5/2}\,ng$
$3d_{3/2}\,n'h'$
$3d_{5/2}\,nh$
ENERGY (cm⁻¹)
221.7 x 10³
202.0 x 10³
201.9 x 10³
201.6 x 10³
SCALE CHANGE
194.7 x 10³
194.6 x 10³
SCALE CHANGE
194.3 x 10³
194.2 x 10³
SCALE CHANGE
182.7 x 10³
182.4 x 10³
182.2 x 10³
182.0 x 10³
194 264.82, 194 261.00 — 6[11/2](5,6)
194 201.57, 194 200.46 — 6[7/2](3,4)
194 169.81, 194 168.25 — 6[9/2](5,4)
182 225.09, 182 224.72 — 5[5/2](3,2)
182 170.72, 182 166.91 — 5[11/2](5,6)
182 068.87, 182 067.49 — 5[7/2](3,4)
182 014.92, 182 013.32 — 5[9/2](5,4)
194 669.10, 194 664.08 — 6[13/2](7,6)
194 656.18 — 6[5/2](3)
194 592.98, 194 592.07 — 6[7/2](3,4)
194 569.80, 194 567.08 — 6[11/2](5,6)
194 556.67 — 6[9/2](4)
182 680.53, 182 680.27 — 5[3/2](1,2)
182 601.96, 182 596.87 — 5[13/2](7,6)
182 587.62, 182 587.12 — 5[5/2](3,2)
182 474.38, 182 473.21 — 5[7/2](3,4)
182 439.62, 182 436.46 — 5[11/2](5,6)
182 405.25, 182 403.64 — 5[9/2](5,4)
201 588.7 — 7[7/2]
201 578.7 — 7[13/2]
201 553.5 — 7[9/2]
201 543.5 — 7[11/2]
194 317.2 — 6[7/2](3,4)
194 301.6 — 6[13/2](6,7)
194 261.8 — 6[9/2](5,6)
194 246.5 — 6[11/2](5,6)
201 981.7 — 7[5/2]
201 967.5 — 7[15/2]
201 958.3 — 7[7/2]
201 937.4 — 7[9/2]
201 932.7 — 7[13/2]
201 926.0 — 7[11/2]
194 717.4 — 6[5/2](2,3)
194 694.6 — 6[15/2](7,8)
194 680.2 — 6[7/2](3,4)
194 646.3 — 6[9/2](4,5)
194 639.2 — 6[13/2](6,7)
194 628.2 — 6[11/2](5,6)
KEY
194 592.98, 194 592.07 — 6[7/2](3,4)
Final J, low to high
j-ℓ Coupling
n
Energy levels (cm⁻¹)
Ionization Limit
221 735 cm⁻¹
(27.491 eV)
[Ti III $3d^2\ ^3F_2$ → Ti IV $3d\ ^2D_{3/2}$]

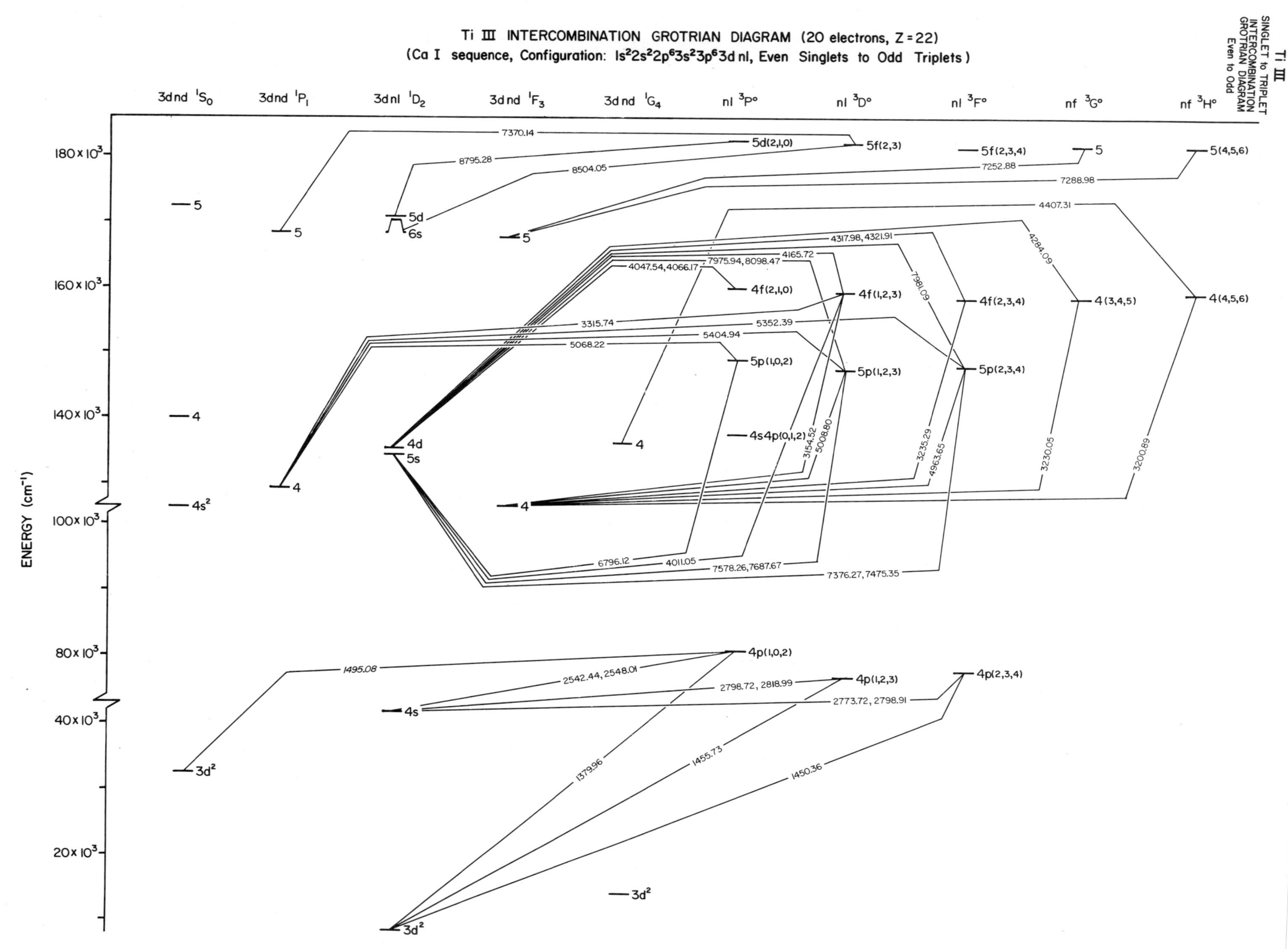
Ti III
SINGLET to TRIPLET
INTERCOMBINATION
GROTRIAN DIAGRAM
Even to Odd
Ti III INTERCOMBINATION GROTRIAN DIAGRAM (20 electrons, Z = 22)
(Ca I sequence, Configuration: $1s^2 2s^2 2p^6 3s^2 3p^6 3d\,nl$, Even Singlets to Odd Triplets)
3dnd $^1S_0$
3dnd $^1P_1$
3dnl $^1D_2$
3dnd $^1F_3$
3dnd $^1G_4$
nl $^3P^o$
nl $^3D^o$
nl $^3F^o$
nf $^3G^o$
nf $^3H^o$
ENERGY (cm^{-1})
180×10^3
160×10^3
140×10^3
100×10^3
80×10^3
40×10^3
20×10^3
5
4
$4s^2$
$3d^2$
5d
6s
4d
5s
4s
5d(2,1,0)
4f(2,1,0)
5p(1,0,2)
4s4p(0,1,2)
4p(1,0,2)
5f(2,3)
4f(1,2,3)
5p(1,2,3)
4p(1,2,3)
5f(2,3,4)
4f(2,3,4)
5p(2,3,4)
4p(2,3,4)
4(3,4,5)
5(4,5,6)
4(4,5,6)
7370.14
8795.28
8504.05
7252.88
7288.98
4407.31
4317.98, 4321.91
4284.09
4165.72
7975.94, 8098.47
4047.54, 4066.17
7981.09
3315.74
5352.39
5404.94
5068.22
3154.52
5008.80
3235.29
4963.65
3230.05
3200.89
6796.12
4011.05
7578.26, 7687.67
7376.27, 7475.35
1495.08
2542.44, 2548.01
2798.72, 2818.99
2773.72, 2798.91
1379.96
1455.73
1450.36

Ti III INTERCOMBINATION GROTRIAN DIAGRAM (20 electrons, Z = 22)
(Ca I sequence, Configuration: $1s^22s^22p^63s^23p^63dnl$, Odd Singlets to Even Triplets)

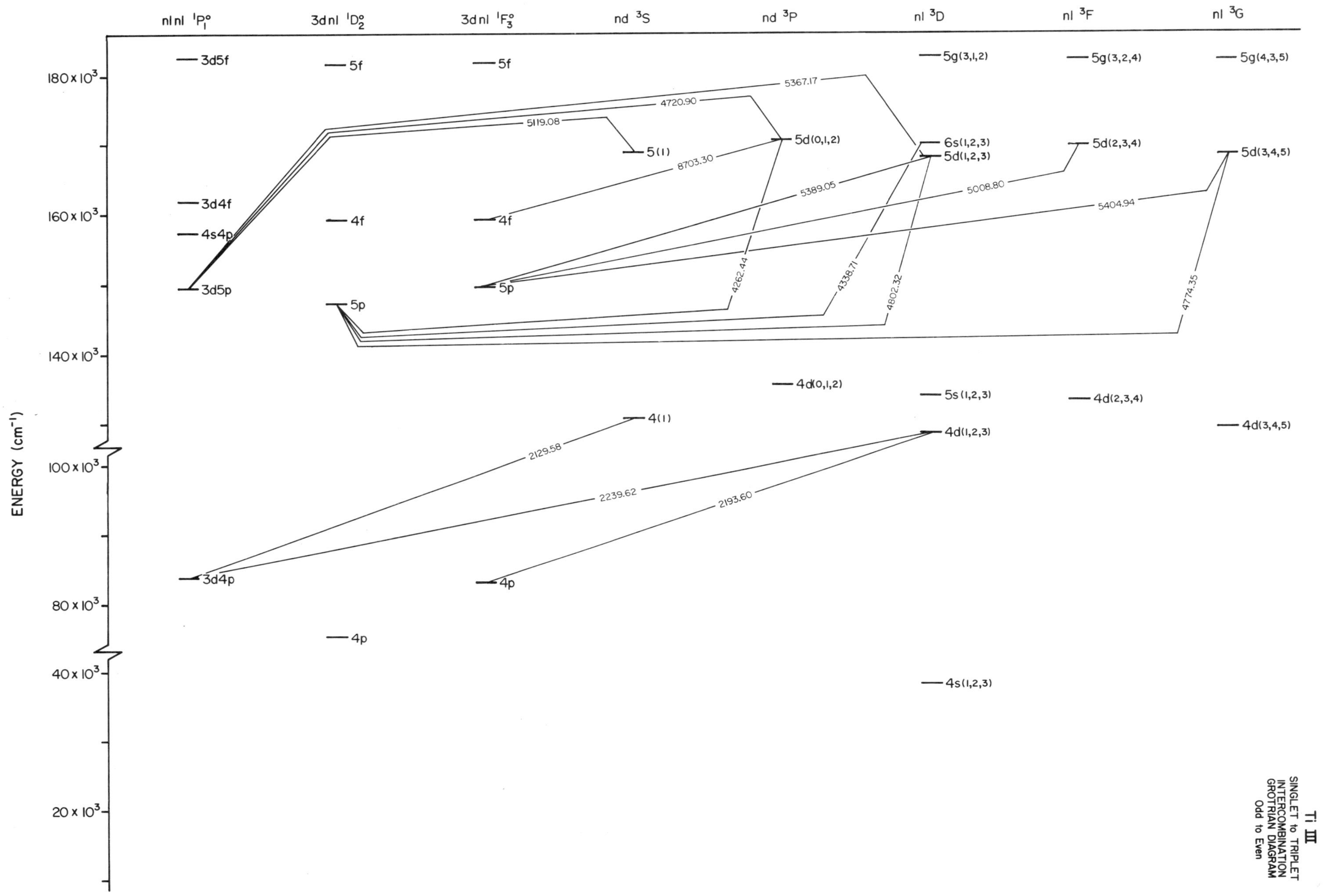

Ti III
SINGLET to TRIPLET
INTERCOMBINATION
GROTRIAN DIAGRAM
Odd to Even

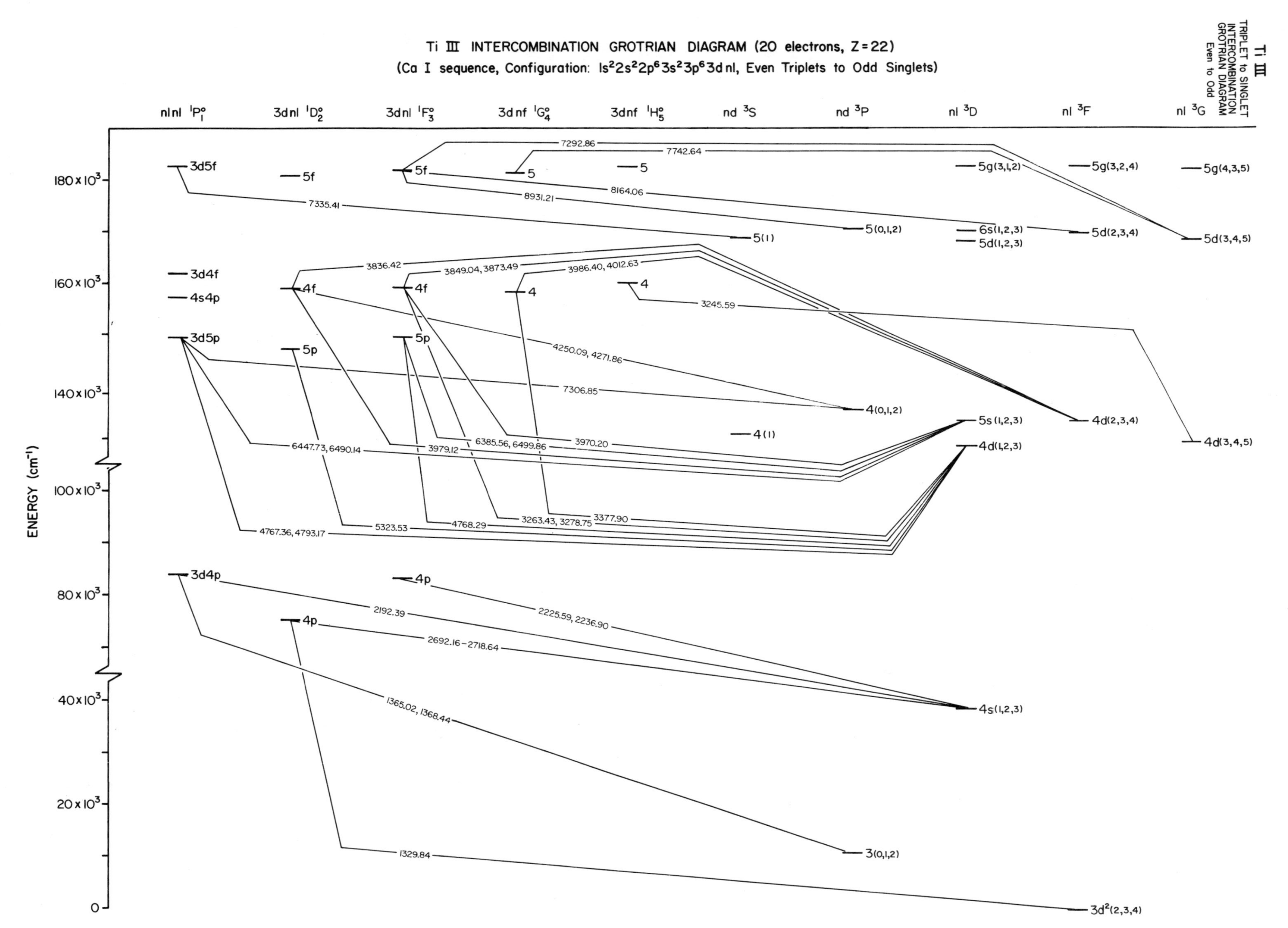
Ti III
TRIPLET to SINGLET
INTERCOMBINATION
GROTRIAN DIAGRAM
Even to Odd
Ti III INTERCOMBINATION GROTRIAN DIAGRAM (20 electrons, Z = 22)
(Ca I sequence, Configuration: $1s^2 2s^2 2p^6 3s^2 3p^6 3d\,nl$, Even Triplets to Odd Singlets)
nlnl $^1P^o_1$
3dnl $^1D^o_2$
3dnl $^1F^o_3$
3dnf $^1G^o_4$
3dnf $^1H^o_5$
nd ^3S
nd ^3P
nl ^3D
nl ^3F
nl ^3G
3d5f
3d4f
4s4p
3d5p
3d4p
5f
4f
5p
4p
5
4
5(1)
4(1)
5(0,1,2)
4(0,1,2)
3(0,1,2)
5g(3,1,2)
6s(1,2,3)
5d(1,2,3)
5s(1,2,3)
4d(1,2,3)
4s(1,2,3)
5g(3,2,4)
5d(2,3,4)
4d(2,3,4)
$3d^2$(2,3,4)
5g(4,3,5)
5d(3,4,5)
4d(3,4,5)
7292.86
7742.64
8164.06
8931.21
7335.41
3836.42
3849.04, 3873.49
3986.40, 4012.63
3245.59
4250.09, 4271.86
7306.85
6385.56, 6499.86
3970.20
3979.12
6447.73, 6490.14
3263.43, 3278.75
3377.90
4768.29
5323.53
4767.36, 4793.17
2192.39
2225.59, 2236.90
2692.16 – 2718.64
1365.02, 1368.44
1329.84
180 x 10^3
160 x 10^3
140 x 10^3
100 x 10^3
80 x 10^3
40 x 10^3
20 x 10^3
0
ENERGY (cm^{-1})

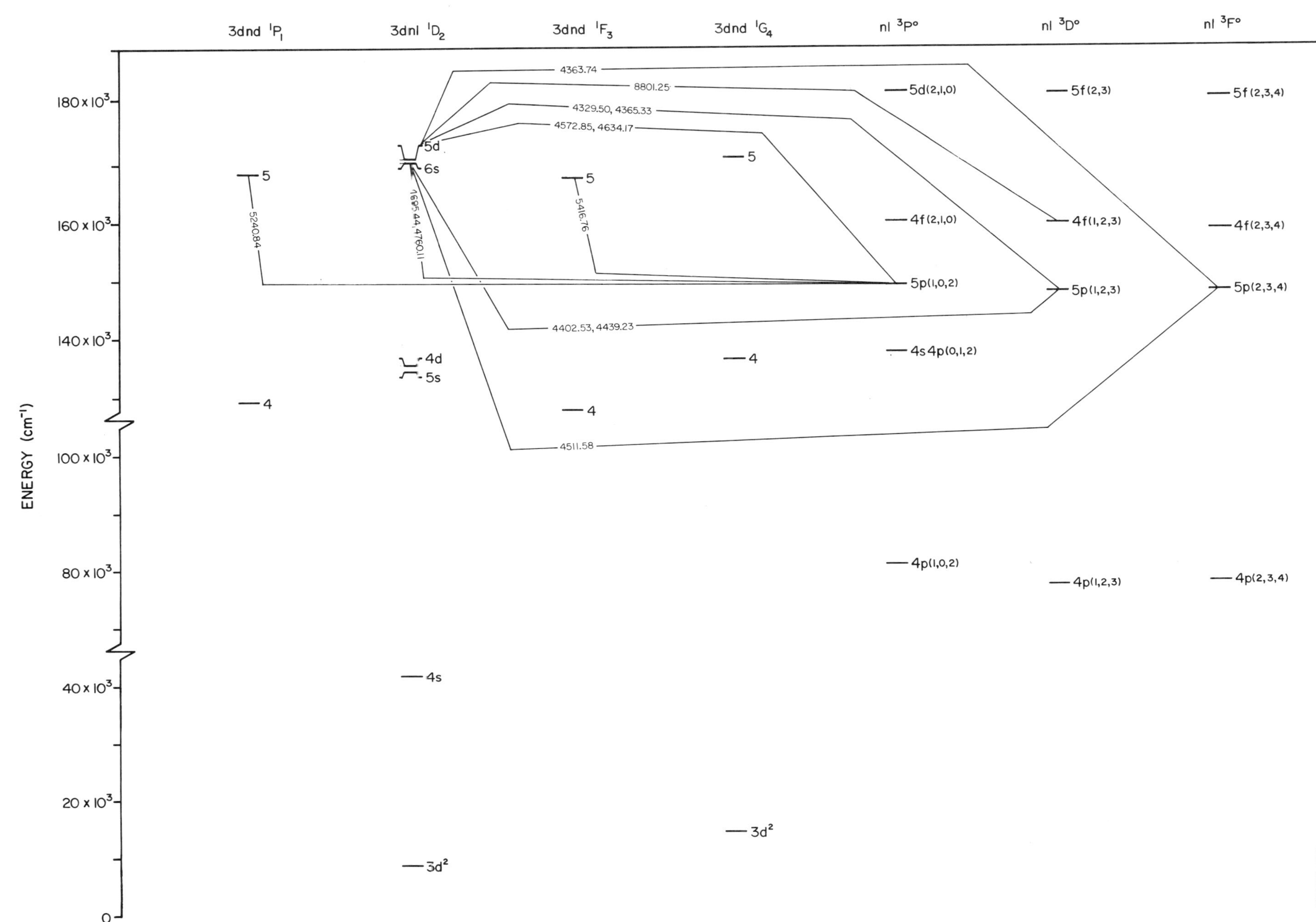
Ti III INTERCOMBINATION GROTRIAN DIAGRAM (20 electrons, Z = 22)
(Ca I sequence, Configuration: 1s²2s²2p⁶3s²3p⁶3dnl, Odd Triplets to Even Singlets)
3dnd ¹P₁
3dnl ¹D₂
3dnd ¹F₃
3dnd ¹G₄
nl ³P°
nl ³D°
nl ³F°
4363.74
8801.25
4329.50, 4365.33
4572.85, 4634.17
5d(2,1,0)
5f(2,3)
5f(2,3,4)
5d
6s
5
5240.84
1695.44, 4760.11
5416.76
4f(2,1,0)
4f(1,2,3)
4f(2,3,4)
5p(1,0,2)
5p(1,2,3)
5p(2,3,4)
4402.53, 4439.23
4s4p(0,1,2)
4d
5s
4
4511.58
4p(1,0,2)
4p(1,2,3)
4p(2,3,4)
4s
3d²
ENERGY (cm⁻¹)
180×10³
160×10³
140×10³
100×10³
80×10³
40×10³
20×10³
0

Ti III
INTERCOMBINATION
GROTRIAN DIAGRAM
SINGLET to
j-l COUPLING

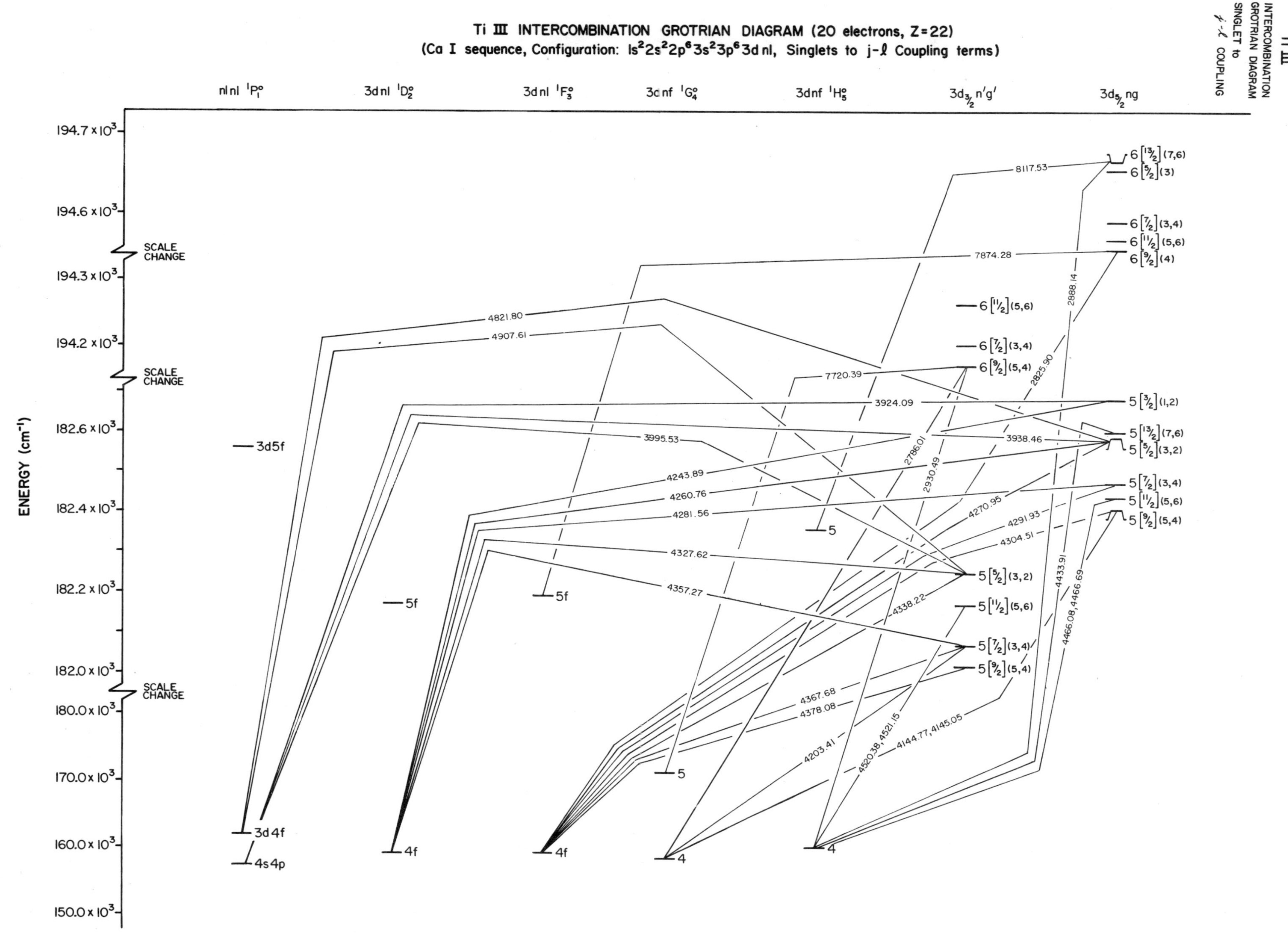

Ti III INTERCOMBINATION GROTRIAN DIAGRAM (20 electrons, Z=22)

(Ca I sequence, Configuration: $1s^2 2s^2 2p^6 3s^2 3p^6 3d\,nl$, Triplets below $180 \times 10^3\ cm^{-1}$ to j-ℓ Coupling)

Ti III
INTERCOMBINATION GROTRIAN DIAGRAM
TRIPLETS
below $180 \times 10^3\ cm^{-1}$ to
j-ℓ COUPLING

nl $^3P^o$
nl $^3D^o$
nl $^3F^o$
nf $^3G^o$
nf $^3H^o$
$3d_{3/2}\,n'g'$
$3d_{5/2}\,ng$

ENERGY (cm^{-1})
194.7×10^3
194.5×10^3
194.3×10^3
194.1×10^3
182.6×10^3
182.4×10^3
182.2×10^3
182.0×10^3
SCALE CHANGE
180.0×10^3
160.0×10^3
140.0×10^3

$6[13/2](7,6)$
$6[5/2](3)$
$6[7/2](3,4)$
$6[11/2](5,6)$
$6[9/2](4)$
$6[11/2](5,6)$
$6[7/2](3,4)$
$6[9/2](5,4)$
$5[3/2](1,2)$
$5[13/2](7,6)$
$5[5/2](3,2)$
$5[7/2](3,4)$
$5[11/2](5,6)$
$5[9/2](5,4)$
$5[5/2](3,2)$
$5[11/2](5,6)$
$5[7/2](3,4)$
$5[9/2](5,4)$
5f (2,1,0)
5f (2,3)
5f (2,3,4)
5
5 (4,5,6)
4f (2,1,0)
4f (1,2,3)
4f (2,3,4)
4 (3,4,5)
4 (4,5,6)

2847.26
2824.45
2812.57, 2820.78
4293.34 – 4309.40
4310.48 – 4326.82
2803.15
2807.20
2802.94, 2803.15
2821.69
4406.20 – 4440.66
4424.40, 4446.56
4333.54, 4348.04
4361.15
4156.67 – 4183.58
4176.54 – 4203.41
4210.13
4192.14, 4215.95
4259.01 – 4285.61
4262.93
4269.84 – 4289.25
4275.53 – 4295.03
4224.58 – 4254.11
4241.29, 4247.62
4212.47 – 4241.29
4214.29
4218.52
4378.94, 4380.73
4220.28, 4224.13
4258.47
4496.51, 4519.42
4257.05
4296.70
4248.54 – 4276.35
4261.90 – 4286.52
4308.39
4319.56 – 4365.33
4295.42 – 4325.93
4348.40, 4368.56

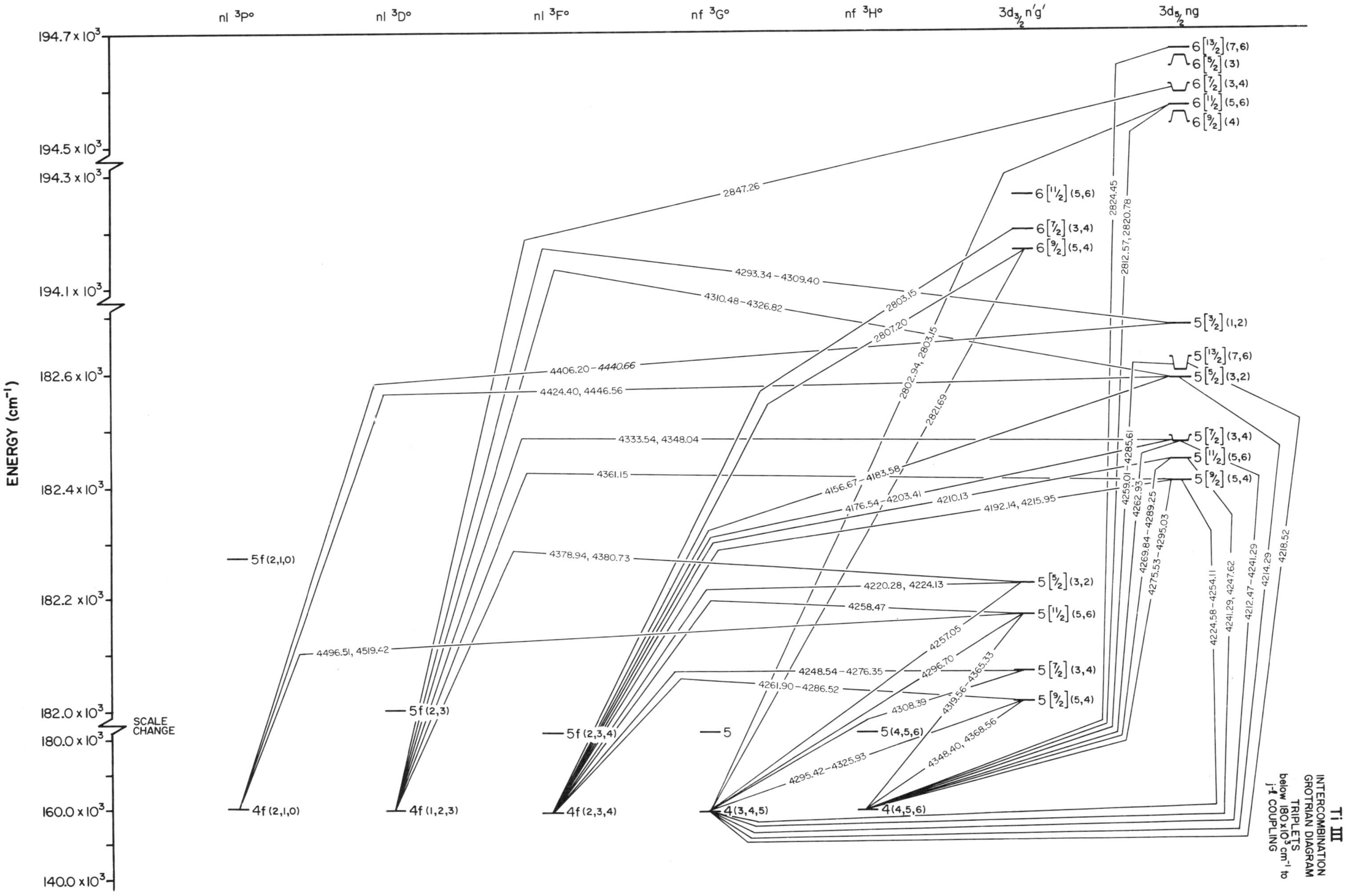

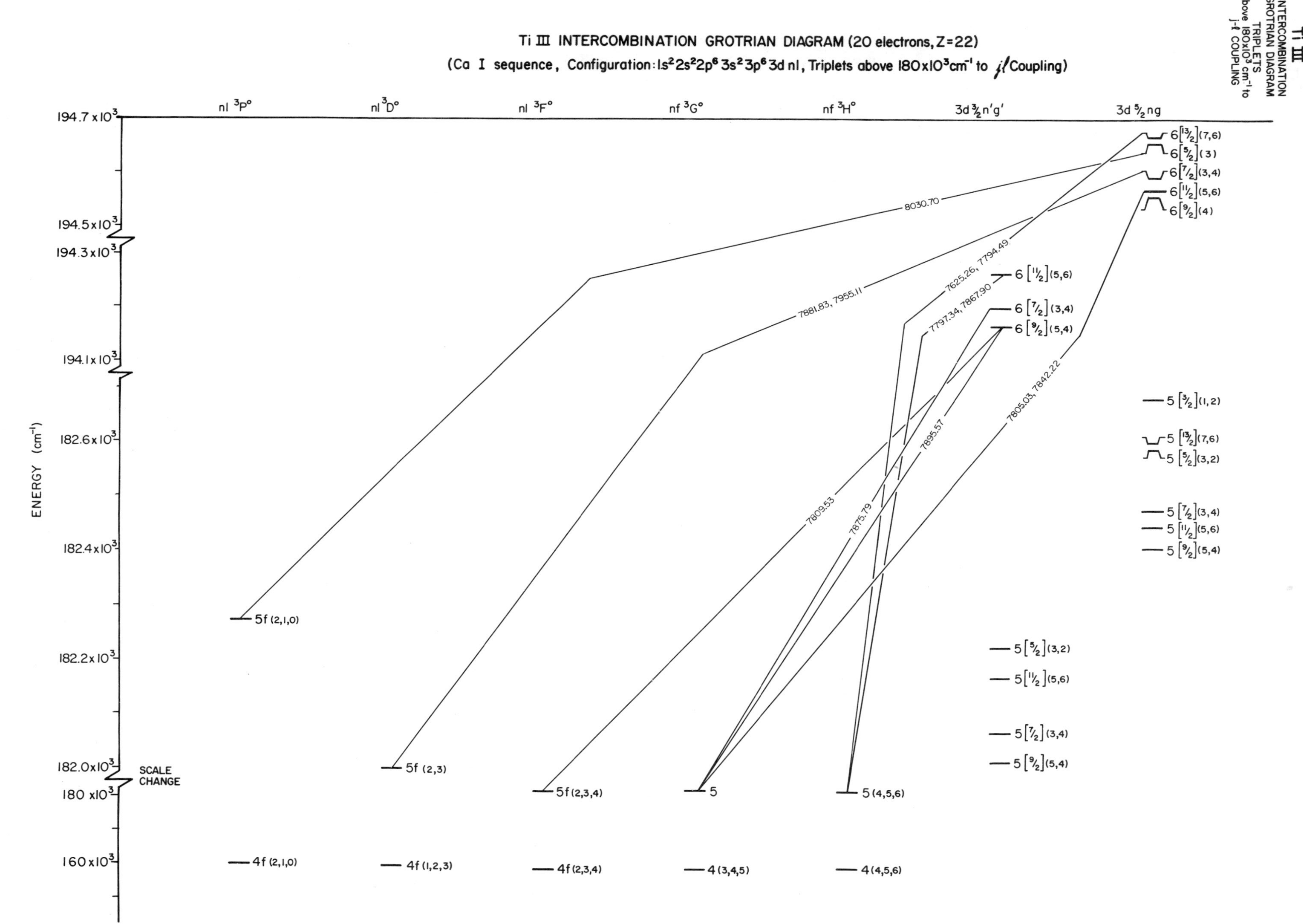

Ti III INTERCOMBINATION GROTRIAN DIAGRAM (20 electrons, Z=22)
(Ca I sequence, Configuration: 1s²2s²2p⁶3s²3p⁶3d nl, Triplets above 180x10³cm⁻¹ to jℓ Coupling)
Ti III
INTERCOMBINATION
GROTRIAN DIAGRAM
TRIPLETS
above 180x10³ cm⁻¹ to
j-ℓ COUPLING
nl ³P°
nl ³D°
nl ³F°
nf ³G°
nf ³H°
3d ³⁄₂ n'g'
3d ⁵⁄₂ ng
ENERGY (cm⁻¹)
194.7 x 10³
194.5 x 10³
194.3 x 10³
194.1 x 10³
182.6 x 10³
182.4 x 10³
182.2 x 10³
182.0 x 10³
180 x 10³
160 x 10³
SCALE CHANGE
6 [13/2] (7,6)
6 [5/2] (3)
6 [7/2] (3,4)
6 [11/2] (5,6)
6 [9/2] (4)
6 [11/2] (5,6)
6 [7/2] (3,4)
6 [9/2] (5,4)
5 [3/2] (1,2)
5 [13/2] (7,6)
5 [5/2] (3,2)
5 [7/2] (3,4)
5 [11/2] (5,6)
5 [9/2] (5,4)
5 [5/2] (3,2)
5 [11/2] (5,6)
5 [7/2] (3,4)
5 [9/2] (5,4)
5f (2,1,0)
5f (2,3)
5f (2,3,4)
5
5 (4,5,6)
4f (2,1,0)
4f (1,2,3)
4f (2,3,4)
4 (3,4,5)
4 (4,5,6)
8030.70
7881.83, 7955.11
7625.26, 7794.49
7797.34, 7867.90
7809.53
7875.79
7895.57
7805.03, 7842.22

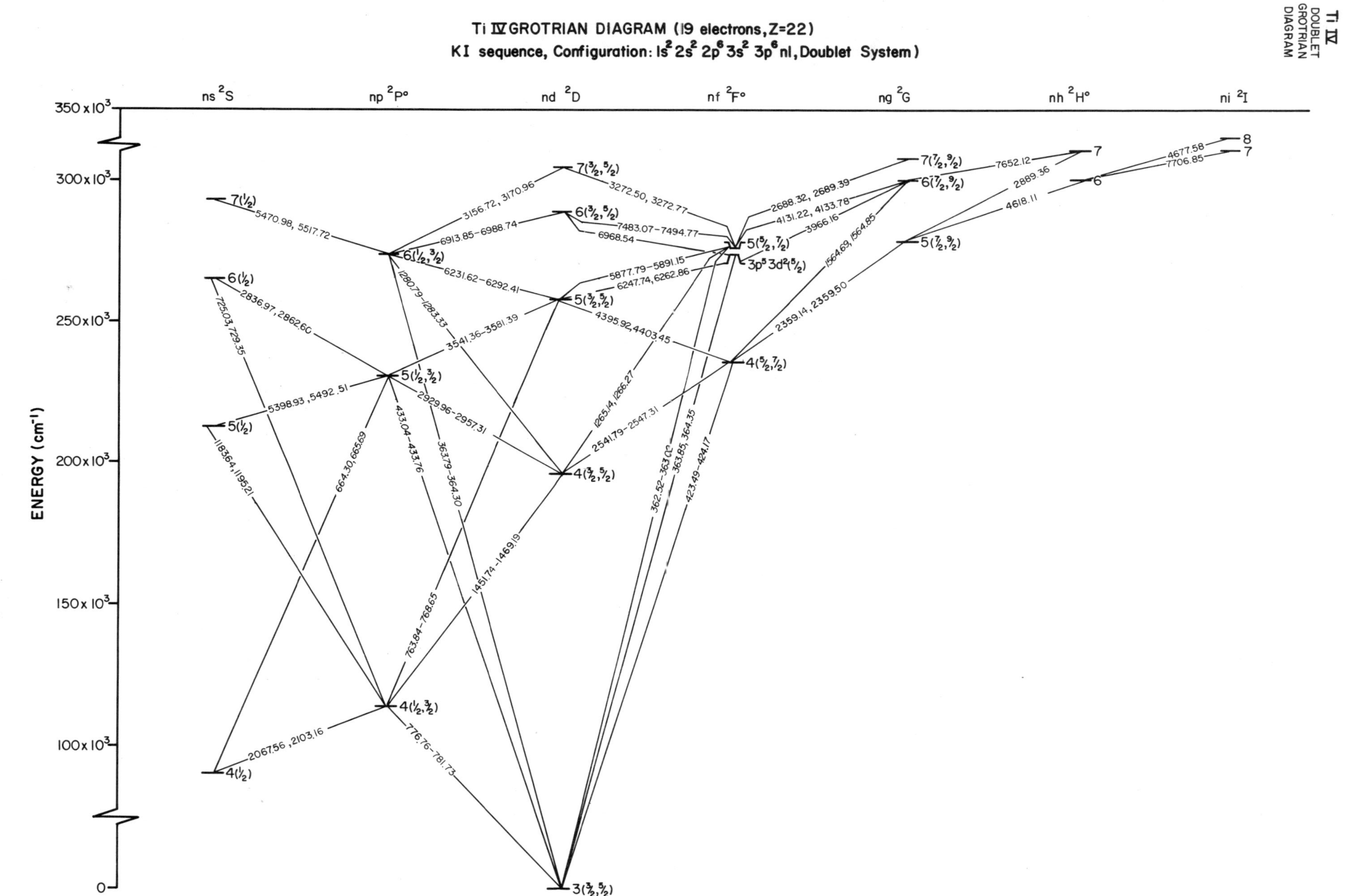
Ti IV GROTRIAN DIAGRAM (19 electrons, Z=22)
KI sequence, Configuration: 1s² 2s² 2p⁶ 3s² 3p⁶ nl, Doublet System)
ns ²S
np ²P°
nd ²D
nf ²F°
ng ²G
nh ²H°
ni ²I
ENERGY (cm⁻¹)
350 x 10³
300 x 10³
250 x 10³
200 x 10³
150 x 10³
100 x 10³
0
7(½)
6(½)
5(½)
4(½)
6(½,³⁄₂)
5(½,³⁄₂)
4(½,³⁄₂)
7(³⁄₂,⁵⁄₂)
6(³⁄₂,⁵⁄₂)
5(³⁄₂,⁵⁄₂)
4(³⁄₂,⁵⁄₂)
3(³⁄₂,⁵⁄₂)
5(⁵⁄₂,⁷⁄₂)
3p⁵3d²(⁵⁄₂)
4(⁵⁄₂,⁷⁄₂)
7(⁷⁄₂,⁹⁄₂)
6(⁷⁄₂,⁹⁄₂)
5(⁷⁄₂,⁹⁄₂)
7
6
8
7
5470.98, 5517.72
2836.97, 2862.60
725.03, 729.35
5398.93, 5492.51
1183.64, 1195.21
664.30, 665.69
2067.56, 2103.16
776.76–781.73
3156.72, 3170.96
6913.85–6988.74
6231.62–6292.41
1280.79–1283.33
3541.36–3581.39
2929.96–2957.31
433.04–433.76
363.79–364.30
763.84–768.65
1451.74–1469.19
3272.50, 3272.77
7483.07–7494.77
6968.54
5877.79–5891.15
6247.74, 6262.86
4395.92, 4403.45
1265.14, 1266.27
2541.79–2547.31
362.52–363.02
363.85, 364.35
423.49–424.17
2688.32, 2689.39
4131.22, 4133.78
3966.16
1564.69, 1564.85
2359.14, 2359.50
7652.12
2889.36
4618.11
4677.58
7706.85

Ti IV ENERGY LEVELS (19 electrons, Z=22)

(K I sequence, Configuration: $1s^2\ 2s^2\ 2p^6\ 3s^2\ 3p^6\ nl$, Doublet System)

ENERGY (cm^{-1})

Axis: 350 x 10^3, 300 x 10^3, 250 x 10^3, 200 x 10^3, 150 x 10^3, 100 x 10^3, 0

ns ^2S

292 999.54 — 7(1/2)
265 847.42 — 6(1/2)
212 407.34 — 5(1/2)
80 388.93 — 4(1/2)

np $^2P^o$

274 881.21, 274 726.29 — 6(1/2, 3/2)
230 924.38, 230 608.89 — 5(1/2, 3/2)
128 739.59, 128 921.36 — 4(1/2, 3/2)

nd ^2D

306 408.30, 306 395.69 — 7(3/2, 5/2)
289 206.93, 289 185.99 — 6(3/2, 5/2)
258 877.08, 258 838.48 — 5(3/2, 5/2)
196 889.96, 196 804.27 — 4(3/2, 5/2)
382.1, 0.0 — 3(3/2, 5/2)

nf $^2F^o$

275 861.94, 275 847.01 — 5(5/2, 7/2)
274 839.82 — $3p^5 3d^2$(5/2)
236 142.30, 236 135.29 — 4(5/2, 7/2)

ng ^2G

313 034.1, 313 033.9 — 7(7/2, 9/2)
300 046.2, 300 045.9 — 6(7/2, 9/2)
278 511.23, 278 510.63 — 5(7/2, 9/2)

nh $^2H^o$

313 110.72 — 7
300 158.76 — 6

ni ^2I

321 531.3 — 8
313 130.66 — 7

KEY

236 142.30, 236 135.29 — 4(5/2, 7/2)
J, low to high
n
Energy Levels (cm^{-1})

Ionization Limit
348 973.3 cm^{-1}
(43.266 eV)
[Ti IV $3p^6\ 3d\ ^2D_{3/2}$ → Ti V $3p^6\ ^1S_0$]

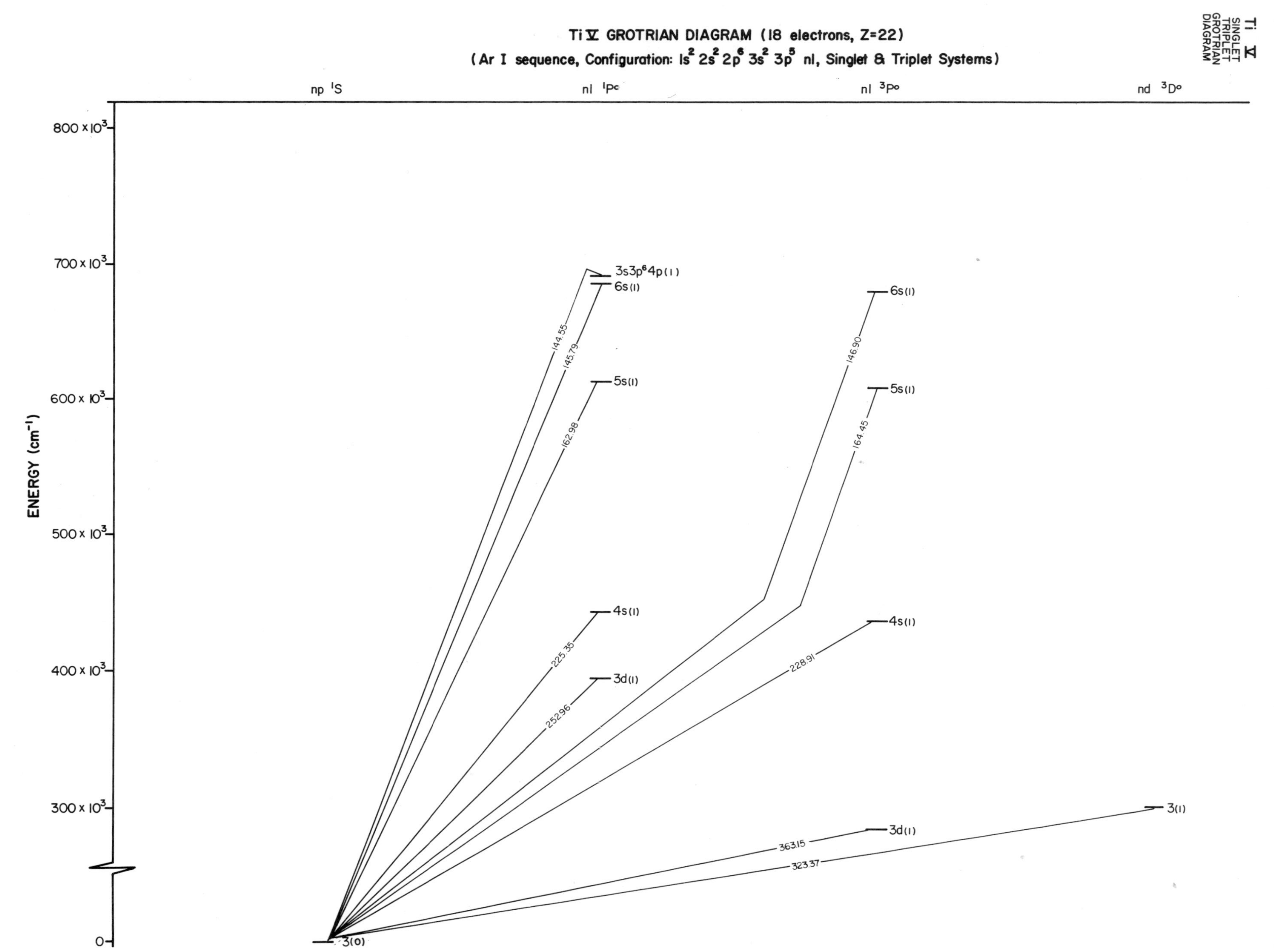
Ti V
SINGLET
TRIPLET
GROTRIAN
DIAGRAM
Ti V GROTRIAN DIAGRAM (18 electrons, Z=22)
(Ar I sequence, Configuration: 1s² 2s² 2p⁶ 3s² 3p⁵ nl, Singlet & Triplet Systems)
np ¹S
nl ¹P°
nl ³P°
nd ³D°
ENERGY (cm⁻¹)
800 x 10³
700 x 10³
600 x 10³
500 x 10³
400 x 10³
300 x 10³
0
3s3p⁶4p(1)
6s(1)
5s(1)
4s(1)
3d(1)
3(0)
3(1)
144.55
145.79
162.98
225.35
252.96
146.90
164.45
228.91
363.15
323.37

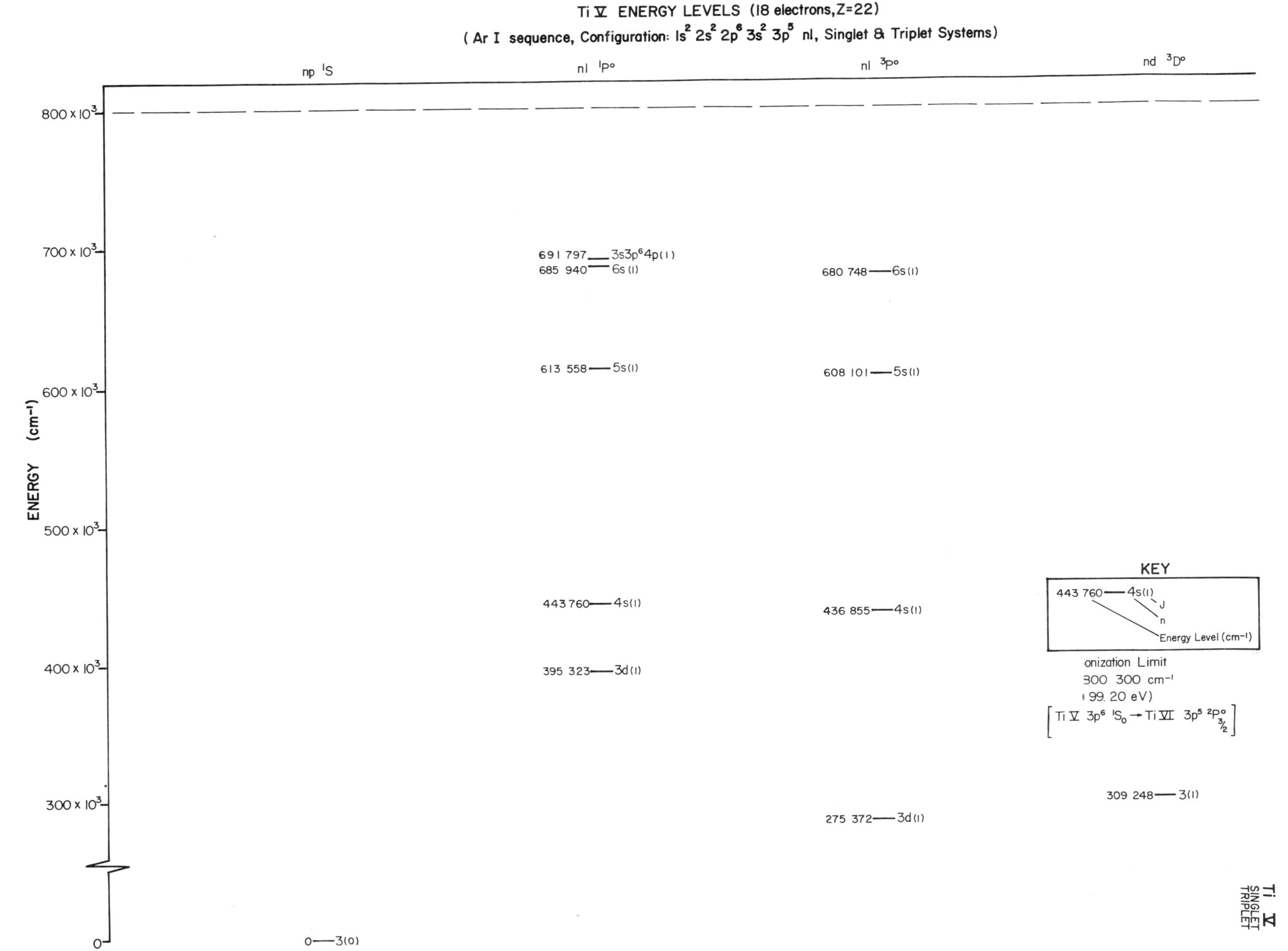

Ti V ENERGY LEVELS (18 electrons, Z=22)
(Ar I sequence, Configuration: $1s^2 2s^2 2p^6 3s^2 3p^5$ nl, Singlet & Triplet Systems)
np ^1S
nl $^1P^o$
nl $^3P^o$
nd $^3D^o$
800 x 10^3
700 x 10^3
600 x 10^3
500 x 10^3
400 x 10^3
300 x 10^3
0
ENERGY (cm^{-1})
691 797 — $3s3p^64p$(1)
685 940 — 6s (1)
613 558 — 5s(1)
443 760 — 4s(1)
395 323 — 3d(1)
680 748 — 6s(1)
608 101 — 5s(1)
436 855 — 4s(1)
275 372 — 3d(1)
309 248 — 3(1)
0 — 3(0)
KEY
443 760 — 4s(1)
J
n
Energy Level (cm^{-1})
onization Limit
300 300 cm^{-1}
(99.20 eV)
[Ti V $3p^6$ $^1S_0$ → Ti VI $3p^5$ $^2P^o_{3/2}$]
Ti V
SINGLET
TRIPLET

Ti VI GROTRIAN DIAGRAM (17 electrons, Z=22)

(Cl I sequence, Configuration: $1s^2\,2s^2\,2p^6\,3s^2\,3p^4\,nl$, Doublet and Quartet Systems)

nl ^2S | nl ^2P | np $^2P^o$ | nl ^2D | nd ^2F | nl ^4P | nd ^4F

ENERGY (cm^{-1})

1000 x 10^3
900 x 10^3
800 x 10^3
700 x 10^3
600 x 10^3
500 x 10^3
SCALE CHANGE
400 x 10^3
200 x 10^3
0

nl ^2S:
4d′(1/2) — 149.56
4s″(1/2) — 182.15, 184.11
3d′(1/2) — 263.25, 267.34
3s3p^6(1/2) — 508.58–524.11

nl ^2P:
5d′(3/2) — 125.69, 126.57
5s(3/2, 1/2) — 139.91–141.06
4d′(3/2, 1/2) — 149.01–150.32
4d(3/2) — 151.90, 153.26
4s(3/2, 1/2) — 197.46 – 201.31
3d′(3/2, 1/2) — 254.03 – 259.23
3d(1/2, 3/2) — 342.60 – 353.88

np $^2P^o$:
3(3/2, 1/2)

nl ^2D:
5d′(5/2, 3/2) — 125.46, 126.–
5d(5/2, 3/2) — 129.15 – 130.11
5s′(3/2, 5/2) — 136.71, 137.81
4d″(3/2, 5/2) — 141.99, 143.18
4d′(5/2, 3/2) — 148.10 – 149.39
4d(5/2, 3/2) — 158.38 – 154.77
4s′(5/2, 3/2) — 192.71 – 194.90
3d′(5/2, 3/2) — 247.45 – 251.07
3d″(3/2, 5/2) — 282.22 – 288.36
3d(3/2, 5/2) — 330.70 – 341.11

nd ^2F:
4d(5/2, 3/2) — 152.34
3d′(5/2) — 301.91

nl ^4P:
5s(3/2) — 141.11
4s(5/2, 3/2, 1/2) — 201.87 – 203.43
3d(1/2) — 331.77

nd ^4F:
5(5/2) — 128.45
4(5/2, 3/2) — 152.96, 154.16

CORES:
nl = $3p^4$ (^3P)
n′l′ = $3p^4$ (^1D)
n″l″ = $3p^4$ (^1S)

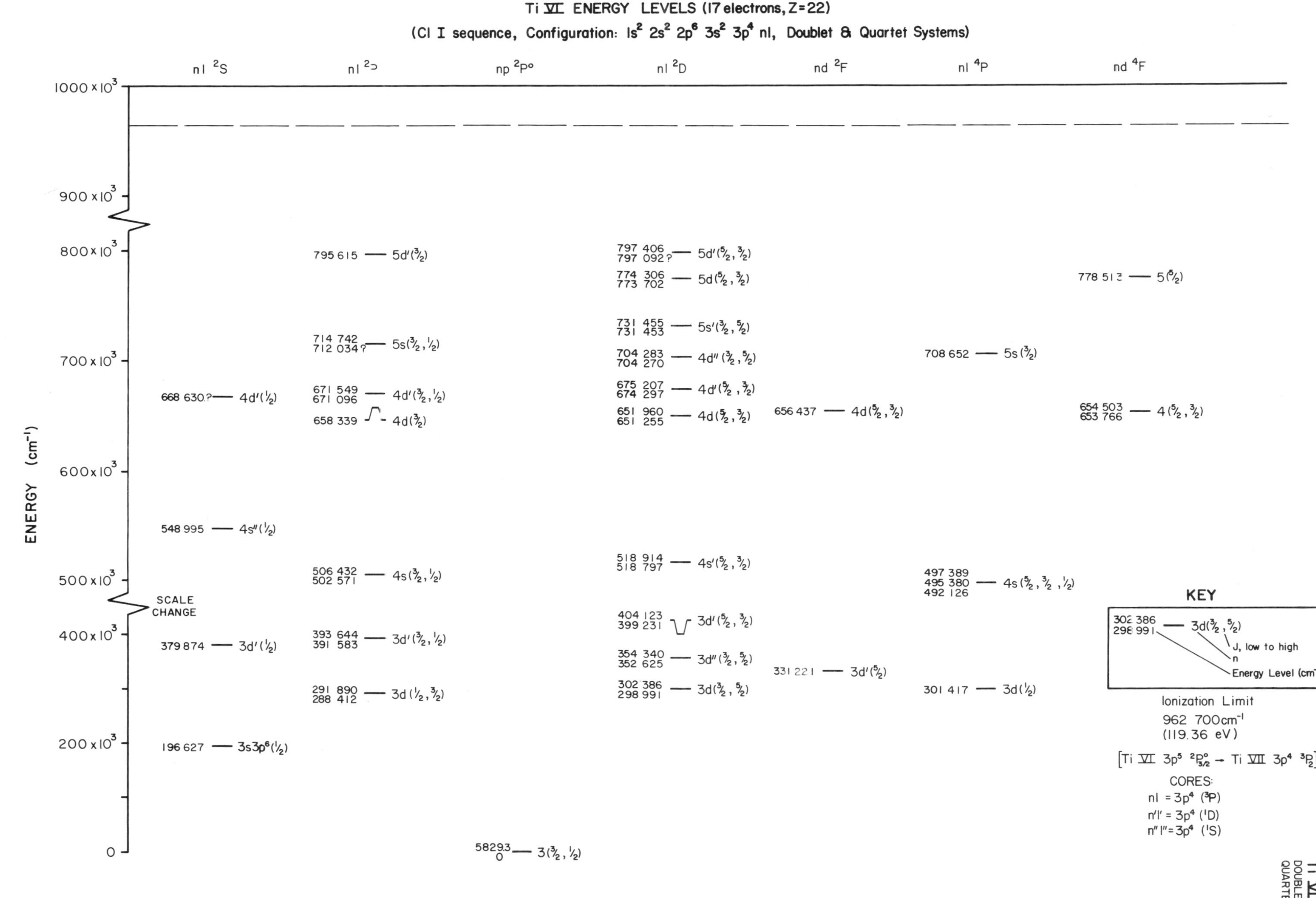

Ti VI ENERGY LEVELS (17 electrons, Z=22)
(Cl I sequence, Configuration: 1s² 2s² 2p⁶ 3s² 3p⁴ nl, Doublet & Quartet Systems)
nl ²S
nl ²P
np ²P°
nl ²D
nd ²F
nl ⁴P
nd ⁴F
ENERGY (cm⁻¹)
1000 x 10³
900 x 10³
800 x 10³
700 x 10³
600 x 10³
500 x 10³
400 x 10³
200 x 10³
0
SCALE CHANGE
668 630? — 4d′(½)
548 995 — 4s″(½)
379 874 — 3d′(½)
196 627 — 3s3p⁶(½)
795 615 — 5d′(3/2)
714 742
712 034? — 5s(3/2, ½)
671 549
671 096 — 4d′(3/2, ½)
658 339 — 4d(3/2)
506 432
502 571 — 4s(3/2, ½)
393 644
391 583 — 3d′(3/2, ½)
291 890
288 412 — 3d(½, 3/2)
58293
0 — 3(3/2, ½)
797 406
797 092? — 5d′(5/2, 3/2)
774 306
773 702 — 5d(5/2, 3/2)
731 455
731 453 — 5s′(3/2, 5/2)
704 283
704 270 — 4d″(3/2, 5/2)
675 207
674 297 — 4d′(5/2, 3/2)
651 960
651 255 — 4d(5/2, 3/2)
518 914
518 797 — 4s′(5/2, 3/2)
404 123
399 231 — 3d′(5/2, 3/2)
354 340
352 625 — 3d″(3/2, 5/2)
302 386
298 991 — 3d(3/2, 5/2)
656 437 — 4d(5/2, 3/2)
331 221 — 3d′(5/2)
708 652 — 5s(3/2)
497 389
495 380
492 126 — 4s(5/2, 3/2, ½)
301 417 — 3d(½)
778 513 — 5(5/2)
654 503
653 766 — 4(5/2, 3/2)
KEY
302 386
298 991 — 3d(3/2, 5/2)
J, low to high
n
Energy Level (cm⁻¹)
Ionization Limit
962 700cm⁻¹
(119.36 eV)
[Ti VI 3p⁵ ²P°3/2 → Ti VII 3p⁴ ³P2]
CORES:
nl = 3p⁴ (³P)
n′l′ = 3p⁴ (¹D)
n″l″ = 3p⁴ (¹S)
Ti VI
DOUBLET
QUARTET

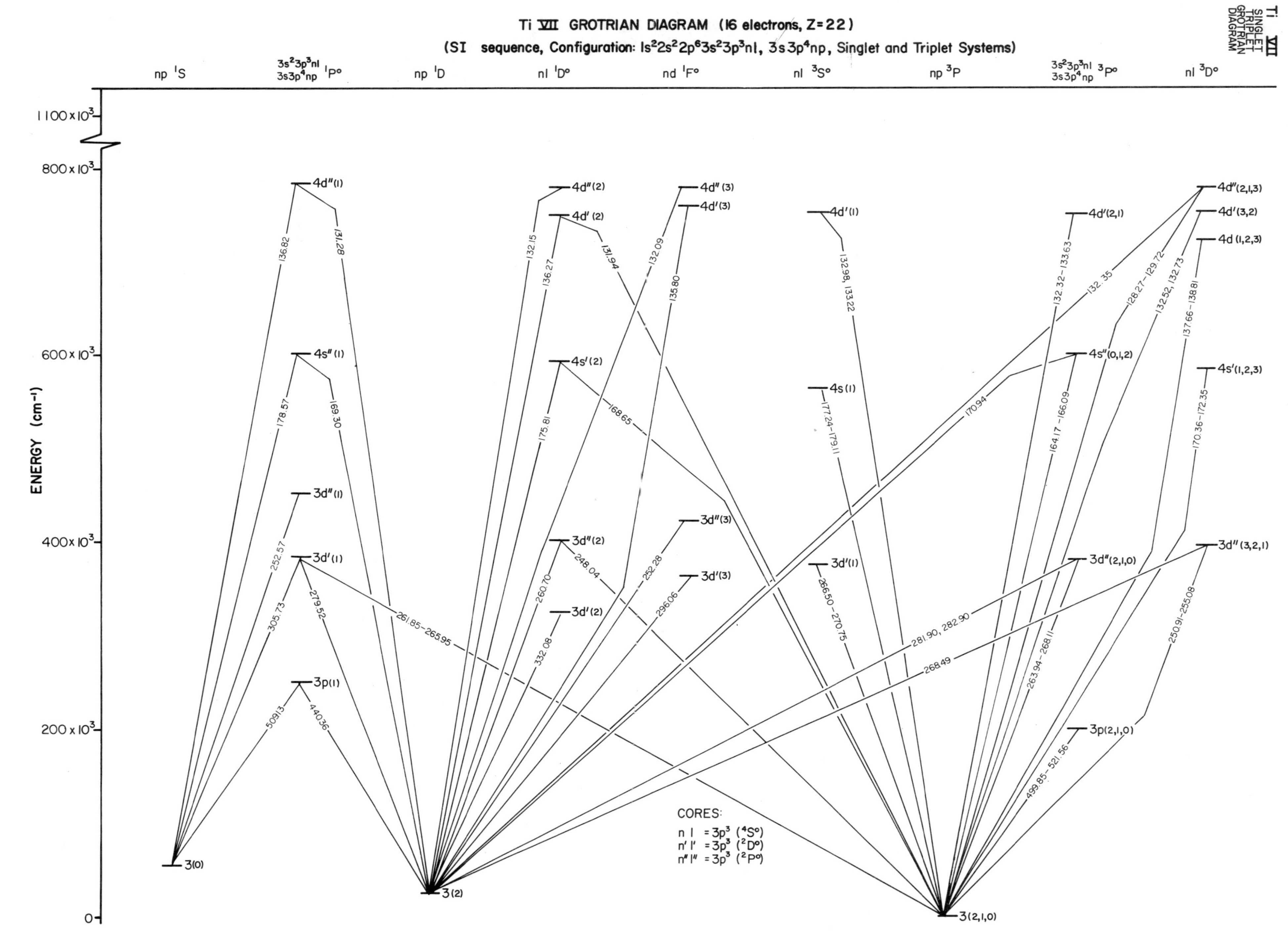

Ti VII
SINGLET
TRIPLET
GROTRIAN
DIAGRAM
Ti VII GROTRIAN DIAGRAM (16 electrons, Z=22)
(SI sequence, Configuration: 1s²2s²2p⁶3s²3p³nl, 3s3p⁴np, Singlet and Triplet Systems)
np ¹S
3s²3p³nl ¹P°
3s3p⁴np
np ¹D
nl ¹D°
nd ¹F°
nl ³S°
np ³P
3s²3p³nl ³P°
3s3p⁴np
nl ³D°
ENERGY (cm⁻¹)
1100×10³
800×10³
600×10³
400×10³
200×10³
0
3(0)
3(2)
3(2,1,0)
4d''(1)
4s''(1)
3d''(1)
3d'(1)
3p(1)
4d''(2)
4d'(2)
4s'(2)
3d''(2)
3d'(2)
4d''(3)
4d'(3)
3d''(3)
3d'(3)
4d'(1)
4s(1)
3d'(1)
4d'(2,1)
4s''(0,1,2)
3d''(2,1,0)
3p(2,1,0)
4d''(2,1,3)
4d'(3,2)
4d(1,2,3)
4s'(1,2,3)
3d''(3,2,1)
136.82
131.28
178.57
169.30
252.57
305.73
279.52
509.13
440.36
261.85–265.95
132.15
136.27
131.94
175.81
168.65
260.70
248.04
332.08
132.09
135.80
252.28
296.06
132.98, 133.22
177.24–179.11
266.50–270.75
132.32–133.63
164.17–166.09
263.94–268.11
499.85–521.56
132.35
170.94
281.90, 282.90
268.49
128.27–129.72
132.52, 132.73
137.66–138.81
170.36–172.35
250.91–255.08
CORES:
n l = 3p³ (⁴S°)
n' l' = 3p³ (²D°)
n'' l'' = 3p³ (²P°)

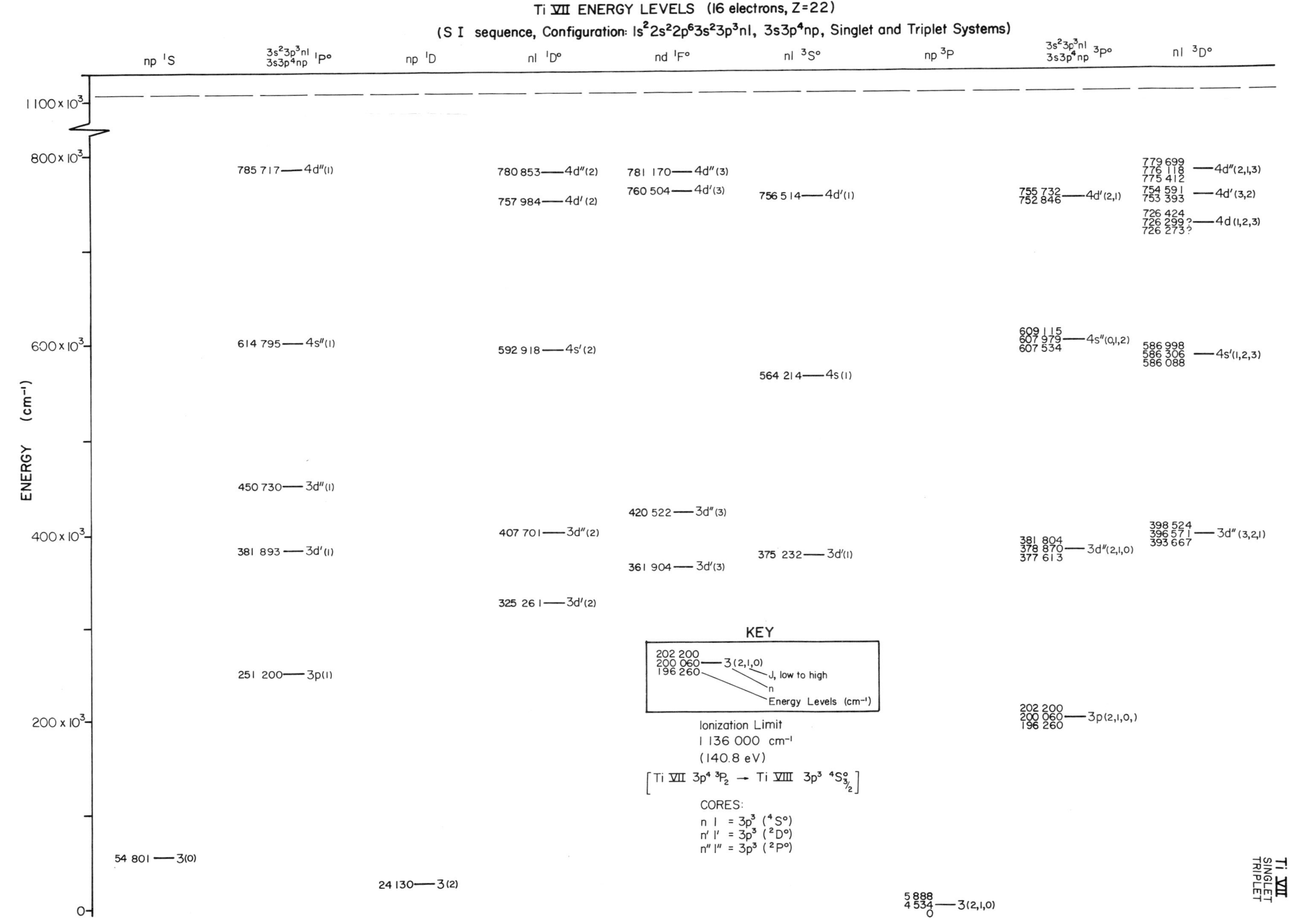
Ti VII ENERGY LEVELS (16 electrons, Z=22)
(S I sequence, Configuration: 1s²2s²2p⁶3s²3p³nl, 3s3p⁴np, Singlet and Triplet Systems)
np ¹S
3s²3p³nl 3s3p⁴np ¹P°
np ¹D
nl ¹D°
nd ¹F°
nl ³S°
np ³P
3s²3p³nl 3s3p⁴np ³P°
nl ³D°
1100×10³
800×10³
600×10³
400×10³
200×10³
0
ENERGY (cm⁻¹)
54 801 — 3(0)
785 717 — 4d″(1)
614 795 — 4s″(1)
450 730 — 3d″(1)
381 893 — 3d′(1)
251 200 — 3p(1)
24 130 — 3(2)
780 853 — 4d″(2)
757 984 — 4d′(2)
592 918 — 4s′(2)
407 701 — 3d″(2)
325 261 — 3d′(2)
781 170 — 4d″(3)
760 504 — 4d′(3)
420 522 — 3d″(3)
361 904 — 3d′(3)
756 514 — 4d′(1)
564 214 — 4s(1)
375 232 — 3d′(1)
5 888 4 534 0 — 3(2,1,0)
755 732 752 846 — 4d′(2,1)
609 115 607 979 607 534 — 4s″(0,1,2)
381 804 378 870 377 613 — 3d″(2,1,0)
202 200 200 060 196 260 — 3p(2,1,0,)
779 699 776 118 775 412 — 4d″(2,1,3)
754 591 753 393 — 4d′(3,2)
726 424 726 299? 726 273? — 4d(1,2,3)
586 998 586 306 586 088 — 4s′(1,2,3)
398 524 396 571 393 667 — 3d″(3,2,1)
KEY
202 200 200 060 196 260 — 3(2,1,0)
J, low to high
n
Energy Levels (cm⁻¹)
Ionization Limit
1 136 000 cm⁻¹
(140.8 eV)
[Ti VII 3p⁴ ³P₂ → Ti VIII 3p³ ⁴S°3/2]
CORES:
n l = 3p³ (⁴S°)
n′ l′ = 3p³ (²D°)
n″ l″ = 3p³ (²P°)
Ti VII
SINGLET
TRIPLET

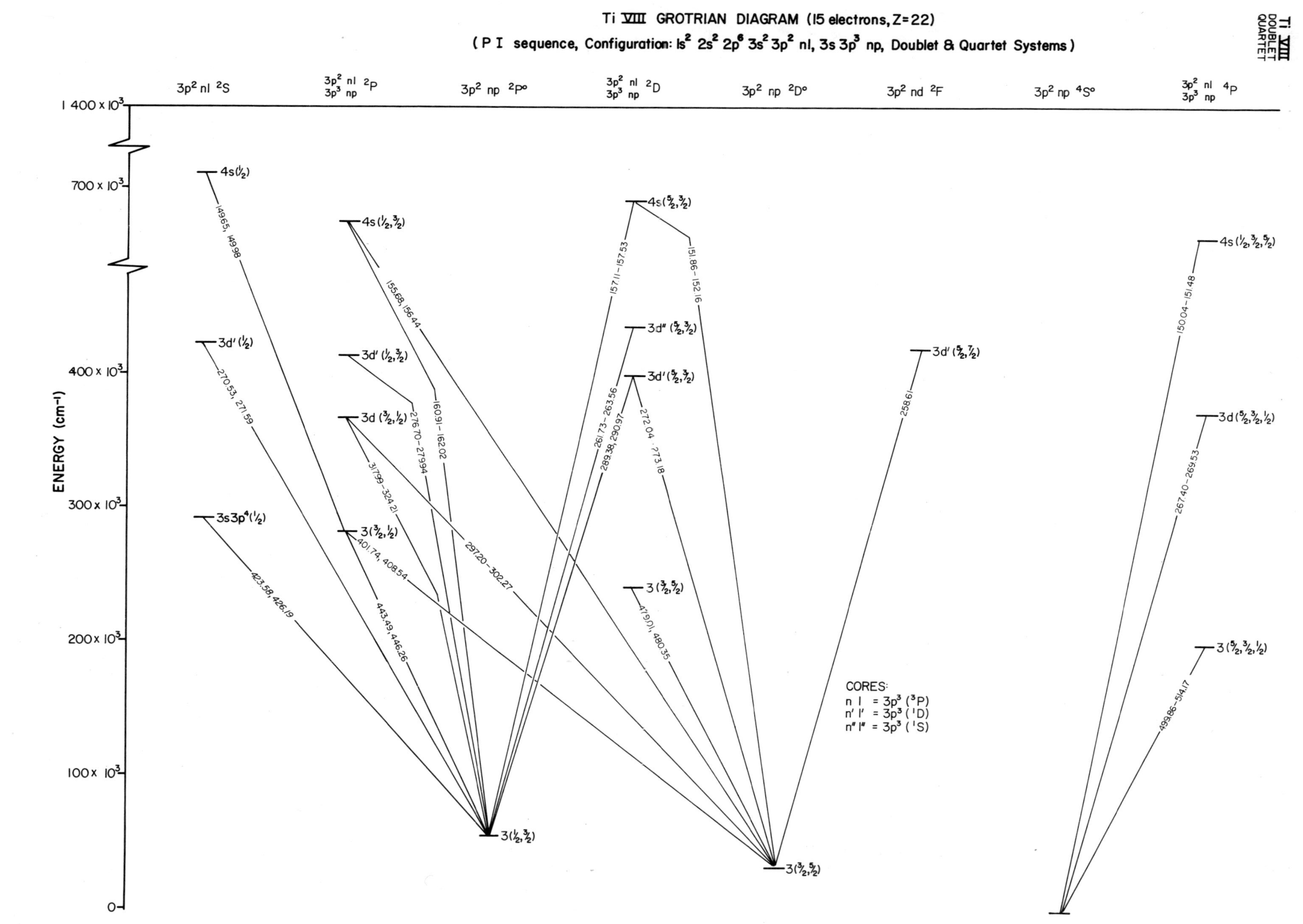

Ti VIII
DOUBLET
QUARTET
Ti VIII GROTRIAN DIAGRAM (15 electrons, Z=22)
(P I sequence, Configuration: 1s² 2s² 2p⁶ 3s² 3p² nl, 3s 3p³ np, Doublet & Quartet Systems)
3p² nl ²S
3p² nl 3p³ np ²P
3p² np ²P°
3p² nl 3p³ np ²D
3p² np ²D°
3p² nd ²F
3p² np ⁴S°
3p² nl 3p³ np ⁴P
ENERGY (cm⁻¹)
1 400 x 10³
700 x 10³
400 x 10³
300 x 10³
200 x 10³
100 x 10³
0
4s(1/2)
3d′(1/2)
3s3p⁴(1/2)
4s(1/2,3/2)
3d′(1/2,3/2)
3d(3/2,1/2)
3(3/2,1/2)
3(1/2,3/2)
4s(5/2,3/2)
3d″(5/2,3/2)
3d′(5/2,3/2)
3(3/2,5/2)
3(3/2,5/2)
3d′(5/2,7/2)
4s(1/2,3/2,5/2)
3d(5/2,3/2,1/2)
3(5/2,3/2,1/2)
149.65, 149.98
270.53, 271.59
423.58, 426.19
155.68, 156.44
160.91 – 162.02
276.70 – 279.94
317.99 – 324.21
297.20 – 302.27
401.74, 408.54
443.49, 446.26
157.11 – 157.53
151.86 – 152.16
261.73 – 263.56
289.38, 290.97
272.04 – 273.18
479.01, 480.35
258.61
150.04 – 151.48
267.40 – 269.53
499.86 – 514.17
CORES:
n l = 3p³ (³P)
n′ l′ = 3p³ (¹D)
n″ l″ = 3p³ (¹S)

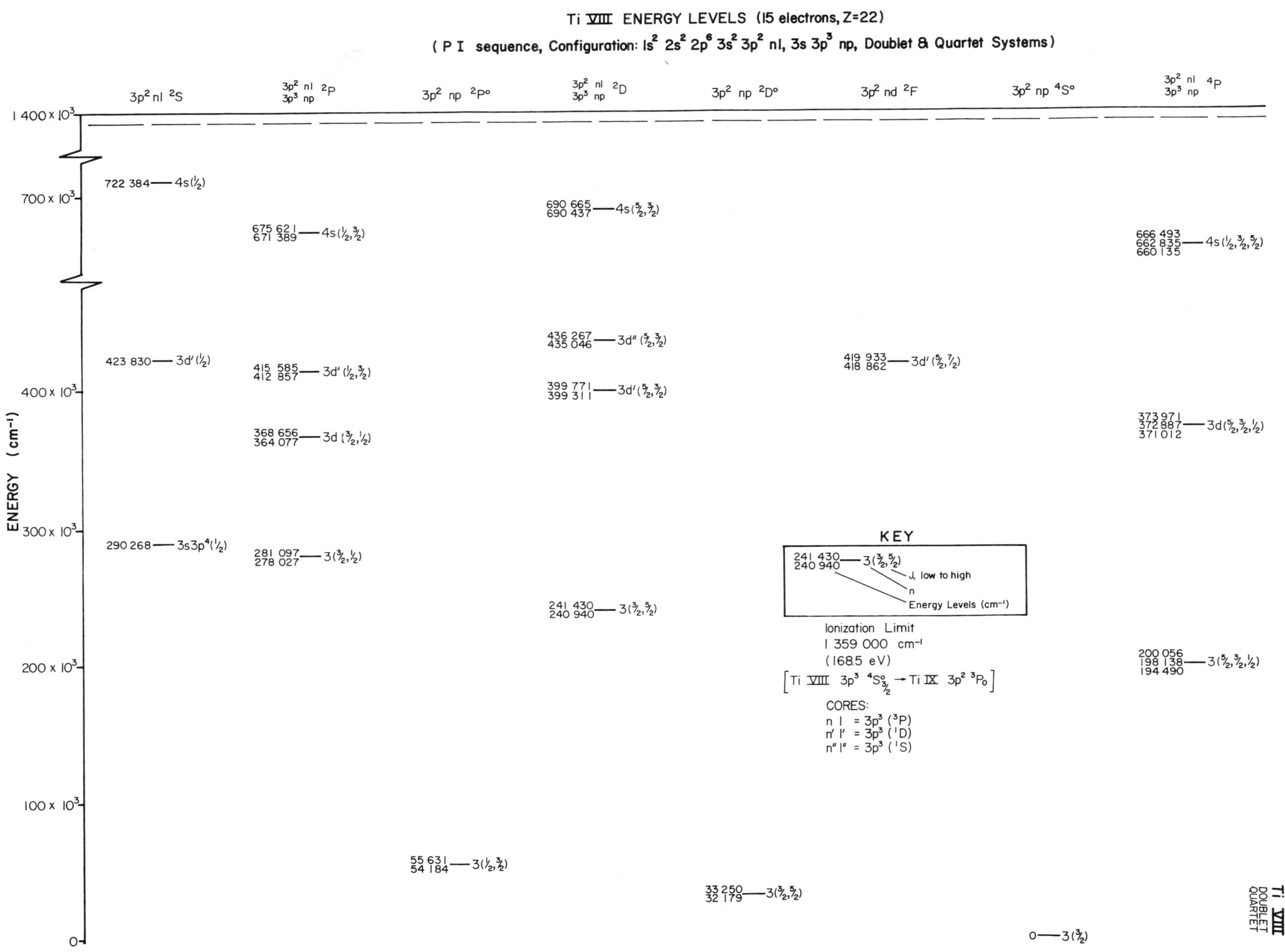

Ti VIII ENERGY LEVELS (15 electrons, Z=22)
(P I sequence, Configuration: $1s^2\ 2s^2\ 2p^6\ 3s^2\ 3p^2$ nl, $3s\ 3p^3$ np, Doublet & Quartet Systems)
$3p^2$ nl ^2S
$3p^2$ nl / $3p^3$ np ^2P
$3p^2$ np $^2P^o$
$3p^2$ nl / $3p^3$ np ^2D
$3p^2$ np $^2D^o$
$3p^2$ nd ^2F
$3p^2$ np $^4S^o$
$3p^2$ nl / $3p^3$ np ^4P
ENERGY (cm^{-1})
1 400 x 10^3
700 x 10^3
400 x 10^3
300 x 10^3
200 x 10^3
100 x 10^3
0
722 384 — 4s(1/2)
423 830 — 3d′(1/2)
290 268 — $3s3p^4$(1/2)
675 621 / 671 389 — 4s(1/2, 3/2)
415 585 / 412 857 — 3d′(1/2, 3/2)
368 656 / 364 077 — 3d (3/2, 1/2)
281 097 / 278 027 — 3(3/2, 1/2)
55 631 / 54 184 — 3(1/2, 3/2)
690 665 / 690 437 — 4s(5/2, 3/2)
436 267 / 435 046 — 3d″(5/2, 3/2)
399 771 / 399 311 — 3d′(5/2, 3/2)
241 430 / 240 940 — 3(3/2, 5/2)
33 250 / 32 179 — 3(3/2, 5/2)
419 933 / 418 862 — 3d′(5/2, 7/2)
0 — 3(3/2)
666 493 / 662 835 / 660 135 — 4s(1/2, 3/2, 5/2)
373 971 / 372 887 / 371 012 — 3d(5/2, 3/2, 1/2)
200 056 / 198 138 / 194 490 — 3(5/2, 3/2, 1/2)
KEY
241 430 / 240 940 — 3(3/2, 5/2)
J, low to high
n
Energy Levels (cm^{-1})
Ionization Limit
1 359 000 cm^{-1}
(168.5 eV)
[Ti VIII $3p^3\ ^4S^o_{3/2}$ → Ti IX $3p^2\ ^3P_0$]
CORES:
n l = $3p^3$ (^3P)
n′ l′ = $3p^3$ (^1D)
n″ l″ = $3p^3$ (^1S)
Ti VIII
DOUBLET
QUARTET

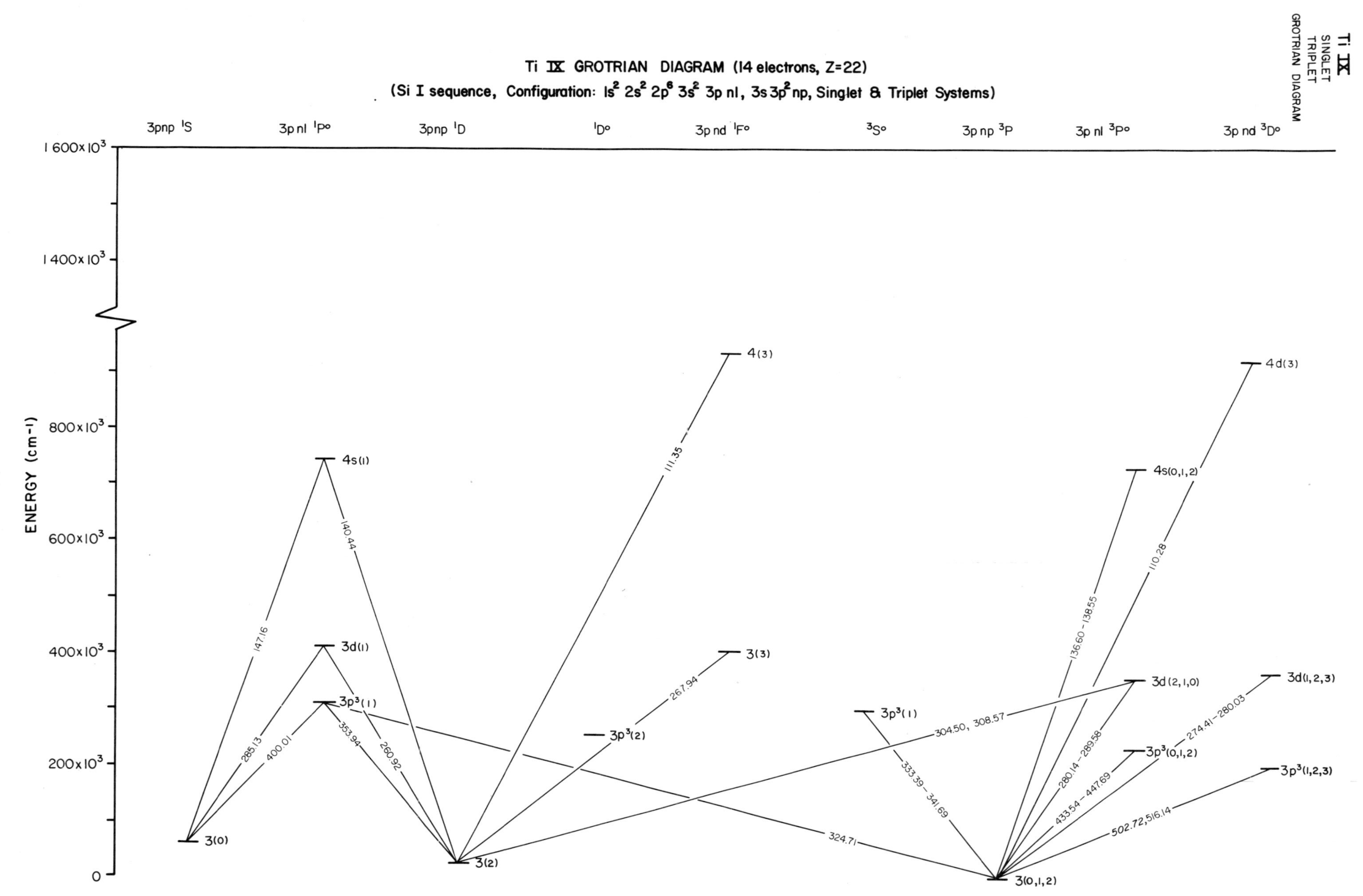
Ti IX GROTRIAN DIAGRAM (14 electrons, Z=22)
(Si I sequence, Configuration: $1s^2 2s^2 2p^6 3s^2$ 3p nl, $3s3p^2$np, Singlet & Triplet Systems)
3pnp ¹S
3p nl ¹P°
3pnp ¹D
¹D°
3p nd ¹F°
³S°
3p np ³P
3p nl ³P°
3p nd ³D°
ENERGY (cm-1)
1600x10³
1400x10³
800x10³
600x10³
400x10³
200x10³
0
3(0)
4s(1)
3d(1)
3p³(1)
3(2)
4(3)
3(3)
3p³(2)
3p³(1)
3(0,1,2)
4s(0,1,2)
4d(3)
3d(2,1,0)
3d(1,2,3)
3p³(0,1,2)
3p³(1,2,3)
147.16
285.13
400.01
140.44
353.94
260.92
111.35
267.94
324.71
304.50, 308.57
333.39 – 341.69
136.60 – 138.55
110.28
280.14 – 289.58
433.54 – 447.69
274.41 – 280.03
502.72, 516.14

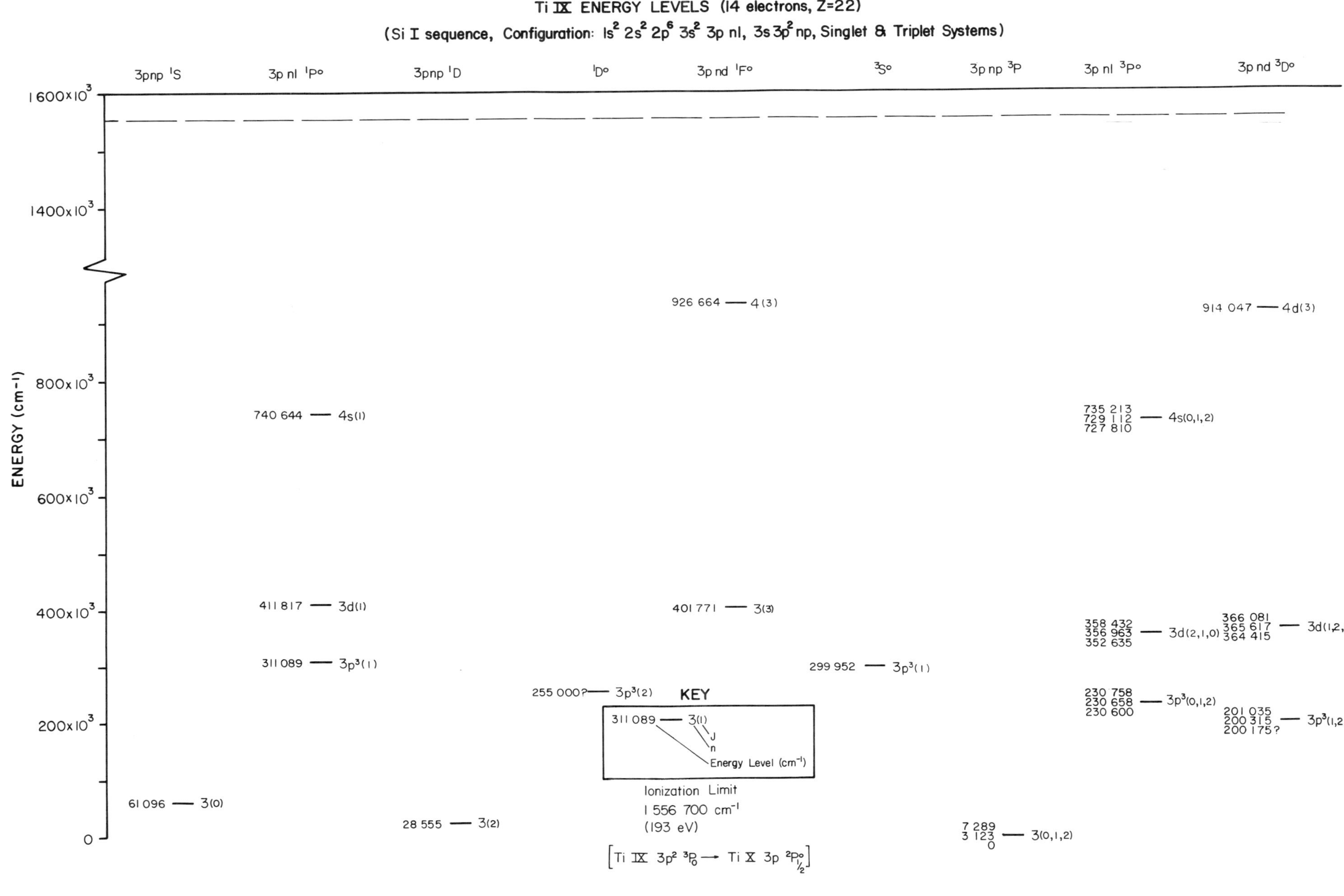
Ti IX ENERGY LEVELS (14 electrons, Z=22)
(Si I sequence, Configuration: 1s² 2s² 2p⁶ 3s² 3p nl, 3s 3p² np, Singlet & Triplet Systems)
3pnp ¹S
3p nl ¹P°
3pnp ¹D
¹D°
3p nd ¹F°
³S°
3p np ³P
3p nl ³P°
3p nd ³D°
1600x10³
1400x10³
800x10³
600x10³
400x10³
200x10³
0
ENERGY (cm⁻¹)
926 664 — 4(3)
914 047 — 4d(3)
740 644 — 4s(1)
735 213
729 112
727 810 — 4s(0,1,2)
411 817 — 3d(1)
401 771 — 3(3)
358 432
356 963
352 635 — 3d(2,1,0)
366 081
365 617
364 415 — 3d(1,2,3)
311 089 — 3p³(1)
299 952 — 3p³(1)
255 000? — 3p³(2)
230 758
230 658
230 600 — 3p³(0,1,2)
201 035
200 315
200 175? — 3p³(1,2,3)
61 096 — 3(0)
28 555 — 3(2)
7 289
3 123
0 — 3(0,1,2)
KEY
311 089 — 3(1)
J
n
Energy Level (cm⁻¹)
Ionization Limit
1 556 700 cm⁻¹
(193 eV)
[Ti IX 3p² ³P₀ → Ti X 3p ²P°½]

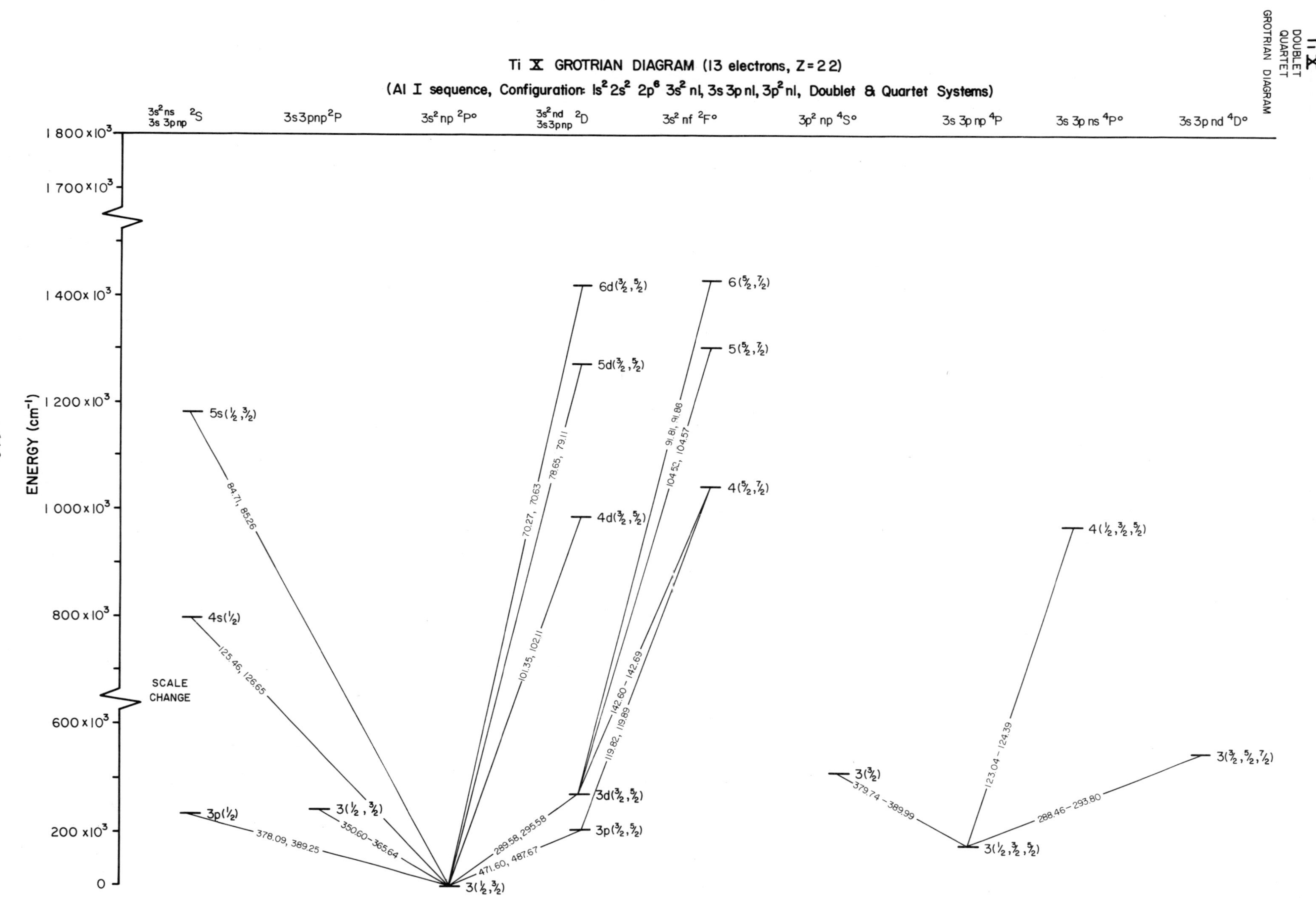

Ti X
DOUBLET
QUARTET
GROTRIAN DIAGRAM
Ti X GROTRIAN DIAGRAM (13 electrons, Z = 22)
(Al I sequence, Configuration: 1s^{2}2s^2 2p^6 3s^2 nl, 3s 3p nl, 3p^2 nl, Doublet & Quartet Systems)
3s^{2}ns 3s 3pnp ^{2}S
3s3pnp ^{2}P
3s^2 np ^{2}P°
3s^{2}nd 3s3pnp ^{2}D
3s^2 nf ^{2}F°
3p^2 np ^{4}S°
3s 3p np ^{4}P
3s 3p ns ^{4}P°
3s 3p nd ^{4}D°
ENERGY (cm^{-1})
1 800 x 10^3
1 700 x 10^3
1 400 x 10^3
1 200 x 10^3
1 000 x 10^3
800 x 10^3
600 x 10^3
200 x 10^3
0
SCALE CHANGE
5s(1/2, 3/2)
4s(1/2)
3p(1/2)
3(1/2, 3/2)
84.71, 85.26
125.46, 126.65
378.09, 389.25
350.60 – 365.64
6d(3/2, 5/2)
5d(3/2, 5/2)
4d(3/2, 5/2)
3d(3/2, 5/2)
3p(3/2, 5/2)
70.27, 70.63
78.65, 79.11
101.35, 102.11
289.58, 295.58
471.60, 487.67
6(5/2, 7/2)
5(5/2, 7/2)
4(5/2, 7/2)
91.81, 91.86
104.52, 104.57
142.60 – 142.69
119.82, 119.89
3(3/2)
379.74 – 389.99
3(1/2, 3/2, 5/2)
4(1/2, 3/2, 5/2)
123.04 – 124.39
288.46 – 293.80
3(3/2, 5/2, 7/2)

Ti X ENERGY LEVELS (13 electrons, Z = 22)

(Al I sequence, Configuration: $1s^2 2s^2 2p^6 3s^2 nl, 3s3p\,nl, 3p^2 nl$, Doublet & Quartet Systems)

ENERGY (cm^{-1})

Axis: 0; 200×10^3; 600×10^3; 800×10^3; 1000×10^3; 1200×10^3; 1400×10^3; 1700×10^3; 1800×10^3

SCALE CHANGE

$3s^2 ns$, $3s3pnp$ ^2S
- 1 180 393 — $5s(\frac{1}{2},\frac{3}{2})$
- 797 112 — $4s(\frac{1}{2})$
- 264 467 — $3p(\frac{1}{2})$

$3s3pnp$ ^2P
- 285 217, 281 044 — $3(\frac{1}{2},\frac{3}{2})$

$3s^2 np$ $^2P°$
- 7 542, 0 — $3(\frac{1}{2},\frac{3}{2})$

$3s^2 nd$, $3s3pnp$ ^2D
- 1 423 471, 1 423 182 — $6d(\frac{3}{2},\frac{5}{2})$
- 1 271 678, 1 271 459 — $5d(\frac{3}{2},\frac{5}{2})$
- 986 918, 986 656 — $4d(\frac{3}{2},\frac{5}{2})$
- 345 856, 345 329 — $3d(\frac{3}{2},\frac{5}{2})$
- 212 598, 212 046 — $3p(\frac{3}{2},\frac{5}{2})$

$3s^2 nf$ $^2F°$
- 1 434 556 — $6(\frac{5}{2},\frac{7}{2})$
- 1 302 172, 1 302 120 — $5(\frac{5}{2},\frac{7}{2})$
- 1 046 690, 1 046 618 — $4(\frac{5}{2},\frac{7}{2})$

$3p^2 np$ $^4S°$
- 421 188 — $3p^3(\frac{3}{2})$

$3s3pnp$ ^4P
- 164 764, 160 655, 157 850 — $3(\frac{1}{2},\frac{3}{2},\frac{5}{2})$

$3s3pns$ $^4P°$
- 973 441, 968 680, 966 176 — $4(\frac{1}{2},\frac{3}{2},\frac{5}{2})$

$3s3pnd$ $^4D°$
- 505 266, 505 134, 504 516 — $3(\frac{3}{2},\frac{5}{2},\frac{7}{2})$

KEY

164764, 160655, 157850 — $3(\frac{1}{2},\frac{3}{2},\frac{5}{2})$
J, low to high
n
Energy Levels (cm^{-1})

Ionization Limit
1 741 500 cm^{-1}
(215.91 eV)
[Ti X $3s^2 3p\ ^2P°_{1/2}$ → Ti XI $3s^2\ ^1S_0$]

Ti XI
SINGLET
GROTRIAN DIAGRAM

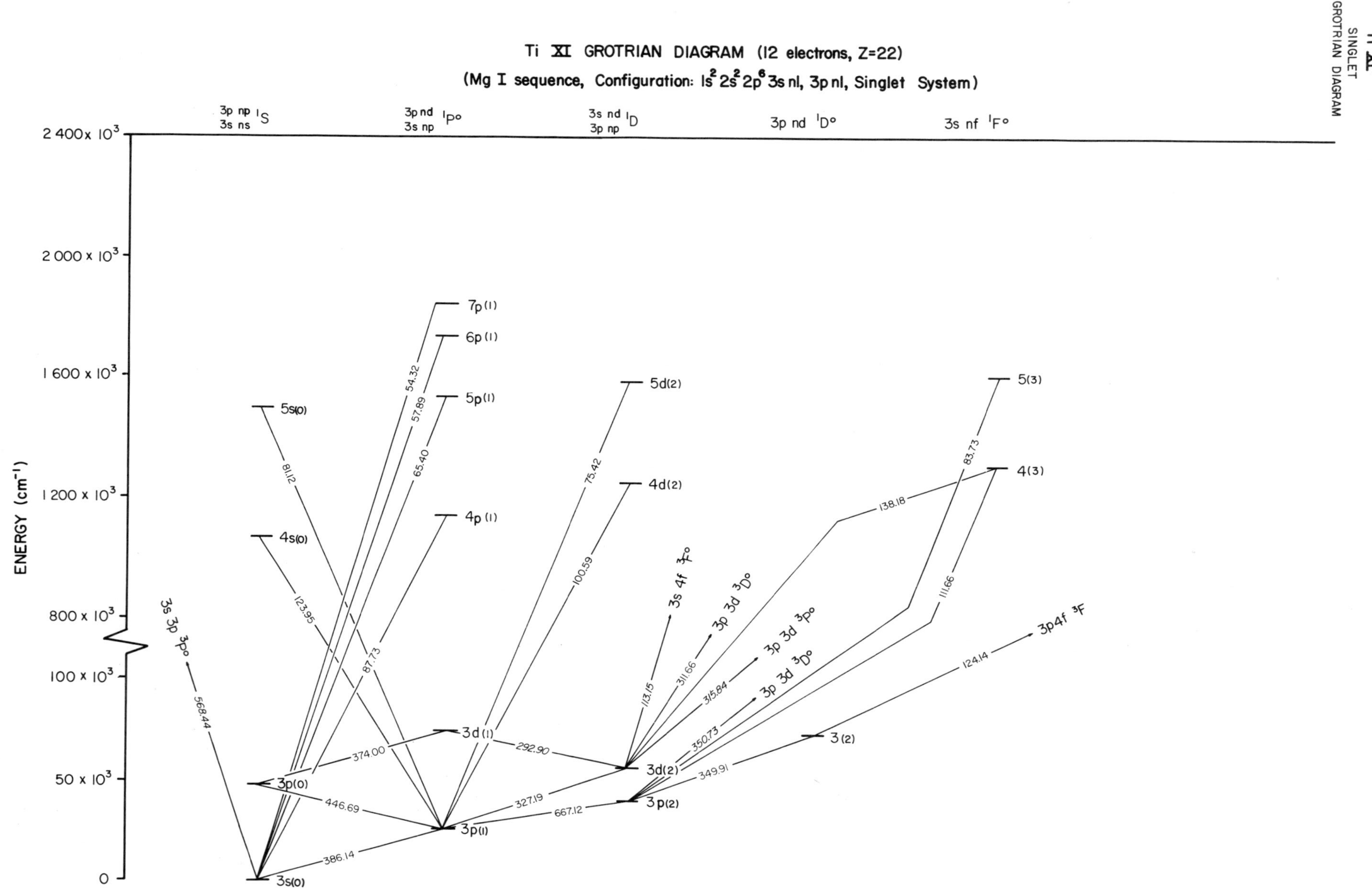

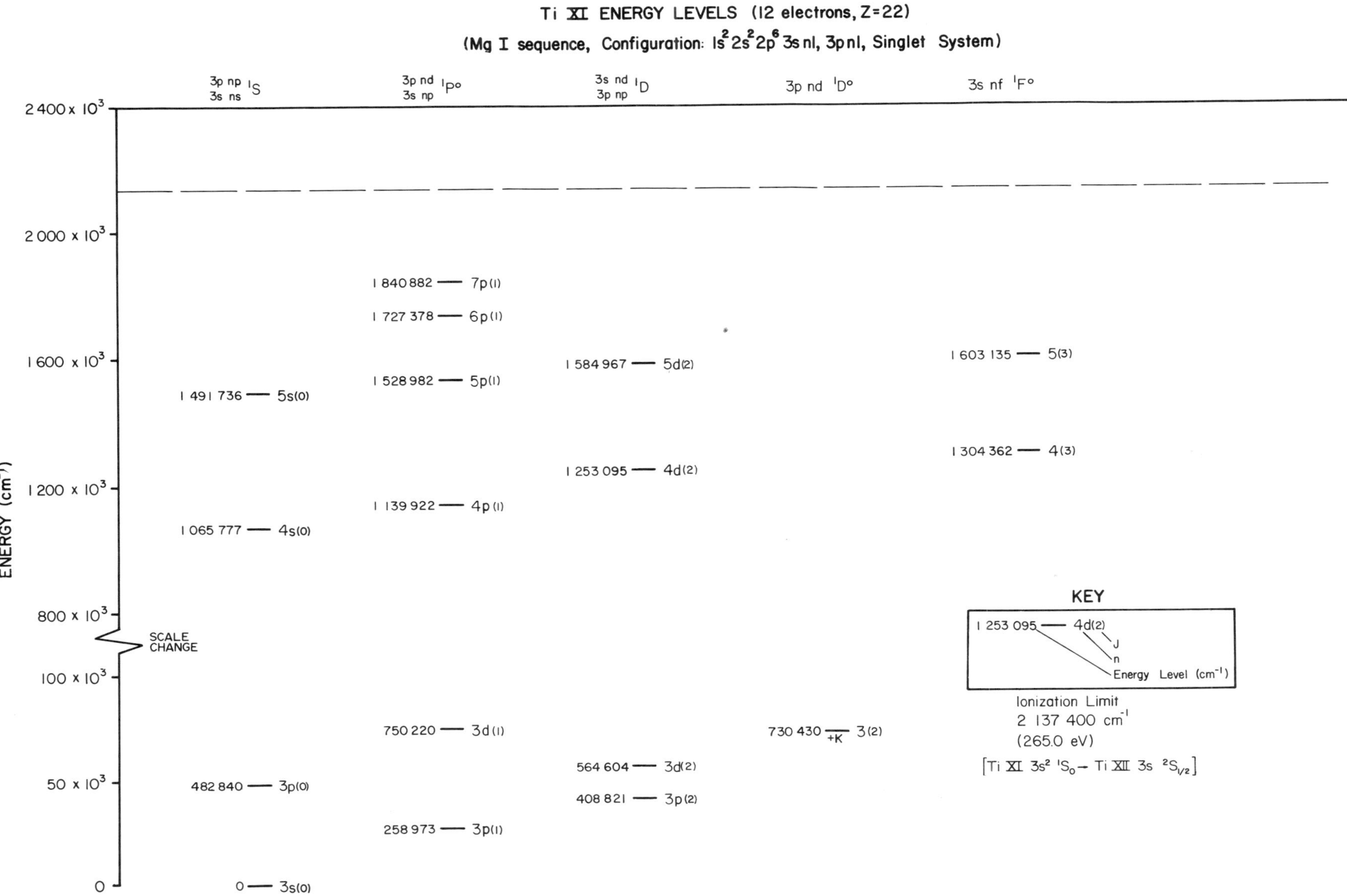
Ti XI ENERGY LEVELS (12 electrons, Z=22)
(Mg I sequence, Configuration: $1s^2 2s^2 2p^6$ 3s nl, 3p nl, Singlet System)
3p np 3s ns ^{1}S
3p nd 3s np ^{1}P°
3s nd 3p np ^{1}D
3p nd ^{1}D°
3s nf ^{1}F°
2 400 x 10^3
2 000 x 10^3
1 600 x 10^3
1 200 x 10^3
800 x 10^3
100 x 10^3
50 x 10^3
0
ENERGY (cm^{-1})
SCALE CHANGE
1 491 736 — 5s(0)
1 065 777 — 4s(0)
482 840 — 3p(0)
0 — 3s(0)
1 840 882 — 7p(1)
1 727 378 — 6p(1)
1 528 982 — 5p(1)
1 139 922 — 4p(1)
750 220 — 3d(1)
258 973 — 3p(1)
1 584 967 — 5d(2)
1 253 095 — 4d(2)
564 604 — 3d(2)
408 821 — 3p(2)
730 430 — 3(2)
+K
1 603 135 — 5(3)
1 304 362 — 4(3)
KEY
1 253 095 — 4d(2)
J
n
Energy Level (cm^{-1})
Ionization Limit
2 137 400 cm^{-1}
(265.0 eV)
[Ti XI $3s^2$ $^1S_0$ – Ti XII 3s $^2S_{1/2}$]

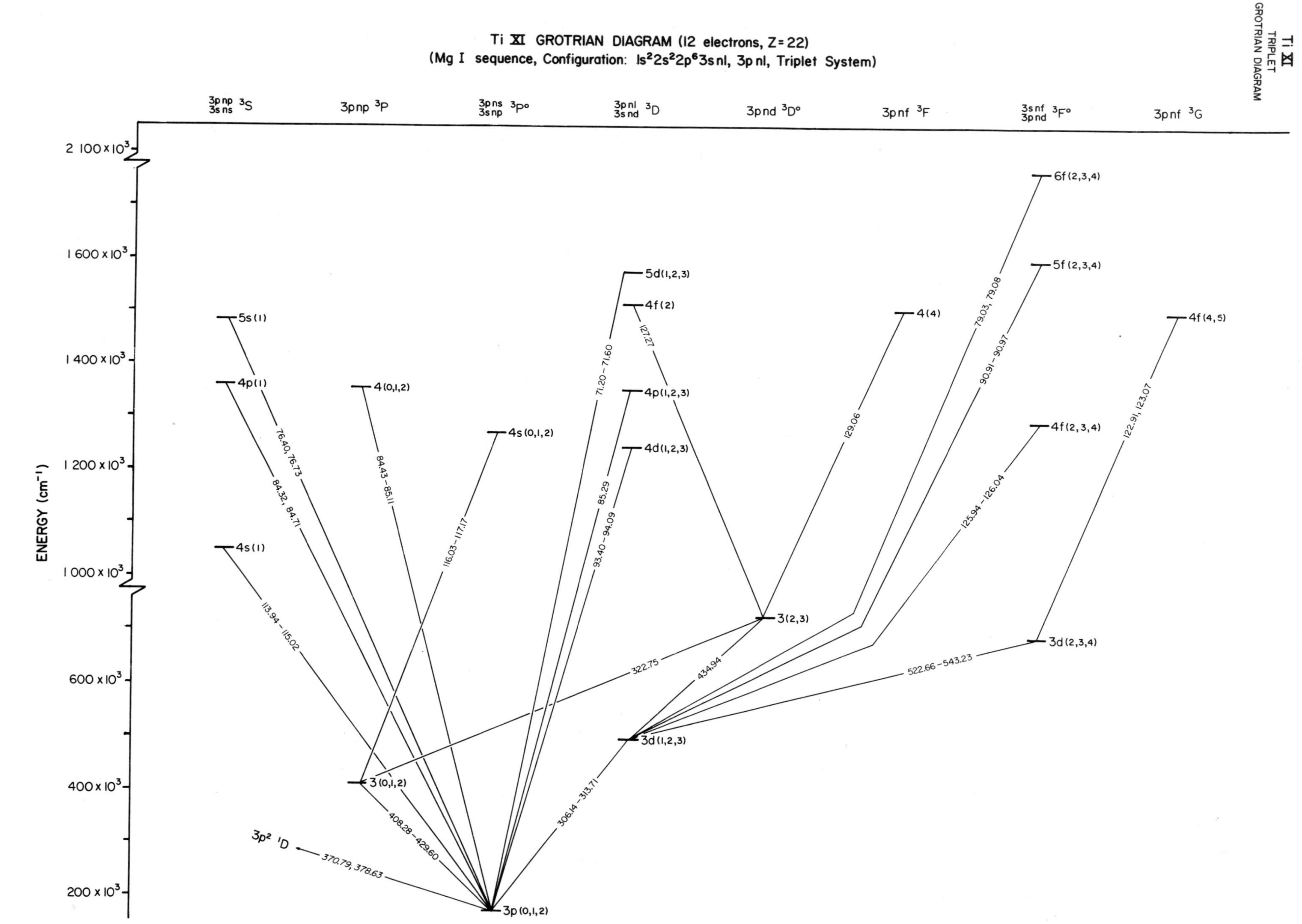

Ti XI
TRIPLET
GROTRIAN DIAGRAM
Ti XI GROTRIAN DIAGRAM (12 electrons, Z=22)
(Mg I sequence, Configuration: 1s²2s²2p⁶3snl, 3pnl, Triplet System)
3pnp 3sns ³S
3pnp ³P
3pns 3snp ³P°
3pnl 3snd ³D
3pnd ³D°
3pnf ³F
3snf 3pnd ³F°
3pnf ³G
2 100 x 10³
1 600 x 10³
1 400 x 10³
1 200 x 10³
1 000 x 10³
600 x 10³
400 x 10³
200 x 10³
ENERGY (cm⁻¹)
5s(1)
4p(1)
4s(1)
4(0,1,2)
4s(0,1,2)
5d(1,2,3)
4f(2)
4p(1,2,3)
4d(1,2,3)
4(4)
6f(2,3,4)
5f(2,3,4)
4f(2,3,4)
4f(4,5)
3(2,3)
3d(2,3,4)
3d(1,2,3)
3(0,1,2)
3p(0,1,2)
3p² ¹D
76.40, 76.73
84.32, 84.71
113.94 – 115.02
84.43 – 85.11
116.03 – 117.17
71.20 – 71.60
127.27
85.29
93.40 – 94.09
129.06
79.03, 79.08
90.91 – 90.97
125.94 – 126.04
122.91, 123.07
322.75
434.94
522.66 – 543.23
306.14 – 313.71
408.28 – 429.60
370.79, 378.63

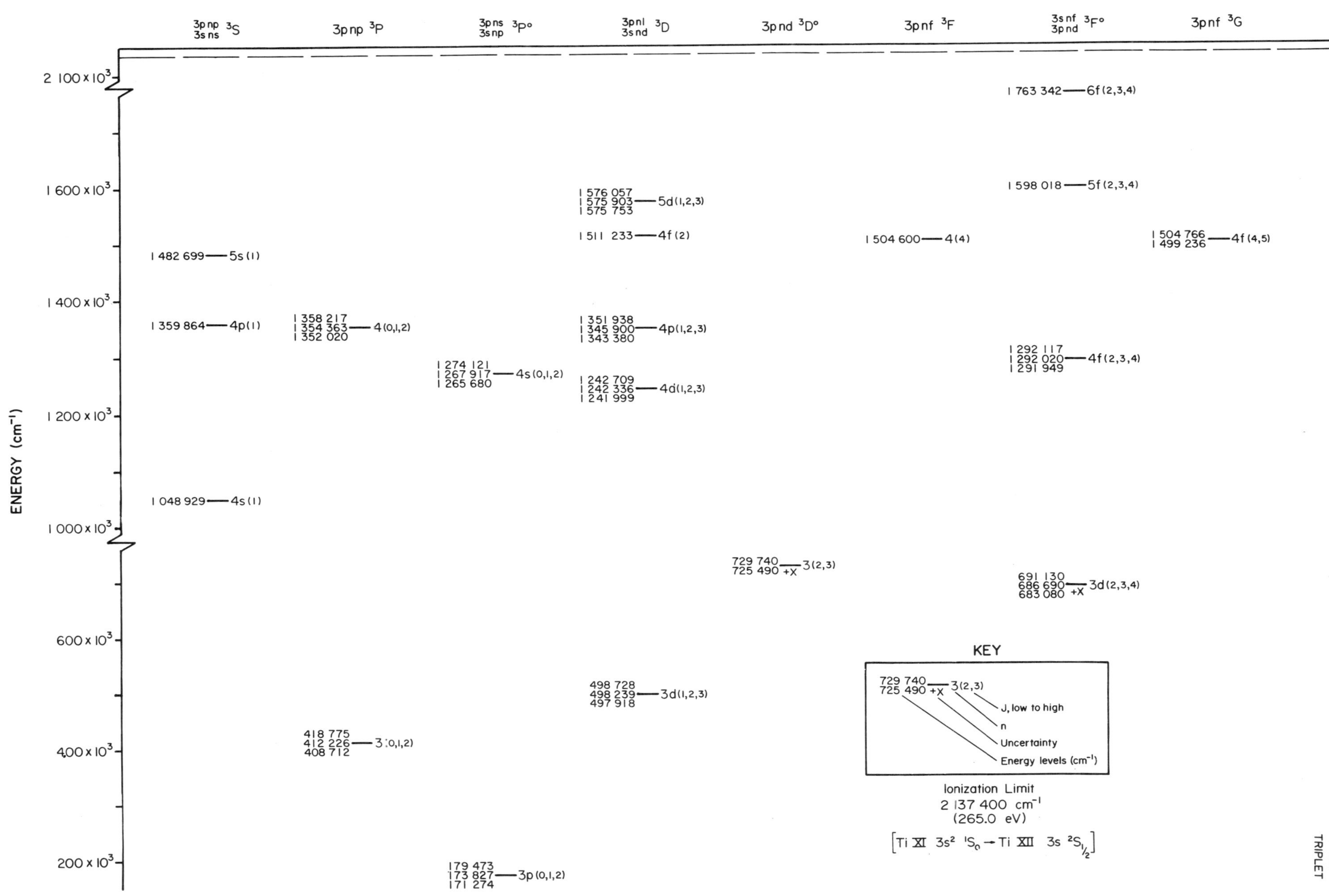

Ti XI
TRIPLET
Ti XI ENERGY LEVELS (12 electrons, Z=22)
(Mg I sequence, Configuration: 1s²2s²2p⁶3snl, 3pnl, Triplet System)
3pnp 3snS ³S
3pnp ³P
3pns 3snp ³P°
3pnl 3snd ³D
3pnd ³D°
3pnf ³F
3snf 3pnd ³F°
3pnf ³G
ENERGY (cm⁻¹)
2 100 x 10³
1 600 x 10³
1 400 x 10³
1 200 x 10³
1 000 x 10³
600 x 10³
400 x 10³
200 x 10³
1 482 699 — 5s (1)
1 359 864 — 4p (1)
1 048 929 — 4s (1)
1 358 217 1 354 363 1 352 020 — 4 (0,1,2)
418 775 412 226 408 712 — 3 (0,1,2)
1 274 121 1 267 917 1 265 680 — 4s (0,1,2)
179 473 173 827 171 274 — 3p (0,1,2)
1 576 057 1 575 903 1 575 753 — 5d (1,2,3)
1 511 233 — 4f (2)
1 351 938 1 345 900 1 343 380 — 4p (1,2,3)
1 242 709 1 242 336 1 241 999 — 4d (1,2,3)
498 728 498 239 497 918 — 3d (1,2,3)
729 740 725 490 +x — 3 (2,3)
1 504 600 — 4 (4)
1 763 342 — 6f (2,3,4)
1 598 018 — 5f (2,3,4)
1 292 117 1 292 020 1 291 949 — 4f (2,3,4)
691 130 686 690 683 080 +x — 3d (2,3,4)
1 504 766 1 499 236 — 4f (4,5)
KEY
729 740 725 490 +x — 3 (2,3)
J, low to high
n
Uncertainty
Energy levels (cm⁻¹)
Ionization Limit
2 137 400 cm⁻¹
(265.0 eV)
[Ti XI 3s² ¹S₀ → Ti XII 3s ²S½]

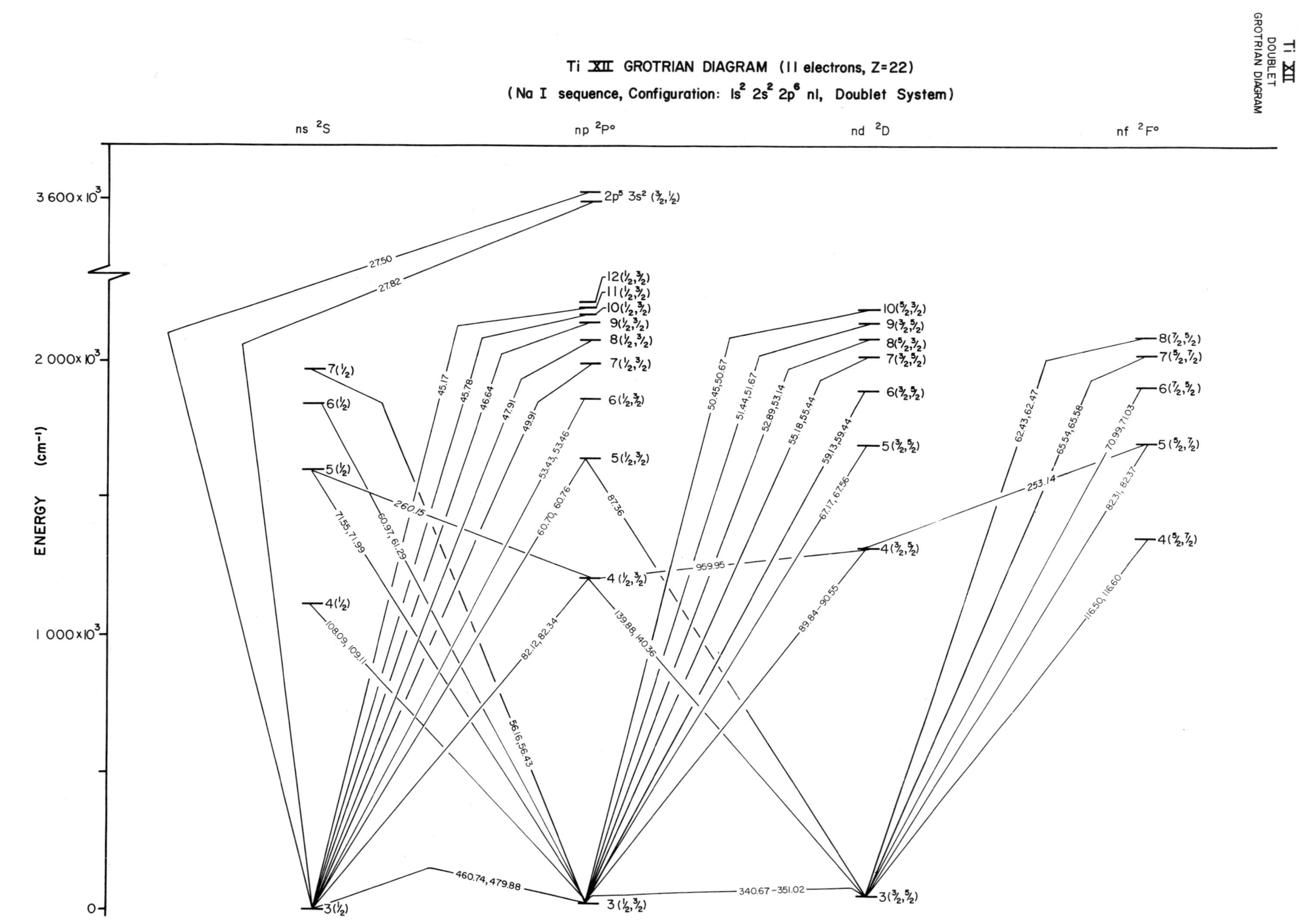
Ti XII GROTRIAN DIAGRAM (11 electrons, Z=22)
(Na I sequence, Configuration: 1s² 2s² 2p⁶ nl, Doublet System)
ns ²S
np ²P°
nd ²D
nf ²F°
ENERGY (cm⁻¹)
3 600 x 10³
2 000 x 10³
1 000 x 10³
0
3(½)
4(½)
5(½)
6(½)
7(½)
3(½,³⁄₂)
4(½,³⁄₂)
5(½,³⁄₂)
6(½,³⁄₂)
7(½,³⁄₂)
8(½,³⁄₂)
9(½,³⁄₂)
10(½,³⁄₂)
11(½,³⁄₂)
12(½,³⁄₂)
2p⁵ 3s² (³⁄₂,½)
3(³⁄₂,⁵⁄₂)
4(³⁄₂,⁵⁄₂)
5(³⁄₂,⁵⁄₂)
6(³⁄₂,⁵⁄₂)
7(³⁄₂,⁵⁄₂)
8(⁵⁄₂,³⁄₂)
9(³⁄₂,⁵⁄₂)
10(⁵⁄₂,³⁄₂)
4(⁵⁄₂,⁷⁄₂)
5(⁵⁄₂,⁷⁄₂)
6(⁷⁄₂,⁵⁄₂)
7(⁵⁄₂,⁷⁄₂)
8(⁷⁄₂,⁵⁄₂)
27.50
27.82
108.09, 109.11
71.55, 71.99
60.97, 61.29
260.15
45.17
45.78
46.64
47.91
49.91
53.43, 53.46
60.70, 60.76
87.36
82.12, 82.34
56.16, 56.43
460.74, 479.88
139.88, 140.36
959.95
340.67 - 351.02
50.45, 50.67
51.44, 51.67
52.89, 53.14
55.18, 55.44
59.13, 59.44
67.17, 67.56
89.84 - 90.55
253.14
62.43, 62.47
65.54, 65.58
70.99, 71.03
82.31, 82.37
116.50, 116.60

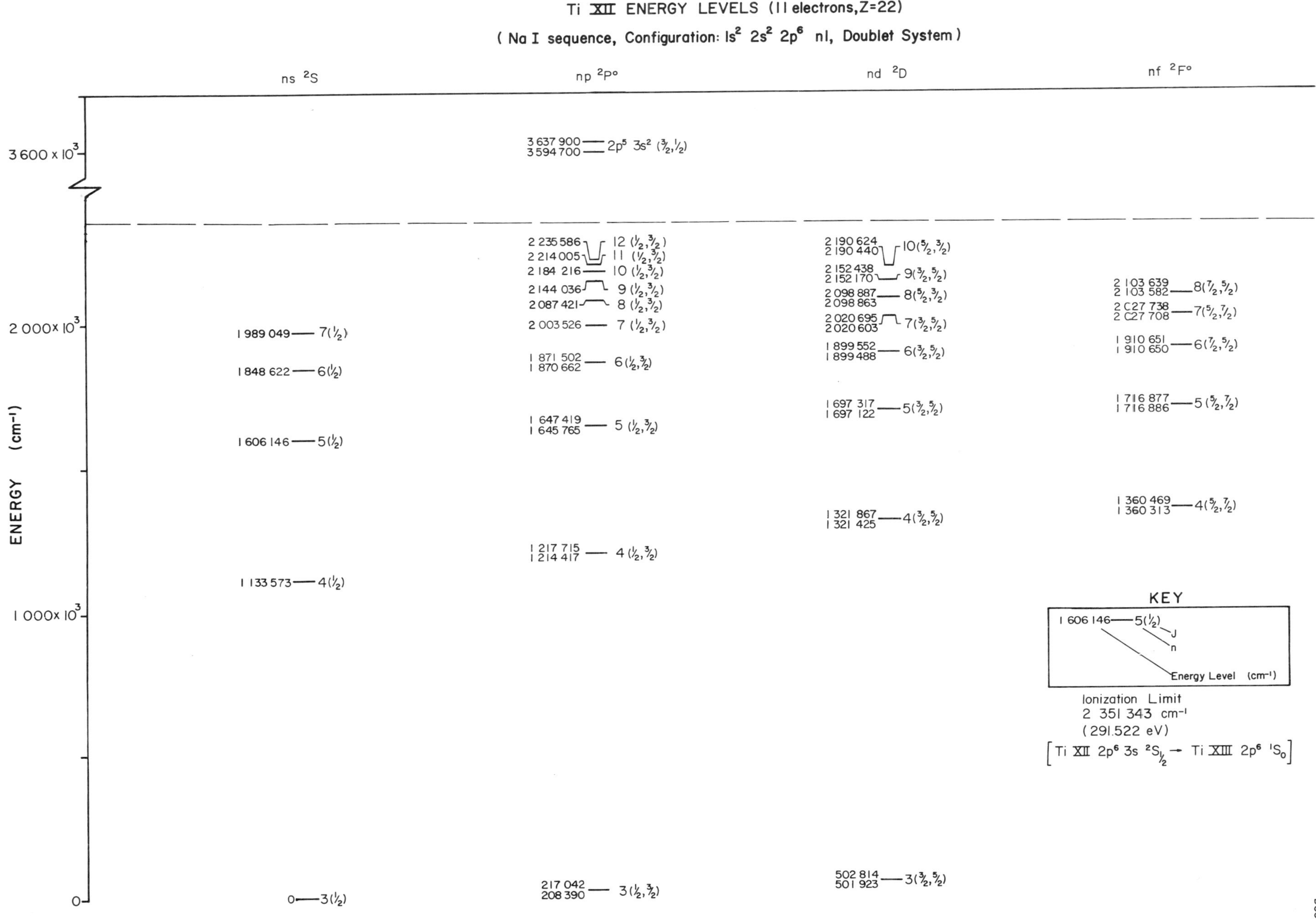
Ti XII ENERGY LEVELS (11 electrons, Z=22)
(Na I sequence, Configuration: 1s² 2s² 2p⁶ nl, Doublet System)
ns ²S
np ²P°
nd ²D
nf ²F°
3 637 900 — 2p⁵ 3s² (3/2, 1/2)
3 594 700
1 989 049 — 7(1/2)
1 848 622 — 6(1/2)
1 606 146 — 5(1/2)
1 133 573 — 4(1/2)
0 — 3(1/2)
2 235 586 — 12 (1/2, 3/2)
2 214 005 — 11 (1/2, 3/2)
2 184 216 — 10 (1/2, 3/2)
2 144 036 — 9 (1/2, 3/2)
2 087 421 — 8 (1/2, 3/2)
2 003 526 — 7 (1/2, 3/2)
1 871 502 — 6 (1/2, 3/2)
1 870 662
1 647 419 — 5 (1/2, 3/2)
1 645 765
1 217 715 — 4 (1/2, 3/2)
1 214 417
217 042 — 3 (1/2, 3/2)
208 390
2 190 624 — 10 (5/2, 3/2)
2 190 440
2 152 438 — 9 (3/2, 5/2)
2 152 170
2 098 887 — 8 (5/2, 3/2)
2 098 863
2 020 695 — 7 (3/2, 5/2)
2 020 603
1 899 552 — 6 (3/2, 5/2)
1 899 488
1 697 317 — 5 (3/2, 5/2)
1 697 122
1 321 867 — 4 (3/2, 5/2)
1 321 425
502 814 — 3 (3/2, 5/2)
501 923
2 103 639 — 8 (7/2, 5/2)
2 103 582
2 C27 738 — 7 (5/2, 7/2)
2 C27 708
1 910 651 — 6 (7/2, 5/2)
1 910 650
1 716 877 — 5 (5/2, 7/2)
1 716 886
1 360 469 — 4 (5/2, 7/2)
1 360 313
KEY
1 606 146 — 5(1/2)
J
n
Energy Level (cm⁻¹)
Ionization Limit
2 351 343 cm⁻¹
(291.522 eV)
[Ti XII 2p⁶ 3s ²S1/2 → Ti XIII 2p⁶ ¹S0]
3 600 x 10³
2 000 x 10³
1 000 x 10³
0
ENERGY (cm⁻¹)

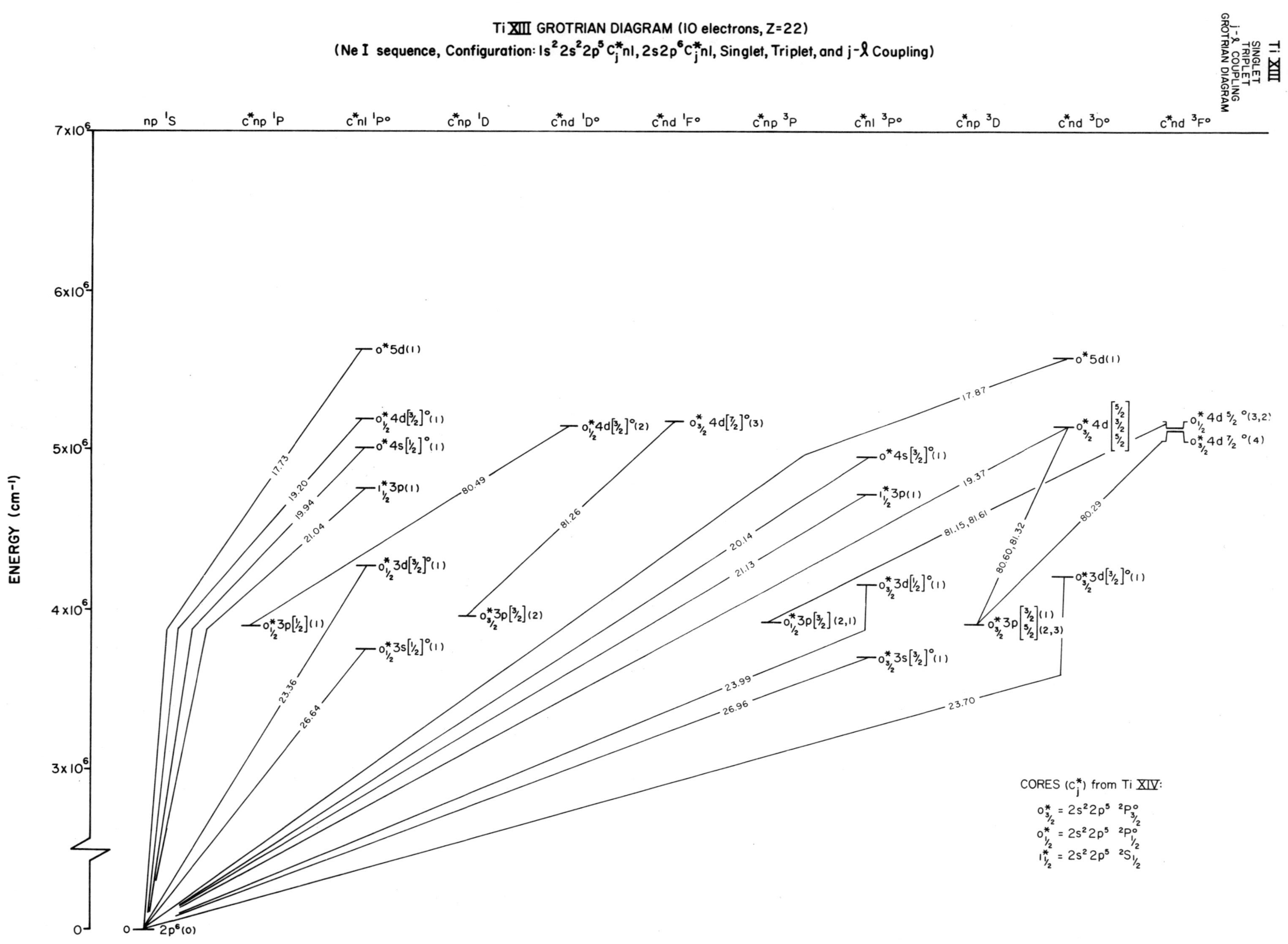
Ti XIII GROTRIAN DIAGRAM (10 electrons, Z=22)
(Ne I sequence, Configuration: 1s^{2}2s^{2}2p^{5}C$^*_j$nl, 2s2p^{6}C$^*_j$nl, Singlet, Triplet, and j-ℓ Coupling)
Ti XIII
SINGLET
TRIPLET
j-ℓ COUPLING
GROTRIAN DIAGRAM
np ^{1}S
c*np ^{1}P
c*nl ^{1}P°
c*np ^{1}D
c*nd ^{1}D°
c*nd ^{1}F°
c*np ^{3}P
c*nl ^{3}P°
c*np ^{3}D
c*nd ^{3}D°
c*nd ^{3}F°
ENERGY (cm^{-1})
7x10^6
6x10^6
5x10^6
4x10^6
3x10^6
0
0
2p^6(0)
o*5d(1)
o$^*_{1/2}$4d[3/2]°(1)
o*4s[1/2]°(1)
1$^*_{1/2}$3p(1)
o$^*_{1/2}$3d[3/2]°(1)
o$^*_{1/2}$3p[1/2](1)
o$^*_{1/2}$3s[1/2]°(1)
o$^*_{1/2}$4d[3/2]°(2)
o$^*_{3/2}$3p[3/2](2)
o$^*_{3/2}$ 4d[7/2]°(3)
o*5d(1)
o*4s[3/2]°(1)
1$^*_{1/2}$3p(1)
o$^*_{3/2}$3d[1/2]°(1)
o$^*_{1/2}$3p[3/2](2,1)
o$^*_{3/2}$3s[3/2]°(1)
o$^*_{3/2}$ 4d [5/2, 3/2, 5/2]
o$^*_{3/2}$3p [3/2](1) [5/2](2,3)
o$^*_{3/2}$3d[3/2]°(1)
o$^*_{1/2}$ 4d 5/2 °(3,2)
o$^*_{3/2}$ 4d 7/2 °(4)
17.73
19.20
19.94
21.04
23.36
26.64
80.49
81.26
17.87
20.14
21.13
19.37
81.15, 81.61
80.60, 81.32
80.29
23.99
26.96
23.70
CORES (C$^*_j$) from Ti XIV:
o$^*_{3/2}$ = 2s^{2}2p^5 ^{2}P°$_{3/2}$
o$^*_{1/2}$ = 2s^{2}2p^5 ^{2}P°$_{1/2}$
1$^*_{1/2}$ = 2s^{2}2p^5 ^{2}S$_{1/2}$

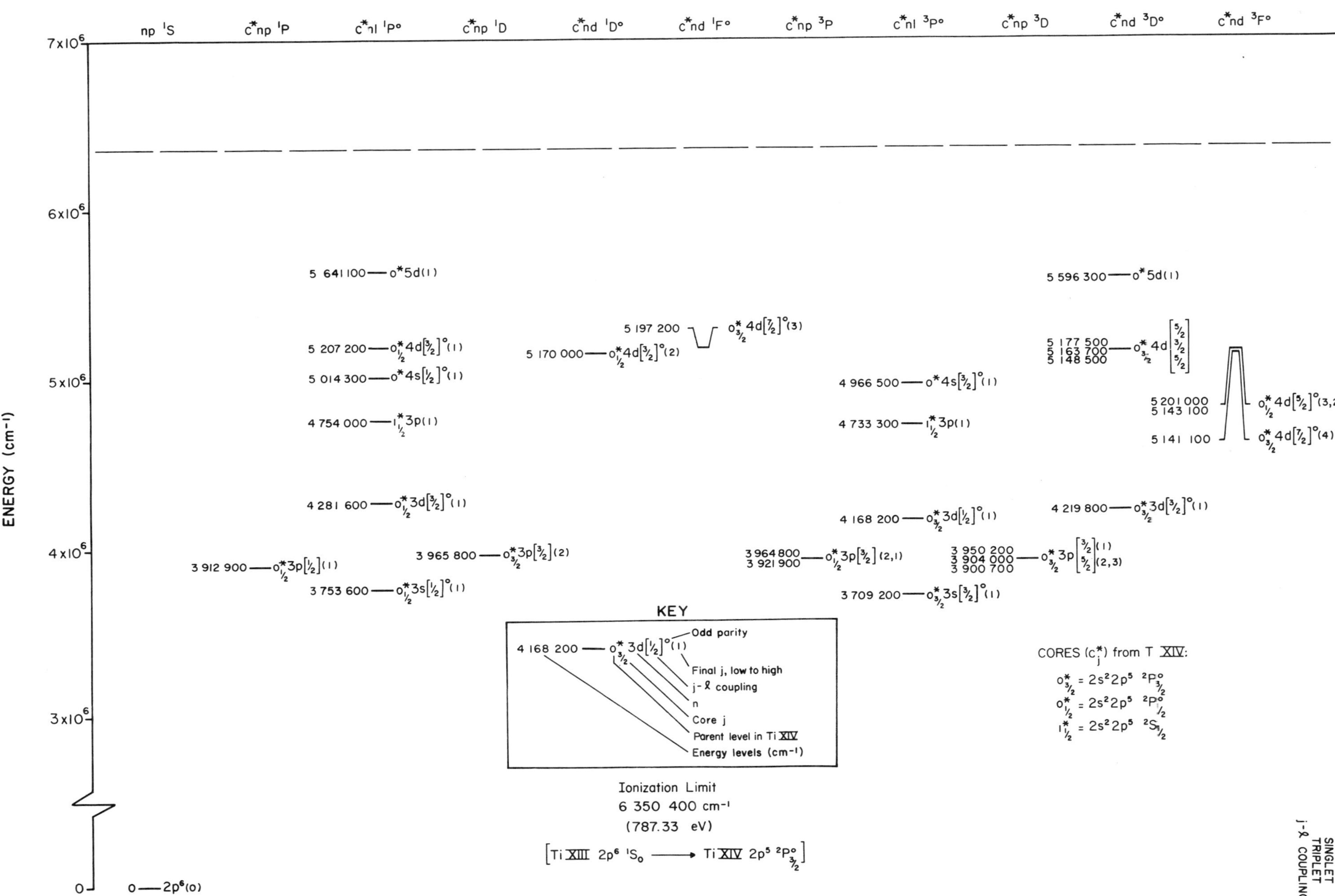

Ti XIII ENERGY LEVELS (10 electrons, Z = 22)
(Ne I sequence, Configuration: 1s²2s²2p⁵C*ⱼnl, 2s2p⁶C*ⱼnl, Singlet, Triplet, and j-ℓ Coupling)
np ¹S
c*np ¹P
c*nl ¹P°
c*np ¹D
c*nd ¹D°
c*nd ¹F°
c*np ³P
c*nl ³P°
c*np ³D
c*nd ³D°
c*nd ³F°
7x10⁶
6x10⁶
5x10⁶
4x10⁶
3x10⁶
0
ENERGY (cm⁻¹)
5 641 100 — o*5d(1)
5 207 200 — o*½4d[3/2]°(1)
5 014 300 — o*4s[½]°(1)
4 754 000 — 1*½3p(1)
4 281 600 — o*½3d[3/2]°(1)
3 912 900 — o*½3p[½](1)
3 753 600 — o*½3s[½]°(1)
3 965 800 — o*3/23p[3/2](2)
5 170 000 — o*½4d[3/2](2)
5 197 200 o*3/24d[7/2]°(3)
3 964 800
3 921 900 — o*½3p[3/2](2,1)
4 966 500 — o*4s[3/2]°(1)
4 733 300 — 1*½3p(1)
4 168 200 — o*3/23d[½]°(1)
3 709 200 — o*3/23s[3/2]°(1)
5 596 300 — o*5d(1)
5 177 500
5 163 700
5 148 500 — o*3/24d [5/2, 3/2, 5/2]
4 219 800 — o*3/23d[3/2]°(1)
3 950 200
3 904 000
3 900 700 — o*3/23p [3/2](1) [5/2](2,3)
5 201 000
5 143 100 o*½4d[5/2]°(3,2)
5 141 100 o*3/24d[7/2]°(4)
0 — 2p⁶(0)
KEY
4 168 200 — o*3/23d[½]°(1)
Odd parity
Final j, low to high
j-ℓ coupling
n
Core j
Parent level in Ti XIV
Energy levels (cm⁻¹)
CORES (c*ⱼ) from T XIV:
o*3/2 = 2s²2p⁵ ²P°3/2
o*½ = 2s²2p⁵ ²P°½
1*½ = 2s²2p⁵ ²S½
Ionization Limit
6 350 400 cm⁻¹
(787.33 eV)
[Ti XIII 2p⁶ ¹S₀ ⟶ Ti XIV 2p⁵ ²P°3/2]
Ti XIII
SINGLET
TRIPLET
j-ℓ COUPLING

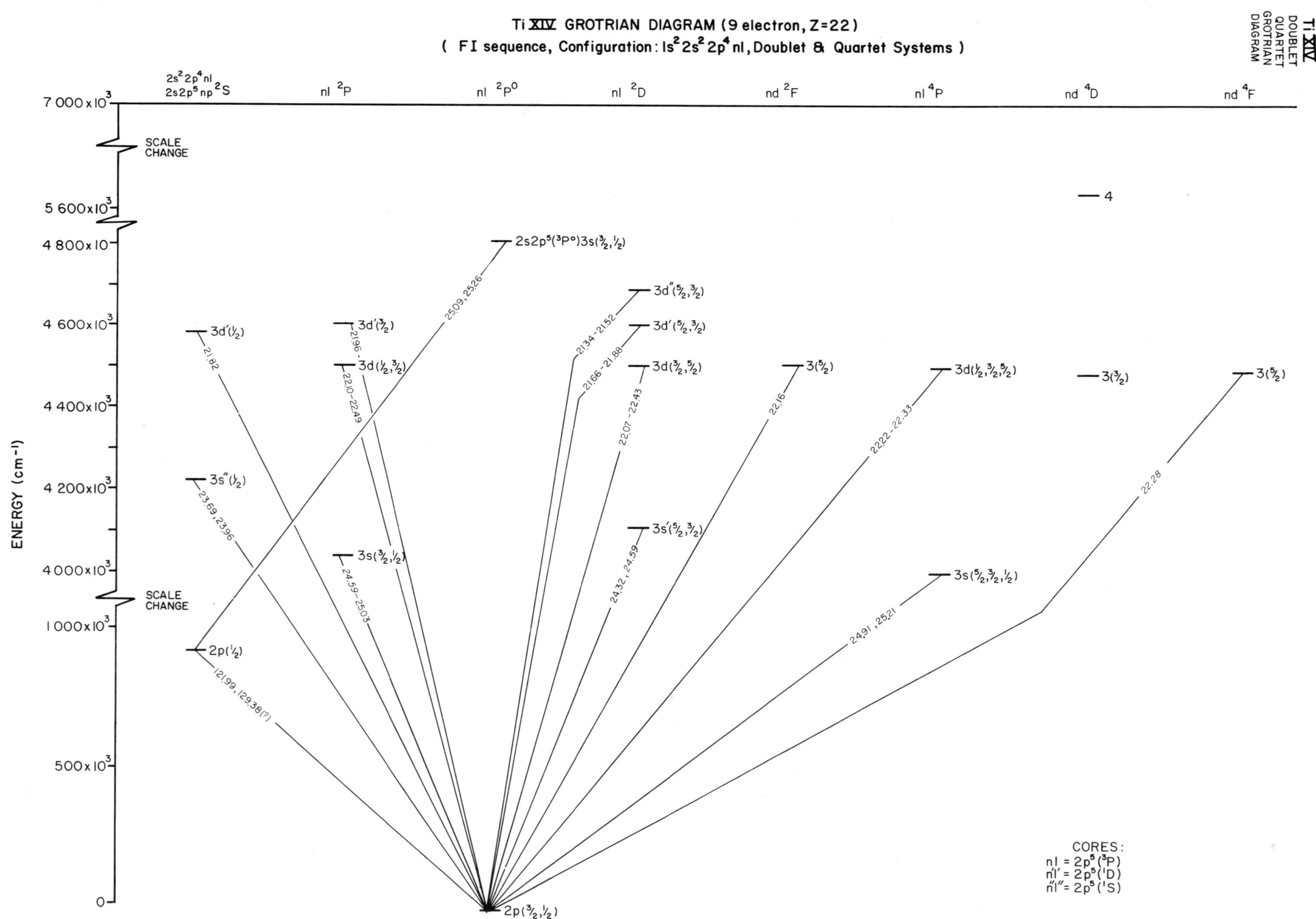

Ti XIV GROTRIAN DIAGRAM (9 electron, Z=22)
(FI sequence, Configuration: 1s²2s²2p⁴nl, Doublet & Quartet Systems)
Ti XIV
DOUBLET
QUARTET
GROTRIAN
DIAGRAM
2s²2p⁴nl
2s2p⁶np ²S
nl ²P
nl ²P°
nl ²D
nd ²F
nl ⁴P
nd ⁴D
nd ⁴F
7000×10³
SCALE CHANGE
5600×10³
4800×10
4600×10³
4400×10³
4200×10³
4000×10³
SCALE CHANGE
1000×10³
500×10³
0
ENERGY (cm⁻¹)
3d′(½)
21.82
3s″(½)
23.69, 23.96
2p(½)
121.99, 129.38(?)
3d′(3/2)
21.96
3d(½,3/2)
22.10-22.49
3s(3/2,½)
24.59-25.03
2s2p⁵(³P°)3s(3/2,½)
25.09, 25.26
3d″(5/2,3/2)
21.34-21.52
3d′(5/2,3/2)
21.66-21.88
3d(3/2,5/2)
22.07-22.43
3s′(5/2,3/2)
24.32, 24.59
3(5/2)
22.16
3d(½,3/2,5/2)
22.22-22.33
3s(5/2,3/2,½)
24.91, 25.21
4
3(3/2)
3(5/2)
22.28
2p(3/2,½)
CORES:
nl = 2p⁵(³P)
nl′ = 2p⁵(¹D)
nl″ = 2p⁵(¹S)

Ti XIV ENERGY LEVELS (9 electrons, Z = 22)

(F I sequence, Configuration: $1s^2\ 2s^2\ 2p^4\ nl$, Doublet & Quartet Systems)

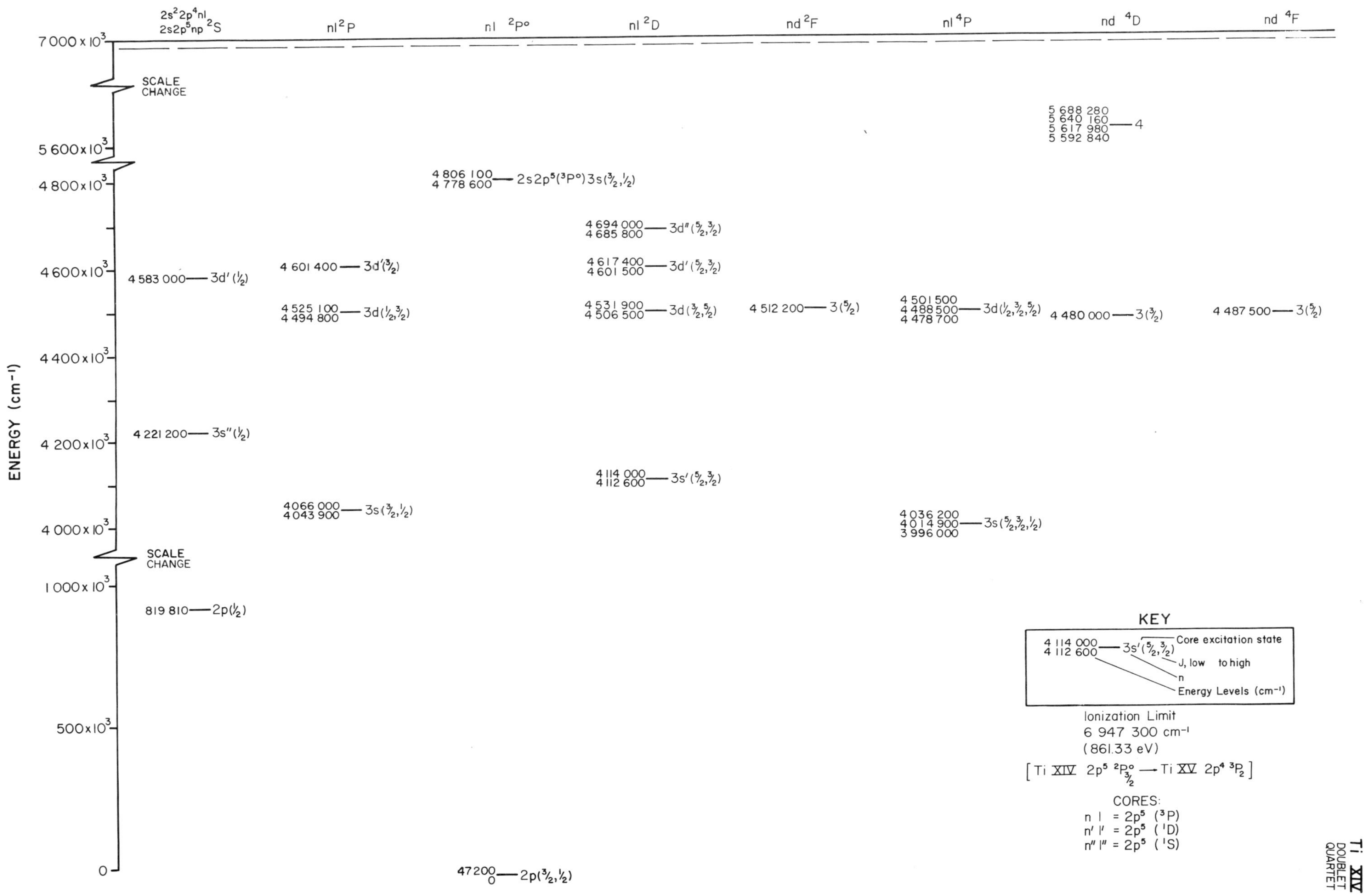

ENERGY (cm^{-1})

Axis: 7000 x 10^3; SCALE CHANGE; 5600 x 10^3; 4800 x 10^3; 4600 x 10^3; 4400 x 10^3; 4200 x 10^3; 4000 x 10^3; SCALE CHANGE; 1000 x 10^3; 500 x 10^3; 0

$2s^2 2p^4 nl$ / $2s2p^5np$ ^2S
- 4 583 000 — 3d′(½)
- 4 221 200 — 3s″(½)
- 819 810 — 2p(½)

$nl\ ^2P$
- 4 601 400 — 3d′(3/2)
- 4 525 100, 4 494 800 — 3d(1/2, 3/2)
- 4066 000, 4043 900 — 3s(3/2, 1/2)

$nl\ ^2P^o$
- 4 806 100, 4 778 600 — $2s2p^5(^3P^o)3s$(3/2, 1/2)
- 47200, 0 — 2p(3/2, 1/2)

$nl\ ^2D$
- 4 694 000, 4 685 800 — 3d″(5/2, 3/2)
- 4 617 400, 4 601 500 — 3d′(5/2, 3/2)
- 4 531 900, 4 506 500 — 3d(3/2, 5/2)
- 4 114 000, 4 112 600 — 3s′(5/2, 3/2)

$nd\ ^2F$
- 4 512 200 — 3(5/2)

$nl\ ^4P$
- 4 501 500, 4 488 500, 4 478 700 — 3d(1/2, 3/2, 5/2)
- 4 036 200, 4 014 900, 3 996 000 — 3s(5/2, 3/2, 1/2)

$nd\ ^4D$
- 5 688 280, 5 640 160, 5 617 980, 5 592 840 — 4
- 4 480 000 — 3(3/2)

$nd\ ^4F$
- 4 487 500 — 3(5/2)

KEY

4 114 000, 4 112 600 — 3s′(5/2, 3/2): Core excitation state; J, low to high; n; Energy Levels (cm^{-1})

Ionization Limit
6 947 300 cm^{-1}
(861.33 eV)

[Ti XIV $2p^5\ ^2P^o_{3/2}$ ⟶ Ti XV $2p^4\ ^3P_2$]

CORES:
- n l = $2p^5$ (^3P)
- n′ l′ = $2p^5$ (^1D)
- n″ l″ = $2p^5$ (^1S)

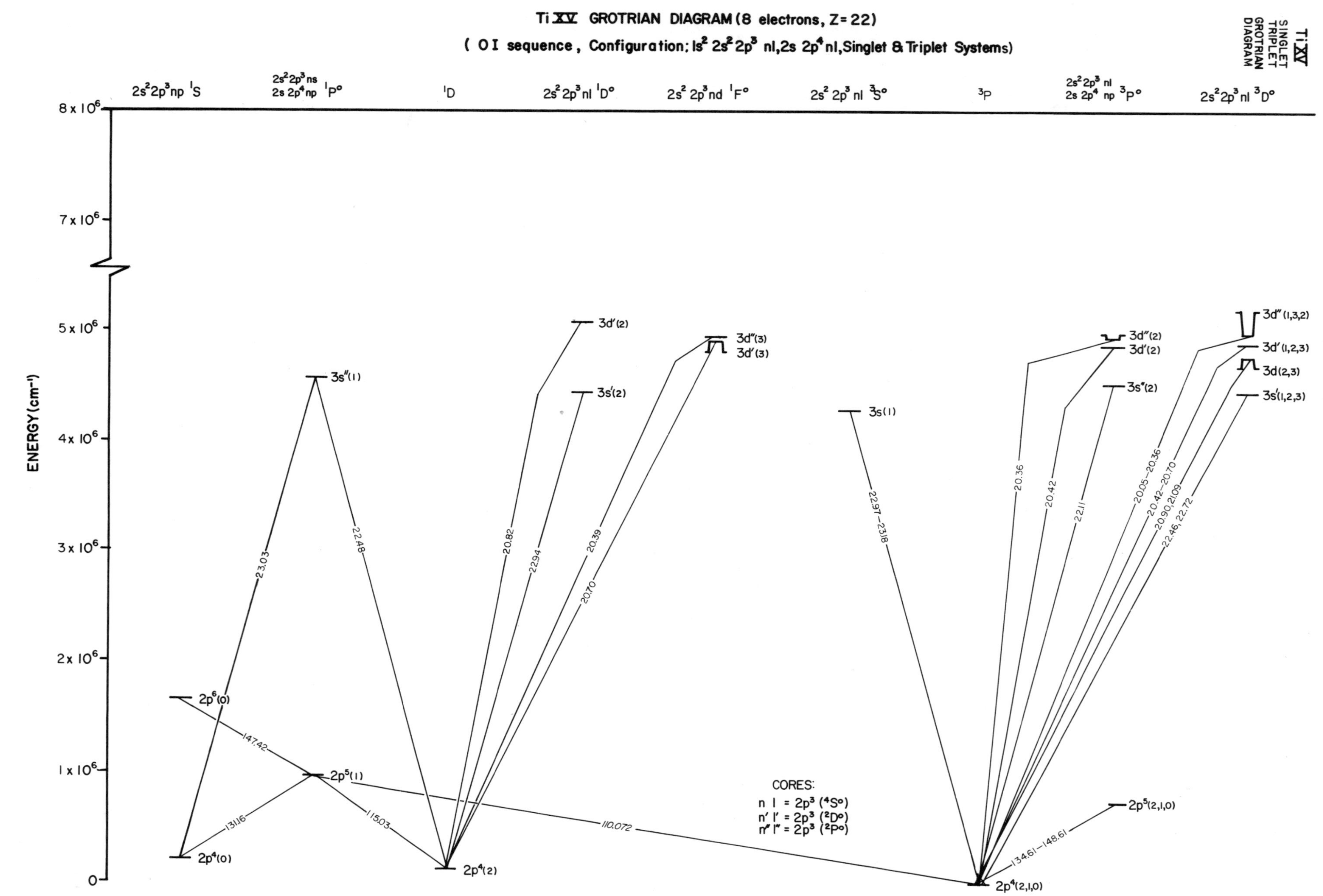
Ti XV GROTRIAN DIAGRAM (8 electrons, Z = 22)
(O I sequence, Configuration: 1s2 2s2 2p3 nl, 2s 2p4 nl, Singlet & Triplet Systems)
Ti XV
SINGLET
TRIPLET
GROTRIAN
DIAGRAM
2s2 2p3 np 1S
2s2 2p3 ns
2s 2p4 np 1P°
1D
2s2 2p3 nl 1D°
2s2 2p3 nd 1F°
2s2 2p3 nl 3S°
3P
2s2 2p3 nl
2s 2p4 np 3P°
2s2 2p3 nl 3D°
ENERGY (cm-1)
8 x 10^6
7 x 10^6
5 x 10^6
4 x 10^6
3 x 10^6
2 x 10^6
1 x 10^6
0
2p6(0)
2p4(0)
3s''(1)
2p5(1)
2p4(2)
3d'(2)
3s'(2)
3d''(3)
3d'(3)
3s(1)
2p4(2,1,0)
2p5(2,1,0)
3d''(2)
3d'(2)
3s''(2)
3d''(1,3,2)
3d'(1,2,3)
3d(2,3)
3s'(1,2,3)
23.03
22.48
147.42
131.16
115.03
110.072
20.82
22.94
20.39
20.70
22.97–23.18
20.36
20.42
22.11
20.05–20.36
20.42–20.70
20.90, 21.09
22.46, 22.72
134.61–148.61
CORES:
n l = 2p3 (4S°)
n' l' = 2p3 (2D°)
n'' l'' = 2p3 (2P°)

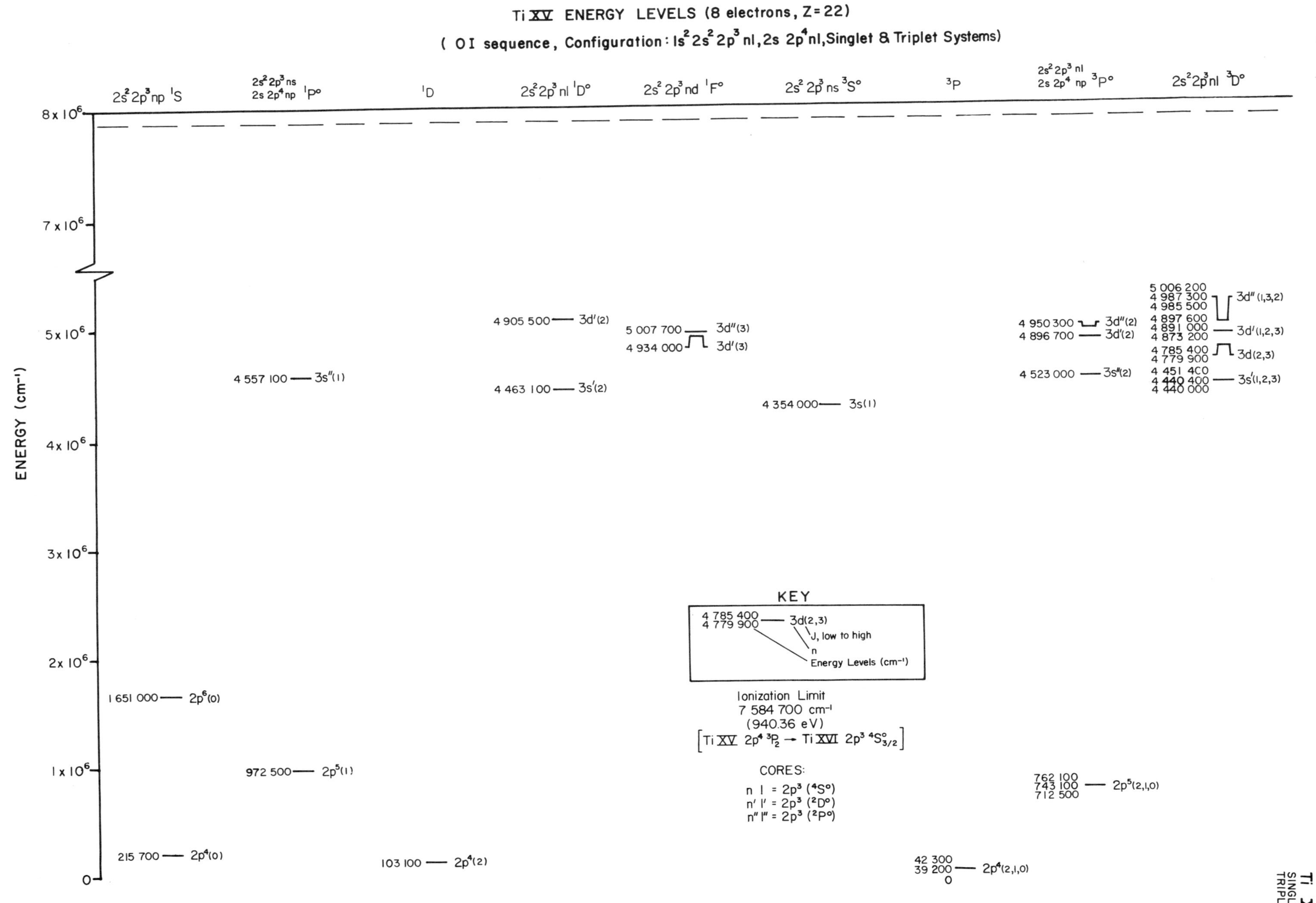
Ti XV ENERGY LEVELS (8 electrons, Z=22)
(O I sequence, Configuration: $1s^2 2s^2 2p^3 nl$, $2s\,2p^4 nl$, Singlet & Triplet Systems)
$2s^2 2p^3 np\ ^1S$
$2s^2 2p^3 ns$ / $2s\,2p^4 np\ ^1P^o$
^1D
$2s^2 2p^3 nl\ ^1D^o$
$2s^2 2p^3 nd\ ^1F^o$
$2s^2 2p^3 ns\ ^3S^o$
^3P
$2s^2 2p^3 nl$ / $2s\,2p^4 np\ ^3P^o$
$2s^2 2p^3 nl\ ^3D^o$
ENERGY (cm^{-1})
8×10^6
7×10^6
5×10^6
4×10^6
3×10^6
2×10^6
1×10^6
0
1 651 000 — $2p^6(0)$
215 700 — $2p^4(0)$
4 557 100 — $3s''(1)$
972 500 — $2p^5(1)$
103 100 — $2p^4(2)$
4 905 500 — $3d'(2)$
4 463 100 — $3s'(2)$
5 007 700 — $3d''(3)$
4 934 000 — $3d'(3)$
4 354 000 — $3s(1)$
42 300
39 200 — $2p^4(2,1,0)$
0
4 950 300 — $3d''(2)$
4 896 700 — $3d'(2)$
4 523 000 — $3s''(2)$
762 100
743 100 — $2p^5(2,1,0)$
712 500
5 006 200
4 987 300 — $3d''(1,3,2)$
4 985 500
4 897 600
4 891 000 — $3d'(1,2,3)$
4 873 200
4 785 400
4 779 900 — $3d(2,3)$
4 451 400
4 440 400 — $3s'(1,2,3)$
4 440 000
KEY
4 785 400
4 779 900 — 3d(2,3)
J, low to high
n
Energy Levels (cm^{-1})
Ionization Limit
7 584 700 cm^{-1}
(940.36 eV)
[Ti XV $2p^4\ ^3P_2$ → Ti XVI $2p^3\ ^4S^o_{3/2}$]
CORES:
$n\ l = 2p^3\ (^4S^o)$
$n'\ l' = 2p^3\ (^2D^o)$
$n''\ l'' = 2p^3\ (^2P^o)$

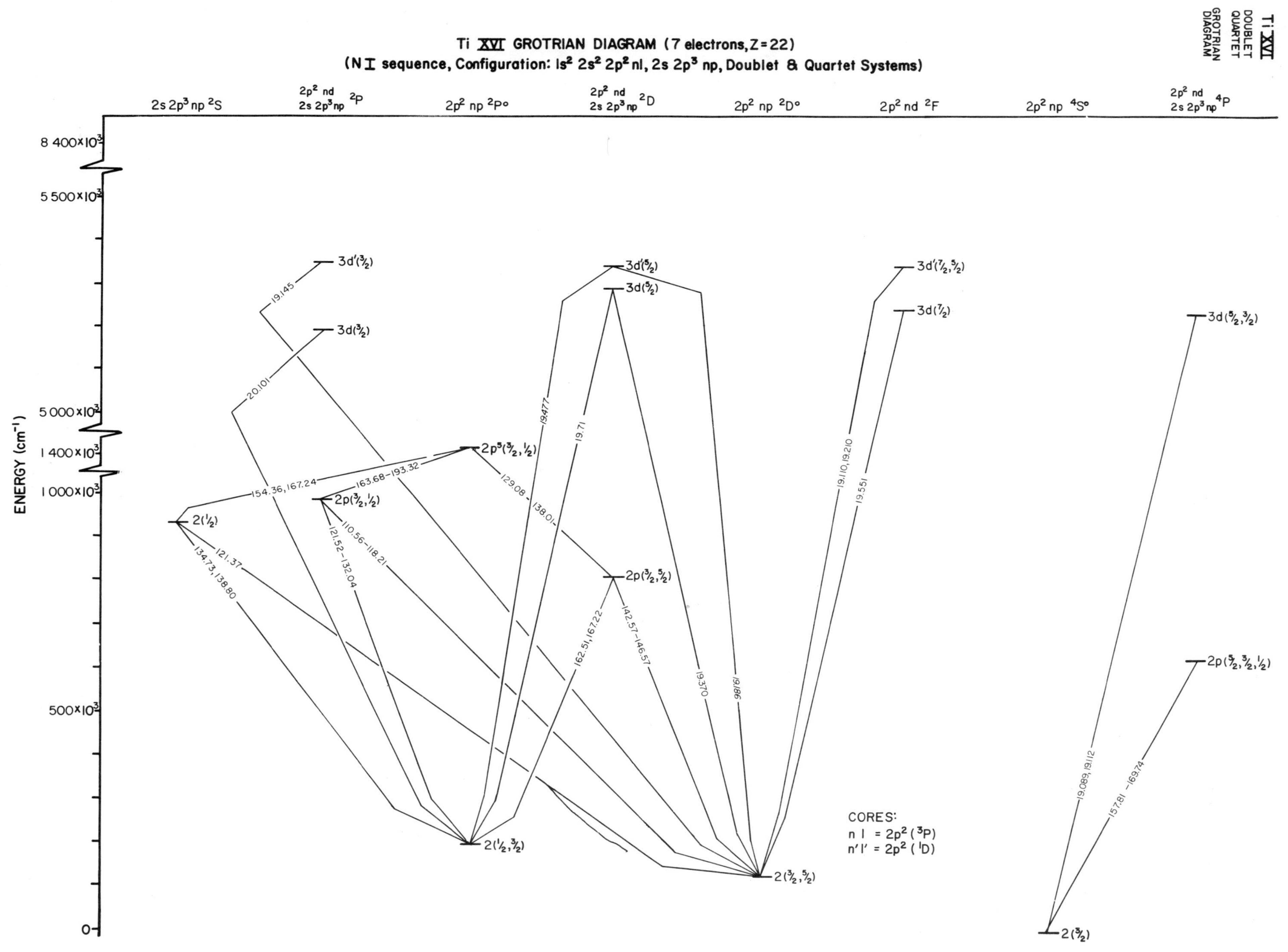

Ti XVI
DOUBLET
QUARTET
GROTRIAN
DIAGRAM
Ti XVI GROTRIAN DIAGRAM (7 electrons, Z=22)
(N I sequence, Configuration: $1s^2\ 2s^2\ 2p^2\ nl$, $2s\ 2p^3\ np$, Doublet & Quartet Systems)
$2s\ 2p^3\ np\ ^2S$
$2p^2\ nd$ / $2s\ 2p^3\ np\ ^2P$
$2p^2\ np\ ^2P^o$
$2p^2\ nd$ / $2s\ 2p^3\ np\ ^2D$
$2p^2\ np\ ^2D^o$
$2p^2\ nd\ ^2F$
$2p^2\ np\ ^4S^o$
$2p^2\ nd$ / $2s\ 2p^3\ np\ ^4P$
ENERGY (cm^{-1})
8 400×10³
5 500×10³
5 000×10³
1 400×10³
1 000×10³
500×10³
0
3d′(3/2)
3d(3/2)
2p(3/2,1/2)
2(1/2)
2p⁵(3/2,1/2)
3d′(5/2)
3d(5/2)
2p(3/2,5/2)
2(1/2,3/2)
3d′(7/2,5/2)
3d(7/2)
2(3/2,5/2)
3d(5/2,3/2)
2p(5/2,3/2,1/2)
2(3/2)
19.145
20.101
154.36, 167.24
163.68–193.32
121.52–132.04
110.56–118.21
121.37
134.73, 138.80
129.08–138.01
19.477
19.71
162.51, 167.22
142.57–146.57
19.370
19.186
19.110, 19.210
19.551
19.089, 19.112
157.81 –169.74
CORES:
n l = $2p^2\ (^3P)$
n′l′ = $2p^2\ (^1D)$

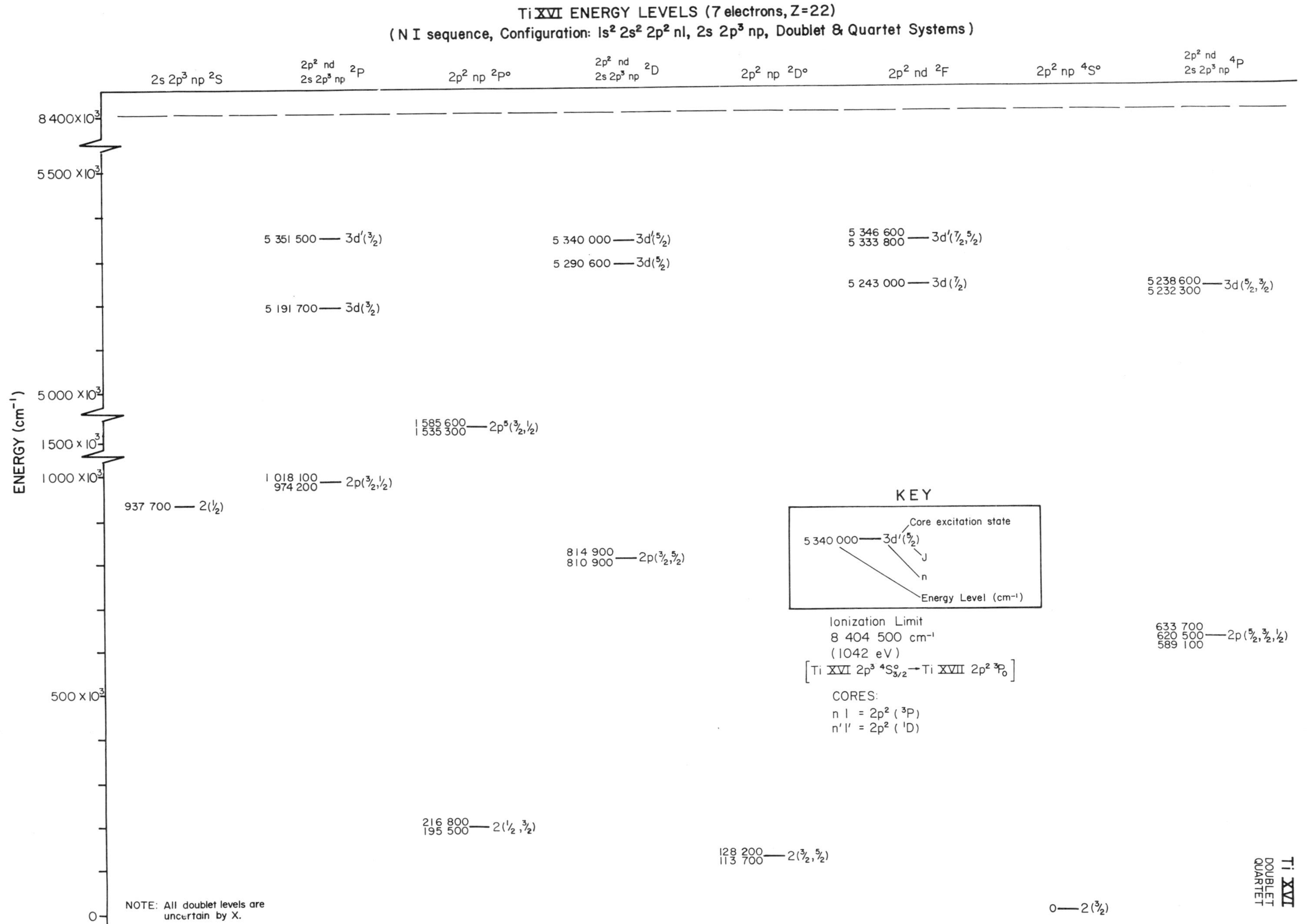

Ti XVI ENERGY LEVELS (7 electrons, Z=22)
(N I sequence, Configuration: 1s² 2s² 2p² nl, 2s 2p³ np, Doublet & Quartet Systems)
2s 2p³ np ²S
2p² nd 2s 2p³ np ²P
2p² np ²P°
2p² nd 2s 2p³ np ²D
2p² np ²D°
2p² nd ²F
2p² np ⁴S°
2p² nd 2s 2p³ np ⁴P
ENERGY (cm⁻¹)
8 400 X10³
5 500 X10³
5 000 X10³
1 500 x 10³
1 000 X10³
500 X10³
0
5 351 500 — 3d′(3/2)
5 191 700 — 3d(3/2)
5 340 000 — 3d′(5/2)
5 290 600 — 3d(5/2)
5 346 600
5 333 800 — 3d′(7/2,5/2)
5 243 000 — 3d(7/2)
5 238 600
5 232 300 — 3d(5/2,3/2)
1 585 600
1 535 300 — 2p⁵(3/2,1/2)
1 018 100
974 200 — 2p(3/2,1/2)
937 700 — 2(1/2)
814 900
810 900 — 2p(3/2,5/2)
KEY
5 340 000 — 3d′(5/2)
Core excitation state
J
n
Energy Level (cm⁻¹)
Ionization Limit
8 404 500 cm⁻¹
(1042 eV)
[Ti XVI 2p³ ⁴S°3/2 → Ti XVII 2p² ³P0]
CORES:
n l = 2p² (³P)
n′l′ = 2p² (¹D)
633 700
620 500
589 100 — 2p(5/2,3/2,1/2)
216 800
195 500 — 2(1/2,3/2)
128 200
113 700 — 2(3/2,5/2)
0 — 2(3/2)
NOTE: All doublet levels are uncertain by X.
Ti XVI
DOUBLET
QUARTET

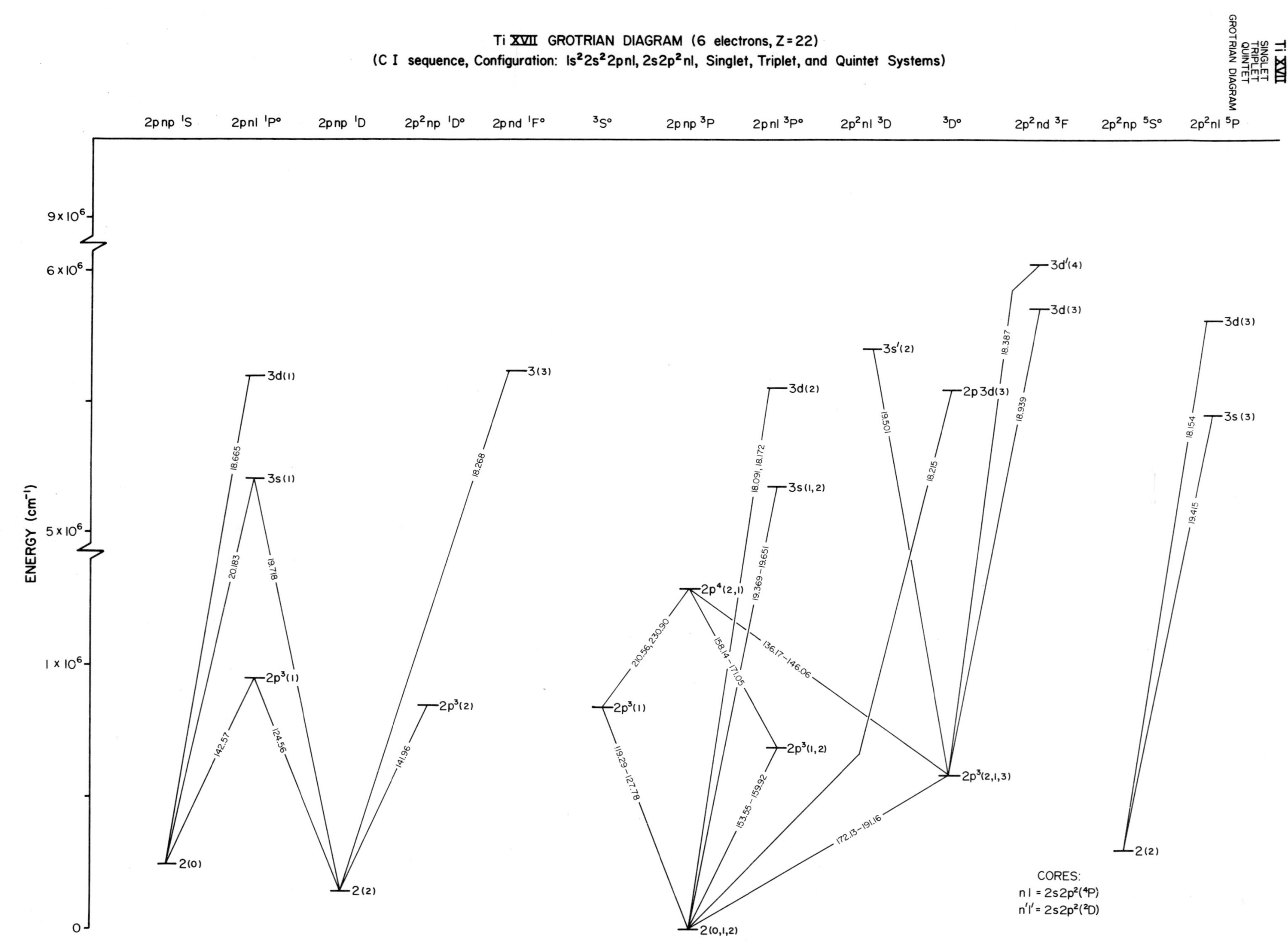

Ti XVII
SINGLET
TRIPLET
QUINTET
GROTRIAN DIAGRAM
Ti XVII GROTRIAN DIAGRAM (6 electrons, Z = 22)
(C I sequence, Configuration: $1s^2 2s^2 2pnl$, $2s2p^2nl$, Singlet, Triplet, and Quintet Systems)
2pnp ^1S
2pnl $^1P°$
2pnp ^1D
$2p^2$np $^1D°$
2pnd $^1F°$
$^3S°$
2pnp ^3P
2pnl $^3P°$
$2p^2$nl ^3D
$^3D°$
$2p^2$nd ^3F
$2p^2$np $^5S°$
$2p^2$nl ^5P
ENERGY (cm^{-1})
9 x 10^6
6 x 10^6
5 x 10^6
1 x 10^6
0
2(0)
3d(1)
3s(1)
$2p^3$(1)
2(2)
$2p^3$(2)
3(3)
18.665
20.183
19.718
142.57
124.56
141.96
18.268
$2p^3$(1)
2(0,1,2)
$2p^4$(2,1)
$2p^3$(1,2)
3d(2)
3s(1,2)
3s′(2)
2p3d(3)
$2p^3$(2,1,3)
3d′(4)
3d(3)
119.29 – 127.78
210.56, 230.90
158.14 – 171.05
153.55 – 159.92
136.17 – 146.06
18.091, 18.172
19.369 – 19.651
19.501
18.215
18.387
18.939
172.13 – 191.16
2(2)
3d(3)
3s(3)
18.154
19.415
CORES:
nl = $2s2p^2(^4P)$
n′l′ = $2s2p^2(^2D)$

Ti XVII ENERGY LEVELS (6 electrons, Z = 22)
(C I sequence, Configuration: $1s^2 2s^2 2p\,nl$, $2s2p^2 nl$, Singlet, Triplet, and Quintet Systems)

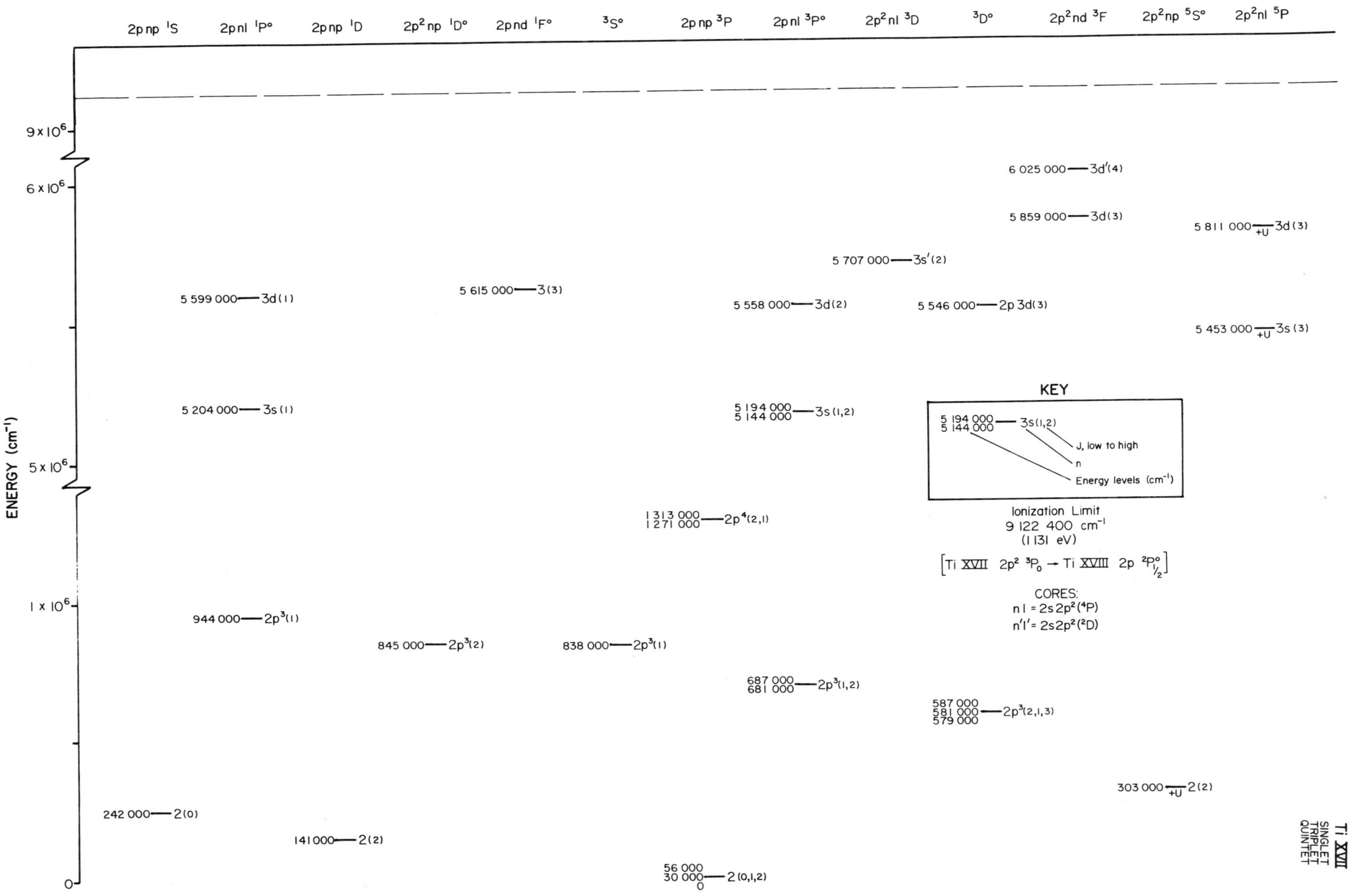

Ti XVII
SINGLET
TRIPLET
QUINTET

ENERGY (cm^{-1})

9×10^6
6×10^6
5×10^6
1×10^6
0

2pnp ^1S
242 000 — 2(0)

2pnl $^1P°$
5 599 000 — 3d(1)
5 204 000 — 3s(1)
944 000 — $2p^3$(1)

2pnp ^1D
141 000 — 2(2)

$2p^2np$ $^1D°$
845 000 — $2p^3$(2)

2pnd $^1F°$
5 615 000 — 3(3)

$^3S°$
838 000 — $2p^3$(1)

2pnp ^3P
1 313 000
1 271 000 — $2p^4$(2,1)
56 000
30 000
0 — 2(0,1,2)

2pnl $^3P°$
5 558 000 — 3d(2)
5 194 000
5 144 000 — 3s(1,2)
687 000
681 000 — $2p^3$(1,2)

$2p^2nl$ ^3D
5 707 000 — 3s′(2)

$^3D°$
5 546 000 — 2p 3d(3)
587 000
581 000
579 000 — $2p^3$(2,1,3)

$2p^2nd$ ^3F
6 025 000 — 3d′(4)
5 859 000 — 3d(3)

$2p^2np$ $^5S°$
303 000 +U — 2(2)

$2p^2nl$ ^5P
5 811 000 +U — 3d(3)
5 453 000 +U — 3s(3)

KEY

5 194 000
5 144 000 — 3s(1,2)
J, low to high
n
Energy levels (cm^{-1})

Ionization Limit
9 122 400 cm^{-1}
(1131 eV)
[Ti XVII $2p^2\ ^3P_0$ → Ti XVIII $2p\ ^2P°_{1/2}$]

CORES:
$nl = 2s2p^2(^4P)$
$n'l' = 2s2p^2(^2D)$

Ti XVIII GROTRIAN DIAGRAM (5 electrons, Z = 22)

(B I sequence, Configuration: $1s^2 2s^2 nl$, $1s^2 2s2p\,nl$, Doublet and Quartet Systems)

ns ^2S
2s2pnl ^2P
$2s^2np$ $^2P^o$
$2s^2nd$ / 2s2pnl ^2D
2s2pnd $^2D^o$
2s2pnd $^2F^o$
$^4S^o$
^4P
2s2pnd $^4P^o$
2s2pnd $^4D^o$

ENERGY (cm^{-1})
10×10^6
8×10^6
7×10^6
6×10^6
5×10^6
4×10^6
3×10^6
2×10^6
1×10^6
0

$4s(1/2)$
$3s(1/2)$
$2p^2(1/2)$
$3p'(1/2,3/2)$
$2p^2(1/2,3/2)$
$3d''(5/2)$
$3p'(3/2,5/2)$
$3d(3/2,5/2)$
$2p(1/2,3/2)$
$3d'(5/2)$
$2p^2(3/2,5/2)$
$3d''(7/2)$
$3d'(5/2,7/2)$
$2p^3(3/2)$
$3d'(5/2)$
$3d'(7/2)$
$2p^2(3/2,5/2)$

13.45
18.38
148.42, 161.94
133.86–147.58
16.90, 16.94
17.59
16.56, 16.62
17.22, 17.37
179.87–200.11
17.92
17.15
17.63, 17.72
153.23–166.22
17.28
17.30

CORES:
$n'l' = 2s2p\,(^3P^o)$
$n''l'' = 2s2p\,(^1P^o)$

Ti XVIII ENERGY LEVELS (5 electrons, Z=22)

(B I sequence, Configuration: $1s^2 2s^2 nl$, $1s^2 2s2p\ nl$, Doublet and Quartet Systems)

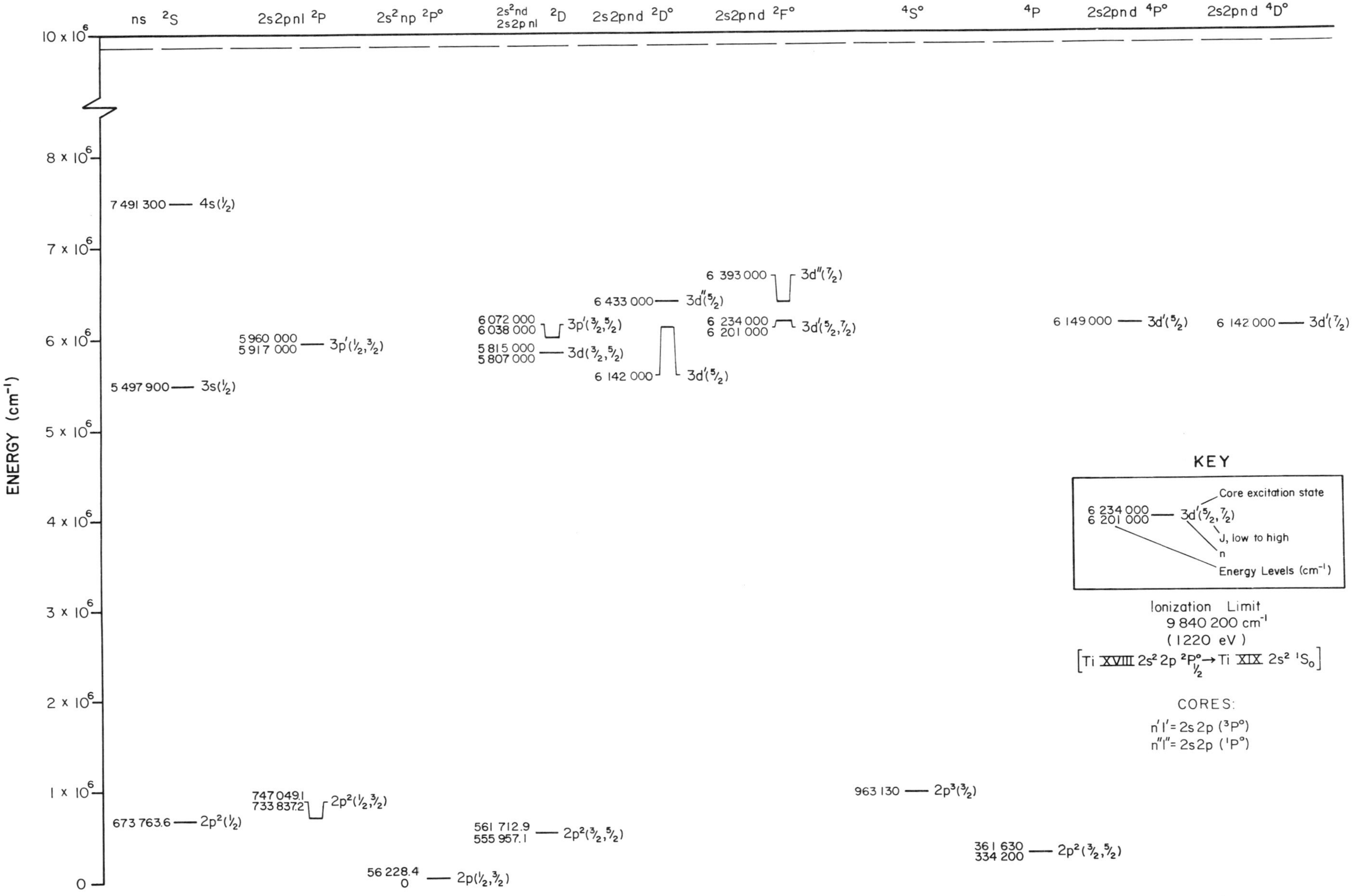

Ionization Limit
9 840 200 cm^{-1}
(1220 eV)
[Ti XVIII $2s^2 2p\ ^2P^o_{1/2}$ → Ti XIX $2s^2\ ^1S_0$]

CORES:
$n'l' = 2s2p\ (^3P^o)$
$n''l'' = 2s2p\ (^1P^o)$

NOTE: All quartets uncertain by +X.

Ti XIX
SINGLET
TRIPLET
GROTRIAN
DIAGRAM

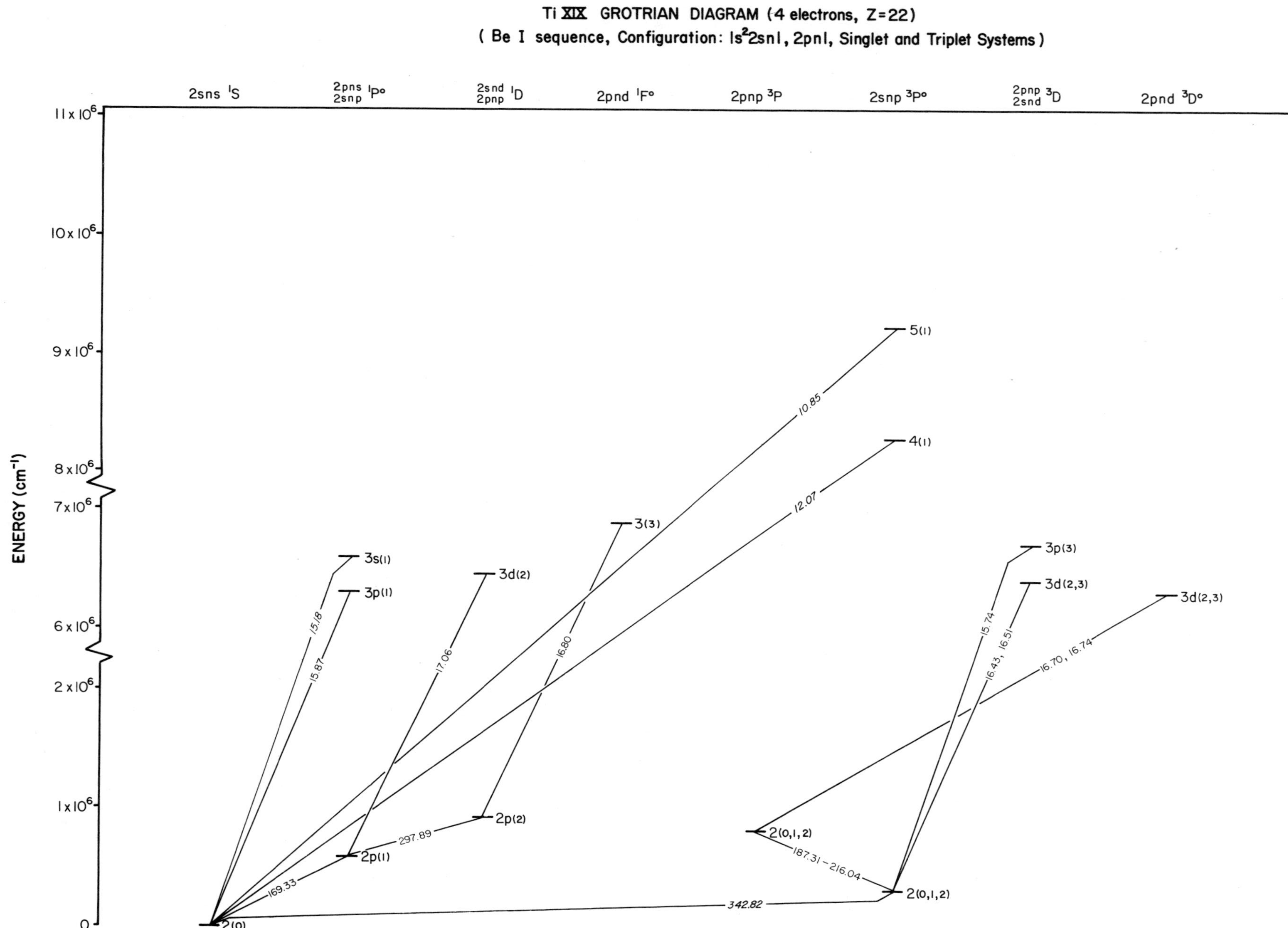

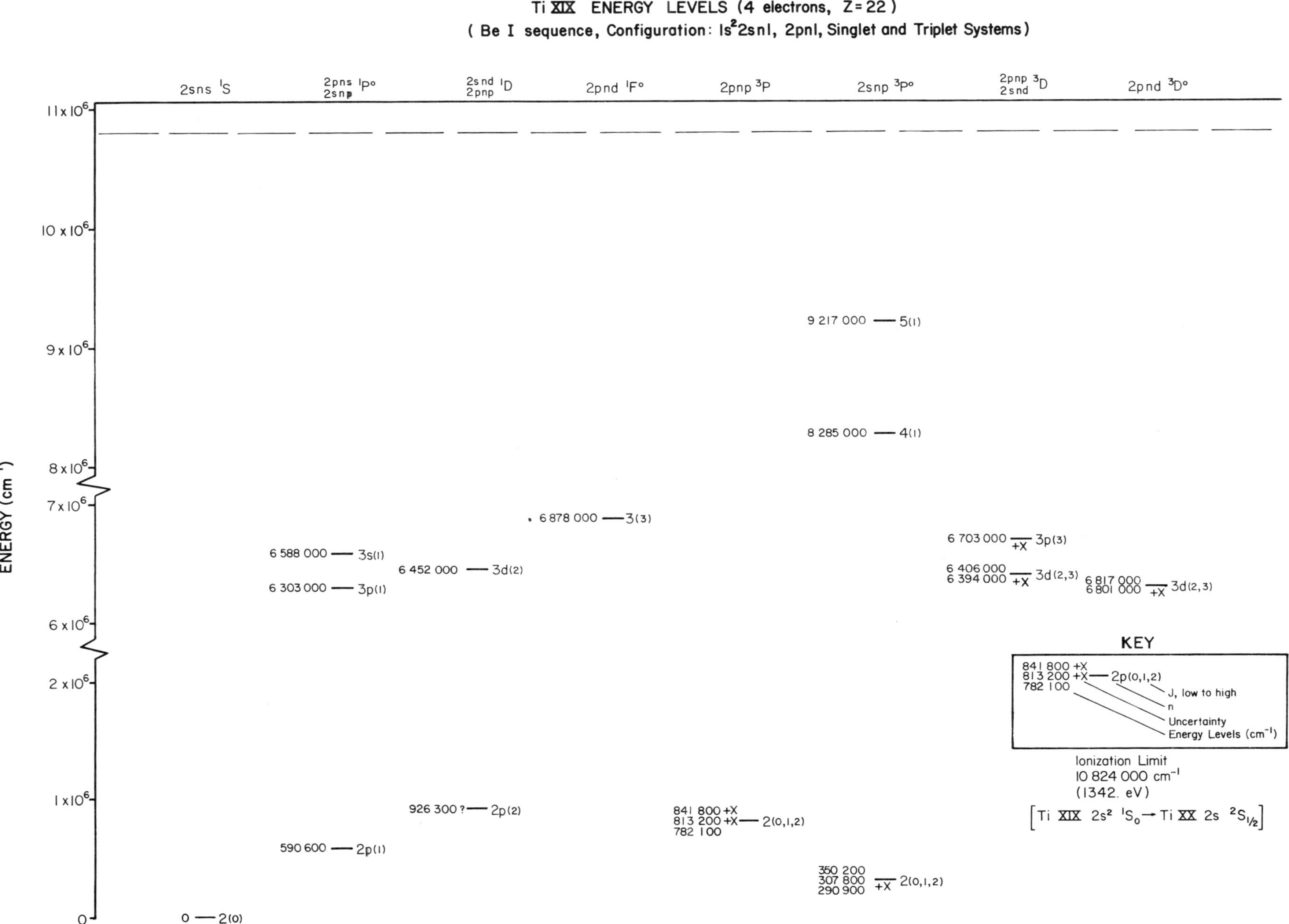
Ti XIX ENERGY LEVELS (4 electrons, Z = 22)
(Be I sequence, Configuration: $1s^2$2snl, 2pnl, Singlet and Triplet Systems)
ENERGY (cm^{-1})
11×10^6
10×10^6
9×10^6
8×10^6
7×10^6
6×10^6
2×10^6
1×10^6
0
2sns ^1S
2pns $^1P^o$ 2snp
2snd ^1D 2pnp
2pnd $^1F^o$
2pnp ^3P
2snp $^3P^o$
2pnp ^3D 2snd
2pnd $^3D^o$
0 — 2(0)
6 588 000 — 3s(1)
6 303 000 — 3p(1)
590 600 — 2p(1)
6 452 000 — 3d(2)
926 300 ? — 2p(2)
6 878 000 — 3(3)
841 800 +X
813 200 +X — 2(0,1,2)
782 100
9 217 000 — 5(1)
8 285 000 — 4(1)
350 200
307 800 +X 2(0,1,2)
290 900
6 703 000 +X 3p(3)
6 406 000
6 394 000 +X 3d(2,3)
6 817 000
6 801 000 +X 3d(2,3)
KEY
841 800 +X
813 200 +X — 2p(0,1,2)
782 100
J, low to high
n
Uncertainty
Energy Levels (cm^{-1})
Ionization Limit
10 824 000 cm^{-1}
(1342. eV)
[Ti XIX $2s^2$ $^1S_0$ → Ti XX 2s $^2S_{1/2}$]

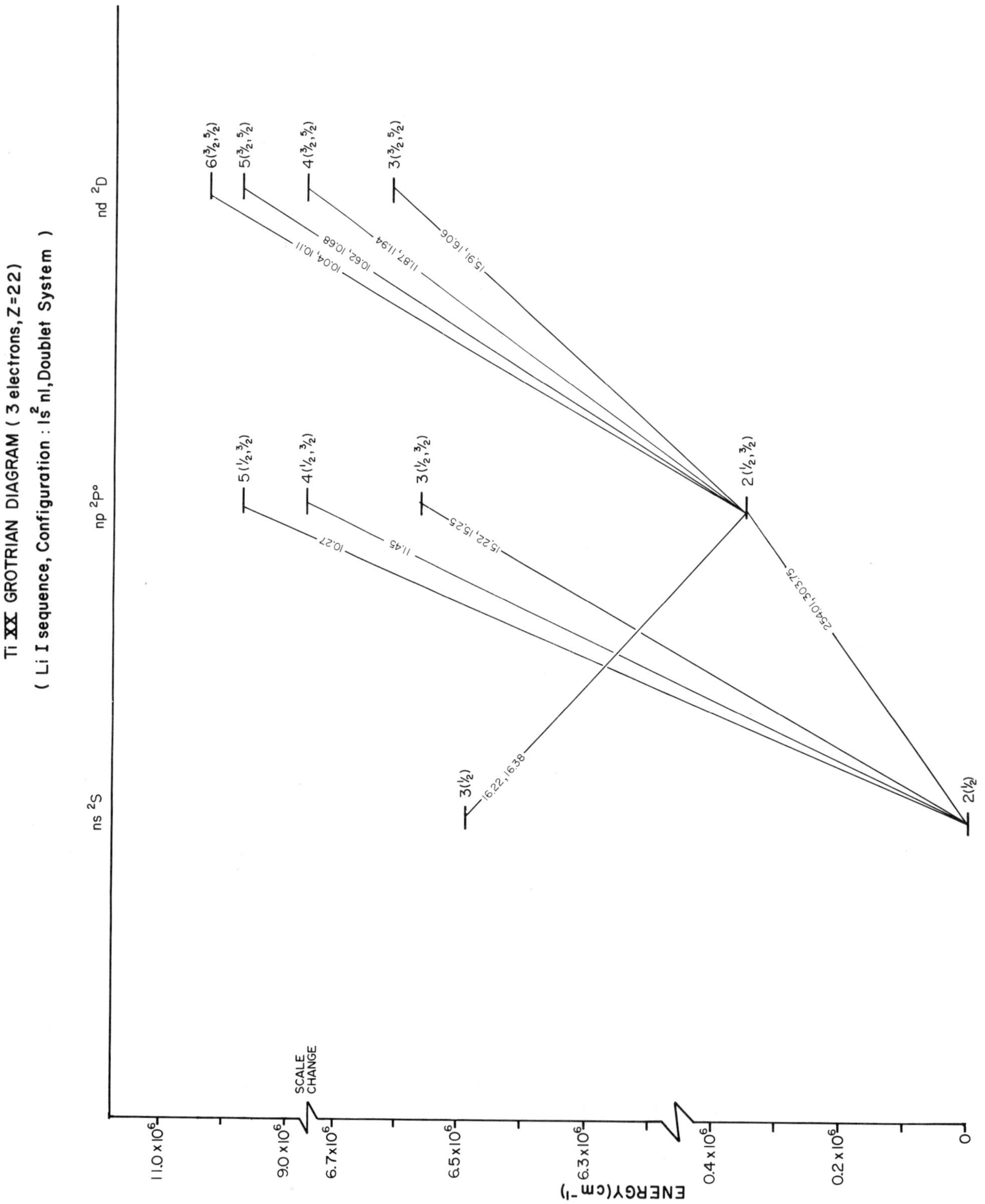

Ti XX GROTRIAN DIAGRAM (3 electrons, Z=22)

(Li I sequence, Configuration : $1s^2 nl$, Doublet System)

ns ^2S

np $^2P^o$

nd ^2D

ENERGY(cm^{-1})

11.0 x 10^6

9.0 x 10^6

SCALE CHANGE

6.7 x 10^6

6.5 x 10^6

6.3 x 10^6

0.4 x 10^6

0.2 x 10^6

0

6(3/2, 5/2)

5(3/2, 5/2)

4(3/2, 5/2)

3(3/2, 5/2)

5(1/2, 3/2)

4(1/2, 3/2)

3(1/2, 3/2)

2(1/2, 3/2)

3(1/2)

2(1/2)

10.27

11.45

15.22, 15.25

16.22, 16.38

254.01, 303.75

10.04, 10.11

10.62, 10.68

11.87, 11.94

15.91, 16.06

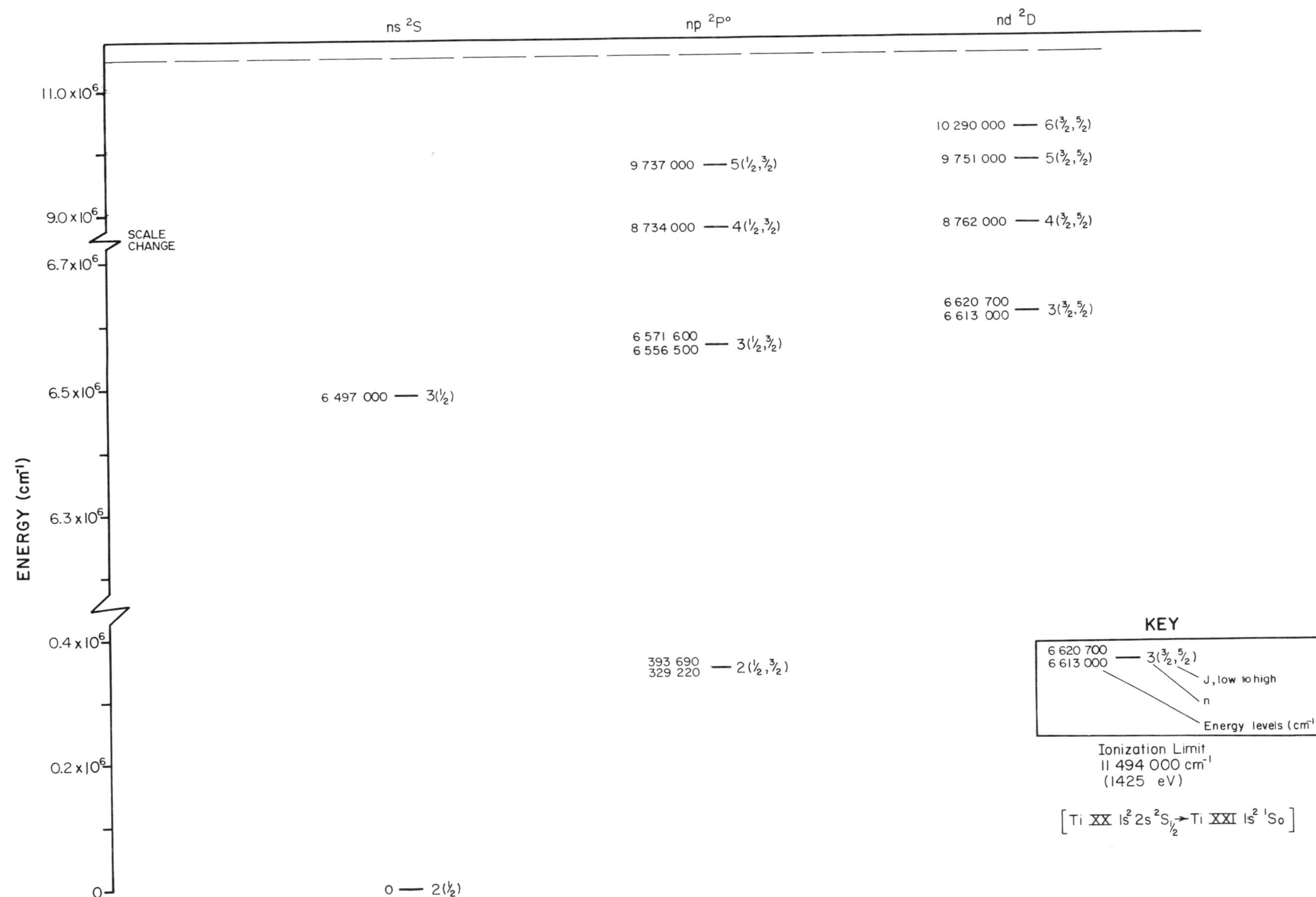
Ti XX ENERGY LEVELS (3electrons, Z=22)
(Li I sequence, Configuration : $1s^2nl$, Doublet System)
ns ^2S
np $^2P^o$
nd ^2D
11.0x10^6
9.0x10^6
SCALE CHANGE
6.7x10^6
6.5x10^6
6.3x10^6
0.4x10^6
0.2x10^6
0
ENERGY (cm^{-1})
10 290 000 — 6(3/2, 5/2)
9 737 000 — 5(1/2, 3/2)
9 751 000 — 5(3/2, 5/2)
8 734 000 — 4(1/2, 3/2)
8 762 000 — 4(3/2, 5/2)
6 620 700
6 613 000 — 3(3/2, 5/2)
6 571 600
6 556 500 — 3(1/2, 3/2)
6 497 000 — 3(1/2)
393 690
329 220 — 2(1/2, 3/2)
0 — 2(1/2)
KEY
6 620 700
6 613 000 — 3(3/2, 5/2)
J, low to high
n
Energy levels (cm^{-1})
Ionization Limit
11 494 000 cm^{-1}
(1425 eV)
[Ti XX $1s^2 2s\ ^2S_{1/2}$ → Ti XXI $1s^2\ ^1S_0$]

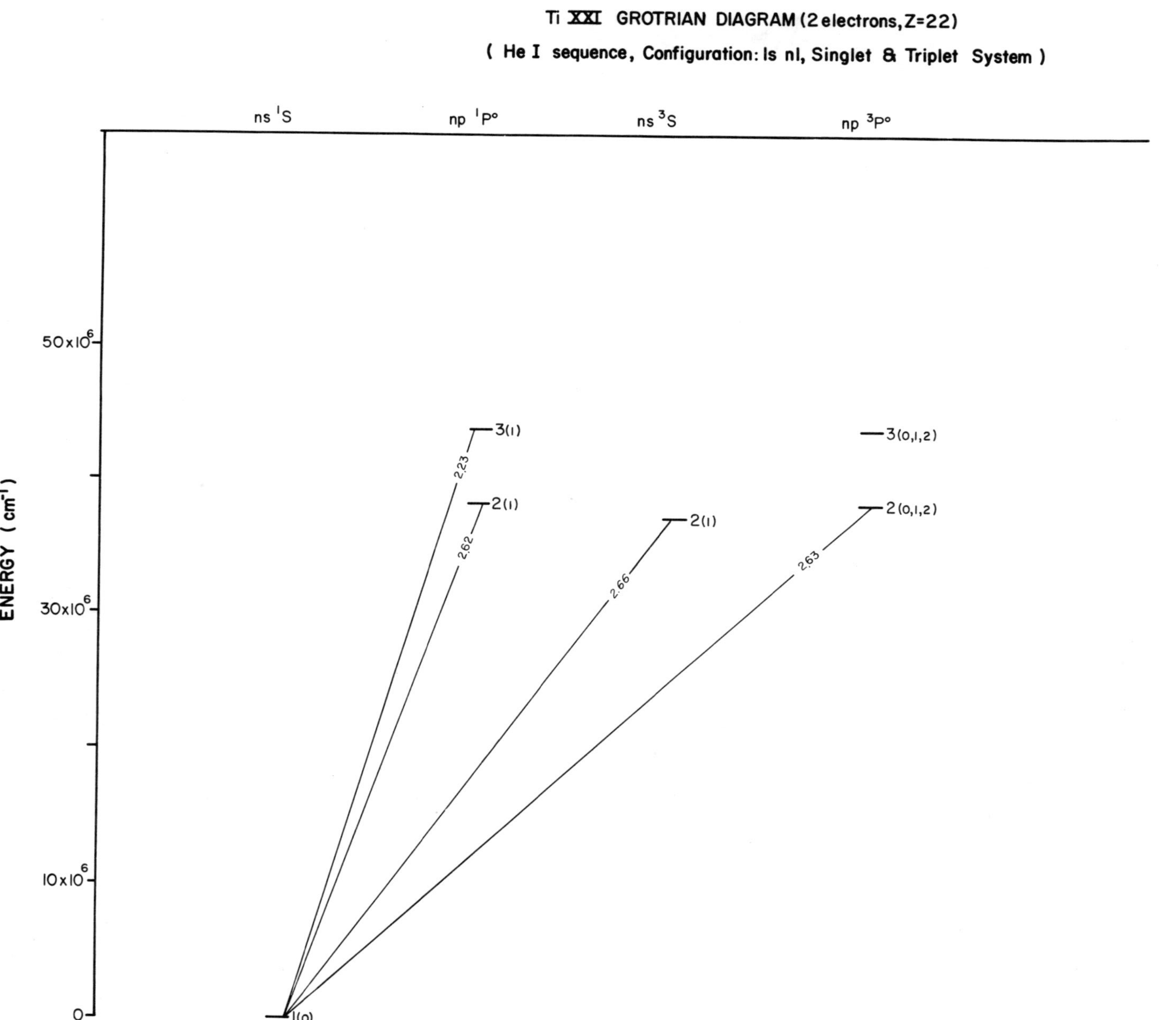
Ti XXI GROTRIAN DIAGRAM (2 electrons, Z=22)
(He I sequence, Configuration: 1s nl, Singlet & Triplet System)
ns 1S
np 1P°
ns 3S
np 3P°
3(1)
2(1)
2(1)
3(0,1,2)
2(0,1,2)
1(0)
2.23
2.62
2.66
2.63
ENERGY (cm-1)
50x10^6
30x10^6
10x10^6
0

Ti XXI ENERGY LEVELS (2 electrons, Z=22)

(He I sequence, Configuration: 1s nl, Singlet & Triplet Systems)

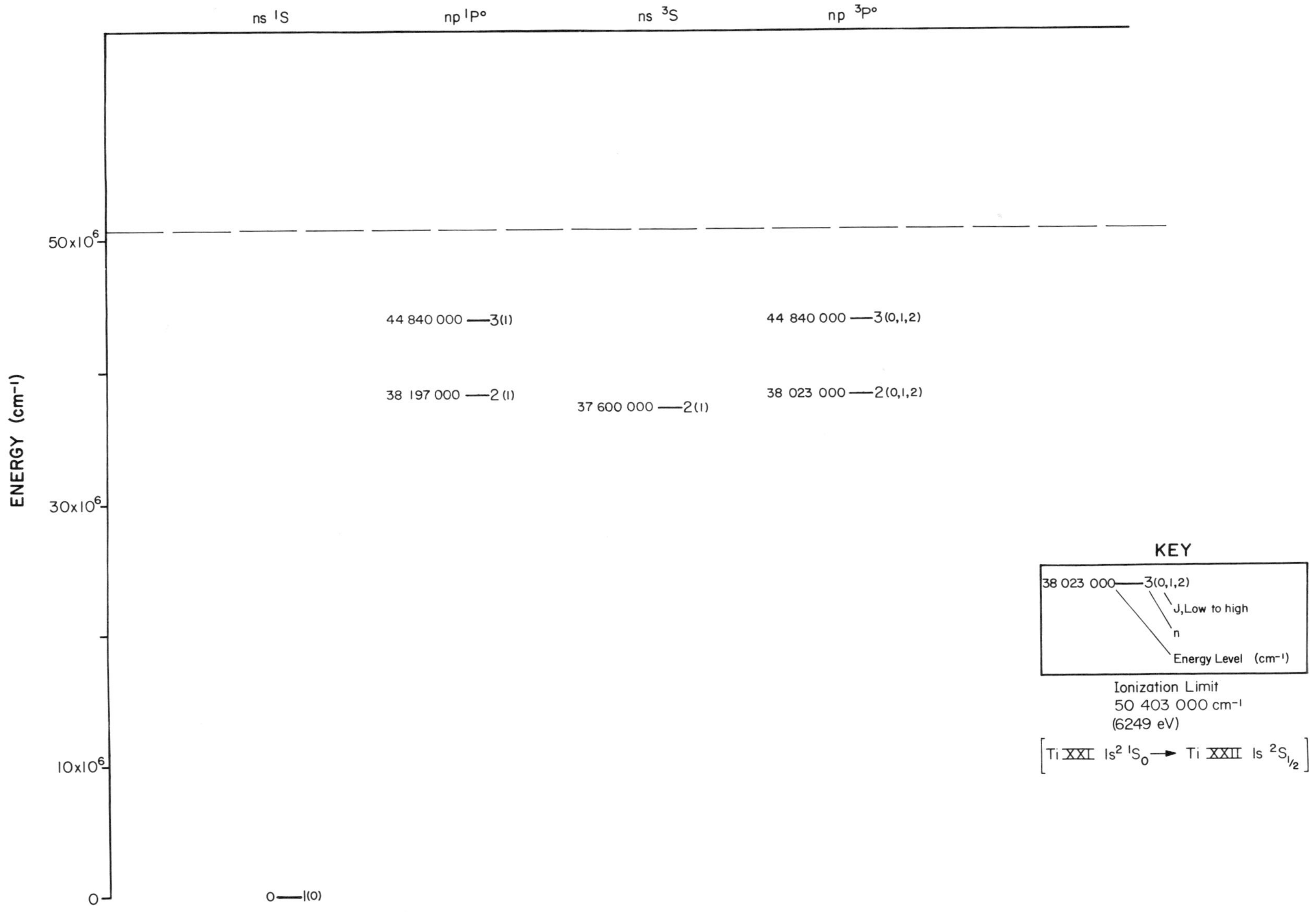

Ionization Limit
50 403 000 cm^{-1}
(6249 eV)

[Ti XXI $1s^2\ ^1S_0$ → Ti XXII $1s\ ^2S_{1/2}$]

Ti XXII GROTRIAN DIAGRAM (1 electron, Z=22)
(H I sequence, Configuration: nl, Doublet System)

ENERGY (cm^{-1})

54×10^6
52×10^6
50×10^6
48×10^6
47×10^6
40×10^6
0

ns ^2S
11(1/2)
10(1/2)
9(1/2)
8(1/2)
7(1/2)
6(1/2)
5(1/2)
4(1/2)
3(1/2)
2(1/2)
1(1/2)

np $^2P^o$
11(1/2)
10(1/2)
9(1/2)
8(1/2)
7(1/2)
6(1/2)
5(1/2,3/2)
4(1/2,3/2)
3(1/2,3/2)
2(1/2,3/2)

nd ^2D
11(3/2,5/2)
10(3/2,5/2)
9(3/2,5/2)
8(3/2,5/2)
7(3/2,5/2)
6(3/2,5/2)
5(3/2,5/2)
4(3/2,5/2)
3(3/2,5/2)

nf $^2F^o$
7(5/2,7/2)
6(5/2,7/2)
5(5/2,7/2)
4(5/2,7/2)

ng ^2G
7(7/2,9/2)
6(7/2,9/2)
5(7/2,9/2)

1.887
1.890
1.895
1.901
1.910
1.925
1.9489, 1.9491
1.995, 1.996
2.104, 2.105
2.491, 2.497
7.72 – 7.78
7.78 – 7.83
7.86 – 7.91
7.97 – 8.02
8.13 – 8.19
8.40 – 8.46
8.882 – 8.956
9.940 – 10.037
13.382 – 13.586
20.61 – 20.72
22.43 – 22.56
26.236 – 26.454
38.261 – 38.805
44.45 – 44.67
53.87 – 54.19
82.73 – 83.87

Ti XXII ENERGY LEVELS (1 electron, Z=22)
(H I sequence, Configuration: nl, Doublet System)

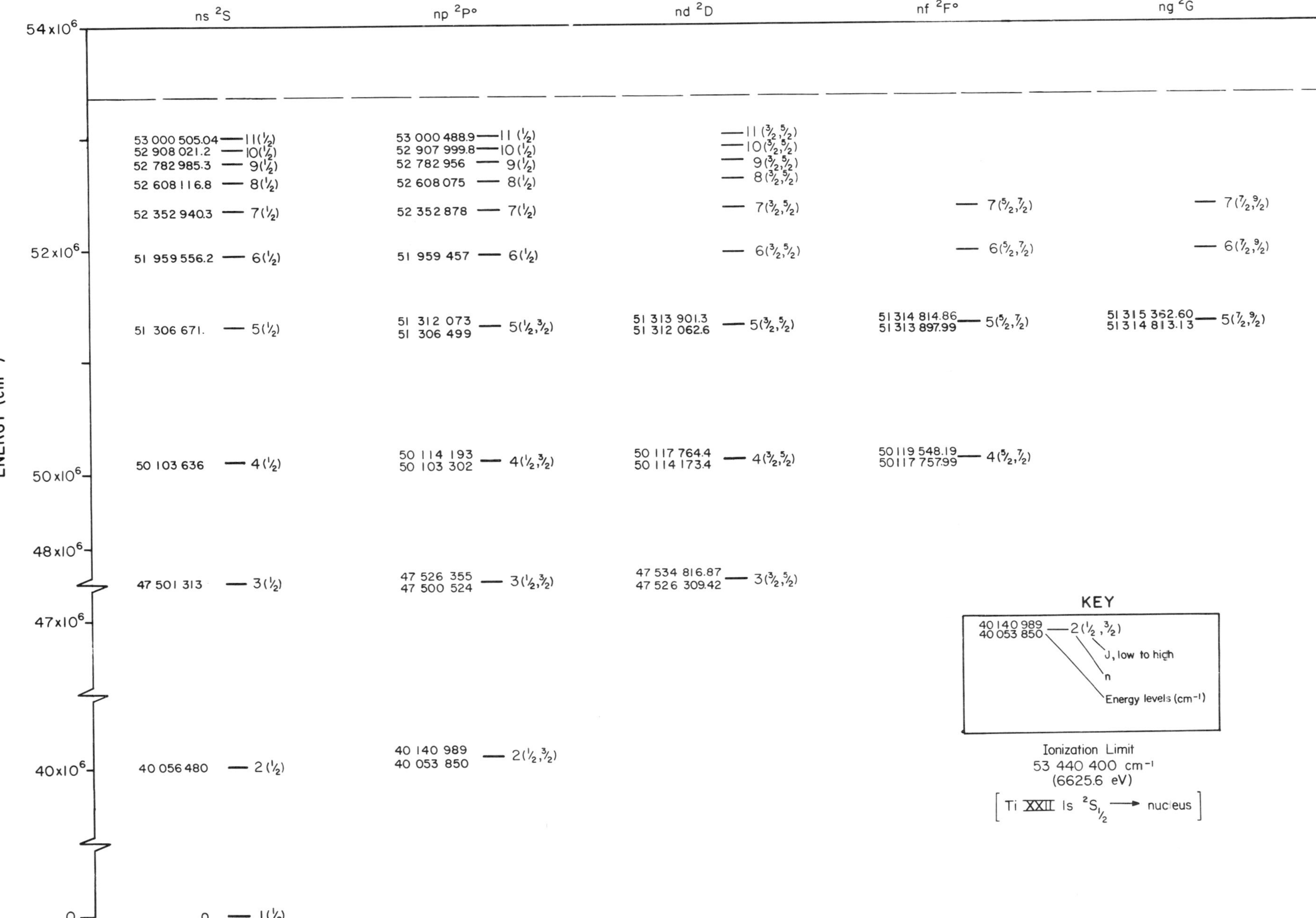

Ti I $Z = 22$ 22 electrons

Configuration interaction involving $(3d+4s)^3$ electrons is treated extensively by Roth[1,2] and by Smith and Siddall[3]. For the odd-parity terms, both papers made substantial revisions in the configurations and some revisions in the designations for the terms given in AEL. Smith and Siddall use statistically weighted averages over the *J*-levels for each term, whereas Roth carries out individual calculations for each *J*-level. The two papers disagree significantly in the parentage coefficients for a number of terms.

Since the configurations are not pure, it is often difficult or even meaningless to describe a term by means of a simple electronic arrangement. In AEL, the configurations were based on the coupling of $3d^2 4s$ electrons to form a system with which the optical electron subsequently coupled. However, Roth shows that it is more likely that there is coupling of the form $d^2(V,S,L)\,sp\,(^{1,3}P)\,SL$, or $d^3(V,S,L)\,pSL$. To illustrate the difference, consider the $^1D_2^\circ$ term at 22 405 cm^{-1}, for which AEL gives $3d^2 4s(^3F)4p\,^1D_2^\circ$, whereas Roth finds $[86\%\ 3d^2(^3F)4s4p(^3P^\circ)+10\%\ 3d^2(^3P)4s4p(^3P^\circ)]\ ^1D_2^\circ$. The situation becomes even more complicated for certain triplets and quintets, for which there are as many as three parents which are substantially different for the component *J*-levels.

In attempting to picture the levels and transitions, it is hopeless to include Roth's configurations on the various levels. The AEL system has the beauty of simplicity, particularly in that the Ti I levels are represented as cores from Ti II levels coupled to the optical electron. We have elected, therefore, to follow the AEL prescriptions, but we advise the user that the detailed and more accurate description of the terms is to be sought in Roth's work. Our diagrams incorporate the AEL letter identifications of the levels.

There is a further difficulty in that there are numerous unclassified lines; these have been left out of our drawings. We might mention that Roth also calculated the energies and configurations for a number of levels which have not been observed; we have omitted those levels from our diagrams. Where there are disagreements as to the level designations, we have adopted those given by Roth. Certain AEL levels rejected by Roth appear in Ref. 3 and so are retained by us. Finally, we caution that Refs. 1 and 2 deal only with the odd-parity levels involving the 4p (or 5p) optical electron.

1. C. Roth, J. Res. N.B.S. 73A, 125 (1969).
2. C. Roth, J. Res. N.B.S. 73A, 159 (1969).
3. G. Smith and J. Siddell, J. Opt. Soc. Am. **59**, 419 (1969).

H.W. Banks, W.R. Bozman, and C.M. Wilson, Georgetown Observatory Monograph #20 (unpublished), Georgetown College Observatory, Washington, D.C. 20007.

These authors present an analysis of this spectrum from 2000 Å to 9000 Å.

M.A. Catalán and R. Velasco, An. Real Soc. Esp. Fis. Quim. (A) **48**, 247 (1952).

These authors analyze series to obtain the ionization limit for this system (among others).

J.F. Giuliani and M.P. Thekaekara, J. Opt. Soc. Am. **54**, 460 (1964).

Authors give a table of interferometrically-measured lines observed in the region 3340–4190 Å.

C.C. Kiess and M.P. Thekaekara, Ap. J. **130**, 1008 (1959).

Authors give a table of term values from empirical data.

A.K. Wardakee, J. Opt. Soc. Am. **45**, 354 (1955).

Author gives a table of lines observed in the interval 5615–5930 Å.

C.M. Wilson and M.P. Thekaekara, J. Opt. Soc. Am. **51**, 289 (1961).

Authors give line and energy levels tables of calculated and observed values from lines observed in the region 2115–3075 Å.

We have taken lines primarily from Wilson and Thekaekara (1961), Giuliani and Thekaekara (1964), and Banks *et al.* Levels are taken from Kelly except where Kiess and Thekaekara (1959) give revised values.

Ti II $Z = 22$ 21 electrons

Here we deal with $3d^2 4p$ and 3d4s4p electrons for the odd terms. For odd levels of 29544 cm^{-1} and higher, Roth has contributed two papers[1,2]. Both papers modify a number of the configurations in AEL, in particular

$$\text{(AEL) } 3d4s(a^2D)4p\ x^2P^\circ_{1/2,3/2} \rightarrow \text{(Roth}^1\text{) } d(^2D)sp(^3P)x^4D^\circ_{1/2,3/2}, \text{ and}$$

$$^4D^\circ_{3/2-7/2} \rightarrow y^4F^\circ_{3/2-7/2}.$$

Ref. 1 also modifies more of the parentage coefficients but the later paper unfortunately omits any discussion of the changes. Ref. 2 also includes more levels than are treated in Ref. 1.

In the interest of simplicity of representation, we have used the AEL configurations and designations.

1. C. Roth, J. Res. N.B.S. **73A**, 125 (1969).
2. C. Roth, J. Res. N.B.S. **73A**, 159 (1969).

M.A. Catalán and R. Velasco, An. Real Soc. Esp. Fis. Quim. **48**, 247 (1952).

These authors derive improved ionization potentials for several systems.

J.F. Giuliani and M.P. Thekaekara, J. Opt. Soc. Am. **54**, 460 (1964).

Authors give a table of lines observed in the region 3340–4190 Å.

H. Mendlowitz, Ap. J. **154**, 1099 (1968).

Author gives a table of calculated and observed energy levels.

H.N. Russell, J. Opt. Soc. Am. **40**, 618 (1950).

This author provides a uniform system for estimating ionization potentials.

N. Sack, Phys. Rev. **102**, 1302 (1956).

Author gives a table of observed and calculated term values.

R.E. Trees, Phys. Rev. **97**, 686 (1955).

Author gives a table of calculated and observed term values.

Ti III $Z = 22$ 20 electrons

In Roth's (1968) calculations of the 3d4p configuration, there is found to be little interaction with sp. A more complete study by Edlén and Swensson shows that configuration mixing is significant for a number of levels. The latter paper greatly extends the information in AEL and corrects more of the designations given by Warner and Kirkpatrick (1969). We base our drawings on Edlén and Swensson (1974) but include only levels listed as observed. Recent calculations of mixing in levels in levels of high excitation are given by Wyart, Physica Scripta **12**, 33 (1975).

Transitions below 2000 Å are taken from Kelly and Palumbo; above 2000 Å they are taken from Striganov. The $5g\,^3G$ term listed by Kelly (unpublished) has been deleted from the *LS*-coupling triplet drawings because its energies correspond to those reported by Edlén and Swensson (1975) and presented here in the j–l-coupling drawings.

I.S. Bowen, Ap. J. **132**, 1 (1960).

Author gives a table of observed forbidden lines.

M.A. Catalán and R. Velasco, An. Real. Soc. Fis. Quim. **48**, 247 (1952).

These authors calculate ionization limits for several spectra.

B. Edlén and J.W. Swensson, Physica Scripta **12**, 21 (1974).

These authors present an extended analysis of this spectrum.

H. Mendlowitz, Ap. J. **158**, 385 (1969).

This author calculates a few wavelengths in this system.

C. Roth, J. Research NBS-A. Phys. and Chem. **72A**, 505 (1968).

Author gives a table of calculated and observed energy levels.

H.N. Russell and R.J. Lang, Ap. J. **66**, 13 (1927).

Authors give a line table, spectrograms, a multiplet array and a term table from lines observed in the interval 1000–4215 Å.

R.E. Trees, Phys. Rev. **97**, 686 (1955).

Author gives a table of average energies.

B. Warner and R.C. Kirkpatrick, Mon. Not. Roy. Astron. Soc. **144**, 397 (1969).

These authors claculate energy levels and forbidden transition probabilities in this system.

Ti IV $Z = 22$ 19 electrons

The levels and transitions have been taken from Swensson and Edlén (1974).

H.N. Russell and R.J. Lang, Ap. J. **66**, 13 (1927).

Authors give a line table, a term table, and spectrograms from lines observed in the region 420–5500 Å.

J.W. Swensson and B. Edlén, Physica Scripta **9**, 335 (1974).

These authors present tables of wavelengths and energies, and a Grotrian diagram.

Ti V $Z = 22$ 18 electrons

Levels are taken from Kelly (unpublished), in turn from Svensson and Ekberg (1968); transitions are taken from Svensson and Ekberg (1968).

R.D. Cowan, J. Opt. Soc. Am. **58**, 924 (1968).

Author gives a table of observed and calculated lines.

B.C. Fawcett, N.J. Peacock, and R.D. Cowan, J. Phys. **B1**, 295 (1968).

Authors give the wavelength of an observed transition.

A.H. Gabriel, B.C. Fawcett, and Carole Jordan, Proc. Phys. Soc. **87**, 825 (1966).

Authors give the wavelength of an observed transition.

P.G. Kruger and S.G. Weissberg, Phys. Rev. **48**, 659 (1935).

Authors give line and term tables from observations.

P.G. Kruger, S.G. Weissberg, and L.W. Phillips, Phys. Rev. **51**, 1090 (1937).

Authors give line and term tables from observation.

L.Å. Svensson and J.O. Ekberg, Ark. Fys. **37**, 65 (1968).

Authors give a table of lines observed in the region 140–365 Å.

L.Å. Svensson and J.O. Ekberg, Ark. Fys. **40**, 145 (1969).

These authors give a table of lines observed in the range 52–425 Å.

Ti VI $Z = 22$ 17 electrons

Levels have been taken from Kelly's unpublished compilation, in turn from Svensson and Ekberg (1968). Transitions are taken from Svensson and Ekberg (1968).

Fawcett *et al.* (1972) give the following transitions which we have omitted from our diagrams

	λ(Å)
$3p^4 3d\ ^3P_{9/2} - 3p^4(^3P)4f\ ^4G^\circ_{11/2}$	235.408
$^3P_{7/2}$ $\quad ^4G^\circ_{9/2}$	235.836
$^3P_{5/2}$ $\quad ^4G^\circ_{7/2}$	235.066
$3p^4(^1D)3d\ ^2G_{9/2} - 3p^4(^1D)4f\ ^2H^\circ_{11/2}$	235.310
$3p^4(^3P)3d\ ^4D_{7/2} - 3p^4(^3P)4f\ ^4F^\circ_{9/2}$	225.561

B. Edlén, Z. Physik **104**, 407 (1937).

Author gives line and term tables from observations in the interval 180–200 Å.

B.C. Fawcett, R.D. Cowan, and R.W. Hayes, J. Phys. B5, 2143 (1972).

These authors list several lines in this spectrum.

B.C. Fawcett and A.H. Gabriel, Proc. Phys. Soc. **88**, 262 (1966).

Authors give a table of lines observed in the region 250–270 Å.

B.C. Fawcett, N.J. Peacock, and R.D. Cowan, J. Phys. B1, 295 (1968).

Authors give a table of lines observed in the region 140–155 Å.

A.H. Gabriel, B.C. Fawcett, and Carole Jordan, Proc. Phys. Soc. **87**, 825 (1966).

Authors give the wavelengths of the observed transitions of two multiplets.

L.Å. Svensson, Physica Scripta **4**, 111 (1971).

This author reports two wavelengths and the separation $3s^2 3p^5\ ^2P^\circ_{3/2-1/2}$ for this system.

L.Å. Svensson and J.O. Ekberg, Ark. Fys. **37**, 65 (1968).

Authors give line and energy level tables from lines observed in the region 125–360 Å.

Ti VII $\quad Z = 22 \quad$ 16 electrons

Fawcett *et al.* list the following transitions omitted from our diagrams:

	λ(Å)		λ(Å)
$3p^3 3d\ ^5D^\circ_4 - 3p^3 4f\ ^5F_5$	193.668	$(^2D)\ ^3F^\circ_4 - (^2D)\ ^3G_5$	192.474
$3p^3 3d\ ^5D^\circ_3 - 3p^3 4f\ ^5F_4$	193.505	$(^2D)\ ^3F^\circ_3 - (^2D)\ ^3G_4$	192.272
$3p^3 3d\ ^5D^\circ_2 - 3p^3 4f\ ^5F_3$	193.534	$(^2D)\ ^3F^\circ_2 - (^2D)\ ^3G_3$	192.102
$3p^3 3d\ ^5D^\circ_1 - 3p^3 4f\ ^5F_2$	193.501		

I.S. Bowen, Ap. J. **132**, 1 (1960).

Author gives a table of observed forbidden transitions.

B. Edlén, Z. Physik **104**, 188 (1936).

Author gives line and term tables and spectrograms from observations in the interval 160–180 Å.

B. Edlén, Phys. Rev. **61**, 434 (1942).

Author gives the wavelength of an observed transition and term values for the $3s^2 3p^4\ ^1S$ and $3s3p^5\ ^1P_1$ configurations.

B.C. Fawcett, R.D. Cowan, and R.W. Hayes, J. Phys. B5, 2143 (1972).

These authors report several lines in this spectrum.

B.C. Fawcett and A.H. Gabriel, Proc. Phys. Soc. **88**, 262 (1966).

Author give the wavelengths of the observed transitions of two multiplets.

B.C. Fawcett and N.J. Peacock, Proc. Phys. Soc. **91**, 973 (1967).

Authors give the wavelengths of the observed transitions of two multiplets.

A.H. Gabriel, B.C. Fawcett, and Carole Jordan, Proc. Phys. Soc. **87**, 825 (1966).

Authors give the wavelengths of the observed transitions of two multiplets.

P.G. Kruger and H.S. Pattin, Phys. Rev. **52**, 621 (1937).

Authors give an array of multiplets and term values from empirical data.

L.Å. Svensson, Physica Scripta **4**, 111 (1971).

This author reports several lines belonging to transitions $3s^2 3p^4$–$3s3p^5$.

L.Å. Svensson and J.O. Ekberg, Ark. Fys. **37**, 65 (1967).

Authors give line and energy level tables from lines observed in the interval 125–340 Å.

Ti VIII $Z = 22$ 15 electrons

The wavelengths for the transitions $3s^2 3p^3\ ^2P°$–$3s3p^4\ ^2S$ were taken from Fawcett (1960). Energy levels were found from these wavelengths.

J.O. Ekberg and L.Å. Svensson, Physica Scripta **2**, 283 (1970).

These authors present a table of classified lines and a table of energy levels for this system.

B.C. Fawcett, J. Phys. B3, 1732 (1970).

This author gives a table of classified lines in this spectrum.

B.C. Fawcett, R.D. Cowan, and R.W. Hayes, J. Phys. B5, 2143 (1972).

These authors classify four lines in this spectrum.

B.C. Fawcett, A.H. Gabriel, and P.A.H. Saunders, Proc. Phys. Soc. **90**, 863 (1967).

Authors give the wavelengths of the transitions of one multiplet.

B.C. Fawcett and N.J. Peacock, Proc. Phys. Soc. **91**, 973 (1967).

Authors give the wavelength of an observed transition.

A.H. Gabriel, B.C. Fawcett, and Carole Jordan, Proc. Phys. Soc. **87**, 825 (1966).

Authors give the wavelength of an observed transition.

P.G. Kruger and H.S. Pattin, Phys. Rev. **52**, 621 (1937).

Authors give an array of multiplets and term values from empirical data.

Ti IX $Z = 22$ 14 electrons

J.O. Ekberg and L.Å. Svensson, Physica Scripta **2**, 283 (1970).

These authors present a table of classified lines and a list of energy levels for this system.

B.C. Fawcett, J. Phys. B3, 1732 (1970).

This author presents several classifications of lines in this spectrum.

B.C. Fawcett, R.D. Cowan, and R.W. Hayes, J. Phys. B5, 2143 (1972).

These authors classify three lines in this spectrum.

B.C. Fawcett, A.H. Gabriel, and P.A.H. Saunders, Proc. Phys. Soc. **90**, 863 (1967).

Authors give the wavelengths of the observed transitions of two multiplets.

B.C. Fawcett and N.J. Peacock, Proc. Phys. Soc. **91**, 973 (1967).

Authors give the wavelengths of the observed transitions of two multiplets.

L.W. Phillips, Phys. Rev. **55**, 708 (1939).

Author gives line and term tables from observations in the region 280–345 Å.

Ti X $Z = 22$ 13 electrons

J.O. Ekberg and L.Å. Svensson, Physica Scripta **2**, 283 (1970).

These authors present tables of classified wavelengths and energy levels for this system.

B.C. Fawcett, J. Phys. B3, 1732 (1970).

This author classifies several wavelengths in this spectrum.

B.C. Fawcett and N.J. Peacock, Proc. Phys. Soc. **91**, 973 (1967).

Authors give a table of lines observed in the interval 300–430 Å.

A.H. Gabriel, B.C. Fawcett, and Carole Jordan, Proc. Phys. Soc. **87**, 825 (1966).

Authors give the wavelengths of the observed transitions of one multiplet.

Ti XI $Z = 22$ 12 electrons

Levels have been taken from Kelly's unpublished tabulation, in turn largely from Ekberg (1971). Lines have been taken from Ekberg (1971).

B. Edlén, Z. Physik **103**, 536 (1936).

Author gives line and term tables from lines observed in the region 70–130 Å.

J.O. Ekberg, Physica Scripta **4**, 101 (1971).

This author presents tables of energy levels and classified wavelengths of this spectrum.

B.C. Fawcett, J. Phys. B3, 1732 (1970).

This author presents several classified lines of this spectrum.

B.C. Fawcett, J. Opt. Soc. Am. **66**, 632 (1976).

This author classifies several additional wavelengths in this system.

B.C. Fawcett and N.J. Peacock, Proc. Phys. Soc. **91**, 973 (1967).

Authors give a table of lines observed in the region 305–430 Å.

Ti XII $Z = 22$ 11 electrons

B. Edlén, Z. Physik **100**, 621 (1936).

Author gives line and term tables from lines observed in the interval 60–120 Å.

J.O. Ekberg and L.Å, Svensson, Physica Scripta **12**, 116 (1975).

These authors analyze this spectrum and present a line list, a table of levels' energies, and a Grotrian diagram.

B.C. Fawcett and N.J. Peacock, Proc. Phys. Soc. **91**, 973 (1967).

These authors give a table of lines observed in the region 340–480 Å.

U. Feldman and L. Cohen, J. Opt. Soc. Am. **57**, 1128 (1967).

Authors give tables of lines from observations.

U. Feldman, G.A. Doschek, D.K. Prinz, and D.J. Nagel, J. Appl. Phys. **47**, 1341 (1976).

These authors observe four wavelengths in this spectrum.

L. Cohen and W.E. Behring, J. Opt. Soc. Am. **66**, 899 (1976).

These authors analyze this spectrum and present a line list and a table of levels' energies.

Ti XIII | $Z = 22$ | 10 electrons

Kastner *et al.* identify their observed levels in both *LS* and *j–l* coupling, and most other levels are also identified in both *LS* (Edlén and Tyrén, Feldman and Cohen) and *j–l* (Moore, Fawcett) coupling; we have used both systems simultaneously on the drawings. Where the *LS* coupling in Edlén and Tyrén disagreed with that in Crance, the identification of Crance has been used.

Energies for $2p^5 3p$ and $2p^5 4d$ terms are largely calculated (Crance); although these are observed levels (Kastner *et al.*), they are not connected to the ground state.

Kelly (private communication, 1975) mentions a $2p^5 5d\ ^3P_1^0$ level at 5 640 000 cm^{-1}, but it is not included here.

M. Crance, Atomic Data **5**, 2 (1973).

Author gives calculated energies for $2p^5 3p$ and $2p^5 4d$ terms not listed by other references. Also corrects *LS*-coupling identifications by Edlén and Tyrén of levels at 4 219 800 and 4 281 600 cm^{-1}.

B. Edlén and F. Tyrén, Z. Physik **101**, 206 (1936).

Authors give line and term tables from observations in the region 20–30 Å, in *LS*-coupling.

B.C. Fawcett, Proc. Phys. Soc. **86**, 1087 (1965).

Author gives *j–l*-coupling information on levels at 5 163 700 and 5 207 200 cm^{-1}.

U. Feldman and L. Cohen, Ap. J. **149**, 265 (1967).

Authors give line and term tables from observations in the region 15–25 Å, in *LS*-coupling.

S.O. Kastner, W.E. Behring, and L. Cohen, Ap. J. **199**, 777 (1975).

These authors identify some transitions classified as $2p^5 3p$–$2p^5 4d$, involving levels not observed by other references (no observed transitions to the ground state).

Ti XIV | $Z = 22$ | 9 electrons

We have taken wavelengths and energies from Feldman *et al.* (1973); in addition, the level $3d\ ^4F_{5/2}$ is given tentatively by Kelly (unpublished).

L. Cohen, U. Feldman, and S.O. Kastner, J. Opt. Soc. Am. **58**, 331 (1968).

Authors give tables of lines in the interval 20–25 Å.

G.A. Doschek, U. Feldman, R.D. Cowan, and L. Cohen, Ap. J. **188**, 417 (1974).

These authors classify two wavelengths observed in this spectrum.

B.C. Fawcett, Proc. Phys. Soc. **86**, 1087 (1965).

Author gives a table of lines observed in the region 20–25 Å.

B.C. Fawcett, J. Phys. **B4**, 981 (1971).

This author classifies six wavelengths in this spectrum.

B.C. Fawcett, Atomic Data and Nuclear Data Tables **16**, 135 (1975).

This author tabulates wavelengths for transitions of the sort $2s^2 2p^n - 2s2p^{n+1}$ and $2s2p^n - 2p^{n+1}$.

B.C. Fawcett, D.D. Burgess, and N.J. Peacock, Proc. Phys. Soc. **91**, 970 (1967).

Authors give the wavelengths of the observed transitions of one multiplet.

U. Feldman, G.A. Doschek, R.D. Cowan, and L. Cohen, J. Opt. Soc. Am. **63**, 1445 (1973).

These authors classify many transitions in this system and provide tables of wavelengths and energy levels.

Ti XV $Z = 22$ 8 electrons

Agreement of wavelength measurements by different authors is typically ±0.01–0.03 Å. We have used the wavelengths given by Doschek *et al.* (1973), Doschek *et al.* (1974) and Fawcett and Hayes (1975), except for the transition $2s2p^5\ ^1P^\circ_1 - 2p^6\ ^1S_0$, for which we have used the wavelength given by Kasyanov *et al.* (1974). The levels below 2×10^6 cm^{-1} are taken from Fawcett (1975), rounded to the nearest 100 cm^{-1}. We have calculated the higher levels from the wavelengths that we selected. The ionization limit is taken from Kelly and Palumbo. Only the tentatively identified transition at 110.072 Å is observed to connect the triplet and singlet systems.

G.A. Doschek, U. Feldman, and L. Cohen, J. Opt. Soc. Am. **63**, 1463 (1973).

These authors provided several wavelengths and energy levels for this system.

G.A. Doschek, U. Feldman, R.D. Cowan and L. Cohen, Ap. J. **188**, 417 (1974).

These authors present several levels and wavelengths in this system.

G.A. Doschek, U. Feldman, J. Davis, and R.D. Cowan, Phys. Rev. **A12**, 980 (1975).

These authors list one wavelength in this spectrum.

B.C. Fawcett, J. Phys. **B4**, 981 (1971).

This author classifies several wavelengths in this spectrum.

B.C. Fawcett, Atomic Data and Nuclear Data Tables **16**, 135 (1975).

This author tabulates wavelengths of the sort $2s^2 2p^n - 2s2p^{n+1}$ and $2s2p^n - 2p^{n+1}$.

B.C. Fawcett and R.W. Hayes, Mon. Not. Roy. Astron. Soc. **170**, 185 (1975).

These authors compare observed and calculated wavelengths for several transitions in this spectrum.

S. Goldsmith, U. Feldman, and L. Cohen. J. Opt. Soc. Am. **61**, 615 (1971).

Authors give line and term tables from lines observed in the region 20–25 Å.

Yu.S. Kasyanov, E.Ya. Kononov, V.V. Korobkin, K.N. Koshelev, A.N. Ryabtsev, R.V. Serov, and E.V. Skokan, Opt. Spectrosc. **36**, 4 (1974).

These authors classify wavelengths and give tables of wavelengths and energy level in this system.

Ti XVI $Z = 22$ 7 electrons

There is fair agreement (typically ±0.03 Å) among wavelengths reported by different investigators since 1974 in this spectrum. We have used the results of Fawcett and Hayes (1975) for wavelengths in the range 19.0–20.2 Å; for longer wavelengths we have taken the most-recent experimental determinations, except that we have used the calculated values for the transitions $2s2p^4\ ^2S$, $^2P - 2p^5\ ^2P^\circ$ (from Fawcett, 1975).

Levels below 1 600 000 cm^{-1} have been taken from Fawcett (1975), rounded to the nearest 100 cm^{-1}. We have calculated the positions of higher levels from the wavelengths reported by Fawcett and Hayes (1975). No intersystem combinations have been identified; hence the position of the doublet levels relative to the quartet levels is known only from the calculations of Fawcett (1975).

G.A. Doschek, U. Feldman, R.D. Cowan, and L. Cohen, Ap. J. **188**, 417 (1974).

These authors tabulate wavelengths observed in this spectrum.

G.A. Doschek, U. Feldman, J. Davis, and R.D. Cowan, Phys. Rev. **A12**, 980 (1975).

These authors list three lines in this spectrum.

B.C. Fawcett, J. Phys. **B4**, 981 (1971).

This author classifies numerous lines in this spectrum.

B.C. Fawcett, Atomic Data and Nuclear Data Tables **16**, 135 (1975).

This author tabulates wavelengths for transitions of the sort $2s^22p^n$–$2s2p^{n+1}$ and $2s2p^n$–$2p^{n+1}$.

B.C. Fawcett and R.W. Hayes, Mon. Not. Roy. Astron. Soc. **170**, 185 (1975).

These authors compare observed and calculated values for several wavelengths in this spectrum.

U. Feldman, G.A. Doschek, R.D. Cowan, and L. Cohen, Ap. J. **196**, 613 (1975).

These authors classify three wavelengths in this spectrum.

Yu.S. Kasyanov, E.Ya. Kononov, V.V. Korobkin, K.N. Koshelev, A.N. Ryabtsev, R.V. Serov, and E.V. Skokan, Opt. Spectrosc. **36**, 4 (1974).

These authors present several wavelengths and energy levels for this system.

Ti XVII — $Z = 22$ — 6 electrons

Wavelength measurements and classifications reported by various authors are not fully consistent. Since the wavelengths reported by Fawcett (1975), Fawcett *et al.* (1974), and Feldman (1975) are consistent, we have taken the levels presented by Fawcett (1975) for the configurations $2s^22p^2$, $2s2p^3$, and $2p^4$, rounded to the nearest 1 000 cm^{-1}. We have then calculated the positions of the levels in the configurations $2s^22p3d$, $2s^22p3s$, $2s2p^23d$, and $2s2p^23s$, and the level $2s2p^3\ {}^5S^\circ_2$, from the wavelengths given by Goldsmith *et al.* (1972). Wavelengths on the Grotrian diagram are from Goldsmith *et al.* (1972) and Fawcett (1975).

B.C. Fawcett, Atomic Data and Nuclear Data Tables **16**, 135 (1975).

This author tabulates wavelengths for transitions of the sort $2s^22p^n$–$2s2p^{n+1}$ and $2s2p^n$–$2p^{n+1}$.

B.C. Fawcett, M. Galanti, and N.J. Peacock, J. Phys. **B7**, 1149 (1974).

These authors classify several wavelengths in this spectrum.

U. Feldman, G.A. Doschek, R.D. Cowan, and L. Cohen, Ap. J. **196**, 613 (1975).

These authors present classifications for six lines in this spectrum.

S. Goldsmith, U. Feldman, A. Crooker, and L. Cohen, J. Opt. Soc. Am. **62**, 260 (1972).

Authors give line and energy level tables from lines observed in the region 18–21 Å.

Yu.S. Kasyanov, E.Ya. Kononov, V.V. Korobkin, K.N. Koshelev, A.N. Ryabtsev, R.V. Serov, and E.V. Skokan, Opt. Spectrosc. **36**, 4 (1974).

These authors give several energy levels and wavelengths for this spectrum.

Ti XVIII — $Z = 22$ — 5 electrons

The levels $2s^23s\ {}^2S_{1/2}$ and $2s^24s\ {}^2S_{1/2}$, and the ionization limit are taken from Kelly's unpublished listing. The $2s2p^2\ {}^4P$ and $2p^3\ {}^4S^o$ levels are taken from Kasyanov *et al.* (1974). The doublet levels built on $2s^2sp$ and $2s2p^2$ are taken from Fawcett (1975), but it is unlikely that the accuracies quoted are as high as the precisions quoted. Wavelengths are taken from Fawcett and Hayes (1975) except for the transitions $2s^22p$–$2s^23s$, $2s^24s$. taken from Kelly and Palumbo. The positions of the levels $2s^23d$, $2s2p3p$ and $2s2pnd$ are estimated using the wavelengths of Fawcett and Hayes (1975).